Scanning Transmission Electron Microscopy

Scanning Transmission Electron Microscopy

Imaging and Analysis

Edited by

Stephen J. Pennycook
Peter D. Nellist

Springer

Editors
Stephen J. Pennycook
Materials Science and Technology Division
Oak Ridge National Laboratory
1 Bethel Valley Road
Oak Ridge, TN 37831-6071, USA
pennycooksj@ornl.gov

Peter D. Nellist
Department of Materials
University of Oxford
Parks Road
Oxford, OX1 3PH, UK
peter.nellist@materials.ox.ac.uk

ISBN 978-1-4419-7199-9 ISBN 978-1-4419-7200-2 (eBook)
DOI 10.1007/978-1-4419-7200-2
Springer New York Dordrecht Heidelberg London

Printed on acid-free paper

Springer is part of Springer Science+Business Media (www.springer.com)

Preface

Over the last two decades, scanning transmission electron microscopy (STEM) has become a very popular and widespread technique, with the number of publications and presentations making use of STEM techniques increasing by about an order of magnitude. Although the strengths of the technique for providing high-resolution structural and analytical information have been known and understood for much longer than that, the key to its more recent popularity has undoubtedly been the availability of STEM modes on instruments available from the major TEM manufacturers. Gone are the days when researchers wanting the unique capabilities of high-resolution STEM had to undertake the task of keeping a VG dedicated STEM instrument operating.

Given the current interest in the technique, we felt that the time was right to review the current state of knowledge about STEM and STEM-related techniques and their application to a range of materials problems. The purpose of this volume is both to educate those who wish to deepen their understanding of STEM and to inform those who are seeking a review of the latest applications and methods associated with STEM. We are delighted that so many of our colleagues accepted our invitation to contribute to this volume, and we are indebted to them for their efforts in creating such excellent contributions. The following chapters illustrate how close STEM has brought us to the ultimate materials characterisation challenge of analysing materials atom by atom.

We hope that the following chapters demonstrate the spectacular results that can be achieved when performing the relatively simple experiment of focusing a beam of electrons down to an atomic scale and measuring the scattering that results.

Stephen J. Pennycook
Peter D. Nellist

Contents

Contributors

Eiji Abe
Department of Materials Science and Engineering, University of Tokyo, Tokyo, Japan

John R. Abelson
Department of Materials Science and Engineering, University of Illinois, Urbana-Champaign, IL, USA

Leslie J. Allen
School of Physics, University of Melbourne, Melbourne, VIC, Australia

Ilke Arslan
Department of Chemical Engineering and Materials Science, University of California-Davis, Davis, CA, USA

Stephanie Bogle
Department of Materials Science and Engineering, University of Illinois, Urbana-Champaign, IL, USA

Gianluigi A. Botton
Department of Materials Science and Engineering, McMaster University, Hamilton, ON, Canada

Nigel D. Browning
Departments of Chemical Engineering and Materials Science, Molecular and Cellular Biology, University of California-Davis, Davis, CA, USA; Physical and Life Sciences Directorate, Lawrence Livermore National Laboratory, Livermore, CA, USA

Matthew F. Chisholm
Materials Science and Technology Division, Oak Ridge National Laboratory, Oak Ridge, TN, USA

Christian Colliex
Laboratoire de Physique des Solides, CNRS/UMR8502, Université Paris-Sud, Orsay, France

Peter A. Crozier
School of Mechanical, Aerospace, Chemical and Materials Engineering, Arizona State University, Tempe, AZ, USA

Niklas Dellby
Nion Co., 1102 8th St., Kirkland, WA, USA

Rolf Erni
Electron Microscopy Center, Empa, Swiss Federal Laboratories for Materials Science and Technology, Dubendorf, Switzerland

Scott D. Findlay
Institute of Engineering Innovation, The University of Tokyo, Tokyo, Japan

Jaume Gazquez
Materials Science and Technology Division, Oak Ridge National Laboratory, Oak Ridge, TN, USA; Departament de Física Aplicada III, University Complutense of Madrid, Madrid, Spain

Yuichi Ikuhara
Institute of Engineering Innovation, The University of Tokyo, Tokyo, Japan

Dmitri O. Klenov
FEI Company, Eindhoven, The Netherlands

Robert F. Klie
Department of Physics, University of Illinois, Chicago, IL, USA

Mathieu Kociak
Laboratoire de Physique des Solides, CNRS/UMR8502, Université Paris-Sud, Orsay, France

Ondrej L. Krivanek
Nion Co., 1102 8th St., Kirkland, WA, USA

James M. LeBeau
Materials Department, University of California, Santa Barbara, CA, USA

Andrew R. Lupini
Materials Science and Technology Division, Oak Ridge National Laboratory, Oak Ridge, TN, USA

Cesar Magen
Materials Science and Technology Division, Oak Ridge National Laboratory, Oak Ridge, TN, USA; Instituto de Nanociencia de Aragon-ARAID and Departamento de Física de la Materia Condensada, Universidad de Zaragoza, Spain

Paul A. Midgley
Department of Materials Science and Metallurgy, University of Cambridge, Cambridge, UK

Matthew F. Murfitt
Nion Co., 1102 8th St., Kirkland, WA, USA

Peter D. Nellist
Department of Materials, University of Oxford, Oxford, UK

Mark P. Oxley
Department of Physics and Astronomy, Vanderbilt University, Nashville, TN, USA; Materials Science and Technology Division, Oak Ridge National Laboratory, Oak Ridge, TN, USA

Stephen J. Pennycook
Materials Science and Technology Division, Oak Ridge National Laboratory, Oak Ridge, TN, USA

Timothy J. Pennycook
Department of Physics and Astronomy, Vanderbilt University, Nashville, TN, USA; Materials Science and Technology Division, Oak Ridge National Laboratory, Oak Ridge, TN, USA

Guillaume Radtke
Institut Matériaux Microélectronique Nanoscience de Provence, UMR CNRS 6242, Université Paul Cézanne Aix-Marseille III, Marseille, France

Bryan W. Reed
Physical and Life Sciences Directorate, Lawrence Livermore National Laboratory, Livermore, CA, USA

Naoya Shibata
Institute of Engineering Innovation, The University of Tokyo, Tokyo, Japan

Susanne Stemmer
Materials Department, University of California, Santa Barbara, CA, USA

Odile Stéphan
Laboratoire de Physique des Solides, CNRS/UMR8502, Université Paris-Sud, Orsay, France

Jing Tao
Brookhaven National Laboratory, Upton, NY, USA

Marcel Tencé
Laboratoire de Physique des Solides, CNRS/UMR8502, Université Paris-Sud, Orsay, France

Maria Varela
Materials Science and Technology Division, Oak Ridge National Laboratory, Oak Ridge, TN, USA; Departament de Física Aplicada III, University Complutense of Madrid, Madrid, Spain

Paul M. Voyles
Department of Materials Science and Engineering, University of Wisconsin, Madison, WI, USA

Weronika Walkosz
Department of Physics, University of Illinois, Chicago, IL, USA

Michael G. Walls
Laboratoire de Physique des Solides, CNRS/UMR8502, Université Paris-Sud, Orsay, France

Masashi Watanabe
Department of Materials Science and Engineering, Lehigh University, Bethlehem, PA, USA

Matthew Weyland
Monash Centre for Electron Microscopy, Monash University, Melbourne, VIC, Australia

Guang Yang
Department of Physics, University of Illinois, Chicago, IL, USA

Yuan Zhao
Department of Physics, University of Illinois, Chicago, IL, USA

Jian-Min Zuo
Department of Materials Science and Engineering and Frederick Seitz Materials Research Laboratory, University of Illinois at Urbana-Champaign, Urbana, IL, USA

1

A Scan Through the History of STEM

Stephen J. Pennycook

1.1 Baron Manfred von Ardenne

The first STEM was designed and constructed by Manfred von Ardenne in Berlin in 1937–1938. In his 1938 paper (submitted for publication on December 25, 1937) he showed an image of ZnO crystals demonstrating a resolution of 40 nm in the scan direction, reproduced in Figure 1–1 (von Ardenne 1938b). He also showed how the detector could be arranged for either bright field or dark field imaging in transmission and for reflection or secondary imaging of solid surfaces. A paper submitted less than 9 months later, on September 7, 1938, demonstrated an impressive fourfold improvement in resolution to 10 nm (von Ardenne 1938a).

These developments took place several years after the development of the TEM by Max Knoll and Ernst Ruska (Knoll and Ruska 1932), and they grew from somewhat different origins and motivations (for accounts in English, see von Ardenne 1985, 1996). The TEM was based on the principles of the light microscope, with the goal of achieving a resolution exceeding that of the optical microscope (Ruska 1987). It was some time, however, before it was realized that the image contrast arises quite differently, from absorption in the light microscope but from scattering in the electron microscope, see Süsskind (1985). The STEM originated through von Ardenne's efforts to develop the scanning electron microscope (SEM) motivated mostly by the development of camera tubes for television (for reviews, see McMullan 1989, 1995, 2004). von Ardenne's first SEM images of solid surfaces were obtained in 1933 but were part of a patent application and did not appear in the

---

Note we use the abbreviation STEM to denote both the instrument (the scanning transmission electron microscope) and the technique (scanning transmission electron microscopy), similarly for TEM.

Notice: This manuscript has been authored by UT-Battelle, LLC, under Contract No. DE-AC05-00OR22725 with the U.S. Department of Energy. The United States Government retains and the publisher, by accepting the article for publication, acknowledges that the United States Government retains a non-exclusive, paid-up, irrevocable, world-wide license to publish or reproduce the published form of this manuscript, or allow others to do so, for United States Government purposes.

S.J. Pennycook, P.D. Nellist (eds.), *Scanning Transmission Electron Microscopy*,
DOI 10.1007/978-1-4419-7200-2_1, Springer Science+Business Media, LLC 2011

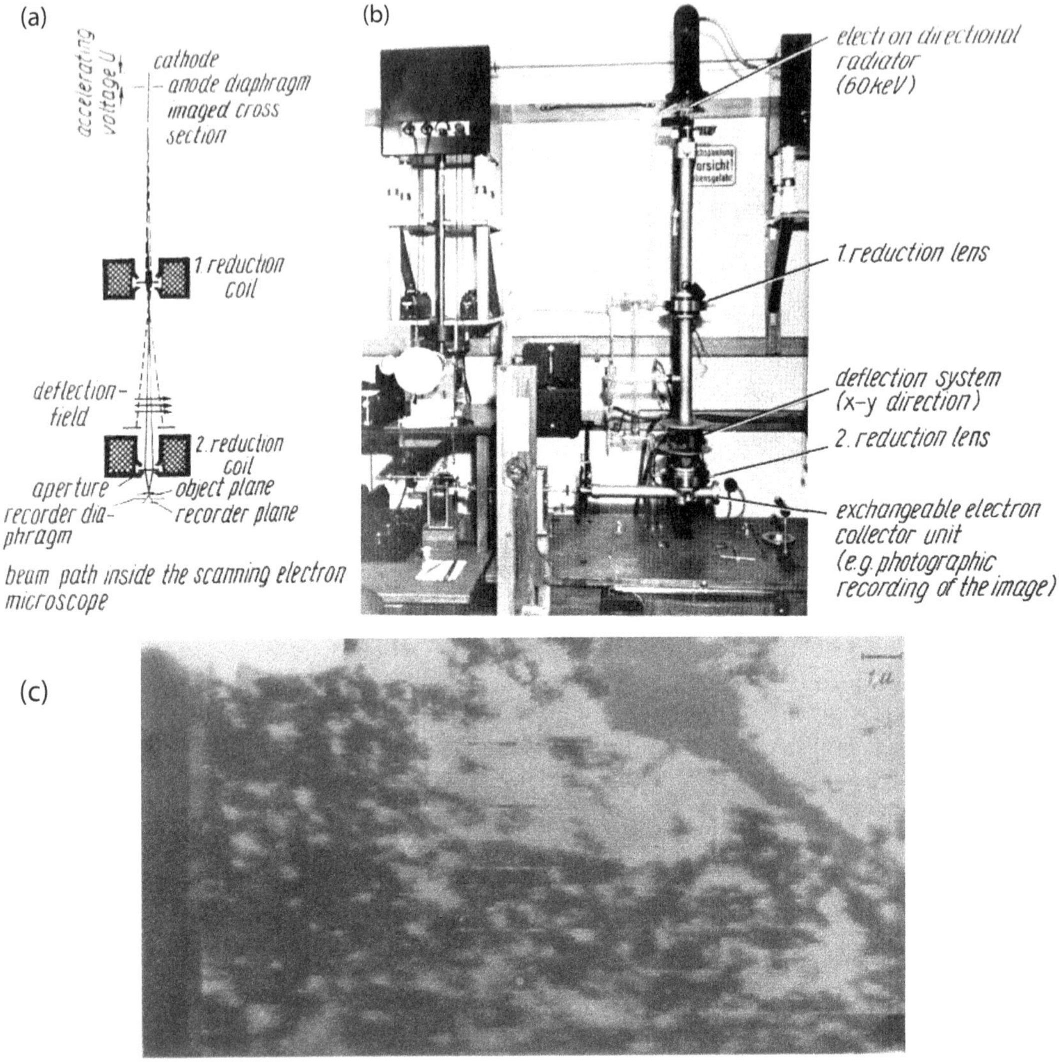

Figure 1–1. (**a**) Schematic diagram of the first STEM built by Manfred von Ardenne. (**b**) Photograph of the microscope. (**c**) Image of ZnO crystals showing a resolution in the scan direction (*horizontal*) of 40 nm. Reproduced from von Ardenne (1938b, 1985) with permission.

open literature. They were later reproduced in von Ardenne (1985). The first published SEM images were by Knoll (1935). von Ardenne's motivation for developing the STEM was his realization that the transmitted electrons would not need to be refocused to form a high-resolution image, merely detected, and hence the resolution of a STEM image would not be degraded by chromatic aberration of the imaging lenses, as was the case with the TEM (von Ardenne 1985).

However, as is clear from Figure 1–1, the major limitation of the STEM was noise, and von Ardenne soon turned his efforts toward developing his universal TEM based on the design of Ruska. It may seem odd that there were no attempts to use a field emission source in any early transmission microscope, especially as both experimental and theoretical work on field emission was being carried out in the

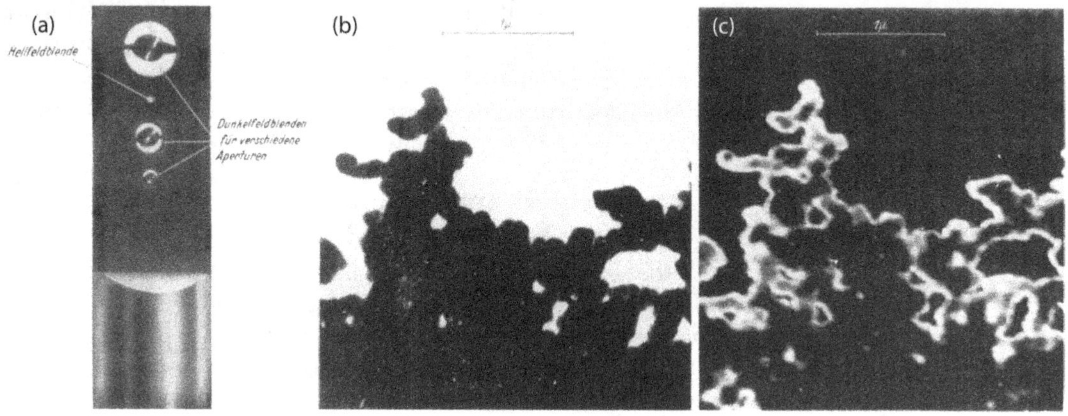

Figure 1–2. (**a**) An annular aperture used as a central beam stop for dark field imaging in von Ardenne's universal TEM. (**b**) The first bright field/dark field pair of images of ZnO crystals, from von Ardenne (1940b) with permission.

same city, Berlin (Fowler and Nordheim 1928, Müller 1936). One was used in a scanning microscope (Zworykin et al. 1942), but to achieve the required high vacuum the source and specimen had to be inside a glass container which was baked and then sealed, not exactly convenient for sample exchanges.

Besides his two papers on scanning microscopy in 1938, von Ardenne published two further papers on transmission microscopy, one on the limits to resolution (von Ardenne 1938c) and the other on questions of intensity and resolution (von Ardenne 1939), which contains the optimistic prediction that "sooner or later the ultramicroscopy technique will be able to reveal single atoms and their distribution in the object plane." von Ardenne's universal TEM produced images in bright field showing clear features 30 Å in diameter and contained several innovations including the first annular aperture for dark field imaging (von Ardenne 1940b). He showed the first bright field/dark field pair using a central beam stop aperture (Figure 1–2), commenting that the dark field image showed more detail than the bright field image, although it actually showed lower resolution because the higher illumination angles introduced more spherical aberration. He also introduced the idea of stereomicroscopy, suggesting it to be "the ultimate tool for future structure investigations." He even published a complete book on electron microscopy (von Ardenne 1940c). In 1944 von Ardenne's STEM equipment was destroyed in an air raid, and a STEM was not developed again until the field emission gun was successfully incorporated over 20 years later by Albert Crewe.

1.2 Development of TEM

Much of the physics of the electron microscope was established relatively quickly, and we mention a few notable points in passing. More detailed accounts of the early development of TEM can be found in the book by P. W. Hawkes (1985), Ruska's Nobel lecture (Ruska 1987), his book (English translation, Ruska 1980), and the articles "Key events in the history of electron microscopy" (Haguenau et al. 2003) and Müller (2009). In particular, the potential resolution of the electron microscope was well appreciated from the earliest time. Ruska, in his Nobel lecture, mentions that when they became aware of de Broglie's wave theory of the electron in the summer of 1931 they were "very heartened" to calculate a resolution limit for their objective aperture of around 2 Å at their accelerating voltage of 75 kV. In 1936 Scherzer published his classic proof that spherical aberration is intrinsically positive for round lenses, with the prescient comment that "the unavoidability of spherical aberration is a technical barrier but not a barrier in principle." He pointed out that, unlike the light microscope, because of spherical aberration the resolution of the electron microscope was limited to around 100 wavelengths, but nevertheless, one day it should be possible to see atoms (Scherzer 1939). Theoretical calculations for single atoms were published by Hillier (1941), based on absorption contrast, concluding that atoms of atomic number (Z) greater than 25 should be visible at 60 kV accelerating voltage. The next year Schiff published an estimate that atoms with Z greater than 7 could be imaged, also at 60 kV (Schiff 1942). His estimate was based on an interference of the elastically scattered beam with the forward scattered beam, phase contrast as it would be called today, producing a contrast of twice the forward scattered amplitude, hence the greater sensitivity to lighter atoms. He also states that this contrast limit disappears if dark field conditions are used, citing von Ardenne for an annular condenser aperture of such a size that the unscattered beam does not pass through the objective aperture.

The first recognition that aberrations need to be combined wave optically was by Glaser (1943). Prior estimates had been made by quadratic minimization as if the defects were independent contributions as opposed to different distortions of the wave front. In this way it is easy to obtain the minimum probe diameter by minimizing the broadening due to spherical aberration and diffraction (see, e.g., Marton (1944)) as

$$d_{\text{min}} = C\lambda^{3/4}C_S^{1/4}, \tag{1}$$

where the constant $C = 1.32$. Glaser's treatment actually produced a very similar value for C since he used only the Gaussian focus (the focal plane for limiting small angles). However, his calculations paved the way toward an exploration of the optimum combination of aberrations, the idea that the intrinsically positive spherical aberration of the objective lens could be balanced to some degree by choosing a negative defocus (weakening the lens to compensate for the overfocus of the rays at higher angles due to spherical aberration). Boersch (1947) considered phase and amplitude contrast, using the lens aberrations as a source of phase contrast. He also considered single atoms and even considered the possibility of energy filtering. Scherzer (1949) examined the optimum conditions in detail, arriving at the values for apertures and defocus which today we refer to as the Scherzer optimum conditions for coherent and incoherent imaging. Scherzer did not refer to them in this way, as the two modes were not in common practice as they are today, he was exploring the physics. He certainly appreciated that for axial illumination of two point scatterers the image amplitudes should be added before taking the intensity, whereas with a wide illumination aperture the intensity from each point object would contribute to the image independently. This is exactly analogous to the situation for light optical imaging described by Lord Rayleigh (1896) and appears to be the first detailed comparison in the context of electron imaging.

The difference from the light optical situation is the significance of spherical aberration, in that with uncorrected electron optics the optimum conditions are not near Gaussian focus but at significant underfocus. Scherzer showed the first contrast transfer function, although he did not use that terminology, and defined an optimum $C \sim 0.6$ for axial (coherent) illumination. For incoherent illumination he estimated $C \sim 0.4$. As in light optics the resolution for incoherent imaging is significantly greater than that for coherent imaging for the simple reason that squaring an amplitude distribution makes it sharper. The prefactors depend on the exact optimization procedure employed but are now generally accepted as 0.66 for coherent imaging (Eisenhandler and Siegel 1966, Cowley 1988) and 0.43 for the case of incoherent imaging (Black and Linfoot 1957, Wall et al. 1974, Beck and Crewe 1975).

In crystalline specimens it was seen from the earliest observations that the contrast did not depend just on the mass thickness but also on the crystal orientation (von Ardenne 1940a, von Borries and Ruska 1940) which was identified as diffraction contrast (Hillier and Baker 1942). Heidenreich (1942) showed how the use of a small objective aperture led to thickness fringes and so began the whole field of diffraction contrast imaging of defects (Heidenreich 1949) and their interpretation through dynamical theory. Dislocations were first identified in an image by Bollmann (1956) and Hirsch et al. (1956), and their image contrast studied by kinematical theory (Hirsch et al. 1960) and dynamical theory (Howie and Whelan 1961, 1962).

The effects of inelastic scattering on image contrast were also investigated at this time. Kamiya and Uyeda (1961) showed that diffraction contrast could be preserved under inelastic scattering. They compared conventional TEM diffraction contrast bright field and dark field images to images obtained by moving the objective aperture so that no diffracted beam contributed to the image. Such images still contained thickness fringes. The following year, thickness fringes were seen in an image formed from plasmon loss electrons using an energy filtering microscope (Watanabe and Uyeda 1962). Theoretical study soon showed that diffraction contrast should be preserved in the inelastic image provided the inelastic wave was scattered through only a small angle (Fujimoto and Kainuma 1963, Fukuhara 1963, Howie 1963). While the electron would be slowed down due to the inelastic excitation, and hence be incoherent with the zero-loss beam, if selected by the spectrometer it would show the same diffraction contrast as the zero-loss beam in the limit of zero angular deflection. The condition that the scattering angle be small was valid for plasmon excitation (Cundy et al.

1966, 1967) and many single electron excitations (Cundy et al. 1969, Humphreys and Whelan 1969). Experiments were extended to core losses by Craven et al. (1978), who found contrast of a stacking fault to be largely preserved at the Si L edge near 100 eV.

For larger scattering angles the contrast was found not to be preserved (Kamiya and Uyeda 1961, Cundy et al. 1969, Kamiya and Nakai, 1971, Melander and Sandstrom 1975). High-angle scattering is dominated by thermal diffuse scattering involving large momentum transfers, which leads to transitions between Bloch states and changes in the image contrast. However, the contribution of phonon scattering was found to be small for the apertures used in TEM at the time. Most of the phonon-scattered electrons were intercepted by the objective aperture, contributing more to absorption effects in the image than directly to image contrast (Hashimoto et al. 1962, Heidenreich 1962, Hall and Hirsch 1965, Humphreys and Hirsch 1968, Humphreys 1979).

The first lattice image to be recorded in an electron microscope was by Menter in 1956. By opening up the objective aperture more than one diffracted beam could reach the image plane where they could interfere and form lattice fringes. Menter achieved a resolution of 12 Å, starting the field that came to be known as high-resolution electron microscopy. For reviews of the achievement of atomic resolution prior to the era of aberration correction, see Herrmann (1978), Cowley and Smith (1987), Smith (1997), and Spence (1999).

In 1965 the first proposal for an annular detector appeared in the literature in a paper entitled "Possibilities and Limitations for the Differentiation of Elements in the Electron Microscope" (Cosslett 1965). The suggestion grew out of the observation that the contrast of thin amorphous films was relatively weakly dependent on atomic number under the usual conditions for bright field imaging. Cosslett pointed out that elemental differentiation would be much improved with an off-axis detector and suggested use of a ring-shaped detector to increase the detected intensity.

1.3 The Crewe Innovations

While serving as Director of Argonne National Laboratory Albert Crewe published an article entitled "Scanning Electron Microscopes: Is High Resolution Possible?" (Crewe 1966) with the subtitle "Use of a field-emission electron source may make it possible to overcome existing limitations on resolution." He advocates the STEM mode of operation because of the fact that the critical resolution-limiting optics are positioned before the beam strikes the specimen, allowing great flexibility in the nature of the detector. However, he does not mention the annular detector, instead he shows many examples of energy loss spectra and energy-filtered images. The resolution achieved with his field emission source is 50 Å, but he predicts that the resolution will improve to the same levels as achieved in the conventional microscope. Two years later the resolution is reported at 30 Å (Crewe et al. 1968).

Crewe left Argonne National Laboratory in 1967 to return to the University of Chicago, Enrico Fermi Institute, where he was a professor. Remarkable progress was made over the next few years (Crewe 2009). In 1970 a new microscope was described (Crewe and Wall 1970, Crewe et al. 1970) with a high-resolution objective lens giving a spot size of around 5 Å. This microscope does now incorporate an annular detector as well as a spectrometer to collect electrons transmitted through its central hole. Due to the fact that elastic scattering is broad in angle whereas inelastic scattering is more sharply peaked in the forward direction, these two detectors were arranged to give signals representing approximately the elastic and inelastic cross-sections of the specimen. Using molecules stained with uranium and thorium atoms, supported on a 2-nm-thick carbon film, images were obtained that indeed showed bright spots having visibilities close to the theoretical values for individual atoms, and, furthermore, they showed the expected geometric patterns for the respective support molecules (pairs or chains). It was concluded that the bright spots were probably due to single atoms, the first observation of single atoms by an electron microscope.

These images were formed by taking the ratio of the elastic signal collected by the annular detector to the inelastic signal collected by the spectrometer. The ratio image showed similar visibility of heavy atoms to the elastic signal alone but suppressed contrast caused by thickness variations in the carbon support film. This ratio signal was termed a "Z" contrast signal, the first use of the term in connection with an image formed from transmitted electrons (Crewe 1971).

A paper by Wall et al. (1974) showed the first annular dark field (ADF) images from small crystallites containing uranium and thorium atoms, identified as oxides or carbides, see Figure 1–3. They referred to the images as elastic dark field, but they represent the first ADF images of crystalline materials. Also, the paper presents the first line traces recorded from individual atoms, demonstrating a full width half maximum of 2.5 ± 0.2 Å at 42.5 kV accelerating voltage, as shown in Figure 1–4. Also in 1974 we find the first reference to aberration correction with regard to STEM, that the STEM is ideally suited to aberration correction since it can be applied to the monochromatic incident beam (Crewe 1974).

Phase contrast lattice images were also obtained, as in the conventional TEM, by using a small axial collector aperture and a defocused lens (Crewe and Wall 1970), see Figure 1–5. It was long appreciated (see, for example, von Ardenne 1996) that the conventional and scanning microscopes were related by reversal of the ray paths. This is one example of the reciprocity principle for elastic scattering amplitudes, see also Cowley (1969) and Zeitler and Thomson (1970).

The ability to image individual atoms spurred theoretical investigations into the expected contrast and the efficiency of various modes of detection. Important for such considerations is the characteristic angle for elastic scattering in relation to the objective aperture and also whether the image contrast arises through phase contrast (interference between scattered and unscattered amplitudes) or scattering contrast

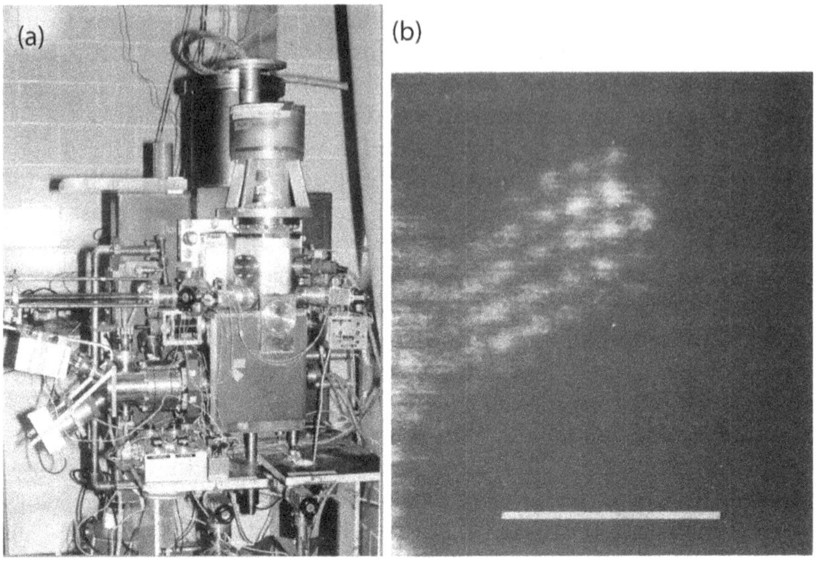

Figure 1–3. (**a**) Image of the Crewe STEM equipped with a cold field emission gun (courtesy O. L. Krivanek). (**b**) ADF image of small crystallites containing uranium and thorium atoms. Scale bar is 20 Å, reproduced from Wall et al. (1974) with permission.

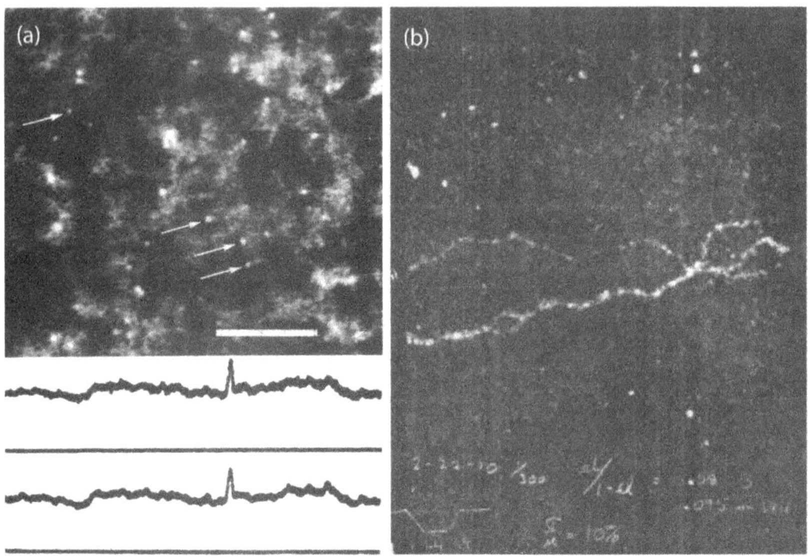

Figure 1–4. ADF image of a sample of mercuric acetate showing individual atoms. Two line traces across the same atom show a full width half maximum of 2.5 ± 0.2 Å. Scale bar is 50 Å, reproduced from Wall et al. (1974) with permission.

(when electrons are blocked from contributing to the image by an aperture, Zeitler and Thomson (1970)). Zeitler and Thomson were also at the Enrico Fermi Institute in Chicago and showed explicitly that for phase contrast to be observed in STEM with an axial detector the scattering angle must be less than the radius of the incident cone, which is equivalent to the reciprocal TEM case where the Bragg reflections must pass

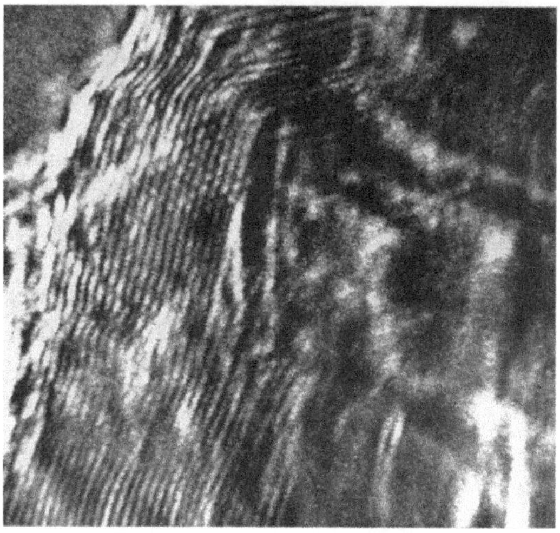

Figure 1–5. STEM bright field phase contrast image taken with a small axial collector aperture showing the 3.4 Å fringes of partially graphitized carbon, reproduced from Crewe and Wall (1970) with permission.

within the objective aperture. They considered both phase contrast and scattering contrast, showing that the TEM and STEM are equivalent, but did not compare detailed noise statistics for the two modes. Maximum phase contrast in uncorrected STEM requires a small axial detector aperture resulting in only a small fraction of the incident beam being detected, and consequently a high noise level. Thomson (1973) treated the noise statistics in detail. He showed that the axial aperture required for phase contrast could be increased to collect about a quarter of the incident beam without substantial loss of contrast. However, the TEM mode would still have around three times better signal to noise ratio even under these conditions. The reverse was true for dark field imaging, since for the apertures of the time (~12 mrad) the annular detector could be made sufficiently wide in angle to intercept the majority of the elastic scattering. The same point was made by Langmore et al. (1973), see Figure 1–6, where they compare the ADF collection efficiency to that of two modes of TEM dark field imaging. Their Hartree–Fock–Slater calculations also reveal the shell structure of the atoms. Both papers also point out that for ultrahigh resolution (below 1 Å) the beam stop

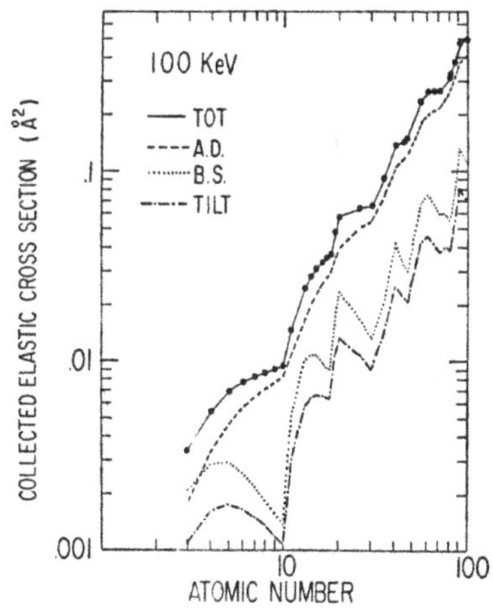

Figure 1–6. Total elastic scattering cross-sections calculated for 100 kV electrons with a Hartree–Fock–Slater atomic model (*solid line*), compared to the cross-sections for elastic scattering into an ADF detector in STEM (*dashed line*), a TEM with dark field beam stop (*dotted line*) and a TEM with tilted illumination (*dash-dotted line*), reproduced from Langmore et al. (1973) with permission.

TEM mode (with the direct beam excluded from the objective aperture) would become the most efficient, since most of the elastic scattering would now occur *within* the angle of the objective aperture instead of outside it. With the success of aberration correction this mode may need to be reexamined, although the TEM cannot provide the simultaneous detection capabilities of the STEM.

A detailed comparison of the TEM and STEM phase contrast modes was given by Rose (1974) during a sabbatical at the Crewe laboratory at the University of Chicago. Rose proposed the use of an annular bright field detector aperture, which markedly improved the efficiency of the STEM phase contrast image compared to the small axial detector, as with the large axial detector considered by Thomson. Rose showed that the highest signal to noise ratio was obtained for an image formed by the difference between the annular bright field signal and the remainder of the bright field disc. The images also show a greatly reduced speckle pattern, reducing the possibility of image artifacts, similar to that found in TEM hollow cone dark field imaging (Thon and Willasch 1972). Surprisingly, the optimum conditions use an aperture and defocus that are larger than those for optimum ADF imaging and therefore give higher resolution, with a prefactor given by $C = 0.36$. Rose also shows calculated single atom images for an aberration-corrected microscope and makes the point that as resolution improves more of the scattered electrons stay within the bright field cone. Encouraging results with annular bright field imaging have indeed been reported recently using an aberration-corrected STEM (Findlay et al. 2009, Okunishi et al. 2009).

Experimental comparison of observed and calculated single atom cross-sections was made by Retsky (1974). The paper shows the first histogram of intensities corresponding to one- and two-atom clusters of U, as shown in Figure 1–7. The second peak is at twice the intensity of the first, providing convincing proof that the first peak is due to single atoms and also implying the incoherent nature of the image (intensities from individual atoms adding linearly). Such studies paved the way for the direct discrimination of elements based on the image contrast. Isaacson et al. (1979) showed the discrimination of Pt atoms

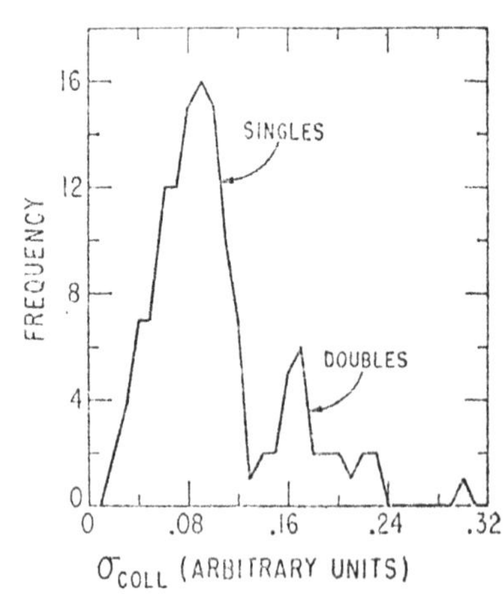

Figure 1–7. A histogram of intensities for 135 bright spots from a uranium specimen showing peaks corresponding to one and two atoms, reproduced from Retsky (1974) with permission.

from Pd atoms and spectacular studies of atomic diffusion (Isaacson et al. 1977). Using a beam energy of only 28.6 keV, the motion was primarily due to thermal diffusion; varying the incident beam current resulted in negligible change in hopping rate. Furthermore, on cooling the entire microscope by 10°C the hopping rate dropped a factor of three (Crewe 1979).

Attempts were also made to observe single atoms in the conventional TEM in dark field (Henkelman and Ottensmeyer 1971, Whiting and Ottensmeyer 1972, Ottensmeyer et al. 1973, 1979). However, because a small off-axis aperture was used there remained significant (coherent) speckle pattern from the carbon support film, unlike in the (incoherent) ADF images obtained by Crewe and coworkers, which raised questions on the interpretation (Dubochet 1979). The speckle pattern of the support could also be removed by using single crystal graphite, and Hashimoto et al. (1971) imaged single Th atoms in this way. Heavy atoms on amorphous carbon were much more difficult to see in bright field images (Baumeister and Hahn 1973, Parsons et al. 1973), but using the graphite support Iijima (1977) was also able to obtain clear images of single heavy atoms in bright field TEM. Single atoms were also imaged using hollow cone illumination, equivalent by reciprocity to an ADF image in the STEM (Thon and Willasch 1972).

Crewe's STEM also gave new impetus to the application of EELS in the microscope. Originally proposed decades earlier (Hillier 1943), it had been used mainly for investigation of the issue of preservation of contrast. Castaing and Henry (1962) introduced spectroscopic imaging in the TEM and showed the location of Al on ZnO via its characteristic plasmon loss. In 1966 Crewe showed identification of C and O K loss edges with his STEM and comments that "It is attractive to consider the possibility of chemical analysis of selected areas of a specimen." EELS in the TEM was also soon demonstrated (Wittry 1969, Wittry et al. 1969, Colliex and Jouffrey 1972, Egerton and Whelan 1974), with spatial resolution being defined by an area selecting aperture. Studies in the STEM began with biological material (Crewe et al. 1971, Isaacson et al. 1973). In 1975 two papers appeared that discussed detailed energy and angular variations of the inelastic scattering necessary for quantitative analysis. Egerton (1975) presented results from the TEM, introduced the power law background subtraction technique, and discussed deconvolution of multiple scattering and instrumental broadening. Isaacson and Johnson (1975) presented results from the STEM, with the first theoretical expressions for the analysis of composition from characteristic losses based on calculated cross-sections and collection efficiencies. They gave expressions for the minimum detectable mass fraction and the minimum detectable mass and discussed the use of fine structure for extracting information on chemical bonding. Isaacson and Johnson predicted that several orders of magnitude improved sensitivity could be achieved by higher collection efficiency spectrometers and parallel detection, which they expected to open up "new vistas in microanalysis beyond that attainable with conventional X-ray detection methods." A flurry of activity ensued (Egerton et al. 1975, Colliex et al. 1976, Leapman and Cosslett 1976, Williams and Edington 1976,

Jeanguillaume et al. 1978, Egerton 1978a, b, Joy and Maher 1978a, b), and detection efficiency was improved (Johnson 1980, Isaacson and Scheinfein 1983). In 1982 the spatial difference technique was first utilized to demonstrate the detection of nitrogen at platelets in diamond (Berger and Pennycook 1982), although the terminology developed later (Mullejans and Bruley 1994, 1995, Scheu 2002). Parallel detection was eventually introduced, and the widespread adoption of the system by Krivanek et al. (1987) paved the way for EELS as we know it today.

1.4 Coherent or Incoherent Imaging?

Much of the early imaging of atoms tacitly assumed that the STEM ADF image produced an incoherent image. The atom size seen in the images was assumed to be the size of the intensity distribution of the scanning spot (Crewe and Wall 1970) and the images did not reverse contrast with focus, both characteristics of an incoherent image. A weak-phase object had been used as a convenient test object for phase contrast imaging theory, since weak-phase variations in the object could be efficiently converted into intensity variations in the image using the lens aberrations to rotate the phase of the scattered beam by 90° (Scherzer 1949, Eisenhandler and Siegel 1966, Heidenreich 1967). Often the phase object was a single test frequency, a sine wave potential variation (Zeitler and Thomson 1970). Cowley (1973) first examined the nature of the ADF image using such a weak-phase test object, and we present the essence of the argument here, as it is key to subsequent controversies.

A phase object is one sufficiently thin that all atoms can be regarded as being located in one two-dimensional plane, then their effect can be regarded as a phase shift of the incident beam, i.e., the object can be represented by a transmission function. A plane incident wave of unity amplitude experiences a pure phase shift of

$$\varphi\left(\mathbf{R}\right) = e^{i\sigma V(\mathbf{R})}, \tag{2}$$

where $\sigma = \pi/\lambda E$ is the interaction constant, λ is the wavelength, E is the accelerating voltage, V is the projected potential, defined as positive when attractive to electrons, and $\mathbf{R}$ is a two-dimensional position vector in the specimen. To see the formation of an ideal phase contrast image, take the case of a weak-phase object, when the exponential can be expanded to give

$$\varphi\left(\mathbf{R}\right) = 1 + i\sigma V\left(\mathbf{R}\right), \tag{3}$$

the sum of an incident wave and a small scattered wave 90° out of phase. Fourier transforming into reciprocal space (with two-dimensional coordinate $\mathbf{K}$) we have

$$\varphi\left(\mathbf{K}\right) = \delta(0) + i\sigma V\left(\mathbf{K}\right). \tag{4}$$

On passing through the objective lens additional phase changes are picked up due to the aberrations:

$$\varphi_i(\mathbf{K}) = [\delta(0) + i\sigma V(\mathbf{K})]\, e^{-i\chi(\mathbf{K})} = [\delta(0) + i\sigma V(\mathbf{K})]\,[\cos\chi(\mathbf{K}) - i\sin\chi(\mathbf{K})]\,, \tag{5}$$

where the aberration term for an uncorrected microscope is dominated by just the terms due to (third-order) spherical aberration, with coefficient C_S and defocus Δf,

$$\chi(K) = \pi\left(\Delta f \lambda K^2 + \frac{1}{2} C_S \lambda^3 K^4\right). \tag{6}$$

To create a phase contrast image we must rotate the phase of the scattered wave another 90° to interfere with the unscattered wave and produce amplitude changes. This can be done by using a negative defocus (reduced lens strength) to give a negative value for χ over a range of K until the positive spherical aberration term dominates at large K. In the ideal case, a passband is created over which $\cos\chi \approx 0$ and $\sin\chi \approx -1$, which produces an amplitude

$$\varphi_i(\mathbf{K}) = \delta(0) - \sigma V(\mathbf{K})\,. \tag{7}$$

In the TEM case the sample is before the objective lens while in the STEM case it is after the lens, but in both cases the combined phase changes are given by Eq. 7. Fourier transforming gives the image amplitude as

$$\varphi_i(\mathbf{R}) = 1 - \sigma V(\mathbf{R})\,, \tag{8}$$

which when squared, for small phase changes, gives a bright field image intensity

$$I_{BF}(\mathbf{R}) = 1 - 2\sigma V(\mathbf{R})\,. \tag{9}$$

The phase changes produced by the specimen have been efficiently converted into intensity variations in the bright field image. For a more detailed discussion of bright field imaging, see Chapter 2. The simplest description of the dark field image is to assume that just the transmitted beam is excluded, in other words that the total scattering is collected to form the image. In this case the image amplitude becomes

$$\varphi_{DF}(\mathbf{R}) = -\sigma V(\mathbf{R}) \tag{10}$$

producing an image intensity

$$I_{DF}(\mathbf{R}) = [\sigma V(\mathbf{R})]^2\,. \tag{11}$$

Cowley points out that a problem arises with a phase grating comprising a sinusoidal potential because squaring the potential introduces double periodicities into the image; however, he also says that for well-isolated heavy atoms in a support film this effect will not be a problem.

If instead of making the approximations for a weak-phase object, Eq. 3, and ideal lens transfer we use the full expressions we can easily show that the bright and dark field images correspond to coherent and incoherent images, respectively. The amplitude transmitted by the lens becomes the Fourier transform of Eq. 2 multiplied by the phase factor due to aberrations.

$$\varphi_i(\mathbf{K}) = e^{i\sigma V(\mathbf{K})} e^{-i\chi(\mathbf{K})}. \tag{12}$$

In the Fourier transform into the image the multiplication becomes a convolution and we have

$$\varphi_i(\mathbf{R}) = \varphi(\mathbf{R}) \otimes \mathrm{FT}[e^{-i\chi(\mathbf{K})}]. \tag{13}$$

In the STEM case, the Fourier transform of the phase changes due to aberrations, $\mathrm{FT}[e^{-i\chi(\mathbf{K})}]$, is the probe amplitude distribution $P(\mathbf{R})$ (in the TEM case it is referred to as the impulse response function),

$$\mathrm{FT}[e^{-i\chi(\mathbf{K})}] = P(\mathbf{R}) = \int e^{2\pi i\mathbf{K}\cdot\mathbf{R}} e^{-i\chi(\mathbf{K})} d\mathbf{K}. \tag{14}$$

Thus squaring Eq. 13 we find the bright field image intensity is the square of a convolution

$$I_{\mathrm{BF}}(\mathbf{R}) = |\varphi(\mathbf{R}) \otimes P(\mathbf{R})|^2, \tag{15}$$

which is the reason that bright field phase contrast images can show positive or negative contrast depending on the phase of the transfer function.

In 1974 two papers appeared that described the imaging of a phase object by a STEM annular detector in detail (Engel et al. 1974, Misell et al. 1974). Both papers showed explicitly that to convert pure *phase* variations in the object to *intensity* variations in the ADF image it is necessary to assume that the annular detector collects all of the scattering. If instead of a plane wave incident on the phase object we have the STEM probe, then the wave function emerging from the phase object is just

$$\psi(\mathbf{R}, \mathbf{R}_0) = \varphi(\mathbf{R}) P(\mathbf{R} - \mathbf{R}_0), \tag{16}$$

where $\mathbf{R}_0$ is a scan coordinate that locates the center of the probe. In the plane of the annular detector we find the Fourier transform of the exit wave

$$\psi(\mathbf{K}_f) = \int e^{-2\pi i\mathbf{K}_f\cdot\mathbf{R}} \varphi(\mathbf{R}) P(\mathbf{R} - \mathbf{R}_0) d\mathbf{R}. \tag{17}$$

Now comes the crucial assumption that the annular detector covers a sufficiently wide angular range to detect *all* of the scattering. In this case we can take the intensity and integrate over all scattering angles $\mathbf{K}_f$

$$I(\mathbf{R}_0) = \int \left| \int e^{-2\pi i\mathbf{K}_f\cdot\mathbf{R}} \varphi(\mathbf{R}) P(\mathbf{R} - \mathbf{R}_0) d\mathbf{R} \right|^2 d\mathbf{K}_f, \tag{18}$$

$$= \iiint \varphi(\mathbf{R}) \varphi^*(\mathbf{R}') P(\mathbf{R} - \mathbf{R}_0) P^*(\mathbf{R}' - \mathbf{R}_0) e^{-2\pi i\mathbf{K}_f\cdot(\mathbf{R}-\mathbf{R}')} d\mathbf{R} d\mathbf{R}' d\mathbf{K}_f. \tag{19}$$

Recognizing that

$$\int e^{-2\pi i\mathbf{K}_f\cdot(\mathbf{R}-\mathbf{R}')} d\mathbf{K}_f = \delta(\mathbf{R} - \mathbf{R}') \tag{20}$$

is just a delta function we can then integrate over $\mathbf{R}'$ to obtain

$$I(\mathbf{R}_0) = \int |\varphi(\mathbf{R})|^2 |P(\mathbf{R} - \mathbf{R}_0)|^2 d\mathbf{R}, \tag{21}$$

which represents a convolution of intensities instead of amplitudes. We thus obtain the fundamental equation for incoherent imaging

$$I(\mathbf{R}_0) = |\varphi(\mathbf{R}_0)|^2 \otimes |P(\mathbf{R}_0)|^2 . \tag{22}$$

A convolution of intensities cannot show a contrast reversal with focus, and the characteristics of the image are like that of a camera. The paper by Engel goes on to compare the annular detector of the STEM with five modes of transmission electron microscopy. He did not make the assumptions leading to Eq. (22) but used the full expression, Eq. (18), limiting the integration to cover the annular detector. Nevertheless, the resulting image calculation showed all the characteristics of an incoherent image. He chose an object of relevance to the biological goals of DNA sequencing (Beer and Moudrianakis 1962, Moudrianakis and Beer 1965, Crewe 1971, Koller et al. 1971, Cole et al. 1977, Crewe 2009), a row of single heavy atoms on a thin carbon support, but chose their spacings cleverly to show the different characteristics of the coherent and incoherent image modes. The two closest atoms were 2.7 Å apart, higher than the resolution limit for incoherent imaging but below that for coherent imaging, and indeed, only in the ADF image simulation are the two closest atoms resolved (see Figure 1–8). Furthermore, the focal series shows the simple dependence on focus expected for an incoherent image, with no contrast reversals. His is the first paper to compare the nature of the contrast obtained in coherent and incoherent modes, and the practical differences between them. He states "An important experimental advantage of the STEM is the lack of interference artifacts, which eliminates the misinterpretation of micrographs and the problem of achieving optimum conditions." This paper predicts almost all the advantages of STEM that we see today. Engel also gives a simple physical interpretation for the incoherence, explicitly stating that the STEM ADF image integrates over the diffraction pattern and therefore destroys the phase information in the object.

Misell et al. (1974) also present a detailed theoretical comparison of the coherent and incoherent modes of operation using a weak-phase object, concluding that since the ADF detector geometry of the time included most of the elastic scattering "the STEM permits imaging of phase objects in an incoherent mode with all its favourable properties." These papers stirred up considerable controversy. In 1976 Burge and Dainty (1976) examined the issue by propagating the mutual intensity to the detector plane and integrating over the detector. Using an object consisting of two sinusoidal phase contributions he found sum and difference frequencies, concluding, like Cowley, that the resulting image intensity was strictly nonlinear. However, he pointed out that for widely separated *point* objects there is no difference between coherent and incoherent imaging. That atoms are indeed point-like sources will become important later in connection with the high-angle annular detector. The paper by Burge also presented the first mathematical consideration of the effects of partial coherence on the image, showing

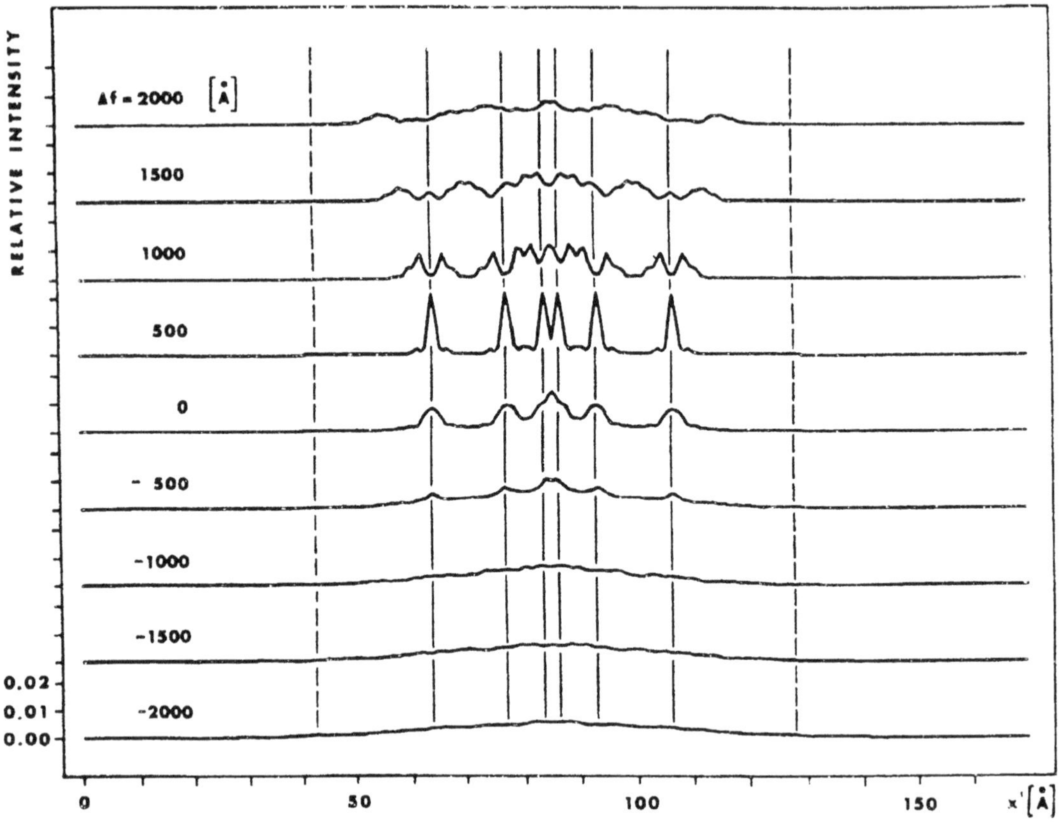

Figure 1–8. Calculated ADF intensity traces for a row of Os atoms as a function of defocus, showing resolution of the closest pairs and the lack of contrast reversals and interference artifacts, reproduced from Engel et al. (1974) with permission.

that the effects of a finite source size were exactly accounted for by convolving the demagnified intensity distribution of the source with the intensity distribution of the perfectly coherent probe (i.e., that obtained assuming no source size contribution, just the geometric aberrations). A thorough discussion of partial coherence in electron optics was given by Hawkes (1978).

One issue that is not captured in a phase object approximation is what happens if instead of a single atom there are multiple atoms lying under the beam, for example, a column of atoms. Such questions are primarily the concern of those who study crystals, and in 1976 Cowley showed that the assumption of independent intensity contributions from each atom does not apply in general with the ADF detector. He pointed out that even if all the scattering were detected, two atoms lying directly over each other would produce twice the phase change of one atom and so produce an intensity variation four times greater. He again pointed out that some of the elastic scattering passes through the hole in the ADF detector and showed quantitative deviations of the image intensity from the incoherent result for atoms spaced closer than the resolution limit. This problem becomes more severe as the objective aperture is increased, i.e., at high resolution, and the following year

Ade (1977) predicted that the problem might become very significant in aberration-corrected STEM.

In 1977 Fertig and Rose examined the mutual intensity for two atoms in various TEM and STEM modes of imaging, but without making the phase object approximation, that is, they could explicitly examine the mutual coherence for atoms displaced not only in the lateral plane but also in the z (beam)-direction. They found that for the STEM annular detector (or hollow cone imaging in TEM) the degree of coherence fell more slowly for atoms separated in the z-direction than in the transverse direction. Hollow cone imaging was also being applied to amorphous materials by Gibson and Howie (1979) in an attempt to suppress the statistical speckle in bright field phase contrast images caused by interference effects. They came to a similar conclusion, that interference effects could be suppressed quite well in the lateral direction but much less effectively along the beam direction. The same year Craven and Colliex (1977) showed the first lattice images of graphitized carbon using the plasmon loss signal as shown in Figure 1–9. The contrast is preserved, though reduced, and they showed that the reduction was consistent with opening up the collector aperture by the characteristic angle of plasmon scattering, which effectively reduced the coherence of the image and hence the fringe visibility.

In 1978 Spence and Cowley showed that lattice contrast in STEM arises from the interference between overlapping convergent beam discs as the probe is scanned. They showed that this is true not only for the coherent bright field phase contrast image but also for the ADF image, since lattice contrast only arises from regions of overlap between diffraction discs. This result gave a simple reciprocal space reason for the improved resolution of the dark field image. For imaging a spacing a with a small axial detector an aperture semiangle of at least λ/a is necessary, whereas applying the Rayleigh criterion would predict that an aperture size of only $0.61\lambda/a$ should be necessary. This smaller size

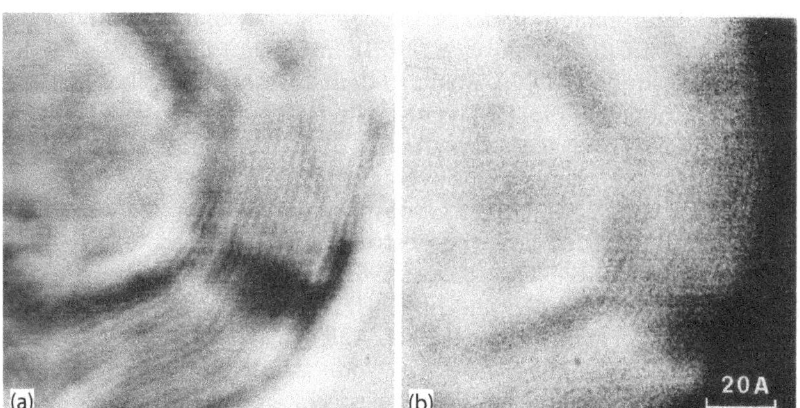

Figure 1–9. Phase contrast lattice images of graphitized carbon (**a**) bright field zero loss and (**b**) plasmon loss showing preservation of contrast, reproduced from Craven and Colliex (1977) with permission.

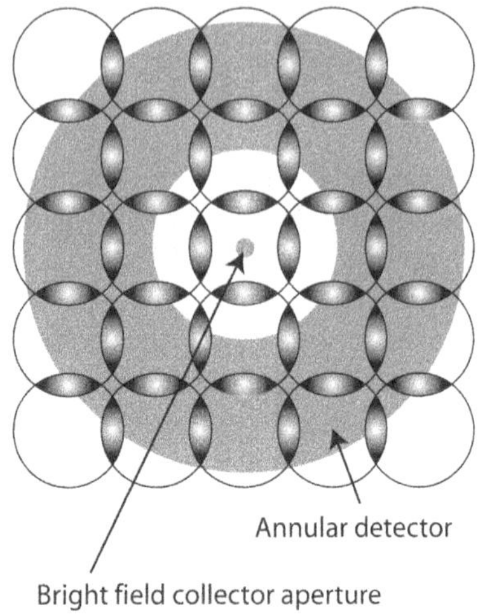

Figure 1–10. Schematic showing overlapping convergent beam discs for a case where the aperture radius is less than the diffraction angle λ/a. No overlaps occur for an axial detector, so no bright field lattice fringes are formed, but overlaps do lie on the annular detector, which makes possible atomic resolution ADF images.

aperture only produces overlapping regions of the diffraction discs on the annular detector, not on the axis, see Figure 1–10. Hence lattice fringes should be seen on the ADF image but not on the bright field image in this case. Of course, fringes would also be seen on an annular bright field image.

It is interesting to note the philosophical difference between those (primarily in biology such as Engel and Misell) who refer to the incoherent image as the ideal linear image of the specimen, as indeed do those in other fields of optics, and those familiar with phase contrast TEM imaging of crystals (as Cowley) who refer to the coherent image as the ideal linear representation of the specimen potential. Such different perspectives persist today. In any case, the STEM did find wide application to biological specimens, where, lacking crystallinity, diffraction contrast was not an issue, and image intensities could be directly interpreted as mass thickness (Engel 1978). A comparison of STEM and phase contrast TEM imaging of biological macromolecules showed the improved interpretability and collection efficiency of the STEM images (Engel et al. 1976, Ohtsuki et al. 1979). Successful imaging of single- and double-stranded DNA was also demonstrated using heavy atom staining (Mory et al. 1981). Mass measurement evolved into mass mapping of protein complexes (Engel et al. 1982, Mastrangelo et al. 1985, Sosinsky et al. 1992) and image averaging came into use, see, for example, Ottensmeyer et al. (1979), Ottensmeyer (1982), and Crewe (1983). Very soon three-dimensional reconstruction was applied (Crewe et al. 1984, Hough et al. 1987, Kapp et al. 1987). Such studies continue today (Müller et al. 1992, Müller and Engel 2001, Xiao et al. 2003, Yuan et al. 2005, de Jonge et al. 2007, Engel 2009, Wall et al. 2009).

Materials applications, however, indeed encountered difficulties because of the crystalline nature of the specimen. The presence of sharp diffracted beams destroys the smooth angular distributions of elastic and inelastic scattering that is characteristic of independently scattering atoms on which the Crewe ratio method relies. Diffraction contrast is *not* removed by taking the ratio of the ADF and energy-filtered images, making interpretation in terms of a simple Z-contrast method difficult or impossible. Donald and Craven (1979) were not able to locate Bi in Cu grain boundaries using the ratio image as residual diffraction contrast was always present making image interpretation ambiguous. The ratio technique was more successful in the case of catalyst particles (Treacy et al. 1978) when even with a crystalline support, small particles comparable to the probe size could be imaged that were not visible in the bright field image. Nevertheless, diffraction from the substrate was still a source of contrast in the images. It also was realized that the thickness independence of the ratio image only applied if the specimen was well below the mean free path for elastic scattering. As specimen thickness increases the inelastic signal into the spectrometer increases initially, but then becomes depleted by the increasing elastic scattering and the collected intensity decreases (Treacy et al. 1978, Egerton 1982, Colliex et al. 1984, Reichelt and Engel 1984).

From the materials science perspective, therefore, by the mid-1980s the interest in STEM was primarily for the purposes of microanalysis, not atomic resolution imaging, see, for example, Brown (1981). High spatial resolution analysis became widely available following the commercial introduction of the field emission gun dedicated STEM by VG Microscopes (for a review of the development, see Wardell and Bovey, 2009). Numerous applications appeared using either the characteristic X-ray (Williams and Edington 1976, Vandersande and Hall 1979, Hall et al. 1981, Michael and Williams 1987, Williams and Romig 1989) or the core loss EELS for elemental identification (Colliex and Trebbia 1982, Colliex et al. 1976, Egerton 1976, Isaacson and Johnson 1975, Jeanguillaume et al. 1978). In addition, the small probe of the STEM allowed diffraction studies from nanometer-sized volumes of materials (Cowley and Spence 1978, 1981, Howie et al. 1982, Cowley 1985). Spence and Lynch (1982) showed how a STEM probe could be located at specific crystallographic sites in the unit cell using the microdiffraction pattern, then core loss EELS could locate specific atomic species at those coordinates. Ba M edges were seen with the probe located over Ba-containing planes, but they were absent when the beam was in-between. The spatial resolution was 5.7 Å. Interestingly, the low-energy Ba N edge did not disappear. Microanalysis has remained of major interest, growing especially as capabilities for atomic resolution were improved.

The issue of preservation of diffraction contrast under inelastic scattering continued to be important for quantitative analysis (Rossouw and Whelan 1981), and remains so today, especially if elemental ratios are being taken using edges with very different energy losses. In such cases diffraction contrast will be preserved to different extents (Bakenfelder et al. 1990, Hofer et al. 1997, Moore et al. 1999), complicating quantitative analysis.

Diffraction contrast was also put to beneficial use at this time to analyze the lattice location of impurities in crystals. The variation of characteristic X-ray (Taftø 1982, Taftø and Spence 1982a, b) or EELS (Krivanek et al. 1982, Taftø and Krivanek 1982) signals was compared as a function of crystal orientation. The method was soon extended from planar channeling conditions to axial channeling (Pennycook and Narayan 1985, Rossouw and Maslen 1987), and the effects of different delocalization of the excitations needed to be quantitatively accounted for (Pennycook 1988, Rossouw et al. 1988a, b, 1989, Spence et al. 1988). The method was later extended to a full statistical analysis of the two-dimensional orientation effect (Rossouw et al. 1996a, b) together with improved account of the effects of delocalization (Allen et al. 1994, 2006, Oxley and Allen 1998, Oxley et al. 1999, Rossouw et al. 2003).

1.5 The High-Angle Annular Dark Field (HAADF) Signal

With the major interest and success of Z-contrast methods being in the biological area the key requirement for the ADF detector was collection efficiency and so the only detector geometry considered was one with an inner angle as close as possible to the objective aperture angle. The first proposal for a high-angle ADF detector was made by Humphreys et al. (1973). They pointed out that much higher contrast from single atoms would result if the inner angle of the ADF detector were increased to a high angle. The Z-dependence of the scattering cross-section would increase to the Z^2-dependence of unscreened Rutherford scattering from the nucleus, although they point out that the reduced signal level would most likely result in a resolution limited by signal to noise ratio rather than beam diameter. Crewe et al. (1975) proposed the use of concentric annular detectors noting that the "outer detector will record electrons scattered from close to the nucleus and thus show an intensity dependence of Z^2 instead of $Z^{3/2}$." The same idea surfaced again in connection with the imaging of catalyst particles on light crystalline supports (Treacy et al. 1978), the motivation being to suppress diffraction contrast effects and improve the visibility of the high-Z particles. They state "a possible method of improving particle contrast in the case of a crystalline substrate would be the use of a detector for collecting the high-angle Rutherford scattered electrons (θ >100 mrad). This signal would be expected to be relatively insensitive to the crystalline nature of the specimen but more sensitive to the atomic number of the scattering elements." Howie (1979) pointed out that at such high scattering angles the Debye–Waller factor would replace the coherent Bragg scattering with diffuse scattering. He states "Thus, unless the annular detector accepts only fairly large scattering angles (θ > 40 mrad), it will collect the coherent (Bragg) scattering as well as the diffuse signal."

The first experimental investigations of HAADF imaging were made by Treacy in the Cavendish Laboratory. He was following up on his idea to collect only high-angle scattering by constructing a solid-state detector that would fit inside the cartridge on the VG Microscopes HB5 STEM. Unfortunately it was damaged on drilling the central hole, and he instead decided to use a scintillator detector in conjunction with a recently constructed cathodoluminescence detector (Pennycook and

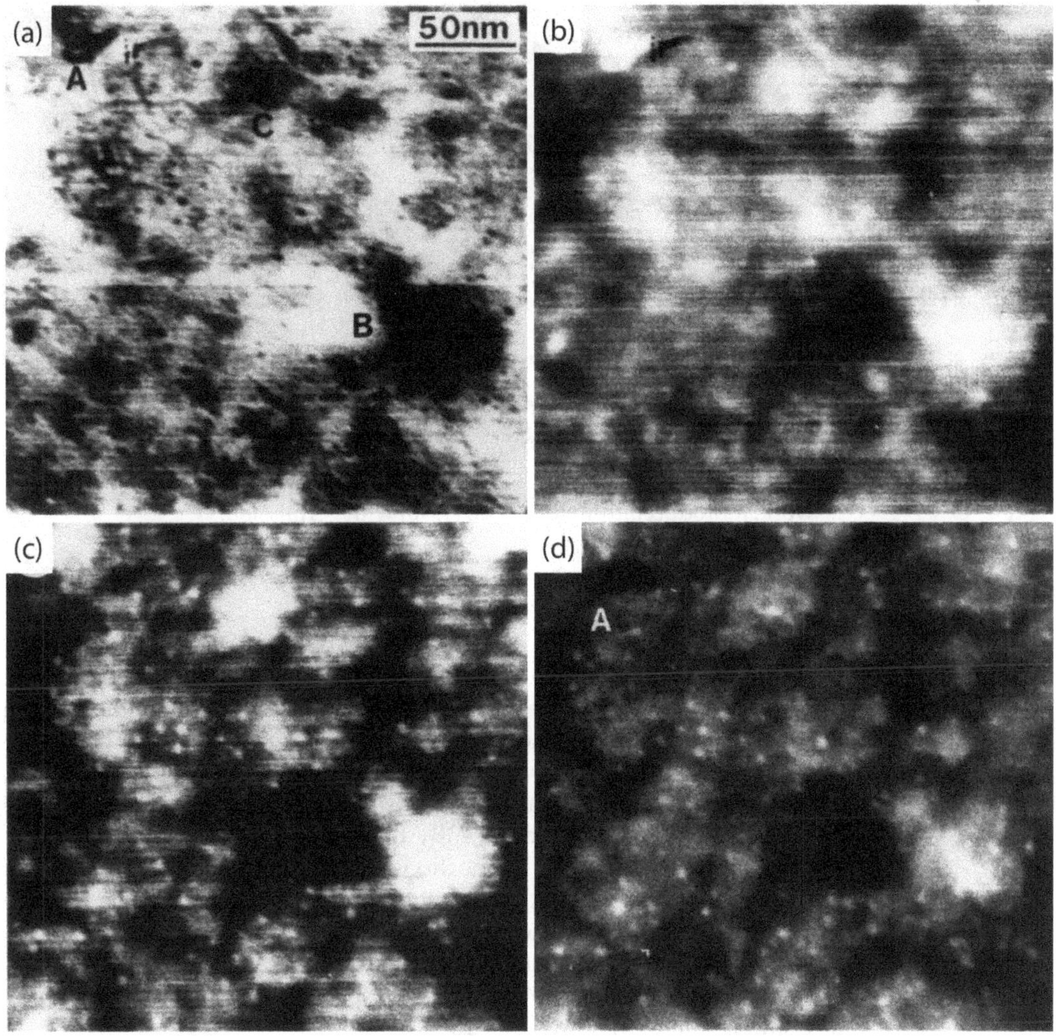

Figure 1–11. Images of a Pt particles on γ-alumina recorded in (**a**) bright field, (**b**) low-angle ADF, (**c**) HAADF, and (**d**) the ratio of high-angle to low-angle ADF signals. Particle contrast is highest in the HAADF image, reproduced from M. M. J. Treacy, PhD thesis, University of Cambridge, 1979, with permission.

Howie 1980, Pennycook et al. 1980). The detector comprised a simple silver tube looking at the specimen and was easily modified to include a scintillator. The first images were reported in Treacy et al. (1980) and a similar set is shown in Figure 1–11. Although the images show a lot of tip instabilities, it is clear that the high-angle detector gives much more contrast than the conventional annular detector. Forming the ratio image is very effective at removing the tip fluctuations and improves the contrast of some particles; however, it also reintroduces some diffraction contrast.

The improved visibility of the small particles was useful in locating them for study by other techniques such as EELS (Pennycook 1981) or

Figure 1–12. HAADF image showing individual Pt atoms in a zeolite framework, reproduced from Rice et al. (1990) with permission.

secondary electrons (Liu and Cowley 1990, 1991). The HAADF method could even image individual Pt atoms in a beam-sensitive zeolite (Rice et al. 1990), although the zeolite framework was damaged and precise atomic locations could not be determined, see Figure 1–12. The framework was more reliably imaged in bright field TEM, but then the Pt atoms could not be seen. It was also realized that if the high-angle signal was generated incoherently then the *integrated* intensity from a small particle should be proportional to the number of atoms and not dependent on the imaging parameters such as resolution. The method was introduced by Treacy and Rice (1989) and extended by Singhal et al. (1997) who measured the number of atoms in a small cluster to ± 2 atoms. Such STEM-based mass spectroscopy techniques remain popular today (Menard et al. 2006).

Quantitative analysis of HAADF images was also used to extract dopant concentrations in ion-implanted Si, as shown in Figure 1–13 (Pennycook and Narayan 1984). The suppression of diffraction contrast in the HAADF image is striking, and the dopant profile agrees with X-ray and Rutherford backscattering spectrometry and also shows better depth resolution than the other techniques. However, as with spectroscopic imaging, diffraction contrast must be avoided for quantitative results and the technique is only sensitive to relatively high concentrations of dopant (Pennycook et al. 1986). More recently, the technique has been applied to delta-doped layers in semiconductors (Vanfleet et al. 2001) and to shallow junctions formed by low-energy implantation (Parisini et al. 2008), where it gives better spatial resolution than available with secondary ion mass spectrometry. It has also been used to image semiconductor quantum wells (Lakner et al. 1991, Otten 1991, Lakner et al. 1996, Liu et al. 1999, Mkhoyan et al. 2003, Mkhoyan et al. 2004) and multilayers (Liu et al. 1992).

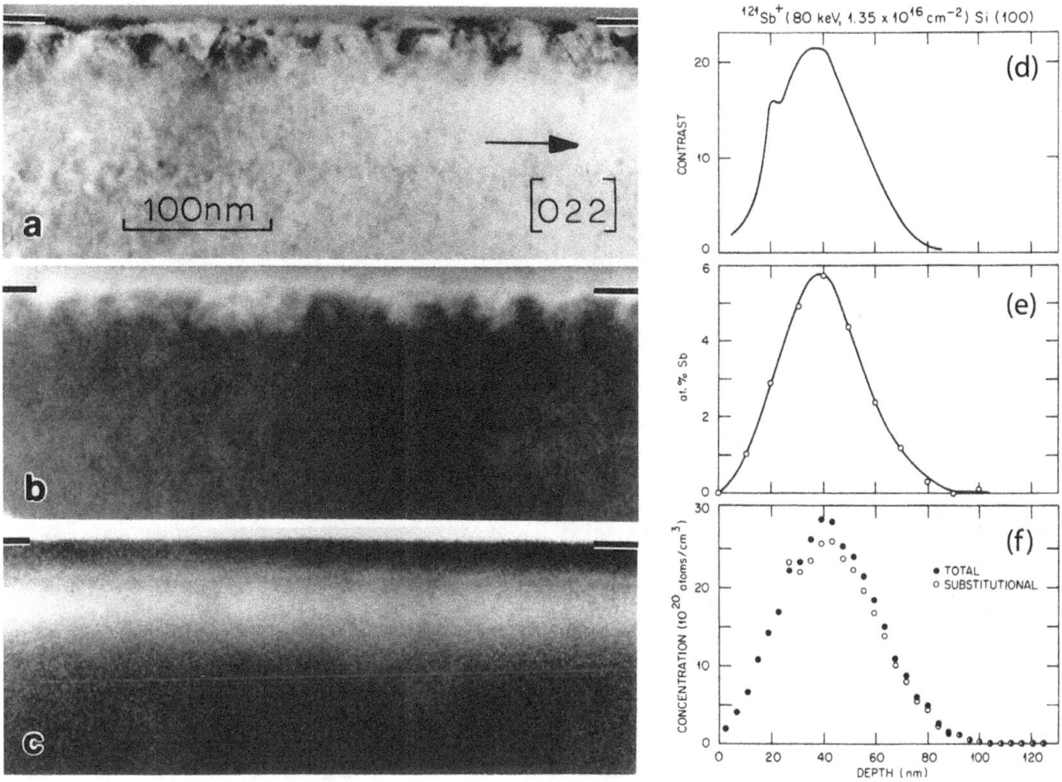

Figure 1–13. Cross-section images of Sb-implanted Si. (**a**) TEM diffraction contrast image showing defects near the surface and end of range damage, (**b**) low-angle ADF image also dominated by diffraction contrast, (**c**) HAADF image revealing the Sb profile (**d**), in agreement with (**e**) X-ray microanalysis and (**f**) Rutherford backscattering spectroscopy, reproduced from Pennycook and Narayan (1984).

1.6 Atomic Resolution Incoherent Imaging of Crystals

In 1984 Cowley published a high-resolution bright field and wide-angle ADF image of $TiNb_{10}O_{29}$, noting the improved resolution in the ADF image compared to the bright field image, see Figure 1–14. He states "atom rows, 3.8 Å apart, are clearly resolved as white spots and the representation of the structure is good," and "It seems clear that when, in the near future, the same ultra-high resolution pole pieces are used in STEM as in TEM instruments, important advances in the imaging of crystals will be possible." However, apart from improved resolution, he made no mention of any other incoherent characteristic, any lack of contrast variation with defocus or specimen thickness, and apparently turned his attention to microdiffraction (Cowley 1986a, Lin and Cowley 1986a) and holography (Lin and Cowley 1986b). A 1987 review article shows the same pair of images and he states "The improvement in resolution over the BF case is roughly the same as given by the approximate theoretical treatment applicable to thin weakly scattering objects (Cowley and Au 1978), but inapplicable here." The prevailing feeling at

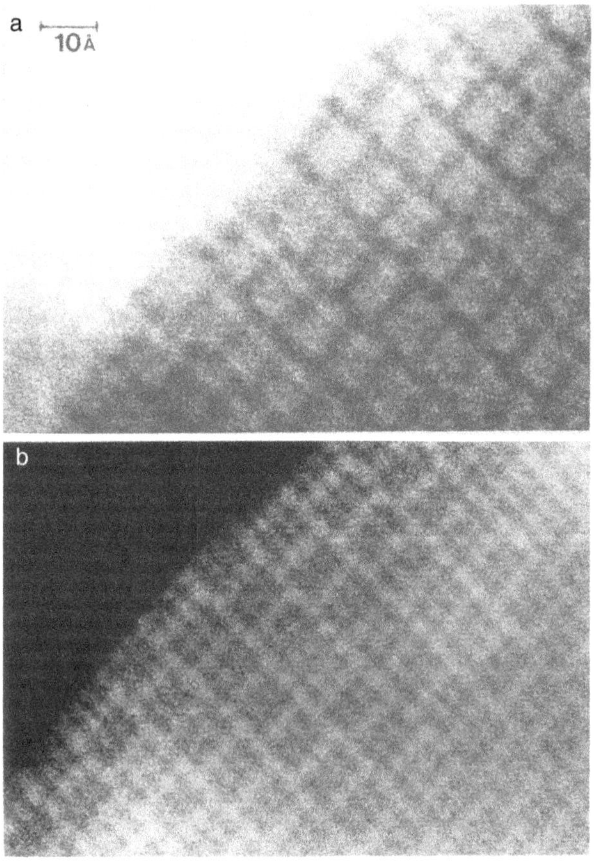

Figure 1–14. (**a**) Bright field and (**b**) ADF images of $Ti_2Nb_{10}O_{29}$ showing improved atomic resolution detail in the dark field image, reproduced from Cowley (1986b) with permission.

the time was that incoherent imaging simply did not apply to the thick crystalline samples typical in materials science.

However, Cowley was primarily interested in coherent diffraction phenomena, and the experience with the HAADF signal suggested a different route toward incoherent imaging. The dopant profiling results shown in Figure 1–13 were obtained using Rutherford scattered electrons, and it was well appreciated that Rutherford scattering is generated close to the atomic nucleus and therefore each atom would generate the scattering independently (Rossouw 1985, Rossouw and Bursill 1985) and in proportion to the intensity close to the nucleus. The situation was similar to that of the generation of secondary excitations such as X-rays (Cherns et al. 1973) or cathodoluminescence (Pennycook and Howie 1980). In other words the image should be an incoherent image, with all the concomitant advantages of freedom from focus variations with thickness or defocus and capable of showing a resolution higher than achievable by phase contrast imaging, just as in the classic light optical case (Rayleigh 1896). In Pennycook et al. (1986) it is stated that "In thin samples of crystalline materials, atomic resolution Z-contrast imaging seems entirely feasible and entirely complementary to conventional high resolution structure imaging."

The first multislice simulations of an ADF image were published the following year (Kirkland et al. 1987, Loane et al. 1988) and predicted single Pt and Au atoms would be visible on a thin crystal of Si using an annular detector that excluded the first-order diffracted beams (30–79 mrad at 100 kV). The simulations included only coherent scattering and therefore the magnitude of the contrast was oscillatory with crystal thickness. The simulations showed contrast not only from the Au atom but for a high-resolution pole piece the Si lattice itself was visible, and the sign of the contrast did not reverse on increasing thickness. Hence even images formed from coherent scattering should show some incoherent characteristics, in agreement with the simple analysis of Eq. (22).

A VG Microscopes HB501UX high-resolution STEM was installed at Oak Ridge National Laboratory in 1988, and the first experimental results were obtained using the high-temperature superconductors $YBa_2Cu_3O_{7-\delta}$ and $ErBa_2Cu_3O_{7-\delta}$ (Pennycook and Boatner 1988, Pennycook 1989b). The specimens were single crystals and observed in a planar geometry as shown in Figure 1–15. The strong Z-contrast is obvious and the images showed no contrast reversals with thickness or defocus. A simple calculation of the expected image intensity was made by convolving a Gaussian probe with the respective high-angle

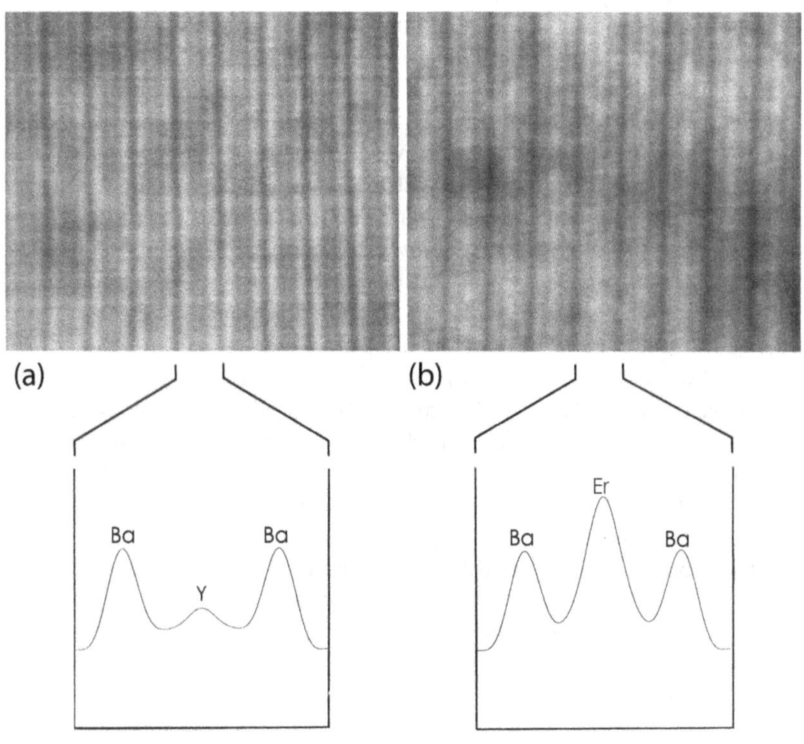

Figure 1–15. HAADF images in a planar channeling condition from (**a**) $YBa_2Cu_3O_{7-\delta}$ and (**b**) $ErBa_2Cu_3O_{7-\delta}$, with calculated intensity profiles across each unit cell. The expected strong Z-contrast closely matches the experimental result, reproduced from Pennycook (1989b).

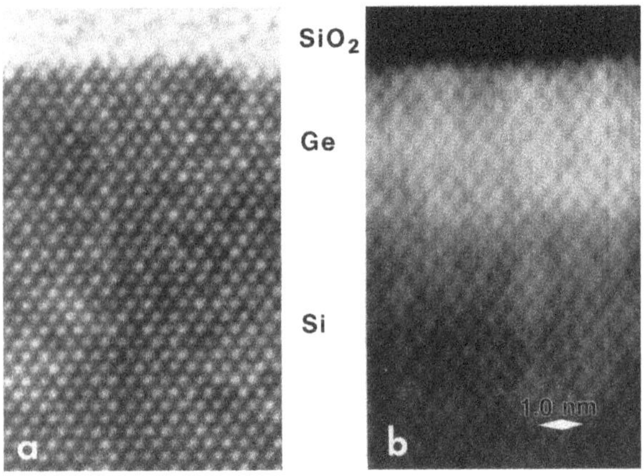

Figure 1–16. Images of a Ge film grown epitaxially on Si by an implantation and oxidation method. (**a**) Conventional TEM image from a JEOL 200CX, (**b**) Z-contrast image obtained with a VG Microscopes HB501UX clearly delineating the Ge layer, reproduced from Pennycook (1989a).

scattering cross-sections, shown in the line traces in Figure 1–15. It was pointed out that this was analogous to the simple phase grating approximation for coherent imaging but should hold to greater thicknesses because of the tendency of the electrons to channel along the atomic planes (Fertig and Rose 1981). Any thickness dependence due to dynamical diffraction should be minimized because the signal is the intensity of Rutherford scattering integrated through the entire sample thickness, just as for X-ray generation.

The first Z-contrast images of semiconductors were published the following year (Pennycook 1989a, Pennycook et al. 1989). Figure 1–16 shows a comparison between TEM phase contrast and STEM HAADF images of a thin film of Ge formed on Si. The Ge layer cannot be distinguished in the phase contrast image but shows strong contrast in the Z-contrast image. Because of this strong dependence on atomic number such images were referred to as Z-contrast images even though no ratio was used as in the case of Crewe's Z-contrast images. Figure 1–17 compares STEM phase contrast and Z-contrast images of a $Si_{0.61}Ge_{0.39}$ alloy layer grown on Si in which the interfacial roughness is clear from the Z-contrast image. The theoretical optimum (Scherzer) resolution for the VG Microscopes high-resolution pole piece used in these experiments was around 2.2 Å, so that the Si dumbbell (1.36 Å separation) is not resolved. Also that year Shin et al. (1989) showed how the transfer function could be extended using an oversized objective aperture and demonstrated a resolution of 1.9 Å in $YBa_2Cu_3O_{7-\delta}$. The following year, with a higher resolution pole piece and an oversized objective aperture, simultaneous ADF and bright field lattice images were demonstrated with 1.92 Å resolution in both modes (Xu et al. 1990).

By a curious coincidence, with the normal pole piece on these microscopes, the optimum objective aperture for the Z-contrast image was around 10 mrad, which excluded the first-order diffracted beams of Si ⟨110⟩. Consequently, there was no lattice image in the bright field signal under these conditions, but instead a set of thickness fringes was observed, while the Z-contrast image did resolve the lattice, as shown in Figure 1–18 (Pennycook and Jesson 1991, Pennycook et al. 1990). This

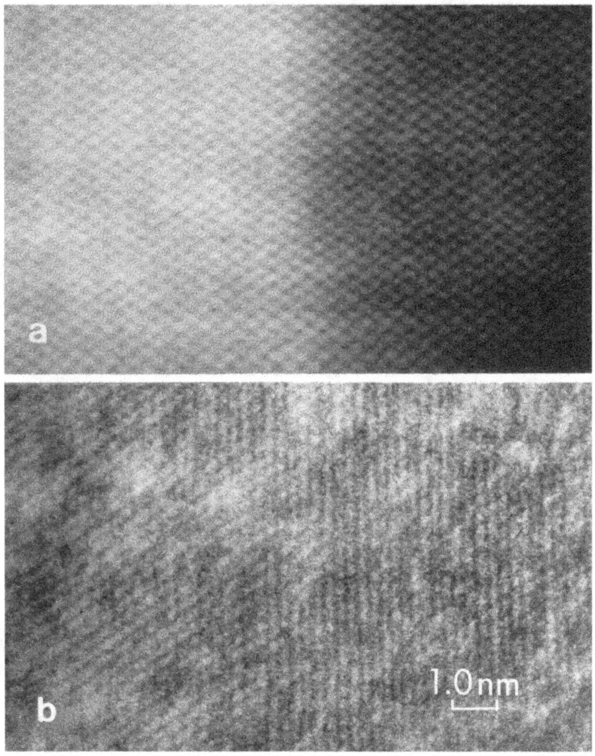

Figure 1–17. (**a**) *Z*-contrast image and (**b**) STEM phase contrast image of a $Si_{0.61}Ge_{0.39}$ alloy layer grown on Si. The interfacial roughness is clear from the *Z*-contrast image, reproduced from Pennycook (1989a).

showed strikingly how the form of the *Z*-contrast image was insensitive to thickness and showed little sign of any effects due to dynamical diffraction.

Theoretical investigation into the reasons for this thickness insensitivity was undertaken using a Bloch wave analysis (Pennycook and Jesson 1990). It was well known that the reason for the thickness dependence of phase contrast zone axis images is the beating between highly excited Bloch states traveling with different wave vectors along the direction of propagation (Kambe 1982). The key to the incoherent nature of the images is the fact that the high-angle Rutherford scattering is generated only very close to the atomic nuclei. Therefore the 1 s Bloch state which has a high intensity in this region is much more effective in generating high-angle scattering than the 2 s Bloch state or any other Bloch states that peak in between the atom columns. For this reason interference between the different Bloch states has little effect on the image intensity. It is almost as if only the 1 s state was present in the specimen, as shown in Figure 1–19(a, b). In addition, the intensity is integrated through the sample thickness, increasing monotonically with thickness, and showing only minor oscillations due to interference with other Bloch states, see Figure 1–19(c). It was also pointed out that since these states are highly localized around the atomic columns there should be minimal interference effects at interfaces and images should be interpretable by deconvolution to higher resolution than the probe size. Images of Si and InP were compared to show that incoherent characteristics applied even

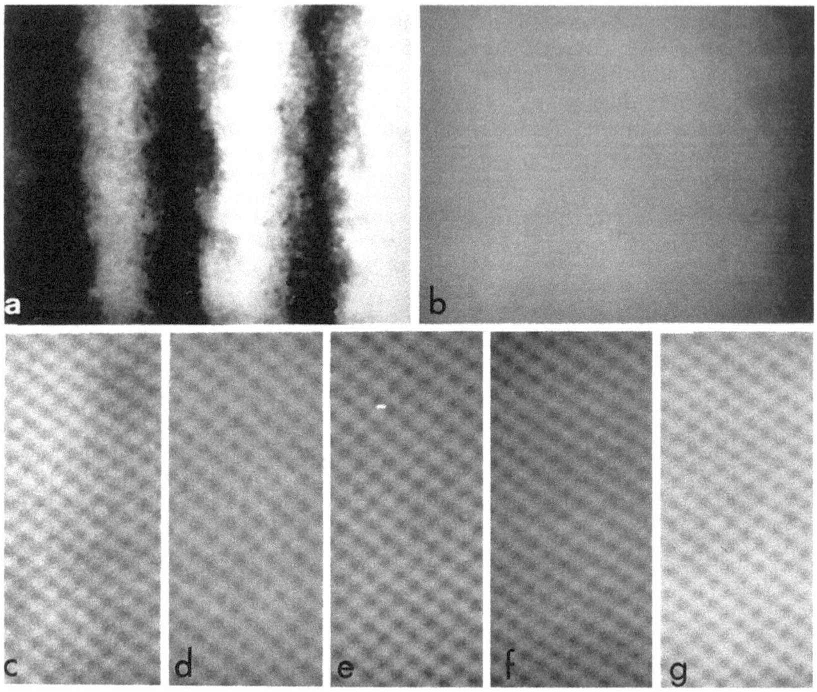

Figure 1–18. Simultaneously recorded (**a**) bright field and (**b**) HAADF images of Si in the ⟨110⟩ zone axis using an objective aperture too small to allow bright field lattice images, as shown schematically in Figure 1–10. In this case the bright field image shows thickness fringes allowing the *Z*-contrast image to be measured as a function of sample thickness (**c–g**), 120,230,350,470 and 610 Å respectively, reproduced from Pennycook et al. (1990), Pennycook and Jesson (1991).

below the resolution limit. In Si the presence of the dumbbell resulted in clearly elongated image features, whereas in InP, the light P column contributed little to the total intensity and the image shows just round features due to the In columns, see Figure 1–20.

The Bloch wave analysis also provided insight into the thickness dependence of the image, or rather the lack of it, as shown for Si in Figure 1–19(c). However, because the 1s Bloch state is peaked near the nucleus it is also the most highly absorbed Bloch state, and with heavier columns it is almost completely depleted after as little as 10 nm or so. The remainder of the crystal generates little additional intensity. With the probe located between the columns a background would be expected from thicker regions as the probe spreads onto neighboring columns, as found experimentally. Hence most of the physical characteristics of the image could be understood from these Bloch wave studies (Pennycook and Jesson 1991, 1992). Also at this time several applications appeared to grain boundaries in superconductors (Chisholm and Pennycook 1991), and insights into growth mechanisms and transport properties were obtained from images of superconductor superlattices (Norton et al. 1991, Pennycook et al. 1991, 1992). Similarly in semiconductor superlattices, new insights were obtained

Figure 1–19. (**a**) HAADF image intensity calculated for a line trace across the dumbbells in Si ⟨110⟩ using all Bloch states (*squares*) and only the 1s Bloch state (*circles*) at an accelerating voltage of (**a**) 100 kV and (**b**) 300 kV. Almost all the image contrast is accounted for by the 1s Bloch states. The *solid line* is a convolution of the thickness integrated 1s state intensity with an effective surface probe that includes the variation in 1s state excitation with angle. (**c**) Calculated thickness dependence of the image at 100 kV for a probe centered on the dumbbell (*upper*) and between the dumbbells (*lower*) using all Bloch states (*solid line*) and only the 1s states (*dashed line*). The points are experimental data. Reproduced from Pennycook and Jesson (1990).

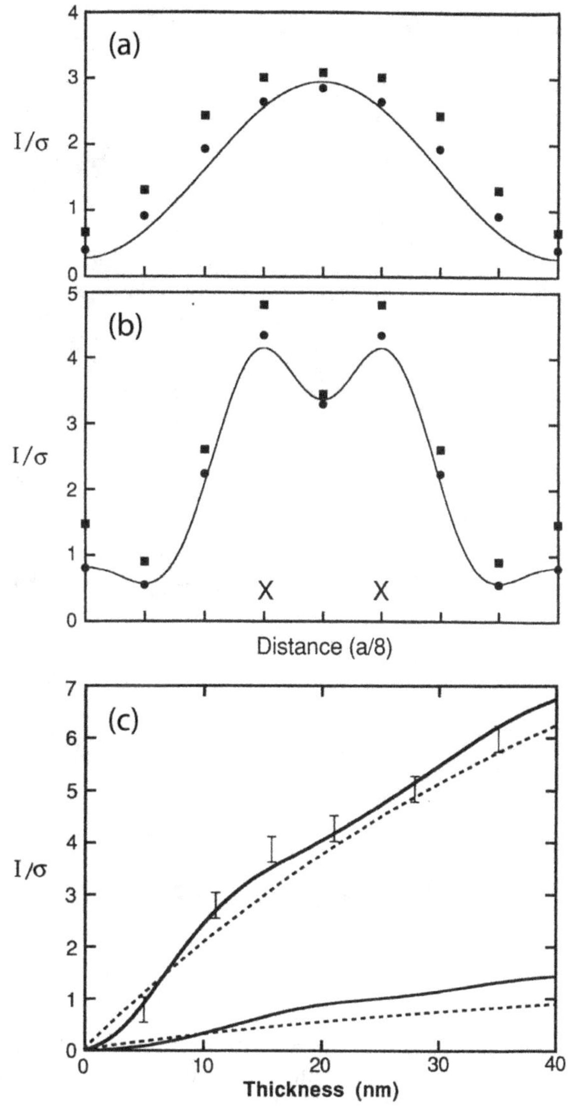

into morphological instabilities during strained layer growth (Jesson et al. 1993a, b).

In 1989 Wang and Cowley introduced a modified multislice algorithm to include the inelastic thermal diffuse scattering (Wang and Cowley 1989, 1990). Only single phonon scattering was included, which resulted in a doughnut-shaped scattering potential, but nevertheless the images appeared similar to those obtained experimentally since the probes at the time were much larger than the doughnut. The multiphonon terms that were neglected dominate close to the nucleus, when the atomic recoil is strong and a superposition of large numbers of phonons is required to describe the atomic displacement. As a result the treatment underestimated the diffuse intensity at high scattering angles (Hall 1965) and overestimated the contribution of coherent scattering

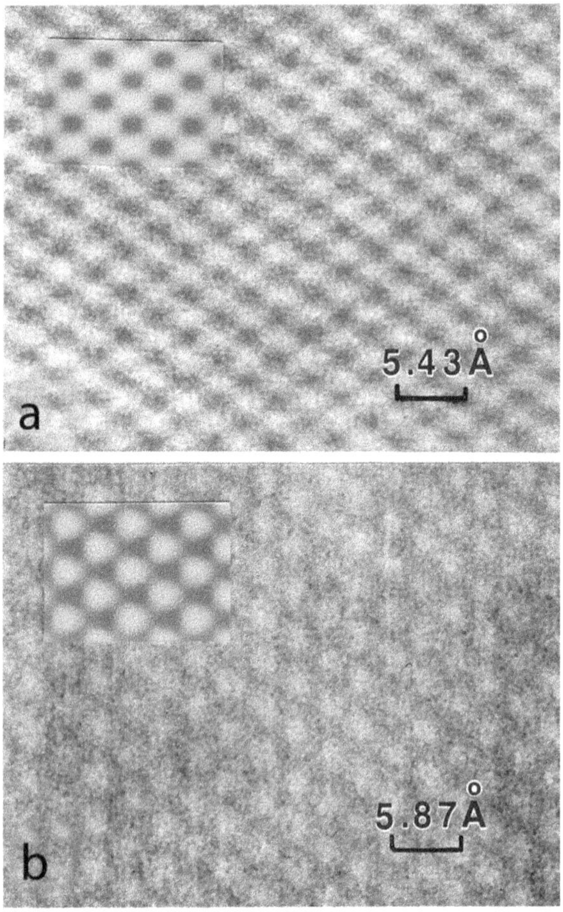

Figure 1–20. *Z*-contrast images from (**a**) Si and (**b**) InP in the ⟨110⟩ zone axis showing the elongated features characteristic of the Si dumbbells are absent from InP where the image is dominated by the heavy In column. Insets show simulated images, reproduced from Pennycook and Jesson (1990).

to the HAADF image. In 1991 the frozen phonon method was introduced into multislice image simulations and successfully reproduced most of the features in convergent beam diffraction patterns (Loane et al. 1991). The method is based on the static lattice concept used in X-ray diffraction (James 1962) and introduced into electron microscopy by Hall and Hirsch (1965). It relies on the fact that the time spent by the electron inside the specimen is much shorter than the period of thermal vibration, and so the vibrating atoms appear frozen in their instantaneous configuration. The implementation ignores the fact that phonon scattering is inelastic and just propagates the beam elastically to the detector, averaging over a number of different configurations of atomic displacements. Normally an Einstein model is used for the atomic displacements as opposed to a full phonon model, but this was later shown to have negligible effect on image simulations (Muller et al. 2001). The following year, a detailed comparison was made between frozen phonon image simulations and the incoherent imaging model, and the match was surprisingly good, mostly within 1% (Loane et al. 1992), see Figure 1–21. Good agreement with experiment was also found provided account was taken of a finite source size. The frozen phonon

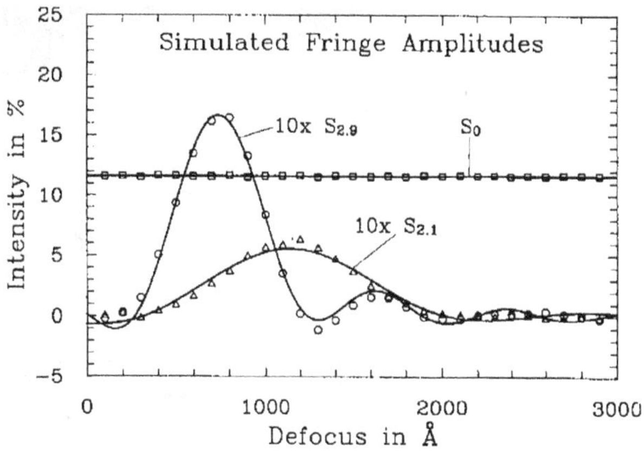

Figure 1–21. Simulated fringe amplitudes for InP ⟨110⟩ as a function of specimen thickness using a frozen phonon algorithm (points) compared to the incoherent imaging prediction (*solid lines*), reproduced from Loane et al. (1992) with permission.

simulations also reproduced the observed thickness dependence and vividly showed the rapid depletion of the channeling peak with thickness along heavy columns (Hillyard and Silcox 1993, Hillyard et al. 1993).

All these results stimulated further investigations into the physical explanation for the image contrast through a reexamination of issues of coherence in ADF images. In connection to the imaging of very thin crystals (phase objects) Jesson and Pennycook (1993) studied the so-called hole-in-the-detector problem from a quantitative viewpoint, again describing the degree of coherence in the ADF image via the mutual intensity but this time explicitly examining the effect of increasing the inner angle of the annular detector. For a pair of atoms separated in the transverse plane by a distance ΔR illuminated by a (larger) probe placed centrally between them, increasing the inner radius to

$$\theta_i = 1.22\lambda/\Delta R \qquad (23)$$

resulted in an image intensity within 5% of the incoherent result. The physical reason, as already stated by Engel, is the averaging of the fringes over the ADF detector. As shown in Figure 1–22(a), the pattern for two point scatterers is a set of Young's fringes, and, as the inner angle is increased, more and more fringes are sampled around the inner cutoff of the detector. With real atoms, the intensity falls off with increasing angle and the total intensity is dominated by the fringes around the inner cutoff. Real atoms are not points but do have a sharp enough potential that the inner angle can be increased without losing too much signal, giving better averaging and the intensity becomes closer to the incoherent result, see Figure 1–22(b). The intensity from a column of atoms was also calculated, as pertaining to the imaging of zone axis crystals. In this case increasing the inner detector angle was much less effective in suppressing coherent interference. With increasing column length atoms at different depths were found to interfere destructively, and the total intensity never increased over that scattered from a thin crystal, see Figure 1–23. This was completely at odds with the experimental results (see Figure 1–18) and highlighted

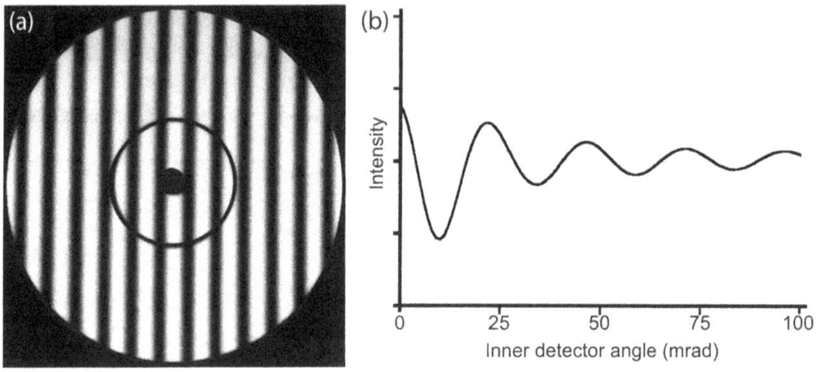

Figure 1–22. (**a**) Intensity distribution in the detector plane for two point scatterers 1.5 Å apart, with a probe centrally located between them. Inner and outer detector angles are 10.3 and 150 mrad, respectively. The circle marks 50 mrad radius and samples many fringes around its perimeter. (**b**) Ratio of the detected intensity to the incoherent scattering prediction for a pair of Si atoms as a function of inner detector angle. Probe is optimum for an uncorrected 100 kV STEM with $C_S = 1.3$ mm, adapted from Jesson and Pennycook (1993).

Figure 1–23. Intensity of coherent scattering reaching a HAADF detector as a function of crystal thickness, showing how the maximum intensity occurs with a thin crystal. Specimen is Si ⟨110⟩ with a 2.2 Å probe located centrally over a dumbbell, reproduced from Jesson and Pennycook (1993).

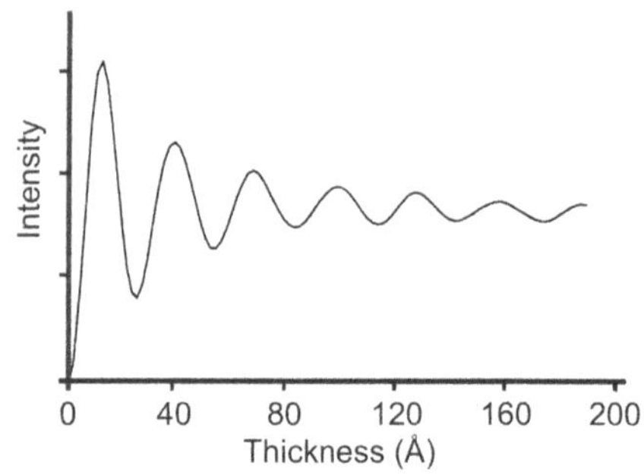

the importance of incoherently generated thermal diffuse scattering to these images.

The same year, Treacy and Gibson (1993) also examined the mutual coherence for ADF or hollow cone imaging, using the term coherence volume to describe its "cigar-like" shape, narrow in the transverse direction but elongated along the beam direction. They showed good agreement with experimental results from wedge-shaped silicon samples. An indication of the effect of thermal vibrations was obtained with an Einstein model, which approximates the degree of correlation between pairs of atoms as e^{-2M}, independent of their separation. A better description of the effect of thermal vibrations is a phonon model to capture the fact that near-neighbor atoms tend to vibrate in phase, and only atoms far apart are correlated by the Einstein value (Warren 1990). This model was applied to a column of atoms by Jesson and Pennycook (1995). Figure 1–24(a) shows the degree of coherence with atomic separation along a column. The physical picture to emerge is that each atom is coherent with a few neighbors above and below, so that a column of n atoms vibrates as a number of independent packets, see Figure 1–24(b), with a resulting scattered intensity that can be above

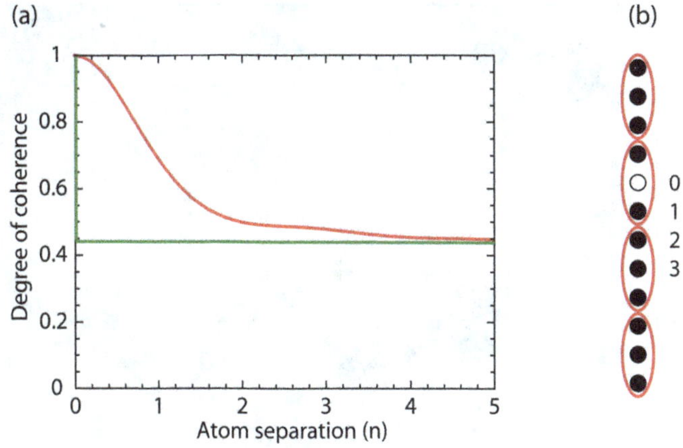

Figure 1–24. (**a**) Degree of coherence along a column of atoms on the Einstein model (*green*) and on a phonon model (*red*), showing how near-neighbor atoms are more highly correlated than in the Einstein model because they tend to vibrate in phase. (**b**) The effect is that the column of *n* atoms behaves as a number of independently vibrating packets, adapted from Jesson and Pennycook (1995).

or below that for incoherent scattering. For the HAADF signal the packets are short and coherence is only important in crystals shorter than the packet length. We thus arrive at the picture that transverse coherence is primarily destroyed by the lateral extent of the detector but *z*-coherence is only destroyed by phonons. We have made no mention of coherent HOLZ lines since their contribution to the ADF image is small (Amali and Rez 1997, Pennycook and Jesson 1991).

Also in the same year Liu and Cowley (1993) introduced a new imaging mode they called large-angle bright field imaging, formed by detecting all electrons, the transmitted cone as well as any diffracted beams, up to an angle comparable to the inner angle of the HAADF detector. By conservation of flux the image would be the complement of the HAADF image, showing the same improved resolution and incoherent characteristics. They also showed the first indications of the resolution of the Si dumbbell in a HAADF image of Si ⟨110⟩, the classic resolution test for phase contrast imaging. Their microscope was equipped with a special high-resolution pole piece ($C_S = 0.8$ mm) and an optical detection system for efficient collection of microdiffraction patterns (Cowley and Spence 1978). Using a quarter coin to block the central region of the diffraction pattern and a slight underfocus of the objective lens they produced images as shown in Figure 1–25. Not all dumbbells show dips due to instabilities but the comparison of the experimental line trace with the calculated trace is convincing.

Numerous applications were also being found during this time. In many cases the interface structures that were seen were much more complicated than previously supposed (Jesson et al. 1991, Pennycook et al. 1993, Chisholm et al. 1994a, b). Misfit dislocations were seen to stand off from the interface by a few lattice spacings and to nucleate preferentially at interface steps (Pennycook et al. 1993, Takasuka

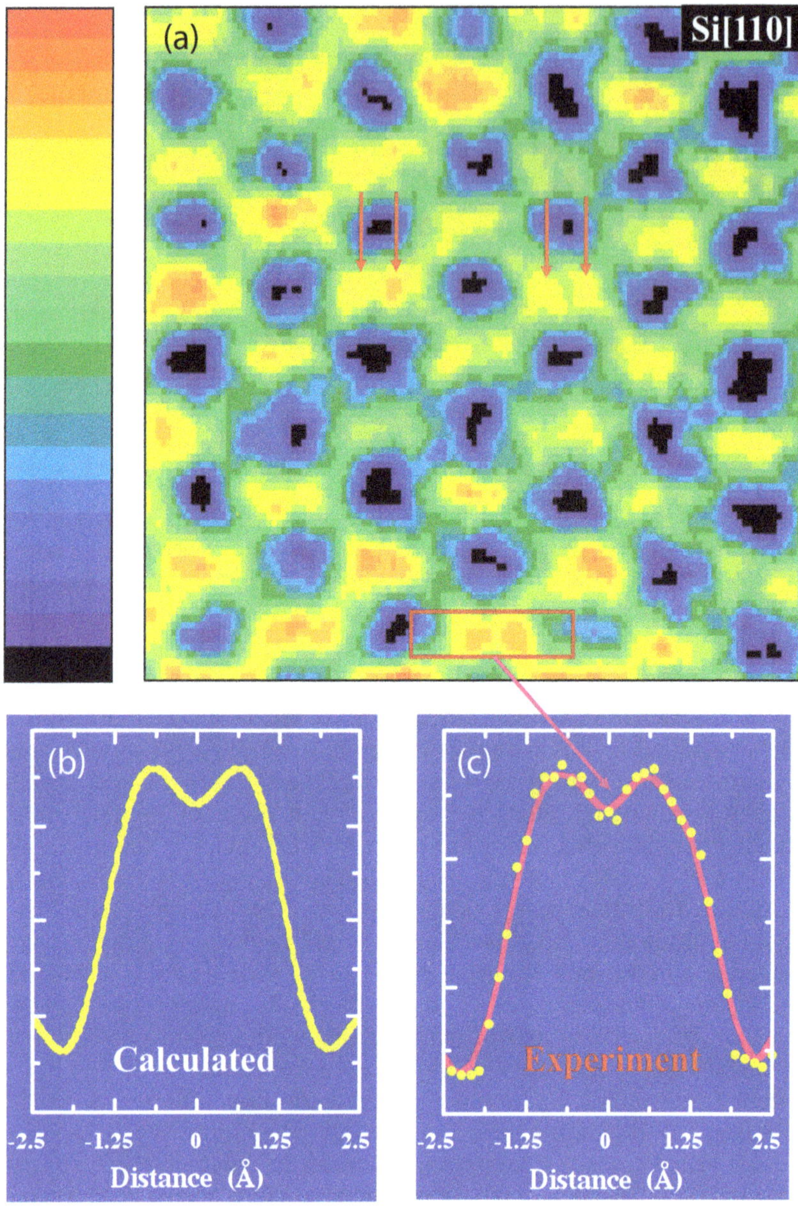

Figure 1–25. (**a**) HAADF image of Si ⟨110⟩ recorded in a VG Microscopes HB5 STEM equipped with an ultrahigh-resolution pole piece, with C_S = 0.8 mm. (**b**) Calculated intensity profile for a defocus of –825 Å. (**c**) Experimental profile across the dumbbell framed in (a), adapted from Liu and Cowley (1993) with permission.

et al. 1992). In 1994 the direct determination of grain boundary structure was demonstrated by combining *Z*-contrast imaging, electron energy loss spectroscopy (EELS), and bond valence sum calculations, see Figure 1–26 (McGibbon et al. 1994). This paper also introduced the maximum entropy method for extracting column positions with

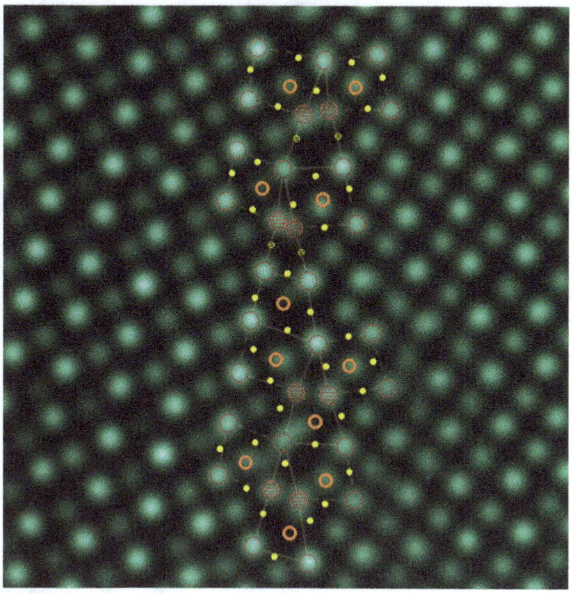

Figure 1–26. Z-contrast image of a 25° symmetric tilt grain boundary in $SrTiO_3$ [001] after maximum entropy processing, with superimposed structure model determined from combined use of the cation coordinates from the maximum entropy analysis, the O coordination from EELS, and the O positions from a bond valence sum analysis. Sr columns are shown as *larger red circles*, TiO columns as *smaller orange circles* and O columns as *yellow dots*, adapted from McGibbon et al. (1994).

an accuracy much exceeding the resolution, estimated at ±0.2 Å. The atomic structure of grain boundary dislocation cores could now be clearly seen, and the perovskite $SrTiO_3$ was shown to follow the structural unit model very closely (Browning et al. 1995, McGibbon et al. 1996).

Another milestone for 1993 was the delivery of a 300 kV STEM to Oak Ridge National Laboratory, a VG Microscopes HB603U, the first of its kind to be equipped with a high-resolution pole piece ($C_S = 1$ mm). This provided a theoretical Scherzer resolution of 1.27 Å, enough to resolve the Si dumbbell at 1.36 Å (Pennycook et al. 1993, von Harrach et al. 1993, von Harrach 1994, 2009). The initial images showed clear indications of a dip between the dumbbells, but there were also significant instabilities which were systematically removed over the next 2 years. Figure 1–27 shows a comparison of images of Si and GaAs, showing the sublattice sensitivity (Pennycook et al. 1996).

The sublattice sensitivity found immediate applications in determining dislocation core structures in compound semiconductor heterostructures (McGibbon et al. 1995) and in GaN (Xin et al. 1998, 2000a). The small probe also provided much better visibility for small particles on supports, and the first images of Pt atoms and clusters on a real catalyst support were obtained (Nellist and Pennycook 1996). This was the first indication that such small clusters might be catalytically important. The improved visibility of grain boundary structure allowed studies to be extended to more complex materials including Ni-ZrO_2 (Dickey et al. 1997) and NiO-ZrO_2 (Dickey et al. 1998). Further development of the maximum entropy technique (Nellist and Pennycook 1998a, McGibbon et al. 1999) allowed grain boundary structures in the high-temperature superconductor $YBa_2Cu_3O_{7-\delta}$ to be determined. They were found to follow the same structural unit model as $SrTiO_3$, and the misorientation

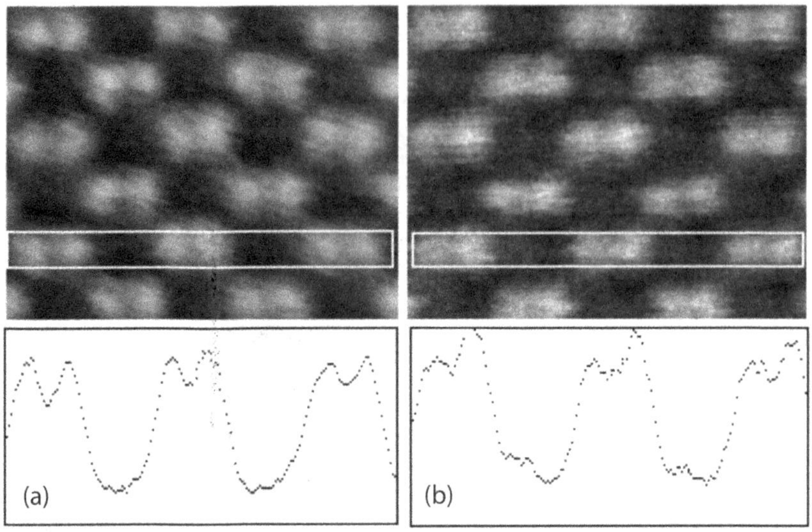

Figure 1–27. Sublattice sensitivity in a 300 kV VG Microscopes HB603U STEM. Z-contrast images of (**a**) Si and (**b**) GaAs ⟨110⟩ with line traces averaged vertically within the *white rectangles*, adapted from Pennycook et al. (1996).

dependence of critical current could be explained at the microscopic level (Browning et al. 1998b, 1999b). Impurities could be imaged at specific sites in grain boundaries for the first time, allowing correlation of experiment to atomistic total energy calculations. Arsenic sites were seen in a Si grain boundary (Chisholm et al. 1998, Maiti et al. 1996), and an impurity-induced structural transformation was seen at an MgO grain boundary (Yan et al. 1998a). For a recent review of grain boundary structure determination, see Chisholm and Pennycook (2006). The HB603U was also influential in the field of quasicrystals, resolving the Al sites for the first time in decagonal $Al_{72}Ni_{20}Co_8$ (Yan et al. 1998b, Yan and Pennycook 2000, 2001, Abe et al. 2003), see Chapter 14. For several years the Oak Ridge HB603U had the world's smallest electron beam.

1.7 Atomic Resolution EELS

The year 1993 was also the year that atomic resolution was demonstrated in EELS. As mentioned above, the inelastic signal was a major motivation for the development of the STEM, both for Crewe and also for the introduction of the commercial STEM by VG Microscopes. It was also well appreciated that being strongly forward peaked a large fraction of the scattering would be quite delocalized reflecting the long-range nature of the Coulomb interaction. This was the reason that the single heavy atoms clearly visible in the ADF image were not visible in the inelastic image (Crewe et al. 1975). For the same reason, low-energy, low-momentum transfer losses largely preserved any image contrast due to elastic scattering mechanisms (Howie 1963). Experimental edge resolution tests were performed by Isaacson et al. (1974), by examining holes in a thin carbon film. They found that the inelastic signal (from 7 to 200 eV loss) was still 6% of its value on the film when the probe was 20 Å from the edge, where the elastic signal was negligible. In 1976 Rose gave a detailed discussion on the nature of the image contrast from inelastic scattering including the effects of delocalization. He showed simulated images of single atoms showing a central peak sitting on top of a long tail due to delocalization. For a carbon atom imaged in a 100 kV microscope with a 3 Å probe he calculated that 50% of the inelastic scattering would be at distances greater than 5.5 Å from the atom. These calculations were motivated by experiments such as by Isaacson in which the total elastic scattering was collected, and the calculations assumed a mean excitation energy for carbon of only 35 eV. Using his simple rule of thumb expression (his Eq. (38)) for the carbon K edge at 285 eV gives a halfwidth of 1.5 Å, much more commensurate with the possibility of atomic resolution. Other rules of thumb subsequently appeared, for example, eq. (16) in Pennycook (1988) based on a root mean square impact parameter also gives 1.5 Å, and Egerton's $L_{50}/2$ is 1.3 Å (Egerton 1996, 2007). While these numbers are of historical interest it must be remembered that for atomic resolution imaging, a single parameter is not useful in predicting image contrast. A full quantum-mechanical treatment is necessary (Kohl 1983, Rose 1984, Kohl and Rose 1985, Muller and Silcox 1995, Oxley and Allen 1999, Cosgriff et al. 2005, Oxley et al., 2007, Oxley and Pennycook, 2008). This issue is discussed fully in Chapter 6.

For a high energy loss the interaction would therefore be expected to be sufficiently localized to allow atomic resolution analysis. Single U atoms imaged with their characteristic $O_{4,5}$ loss at 105 eV were visible, although line scans showed both the resolution and contrast to be significantly degraded (Colliex 1985). Scheinfein et al. (1985) scanned a similar 5 Å probe across a Si(100)/CaF_2 interface and plotted intensities of the Si L_{23} edge at 98 eV and the Ca L_{23} edge at 343 eV, taking spectra every 4 Å. They concluded that the width of the interface was about 5 Å, consistent with an atomically abrupt interface. Batson (1991) observed changes in pre-edge features at the Si L_{23} edge on moving the probe to within 6 Å of a Si(111)/Al interface. Near the interface he saw changes on moving the probe by only 2 Å, but he did not correlate the data to an atomic resolution image at this time. Also in 1991, Mory et al. concluded that an upper limit for delocalization at around 100 eV energy loss was 3–4 Å. More recently Suenaga et al. (2000) imaged single Gd atoms inside fullerenes inside a single-wall carbon nanotube, again using a beam of around 5 Å diameter. Some beam-induced migration and coalescence of Gd was seen, but single atoms could be identified based on the number of counts in the Gd N edge at 150 eV.

The first attempt to perform core loss EELS with atomic resolution used a Si(111)/$CoSi_2$ interface, well known to be atomically abrupt. Spectra were recorded with the sample aligned to a zone axis and while scanning the beam in a line parallel to the interface. By monitoring the Z-contrast image intensity the probe could be accurately maintained over each plane of interest. This minimized beam damage while maintaining the possibility of atomic resolution perpendicular to the

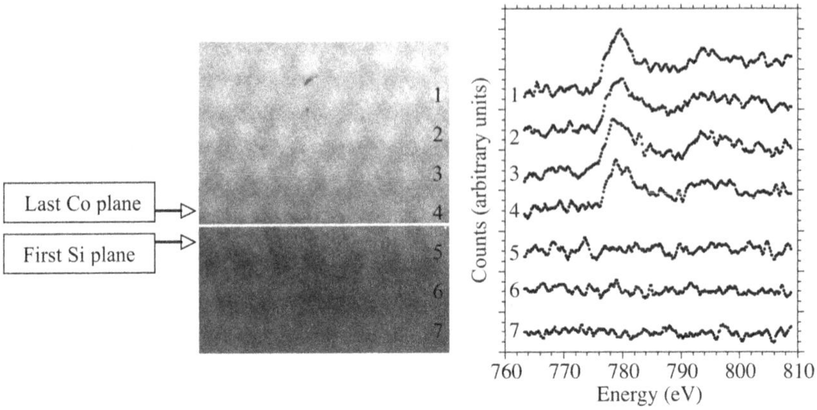

Figure 1–28. Z-contrast image of a $CoSi_2/Si$ {111} interface with spectra obtained plane by plane across the interface, which is marked with a *white line*. The first Si plane shows dumbbells in a twin orientation resulting in a separation between the last Co plane and the first Si plane of 2.7 Å. Spectra 1–4 were background subtracted by the usual power law fit, spectra 5–6 were obtained using the spatial difference method, using a reference spectrum from Si far from the interface, adapted from Browning et al. (1993a, b, 2006).

interface. The results are shown in Figure 1–28 (Browning et al. 1993a, b, 2006). The Co L_{23} edge shows a substantial drop between the last Co plane and the first Si plane. The magnitude of the drop exceeds that required to demonstrate atomic resolution and is consistent with recent EELS simulations for a thin specimen (Pennycook et al. 2009a).

That same year Batson (1993) demonstrated changes in the Si L_{23} edge fine structure at a $Si(100)/SiO_2$ interface oriented to the ⟨110⟩ zone axis. Now the Z-contrast image was used to locate the probe on particular atomic columns (Si dumbbells). Moving the probe from the last Si dumbbell into the SiO_2 gave additional small peaks in the spectrum, see Figure 1–29. Also in 1993, Muller et al. demonstrated two-dimensional mapping with EELS fine structure, using the $\pi*$ and $\sigma*$ peaks at the C K edge to map sp^2 and sp^3 bonded carbon with sub-nanometer resolution. These capabilities found many applications to grain boundaries and interfaces (Browning et al. 1993c, Pennycook et al. 1993, Muller et al. 1996, 1998, 1999, Wallis et al. 1997a, b). The first atomic resolution spectroscopic identification of impurity valence was achieved in 1998 using a Mn-doped $SrTiO_3$ grain boundary (Duscher et al. 1998a), shown in Figure 1–30. For a recent review of the history of atomic resolution EELS see Pennycook et al. (2009a).

1.8 Atomic Resolution with TEM/STEM Instruments

Atomic resolution imaging and spectroscopy was not widely taken up due to the lack of instruments capable of achieving such a small, stable probe. There were only ever four 300 kV STEMs built by VG Microscopes, and the only one with a high-resolution pole piece was

Figure 1–29. Column-by-column spectroscopy at the Si/SiO_2 interface showing a pre-edge feature at the interface which is not seen from spectrum #1 approximately 2 Å away. Reproduced from Batson (1993) with permission.

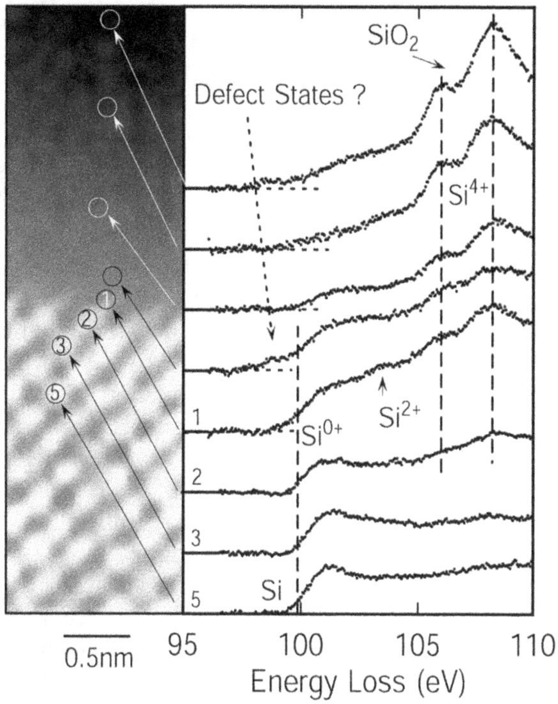

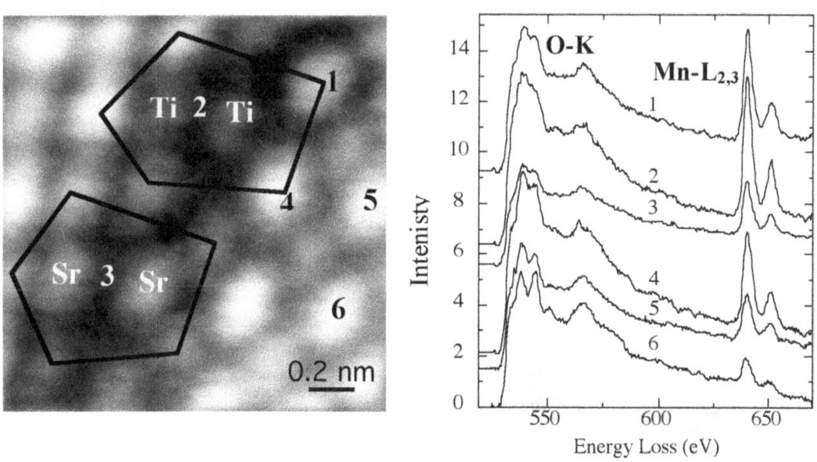

Figure 1–30. (**a**) Z-contrast image from a Mn-doped 36° $SrTiO_3$ grain boundary recorded on the uncorrected 300 kV VG Microscopes HB603U STEM. (**b**) EELS spectra recorded by stopping the probe on the corresponding atomic columns in an uncorrected 100 kV VG Microscopes HB501UX STEM, revealing differences in Mn concentration and valence at different sites. Reproduced from Duscher et al. (1998a).

at Oak Ridge. While VG Microscopes were supplying dedicated STEMs there was little effort from manufacturers of conventional TEM columns to compete with their high-resolution performance. While many more 100 kV VG machines were installed, none had quite the same EELS capability as the Oak Ridge machine. Many had photodiode arrays which had too high a dark current for the low-signal levels generated by the small probes necessary for atomic resolution. The Oak Ridge system was based on a design by McMullan et al. (1990) at the Cavendish Laboratory, Cambridge, but was the first to be designed specifically for column-by-column spectroscopy. It used a thinned, multi-phase-pinned charge-coupled device for the highest sensitivity and lowest dark count and its optical coupling avoided channel to channel gain variations (Pennycook et al. 2009a).

Ironically, it seems to have been the demise of VG Microscopes that stimulated the other manufacturers to improve their STEM performance. In 1995 VG Microscopes was acquired by new owners and ceased production the following year. Nigel Browning was setting up his group at the University of Illinois at the time and found himself with money to buy a dedicated STEM but no supplier. He instead purchased a JEOL 2010F and worked with the manufacturer to achieve atomic resolution capability (James et al. 1998, James and Browning 1999), as shown in Figure 1–31. Due to the lower C_S of the pole piece the resolution achieved at 200 kV was very comparable to that achieved at 300 kV in the VG Microscopes HB603U. The EELS performance was not comparable, however, due to the use of a Schottky source with lower brightness and about double the energy spread compared to the cold field emission source used by VG. In addition, high-resolution STEMs are very sensitive to environmental factors (Muller and Grazul 2001). Nevertheless, whereas VG Microscopes had installed about 70 STEMs in their 22-year history, the number of STEMs in service that were capable of atomic resolution imaging doubled within just a few years.

Rapid progress ensued, with new applications and new approaches to image simulation and quantitative compositional profiling. More detailed studies of the role of phonons on image contrast were carried out by Dinges et al. (1995) and Hartel et al. (1996). They extended

Figure 1–31. Z-contrast image of Si ⟨110⟩ recorded in a JEOL 2010F TEM/STEM, reproduced from James and Browning (1999) with permission.

the mutual coherence function approach introduced earlier by Rose and coworkers to the consideration of phonon scattering and introduced a modified multislice approach for image simulation. This summed over statistical phases introduced to ensure incoherence between different inelastic excitations propagating to the detector. In 1997 Anderson et al. (1997) presented a method for the quantitative analysis of composition using a modified multislice method based on matching image intensities within a two-dimensional unit cell. Using a GaAs/$Al_{0.6}Ga_{0.4}As$ interface, they found good agreement with the method of Ourmazd et al. (1989, 1990), which is based on the chemical sensitivity of the {200} reflection in such materials. Amali and Rez (1997) used a Bloch wave expression to show that even with multiphonon scattering, which dominates the HAADF image, and dynamical diffraction conditions, the criterion for lattice resolution remained that the Bragg angle must be less than the probe convergence angle, as noted before for a phase object (Spence and Cowley 1978). Nakamura et al. (1997) developed a multislice formulation that used a complex potential to calculate the diffuse scattering over the annular detector, thereby avoiding the need to average over many vibrational snapshots. They pointed out that because the 1s state intensity was strongly depth dependent, the visibility of a single heavy atom in a crystal would also be very dependent on the depth of the atom, showing maximum visibility at the depth of the first strong channeling peak. Plamann and Hÿtch (1999) pointed out that the depletion of 1 s states on heavy columns could also be aided by capture of flux by adjacent lighter columns and showed that the channeling effect was strongly affected by correlated displacements such as strain fields.

In 1988 a significant advance in the understanding of the frozen phonon method was made by Wang (1998a). The frozen lattice approximation is a semi-classical means of including the effects of thermal diffuse scattering in image simulations where the scattered wave remains coherent with the unscattered wave. The relation with the quantum theory, the excitation and annihilation of phonons where the phonon-scattered electrons are incoherent with the unscattered wave, had not so far been established. Wang showed that the two theories are equivalent as long as the mixed dynamic form factor is used to describe the phonon scattering, that is, a full non-local description is necessary. He showed how this could be formulated within a multislice Bloch wave method (Wang 1998b).

In 1998 Nellist pushed the performance of the HB603U into the sub-angstrom regime by using an oversized objective aperture and an underfocused lens to resolve the 0.93 Å separation of the Cd and Te columns in CdTe ⟨110⟩ (Nellist and Pennycook 1998b). Figure 1–32 compares the result with the normal Scherzer condition in which the dumbbells are unresolved. Although the image is extremely noisy there are indications that the dumbbell is split in the line scan, and the enhanced transfer in the underfocus condition is seen by the presence of a strong {444} spot in the Fourier transform. With Si {112} information transfer was obtained to 0.78 Å. The paper also pointed out that the conventional information limit does not apply to ADF images. Image contrast arises from the regions of overlap between the diffraction discs,

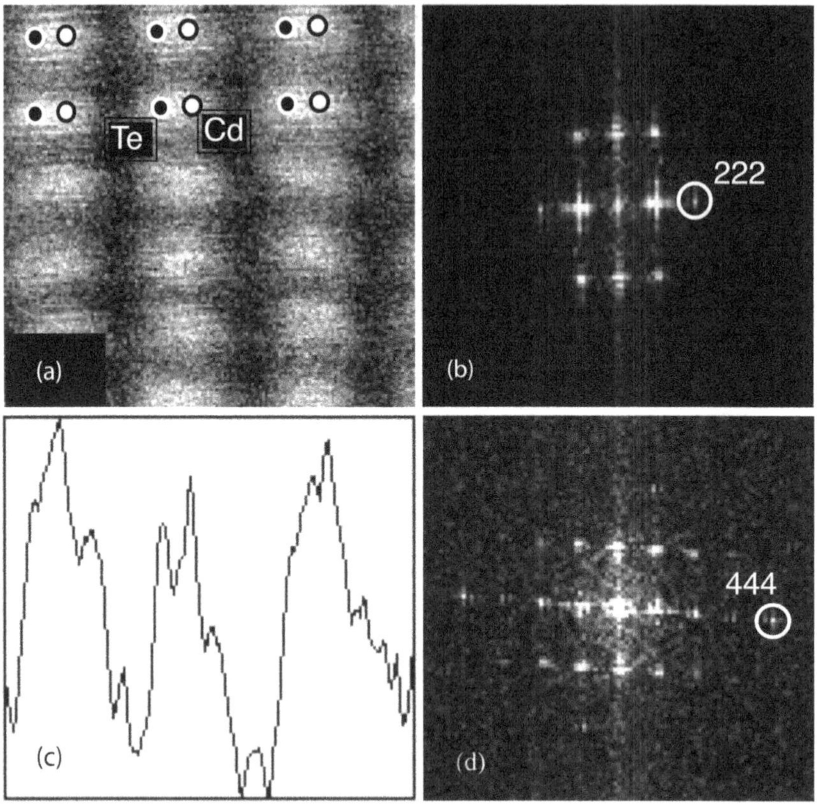

Figure 1–32. (**a**) Z-contrast image of CdTe ⟨112⟩ taken with a 300 kV STEM under Scherzer conditions when the resolution of 1.36 Å is insufficient to resolve the CdTe dumbbell spacing of 0.93 Å. (**b**) Fourier transform showing information transferred to the 1.86 Å {222} spacing but not to the 0.93 Å {444} spacing. (**c**) Profile plot obtained by summing vertically over 200 pixels of an image of CdTe ⟨112⟩ taken with an oversized objective aperture and higher defocus showing the {444} fringes and (**d**) their corresponding spots in the Fourier transform. Reproduced from Nellist and Pennycook (1998b).

and the centers of these overlaps are symmetrical about the optic axis and as a result are insensitive to small changes in energy. Energy spread does not represent the information limit in incoherent imaging as it does for axial phase contrast imaging.

The following year Nellist and Pennycook (1999) developed a fully reciprocal space expression for the coherent ADF image intensity in a Bloch wave formulation that allowed the contribution of different states to be calculated. The results confirmed that incoherent images would be obtained under dynamical conditions even if only coherent scattering contributed to the image and again highlighted the role of the 1s state in generating the high-angle scattering. They showed that in ⟨110⟩ GaAs, although the 2s state is the greatest contributor to the intensity inside the crystal it is the much more weakly excited 1s state that dominates the high-angle scattering. Plots of the intensity inside the crystal may not therefore be a good indicator of contributions to the HAADF image.

Applications of atomic and near-atomic resolution imaging and EELS continued to grow. In 1999 Batson reported the electronic structure of a dissociated misfit dislocation in a Si/Ge_xSi_{1-x} heterostructure (Batson 1999a, b). There were further developments in the interpretation of EELS fine structure at grain boundaries (Browning et al. 1998a, 1999a, Muller 1999, Shashkov et al. 1999, Titchmarsh 1999) and further applications to semiconductors (Lakner et al. 1999, Muller and Mills 1999, Muller et al. 1999, Kim et al. 2000, Xin et al. 2000b, Yamazaki et al. 2000a), ceramics (Yan et al. 1999, Duscher et al. 2000, Klie and Browning 2000, Stemmer et al. 2000, Xu et al. 2000, Yamazaki et al. 2000b), precipitates in metal alloys (Hutchinson et al. 2001, Mitsuishi et al. 1999), and nanomaterials (Grigorian et al. 1998, Fan et al. 1999, 2000).

With increasing applications there came increasing demand for rapid image simulations. In the frozen phonon method of Loane et al. (1991), or the method of statistical phases introduced by Dinges et al. (1995), the incoherence of thermally scattered electrons is maintained by averaging over many configurations, requiring many multislice simulations per image point. Furthermore, the intensity needs to be accurately tracked to the detector, over which it is then integrated, so losing all the details of the distribution just calculated. The absorptive potential approach simulates only the total scattering onto the detector and so is much faster. In addition, for accurate simulations both the coherent and the incoherent scattering reaching the detector should be calculated, especially with low inner detector angles. In 2001 two groups extended the Bloch wave method to include the coherent scattering. Mitsuishi et al. (2001) used a delta-function approximation for the absorptive potential, appropriate for high-angle scattering, whereas Watanabe et al. (2001) did not use the delta-function approximation but instead used an approximation based on two optical potentials. In their formulation the intensity falling onto the ADF detector is not included in the absorption of the wave function inside the crystal, which is presumably only accurate when the detected intensity is a small fraction of the total absorption, i.e., for a high-angle detector. The same year Ishizuka incorporated an optical potential in a multislice code and did not make a delta-function approximation, hence this remains accurate for low-angle scattering (Ishizuka 2001, 2002). He points out that since the multislice formulation treats the entire incident cone simultaneously, it should be more efficient in the case of aberration-corrected probes than the Bloch wave method that so far had summed individual plane wave components in the incident probe.

The same year, the approach of Nellist and Pennycook (1999) was also extended, with the goal of probing the physical mechanisms contributing to the HAADF image rather than in a quest for accurate image simulations. Rafferty et al. (2001) pointed out that the phonon scattering acted only to blur the distribution on the detector and that therefore the coherent Bloch wave formulation would provide an accurate measure of the contribution of various Bloch states to the image. They investigated the issue of cross talk by removing half or all the In column in ⟨110⟩ InAs, finding negligible effect on the As column intensity. Interestingly they also found that the Z-dependence of the

1 s state intensity for a high-angle detector becomes identical to the Z-dependence for screened Rutherford scattering from isolated atoms, although with much higher intensity due to the channeling effect. This supports the original picture of the electron flux being concentrated onto the atomic columns and each atom acting as an independent generator of high-angle scattering.

The incoherent nature of the HAADF image means that it is a much better approximation to a mass thickness image than a bright field image, even if diffraction effects are not completely avoidable. This characteristic was used to achieve a three-dimensional tomographic reconstruction with a resolution of 1 nm in all directions (Midgley et al. 2001, Weyland et al. 2001). The technique has become very widely applied, especially to catalytic materials (Midgley and Weyland 2003, Midgley et al. 2004) and embedded nanostructures (Ozasa et al. 2003, Arslan et al. 2005), and a full account is presented in Chapter 8. A scanning confocal mode of imaging was introduced by Frigo et al. (2002) as a means to image buried structures in integrated circuits.

The number of installed TEM/STEM instruments capable of atomic resolution Z-contrast imaging and EELS continued to grow, as did their applications. Studies of the Si/SiO_2 interface were continuing (Muller et al. 1995, Duscher et al. 1998b, Muller 2001, Muller and Wilk 2001) and in 2003, the incoherent Z-contrast imaging was recommended as the preferred method for determining the thickness of thin dielectric films, due to its relative simplicity and insensitivity to Fresnel fringe effects (Diebold et al. 2003). A large number of studies have appeared in this context, see Chapter 12 for more details. The first applications to complex oxides also appeared (Verbeeck et al. 2001, Ohtomo et al. 2002a, b, Varela et al. 2002, 2003), see Chapter 10. Also notable are the insights into the growth of crystalline oxides on Si (McKee et al. 1998, 2001). These areas are ideal examples of how the ability of the STEM to correlate local electronic structure, composition, and bonding through simultaneous EELS and HAADF imaging can provide fundamental insights into the origin of interfacial properties. This goal was enormously advanced with the advent of aberration correction.

1.9 The Successful Correction of Lens Aberrations

It has often been stated that history is littered with unsuccessful attempts at aberration correction. The physics was well established in the last century, with specific proposals first introduced by Scherzer (1947). For a review of the history of aberration correction in electron optics, see Rose (2008), Krivanek et al. (2009a), and Hawkes (2009). It was only in the era of computers and charge-coupled device detectors that it became possible to design a system to diagnose its aberrations and control the large number of optical elements to the necessary accuracy to improve the resolution. The first success was with the scanning electron microscope (Zach and Haider 1995) using an electrostatic-electromagnetic 4-layer quadrupole-octupole corrector of spherical and chromatic aberration in a low-voltage microscope. The electrostatic designs are more difficult to operate at higher voltages and so this was not the approach adopted for the TEM. Successful correction of aberrations in the TEM was first demonstrated using a hexapole corrector to improve the resolution from 2.2 to about 1.3 Å (Haider et al. 1998a, b, c). Along with improved spatial resolution there was a substantial decrease in the delocalization of the phase contrast image.

The first successful aberration correction in the STEM was achieved at about the same time using an old VG Microscopes HB5 and a quadrupole/octupole corrector (Krivanek et al. 1997, 1999). The resolution was limited by microscope instabilities to between 2.3 and 3.4 Å. An improved design was incorporated into a VG Microscopes HB501 STEM and demonstrated resolution of the dumbbells in Si ⟨110⟩ at 1.36 Å (Dellby et al. 2001). This performance was much better than the theoretical Scherzer resolution limit of 2.2 Å for the uncorrected lens and represented a new resolution limit for the accelerating voltage used, just 100 kV. The current in the probe was also high, about 160 pA, indicating that with more source demagnification the electron optical limit to resolution should be in the range of 0.8 Å. The following year this level of performance was demonstrated at IBM by pushing the accelerating voltage to 120 kV (Batson et al. 2002). Line profiles across single Au atoms showed that sub-angstrom resolution had finally been achieved in electron microscopy. A similar corrector was installed in the VG Microscopes HB501UX at Oak Ridge in March 2001 and soon achieved resolution of the Si ⟨110⟩ dumbbells (Pennycook 2002). Figure 1–33 shows the imaging of individual Bi atoms within the Si lattice (Lupini and Pennycook 2003, Pennycook et al. 2003b). The

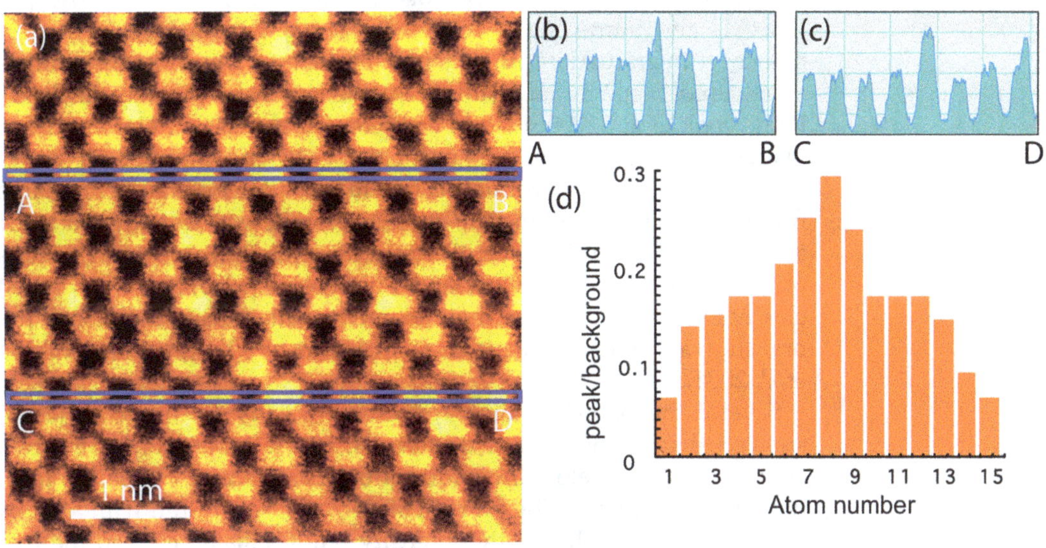

Figure 1–33. Z-contrast image of Bi-doped Si ⟨110⟩ taken with a VG Microscopes HB501UX with Nion aberration corrector operating at 100 kV, revealing columns containing individual Bi atoms. The upper intensity profile shows a Bi atom on the *right-hand column* of a Si dumbbell and the lower profile shows a Bi atom in each of the two columns of a dumbbell. Reproduced from Lupini and Pennycook (2003) and Pennycook et al. (2003b).

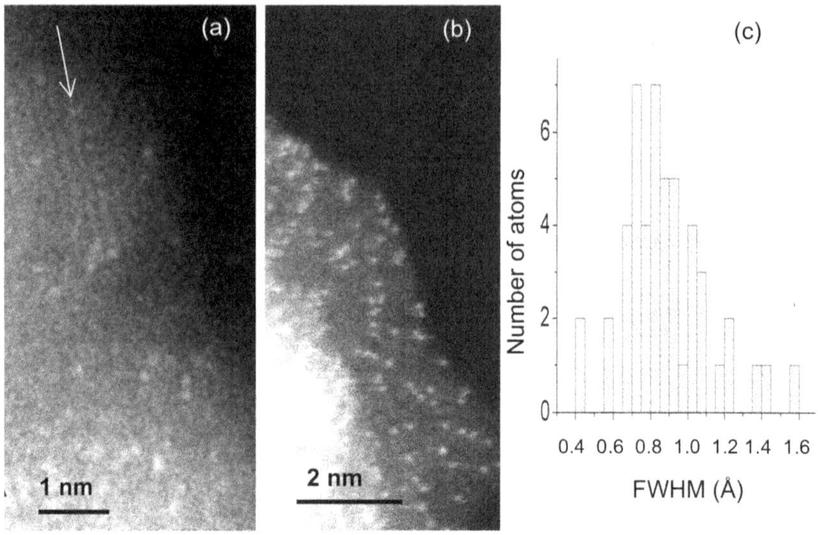

Figure 1–34. Images of La-doped γ-alumina obtained with a VG Microscopes HB603U operating at 300 kV before aberration correction (**a**) showing faint fringes from the γ-alumina (*arrow*) and some individual La atoms. After correction (**b**) there is a significant improvement in contrast, resolution, and signal to noise ratio. Demonstration that the probe size is sub-angstrom comes from (**c**) a histogram showing the full width half maximum (FWHM) of intensity profiles across single atoms. Reproduced from Pennycook et al. (2003a).

signal to noise ratio is much improved compared to images obtained in an uncorrected microscope at 200 kV (Voyles et al. 2002, 2003, 2004).

In 2002 a similar corrector was installed in the VG Microscopes HB603U and immediately achieved a sub-angstrom level of performance, with gradual improvement over the next 2 years as instabilities were cured. Some initial results are presented in Krivanek et al. (2003), Pennycook et al. (2003b), and Pennycook et al. (2003c). The improved visibility of single atoms was particularly striking. Figure 1–34 shows images of La atoms on γ-alumina before and after correction. From line traces across single atoms the corrected probe was determined to be about 0.7–0.8 Å in diameter (Pennycook et al. 2003a). The Pt trimers originally imaged by Nellist and Pennycook (1996) were now seen clearly enough to correlate their geometry with density functional theory, showing that they were in fact capped by OH groups (Sohlberg et al. 2004).

Several theoretical studies appeared in 2003. Anstis et al. (2003) showed that in the case of dumbbells, as the separation of the two columns reduces below the width of the 1s state they overlap to form bonding and antibonding pairs of states, and, as a result, with a probe placed over one column, the intensity will oscillate between the two columns with a depth periodicity depending on the degree of overlap. A similar effect was found by Dwyer and Etheridge (2003) using frozen phonon simulations. They showed details of the probe broadening for probes of various sizes incident on both ⟨110⟩ and ⟨100⟩ Si, finding a more rapid broadening of the smaller probes. However, in general locating the probe directly over a column results in more intensity on that column than the total on all other columns for quite substantial thicknesses. Voyles et al. (2004), however, showed that in thick crystals, due to this transfer of intensity from one column of a dumbbell to the other, it was possible for a single heavy atom to contribute to the image intensity on the wrong column. Also, Yu et al. (2003) showed through simulations how use of an experimental black level could introduce artifacts in the images. Clipping could introduce frequencies into the image that were beyond the diffraction limit, hence care should be taken to avoid clipping when looking at the Fourier transform to assess information limit.

Also in 2003, a major advance in image simulation was published (Allen et al. 2003b, Findlay et al. 2003). These papers showed how any inelastic signal, thermal diffuse, X-ray, or EELS could be simulated accurately through either a multislice or Bloch wave methodology. They showed that it was much more efficient to consider the excitation of Bloch states by the whole probe as opposed to the previous treatments where each plane wave component was individually expanded into a set of phase-linked Bloch states and showed the equivalence of the two formulations. Their simulations of ADF, X-ray, and EELS images are shown in Figure 1–35 and show the expected slight degradation of resolution, in that order, due to increasing ionization delocalization. The ability to perform accurate simulations for inelastic scattering allowed the first comparison with an experimental EELS line trace (Allen et al. 2003a). For more details and recent comparisons between theory and experiment see Chapter 6.

In 2004, the first sub-angstrom resolution image of a crystal lattice was published, as shown in Figure 1–36 (Nellist et al. 2004). Every dumbbell in ⟨112⟩ Si shows a clear dip in the middle, indicating a resolution of 0.78 Å. The Fourier transform of the image intensity indicated an information limit of 0.63 Å (avoiding clipping). Later more detailed analysis allowed the microscope and specimen parameters to be extracted through comparison to full image simulations (Peng et al. 2008).

Also in 2004 another milestone was achieved in energy loss spectroscopy, the spectroscopic identification of a single atom, the smallest quantity of matter, as shown in Figure 1–37 (Varela et al. 2004). The small signal from the adjacent columns is due to beam broadening and EELS image simulations showed the depth of the atom to be approximately 100 Å below the surface.

The same year it was realized that the larger objective aperture made possible by aberration correction improved not only the lateral resolution but also the depth resolution, and it became feasible to obtain three-dimensional information through depth slicing. A focal series had become a depth sequence of images, at least for non-channeling materials. Figure 1–38 shows an example of the imaging of a Pt_2Ru_4 catalyst supported on γ-alumina (Pennycook et al. 2004b). The Bloch wave analysis of HAADF image formation was extended into the sub-angstrom regime (Peng et al. 2004) and found increasing contributions from plane wave-like Bloch states around the periphery of the aperture.

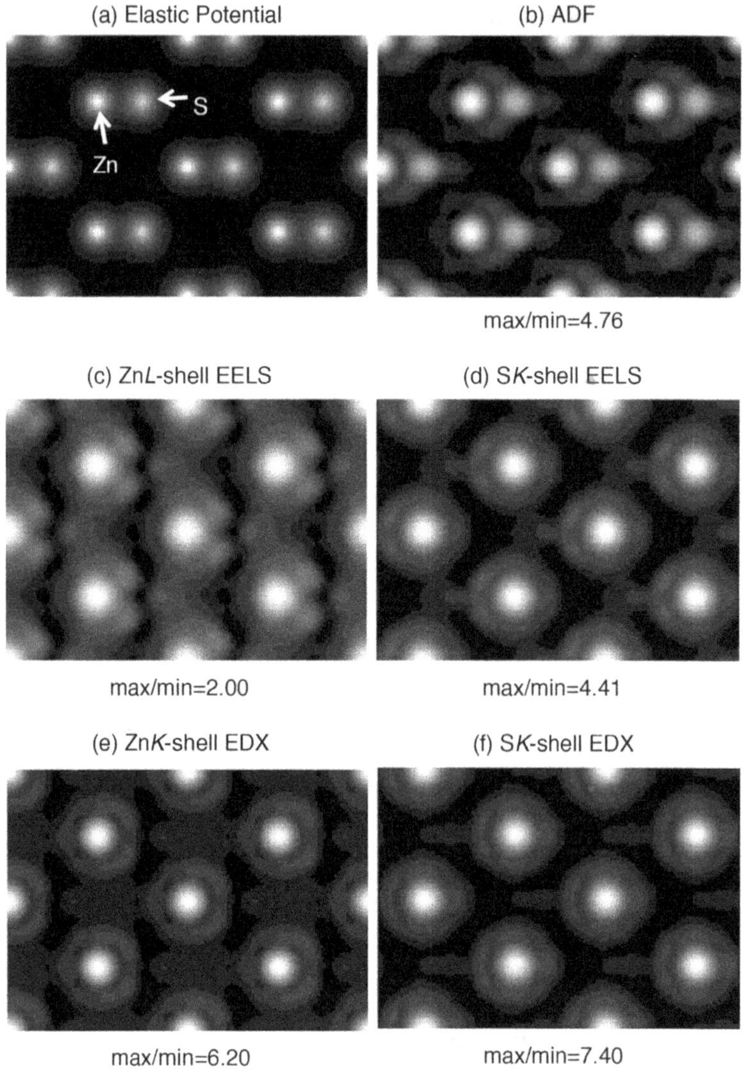

Figure 1–35. (**a**) Projected potential down the ZnS ⟨110⟩ zone axis. (**b**) Simulated ADF lattice images (acceptance angles 60–160 mrad). (**c**) Lattice image from L-shell ionization of Zn. (**d**) Lattice image for K-shell ionization of S. (**e**) Lattice image from K-shell ionization of Zn for EDX. (**f**) Lattice image for K-shell ionization of S for EDX. Beam energy is 200 kV with a probe size of 0.5 Å, reproduced from Allen et al. (2003b).

These beams were so far from the zone axis that they propagated essentially as in free space. The probe could be therefore be thought of as the superposition of a channeling component near the zone axis and a free-space-like component that would come to a focus at a specific depth, and it was shown how the channeling peak could be pushed down the column by a change in focus.

Experimental verification was first made with La-stabilized γ-alumina, a support material for metal nanoparticles (Wang et al. 2004).

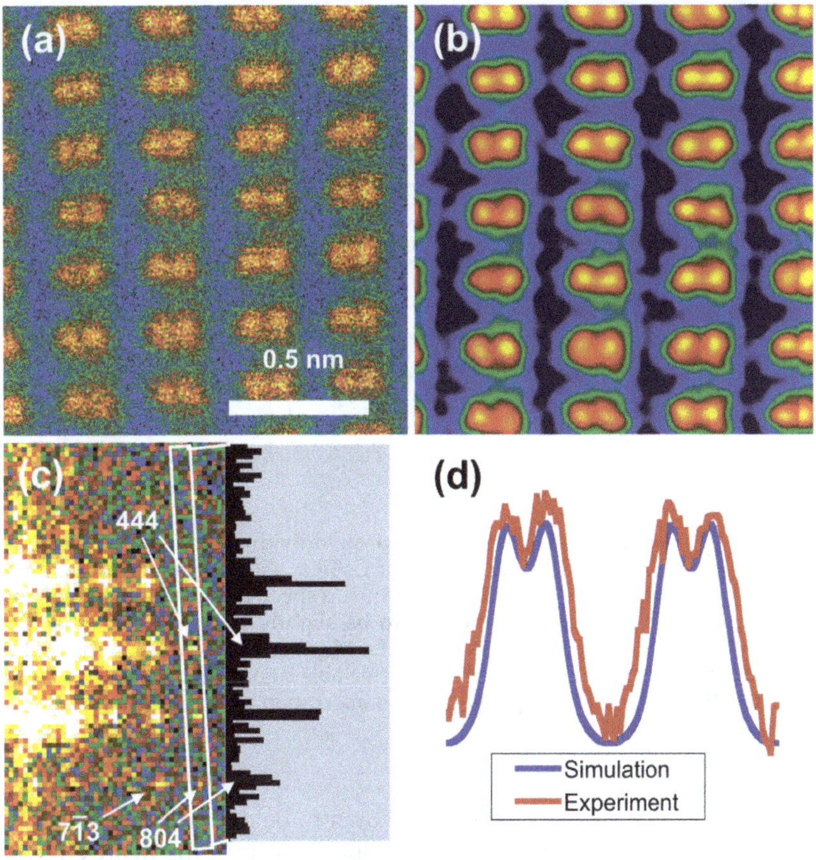

Figure 1–36. (**a**) Image of ⟨112⟩ Si recorded using a VG Microscopes HB603U with Nion aberration corrector operating at 300 kV. (**b**) Image after low-pass filtering and unwarping. (**c**) Modulus of the Fourier transform and a profile of the region shown in the white rectangle, showing the (444) 78 pm spacing of the dumbbell and evidence of information transfer at the (804) 61 pm spacing. (**d**) An intensity profile through two column pairs in (a) formed by summing over a width of 10 pixels with a simulated profile. Reproduced from Nellist et al. (2004).

A focal series showed the La atoms to be located on the surface, in agreement with density functional theory predictions. Image simulations showed La atoms at the exit surface of an aligned crystal showed brighter since the probe was focused into a sharp channeling peak by the crystal. The resolution was identical nevertheless, since resolution is determined by the change in scattered intensity as the probe is scanned, not by the width of the channeling peak for a single probe position. A spectacular application to semiconductor devices was the three-dimensional mapping of stray Hf atoms within the nanometer thick gate dielectric of a semiconductor device (van Benthem et al. 2005). A focal series of images showed individual Hf atoms to appear and disappear allowing their three-dimensional coordinates to be determined. The change in focus over which each Hf atom could be seen was much smaller than expected from the theoretical depth of focus,

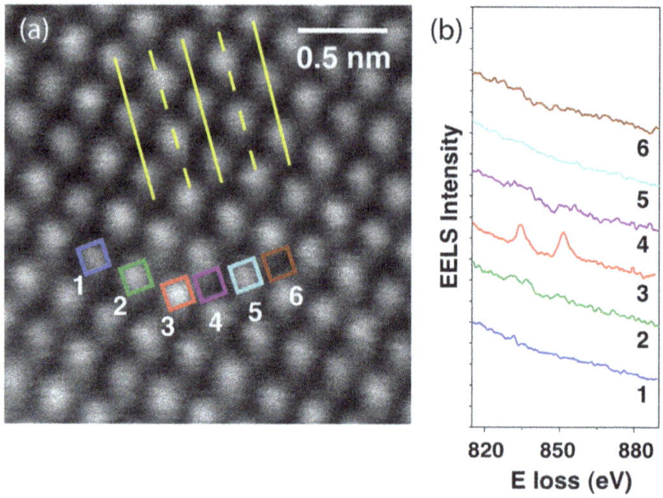

Figure 1–37. Spectroscopic identification of an individual atom in its bulk environment by EELS. (**a**) Z-contrast image of $CaTiO_3$ showing traces of the CaO and TiO_2 {100} planes as *solid* and *dashed lines*, respectively. A single La dopant atom in column 3 causes this column to be slightly brighter than other Ca columns, and EELS from it shows a clear La $M_{4,5}$ signal (**b**). Moving the probe to adjacent columns gives reduced or undetectable signals. Data recorded using the VG Microscopes HB501UX with Nion aberration corrector, adapted from Varela et al. (2004).

as shown in Figure 1–39. This was shown to be due to the high background signal from out of focus contributions from the nearby HfO_2 (van Benthem et al. 2006), since images of Pt atoms on a thin carbon film did show the expected variation with focus, as seen in Figure 1–38(d). Theoretical studies for aligned crystals showed that Bi atoms in ⟨110⟩ Si could be easily located in depth with a 35 mrad probe angle, but heavier materials would channel stronger and a higher probe-forming angle would be necessary (Borisevich et al. 2006b). Recently, depth-sensitive imaging of Bi atoms in ⟨100⟩ Si has been obtained with reduced probe angle (Lupini et al. 2009).

Several new areas of application were opened up by the new capabilities. The ability to resolve and distinguish the sublattice in compound semiconductors allowed nanocrystal shape and polarity to be determined from a single image, providing insight into growth mechanisms (McBride et al. 2004). Subsequently the technique has been used to understand the growth and optical efficiency of core-shell nanocrystals (McBride et al. 2006), nanowires (Heo et al. 2004), and white light-emitting nanocrystals (Bowers et al. 2009). For further details, see the recent reviews by Rosenthal et al. (2007) and Pennycook et al. (2010). The size-dependent energy gap of individual quantum dots was measured by low-loss EELS (Erni and Browning 2007), the variation being different from that obtained by bulk measurements which necessarily average over the particle size distribution. The field of structural ceramics also saw a significant advance in 2004, when rare earth dopants were imaged for the first time in the intergranular phase of a

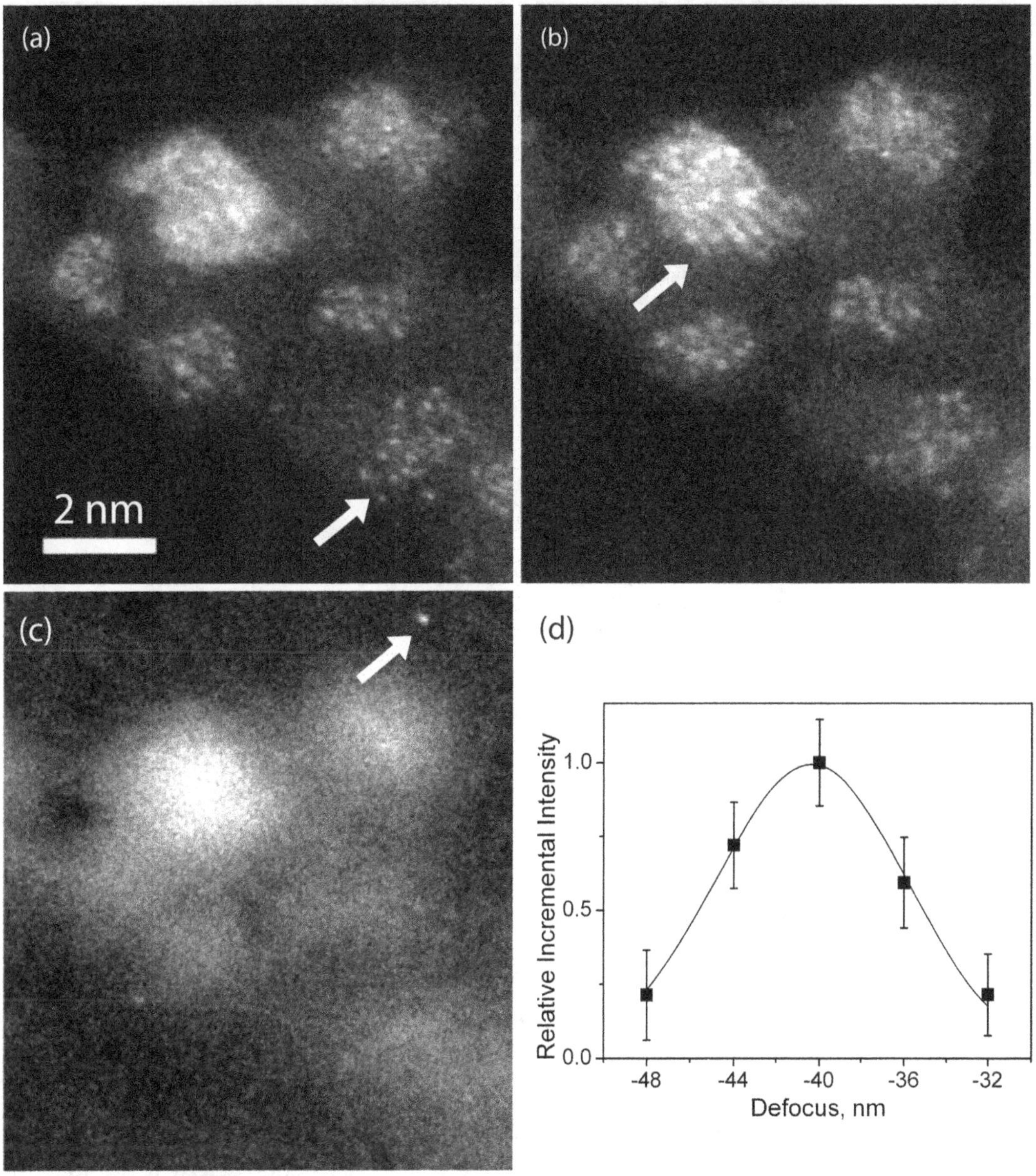

Figure 1–38. Three frames from a through-focal series of *Z*-contrast images from a Pt/Ru catalyst on a γ-alumina support, at a defocus of (**a**) –12 nm, (**b**) –16 nm, (**c**) –40 nm from initial setting. *Arrows* point to regions in focus. (**c**) A single atom is in focus on the carbon support film. (**d**) Integrated intensity of the Pt atom in (**c**) as a function of defocus, compared to a Gaussian fit. The FWHM of the fit is 12 nm but the precision of the location of the peak intensity is 0.2 nm with 95% confidence. Results obtained with a VG Microscopes HB603U operating at 300 kV with a Nion aberration corrector, reproduced from Borisevich et al. (2006a).

Si_3N_4 ceramic (Shibata et al. 2004). Numerous other studies soon followed (Winkelman et al. 2004, Ziegler et al. 2004, Shibata et al. 2005, Winkelman et al. 2005, Becher et al. 2006, Buban et al. 2006, Dwyer et al. 2006, Sato et al. 2006, Shibata et al. 2006), and further details are given in Chapter 11.

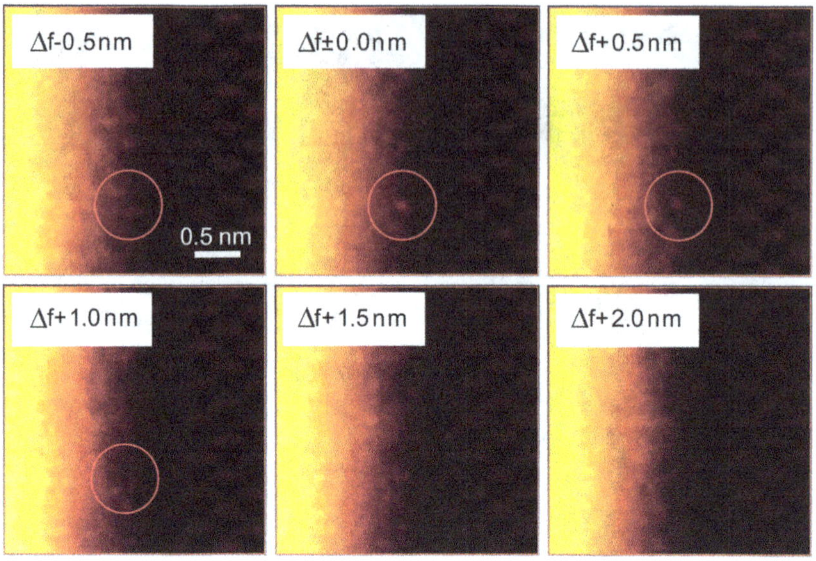

Figure 1–39. A sequence of frames from a through-focal series of *Z*-contrast images of a $Si/SiO_2/HfO_2$ high dielectric constant device structure showing an individual Hf atom coming in and out of focus (*circled*). The HfO_2 is seen brightly on the *left*, the Si lattice dimly on the *right*, and the Hf atoms are in the SiO_2 gate oxide region. Results obtained with a VG Microscopes HB603U operating at 300 kV with a Nion aberration corrector, reproduced from van Benthem et al. (2006).

New facilities also appeared. The Daresbury SuperSTEM facility opened in 2003 and new insights were obtained into $CoSi_2/Si$ interfaces (Falke et al. 2004, 2005), dislocation core structures in GaAs (Xu et al. 2005), and numerous other materials. In 2004 the Ernst Ruska Center opened at Jülich equipped with two aberration-corrected instruments: one for TEM and one for STEM. In the TEM, the reduced delocalization and enhanced signal to noise ratio allowed oxygen columns to be clearly seen for the first time in perovskites and related materials. Their positions and intensities could be quantitatively analyzed (Jia et al. 2003, Jia and Urban 2004), allowing the direct mapping of ferroelectric distortions at the unit cell level (Jia et al. 2006). Using exit wave reconstruction the detailed atomic reconstruction at a twin boundary in $YBa_2Cu_3O_{7-\delta}$ was able to be determined (Houben et al. 2006).

These TEM results stimulated a reexamination of phase contrast imaging in the aberration-corrected STEM, and it became apparent that the large flat phase region on axis in the detector plane should allow the bright field collector aperture to be enlarged, thus improving the efficiency of the phase contrast image. Contrast transfer functions indicated that the collector aperture diameter could be increased tenfold, without losing coherence, as shown in Figure 1–40. STEM phase contrast imaging had finally become a useful technique, with accuracies comparable to the TEM method but with the advantage of the availability of simultaneous *Z*-contrast imaging. A comparison of the two modes of imaging is shown in Figure 1–41 (Pennycook et al. 2004a). Note that the optimum tuning is performed for the *Z*-contrast image

Figure 1–40. (**a**) Contrast transfer functions for an uncorrected 300 kV microscope, *gray line* (C_S = 1.0 mm, Δf = –44 nm), with the damping envelopes introduced by a beam divergence of 0.25 mrad (*dotted line*) and an energy spread of 0.6 eV (*dashed line*). (**b**) Transfer for a corrected 300 kV microscope, *solid line* (C_S = −37µm, C_5 = 100 mm, Δf = 5 nm) , with the damping envelopes introduced by a beam divergence of 2.5 mrad (*dotted line*) and an energy spread of 0.3 eV (*dashed line*). Reproduced from Pennycook et al. (2007).

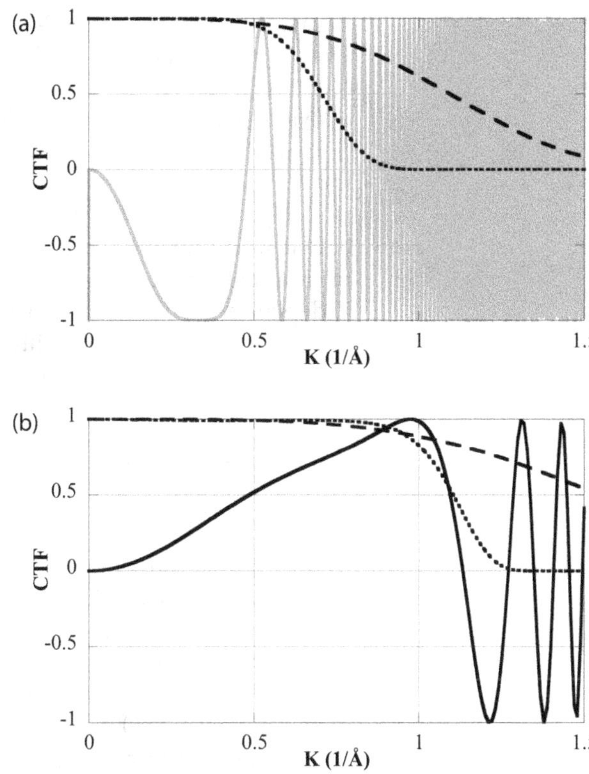

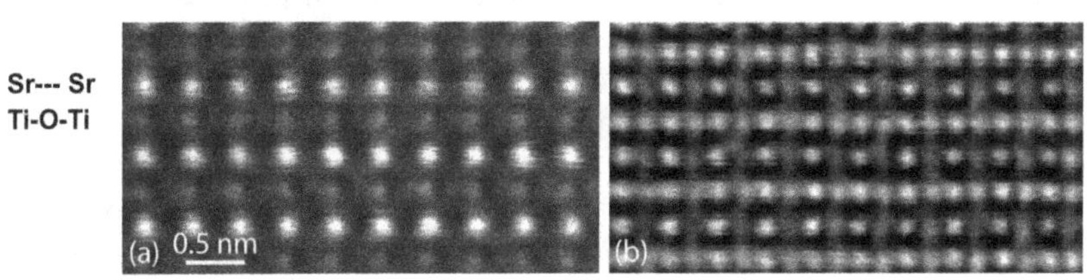

Figure 1–41. (**a**) Z-contrast and (**b**) phase contrast images of ⟨110⟩ $SrTiO_3$ taken with a VG Microscopes HB603U with Nion aberration corrector operating at 300 kV, using a defocus of +2 and +6 nm, respectively (raw data). Reproduced from Pennycook et al. (2004a).

(for the optimum probe) and uses a slightly negative C_S to balance the positive fifth-order spherical aberration. Hence the conditions are close to those used by Jia et al., although the optimum focus for the two images is slightly different and a focal series is useful (Pennycook 2006).

Besides higher resolution, aberration correction also brings the possibility of much higher currents, allowing roughly an order of magnitude more current to be focused into a probe the same size as before correction (Krivanek et al. 2003). This is very useful for low-intensity signals such as elemental mapping with X-rays (Watanabe et al. 2006), see Chapter 7.

Uncorrected STEMs were also bringing advances into a wide range of materials. To give just a few examples, Bi segregation sites were imaged in a Cu grain boundary and linked to embrittlement (Duscher et al. 2004); Sb segregation sites were seen at inversion domain boundaries in ZnO (Yamazaki et al. 2004); clustering of substitutional K was seen in $PbZrO_3$ (Viehland et al. 2004); and new shell-like and planar arrangements of Ag atoms were seen in the early stages of aging of Al–Ag alloys (Konno et al. 2004). An interesting approach to image quantification in this system was presented by Erni et al. (2003). They measured the number of Ag atoms in columns of Al directly from the image intensity. The method relies on detecting the difference in image intensity between columns that differ in Ag content by one atom as determined from a histogram of column intensities. Then, assuming a power law Z-dependence the specimen thickness could be determined and was found to be in agreement with the thickness estimated by EELS. Clearly the method assumes that the HAADF image is a true incoherent Z-contrast image and cannot be expected to be as accurate as a full simulation that includes all depth-dependent dynamical diffraction and thermal diffuse scattering, but it has the advantage of simplicity and appears to work well for heavy atoms in a light matrix.

Another area of major interest was that of semiconductor quantum dots and quantum wells for optoelectronic applications. Initial investigations with HAADF STEM showed evidence for compositional fluctuations (Duxbury et al. 2000) and led to attempts to map compositional profiles either without resolving the lattice (Crozier et al. 2003, Fewster et al. 2003, Tey et al. 2005) or from atomic column intensities (Takeguchi et al. 2004, Wallis et al. 2005). Monolayer fluctuations in well width were proposed to be the origin of the exciton localization in InGaN/GaN quantum-well structures (Graham et al. 2005). Wang et al. (2006) used both the HAADF intensity and the EELS composition mapping to obtain the shape and composition of InAs quantum dots in a GaAs matrix. The size and shape of the dot was obtained from the Z-contrast image, then the In concentration was obtained by mapping the As EELS signal. With an aberration-corrected STEM the method was extended to quaternary alloys (Molina et al. 2007a) and combined with finite element analysis to produce strain maps (Molina et al. 2006). They showed that an asymmetric stress distribution meant that nanowire arrays would grow with a slight angle to the growth direction and found good agreement between calculations and observations. The origin of the asymmetry was traced to the nucleation of the nanowires at step edges (Molina et al. 2007b), the preferential nucleation site being the upper terrace in a strained system. The method was later extended to use compositions extracted from Z-contrast image intensities (Molina et al. 2008, 2009).

Van Aert et al. (2009) have recently presented another method for quantification of column composition directly from the HAADF image intensities. In their method they parameterize atomic columns of known composition, not only their scattering strength but also their width, modeled as a Gaussian. This allows the apparent background in between the columns to vary, as it is known to do experimentally. Figure 1–42 shows a histogram of known column intensities from a $La_{0.7}Sr_{0.3}MnO_3$–$SrTiO_3$ multilayer structure, where all the different column types are well separated. Columns from the interface region showing intermediate intensity could then be identified as mixed, as

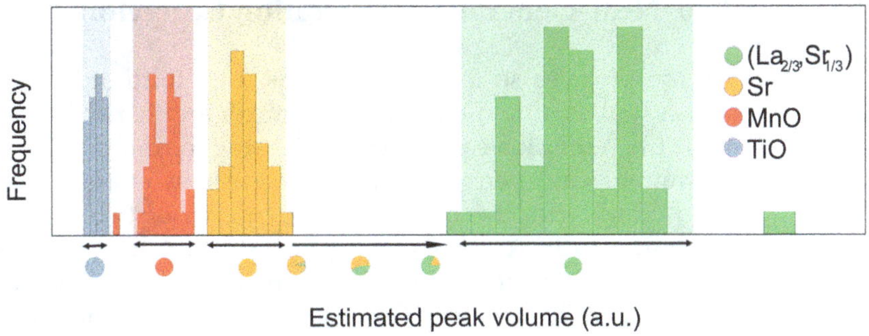

Figure 1–42. Histograms of the estimated peak volumes of known columns in a $La_{0.7}Sr_{0.3}MnO_3$–$SrTiO_3$ multilayer structure. The colored vertical bands represent the corresponding tolerance intervals. Unknown columns near the interfaces can be identified by comparing their estimated peak volumes with these tolerance intervals. Reproduced from Van Aert et al. (2009) with permission.

shown in Figure 1–43. Such parameterization methods are much faster than full simulations. Based on intensity measurements, Luysberg et al. (2009) have observed an interesting 2×1 interfacial reconstruction at a $SrTiO_3/DyScO_3$ interface. The presence of a half plane of Dy at the interface presumably helps to disperse the valence mismatch between a full $(DyO)^{+1}$ cation layer and the neutral $(TiO_2)^0$ layer. An approximation to the frozen phonon method was presented by Croitoru et al. (2006), which gave reasonable agreement with the full simulations but an increase in speed by a factor of 3–5.

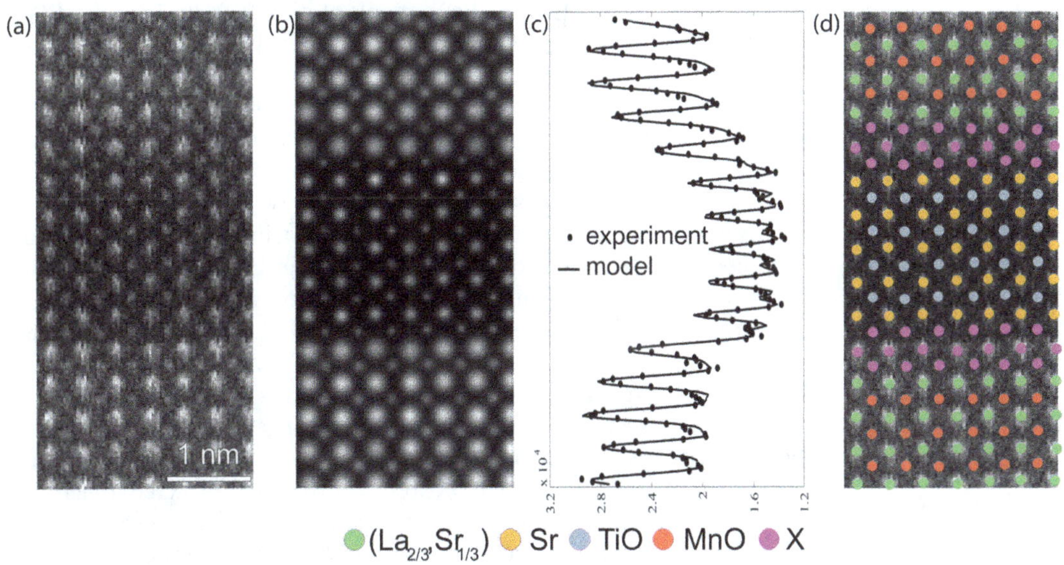

Figure 1–43. (**a**) HAADF STEM image of a $La_{0.7}Sr_{0.3}MnO_3$–$SrTiO_3$ multilayer structure along the [001] zone axis taken using a FEI Titan 80-300 microscope operated at 300 kV. (**b**) Refined model. (**c**) Experimental data (**a**) and refined model (**b**) averaged along the horizontal direction. (**d**) Overlay indicating the estimated positions of the columns together with their atomic column types. The columns whose composition is unknown are indicated by the symbol "X." Reproduced from Van Aert et al. (2009).

1.10 Next-Generation Aberration Correctors

All the progress so far described has used third-order correctors, in which the resolution-limiting aberrations are therefore of fifth order. With 1 Å resolution easily surpassed using third-order correctors, consideration was soon given to the correction of these higher order aberrations, with the goal of another factor of two improvement in resolution, to 0.5 Å. Haider et al. (2000) showed that 0.5 Å was theoretically achievable with a fifth-order corrector and an accelerating voltage greater than 200 kV. In 2003 Krivanek et al. presented a new design for a quadrupole/octapole-based corrector, anticipated to be capable of achieving 0.5 Å resolution at 200 kV accelerating voltage. In 2006 Müller et al. showed how the sextupole design could be improved to allow fifth-order correction of geometric aberrations, again with the conclusion that 0.5 Å could be achieved electron-optically with sufficient reduction of parasitic aberrations and instabilities.

A number of projects were initiated around the world to attempt to realize these alluring goals. In the USA, the Department of Energy Transmission Electron Aberration-Corrected Microscope (TEAM) project was begun in collaboration with FEI and CEOS, aiming to achieve 0.5 Å resolution in both TEM and STEM, and in Japan the Core Research for Evolutional Science and Technology (CREST) project started development of the R005 (resolution 0.05 nm) instrument with JEOL. In 2006 a Titan 80-300 was delivered to Oak Ridge National Laboratory equipped with a CEOS third-order corrector and Schottky gun and achieved sub-angstrom resolution at 300 kV accelerating voltage with ⟨112⟩ Ge (Pennycook et al. 2010). In 2007 the R005 project demonstrated a resolution of 0.63 Å in [211] GaN using their 300 kV cold field emission system (Sawada et al. 2007). The same year, the Oak Ridge machine was upgraded with a prototype high brightness Schottky gun and the improved CEOS fifth-order corrector (Müller et al. 2006), also eventually achieving 0.63 Å resolution in [211] GaN (Lupini et al. 2009). Meanwhile, an improved column was under development for Lawrence Berkeley National Laboratory, using a higher resolution pole piece and a better column suspension and isolation system and demonstrated 0.63 Å resolution in 2008 (Kisielowski et al. 2008). In 2009, both projects achieved the 0.5 Å goal with a HAADF image of dumbbells in ⟨114⟩ Ge spaced just 0.47 Å apart (Erni et al. 2009, Sawada et al., 2009), see Figure 1–44.

In terms of resolution, the STEM has clearly benefited more from aberration correction than the TEM, since we have seen more than a factor of two improvement in resolution over uncorrected machines making new applications possible in many fields (for a recent review of applications, see Pennycook et al., 2009b). Aberration correction in STEM has finally overcome the historic limitations of noise, and for the first time the STEM has held the record for resolution over the TEM, as physics says it should, incoherent imaging having higher theoretical resolution (Rayleigh 1896, Scherzer 1949). In TEM the benefits have been more precision in quantification rather than image resolution, as mentioned before, allowing quantitative analysis of O concentration in

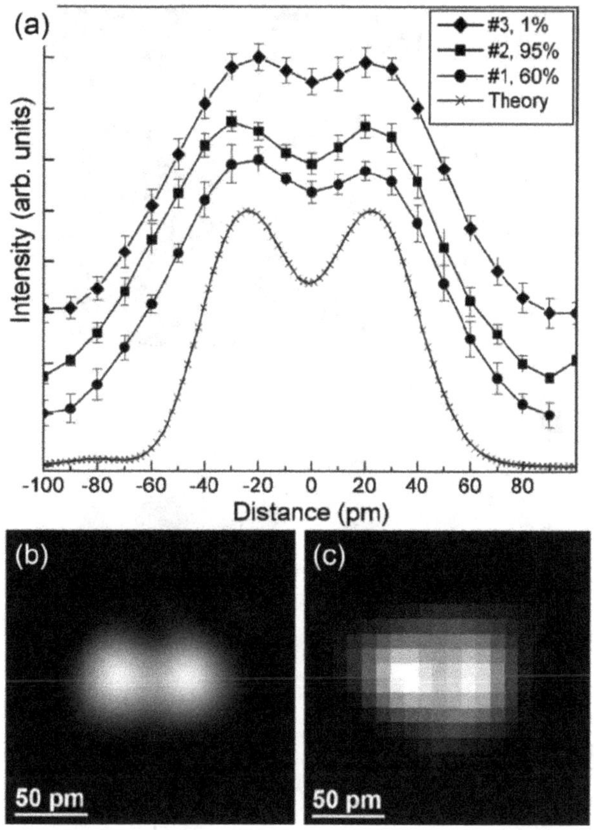

Figure 1–44. (**a**) Representative intensity profiles averaged over 12–13 dumbbells of a Z-contrast image from ⟨114⟩ Ge taken with the TEAM 0.5 microscope showing resolution of the 47 pm spacing. *Solid line* is the theoretical curve. (**b**) Calculated dumbbell image. (**c**) Averaged experimental image. Reproduced from Erni et al. (2009).

dislocation cores (Jia et al. 2005, Jia 2006), the effect of a dislocation on local ferroelectric polarization (Jia et al. 2009b), and the tracking of octahedral reconstruction across a complex oxide interface (Jia et al. 2009a). Such measurements are the key to structure property correlation in these materials (Urban 2008).

Similar precision is achievable in STEM, either with the phase contrast bright field image or with the HAADF image (Saito et al. 2009), with the advantage that the HAADF image is available in thicker regions of sample so that potential issues of surface relaxation or damage are less of a concern. Figure 1–45 shows how oxygen octahedral rotation angles in $BiFeO_3$ can be measured to a high accuracy by STEM bright field imaging (Borisevich et al. 2010). The reduced noise in STEM has also made possible the imaging of individual light atoms in a Z-contrast image, allowing the resolution and identification of B, C, N, and O atoms in monolayer BN, with the identification of substitutional and adatom defects (Krivanek et al. 2010), see Chapter 15. Single dopant atoms have also been imaged at a buried interface, allowing their distribution in the interface plane to be directly observed (Shibata et al. 2009), see Chapter 11.

We have also seen the first imaging of point defect configurations (Oh et al. 2008). Figure 1–46 shows images of single gold atoms inside Si in substitutional and several interstitial configurations, obtained with

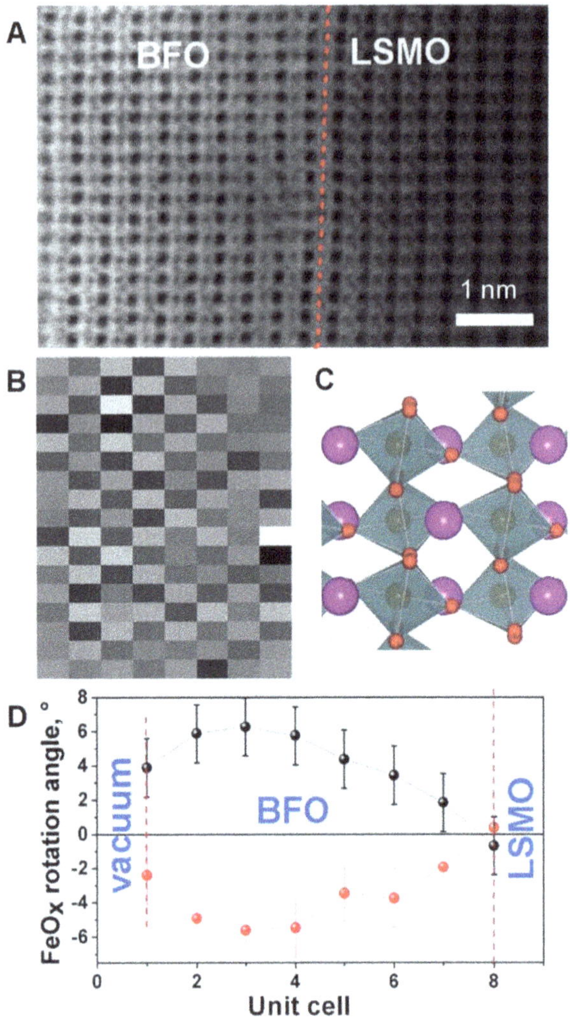

Figure 1–45. Quantitative measurement of oxygen octahedral rotation angles using a VG Microscopes HB603U with Nion aberration corrector operating at 300 kV. (**a**) Bright field image of a $BiFeO_3/La_{0.7}Sr_{0.3}MnO_3$ ultrathin film on $SrTiO_3$. (**b**) Corresponding two-dimensional map of in-plane octahedral rotation angles in BFO showing checkerboard ordering. (**c**) BFO structure in the rhombohedral (001) orientation showing the tilt pattern. (**d**) Line profile obtained from the map in (b), which was corrected for local Bi–Bi angle variations due to drift. Error bars in (d) are set equal to the standard deviation of the local Bi–Bi angles. Reproduced from Borisevich et al. (2010).

a 300 kV STEM and third-order corrector. It was assumed that the Au atoms inside the Si were most likely injected there by the electron beam; however, Allen et al. (2008) also imaged gold atoms inside Si nanowires, finding them to segregate to a twin boundary. They concluded from the lack of concentration gradient along the nanowire that the Au atoms were incorporated during growth. The most sensitive detection of a point defect is with a third-order-corrected TEM, the detection of self interstitials in Ge (Alloyeau et al. 2009). O interstitial impurities in α-Si_3N_4 have been imaged and identified spectroscopically by EELS (Idrobo et al. 2009), see Figure 1–47, antisite defects have been seen in $LiFePO_4$ (Chung et al. 2008, 2009), and individual Eu dopant atoms have been imaged in a β-SiAlON phosphor (Kimoto et al. 2009).

EELS performance is enormously enhanced, with better collection efficiency and more available current if required. The first two-dimensional EELS maps were reported using just a third-order corrector (Bosman et al. 2007, see Figure 6–12), achieving much higher

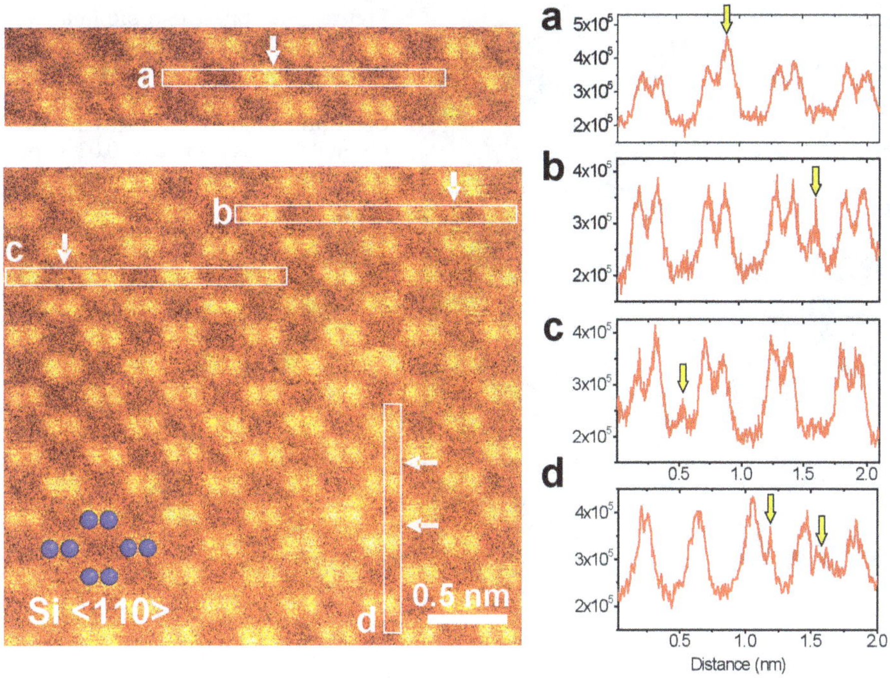

Figure 1–46. HAADF images of a Si nanowire in ⟨110⟩ zone axis orientation (*left panel*) taken with a VG Microscopes HB603U with Nion aberration corrector operating at 300 kV. *Boxes* show the regions used for intensity profiles, with Au atoms in various configurations *arrowed*; (**a**) substitutional; (**b**) tetrahedral; (**c**) hexagonal; (**d**) buckled Si–Au–Si chain configurations. The intensity profiles across the Si dumbbells correspond to a width of 18 pixels. Individual Au atoms are *arrowed*, adapted from Oh et al. (2008).

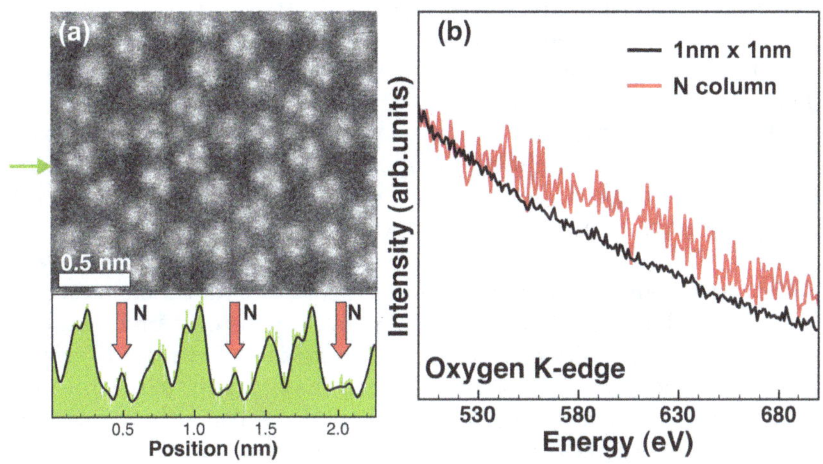

Figure 1–47. (**a**) *Z*-contrast image of α-Si_3N_4 with a line trace across the arrowed position showing unexpectedly strong intensity from particular N columns. (**b**) EELS from these columns reveals the presence of O (*red trace*). The *black trace* is the spectrum obtained from scanning a larger region, when no O is detectable. Reproduced from Idrobo et al. (2009).

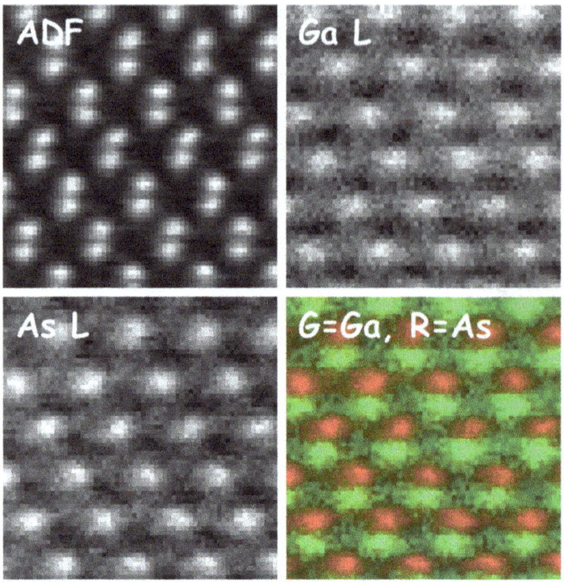

Figure 1–48. Spectroscopic imaging of GaAs in the ⟨110⟩ projection comparing the ADF image to the Ga- and As L spectroscopic images, obtained on the Nion UltraSTEM operating at 100 kV. Images are 64 × 64 pixels, with collection time 0.02 s/pixel and a beam current of approximately 100 pA, after noise reduction by principle component analysis (Varela et al. 2009), recorded by M. Varela.

efficiency than an uncorrected microscope could provide (Kimoto et al. 2007). With the fifth-order corrector further enhancement in collection efficiency was seen, giving recognizable atomic resolution images using just a small energy window (Muller et al. 2008), see Figure 1–17. Such capability should greatly facilitate atomic resolution mapping of fine structure (Varela et al. 2009), as discussed in Chapter 10. An impressive example of the spectroscopic imaging of GaAs is given in Figure 1–48.

Major progress has also been achieved in obtaining agreement between theoretical image simulations and experiment. The so-called Stobbs factor (Hÿtch and Stobbs 1994), the ratio between theoretical and experimental image contrast, was shown not to exist for HAADF STEM (LeBeau and Stemmer 2008, LeBeau et al. 2008). Absolute intensities in experimental images were compared to both Bloch wave and frozen phonon simulations, as shown in Figure 6–11. Quantitative agreement was obtained in both methods for thin crystals. For thicker crystals the frozen phonon method is more accurate since it does not lose the "absorbed" electrons as does the Bloch wave method. Accurate Debye–Waller factors are required, which may be different for different atomic columns (LeBeau et al. 2009b). This solves a historic problem and shows that the physics of electron scattering is sufficiently described by present methods. Although the comparison used an uncorrected microscope it is reasonable to assume that similar agreement with aberration-corrected data would be obtained provided the aberrations were well characterized.

As part of this work a new method for the accurate determination of thickness was developed, position-averaged convergent beam electron diffraction, which is less sensitive to surface effects than EELS methods and is accurate to ±2–4 nm (LeBeau et al. 2009b). In 2009 the Stobbs factor for the TEM phase contrast image was traced to an underestimate of the effect of the point spread function in the detector (Thust 2009). STEM bright field images were also shown to be free of a Stobbs factor (LeBeau et al. 2009a). It has also been shown that the established Young's fringe method for determining the information limit of phase contrast images is not valid (Barthel and Thust 2008). Inserting an objective aperture to limit the frequencies transferred to the image, the resulting Young's fringes extended significantly beyond the aperture cutoff, indicating false detail due to nonlinearities in the imaging. Also in 2009 a new derivation appeared that showed in rigorous but transparent way how the frozen phonon model is equivalent to a full quantum-mechanical treatment of the inelastic phonon scattering process (Van Dyck 2009).

A significant development in three-dimensional imaging was also achieved in recent years with the introduction of a true confocal mode. As initially used on an uncorrected microscope (Frigo et al. 2002), the pinhole detector did not allow depth sectioning by changing objective lens focus. For this reason Takeguchi et al. (2008) introduced a stage scanning system and demonstrated atomic resolution in the lateral plane. Meanwhile, in 2005 the first double-corrected microscope was installed at the University of Oxford (Hutchison et al. 2005, Sawada et al. 2005) and was soon used in a confocal mode (Nellist et al. 2006, 2008). The problem with the simple focal series method of depth sectioning is that while it works reasonably well for point objects, for larger objects the depth resolution is substantially degraded to a value of d/α, where d is the lateral size of the object and α is the semiangle of the probe-forming aperture. This can be 100 nm or more for a particle just a few nanometers in diameter. The physical reason is that the probe needs to be defocused until its lateral extent is comparable to the size of the object before a significant change in scattered intensity will be seen. The confocal mode of operation overcomes this limitation, filling in the missing wedge in the otherwise propeller-shaped transfer function (D'Alfonso et al. 2008, Xin and Muller 2009). Image simulations suggested that 1 nm depth resolution could be achieved with fifth-order correctors (Einspahr and Voyles 2006). Similar considerations apply to EELS, and theoretical studies showed how optical sectioning would give much improved elemental selectivity (D'Alfonso et al. 2007). However, again the confocal mode provides better localization of the signal (D'Alfonso et al. 2008). A number of theoretical analyses have also appeared (Cosgriff and Nellist 2007, Cosgriff et al. 2008), see Chapter 2.

In the absence of a confocal mode, there have been attempts to use deconvolution to reconstruct the missing wedge in depth sectioning. Behan et al. (2009) have shown that use of some a priori knowledge can be effective, such as the assumption of sharp edged spherical particles. One disadvantage of the confocal mode compared to a simple optical sectioning mode is its relative inefficiency in the use of electrons. This is an important consideration for biological material and makes deconvolution more attractive to minimize beam exposure. De Jonge et al. (2010) showed that deconvolution could reduce the FWHM in the depth

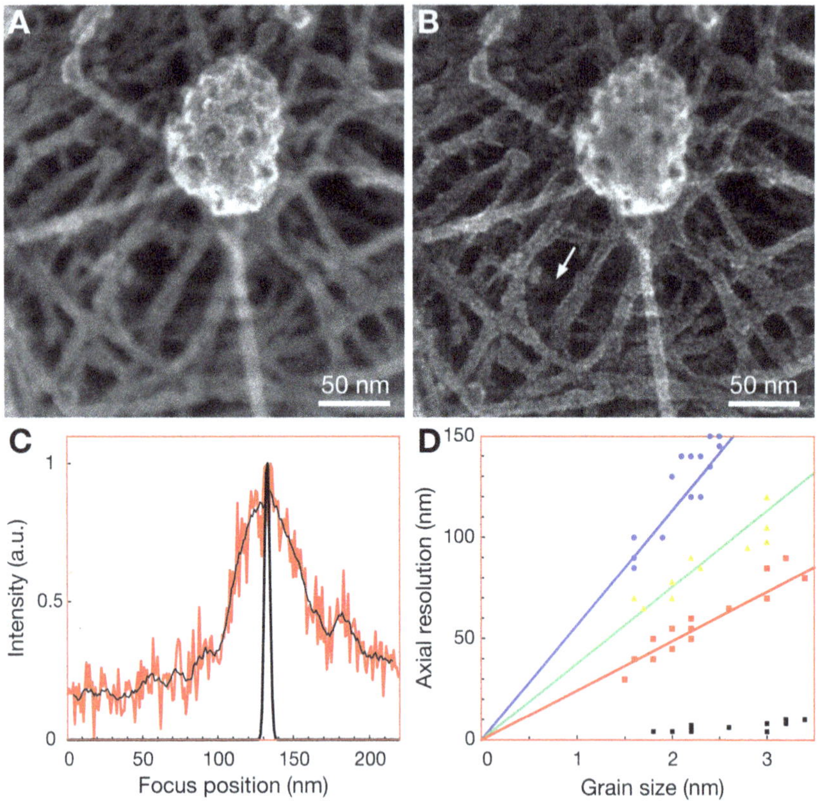

Figure 1–49. STEM images of a clathrin-coated pit and parts of the cytoskeleton of mammalian cells. (**a, b**) are from a through-focal series differing 67 nm in focus (*vertical*) position. (**c, d**) Horizontal slices from a data set after deconvolution of the point spread function differing by 20 nm in vertical position. (**e**) Line scan in the vertical direction over the grain indicated by the *arrow* in (b) (*red*). The line scan at the same position in the data set after deconvolution is shown as a *thick black line*, showing a FWHM of the depth profile reduced by an order of magnitude. (**f**) The vertical resolution was determined as a function of the grain size for several particles and for three different beam semi-angles, 41 mrad (*red*), 26.5 mrad (*green*), and 17.7 mrad (*blue*). The theoretical prediction is shown as lines of the corresponding colors. The vertical FWHM measured on the same grains but after deconvolution is shown as *black squares*. Adapted from de Jonge et al. (2010).

profile of a 2.2 nm sized particle from 55 nm to just over 5 nm as shown in Figure 1–49.

1.11 Outlook

It seems unlikely that we will see the implementation of seventh-order aberration correctors due to the increasing complexity and diminishing returns on resolution. Furthermore, especially for the lower beam voltages popular to avoid knock-on damage, chromatic aberration becomes the limiting factor even with today's available objective apertures. Chromatic aberration pushes current from the central peak into the probe tails and therefore tends to limit contrast more than it does resolution (Krivanek et al. 2003, Intaraprasonk et al. 2008). In TEM, use of a monochromator can successfully improve the information limit (Freitag et al. 2005), but this is not an attractive route in STEM since it reduces the probe current and reintroduces the problems of noise. More attractive is the correction of chromatic aberration, recently successfully demonstrated in TEM (Kabius et al. 2009), and there are two proposals for a combined spherical and chromatic aberration corrector for the STEM (Haider et al. 2009, Krivanek et al. 2009b).

We are likely therefore to see a trend toward reducing beam voltage, in an attempt to reduce knock-on damage. However, ionization damage increases with lower accelerating voltage, and the optimum voltage is very material dependent. We are also likely to see more developments for stable, atomic resolution in situ observation, superconductors at liquid helium temperature, for example. Another obvious development is the use of alternative imaging signals, such as cathodoluminescence for correlating defect atomic structure and impurity segregation with optical properties (Pennycook 2008) or electron beam-induced current for mapping solar cell efficiencies at the nanoscale. We can anticipate STEM combined with scanning tunneling microscopy, applying electric fields to watch transformation processes develop, such as the nucleation of domain boundaries in ferroelectrics or ionic migration mechanisms in energy storage materials. Many scanning probe techniques exist that map functionality, but the link between functionality and defects can best be established by correlating with STEM observations.

There is one thing, however, that can be said with certainty – the future of STEM has never been brighter.

Acknowledgments The author would like to express his deep appreciation to L. A. Allen, O. L. Krivanek, and P. W. Hawkes for valuable comments on the chapter, to P. W. Hawkes and O. L. Krivanek for photographs of the microscopes in Figures 1–1 and 1–3, respectively, and to his many colleagues who have contributed to the work presented here, especially, L. A. Boatner, K. van Benthem, N. D. Browning, A. Y. Borisevich, M. F. Chisholm, H. M. Christen, N. Dellby, V. P. Dravid, G. Duscher, J. C. Idrobo, D.E. Jesson, N. de Jonge, O. L. Krivanek, J. T. Luck, A. R. Lupini, A. J. McGibbon, M. M. McGibbon, S. I. Molina, M. F. Murfitt, J. Narayan, P. D. Nellist, S. H. Oh, M. P. Oxley, S.T. Pantelides, Y. Peng, W. H. Sides, Z. S. Szilagyi, and M. Varela, which was supported largely by the Materials Sciences and Engineering Division, US Department of Energy.

References

E. Abe, S.J. Pennycook, A.P. Tsai, Direct observation of a local thermal vibration anomaly in a quasicrystal. Nature **421**, 347–350 (2003)

G. Ade, On the incoherent imaging in the scanning transmission electron microscope (STEM). Optik **49**, 113–116 (1977)

J.E. Allen, E.R. Hemesath, D.E. Perea, J.L. Lensch-Falk, Z.Y. Li, F. Yin, M.H. Gass, P. Wang, A.L. Bleloch, R.E. Palmer, L.J. Lauhon, High-resolution detection of Au catalyst atoms in Si nanowires. Nat. Nanotechnol. **3**, 168–173 (2008)

L.J. Allen, S.D. Findlay, A.R. Lupini, M.P. Oxley, S.J. Pennycook, Atomic-resolution electron energy loss spectroscopy imaging in aberration corrected scanning transmission electron microscopy. Phys. Rev. Lett. **91**, 105503 (2003a)

L.J. Allen, S.D. Findlay, M.P. Oxley, C.J. Rossouw, Lattice-resolution contrast from a focused coherent electron probe. Part I. Ultramicroscopy **96**, 47–63 (2003b)

L.J. Allen, S.D. Findlay, M.P. Oxley, C. Witte, N.J. Zaluzec, Channeling effects in high-angular-resolution electron spectroscopy. Phys. Rev. B **73**, 094104 (2006)

L.J. Allen, T.W. Josefsson, C.J. Rossouw, Interaction delocalization in characteristic x-ray-emission from light-elements. Ultramicroscopy **55**, 258–267 (1994)

D. Alloyeau, B. Freitag, S. Dag, L.W. Wang, C. Kisielowski, Atomic-resolution three-dimensional imaging of germanium self-interstitials near a surface: Aberration-corrected transmission electron microscopy. Phys. Rev. B **80**, 014114 (2009)

A. Amali, P. Rez, Theory of lattice resolution in high-angle annular dark-field images. Microsc. Microanal. **3**, 28–46 (1997)

S.C. Anderson, C.R. Birkeland, G.R. Anstis, D.J.H. Cockayne, An approach to quantitative compositional profiling at near-atomic resolution using high-angle annular dark field imaging. Ultramicroscopy **69**, 83–103 (1997)

G.R. Anstis, D.Q. Cai, D.J.H. Cockayne, Limitations on the s-state approach to the interpretation of sub-angstrom resolution electron microscope images and microanalysis. Ultramicroscopy **94**, 309–327 (2003)

I. Arslan, T.J.V. Yates, N.D. Browning, P.A. Midgley, Embedded nanostructures revealed in three dimensions. Science **309**, 2195–2198 (2005)

A. Bakenfelder, I. Fromm, L. Reimer, R. Rennekamp, Contrast in the electron spectroscopic imaging mode of a TEM.3. Bragg contrast of crystalline specimens. J. Microsc.-Oxford **159**, 161–177 (1990)

J. Barthel, A. Thust, Quantification of the information limit of transmission electron microscopes. Phys. Rev. Lett. **101**, 200801 (2008)

P.E. Batson Silicon $L_{2,3}$ core absorption obtained at the buried Al/Si(111) interface. Phys. Rev. B **44**, 5556–5561 (1991)

P.E. Batson, Simultaneous STEM imaging and electron energy-loss spectroscopy with atomic-column sensitivity. Nature **366**, 727–728 (1993)

P.E. Batson, Atomic and electronic structure of a dissociated 60 degrees misfit dislocation in $Ge_xSi_{(1-x)}$. Phys. Rev. Lett. **83**, 4409–4412 (1999a)

P.E. Batson, Atomic resolution EELS analysis of a misfit dislocation at a GeSi/Si interface. Physica B **274**, 593–597 (1999b)

P.E. Batson, N. Dellby, O.L. Krivanek, Sub-angstrom resolution using aberration corrected electron optics. Nature **418**, 617–620 (2002)

W. Baumeister, M.H. Hahn, Electron microscopy of monomolecular layers of thorium atoms. Nature **241**, 445–447 (1973)

P.F. Becher, G.S. Painter, N. Shibata, R.L. Satet, M.J. Hoffmann, S.J. Pennycook, Influence of additives on anisotropic grain growth in silicon nitride ceramics. Mater. Sci. Eng. A **422**, 85–91 (2006)

V. Beck, A.V. Crewe, High-resolution imaging properties of STEM. Ultramicroscopy **1**, 137–144 (1975)

M. Beer, E. Moudrianakis, Determination of base sequence in nucleic acids with the electron microscope: Visibility of a marker. Proc. Natl. Acad. Sci. USA. **48**, 409–416 (1962)

G. Behan, E.C. Cosgriff, A.I. Kirkland, P.D. Nellist, Three-dimensional imaging by optical sectioning in the aberration-corrected scanning transmission electron microscope. Philos. Trans. R. Soc. A **367**, 3825–3844 (2009)

S.D. Berger, S.J. Pennycook, Detection of nitrogen at (100) platelets in diamond. Nature **298**, 635–637 (1982)

G. Black, E.H. Linfoot, Spherical aberration and the information content of optical images. Proc. R. Soc. London A **239**, 522–540 (1957)

H. Boersch, Über die Kontraste von Atomen im Elektronenmikroskop. Z. Naturforschung **2a**, 615–633 (1947)

W. Bollmann, Interference effects in the electron microscopy of thin crystal foils. Phys. Rev. **103**, 1588–1589 (1956)

A.Y. Borisevich, H.J. Chang, M. Huijben, M.P. Oxley, S. Okamoto, M.K. Niranjan, J.D. Burton, E.Y. Tsymbal, Y.H. Chu, P. Yu, R. Ramesh, S.V. Kalinin, S.J. Pennycook, Suppression of octahedral tilts and associated changes in electronic properties at epitaxial oxide heterostructure interfaces. Phys. Rev. Lett., **105**, 087204 (2010)

A.Y. Borisevich, A.R. Lupini, S.J. Pennycook, Depth sectioning with the aberration-corrected scanning transmission electron microscope. Proc. Natl. Acad. Sci. USA **103**, 3044–3048 (2006a)

A.Y. Borisevich, A.R. Lupini, S. Travaglini, S.J. Pennycook, Depth sectioning of aligned crystals with the aberration-corrected scanning transmission electron microscope. J. Electron. Microsc. **55**, 7–12 (2006b)

M. Bosman, V.J. Keast, J.L. Garcia-Munoz, D'A.J. Alfonso, S.D. Findlay, L.J. Allen, Two-dimensional mapping of chemical information at atomic resolution. Phys. Rev. Lett. **99**, 086102 (2007)

M.J. Bowers, J.R. McBride, M.D. Garrett, J.A. Sammons, A.D. Dukes, M.A. Schreuder, T.L. Watt, A.R. Lupini, S.J. Pennycook, S.J. Rosenthal, Structure and ultrafast dynamics of white-light-emitting CdSe nanocrystals. J. Am. Chem. Soc. **131**, 5730–5731 (2009)

M. Brown, Scanning transmission electron microscopy: Microanalysis for the microelectronic age. J. Phys. F **11**, 1–26 (1981)

N. D. Browning, J. P. Buban, H. O. Moltaji, S. J. Pennycook, G. Duscher, K. D. Johnson, R. P. Rodrigues, V. P. Dravid, (1999a) The influence of atomic structure on the formation of electrical barriers at grain boundaries in $SrTiO_3$. Appl. Phys. Lett. **74**, 2638–2640

N.D. Browning, J.P. Buban, P.D. Nellist, D.P. Norton, M.F. Chisholm, S.J. Pennycook, The atomic origins of reduced critical currents at [001] tilt grain boundaries in $YBa_2Cu_3O_{7-\delta}$ thin films. Physica C **294**, 183–193 (1998b)

N.D. Browning, J.P. Buban, C. Prouteau, G. Duscher, S.J. Pennycook, Investigating the atomic scale structure and chemistry of grain boundaries in high-T_c superconductors. Micron **30**, 425–436 (1999b)

N.D. Browning, M.F. Chisholm, S.J. Pennycook, Atomic-resolution chemical-analysis using a scanning-transmission electron-microscope. Nature **366**, 143–146 (1993a)

N.D. Browning, M.F. Chisholm, S.J. Pennycook, Atomic-resolution electron energy loss spectroscopy in the scanning transmission electron microscope, *27th Annual Proceedings of Microbeam Analysis Society* (VCH, Los Angeles, 1993b), pp. S270–S271

N.D. Browning, M.F. Chisholm, S.J. Pennycook, Atomic-resolution chemical analysis using a scanning transmission electron microscope (**366**, 143, 1993). Nature **444**, 235–235 (2006)

N.D. Browning, M.F. Chisholm, S.J. Pennycook, D.P. Norton, D.H. Lowndes, Correlation between hole depletion and atomic-structure at high-angle grain-boundaries in $YBa_2Cu_3O_{7-\delta}$ Physica C **212**, 185–190 (1993c)

N.D. Browning, H.O. Moltaji, and J.P. Buban, Investigation of three-dimensional grain-boundary structures in oxides through multiple-scattering analysis of spatially resolved electron-energy-loss spectra. Phys. Rev. B **58**, 8289–8300 (1998a)

N.D. Browning, S.J. Pennycook, M.F. Chisholm, M.M. McGibbon, A.J. McGibbon, Observation of structural units at symmetric [001] tilt boundaries in $SrTiO_3$. Interface Sci. **2**, 397–423 (1995)

J. Buban, K. Matsunaga, J. Chen, N. Shibata, W. Ching, T. Yamamoto, Y. Ikuhara, Grain boundary strengthening in alumina by rare earth impurities. Science **311**, 212 (2006)

R.E. Burge, J.C. Dainty, Partially coherent image-formation in scanning-transmission electron-microscope (STEM). Optik **46**, 229–240 (1976)

R. Castaing, L. Henry, Filtrage magnetique des vitesses en microscopie electronique. Comptes Rendus Hebdomadaires Des Seances De L'Academie Des Sciences Serie B **255**, 76–78 (1962)

D. Cherns, A. Howie, M.H. Jacobs, Characteristic x-ray production in thin crystals. Z. Naturforsch. A **28**, 565–571 (1973)

M.F. Chisholm, N.D. Browning, S.J. Pennycook, R. Jebasinski, S. Mantl, Z-contrast investigation of the ordered atomic interface of $CoSi_2$/Si(001) layers. Appl. Phys. Lett. **64**, 3608–3610 (1994a)

M.F. Chisholm, A. Maiti, S.J. Pennycook, S.T. Pantelides, Atomic configurations and energetics of arsenic impurities in a silicon grain boundary. Phys. Rev. Lett. **81**, 132–135 (1998)

M.F. Chisholm, S.J. Pennycook, Structural origin of reduced critical currents at $YBa_2Cu_3O_{7-\Delta}$ grain-boundaries. Nature **351**, 47–49 (1991)

M.F. Chisholm, S.J. Pennycook, Direct imaging of dislocation core structures by Z-contrast STEM. Philos. Mag. **86**, 4699–4725 (2006)

M.F. Chisholm, S.J. Pennycook, R. Jebasinski, S. Mantl, New interface structure for A-type $CoSi_2$/Si(111). Appl. Phys. Lett. **64**, 2409–2411 (1994b)

S.Y. Chung, S.Y. Choi, T. Yamamoto, Y. Ikuhara, Atomic-scale visualization of antisite defects in $LiFePO_4$. Phys. Rev. Lett. **100**, 125502 (2008)

S.Y. Chung, S.Y. Choi, T. Yamamoto, Y. Ikuhara, Orientation-dependent arrangement of antisite defects in lithium iron(II) phosphate crystals. Ang. Chem.-Int. Edn. **48**, 543–546 (2009)

M.D. Cole, J.W. Wiggins, M. Beer, Molecular microscopy of labeled polynucleotides: stability of osmium atoms. J. Mol. Biol. **117**, 387–400 (1977)

C. Colliex, An illustrated review of various factors governing the high spatial-resolution capabilities in EELS microanalysis. Ultramicroscopy **18**, 131–150 (1985)

C. Colliex, V.E. Cosslett, R.D. Leapman, P. Trebbia, Contribution of electron energy loss spectroscopy to the development of analytical electron microscopy. Ultramicroscopy **1**, 301–315 (1976)

C. Colliex, C. Jeanguillaume, C. Mory, Unconventional modes for STEM imaging of biological structures. J. Ultrastruct. Res. **88**, 177–206 (1984)

C. Colliex, B. Jouffrey, Diffusion inelastique des electrons dans un solide par excitation de niveaux atomiques profonds – I. Spectres de pertes energie. Philos. Mag. **25**, 491–511 (1972)

C. Colliex, P. Trebbia, Performance and applications of electron energy loss spectroscopy in STEM. Ultramicroscopy **9**, 259–266 (1982)

E.C. Cosgriff, A.J. D'Alfonso, L.J. Allen, S.D. Findlay, A.I. Kirkland, P.D. Nellist, Three-dimensional imaging in double aberration-corrected scanning confocal electron microscopy, part I: Elastic scattering. Ultramicroscopy **108**, 1558–1566 (2008)

E.C. Cosgriff, P.D. Nellist, A Bloch wave analysis of optical sectioning in aberration-corrected STEM. Ultramicroscopy **107**, 626–634 (2007)

E.C. Cosgriff, M.P. Oxley, L.J. Allen, S.J. Pennycook, The spatial resolution of imaging using core-loss spectroscopy in the scanning transmission electron microscope. Ultramicroscopy **102**, 317–326 (2005)

V.E. Cosslett, Possibilities and limitations for the differentiation of elements in the electron microscope. Lab. Invest. **14**, 1009–1019 (1965)
J.M. Cowley, Image contrast in a transmission scanning electron microscope. Appl. Phys. Lett. **15**, 58–59 (1969)
J.M. Cowley, High-resolution dark-field electron-microscopy. 1. Useful approximations. Acta Cryst. A, A **29**, 529–536 (1973)
J.M. Cowley, Scanning-transmission electron-microscopy of thin specimens. Ultramicroscopy **2**, 3–16 (1976)
J.M. Cowley, Scanning transmission electron microscopy and microdiffraction techniques. Bull. Mater. Sci. **6**, 477–490 (1984)
J.M. Cowley, High-resolution electron-microscopy and microdiffraction. Ultramicroscopy **18**, 11–17 (1985)
J.M. Cowley, Electron-diffraction phenomena observed with a high-resolution STEM instrument. J. Electron Microsc. Tech. **3**, 25–44 (1986a)
J.M. Cowley, in *Principles of Analytical Electron Microscopy*, ed. by J.J. Hren, J.I. Goldstein, D.C. Joy (Plenum, New York, NY, 1986b), pp. 343–368
J.M. Cowley, in *High Resolution Electron Microscopy and Associated Techniques*, 1 st edn, ed. by P.R. Buseck, J.M. Cowley, L. Eyring (Oxford University Press, New York, NY, 1988), pp. 38–57
J.M. Cowley, A.Y. Au, Diffraction by crystals with planar faults. 3. Structure analyses using microtwins. Acta Cryst. A **34**, 738–743 (1978)
J.M. Cowley, D.J. Smith, The present and future of high-resolution electron-microscopy. Acta Cryst. A **43**, 737–751 (1987)
J.M. Cowley, J.C.H. Spence, Innovative imaging and micro-diffraction in STEM. Ultramicroscopy **3**, 433–438 (1978)
J.M. Cowley, J.C.H. Spence, Convergent beam electron micro-diffraction from small crystals. Ultramicroscopy **6**, 359–366 (1981)
A.J. Craven, C. Colliex, High-resolution energy filtered images in STEM. J. Microsc. Spect. Elec. **2**, 511–522 (1977)
A.J. Craven, J.M. Gibson, A. Howie, D.R. Spalding, Study of single-electron excitations by electron-microscopy. 1. Image-contrast from delocalized excitations. Philos. Mag. A **38**, 519–527 (1978)
Core Research for Evolutional Science and Technology. http://www.busshitu.jst.go.jp/en/index.html (accessed Nov. 3, 2010)
A.V. Crewe, Scanning electron microscopes – is high resolution possible. Science **154**, 729–738 (1966)
A.V. Crewe, High resolution scanning microscopy of biological specimens. Philos. Trans. R. Soc. B **261**, 61–70 (1971)
A.V. Crewe, Scanning-transmission electron-microscopy. J. Microsc.-Oxford **100**, 247–259 (1974)
A.V. Crewe, Direct imaging of single atoms and molecules using the STEM. Chemica Scripta **14**, 17–20 (1979)
A.V. Crewe, High-resolution scanning-transmission electron-microscopy. Science **221**, 325–330 (1983)
A.V. Crewe, in *Advances in Imaging and Electron Physics*, ed. by P.W. Hawkes (Elsevier, Amsterdam, The Netherlands, 2009), pp. 1–61
A.V. Crewe, D.A. Crewe, O.H. Kapp, Inexact 3-dimensional reconstruction of a biological macromolecule from a restricted number of projections. Ultramicroscopy **13**, 365–371 (1984)
A.V. Crewe, M. Isaacson, D. Johnson, Electron energy loss spectra of nucleic acid bases. Nature **231**, 262–263 (1971)
A.V. Crewe, J.P. Langmore, M.S. Isaacson, in *Physical Aspects of Electron Microscopy and Microbeam Analysis*, ed. by B.M. Siegel, D.R. Beaman (Wiley, New York, NY, 1975), pp. 47–62

A.V. Crewe, J. Wall, A scanning microscope with 5 Å resolution. J. Mol. Biol. **48**, 375–393 (1970)

A.V. Crewe, J. Wall, J. Langmore, Visibility of single atoms. Science **168**, 1338–1340 (1970)

A.V. Crewe, J. Wall, L.M. Welter, A high-resolution scanning transmission electron microscope. J. Appl. Phys. **39**, 5861–5868 (1968)

M.D. Croitoru, D. Van Dyck, S. Van Aert, S. Bals and J. Verbeeck, An efficient way of including thermal diffuse scattering in simulation of scanning transmission electron microscopic images, Ultramicroscopy, **106**, 933–940 (2006)

P.A. Crozier, M. Catalano, R. Cingolani, A modeling and convolution method to measure compositional variations in strained alloy quantum dots. Ultramicroscopy **94**, 1–18 (2003)

S.L. Cundy, A. Howie, U. Valdrè, Preservation of electron microscope image contrast after inelastic scattering. Philos. Mag. **20**, 147–163 (1969)

S.L. Cundy, A.J. Metherel, M.J. Whelan, An energy analysing electron microscope. J. Sci. Instrum. **43**, 712–715 (1966)

S.L. Cundy, A.J. Metherel, M.J. Whelan, Contrast preserved by elastic and quasi-elastic scattering of fast electrons near Bragg beams. Philos. Mag. **15**, 623–630 (1967)

A.J. D'Alfonso, E.C. Cosgriff, S.D. Findlay, G. Behan, A.I. Kirkland, P.D. Nellist, L.J. Allen, Three-dimensional imaging in double aberration-corrected scanning confocal electron microscopy, part II: Inelastic scattering. Ultramicroscopy **108**, 1567–1578 (2008)

A.J. D'Alfonso, S.D. Findlay, M.P. Oxley, S.J. Pennycook, K. van Benthem, L.J. Allen, Depth sectioning in scanning transmission electron microscopy based on core-loss spectroscopy. Ultramicroscopy **108**, 17–28 (2007)

N. De Jonge, R. Sougrat, B.M. Northan, S.J. Pennycook, Three-dimensional scanning transmission electron microscopy of biological specimens. Microsc. Microanal. **16**, 54–63 (2010)

N. de Jonge, R. Sougret, D.B. Peckys, A.R. Lupini, S.J. Pennycook, in *Nanotechnology in Biology and Medicine*, ed. by T. Vo-Dinh (Taylor and Francis, Boca Raton, FL, 2007), pp. 13.11–13.27

N. Dellby, O.L. Krivanek, P.D. Nellist, P.E. Batson, A.R. Lupini, Progress in aberration-corrected scanning transmission electron microscopy. J. Electron. Microsc. **50**, 177–185 (2001)

E.C. Dickey, V.P. Dravid, P.D. Nellist, D.J. Wallis, S.J. Pennycook, Three-dimensional atomic structure of NiO-ZrO_2(cubic) interfaces. Acta Mater. **46**, 1801–1816 (1998)

E.C. Dickey, V.P. Dravid, P.D. Nellist, D.J. Wallis, S.J. Pennycook, A. Revcolevschi, Structure and bonding at Ni-ZrO_2 (cubic) interfaces formed by the reduction of a NiO-ZrO_2 (cubic) composite. Microsc. Microanal. **3**, 443–450 (1997)

A.C. Diebold, B. Foran, C. Kisielowski, D.A. Muller, S.J. Pennycook, E. Principe, S. Stemmer, Thin dielectric film thickness determination by advanced transmission electron microscopy. Microsc. Microanal. **9**, 493–508 (2003)

C. Dinges, A. Berger, H. Rose, Simulation of TEM images considering phonon and electronic excitations. Ultramicroscopy **60**, 49–70 (1995)

A.M. Donald, A.J. Craven, Study of grain-boundary segregation in Cu-Bi alloys using STEM. Philos. Mag. A **39**, 1–11 (1979)

J. Dubochet, Appendix to: Klug, A. Direct imaging of atoms in crystals and molecules – status and prospects for biological sciences. Chemica Scripta **14**, 291–293 (1979)

G. Duscher, N.D. Browning, S.J. Pennycook, Atomic column resolved electron energy-loss spectroscopy. Phys. Stat. Solidi A **166**, 327–342 (1998a)

G. Duscher, J.P. Buban, N.D. Browning, M.F. Chisholm, S.J. Pennycook, The electronic structure of pristine and doped (100) tilt grain boundaries in $SrTiO_3$. Interface Sci. **8**, 199–208 (2000)

G. Duscher, M.F. Chisholm, U. Alber, M. Ruhle, Bismuth-induced embrittlement of copper grain boundaries. Nat. Mat. **3**, 621–626 (2004)

G. Duscher, S.J. Pennycook, N.D. Browning, R. Rupangudi, C. Takoudis, H.J. Gao, R. Singh, in *Characterization and Metrology for ULSI Technology*: 1998 International Conference ed. by D.G. Seiler, A.C. Diebold, W.M. Bullis, T.J. Shaffner, R. Mcdonald, E.J. Walters, American Institute of Physics, Melville, NY, v.**449**, (1998b), pp. 191–195

N. Duxbury, U. Bangert, P. Dawson, E.J. Thrush, W. Van der Stricht, K. Jacobs, I. Moerman, Indium segregation in InGaN quantum-well structures. Appl. Phys. Lett. **76**, 1600 (2000)

C. Dwyer, J. Etheridge, Scattering of Å-scale electron probes in silicon. Ultramicroscopy **96**, 343–360 (2003)

C. Dwyer, A. Ziegler, N. Shibata, G.B. Winkelman, R.L. Satet, M.J. Hoffmann, M.K. Cinibulk, P.F. Becher, G.S. Painter, N.D. Browning, D.J.H. Cockayne, R.O. Ritchie, S.J. Pennycook, Interfacial structure in silicon nitride sintered with lanthanide oxide. J. Mat. Sci. **41**, 4405–4412 (2006)

R.F. Egerton, Inelastic-scattering and energy filtering in transmission electron-microscope. Philos. Mag. **34**, 49–65 (1976)

R.F. Egerton, Inelastic-scattering of 80 keV electrons in amorphous carbon. Philos. Mag. **31**, 199–215 (1975)

R.F. Egerton, Formulas for light-element microanalysis by electron energy-loss spectrometry. Ultramicroscopy **3**, 243–251 (1978a)

R.F. Egerton, Simple electron spectrometer for energy analysis in transmission microscope. Ultramicroscopy **3**, 39–47 (1978b)

R.F. Egerton, Thickness dependence of the STEM ratio image. Ultramicroscopy **10**, 297–299 (1982)

R.F. Egerton, *Electron Energy-Loss Spectroscopy in the Electron Microscope* (Plenum Press, New York, NY, 1996)

R.F. Egerton, Limits to the spatial, energy and momentum resolution of electron energy-loss spectroscopy. Ultramicroscopy **107**, 575–586 (2007)

R.F. Egerton, J.G. Philip, P.S. Turner, M.J. Whelan, Modification of a transmission electron-microscope to give energy-filtered images and diffraction patterns, and electron-energy loss spectra. J. Phys. E. **8**, 1033–1037 (1975)

R.F. Egerton, M.J. Whelan, Electron-energy loss spectrum and band-structure of diamond. Philos. Mag. **30**, 739–749 (1974)

J.J. Einspahr, P.M. Voyles, Prospects for 3D, nanometer-resolution imaging by confocal STEM. Ultramicroscopy **106**, 1041–1052 (2006)

C.B. Eisenhandler, B.M. Siegel, Imaging of single atoms with the electron microscope by phase contrast. J. Appl. Phys. **37**, 1613–1620 (1966)

A. Engel, Molecular weight determination by scanning transmission electron microscopy. Ultramicroscopy **3**, 273 (1978)

A. Engel, in *Advances in Imaging and Electron Physics*, ed. by P.W. Hawkes (Elsevier, Amsterdam, The Netherlands, 2009), pp. 357–386

A. Engel, W. Baumeister, W.O. Saxton, Mass mapping of a protein complex with the scanning transmission electron microscope. Proc. Natl. Acad. Sci. **79**, 4050 (1982)

A. Engel, J. Dubochet, E. Kellenberger, Some progress in the use of a scanning transmission electron microscope for the observation of biomacromolecules. J. Ultrastruct. Res. **57**, 322 (1976)

A. Engel, J.W. Wiggins, D.C. Woodruff, Comparison of calculated images generated by 6 modes of transmission electron-microscopy. J. Appl. Phys. **45**, 2739–2747 (1974)

R. Erni, N.D. Browning, Quantification of the size-dependent energy gap of individual CdSe quantum dots by valence electron energy-loss spectroscopy. Ultramicroscopy **107**, 267–273 (2007)

R. Erni, H. Heinrich, G. Kostorz, Quantitative characterisation of chemical inhomogeneities in Al-Ag using high-resolution Z-contrast STEM. Ultramicroscopy **94**, 125–133 (2003)

R. Erni, M.D. Rossell, C. Kisielowski, U. Dahmen, Atomic-resolution imaging with a sub-50-pm electron probe. Phys. Rev. Lett. **102**, 096101 (2009)

U. Falke, A. Bleloch, M. Falke, Teichert, S. Atomic structure of a (2 × 1) reconstructed $NiSi_2$/Si(001) interface. Phys. Rev. Lett. **92**, 116103 (2004)

M. Falke, U. Falke, A. Bleloch, S. Teichert, G. Beddies, H.J. Hinneberg, Real structure of the $CoSi_2$/Si(001) interface studied by dedicated aberration-corrected scanning transmission electron microscopy. Appl. Phys. Lett. **86**, 203103 (2005)

X. Fan, E.C. Dickey, P.C. Eklund, K.A. Williams, L. Grigorian, R. Buczko, S.T. Pantelides, S.J. Pennycook, Atomic arrangement of iodine atoms inside single-walled carbon nanotubes. Phys. Rev. Lett. **84**, 4621–4624 (2000)

X. Fan, E.C. Dickey, S.J. Pennycook, M.K. Sunkara, Z-contrast imaging and electron energy-loss spectroscopy analysis of chromium-doped diamond-like carbon films. Appl. Phys. Lett. **75**, 2740–2742 (1999)

J. Fertig, H. Rose, A reflection on partial coherence in electron microscopy. Ultramicroscopy **2**, 269–279 (1977)

J. Fertig, H. Rose, Resolution and contrast of crystalline objects in high-resolution scanning-transmission electron-microscopy. Optik **59**, 407–429 (1981)

P.F. Fewster, V. Holy, D. Zhi, Composition determination in quantum dots with in-plane scattering compared with STEM and EDX analysis. J. Phys. D. **36**, A217-A221 (2003)

S.D. Findlay, L.J. Allen, M.P. Oxley, C.J. Rossouw, Lattice-resolution contrast from a focused coherent electron probe. Part II. Ultramicroscopy **96**, 65–81 (2003)

S.D. Findlay, N. Shibata, H. Sawada, E. Okunishi, Y. Kondo, T. Yamamoto, Y. Ikuhara, Robust atomic resolution imaging of light elements using scanning transmission electron microscopy. Appl. Phys. Lett. **95**, 191913 (2009)

R. Fowler, L. Nordheim, Electron emission in intense electric fields. Proc. R. Soc. A. **119**, 173–181 (1928)

B. Freitag, S. Kujawa, P. Mul, J. Ringnalda, P. Tiemeijer, Breaking the spherical and chromatic aberration barrier in transmission electron microscopy. Ultramicroscopy **102**, 209–214 (2005)

S. Frigo, Z. Levine, N.J. Zaluzec, Submicron imaging of buried integrated circuit structures using scanning confocal electron microscopy. Appl. Phys. Lett. **81**, 2112 (2002)

F. Fujimoto, Y. Kainuma, Inelastic scattering of fast electrons by thin crystals. J. Phys. Soc. Jpn. **18**, 1792 (1963)

A. Fukuhara, Inelastic scattering of electrons with a definite momentum transfer in a crystal. J. Phys. Soc. Jpn. **18**, 496 (1963)

J.M. Gibson, A. Howie, Investigation of local-structure and composition in amorphous solids by high-resolution electron-microscopy. Chemica Scripta **14**, 109–116 (1979)

W. Glaser, Bildentstehung und Auflösungsvermögen des Elektronenmikroskops vom Standpunkt der Wellenmechanik. Z. Phys. A. **121**, 647–666 (1943)

D.M. Graham, A. Soltani-Vala, P. Dawson, M.J. Godfrey, T.M. Smeeton, J.S. Barnard, M.J. Kappers, C.J. Humphreys, E.J. Thrush, Optical and microstructural studies of InGaN/GaN single-quantum-well structures. J. Appl. Phys. **97**, 103508 (2005)

L. Grigorian, K.A. Williams, S. Fang, G.U. Sumanasekera, A.L. Loper, E.C. Dickey, S.J. Pennycook, P.C. Eklund, Reversible intercalation of charged iodine chains into carbon nanotube ropes. Phys. Rev. Lett. **80**, 5560–5563 (1998)

F. Haguenau, P.W. Hawkes, J.L. Hutchison, B. Satiat-Jeunemaitre, G.T. Simon, D.B. Williams, Key events in the history of electron microscopy. Microsc. Microanal. **9**, 96–138 (2003)

M. Haider, P. Hartel, H. Müller, Current and future aberration correctors for the improvement of resolution in electron microscopy. Philos. Trans. R. Soc. A **367**, 3665–3682 (2009)

M. Haider, H. Rose, S. Uhlemann, B. Kabius, K. Urban, Towards 0.1 nm resolution with the first spherically corrected transmission electron microscope. J. Electron. Microsc. **47**, 395–405 (1998a)

M. Haider, H. Rose, S. Uhlemann, E. Schwan, B. Kabius, K. Urban, A spherical-aberration-corrected 200 kV transmission electron microscope. Ultramicroscopy **75**, 53–60 (1998b)

M. Haider, S. Uhlemann, E. Schwan, H. Rose, B. Kabius, K. Urban, Electron microscopy image enhanced. Nature **392**, 768–769 (1998c)

M. Haider, S. Uhlemann, J. Zach, Upper limits for the residual aberrations of a high-resolution aberration-corrected STEM. Ultramicroscopy **81**, 163–175 (2000)

C.R. Hall, Scattering of high energy electrons by thermal vibrations of crystals. Philos. Mag. **12**, 815–826 (1965)

C.R. Hall, P.B. Hirsch, Effect of thermal diffuse scattering on propagation of high energy electrons through crystals. Proc. R. Soc. London A **286**, 158–177 (1965)

E.L. Hall, D. Imeson, J.B. Vander Sande, On producing high-spatial-resolution composition profiles via scanning transmission electron microscopy. Philos. Mag. A **43**, 1569–1585 (1981)

P. Hartel, H. Rose, C. Dinges, Conditions and reasons for incoherent imaging in STEM. Ultramicroscopy **63**, 93–114 (1996)

H. Hashimoto, A. Howie, M.J. Whelan, Anomalous electron absorption effects in metal foils: Theory and comparison with experiment. Proc. R. Soc. London A **269**, 80–103 (1962)

H. Hashimoto, A. Kumao, K. Hino, H. Yotsumoto, A. Ono, Images of thorium atoms in transmission electron microscopy. Jap. J. Appl. Phys. **10**, 1115–1116 (1971)

P.W. Hawkes, Coherence in electron optics. Adv. Opt. Electron. Microsc. **7**, 101–184 (1978)

P.W. Hawkes (ed.), in *The Beginnings of Electron-Microscopy*. Advances in Imaging and Electron Physics, Suppl 16 (Academic, San Diego, CA, 1985)

P.W. Hawkes, Aberration correction past and present. Philos. Trans. R. Soc. A **367**, 3637–3664 (2009)

R.D. Heidenreich, Electron reflections in MgO crystals with the electron microscope. Phys. Rev. **62**, 291–292 (1942)

R.D. Heidenreich, Electron microscope and diffraction study of metal crystal textures by means of thin sections. J. Appl. Phys. **20**, 993–1010 (1949)

R.D. Heidenreich, Attenuation of fast electrons in crystals and anomalous transmission. J. Appl. Phys. **33**, 2321–2333 (1962)

R.D. Heidenreich, Electron phase contrast images of molecular detail. J. Electron. Microsc. **16**, 23–38 (1967)

R.M. Henkelman, F.P. Ottensmeyer, Visualization of single heavy atoms by dark field electron microscopy. Proc. Natl. Acad. Sci. USA. **68**, 3000–3004 (1971)

Y.W. Heo, C.R. Abernathy, K. Pruessner, W. Sigmund, D.P. Norton, M. Overberg, F. Ren, M.F. Chisholm, Structure and optical properties of cored wurtzite (Zn,Mg)O heteroepitaxial nanowires. J. Appl. Phys. **96**, 3424–3428 (2004)

K.H. Herrmann, Present state of instrumentation in high-resolution electron-microscopy. J. Phys. E. **11**, 1076–1091 (1978)

J. Hillier, A discussion of the fundamental limit of performance of an electron microscope. Phys. Rev. **60**, 743–745 (1941)

J. Hillier, On microanalysis by electrons. Phys. Rev. **64**, 318–319 (1943)

J. Hillier, R.F. Baker, The observation of crystalline reflections in electron microscope images. Phys. Rev. **61**, 722–723 (1942)

S. Hillyard, R.F. Loane, J. Silcox, Annular dark-field imaging – resolution and thickness effects. Ultramicroscopy **49**, 14–25 (1993)

S. Hillyard, J. Silcox, Thickness effects in ADF STEM zone-axis images. Ultramicroscopy **52**, 325–334 (1993)

P.B. Hirsch, R.W. Horne, M.J. Whelan, Direct observations of the arrangement and motion of dislocations in aluminium. Philos. Mag. **1**, 677–684 (1956)

P.B. Hirsch, A. Howie, M.J. Whelan, A kinematical theory of diffraction contrast of electron transmission microscope images of dislocations and other defects. Philos. Trans. R. Soc. A. **252**, 499–529 (1960)

F. Hofer, W. Grogger, G. Kothleitner, P. Warbichler, Quantitative analysis of EFTEM elemental distribution images. Ultramicroscopy **67**, 83–103 (1997)

L. Houben, A. Thust, K. Urban, Atomic-precision determination of the reconstruction of a 90° tilt boundary in $YBa_2Cu_3O_{7-\delta}$ by aberration corrected HRTEM. Ultramicroscopy **106**, 200–214 (2006)

P.V.C. Hough, I.A. Mastrangelo, J.S. Wall, J.F. Hainfeld, M. Sawadogo, R.G. Roeder, The gene-specific initiation factor USF (upstream stimulatory factor) bound at the adenovirus type 2 major late promoter: Mass and three-dimensional structure. Proc. Natl. Acad. Sci. USA. **84**, 4826–4830 (1987)

A. Howie, Inelastic scattering of electrons by crystals I. The theory of small-angle inelastic scattering. Proc. R. Soc. Lond. A **271**, 268–287 (1963)

A. Howie, Image-contrast and localized signal selection techniques. J. Microsc.-Oxford **117**, 11–23 (1979)

A. Howie, L.D. Marks, S.J. Pennycook, New imaging methods for catalyst particles. Ultramicroscopy **8**, 163–174 (1982)

A. Howie, M.J. Whelan, Diffraction contrast of electron microscope images of crystal lattice defects. II. The development of a dynamical theory. Proc. Roy. Soc. London A **263**, 217–237 (1961)

A. Howie, M.J. Whelan, Diffraction contrast of electron microscope images of crystal lattice defects. III. Results and experimental confirmation of the dynamical theory of dislocation image contrast. Proc. R. Soc. Lond. A **267**, 206–230 (1962)

C.J. Humphreys, Scattering of fast electrons by crystals. Rep. Prog. Phys. **42**, 1825–1887 (1979)

C.J. Humphreys, P.B. Hirsch, Absorption parameters in electron diffraction theory. Philos. Mag. **18**, 115–122 (1968)

C.J. Humphreys, R. Sandstrom, J.P. Spencer, in *Scanning Electron Microscopy/1973 (Part II)*, ed. by O. Johari (IIT Research Institue, Chicago, IL, 1973), pp. 233–239

C.J. Humphreys, M.J. Whelan, Inelastic scattering of fast electrons by crystals – I. Single electron excitations. Philos. Mag. **20**, 165–172 (1969)

C.R. Hutchinson, X. Fan, S.J. Pennycook, G.J. Shiflet, On the origin of the high coarsening resistance of omega plates in Al-Cu-Mg-Ag alloys. Acta Mater. **49**, 2827–2841 (2001)

J.L. Hutchison, J.M. Titchmarsh, D.J.H. Cockayne, R.C. Doole, C.J.D. Hetherington, A.I. Kirkland, H. Sawada, A versatile double aberration-corrected, energy filtered HREM/STEM for materials science. Ultramicroscopy **103**, 7–15 (2005)

M.J. Hÿtch, W.M. Stobbs, Quantitative comparison of high-resolution TEM images with image simulations. Ultramicroscopy **53**, 191–203 (1994)

J.C. Idrobo, M.P. Oxley, W. Walkosz, R.F. Klie, S. Ogut, B. Mikijelj, S.J. Pennycook, S.T. Pantelides, Identification and lattice location of oxygen impurities in alpha-Si_3N_4. Appl. Phys. Lett. **95**, 164101 (2009)

S. Iijima, Observation of single and clusters of atoms in bright field electron-microscopy. Optik **48**, 193–214 (1977)

V. Intaraprasonk, H.L. Xin, D.A. Muller, Analytic derivation of optimal imaging conditions for incoherent imaging in aberration-corrected electron microscopes. Ultramicroscopy **108**, 1454–1466 (2008)

M. Isaacson, D. Johnson, Microanalysis of light-elements using transmitted energy-loss electrons. Ultramicroscopy **1**, 33–52 (1975)

M. Isaacson, D. Johnson, A.V. Crewe, Electron-beam excitation and damage of biological molecules – its implications for specimen damage in electron microscopy. Radiat. Res. **55**, 205–224 (1973)

M. Isaacson, D. Kopf, M. Ohtsuki, M. Utlaut, Atomic imaging using the dark-field annular detector in the STEM. Ultramicroscopy **4**, 101–104 (1979)

M. Isaacson, D. Kopf, M. Utlaut, N.W. Parker, A.V. Crewe, Direct observations of atomic diffusion by scanning-transmission electron-microscopy. Proc. Natl. Acad. Sci. USA. **74**, 1802–1806 (1977)

M. Isaacson, J.P. Langmore, H. Rose, Determination of nonlocalization of inelastic-scattering of electrons by electron-microscopy. Optik **41**, 92–96 (1974)

M. Isaacson, M. Scheinfein, A high-performance electron-energy loss spectrometer for use with a dedicated STEM. J. Vac. Sci. Technol. B **1**, 1338–1343 (1983)

K. Ishizuka Prospects of atomic resolution imaging with an aberration-corrected STEM. J. Electron Microsc. **50**, 291–305 (2001)

K. Ishizuka, A practical approach for STEM image simulation based on the FFT multislice method. Ultramicroscopy **90**, 71–83 (2002)

E.M. James, N.D. Browning, Practical aspects of atomic resolution imaging and analysis in STEM. Ultramicroscopy **78**, 125–139 (1999)

E.M. James, N.D. Browning, A.W. Nicholls, M. Kawasaki, Y. Xin, S. Stemmer, Demonstration of atomic resolution Z-contrast imaging by a JEOL JEM-2010F scanning transmission electron microscope. J. Electron Microsc. **47**, 561–574 (1998)

R.W. James, *The Optical Principles of the Diffraction of X-Rays* (Bell and Sons, London, 1962)

C. Jeanguillaume, P. Trebbia, C. Colliex, About use of electron energy-loss spectroscopy for chemical mapping of thin foils with high spatial-resolution. Ultramicroscopy **3**, 237–242 (1978)

D.E. Jesson, S.J. Pennycook, Incoherent imaging of thin specimens using coherently scattered electrons. Proc. R. Soc. Lond. A **441**, 261–281 (1993)

D.E. Jesson, S.J. Pennycook, Incoherent imaging of crystals using thermally scattered electrons. Proc. R. Soc. Lond. A **449**, 273–293 (1995)

D.E. Jesson, S.J. Pennycook, J.M. Baribeau, Direct imaging of interfacial ordering in ultrathin $(Si_mGe_n)_p$ superlattices. Phys. Rev. Lett. **66**, 750–753 (1991)

D.E. Jesson, S.J. Pennycook, J.M. Baribeau, D.C. Houghton, Direct imaging of surface cusp evolution during strained-layer epitaxy and implications for strain relaxation. Phys. Rev. Lett. **71**, 1744–1747 (1993a)

D.E. Jesson, S.J. Pennycook, J.Z. Tischler, J.D. Budai, J.M. Baribeau, D.C. Houghton, Interplay between evolving surface-morphology, atomic-scale growth modes, and ordering during Si_xGe_{1-X} epitaxy. Phys. Rev. Lett. **70**, 2293–2296 (1993b)

C.L. Jia, Atom vacancies at a screw dislocation core in $SrTiO_3$. Philos. Mag. Lett. **86**, 683–690 (2006)

C.L. Jia, M. Lentzen, K. Urban, Atomic-resolution imaging of oxygen in perovskite ceramics. Science **299**, 870–873 (2003)

C.L. Jia, S.B. Mi, M. Faley, U. Poppe, J. Schubert, K. Urban, Oxygen octahedron reconstruction in the $SrTiO_3/LaAlO_3$ heterointerfaces investigated using aberration-corrected ultrahigh-resolution transmission electron microscopy. Phys. Rev. B **79**, 081405 (2009a)

C.L. Jia, S.B. Mi, K. Urban, I. Vrejoiu, M. Alexe, D. Hesse, Effect of a single dislocation in a heterostructure layer on the local polarization of a ferroelectric layer. Phys. Rev. Lett. **102**, 117601 (2009b)

C. Jia, V. Nagarajan, J. He, L. Houben, T. Zhao, R. Ramesh, K. Urban, R. Waser, Unit-cell scale mapping of ferroelectricity and tetragonality in epitaxial ultrathin ferroelectric films. Nat. Mat. **6**, 64–69 (2006)

C.L. Jia, A. Thust, K. Urban, Atomic-scale analysis of the oxygen configuration at a $SrTiO_3$ dislocation core. Phys. Rev. Lett. **95**, 225506 (2005)

C.L. Jia, K. Urban, Atomic-resolution measurement of oxygen concentration in oxide materials. Science **303**, 2001–2004 (2004)

D.E. Johnson, Electron-energy loss microanalysis system with high collection efficiency. Rev. Sci. Instrum. **51**, 705–709 (1980)

D.C. Joy, D.M. Maher, Choice of operating parameters for microanalysis by electron energy-loss spectroscopy. Ultramicroscopy **3**, 69–74 (1978a)

D.C. Joy, D.M. Maher, Practical electron spectrometer for chemical-analysis. J. Microsc. Oxford **114**, 117–129 (1978b)

B. Kabius, P. Hartel, M. Haider, H. Muller, S. Uhlemann, U. Loebau, J. Zach, H. Rose, First application of Cc-corrected imaging for high-resolution and energy-filtered TEM. J. Electron. Microsc. **58**, 147–155 (2009)

K. Kambe, Visualization of Bloch waves of high-energy electrons in high-resolution electron-microscopy. Ultramicroscopy **10**, 223–227 (1982)

Y. Kamiya, Y. Nakai, Diffraction contrast effect of electrons scattered inelastically through large angles. J. Phys. Soc. Jpn. **31**, 195–203 (1971)

Y. Kamiya, R. Uyeda, Effect of incoherent waves on the electron microscopic images of crystals. J. Phys. Soc. Jpn. **16**, 1361–1366 (1961)

O.H. Kapp, M.G. Mainwaring, S.N. Vinogradov, A.V. Crewe, Scanning-transmission electron-microscopic examination of the hexagonal bilayer structures formed by the reassociation of 3 of the 4 subunits of the extracellular hemoglobin of lumbricus-terrestris. Proc. Natl. Acad. Sci. U.S.A. **84**, 7532–7536 (1987)

C.S. Kim, M. Kim, J.K. Furdyna, M. Dobrowolska, S. Lee, H. Rho, L.M. Smith, H.E. Jackson, E.M. James, Y. Xin, Evidence for 2D precursors and interdiffusion in the evolution of self-assembled CdSe quantum dots on ZnSe. Phys. Rev. Lett. **85**, 1124–1127 (2000)

K. Kimoto, T. Asaka, T. Nagai, M. Saito, Y. Matsui, K. Ishizuka, Element-selective imaging of atomic columns in a crystal using STEM and EELS. Nature **450**, 702–704 (2007)

K. Kimoto, R. Xie, Y. Matsui, K. Ishizuka, N. Hirosaki, Direct observation of single dopant atom in light-emitting phosphor of β-SiAlON: Eu. Appl. Phys. Lett. **94**, 041908 (2009)

E.J. Kirkland, R.F. Loane, J. Silcox, Simulation of annular dark field STEM images using a modified multislice method. Ultramicroscopy **23**, 77–96 (1987)

C. Kisielowski, B. Freitag, M. Bischoff, van H. Lin, S. Lazar, G. Knippels, P. Tiemeijer, M. van der Stam, S. von Harrach, M. Stekelenburg, M. Haider, S. Uhlemann, H. Muller, P. Hartel, B. Kabius, D. Miller, I. Petrov, E.A. Olson, T. Donchev, E.A. Kenik, A.R. Lupini, J. Bentley, S.J. Pennycook, I.M. Anderson, A.M. Minor, A.K. Schmid, T. Duden, V. Radmilovic, Q.M. Ramasse, M. Watanabe, R. Erni, E.A. Stach, P. Denes, U. Dahmen, Detection of single atoms and buried defects in three dimensions by aberration-corrected electron microscope with 0.5-angstrom information limit. Microsc. Microanal. **14**, 469–477 (2008)

R.F. Klie, N.D. Browning, Atomic scale characterization of oxygen vacancy segregation at $SrTiO_3$ grain boundaries. Appl. Phys. Lett. **77**, 3737–3739 (2000)

M. Knoll, Aufladepotentiel und Sekundäremission elektronenbestrahlter Körper. Z. Tech. Phys. **16**, 467–475 (1935)

M. Knoll, E. Ruska, Das Elektronenmikroskop. Z. Phys. A. **78**, 318–339 (1932)

H. Kohl, Image formation by inelastically scattered electrons: Image of a surface plasmon. Ultramicroscopy **11**, 53–65 (1983)

H. Kohl, H. Rose, Theory of image-formation by inelastically scattered electrons in the electron-microscope. Adv. Imag. Electron. Phys. **65**, 173–227 (1985)

T. Koller, M. Beer, M. Muller, K. Muhletha, Electron-microscopy of selectively stained molecules. Cytobiologie **4**, 369–408 (1971)

T.J. Konno, E. Okunishi, T. Ohsuna, K. Hiraga, HAADF-STEM study on the early stage of precipitation in aged Al-Ag alloys. J. Electron. Microsc. **53**, 611–616 (2004)

O.L. Krivanek, C.C. Ahn, R.B. Keeney, Parallel detection electron spectrometer using quadrupole lenses. Ultramicroscopy **22**, 103–115 (1987)

O.L. Krivanek, N. Dellby, A.J. Spence, R.A. Camps, L.M. Brown, in: *Institute of Physics Conference Series No. 153*, ed. by J.M. Rodenburg (1997), pp. 35–40

O.L. Krivanek, N. Dellby, A.R. Lupini, Towards sub-Å electron beams. Ultramicroscopy **78**, 1–11 (1999)

O.L. Krivanek, N. Dellby, M.F. Murfitt, in *Handbook of Charged Particle Optics*, 2nd edn. ed. by J. Orloff (CRC Press, Boca Raton, 2009a), pp. 601–640

O.L. Krivanek, M.M. Disko, J. Taftø, J.C.H. Spence, Electron energy loss spectroscopy as a probe of the local atomic environment. Ultramicroscopy **9**, 249–254 (1982)

O.L. Krivanek, P.D. Nellist, N. Dellby, M.F. Murfitt, Z.S. Szilagyi, Towards sub-0.5 Å electron beams. Ultramicroscopy **96**, 229–237 (2003)

O.L. Krivanek, J.P. Ursin, N.J. Bacon, G.J. Corbin, N. Dellby, P. Hrncirik, M.F. Murfitt, C.S. Own, Z.S. Szilagyi High-energy-resolution monochromator for aberration-corrected scanning transmission electron microscopy/electron energy-loss spectroscopy. Philos. Trans. R. Soc. A **367**, 3683–3697 (2009b)

O.L. Krivanek, M.F. Chisholm, V. Nicolosi, T.J. Pennycook, G.J. Corbin, N. Dellby, M.F. Murfitt, C.S. Own, Z.S. Szilagyi, M.P. Oxley, S.T. Pantelides, S.J. Pennycook, Atom-by-atom structural and chemical analysis by annular dark-field electron microscopy. Nature **464**, 571–574 (2010)

H. Lakner, B. Bollig, E. Kubalek, M. Heuken, K. Heime, F. Scheffer, F.E.G. Guimaraes, Scanning-transmission electron-microscopy of heterointerfaces grown by metalorganic vapor-phase epitaxy (MOVPE). J. Cryst. Growth **107**, 452–457 (1991)

H. Lakner, B. Bollig, S. Ungerechts, E. Kubalek, Characterization of III–V semiconductor interfaces by Z-contrast imaging, EELS and CBED. J. Phys. D. **29**, 1767–1778 (1996)

H. Lakner, B. Rafferty, G. Brockt, Electronic structure analysis of (In,Ga,Al)n heterostructures on the nanometre scale using EELS. J. Microsc.-Oxford **194**, 79–83 (1999)

J.P. Langmore, J. Wall, M.S. Isaacson, Collection of scattered electrons in dark field electron-microscopy. 1. Elastic-scattering. Optik **38**, 335–350 (1973)

R.D. Leapman, V.E. Cosslett, Extended fine-structure above x-ray edge in electron-energy loss spectra. J. Phys. D. **9**, L29–L32 (1976)

J.M. LeBeau, A.J. D'Alfonso, S.D. Findlay, S. Stemmer, L.J. Allen, Quantitative comparisons of contrast in experimental and simulated bright-field scanning transmission electron microscopy images. Phys. Rev. B **80**, 214110 (2009a)

J.M. LeBeau, S.D. Findlay, L.J. Allen, S. Stemmer, Quantitative atomic resolution scanning transmission electron microscopy. Phys. Rev. Lett.**100**, 206101 (2008)

J.M. LeBeau, S.D. Findlay, X.Q. Wang, A.J. Jacobson, L.J. Allen, S. Stemmer, High-angle scattering of fast electrons from crystals containing heavy elements: Simulation and experiment. Phys. Rev. B **79**, 214110 (2009b)

J.M. LeBeau, S. Stemmer, Experimental quantification of annular dark-field images in scanning transmission electron microscopy. Ultramicroscopy **108**, 1653–1658 (2008)

J.A. Lin, J.M. Cowley, Calibration of the operating parameters for an HB5 STEM instrument. Ultramicroscopy **19**, 31–42 (1986a)

J.A. Lin, J.M. Cowley, Reconstruction from in-line electron holograms by digital processing. Ultramicroscopy **19**, 179–190 (1986b)

C.P. Liu, A.R. Preston, C.B. Boothroyd, C.J. Humphreys, Quantitative analysis of ultrathin doping layers in semiconductors using high-angle annular dark field images. J. Microsc. Oxford **194**, 171–182 (1999)

J. Liu, Y. Cheng, J.M. Cowley, M.B. Stearns, High-angle annular dark-field microscopy of Mo/Si multilayer structures. Ultramicroscopy **40**, 352–364 (1992)

J. Liu, J.M. Cowley, High-angle ADF and high-resolution SE imaging of supported catalyst clusters. Ultramicroscopy **34**, 119–128 (1990)

J. Liu, J.M. Cowley, Imaging with high-angle scattered electrons and secondary electrons in the STEM. Ultramicroscopy **37**, 50–71 (1991)

J. Liu, J.M. Cowley, High-resolution scanning-transmission electron-microscopy. Ultramicroscopy **52**, 335–346 (1993)

R.F. Loane, E.J. Kirkland, J. Silcox, Visibility of single heavy-atoms on thin crystalline silicon in simulated annular dark-field STEM images. Acta Cryst. A **44**, 912–927 (1988)

R.F. Loane, P. Xu, J. Silcox, Incoherent imaging of zone axis crystals with ADF STEM. Ultramicroscopy **40**, 121–138 (1992)

R.F. Loane, P. Xu, J. Silcox, Thermal vibrations in convergent-beam electron-diffraction. Acta Cryst. A **47**, 267–278 (1991)

A.R. Lupini, A.Y. Borisevich, J.C. Idrobo, H.M. Christen, M. Biegalski, S.J. Pennycook, Characterizing the two- and three-dimensional resolution of an improved aberration-corrected STEM. Microsc. Microanal. **15**, 441–453 (2009)

A.R. Lupini, S.J. Pennycook, Localization in elastic and inelastic scattering. Ultramicroscopy **96**, 313–322 (2003)

M. Luysberg, M. Heidelmann, L. Houben, M. Boese, T. Heeg, J. Schubert, M. Roeckerath, Intermixing and charge neutrality at $DyScO_3/SrTiO_3$ interfaces. Acta Mater. **57**, 3192–3198 (2009)

A. Maiti, M.F. Chisholm, S.J. Pennycook, S.T. Pantelides, Dopant segregation at semiconductor grain boundaries through cooperative chemical rebonding. Phys. Rev. Lett. **77**, 1306–1309 (1996)

L. Marton, Electron microscopy. Rep. Prog. Phys. **10**, 204–252 (1944)
I.A. Mastrangelo, P.V.C. Hough, Van G. Wilson, J.S. Wall, J.F. Hainfeeld, P. Tegtmeyer, Monomers through trimers of large tumor antigen bind in region I and monomers through tetramers bind in region II of simian virus 40 origin of replication DNA as stable structures in solution. Proc. Natl. Acad. Sci. USA. **82**, 3626–3630 (1985)
J. McBride, J. Treadway, L.C. Feldman, S.J. Pennycook, S.J. Rosenthal, Structural basis for near unity quantum yield core/shell nanostructures. Nano Lett. **6**, 1496–1501 (2006)
J.R. McBride, T.C. Kippeny, S.J. Pennycook, S.J. Rosenthal, Aberration-corrected Z-contrast scanning transmission electron microscopy of CdSe nanocrystals. Nano Lett. **4**, 1279–1283 (2004)
A.J. McGibbon, S.J. Pennycook, J.E. Angelo, Direct observation of dislocation core structures in CdTe/GaAs(001). Science **269**, 519–521 (1995)
A.J. McGibbon, S.J. Pennycook, D.E. Jesson, Crystal structure retrieval by maximum entropy analysis of atomic resolution incoherent images. J. Microsc.-Oxford **195**, 44–57 (1999)
M.M. McGibbon, N.D. Browning, M.F. Chisholm, A.J. McGibbon, S.J. Pennycook, V. Ravikumar, V.P. Dravid, Direct determination of grain-boundary atomic-structure in $SrTiO_3$. Science **266**, 102–104 (1994)
M.M. McGibbon, N.D. Browning, A.J. McGibbon, S.J. Pennycook, The atomic structure of asymmetric [001] tilt boundaries in $SrTiO_3$. Philos. Mag. A **73**, 625–641 (1996)
R.A. McKee, F.J. Walker, M.F. Chisholm, Crystalline oxides on silicon: The first five monolayers. Phys. Rev. Lett. **81**, 3014–3017 (1998)
R.A. McKee, F.J. Walker, M.F. Chisholm, Physical structure and inversion charge at a semiconductor interface with a crystalline oxide. Science **293**, 468–471 (2001)
D. McMullan, SEM past, present and future. J. Microsc.-Oxford **155**, 373–392 (1989)
D. McMullan, Scanning electron-microscopy 1928–1965. Scanning **17**, 175–185 (1995)
D. McMullan, A history of the scanning electron microscope, 1928–1965. Adv. Imag. Elect. Phys. **133**, 523–545 (2004)
D. McMullan, J.M. Rodenburg, Y. Murooka, A.J. McGibbon, Parallel EELS CCD detector for a VG HB501 STEM. Inst. Phys. Conf. Ser. No. **98**, 55–58 (1990)
A. Melander, R. Sandstrom, Stacking-fault contrast from phonon-scattered electrons. J. Phys. C. **8**, 767–779 (1975)
L.D. Menard, S.P. Gao, H.P. Xu, R.D. Twesten, A.S. Harper, Y. Song, G.L. Wang, A.D. Douglas, J.C. Yang, A.I. Frenkel, R.G. Nuzzo, R.W. Murray, Sub-nanometer Au monolayer-protected clusters exhibiting molecule-like electronic behavior: Quantitative high-angle annular dark-field scanning transmission electron microscopy and electrochemical characterization of clusters with precise atomic stoichiometry. J. Phys. Chem. B **110**, 12874–12883 (2006)
J. Menter, The direct study by electron microscopy of crystal lattices and their imperfections. Proc. R. Soc. Lond. A **236**, 119–135 (1956)
J.R. Michael, D.B. Williams, A consistent definition of probe size and spatial-resolution in the analytical electron-microscope. J. Microsc.-Oxford **147**, 289–303 (1987)
P.A. Midgley, J.M. Thomas, L. Laffont, M. Weyland, R. Raja, B.F.G. Johnson, T. Khimyak, High-resolution scanning transmission electron tomography and elemental analysis of zeptogram quantities of heterogeneous catalyst. J. Phys. Chem. B **108**, 4590–4592 (2004)

P.A. Midgley, M. Weyland, 3D electron microscopy in the physical sciences: The development of Z-contrast and EFTEM tomography. Ultramicroscopy **96**, 413 (2003)

P.A. Midgley, M. Weyland, J.M. Thomas, B.F.G. Johnson, Z-contrast tomography: A technique in three-dimensional nanostructural analysis based on Rutherford scattering. Chem. Commun. **10**, 907–908 (2001)

D.L. Misell, G.W. Stroke, M. Halioua, Coherent and incoherent imaging in scanning-transmission electron-microscope. J. Phys. D. **7**, L113–L117 (1974)

K. Mitsuishi, M. Kawasaki, M. Takeguchi, K. Furuya, Determination of atomic positions in a solid Xe precipitate embedded in an Al matrix. Phys. Rev. Lett. **82**, 3082–3084 (1999)

K. Mitsuishi, M. Takeguchi, H. Yasuda, K. Furuya, New scheme for calculation of annular dark-field STEM image including both elastically diffracted and TDS waves. J. Electron Microsc. **50**, 157–162 (2001)

K.A. Mkhoyan, E.J. Kirkland, J. Silcox, E.S. Alldredge, Atomic level scanning transmission electron microscopy characterization of GaN/AlN quantum wells. J. Appl. Phys. **96**, 738–746 (2004)

K.A. Mkhoyan, J. Silcox, H. Wu, W.J. Schaff, L.F. Eastman, Nonuniformities in GaN/AlN quantum wells. Appl. Phys. Lett. **83**, 2668–2670 (2003)

S.I. Molina, T. Ben, D.L. Sales, J. Pizarro, P.L. Galindo, M. Varela, S.J. Pennycook, D. Fuster, Y. Gonzalez, L. Gonzalez, Determination of the strain generated in InAs/InP quantum wires: Prediction of nucleation sites. Nanotechnology **17**, 5652–5658 (2006)

S.I. Molina, A.M. Sanchez, A.M. Beltran, D.L. Sales, T. Ben, M.F. Chisholm, M. Varela, S.J. Pennycook, P.L. Galindo, A.J. Papworth, P.J. Goodhew, and J.M. Ripalda, Incorporation of Sb in InAs/GaAs quantum dots. Appl. Phys. Lett. **91**, 263105 (2007a)

S.I. Molina, D.L. Sales, P.L. Galindo, D. Fuster, Y. Gonzalez, B. Alen, L. Gonzalez, M. Varela, S.J. Pennycook, Column-by-column compositional mapping by Z-contrast imaging. Ultramicroscopy **109**, 172–176 (2009)

S.I. Molina, M. Varela, T. Ben, D.L. Sales, J. Pizarro, P.L. Galindo, D. Fuster, Y. Gonzalez, L. Gonzalez, S.J. Pennycook, A method to determine the strain and nucleation sites of stacked nano-objects. J. Nanosci. Nanotech. **8**, 3422–3426 (2008)

S.I. Molina, M. Varela, D.L. Sales, T. Ben, J. Pizarro, P.L. Galindo, D. Fuster, Y. Gonzalez, L. Gonzalez, S.J. Pennycook, Direct imaging of quantum wires nucleated at diatomic steps. Appl. Phys. Lett. **91**, 143112 (2007b)

K.T. Moore, J.M. Howe, D.C. Elbert, Analysis of diffraction contrast as a function of energy loss in energy-filtered transmission electron microscope imaging. Ultramicroscopy **80**, 203–219 (1999)

C. Mory, C. Colliex, B. Revet, E. Delain, Improved visualization of single-stranded and double-stranded nucleic-acids by STEM. Ultramicroscopy **7**, 161–168 (1981)

C. Mory, H. Kohl, M. Tence, C. Colliex, Experimental investigation of the ultimate EELS spatial-resolution. Ultramicroscopy **37**, 191–201 (1991)

E.N. Moudrianakis, M. Beer, Base sequence determination in nucleic acids with the electron microscope. III. Chemistry and microscopy of guanine-labeled DNA. Proc. Natl. Acad. Sci. USA. **53**, 564–571 (1965)

H. Mullejans, J. Bruley, Improvements in detection sensitivity by spatial difference electron-energy-loss spectroscopy at interfaces in ceramics. Ultramicroscopy **53**, 351–360 (1994)

H. Mullejans, J. Bruley, Electron-energy-loss near-edge structure of internal interfaces by spatial difference spectroscopy. J. Microsc. Oxford **180**, 12–21 (1995)

D.A. Muller, Why changes in bond lengths and cohesion lead to core-level shifts in metals, and consequences for the spatial difference method. Ultramicroscopy **78**, 163–174 (1999)
D.A. Muller, M.J. Mills, Electron microscopy: Probing the atomic structure and chemistry of grain boundaries, interfaces and defects. Mater. Sci. Eng. A **260**, 12–28 (1999)
D.A. Muller, G. Wilk, Atomic scale measurements of the interfacial electronic structure and chemistry of zirconium silicate gate dielectrics. Appl. Phys. Lett. **79**, 4195 (2001)
D.A. Muller, in *Characterization and Metrology for ULSI Technology: 2000 International Conference*, ed. by D.G. Seller, A.C. Diebold, T.J. Shaffner, R. Mcdonald, W.M. Bullis, P.J. Smith, E.M. Secula (American Institute of Physics, Melville, NY, 2001), pp. 500–505
D.A. Muller, B. Edwards, E.J. Kirkland, J. Silcox, Simulation of thermal diffuse scattering including a detailed phonon dispersion curve. Ultramicroscopy **86**, 371–380 (2001)
D.A. Muller, J. Grazul, Optimizing the environment for sub-0.2 nm scanning transmission electron microscopy. J. Electron Microsc. **50**, 219–226 (2001)
D.A. Muller, L.F. Kourkoutis, M. Murfitt, J.H. Song, H.Y. Hwang, J. Silcox, N. Dellby, O.L. Krivanek, Atomic-scale chemical imaging of composition and bonding by aberration-corrected microscopy. Science **319**, 1073–1076 (2008)
D.A. Muller, D.A. Shashkov, R. Benedek, L.H. Yang, J. Silcox, D.N. Seidman, Atomic scale observations of metal-induced gap states at {222}MgO/Cu interfaces. Phys. Rev. Lett. **80**, 4741–4744 (1998)
D.A. Muller, J. Silcox, Delocalization in inelastic scattering. Ultramicroscopy **59**, 195–213 (1995)
D.A. Muller, T. Sorsch, S. Moccio, F.H. Baumann, K. Evans-Lutterodt, G. Timp, The electronic structure at the atomic scale of ultrathin gate oxides. Nature **399**, 758–761 (1999)
D.A. Muller, S. Subramanian, P.E. Batson, S.L. Sass, J. Silcox, Near atomic scale studies of electronic structure at grain boundaries in Ni_3Al. Phys. Rev. Lett. **75**, 4744–4747 (1995)
D.A. Muller, S. Subramanian, P.E. Batson, J. Silcox, S.L. Sass, Structure, chemistry and bonding at grain boundaries in Ni_3Al-1. The role of boron in ductilizing grain boundaries. Acta Mater. **44**, 1637–1645 (1996)
D.A. Muller, Y. Tzou, R. Raj, J. Silcox, Mapping sp(2) and sp(3) states of carbon at subnanometer spatial-resolution. Nature **366**, 725–727 (1993)
E.W. Müller, Die Abhängigkeit der Feldelektronenemission von der Austrittsarbeit. Z. Phys. A. **102**, 734–761 (1936)
F. Müller, in *Scientific Research in World War II. What Scientists did in the War*, ed. by A. Mass, H. Hooijmaijers (Routledge, Abingdon and New York, 2009), pp. 121–146
S. Müller, A. Engel, Structure and mass analysis by scanning transmission electron microscopy. Micron **32**, 21–31 (2001)
H. Müller, S. Uhlemann, P. Hartel, M. Haider, Advancing the hexapole Cs-corrector for the scanning transmission electron microscope. Microsc. Microanal. **12**, 442–455 (2006)
S. Müller, K. Goldie, R. Bürki, R. Häring, A. Engel, Factors influencing the precision of quantitative scanning transmission electron microscopy. Ultramicroscopy **46**, 317–334 (1992)
K. Nakamura, H. KaKibayashi, K. Kanehori, N. Tanaka, Position dependence of the visibility of a single gold atom in silicon crystals in HAADF-STEM image simulation. J. Electron. Microsc. **46**, 33–43 (1997)

P.D. Nellist, G. Behan, A.I. Kirkland, C.J.D. Hetherington, Confocal operation of a transmission electron microscope with two aberration correctors. Appl. Phys. Lett. **89**, 124105 (2006)

P.D. Nellist, M.F. Chisholm, N. Dellby, O.L. Krivanek, M.F. Murfitt, Z.S. Szilagyi, A.R. Lupini, A. Borisevich, W.H. Sides, S.J. Pennycook, Direct sub-angstrom imaging of a crystal lattice. Science **305**, 1741–1741 (2004)

P.D. Nellist, E.C. Cosgriff, G. Behan, A.I. Kirkland, Imaging modes for scanning confocal electron microscopy in a double aberration-corrected transmission electron microscope. Microsc. Microanal. **14**, 82–88 (2008)

P.D. Nellist, S.J. Pennycook, Direct imaging of the atomic configuration of ultradispersed catalysts. Science **274**, 413–415 (1996)

P.D. Nellist, S.J. Pennycook, Accurate structure determination from image reconstruction in ADF STEM. J. Microsc. Oxford **190**, 159–170 (1998a)

P.D. Nellist, S.J. Pennycook, Subangstrom resolution by underfocused incoherent transmission electron microscopy. Phys. Rev. Lett. **81**, 4156–4159 (1998b)

P.D. Nellist, S.J. Pennycook, Incoherent imaging using dynamically scattered coherent electrons. Ultramicroscopy **78**, 111–124 (1999)

D.P. Norton, D.H. Lowndes, S.J. Pennycook, J.D. Budai, Depression and broadening of the superconducting transition in superlattices based on $YBa_2Cu_3O_{7-\delta}$ – influence of the barrier layers. Phys. Rev. Lett. **67**, 1358–1361 (1991)

S.H. Oh, K. van Benthem, S.I. Molina, A.Y. Borisevich, W.D. Luo, P. Werner, N.D. Zakharov, D. Kumar, S.T. Pantelides, S.J. Pennycook, Point defect configurations of supersaturated Au atoms inside Si nanowires. Nano Lett. **8**, 1016–1019 (2008)

A. Ohtomo, D.A. Muller, J.L. Grazul, H.Y. Hwang, Epitaxial growth and electronic structure of $LaTiO_x$ films. Appl. Phys. Lett. **80**, 3922–3924 (2002a)

A. Ohtomo, D.A. Muller, J.L. Grazul, H.Y. Hwang, Artificial charge-modulation in atomic-scale perovskite titanate superlattices. Nature **419**, 378–380 (2002b)

M. Ohtsuki, M.S. Isaacson, A.V. Crewe, Dark field imaging of biological macromolecules with the scanning-transmission electron-microscope. Proc. Natl. Acad. Sci. USA. **76**, 1228–1232 (1979)

E. Okunishi, I. Ishikawa, H. Sawada, F. Hosokawa, M. Hori, Y. Kondo, Visualization of light elements at ultrahigh resolution by STEM annular bright field microscopy. Microsc. Microanal. **15**, 164–165 (2009)

M.T. Otten, High-angle annular dark-field imaging on a TEM/STEM system. J. Electron Microsc. Tech. **17**, 221–230 (1991)

F.P. Ottensmeyer, Scattered electrons in microscopy and microanalysis. Science **215**, 461–466 (1982)

F.P. Ottensmeyer, D.P. Bazettjones, R.M. Henkelman, A.P. Korn, R.F. Whiting, Imaging of atoms – its application to the structure determination of biological macromolecules. Chemica Scripta **14**, 257–262 (1979)

F.P. Ottensmeyer, E.E. Schmidt, A.J. Olbrecht, Image of a sulfur atom. Science **179**, 175–176 (1973)

A. Ourmazd, F.H. Baumann, M. Bode, Y. Kim, Quantitative chemical lattice imaging – theory and practice. Ultramicroscopy **34**, 237–255 (1990)

A. Ourmazd, D.W. Taylor, J. Cunningham, Chemical mapping of semiconductor interfaces at near-atomic resolution. Phys. Rev. Lett. **62**, 933–936 (1989)

M.P. Oxley, L.J. Allen, Delocalization of the effective interaction for inner-shell ionization in crystals. Phys. Rev. B **57**, 3273–3282 (1998)

M.P. Oxley, L.J. Allen, Impact parameters for ionization by high-energy electrons. Ultramicroscopy **80**, 125–131 (1999)

M.P. Oxley, L.J. Allen, C.J. Rossouw, Correction terms and approximations for atom location by channelling enhanced microanalysis. Ultramicroscopy **80**, 109–124 (1999)

M.P. Oxley, S.J. Pennycook, Image simulation for electron energy loss spectroscopy. Micron. **39**, 676 (2008)

M.P. Oxley, M. Varela, T.J. Pennycook, K. van Benthem, S.D. Findlay, A.J. D'Alfonso, L.J. Allen, S.J. Pennycook, Interpreting atomic-resolution spectroscopic images. Phys. Rev. B. **76**, 064303 (2007)

K. Ozasa, Y. Aoyagi, M. Iwaki, H. Kurata, Facets, indium distribution, and lattice distortion of InGaAs/GaAs quantum dots observed by three-dimensional scanning transmission electron microscope. J. Appl. Phys. **94**, 313–317 (2003)

A. Parisini, V. Morandi, S. Solmi, P.G. Merli, D. Giubertoni, M. Bersani, J.A. van den Berg, Quantitative determination of the dopant distribution in Si ultra-shallow junctions by tilted sample annular dark field scanning transmission electron microscopy. Appl. Phys. Lett. **92**, 261907 (2008)

J.R. Parsons, H.M. Johnson, C.W. Hoelke, R.R. Hosbons, Imaging of uranium atoms with the electron microscope by phase contrast. Philos. Mag. **27**, 1359–1368 (1973)

Y. Peng, M.P. Oxley, A.R. Lupini, M.F. Chisholm, S.J. Pennycook, Spatial resolution and information transfer in scanning transmission electron microscopy. Microsc. Microanal. **14**, 36–47 (2008)

Y.P. Peng, P.D. Nellist, S.J. Pennycook, HAADF-STEM imaging with sub-angstrom probes: A full Bloch wave analysis. J. Electron Microsc. **53**, 257–266 (2004)

S.J. Pennycook, Study of supported ruthenium catalysts by STEM. J. Microsc.-Oxford **124**, 15–22 (1981)

S.J. Pennycook, Delocalization corrections for electron channeling analysis. Ultramicroscopy **26**, 239–248 (1988)

S.J. Pennycook, High-resolution imaging with large-angle elastically scattered electrons. EMSA Bull. **19**, 67–74 (1989a)

S.J. Pennycook, Z-contrast STEM for materials science. Ultramicroscopy **30**, 58–69 (1989b)

S.J. Pennycook, in *Advances in Imaging and Electron Physics*, vol. 123, ed. by P.G. Merli, G. Calestani, M. Vittori-Antisari, (Academic, San Diego, 2002), pp. 173–206

S.J. Pennycook, in *Encyclopedia of Condensed Matter Physics*, ed. by F. Bassani, J. Liedl, P. Wyder, (Elsevier Science, Kidlington, Oxford, 2006), pp. 240–247

S.J. Pennycook, Investigating the optical properties of dislocations by scanning transmission electron microscopy. Scanning **30**, 287–298 (2008)

S.J. Pennycook, S.D. Berger, R.J. Culbertson, Elemental mapping with elastically scattered electrons. J. Microsc. Oxford **144**, 229–249 (1986)

S.J. Pennycook, L.A. Boatner, Chemically sensitive structure-imaging with a scanning-transmission electron-microscope. Nature **336**, 565–567 (1988)

S.J. Pennycook, L.M. Brown, A.J. Craven, Observation of cathodoluminescence at single dislocations by STEM. Philos. Mag. A **41**, 589–600 (1980)

S.J. Pennycook, N.D. Browning, D.E. Jesson, M.F. Chisholm, A.J. McGibbon, Atomic-resolution imaging and spectroscopy of semiconductor interfaces. Appl. Phys. A **57**, 385–391 (1993)

S.J. Pennycook, N.D. Browning, M.M. McGibbon, A.J. McGibbon, D.E. Jesson, M.F. Chisholm, Direct determination of interface structure and bonding with the scanning transmission electron microscope. Philos. Trans. R. Soc. A. **354**, 2619–2634 (1996)

S.J. Pennycook, M.F. Chisholm, D.E. Jesson, R. Feenstra, S. Zhu, X.Y. Zheng, D.J. Lowndes, Growth and relaxation mechanisms of $YBa_2Cu_3O_{7-x}$ films. Physica C **202**, 1–11 (1992)
S.J. Pennycook, M.F. Chisholm, D.E. Jesson, D.P. Norton, D.H. Lowndes, R. Feenstra, H.R. Kerchner, J.O. Thomson, Interdiffusion, growth mechanisms, and critical currents in $YBa_2Cu_3O_{7-x}/PrBa_2Cu_3O_{7-x}$ superlattices. Phys. Rev. Lett. **67**, 765–768 (1991)
S.J. Pennycook, M.F. Chisholm, A.R. Lupini, M. Varela, van K. Benthem, A.Y. Borisevich, M.P. Oxley, W. Luo, S.T. Pantelides, in *Aberration-Corrected Electron Microscopy*, ed. by P.W. Hawkes (Academic, Oxford, 2008), pp. 327–384
S.J. Pennycook, M.F. Chisholm, A.R. Lupini, M. Varela, A.Y. Borisevich, M.P. Oxley, W.D. Luo, van K. Benthem, S.H. Oh, D.L. Sales, S.I. Molina, J. Garcia-Barriocanal, C. Leon, J. Santamaria, S.N. Rashkeev, S.T. Pantelides, Aberration-corrected scanning transmission electron microscopy: From atomic imaging and analysis to solving energy problems. Philos. Trans. R. Soc. A. **367**, 3709–3733 (2009b)
S.J. Pennycook, M.F. Chisholm, M. Varela, A.R. Lupini, A. Borisevich, Y. Peng, van K. Benthem, N. Shibata, V.P. Dravid, P. Prabhumirashi, S.D. Findlay, M.P. Oxley, L.J. Allen, N. Dellby, P.D. Nellist, Z.S. Szilagyi, O.L. Krivanek, Materials applications of aberration-corrected STEM, *2004 Focused Interest Group Pre-Congress Meeting "Materials Research in an Aberration Free Environment"* (2004a)
S.J. Pennycook, A. Howie, Study of single-electron excitations by electron-microscopy. 2. Cathodoluminescence image-contrast from localized energy transfers. Philos. Mag. A **41**, 809–827 (1980)
S.J. Pennycook, D.E. Jesson, High-resolution incoherent imaging of crystals. Phys. Rev. Lett. **64**, 938–941 (1990)
S.J. Pennycook, D.E. Jesson, High-resolution Z-contrast imaging of crystals. Ultramicroscopy **37**, 14–38 (1991)
S.J. Pennycook, D.E. Jesson, Atomic resolution Z-contrast imaging of interfaces. Acta Metall. Mater. **40**, S149–S159 (1992)
S.J. Pennycook, D.E. Jesson, M.F. Chisholm, in *Microscopy of Semiconducting Materials 1989*, ed. by A.G. Cullis, J.L. Hutchison, Institute of Physics Conference Series No. 100, (1989), pp. 51–58
S.J. Pennycook, D.E. Jesson, M.F. Chisholm, in *High Resolution Electron Microscopy of Defects in Materials*, ed. by R. Sinclair, D.J. Smith, U. Dahmen (Materials Research Society, Warrendale, PA, 1990), pp. 211–222
S.J. Pennycook, A.R. Lupini, A. Borisevich, Y. Peng, N. Shibata, 3D atomic resolution imaging through aberration-corrected STEM. Microsc. Microanal. **10**(Suppl. 2), 1172–1173 (2004b)
S.J. Pennycook, A.R. Lupini, M. Varela, A.Y. Borisevich, Y. Peng, M.F. Chisholm, N. Dellby, O. Krivanek, P.D. Nellist, Z.S. Szlagyi, G. Duscher, Sub-angstrom resolution through aberration-corrected STEM. Microsc. Microanal. **9**, 926–927 (2003a)
S.J. Pennycook, A.R. Lupini, A. Kadavanich, J.R. McBride, S.J. Rosenthal, R.C. Puetter, A. Yahil, O.L. Krivanek, N. Dellby, P.D. Nellist, G. Duscher, L.G. Wang, S.T. Pantelides, Aberration-corrected scanning transmission electron microscopy: The potential for nano- and interface science. Z. Metalkd. **94**, 350–357 (2003b)
S.J. Pennycook, A.R. Lupini, M. Varela, A. Borisevich, M.F. Chisholm, E. Abe, N. Dellby, O.L. Krivanek, P.D. Nellist, L.G. Wang, R. Buczko, X. Fan, S.T. Pantelides, in *Spatially Resolved Characterization of Local Phenomena in Materials and Nanostructures*, ed. by D.A. Bonnell, A.J. Piqueras, P. Shreve, F. Zypman (Materials Research Society, Warrendale, PA, G1.1, 2003c)

S.J. Pennycook, A.R. Lupini, M. Varela, A.Y. Borisevich, Y. Peng, M.P. Oxley, K. van Benthem, M.F. Chisholm, in *Scanning Microscopy for Nanotechnology: Techniques and Applications*, ed. by W. Zhou, Z.L. Wang (Springer, New York, 2007), pp. 152–191

S.J. Pennycook, J. Narayan, Direct imaging of dopant distributions in silicon by scanning-transmission electron-microscopy. Appl. Phys. Lett. **45**, 385–387 (1984)

S.J. Pennycook, J. Narayan, Atom location by axial-electron-channeling analysis. Phys. Rev. Lett. **54**, 1543–1546 (1985)

S.J. Pennycook, M. Varela, A.R. Lupini, M.P. Oxley, M.F. Chisholm, Atomic-resolution spectroscopic imaging: Past, present and future. J. Electron Microsc. **58**, 87–97 (2009a)

S.J. Pennycook, M. Varela, M.F. Chisholm, A.Y. Borisevich, A.R. Lupini, K. van Benthem, M.P. Oxley, W. Luo, J.R. McBride, S.J. Rosenthal, S.H. Oh, D.L. Sales, S.I. Molina, K. Sohlberg, S.T. Pantelides, in *Oxford Handbook of Nanoscience and Nanotechnology*, ed. by A.V. Narlikar, Y.Y. Fu (Oxford University Press, Oxford, 2010), pp. 205–248

T. Plamann, M.J. Hÿtch, Tests on the validity of the atomic column approximation for STEM probe propagation. Ultramicroscopy **78**, 153–161 (1999)

B. Rafferty, P.D. Nellist, S.J. Pennycook, On the origin of transverse incoherence in Z-contrast STEM. J. Electron Microsc. **50**, 227–233 (2001)

Rayleigh, Lord, On the theory of optical images with special reference to the microscope. Philos. Mag. **42**(5), 167–195 (1896)

R. Reichelt, A. Engel, Monte carlo calculations of elastic and inelastic electron scattering in biological and plastic materials. Ultramicroscopy **13**, 279–293 (1984)

M. Retsky, Observed single atom elastic cross-sections in a scanning electron-microscope. Optik **41**, 127–142 (1974)

S.B. Rice, J.Y. Koo, M.M. Disko, M.M.J. Treacy, On the imaging of Pt atoms in zeolite frameworks. Ultramicroscopy **34**, 108–118 (1990)

H. Rose, Phase-contrast in scanning-transmission electron-microscopy. Optik **39**, 416–436 (1974)

H. Rose, Image formation by inelastically scattered electrons in electron microscopy. Optik **45**, 139–158 and 187–208 (1976)

H. Rose, Information transfer in transmission electron microscopy. Ultramicroscopy **15**, 173–191 (1984)

H. Rose, in *Advances in Imaging and Electron Physics*, vol. 153, ed. by P.W. Hawkes (Elsevier, Amsterdam, The Netherlands, 2008), pp. 3–39

S.J. Rosenthal, J. McBride, S.J. Pennycook, L.C. Feldman, Synthesis, surface studies, composition and structural characterization of CdSe, core/shell and biologically active nanocrystals. Surf. Sci. Rep. **62**, 111–157 (2007)

C.J. Rossouw, Coherence in inelastic electron-scattering. Ultramicroscopy **16**, 241–254 (1985)

C.J. Rossouw, L.A. Bursill, Diffuse scattering of fast electrons in barium titanate. Acta Crystallogr **B41**, 248–254 (1985)

C.J. Rossouw, P.S. Turner, T.J. White, Axial electron-channelling analysis of perovskite. I: Theory and experiment for $CaTiO_3$. Philos. Mag. B. **57**, 209–225 (1988a)

C.J. Rossouw, P.S. Turner, T.J. White, Axial electron-channelling analysis of perovskite. II: Site identification of Sr, Zr and U impurities. Philos. Mag. B. **57**, 227–241 (1988b)

C.J. Rossouw, M.J. Whelan, Diffraction contrast retained by plasmon and K-loss electrons. Ultramicroscopy **6**, 53–66 (1981)

C.J. Rossouw, L.J. Allen, S.D. Findlay, M.P. Oxley, Channelling effects in atomic resolution STEM. Ultramicroscopy **96**, 299–312 (2003)

C.J. Rossouw, C.T. Forwood, M.A. Gibson, P.R. Miller, Statistical ALCHEMI: General formulation and method with application to Ti-Al ternary alloys. Philos. Mag. A **74**, 57–76 (1996a)

C.J. Rossouw, C.T. Forwood, M.A. Gibson, P.R. Miller, Zone-axis convergent-beam electron diffraction and ALCHEMI analysis of Ti-Al alloys with ternary additions. Philos. Mag. A, **74**, 77–102 (1996b)

C.J. Rossouw, V.W. Maslen, Localization and ALCHEMI for zone axis orientations. Ultramicroscopy **21**, 277–289 (1987)

C.J. Rossouw, P.S. Turner, T.J. White, A.J. Oconnor, Statistical-analysis of electron channeling microanalytical data for the determination of site occupancies of impurities. Philos. Mag. Lett. **60**, 225–232 (1989)

E. Ruska, *The Early Development of Electron Lenses and Electron Microscopy* (Stuttgart, Hirzel, 1980)

E. Ruska, The development of the electron-microscope and of electron-microscopy. Rev. Mod. Phys. **59**, 627–638 (1987)

M. Saito, K. Kimoto, T. Nagai, S. Fukushima, D. Akahoshi, H. Kuwahara, Y. Matsui, K. Ishizuka, Local crystal structure analysis with 10-pm accuracy using scanning transmission electron microscopy. J. Electron. Microsc. **58**, 131–136 (2009)

Y. Sato, J.P. Buban, T. Mizoguchi, N. Shibata, M. Yodogawa, T. Yamamoto, Y. Ikuhara, Role of Pr segregation in acceptor-state formation at ZnO grain boundaries. Phys. Rev. Lett. **97**, 106802 (2006)

H. Sawada, F. Hosokawai, T. Kaneyama, T. Ishizawa, M. Terao, M. Kawazoe, T. Sannomiya, T. Tomita, Y. Kondo, T. Tanaka, Y. Oshima, Y. Tanishiro, N. Yamamoto, K. Takayanagi, Achieving 63 pm resolution in scanning transmission electron microscope with spherical aberration corrector. Jpn. J. Appl. Phys. **46**, L568–L570 (2007)

H. Sawada, Y. Tanishiro, N. Ohashi, T. Tomita, F. Hosokawa, T. Kaneyama, Y. Kondo, K. Takayanagi, STEM imaging of 47-pm-separated atomic columns by a spherical aberration-corrected electron microscope with a 300-kV cold field emission gun. J. Electron Microsc. **58**, 357–361 (2009)

H. Sawada, T. Tomita, M. Naruse, T. Honda, P. Hambridge, P. Hartel, M. Haider, C. Hetherington, R. Doole, A. Kirkland, J. Hutchison, J. Titchmarsh, D. Cockayne, Experimental evaluation of a spherical aberration-corrected TEM and STEM. J. Electron Microsc. **54**, 119–121 (2005)

M. Scheinfein, A .Muray, M. Isaacson, Electron-energy loss spectroscopy across a metal-insulator interface at sub-nanometer spatial-resolution. Ultramicroscopy **16**, 233–239 (1985)

O. Scherzer, Uber einige Fehler von Elektronenlinsen. Z. Physik **101**, 114–132 (1936)

O. Scherzer, Das theoretisch erreichbare Auflösungsvermögen des Elektronenmikroskops. Z. Physik **114**, 427–434 (1939)

O. Scherzer, Sparische und chromatische Korrektur von Electronen-Linsen. Optik **2**, 114–132 (1947)

O. Scherzer, The theoretical resolution limit of the electron microscope. J. Appl. Phys. **20**, 20–29 (1949)

C. Scheu, Electron energy-loss near-edge structure studies at the atomic level: Reliability of the spatial difference technique. J. Microsc.-Oxford **207**, 52–57 (2002)

L.I. Schiff, Ultimate resolving power of the electron miscroscope. Phys. Rev. **61**, 721 (1942)

D.A. Shashkov, D.A. Muller, D.N. Seidman, Atomic-scale structure and chemistry of ceramic/metal interfaces–II. Solute segregation at MgO/Cu (Ag) and CdO/Ag (Au) interfaces. Acta Mater. **47**, 3953–3963 (1999)

N. Shibata, S.D. Findlay, S. Azuma, T. Mizoguchi, T. Yamamoto, Y. Ikuhara, Atomic-scale imaging of individual dopant atoms in a buried interface. Nat. Mater. **8**, 654–658 (2009)

N. Shibata, G.S. Painter, P.F. Becher, S.J. Pennycook, Atomic ordering at an amorphous/crystal interface. Appl. Phys. Lett. **89**, 051908 (2006)

N. Shibata, G.S. Painter, R.L. Satet, M.J. Hoffmann, S.J. Pennycook, P.F. Becher, Rare-earth adsorption at intergranular interfaces in silicon nitride ceramics: Subnanometer observations and theory. Phys. Rev. B **72**, 140101 (2005)

N. Shibata, S.J. Pennycook, T.R. Gosnell, G.S. Painter, W.A. Shelton, P.F. Becher, Observation of rare-earth segregation in silicon nitride ceramics at subnanometre dimensions. Nature **428**, 730–733 (2004)

D.H. Shin, E.J. Kirkland, J. Silcox, Annular dark field electron-microscope images with better than 2 Å resolution at 100 kV. Appl. Phys. Lett. **55**, 2456–2458 (1989)

A. Singhal, J.C. Yang, J.M. Gibson, STEM-based mass spectroscopy of supported Re clusters. Ultramicroscopy **67**, 191–206 (1997)

D.J. Smith, The realization of atomic resolution with the electron microscope. Rep. Prog. Phys. **60**, 1513–1580 (1997)

K. Sohlberg, S. Rashkeev, A.Y. Borisevich, S.J. Pennycook, S.T. Pantelides, Origin of anomalous Pt-Pt distances in the Pt/alumina catalytic system. Chemphyschem **5**, 1893–1897 (2004)

G.E. Sosinsky, N.R. Francis, D.J. Derosier, J.S. Wall, M.N. Simon, J. Hainfeld, Mass determination and estimation of subunit stoichiometry of the bacterial hook basal body flagellar complex of salmonella-typhimurium by scanning-transmission electron-microscopy. Proc. Natl. Acad. Sci. U.S.A. **89**, 4801–4805 (1992)

J.C.H. Spence, The future of atomic resolution electron microscopy for materials science. Mater. Sci. Eng. R **26**, 1–49 (1999)

J.C.H. Spence, J.M. Cowley, Lattice imaging in STEM. Optik **50**, 129–142 (1978)

J.C.H. Spence, J. Lynch, STEM microanalysis by transmission electron-energy loss spectroscopy in crystals. Ultramicroscopy **9**, 267–276 (1982)

J.C.H. Spence, M. Kuwabara, Y. Kim, Localization effects on quantification in axial and planar ALCHEMI. Ultramicroscopy **26**, 103–112 (1988)

S. Stemmer, A. Sane, N.D. Browning, T.J. Mazanec, Characterization of oxygen-deficient $SrCoO_{3-\delta}$ by electron energy-loss spectroscopy and Z-contrast imaging. Solid State Ionics **130**, 71–80 (2000)

K. Suenaga, T. Tence, C. Mory, C. Colliex, H. Kato, T. Okazaki, H. Shinohara, K. Hirahara, S. Bandow, S. Iijima, Element-selective single atom imaging. Science **290**, 2280–2282 (2000)

C. Süsskind, in *The Beginnings of Electron-Microscopy*, ed. by P.W. Hawkes, Advances in Imaging and Electron Physics, suppl. 16 (Academic, San Diego, 1985), pp. 501–523

J. Taftø, The cation-atom distribution in a $(Cr, Fe, Al, Mg)_3O_4$ spinel as revealed from the channelling effect in electron-induced X-ray emission. J. Appl. Crystallogr. **15**, 378–381 (1982)

J. Taftø, O. Krivanek, Site-specific valence determination by electron energy-loss spectroscopy. Phys. Rev. Lett. **48**, 560–563 (1982)

J. Taftø, J.C.H. Spence, Crystal site location of iron and trace elements in a magnesium-iron olivine by a new crystallographic technique. Science **218**, 49–51 (1982a)

J. Taftø, J.C.H. Spence, Atomic site determination using the channeling effect in electron-induced X-ray emission. Ultramicroscopy **9**, 243–248 (1982b)

E. Takasuka, K. Asai, K. Fujita, M.F. Chisholm, S.J. Pennycook, High-resolution Z-contrast observation of GaAs/Si heterointerfaces through scanning-transmission electron-microscope. Jpn. J. Appl. Phys. **31**, L1788-L1790 (1992)

M. Takeguchi, A. Hashimoto, M. Shimojo, K. Mitsuishi, K. Furuya, Development of a stage-scanning system for high-resolution confocal STEM. J. Electron Microsc. **57**, 123–127 (2008)

M. Takeguchi, M.R. McCartney, D.J. Smith, Mapping In concentration, strain, and internal electric field in InGaN/GaN quantum well structure. Appl. Phys. Lett. **84**, 2103–2105 (2004)

Transmission Electron Aberration-Corrected Microscope Project. http://ncem.lbl.gov/TEAM-project (accessed Nov. 3, 2010)

C.M. Tey, H.Y. Liu, A.G. Cullis, I.M. Ross, M. Hopkinson, Structural studies of a combined InAlAs/InGaAs capping layer on 1.3-μm InAs/GaAs quantum dots. J Cryst Growth **285**, 17–23 (2005)

M.G.R. Thomson, Resolution and contrast in conventional and scanning high-resolution transmission electron-microscopes. Optik **39**, 15–38 (1973)

F. Thon, D. Willasch, Imaging of heavy atoms in darkfield electron-microscopy using hollow cone illumination. Optik **36**, 55–58 (1972)

A. Thust, High-resolution transmission electron microscopy on an absolute contrast scale. Phys. Rev. Lett. **102**, 220801 (2009)

J.M. Titchmarsh, Detection of electron energy-loss edge shifts and fine structure variations at grain boundaries and interfaces. Ultramicroscopy **78**, 241–250 (1999)

M.M.J. Treacy, J.M. Gibson, Coherence and multiple-scattering in "Z-contrast" images. Ultramicroscopy **52**, 31–53 (1993)

M.M.J. Treacy, A. Howie, S.J. Pennycook, in *Electron Microscopy and Analysis 1979* ed. by T. Mulvey (Institute of Physics, Brighton, 1980), pp. 261–266

M.M.J. Treacy, A. Howie, C.J. Wilson, Z-contrast of platinum and palladium catalysts. Philos. Mag. A **38**, 569–585 (1978)

M.M.J. Treacy, S.B. Rice, Catalyst particle sizes from Rutherford scattered intensities. J. Microsc.-Oxford **156**, 211–234 (1989)

K. Urban, Studying atomic structures by aberration-corrected transmission electron microscopy. Science **321**, 506–510 (2008)

S. Van Aert, J. Verbeeck, R. Erni, S. Bals, M. Luysberg, D. Van Dyck, G. Van Tendeloo, Quantitative atomic resolution mapping using high-angle annular dark field scanning transmission electron microscopy. Ultramicroscopy **109**, 1236–1244 (2009)

K. van Benthem, A.R. Lupini, M. Kim, H.S. Baik, S. Doh, J.H. Lee, M.P. Oxley, S.D. Findlay, L.J. Allen, J.T. Luck, S.J. Pennycook, Three-dimensional imaging of individual hafnium atoms inside a semiconductor device. Appl. Phys. Lett. **87**, 034104 (2005)

K. van Benthem, A.R. Lupini, M.P. Oxley, S.D. Findlay, L.J. Allen, S.J. Pennycook, Three-dimensional ADF imaging of individual atoms by through-focal series scanning transmission electron microscopy. Ultramicroscopy **106**, 1062–1068 (2006)

D. Van Dyck, Is the frozen phonon model adequate to describe inelastic phonon scattering? Ultramicroscopy **109**, 677–682 (2009)

J.B. Vandersande, E.L. Hall, Applications of dedicated scanning-transmission electron-microscopy to non-metallic materials. J. Am. Ceram. Soc. **62**, 246–254 (1979)

R. Vanfleet, D.A. Muller, H.J. Gossmann, P.H. Citrin, J. Silcox, in *Advances in Materials Problem Solving with the Electron Microscope*, ed. by J. Bentley, C. Allen, U. Dahmen, I. Petrov (Materials Research Society, Warrendale, PA, 2001), pp. 173–178

M. Varela, D. Arias, Z. Sefrioui, C. Leon, C. Ballesteros, S.J. Pennycook, J. Santamaria, Direct correlation between T_c and CuO_2 bilayer spacing in $YBa_2Cu_3O_{7-x}$. Phys. Rev. B **66**, 134517 (2002)

M. Varela, S.D. Findlay, A.R. Lupini, H.M. Christen, A.Y. Borisevich, N. Dellby, O.L. Krivanek, P.D. Nellist, M.P. Oxley, L.J. Allen, S.J. Pennycook, Spectroscopic imaging of single atoms within a bulk solid. Phys. Rev. Lett. **92**, 095502 (2004)

M. Varela, A.R. Lupini, S.J. Pennycook, Z. Sefrioui, J. Santamaria, Nanoscale analysis of $YBa_2Cu_3O_{7-x}/La_{0.67}Ca_{0.33}MnO_3$ interfaces. Solid State Electron. **47**, 2245–2248 (2003)

M. Varela, M.P. Oxley, W. Luo, J. Tao, M. Watanabe, A.R. Lupini, S.T. Pantelides, S.J. Pennycook, Atomic-resolution imaging of oxidation states in manganites. Phys. Rev. B **79**, 085117 (2009)

J. Verbeeck, O.I. Lebedev, G. Van Tendeloo, J. Silcox, B. Mercey, M. Hervieu, A.M. Haghiri-Gosnet, Electron energy-loss spectroscopy study of a $(LaMnO_3)(8)(SrMnO_3)(4)$ heterostructure. Appl. Phys. Lett. **79**, 2037–2039 (2001)

D. Viehland, J. Li, Z. Xu, Direct evidence of substituent clustering in lower valent K+ modified $PbZrO_3$ by high-resolution Z-contrast imaging. Appl. Phys. A: Mater. Sci. Process. **79**, 1955–1958 (2004)

M. von Ardenne, Das Elektronen-Rastermikroskop. Praktische Ausführung. Z. Tech. Phys. **19**, 407–416 (1938a)

M. von Ardenne, Das Elektronen-Rastermikroskop. Theoretische Grundlagen. Z. Physik **109**, 553–572 (1938a)

M. von Ardenne, Die Grenzen für das Auflösungsvermögen des Elektronenmikroskops. Z. Physik **108**, 338–353 (1938c)

M. von Ardenne, Intensitätsfragen und Auflösungsvermögen des Elektronenmikroskops. Z. Physik **112**, 744–752 (1939)

M. von Ardenne, Über das Auftreten von Schwärzungslinien bei der elektronenmikroskopischen Abbildung kristalliner Lamellen. Z. Physik **116**, 736 (1940a)

M. von Ardenne, Über ein Universal-Elektronenmikroskop für Hellfeld-, Dunkelfeld- und Stereobild-Betrieb. Z. Physik **115**, 339–368 (1940b)

M. von Ardenne, *Elektronen-Übermikroskopie* (Springer, Berlin, 1940c)

M. von Ardenne, in *The Beginnings of Electron-Microscopy*, ed. by P.W. Hawkes, Advances in Imaging and Electron Physics, Suppl 16. (Academic, Orlando, FL, 1985), pp. 1–21

M. von Ardenne, in *The Growth of Electron Microscopy*, ed. by T. Mulvey. Advances in Imaging and Electron Physics, vol. 96 (Academic, San Diego, 1996), pp. 635–652

B. von Borries, E. Ruska, Der Einfluß von Elektroneninterferenzen auf die Abbildung von Kristallen im Übermikroskop. Naturwiss. **23**, 366–367 (1940)

H.S. von Harrach, in *Advances in Imaging and Electron Physics*, vol. 159. ed. by P.W. Hawkes (Elsevier, Amsterdam, The Netherlands, 2009), pp. 287–323

H.S. von Harrach, Medium-voltage field-emission STEM – the ultimate AEM. Microsc. Microanal. Microstruct. **5**, 153–164 (1994)

H.S. von Harrach, A.W. Nicholls, D.E. Jesson, S.J. Pennycook, in *Electron Microscopy and Analysis 1993*, ed. by A.J. Craven, Institute of Physics Conference Series No. 138. (Institute of Physics, Bristol, 1993), pp. 499–502

P. Voyles, D. Muller, E. Kirkland, Depth-dependent imaging of individual dopant atoms in silicon. Microsc. Microanal. **10**, 291–300 (2004)

P.M. Voyles, J.L. Grazul, D.A. Muller, Imaging individual atoms inside crystals with ADF-STEM. Ultramicroscopy **96**, 251–273 (2003)

P.M. Voyles, D.A. Muller, J.L. Grazul, P.H. Citrin, H.J.L. Gossmann, Atomic-scale imaging of individual dopant atoms and clusters in highly n-type bulk Si. Nature **416**, 826–829 (2002)

J. Wall, J. Langmore, M. Isaacson, A.V. Crewe, Scanning-transmission electron-microscopy at high-resolution. Proc. Natl. Acad. Sci. USA. **71**, 1–5 (1974)
J.S. Wall, M.N. Simon, J.F. Hainfeld, in *Advances in Imaging and Electron Physics*, vol. 159, ed. by P.W. Hawkes (Elsevier, Amsterdam, The Netherlands, 2009), pp. 101–121
D.J. Wallis, R.S. Balmer, A.M. Keir, T. Martin, Z-contrast imaging of AlN exclusion layers in GaN field-effect transistors. Appl. Phys. Lett. **87**, 042101 (2005)
D.J. Wallis, N.D. Browning, P.D. Nellist, S.J. Pennycook, I. Majid, Y. Liu, J.B. Van der Sande, Atomic structure of a 36.8 (210) tilt grain boundary in TiO_2. J. Am. Ceram. Soc. **80**, 499–502 (1997a)
D.J. Wallis, N.D. Browning, S. Sivananthan, P.D. Nellist, S.J. Pennycook, Atomic layer graphoepitaxy for single crystal heterostructures. Appl. Phys. Lett. **70**, 3113–3115 (1997b)
P. Wang, A.L. Bleloch, M. Falke, P.J. Goodhew, J. Ng, M. Missous, Direct measurement of composition of buried quantum dots using aberration-corrected scanning transmission electron microscopy. Appl. Phys. Lett. **89**, 072111 (2006)
S.W. Wang, A.Y. Borisevich, S.N. Rashkeev, M.V. Glazoff, K. Sohlberg, S.J. Pennycook, S.T. Pantelides, Dopants adsorbed as single atoms prevent degradation of catalysts. Nat. Mater. **3**, 274–274 (2004)
Z.L. Wang, The 'frozen-lattice' approach for incoherent phonon excitation in electron scattering. How accurate is it? Acta Cryst. A **54**, 460–467 (1998a)
Z.L. Wang, An optical potential approach to incoherent multiple thermal diffuse scattering in quantitative HRTEM. Ultramicroscopy **74**, 7–26 (1998b)
Z.L. Wang, J.M. Cowley, Simulating high-angle annular dark-field STEM images including inelastic thermal diffuse-scattering. Ultramicroscopy **31**, 437–454 (1989)
Z.L. Wang, J.M. Cowley, Dynamic theory of high-angle annular-dark-field STEM lattice images for a Ge/Si interface. Ultramicroscopy **32**, 275–289 (1990)
I.R.M. Wardell, P.E. Bovey, in *Advances in Imaging and Electron Physics*, vol. 159, ed. by P.W. Hawkes (Elsevier, 2009), pp. 221–285
B.E. Warren, *X-ray Diffraction* (Dover, New York, NY, 1990)
H. Watanabe, R. Uyeda, Energy-selecting electron microscope. J. Phys. Soc. Jpn. **17**, 569–570 (1962)
K. Watanabe, T. Yamazaki, I. Hashimoto, M. Shiojiri, Atomic-resolution annular dark-field STEM image calculations. Phys. Rev. B **64**, 115432 (2001)
M. Watanabe, D.W. Ackland, A. Burrows, C.J. Kiely, D.B. Williams, O.L. Krivanek, N. Dellby, M.F. Murfitt, Z.S. Szilagyi, Improvements in the X-ray analytical capabilities of a scanning transmission electron microscope by spherical-aberration correction. Microsc. Microanal. **12**, 515–526 (2006)
M. Weyland, P.A. Midgley, J.M. Thomas, Electron tomography of nanoparticle catalysts on porous supports: A new technique based on Rutherford scattering. J. Phys. Chem. B **105**, 7882–7886 (2001)
R.F. Whiting, F.P. Ottensmeyer, Heavy atoms in model compounds and nucleic acids imaged by dark field transmission electron microscopy. J. Mol. Biol. **67**, 173–176 (1972)
D.B. Williams, J.W. Edington, High-resolution microanalysis in materials science using electron-energy loss measurements. J. Microsc.-Oxford **108**, 113–145 (1976)
D.B. Williams, A.D. Romig, Studies of interfacial segregation in the analytical electron-microscope – a brief review. Ultramicroscopy **30**, 38–51 (1989)

G.B. Winkelman, C. Dwyer, T.S. Hudson, Nguyen-D. Manh, M. Doblinger, R.L. Satet, M.J. Hoffmann, D.J.H. Cockayne, Arrangement of rare-earth elements at prismatic grain boundaries in silicon nitride. Philos. Mag. Lett. **84**, 755–762 (2004)

G.B. Winkelman, C. Dwyer, T.S. Hudson, Nguyen-D. Manh, M. Doblinger, R.L. Satet, M.J. Hoffmann, D.J.H. Cockayne, Three-dimensional organization of rare-earth atoms at grain boundaries in silicon nitride. Appl. Phys. Lett. **87**, 061911 (2005)

D.B. Wittry, An electron spectrometer for use with transmission electron microscope. J. Phys. D. **2**, 1757–1766 (1969)

D.B. Wittry, R.P. Ferrier, V.E. Cosslett, Selected-area electron spectrometry in transmission electron microscope. J. Phys. D. **2**, 1767–1773 (1969)

Y. Xiao, F. Patolsky, E. Katz, J.F. Hainfeld, I. Willner, "Plugging into enzymes": Nanowiring of redox enzymes by a gold nanoparticle. Science **299**, 1877–1881 (2003)

H.L. Xin, D.A. Muller, Aberration-corrected ADF-STEM depth sectioning and prospects for reliable 3D imaging in S/TEM. J. Electron Microsc. **58**, 157–165 (2009)

Y. Xin, E.M. James, I. Arslan, S. Sivananthan, N.D. Browning, S.J. Pennycook, F. Omnes, B. Beaumont, J.P. Faurie, P. Gibart, Direct experimental observation of the local electronic structure at threading dislocations in metalorganic vapor phase epitaxy grown wurtzite GaN thin films. Appl. Phys. Lett. **76**, 466–468 (2000a)

Y. Xin, E.M. James, N.D. Browning, S.J. Pennycook, Atomic resolution Z-contrast imaging of semiconductors. J. Electron Microsc. **49**, 231–244 (2000b)

Y. Xin, S.J. Pennycook, N.D. Browning, P.D. Nellist, S. Sivananthan, F. Omnes, B. Beaumont, J.P. Faurie, P. Gibart, Direct observation of the core structures of threading dislocations in GaN. Appl. Phys. Lett. **72**, 2680–2682 (1998)

P. Xu, E.J. Kirkland, J. Silcox, R. Keyse, High-resolution imaging of silicon (111) using a 100 keV STEM. Ultramicroscopy **32**, 93–102 (1990)

X. Xu, S.P. Beckman, P. Specht, E.R. Weber, D.C. Chrzan, R.P. Erni, I. Arslan, N. Browning, A. Bleloch, C. Kisielowski, Distortion and segregation in a dislocation core region at atomic resolution. Phys. Rev. Lett. **95**, 145501 (2005)

Z.K. Xu, S.M. Gupta, D. Viehland, Y.F. Yan, S.J. Pennycook, Direct imaging of atomic ordering in undoped and La-doped $Pb(Mg_{1/3}Nb_{2/3})O_3$. J. Am. Ceram. Soc. **83**, 181–188 (2000)

T. Yamazaki, N. Nakanishi, A. Rečnik, M. Kawasaki, K. Watanabe, M. Čeh, M. Shiojiri, Quantitative high-resolution HAADF-STEM analysis of inversion boundaries in Sb_2O_3-doped zinc oxide. Ultramicroscopy **98**, 305–316 (2004)

T. Yamazaki, K. Watanabe, Y. Kikuchi, M. Kawasaki, I. Hashimoto, M. Shiojiri, Two-dimensional distribution of As atoms doped in a Si crystal by atomic-resolution high-angle annular dark field STEM. Phys. Rev. B **61**, 13833–13839 (2000a)

T. Yamazaki, K. Watanabe, A. Recnik, M. Čeh, M. Kawasaki, M. Shiojiri, Simulation of atomic-scale high-angle annular dark field scanning transmission electron microscopy images. J. Electron Microsc. **49**, 753–759 (2000b)

Y. Yan, M.F. Chisholm, G. Duscher, A. Maiti, S.J. Pennycook, S.T. Pantelides, Impurity-induced structural transformation of a MgO grain boundary. Phys. Rev. Lett. **81**, 3675–3678 (1998a)

Y. Yan, S.J. Pennycook, A.P. Tsai, Direct imaging of local chemical disorder and columnar vacancies in ideal decagonal Al-Ni-Co quasicrystals. Phys. Rev. Lett. **81**, 5145–5148 (1998b)

Y.F. Yan, S.J. Pennycook, Alloys – atomic structure of the quasicrystal $Al_{72}Ni_{20}Co_8$. Nature **403**, 266–267 (2000)
Y.F. Yan, S.J. Pennycook, Chemical ordering in $Al_{72}Ni_{20}Co_8$ decagonal quasicrystals. Phys. Rev. Lett. **86**, 1542–1545 (2001)
Y.F. Yan, S.J. Pennycook, M. Terauchi, M. Tanaka, Atomic structures of oxygen-associated defects in sintered aluminum nitride ceramics. Microsc. Microanal. **5**, 352–357 (1999)
Z. Yu, P.E. Batson, J. Silcox, Artifacts in aberration-corrected ADF-STEM imaging. Ultramicroscopy **96**, 275–284 (2003)
J.F. Yuan, D.R. Beniac, G. Chaconas, F.P. Ottensmeyer, 3D reconstruction of the Mu transposase and the type 1 transpososome: A structural framework for Mu DNA transposition. Genes Develop. **19**, 840–852 (2005)
J. Zach, M. Haider, Aberration correction in a low voltage SEM by a multipole corrector. Nucl. Inst. Meth. A **363**, 316–325 (1995)
E. Zeitler, M.G.R. Thomson, Scanning transmission electron microscopy. Optik **31**, 258–280 and 359–366 (1970)
A. Ziegler, J.C. Idrobo, M.K. Cinibulk, C. Kisielowski, N.D. Browning, R.O. Ritchie, Interface structure and atomic bonding characteristics in silicon nitride ceramics. Science **306**, 1768–1770 (2004)
V.A. Zworykin, J. Hillier, R.L. Snyder, A scanning electron microscope. ASTM Bull. **117**, 15–23 (1942)

2

The Principles of STEM Imaging

Peter D. Nellist

2.1 Introduction

The purpose of this chapter is to review the principles underlying imaging in the scanning transmission electron microscope (STEM). Consideration of interference between parts of the convergent illuminating beam will be used to provide a common framework which allows contrast in various modes to be considered, and serves to allow the resolution limits of imaging to be determined. Several of the other chapters in this volume deal with specific imaging modes, so we do not seek to provide a detailed analysis of all those modes here, rather we will point out how these imaging modes may be considered in similar ways.

Figure 2–1 shows a schematic of the STEM optical configuration. A series of lenses focuses a beam to form a small spot, or probe, incident upon a thin, electron-transparent sample. Except for the final focusing lens, which is referred to as the objective, the other pre-sample lenses are referred to as condenser lenses. The aim of the lens system is to provide enough demagnification of the finite-sized electron source in order to form an atomic-scale probe at the sample. The objective lens provides the final, and largest, demagnification step. It is the aberrations of this lens that dominate the optical system. An objective aperture is used to restrict its numerical aperture to a size where the aberrations do not lead to significant blurring of the probe. The requirement of an objective aperture has two important consequences: (i) it imposes a diffraction limit to the smallest probe diameter that may be formed and (ii) electrons that do not pass through the aperture are lost, and therefore the aperture restricts the amount of beam current available.

Scan coils are arranged to scan the probe over the sample in a raster, and a variety of scattered signals can be detected and plotted as a function of probe position to form a magnified image. There is a wide range of possible signals available in the STEM, but the commonly collected ones are the following

(i) Transmitted electrons that leave the sample at relatively low angles with respect to the optic axis (smaller than the incident beam convergence angle). This mode is referred to as bright field (BF).

S.J. Pennycook, P.D. Nellist (eds.), *Scanning Transmission Electron Microscopy*, DOI 10.1007/978-1-4419-7200-2_2, © Springer Science+Business Media, LLC 2011

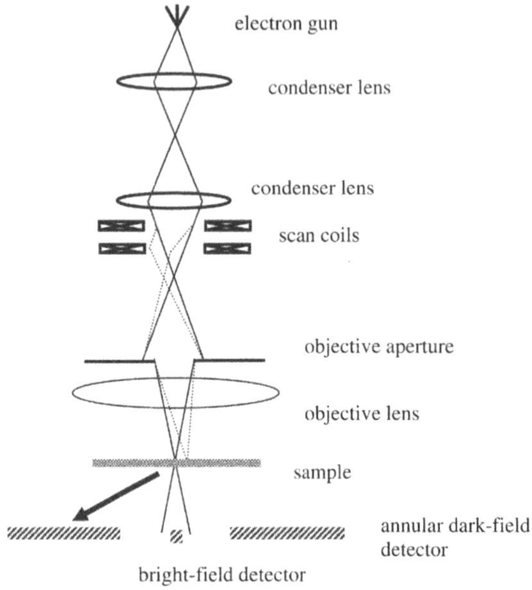

Figure 2–1. A schematic diagram of a STEM instrument showing the elements discussed in this chapter.

(ii) Transmitted electrons that leave the sample at relatively high angles with respect to the optic axis (usually at an angle several times the incident beam convergence angle). This mode is referred to as annular dark field (ADF).
(iii) Transmitted electrons that have lost a measurable amount of energy as they pass through the sample. Forming a spectrum of these electrons as a function of the energy lost leads to electron energy loss spectroscopy (EELS).
(iv) X-rays generated from electron excitations in the sample (EDX).

Post-specimen optics may also be present to control the angles subtended by some of these detectors, but such optics play no part in the image formation process and will not be considered here.

In this chapter we will mainly consider the first two detection modes on the above list. Chapter 6 deals more extensively with quantitative ADF imaging calculations and imaging using inelastically scattered electrons and Chapter 7 deals with EDX mapping.

2.2 The Principle of Reciprocity

Before embarking on a discussion of the origins of contrast and resolution limits in STEM imaging, it is first important to consider the implications of the *principle of reciprocity*. Consider elastic scattering so that all the electron waves in the microscope have the same energy. Under these conditions, the propagation of the electrons is time reversible. Points in the original detector plane could be replaced with electron sources, and the original source replaced with a detector, and a similar

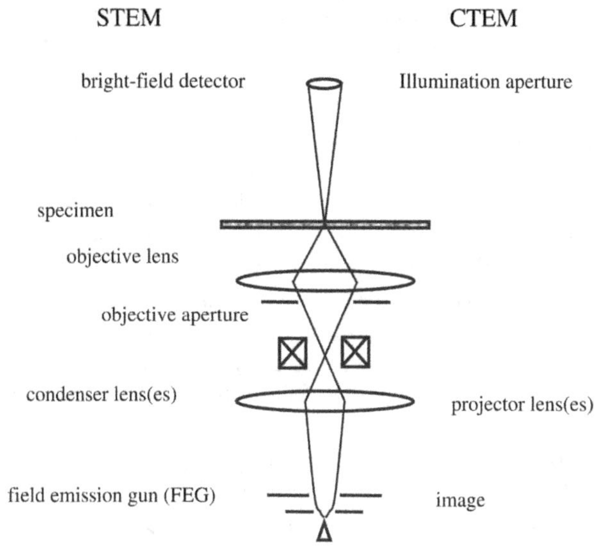

Figure 2–2. A schematic diagram showing the equivalence between bright-field STEM and HRTEM imaging making use of the principle of reciprocity.

intensity would be seen. Applying this concept to STEM (Cowley 1969, Zeitler and Thomson 1970), it becomes clear that the STEM imaging optics (before the sample) are equivalent to the imaging optics (after the sample) in the conventional TEM (CTEM). Similarly, the detector plane in STEM plays a similar role to the illumination configuration in CTEM.

We will see later that many of the concepts relating to coherence derived for CTEM can be transferred to STEM making use of the principle of reciprocity. As an immediate illustration of reciprocity, consider simple bright-field (BF) imaging. In the CTEM the ideal situation is that the sample is illuminated by perfectly coherent plane-wave illumination, and post-specimen optics form a highly magnified image of the wave that is transmitted by the sample. Now reverse this process to reveal the BF configuration for STEM (Figure 2–2). The electron source in the STEM plays an equivalent role to an image pixel in CTEM. The STEM imaging optics form a highly demagnified image of the source at the sample, and that can be scanned over the sample. Plane-wave transmission is then detected, usually with a small detector placed on the optic axis in the far field, and plotted as a function of probe position. The principle of reciprocity suggests that the image contrast will have the same form in both the CTEM and STEM cases, and this is observed experimentally (Crewe and Wall 1970) (see Figure 1–5). In the rest of this chapter, we will derive the imaging attributes from the STEM point of view but make the connection to CTEM where appropriate.

2.3 Interference Between Overlapping Discs

The origins of contrast in STEM arise from the interference between partial plane waves in the convergent beam that form the probe. Many such beams can interfere as they are scattered into the final beam that propagates to the detector, leading to a change in the intensity of this final beam as the probe is moved and hence image contrast (Spence and Cowley 1978). To understand this process, it is instructive to first consider lattice imaging of a simple sample that only scatters to reciprocal lattice vectors **g** and –**g** in addition to transmitting an unscattered beam. Plane-wave illumination of such a sample would lead to three spots: the direct beam and the two scattered beams. In STEM we have a coherent convergent beam illuminating the sample, and so the diffracted beams broaden to form discs. Where these diffracted discs overlap, interference features will be seen, and it is these interference features that lead to image contrast in STEM (Figure 2–3). To explain the form of these interference features, we need to follow the wavefunction through the microscope.

We start by assuming that the front focal plane of the objective lens is coherently illuminated. We assume that the effects of aberrations can be treated as a phase shift χ that has the form

$$\chi(\mathbf{K}) = \left(\pi C_{1,0}\lambda|\mathbf{K}|^2 + \frac{1}{2}\pi C_{3,0}\lambda^3|\mathbf{K}|^4\right), \tag{1}$$

where we have considered only defocus $C_{1,0}$ and spherical aberration $C_{3,0}$ as being present, and $\mathbf{K}$ is the transverse component of the wavevector at that position in the front focal plane. In an aberration-corrected microscope, the instrument will not be limited by $C_{3,0}$, and the general aberration phase surface is given in Chapters 3 and 7. To limit the influence of aberrations, an aperture is used, allowing beams to contribute up to a maximum transverse wavevector $K_{\text{max}} = \lambda/\alpha$; thus the

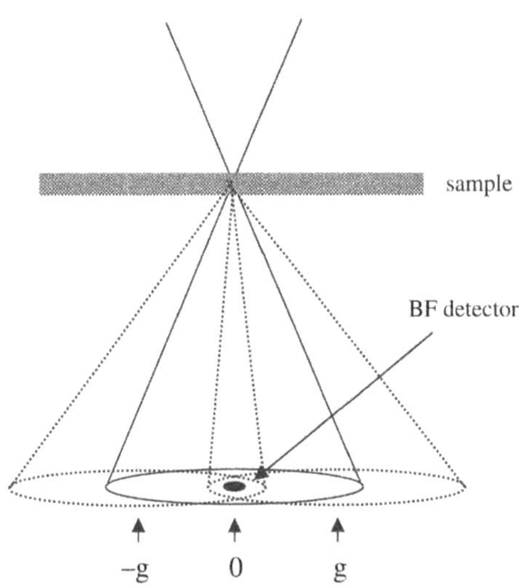

Figure 2–3. Diffraction of the coherent convergent beam by a specimen leads to diffracted discs. Where these discs overlap, interference will be seen. The bright-field detector is sensitive to interference between the direct beam and the two opposite diffracted beams.

overall wave at the front focal plane is given by the lens transmission function

$$T(\mathbf{K}) = A(\mathbf{K}) \exp\left[-i\chi(\mathbf{K})\right], \tag{2}$$

where A is a function that describes the size of the objective aperture, having a value of 1 for $|\mathbf{K}| \leq K_{max}$ and 0 elsewhere.

The electron probe can now be calculated by simply taking the inverse Fourier transform of the wave at the front focal plane, thus

$$P(\mathbf{R}) = \int T(\mathbf{K}) \exp(i2\pi\mathbf{K}\cdot\mathbf{R})\,d\mathbf{K}. \tag{3}$$

To express the ability of the STEM to move the probe over the sample, we can include a shift term in Eq. (3) to give

$$P(\mathbf{R}-\mathbf{R}_0) = \int T(\mathbf{K}) \exp(i2\pi\mathbf{K}\cdot\mathbf{R}) \exp(-i2\pi\mathbf{K}\cdot\mathbf{R}_0)\,d\mathbf{K}, \tag{4}$$

where $\mathbf{R}_0$ is the probe position. Moving the probe is therefore equivalent to adding a linear ramp to the phase variation across the front focal plane, which is exactly what the scan coils do.

Now consider diffraction by a sample. If we assume a thin sample that can be treated as being a thin, multiplicative transmission function, ϕ, then the wave exiting the sample can be written as

$$\psi(\mathbf{R},\mathbf{R}_0) = P(\mathbf{R}-\mathbf{R}_0)\phi(\mathbf{R}). \tag{5}$$

To calculate the wave at the detector plane, we take the Fourier transform of Eq. (5). Because Eq. (5) is a product, its Fourier transform becomes a convolution and can be written as

$$\psi(\mathbf{K}_f,\mathbf{R}_0) = \int \phi(\mathbf{K}_f-\mathbf{K})T(\mathbf{K})\exp(-i2\pi\mathbf{K}\cdot\mathbf{R}_0), \tag{6}$$

where changes in the argument of a function to reciprocal space vectors indicate that the Fourier transform has been taken. This equation has a relatively simple interpretation. The detector is in diffraction space, and the wave incident upon the detector at a position corresponding to a transverse wavevector $\mathbf{K}_f$, is the sum of all waves incident upon the sample, with transverse wavevectors $\mathbf{K}$, that are scattered by the object to $\mathbf{K}_f$. Now consider a sample that transmits a direct beam and scatters into $+\mathbf{g}$ and $-\mathbf{g}$ beams, i.e. it contains only a simple sinusoidal variation, either in amplitude or phase. The Fourier transform of the sample transmission function will contain Dirac delta functions at $\mathbf{0}$, $-\mathbf{g}$ and $+\mathbf{g}$. Substituting this form into Eq. (6) gives

$$\begin{aligned}\psi(\mathbf{K}_f,\mathbf{R}_0) = {} & T(\mathbf{K})\exp\left[-i2\pi\mathbf{K}\cdot\mathbf{R}_0\right] + \phi_{\mathbf{g}}T(\mathbf{K}-\mathbf{g})\exp\left[-i2\pi\left(\mathbf{K}-\mathbf{g}\right)\cdot\mathbf{R}_0\right] \\ & + \phi_{-\mathbf{g}}T(\mathbf{K}+\mathbf{g})\exp\left[-i2\pi\left(\mathbf{K}+\mathbf{g}\right)\cdot\mathbf{R}_0\right],\end{aligned} \tag{7}$$

where $\phi_{\mathbf{g}}$ represent the complex amplitude (amplitude and phase) of the beam scattered to $+\mathbf{g}$. Because T has an amplitude that is disc shaped (being controlled by the shape and size of the objective aperture), the form of the diffraction pattern will be three discs. If the objective aperture is large enough, the discs will overlap, as shown for example in Figure 2–3. Where the discs overlap, coherent interference can occur (Cowley 1979 1981, Spence 1992). To examine the form of the interference in the region where only the **0** and **+g** discs overlap, we need to calculate the intensity in this region. Taking the modulus squared of Eq. (7) and only considering the **0** and +**g** terms, which are the only ones contributing in this region, gives

$$I(\mathbf{K}_f, \mathbf{R}_0) = 1 + |\phi_\mathbf{g}|^2 + 2|\phi_\mathbf{g}|\cos\left[-\chi(\mathbf{K}_f) + \chi(\mathbf{K}_f - g) + 2\pi\mathbf{g}\cdot\mathbf{R}_0 + \angle\phi_\mathbf{g}\right], \tag{8}$$

where $\angle\phi_\mathbf{g}$ is the phase of the beam diffracted to **+g**.

Equation (8) reveals features of the interference that are important for understanding STEM imaging:

(i) The intensity in the overlap region varies sinusoidally as the probe is scanned. If a point detector was placed in this region and used to form a STEM image, fringes would be seen in the image corresponding to the spacing of the sample, and their geometric position is controlled by the phase relationship of the interfering beams.

(ii) Lens aberration can also affect the form in this overlap region. Consider just defocus (i.e. ignoring all other aberrations). Using Eq. (2) it is possible to evaluate the quantity

$$-\chi(\mathbf{K}_f) + \chi(\mathbf{K}_f - g) = \pi z\lambda\left[-\mathbf{K}_f^2 + (\mathbf{K}_f - \mathbf{g})^2\right] = \pi z\lambda\left[-2\mathbf{K}_f\cdot\mathbf{g} + |\mathbf{g}|^2\right]. \tag{9}$$

This quantity is linear in $\mathbf{K}_f$, and so substituting it into Eq. (8) reveals that a uniform set of fringes will be seen running perpendicular to the **g** vector. Such a set of interference fringes are seen in Figure 2–4. Although these fringes exist in diffraction space, their spacing, as specified in diffraction angle, corresponds to the spacing in the sample divided by the value of the defocus. Thus they can be thought of as a shadow image of the lattice in the sample. This illustrates how the detector plane in STEM, albeit nominally in diffraction space, can show real-space information. Removing the aperture completely gives an electron Ronchigram (see Chapter 3). As the defocus is reduced to zero, the apparent magnification of the shadow increases until at zero defocus the shadow has infinite magnification, and the disc overlap region contains a uniform intensity.

If we now include higher order aberrations rather than just defocus, such as spherical aberration, the form of the interference features will become more complicated. The fringes will distort, and it will not be possible to fill the overlap region with a uniform intensity.

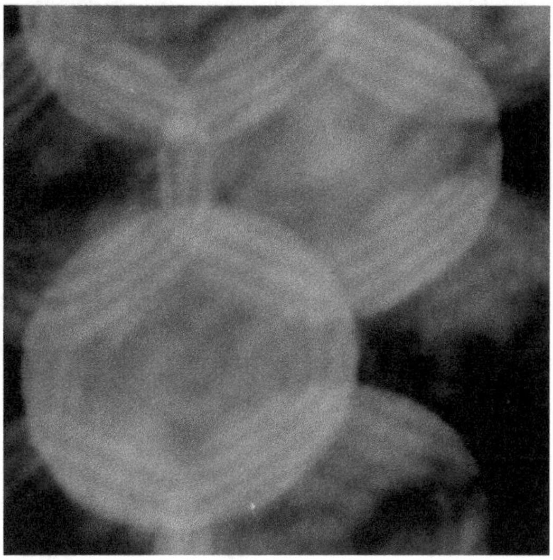

Figure 2–4. Overlapping diffracted discs in a coherent convergent-beam electron diffraction pattern. The probe has been defocused leading to relatively fine interference features in the disc overlap regions.

2.4 Bright-Field Imaging

As mentioned previously, reciprocity shows that the STEM equivalent of CTEM imaging is to use a small detector on the optic axis. From Figure 2–3 it can be seen that such a detector makes use of the intensity in a triple overlap region where the direct **0** beam and the +**g** and –**g** beams overlap. The wavefunction at this point is given by

$$\begin{aligned}\Psi(\mathbf{K}_\mathrm{f}=\mathbf{0},\mathbf{R}_0) = 1 &+ \phi_\mathbf{g}\exp\left[-\mathrm{i}\chi(-\mathbf{g}) - \mathrm{i}2\pi\mathbf{g}\cdot\mathbf{R}_0\right]\\ &+\phi_{-\mathbf{g}}\exp\left[-\mathrm{i}\chi(\mathbf{g}) + \mathrm{i}2\pi\mathbf{g}\cdot\mathbf{R}_0\right].\end{aligned} \tag{10}$$

Taking the modulus squared of Eq. (10) and neglecting terms of higher order than linear gives

$$\begin{aligned}I(\mathbf{K}_\mathrm{f}=\mathbf{0},\mathbf{R}_0) = 1 &+ \phi_\mathbf{g}\exp\left[-\mathrm{i}\chi(-\mathbf{g}) - \mathrm{i}2\pi\mathbf{g}\cdot\mathbf{R}_0\right]\\ &+\phi_{-\mathbf{g}}\exp\left[-\mathrm{i}\chi(\mathbf{g}) + \mathrm{i}2\pi\mathbf{g}\cdot\mathbf{R}_0\right]\\ &+\phi^*_\mathbf{g}\exp\left[\mathrm{i}\chi(-\mathbf{g}) + \mathrm{i}2\pi\mathbf{g}\cdot\mathbf{R}_0\right]\\ &+\phi^*_{-\mathbf{g}}\exp\left[i\chi(\mathbf{g}) - i2\pi\mathbf{g}\cdot\mathbf{R}_0\right].\end{aligned} \tag{11}$$

Now consider a weak-phase object where we can write

$$\phi_\mathbf{g} = \mathrm{i}\sigma V_\mathbf{g} \tag{12}$$

where $V_\mathbf{g}$ is the **g**th Fourier component of the specimen potential. Because the potential is real,

$$V^*_\mathbf{g} = V_{-\mathbf{g}}, \tag{13}$$

and therefore

$$
\begin{aligned}
I(\mathbf{K}_\mathrm{f}=0,\mathbf{R}_\mathrm{p}) = 1 &+ \mathrm{i}\sigma V_\mathbf{g} \exp \mathrm{i}\left[-\chi(-\mathbf{g}) - 2\pi \mathbf{g}\cdot\mathbf{R}_0\right] \\
&+ \mathrm{i}\sigma V^*_\mathbf{g} \exp \mathrm{i}\left[-\chi(\mathbf{g}) + 2\pi \mathbf{g}\cdot\mathbf{R}_0\right] \\
&- \mathrm{i}\sigma V^*_\mathbf{g} \exp \mathrm{i}\left[\chi(-\mathbf{g}) + 2\pi \mathbf{g}\cdot\mathbf{R}_0\right] \\
&- \mathrm{i}\sigma V_\mathbf{g} \exp \mathrm{i}\left[\chi(\mathbf{g}) - 2\pi \mathbf{g}\cdot\mathbf{R}_0\right].
\end{aligned} \tag{14}
$$

Collecting terms and assuming that χ is a symmetric function,

$$
\begin{aligned}
I_\mathrm{BF}(\mathbf{R}_\mathrm{p}) &= 1 + \mathrm{i}\left(\exp\left[-\mathrm{i}\chi(\mathbf{g})\right] - \exp\left[\mathrm{i}\chi(\mathbf{g})\right]\right) \\
&\quad \times \left(\sigma V_\mathbf{g} \exp\left[-\mathrm{i}2\pi\mathbf{g}\cdot\mathbf{R}_0\right] + \sigma V^*_\mathbf{g} \exp\left[\mathrm{i}2\pi\mathbf{g}\cdot\mathbf{R}_0\right]\right) \\
&= 1 + 2\sigma \sin\left(\chi(\mathbf{g})\right)\left[\left|V_\mathbf{g}\right| \exp\left(-\mathrm{i}2\pi\mathbf{g}\cdot\mathbf{R}_0 + \angle V_\mathbf{g}\right)\right. \\
&\quad \left. + \left|V_\mathbf{g}\right| \exp\left[\mathrm{i}2\pi\mathbf{g}\cdot\mathbf{R}_0 - \angle V_\mathbf{g}\right]\right],
\end{aligned} \tag{15}
$$

which simplifies to give

$$
I_\mathrm{BF}\left(\mathbf{R}_0\right) = 1 + 4\left|\sigma V_\mathbf{g}\right| \cos\left(2\pi\mathbf{g}\cdot\mathbf{R}_0 - \angle V_\mathbf{g}\right) \sin\chi(\mathbf{g}), \tag{16}
$$

which is the standard form of phase contrast imaging in the electron microscope (Spence 1988), with the phase contrast transfer function being given by $\sin(\chi)$.

Thus BF imaging in STEM shows the usual phase contrast imaging, with a phase contrast transfer function that is controlled by the lens aberrations, in a similar way to phase contrast imaging in CTEM. The principle of reciprocity is thereby confirmed. It should be pointed out, however, that BF imaging in STEM is much less efficient of electrons than that in CTEM because the small detector does not collect the majority of the electrons in the detector plane.

2.5 Resolution Limits

Figure 2–3 shows that triple overlap conditions can occur only if the magnitude of **g** is less than the radius of the aperture. The aperture itself is used to prevent highly aberrated rays contributing to the image (which in the bright-field model would correspond to the oscillatory region of the phase contrast transfer function). If the magnitude of **g** has a value lying between the aperture radius and the aperture diameter, there will still be interference in the single overlap regions (see Figure 2–5). Thus information at this resolution can be recorded in a STEM, but not using an axial detector. An off-axis detector needs to be used to record this so-called single sideband interference. By reciprocity, the equivalent approach in HRTEM is to use tilted illumination, which has been shown to improve image resolution (Haigh et al. 2009).

Ideas for making use of this single sideband interference include differential phase contrast detectors (Dekkers and de Lang 1974) and annular bright-field detectors (Rose 1974). It can also be seen that an

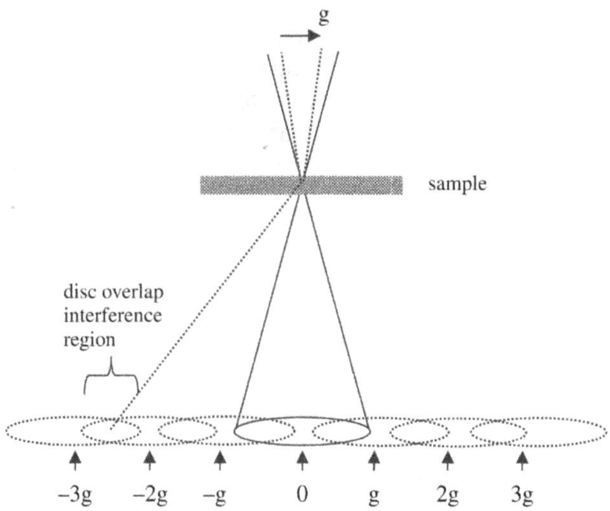

Figure 2–5. For smaller lattice spacings, the triple overlap regions necessary for bright-field STEM may not exist and no contrast will be seen. Such spacings can be resolved, however, using non-axial detector geometries including annular dark field.

annular dark-field detector would also detect single overlap interference, though at the angles usually detected, discrete discs are no longer observable because of the effects of thermal diffuse scattering.

A broad statement for STEM resolution is that for a spatial frequency **Q** to show up in the image, two beams incident on the sample separated by **Q** must be scattered by the sample so that they end up in the same final wavevector $\mathbf{K}_f$ where they can interfere. This model of STEM imaging is applicable to any imaging mode, even when TDS or inelastic scattering is included. We can immediately conclude that STEM is unable to resolve any spacing smaller than that allowed by the diameter of the objective aperture, no matter which imaging mode is used.

2.6 Partial Coherence in STEM Imaging and the Need for Brightness

The models so far presented have assumed that the illuminating electron beam emanates from a point source (has perfect spatial coherence), is perfectly monochromatic (has perfect temporal coherence) and that the BF detector is infinitesimal. Coherence is used to model the degree to which different beams can interfere, therefore the effects of partial coherence can strongly influence the form of STEM images. Let us consider each in turn.

2.6.1 Source Spatial Coherence and Brightness

Any electron gun emits radiation from a finite-size source, which is regarded to be self-luminous. Radiation emitted from one point is assumed to be unable to interfere with the radiation from any neighbouring point. To model this coherence, we can treat each point in the electron source as giving rise to its own illuminating probe at the sample. For each desired probe position, corresponding to a pixel in the image, the detected image intensity arises from a range of actual probe positions that are then added. This can be described by a convolution, and it can be written as

$$I_{\mathrm{src}}(\mathbf{R}_0) = I(\mathbf{R}_0) \otimes S(\mathbf{R}_0), \tag{17}$$

where S is the source intensity distribution as measured at the sample plane, i.e. after taking source demagnification into account.

It is immaterial how the nominally coherent image $I(\mathbf{R}_0)$ is formed, the effects of partial source coherence can always be modelled as a simple convolution of the image with the effective source size. The purpose of condenser lenses is to demagnify the electron source as much as possible to reduce the deleterious effects of partial source coherence. The more the source is demagnified, the lower the current in the probe, as shown in Figure 2–6. The crucial quantity is brightness B, which is defined as the current per unit area per unit solid angle subtended by the beam. Brightness is conserved in an optical system, and so knowledge of the brightness of the electron source allows calculation of the current available in the STEM probe. Given that the solid angle subtended by the incident beam is controlled by the size of the objective aperture, it is possible to write the current available in the probe J in terms of the probe diameter d and the brightness B:

$$J = B\pi^2\alpha^2 d^2/4. \tag{18}$$

Thus the smaller the STEM probe, the lower the current available and the higher the brightness needed to provide a reasonable current. It is for this reason that the development of the modern STEM required the development of a high-brightness gun (Crewe et al. 1968).

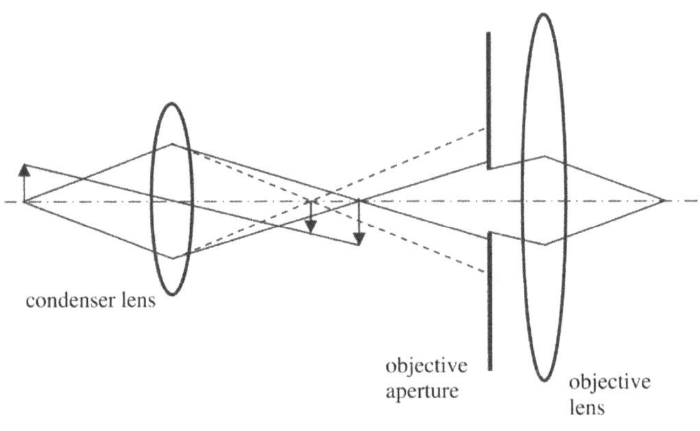

Figure 2–6. Increasing the strength of the condenser lens to provide greater source demagnification leads to greater loss of current at the objective aperture and less probe current.

2.6.2 Partial Detector Spatial Coherence

It might be considered strange to think of the effects of finite detector size as being regarded as a partial coherence. Clearly detectors do not affect the beam. However, a finite-sized detector might not detect very small interference features, and coherence refers to the ability to observe interference effects. Furthermore, by reciprocity a finite-sized detector in STEM is equivalent to a finite source in CTEM, and the latter would be conventionally regarded as a source of partial coherence.

The effects of partial detector coherence depend very much on the STEM imaging mode. For BF imaging, it leads to a coherence envelope similar to that seen for partial source coherence in CTEM (Nellist and Rodenburg 1994). It has a dependence on the slope of the aberration function χ and the reason for this becomes clear when one considers the interference in disc overlap regions. Aberrations will lead to smaller interference features in the overlap region and may therefore not be detected by a finite-sized detector.

It might be assumed that as the detector becomes larger, the effects of decreased coherence lead to weaker image contrast. Although it is indeed the case that the imaging process does become incoherent, the image contrast can be maintained, which brings us to the concept of incoherent imaging using an annular dark-field detector.

2.6.3 Partial Temporal Coherence

One of the important advantages of STEM is that all the imaging optics are placed before the sample, and optics after the sample do not influence the imaging process except for allowing the collection angles of detectors to be varied (essentially by changing the camera length of the post-specimen diffraction). The effects of temporal coherence arise from the finite energy spread of the beam, and the chromatic aberrations of the lenses. In CTEM, the energy spread can arise from inelastic scattering in the sample and can be broad. In STEM, partial temporal coherence can arise only because of the spread in energies of the illuminating beam, which is likely to be relatively low given that field emission sources are used.

Again, the exact effect of partial temporal coherence depends on the imaging mode being used. For BF imaging, the effect is similar to that for CTEM by reciprocity (Nellist and Rodenburg 1994), but for incoherent imaging modes, the effect of partial temporal coherence is not as severe (Nellist and Pennycook 1998).

2.7 Annular Dark-Field Imaging

The use of an annular dark-field (ADF) detector gave rise to one of the first detection modes used by Crewe and co-workers during the initial development of the modern STEM (Crewe 1980). The detector consists of an annular sensitive region that detects electrons scattered over an angular range with an inner radius that may be a few tens of milliradians up to perhaps 100 mrad and an outer radius of several hundred milliradians. It has remained by far the most popular STEM imaging mode. It was later proposed that high scattering angles (~100 mrad) would enhance the compositional contrast (Treacy et al. 1978) and that the coherent effects of elastic scattering could be neglected because the scattering was almost entirely thermally diffuse (Howie 1979). This idea led to the use of the high-angle annular dark-field detector (HAADF). In this chapter, we will consider scattering over all angular ranges and will refer to the technique generally as ADF STEM.

It is indeed the case that for typical ADF detector angles, the scattering predominantly detected will be TDS. To understand the nature of incoherent imaging, and the resolution limits that apply, it is useful to first consider a lattice with no thermal vibrations so that the overlapping disc model used earlier applies. Figure 2–5 shows that an ADF detector will not only sum the intensity over entire disc overlap regions but also sum the intensity over many of such overlap regions. It might be expected that such an approach would generally wash out most of the available image contrast, but somewhat surprisingly this is not the case. The approach we take below follows very closely previous approaches (Jesson and Pennycook 1993, Loane et al. 1992, Nellist and Pennycook 1998).

Consider a sample that is continuous in Fourier space. An equivalent to Eq. (6) can be formed, the modulus squared taken to form an intensity, and that intensity then integrated over a detector function:

$$I_{\mathrm{ADF}}(\mathbf{R}_0) = \int D_{\mathrm{ADF}}(\mathbf{K}_\mathrm{f}) \times \left| \int \phi(\mathbf{K}_\mathrm{f} - \mathbf{K}) T(\mathbf{K}) \exp(-\mathrm{i}2\pi \mathbf{K} \cdot \mathbf{R}_0)\,\mathrm{d}\mathbf{K} \right|^2 \mathrm{d}\mathbf{K}_\mathrm{f}. \tag{19}$$

Taking the Fourier transform, after expanding the modulus squared, gives

$$\begin{aligned} I_{\mathrm{ADF}}(\mathbf{Q}) = \int \exp(-\mathrm{i}2\pi \mathbf{Q} \cdot \mathbf{R}_0) \int D_{\mathrm{ADF}}(\mathbf{K}_\mathrm{f}) \\ \times \left\{ \int \phi(\mathbf{K}_\mathrm{f} - \mathbf{K}) T(\mathbf{K}) \exp(-\mathrm{i}2\pi \mathbf{K} \cdot \mathbf{R}_0)\,\mathrm{d}\mathbf{K} \right\} \\ \times \left\{ \int \phi^*(\mathbf{K}_\mathrm{f} - \mathbf{K}') T^*(\mathbf{K}') \exp(\mathrm{i}2\pi \mathbf{K}' \cdot \mathbf{R}_0)\,\mathrm{d}\mathbf{K}' \right\} \mathrm{d}\mathbf{K}_\mathrm{f}\,\mathrm{d}\mathbf{R}_0. \end{aligned} \tag{20}$$

Performing the $\mathbf{R}_0$ integral first results in a Dirac δ function:

$$I_{\mathrm{ADF}}(\mathbf{Q}) = \iiint D_{\mathrm{ADF}}(\mathbf{K}_\mathrm{f}) \phi(\mathbf{K}_\mathrm{f} - \mathbf{K}) T(\mathbf{K}) \phi^*(\mathbf{K}_\mathrm{f} - \mathbf{K}') \times T^*(\mathbf{K}') \delta(\mathbf{Q} + \mathbf{K} - \mathbf{K}')\,\mathrm{d}\mathbf{K}_\mathrm{f}\,\mathrm{d}\mathbf{K}\,\mathrm{d}\mathbf{K}', \tag{21}$$

which allows simplification by performing the $\mathbf{K}'$ integral:

$$I_{\mathrm{ADF}}(\mathbf{Q}) = \iint D_{\mathrm{ADF}}(\mathbf{K}_\mathrm{f}) T(\mathbf{K}) T^*(\mathbf{K} + \mathbf{Q}) \times \phi(\mathbf{K}_\mathrm{f} - \mathbf{K}) \phi^*(\mathbf{K}_\mathrm{f} - \mathbf{K} - \mathbf{Q})\,\mathrm{d}\mathbf{K}_\mathrm{f}\,\mathrm{d}\mathbf{K}. \tag{22}$$

Equation (22) is straightforward to interpret in terms of interference between diffracted discs (Figure 2–5). The integral over $\mathbf{K}$ is a convolution so that Eq. (22) could be written as

$$I_{\mathrm{ADF}}(\mathbf{Q}) = \int D_{\mathrm{ADF}}(\mathbf{K})\{[T(\mathbf{K})T^*(\mathbf{K}+\mathbf{Q})] \otimes_{\mathbf{K}} [\phi(\mathbf{K})\phi^*(\mathbf{K}-\mathbf{Q})]\}\,\mathrm{d}\mathbf{K}. \tag{23}$$

The first bracket of the convolution is the overlap product of two apertures, and this is then convolved with a term that encodes the interference between scattered waves separated by the image spatial frequency $\mathbf{Q}$. For a crystalline sample, $\phi(\mathbf{K})$ will only have values for discrete $\mathbf{K}$ values corresponding to the diffracted spots. In this case Eq. (23) is easily interpretable as the sum over many different disc overlap features that are within the detector function.

We can expect that the aperture overlap region is small compared with the physical size of the ADF detector. In terms of Eq. (22) we can say the domain of the $\mathbf{K}$ integral (limited to the disc overlap region) is small compared with the domain of the $\mathbf{K}_{\mathrm{f}}$ integral, and we can make the approximation:

$$I_{\mathrm{ADF}}(\mathbf{Q}) = \int T(\mathbf{K})T^*(\mathbf{K}+\mathbf{Q})\mathrm{d}\mathbf{K} \times \int D_{\mathrm{ADF}}(\mathbf{K}_{\mathrm{f}})\phi(\mathbf{K}_{\mathrm{f}})\phi^*(\mathbf{K}_{\mathrm{f}}-\mathbf{Q})\mathrm{d}\mathbf{K}_{\mathrm{f}}. \tag{24}$$

In making this approximation we have assumed that the contribution of any overlap regions that are partially detected by the ADF detector is small compared with the total signal detected. The integral containing the aperture functions is actually the autocorrelation of the aperture function. The Fourier transform of the probe intensity is the autocorrelation of T, thus Fourier transforming Eq. (24) to give the image results in

$$I(\mathbf{R}_0) = |P(\mathbf{R}_0)| \otimes O(\mathbf{R}_0), \tag{25}$$

where $O(\mathbf{R}_0)$ is the inverse Fourier transform with respect to $\mathbf{Q}$ of the integral over $\mathbf{K}_{\mathrm{f}}$ in Eq. (24).

Equation (25) is the definition of incoherent imaging. The image is regarded as being formed from an object function that is then convolved with a real-positive intensity point-spread function. The Fourier transform of the image will therefore be a product of the Fourier transform of the probe intensity and the Fourier transform of the object function. The Fourier transform of the probe intensity is known as the optical transfer function (OTF) and its typical form in shown in Figure 2–7. Unlike the phase contrast transfer function for BF imaging, it shows no contrast reversals and decays monotonically as a function of spatial frequency.

It is fair to say that the majority of imaging across all radiations can be regarded as incoherent. Generally, an imaged object can be regarded as being effectively self-luminous, which leads directly to an incoherent imaging model (Rayleigh 1896). In this case, the object is not self-luminous, and the illuminating probe is coherent. We noted earlier that the detector geometry can control coherence, and that is exactly what is happening here. Furthermore, by reciprocity, the large annular detector is equivalent to a large (and therefore incoherent) illuminating source, and large sources are another route to ensuring that an imaging process is incoherent.

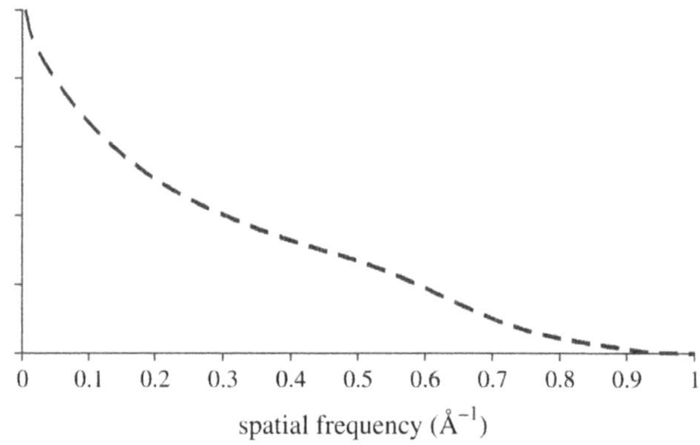

Figure 2–7. A typical optical transfer function (OTF) for incoherent imaging in STEM. This OTF has been calculated for a 300-kV STEM with spherical aberration $C_S = 1$ mm.

Incoherent imaging leads to data that is much easier to interpret. The contrast reversals and delocalization usually associated with HRTEM images are absent, and generally bright features in an ADF image can be associated with the presence of atoms or atomic columns in an aligned crystal. Combined with the strong Z contrast that arises from the high-angle scattering (see Chapter 1) this leads to a high-contrast, chemically sensitive imaging mode. Optimising the conditions for incoherent imaging in STEM is simply a matter of getting the smallest, most intense probe possible. Use of aberrations to generate contrast (as seen in BF imaging) is not required.

As pointed out in Chapter 1, the early investigations suggested that ADF imaging could be regarded as being incoherent only if the all the electrons in the detector plane were summed over, but that this mode would lead to no-image contrast (Ade 1977, Treacy and Gibson 1995). The hole in the ADF detector is therefore crucial to generate contrast, and it is useful to examine its influence on the detector function. By assuming that the maximum image spatial frequency **Q** vector is small compared to the geometry of the detector and noting that the detector function is either unity or zero, we can write the Fourier transform of the object function as

$$O(\mathbf{Q}) = \int D_{\mathrm{ADF}}(\mathbf{K}_f)\varphi(\mathbf{K}_f)D_{\mathrm{ADF}}(\mathbf{K}_f - \mathbf{Q})\varphi^*(\mathbf{K}_f - \mathbf{Q})\mathrm{d}\mathbf{K}_f. \tag{26}$$

This equation is just the autocorrelation of $D(\mathbf{K})\varphi(\mathbf{K})$, and so the object function is

$$O(\mathbf{R}_0) = |D(\mathbf{R}_0) \otimes \phi(\mathbf{R}_0)|^2 . \tag{27}$$

Neglecting the outer radius of the detector, where we can assume the strength of the scattering has become negligible, $D(\mathbf{K})$ can be thought of as a sharp high-pass filter. The object function is therefore the modulus squared of the high-pass filtered specimen transmission function.

Nellist and Pennycook (2000) have taken this analysis further by making the weak-phase object approximation, under which condition the object function becomes

$$O(\mathbf{R}_0) = \int_{\text{half plane}} \frac{J_1(2\pi k_{\text{inner}}\,|\mathbf{R}|)}{2\pi\,|\mathbf{R}|} \times [\sigma V(\mathbf{R}_0 + \mathbf{R}/2) - \sigma V(\mathbf{R}_0 - \mathbf{R}/2]^2\,\mathrm{d}\mathbf{R}, \tag{28}$$

where k_{inner} is the spatial frequency corresponding to the inner radius of the ADF detector, and J_1 is a first-order Bessel function of the first kind. This is essentially the result derived by Jesson and Pennycook (1993). The coherence envelope expected from the Van Cittert–Zernicke theorem is now seen in Eq. (28) as the Airy function involving the Bessel function. If the potential is slowly varying within this coherence envelope, the value of $O(\mathbf{R}_0)$ is small. For $O(\mathbf{R}_0)$ to have significant value, the potential must vary quickly within the coherence envelope. A coherence envelope that is broad enough to include more than one atom in the sample (arising from a small hole in the ADF), however, will show unwanted interference effects between the atoms. Making the coherence envelope too narrow by increasing the inner radius, on the other hand, will lead to too small a variation in the potential within the envelope, and therefore no signal. If there is no hole in the ADF detector, then $D(\mathbf{K}) = 1$ everywhere, and its Fourier transform will be a delta function. Equation (27) then becomes the modulus squared of ϕ, and there will be no contrast. To get signal in an ADF image, we require a hole in the detector, leading to a coherence envelope that is narrow enough to destroy coherence from neighbouring atoms but broad enough to allow enough interference in the scattering from a single atom. In practice, there are further factors that can influence the choice of inner radius, such as the presence of strain contrast. A typical choice for incoherent imaging is that the ADF inner radius should be about three times the objective aperture radius (Hartel et al. 1996), which ensures that the coherence envelope is significantly narrower than the probe.

2.7.1 Incoherent Imaging with Dynamical Diffraction

If one can assume ADF imaging to be incoherent, then it is reasonable to expect that the total scattered intensity would be simply proportional to the number of atoms illuminated by the probe. Early applications of ADF imaging showed that diffraction of the electron beam in the sample could still influence the intensity seen in ADF images (Donald and Craven 1979). Specifically, when a crystal is aligned with a low-order zone axis parallel to the beam, strong channelling conditions which enhance the strength of the scattering to high angles are established. To explain this, we need to examine the influence of dynamical diffraction.

The analysis performed above has assumed that the scattering by the sample can be treated as being a simple, multiplicative transmission function, i.e. the sample is thin. Under dynamical diffraction conditions, the multiplicative transmission function approximation cannot be made. If we continue to neglect thermal diffuse scattering (which we include in Section 2.7.3), then it is possible to include dynamical diffraction by making use of the Bloch wave model. In Eq. (22), the Fourier transform of the object function gives the strength of the scattering from an incoming partial plane wave to an outgoing one. The effect of dynamical diffraction is that the strength of the scattering is no longer simply dependent on the change of wavevector but on the incoming and outgoing wavevectors independently, thus

$$\begin{aligned} I_{\mathrm{ADF}}(\mathbf{Q},z) = \iint D_{\mathrm{ADF}}(\mathbf{K}_\mathrm{f})T(\mathbf{K})T^*(\mathbf{K}+\mathbf{Q}) \\ \times\phi(\mathbf{K}_\mathrm{f},\mathbf{K},z)\phi^*(\mathbf{K}_\mathrm{f},\mathbf{K}-\mathbf{Q},z)\mathrm{d}\mathbf{K}_\mathrm{f}\,\mathrm{d}\mathbf{K}. \end{aligned} \tag{29}$$

To include the effects of dynamical scattering, in a perfect crystal that only contains spatial frequencies corresponding to reciprocal lattice points it is possible to follow the approach of Nellist and Pennycook (1999) and write the scattering as a sum over Bloch waves (see, for example, Humphreys and Bithell 1992):

$$\begin{aligned} \tilde{I}_{\mathrm{ADF}}(\mathbf{Q},z) = \sum_{\mathbf{g}} D_{\mathrm{ADF}}(\mathbf{g}) \int T(\mathbf{K})T^*(\mathbf{K}+\mathbf{Q}) \\ \times \sum_j \Phi_{\mathbf{0}}^{(j)*}(\mathbf{K})\Phi_{\mathbf{g}}^{(j)}(\mathbf{K})\exp\left[-\mathrm{i}2\pi z k_z^{(j)}(\mathbf{K})\right] \\ \times \sum_k \Phi_{\mathbf{Q}}^{(k)}(\mathbf{K})\Phi_{\mathbf{g}}^{(k)*}(\mathbf{K})\exp\left[\mathrm{i}2\pi z k_z^{(k)}(\mathbf{K})\right]\mathrm{d}\mathbf{K}, \end{aligned} \tag{30}$$

where $\Phi_{\mathbf{g}}^{(j)}(\mathbf{K})$ is the $\mathbf{g}$th Fourier component of the jth Bloch wave for an incoming beam with transverse wavevector $\mathbf{K}$. By performing the $\mathbf{g}$ summation first, which plays an equivalent role in the sum over the detector in Eq. (22), it is possible to look at the degree of coherent interference between different Bloch waves, thus

$$C_{ij}(\mathbf{K}) = \sum_{\mathbf{g}} D_{\mathrm{ADF}}(\mathbf{g})\Phi_{\mathbf{g}}^{(j)}(\mathbf{K})\Phi_{\mathbf{g}}^{(k)*}(\mathbf{K}), \tag{31}$$

which, in a similar fashion to the approach in Eq. (28), can be written in terms of the hole in the detector:

$$C_{ij}(\mathbf{K}') = \delta_{ij} - \int \frac{J_1(2\pi u_{\mathrm{in}}\,|\mathbf{B}|)}{2\pi\,|\mathbf{B}|} \int \Phi^{(j)}(\mathbf{C},\mathbf{K})\Phi^{(k)*}(\mathbf{C}+\mathbf{B},\mathbf{K})\mathrm{d}\mathbf{C}\,\mathrm{d}\mathbf{B}, \tag{32}$$

where $\mathbf{B}$ and $\mathbf{C}$ are dummy real-space variables of integration, and the Bloch waves have been written as real-space functions for a given incident beam transverse wavevector $\mathbf{K}$. As we saw before, the hole in the detector is imposing a coherence envelope. Thus $C_{jk}(\mathbf{K})$ allows only interference effects to show up in the ADF image between Bloch states that are sharply peaked and whose peaks are physically close such that they lie within a few tenths of an angstrom of each other. A physical interpretation is that the high-angle ADF detector is acting like a high-pass filter (as it has been seen to do for thin specimens – see Section 2.7.1) acting on the exit-surface wavefunction. Only when the probe excites sharply peaked Bloch states will the electron density be sharply peaked.

2.7.2 The Effect of Thermal Diffuse Scattering

Early analyses of ADF imaging took the approach that at high enough scattering angles, the thermal diffuse scattering (TDS) arising from phonons would dominate the image contrast (Howie 1979). In the Einstein approximation, this scattering is completely uncorrelated between atoms, and therefore there could be no coherent interference effects between the scattering from different atoms. In this approach the intensity of the wavefunction at each site needs to be computed using a dynamical elastic scattering model and then the TDS from each atom summed (Allen et al. 2003, Pennycook and Jesson 1990). When the probe is located over an atomic column in the crystal, the most bound, least dispersive states (usually 1s or 2s-like) are predominantly excited and the electron intensity "channels" down the column. This channelling effect reduces the spreading of the probe as it propagates, which is useful for thicker samples, though spreading can still be seen, especially for aberration-corrected instruments with larger convergence angles (Dwyer and Etheridge 2003). When the probe is not located over a column, it excites more dispersive, less bound states and spreads leading to reduced intensity at the atom sites and a lower ADF signal. Both the Bloch wave (for example Amali and Rez 1997, Findlay et al. 2003, Mitsuishi et al. 2001, Pennycook 1989) and multislice (for example Allen et al. 2003, Dinges et al. 1995, Kirkland et al. 1987, Loane et al. 1991) methods have been used for simulating the TDS scattering to the ADF detector. Details of the way TDS is incorporated into image calculations can be found in Chapter 6.

It is possible to see the incoherence due to the detector geometry and the incoherence due to TDS in a similar framework. In the analyses presented here, the key to incoherent imaging has been the sum over the many final wavevectors that are incident upon the detector. One way of explaining the diffuse nature of thermal scattering is to consider that, in addition to the transverse momentum imparted by the elastic scattering from the crystal, additional momentum is imparted by scattering from a phonon. Phonon momenta will be comparable to reciprocal lattice vectors, and the range of phonon momenta present in a crystal will therefore blur the elastic diffraction pattern. Furthermore, each phonon will impart a slightly different energy to others, and therefore scattering by different phonon momenta will lead to waves that are mutually incoherent. If we consider a single detection point, many beams elastically scattered to different final wavevectors will be additionally scattered by phonons to the detector, leading to a sum in intensity over final elastic wavevectors – exactly what is required for incoherent imaging. It is fair to say, however, that the geometry of the ADF detector will always be larger than typical phonon momenta (not least because longer wavelength phonons are usually more common) and that transverse incoherence is ensured by the use of a large detector. It is interesting to speculate whether a small detector at high angle

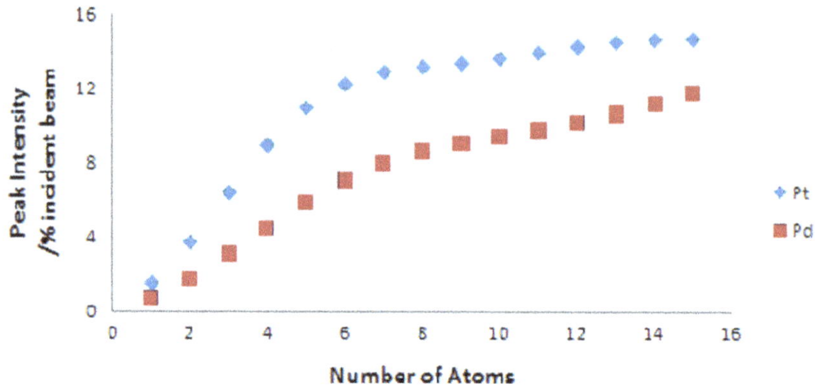

Figure 2–8. The peak ADF image intensity (expressed as a fraction of the incident beam current) for isolated Pt and Pd columns expressed as a function of number of atoms in a column (graph courtesy of H. E).

would give a strongly incoherent signal, relying as it would purely on the sum over phonon momenta to give the necessary integral to destroy the coherence.

The combined effects of channelling and TDS give rise to a dependence of ADF image peak intensity on sample thickness typically of the form shown in Figure 2–8. The ADF signal rises monotonically with thickness, but is clearly non-linear, and so is not proportional to the number of illuminated atoms. Changes in the slope of the graph are caused by variations in the strength of the electron beam channelling along the column, arising from both channelling oscillations and absorption. It is therefore clear that quantitative interpretation of ADF images does require matching to simulations.

Almost all the simulations currently performed assume an Einstein phonon dispersion model. Other, more realistic, dispersions have been considered (Jesson and Pennycook 1995). Although the detector geometry is highly effective for destroying coherence perpendicular to the beam direction, phonons play a much more important role in controlling the coherence parallel to the beam direction. Jesson and Pennycook (1995) showed that a realistic phonon dispersion could give rise to short-range coherence envelopes in the depth direction. Detailed multislice simulations (Muller et al. 2001) suggest that the effect of a realistic phonon dispersion on the ADF intensities for a perfect crystal is small.

The combination of channelling and absorption can also lead to some unexpected effects when the displacement of atoms varies along a column, referred to as strain contrast. Strain can lead to either a depletion or an enhancement of ADF intensities depending on the inner radius of the detector (Yu et al. 2004). This phenomenon has been ascribed to the strain causing interband scattering between Bloch waves (Perovic et al. 1993). A channelling wave that has been strongly absorbed may be replenished by interband scattering, thereby leading to increased intensity.

2.8 Imaging Using Inelastic Electrons

Using the STEM to image at, or close to, atomic resolution using inelastically scattered electrons is a powerful experimental mode. Remarkable progress has been made since it was first demonstrated (Browning et al. 1993) and the development of aberration-corrected STEM has allowed impressive atomic-resolution mapping to be demonstrated. Only core-loss inelastic scattering provides a sufficiently localized signal to allow atomic resolution, and because such scattering involves the excitation of an atomic core state to a final state, it is clear that such scattering will be independent of neighbouring atoms and that no interference between the scattering from neighbouring atoms can be expected.

The inclusion of inelastic scattering is discussed extensively in Chapter 6. As shown there, scattering from a specific initial state to a specific final state can be treated by a simple, multiplicative scattering function (see Eq. (11) of Chapter 6). The final image will be a sum in intensity over many of such scattering functions because for any experiment with finite energy resolution, a significant number of final states must be included. Because each final state differs slightly in energy, a sum in intensity is required, thereby breaking the coherence in the imaging process. As noted in Chapter 6, however, this summation is often not sufficient to prevent partial coherence effects from being observed, and the use of a large collector aperture is further required to ensure incoherent imaging. A large collector aperture destroys coherence in exactly the same way as the large ADF detector does for elastic or quasi-elastic scattering.

2.9 Optical Depth Sectioning and Confocal Microscopy

So far we have considered only two-dimensional imaging. The development of aberration correctors in STEM has led to dramatic improvements in lateral resolution due to the larger objective lens numerical aperture allowed. Whilst the lateral resolution varies as the inverse of the numerical aperture, the depth of focus is inversely proportional to the square of the numerical aperture. In a state-of-the-art, aberration-corrected STEM, the depth of focus may fall to just a few nanometres, which is less than the typical thickness of TEM samples. The full width at half-maximum of the probe intensity along the optic axis is given by

$$\Delta z = 1.77\frac{\lambda}{\alpha^2}. \tag{33}$$

Whilst this raises concerns about interpreting high-resolution images from thicker samples, it does raise the possibility of using this reduced depth of focus to retrieve depth information.

The simplest approach to measuring such 3D information in STEM is to record a focal series of images, thereby forming a 3D stack. Clearly we want an incoherent imaging mode where the scattering is simply dependent on the 3D probe intensity distribution, and ADF imaging is therefore suitable. Such an approach has been used for the 3D imaging of Hf atoms in a transistor gate oxide stack (Van Benthem et al. 2005). Applications to the mapping of nanoparticle locations in heterogeneous catalysts (Borisevich et al. 2006) showed significant elongation in the depth direction. This elongation was subsequently investigated by Behan et al. (2009) and was seen to arise from the form of the OTF in three dimensions. Figure 2–7 has already shown the form of the OTF in 2D, and a 3D OTF can simply be formed by taking the Fourier transform of the 3D probe intensity distribution. As seen in Figure 2–9, the OTF is approximately of a donut shape and has a large missing region. This missing region is known from light optics (Frieden 1967) and has an opening angle that is given by 90°-α, where α is the acceptance angle of the lens. In light optics, α can approach close to 90°, whereas even in an aberration-corrected STEM, α is less than 2°, leading to a large missing region in the OTF. For laterally extended objects that are dominated by low transverse spatial frequencies, only low longitudinal spatial frequencies will be transferred, leading to longitudinal elongation. The depth resolution for an extended object can be approximated as

$$\Delta z = \frac{d}{\alpha}, \tag{34}$$

where d is the lateral extent of the object. Even for a 5-nm particle, the depth resolution in an aberration-corrected STEM would be typically 200 nm. Methods to use deconvolution to overcome this problem have been investigated (Behan et al. 2009, de Jonge et al. 2010), but it must be

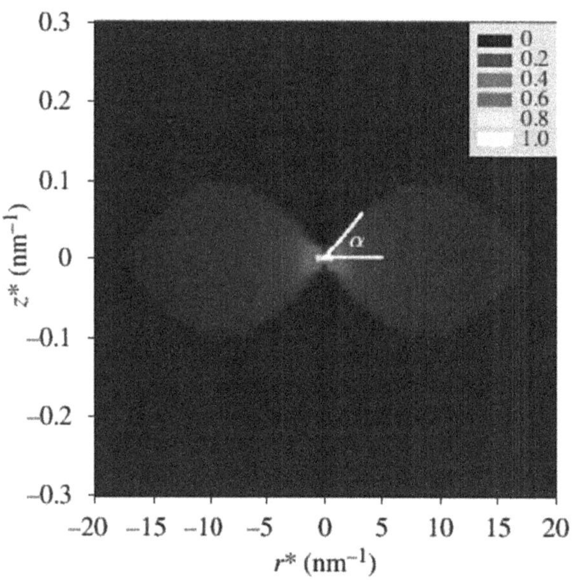

Figure 2–9. A cross section through the 3D OTF for incoherent imaging. Note the missing cone region. The longitudinal (z^*) and lateral (r^*) axes have different scales. A 200-kV microscope with $\alpha = 22$ mrad has been assumed. Reproduced from Behan et al. (2009).

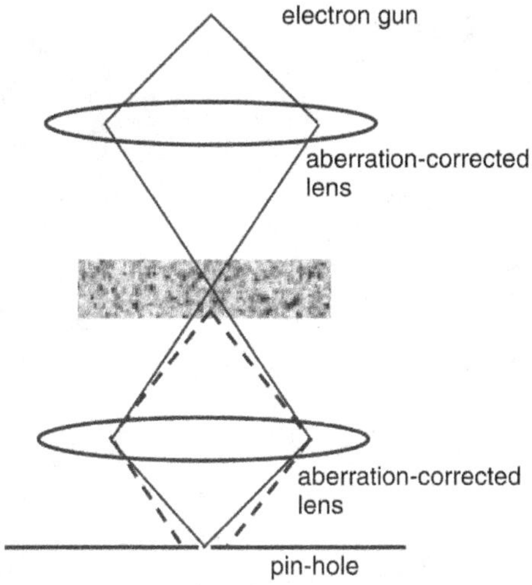

Figure 2–10. A schematic of the scanning confocal electron microscope. Scattering from regions of the sample away from the confocal point (*dashed lines*) is neither strongly illuminated nor focused at the detector pinhole.

remembered that it is not possible to reconstruct the information in the missing cone unless prior information is included.

It has recently been shown that it is possible to use a microscope fitted with aberration correctors both before and after the sample in a confocal geometry (Nellist et al. 2006), similar to the confocal scanning optical microscope that is widely used in light optics (Figure 2–10). The advantage of such a configuration is that the second lens provides additional depth resolution and selectivity. Further detailed analysis of SCEM image contrast has been performed for both elastic (Cosgriff et al. 2008) and inelastic (D'Alfonso et al. 2008) scattering. For elastic scattering, there is no first-order phase contrast transfer, and so the contrast is weak and relies on multiple scattering. Collection of inelastic scattering, in the energy-filtered SCEM (EFSCEM) mode, is much more promising (see also Chapter 6). There is no missing cone in the transfer function (Figure 2–11) and recent results suggest that nanoscale depth resolutions are achievable from laterally extended objects (Wang et al. 2010).

2.10 Conclusions

In this chapter we have reviewed imaging in the STEM, with particular focus on BF and ADF imaging. A key strength of ADF imaging is its incoherent nature, which it shares with many other STEM signals such as EELS and EDX. Unlike conventional high-resolution TEM, the main requirement for STEM is to minimize the aberrations so that a small, intense probe is formed.

In this chapter we concentrated on single signals (e.g. BF or ADF) that are recorded as a function of probe position. Large areas of the

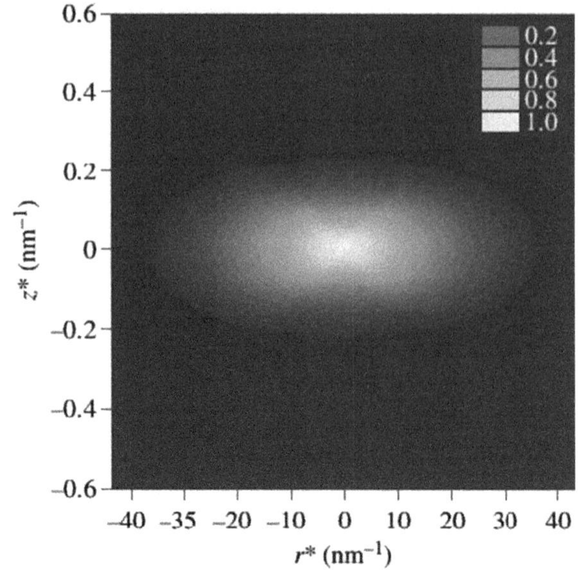

Figure 2–11. A cross section through the transfer function for incoherent SCEM imaging. A 200-kV microscope with both the pre- and post-specimen optics subtending 22 mrad has been assumed. Reproduced from Behan et al. (2009).

detector plane are summed over to record these signals, thus discarding significant amounts of information. Attempts have been made to use position-sensitive detectors in STEM imaging (see, for example, Nellist et al. 1995, Rodenburg and Bates 1992) but have been limited to rather small fields of view because of the problems of acquiring and handling the vast amounts of data. With improved detectors and information technology, we may well see a re-emergence of the idea of collecting the entire detector plane as a function of each probe position (for some recent ideas, see Faulkner and Rodenburg 2004). All possible detector geometries can then be synthesized, or the entire 4D data set used to retrieve information about the sample.

Acknowledgements The author would like to thank the many colleagues and collaborators that have been involved in furthering our understanding of STEM imaging. P.D.N. acknowledges support from the Leverhulme Trust (F/08749/B), Intel Ireland, and the Engineering and Physical Sciences Research Council (EP/F048009/1).

References

G. Ade, On the incoherent imaging in the scanning transmission electron microscope. Optik **49**, 113–116 (1977)

L.J. Allen, S.D. Findlay, M.P. Oxley, C.J. Rossouw, Lattice-resolution contrast from a focused coherent electron probe. Part I. Ultramicroscopy **96**, 47–63 (2003)

A. Amali, P. Rez, Theory of lattice resolution in high-angle annular dark-field images. Microsc. Microanal. **3**, 28–46 (1997)

G. Behan, E.C. Cosgriff, A.I. Kirkland, P.D. Nellist, Three-dimensional imaging by optical sectioning in the aberration-corrected scanning transmission electron microscope. Philos. Trans. R. Soc. Lond. A **367**, 3825–3844 (2009)

A.Y. Borisevich, A.R. Lupini, S.J. Pennycook, Depth sectioning with the aberration-corrected scanning transmission electron microscope. Proc. Natl. Acad. Sci. **103**, 3044–3048 (2006)

N.D. Browning, M.F. Chisholm, S.J. Pennycook, Atomic-resolution chemical analysis using a scanning transmission electron microscope. Nature **366**, 143–146 (1993)

E.C. Cosgriff, A.J. D'Alfonso, L.J. Allen, S.D. Findlay, A.I. Kirkland, P.D. Nellist, Three dimensional imaging in double aberration-corrected scanning confocal electron microscopy. Part I: Elastic scattering. Ultramicroscopy **108**, 1558–1566 (2008)

J.M. Cowley, Image contrast in a transmission scanning electron microscope. Appl. Phys. Lett. **15**, 58–59 (1969)

J.M. Cowley, Coherent interference in convergent-beam electron diffraction & shadow imaging. Ultramicroscopy **4**, 435–450 (1979)

J.M. Cowley, Coherent interference effects in SIEM and CBED. Ultramicroscopy **7**, 19–26 (1981)

A.V. Crewe, The physics of the high-resolution STEM. Rep. Progr. Phys. **43**, 621–639 (1980)

A.V. Crewe, D.N. Eggenberger, J. Wall, L.M. Welter, Electron gun using a field emission source. Rev. Sci. Instrum. **39**, 576–583 (1968)

A.V. Crewe, J. Wall, A scanning microscope with 5 Å resolution. J. Mol. Biol. **48**, 375–393 (1970)

A.J. D'Alfonso, E.C. Cosgriff, S.D. Findlay, G. Behan, A.I. Kirkland, P.D. Nellist, L.J. Allen, Three dimensional imaging in double aberration-corrected scanning confocal electron microscopy. Part II: Inelastic scattering. Ultramicroscopy **108**, 1567–1578 (2008)

N. de Jonge, R. Sougrat, B.M. Northan, S.J. Pennycook, Three-dimensional scanning transmission electron microscopy of biological specimens. Microsc. Microanal. **16**, 54–63 (2010)

N.H. Dekkers, H. de Lang, Differential phase contrast in a STEM. Optik **41**, 452–456 (1974)

C. Dinges, A. Berger, H. Rose, Simulation of TEM images considering phonon and electron excitations. Ultramicroscopy **60**, 49–70 (1995)

A.M. Donald, A.J. Craven, A study of grain boundary segregation in Cu–Bi alloys using STEM. Philos. Mag. A **39**, 1–11 (1979)

C. Dwyer, J. Etheridge, Scattering of Å-scale electron probes in silicon. Ultramicroscopy **96**, 343–360 (2003)

H.M.L. Faulkner, J.M. Rodenburg, Moveable aperture lensless transmission microscopy: A novel phase retrieval algorithm. Phys. Rev. Lett. **93**, 023903 (2004)

S.D. Findlay, L.J. Allen, M.P. Oxley, C.J. Rossouw, Lattice-resolution contrast from a focused coherent electron probe. Part II. Ultramicroscopy **96**, 65–81 (2003)

B.R. Frieden, Optical transfer of the three-dimensional object. J. Opt. Soc. Am. **57**, 36–41 (1967)

S.J. Haigh, H. Sawada, A.I. Kirkland, Atomic structure imaging beyond conventional resolution limits in the transmission electron microscope. Phys. Rev. Lett. **103**, 126101 (2009)

P. Hartel, H. Rose, C. Dinges, Conditions and reasons for incoherent imaging in STEM. Ultramicroscopy **63**, 93–114 (1996)

A. Howie, Image contrast and localised signal selection techniques. J. Microsc. **117**, 11–23 (1979)

C.J. Humphreys, E.G. Bithell, in *Electron Diffraction Techniques*, vol. 1, ed. by J.M. Cowley (OUP, New York, NY, 1992), pp. 75–151

D.E. Jesson, S.J. Pennycook, Incoherent imaging of thin specimens using coherently scattered electrons. Proc. R. Soc. (Lond.) Ser. A **441**, 261–281 (1993)

D.E. Jesson, S.J. Pennycook, Incoherent imaging of crystals using thermally scattered electrons. Proc. Roy. Soc. (Lond.) Ser. A **449**, 273–293 (1995)

E.J. Kirkland, R.F. Loane, J. Silcox, Simulation of annular dark field STEM images using a modified multislice method. Ultramicroscopy **23**, 77–96 (1987)

R.F. Loane, P. Xu, J. Silcox, Thermal vibrations in convergent-beam electron diffraction. Acta Crystallogr. A **47**, 267–278 (1991)

R.F. Loane, P. Xu, J. Silcox, Incoherent imaging of zone axis crystals with ADF STEM. Ultramicroscopy **40**, 121–138 (1992)

K. Mitsuishi, M. Takeguchi, H. Yasuda, K. Furuya, New scheme for calculation of annular dark-field STEM image including both elastically diffracted and TDS wave. J. Electron Microsc. **50**, 157–162 (2001)

D.A. Muller, B. Edwards, E.J. Kirkland, J. Silcox, Simulation of thermal diffuse scattering including a detailed phonon dispersion curve. Ultramicroscopy **86**, 371–380 (2001)

P.D. Nellist, G. Behan, A.I. Kirkland, C.J.D. Hetherington, Confocal operation of a transmission electron microscope with two aberration correctors. Appl. Phys. Lett. **89**, 124105 (2006)

P.D. Nellist, B.C. McCallum, J.M. Rodenburg, Resolution beyond the 'information limit' in transmission electron microscopy. Nature **374**, 630–632 (1995)

P.D. Nellist, S.J. Pennycook, Accurate structure determination from image reconstruction in ADF STEM. J. Microsc. **190**, 159–170 (1998)

P.D. Nellist, S.J. Pennycook, Subangstrom resolution by underfocussed incoherent transmission electron microscopy. Phys. Rev. Lett. **81**, 4156–4159 (1998)

P.D. Nellist, S.J. Pennycook, Incoherent imaging using dynamically scattered coherent electrons. Ultramicroscopy **78**, 111–124 (1999)

P.D. Nellist, S.J. Pennycook, The principles and interpretation of annular dark-field Z-contrast imaging. Adv. Imag. Electron Phys. **113**, 148–203 (2000)

P.D. Nellist, J.M. Rodenburg, Beyond the conventional information limit: the relevant coherence function. Ultramicroscopy **54**, 61–74 (1994)

S.J. Pennycook, Z-contrast STEM for materials science. Ultramicroscopy **30**, 58–69 (1989)

S.J. Pennycook, D.E. Jesson, High-resolution incoherent imaging of crystals. Phys. Rev. Lett. **64**, 938–941 (1990)

D.D. Perovic, C.J. Rossouw, A. Howie, Imaging elastic strain in high-angle annular dark-field scanning transmission electron microscopy. Ultramicroscopy **52**, 353–359 (1993)

Lord Rayleigh, On the theory of optical images with special reference to the microscope. Philos. Mag. **42**(5), 167–195 (1896)

J.M. Rodenburg, R.H.T. Bates, The theory of super-resolution electron microscopy via Wigner-distribution deconvolution. Philos. Trans. R. Soc. Lond. A **339**, 521–553 (1992)

H. Rose, Phase contrast in scanning transmission electron microscopy. Optik **39**, 416–436 (1974)

J.C.H. Spence, *Experimental High-Resolution Electron Microscopy* (OUP, New York, NY, 1988)

J.C.H. Spence, Convergent-beam nanodiffraction, in-line holography and coherent shadow imaging. Optik **92**, 57–68 (1992)

J.C.H. Spence, J.M. Cowley, Lattice imaging in STEM. Optik **50**, 129–142 (1978)
M.M.J. Treacy, J.M. Gibson, Atomic contrast transfer in annular dark-field images. J. Microsc. **180**, 2–11 (1995)
M.M.J. Treacy, A. Howie, C.J. Wilson, Z contrast imaging of platinum and palladium catalysts. Philos. Mag. A **38**, 569–585 (1978)
K. Van Benthem, A.R. Lupini, M. Kim, H.S. Baik, S. Doh, J.-H. Lee, M.P. Oxley, S.D. Findlay, L.J. Allen, J.T. Luck, S.J. Pennycook, Three-dimensional imaging of individual hafnium atoms inside a semiconductor device. Appl. Phys. Lett. **87**, 034104 (2005)
P. Wang, G. Behan, M. Takeguchi, A. Hashimoto, K. Mitsuishi, M. Shimojo, A.I. Kirkland, P.D. Nellist, Nanoscale energy-filtered scanning confocal electron microscopy using a double-aberration-corrected transmission electron microscope. Phys. Rev. Lett. **104**, 200801 (2010)
Z. Yu, D.A. Muller, J. Silcox, Study of strain fields at a-Si/c-Si interface. J. Appl. Phys. **95**, 3362–3371 (2004)
E. Zeitler, M.G.R. Thomson, Scanning transmission electron microscopy. Optik **31**, 258–280 and 359–366 (1970)

3

The Electron Ronchigram

Andrew R. Lupini

3.1 Introduction

The electron Ronchigram is an extremely interesting form of image that can be obtained in a transmission electron microscope (TEM). The main reason why this imaging mode is so interesting is that it contains a mixture of *both* real space (direct image) and reciprocal space (diffraction) information. Moreover, an electron Ronchigram is also an inline hologram recorded in almost exactly the configuration used by Gabor in his Nobel Prize winning invention of holography (Gabor 1948). Gabor himself pointed out that he chose the name "hologram" from the Greek word "holos" because it contains "the whole information" about the sample (Gabor 1992). As well as being interesting, this image is also useful from a practical point of view, because it allows measurements that would be more difficult to perform purely from real images or diffraction patterns. The name "Ronchigram" is used because the optical arrangement is essentially similar to the light-optical Ronchigram pioneered as a lens test in conventional optics by Ronchi (Ronchi 1964). Furthermore, a Ronchigram is also sometimes known as a shadow image, because it is literally the transmitted shadow of the sample.

One of the major uses of the electron Ronchigram is to provide a convenient route toward aligning a scanning transmission electron microscope (STEM). At the present time, all of the aberration correctors available for a STEM use the Ronchigram as part of their alignment procedures. The Nion system uses a software routine to automatically analyze and correct aberrations directly from the Ronchigram (Dellby et al. 2001, Krivanek et al. 1999). The CEOS system relies on the user (assisted by software tools) performing manual adjustments to the corrector settings. The more recently developed STEM alignment system

---

Notice: This manuscript has been authored by UT-Battelle, LLC, under Contract No. DE-AC05-00OR22725 with the U.S. Department of Energy. The United States Government retains and the publisher, by accepting the article for publication, acknowledges that the United States Government retains a non-exclusive, paid-up, irrevocable, world-wide license to publish or reproduce the published form of this manuscript, or allow others to do so, for United States Government purposes.

S.J. Pennycook, P.D. Nellist (eds.), *Scanning Transmission Electron Microscopy*,
DOI 10.1007/978-1-4419-7200-2_3, Springer Science+Business Media, LLC 2011

by JEOL also uses automated analysis of the Ronchigram (Sawada et al. 2008).

There are several key differences between a conventional TEM or STEM image and a Ronchigram. First, just like in a STEM, the most important optical elements are before the sample. However, since the Ronchigram is recorded as a function of angle and the beam of electrons is converging, the position of a feature in the Ronchigram depends on angle as well as probe position. Because the aberration function changes with angle, this dependence means that the appearance of the particular feature in the Ronchigram may vary dramatically at different places.

In this chapter we will assume that the sample is thin enough that it can be approximated as a single plane. We will not consider the interaction of the electron beam with the sample, thickness, or limited coherence in detail. Even with these simplifications, there are two rather different ways of considering the Ronchigram and we will make the slightly artificial distinction between a wave-optical and a geometrical approach. The relationship between these two approaches can be understood by recalling that a ray travels perpendicular to a wavefront (surface of constant phase).

3.2 A Brief Summary of Aberrations

In a "perfect" imaging system, all rays with the same origin would be focused at the same point and the resolution would only be limited by diffraction. In practice, the resolving power of a real lens depends on the position, angle, and energy of the incoming rays. The aberration function provides a mathematical description of these imperfections.

For a STEM, it is usual to neglect the off-axis aberrations, treat the energy dependence separately, and assume that the aberration function depends only upon the angle that a particular ray makes to the optic axis. While these assumptions are not completely accurate, doing so makes the notation rather simpler. Some justification can be obtained by noting that the energy spread is rather small and that the number of pixels in a digital image is limited. Thus as the magnification is increased to the point that variations with position can be seen, the field of view decreases, reducing these same variations. In newer instruments, the field of view available at high resolution is increasing and the chromatic aberration is also adjustable. Thus it should be remembered that although this description is adequate for most users, it is only approximate and for some details it is necessary to consider a more general description such as by Born and Wolf 1959 or Hawkes and Kasper 1989.

Here we use notation based on that of Krivanek to describe the aberration function (Krivanek et al. 1999). Where possible we will attempt to use a general notation so that the analysis is valid to arbitrary order. The slight difference to the rest of this book is that we will take the aberration function as the *distance* rather than the *phase*, which makes the notation simpler in the geometrical optics section. Thus there is a difference of $2\pi/\lambda$ in the definition here to other chapters. To third order in Cartesian coordinates, the aberration function is explicitly given by

$$\begin{aligned}\chi(u,v) = {} & C_{01a}u + C_{01b}v + \tfrac{1}{2}\left(C_1\left(u^2+v^2\right) + C_{12a}\left(u^2-v^2\right) + 2C_{12b}uv\right) \\ & + \tfrac{1}{3}\left(C_{23a}\left(u^3-3uv^2\right) + C_{23b}\left(3u^2v-v^3\right) + C_{21a}\left(u^3+uv^2\right)\right. \\ & \left. + C_{21b}\left(v^3+u^2v\right)\right) + \tfrac{1}{4}\left(C_3\left(u^4+v^4+2u^2v^2\right) + C_{34a}\left(u^4-6u^2v^2+v^4\right)\right. \\ & \left. + C_{34b}\left(4u^3v-4uv^3\right) + C_{32a}\left(u^4-v^4\right) + C_{32b}\left(2u^3v+2uv^3\right)\right), \end{aligned} \tag{1}$$

where each term has the form, C_{NSA}. The subscript N indicates the order of the aberration, S the symmetry or multiplicity, and A the orientation, a or b. The order reveals how rapidly the aberration increases off-axis, the symmetry indicates the number of times that the aberration repeats upon rotation about the optic axis, and the orientation indicates whether it is a sine or a cosine term (which are equivalent but rotated). The orientation is omitted when redundant (such as for round terms), commas may be included for clarity, and zeroes can optionally be omitted. Thus the round third-order spherical aberration, C_s, could be written as C_3 or C_{30} or $C_{3,0}$ in this notation. For completeness, we have included the probe position above (C_{01a}, C_{01b}), while for most of the rest of this chapter we will explicitly include as a separate term **R**, since it is a common usage to define an image as a function of probe position.

A more detailed summary of this notation can be found in this book and elsewhere (Krivanek et al. 2008, Lupini 2001). Some of the limitations of this description have been discussed above, but one advantage of this notation is that it can be extended to arbitrary order without ambiguity over the numerical pre-factor, which is always defined to be $1/(N{+}1)$ or needing to memorize an arbitrary series of names. Other notations can also be used (Kirkland et al. 2006, Uhlemann and Haider 1998) and we will attempt to keep the following derivations general enough that they do not rely strongly on this particular naming convention. The symmetry and radial behaviors are similar in most of these notations; the main differences are in the naming schemes and the choice of numerical pre-factors. As a result there can be numerical factor differences in some of the aberration terms from these different schemes (Ishizuka 1994, Krivanek 1994).

Another cautionary note for the reader is that we write the aberration function as a function of angle, whereas the sine or the tangent of the angle is sometimes used. Of course for small angles, all of these definitions are equivalent to leading order. We will tend to use the term "angle" interchangeably. The main consequence to beware of is that combining numerical results from different software packages can occasionally lead to errors.

Figure 3–1 is based upon a diagram by Scherzer and illustrates the relationship between wavefronts and ray aberrations (Scherzer 1949). From the figure, it is apparent that ray deviations depend upon the gradient of the aberration function. A more complete derivation including a quantification of the error is given by Born and Wolf. Equally, however, we could take this definition of the ray deviation as the starting point and define a geometrical aberration function such that the ray deviation is given by the gradient of that function.

Figure 3–1. Relationship between ray deviation and the aberration function in one dimension. We can see that $\delta \approx d\chi/dr$. Based on Scherzer (1949).

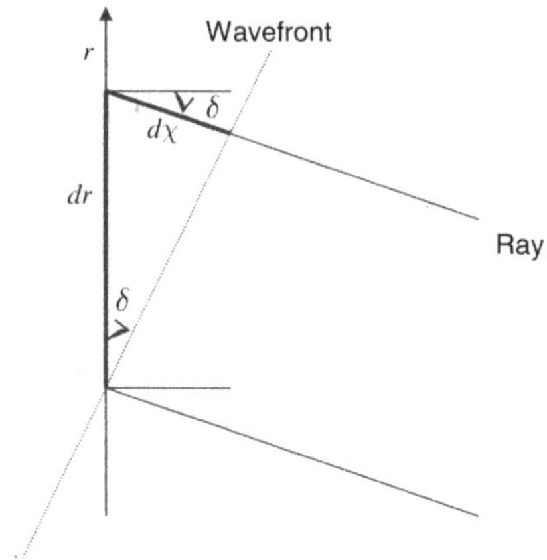

3.3 Formation of the Ronchigram

A source emits electrons that are accelerated by a high voltage. Apertures are used to limit the beam of electrons, which is focused by a series of condenser lenses. Finally an objective lens is used to form a probe. Suitable choice of apertures and lens settings allows the magnification of the electron source and the range of angles incident upon the specimen to be adjusted. Since the angles in the condenser system are small, it is usual to assume that the contribution to the aberrations from the gun and condensers can be neglected or added into the objective lens aberrations. Additionally an aberration corrector may be used to provide an adjustable method to cancel the lens aberrations.

Figure 3–2 illustrates the formation of an electron Ronchigram. An aperture is used to define a range of angles about an optical axis and the

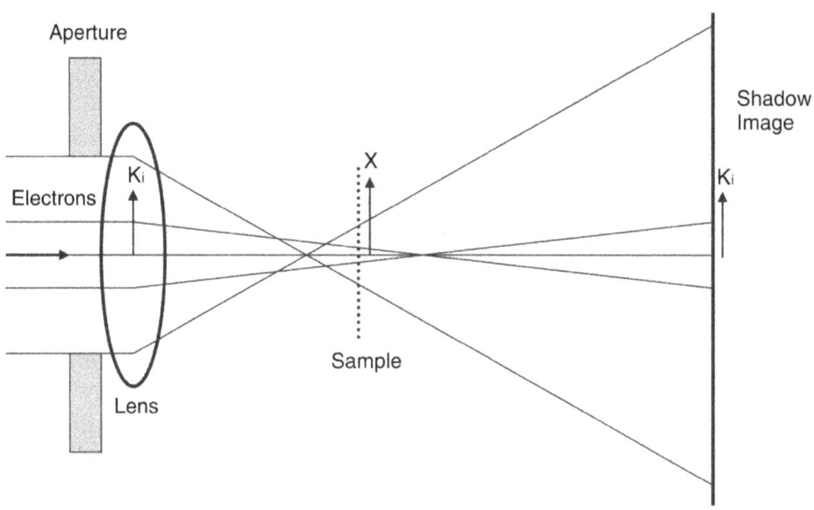

Figure 3–2. Schematic for the formation of a Ronchigram. In this model, the Ronchigram is the geometrical shadow of the sample. We can see that the image will be distorted by the aberrations that cause the rays passing through the sample to deviate.

electron beam is brought to a focus at or near the plane of a thin sample. The image is recorded on a CCD camera (Krivanek and Fan 1992) as a function of angle. From this figure, the picture of a Ronchigram as the geometrical shadow of a suitably thin sample should be apparent. We will consider this geometrical description of a Ronchigram in the next section.

From Figure 3–2, the relationship to STEM imaging should also be clear. In a STEM the probe is focused at the plane of the sample and the image is formed by recording the scattering to different detectors as a function of probe position. A detector effectively integrates over a range of scattering angles. Thus one of the most common uses for the Ronchigram, even before aberration correction, is to correctly align a STEM column to obtain high-quality images (Cowley 1979, James and Browning 1999, Rodenburg and Macak 2002).

3.4 A Geometrical Optics Approach to the Ronchigram

Many of the interesting features of the Ronchigram can be understood from a geometrical viewpoint. This approximation is most useful when the sample is amorphous. However, we should remember that important diffraction effects are neglected. In this limit we consider the Ronchigram as the geometrical shadow of a semi-transparent object (Cowley 1979). Rays propagate through the sample and the shadow image is the intensity in the far field.

As discussed earlier, the position at which a ray goes through the sample depends on the gradient of the aberration function. Thus a particular ray at angle $\mathbf{K}_1 = (u,v)$ goes through the sample at a position $\mathbf{X}_1$. Note that we take $\mathbf{K}_1$ as an angle meaning that there is a factor of λ difference to the wavevector. We shall explicitly include a term for the probe position $\mathbf{R}_1$, meaning that the ray position is

$$\mathbf{X}_1 = \nabla\chi\,(\mathbf{K}_1) + \mathbf{R}_1 \tag{2}$$

We then assume that there is a sufficiently large magnification between the sample plane and the detector plane, such that the position at which a ray at angle $\mathbf{K}$ hits the detector is simply proportional to the angle. Here we will ignore the overall magnification factor and assume that all coordinates can be suitably scaled. When working with the Ronchigram experimentally it is important to calibrate this relationship quite carefully. Note that post-sample distortions, which we shall ignore, can be relevant. (For example, we have observed significant post-sample astigmatism.) It is convenient to measure a calibration in terms of radians per pixel. A useful method to calibrate this value is to adjust the condenser lenses such that a diffraction pattern is formed from a known sample.

It is also important to recall that the objective lens post-field will affect this calibration. Thus changing the objective lens current to refocus at a different sample height can change this calibration. In a microscope with a z-stage, where the sample is kept at the eucentric height, these changes should be small. Finally the reader should also recall that round electron lenses cause the electrons to rotate about the axis, so the rotations between the different elements (such as sample to detector) also need to be considered.

Because the position at which a ray goes through the sample can be a non-linear function, it should be apparent that we need to think carefully about what exactly magnification means when aberrations are present. Clearly a constant (scalar) does not completely describe the magnification encountered here. We can see that the above equations define a form of absolute magnification, relating where a ray goes through the sample to where it hits the detector. This measure is sometimes useful but also highlights a potential difficulty: The magnification will change with position when the aberrations are non-zero (Cowley 1979, Lin and Cowley 1986b).

To further investigate the magnification, we consider a second ray at an angle $\mathbf{K}_2$ (and assume that the aberration function is unchanged). That ray would go through the sample at a position

$$\mathbf{X}_2 = \nabla\chi\,(\mathbf{K}_2) + \mathbf{R}_2. \tag{3}$$

We can then examine the separation between the points at which the rays at $\mathbf{K}_2$ and $\mathbf{K}_1$ went through the sample and find

$$\mathbf{X}_2 - \mathbf{X}_1 = \nabla\chi\,(\mathbf{K}_2) - \nabla\chi\,(\mathbf{K}_1) + \mathbf{R}_2 - \mathbf{R}_1. \tag{4}$$

Now this definition is useful, but rather difficult to consider since it describes a kind of multidimensional magnification. It seems more intuitive to assume that the term "magnification" gives the relationship between a vector $d\mathbf{X}$ on the sample and how that same vector appears at the detector $d\mathbf{K}$, where $d\mathbf{X}=\mathbf{X}_2–\mathbf{X}_1$ and $d\mathbf{K}=\mathbf{K}_2–\mathbf{K}_1$. We can achieve this simplification by considering how a very small vector at the sample would appear at the detector plane. We are using the words "very small" in the differential sense, such that $d\mathbf{K}$ is sufficiently small that terms in $d\mathbf{K}^2$ can be neglected. Thus we can Taylor expand using

$$\mathbf{K}_2 = \mathbf{K}_1 + d\mathbf{K} \text{ to give } \nabla\chi\,(\mathbf{K}_1 + d\mathbf{K}) \approx \nabla\chi\,(\mathbf{K}_1) + d\mathbf{K}.\nabla\,(\nabla\chi\,(\mathbf{K}_1))\,. \tag{5}$$

The double differential quantity in the above equation is best represented as a matrix of the second derivatives of the aberration function evaluated at $\mathbf{K}_1$, which we call $\underline{\mathbf{H}}_{\mathbf{K}_1}$ so that we can write

$$d\mathbf{X} = \underline{\mathbf{H}}_{\mathbf{K}_1} d\mathbf{K} + d\mathbf{R}, \tag{6}$$

where we have used $d\mathbf{R}=\mathbf{R}_2–\mathbf{R}_1$ to allow for a possible probe shift. We have defined the matrix of second derivatives $\underline{\mathbf{H}}_{\mathbf{K}}$ evaluated at an angle $\mathbf{K}$ as

$$\underline{\mathbf{H}}_{\mathbf{K}} = \begin{pmatrix} \frac{\partial^2\chi}{\partial u^2} & \frac{\partial^2\chi}{\partial u \partial v} \\ \frac{\partial^2\chi}{\partial v \partial u} & \frac{\partial^2\chi}{\partial v^2} \end{pmatrix}_{\mathbf{K}}. \tag{7}$$

This matrix is rather useful and will be seen several times. If we want to consider how the aberrations magnify a distance $d\mathbf{X}$ on the sample to the distance in the image $d\mathbf{K}$ then we need the inverse of this matrix such that

$$d\mathbf{K} = \underline{\mathbf{H}}_{\mathbf{K}_1}^{-1}\,(d\mathbf{X} - d\mathbf{R})\,. \tag{8}$$

Two interesting properties are easily seen from this formulation.

(1) The matrix has no inverse where its determinant is equal to 0. The magnification at those locations is undefined, or infinite. Many of the approximations used within this chapter will fail near such loci.
(2) The magnification matrix at a point in the Ronchigram is related to the second derivatives of the aberration function at that point (obviously the appropriate calibrations should be included). If the magnification can somehow be measured then the local second derivatives can be determined. Repeating this measurement at several points in the Ronchigram therefore allows us to fit for the whole aberration function from its second derivatives. Thus if we can somehow determine the magnification at multiple points within the Ronchigram, we can measure the microscope aberrations.

We will return to both of these properties below. Note that we will not include the transpose operator and will freely change the order in dot products involving vectors to preserve readability.

We repeat that the important conclusion from this section is that the magnification of the Ronchigram depends on the (second derivatives of the) aberration function. The experienced user will therefore quickly be able to gain insight into the symmetry and approximate magnitude of the aberrations by examining the appearance of a Ronchigram.

3.5 Where This Description Fails

One experimental test of this description of the magnification is to identify the loci of infinite magnification within the Ronchigram. Cowley has presented a comprehensive description of the circles of infinite magnification (Cowley 1979, Cowley and Disko 1980, Lin and Cowley 1986a). However, the expressions given by Cowley were derived for the case when the loci of infinite magnification are (at least approximately) circular. The answer for non-round aberrations has been derived for the purely geometric case and given previously (Lupini 2001). We have seen that the magnification in the Ronchigram depends on the inverse of the matrix $\underline{\mathbf{H}}_{\mathbf{K}}$. When the determinant of a matrix is zero, the inverse does not exist. We could write the matrix $\underline{\mathbf{H}}_{\mathbf{K}}$, its inverse, and determinant as

$$\underline{\mathbf{H}}_{\mathbf{K}} = \begin{pmatrix} \chi_{uu} & \chi_{uv} \\ \chi_{vu} & \chi_{vv} \end{pmatrix}_{\mathbf{K}}, \qquad \underline{\mathbf{H}}_{\mathbf{K}}^{-1} = \frac{1}{\det(\underline{\mathbf{H}}_{\mathbf{K}})} \begin{pmatrix} \chi_{vv} & -\chi_{uv} \\ -\chi_{vu} & \chi_{uu} \end{pmatrix}_{\mathbf{K}}, \text{ and} \tag{9}$$
$$\det\left(\underline{\mathbf{H}}_{\mathbf{K}}\right) = \chi_{uu}\chi_{vv} - \chi_{uv}\chi_{vu},$$

respectively, where $\chi_{uu} = \frac{\partial^2 \chi}{\partial u^2}$ and so on, and all of the derivatives are evaluated at $\mathbf{K}$. Thus the magnification tends to infinity where

$$\det\left(\underline{\mathbf{H}}_{\mathbf{K}}\right) \rightarrow 0. \tag{10}$$

It is instructive to consider the determinant of this matrix for C_3 and defocus only. In this case, we have

$$\underline{\mathbf{H}}_{\mathbf{K},C1\&C3} = \begin{pmatrix} C_1 + C_3\left(3u^2 + v^2\right) & C_3 2uv \\ C_3 2uv & C_1 + C_3\left(u^2 + 3v^2\right) \end{pmatrix}. \tag{11}$$

Taking the determinant and rearranging terms, we find that the magnification tends to infinity where

$$\det\left(\underline{\mathbf{H}}_{\mathbf{K}}\right) = \left(C_1 + 3C_3\left(u^2 + v^2\right)\right)\left(C_1 + C_3\left(u^2 + v^2\right)\right) \rightarrow 0. \tag{12}$$

Thus there are two circles at an angular radius r where $r^2 = u^2 + v^2$, given by

$$r_{\text{radial}} = \sqrt{\frac{C_1}{3C_3}} \text{ and } r_{\text{azimuthal}} = \sqrt{\frac{C_1}{C_3}}, \tag{13}$$

where the subscript indicates the direction in which the magnification tends to infinity, corresponding to the circles of infinite radial or azimuthal magnification (Cowley and Disko 1980). There is not necessarily a clear distinction to be made between these two circles, other than in the particular case of round aberrations. In this case, the azimuthal circle of infinite magnification can be defined as the locus where all rays at the same radius go through the same point on the sample, irrespective of azimuthal angle. Thus a single point on the sample will give a ring in the Ronchigram. For round terms, the radius $|\mathbf{X}|$ at which a ray goes through the sample is given by

$$|\mathbf{X}| = \frac{\partial \chi}{\partial r}. \tag{14}$$

For non-zero aberrations, each ray at a different azimuthal angle goes through the sample at a different position apart from where

$$\left.\frac{\partial \chi}{\partial r}\right|_{r_{\text{radial}}} = 0. \tag{15}$$

Still considering only round aberrations, the radial circle of infinite magnification is defined as the locus at which radially adjacent rays pass through the same point:

$$\left.\frac{\partial^2 \chi}{\partial r^2}\right|_{r_{\text{azimuthal}}} = 0. \tag{16}$$

Evaluating the differentials in equations (15) and (16) gives exactly the same solutions seen in equation (13). Measuring aberrations from the locations of these rings has been proposed (Hanai et al. 1986, Rodenburg and Lupini 1997), although we note that this approach is likely to break down when higher order aberrations or significant departures from rotational symmetry are present. For rotationally symmetric aberrations both approaches give the same answer, but this is a special case and so we believe that the two-dimensional model used above is more general than this simplified one-dimensional model. Although purely examining these circles is unlikely to provide a completely general method, it is an extremely useful manual method to measure aberrations, since these rings of infinite magnification are often easily observed by eye (Figure 3–3). Furthermore, these loci of infinite magnification are important for this chapter because any method to determine the magnification in the Ronchigram might have significant problems at locations where the determinant is small.

Both circles of infinite magnification are apparent in Figure 3–3, where we use a very simple model to simulate Ronchigrams for a thin sample. We form a probe using the wave-optical equations given later, multiply by a sample function, then Fourier transform to the Ronchigram plane, and take the intensity. Here we generate the sample function for an amorphous sample by generating a random phase and applying a Gaussian filter to the Fourier transform. A crystalline sample can be generated by creating a series of delta functions in the diffraction plane and use of the Fourier transform. Although the simulations in this chapter are not intended to be physically realistic models

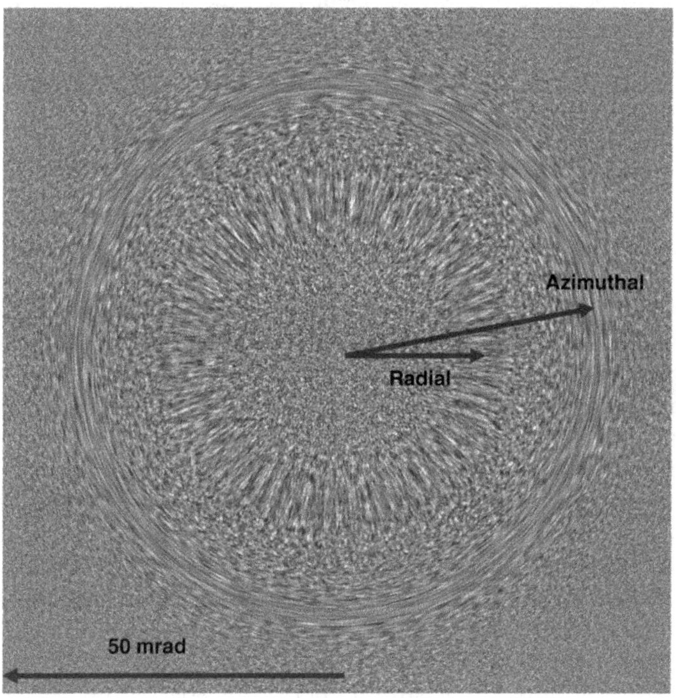

Figure 3–3. Amorphous Ronchigram simulation with half-angle 50 mrad, C_3=1 mm C_1=–1500 nm, 300 kV. The *circles* of infinite radial and azimuthal magnification are marked.

of the samples, they capture many of the essential features seen experimentally. It is also interesting that the geometrical approach applies so well to the purely wave-optical simulations presented here.

3.6 The Nion Method for Measuring Aberrations

As indicated above, one of the primary reasons for interest in the electron Ronchigram is that it can be used to measure the aberration function quickly. We have seen in the previous sections that the magnification depends upon the derivatives of the aberration function. Thus we were able to deduce that methods that allow the magnification to be determined will allow us to fit an aberration function. Obviously, several routes exist to determine the magnification, which may be appropriate for different samples. It is likely that there are further methods that have not yet been considered.

For samples with unknown but recognizable features one way to determine the magnification is to shift the sample by a known amount and measure the resulting apparent shifts of different features. In practice, this is usually achieved by shifting the probe by a known amount (although a precisely calibrated piezo stage or similar method might also be suitable). Cross-correlation of small patches of the Ronchigrams is used to measure the shifts. This method forms the basis of the Nion method to measure aberrations (Dellby et al. 2001, Krivanek et al. 1999). We take equation (6) and locate the same feature, which is equivalent to setting $\mathbf{X}_1 = \mathbf{X}_2$ to give

$$d\mathbf{R} = -\underline{\mathbf{H}}_{\mathbf{K}1} d\mathbf{K}. \tag{17}$$

(The sign is rather arbitrary, since we could just define a different axis at the sample $\mathbf{R}' = -\mathbf{R}$.) In other words, the apparent shift ($d\mathbf{K}$) of features within a "small patch" at $\mathbf{K}_1$ depends upon the known probe shift $d\mathbf{R}$ and on the matrix of second derivatives of the aberration function evaluated at that patch. Thus we can see that one reason why the patch has to be "small" is because the apparent shifts will vary over the Ronchigram. These changes make the fitting more difficult, since the appearance of a particular feature may change as it appears at different angles in the Ronchigram. (Think of stretching a feature by, say, moving the left parts further than the right.) However, these changes illustrate the very effect that allows us to determine the aberrations higher than first order. Since the aberration function changes with angle, patches taken at different angles will have different shifts. Thus each patch gives us a measure of the *local* second derivatives of the aberration function and we fit an aberration function to these measurements.

An important result is that, at least in principle, we do not need more Ronchigrams to measure higher order aberrations. We are able to vary the order of the measurement by taking more patches at different angles. The subdivision of the Ronchigram into a series of patches is schematically illustrated in Figure 3–4, along with a very simple visual picture of some of the approximations that will be made. Clearly the

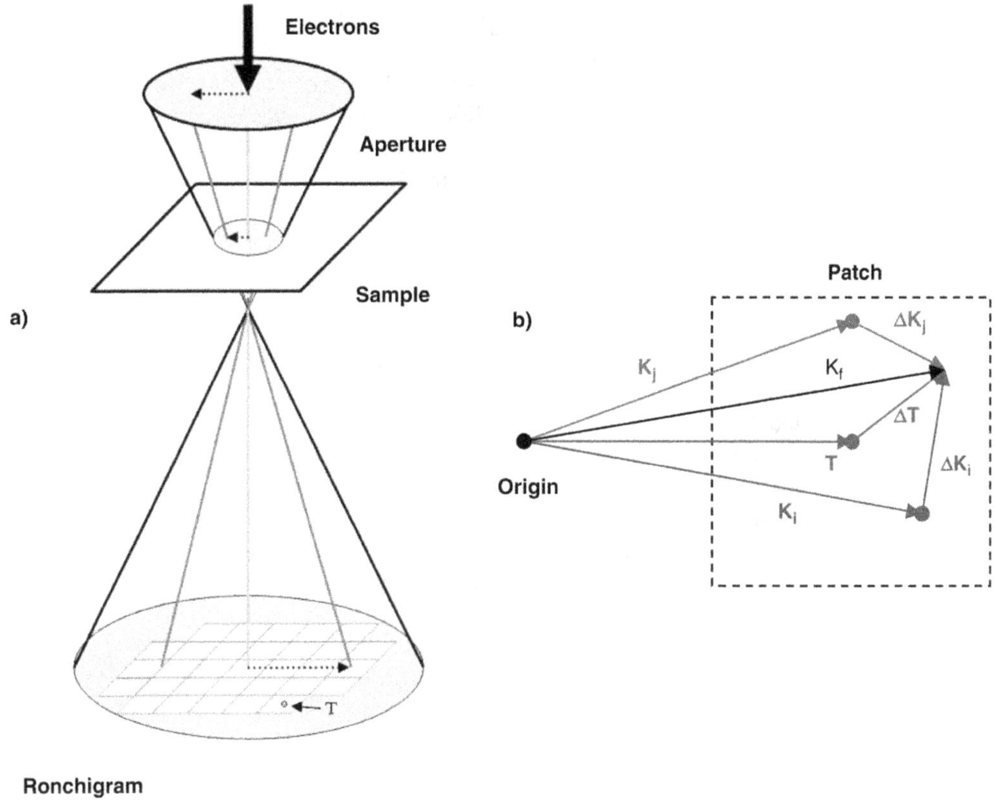

Figure 3–4. (**a**) Schematic formation of an electron Ronchigram or "shadow image" and the division of the Ronchigram into small patches. Each *patch* corresponds to a different scattering angle (e.g., Patch **T**). (**b**) Schematic diagram showing scattering from incident angles $\mathbf{K}_i$ and $\mathbf{K}_j$ into a particular final angle $\mathbf{K}_f$ that is within a patch centered at **T**. Although none of these angles are necessarily small, with an amorphous specimen the scattering decreases with angle, so that the only significant contributions (denoted Δ) involve small angle scattering. From Lupini et al. (2010).

choice of patch position and size will have to depend on the aberrations present. In practice this mostly means that the defocus and shift should be chosen appropriately to allow the measurement of high-order terms.

In practice, the probe shift can often be chosen based on a guess for what we expect the apparent shifts to be. Since we are fitting to the second derivatives the aberration function will have undetermined constant and linear terms, which are unimportant here. It should be clear that for a single shift this equation is still underdetermined (there are more unknowns than measurements for each patch). The obvious solution is to employ shifts in two (or more) different directions.

In previous work (Lupini 2001), we originally formed matrices each consisting of two shift vectors ($\underline{d\mathbf{R}}$ for the known shifts and $\underline{d\mathbf{K}}$ for the apparent shifts in the Ronchigram), allowing us to directly invert the equation at each patch as

$$\underline{\mathbf{H}}_{\mathbf{K}1} = -\underline{d\mathbf{R}}.\underline{d\mathbf{K}}^{-1}. \tag{18}$$

This equation is useful to consider, because it shows immediately why we need to use non-collinear shifts in order to invert $\underline{d\mathbf{K}}$. However for noisy experimental data, inverting this matrix at each patch tends to be numerically unstable. In practice, therefore we do not actually use this inversion and instead use multiple shifts in the form above and put them all into a single least-squares fit

$$\underline{d\mathbf{R}} = -\underline{\mathbf{H}}_{\mathbf{K}1}\underline{d\mathbf{K}}, \tag{19}$$

where the matrices $\underline{d\mathbf{R}}$ and $\underline{d\mathbf{K}}$ can now contain arbitrarily large numbers of known and measured shifts. Using a technique such as a least-squares fit allows reliable estimates of the aberrations from multiple measurements. Typically five Ronchigrams are used. This approach also allows a linear drift measurement and a rotation term to be included.

As a simple example to see how the fitting from second derivatives works, consider the case where the only aberration is defocus:

$$\chi\,(\mathbf{K}) = \frac{1}{2}C_1\mathbf{K}^2, \tag{20}$$

which gives

$$d\mathbf{R} = -C_1 d\mathbf{K}. \tag{21}$$

Thus the shifts in the Ronchigram will be proportional to the probe shifts and the defocus can be measured by comparing the apparent to the known values. Similarly if only first-order aberrations are present we arrive at

$$d\mathbf{R} = -\begin{pmatrix} C_1 + C_{12a} & C_{12b} \\ C_{12b} & C_1 - C_{12b} \end{pmatrix} d\mathbf{K}. \tag{22}$$

Thus the apparent shifts will be an appropriately scaled and rotated version of the probe movement. We see that to solve for the four unknown elements in the matrix, we need multiple shifts (at least two, since each shift vector has two components). For this first-order case, we can see that the shifts are constant over the whole frame. If higher order aberrations are present, then the apparent shifts will vary with the position at which a patch is cut from the Ronchigram. Thus as the number of aberration terms to be measured is increased we need to increase the number of patches and cut them from higher angles.

As was mentioned earlier, when higher order terms are present, the shifts will vary *within* each patch, which causes the appearance of a particular feature to distort. This effect will limit the precision of the cross-correlation (since the features are changed), which causes errors in the aberration measurement. However, if the aberration measurement is roughly correct, then the measurement can be used to provide an estimate for the shifts, calculated for each pixel, which will include this distorting effect. Applying this varying shift to the Ronchigrams being examined means that the aberrations can be "unwarped." The resulting "unwarped" image can then be used to repeat the cross-correlation. If the "unwarping" process was successful, the processed images will be more similar to each other than the originals, meaning that the cross-correlation should be more precise. Thus the measurement of the aberration function should also be more precise and the output from the cross-correlation should reveal the error in the "unwarping." This process can be iterated several times with, hopefully, an improvement in accuracy in each iteration. An extremely useful consequence is that if the process is converging to a realistic value, then the corrections should be smaller and the cross-correlations better with each successive step until some noise limit is reached. If the corrections diverge or rattle too much, then the measurement can be automatically flagged as unreliable.

3.7 Related Methods

We can generalize some of the above results. Within this section we will include the probe position **R** into the aberration function instead of writing it as a separate term. We then imagine a change to the aberration function that causes a specific feature in the Ronchigram at $\mathbf{K}_1$ to move to $\mathbf{K}_2$. Since it is the same feature on the sample then $\mathbf{X}_1 = \mathbf{X}_2$ giving

$$\nabla \chi_{\text{changed}} (\mathbf{K}_2) - \nabla \chi (\mathbf{K}_1) = 0. \tag{23}$$

We consider a change $\Omega (\mathbf{K})$ in the aberration function, so that

$$\chi_{\text{Changed}} (\mathbf{K}) = \chi (\mathbf{K}) + \Omega (\mathbf{K}) . \tag{24}$$

Thus

$$\nabla \chi (\mathbf{K}_1) - \nabla \chi (\mathbf{K}_2) = \nabla \Omega (\mathbf{K}_2) . \tag{25}$$

We therefore have an equation that relates the change that we make to the aberration function to the shift of the feature in the Ronchigram. This result summarizes a class of methods and we will consider it in detail.

If the change is a probe shift as before, we would have

$$\Omega (\mathbf{K}) = -d\mathbf{R}.\mathbf{K}. \tag{26}$$

Giving

$$\nabla \Omega (\mathbf{K}) = -d\mathbf{R}. \tag{27}$$

Therefore substituting equation (27) back into equation (25) gives

$$\nabla \chi (\mathbf{K}_1) - \nabla \chi (\mathbf{K}_2) = -d\mathbf{R}, \tag{28}$$

which is, rather reassuringly, just the same as derived earlier when we treated the probe shift separately. We will see that we can expand this equation to the second derivatives just as before to obtain the Nion method (equation (17)).

The first generalization we make here is that we can fit to this function directly. Thus instead of fitting to the second derivatives of the aberration function, we would be fitting to a discrete difference in the first derivative. However, the second derivative method seems to work well with a cross-correlation algorithm, whereas this method would be better suited to some kind of feature identifying and tracking algorithm. We have not yet extensively investigated such methods, but we note that the fitting is closely related to the method for crystalline samples that will be described later.

For now, we proceed as before by assuming that the change $d\mathbf{K}$ is small and neglect higher order changes in equation (25) to arrive at

$$\underline{\mathbf{H}}_{\mathbf{K}2} d\mathbf{K} = \nabla\Omega\left(\mathbf{K}_2\right). \tag{29}$$

In other words, if we make a change to the aberration function, the resulting shifts in the Ronchigram will reveal the second derivatives of the aberration function. We can think of this as a generalization of the probe-shifting method. The principle is unchanged, because we are still using the magnification to give the second derivatives of the aberration function. Furthermore, we can see why this formulation is more general: We could potentially use any change in the aberration function to determine the magnification. (Note that this indicates the slightly ambiguous nature of writing the probe position as a separate vector $\mathbf{R}$ when we could have left it inside the function.)

A method to perform this measurement using focus changes has previously been suggested, but was found not to be invertible for individual patches (Lupini 2001). For a focus change z the change in the aberration function would be

$$\Omega\left(\mathbf{K}_2\right) = \frac{1}{2} z\mathbf{K}_2^2, \text{ giving } \nabla\Omega\left(\mathbf{K}_2\right) = z\mathbf{K}_2. \tag{30}$$

Thus

$$\underline{\mathbf{H}}_{\mathbf{K}2} d\mathbf{K} = z\mathbf{K}_2. \tag{31}$$

This equation reveals that this method is different from the probe-shifting method, in that we are not able to use orthogonal directions to form an invertible matrix for each patch. However, as discussed above, it appears that avoiding the explicit inversion and fitting directly to equation (31) offers a route to measure aberrations. We can qualitatively understand this result by noting that the vector $\mathbf{K}_2$ is different for each patch, whereas the shift $d\mathbf{R}$ was a constant. Thus by combining information from different patches this method does indeed allow us to probe all aberrations. This method requires pairs of Ronchigrams, where the probe-shifting required a minimum of three Ronchigrams. Initial tests suggest that this method suffers a little because the magnification is different after the focus change, so the cross-correlations will usually be poorer than for the probe-shifting method. (This observation suggests that an "unwarping" step might be essential here.) We also found that such methods are very sensitive to drifts and imperfect centering of the optic axis on the CCD.

A further generalization we can make is that the exact nature of the change could be more complicated than a simple shift or defocus change. For example, we have successfully tested a method, on simulated images, that uses astigmatism changes. Admittedly it is not quite so clear how useful such methods might be, since it might not be desirable to change the aberration function in this way. Also most aberration changes tend not to be pure and include a parasitic shift. Potentially therefore this method might provide a route to control purification: The first step in removing unwanted parasitic changes is to be able to measure them.

Finally, we note that there is a common theme in the methods to measure aberrations in this section; We related the magnification to the derivatives, or second derivatives, of the aberration function. In the Nion method, we determine the magnification from the shifts of small parts of the Ronchigram. Although this tracking is automated in the aberration measurement software, a useful check is that the shifts should be visible by eye; if they are not, then the software will also struggle.

Thus it should be apparent that the important aspect in these methods is the determination of the magnification. This implies that if we can measure the magnification from single frames, we should be able to fit the whole aberration function from single Ronchigrams. In some sense it can be seen that these methods are descendents of Cowley's original suggestion of determining magnification in the Ronchigram by comparing to bright-field (BF) images (Lin and Cowley 1986a). If the magnification can be determined by some other way then that would allow an alternative method to be developed. For example, knowing exactly what the sample looks like or using many Ronchigrams while shifting the same feature to many different positions. There is thus a relationship to other methods, such as that of Sawada that looks at the size of the autocorrelation function (Sawada et al. 2008), or our own work (Lupini and Pennycook 2008, Lupini et al. 2010) that examines the Fourier transforms of small patches, since we might expect those properties to depend on the magnification in some manner. Similarly other techniques such as examining the variance or a characteristic feature size might provide methods to perform this measurement for single frames. We will examine some of these possibilities in the next sections.

3.8 A Wave-Optical Approach

We have seen in the previous sections how a geometrical optics approach can be used to understand many of the striking features seen experimentally in the electron Ronchigram. However, this approach has neglected diffraction effects, which we can now include.

Consider the following question: How did a particular feature in the Ronchigram get to that position? We can arrive at two limiting cases for the answer. First that the feature is the geometrical shadow of the corresponding part of the sample, as laid out in the previous section, or alternatively that the feature is a diffraction effect with contributions from the whole illuminated area.

We can visualize the formation of the Ronchigram with the following thought experiment. If we illuminate the specimen through a very small aperture then the beam incident upon the specimen will be almost parallel. Because of diffraction from the aperture (equivalent to the uncertainty principle) a very small angle must illuminate a large area of the specimen. Depending upon the arrangement of atoms within the material, at some angles the scattered wavefronts will interfere constructively, at others destructively, leading to the formation of a diffraction pattern. In the limit of a very small aperture, and near parallel illumination, a Ronchigram will therefore be a diffraction pattern.

We next imagine increasing the angular size of the aperture. We can imagine each part of the illumination aperture to be giving rise to a diffraction pattern, with each pattern at a slightly different angle (due to the different incident angle). Thus, the diffraction spots are broadened into disks and we will see a convolution of the diffraction pattern with the aperture function. As the aperture size is increased still further, these disks overlap and, assuming suitably coherent conditions, they interfere with each other. Therefore in the overlap regions we see a series of fringes that must in some way depend on the lattice spacings, the aberrations, probe position, and so on.

Finally we imagine using a very large aperture. As the disks become larger more of the aforementioned disks will interfere and the resulting interference pattern looks more and more like a real image of the specimen rather than a diffraction pattern. However, the dilemma remains: Our feature of interest could be either a geometrical shadow of a particular piece of the specimen or a diffraction effect due to a particular spacing. Conceptually we can separate specimens into two limiting cases: amorphous, where the diffraction is negligible and our geometrical model holds reasonably well; and crystalline, where diffraction dominates. This dichotomy is why Ronchigrams of mixed specimens, where effects from both models are visible, can look rather odd.

3.9 Probe Formation

We assume that the objective aperture is uniformly and coherently illuminated by a monochromatic beam of electrons. Later we will see that the effects of limited coherence can be included, meaning that these assumptions are adopted to simplify the derivation, rather than as strict requirements. The probe is given by the (reverse) Fourier transform of the wavefront across the objective aperture as

$$P(\mathbf{R},\mathbf{R}_0) = \int A(\mathbf{K}_i)\, e^{-2\pi i\lambda^{-1}\mathbf{K}_i.(\mathbf{R}-\mathbf{R}_0)} d\mathbf{K}_i, \tag{32}$$

where $\mathbf{K}_i$ is the incident wavevector expressed as an angle, λ is the electron wavelength, $\mathbf{R}$ is a coordinate on the sample (a dummy variable here), and $\mathbf{R}_0$ is the probe position; all vectors are two dimensional. The aperture function, A, is a product of a top-hat function and phase change due to the aberration function. Remember that the notation used here is slightly different from the rest of this book because we use the aberration function as a distance, rather than a phase, and also that we are taking $\mathbf{K}_i$ as an angle, which gives a factor of λ difference. To further simplify the notation, we will only consider points inside the aperture and omit the aperture cut-off. We assume that the aperture is large enough that the relevant waves are able to interfere.

We further assume that the sample is thin and that the effect it has on the beam is purely multiplicative. Thus subject to these conditions, the exit surface wave function (EWF) ψ is the product of the probe function and the specimen function φ:

$$\psi\left(\mathbf{R}, \mathbf{R}_0\right) = \int \varphi\left(\mathbf{R}\right) e^{2\pi i\lambda^{-1}\chi(\mathbf{K}_i)-2\pi i\lambda^{-1}\mathbf{K}_i.(\mathbf{R}-\mathbf{R}_0)} d\mathbf{K}_i. \tag{33}$$

Since the Ronchigram is recorded as a function of angle, we Fourier transform the EWF to

$$\psi\left(\mathbf{K}_f, \mathbf{R}_0\right) = \int \varphi\left(\mathbf{K}_f - \mathbf{K}_i\right) e^{2\pi i\lambda^{-1}\chi(\mathbf{K}_i)+2\pi i\lambda^{-1}\mathbf{R}_0.\mathbf{K}_i} d\mathbf{K}_i. \tag{34}$$

By a suitable change of variable, we could also write

$$\psi\left(\mathbf{K}_f, \mathbf{R}_0\right) = \int \varphi\left(\mathbf{K}_1\right) e^{2\pi i\lambda^{-1}\chi\left(\mathbf{K}_f-\mathbf{K}_1\right)+2\pi i\lambda^{-1}\mathbf{R}_0.\left(\mathbf{K}_f-\mathbf{K}_1\right)} d\mathbf{K}_1. \tag{35}$$

Note that this means that we can write the intensity in slightly different forms, depending on which is most convenient. At this point, we should recall that the electron wavefunction propagates through the specimen and interferes with itself coherently, and that the observable is the intensity. This leads to the familiar phase problem (Rodenburg 1989). Much of the interesting information about the sample is contained in the phase of the wavefunction, but we can only record the intensity. Therefore, we now consider the intensity, but in the following discussion, it should be remembered that we would like to recover the complex amplitude. Since the intensity is recorded on a CCD camera in the far-field, we write it as a function of detection angle $\mathbf{K}_f$ and probe position. We can do this by taking the modulus squared of the equation for the amplitude:

$$I\left(\mathbf{K}_f, \mathbf{R}_0\right) = \iint \varphi\left(\mathbf{K}_f - \mathbf{K}_i\right)\varphi^*\left(\mathbf{K}_f - \mathbf{K}_j\right) e^{2\pi i\lambda^{-1}\left[\chi(\mathbf{K}_i)-\chi(\mathbf{K}_j)+\mathbf{R}_0.(\mathbf{K}_i-\mathbf{K}_j)\right]} d\mathbf{K}_i d\mathbf{K}_j. \tag{36}$$

One way to determine the phase of the complex wavefunction is to interfere it with a known reference. Since the phase of the reference is known, it may be possible to determine the phase of the scattered wavefunction. We note that this situation describes holography as invented by Gabor in 1948. Most of the central beam passes almost unchanged through the specimen, and thus provides a convenient reference. For a thin sample, a small fraction of the incident beam is scattered and interferes with this reference, and the intensity can be recorded. In practice nowadays for holography a "skew" reference beam is used, which facilitates reconstruction of a unique image. However, it should be clear that the phase information is contained in the electron Ronchigram and can be extracted via suitable methods, such as ptychography (Nellist and Rodenburg 1994, Rodenburg and Bates 1992, Rodenburg et al. 1993).

3.10 An Aside on Bright-Field Imaging

We will see that the electron Ronchigram is, in many ways, very similar to a conventional bright-field (BF) image and so to help understand the Ronchigram we first briefly review the formation of the BF image. We Fourier transform the intensity over the probe position to give the (Fourier transform of the) intensity as a function of spatial frequency as

$$I\left(\mathbf{K}_f, \boldsymbol{\rho}\right) = \int \varphi\left(\mathbf{K}_f - \mathbf{K}_i\right) \varphi^*\left(\mathbf{K}_f - \mathbf{K}_i - \boldsymbol{\rho}\right) e^{2\pi i \lambda^{-1}\chi(\mathbf{K}_i) - 2\pi i\lambda^{-1}\chi(\mathbf{K}_i+\boldsymbol{\rho})} d\mathbf{K}_i. \tag{37}$$

We next apply the weak phase object approximation (WPOA). In this approximation, it is assumed that the sample only changes the phase of the wavefront passing through, and we further assume that the phase change is small, such that we can write the sample function as

$$\varphi(\mathbf{R}) = e^{-i\sigma V(\mathbf{R})} \approx 1 - i\sigma V(\mathbf{R}). \tag{38}$$

The Fourier transform of the sample function can then be written as

$$\varphi(\mathbf{K}) \approx \delta(\mathbf{K}) - i\sigma V(\mathbf{K}). \tag{39}$$

We neglect constants and higher order terms to give

$$\begin{aligned} \varphi\left(\mathbf{K}_f - \mathbf{K}_i\right) \varphi^*\left(\mathbf{K}_f - \mathbf{K}_j\right) \approx\ & i\sigma V^*\left(\mathbf{K}_f - \mathbf{K}_j\right) \delta\left(\mathbf{K}_f - \mathbf{K}_i\right) \\ & - i\sigma V\left(\mathbf{K}_f - \mathbf{K}_i\right) \delta\left(\mathbf{K}_f - \mathbf{K}_j\right). \end{aligned} \tag{40}$$

This approximation means that

$$\begin{aligned} I\left(\mathbf{K}_f, \boldsymbol{\rho}\right) = i\sigma \Big(& V^*(-\boldsymbol{\rho})\, e^{2\pi i\lambda^{-1}\chi\left(\mathbf{K}_f\right) - 2\pi i\lambda^{-1}\chi\left(\mathbf{K}_f+\boldsymbol{\rho}\right)} \\ & - V(\boldsymbol{\rho})\, e^{2\pi i\lambda^{-1}\chi\left(\mathbf{K}_f-\boldsymbol{\rho}\right) - 2\pi i\lambda^{-1}\chi\left(\mathbf{K}_f\right)} \Big). \end{aligned} \tag{41}$$

We can simplify the notation somewhat by making the following definition:

$$\chi_{\mathbf{T}}(\Delta\mathbf{T}) := \chi(\mathbf{T} + \Delta\mathbf{T}) - \chi(\mathbf{T}). \tag{42}$$

We further know that if $V(\mathbf{R})$ is real then after Fourier transforming $V(\boldsymbol{\rho}) = V^*(-\boldsymbol{\rho})$. Thus

$$I\left(\mathbf{K}_f,\boldsymbol{\rho}\right)=i\sigma V\left(\boldsymbol{\rho}\right)\left(e^{-2\pi i\lambda^{-1}\chi_{\mathbf{K}_f}(\boldsymbol{\rho})}-e^{2\pi i\lambda^{-1}\chi_{\mathbf{K}_f}(-\boldsymbol{\rho})}\right). \tag{43}$$

Finally, by writing the aberration function as a sum of even E and odd O parts, we arrive at

$$I\left(\mathbf{K}_f,\boldsymbol{\rho}\right)=2\sigma V\left(\boldsymbol{\rho}\right)\sin\left(2\pi i\lambda^{-1}E_{\mathbf{K}_f}\left(\boldsymbol{\rho}\right)\right)e^{-2\pi i\lambda^{-1}O_{\mathbf{K}_f}(\boldsymbol{\rho})}. \tag{44}$$

Thus we have derived the usual BF contrast transfer function (CTF) (including the changes for tilted illumination) that is frequently seen in TEM (de Jong and Koster 1992, Krivanek and Fan 1992, Meyer et al. 2002, 2004, Saxton 2000, Zemlin et al. 1978). Although this derivation is for STEM, it is essentially the same as the result for BF TEM with the appropriate changes of angles, as might be expected from the principle of reciprocity (Cowley 1969, Zeitler and Thomson 1970). If the detector is a delta function on axis, this reproduces the familiar form of the CTF.

It should be clear that the aberration function will change off-axis. For zero tilt, this gives the usual axial aberration function, but off-axis, the apparent aberrations will change. This change causes the BF images off-axis to be shifted relative to the on-axis image and the focus and astigmatism will change noticeably. These changes are of course analogous to the changes in conventional BF TEM with tilted illumination. Measuring either the induced shift or the induced defocus and astigmatism allows the aberration function to be measured. We can Taylor series expand the aberration function as

$$\begin{aligned}\chi_{\mathbf{T}}\left(\Delta\mathbf{T}\right)&=\Delta\mathbf{T}\cdot\nabla\chi_{\mathbf{T}}\left(\mathbf{T}\right)+\tfrac{1}{2}\Delta\mathbf{T}\cdot\underline{\mathbf{H}}_{\mathbf{T}}\cdot\Delta\mathbf{T}+O\left(\Delta\mathbf{T}^3\right)\\&=\Delta\mathbf{T}.\left(C_{01a}^{\mathbf{T}},C_{01b}^{\mathbf{T}}\right)+\tfrac{1}{2}\Delta\mathbf{T}\begin{pmatrix}C_1^{\mathbf{T}}+C_{12a}^{\mathbf{T}} & C_{12b}^{\mathbf{T}}\\ C_{12b}^{\mathbf{T}} & C_1^{\mathbf{T}}-C_{12a}^{\mathbf{T}}\end{pmatrix}\Delta\mathbf{T}+O\left(\Delta\mathbf{T}^3\right),\end{aligned} \tag{45}$$

where $\underline{\mathbf{H}}_{\mathbf{T}}$ is the Hessian matrix of second derivatives of the aberration function evaluated at $\mathbf{T}$. We note that $\chi_{\mathbf{T}}\left(\Delta\mathbf{T}\right)$ has the same form as $\chi\left(\mathbf{K}\right)$ although the values of the coefficients are changed. Thus we see that the shifts will depend on the derivatives and the apparent aberration will depend on the second derivatives of the aberration function. Here we have used superscripts to indicate apparent aberration coefficients $C^{\mathbf{T}}$ evaluated at a tilt angle $\mathbf{T}$ since they differ from the axial values. As an example consider the case where only round aberrations are present, defocus C_1 and spherical aberration C_3. In this case we can evaluate the second derivatives to give

$$\underline{\mathbf{H}}_{\mathbf{T},C1\&C3}=\begin{pmatrix}C_1+C_3\left(3u^2+v^2\right) & C_3 2uv\\ C_3 2uv & C_1+C_3\left(u^2+3v^2\right)\end{pmatrix}. \tag{46}$$

Thus, on-axis (at u=v=0) the apparent defocus is the true focus and the astigmatism is 0 as might be expected. Off-axis, the defocus and astigmatism will vary with the tilt angle used. Thus simply by measuring the effective defocus and astigmatism (and/or the shift) from a series of images acquired at different tilts we are able to measure the whole aberration function. Clearly to include more and higher order terms we would need to use more images. Furthermore, one potential benefit of the STEM geometry is that it is possible to obtain multiple tilted BF images simultaneously, thus mitigating problems from sample drift, and therefore exploring the use of multiple detectors would be a fruitful area for development. Note that the higher order apparent aberrations will also change, and thus more terms could also potentially be used to measure aberrations. See for example the method for measuring spherical aberration from a single BF image (Krivanek 1976). We note that odd terms will still require more than one image, and that it might be difficult to accurately fit higher-order terms.

3.11 Small Patches

The Ronchigram is rather complicated because it does not separate into a simple contrast transfer function in the same way as we have just seen for bright-field images. Cowley derived a CTF, which is rather complicated because it is non-isoplanatic, which is to say that it changes over the field of view (Cowley and Lin 1986). One approach to simplifying the Ronchigram is suggested by the previous section, where we considered many images at different angles. We divide the Ronchigram into a series of small patches with each patch small enough that it can be regarded, at least approximately, as isoplanatic (i.e., that we can neglect changes within the patch). This is the same approach used in the Nion method (Dellby et al. 2001, Lupini 2001) and it is an important way of dealing with expressions that can otherwise become too complicated for simple consideration. This process is schematically illustrated in Figure 3–4.

We therefore consider a *small patch* of the Ronchigram at a particular detection (tilt) angle $\mathbf{T}$, such that $\mathbf{K}_f = \mathbf{T} + \mathbf{K}_\mathbf{T}$ for small $\mathbf{K}_\mathbf{T}$, and Taylor series expand

$$\begin{aligned}\chi\left(\mathbf{K}_f - \mathbf{K}_i\right) - \chi\left(\mathbf{K}_f - \mathbf{K}_j\right) \approx \chi\left(\mathbf{T} - \mathbf{K}_i\right) - \chi\left(\mathbf{T} - \mathbf{K}_j\right) \\ + \mathbf{K}_\mathbf{T}.\nabla\left(\chi\left(\mathbf{T} - \mathbf{K}_i\right) - \chi\left(\mathbf{T} - \mathbf{K}_j\right)\right) + \cdots\end{aligned} \tag{47}$$

Thus the condition for a "small" patch is that it is reasonable to neglect terms of order $\mathbf{K}_\mathbf{T}^2$ and higher.

The next step is to consider the sample. We will consider two limiting cases, a perfectly amorphous material and a perfectly crystalline material. In practice, real materials are somewhere between these two limits. For example, most microscopists are probably familiar with the thin amorphous layers that can be found on top of many crystalline TEM samples, resulting from damage during preparation and contamination.

3.12 Amorphous Materials

In the Ronchigram geometry, the distance at which rays go through the sample will depend on their angular separation. For an amorphous material we assume that there is no long-range ordering of the atoms.

This means that there is no long range ordering of the scattering from different parts of the specimen. Thus, when all the scattering from a finite area on an amorphous sample is integrated, we expect that the total will be dominated by small angle scattering. Note that this is different from the crystalline samples where, by definition, there will be long-range order. (We note that the Ronchigram provides a route by which the degree of ordering can be probed (Rodenburg 1988)).

Obviously, this small-angle approximation is very similar to the small patch approximation (considering points near a particular angle $\mathbf{T}$). For an amorphous material, we can therefore write the differences between the incident and scattered wavevectors as $\Delta\mathbf{K}_i = \mathbf{K}_f - \mathbf{K}_i$ and $\Delta\mathbf{K}_j = \mathbf{K}_f - \mathbf{K}$, respectively where both $\Delta\mathbf{K}_i$ and $\Delta\mathbf{K}_j$ are small. This approximation is important because it means that we can neglect higher order terms. We therefore expand the difference between the interfering waves as (Lupini et al. 2010)

$$\begin{aligned}\chi(\mathbf{K}_i) - \chi(\mathbf{K}_j) &= \chi(\mathbf{T}+\Delta\mathbf{T}-\Delta\mathbf{K}_i) - \chi(\mathbf{T}+\Delta\mathbf{T}-\Delta\mathbf{K}_j)\\ &\approx \chi(\mathbf{T}-\Delta\mathbf{K}_i) - \chi(\mathbf{T}-\Delta\mathbf{K}_j) + \Delta\mathbf{T}\cdot(\nabla\chi(\mathbf{T}-\Delta\mathbf{K}_i)\\ &\quad -\nabla\chi(\mathbf{T}-\Delta\mathbf{K}_j)) \approx \chi(\mathbf{T}-\Delta\mathbf{K}_i) - \chi(\mathbf{T}-\Delta\mathbf{K}_j)\\ &\quad -\Delta\mathbf{T}\cdot\underline{\mathbf{H}}_{\mathbf{T}}\cdot(\Delta\mathbf{K}_i-\Delta\mathbf{K}_j).\end{aligned} \tag{48}$$

We had to be careful to expand enough terms to give a result that is relatively easy to integrate in subsequent steps, but without throwing away all of the significant terms. (See Figure 3–4 for a visual interpretation.) We remain hopeful that a more exact treatment might be derived in the future. We substitute this approximation into our earlier equation for the intensity and neglect the dependence upon probe position to give

$$I(\mathbf{T},\Delta\mathbf{T}) = \iint \varphi(\Delta\mathbf{K}_i)\varphi^*(\Delta\mathbf{K}_j)\cdot e^{-2\pi\lambda^{-1}i[\chi(\mathbf{T}+\Delta\mathbf{T}-\Delta\mathbf{K}_i)-\chi(\mathbf{T}+\Delta\mathbf{T}-\Delta\mathbf{K}_j)]}d\Delta\mathbf{K}_i d\Delta\mathbf{K}_j. \tag{49}$$

Thus the advantage of our approximate treatment is that it lets us take the Fourier transform just like we did in the BF derivation. However, note that for the BF image we were Fourier transforming over probe position, whereas here we are keeping the probe position fixed and instead transform over the range of angles $\Delta\mathbf{T}$ inside our small patch at $\mathbf{T}$. Thus

$$I(\mathbf{T},\boldsymbol{\tau}) = \iint\int \varphi(\Delta\mathbf{K}_i)\,\varphi^*(\Delta\mathbf{K}_j)\; e^{-2\pi\lambda^{-1}i\left[\chi(\mathbf{T}-\Delta\mathbf{K}_i)-\chi\left(\mathbf{T}-\Delta\mathbf{K}_j\right)-\Delta\mathbf{T}\cdot\underline{\mathbf{H}}_{\mathbf{T}}\cdot\left(\Delta\mathbf{K}_i-\Delta\mathbf{K}_j\right)-\Delta\mathbf{T}.\boldsymbol{\tau}\right]}d\Delta\mathbf{K}_i d\Delta\mathbf{K}_j d\Delta\mathbf{T} \tag{50}$$

Where $\boldsymbol{\tau}$ is the conjugate variable to $\Delta\mathbf{T}$. We note that the integral over $\Delta\mathbf{T}$ gives a delta function. At regions where $\underline{\mathbf{H}}_{\mathbf{T}}$ is invertible, we assume that we can write

$$\int e^{-2\pi\lambda^{-1}i\left(\Delta\mathbf{T}\cdot\underline{\mathbf{H}}_{\mathbf{T}}\cdot(\Delta\mathbf{K}_i-\Delta\mathbf{K}_j)+\Delta\mathbf{T}\cdot\boldsymbol{\tau}\right)}d\Delta\mathbf{T} = \delta\left(\underline{\mathbf{H}}_{\mathbf{T}}.\Delta\mathbf{K}_j - \underline{\mathbf{H}}_{\mathbf{T}}.\Delta\mathbf{K}_i - \boldsymbol{\tau}\right) = \delta\left(\Delta\mathbf{K}_j - \Delta\mathbf{K}_i - \underline{\mathbf{H}}_{\mathbf{T}}^{-1}.\boldsymbol{\tau}\right). \tag{51}$$

Obviously there are some regions where $\underline{\mathbf{H}}_{\mathbf{T}}$ is not invertible (such as the loci of infinite magnification seen earlier) and regions where the approximations that we have used are not particularly accurate. It is therefore appropriate to realize that this result will be most accurate when the aberration function is dominated by first-order aberrations such as defocus. This delta function allows us to perform the integral over $\mathbf{K}_j$ to obtain

$$I(\mathbf{T},\boldsymbol{\tau})=\int\varphi\left(\Delta\mathbf{K}_i\right)\varphi^*\left(\Delta\mathbf{K}_i+\underline{\mathbf{H}}_{\mathbf{T}}^{-1}.\boldsymbol{\tau}\right)e^{-2\pi i\lambda^{-1}\left[\chi(\mathbf{T}-\Delta\mathbf{K}_i)-\chi\left(\mathbf{T}-\Delta\mathbf{K}_i-\underline{\mathbf{H}}_{\mathbf{T}}^{-1}.\boldsymbol{\tau}\right)\right]}d\Delta\mathbf{K}_i. \tag{52}$$

We next apply the weak phase object approximation again, such that we can write

$$\varphi\left(\Delta\mathbf{K}_i\right)\varphi^*\left(\Delta\mathbf{K}_i+\underline{\mathbf{H}}_{\mathbf{T}}^{-1}.\boldsymbol{\tau}\right)=\left(\delta\left(\Delta\mathbf{K}_i\right)-i\sigma V\left(\Delta\mathbf{K}_i\right)\right)\left(\delta\left(\Delta\mathbf{K}_i+\underline{\mathbf{H}}_{\mathbf{T}}^{-1}.\boldsymbol{\tau}\right)+i\sigma V^*\left(\Delta\mathbf{K}_i+\underline{\mathbf{H}}_{\mathbf{T}}^{-1}.\boldsymbol{\tau}\right)\right). \tag{53}$$

Using this sample, and ignoring the constants and terms in σ^2 we have

$$I\left(\mathbf{T},\boldsymbol{\tau}\right)=i\sigma\int\left(\delta\left(\Delta\mathbf{K}_i\right)V^*\left(\Delta\mathbf{K}_i+\underline{\mathbf{H}}_{\mathbf{T}}^{-1}.\boldsymbol{\tau}\right)-V\left(\Delta\mathbf{K}_i\right)\delta\left(\Delta\mathbf{K}_i+\underline{\mathbf{H}}_{\mathbf{T}}^{-1}.\boldsymbol{\tau}\right)\right) e^{-2\pi i\lambda^{-1}\left[\chi(\mathbf{T}-\Delta\mathbf{K}_i)-\chi\left(\mathbf{T}-\Delta\mathbf{K}_i-\underline{\mathbf{H}}_{\mathbf{T}}^{-1}.\boldsymbol{\tau}\right)\right]}d\Delta\mathbf{K}_i. \tag{54}$$

The integral over the small patch gives

$$I\left(\mathbf{T},\boldsymbol{\tau}\right)=i\sigma\left(V^*\left(\underline{\mathbf{H}}_{\mathbf{T}}^{-1}.\boldsymbol{\tau}\right)e^{-2\pi i\lambda^{-1}\left[\chi(\mathbf{T})-\chi\left(\mathbf{T}-\underline{\mathbf{H}}_{\mathbf{T}}^{-1}.\boldsymbol{\tau}\right)\right]} - V\left(-\underline{\mathbf{H}}_{\mathbf{T}}^{-1}.\boldsymbol{\tau}\right)e^{-2\pi i\lambda^{-1}\left[\chi\left(\mathbf{T}+\underline{\mathbf{H}}_{\mathbf{T}}^{-1}.\boldsymbol{\tau}\right)-\chi(\mathbf{T})\right]}\right). \tag{55}$$

We again use the relation that if $V(\mathbf{R})$ is real then $V^*\left(\mathbf{K}\right)=V\left(-\mathbf{K}\right)$:

$$I\left(\mathbf{T},\boldsymbol{\tau}\right)=i\sigma V\left(-\underline{\mathbf{H}}_{\mathbf{T}}^{-1}.\boldsymbol{\tau}\right)\left(e^{-2\pi i\lambda^{-1}\left[\chi(\mathbf{T})-\chi\left(\mathbf{T}-\underline{\mathbf{H}}_{\mathbf{T}}^{-1}.\boldsymbol{\tau}\right)\right]} - e^{-2\pi i\lambda^{-1}\left[\chi\left(\mathbf{T}+\underline{\mathbf{H}}_{\mathbf{T}}^{-1}.\boldsymbol{\tau}\right)-\chi(\mathbf{T})\right]}\right). \tag{56}$$

Using our earlier notation for $\chi_{\mathbf{T}}$ gives

$$I\left(\mathbf{T},\boldsymbol{\tau}\right)=i\sigma V\left(-\underline{\mathbf{H}}_{\mathbf{T}}^{-1}.\boldsymbol{\tau}\right)\left(e^{2\pi i\lambda^{-1}\left[\chi_{\mathbf{T}}\left(-\underline{\mathbf{H}}_{\mathbf{T}}^{-1}.\boldsymbol{\tau}\right)\right]} - e^{-2\pi i\lambda^{-1}\left[\chi_{\mathbf{T}}\left(\underline{\mathbf{H}}_{\mathbf{T}}^{-1}.\boldsymbol{\tau}\right)\right]}\right). \tag{57}$$

Since we can write a function as a sum of odd and even parts $\chi_{\mathbf{T}} := E_{\mathbf{T}} + O_{\mathbf{T}}$ such that $\chi_{\mathbf{T}}\left(-\mathbf{K}\right) = E_{\mathbf{T}}\left(\mathbf{K}\right) - O_{\mathbf{T}}\left(\mathbf{K}\right)$ we find

$$I(\mathbf{T},\boldsymbol{\tau}) = i\sigma V\left(-\underline{\mathbf{H}}_{\mathrm{T}}^{-1}.\boldsymbol{\tau}\right)\left(e^{2\pi i\lambda^{-1}\left[E_{\mathrm{T}}\left(\underline{\mathbf{H}}_{\mathrm{T}}^{-1}.\boldsymbol{\tau}\right)-O_{\mathrm{T}}\left(\underline{\mathbf{H}}_{\mathrm{T}}^{-1}.\boldsymbol{\tau}\right)\right]}\right.$$
$$\left.-e^{-2\pi i\lambda^{-1}\left[E_{\mathrm{T}}\left(\underline{\mathbf{H}}_{\mathrm{T}}^{-1}.\boldsymbol{\tau}\right)+O_{\mathrm{T}}\left(\underline{\mathbf{H}}_{\mathrm{T}}^{-1}.\boldsymbol{\tau}\right)\right]}\right). \tag{58}$$

To finally give

$$I(\mathbf{T},\boldsymbol{\tau}) \approx 2\sigma V\left(-\underline{\mathbf{H}}_{\mathrm{T}}^{-1}\boldsymbol{\tau}\right)\sin\left[2\pi\lambda^{-1}\chi_{\mathrm{T}}^{even}\left(-\underline{\mathbf{H}}_{\mathrm{T}}^{-1}\boldsymbol{\tau}\right)\right].e^{-2\pi i\lambda^{-1}\chi_{T}^{odd}\left(-\underline{\mathbf{H}}_{\mathrm{T}}^{-1}\boldsymbol{\tau}\right)}. \tag{59}$$

Thus we arrive at almost exactly the same contrast transfer function as for tilted BF images. The (simulated) CTF in Figure 3–5 closely resembles the usual BF CTF. The difference is that instead of spatial frequency, we have a distorted position magnified by the inverse matrix of the second derivatives. This result should not be surprising, since we have identified this matrix as corresponding to magnification in the geometrical optics approach. We can gain some further confidence in this result by again noting the similarity to Cowley's method to measure aberrations by comparing the magnification of the Ronchigram to that of a BF image of the same area taken at a known magnification (Lin and Cowley 1986a).

The importance of this result is that it shows that, in principle, we can measure all geometrical aberrations to arbitrary order from a single Ronchigram. In practice, this will rely on a suitable choice of conditions, since we will have to worry about limited coherence, sampling (many pixels might be required), and sources of noise.

We can simplify this equation by assuming that the aberration function is dominated by the first-order terms:

$$\chi_{\mathrm{T}}^{\mathrm{even}}(\boldsymbol{\tau}) \approx \frac{1}{2}\boldsymbol{\tau}\underline{\mathbf{H}}_{\mathrm{T}}\boldsymbol{\tau} + \ldots. \tag{60}$$

Thus

$$CTF_{\mathrm{Ronchi}}(\mathbf{T},\boldsymbol{\tau}) \approx \sin\left(\pi\lambda^{-1}\boldsymbol{\tau}\underline{\mathbf{H}}_{\mathrm{T}}^{-1}\boldsymbol{\tau}\right). \tag{61}$$

Applying the same approximation to the BF CTF (equation (44)) gives the CTF of a BF image with a tilted detector as

$$CTF_{\mathrm{BF}}(\mathbf{T},\boldsymbol{\rho}) \approx \sin\left(\pi\lambda^{-1}\boldsymbol{\rho}\underline{\mathbf{H}}_{\mathrm{T}}\boldsymbol{\rho}\right). \tag{62}$$

This last equation demonstrates the familiar result that the Fourier transform of a BF image will be a series of light and dark rings. The radii and shape of those rings will depend upon the effective aberrations. Thus a series of BF images recorded with different tilt angles will allow the aberration function to be determined via a Zemlin tableau (Zemlin et al. 1978). However, the important point for the present work is that we have derived almost exactly the same result for the CTF of the Ronchigram. An example plot is shown in Figure 3–5.

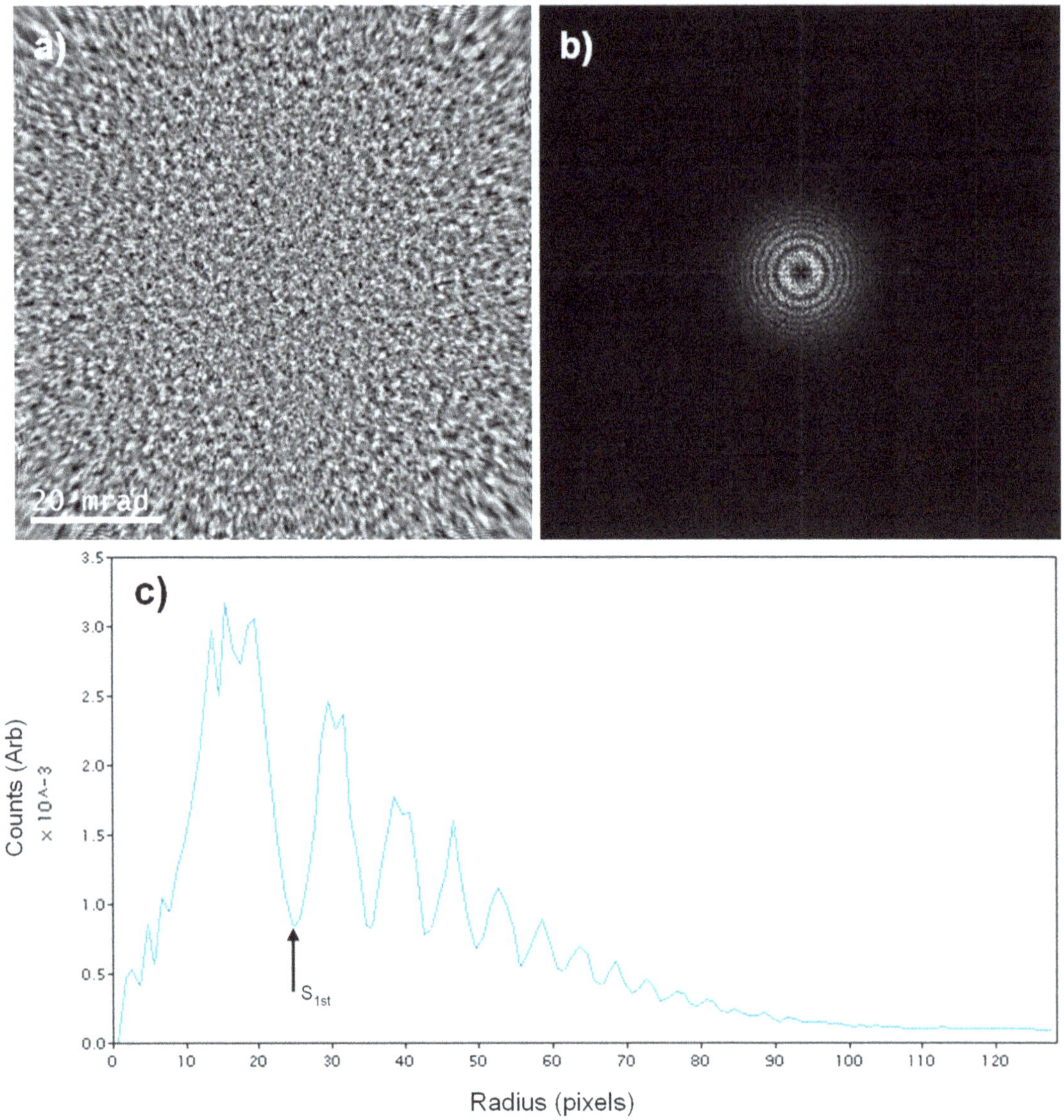

Figure 3–5. (**a**) Simulated Ronchigram for a defocus, C_{10}=–1000 nm and spherical aberration, C_{30}=0.1 mm and with other aberrations equal to 0 at 200 kV. Note how the magnification changes toward the corners of the image. (**b**) Fourier transform of the 512 × 512 pixel central patch showing a pattern of *light and dark rings*. (**c**) The radial profile of the Fourier transform in (b). The *arrow* indicates the "first zero" in the CTF oscillation. From Lupini et al. (2010).

Thus we see that a convenient route to measure aberrations is to fit the apparent defocus and astigmatism to a series of patches cut from the Ronchigram at different angles and again to fit the aberration function from its second derivatives. The only important difference between the BF and the Ronchigram description is the matrix inversion in the expression for the Ronchigram. However, the Ronchigram-based method offers a significant advantage in that we can cut many patches

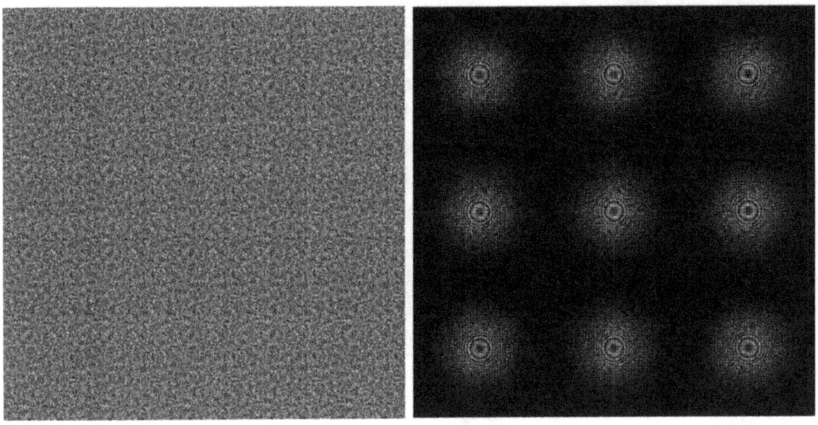

Figure 3–6. (*Left*) Amorphous Ronchigram simulation with half-angle 50 mrad, C_1=–1000 nm, 300 kV. (*Right*) The Fourier transforms of a series of patches cut from the Ronchigram at a corresponding angle. Each Fourier transform is essentially identical.

from a single Ronchigram, unlike the BF case, where multiple images are required.

Figure 3–6 shows an example simulation (using our simple thin-sample model) with no aberrations apart from defocus. Each patch gives a very similar diffractogram, since the matrix $\underline{\mathbf{H}}_\mathrm{T}$ is a constant over the field of view in this case. Figure 3–7 shows a simulation under very similar conditions, but with a small amount of spherical aberration, which causes the magnification and the matrix $\underline{\mathbf{H}}_\mathrm{T}$ to vary over the field of view. Thus the diffractograms of small patches show corresponding changes in apparent defocus and astigmatism, meaning that the rings in the diffractograms become elliptical.

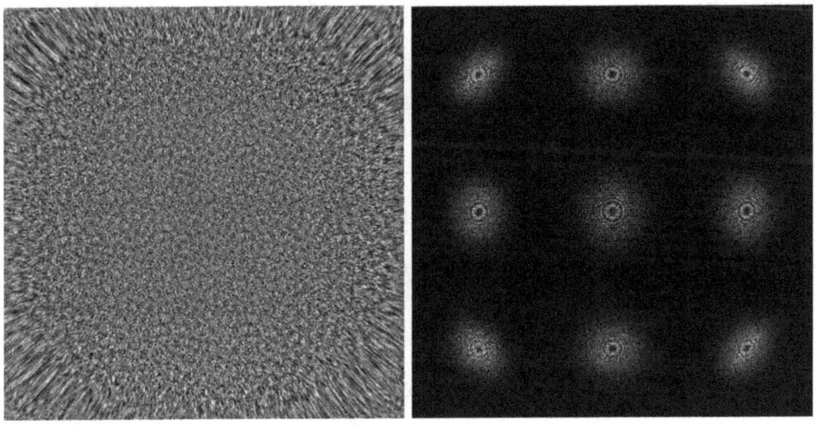

Figure 3–7. (*Left*) Amorphous Ronchigram simulation with half-angle 50 mrad, C_3=0.1 mm, C_1=–1000 nm, 300 kV. (*Right*) The Fourier transforms of a series of patches cut from the Ronchigram at a corresponding angle. Note that the off-axis Fourier transforms are distorted due to the spherical aberration.

Using the analytic description derived above, we have recently developed a method capable of fitting to these diffractograms and shown that it is capable of measuring aberrations (Lupini et al. 2010). The disadvantage of this method is that it can be difficult to select a patch that is large enough to provide enough features to give a good Fourier transform and yet is also small enough that the magnification is roughly constant over the patch. Therefore one of the main limitations was the difficulty of fitting to the Fourier transformed patches. We had to develop an iterative fitting method and include damping envelopes due to the limited coherence. This is interesting because it suggests that more accurate and faster methods to fit to these patterns could certainly be developed, and also because it is necessary to consider the effect of incoherence.

3.13 Coherence

There are two main source of incoherence that we will consider in the Ronchigram and further consideration can be found elsewhere (Cowley 1995, Cowley and Spence 1981, James and Rodenburg 1997, Nellist et al. 1995, Rodenburg 1988). Our approach here is a summary of Lupini et al. (2010).

Spatial coherence is limited because the source of electrons has a finite size. The gun and condenser optics are used to adjust the effective size that the source appears to be when projected to the specimen plane. This size will only be zero when there is zero current, but can be quite large when supplying enough current for microanalysis. Thus in practice these lenses are usually optimized so that the effective size is just a little smaller than the imaging resolution required. Note that we can also include broadening from instabilities or sample movement in this effective size. The resulting intensity I_s will be given by the integration of intensities over the effective source size. Applying the small scattering approximation and then the same delta function approximation used earlier gives the probe position dependence in the intensity as

$$e^{2\pi i\lambda^{-1}\left(\Delta\mathbf{K}_i-\Delta\mathbf{K}_j\right).\mathbf{R}_0} \text{ and thus } e^{2\pi i\lambda^{-1}\underline{\mathbf{H}}_{\mathrm{T}}^{-1}\cdot\boldsymbol{\tau}\cdot\mathbf{R}_0}. \tag{63}$$

Therefore, integration over a Gaussian distribution of intensities $e^{-\alpha\mathbf{R}_0^2}$ with $\alpha = 1/w^2$ and neglecting constants gives

$$I_S(\mathbf{T},\boldsymbol{\tau}) = \int I(\mathbf{T},\boldsymbol{\tau})\, e^{-\alpha\mathbf{R}_0^2} e^{2\pi i\lambda^{-1}\underline{\mathbf{H}}_{\mathrm{T}}^{-1}\cdot\boldsymbol{\tau}\cdot\mathbf{R}_0} d\mathbf{R}_0, \tag{64}$$

where we assumed a $1/e$ width for the distribution of w. Using the standard integral

$$\int e^{-\alpha x^2} e^{-2\pi i x\xi}\, dx = \sqrt{\frac{\pi}{\alpha}} e^{-\frac{(\pi\xi)^2}{\alpha}}, \tag{65}$$

gives a damping envelope,

$$\propto e^{-\left(w\pi\lambda^{-1}\underline{\mathbf{H}}_{\mathrm{T}}^{-1}\cdot\boldsymbol{\tau}\right)^2}. \tag{66}$$

Thus we find that a large effective source size will effectively damp contrast transfer into the Ronchigram in a very similar way to the damping in BF images. The difference is that damping here also depends on the magnification matrix.

This result suggests that we can measure the effective source size when we perform the aberration measurement. As well as fitting to the ring positions, we can fit to the damping envelope. Of course we would need to worry about other damping and sample effects to be sure that we are measuring a property of the source.

Alternatively, we note that the damping envelopes could be used to determine that aberration function directly if the effective source size is calibrated, thus providing an alternative method to measure the aberration function. The calibration could be obtained by taking measurements at a known condition where the matrix is dominated by defocus. Sawada et al. have recently demonstrated a method where the autocorrelation function for a series of patches is measured and the aberration function fitted (Sawada et al. 2008). Those authors attribute the width of the autocorrelation function (ACF) to the effective source size and the aberrations. We can see that work is consistent with our present derivation by noting that there will be an inverse relationship between the width of the ACF and the CTF damping envelope.

Additionally, the electrons emitted from the source will have different energies and, since the lenses suffer from chromatic aberration, this will cause a change in focus, which can be included as a temporal coherence damping envelope. We include the focus change in a similar manner to the method used for the source size by integrating over the intensities from a range of focus values.

We integrate the last term in equation (55) over a focus change z using a Gaussian distribution with $\alpha=1/\Delta^2$, again neglecting constants to give

$$I_{\Delta f}(\mathbf{T},\boldsymbol{\tau}) = \int I(\mathbf{T},\boldsymbol{\tau})\, e^{2\pi i \frac{z}{2\lambda}\left(2\mathbf{T}\cdot\underline{\mathbf{H}}_{\mathrm{T}}^{-1}\cdot\boldsymbol{\tau}+\left(\underline{\mathbf{H}}_{\mathrm{T}}^{-1}\cdot\boldsymbol{\tau}\right)^2\right)} e^{-\alpha z^2} dz. \tag{67}$$

We use the same standard integral to arrive at a damping envelope of the form,

$$\propto e^{-\left(\left(2\mathbf{T}\cdot\mathbf{H}_{\mathrm{T}}^{-1}\cdot\boldsymbol{\tau}+\left(\mathbf{H}_{\mathrm{T}}^{-1}\cdot\boldsymbol{\tau}\right)^2\right)\frac{\pi\Delta}{2\lambda}\right)^2}, \tag{68}$$

for a $1/e$ focus spread of Δ. This closely resembles previously derived damping envelopes, e.g., Nellist and Rodenburg (1994). However, using the same process for the first term in equation (55) gives a similar, but slightly different, term,

$$\propto e^{-\left(\left(-2\mathbf{T}\cdot\mathbf{H}_{\mathrm{T}}^{-1}\cdot\boldsymbol{\tau}+\left(\mathbf{H}_{\mathrm{T}}^{-1}\cdot\boldsymbol{\tau}\right)^2\right)\frac{\pi\Delta}{2\lambda}\right)^2}. \tag{69}$$

This result is therefore slightly more complicated than the more usual form of the on-axis damping envelope. However, we note that a closely related result has recently been derived (Bartel and Thust 2008) for off-axis BF contrast damping envelopes.

In many situations, it seems reasonable to ignore the antisymmetrical (odd) part of the damping to arrive at a symmetrical (even) damping envelope of the form,

$$\propto e^{-\left(\frac{\pi\Delta}{2\lambda}\right)^2\left(4\left(\mathbf{T}\cdot\mathbf{H}_\mathrm{T}^{-1}\cdot\boldsymbol{\tau}\right)^2+\left(\mathbf{H}_\mathrm{T}^{-1}\cdot\boldsymbol{\tau}\right)^4\right)}. \tag{70}$$

When we combine this result with the previous damping envelope, we see an interesting behavior. Near to the optic axis, this term will be less severe than the source-size effect, meaning that the assumption that the damping will be dominated by the source size is valid. However, off-axis the effect of chromatic aberration increases. This transition should occur when the product of the tilt angle multiplied by the focus spread is about the same as the effective source size: $|\mathbf{T}|\,\Delta \approx w$.

There is an interesting parallel to conventional imaging, where chromatic aberration can provide a significant resolution limit at the large aperture angles permitted by newly developed aberration-correctors. Thus it may be the chromatic aberration rather than the geometrical aberrations that limit the choice of aperture in a fifth-order corrected machine (Lupini et al. 2009). This limitation has spurred recent development of monochromators (Krivanek et al. 2009, Mitterbauer et al. 2003) and chromatic aberration-correctors (Kabius et al. 2009). Analysis of the off-axis damping envelope could potentially provide a useful diagnostic in such systems.

Experimentally we have seen that a large high voltage instability will cause blurring toward the edges of the Ronchigram (see Figures 3–8 and 3–9). Even for purely round aberrations, we also note that the chromatic term gives a direction-dependent damping envelope. Thus the intriguing possibility is that we can measure the aberrations from the rings

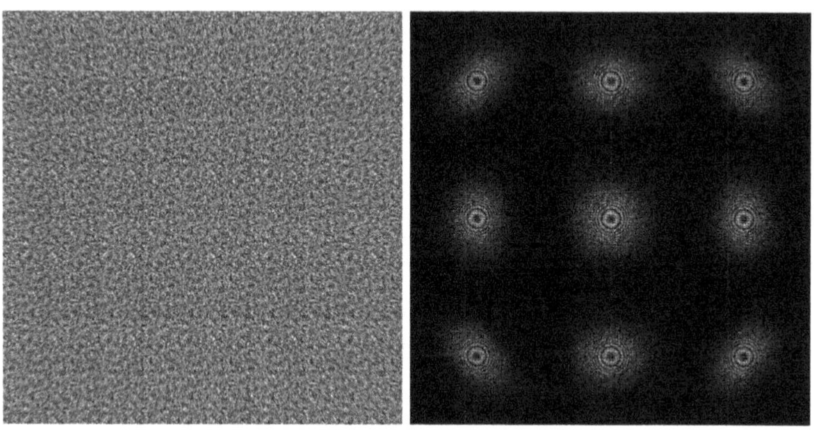

Figure 3–8. (*Left*) Amorphous Ronchigram simulation with half-angle 50 mrad, C_1=–1000 nm, 300 kV, 10 nm focal spread (uniformly weighted frames). (*Right*) The Fourier transforms of a series of patches cut from the Ronchigram at a corresponding angle. The effect of the contrast damping from the chromatic focus spread can be seen.

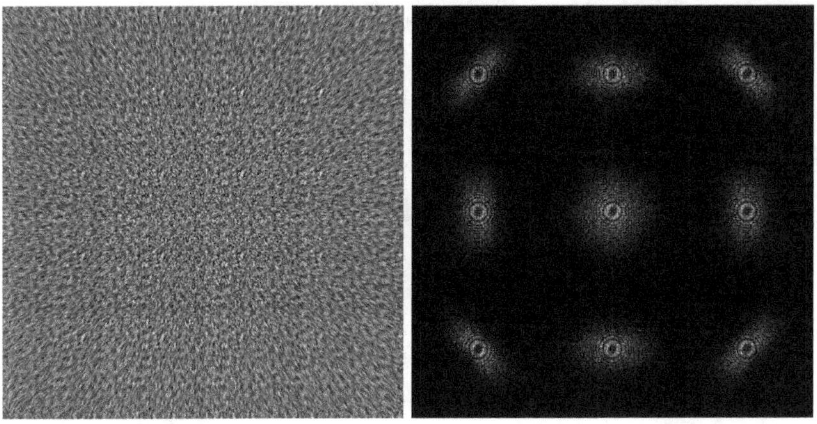

Figure 3–9. (*Left*) Amorphous Ronchigram simulation with half-angle 50 mrad, C_1=–1000 nm, 300 kV, 20 nm focal spread (uniformly weighted frames). (*Right*) The Fourier transforms of a series of patches cut from the Ronchigram at a corresponding angle. The effect of the contrast damping from the chromatic focus spread is clear.

in the CTF, fit an effective source size for the damping envelopes of patches near to the axis, and fit the chromatic aberration from patches at larger angles. Obviously the conditions to allow this measurement might have to be carefully chosen, but it is remarkable that so many of the important microscope parameters can be measured from a single image.

Note that we have used the 1/e widths, which will lead to differences in numerical factors when compared to derivations that use the standard deviation. It seems reasonable that we might consider potential damping envelopes caused by patch size or finite pixel size in a similar manner. Note that we have neglected sample effects (such as the atomic scattering factors) and are assuming a perfectly thin amorphous sample. We might expect a slightly different result for crystals, since we will see that we cannot make all of the same approximations in that case.

3.14 Crystalline Materials

The crucial difference when considering a crystalline sample is that there is long range ordering of the atoms. Thus the diffraction pattern from an infinite perfect crystal will consist of a series of delta function spots. The position of these spots will depend on the crystal lattice. Larger spacings in real space will give smaller spacings in this reciprocal space image, and vice-versa. Again we will be using a large aperture such that these spots are broadened into disks, which overlap and interfere. For a real sample, there will be features inside the disks due to dynamical effects (such as channeling as the beam propagates through a thick sample, Kikuchi lines, and so on), which we will neglect, so that we have featureless disks. In the interference regions, there will be a

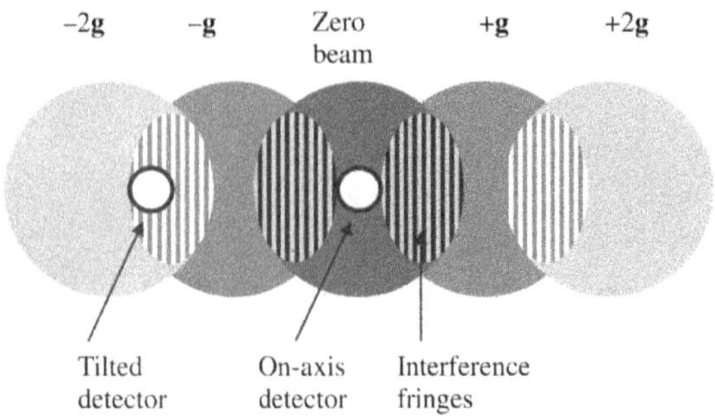

Figure 3–10. Schematic illustration of the Ronchigram from a 5-beam sample.

series of fine fringes. We will see that these are the cosine of the phase difference between the interfering beams, and that they depend upon the probe position and aberrations (Boothroyd 1997, Cowley and Lin 1985, Kuramochi et al. 2007, Lupini and Pennycook 2008, Nellist and Rodenburg 1994, Yamazaki et al. 2006).

In the description of imaging given by Cowley, we can imagine a small (bright-field) detector. For the case shown in Figure 3–10, if the detector was on-axis, it would not detect any interference and so the image would not reveal any lattice fringes as the probe is moved. If however, the detector is positioned such that it does see some fringes, then as the probe is moved the fringes move over the detector a varying signal is recorded. Thus we see that this simple model contains the diffraction limit – we could make the fringes overlap on-axis by using a larger aperture. This model is also a useful way to show how by tilting the detector off-axis such that it sits on the interference fringes, we would be able to record fringes that are not detectable with an on-axis detector, at the expense of a directional bias to the resolution. We can also see that a larger annular detector would be able to record this interference, suggesting that higher resolution might potentially be available in modes that use an annular detector (Cowley 1976).

This model also gives a simple model for coherence. A very small detector might be able to record very fine fringes faithfully, although it might be too small to collect a significant signal. If the detector size is increased, the detector would not be able to record single fringes and would average over several periods, which would reduce the contrast. If the detector was too large, then the fringe contrast would entirely average out. This effect produces a spatial coherence damping envelope. We can also include the effect of a finite source size. Since the fringe spacings depend on the position of the probe upon the sample, we can model a finite effective source size as a summation over several probe positions, again with a corresponding decrease in fringe contrast. In a similar way the defocus change due to chromatic aberration and the finite energy spread of the tip will generate a sum of fringes with slightly different spacings to provide a temporal coherence damping envelope.

Taking our example in Figure 3–10, we can imagine a case where increasing the objective aperture size to the point where the 0-beam interferes with the **2g** might result in a situation where we are unable to detect coherent interference. For our toy example of a perfectly thin sample, we would still like to know the phase difference between the 0 and 2**g** beams. However, in this example only coherent interference between the 0 and **g** beams and between the **g** and 2**g** beams is observable. Although a single point detector would be unable to resolve this spacing, we can imagine recording the whole Ronchigram as a function of probe position. We might be able to devise a method to determine the phase of the **g** beam relative to the 0 beam and the 2**g** beam relative to the **g** beam and so on. Combining this information allows us to determine the phase of the 2**g** beam relative to the 0 beam and could in a similar way be extended to all other beams. This model forms a very simple way to consider the super resolution methods, which have been further extended to more complicated structures and amorphous materials (Nellist and Rodenburg 1994, Rodenburg and Bates 1992, Rodenburg et al. 1993).

By way of contrast we note that the fine fringes that allow us to fit the aberration function also mean that the resulting pattern is sensitive to the exact position of the probe and rather complicated to interpret. A clever technique whereby the probe position is varied, causing these coherent effects to be averaged out, has recently been developed (Lebeau et al. 2010). The resulting position-averaged patterns can then be matched to simulations to accurately measure the sample thickness. The advantage is that it is not necessary to match the aberrations or incoherent damping envelopes. We note that this is almost the opposite approach to that taken here, where we ignored the details of the sample scattering function, and it therefore provides complementary information.

We now consider how to measure aberrations using a crystalline Ronchigram. Although we can still define a suitably sized patch that the small-patch approximation will hold, we cannot make the same assumption of small angle scattering that we were able to make for amorphous samples. Thus the result for a crystal will be closely related to, but different from, the amorphous result. In fact the final results are sufficiently similar that for measuring aberrations, this does not always produce a large numerical error (see the appendix in (Lupini and Pennycook 2008)) but the difference is conceptually important.

We assume that a crystal can be represented in reciprocal space as a series of delta functions at positions $\mathbf{g}_i$ with complex amplitudes a_i:

$$\varphi(\mathbf{K}) = \sum_i a_i \delta\left(\mathbf{K} - \mathbf{g}_i\right). \tag{71}$$

This is similar to the approach used by Cowley (Lin and Cowley 1986a). (Those authors used an amplitude *ia* instead, which shifts some cosine terms to sine, a change that can just be included in the phase). Thus a 2-beam sample with a unity amplitude beam at zero and another beam at **g** would be written as

$$\varphi(\mathbf{K}) = \delta(\mathbf{K}) + a\delta\left(\mathbf{K} - \mathbf{g}\right). \tag{72}$$

Substituting this sample into our exit wavefunction equation gives

$$\psi\left(\mathbf{K}_f,\mathbf{R}_0\right)=\int\left(\delta\left(\mathbf{K}\right)+a\delta\left(\mathbf{K}-\mathbf{g}\right)\right)e^{\alpha\chi\left(\mathbf{K}_f-\mathbf{K}_1\right)+\alpha\mathbf{R}_0.\left(\mathbf{K}_f-\mathbf{K}_1\right)}d\mathbf{K}_1. \quad (73)$$

So, as expected we see that the resulting pattern is our zero disk convolved with the diffraction spots. Performing the convolution integral gives

$$\psi\left(\mathbf{K}_f,\mathbf{R}_0\right)=e^{\alpha\chi\left(\mathbf{K}_f\right)+\alpha\mathbf{R}_0.\mathbf{K}_f}+a_ie^{\alpha\chi\left(\mathbf{K}_f-\mathbf{g}\right)+\alpha\mathbf{R}_0.\left(\mathbf{K}_f-\mathbf{g}\right)}. \quad (74)$$

Thus the intensity becomes

$$I\left(\mathbf{K}_f,\mathbf{R}_0\right)=1+a^2+ae^{\alpha\chi\left(\mathbf{K}_f-\mathbf{g}\right)-\alpha\chi\left(\mathbf{K}_f\right)-\alpha\mathbf{R}_0.\mathbf{g}}+a^*e^{\alpha\chi\left(\mathbf{K}_f\right)-\alpha\chi\left(\mathbf{K}_f-\mathbf{g}\right)+\alpha\mathbf{R}_0.\mathbf{g}}, \quad (75)$$

which can be written as

$$I\left(\mathbf{K}_f,\mathbf{R}_0\right)=1+a^2+2a\cos\left(\alpha\chi\left(\mathbf{K}_f-\mathbf{g}\right)-\alpha\chi\left(\mathbf{K}_f\right)-\alpha\mathbf{R}_0.\mathbf{g}+\vartheta_{\mathbf{g}}\right). \quad (76)$$

Thus we see that in the overlap region we obtain cosine fringes that depend on the aberrations, the probe position, and the relative phase of the two beams $\vartheta_{\mathbf{g}}$. Figures 3–11 and 3–12 show example Ronchigram simulations using a thin sample model with delta function diffraction spots generated in the same way as the earlier amorphous Ronchigrams. As the defocus is changed, a clear difference in the fringe spacing can be seen. It might therefore be expected that if we can accurately measure these fringes, we will be able to determine the aberrations.

As an aside we note that when $\mathbf{g}^2-2\mathbf{K}_f.\mathbf{g}=0$ then this equation has no dependence upon round aberrations. We therefore see achromatic lines (Nellist and Rodenburg 1994) where there is no dependence upon chromatic aberration. (This should be compared to our earlier result for the chromatic aberration damping envelope for amorphous materials.)

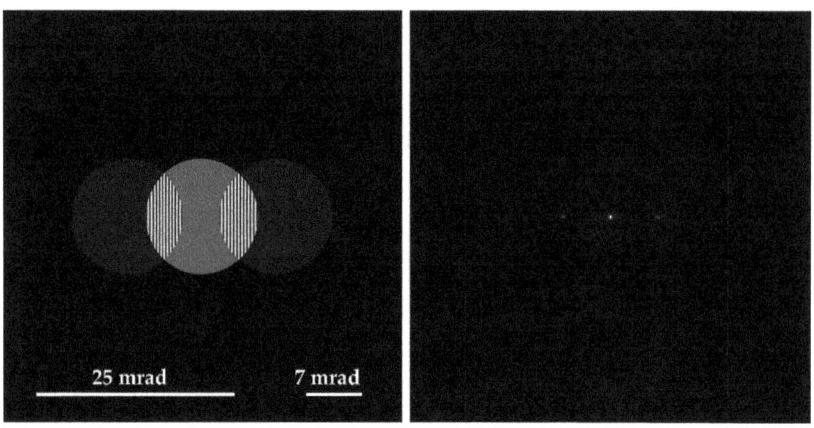

Figure 3–11. (*Left*) Crystalline Ronchigram simulation for a 3-beam model sample with half-angle 25 mrad, C_3=1 mm, 100 kV, C_1=–1000 nm, 7 mrad aperture half-angle (0.4 nm lattice spacing). (*Right*) The Fourier transform shows two delta functions.

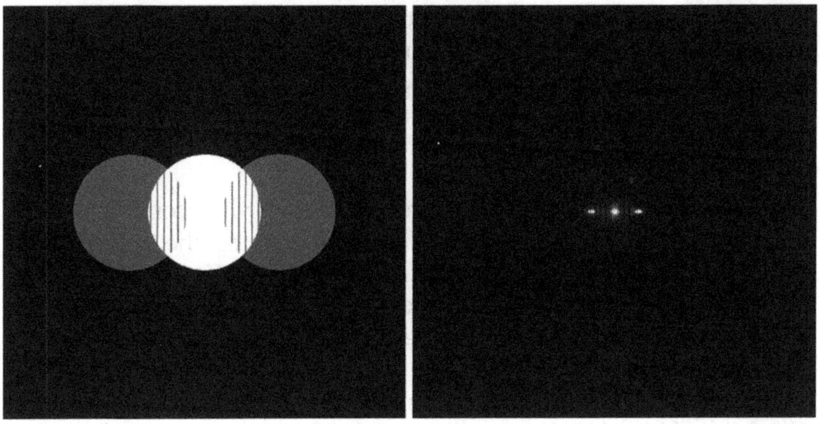

Figure 3–12. (*Left*) Crystalline Ronchigram simulation for a 3-beam model sample with half-angle 25 mrad, C_3=1 mm, 100 kV, C_1=−500 nm, 7 mrad aperture half-angle (0.4 nm lattice spacing). (*Right*) The Fourier transform shows two delta functions. Note that they have moved because of the focus change.

Ramasse and Bleloch use an ingenious method to measure aberrations that relies upon the fact that there is no dependence on round aberrations along loci that meet this criterion, and by changing the focus they are able to directly measure all non-round aberrations (Ramasse and Bleloch 2005).

In the following, we will neglect the dependence upon probe position and relative phases, since those just shift the pattern, while noting that this shifting again suggests the possibility to determine the relative phases by obtaining multiple Ronchigrams as the probe is moved. One further complication is that we have neglected multiple beams. We therefore consider the 3-beam case with an additional beam at –**g**:

$$\varphi(\mathbf{K}) = \delta(\mathbf{K}) + a\delta(\mathbf{K}-\mathbf{g}) + a\delta(\mathbf{K}+\mathbf{g}). \tag{77}$$

We neglect the phase difference $\vartheta_{\mathbf{g}}$, the probe position, and constant terms to derive

$$\begin{aligned} I(\mathbf{K}_f,\mathbf{R}_0) = {} & 2a\cos(\alpha\chi(\mathbf{K}_f-\mathbf{g}) - \alpha\chi(\mathbf{K}_f)) + 2a\cos(\alpha\chi(\mathbf{K}_f+\mathbf{g}) \\ & -\alpha\chi(\mathbf{K}_f)) + 2a^2\cos(\alpha\chi(\mathbf{K}_f+\mathbf{g}) - \alpha\chi(\mathbf{K}_f-\mathbf{g})). \end{aligned} \tag{78}$$

The term in a^2 gives a similar result to what we have just seen in the 2-beam case and is likely to be weaker in a thin crystal. More interestingly, we find that the first two terms are similar to each other but shifted. This means that we get interference regions where the 0 to +**g** terms cancel out the -**g** to 0 terms (Boothroyd 1997, Ishizuka et al. 2009, Lin and Cowley 1986a, Lupini and Pennycook 2008).

We can rearrange the first two terms as

$$\begin{aligned} I(\mathbf{K}_f,\mathbf{R}_0) = {} & 4a\cos(\pi\lambda^{-1}(\chi(\mathbf{K}_f+\mathbf{g}) - \chi(\mathbf{K}_f-\mathbf{g}))) \\ & \times\cos(\pi\lambda^{-1}(2\chi(\mathbf{K}_f) - \chi(\mathbf{K}_f-\mathbf{g}) - \chi(\mathbf{K}_f+\mathbf{g}))). \end{aligned} \tag{79}$$

Therefore the 3-beam case is essentially the same as the 2-beam case, but modulated by an extra term. This term gives loci of contrast reversals and it is possible to use those loci to measure spherical aberration (Cowley and Lin 1985, Lin and Cowley 1986a). The expressions for additional beams follow in a similar manner, so we will not add further beams.

We also note that the 3-beam interference further complicates the interpretation of the Fourier transform. Interfering the pattern after shifting by **g** causes interference fringes $\cos\left(2\pi\mathbf{g}.\mathbf{K}_f\right)$ in the Fourier transform (Boothroyd 1997, Ishizuka et al. 2009). It has been proposed that this cosine modulation can be used to determine the vectors $\mathbf{g}_i$ (Kuramochi et al. 2007).

How should we analyze these fringes? There are several possibilities (Cowley and Lin 1985, Kuramochi et al. 2007, Yamazaki et al. 2006). One particularly promising method, suggested by Boothroyd (Boothroyd 1997) is the Fourier transform, since the Fourier transform of a cosine function at a constant frequency gives a delta function. However, the complication here is that for non-zero aberrations the frequency of the fringes is not constant. Taking the Fourier transform reveals comet-like patterns. As pointed out by Boothroyd the angle of the comet tails is characteristic of the third-order spherical aberration (Boothroyd 1997). A mathematical derivation of this result was given by Ishizuka and we will return to this issue later (Ishizuka et al. 2009).

One solution to the difficulties caused by the non-constant fringe spacings is again to consider a small patch. We attempt to choose a small enough patch that the fringe spacing is roughly constant over the patch. As before we assume that a scattered vector $\mathbf{K}_f$, which can be arbitrarily large, is a small distance $d\mathbf{T}$ away from the center of a patch at **T**, which was deliberately chosen so that the distance $d\mathbf{T}$ is small enough that terms in $(d\mathbf{T})^2$ can be neglected. We can then approximate the difference between the two beams at **g** and **h** as

$$\begin{aligned}\chi\left(\mathbf{K}_f-\mathbf{h}\right)-\chi\left(\mathbf{K}_f-\mathbf{g}\right)\approx\ &\chi\left(\mathbf{T}-\mathbf{h}\right)-\chi\left(\mathbf{T}-\mathbf{g}\right)\\ &-d\mathbf{T}.\nabla\chi\left(\mathbf{T}-\mathbf{h}\right)-\chi\left(\mathbf{T}-\mathbf{g}\right).\end{aligned} \tag{80}$$

The advantage of this expansion is that it allows us to Fourier transform the intensity within a small patch over $d\mathbf{T}$. We then find that the coordinates of the delta functions (from transforming the cosine function) will be at

$$S_{\mathbf{T}}=\pm\nabla\left(\chi\left(\mathbf{T}-\mathbf{h}\right)-\chi\left(\mathbf{T}-\mathbf{g}\right)\right). \tag{81}$$

This result is the key to fitting the aberration function from a crystalline Ronchigram, because it relates the coordinates of the delta functions in the Fourier transform of a patch at a particular angle **T** to the lattice vectors and aberrations (see Lupini and Pennycook (2008) for further discussion and limitations).

We can see that there will be limits on the choice of patch, since it is necessary to have enough fringes that they can be adequately sampled with finite size detector pixels. In practice, no patch choice is perfect, and we find that even small patches will give comet-like Fourier

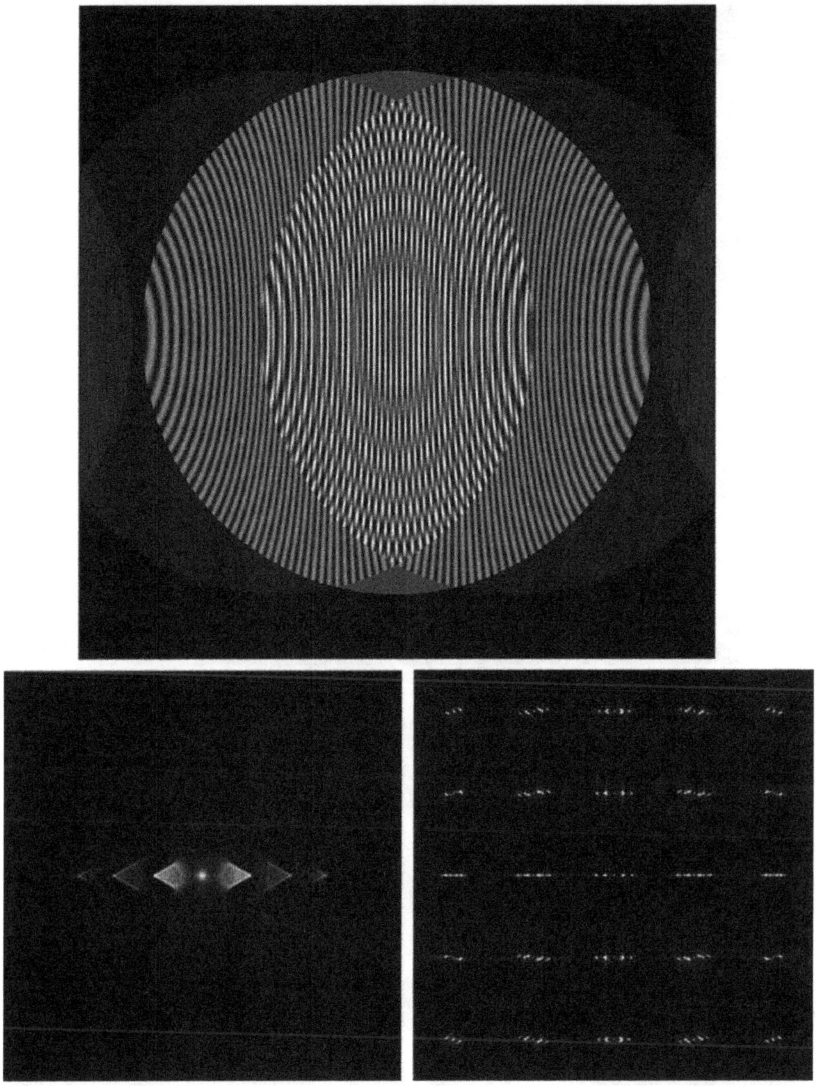

Figure 3–13. Crystalline Ronchigram simulation for a 3-beam model sample with half-angle 25 mrad, C_3=1 mm, 100 kV, C_1=−1000 nm, 20 mrad aperture half-angle (0.4 nm lattice spacing). Note that contrast reversals are visible in the fringes in the 3-beam interference region. (*Bottom left*) The Fourier transform shows comets with interference fringes visible. Note that the outermost spots are due to the periodicity of the Fourier transform. (*Bottom right*) An array of 5×5 Fourier transforms of patches cut from the Ronchigram at the corresponding positions. The Fourier transforms of small patches are nearly delta functions and the positions can be seen to change when cutting patches from different parts of the Ronchigram.

transforms. An algorithm that examines these should be designed such that it gives a reasonable estimate for the position of the head despite the 3-beam interference (which gives fringes in the Fourier transform) and the finite comet tails. Figures 3–13 and 3–14 show the division of the Ronchigram into small patches and the resulting change in

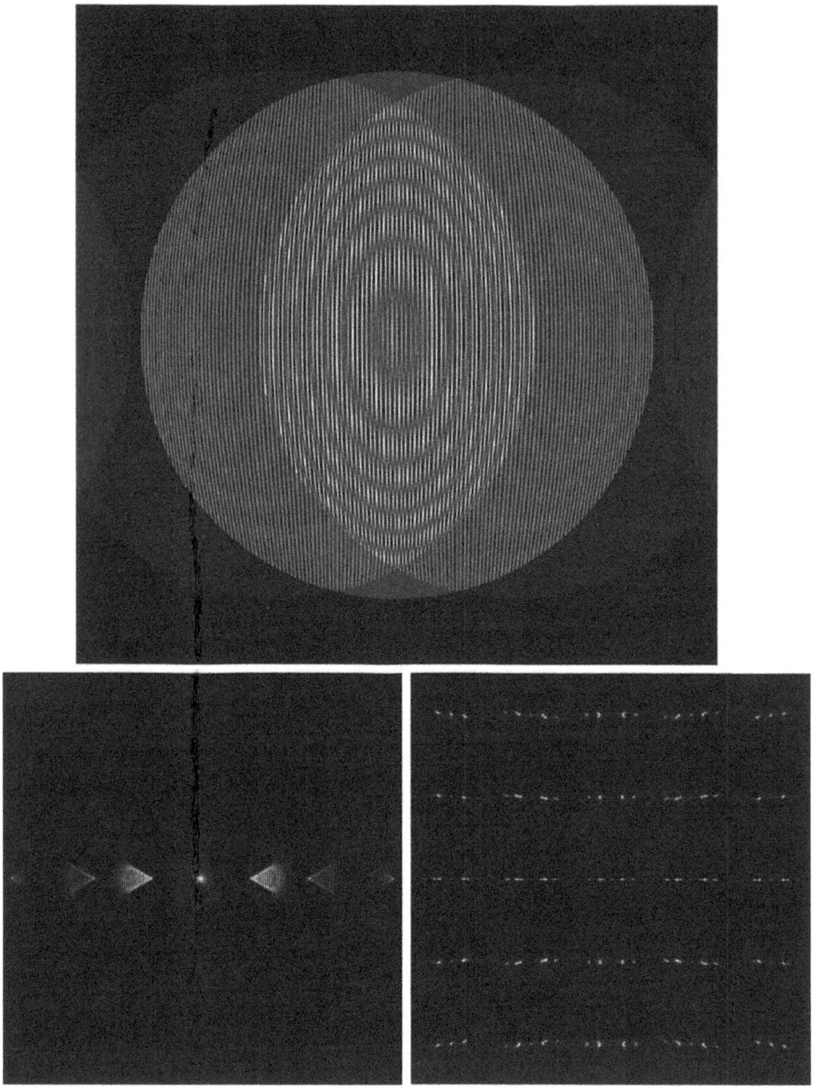

Figure 3–14. Crystalline Ronchigram simulation for a 3-beam model sample with the same parameters as Figure 3–13 except for the opposite sign of defocus. (*Bottom left*) The Fourier transform shows comets in the opposite direction. (Again the outermost spots are due to the periodicity of the Fourier transform.) (*Bottom right*) The Fourier transforms of small patches are again nearly delta functions and the positions again change when cutting patches from different parts of the Ronchigram.

spot positions due to the different effective aberrations within each patch.

A possible algorithm then proceeds as follows: A Ronchigram is divided into a series of patches and each patch is Fourier transformed. We locate the heads of the comets within each patch. If we assume that the vectors are known, then we can combine all of the measurements into a least-squares fit that provides the estimates of the aberration function from these equations. Obviously to sample all components of the function we will require multiple spots in different directions.

This method works well for simulations, but is a little more difficult in practice where the sample and its orientation might be unknown. How should we determine the **g**-vectors? Consider the spot position dependence upon defocus only:

$$\mathbf{S1}_{\mathbf{T}} = C_1\mathbf{g} + ..., \tag{82}$$

where we have neglected the dependence upon other terms. We can then change defocus by an amount dC_1 such that the spot moves to **S2** giving

$$\mathbf{S2}_{\mathbf{T}} = (C_1 + dC_1)\,\mathbf{g} + ... \tag{83}$$

If the focus change is pure, meaning that no other terms in the function were changed, then all those other terms will cancel out. Thus we can determine the vector from

$$\mathbf{g} = (\mathbf{S2}_{\mathbf{T}} - \mathbf{S1}_{\mathbf{T}})\,/dC_1. \tag{84}$$

In other words, although the spot position depends upon all of the aberrations, if a pure focus change is applied then the spot moves along a vector that depends on the reciprocal lattice vector **g**.

Thus we can determine the relevant vector for each comet being considered. In practice at least two (non-collinear) vectors are required to measure all aberrations, although more can be used. It is really just a book-keeping task to keep track of each spot separately and for the algorithm to estimate reasonable constraints that prevent it from losing track of the spots. The keys to getting this method to work were found to be devising a method that could locate the heads of the comets accurately but was reasonably tolerant of the 3-beam interference and comet tails. Even so, the defocus and patch size have to be chosen appropriately that there are enough fringes with roughly constant spacing to give a good approximation to a delta function. This method was shown to be accurate for simulated data and be able to work with experimental data (Lupini and Pennycook 2008), although the values of the aberrations had to lie within a limited range for the algorithm to be able to correctly identify suitable comets. It appears that the limitations in this method are due to the pattern matching. As acquisition devices and processing speeds improve, these limitations should be reduced. We also want to emphasize that the particular comet-finding routine used here was rather primitive; there is clearly room for superior algorithms to be developed.

Finally, we note that the wavelet transform seems ideally suited to this problem (Ramasse 2005).

3.15 Comets

Finally we can return to the question of the Fourier transforms of Ronchigrams from crystals look like "comets" with a 60 degree angle in the presence of C_S. Although defocus and astigmatism can change the comet position they do not change the tail angle (see Figure 3–15). Higher order aberrations can change the comet shapes (Figure 3–16). The comets due to spherical aberration were reported by (Boothroyd 1997) and the shape derived mathematically by (Ishizuka et al. 2009) for the Fourier transform of the whole pattern, but it is worth examining these in our small-patch approximation. If the fringe spacing were a constant, we would get a delta function. However, the magnification varies over the patch, meaning that we can imagine the superposition of a series of delta functions. (In reality the situation is a little more complicated, as they are not truly delta functions since our integrals or apertures do not run to infinity and effects add coherently.) In the case for C_S and defocus only, the delta functions will depend on the derivatives

$$\chi = \frac{1}{2}C_1\left(u^2+v^2\right)+\frac{1}{4}C_3\left(u^4+2u^2v^2+v^4\right), \tag{85}$$

$$\chi_u = C_1u + C_3\left(u^3+uv^2\right) \text{ and } \chi_v = C_1v + C_3\left(v^3+vu^2\right), \tag{86}$$

$$\chi_{uu} = C_1 + C_3\left(3u^2+v^2\right), \chi_{uv} = C_3 2uv, \text{ and } \chi_{vv} = C_1 + C_3\left(3v^2+u^2\right). \tag{87}$$

Interestingly, by using the second derivative approximation to equation (81) we would obtain the spot coordinates (x,y) for a unit g-vector (1,0) as

$$x = C_1 + C_3\left(3u^2+v^2\right) \text{ and } y = C_3 2uv. \tag{88}$$

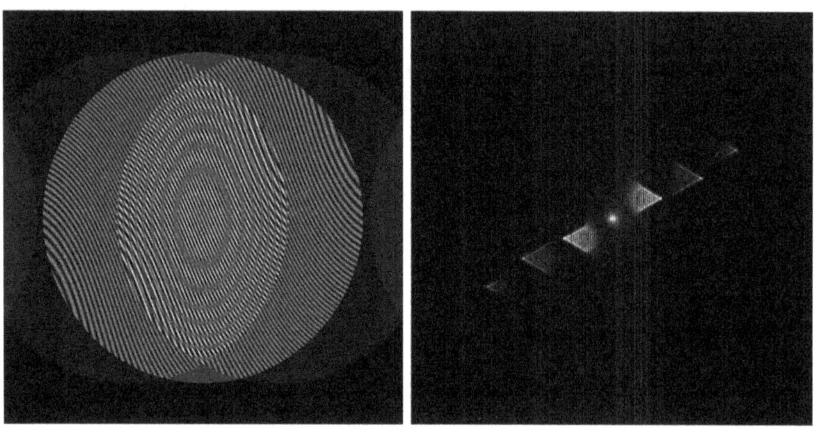

Figure 3–15. (*Left*) Crystalline Ronchigram simulation for a 3-beam model sample with half-angle 25 mrad, C_3=1 mm, 100 kV, C_1=1000 nm, C_{12b}=400 nm, 20 mrad aperture half-angle (0.4 nm lattice spacing). (*Right*) The Fourier transform. Note the shift in the comet positions.

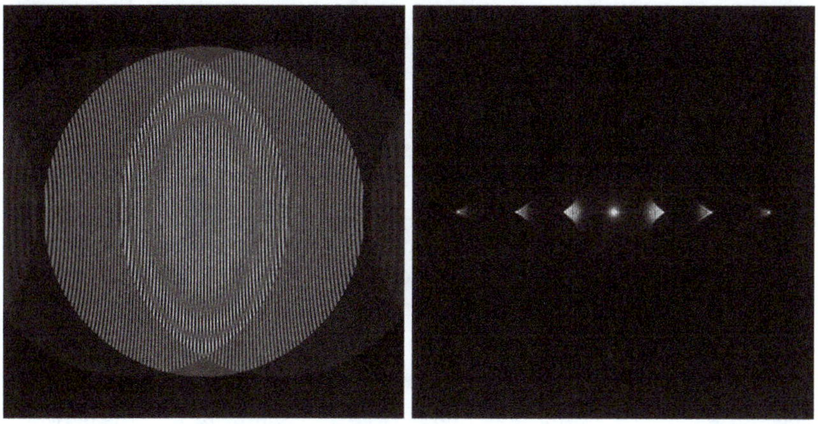

Figure 3–16. (*Left*) Crystalline Ronchigram simulation for a 3-beam model sample with half-angle 25 mrad, C_5=1 m, 100 kV, C_1=1000 nm, 20 mrad aperture half-angle (0.4 nm lattice spacing). (*Right*) The Fourier transform. Note that the comet tails are changed.

Plotting this equation as a function of both u and v gives comets with a 60 degree angle, precisely as described by (Boothroyd 1997) (Figure 3–17). Note that the second derivative approximation gives the correct shape in this case, but with an offset due to neglecting significant terms, meaning that we would need to use equation (81) to obtain the correct spot coordinates, which have been given by (Ishizuka et al. 2009). As shown in Figure 3–16, other aberrations can change the shape of the tails.

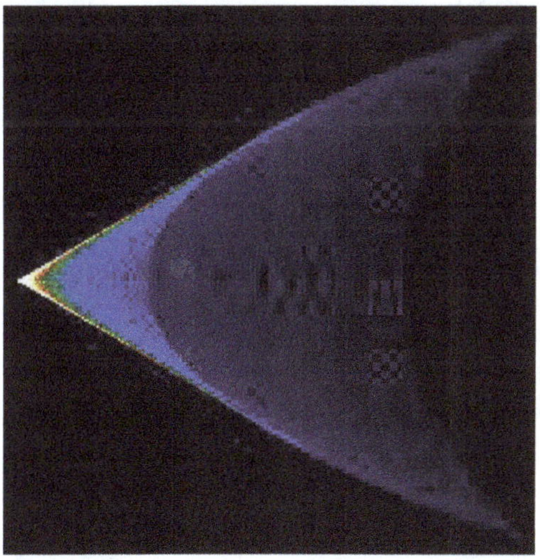

Figure 3–17. Illustration of the parametric function $x = C_1 + C_3\left(3u^2 + v^2\right)$ and $y = C_3 2uv$.

3.16 Rapid Methods to Measure Defocus in the Ronchigram

It is often useful to obtain a rapid measurement of focus in the Ronchigram and this also provides a method to calibrate the microscope focus, as well as a useful test of some of the preceding theory. Here we assume that the Ronchigram is dominated by defocus only and that the camera length has been accurately calibrated. For crystalline samples with a lattice spacing d we can rewrite our earlier result to find that the delta function spot should be at a radius S where

$$C_1 = Sd. \tag{89}$$

If the head of a comet is r_s pixels from the center of a Fourier transform with n pixels cut from a Ronchigram with c mrad/pixel, then

$$S = r_s/nc. \tag{90}$$

Meaning that

$$C_1 = r_s d/nc. \tag{91}$$

(See Figure 3–11, where $r_s \approx 123$ pixels, $d \approx 0.4$ nm, $n \approx 1024$ pixels, $c \approx 4.88\text{e}^{-5}$ rad/pixel.) We can compare this to the result for amorphous materials where r_a is the radius of the first zero of the CTF:

$$C_1 = \left(\frac{r_a}{nc}\right)^2 \lambda. \tag{92}$$

(See Figure 3–5 where $r_a \approx 25$ pixels, $n \approx 512$ pixels, $c \approx 7.81e^{-5}$ rad/pixel, $\lambda \approx 2.51\text{e}^{-3}$ nm.) Note that neither of these methods that examined the diffractogram were able to give the sign of focus, which must be determined by some other route (such as changing focus).

We can also compare to the probe-shifting method (Dellby et al. 2001) for a probe shift dR and an apparent shift dK (where dK is the number of pixels moved multiplied by the calibration c in rad/pixel)

$$C_1 = dR/dK. \tag{93}$$

One useful observation is that the amorphous method has a different dependence on the calibration than the other two methods used here. Since all three methods should measure the same defocus, this provides a route by which the calibrations can be verified. If the mrad/pixel calibration (which depends on the camera length) is not accurate we might expect rather large errors in the measured coefficients. Similarly the probe-shifting method depends upon accurately calibrated shifters.

Another potential use for a rapid measurement of focus is that once this value is known, the equations for the circles of infinite magnification allow a fast estimate of higher order aberrations to be obtained. This rough measurement can often be useful in diagnosing unusual hardware problems.

Finally we note that post-sample aberrations are potentially significant. For a really accurate measurement of the pre-sample aberrations, we need to be able to quantify the effect of post-sample aberrations. The different dependence of the methods upon the post-sample calibration might provide one route to achieve this. Another route would be to obtain repeated measurements as a function of defocus. For example, at very large defocus values the effect of a small amount of pre-sample astigmatism should be negligible and thus the post-sample astigmatism should dominate.

3.17 Summary

Aberration-corrected STEM appears to be one of the leading techniques to investigate the structure of a wide variety of materials, as discussed in detail in other chapters. We recall that Gabor originally proposed the Ronchigram as a method to examine materials at high resolution by overcoming spherical aberration, incidentally inventing holography. It is therefore interesting to note that all aberration-corrected STEMs available at the moment use the Ronchigram at some stage in their alignment. (Although this is not necessarily a requirement, it does greatly assist the user.) While the measurement of aberrations might not be the ultimate goal, it is an important part of obtaining higher resolution images and better quality data from these instruments. There is thus considerable demand for methods that are able to measure the aberration function from the Ronchigram.

We have seen that a geometrical optics approach relates the local magnification in the Ronchigram to the second derivatives of the aberration function, which are most compactly expressed in matrix form. This form incidentally allows the inverse to be considered and allows us to perform least-squares fits rather easily. We have seen how a very promising route to analyze the Ronchigram is to divide it up into small patches that make some of the difficult expressions rather simpler to handle.

We derived an expression for the Ronchigrams of amorphous materials from wave-optical considerations and found that the model of a Ronchigram as a distorted bright-field image is surprisingly good. This model allows us to include and to measure some of the effects of limited coherence. Finally we were able to extend this treatment to Ronchigrams of crystalline materials. Although the appearance of these Ronchigrams is very different, we saw that the underlying description is closely related to the amorphous result, with the primary difference being that we could not make the same small angle scattering approximation.

It is perhaps somewhat disappointing that the methods used to measure aberrations in (S)TEM are still slower than the closely related techniques available in active or adaptive optics. However, we hope that the methods reviewed here might lead to faster measurements in future. In particular we note that both we and other researchers have derived methods that are capable of measuring all aberrations from a single Ronchigram. In principle at least, these measurements could be performed multiple times per second. We also note that the current trend in computer development seems to be toward more parallel processors, which are entirely compatible with our approach of dividing the Ronchigram up into small, nearly independent, patches.

It is also exciting how many parameters can be derived from a single Ronchigram. In this chapter, we have seen how all the coherent aberrations can be extracted and have begun to perform measurements of incoherence as have other workers (James and Rodenburg 1997, Dwyer et al. 2010). It is therefore notable that Ronchigrams also provide methods to extract super-resolution information about the sample (Nellist and Rodenburg 1994, Rodenburg and Bates 1992, Rodenburg et al. 1993) and that a position averaged Ronchigram allows the accurate measurement of sample thickness and polarity (Lebeau et al. 2010), which are areas that we have regretfully skipped over in the models here. It certainly appears that Gabor's choice of the name "hologram" to describe a type of image containing the "whole information" was appropriate.

Acknowledgments Work funded by the Materials Science and Engineering Division of the U.S. Department of Energy. This chapter is a summary of work that has been performed and discussed with several other workers over many years. Fruitful discussions with Drs. A.L. Bleloch, O.L. Krivanek, L.M. Brown, P.D. Nellist, J.M. Rodenburg, A.I. Kirkland, P. Wang, and of course S.J. Pennycook, as well as others are gratefully acknowledged.

References

J. Bartel, A. Thust, Quantification of the information limit of transmission electron microscopes. Phys. Rev. Lett. **101**, 200801 (2008)

C.B. Boothroyd, Quantification of energy filtered lattice images and coherent convergent beam patterns. Scan. Microsc. **11**, 31–42 (1997)

M. Born, E. Wolf, *Principles of Optics* (Pergamon Press, London, 1959)

J.M. Cowley, Image contrast in a transmission scanning electron microscope. Appl. Phys. Lett. **15**, 58–59 (1969)

J.M. Cowley, Scanning transmission electron microscopy of thin specimens. Ultramicroscopy **2**, 3–16 (1976)

J.M. Cowley, Adjustment of a STEM instrument by use of shadow images. Ultramicroscopy **4**, 413–418 (1979)

J.M. Cowley, Chromatic coherence and inelastic scattering in electron holography. Ultramicroscopy **57**, 327–331 (1995)

J.M. Cowley, M.M. Disko, Fresnel diffraction in a coherent convergent electron beam. Ultramicroscopy **5**, 469–477 (1980)

J.M. Cowley, J.A. Lin, Calibration of the operating parameters for an HB5 STEM instrument. Ultramicroscopy **19**, 31–42 (1985)

J.M. Cowley, J.A. Lin, Reconstruction from in-line electron holograms by digital processing. Ultramicroscopy **19**, 179–190 (1986)

J.M. Cowley, J.C.H. Spence, Convergent beam electron micro-diffraction from small crystals. Ultramicroscopy **6**, 359–366 (1981)

A.F. de Jong, A.J. Koster, Measurement of electron-optical parameters for high-resolution electron microscopy image interpretation. Scan. Microsc. Suppl **6**, 95–103 (1992)

N. Dellby, O.L. Krivanek, P.D. Nellist, P.E. Batson, A.R. Lupini, Progress in aberration-corrected scanning transmission electron microscopy. Microsc. Microanal. **50**, 177–185 (2001)

C. Dwyer, R. Erni, J. Etheridge, Measurement of effective source distribution and its importance for quantitative interpretation of STEM images. Ultramicroscopy, **110**, 952–957 (2010)

D. Gabor, A new microscopic principle. Nature **161**, 777–778 (1948)

D. Gabor (ed.), *Nobel Lectures* (World Scientific Publishing, Singapore, 1992)

T. Hanai, M. Hibino, S. Maruse, Measurement of axial geometrical aberrations of the probe-forming lens by means of the shadow image of fine particles. Ultramicroscopy **20**, 329–336 (1986)

P.W. Hawkes, E. Kasper, *Principles of Electron Optics* (Academic Press, New York, 1989)

K. Ishizuka, Coma-free alignment of a high-resolution electron microscope with three-fold astigmatism. Ultramicroscopy **55**, 407–418 (1994)

K. Ishizuka, K. Kimoto, Y. Bando, Fourier analysis of Ronchigram and aberration assessment. Microscopy and Microanalysis **15**, 1094–1095 (Cambridge University Press, 2009)

E.M. James, J.M. Rodenburg, A method for measuring the effective source coherence in a field emission transmission electron microscope. Appl. Surface Sci. **111**, 174–179 (1997)

E.M. James, N.D. Browning, Practical aspects of atomic resolution imaging and analysis in STEM. Ultramicroscopy **78**, 125–139 (1999)

B. Kabius, P. Hartel, M. Haider, H. Muller, S. Uhlemann, U. Loebau, J. Zach, H. Rose, First application of Cc-corrected imaging for high-resolution and energy-filtered TEM. J. Electron. Microsc. (Tokyo) **58**, 147–155 (2009)

A.I. Kirkland, R.R. Meyer, L.-Y. Chang, Local measurement and computational refinement of aberrations for HRTEM. Microsc. Microanal. **12**, 461–468 (2006)

O.L. Krivanek, A method for determining the coefficient of spherical aberration from a single electron micrograph. Optik **1**, 97–101 (1976)

O.L. Krivanek, Three-fold astigmatism in high-resolution transmission electron microscopy. Ultramicroscopy **55**, 419–433 (1994)

O.L. Krivanek, N. Dellby, R.J. Keyse, M.F. Murfitt, C.S. Own, Z.S. Szilagyi, in *Aberration-Corrected Electron Microscopy*, ed. by P.W. Hawkes (Academic Press, New York, 2008), pp. 121–160

O.L. Krivanek, N. Dellby, A.R. Lupini, Towards sub-angstrom electron beams. Ultramicroscopy **78**, 1–11 (1999)

O.L. Krivanek, G.Y. Fan, Application of slow-scan charge-coupled device (CCD) cameras to on-line microscope control. Scan. Microsc. (Suppl. 6), 105–114 (1992)

O.L. Krivanek, J.P. Ursin, N.J. Bacon, G.C. Corbin, N. Dellby, P. Hrncirij, M.F. Murfitt, C.S. Own, Z.S. Szilagyi, High-energy-resolution monochromator for aberration-corrected scanning transmission electron microscopy/electron energy-loss spectroscopy. Phil. Trans. R. Soc. A **367**, 3683–3697 (2009)

K. Kuramochi, T. Yamazaki, Y. Kotaka, Y. Kikuchi, I. Hashimoto, K. Watanabe, Measurement of twofold astigmatism of probe-forming lens using low-order zone-azis Ronchigram. Ultramicroscopy **108**(4), 339–345 (2007)

J.M. Lebeau, S.D. Findlay, L.J.S. Allen, Position averaged convergent beam electron diffraction: Theory and applications. Ultramicroscopy **110**, 118–125 (2010)

J.A. Lin, J.M. Cowley, Calibration of the operating parameters for an HB5 STEM instrument. Ultramicroscopy **19**, 31–42 (1986a)

J.A. Lin, J.M. Cowley, Reconstruction from in-line electron holograms by digital processing. Ultramicroscopy **19**, 179–190 (1986b)

A.R. Lupini, *Aberration Correction in STEM*, PhD Thesis (Cambridge University, Cambridge, UK, 2001)
A.R. Lupini, A. Borisevich, J.C. Idrobo, H.M. Christen, M. Biegalski, S.J. Pennycook, Characterizing the two- and three-dimensional resolution of an improved aberration-corrected STEM. Microsc. Microanal. **15**, 441–453 (2009)
A.R. Lupini, S.J. Pennycook, Rapid autotuning for crystalline specimens from an inline hologram. J. Electron. Microsc. (Tokyo) **57**, 195–201 (2008)
A.R. Lupini, P. Wang, P.D. Nellist, A.I. Kirkland, S.J. Pennycook, Aberration measurement using the Ronchigram contrast transfer function. Ultramicroscopy **110**(7), 891–898 (2010)
R.R. Meyer, A.I. Kirkland, W.O. Saxton, A new method for the determination of the wave aberration function for high resolution TEM. Ultramicroscopy **92**, 89–109 (2002)
R.R. Meyer, A.I. Kirkland, W.O. Saxton, A new method for the determination of the wave aberration function for high-resolution TEM. Ultramicroscopy **99**, 115–123 (2004)
C. Mitterbauer, G. Kothleitner, W. Grogger, H. Zandbergen, B. Freitag, P. Tiemeijer, F. Hofer, Electron energy-loss near-edge structures of 3d transition metal oxides recorded at high-energy resolution. Ultramicroscopy **96**, 469–480 (2003)
P.D. Nellist, B.C. McCallum, J.M. Rodenburg, Resolution beyond the information limit in transmission electron-microscopy. Nature **374**, 630–632 (1995)
P.D. Nellist, J.M. Rodenburg, Beyond the conventional information limit: The relevant coherence function. Ultramicroscopy **54**, 61–74 (1994)
Q.M. Ramasse, *Diagnosis of Aberrations in Scanning Transmission Electron Microscopy*, PhD thesis (Cambridge University, Cambridge, UK, 2005)
Q.M. Ramasse, A.L. Bleloch, Diagnosis of aberrations from crystalline samples in scanning transmission electron microscopy. Ultramicroscopy **106**, 37–56 (2005)
J.M. Rodenburg, Properties of electron microdiffraction patterns from amorphous materials. Ultramicroscopy **25**, 329–344 (1988)
J.M. Rodenburg, The phase problem, microdiffraction and wavelength-limited resolution – a discussion. Ultramicroscopy **27**, 413–422 (1989)
J.M. Rodenburg, R.H.T. Bates, The theory of super-resolution electron microscopy via Wigner-distribution deconvolution. Philos. Trans.: Phys. Sci. Eng. **339**, 521–553 (1992)
J.M. Rodenburg, A.R. Lupini, Measuring lens parameters from coherent ronchigrams in STEM. Inst. Phys. Conf. Ser. No 161: Section 7. Proc. EMAG99, 339–342 (1997)
J.M. Rodenburg, B.C. McCallum, P.D. Nellist, Experimental tests on double-resolution coherent imaging via STEM. Ultramicroscopy **48**, 304–314 (1993)
J.M. Rodenburg, E.B. Macak, Optimising the Resolution of TEM/STEM with the Electron Ronchigram, *Microscopy and Analysis* **90**, 5–7 (2002)
V. Ronchi, Forty years of history of a grating interferometer. Appl. Opt. **3**, 437–451 (1964)
H. Sawada, T. Sannomiya, F. Hosokawa, T. Nakamichi, T. Kaneyama, T. Tomita, Y. Kondo, T. Tanaka, Y. OshimaY. Tanishiro, K. Takayanagi, Measurement method of aberration from Ronchigram by autocorrelation function. Ultramicroscopy **108**, 1467–1475 (2008)
W.O. Saxton, A new way of measuring microscope aberrations. Ultramicroscopy **81**, 41–45 (2000)
O. Scherzer, The theoretical resolution limit of the electron microscope. J. Appl. Phys. **20**, 20–29 (1949)

S. Uhlemann, M. Haider, Residual wave aberrations in the first spherical aberration corrected transmission electron microscope. Ultramicroscopy **72**, 109–119 (1998)

T. Yamazaki, Y. Kotaka, Y. Kikuchi, K. Watanabe, Precise measurement of third-order spherical aberration using low-order zone-axis Ronchigram. Ultramicroscopy **106**, 153–163 (2006)

E. Zeitler, M.G.R. Thomson, Scanning Transmission Electron Microscopy. Optik **31**, 258–280 (1970)

F. Zemlin, K. Weiss, P. Schiske, W. Kunath, K.H. Herrmann, Coma-free alignment of high resolution electron microscopes with the aid of optical diffractograms. Ultramicroscopy **3**, 49–60 (1978)

4

Spatially Resolved EELS: The Spectrum-Imaging Technique and Its Applications

Mathieu Kociak, Odile Stéphan, Michael G. Walls, Marcel Tencé and Christian Colliex

4.1 Introduction

The energy lost by a fast electron through its interaction with a sample gives a wide range of information on this sample. Electron energy-loss spectroscopy (EELS) is thus very useful for studying chemical, physical and even optical properties of materials. However, it realises its full potential when it is combined with the very high spatial resolution furnished by the fast electron in a transmission electron microscope (TEM). Two configurations exist: one in which the beam is fixed and broad, and in which filtered images are acquired, and another in which a focused beam scans a sample and an EELS spectrum is taken at each pixel of the scan. The latter technique, called spectrum imaging (SPIM) (Jeanguilaume and Colliex 1989), and experimentally demonstrated more than 15 years ago (Hunt and Williams 1991), is the object of this chapter. By allowing the parallel acquisition of the elastic and inelastic signals with a resolution now reaching the single atomic column level, it gives invaluable and unambiguous information about the chemical and physical properties of nano-objects in parallel with structural information. As the primary detected and optimised signal is a spectrum, this technique benefits from all the technical, analytical and theoretical background of spectroscopies and thus yields a deep understanding of the material properties down to the atomic scale.

This chapter contains successively

- A short summary of the excitations measured in EELS and how they can be interpreted in the context of spectrum imaging
- A description of the instrumentation and analysis techniques relevant to the spectrum-imaging technique
- A selection of applications using signals across the broad range of measured energy losses, typically between 1 and 1000 eV

S.J. Pennycook, P.D. Nellist (eds.), *Scanning Transmission Electron Microscopy*, DOI 10.1007/978-1-4419-7200-2_4, © Springer Science+Business Media, LLC 2011

4.2 EELS and the Datacube

4.2.1 Anatomy of an EELS Spectrum

In this section, we quickly review the characteristic features and typical physical signals present in an EELS spectrum that are of interest in this chapter. Comprehensive information can be found elsewhere in the present book, especially in Chapter 5.

An EELS spectrum is composed of three main parts, as shown in Figure 4–1. They are the zero-loss (ZL), the low-loss (LL) and the core-loss (CL) regions. Note the high dynamic range across the spectrum.

4.2.1.1 The Zero-Loss Peak

The ZL peak is the signature of all the electrons that have not been significantly inelastically scattered by the sample. It has therefore no measurable spectroscopic information. In addition, it usually has tails that swamp any signal in the lowest energy-loss region, which unfortunately corresponds to the near-infrared (NIR)/visible (vis) range of the optical spectrum and therefore contains valuable information. Hardware and software workarounds for these issues are presented in Section 4.3.2 and in Chapter 16. There are however two useful applications of the ZL peak, especially in the context of SPIM. The first is that its position provides the reference position of the zero energy for the rest of the spectrum, which is difficult to evaluate otherwise. The second is that the Neperian logarithm of the ratio of the integrated low-loss region to the ZL, often referred as the "t over λ ratio", is a good indication of the thickness *t* of the sample in units of the electron mean free path

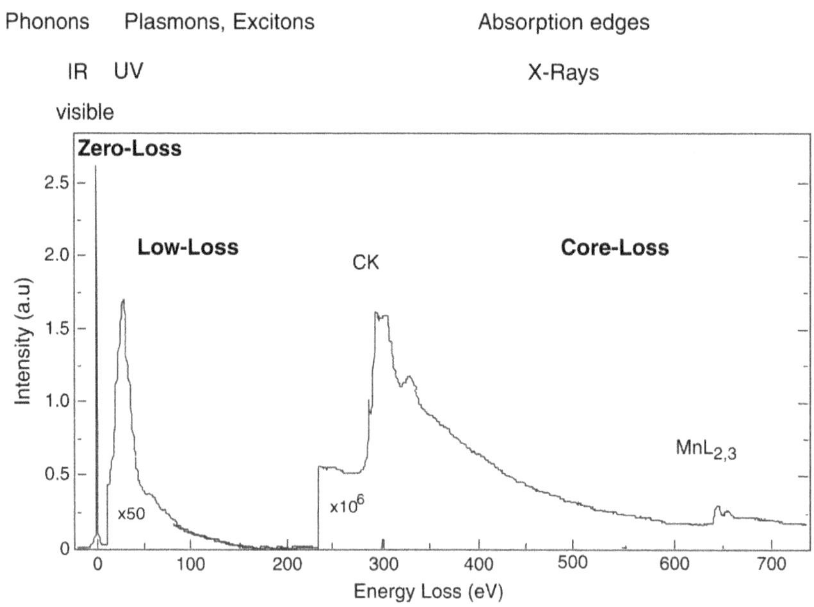

Figure 4–1. A typical EELS spectrum.

for inelastic scattering λ, at the position where it has been measured. Applications will be presented in Section 4.4.

4.2.1.2 *The Low-Loss Region*

The low-loss region has an intimate link with the optical properties of the sample. The reason for this can be intuitively understood in the following manner (Nelayah et al. 2007): the field generated by a fast electron is, to a good approximation, a Coulombic electric field pointing from the electron to the point of interest. Only when the electron is close to this point will the electric field be large. At that time, it will be polarised approximately perpendicularly to the electron trajectory. Also, because the electron is passing fast, the point of interest will feel an impulse of field or, in other words, a field containing a broad range of time frequencies. Therefore, any point in the sample feels an almost transversely polarised wave, containing many frequencies, i.e. a white source of light.

Due to this, the low-loss spectrum depends on the optical properties of the sample, which themselves usually depend on the dielectric function of the material(s), generally through intricate functions. Some electronic properties of the material can thus be recovered from the spectrum, since electronic transitions between the occupied and unoccupied states are involved. However, we have to bear in mind that the excitations measured are essentially optical, in the sense that the number of charge carriers is constant during the transition. Thus, care must be taken when comparing EELS results with one-particle experiments (scanning tunnelling spectroscopy, for example).

The information extracted from the low-loss spectrum is usually separated into three groups: bulk, surface (or interface) and the so-called begrenzung. All three components add up to form the measured EELS signal.

Bulk

The bulk response is that of a non-bounded medium and is, for low enough electron speeds, proportional to the so-called loss function $\mathrm{Im}(-1/\varepsilon)$ where ε is the dielectric function of the medium. In the case of electrons having a speed larger than the speed of light in the medium, Cherenkov emission may be triggered, and the loss function must be modified (see Chapter 16). The real (ε_1) and imaginary (ε_2) parts of ε can be retrieved from this function, taking care to remove surface and relativistic effects.

It is worth emphasising that in the ideal case of a dielectric function of vanishingly small imaginary part, the bulk loss function goes to infinity when

$$\mathrm{Re}(\varepsilon) = 0. \tag{1}$$

This corresponds to the case where an arbitrarily large electric field $\mathbf{E} = \mathbf{D}/\varepsilon$ can exist even with a vanishingly small displacement field $\mathbf{D}$. This situation corresponds to the collective motion of electrons, i.e.

plasmon waves. This makes the bulk loss function highly sensitive to volume plasmons.

Far from the plasmon resonance, the loss function can be very roughly approximated to $\mathrm{Im}(-1/\varepsilon) = \frac{\varepsilon_2}{\varepsilon_2^2+\varepsilon_1^2} \approx \varepsilon_2$ and thus reflects gaps and interband transitions in the material. But the exact energy position and relative intensities may differ from those found in the imaginary part of the dielectric function, because ε_1 can vary in the low-loss range.

A typical bulk spectrum (see Figure 4–2) exhibits different kinds of excitations: optical gap transitions, bulk plasmon(s), semi-core losses[1] and other, less well-defined excitations. A Kramers–Kronig analysis retrieves the real and imaginary parts of the dielectric function, which may give more insight into the underlying physics of the material of interest.

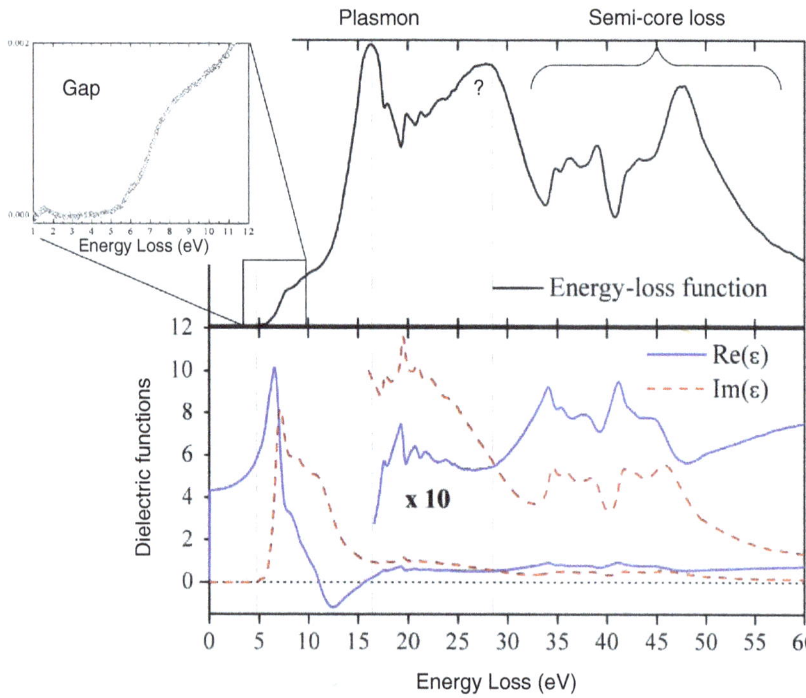

Figure 4–2. A low-loss spectrum (HfO_2) exhibiting different kinds of excitations. Note the question mark for an excitation whose character is uncertain. Adapted from Couillard et al. (2007). Courtesy of M. Couillard.

[1] Although the terminology is not consistent in the literature, we will call interband transitions arising in the 13.5–100 eV range "semi-core losses". The distinction with the core losses is blurred somewhat, but one practical difference is that the real part of ε is very close to unity for core losses and may depart from it significantly for semi-core losses.

Boundary Effects

Both surface and begrenzung effects are boundary effects, but they arise from slightly different physical reasons. The surface term is the signature of extra excitations arising from the presence of surfaces, while the begrenzung terms arise from the screening, in the bulk, of the fields generated by the surface excitations. For the sake of simplicity, relativistic effects involving surfaces (transition radiation, surface Cerenkov) will not be detailed here. Their effect will be briefly commented in Section 4.4.1, and the interested reader might want to read a comprehensive review on the subject (De Abajo et al. 2004).

Surface and begrenzung effects are position dependent, decreasing quickly as a function of distance from the surfaces/interfaces under consideration. The decrease is usually roughly close to an exponential

$$\approx \exp(-\omega b/v), \tag{2}$$

with ω, b and v being, respectively, the excitation energy, the distance to the interface and the speed of the electron.

Begrenzung

The begrenzung effect has the same functional form as the bulk loss function (i.e. $\mathrm{Im}(1/\varepsilon(\omega))$), but with a negative sign and a position-dependent pre-factor (vanishing at large distances from the interface(s)). This means that the apparent (bulk plus begrenzung) bulk signal will be weakened close to the interface. It is worth noting that the begrenzung effect is not restricted to plasmons, but affects the measurement of all kinds of excitations present in the bulk spectrum (plasmons, gap, semi-core losses and even Cerenkov excitations) (Couillard et al. 2007). Not taking it into account may lead to serious quantification errors in the case of semi-core losses (Taverna et al. 2008).

Surface

The expression for the surface loss function is highly dependent on the sample geometry. Unlike in the case of the bulk signal, which is always analytically known, the expression for the surface loss is only known in a few cases, i.e. cases that can be reduced to a 1D, highly symmetric case (spheres, interfaces, cylinders and multilayered structures of similar symmetries, see for example (Bolton and Chen 1995b, Lucas and Vigneron 1984, Moreau et al. 1997, Taverna et al. 2002, Zabala et al. 1997, 1989)). The extraction of electronic information from surface spectra is thus usually very difficult. However, any excitation available in the bulk has a surface counterpart. Indeed, discarding for simplicity the case of relativistic modes and limiting our explanation to the case of an arbitrarily shaped object in vacuum, one can show that the loss function accepts the following modal decomposition (Garcia et al. 1997, Aizpurua et al. 1999):

$$P(\omega) \approx \sum_i A_i \mathrm{Im}(g_i), \tag{3}$$

where i is an integer indexing the excited modes, A_i is a weighting prefactor (dependent on the probe position) and g_i is the contribution to the EELS of the ith mode. g_i reads

$$g_i = \frac{2}{\varepsilon(1+\lambda_i) + (1-\lambda_i)}, \tag{4}$$

with λ_i a real number depending solely on the shape of the interface between the two media. The exact value of the λ_i is usually not known except in the cases mentioned above (as an example, in the case of a semi-infinite plane in vacuum, $\lambda_i = 0$ for all i). However, the knowledge of their exact values is not necessary. When $\varepsilon = \frac{\lambda_i - 1}{1+\lambda_i}$, which is the surface counterpart of Equation (1), the surface function diverges, corresponding to the existence of a surface plasmon. The existence of surface plasmons is directly dependent on the existence of a volume plasmon, as condition (4) diverges for a certain energy only if $\mathrm{Re}(\varepsilon) = 0$ for some other energy. It is worth noting that the surface plasmon energies are very different from those of the bulk plasmons.

However, surface excitations are not limited to surface plasmons. Indeed, any other type of surface excitation, related to a bulk excitation (exciton, band gap transitions) may be measured. This can be directly seen by noting that far from a pole of g_i

$$\mathrm{Im}(g_i) \approx \varepsilon_2. \tag{5}$$

In this case, the energies of the surface excitations are very close to those of the bulk. This can lead to significant errors in the quantification of semi-core loss signals (Taverna et al. 2008).

A more advanced description of spatially resolved low-loss EELS has been given (Garcia de Abajo and Kociak 2008), as described in Section 4.4.1.

4.2.1.3 *The Core Loss*

In the core-loss region (typically beyond 100 eV), the loss function is still $\propto \mathrm{Im}(-1/\varepsilon)$. However, two main differences arise with respect to the LL case:

- As the boundary effects exponentially decrease with increasing energies, the core-loss signals are virtually independent of them (except in the case of atomically resolved EELS, see Section 4.4.2).
- In this range $\varepsilon_1 \approx 1 \gg \varepsilon_2$, $\mathrm{Im}(-1/\varepsilon) = \varepsilon_2$ to a very good approximation. The core-loss signal can thus be directly interpreted in terms of electronic transitions.

ε_2 is a measure of the electronic transitions. The transitions arise between many-body states of the object of interest. However, the core states being usually well-defined as belonging to given atoms, the transitions can be classified following the starting core level, as for atomic physics.

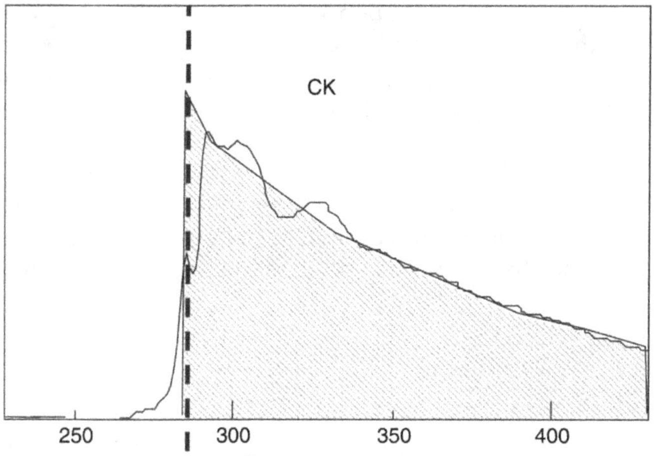

Figure 4–3. Carbon *K*-edge. The relevant information that can be extracted from an EELS spectrum is exemplified here. The onset of the edge is a good indication of the chemical species under consideration. The area under the edge, compared to known cross-sections (schematised as a *shaded area*), gives the concentration of the given element. Finally, the fine structures of the edge give information on the electronic structure in the material.

Much effort has been devoted to computing the cross-sections for such transitions, and we refer the reader to Egerton (1986) and Chapter 5 for more details on the subject. We will instead concentrate on the basic but important information that can be extracted from such spectra: elemental identification, concentration measurements and electronic structure. This is exemplified in the case of the carbon *K*-edge in Figure 4–3.

Elemental Analysis

As a crude approximation, the transitions appear as edges in the spectra, the energy of which is close to the ionisation energy and thus an excellent signature of the chemical species' identity. The exact value of the transition energy is however highly dependent on the environment in the solid or, in other words, on the exact electronic structure, leading to shifts which can be as large as a few electron volts.

Quantification

The area under the edge is proportional to the number of analysed atoms of the given chemical species per unit area. Knowing the geometrical conditions during the experiments (in particular the investigated volume), and the cross-sections for the elements of interest, the absolute number of atoms can in principle be determined. However, in many cases, only the relative concentrations between species can be retrieved, as the exact thickness is usually difficult to determine, especially for nano-structured materials.

Even in this case, the quantification is not simple. Obtaining good cross-sections, either through simulation or standards, may indeed not be straightforward. First, the presence of fine structures (see below) makes it difficult to precisely model (or to obtain calibrated samples) for each chemical species in every configuration. Second, for cross-section calculations, different models of increasing complexity exist, going from simple hydrogenic models to ab initio, multiple scattering or multiplet calculations. However, as the simulations improve in reproducing the fine structures, they become less and less relevant for quantification.

Electronic Structure

Finally, the first few electron volts after the edge contain a lot of information about the electronic structure, bonding and valence. First of all, the overall shape of the edge is a signature of the transition: for example, *K*-edges stemming from 1*s* states have the typical sawtooth profile, while $L_{2,3}$-edges have a delayed maximum but can contain intense narrow peaks at the onset, known as "white lines", corresponding to transitions to narrow *d* bands. See Chapter 5 for more details. However, even more information can be gained by inspection of the fine structures themselves. Indeed, it can be shown that the EELS signal close to an edge is proportional to the local density of electronic states projected onto the atom (Egerton 1986). As such, the analysis of this region, preferably including comparison with the relevant simulations, can yield information on the local bonding, valence state, charge transfer, crystal field, etc.

A more advanced description of the core-loss signal in the case of atomically resolved EELS will be summarised in Section 4.4.2 and described in detail in Chapter 6.

4.2.2 The Datacube

All this information is especially useful if it is obtained at the nanometre scale. There are basically two different methods of achieving this: either by rastering an electron probe on the sample of interest while measuring a spectrum at each point (spectrum imaging, SPIM) or by illuminating a large region and recording a set of filtered images (energy filtered transmission electron microscopy, EFTEM). However, they both aim at recording the same set of data of interest, the so-called datacube.

As shown in Figure 4–4, the datacube is a 3D matrix containing an intensity value at each point (x, y, E), where x and y are two spatial coordinates and E is an energy coordinate. A SPIM will be taken by filling the datacube at each spatial position (x, y) with a whole spectrum, while an EFTEM experiment will consist in acquiring the datacube by sampling the constant E planes. In contrast with the SPIM technique, where the accuracy of the datacube suffers mainly from spatial drift which can in principle be corrected thanks to the parallel acquisition

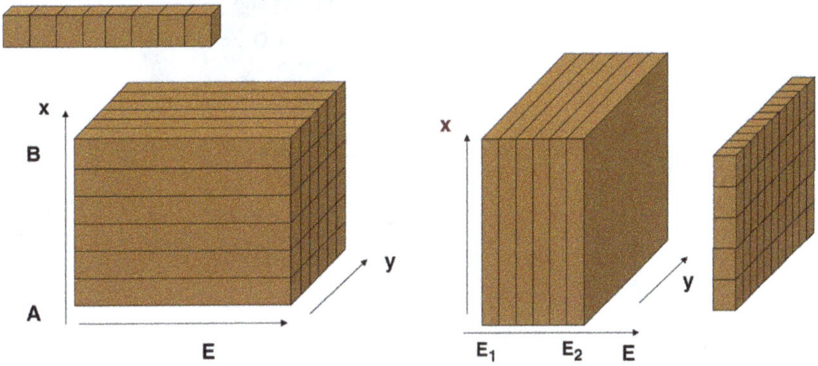

Figure 4–4. The datacube. *Left*: Acquired in the SPIM mode. *Right*: Acquired in the EFTEM mode. The primary data sets (spectra for SPIM and filtered images for EFTEM) are indicated.

of the high-angle annular dark-field (HAADF) image, the reconstruction of the datacube from image series requires much more effort. It is however possible, and we refer the interested reader to the relevant reference (Schaffer et al. 2006).

Finally, it is worth noting that in many cases, 1D spatial information is sufficient and/or easier to acquire. In such cases, one can acquire a spectrum line (SPLI) giving a reduced data set with only two dimensions (x, E).

4.3 Instrumentation and Data Analysis

In the following, we will describe the instrumentation and analysis details relevant for the datacube acquisition and analysis. Although many aspects are relevant for both techniques (EFTEM and SPIM), we will mainly illustrate them via the SPIM mode, and the reader interested in EFTEM is invited to read, for example, Egerton (1986) and Schaffer et al. (2006).

4.3.1 EELS Datacube Acquisition

Figure 4–5 illustrates the typical system for acquiring a datacube in the SPIM mode.

Three main sections have to be considered: the pre-specimen condenser section, the post-specimen projector section and the spectrometer section. Note that these sections do not correspond directly to relevant parts of the real microscope, as, for example, the objective lens (OL) belongs to both the pre- and post-specimen sections.

To form an electron probe, a high brightness electron gun acts as a source of electrons for the condenser system. The condenser system is a combination of typically three condensers and the condenser part of the objective lens. It is the condenser part of the OL that is responsible for most of the demagnification of the source. As such, this is the strongest

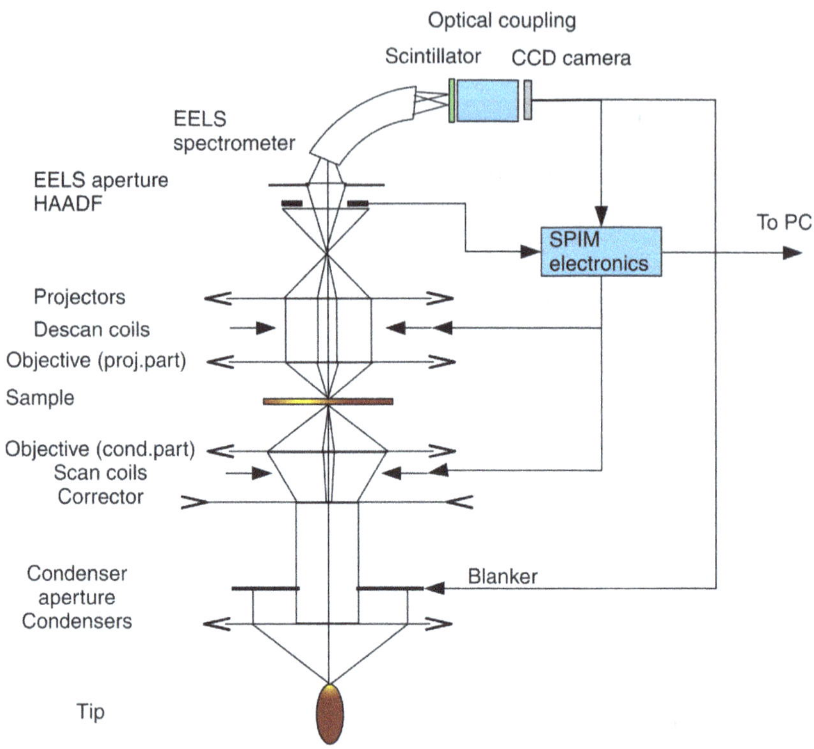

Figure 4–5. Datacube acquisition chain in the SPIM mode for a Cs-corrected probe system.

lens in the condenser system and the one that accepts the large angular range needed to form a small probe – indeed, if we restrict our discussion to the case of a diffraction-limited probe, the larger the angle, the smaller the probe.[2] As such, this lens is the primary lens needing correction for aberrations, which explains why sub-angström probe formation requires a Cs corrector placed before the OL (Krivanek et al. 2008) (see Chapter 15). A monochromator can be inserted before or in the condenser system. Ideally for SPIM, a blanker, preferentially electrostatic for optimised blanking speed, has to be placed in a plane *before* the sample. Doing this ensures that the sample is not irradiated during readout times, which might be an issue to consider for radiation-sensitive materials. Also, scanning coils are inserted somewhere before the condenser part of the OL.

The projector section is not essential. Indeed, in the old-fashioned dedicated STEM, the best example of which is probably the VG 5XX/6XX series, the projector part of the pole piece is much weaker

[2] When very high current is required, the first condenser can also be strongly excited, resulting in an additional contribution to the spherical aberration Cs, which has to be corrected by the Cs corrector. Note that the Cs corrector could be put *after* any of the lenses it has to correct, but obviously is put before the OL due to space limitations.

than the condenser part, and no other projectors are present. Only a set of deflection dipoles is used to centre the beam at the entrance of the spectrometer or on the bright field (BF)/HAADF detectors. Such a configuration is especially useful, as it prevents any Cc or Cs aberrations arising from the *projector section* that might affect the EELS spectra. However, it does not allow for camera length change.

In a typical projector system, descan coils are put after the projector part of the OL and synchronised with the scanning coils so that the EELS spectrum does not shift or broaden when scanning (see below), which would otherwise happen for large scan fields (say, a few hundreds of nanometres). Then, a set of projectors is used to transfer a diffraction pattern to the plane of the spectrometer aperture[3] at various camera lengths. In this plane is also placed the HAADF detector.

The last part consists of the spectrometer section.

The spectrometer itself is a magnetic prism that disperses the electron trajectories along one direction (the so-called dispersion direction) as a function of energy. In the usual single focusing geometry, the transverse component of the electron trajectory is essentially not affected by the presence of the prism. Thus, the output spectrum is essentially 2D, with the direction perpendicular to the dispersion axis containing angle information.

The spectrometer is fitted with various dipole, quadrupole and sextupole lenses. These are used to align, focus and correct aberrations in the spectrum in the detector plane, as well as changing the effective dispersion in this plane. The energy offset of the spectrum can be varied by several means, in particular by applying a voltage to a tube, the so-called drift tube, in the inner part of the spectrometer.

Finally, a scintillator is placed in the imaging plane of the spectrometer, and the electron-generated photons are transferred onto a 2D camera, either through an optical fibre system or by lens coupling. Typically, the camera is a rectangular CCD camera (say $\sim$1000$*$100 pixels) whose readout must be synchronised with the probe movements during the acquisition of a SPIM.

Along the dispersion axis, the spectrometer acts both as a prism *and* as a lens. The consequence is worth emphasising. Ideally, all the electrons with a given energy emerging at various angles from a point source situated in the object plane of the spectrometer are focused at a particular location in the image plane of the spectrometer – a line perpendicular to the dispersion direction in the single focusing geometry. Electrons having suffered various energy losses but all coming from the same point will then form a well-focused spectrum. However, if the object point is so extended that its image for a given energy loss is larger than a camera pixel, then the chromatic images of the source

[3] In some designs, such as the NION STEM (Krivanek et al. 2008), a coupling module can be added between the spectrometer and the projector system, serving various purposes: adapting the energy dispersion-dependent object point of the spectrometer, changing the camera length while keeping a constant HAADF angle (if the HAADF detector is between the projectors and the coupling module) and correcting third-order aberrations of the spectrometer.

point overlap, leading to a loss of energy resolution. It seems at first sight that such a situation may never happen in a typical SPIM experiment, as the size of the image of the probe ($\lesssim 10$ nm) is always much smaller than the typical pixel size ($\approx 10\,\mu$m). It can however arise if the source point is not in focus at the spectrometer, which is likely to happen when the object is changing focus. Also, if no descan is present, the EELS spectrum may considerably shift along the dispersion axis while performing large (typically more than 100 nm) scans.

A typical SPIM acquisition works approximatively as follows:

1. Beam is positioned at position N
2. Beam is unblanked
3. HAADF detector is continuously exposed and read and CCD camera is exposed for a time τ_e
4. Beam is blanked
5. CCD camera is read during time τ_r
6. If the drift correction is active, the beam is unblanked, drift correction images are acquired, beam is blanked, drift images are processed and the drift corrections are sent to the scanning coils.
7. Beam is positioned at position $N+1$, with drift corrections included if necessary.

4.3.2 Technical Issues

At this point, it is worth emphasising some general and technical considerations.

4.3.2.1 *Need for High Brightness*

The brightness ($B = I/(\Omega A)$) with I, Ω and A being the current, the solid angle and the cross-sectional area of the electron beam) is conserved along the electron trajectory in a microscope, discarding the effect of aberrations for simplicity. A high brightness gives the highest current in a given probe size, all other things being equal. Preserving the brightness by increasing the mechanical and electrical stability and aberration correction is thus crucial for fast, high signal-to-noise ratio (SNR) imaging. Also, the use of monochromation, as it decreases the brightness, has to be balanced with SNR and speed issues. Finally, the use of a Schottky gun (typical brightness $= 5 \times 10^8 A/\text{srd}/\text{cm}^2$ at 100 keV) is the minimum needed for an efficient SPIM acquisition, the best gun being, at the time of writing, the cold FEG (typical brightness of the order of $2 \times 10^9 A/\text{srd}/\text{cm}^2$ at 100 keV).

4.3.2.2 *Data Size and Time Issues*

Using the 2D camera unbinned or with low binning[4] levels can be of interest for better deconvolution (Gloter et al. 2003) (see below) and/or increased dynamics and/or recording the full zero loss without saturation for spectra alignment. Also, in some cases, it might be required

[4] To bin is the action of hardware summing different pixels at the time - then gaining readout speed, readout noise, but losing dynamics. In EELS, this is mostly done along the non-dispersing axis.

to record large areas with good sampling, but without too much drift. Data size and time issues thus have to be considered when setting acquisition conditions.

It is worth commenting on the size of a 4D datacube (two spatial dimensions and two other for the CCD camera. Other 4D/5D datacubes will be discussed later). In SPIM mode, the size of a datacube is typically, in the energy direction, of the order of 1024 channels multiplied by 1–100 rows on the CCD camera, each being real-valued[5] (typically 4 bytes). The spatial dimensions are typically from 10∗10=100 pixels to 200∗200=40,000. This gives an overall memory footprint ranging between 100∗1024∗1∗4 ≈ 400 Kb and 40000∗1024∗100∗4 = 16 Gb. While the first value is negligible, the second is only available on recent, 64-bit computers, which means that the user must choose the acquisition conditions carefully.

Several time values have to be considered here: blanking time τ_b, readout time (τ_r), probe settling time (τ_s) and, of course, exposure time τ_e. τ_e will depend on the experimental conditions, but will span a range from 0.1 ∼ 1 ms for fast low-loss applications to 1 ∼ 2 s for slow, fine structure-resolved, core losses. τ_b essentially depends on the type of blanker used: typically milliseconds for magnetic-type blankers and microseconds for electrostatic ones. Probe settling times are dependent on the given scan steps and speed, but are typically negligible whatever the acquisition time. The readout time depends on the CCD camera binning. The readout time of a camera pixel is of the order of 1–10 μs and thus 1–10 ms for a line. At full-binning, the time for the whole camera readout is roughly the same as for a single line, while it goes up to 0.1–1 s for a non-binned CCD. τ_r may become the limiting time when acquiring a SPIM especially in the LL region and/or with the high currents provided by the Cs-corrected machines. For the same reasons, it is worth considering the use of an electrostatic blanker when reaching the fast acquisition limit.

4.3.2.3 *Dose and Time Issues: A Comparison with EFTEM*

It is interesting to compare the typical time and dose required to access a given amount of information by SPIM and EFTEM.

Let us define some useful quantities for this comparison. Let B_i (i being STEM or TEM for the case where SPIM and EFTEM experiments are performed with different microscopes) be the brightness of the microscope, N and M the number of pixels along the two directions, n the number of channels in the spectrum, ΔS the area of an image pixel, and Ω_i the solid incidence angle. Let Δt_j, with j being the image or spectrum, be the time required to acquire an image (EFTEM mode) or a spectrum (STEM mode).

In EFTEM, the dose experienced by the sample during an image acquisition is

$$D_{1\text{image}} = B_{\text{TEM}} N M \Delta S \Omega_{\text{TEM}} \Delta t_{\text{image}} \tag{6}$$

[5] Although the acquisition format is of course non-negative integer, any subsequent treatment is likely to transform the data into real (possibly negative) numbers, so it is advisable to stick to a real number format.

and to collect a datacube

$$D_{\mathrm{EFTEM}} = nB_{\mathrm{TEM}} NM \Delta S \Omega_{\mathrm{TEM}} \Delta t_{\mathrm{image}}, \tag{7}$$

while the dose for a spectrum acquisition in the STEM mode will be

$$D_{1\mathrm{spectrum}} = B_{\mathrm{STEM}} \Delta S \Omega_{\mathrm{STEM}} \Delta t_{\mathrm{spectrum}} \tag{8}$$

and so the dose to collect the datacube

$$D_{\mathrm{SPIM}} = NMB_{\mathrm{STEM}} \Delta S \Omega_{\mathrm{STEM}} \Delta t_{\mathrm{spectrum}}. \tag{9}$$

The dose ratio between the two modes is

$$r = \frac{D_{\mathrm{SPIM}}}{D_{\mathrm{EFTEM}}} = \frac{1}{n} \frac{B_{\mathrm{STEM}} \Omega_{\mathrm{STEM}} \Delta t_{\mathrm{spectrum}}}{B_{\mathrm{TEM}} \Omega_{\mathrm{TEM}} \Delta t_{\mathrm{image}}}, \tag{10}$$

which is the ratio of the current densities multiplied by the ratio between the acquisition time *per spectrum* and the acquisition time *per image* (not the acquisition time of the whole datacube) divided by the number of channels.

Now, we want the gathered signal[6] per voxel (volume element in the x, y, E space) to be the same in both modes, i.e. the acquisition times are such that

$$B_{\mathrm{STEM}} \Omega_{\mathrm{STEM}} \Delta t_{\mathrm{spectrum}} = B_{\mathrm{TEM}} \Omega_{\mathrm{TEM}} \Delta t_{\mathrm{image}}. \tag{11}$$

In such conditions, the dose ratio is given by

$$r = 1/n. \tag{12}$$

In terms of dose, it is thus always more efficient to use the SPIM mode rather than EFTEM. This can be easily understood as a whole image must be recorded for each energy channel in EFTEM. The information provided by the electrons in all the remaining $n-1$ channels is thus lost at each image acquisition, while all the electrons are used in the SPIM mode.

Now, let us compare the typical acquisition times required to acquire a datacube of $N*M$ pixels and n channels. For the EFTEM mode, this time is

$$T_{\mathrm{EFTEM}} = n \Delta t_{1\mathrm{image}}, \tag{13}$$

while for the SPIM it is

$$T_{\mathrm{SPIM}} = NM \Delta t_{1\mathrm{spectrum}}. \tag{14}$$

As we still want to obtain the same amount of information in a voxel, we make use of Equation (11) and write the ratio of total acquisition times as

$$\frac{T_{\mathrm{EFTEM}}}{T_{\mathrm{SPIM}}} = \frac{n}{NM} \frac{\Delta t_{1\mathrm{image}}}{\Delta t_{1\mathrm{spectrum}}} = \frac{n}{NM} \frac{B_{\mathrm{STEM}} \Omega_{\mathrm{STEM}}}{B_{\mathrm{TEM}} \Omega_{\mathrm{TEM}}}. \tag{15}$$

[6] We consider for simplicity here that all the inelastic signals are gathered in both types of experiment, which may of course not be the case in real experiments, as discussed at various places in this chapter.

The ratio of the two acquisition times is thus the ratio between the number of EELS channels and the number of pixels multiplied by the current densities. Assuming to begin with a similar brightness for both microscopes, the ratio of current densities is equal to the ratio of incident solid angles. In EFTEM, let us assume a 1 mrd incident semi-angle, and in SPIM a value ranging from 6 (uncorrected) to 25 mrd (C_3 corrected[7]). This leads to a solid angle ratio, and so to a current density ratio of the order of 40–600. Also, if the STEM is fitted with a cold field emission gun, as is the case for a dedicated STEM, and the TEM is fitted with a Schottky gun, the ratio of brightness is of the order of 10. The current density ratio can thus be of the order of 400–6000.

It is thus obvious that the SPIM technique is worth considering for datacubes with a small number of pixels. However, even with such a large current density ratio, EFTEM becomes preferable whenever large area chemical maps are needed. Indeed, in such a case, where often only three channels are necessary, even a 100∗100 pixel map is worth performing in the EFTEM mode, as the ratio in the most extreme case is

$$6000 * 3/10,000 \approx 1. \tag{16}$$

4.3.3 Novel Acquisition Modes

The new possibilities offered by recent hardware and software developments such as monochromation, Cs-correction and deconvolution trigger the needs for faster detection, higher dynamics and higher energy range. Thus, these new developments must be accompanied in particular with new SPIM and CCD detection schemes. For example, a comprehensive experiment would consist in recording at each point of the SPIM both the low-loss and the core-loss spectra, or perhaps even two different core-loss regions, with, of course, different acquisition times for all of them. The resulting data set should include the following:

- a non-saturated ZL peak – for energy realignment, probe intensity variation measurement and deconvolution purposes, yet with enough dynamics to preserve a sufficient SNR.
- one or several core-loss regions with different acquisition times – and hopefully different acquisition conditions (binning).

Some recent developments in this direction will now be discussed.

4.3.3.1 Chrono-SPIM

The information contained in a CCD camera pixel is limited in dynamics, typically 2^{16} counts for fully binned cameras. This means that a feature in the visible part of the spectrum, which may have less than 0.01% of the ZL intensity, will yield less than say six counts. Assuming

[7] It is worth noting the gain in current density available in a C_5-corrected machine, where the incident semi-angle can be as large as 50 mrd while maintaining sub nanometre spatial resolution.

Poisson statistics, this leads to a SNR of around $\sqrt{6} = 2.4$, and thus an almost undetectable signal. Also, mains fluctuations lead to a broadening of the spectrum and a subsequent loss in resolution. Artificially increasing the dynamics by taking unbinned data might not be wise in many situations, as the readout noise will increase significantly, as well as the readout time. A possible workaround is to record a so-called chrono-SPIM (Nelayah et al. 2007), which is a SPIM with a series of N consecutive spectra contained in each pixel. After realignment (see next section) the energy resolution improves (see Figure 4–6) and the statistics significantly improve (by a factor of $\sqrt{N}$). The memory footprint and acquisition time[8] however increase significantly. This approach is of less interest for core losses, as the realignment requires

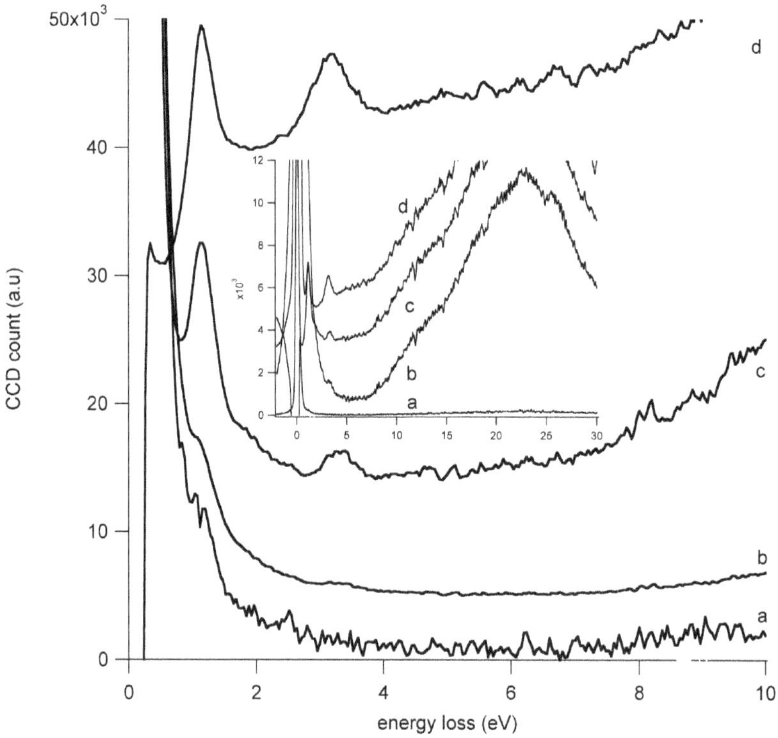

Figure 4–6. Effects of realignment and deconvolution: the case of a chrono-SPIM (50 ms acquisition per spectrum, 50 spectra per point). Spectra taken at the same given position in the chrono-SPIM on a silver nanoprism. The spectra are offset and scaled for clarity. (**a**). One individual 50 ms spectrum. (**b**). The corresponding spectrally realigned and summed spectrum. (**c**). The corresponding deconvoluted spectrum. (**d**). Corresponding ZL subtracted deconvoluted spectrum. The increase in SNR and peak visibility due to realignment and deconvolution is obvious. Inset: same spectra over a wider energy range and with no vertical scaling (offset for clarity). Data courtesy of J. Nelayah.

---

[8] With available currents in low-loss experiments increasing, CCD detectors might saturate so quickly that the limitation becomes the readout time rather than the acquisition time.

acquisition times smaller than mains fluctuations – and therefore very limited signal for the low cross-section core-loss signals. It also requires a well-defined peak in the spectrum, which is not always present.

4.3.3.2 Multiple Window SPIM

An alternative technique consists in acquiring different parts of the spectrum with different acquisition times and binnings – typically a fast, non-binned low loss and a slower, binned CL. This allows for efficient thickness deconvolution as well as CL energy shift calibration and allows a huge dynamic range (typically 100∗1000 times more than with a conventional SPIM if the acquisition time ratio between the LL and CL regions is 1000). Two schemes have been proposed so far (see Figure 4–7)

1. Using a standard camera and switching the drift tube voltage to alternate between the LL and the CL regions on the detector (Scott et al. 2008).
2. Using a dedicated camera divided into at least two sections, typically one for the low loss and one for the core loss (Tencé et al. 2006). At each point of a SPIM, spectra are acquired at two different drift tube voltages and thus over two different energy ranges. Simultaneously, a dipole in the spectrometer arrangement is excited so as to put the spectrum in a different location along the non-dispersive axis. Using two independent sets of readout electronics, the two parts of the spectrum can be read independently. In practice, one is read while the other is exposed, eliminating a large part of the dead time.

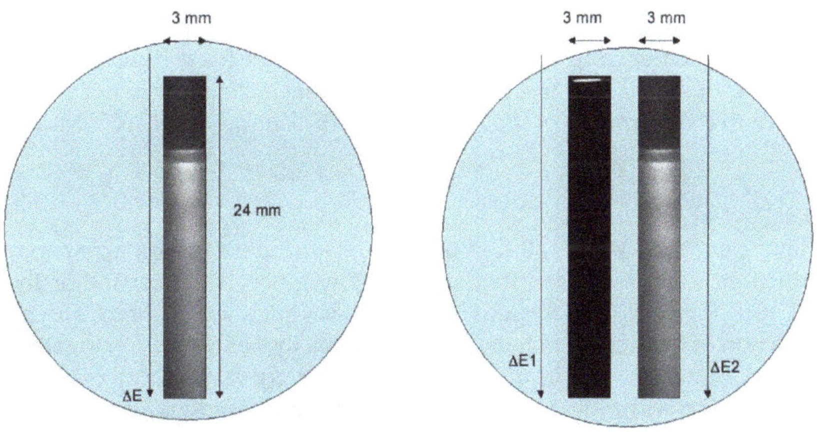

Figure 4–7. Multiple window camera. *Left*: A standard CCD camera, the centre of which can be alternatively illuminated by a core-loss and low-loss region (Scott et al. 2008). This allows for near simultaneous multiple energy range measurements. *Right*: Multi-section CCD camera. Two different parts of the camera are exposed and read sequentially. One of the improvements with respect to the previous situation is the absence of remanences in the core-loss measurement due to zero-loss exposure (Tencé et al. 2006).

We note that each time the energy range is changed, the spectrometer settings (focus, second-order aberrations) have to be changed, as their fine-tuning depends on the energy.

The second alternative has a huge practical advantage over the first one, as the two parts of the spectra are not read at the same physical position, so the unavoidable remanences due to the ZL do not affect the CL region.

4.3.4 Data Analysis

We will now describe some of the routine data analysis techniques used in the treatment of spectrum images.

4.3.4.1 Spatial and Spectral Alignments

As the spectra are acquired sequentially and images reconstituted afterwards, there can be problems with energy and spatial drift. The zero-loss peak position may change during the acquisition, as well as the intensity of the incoming electrons. These effects can be easily a posteriori corrected by automatic peak detection, realignment and normalisation of the integrated spectrum weight (see Figure 4–6). This, of course, requires a non-saturated ZL or an edge of constant energy present throughout the entire scan. In this context, the use of a high-dynamics, high-energy range system as described in Section 4.3.2 is particularly advantageous.

On the other hand, correction of the spatial drift is more difficult. In the case of a known geometry (typically an interface), the SPIM can be spatially realigned (see the example of an interface in Figure 4–8). Otherwise, an online procedure has to be followed. Regularly during the acquisition of the SPIM, the acquisition stops and fast HAADF images of a reference region of the specimen are taken. They are then cross-correlated to the previous images. One can thus determine the shift between each image acquisition and correct for it directly in the scan input signal. It is worth emphasising that such a technique works down to the atomic level (Bosman et al. 2007, Kimoto et al. 2007, Muller et al. 2008).

4.3.4.2 Deconvolution

Raw spectra are a convolution of the signal of interest with a response function. This response function is mainly given by the shape of the ZL, but includes also various contributions such as the point spread function of the detector chain. This induces a loss of resolution together with a huge increase in the background at very low energy loss due to the finite width of the ZL. Therefore, deconvolution procedures are generally needed to increase the energy resolution, decrease ZL tail effects or remove the effects of multiple scattering (so-called Fourier log or Fourier ratio deconvolution (Egerton 1986)). Such procedures are of importance for accurate quantification, accurate detection of fine structures and access to the near-infrared (less than 1 eV) region when using non-monochromated guns. However, they all rely on the acquisition of a kernel spectrum with which the spectrum of interest has to be

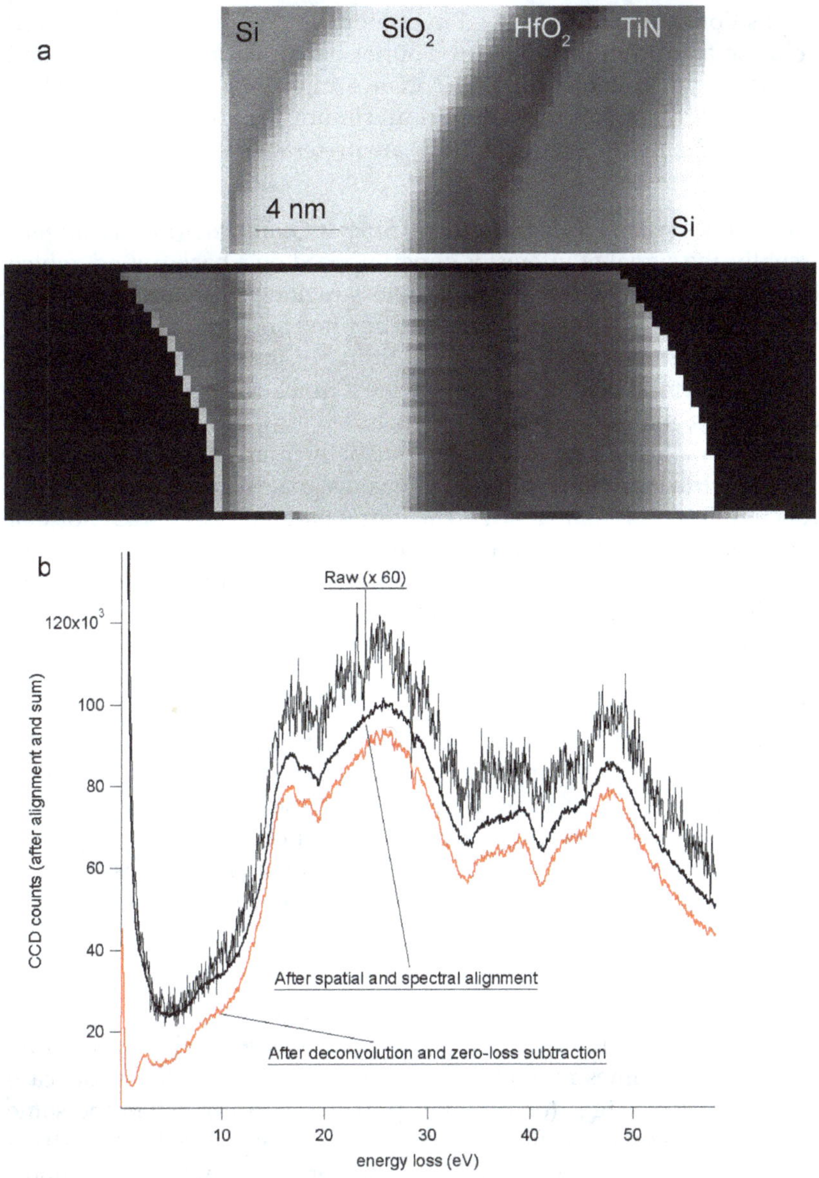

Figure 4–8. Effect of spatial realignment and deconvolution: the case of 1D materials. (**a**). ZL-filtered image of a stack of various materials before (*top*) and after (*bottom*) spatial realignment. The drift in the top image is obvious. (**b**). Spectrum taken within the HfO_2 layer, before and after spectral realignment and spatial sum along plane, and after deconvolution and subtraction. Data courtesy of M. Couillard.

deconvoluted. This can be either a reference zero-loss spectrum (simple deconvolution), which can be acquired before or after the SPIM acquisition, or a full low-loss spectrum (multiple scattering deconvolution), in which case it must be acquired at each pixel of the SPIM. Again, in the later case, one can appreciate the interest of using a multi-window acquisition system as described in Section 3.2 . We note that the deconvolution algorithm is important. Fourier-based algorithms usually fail as they tend systematically to enhance high-frequency noise. More sophisticated algorithms based on maximum likelihood or maximum entropy schemes (Gloter et al. 2003) are preferable.

4.3.4.3 Background Subtraction and Simple Quantification Technique

Usually, the signal of interest is superimposed on a background, which can be due to the ZL peak tail (low-loss region), plasmon tail and/or core-loss tail. This background has to be fitted and removed. While this procedure can be included in a full fitting of the spectrum, it is more often performed beforehand to provide a quick check of the data with enhanced signal-to-background ratio.

In the low-loss region, the ZL tail is usually fitted either by a reference ZL taken in a region in which the relevant signal is absent or by a fitting function (e.g. polynomial law). The fitting parameters, i.e. the choice of the window in which to fit the ZL, have to be carefully chosen; otherwise, the high dynamic range between the ZL and the low-loss region may invalidate the fit in the region of interest.

In the core-loss region, the background due to the plasmon tail is usually removed by a power-law fit. Again, the fitting window has to be chosen carefully.

The data can then be analysed more easily. In particular, quick elemental mapping or plasmon intensity maps can be made by computing images in which each pixel corresponds to the signal integrated over the energy window of interest. In the case of chemical mapping, each chemical signal can be normalised by the corresponding cross-section, giving rise to a more quantitative map. At this level, SPIM mapping is roughly equivalent to EFTEM mapping.

4.3.4.4 Model Based Quantification Techniques

Despite its intrinsic simplicity and efficiency, the previous procedure can (and sometimes must) be improved. This is in particular the case when different edges (or plasmon peaks) overlap or when the same element is present in different chemical forms (and thus the fine structures will vary for a given edge in an intricate way). In these situations, the obvious improvement with respect to the simple procedure is to fit the whole spectrum of interest. Different kinds of models can be used, but it is worth emphasising that the spectrum must be deconvoluted to remove multiple scattering for the fit to be meaningful. The most simple case is when the spectrum can be supposed to be a linear combination of known reference spectra. Then, a simple linear fit (often called multiple least square fit, or MLS) is sufficient to determine the weight of each component in the experimental spectrum. More sophisticated models are non-linear in nature, for example, when adding the possibility of shifting the spectra or when the reference functions are not linear (Gaussians). Within the SPIM context, after applying the procedure at each pixel, maps of the weight of each component in the model can be extracted in the case of linear fits, while subtler features can be mapped in the case of non-linear fits (peak position, width, intensity, for example).

All the above fitting procedures are highly dependent on the chosen model, for the background as well as for the edges or peaks. It is thus essential to determine how appropriate it is to use a given model to fit the data of interest. This can be done by estimating the minimum measurable error available for a given model (Manoubi et al. 1990, Verbeeck and van Aert 2004).

We note that due to the fact that the SPIM approach provides a direct acquisition of the datacube in the form of a series of spectra, the above-mentioned operations are straightforwardly performed, while EFTEM requires additional pre-processing in order to reconstruct the spectra (Schaffer et al. 2006).

4.3.4.5 *Thresholding*

The redundancy of the information in a SPIM is very high, with many spectra usually containing similar information. A simple use of this fact is to sum several spectra taken at different equivalent positions in order to enhance the signal-to-noise ratio and/or avoid radiation effects. This can be done not only when the geometry of the object is known a priori (as for an interface, see Section 4.3.3), but also by using the information extracted from HAADF images or chemical maps, for example. For instance, one can sum all spectra related to pixels associated with a given intensity level in an elemental map, in order to extract an optimised spectral signature of that element.

4.3.4.6 *Principal Component Analysis*

More sophisticated analyses can take advantage of the information redundancy of a SPIM. One can use principal component analysis (PCA) (Bonnet et al. 1999), which is theoretically used to extract redundant spectral features in a SPIM and in practice is very useful for noise removal. In principle, PCA requires the diagonalisation of the covariance matrix of the datacube. The eigenvectors (spectra, for example) can be ranked as a function of their associated eigenvalues. Only the first few are usually significant, and by skipping most of the eigenvectors to reconstruct individual spectra of the datacube, a huge gain in signal-to-noise ratio is observed. Going beyond simple denoising, i.e. retrieving unknown spectral features, is in practice however much more difficult (Bonnet et al. 1999).

4.4 Applications

In the following, we will present some applications of the SPIM technique to different physical, chemical and materials science problems. In many cases, an entire SPIM is not required to solve a problem. Sometimes, only a spectrum line or even a single spectrum is sufficient. However, in these cases, the accurate positioning of the probe, its small size and the parallel acquisition of the HAADF signal are crucial to solve the problem.

4.4.1 Low Loss

In the low-loss region, the quantities of interest related to the material (electron density, gaps) are hidden by electromagnetic effects (Cerenkov radiation, long-range surface plasmons). We will present in the following some applications and limits of low-loss SPIM for material research. However, the low-loss SPIM is actually not well-suited for this purpose, mainly because the information from the material is delocalised. On the other hand, the information concerning the electromagnetic fields themselves is quite localised, so we end this section with a discussion of the usefulness of the low-loss SPIM for the study of optical excitations at the nanometre scale. We anticipate that electromagnetic field mapping at the nanometre scale will be the main application of low-loss SPIM, as chemical mapping is for the core-loss SPIM.

4.4.1.1 Surface and Interface Plasmons

Interface plasmons have been described already in Section 4.2.1. They may arise at the interface between different materials and in different geometries. Comprehensive reviews can be found elsewhere (Wang 1997, Rivacoba et al. 2000). We will illustrate the use of the SPIM for their study in two simple cases, although plasmons have been studied with the SPIM techniques in several other simple geometries: spheres (Ugarte et al. 1992), cylinders (Kociak et al. 2000, Stephan et al. 2001, Taverna et al. 2003, Bursill et al. 1994, Arenal et al. 2005, Stockli et al. 2000, Seepujak et al. 2006) and thin films or multilayers (Moreau et al. 1997, Couillard et al. 2007, 2008, Eberlein 2008), for example. More complex geometries will be discussed in Section 4.4.1.4.

Si/SiO_2 Interface Plasmon

The case of the Si/SiO_2 interface plasmon (IP) is an interesting textbook case as Si/SiO_2 interfaces are relatively easy to obtain in the form of thin cross-sections and the IP energy value (close to 8 eV) is far from the ZL tail. Moreau et al. (1997) have studied such an interface by using a sampling of 0.3 nm and a beam size of 0.7 nm. Figure 4–9 shows the interface plasmon features as the beam is scanned from the silicon to the silica. The first point to be noticed is that the IP signal can be detected far away from the interface, a behaviour loosely referred to as delocalisation in the literature. While the peak is itself well-defined for a given probe position, a noticeable energy shift is evidenced as the probe moves. This effect can be understood by remembering that the IP energy disperses towards higher energies with the momentum transfer, and that the number of momenta values needed to describe the EELS spectra sharply increases as the beam approaches the interface. Thus, as the probe approaches closer and closer to the interface, the momentum transfer participating in the energy loss increases, as does the energy.[9] This explains the discrepancies reported between previous

[9] In the case of a planar system, the fact that the energy depends on the momentum is a relativistic effect. However, in other geometries (spheres

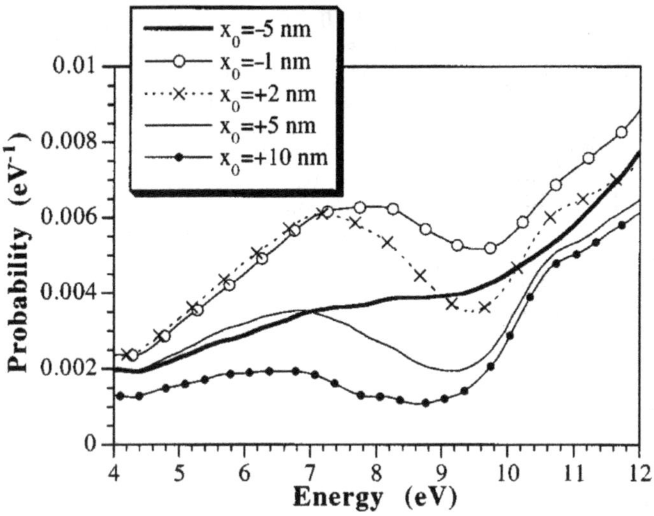

Figure 4–9. Experimental interface plasmon spectra for various impact parameters x_0. Negative values are for the probe in the silicon, positive for the probe in the silica. After Moreau et al. (1997). Courtesy of P. Moreau.

non-spatially resolved EELS experiments and shows the interest of such measurements. Similar effects have been reported in different geometries (Kociak et al. 2000, Seepujak et al. 2006, Stephan et al. 2002).

Silica-Coated Silicon Spheres

A related problem is that of silica-coated silicon spheres (Ugarte et al. 1992). The silicon nanospheres described in this study are 10–300 nm in diameter and naturally covered by an SiO_2 shell due to air oxidation. Figure 4–10 (left) shows the intensity profile of the main detected excitations: (i) 17 eV bulk Si plasmon, (ii) 23 eV bulk SiO_2 plasmon, (iii) 9 eV Si/SiO_2 interface plasmon, and (iv) a feature around 3–4 eV that we will now discuss.

As this last feature could not be described easily in a core–shell approximation, the authors assumed that the SiO_2 shell was surrounded by another Si shell, probably due to electron-induced oxygen desorption. The lower energy mode was supposed to be essentially due to the external (amorphous) silicon/vacuum interface plasmon. This has been confirmed with simulations shown in Figure 4–10. This example nicely shows the power of the dielectric model usually used in simulating spatially resolved experiments (Couillard et al. 2007, 2008, Taverna et al. 2002, Ugarte et al. 1992, Kociak et al. 2001). It also demonstrates the interest of using nanometric spatial resolution, i.e. better than the typical delocalisation length of the problem, to discriminate

(Ugarte et al. 1992) and cylinders (Kociak et al. 2000, for example), the energy is also momentum dependent in classical schemes.

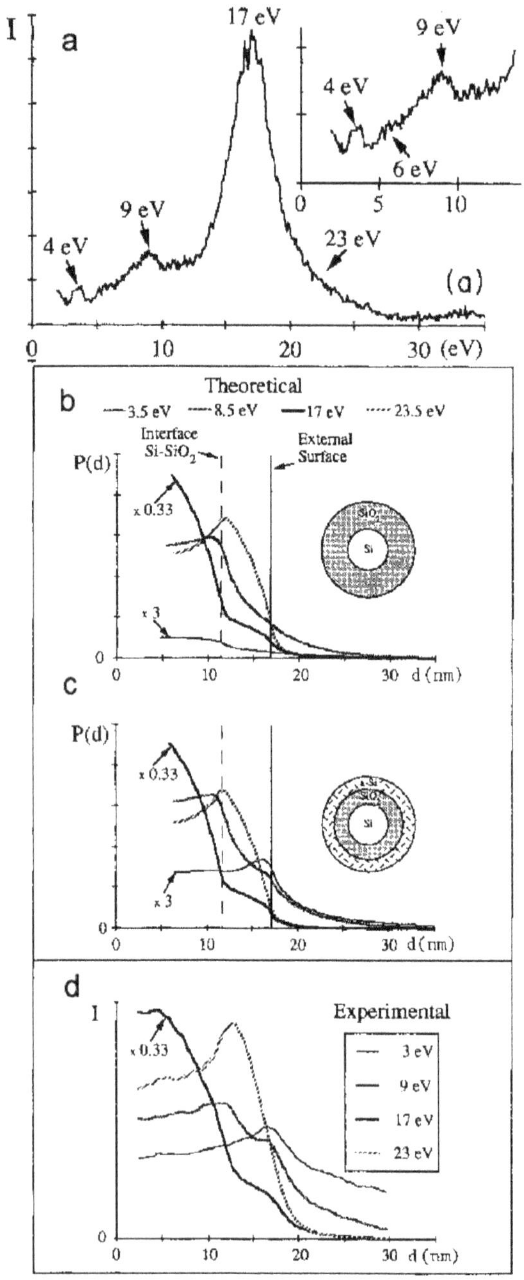

Figure 4–10. (**a**) Typical spectrum taken when the beam is sent through the centre of a 50 nm diameter Si/SiO_2/amorphous Si sphere. (**b**). Theoretical line profiles of the main mode intensities for a 50 nm Si/SiO_2 diameter sphere. (**c**). As (b) but for a Si/SiO_2/amorphous Si 50 nm diameter sphere. (**d**). Experimental line profiles of the main mode intensities of a Si/SiO_2/amorphous Si 50 nm diameter sphere. Adapted from Ugarte et al. (1992). Courtesy of D. Ugarte.

between different morphologies (note that some signal at ≈3.5 eV is theoretically predicted even in the Si/SiO_2 model, but with a totally different spatial behaviour from the experimental one). Finally, this is an example where the low loss can help retrieve material information (i.e. determine which model is better, the Si/SiO_2 or Si/SiO_2/amorphous Si model).

4.4.1.2 *Gaps and Excitons*

As already described, LL EELS is sensitive to all optical excitations, including gaps and excitons.[10].

Measurements of gaps have been shown to be quite successful in bulk material (Rafferty and Brown 1998, Schamm and Zanchi 2003), making the technique very appealing for band gap measurement in nano-structured systems, such as multilayers and nanoparticles. Spatially resolved band gap measurements have been pioneered by Batson et al. on misfit dislocations at the GaAs/GaInAs interface (Batson et al. 1986). The gap can be measured when the electron beam is impinging the object of interest or in an aloof geometry. For very small objects (nanotubes (Kociak et al. 2001), very thin layers (Couillard et al. 2008), however, the begrenzung effect is so strong that only the surface signal is measured. As already seen in Section 4.2.1, this signal is still usable to retrieve gap information. Such measurements have been performed on various nano-objects, including SiO_2 nanospheres (Abe et al. 2000) and boron nitride nanotubes (Arenal et al. 2005). The case of artefacts in the determination of gaps due to Cerenkov emission has been clearly explained in De Abajo et al. (2004), in particular in the case of SiO_2 nanospheres (Abe et al. 2000).

In the case of BN nanotubes (see Figure 4–11), it has been shown that the exciton line position is not dependent on structural parameters (diameters, number of walls) and is very close to the bulk value. For such nanostructures, it has been shown that the surface response is closer to ε_2 than to $\mathrm{Im}(-1/\varepsilon)$ (Taverna et al. 2003, Kociak et al. 2001). This result can be understood within the dielectric model, which has shown its impressive predictive power down to the monoatomic layer level (Arenal et al. 2005, Stephan et al. 2002, Kociak et al. 2001).

It thus seems straightforward to generalise such measurements to any nano-structured material. This is however impossible in general, as additional signals, due to Cerenkov emission and/or surface plasmons, may arise in the band gap of nanomaterials. The effect of (surface) Cerenkov losses has been shown in SiO_2 nanospheres (De Abajo et al. 2004, Abe et al. 2000), while the combined effect of Cerenkov and surface plasmons has been extensively examined in the case of a HfO_2-based multilayer (Couillard et al. 2007). In the latter case, a minute comparison of the spatially resolved experimental data with the dielectric model predictions (Bolton and Chen 1995a) showed impressive agreement, allowing for an accurate assignment of the physical origin to the different features. Apart from the above-mentioned physical restrictions, it is worth emphasising the technical challenge of measuring a band gap. Indeed, the onset of the gap is usually ill-defined, and the intensity of the maximum is usually much smaller than the following plasmons (see e.g. (Couillard et al. 2007).

---

[10] In an EELS experiment, there is practically no way to distinguish between a gap and an exciton. What is experimentally measured is – at best! – the onset of a peak in the low-loss region. Whether it is a pure electronic gap or an exciton has to be determined through additional theoretical work.

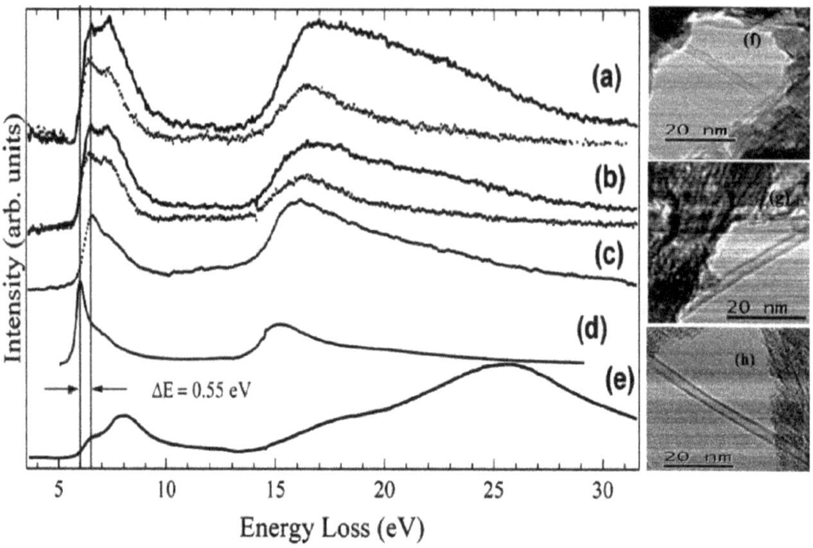

Figure 4–11. EELS spectra of BN nanotubes. From *top* to *bottom*: Experimental deconvoluted spectra for (**a**) a triple-walled nanotube, in intersecting geometry and at grazing incidence, (**b**) a double-walled nanotube in intersecting geometry and at grazing incidence, (**c**) a single-walled nanotube in the rastered mode, (**d**) $\varepsilon_{2,\perp}$ (imaginary part of the component of the dielectric tensor perpendicular to the anisotropy axis of the BN sheets) as extracted from the bulk-loss spectra in (**e**). (**f**), (**g**), (**h**) Bright field images of the tubes (**a**), (**b**), (**c**). From Arenal et al. (2005). Courtesy of R. Arenal.

4.4.1.3 Semi-Core Loss

As described at some length in Section 4.4.2, core losses are routinely used to perform nanometre-scale concentration measurements. The spatial resolution can be high enough to provide atomic resolution (Browning et al. 1993, Muller et al. 1993), mainly because the interaction distance with the electron probe decreases as the edge energy increases. Nonetheless, many elements display edges at low energies, and it is useful to understand the consequence of this for elemental quantification at this scale. A typical example is represented by the He *K*-edge arising from the $1s \rightarrow 2p$ excitation of the He atom. In He gas, the transition is around 21.218 eV, but is blue-shifted in high-density He fluids due to Pauli repulsion (Lucas et al. 1983). This has been confirmed experimentally by spatially resolved EELS in the spot mode (McGibbon 1991, Walsh et al. 2000). However, two additional effects could not be measured without the SPIM technique (Taverna et al. 2008): as exemplified in Figure 4–12, the position of the helium energy peak is red-shifted close to the bubble interface, and the apparent density decreases at the interface. Both effects are related to surface and begrenzung effects, which diminish the measured intensities (as compared to the bulk situation) and blue-shift the apparent transition. Such effects, not detectable without accurate SPIM, may lead to a systematic error in the density measurements. Surface effects have also been demonstrated on more complex core losses (Couillard et al. 2007). Here, the interest of the SPIM technique (in particular with respect to EFTEM) is clear as it allows one

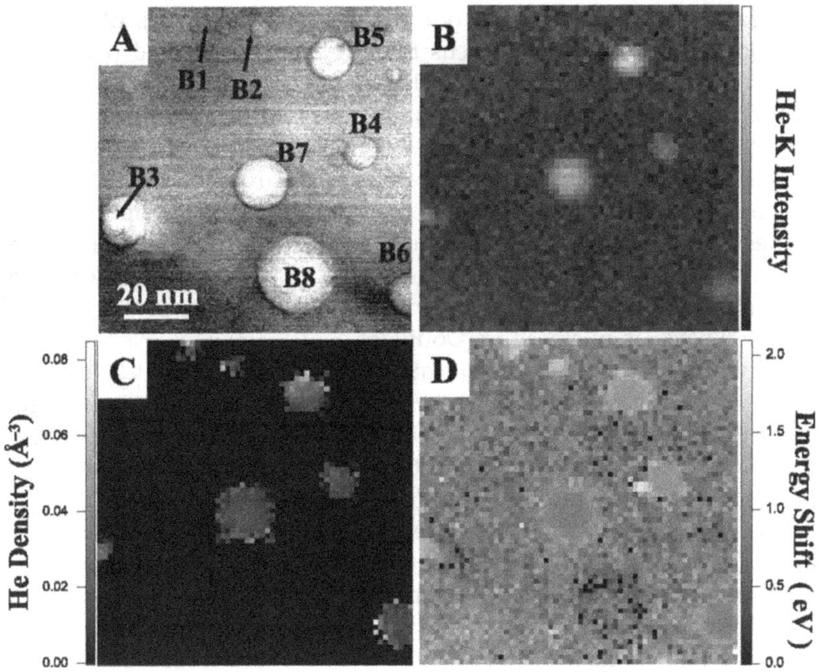

Figure 4–12. Maps extracted from a spectrum image of a selected area of a sample made up of He nanobubbles embedded in a metallic Pd_{90} Pt_{10} matrix. (**a**) Bright field image of the analysed area. Bubbles showing He signal are evidenced. (**b**) Helium chemical map. (**c**) Map of the He density inside the He-filled bubbles. (**d**) Map of the energy shift of the He K-line. The reference energy is chosen as that of the atomic He (21.218 eV). From Taverna et al. (2008). Courtesy of D. Taverna.

to map energy shifts as a function of the probe position. This is particularly useful for retrieving subtle changes in the physical properties of nano-objects at very high spatial resolution.

4.4.1.4 Sub-optical Wavelength Mapping

Up to now, we have only considered LL (i) in the UV (i.e. more than 4 eV) range and (ii) for highly symmetrical cases. Indeed, most studies have concentrated on these situations, due to a lack of combined high spatial/high energy resolution and of relevant theories or simulation tools for handling arbitrarily shaped nanostructures. Recently, however, both limitations have been overcome. This is of primary importance because it enables the study of the optical properties of metallic nanoparticles at the nanometre scale. It is well known that these optical properties depend on the nanoparticle surface plasmons, which themselves depend drastically on the particles' shape, size and dielectric environment, whenever the nanoparticles size is of the order of or smaller than the free-space wavelength of light. For noble metals, the typical wavelengths corresponding to the surface plasmon energies fall in the near-infrared/visible range (roughly 1 μm to 300 nm). Such nanoparticles have many potential applications, ranging from cancer cell therapy to efficient surface-enhanced Raman scattering (SERS) substrate fabrication. Indeed, one of the main properties of surface plasmons is their ability to focus the electromagnetic field over very small distances – much less than the wavelength of light, and thus to store/transport energy on a very small scale. Most, if not all, of these applications depend on this ability. It is thus of great interest to map the subwavelength physical properties of the electromagnetic field. This information cannot, however, be retrieved with standard optical techniques. Some cutting-edge techniques can access part of the information (different kinds of scanning near-field optical microscopes (Imura and Okamoto 2008), cathodoluminescence (Yamamoto et al. 2001) or photo emission electron microscopy (Douillard et al. 2008). However, they are usually quite limited either in terms of SNR, spatial resolution (hardly better than 20 nm in the best cases) and/or spectral information (no datacube is available). The EELS SPIM technique thus seems promising, even if, as we will see, many technical, theoretical and numerical problems have only recently been overcome to reach this goal.

The field of optical subwavelength mapping was pioneered by Batson, who showed, in spot mode and close to the UV regime, surface plasmon coupling effects between nanospheres (Batson 1982) and later modifications of the SP properties at different locations on a hemisphere (Ouyang et al. 1992). More recently, thanks to a monochromatised beam, Khan and co-workers (Khan et al. 2006) obtained visible range information with a ca. 20 nm resolution in the spot mode. It was however difficult to draw conclusions on the nature of the measured signal without SPIM information. Bosman et al. (2007) and Nelayah et al. (2007) independently demonstrated subwavelength mapping with resolution better than 10 nm. Bosman et al. studied different gold nanoparticles, singly or in clusters, while Nelayah et al. studied individual silver nanoprisms. As neither of them used monochromators, they relied on the weak energy spread of a cold FEG, in combination with a deconvolution procedure (Nelayah et al. 2007) or PCA (Bosman et al. 2007).

Figure 4–13 shows intensity maps obtained on a 78 nm side, 10 nm thick, silver triangular nanoprism at three energies spanning the NIR/visible/UV range, corresponding to plasmon resonances in the spectra. To each energy corresponds a given intensity distribution.

These data can be simulated by relevant boundary element method simulation techniques (Nelayah et al. 2007, Garcia et al. 1997) or compared to optical results, as in Bosman et al. (2007). The mapped quantities are related to plasmons, and the energies of the excitations are close to those measured or simulated by optical means (see (Sherry et al. 2006), for example). However, the exact link with optical measurements should be explained. It can be shown (Garcia de Abajo and Kociak 2008) that spatially resolved EELS is, to a very good approximation, a measure of the electromagnetic local density of states (EMLDOS). This quantity, well-known in the near-field optician community (Dereux et al. 2000), is to the Maxwell equation what the electronic Local Density of States is to the Schrödinger equation: a measurement of the probability of finding an optical excitation, photon or plasmon, at a given position (for the EMLDOS), instead of an electron (for the eLDOS).

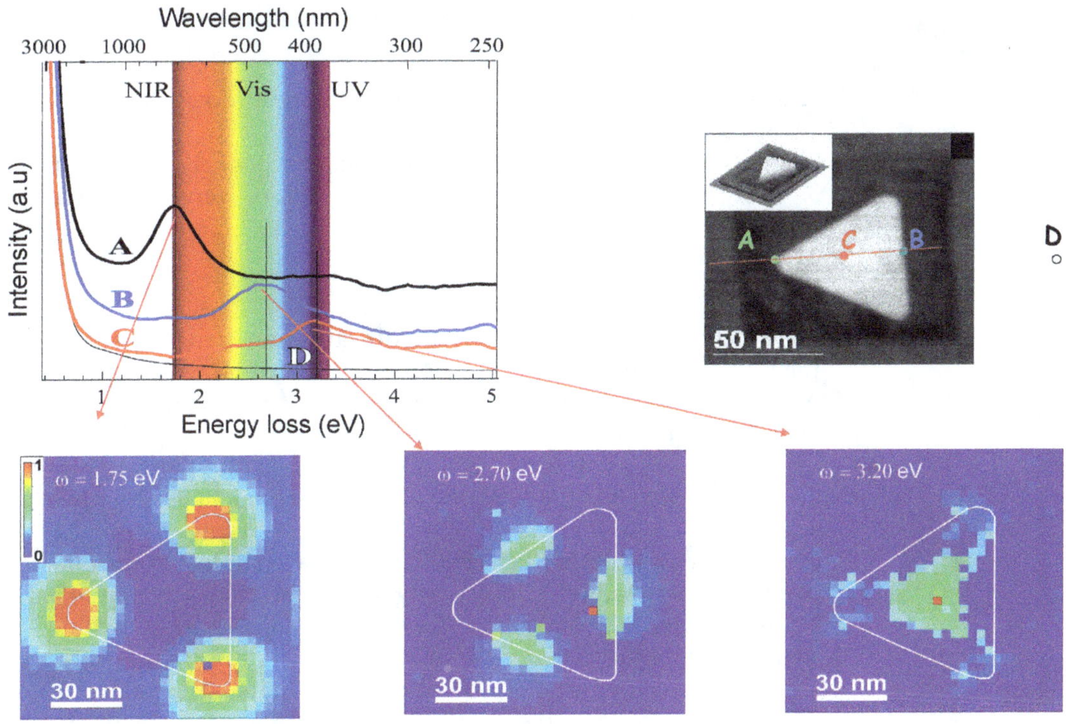

Figure 4–13 Spectrum image on a silver nanoprism. *Left, top*: four different spectra taken at the position indicated on the HAADF image on the *right*. *Bottom*: fitted intensity of the three most important modes. Data courtesy of J. Nelayah.

More precisely, in the case of surface plasmons, it was shown in (Garcia de Abajo and Kociak 2008) that the EELS signal is proportional to the square of the electric field associated with the surface plasmon charge distribution, polarised along the direction of electron propagation. Note the change of point of view here. The standard point of view always tried to link spatially resolved low-loss EELS to some physical quantity, itself linked to the property of the (nano)object of interest (e.g. the dielectric function). In that sense, the EELS signal had to be considered as delocalised. The present point of view emphasises the local measurement of electromagnetic fields. Following this approach, it is interesting to note that the material's physical properties, by themselves, are not of great importance (the dielectric function of gold has been known for many years). It is rather the field distribution as affected by the presence of a nano-object that is the quantity of interest in plasmonics and photonics. This is exactly what LL EELS SPIM is mapping.

4.4.2 Core-Loss

The primary use of the SPIM technique is elemental and concentration mapping and, increasingly, bonding and valence state mapping. However, following the discussion in Section 4.3.2, the EFTEM technique can be a very good competitor when large areas and few energy values are required for extracting the required information. We will thus limit our discussion to the cases where the SPIM is preferable:

1. small areas and very high spatial resolution (down to the single atomic column level, see Section 4.4.2)
2. multiple edges and thus a large number of energy channels required
3. background-sensitive edge extraction and/or very low chemical signal
4. overlapping edges
5. fine structure mapping

Of course, the last four items are directly connected to the fact that the primary information in a SPIM is a spectrum and not an image. We note that to deal most efficiently with the last two points, fitting and PCA techniques are essential.

4.4.2.1 Mapping and Quantifying Chemical Signals at the Nanometre Scale

As already mentioned, the acquisition of a complete spectrum per pixel enables a posteriori processing which can generate a large variety of spectroscopic images. Furthermore, image analysis techniques with segmentation of regions of interest can be introduced: spectra corresponding to a given area can then be summed. This allows for a posteriori checking of the actual presence of a given element at chosen locations in the maps in the case of low signals. This leads to an improvement in accuracy and detection limits. It also allows for a detailed analysis of the characteristic spectral features associated with these specific locations, enabling an investigation of the local electronic properties. More specifically, it improves the background and characteristic signal modelling and fitting. This is highly useful in the case of low characteristic signal-to-background ratios, low signal-to-noise ratios, extraction of edges in the low-loss energy region and finally in the case of overlapping edges. All these possibilities lead to a greater chemical sensitivity as illustrated in the following examples. Maps of nanometre-scale chemical heterogeneities in nanomaterials and nanostructures appear in the literature in various application domains: heterogeneous catalysis (Morales et al. 2005), nanotubes (Stephan et al. 1994, Ajayan et al. 1995, Suenaga et al. 1997, Zhang et al. 1998), or semiconductors (Schamm et al. 2008, Gass et al. 2006, Bruley et al. 1999, Laquerriere et al. 2002, Hunt et al. 1995, Leapman and Ornberg 1988).

One aspect to be emphasised is the high sensitivity and robustness of the chemical maps when deduced from a spectrum image. The example of the mapping of small BN nanodomains substituting the graphene network of single-walled carbon nanotubes (SWCNTs) (Enouz et al. 2007) is very illustrative. Mapping nitrogen in carbon is a difficult case, as the low cross-section *N*-*K*–edge is sitting on the extended fine structures of the carbon *K*-edge. Special care has then to be taken for background modelling and N signal extraction. In the case of the hybrid nanotubes investigated, the challenge was to detect the presence of a very small number of N and B atoms (typically 5–10 atoms within the nanometre size analysis volume) substituting in the carbon network

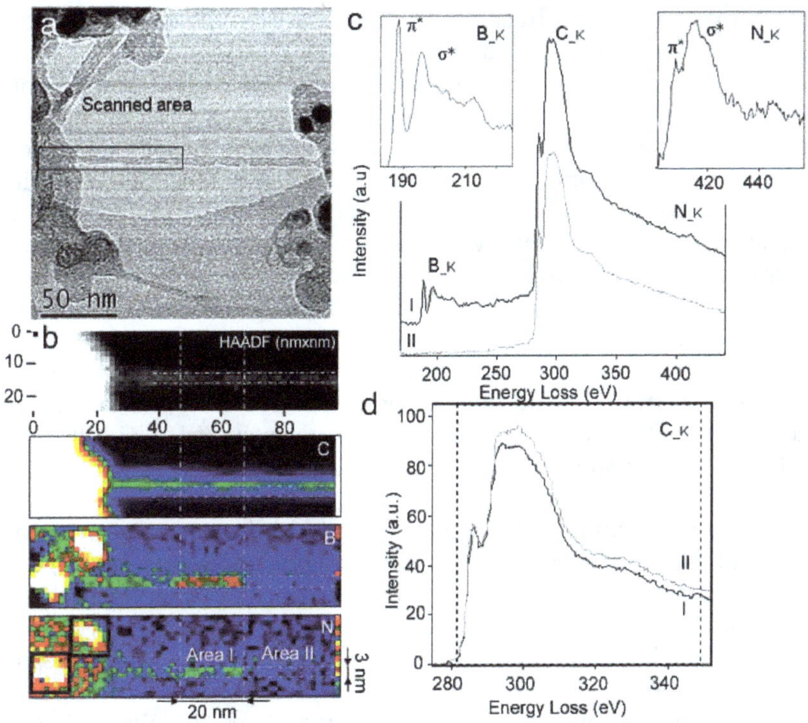

Figure 4–14. (**a**) Bright field image of a nanotube rope 4–5 nm wide. The *rectangle* shows the region analysed by EELS and corresponds to a 64×16 pixel spectrum image. (**b**) HAADF image of the scanned area (98 × 24 nm^2) and the relative intensity chemical maps of C, B, and N elements. The intensity of the signal varies from *dark/blue* (poor signal) to *white/red* (high signal). Areas marked Area I and Area II each measure 20 × 3 nm^2 and are defined at the positions 50–70 and 70–90 nm along the tube axis, respectively. (**c**) EEL spectra I and II are defined as the sum of Areas I and II respectively. Each spectrum is the sum of 30 spectra. Near-edge fine structures of *B*- and *N–K*,-edges are shown in the inset. (**d**) Background subtracted C–*K*-edges in areas I and II. From Enouz et al. (2007), courtesy S. Enouz et al., reproduced with permission.

of a small rope of 5–6 SWCNTs. The nitrogen and boron maps displayed in Figure 4–14 and extracted from a 64 × 16 pixel spectrum image clearly show the presence of B- and N-rich areas 10–20 nm in length. Elemental quantification indicates that these areas arise from the substitution of 20% of the C atoms by B and N in an approximate B/N =1 stoichiometry. This corresponds to the formation of 1 nm^2 BN nanopatches distributed randomly within these B, C, and N regions. An a posteriori check of the B and N signal fine structures associated with these nanodomain regions confirms the incorporation of B and N atoms as BN entities within the C network.

The first successful detection of VN inclusions as small as 1 nm in steel is another example (Mackenzie et al. 2006, Craven et al. 2008). A validation of the noisy V and N maps was obtained by checking the spectra a posteriori to verify that VN precipitates were actually present. Moreover, MLS fitting was used to improve the signal-to-noise ratio in the V maps and to subtract any contribution from the oxygen *K*-edge superimposed on the *V–L*-edge.

The use of a variant of MLS fitting was reported earlier for the characterisation of Au/Ni MBE-grown multilayers (nickel layers, 2–6 atomic planes thick, embedded between 6 nm thick Au spacers) (Tencé et al. 1995). This practical situation combines all the difficulties in terms of required spatial resolution and improved data processing techniques for the quantitative evaluation of strongly overlapping featureless edges. The fitting was performed on the second derivative signals of the Au-$O_{2,3}$-, Au-$N_{6,7}$- and Ni-$M_{2,3}$-edges at 54, 83 and 68 eV, respectively. These edges, although strongly overlapping, were chosen for their high cross-section, a prerequisite for the extraction of information from a small number of atoms. The resulting calculated compositional profiles were accurate enough to evidence Ni diffusion into Au being responsible for asymmetrical profiles at the nanometre scale.

Because it optimises the collected spectral information per unit of incident electron dose, the spectrum-imaging mode is also a very appropriate tool in the case of radiation-sensitive samples such as biological or other soft materials.

For example, it has been shown recently that the dose-limited spatial resolution in spectrum images of solvated polymers can reach a few nanometres, opening interesting perspectives to solve polymer application-oriented problems (Yakovlev and Libera 2008). The authors used MLS fits to map the spatial distribution of different polymer phases associated with specific low-loss fingerprints and to investigate the nature of the interfaces. Typically, 10 nanometre resolution maps were obtained with a reduced incident electron dose of 1200 e/nm^2.

Another advantage is that the use of a focused probe optimises the detection of small numbers of atoms: reducing the probe size improves both the signal-to-background and signal-to-noise ratio.

Pioneering work on biological samples was performed by R. Leapmann. He has demonstrated that the 1 nm probe of a VG-STEM offers the best compromise in terms of detection limit optimisation versus radiation damage minimisation for the detection of phosphorus in macromolecular assemblies. In particular, he has shown the feasibility of mapping base pairs in DNA plasmids using an electron dose of approximately 109 e/nm^2 at a temperature of −160°C. The detection limit of about ten phosphorus atoms associated with a 3 nm spatial resolution due to a necessary undersampling to reduce radiation damage was decreased to the single atom limit with a 1 nm spatial resolution in the case of the moderately e-beam-sensitive tobacco mosaic virus (Leapman and Rizzo 1999). Similarly, the chemical mapping of individual Ca and Fe atoms in appropriately thin biological specimens has also been demonstrated (Leapman 2003).

4.4.2.2 Bond Mapping

A step further in the analysis is the extraction of bonding information from local changes in edges of interest. This can be done by extending the use of MLS and/or PCA techniques (Muller et al. 1993, Arenal et al. 2008) from overlapping edges to the case where the measured edge is a linear combination of edges of the same element in different chemical/bonding states. When the different chemical states of the species of interest are known a priori and when good quality spectra of the species in these different states are available, MLS is a much easier method than PCA, from which the output is not directly physical. If fingerprint spectra are not available and/or edge fine structures are not known a priori (see e.g. Muller et al. 1998, Samet et al. 2003) and/or noise is an important issue (see e.g. Bosman et al. 2007, Samet et al. 2003), PCA has to be used as a first analysis step. We illustrate the use of SPIM to recover bonding information in the case of boron nitride nanotube (BNNT) samples containing a large number of boron-based nanoparticles, in which boron can appear in different forms:

- pure boron
- boron oxide
- hexagonal boron nitride (h-BN). This latter being anisotropic, the orientation of the σ and π bonding with respect to the electron beam has to be considered.

This is well exemplified in Figure 4–15, in which spectra taken at different locations in the sample display very different boron *K*-edge fine structures. Applying an MLS fit at each pixel of the SPIM using reference spectra (Figure 4–15h), the different chemical and bonding states can be disentangled (Figure 4–15 bottom). The hollow anisotropic BN shells are clearly identified, with the $q \perp c$ (i.e. mean transferred momentum vector perpendicular to the anisotropic c axis of the h-BN sheets) map being uniform on a shell, while the $q \parallel c$ (i.e. mean transferred momentum vector parallel to the anisotropic axis c of the h-BN sheets) map is peaked at the edges of the shell.

In the same vein, valence state information, which can be extracted from the EELS spectra (Chen et al. 2009, Gloter et al. 2004, Goulhen et al 2006) as described in Section 4.4.2, and can be mapped (Samet et al. 2003, Kourkoutis et al. 2006). Here again, the use of the SPIM technique (or, at least, the SPLI) is of primary importance. Indeed, changes in valence states/electronic structure across interfaces have huge consequences in multilayers for electronic (Muller et al. 1999) or spintronic (Samet et al. 2003, Maurice et al. 2006, Estrade et al. 2008) properties. Correlating spectroscopic information from EELS and structural information from the HAADF is thus crucial for the in-depth study of the electronic/magnetic properties of such multilayers. This is best illustrated in the case of the vanadium oxidation state at the interface between $LaV^{3+}O_3$ (a Mott insulator) and $LaV^{5+}O_4$ (a band insulator) (Kourkoutis et al. 2006), see Figure 4–16. The question arises as to how the vanadium oxidation state changes across the interface, and in particular whether some of the *V* atoms can be in a 3^+ state (a situation that does not exist in the bulk for LaVO systems).

A spectrum line, in combination with an MLS analysis, clearly shows that vanadium in the 3^+ oxidation state exists at the interface.

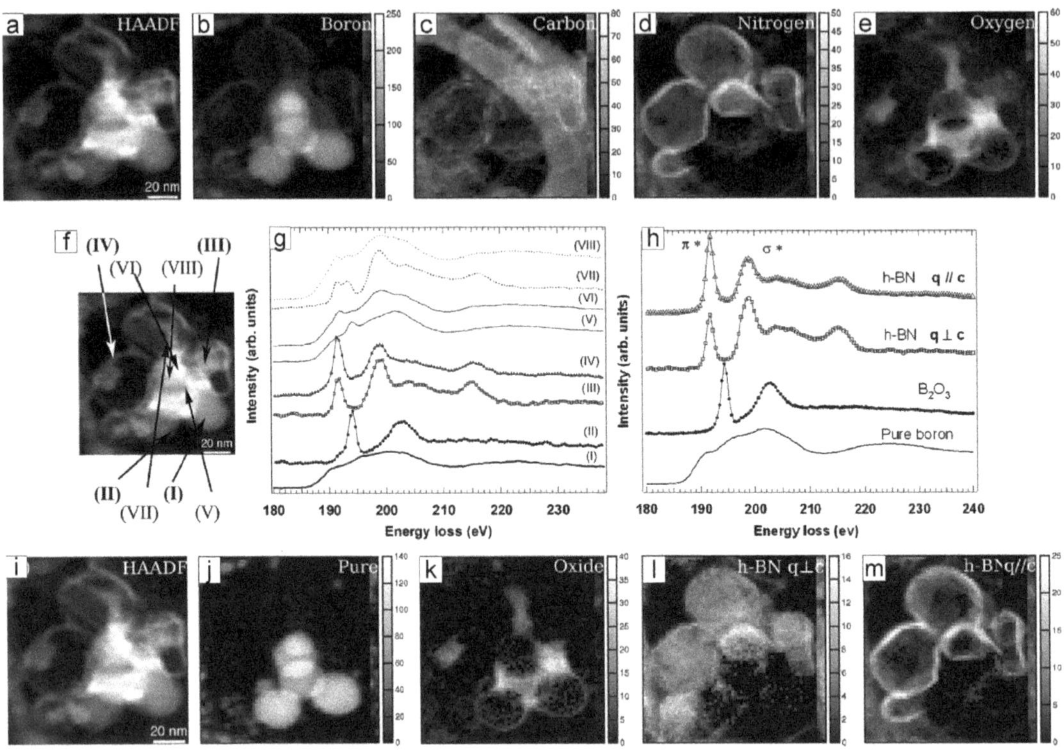

Figure 4–15. MLS analysis of a BNNT sample. (**a**) HAADF image of the sample. (**b–e**) Chemical maps of the different elements contained in the sample. (**f**). Positions where the spectra in (**g**) have been extracted. (**h**) reference spectra. (**i**) HAADF image of the sample. **(i–m)** MLS maps of the different chemical forms of boron. From Arenal et al. (2008). Courtesy of R. Arenal, reprinted with permission.

4.4.2.3 Atomic Column Chemical and Spectroscopic Imaging

Now that new generation microscopes with high-brightness, sub-angström probe formation capabilities and high electrical and mechanical stabilities are available (see Chapter 15), it seems straightforward to use the techniques described in the previous sections to realise chemical and spectroscopic analysis of one individual atomic column. One must not, however, forget the true quantum natures of both the target (the sample) and the probe (the incoming electron) used in the present type of system, which lead to a delocalisation of the information. Let us summarise the conceptual challenges, which are extensively described in Chapter 6.

1. Dechannelling: part of the intensity in an electron probe, originally focused on a given column, is transferred to an adjacent column – this happens usually when the sample is thick and/or when the convergence angle is large.
2. Delocalisation of the inelastic scattering: when the probe becomes smaller than an interatomic distance, it is worth considering the effective size of the scatterer. For inelastic scattering, the relevant scatterer size is not the extension of the electronic wavefunction of the atom under consideration, but is the extension of the (screened)

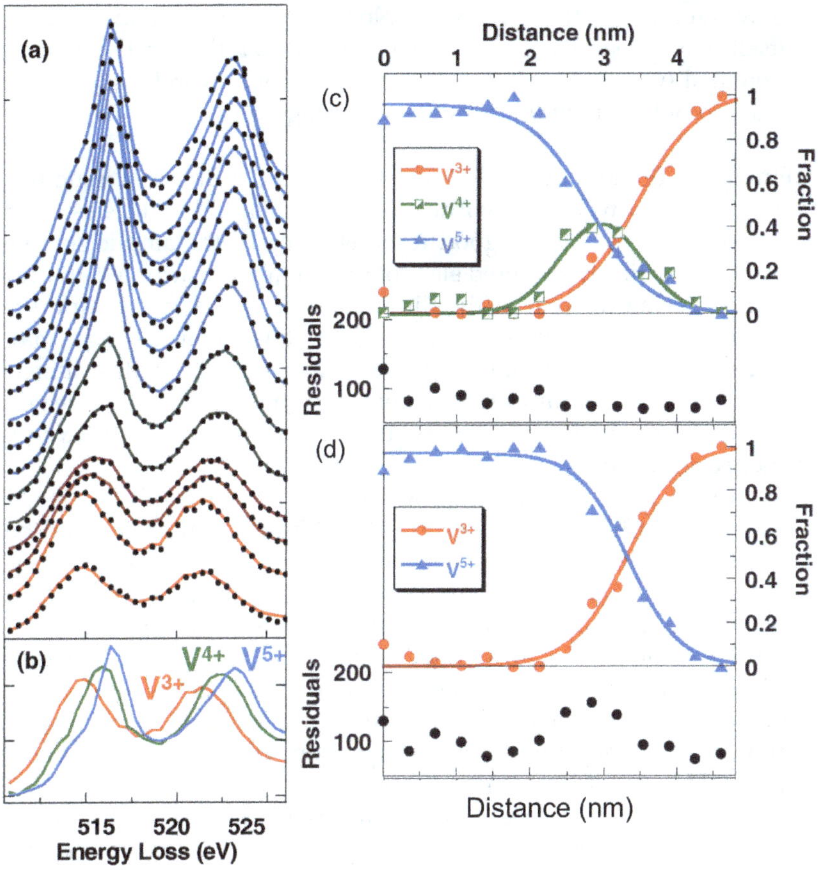

Figure 4–16. (**a**) V–$L_{2,3}$-edge across an $LaV^{3+}O_3/LaV^{5+}O_4$ interface. *Dots* are experimental data, and *solid lines* are fits using the reference spectra in (**b**). (**b**) Reference spectra for the three possible valence states of V. (**c**) Fraction of V^{3+}, V^{4+}, V^{5+} deduced from a fit of the experimental edges. (**d**) Same as (**c**), but using only two reference spectra, V^{3+} and V^{5+}. The high residual in the latter case is strong evidence that the vanadium oxidation is partly 4^+. Adapted from Kourkoutis et al. (2006). Courtesy of D. Muller, reprinted with permission.

Coulomb potential generated by the atom, in very much the same way as discussed in Section 4.4.1 for low-loss scattering. The larger the energy loss, the less pronounced is the effect, but the typical delocalisation length is larger than 1 Å for losses less than 1 keV (Egerton 1986, Kimoto et al. 2007, Browning et al. 1993). This is the most limiting effect for high-resolution imaging, as the signal from high-energy edges, when available, is limited by their small cross-section.

3. Non-locality of the inelastic scattering: as clearly evidenced by Oxley et al. (2005), an electron probe, the intensity of which is localised on a given atomic column, may experience inelastic scattering linked to scatterers well away from the given column. This is due to the fact that the relevant quantity to consider is the wavefunction, not the intensity of the probe. Different parts of the probe wavefunction

may interact together via the non-local scattering potential, which itself may extend over regions even larger than the local scattering potential (Oxley et al. 2005). However, this effect can be effectively washed out by integrating all the scattering events.

Finally, as pointed out in Muller et al. (2008), a practical issue has to be considered when one wishes to obtain chemical images as close as physically possible to a good representation of the atomic species position. Inelastically scattered electrons are sharply distributed around elastic scattering directions (the diffraction directions in case of crystalline systems). If a sample is diffracting strongly, chemical images may reflect elastic contrast if care has not been taken to collect all scattered electrons. To do so requires large acceptance angles and well-corrected spectrometer pre-optics in order to collect all the signal without any loss of energy resolution.

The quest for such imaging, which has been recently achieved (Bosman et al. 2007, Kimoto et al. 2007, Muller et al. 2008), can be traced back to the work described in (Browning et al. 1993, Muller et al. 1993 and Batson 1993), which demonstrated the sensitivity for compositional or bonding change identification across interfaces with a resolution of the order of an atomic column. However, it took 14 years to obtain spectrum images with sub-interatomic distance resolution. The SPIM technique is crucial for practical applications involving atomic column spectroscopy. Indeed, the comparison with the simultaneously acquired HAADF image makes it possible to identify the atomic column position, and therefore to identify any potential non-locality effect. Also, the signal is quite delocalised (see Figure 1–37 for an example of an edge detected one atomic column away from the position of the atom responsible for it). Thus, however small the probe, the signal of a given edge between two identical atoms tends not to drop to zero. The identification of the atom position thus requires clear 2D mapping.

The preceding discussion is well illustrated by the above-mentioned three publications summarised in Table 4–1. Bosman et al. (2007) demonstrated chemical mapping on $Bi_{0.5}Sr_{0.5}MnO_3$ samples with a VG-STEM fitted with a NION Cs corrector. Typical acquisition times were 10 min for an area of $\sim 1.4 * 1.6$ nm∗nm and 3000 spectra (sampling of ca. 0.3 Å). As the detected signal was noisy, the authors used a PCA decomposition in order to reduce the noise. Clear maps of Mn and O were obtained, although in one direction dechannelling from Bi/Sr/O columns to O columns was observed. No non-local effects were evidenced despite the fact that the convergence and collection angle were the same.

Kimoto et al. (2007) demonstrated chemical mapping on an $La_{1.2}Sr_{1.8}Mn_2O_7$ layer oriented along the [010] direction, with a non-corrected Hitachi HD 2300C. Some defects were present in the mapped area. Using two edges of the same element, one at low energy (La $N_{4,5}$) and the other at higher energy (La $M_{5,4}$), they clearly showed the negative effect of delocalisation on image resolution, as the map with

Table 4–1 Summary of three atomically resolved chemical mapping experiments.

Ref.	System	Time per spectrum	Total time	area (nm^2)	Sampling (Å)	α (mrd)	β (mrd)	Chemical imaging	Spectro. imaging
Bosman et al. (2007)	Crystal	0.2 s	10 min	1.4*1.6	0.3	24	24	PCA and BR	–
Kimoto et al. (2007)	Planar defect	2 s	61 min	0.84*2.13	0.35	15	31	BR	–
Muller et al. (2008)	Trilayer	7 ms	30 s	3.1*3.1	0.48	40	60	BR	PCA

α and β stand for the incident and acceptance semi-angles, respectively. "Spectro.imaging" stands for spectroscopic imaging. "BR" means background removal using a power-law fit

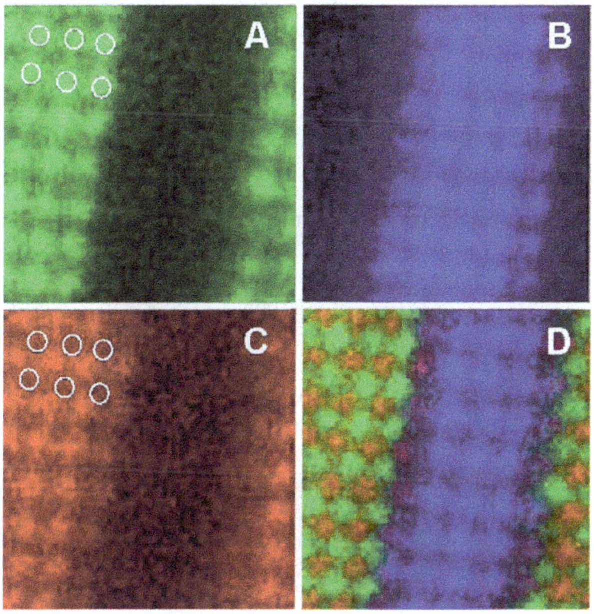

Figure 4–17. Colour-coded elemental maps of a $La_{0.7}Sr_{0.3}MnO_3/SrTiO_3$ multilayer. (**a**) La; (**b**) Ti; (**c**) Mn; and (**d**) RGB representation of the sample obtained by combining the individual maps. From Muller et al. (2008). Courtesy of D. Muller, reprinted with permission.

the $N_{4,5}$-edge was not atomically resolved while the one with the $M_{5,4}$-edge was.

Finally, Muller et al. (2008) demonstrated chemical and spectroscopic mapping at the single column level on a $La_{0.7}Sr_{0.3}MnO_3/SrTiO_3$ sample. Using a C_5-corrected machine (NION UltraSTEM) together with a high acceptance angle, most of the drawbacks described at the beginning of this section were avoided, and ultrafast and unambiguous chemical maps were obtained. They clearly showed the intermixing of

Ti and Mn at the multilayer interfaces, to different degrees at the two interfaces with the $SrTiO_3$ layer. Using a PCA, changes in the Ti fine structures between the nominal $La_{0.7}Sr_{0.3}MnO_3$ and $SrTiO_3$ part were shown – the fine structure being typical of local disorder in the former case.

4.5 Conclusion

We have seen how powerful a technique the SPIM is. In the domain of nano-optics and atomically resolved chemical and spectroscopic imaging, this technique is only in its very early stages. The interpretation of future experiments made possible by the spread of this technique and the rise of Cs-corrected and monochromated machines will be an exciting challenge. Finally, new developments are to be expected, such as the 3D chemical mapping made possible by C_5 and higher order corrected machines.

Acknowledgement We wish to thank J. Nelayah and M. Couillard for providing us with the raw data needed for some figures.

Reference

H. Abe, H. Kurata, K. Hojou, Spatially resolved electron energy-loss spectroscopy of the surface excitations on the insulating fine particle of aluminum oxide. J. Phys. Soc. Jpn. **69**, 1553–1557 (2000)

J. Aizpurua, A. Howie, F.J.G. De Abajo, Valence-electron energy loss near edges, truncated slabs, and junctions. Phys. Rev. B **60**, 11149–11162 (1999)

P.M. Ajayan, O. Stephan, P. Redlich, C. Colliex, Carbon nanotubes as removable templates for metal-oxide nanocomposites and nanostructures. Nature **375**, 564–567 (1995)

R. Arenal, F. De La Peña, O. Stéphan, M. Walls, M. Tencé, A. Loiseau, C. Colliex, Extending the analysis of EELS spectrum-imaging data, from elemental to bond mapping in complex nanostructures. Ultramicroscopy **109**, 32–38 (2008).

R. Arenal, O. Stephan, M. Kociak, D. Taverna, A. Loiseau, C. Colliex, Electron energy loss spectroscopy measurement of the optical gaps on individual boron nitride single-walled and multiwalled nanotubes. Phys. Rev. Lett. **95**, 127601 (2005)

P.E. Batson, Simultaneous STEM imaging and electron-energy-loss spectroscopy with atomic-column sensitivity. Nature **366**, 727–728 (1993)

P.E. Batson, Surface plasmon coupling in clusters of small spheres. Phys. Rev. Lett. **49**, 936–940 (1982)

P.E. Batson, K.L. Kavanagh, J.M. Woodall, J.W. Mayer, Electron-energy-loss scattering near a single misfit dislocation at the GaAs/GaInAs interface. Phys. Rev. Lett. **57**, 2729–2732 (1986)

J.P.R. Bolton, M. Chen, Electron-energy-loss in multilayered slabs .2. parallel incidence. J. Phys. Condens. Matter **7**, 3389–3403 (1995a)

J.P.R. Bolton, M. Chen, Electron energy loss in multilayered slabs. Ultramicroscopy **60**, 247–263 (1995b)

N. Bonnet, N. Brun, C. Colliex, Extracting information from sequences of spatially resolved EELS spectra using multivariate statistical analysis. Ultramicroscopy **77**, 97–112 (1999)

M. Bosman, V.J. Keast, J.L. Garcia-Munoz, A.J. D'alfonso, S.D. Findlay, L.J. Allen, Two-dimensional mapping of chemical information at atomic resolution. Phys. Rev. Lett. **99**, 086102 (2007)

M. Bosman, V.J. Keast, M. Watanabe, A.I. Maaroof, M.B. Cortie, Mapping surface plasmons at the nanometre scale with an electron beam. Nanotechnology **18**, 165505 (2007)

N.D. Browning, M.F. Chisholm, S.J. Pennycook, Atomic-resolution chemical-analysis using a scanning-transmission electron-microscope. Nature **366**, 143–146 (1993)

J. Bruley, J. Cho, H.M. Chan, M.P. Harmer, J.M. Rickman, Scanning transmission electron microscopy analysis of grain boundaries in creep-resistant yttrium- and lanthanum-doped alumina microstructures. J. Am. Ceram. Soc. **82**, 2865–2870 (1999)

L. A. Bursill, P. A. Stadelmann, J. L. Peng, and S. Prawer. Surface plasmon observed for carbon nanotubes. Phys. Rev. B, 49:2882–2887, 1994.

M. Couillard, M. Kociak, O. Stephan, G.A. Botton, C. Colliex, Multiple-interface coupling effects in local electron-energy-loss measurements of band gap energies. Phys. Rev. B **76**, 165131 (2007)

M. Couillard, A. Yurtsever, D.A. Muller, Competition between bulk and interface plasmonic modes in valence electron energy-loss spectroscopy of ultrathin SiO_2 gate stacks. Phys. Rev. B **77**, 085318 (2008)

A.J. Craven, M. Mackenzie, A. Cerezo, T. Godfrey, P.H. Clifton, Spectrum imaging and three-dimensional atom probe studies of fine particles in a vanadium micro-alloyed steel. Mater. Sci. Technol. **24**, 641–650 (2008)

FJG. De Abajo, A. Rivacoba, N. Zabala, N. Yamamoto, Boundary effects in Cherenkov radiation. Phys. Rev. B **69**, 155420 (2004)

A. Dereux, C. Girard, J.C. Weeber, Theoretical principles of near-field optical microscopies and spectroscopies. J. Chem. Phys. **112**, 7775–7789 (2000)

L. Douillard, F. Charra, Z. Korczak, R. Bachelot, S. Kostcheev, G. Lerondel, P. M. Adam, P. Royer, Short range plasmon resonators probed by photoemission electron microscopy. Nano Lett. **8**, 935–940 (2008)

T. Eberlein, U. Bangert, R.R. Nair, R. Jones, M. Gass, A.L. Bleloch, K.S. Novoselov, A. Geim, P.R. Briddon, Plasmon spectroscopy of free-standing graphene films. Phys. Rev. B **77**, 233406 (2008)

R.F. Egerton, *Electron Energy Loss Spectroscopy in the Electron Microscope* (Plenum, New York, NY, 1986)

S. Enouz, O. Stephan, J.L. Cochon, C. Colliex, A. Loiseau, C-BN patterned single-walled nanotubes synthesized by laser vaporization. Nano Lett. **7**, 1856–1862 (2007)

S. Estrade, J. Arbiol, F. Peiro, I.C. Infante, F. Sanchez, J. Fontcuberta, F. De La Pena, M. Walls, C. Colliex, Cationic and charge segregation in $La_{2/3}Ca_{1/3}MnO_3$ thin films grown on (001) and (110) $SrTiO_3$. Appl. Phys. Lett. **93**, 112505 (2008)

F. J. García De Abajo, J. Aizpurua, Numerical simulation of electron energy loss near inhomogeneous dielectrics. Phys. Rev. B **56**, 15873–15884 (1997)

F.J.G. Garcia de Abajo, M. Kociak, Probing the photonic local density of states with electron energy loss spectroscopy. Phys. Rev. Lett. **100**, 106804 (2008)

M.H. Gass, A.J. Papworth, R. Beanland, T.J. Bullough, P.R. Chalker, Mapping the effective mass of electrons in III-V semiconductor quantum confined structures. Phys. Rev. B **73**, 035312 (2006)

A. Gloter, A. Douiri, M. Tence, C. Colliex, Improving energy resolution of EELS spectra: an alternative to the monochromator solution. Ultramicroscopy **96**, 385–400 (2003)

A. Gloter, M. Zbinden, F. Guyot, F. Gaill, C. Colliex, TEM-EELS study of natural ferrihydrite from geological-biological interactions in hydrothermal systems. Earth Planet. Sci. Lett. **222**, 947–957 (2004)

F. Goulhen, A. Gloter, F. Guyot, M. Bruschi, Cr(vi) detoxification by *Desulfovibrio vulgaris* strain Hildenborough: Microbe-metal interactions studies. Appl. Microbiol. Biotechnol. **71**, 892–897 (2006)

J.A. Hunt, D.B. Williams, Electron energy-loss spectrum-imaging. Ultramicroscopy **38**, 47–73 (1991)

J.A. Hunt, M.M. Disko, S.K. Behal, R.D. Leapman, Electron-energy-loss chemical imaging of polymer phases. Ultramicroscopy **58**, 55–64 (1995)

C. Jeanguillaume, C. Colliex, Spectrum-image - the next step in EELS digital acquisition and processing. Ultramicroscopy **28**, 252 (1989)

I.R. Khan, D. Cunningham, S. Lazar, D. Graham, W.E. Smith, D.W. McComb, A tem and electron energy loss spectroscopy (EELS) investigation of active and inactive silver particles for surface enhanced resonance Raman spectroscopy (SERRS). Faraday Discuss. **132**, 171–178 (2006)

K. Kimoto, T. Asaka, T. Nagai, M. Saito, Y. Matsui, K. Ishizuka, Element-selective imaging of atomic columns in a crystal using STEM and EELS. Nature **450**, 702–704 (2007)

M. Kociak, L. Henrard, O. Stephan, K. Suenaga, C. Colliex, Plasmons in layered nanospheres and nanotubes investigated by spatially resolved electron energy-loss spectroscopy. Phys. Rev. B **61**, 13936–13944 (2000)

M. Kociak, O. Stephan, L. Henrard, V. Charbois, A. Rothschild, R. Tenne, C. Colliex, Experimental evidence of surface-plasmon coupling in anisotropic hollow nanoparticles. Phys. Rev. Lett. **8707**, 075501 (2001)

K. Imura, H. Okamoto, Development of novel near-field microspectroscopy and imaging of local excitations and wave functions of nanomaterials. Bull. Chem. Soc. Jpn. **81**, 659–675 (2008)

L.F. Kourkoutis, Y. Hotta, T. Susaki, H.Y. Hwang, D.A. Muller, Nanometer scale electronic reconstruction at the interface between $LaVO_3$ and $LaVO_4$. Phys. Rev. Lett. **97**, 256803 (2006)

O.L. Krivanek, G.J. Corbin, N. Dellby, B.F. Elston, R.J. Keyse, M.F. Murfitt, C.S. Own, Z.S. Szilagyi, J.W. Woodruff. An electron microscope for the aberration-corrected era. Ultramicroscopy **108**, 179–195 (2008)

P. Laquerriere, J. Michel, G. Balossier, and B. Chenais. Light elements quantification in stimulated cells cryosections studied by electron probe microanalysis. Micron, **33**, 597–603, 2002.

R.D. Leapman, Detecting single atoms of calcium and iron in biological structures by electron energy-loss spectrum-imaging. J. Microsc.-Oxf. **210**, 5–15 (2003)

R.D. Leapman, N.W. Rizzo, Towards single atom analysis of biological structures. Ultramicroscopy **78**, 251–268 (1999)

R.D. Leapman, R.L. Ornberg, Quantitative electron-energy loss spectroscopy in biology. Ultramicroscopy **24**, 251–268 (1988)

A.A. Lucas, J.P. Vigneron, Theory of electron energy loss spectroscopy from surfaces of anisotropic materials. Solid State Commun. **49**, 327–330 (1984)

A.A. Lucas, J.P. Vigneron, S.E. Donnelly, J.C. Rife, Theoretical interpretation of the vacuum ultraviolet reflectance of liquid-helium and of the absorption-spectra of helium microbubbles in aluminum. Phys. Rev. B **28**, 2485–2496 (1983)

M. Mackenzie, A.J. Craven, C.L. Collins, Nanoanalysis of very fine VN precipitates in steel. Scripta Mater. **54**, 1–5 (2006)

T. Manoubi, M. Tence, M.G. Walls, C. Colliex, Curve fitting methods for quantitative-analysis in electron-energy loss spectroscopy. Microsc. Microanal. Microstruct. **1**, 23–39 (1990)

J.L. Maurice, D. Imhoff, J.P. Contoury, C. Colliex, Interfaces in 100 epitaxial heterostructures of perovskite oxides. Philos. Mag. **86**, 2127–2146 (2006)

A.J. McGibbon, Inst. Phys. Conf. Ser. **119**, 109 (1991)

F. Morales, F.M.F. De Groot, O.L.J. Gijzeman, A. Mens, O. Stephan, B.M. Weckhuysen, Mn promotion effects in CO/TiO_2 Fischer-Tropsch catalysts as investigated by XPS and STEM-EELS. J. Catal. **230**, 301–308 (2005)

P. Moreau, N. Brun, C.A. Walsh, C. Colliex, A. Howie, Relativistic effects in electron-energy-loss-spectroscopy observations of the Si/SiO_2 interface plasmon peak. Phys. Rev. B **56**, 6774–6781 (1997)

D.A. Muller, D.A. Shashkov, R. Benedek, L.H. Yang, J. Silcox, D.N. Seidman, Atomic scale observations of metal-induced gap states at 222MgO/Cu interfaces. Phys. Rev. Lett. **80**, 4741–4744 (1998)

D.A. Muller, L.F. Kourkoutis, M. Murfitt, J.H. Song, H.Y. Hwang, J. Silcox, N. Dellby, O.L. Krivanek, Atomic-scale chemical imaging of composition and bonding by aberration-corrected microscopy. Science **319**, 1073–1076 (2008)

D.A. Muller, T. Sorsch, S. Moccio, F. H. Baumann, K. Evans-Lutterodt, G. Timp. The electronic structure at the atomic scale of ultrathin gate oxides. Nature **399**, 758–761 (1999)

D.A. Muller, Y. Tzou, R. Raj, J. Silcox, Mapping sp(2) and sp(3) states of carbon at subnanometer spatial-resolution. Nature **366**, 725–727 (1993)

J. Nelayah, M. Kociak, O. Stephan, F.J.G. De Abajo, M. Tence, L. Henrard, D. Taverna, I. Pastoriza-Santos, L.M. Liz-Marzan, C. Colliex, Mapping surface plasmons on a single metallic nanoparticle. Nat. Phys. **3**, 348–353 (2007)

F. Ouyang, P.E. Batson, M. Isaacson, Quantum size effects in the surface-plasmon excitation of small metallic particles by electron-energy-loss spectroscopy. Phys. Rev. B **46**, 15421–15425 (1992)

M.P. Oxley, E.C. Cosgriff, L.J. Allen, Nonlocality in imaging. Phys. Rev. Lett. **94**, 203906 (2005)

B. Rafferty, L.M. Brown, Direct and indirect transitions in the region of the band gap using electron-energy-loss spectroscopy. Phys. Rev. B **58**, 10326–10337 (1998)

A. Rivacoba, N. Zabala, J. Aizpurua, Image potential in scanning transmission electron microscopy. Prog. Surf. Sci. **65**, 1–64 (2000)

L. Samet, D. Imhoff, J.L. Maurice, J.P. Contour, A. Gloter, T. Manoubi, A. Fert, C. Colliex, Eels study of interfaces in magnetoresistive LSMO/STO/LSMO tunnel junctions. Eur. Phys. J. B **34**, 179–192 (2003)

B. Schaffer, G. Kothleitner, W. Grogger, EFTEM spectrum imaging at high-energy resolution. Ultramicroscopy **106**, 1129–1138 (2006)

S. Schamm, C. Bonafos, H. Coffin, N. Cherkashin, M. Carrada, G. Ben Assayag, A. Claverie, M. Tence, C. Colliex, Imaging Si nanoparticles embedded in SiO_2 layers by (S)TEM-EELS. Ultramicroscopy **108**, 346–357 (2008)

S. Schamm, G. Zanchi, Study of the dielectric properties near the band gap by VEELS: Gap measurement in bulk materials. Ultramicroscopy **96**, 559–564 (2003)

J. Scott, P.J. Thomas, M. Mackenzie, S. McFadzean, J. Wilbrink, A.J. Craven, W.A.P. Nicholson, Ultramicroscopy **108**, 1586 (2008)

A. Seepujak, U. Bangert, A.J. Harvey, P.M.F.J. Costa, M.L.H. Green, Redshift and optical anisotropy of collective pi-volume modes in multiwalled carbon nanotubes. Phys. Rev. B **74**, 075402 (2006)

L.J. Sherry, R.C. Jin, C.A. Mirkin, G.C. Schatz, R.P. Van Duyne, Localized surface plasmon resonance spectroscopy of single silver triangular nanoprisms. Nano Lett. **6**, 2060–2065 (2006)

S.-Y. Chen, A. Gloter, A. Zobelli, L. Wang, C.-H. Chen, C. Colliex, Electron energy loss spectroscopy and ab initio investigation of iron oxide nanomaterials grown by a hydrothermal process. Phys. Rev. B **79**, 104103 (2009)

O. Stephan, D. Taverna, M. Kociak, K. Suenaga, L. Henrard, C. Colliex, Dielectric response of isolated carbon nanotubes investigated by spatially resolved electron energy-loss spectroscopy: from multiwalled to single-walled nanotubes. Phys. Rev. B **66**, 155422 (2002)

O. Stephan, M. Kociak, L. Henrard, K. Suenaga, A. Gloter, M. Tence, E. Sandre, C. Colliex, Electron energy-loss spectroscopy on individual nanotubes. J. Electron Spectrosc. Relat. Phenom. **114**, 209–217 (2001)

O. Stephan, P.M. Ajayan, C. Colliex, P. Redlich, J.M. Lambert, P. Bernier, P. Lefin, Doping graphitic and carbon nanotube structures with boron and nitrogen. Science **266**, 1683–1685 (1994)

T. Stockli, J.M. Bonard, A. Chatelain, Z.L. Wang, P. Stadelmann, Interference and interactions in multiwall nanotubes. Physica-B **280**, 4844–4847 (2000)

K. Suenaga, C. Colliex, N. Demoncy, A. Loiseau, H. Pascard, F. Willaime, Synthesis of nanoparticles and nanotubes with well-separated layers of boron nitride and carbon. Science **278**, 653–655 (1997)

D. Taverna, M. Kociak, O. Stephan, A. Fabre, E. Finot, B. Decamps, C. Colliex. Probing physical properties of confined fluids within individual nanobubbles. Phys. Rev. Lett. **1**, 035301 (2008)

D. Taverna, M. Kociak, V. Charbois, L. Henrard, Electron energy-loss spectrum of an electron passing near a locally anisotropic nanotube. Phys. Rev. B **66**, 235419 (2002)

D. Taverna, M. Kociak, V. Charbois, L. Henrard, O. Stephan, C. Colliex, Simulations of electron energy-loss spectra of an electron passing near a locally anisotropic nanotube. J. Electron Spectrosc. Relat. Phenom. **129**, 293–298 (2003)

M. Tencé, H. Pinna, T. Birou, L. Guiraud, A. Mayet, C. Pertel, V. Serin, C. Colliex, in *Proceedings of IMC 16*, H. Ichinose, T. Sasaki, ed. (2006) (Publication Committee of IMC16) pp. 824–825

M. Tence, M. Quartuccio, C. Colliex, PEELS compositional profiling and mapping at nanometer spatial-resolution. Ultramicroscopy **58**, 42–54 (1995)

D. Ugarte, C. Colliex, P. Trebbia, Surface-plasmon and interface-plasmon modes on small semiconducting spheres. Phys. Rev. B **45**, 4332–4343 (1992)

M. Varela, S.D. Findlay, A.R. Lupini, H.M. Christen, A.Y. Borisevich, N. Dellby, O.L. Krivanek, P.D. Nellist, M.P. Oxley, L.J. Allen, S.J. Pennycook, Spectroscopic imaging of single atoms within a bulk solid. Phys. Rev. Lett. **92**, 095502 (2004)

J. Verbeeck, S. Van Aert, Model based quantification of EELS spectra. Ultramicroscopy **101**, 207–224 (2004)

C.A. Walsh, J. Yuan, L.M. Brown, A procedure for measuring the helium density and pressure in nanometre-sized bubbles in irradiated materials using electron-energy-loss spectroscopy. Philos. Mag. A **80**, 1507–1543 (2000)

Z.L. Wang, Valence electron excitations and plasmon oscillations in thin films, surfaces, interfaces and small particles (vol 27, p 265, 1996). Micron **28**, 505–506 (1997)

S. Yakovlev, M. Libera, Dose-limited spectroscopic imaging of soft materials by low-loss EELS in the scanning transmission electron microscope. Micron **39**, 734–740 (2008)

N. Yamamoto, K. Araya, F.J.G. De Abajo, Photon emission from silver particles induced by a high-energy electron beam. Phys. Rev. B **6420**, 205419 (2001)

N. Zabala, A. Rivacoba, P.M. Echenique, Coupling effects in the excitations by an external electron beam near close particles. Phys. Rev. B **56**, 7623–7635 (1997)

N. Zabala, A. Rivacoba, P.M. Echenique, Energy loss of electrons travelling through cylindrical holes. Surf. Sci. **209**, 465–480 (1989)

Y. Zhang, K. Suenaga, C. Colliex, S. Iijima, Coaxial nanocable: Silicon carbide and silicon oxide sheathed with boron nitride and carbon. Science **281**, 973–975 (1998)

5

Energy Loss Near-Edge Structures

Guillaume Radtke and Gianluigi A. Botton

5.1 General Concepts

5.1.1 Introduction

One of the main advantages of scanning transmission electron microscopy (STEM) is the capability of recording a number of signals at the location of the electron beam, including characteristic X-rays and the measurement of the distribution of energy lost by the primary electron beam. Due to their importance in materials research, the use of these two techniques, known in general as "analytical electron microscopy," has been the topic of extended reviews and monographs (Botton 2007, Joy et al. 1986, Sigle 2005, Williams and Carter 1996). In general these techniques are used, primarily, to extract local information on the composition of the sample with a resolution limited in part by the delocalization of the signal due to the long-range interaction discussed in Chapter 6 of this book and in part by the broadening of the beam due to the sample thickness. In this chapter, we will focus on the particular subset of analytical signals that allow the extraction of information on the chemical environment of the atoms probed by the fast primary electron beam. One might wonder why use the transmission electron microscope for such studies? Indeed, a range of techniques are currently used to extract such information. For example, X-ray photoelectron spectroscopy (XPS) is routinely used to extract information on the chemical state of elements in thin films; X-ray absorption spectroscopy (XAS) is also used in most synchrotron facilities to extract valence and coordination; Fourier transform infrared spectroscopy, Raman spectroscopy, and nuclear magnetic resonance are available in most laboratories working on synthesis of new materials or compounds. What is special about chemical environment data extracted in STEM is the potential of obtaining the same or complementary information as XPS and XAS with near, or effectively the same, energy resolution as XAS but with significantly higher spatial resolution, approaching today, with modern aberration-corrected microscopes, the interatomic spacing in crystals.

When the incident primary electrons of the highly focused beam interact with the atoms in a sample, core and valence electrons ejected

S.J. Pennycook, P.D. Nellist (eds.), *Scanning Transmission Electron Microscopy*,
DOI 10.1007/978-1-4419-7200-2_5,

from their initial energy level can subsequently scatter with the potential of the crystal or exit the sample depending on their energy. The minimum energy required by the primary electron to eject the core electrons corresponds, to a first approximation, to the ionization potential of the excited atom in the specimen. However, the ejected core electron can also probe the first unoccupied bound states of the crystal. Therefore, its final state and allowed energy levels will strongly depend on the overall electronic structure and thus structural and chemical environment of the atom excited by the incident electrons (Figure 5–1).

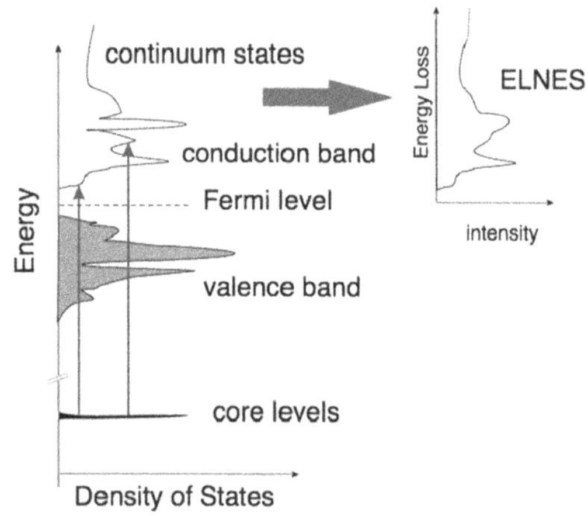

Figure 5–1. Schematic diagram representing the electronic transitions involved in ELNES in a solid. Occupied states are *shaded*.

Primary electrons can, of course, also provide higher energy to the core electrons that can therefore probe much higher energy states of the samples, constituting the continuum. The final state of the ejected electron is ultimately reflected on the probability of energy loss by the primary electron and is thus visible on the electron energy loss spectrum and in particular in the electron energy loss near-edge structure (ELNES) that modulates the first few electron volts from the absorption edge threshold (Figure 5–1).

Such modulations reflect changes in the chemical bonding environment including the type of coordination and the valence (in broader terms of "the electronic structure and bonding"), all of which depend on the crystalline structure surrounding the excited atom. Modulations at higher energies from the edge threshold (Figure 5–2) are also present in the energy loss spectrum, and these reflect the scattering of ejected electrons with increasingly higher kinetic energies in the potential of the crystal. These modulations are known as extended energy loss fine structures (EXELFS) and provide information on the interatomic distances and coordination number, which are also strongly dependent on the crystalline structure surrounding the excited atoms. Forming a continuum, these two parts of the energy loss fine structure represent excited electrons probing the bound states, near the edge threshold and to the continuum states when the excited electron has enough kinetic energy to be essentially free.

Due to the similarities in the scattering and excitation processes, it is clear that the ELNES provides very similar information to XAS (in

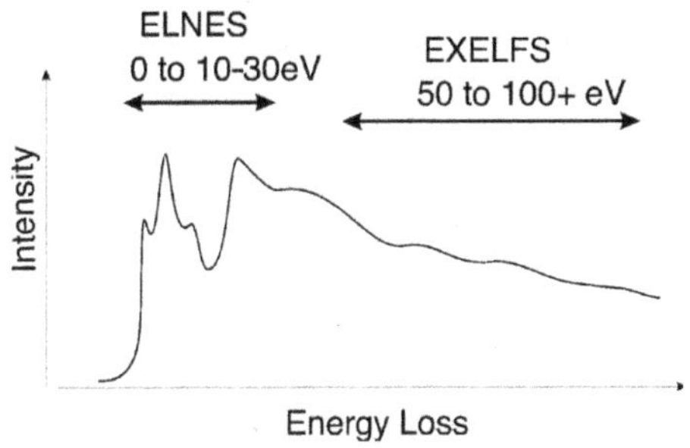

Figure 5–2. Schematic energy loss spectrum corresponding to an ionization edge with two regions showing different types of fine structures.

particular X-ray absorption near-edge structures – XANES) while the EXELFS are equivalent to extended X-ray absorption fine structure – EXAFS. The main difference remains excitation by high-energy electrons (hence the capability of obtaining information from much smaller areas) rather than by photons and the fact that momentum transfer can be precisely tuned depending on the experimental collection conditions (aperture size, position in the diffraction pattern).

In order to give the reader the necessary background to understand the potential of the technique and some examples of spatially resolved ELNES, we will first cover the fundamental aspects of electron scattering so that the reader can relate the "electronic structure and bonding" of the crystal to the features observed in the energy loss spectrum. The origins of the various formalisms and approximations used to understand the fine structures are reviewed so that the reader can identify the most appropriate method to model, if necessary, the spectra and extract quantitative information on structural and chemical parameters such as crystal field, valence, and spin state.

5.1.2 Bethe's Theory of Inelastic Scattering

The modeling of ELNES is intimately associated with the calculation of a fundamental quantity known as the inelastic *double differential scattering cross section* (DDSCS). The DDSCS measures the probability per unit of time, energy, solid angle, and incident electron density for a fast incident electron to be scattered inelastically when propagating through the sample. Among the large number of physical processes that possibly result in a loss of energy of the incident fast electron, energy loss near-edge structures are related to processes involving the creation of only *one* electron–hole pair in the solid and even to the more restricted case where the hole is localized on an atomic core state.

To solve this problem, one has to derive an expression for the probability of an incident fast electron in the initial state $\mathbf{k}_0$ of energy E_0 to be scattered in the final state $\mathbf{k}'$ of energy E' by inducing a transition of an electron of the solid from an atomic core state to an unoccupied conduction state. The geometry is illustrated in Figure 5–3. This scattering process is therefore associated with both a momentum transfer $\mathbf{q} = \mathbf{k}' - \mathbf{k}_0$ and an energy transfer E from the fast electron to the core electron. This probability can be calculated through the first-order time-dependent perturbation theory assuming a Coulomb interaction

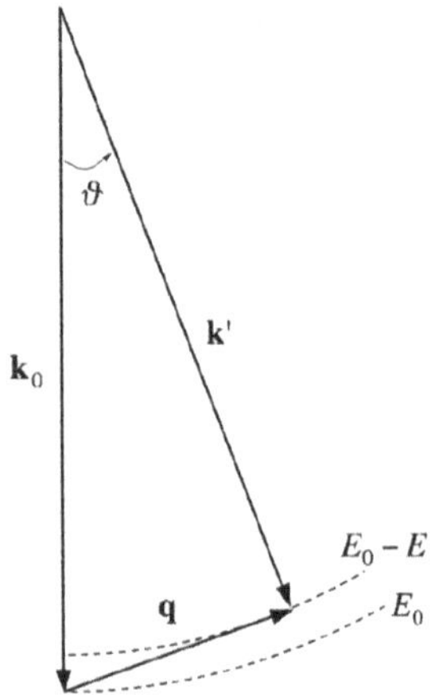

Figure 5–3. Geometry of inelastic scattering.

potential V between the electrons of the solid of coordinates $\mathbf{r}_i$ and the fast electron of coordinate $\mathbf{r}$ given by

$$V = -\sum_i \frac{e}{\|\mathbf{r} - \mathbf{r}_i\|}. \tag{1}$$

The first quantum mechanical expression for the cross section of the inelastic scattering of a charged particle on matter was derived by H.A. Bethe in the early 1930s (Bethe 1930). Applying the first-order plane-wave Born approximation to this problem, Bethe's theory leads to the following expression for DDSCS:

$$\frac{\partial^2 \sigma}{\partial E \partial \Omega} = \frac{4\gamma^2}{a_0^2 q^4} \frac{k'}{k} S(\mathbf{q}, E), \tag{2}$$

where the quantity $S(\mathbf{q}, E)$ is known as the dynamic form factor (DFF):

$$S(\mathbf{q}, E) = \sum_{c,v} \left| \langle \phi_v | \, e^{i\mathbf{qr}} \, | \phi_c \rangle \right|^2 \delta(\varepsilon_v - \varepsilon_c - E), \tag{3}$$

and where $a_0 = \hbar^2/me^2$ is the Bohr radius, $\gamma = 1/\sqrt{1 - v^2/c^2}$ is the relativistic factor, $|\phi_c\rangle$ is the initial core state of energy ε_c, and $|\phi_v\rangle$ is the final conduction state of energy ε_v for the solid electron. In this expression, the Dirac distribution ensures the overall energy conservation of the closed system solid + fast electron. The particular form of the matrix element appearing in this expression has important consequences on the physical interpretation of the near-edge fine structure. Indeed, we should first stress the fact that the initial one-electron state is an atomic core state localized on the nucleus of the excited atom. As a consequence, the evaluation of this matrix element is obtained by integrating the spatial coordinates over a very sharp region (with a spatial extension of order of the Bohr radius a_0) centered on the nucleus. For medium acceleration voltages < 400 keV, the observable momentum transfers fall in the range where $qa_0 << 1$, and therefore the operator $e^{i\mathbf{qr}}$ can be accurately approximated by its first-order series expansion:

$$e^{i\mathbf{qr}} \approx 1 + i\mathbf{qr} + \cdots . \tag{4}$$

This approximation, known as the *dipole approximation*, leads to a simplified version of the dynamic form factor:

$$S_{\text{dip}}(\mathbf{q}, E) = \sum_{c,v} \left| \langle \phi_v | \, \mathbf{qr} \, | \phi_c \rangle \right|^2 \delta(\varepsilon_v - \varepsilon_c - E), \tag{5}$$

where the constant term in the series expansion (4) does not contribute to the matrix element due to the orthogonality of initial and final states.

5.1.3 Probing the Local Density of States

Equation (5) bears similarities to the expression of the density of states (DOS), familiar to solid-state physicists:

$$\rho(E) = \sum_{v} \delta(\varepsilon_v - E), \tag{6}$$

where the sum is taken over the *unoccupied* electronic states of the solid, i.e., states above the Fermi level ($\varepsilon_v > \varepsilon_{\text{F}}$). The dynamic form factor is therefore often further approximated under the form

$$S_{\text{dip}}(q, E) = \left| M_{vc}(q, E) \right|^2 \rho(E), \tag{7}$$

where $M_{vc}(\mathbf{q}, E)$ represents the dipole matrix element and $\rho(E)$ is the total density of unoccupied states of the solid. Evaluation of the angular part of the dipole matrix element appearing in Eq. (7) leads to the very important *dipole selection rule*. This rule states that only unoccupied states characterized by a well-defined orbital momentum ℓ that differs from the orbital momentum ℓ_c of the initial core electron by unity are accessible to the atomic electron, i.e.,

$$\ell = \ell_c \pm 1. \tag{8}$$

Incorporating this result in Eq. (7) gives finally

$$S_{\text{dip}}(q, E) = \left| M_{l_c-1}(q, E) \right|^2 \rho_{l_c-1}(E) + \left| M_{l_c+1}(q, E) \right|^2 \rho_{l_c+1}(E). \tag{9}$$

The dynamic form factor may therefore be interpreted in the dipole approximation as a weighted local density of states (LDOS), i.e., as an image of a site- and symmetry-projected DOS. This "local" character comes from the fact that the dominant contributions to the dipole matrix element in Eq. (9) correspond to the region of space where the overlap between the initial and the final state wave functions is non-negligible, i.e., essentially from a sharp region centered on the excited atom. The symmetry of the initial core state therefore dictates the symmetry of the probed LDOS: the excitation of a 1s core electron involves transitions to unoccupied p-LDOS, the excitation of a 2p core electron involves transitions to unoccupied s- and d-LDOS, and so on. The nomenclature commonly used in core-level spectroscopies is based on the following rules: one uses the letters K, L, M, ..., for the principal quantum number of the core state $n = 1, 2, 3, \ldots$, respectively, and the

Table 5–1. Nomenclature used in X-ray and electron spectroscopies. Edges for which the initial core state has an angular momentum $\ell_c \neq 0$ are split by the spin–orbit coupling of the core hole.

Core state	Spectroscopic name	Symmetry of the probed LDOS
$1s$	K	p
$2s$	L_1	p
$2p_{1/2}$ and $2p_{3/2}$	L_{23}	s + d
$3s$	M_1	p
$3p_{1/2}$ and $3p_{3/2}$	M_{23}	s + d
$3d_{3/2}$ and $3d_{5/2}$	M_{45}	p + f
$4s$	N_1	p
$4p_{1/2}$ and $4p_{3/2}$	N_{23}	s + d
$4d_{3/2}$ and $4d_{5/2}$	N_{45}	p + f
$4f_{5/2}$ and $4f_{7/2}$	N_{67}	d + g

subscripts 1, 2, 3, ..., instead of the orbital angular momentum s, p, d, ..., to label the different peaks from high to low energy within a shell of quantum number n. These spectroscopic notations used in electron and X-ray spectroscopies are summarized in Table 5–1. Edges for which the initial core state has an angular momentum $\ell_c \neq 0$ are grouped by pairs. In these cases, the hole remaining on the core state after the ejection of the core electron indeed experiences a (usually strong) spin–orbit splitting. The spin–orbit interaction is a relativistic effect that couples the orbital and the spin angular momenta of the core hole and splits the resulting states according to the value of their total angular momentum j. In the case of a single core hole, the spin angular momentum s is either $-1/2$ or $+1/2$ in $\hbar$ units, and the total angular momentum j is therefore given by $\ell_c - 1/2$ or $\ell_c + 1/2$. Let us take the example of a $2p$ excitation: according to the previous discussion, two edges are observed (as the core state angular momentum is $\ell_c = 1$), namely the L_3 edge at lower energy and the L_2 edge at higher energy. These edges are split by the core-hole spin–orbit coupling, the L_3 edge involving transitions from the $2p_{3/2}$ and the L_2 edge involving transitions from the $2p_{1/2}$ core states.

Although Eq. (9) provides a simple physical picture for the interpretation of the near-edge fine structure in terms of site- and symmetry-projected LDOS, it is important to keep in mind that, strictly speaking, this proportionality between the dynamic form factor and the LDOS is exact *only* when dynamical diffraction is neglected and in the absence of $\ell + 1/\ell - 1$ cross terms (Nelhiebel et al. 1999a, b).

5.1.4 Key Approximations

At this point of our discussion, it is interesting to highlight some of the underlying approximations that have been used to establish all these results.

First of all, we assumed that the electron inelastic scattering corresponds to a *one-electron process*. This approximation allowed us to write the initial and final state wave functions as, respectively, a single-electron atomic core state wave function and a single-electron valence state wave function delocalized over the whole solid. Practically, this one-electron approach of the near-edge fine structure is very efficient for interpreting K edges as well as certain L_{23} edges but definitely fails in other cases, such as transition metal L_{23} or M_{23} edges and rare earth M_{45} or N_{45} edges. The fundamental reason for this discrepancy is that one does *not* observe a local density of unoccupied electronic states of the solid any more in these specific edges. Instead, strong atomic effects dominate the edge shape. Whereas these effects are very weak and barely visible in K edges, where an electronic transition occurs from the 1*s* core state, they become prominent, for example, in the case of transition metal L_{23} edges, essentially because of the strong overlap of the core (2*p*) and valence (mainly 3*d*) wave functions (de Groot and Kotani 2008). This overlap is responsible for the creation of final states that can be determined by coupling the different orbital and spin momenta of the partly filled electronic shells of the excited atom. These *atomic multiplet effects* will be presented in Section 5.2.2.

A second important approximation lies in the non-relativistic first-order Born approximation used to develop this theory. More precisely, it was thought for a long time that a fully relativistic approach of the problem was not necessary for the usual acceleration voltages used in transmission electron microscopy. Therefore, a simple correction accounting for the relativistic increase of mass of the incident electron was often considered sufficient. As can be seen in Table 5–2, the velocity of the probe electrons reaches 0.69*c* for a 200 keV microscope and 0.77*c* for a 300 keV microscope. At these velocities, the mutual interaction between the incident electrons and the electrons of the solid can no longer be described using the Coulomb potential (1). Instead, a fully relativistic treatment of Bethe's theory in the dipole approximation should be used. Application of this relativistic theory leads to a modified version of the double differential scattering cross section (Schattschneider 2005):

$$\frac{\partial^2\sigma}{\partial E \partial \Omega} = \frac{4\gamma^2}{a_0^2 Q^2} \frac{k'}{k} S_{\text{dip}}(\mathbf{q}', E), \tag{10}$$

where $Q = q^2 - q_z^2\beta^2$ and where the dynamic form factor has now the form

$$S_{\text{dip}}(\mathbf{q}', E) = \sum_{c,v} \left| \langle \phi_v | \, \mathbf{q}'\mathbf{r} \, | \phi_c \rangle \right|^2 \delta(\varepsilon_v - \varepsilon_c - E). \tag{11}$$

In this equation, we used $\mathbf{q}' = \mathbf{q} - q_z\beta^2\hat{\mathbf{z}}$ where the *z*-axis is oriented along the propagation direction of the incident fast electron.

An immediate consequence of this relativistic treatment is the modification of the angular distribution of the scattered intensity. For isotropic systems in the non-relativistic case, $S_{\text{dip}} \propto q^2$ (Eq. (5)) and one can easily obtain

Table 5–2. Velocity, relativistic factor, and wavelength of a fast electron for different primary energies.

Primary energy E_0 (in keV)	Normalized velocity $\beta = \frac{v}{c}$	Relativistic factor $\gamma = \sqrt{1-\beta^2}$	Wavelength λ (in Å)
10	0.1950	1.0196	0.1220
80	0.5024	1.1566	0.0418
100	0.5482	1.1957	0.0370
120	0.5867	1.2348	0.0335
200	0.6953	1.3914	0.0251
300	0.7765	1.5871	0.0197
500	0.8629	1.9785	0.0142
1000	0.9411	2.9569	0.0087

$$\frac{\partial^2\sigma}{\partial E \partial\Omega} \propto \frac{1}{\vartheta^2 + \vartheta_E^2}, \tag{12}$$

where ϑ is the scattering angle (Figure 5–3) and ϑ_E is the *characteristic* angle defined as $\vartheta_E = E/2\gamma T$. In this expression, T is the classical kinetic energy of the probe electron. This is the well-known Lorentzian distribution of the intensity centered on the forward scattering direction. When an appropriate relativistic treatment is applied to the problem, we see that $S_{\text{dip}} \propto \mathbf{q}'^2$ and according to Eq. (10)

$$\frac{\partial^2\sigma}{\partial E \partial\Omega} \propto \frac{\vartheta^2 + \vartheta_E^2/\gamma^4}{(\vartheta^2 + \vartheta_E^2/\gamma^2)^2}. \tag{13}$$

The relativistic effects are particularly visible for high acceleration voltages where the DDSCS shows a large increase for $\vartheta > 0$ (Kurata et al. 1997). Of course, Eq. (13) reduces to Eq. (12) in the non-relativistic case, i.e., in the limit where $\gamma \to 1$.

Finally, the channeling or dynamical scattering of the fast electron prior to and after the inelastic event was totally ignored in the formalism presented in Section 5.1.2. Indeed, the fast electron channeling through the crystal before and after the ionization of one of its atoms cannot be simply described with a single plane wave but instead by many of them (Schattschneider et al. 1996). A number of theoretical works have been devoted to the study of channeling effects on the inelastic scattering cross section using either Bloch state (Allen and Josefsson 1995, Oxley and Allen 1998) or multislice (Dwyer 2005) formulations. The reader is referred to Chapter 6 for a detailed presentation of this problem.

5.1.5 Anisotropy and the Experimental Conditions

Inspection of Eq. (5) or (11) of the DFF reveals that the scattering vector $\mathbf{q}$ in an EELS experiment plays the same role as the polarization vector **ε** does in X-ray absorption spectroscopy (XAS). In the case of anisotropic crystals, the selection of a particular direction of the scattering vector with respect to the reference system of the sample opens the possibility to probe a particular symmetry of the unoccupied electronic states and therefore to study their spatial orientation. Let us take the example of a uniaxial material such as graphite. The C K edge onset in this compound is dominated by two prominent features corresponding to transitions from the 1*s* core state to antibonding π^* orbitals at 284–285 eV and σ^* orbitals at 292–296 eV, as shown in Figure 5–4. These unoccupied states have a well-defined spatial orientation, the former ones being built from the C p_z orbitals oriented *perpendicularly* to the graphene planes and the latter ones being essentially oriented *within* the planes (Leapman et al. 1983). Now let us assume that a graphite sample is observed in the [0001] zone axis under a perfectly parallel illumination in a 200 keV microscope. Under these experimental conditions, the incident electron wave vector $\mathbf{k}_0$ is parallel to the *c*-axis of the crystal. If we look, in the reciprocal space (diffraction plane), at the electrons scattered inelastically after ionization of the C 1*s* shell, we obtain the intensity distribution shown in Figure 5–5(a) given by expression (13).

If now we only consider electrons responsible for the $1s \rightarrow \pi^*$ electronic transition, we obtain the much sharper distribution shown in

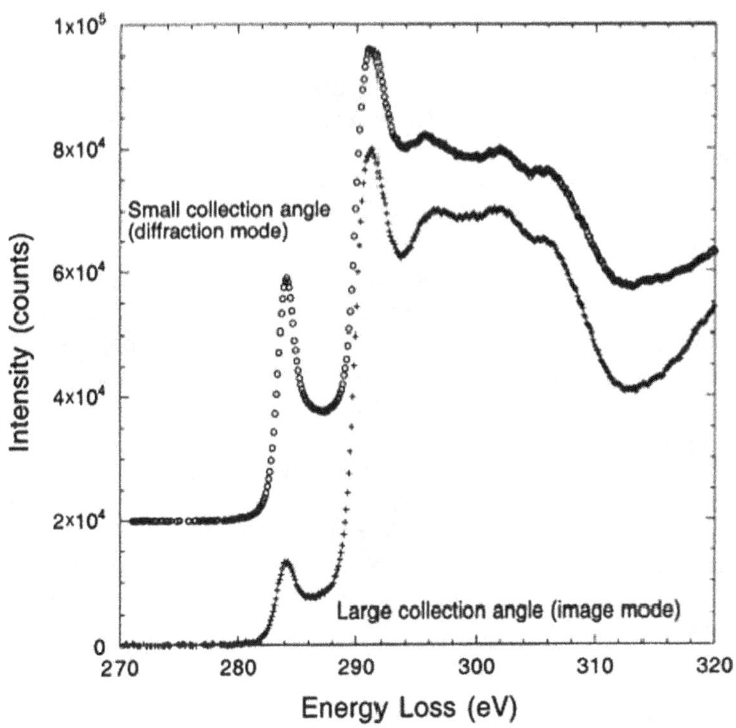

Figure 5–4. C K edge recorded in graphite oriented along the [0001] zone axis under parallel illumination and with two different collection angles (reproduced from Botton 2005).

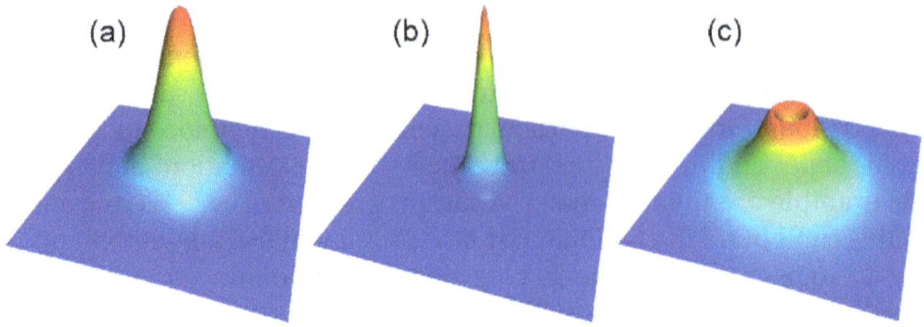

Figure 5–5. Representation of the intensity distribution of electrons scattered inelastically after ionization of the C-1s shell in a graphite sample oriented in the [0001] zone axis under parallel illumination at 200 keV. The area represented is centered on the optical axis with the dimensions $10 \times 10\ \vartheta_E$ ($\vartheta_E \sim 0.83$ mrad for the C K edge at 200 keV): (a) intensity distribution of the overall edge, (b) intensity distribution corresponding to the $1s \rightarrow \pi^*$ transition only, and (c) intensity distribution corresponding to the $1s \rightarrow \sigma^*$ transition only.

5–5 (b). In this case, the intensity is essentially concentrated on the optical axis (Botton 2005). Under these particular experimental conditions, the forward scattered electrons have a momentum transfer **q** parallel to the c-axis, i.e., aligned with the C p_z orbitals of the π^* states. On the contrary, electrons responsible for the $1s \rightarrow \sigma^*$ electronic transition, shown in Figure 5–5(c), are distributed in a totally different manner. The intensity presents a broad maximum located on a ring of angle ϑ_E/γ and falls to zero on the optical axis. Transitions to the p_z orbitals need a component of **q** parallel to the c-axis (in order to have an active electric dipole along this direction). These results may be understood as follows: as the angle between **q** and c increases, the transition probability to the p_z orbitals decreases. The opposite is true for the transitions to the p_x and p_y orbitals belonging to the in-plane σ^* states. One therefore easily understands that the relative intensities of the π^* and σ^* peaks will largely depend on the exact size and position of the spectrometer entrance aperture used to record the experimental spectrum. As can be seen in Figure 5–4, in the experimental conditions described above, the use of small collection angles favors the π^* transitions whereas large collection angles clearly tend to reduce their relative weight.

In a practical CTEM or STEM experiment, such a selection of a particular orientation of the scattering vector with respect to the reference system of the crystal is practically impossible due to the use of finite convergence and collection semi-angles. Instead, an averaged spectrum with a reduced momentum resolution is obtained, resulting from the collection of electrons with different incident and scattered wave vectors. Despite these experimental difficulties, different setups have been proposed in the literature to enhance the orientation sensitivity of ELNES (Botton et al. 1995, Browning et al. 1991, Keast et al. 2001). An important body of work has been devoted to the study of the influence of the experimental conditions – the convergence (α) and collection (β) semi-angles as well as the specimen orientation – on the near-edge fine structure (Hébert et al. 2006, Le Bossé et al. 2006). In particular, these works demonstrate the existence and allow the determination of specific couples (α, β) known as the *magic angle conditions* for which the near-edge structure of anisotropic materials becomes independent of the specimen orientation.

5.2 Quantitative Interpretation of Near-Edge Fine Structures

5.2.1 K Edges and the One-Electron Picture

5.2.1.1 Introduction

We demonstrated in the previous section that the central quantity appearing in the theory of inelastic scattering is the dynamic form factor. The main challenge that one faces when trying to model the near-edge fine structure is therefore to calculate this quantity as accurately as possible. This problem essentially reduces to the calculation of the initial and final states $|\phi_c\rangle$ and $|\phi_v\rangle$, respectively, appearing in Eq. (11) of the dynamic form factor. If the former is relatively insensitive to the details of the electronic structure of the sample under investigation, the latter provides a *fingerprint* of the "chemical bonds" between the atoms in the solid state.

The influence of the core state is essentially limited to two dominant aspects. First, it determines in large part the absolute energy of the edge onset. Indeed, as a first approximation, the minimum energy needed to eject an electron from a core state to an unoccupied state located just above the Fermi level is directly related to the binding energy of the core electron. Second, the core level dictates the symmetry of the probed final states by application of the dipole selection rule. On the other hand, the details of the fine structure are associated with the energy distribution of the unoccupied states, i.e., with the structure of the conduction states.

Different approaches may be used to tackle the problem of calculating the unoccupied states ranging from simple molecular orbital calculations to advanced direct-space multiple scattering (Moreno et al. 2007) and reciprocal-space band structure methods (Hébert 2007). Most of the popular codes currently used to calculate near-edge fine structure are based on density functional theory (DFT) (Jones and Gunnarsson 1989) applied practically within the local spin density approximation (LSDA). Its implementation for calculating the electronic structure of solids may be based on a wide range of methods including the orthogonalized linear combination of atomic orbitals (Tanaka et al. 2005), the linearized muffin-tin orbitals (LMTO) (Paxton 2005), the linearized augmented plane waves (LAPW) (Hébert 2007), the augmented spherical waves (ASW) (de Groot et al. 1993), the plane waves and pseudopotentials (PP) (Elsässer and Köstlmeier 2001, Jayawardane et al. 2001), or real space Korringa-Kohn-Rostoker (KKR) method (Rez et al. 1998). Although LSDA calculations usually provide an excellent description of the experimental near-edge fine structure, other methods have been developed to go beyond this approximation. For example, the LSDA+U

method (U is the Hubbard parameter) is widely used to improve the electronic structure and band gap description of strongly correlated systems and has been employed successfully to model ELNES in compounds such as NiO (Dudarev et al. 1998), $SrCu_2(BO_3)_2$ (Radtke et al. 2008), or La_2CuO_4 (Czyzyk and Sawatzky 1994). Methods based on the solution of the electron-core-hole Bethe–Salpeter equation (Shirley 1998) also clearly improve the theoretical description of the experimental data in certain cases. In the following discussion, however, we will focus on the common LSDA description of the spectra.

5.2.1.2 *A Qualitative Description of the O K Edge in* TiO_2

In order to analyze the different steps involved in the modeling of a K edge using state-of-the-art band structure calculations and to illustrate the deep relationship existing between the fine structure and the chemistry taking place in the solid state, we will discuss in detail the example of the O K edge of *rutile* titanium dioxide. The rutile structure is described in a tetragonal unit cell (space group $P4_2/mnm$) containing two inequivalent atoms, a Ti atom at coordinates (0,0,0) in position 2*a* and an O atom at coordinates $(u, u, 0)$ with $u = 0.3048$ in position 4*f*. The tetragonal unit cell contains therefore two molecular units as shown in Figure 5–6(a). In this structure as in many transition metal oxides, each Ti atom is surrounded by a coordination octahedron built from six oxygen atoms. In this particular case, the octahedron is distorted leading to a lower symmetry of the Ti site (D_{2h} point group). These octahedra may be thought of as basic structural units assembled in a three-dimensional lattice as shown in Figure 5–6(b): columns of edges sharing octahedra are oriented along the *z*-axis of the crystal and held together by sharing corners.

A good starting point for our investigation of the electronic structure of this compound is therefore to study the structure of the molecular orbitals in a perfect (O_h symmetry) TiO_6 octahedron first. If we consider TiO_2 as a purely ionic compound, the actual formal valences of Ti and O are close to Ti^{4+} and O^{2-}, the electrons transferred from the Ti atom filling completely the O 2*p* shell and leaving the Ti 3*d* orbitals

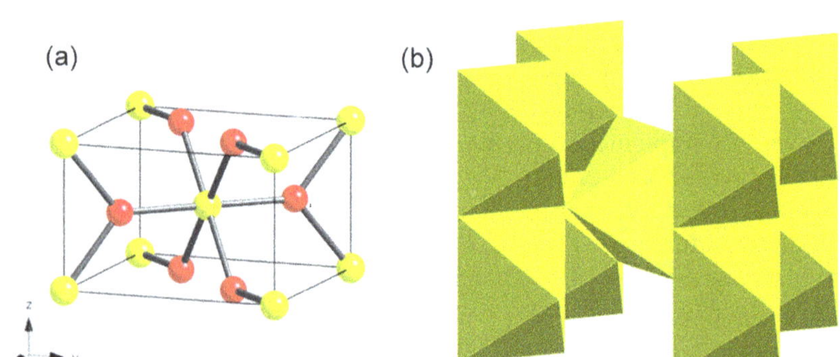

Figure 5–6. (a) Tetragonal unit cell of rutile TiO_2. The Ti atoms are represented in *yellow* and the O atoms are represented in *red*. (b) Arrangement of the coordination octahedra around the Ti atoms.

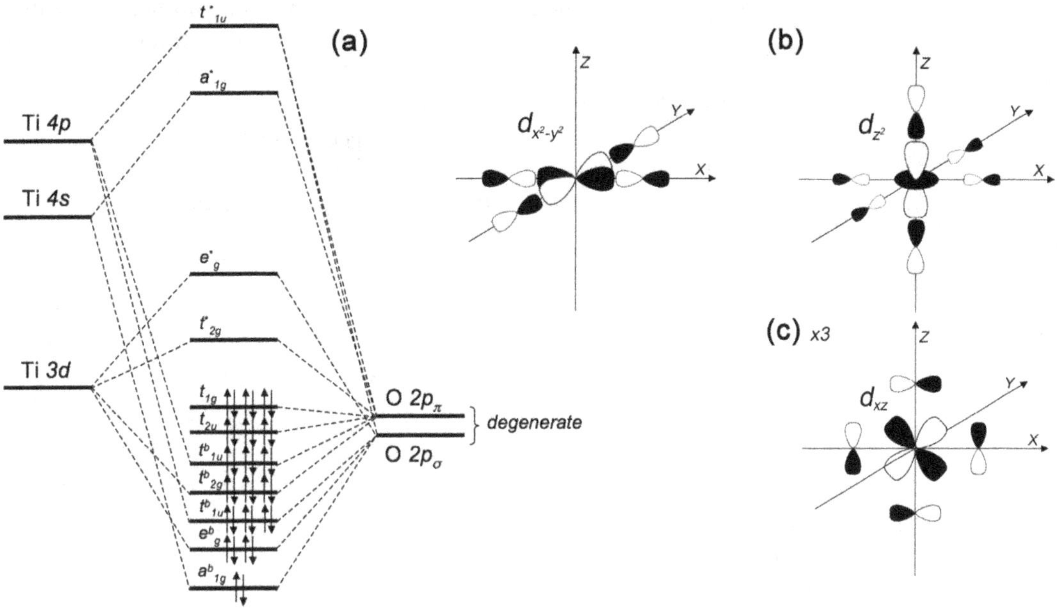

Figure 5–7. (a) Molecular orbital structure of a $[TiO_6]^{8-}$ cluster. The bonding character of the molecular orbitals is indicated by the *superscript b* whereas a *star* symbolizes the antibonding character. (b) Schematic representation of the e_g^* antibonding molecular orbitals. (c) Schematic representation of the t_{2g}^* molecular orbital involving the Ti $3d_{xz}$ state. The other two molecular orbitals may be constructed in the same manner in the (X,Y) and (Y,Z) planes using the Ti $3d_{xy}$ and $3d_{yz}$ states.

unoccupied. In this model, the cohesive energy of the solid comes essentially from the Coulomb attraction between positively and negatively charged ions. This picture is, however, far from accurate since, as discussed below, a large covalent interaction exists between Ti and O orbitals. Nevertheless, we will consider a $[TiO_6]^{8-}$ octahedron in the following discussion. The schematic molecular orbital structure of this cluster is represented in Figure 5–7(a). In this molecular picture, we only consider a restricted number of orbitals playing an important role around the Fermi level and in the first unoccupied states, i.e., the Ti 3*d*, 4*s*, and 4*p* and the O 2*p* states. We also labeled the molecular orbitals resulting from the hybridization of these states according to their symmetry properties by using the names of the irreducible representations of the O_h group they span.

On the right-hand side of the figure, the O 2*p* states are classified in two subgroups, namely O $2p_\sigma$ and O $2p_\pi$, according to the type of overlap they experience with the Ti orbitals. The first group corresponds to states constructed as linear combinations of O 2*p* atomic orbitals that are symmetric about the line joining the ligands and the Ti atoms and therefore leading to σ-type bonds. The second group, on the other hand, corresponds to states built from orbitals oriented perpendicularly to the Ti–O lines and leading to the formation of π-type bonds. As a general principle, each of these symmetry-adapted linear combinations of ligand 2*p* orbitals will interact with the transition metal orbitals of the same symmetry to create bonding and antibonding molecular states. For example, the large π overlap between the O $2p_\sigma$ and the Ti $3d$ states of e_g symmetry leads to the formation of bonding e_g^b and antibonding e_g^* molecular orbitals. Due to the larger electronegativity of the O atoms, or equivalently to the lower energy of the O $2p$ states, the former exhibit a large O character whereas the latter have a dominant Ti component. A schematic representation of the two e_g^* molecular orbitals involving, respectively, the $3d_{x^2-y^2}$ and $3d_{z^2}$ crystalline orbitals of the Ti atoms is given in Figure 5–7(b).

The same type of overlap occurs between the O $2p_\sigma$ and the high-lying Ti $4s$ and $4p$ states stabilizing a_{1g}^b and t_{1u}^borbitals and pushing antibonding a_{1g}^* and t_{1u}^* orbitals toward higher energies. The weaker π overlap between O $2p_\sigma$ and Ti $3d$ orbitals of t_{2g} symmetry leads to a smaller energy splitting between the bonding t_{2g}^b and the antibonding t_{2g}^* molecular orbitals. A representation of the t_{2g}^* molecular orbital involving the Ti $3d_{xz}$ is shown in Figure 5–7(c). The two remaining molecular orbitals present a similar shape but are oriented in the (X,Y) and (Y,Z) planes. The resulting energy separation between antibonding t_{2g}^* and e_g^* orbitals may be identified to the well-known crystal field splitting "10Dq" used in crystal field theory.

It should be noted here that, due to their specific symmetries, some of the O $2p_\pi$ molecular orbitals (labeled t_{1g} and t_{2u}) are non-bonding. To complete this approach, we can now fill out the lowest orbitals with the 36 electrons available following the 0 K Fermi statistics, as indicated in Figure 5–7(a). A HOMO-LUMO gap appears between non-bonding orbitals of dominant O character and the slightly antibonding t_{2g}^* Ti $3d$ states. This simplified molecular orbital model contains most of the basic features of the electronic structure of TiO_2 and provides a first idea of the near-edge fine structure observable on the O K edge in this compound: one expects the first unoccupied states to be dominated by the Ti $3d$ states split by the octahedral crystal field into states of t_{2g} and e_g symmetries and followed at much higher energies by the Ti $4sp$ states.

Strictly speaking, these states are of *dominant* Ti character but are not purely built from the Ti states, as one would expect in the case of a pure ionic bond. On the contrary, these orbitals hybridize strongly with the O $2p$ orbitals. This covalence explains why these structures predominantly associated with the Ti states are visible on the O K edge. Molecular orbital calculations usually provide a simple picture of the unoccupied states useful to analyze the experimental data. However, these results remain largely qualitative, as the real structure of the solid is not taken into account. A much more accurate approach allowing a quantitative modeling of the experimental data is provided by ab initio band structure calculations.

5.2.1.3 Ab Initio Calculation of the O K Edge in TiO_2

Density functional calculations conducted for TiO_2 within the local density approximation (Blaha et al. 2009) are presented in Figure 5–8(a). It

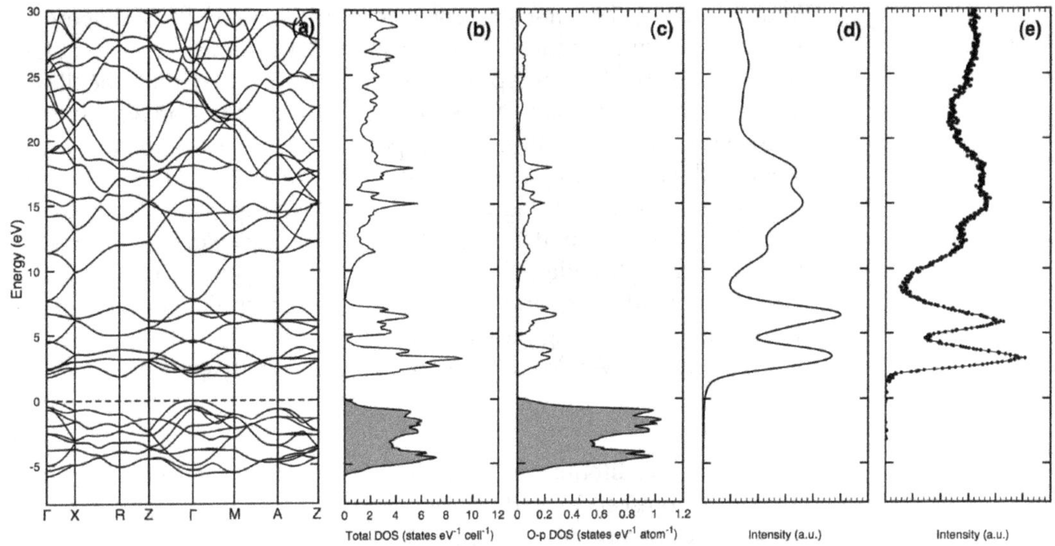

Figure 5–8. First-principle calculation of the TiO_2 electronic structure within the local density approximation. (a) The band structure of TiO_2 (energy relative to the Fermi level). (b) The total density of states. (c) The O p projected density of states. (d) The theoretical O K edge calculated in the dipole approximation. (e) The experimental O K edge. These calculations were performed using the Wien2k (Blaha et al. 2009) code. *Shaded* states are occupied.

is possible to understand the basic features of the complex band structure shown in Figure 5–8(a) in light of the simple model we analyzed in the previous section. First of all, we notice the presence of a 5-eV-wide valence band made of 12 individual bands. These bands are directly built from the 12 distinct O p orbitals present in the unit cell (corresponding to three p orbitals per atom × two atoms per molecular unit × two molecular units per unit cell). They may be interpreted as the solid-state equivalent to the occupied molecular orbitals of dominant O p character of Figure 5–7(a). This valence band is separated from the first unoccupied states of Ti d character by a band gap of 1.7 eV, equivalent for an infinite solid to the HOMO-LUMO gap observed in the molecular orbital model. The effect of the crystal field on the Ti 3d states is also clearly visible in Figure 5–8(a) where two groups of bands correspond, respectively, to the Ti 3d t_{2g} and e_g states. The first group is indeed constituted by six bands (three t_{2g} orbitals per Ti atom × two Ti atoms per unit cell) and the second group by four bands (corresponding to two e_g orbitals per Ti atom × two Ti atoms per unit cell). At higher energies, the Ti 4sp states are mixed with other high-lying orbitals to form a complex network of bands. The information missing in the molecular orbital model concerning the extended crystal structure of TiO_2 is now explicitly included in the calculations and appears in Figure 5–8(a) under the form of the different $E_n(\mathbf{k})$ bands.

Keeping in mind this basic description of the electronic structure of TiO_2, we will now focus our attention on the calculation of the near-edge fine structure of the O K edge. The results from the different steps required to calculate the dynamic form factor are summarized in

Figure 5–8(b–e). The first step consists in integrating all the dispersion relations drawn in Figure 5–8(a) over the first Brillouin zone to obtain the total density of states represented in Figure 5–8(b). Under the dipole approximation, this total DOS should be projected out on a local basis set to extract its dipole-allowed component. In the case of the O K edge, $\ell_c = 0$ and we only keep the $\ell = 1$ component, i.e., the O p local density of states represented in Figure 5–8(c). In a last step, the modulation of the LDOS due to the matrix element is included (see Eq. (9)). The resulting spectrum is broadened to account for the experimental resolution (Gaussian broadening) and the finite lifetimes of initial and final states (Lorentzian broadening), as shown in Figure 5–8(d). The full width at half-maximum (FWHM) of the Lorentzian is usually expressed as

$$\Gamma = \Gamma_c + \Gamma_v(E). \tag{14}$$

The core-hole lifetime, of the order of $\tau_c \approx 10^{-15}$ s, is essentially limited by fluorescence and Auger decay processes. Applying the Heisenberg uncertainty principle $\Gamma_c\tau_c \approx \hbar$, one obtains a core-hole broadening of the order of a few tenths of electronvolts. The precise values for various edges can be obtained in the literature (Fuggle and Inglesfield 1992). The lifetime of the excited electron is limited by its strong interactions with the other electrons of the solid. This electron will ultimately fall to the Fermi level (E_F) after creation of plasmons and electron–hole pairs. The energy-dependent broadening associated with this "quasi-particle" finite lifetime is less well characterized, and its treatment has been subject to various approximations in the literature. Weijs et al. (1990) proposed a linear dependence

$$\Gamma_v(E) \approx 0.1(E - E_F) \tag{15}$$

based on empirical arguments whereas Paxton (2005) and Muller et al. (1998) employed the quadratic expression derived from the random phase approximation (RPA) treatment of jellium:

$$\Gamma_v(E) = \frac{\sigma^2\sqrt{3}}{128} E_p \frac{(E - E_F)^2}{(E_F - E_0)}, \tag{16}$$

where E_p is the plasmon energy and E_0 is the energy of the bottom of the valence band. Alternatively, Moreau et al. (2006) evaluated $\Gamma_v(E)$ from the universal curve describing the inelastic mean free path of the ejected electron as a function of its kinetic energy. Finally, it has been shown that the excited state broadening can be extracted from the low-loss region of the EELS spectrum of the specimen itself (Hébert et al. 2000).

The resulting theoretical spectrum, directly extracted from the original band structure calculation, can be compared with the experimental data acquired in electron energy loss spectroscopy displayed in Figure 5–8(e).

Through this example, we can easily understand that the detailed structure of the ionization edge is representative of the specific crystal structure of the sample. Therefore, two polytypes of the same compound can be easily recognized from the slight differences present in

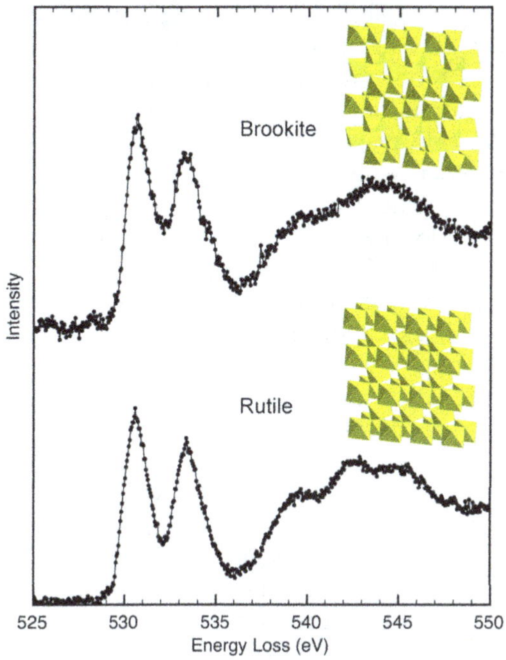

Figure 5–9. Experimental O K edge recorded in two polytypes of TiO_2: rutile and brookite. The arrangements of the coordination octahedra around the Ti atoms in these two structures are also given.

their near-edge fine structure. As we already mentioned, TiO_2 exists under different crystallographic structures among which are the so-called *rutile* and *brookite* forms. The O K edges recorded in these two structures are presented in Figure 5–9 together with the respective long-range arrangements of the TiO_6 coordination octahedra observed in these polytypes. The basic shape of the edge remains very similar with two prominent peaks associated with the Ti 3*d* states followed at higher energies by the 4*sp*-related broader structures. However, slight variations in the intrinsic distortions of the coordination octahedra as well as in their long-range arrangements induce small differences especially visible in the high-lying structures and characteristic of each phase.

5.2.1.4 Core-Hole Effects

Even though ground-state DOS provide a good starting point for the description of the near-edge fine structure in many cases (de Groot et al. 1993), the presence of the core hole in the final state can lead to a deep modification of the spectral shape. Indeed, the removal of a core electron during the excitation process that initially participated to the shielding of the nuclear charge produces a locally more attractive potential. The perturbation of the potential not only strongly modifies the dynamics of the other core electrons of the excited atom but also polarizes the surrounding valence electrons that tend to screen the additional positive charge of the core hole (Elsässer and Köstlmeier 2001, van Benthem et al. 2003). However, the important effect of the core-hole potential in ELNES is related to the modification of local unoccupied states. Under the presence of the core hole, the amplitude of the wave function $\phi_v(\mathbf{r})$ describing the ejected electron will be larger in the surrounding of the ionized atom than around other atoms of the same type but without a core hole. As the amplitude of the matrix element of the DFF in Eq. (11) is related to the amplitude of $\phi_v(\mathbf{r})$, one can expect the presence of the core hole to increase the scattering cross section. This effect is stronger for low-energy unoccupied states and therefore induces larger modifications of the fine structure close to the edge onset. A beautiful illustration of this "localization" of the first unoccupied states is given by Tanaka et al. (2005) in the case of MgO.

Ultimately, a strong Coulomb interaction between the core hole and the ejected electron leads to the formation of a bound electron–hole pair known as a core exciton. Figure 5–10 reproduces the experimental C K edge recorded in diamond (Batson and Bruley 1991) where such an excitonic resonance is clearly observed below the continuum of states of the conduction band.

Inclusion of the core-hole effects in band structure calculations has been very successful in a large number of cases (Elsässer and Köstlmeier 2001, Mizoguchi et al. 2000, 2004, van Benthem 2003). Practically, approaches based on reciprocal space band structure require the use of a sufficiently large supercell in which a core hole is included on one atom. The use of supercells is necessary to enable a large spatial separation between excited atoms and, therefore, to avoid their artificial interaction arising through the application of periodic boundary conditions. It should be noted that real space methods such as first-principle cluster calculations or full multiple scattering offer a more appropriate framework to treat these effects. Two other approximations have been employed to include these core-hole effects: (i) the $Z+1$ *approximation* models the core hole by adding an extra nuclear charge to the excited atom (Serin et al. 1998), i.e., by replacing it by the next atom in the periodic table or (ii) the *transition state* due to Slater which consists in a

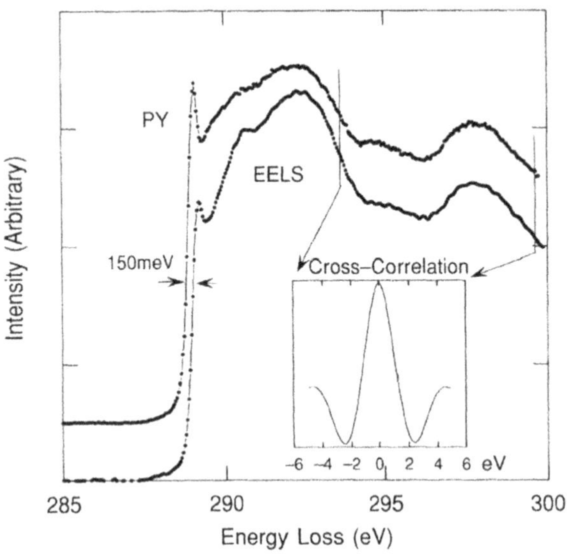

Figure 5–10. Experimental C K edge in diamond recorded in electron energy loss spectroscopy (EELS) and X-ray absorption (PY) (reproduced from Batson and Bruley 1991). A clear excitonic resonance is observed at the edge onset, below the continuum of the conduction band.

calculation where half of the core electron is kept on the atomic core state and the second half occupies the lowest conduction state of the solid (Keast et al. 2002, Paxton et al. 2000). A discussion of the various criteria governing the strength of the core hole at a given edge can be found in Mauchamp et al. (2009).

5.2.2 Transition Metal L_{23} and Rare Earth M_{45} Edges and the Multiplet Effects

We have seen in the previous section that a one-electron approach can be very successful in describing K edges, and it could be tempting to extend the same treatment to other types of edges recorded in EELS. However, it turns out that this approach is only applicable to a limited number of cases including K edges as well as certain L_{23} edges but definitely fails to reproduce the experimental signatures in other cases, mainly for atoms with partially filled *3d* and *4f* shells. In the latter cases, the excitation process should be thought of as resulting from a quasi-atomic process where the core electron is transferred into a bound atomic-like state. This "atomic character" is essentially due to the strong coupling between the *2p* core hole and the *3d* outer electrons in the case of an L_{23} edge or, equivalently, between the *3d* core hole and the *4f* electrons in the case of an M_{45} edge.

5.2.2.1 *Atomic Multiplet Theory*

To illustrate this point, let us start with one of the simplest signatures that can be observed in EELS: the Ba^{2+} M_{45} edge, as can be found in the well-known perovskite $BaTiO_3$. The ground-state electronic configuration of this ion is $[Xe]6s^0$, corresponding to filled *5s* and *5p* outer shells but with empty *4f* orbitals. According to the spectroscopic nomenclature presented in Section 5.1.3, an M_{45} edge corresponds to electronic transitions from the *3d* core states to unoccupied states of *p* and *f* symmetries. In this case, the large spin–orbit coupling of the *3d* core hole present in the final state leads to the observation of two distinct edges, namely the M_5 and M_4 edges. These edges correspond to transitions occurring, respectively, from the $3d_{5/2}$ and $3d_{3/2}$ core states, split in first approximation by $5/2\ \zeta_{3d}$, where ζ_{3d} is the radial factor of the spin–orbit interaction. From the location of Ba in the periodic table just before the lanthanides, one can also speculate on the basic shapes of the different Ba-LDOS probed in this case.

The very localized *f* orbitals are indeed expected to form narrow bands characterized by a high density of states whereas the *p* states are spatially more extended and form flatter bands. The dominant transitions are therefore the $3d_{5/2} \rightarrow 4f$ and $3d_{3/2} \rightarrow 4f$ as we indeed see in the experimental spectrum of Figure 5–11. This one-electron description is still satisfactory as long as we only describe qualitatively the dominant features of this edge. However, a closer look at the experimental spectrum reveals important discrepancies. For example, the M_5 to M_4 intensity ratio should be closely related to the degeneracy of the initial states, as the final LDOS probed in both cases remains the same. One therefore expects a 3:2 ratio arising from the fact that the $j = 5/2$

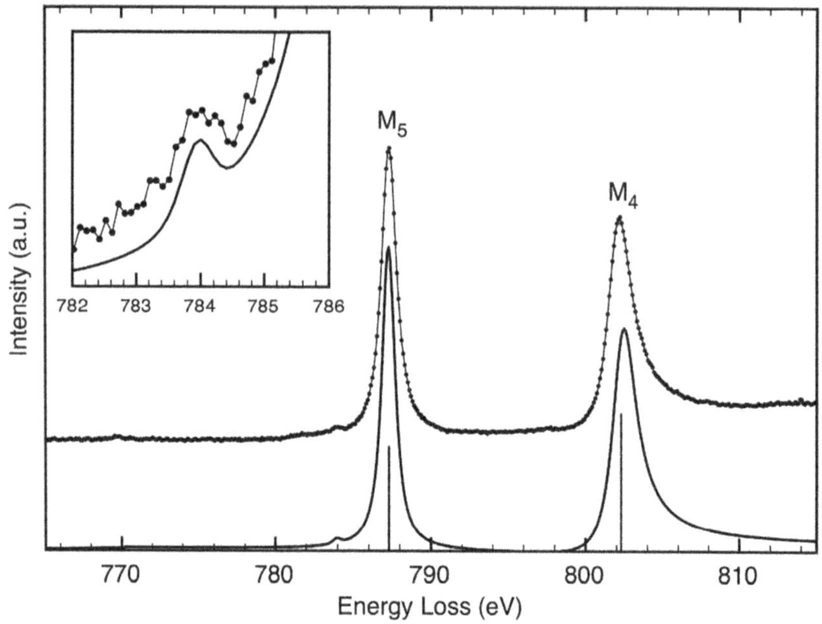

Figure 5–11. Experimental and theoretical Ba M_{45} edge recorded in $BaTiO_3$. The *inset* zooms on the low-intensity peak present at around 784 eV. The calculation has been performed using the TT Multiplets programs (de Groot and Kotani 2008).

configuration consists of $2 \times 5/2 + 1 = 6$ states whereas the $j = 3/2$ configuration consists of $2 \times 3/2 + 1 = 4$ states. This prediction clearly breaks down here, indicating that the spin–orbit split states are actually mixed by other interactions. A detailed inspection of the edge onset also reveals the presence of a third peak of very low intensity at around 784 eV, as shown in the inset of Figure 5–11. This additional structure cannot be interpreted any more in terms of one-electron DOS. Instead, we need to apply the atomic multiplet theory to the Ba^{2+} ion to explain the presence of this additional peak.

In this new framework, the excitation process is considered as an atomic effect and to a good approximation the solid surrounding the Ba^{2+} ion can be neglected. The observation of an M_{45} edge is essentially associated with a transition from a ground-state $3d^{10}4f^0$ to an excited $3d^9 4f^1$ electronic configuration. In order to predict the number of possible transitions observed for this edge, we first need to determine the possible states of the atom for each of these electronic configurations. For simplicity we will work within the *LS* (or Russell–Saunders) coupling scheme in the following discussion. For a multi-electron configuration with quantum numbers *L*, *S*, and *J* associated, respectively, with the orbital, spin, and total angular momenta of the atom, a term symbol is generally written $^{2S+1}L_J$. The values of *L* and *S* are found after coupling the individual orbital and spin angular momenta of the electrons (or holes) present in the partly filled shells of the atom.

The resulting total orbital and spin angular momenta are then coupled together to determine the values of J. In its ground-state $3d^{10}4f^0$ electronic configuration, the Ba^{2+} ion has only totally filled or empty electronic shells and the determination of the term symbol is trivial: with $L = 0$, $S = 0$, and $J = 0$, we obtain a ${}^1S_0$ symmetry. In the excited $3d^94f^1$ electronic configuration, we have to couple the individual momenta of a $3d$ hole ($\ell = 2$ and $s = 1/2$) and a $4f$ electron ($\ell = 3$ and $s = 1/2$). For the spin part, we obtain a singlet $S = 0$ and a triplet $S = 1$ states. For the angular part, L can take any of the values 1, 2, 3, 4, or 5. We therefore obtain 20 terms: ${}^1P_1$, ${}^1D_2$, ${}^1F_3$, ${}^1G_4$, ${}^1H_5$ and ${}^3P_0$, ${}^3P_1$, ${}^3P_2$, ${}^3D_1$, ${}^3D_2$, ${}^3D_3$, ${}^3F_2$, ${}^3F_3$, ${}^3F_4$, ${}^3G_3$, ${}^3G_4$, ${}^3G_5$, ${}^3H_4$, ${}^3H_5$, ${}^3H_6$. Each term corresponds to $2J+1$ states of the atom, the total number of states found after coupling adds up to 140. This indeed corresponds to the $10 \times 14 = 140$ number of different ways to arrange a hole on the $3d$ shell (10 distinct states) together with an electron on the $4f$ shell (14 distinct states).

The total Hamiltonian used to determine the eigenstates of the atom is given by

$$H_{\mathrm{ATOM}} = \sum_N \frac{\mathbf{p}_i^2}{2m} + \sum_N \frac{-Ze^2}{\mathbf{r}_i} + \sum_{i \neq j} \frac{e^2}{\mathbf{r}_{ij}} + \sum_N \zeta(\mathbf{r}_i)\mathbf{l}_i \cdot \mathbf{s}_i, \qquad (17)$$

where the four terms correspond, respectively, to the kinetic energy of the electrons, the Coulomb interaction of the electrons with the nucleus, the electron–electron repulsion, and finally the spin–orbit coupling for each electron. The first two terms in this expression only contribute to the average energy of the electronic configuration. The last two terms on the other hand are responsible for the splitting of the different terms, giving rise to the multiplet structure of the excited atom. Besides the strong $3d$ core-hole spin–orbit coupling which we already discussed, the electron–hole Coulomb interaction also plays a dominant role in the determination of the final state multiplet structure. In our present case of Ba^{2+} M_{45} edge, it is expressed through the two-electron Coulomb $\langle 3d4f | 1/\mathbf{r}_{12} | 3d4f \rangle$ and exchange $\langle 3d4f | 1/\mathbf{r}_{12} | 4f3d \rangle$ integrals. These integrals are usually expanded in a series of Legendre polynomials. For the df Coulomb interaction, one obtains three terms denoted by the Slater integrals F^0, F^2, and F^4 whereas the df exchange interaction yields the G^1, G^3, and G^5 Slater integrals. The F^0 term represents the direct potential of the core hole and actually contributes to the average energy of the final state electronic configuration. The other terms F^2, F^4, G^1, G^3, and G^5 participate in the energy splitting of the terms.

In the absence of spin–orbit coupling, i.e., in the presence of a purely "Coulombic atom," the use of the LS coupling would be fully justified and the dipole selection rule would simply be expressed through the conditions $\Delta L = 1$ and $\Delta S = 0$. We would only observe one transition in this case, from the ${}^1S$ ground state to the ${}^1P$ final state. However, such an atom does not exist and neither the electron–electron interaction nor the spin–orbit coupling can be totally neglected in practical cases. The first consequence is that atomic multiplet calculations are always performed in an intermediate coupling scheme (Cowan 1981) and the terms that we determined above are not eigenstates of the Hamiltonian (17). It therefore makes the *LS* nomenclature inappropriate for such a problem. It will not affect our study, however, providing that we express the dipole selection rule within the framework of the intermediate coupling scheme. In this case, only the value of J is important and the rule may be expressed as $\Delta J = 0, \pm 1$ with one exception, however: if J is equal to zero in the initial state, then J' cannot be zero in the final state. In the case of the Ba^{2+} M_{45} edge, $J = 0$ in the initial state and only three transitions are then expected to the $^1P_1$, $^3P_1$, and $^3D_1$ final states as observed experimentally. As mentioned above, strictly speaking one actually observes transitions to states resulting from the mixing of the different $^1P_1$, $^3P_1$, and $^3D_1$ pure states via the spin–orbit interaction.

This simple example illustrates how the one-electron approach breaks down when atomic multiplet effects dominate the spectral shape. This approach has been successfully applied to more complex cases in the series of lanthanides (Thole et al. 1985). As an example, Figure 5–12 shows the case of the M_{45} edge of trivalent Tb recorded in the pyrochlore $Tb_2Ti_2O_7$. This ion has a $4f^8$ electronic configuration in the ground state with a $^7F_6$ symmetry. The M_{45} edge therefore corresponds to the $3d^{10}4f^8 \rightarrow 3d^94f^9$ transition, giving rise to a spectrum made of 255 distinct lines.

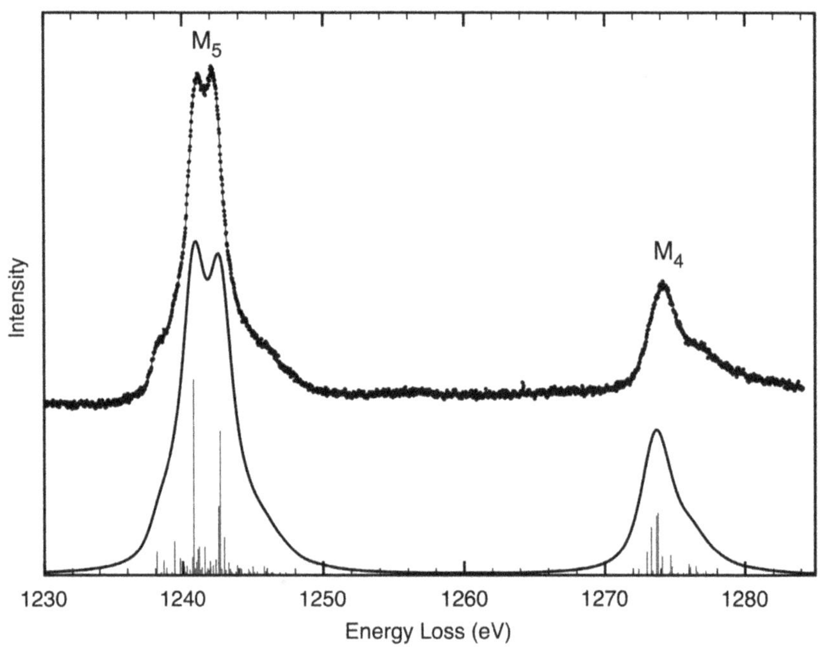

Figure 5–12. Experimental and theoretical Tb M_{45} edge recorded in $Tb_2Ti_2O_7$. The calculation has been performed using the TT Multiplets programs (de Groot and Kotani 2008).

5.2.2.2 *Crystal Field Effects*

The main reason for the success of atomic multiplet theory in providing the framework for an accurate description of rare earth M_{45} edges lies essentially in the very localized nature of their $4f$ states that experience only a weak interaction with the immediate environment of the ion. The situation is quite different in the case of transition metal $3d$ states, which strongly hybridize with the orbitals of neighboring atoms, as discussed in Section 5.2.1.2. A possible way to incorporate the influence of this local environment into the atomic multiplet calculation is to add to the Hamiltonian (17) a term $H_{\mathrm{CF}} = -e\Phi(\mathbf{r})$, which accounts for the local environment surrounding the excited atom. This term appears in the form of an electrostatic potential $\Phi(\mathbf{r})$ with the point group symmetry of the site of the excited atom and which mimics the influence of the neighboring atoms as if they were reduced to point charges. In this formalism, H_{CF} is essentially treated as a perturbation of the atomic results. Let us come back to the example used in Section 5.2.1.2 of a Ti^{4+} ion surrounded by six ligands forming a perfect coordination octahedron. The original SO_3 symmetry (full rotation group) of the isolated ion is now reduced, and the overall system belongs to the O_h point group. We pointed out that the main effect of this symmetry reduction is to split the $3d$ orbitals of the Ti ion into threefold degenerate t_{2g} states at low energy and twofold degenerate e_g states at higher energy. The energy difference between these states defines the crystal field parameter 10Dq. It is now interesting to investigate the influence of this crystal field on the shape of the Ti L_{23} edge. The corresponding $2p^6 3d^0 \rightarrow 2p^5 3d^1$ electronic transition can be studied within the *LS* coupling scheme in a very similar manner as we did for the Ba^{2+} M_{45} edge. The symmetry of the initial state is still $^1S_0$, and the different terms obtained after coupling the orbital and spin angular momenta of the $2p$ core hole ($\ell = 1$ and $s = 1/2$) and a $3d$ electron ($\ell = 2$ and $s = 1/2$) for the final state are the singlets $^1P_1$, $^1D_2$, $^1F_3$ and the triplets $^3P_0$, $^3P_1$, $^3P_2$, $^3D_1$, $^3D_2$, $^3D_3$, and $^3F_2$, $^3F_3$, $^3F_4$. In the atomic approximation, three allowed transitions are therefore expected from the $^1S_0$ initial state to the $^1P_1$, $^3P_1$, $^3D_1$ final states. However, as in the case of the one-electron $3d$ orbitals, the reduced symmetry of the system formed by the transition metal ion and its coordination octahedron will lift the degeneracy of certain of these terms through the crystal field. Group theory (Tinkham 1964) predicts the following branching of the states of total angular momentum J from the SO_3 onto the O_h irreducible representations: $J = 0 \rightarrow A_1, J = 1 \rightarrow T_1$, $J = 2 \rightarrow E + T_2$, $J = 3 \rightarrow A_2 + T_1 + T_2$, and $J = 4 \rightarrow A_1 + E + T_1 + T_2$. As a consequence, the 12 terms associated with the $2p^5 3d^1$ electronic configuration are split into 25 terms in octahedral symmetry: $2 \times A_1, 3 \times A_2, 5 \times E, 7 \times T_1$, and $8 \times T_2$. The initial state symmetry is A_1. We now have to determine which transitions are allowed under the dipole approximation. A well-known result of group theory tells us that the matrix element $\langle \Gamma_{j'} | \Gamma_H | \Gamma_j \rangle$ of an operator belonging to the irreducible representation Γ_H between two states belonging, respectively, to the irreducible representations $\Gamma_{j'}$ and Γ_j is zero unless the direct product $\Gamma_{j'} \otimes \Gamma_H \otimes \Gamma_j$ contains at least once the fully symmetric

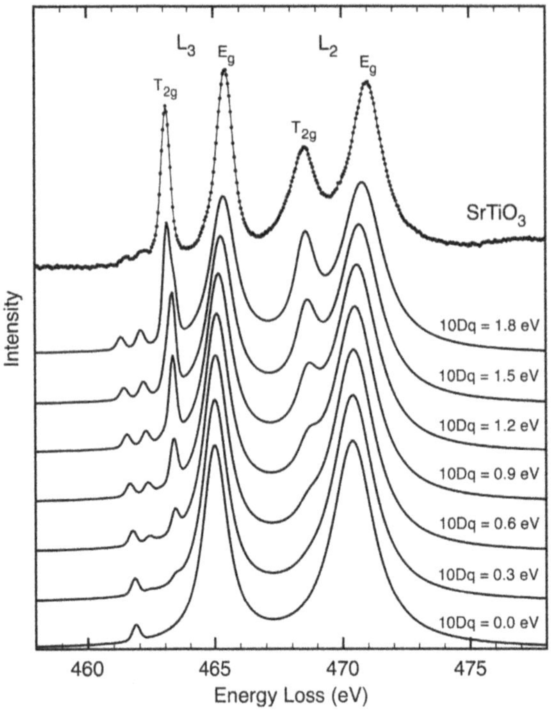

Figure 5–13. Theoretical Ti^{4+} L_{23} edge calculated in octahedral symmetry for increasing crystal field strengths starting from the atomic case (10Dq = 0.0 eV). The theoretical spectra have been performed using the TT Multiplets programs (de Groot and Kotani 2008) and are compared to the experimental spectrum acquired in $SrTiO_3$.

representation A_1. As the dipole operator transforms according to T_1 and the initial state has the A_1 symmetry, the only accessible states are those belonging to the T_1 irreducible representation. From three transitions in the case of the free ion, we now expect a maximum of seven transitions for the L_{23} edge of Ti^{4+} in octahedral symmetry. This effect of the local environment on the spectral shape is illustrated in Figure 5–13. The Ti L_{23} edge has been calculated within the framework of the crystal field multiplet theory for different values of the 10Dq parameter and starting with the atomic calculation at the bottom of the figure.

In this simple case again, the dominant features of the spectrum can be approximately described in terms of one-electron transitions. The dominant interaction in the final state is still the core-hole spin–orbit coupling that splits the lines in two groups corresponding to the L_3 (transitions from the $2p_{3/2}$ core state) and the L_2 (transitions from the $2p_{1/2}$ core state) edges. Each of these two edges experiences a subsequent splitting associated with the octahedral crystal field that lift the degeneracy of the 3*d* states into t_{2g} and e_g components. Multi-electronic effects are, however, clearly visible through the presence of the low-intensity peaks located at the edge onset and the strong deviation from the 6:4:3:2 intensity ratio expected from the degeneracies of the states associated with the $2p_{3/2} \rightarrow t_{2g}$, $2p_{3/2} \rightarrow e_g$, $2p_{1/2} \rightarrow t_{2g}$, and $2p_{1/2} \rightarrow e_g$ transitions. It should be pointed out here that the experimental L_3 (or L_2) splitting is therefore not strictly equal to the 10Dq parameter introduced in crystal field theory. Based on the same arguments, additional splittings should be visible on spectra acquired in compounds where

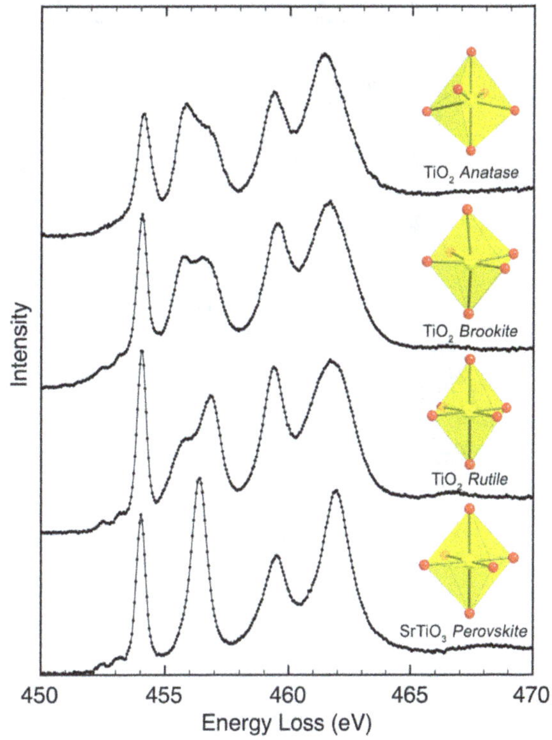

Figure 5–14. Experimental Ti^{4+} L_{23} edges recorded in the three polytypes of TiO_2 (*rutile, anatase,* and *brookite*) and compared to the reference compound $SrTiO_3$. The coordination *octahedron* around the Ti atoms is given on the *right-hand* side for each of these compounds.

the local symmetry of the Ti sites is lowered from O_h to one of its subgroups. These effects are illustrated in Figure 5–14 where the Ti L_{23} edges recorded in the different polytypes of TiO_2 are compared to the reference spectrum of $SrTiO_3$.

The symmetry of the Ti sites in *rutile, brookite,* and *anatase* are, respectively, reduced to D_{2h}, C_1, and D_{2d} point groups. In *rutile* and *anatase,* the dominant distortion consists of an elongation of one of the axes of the octahedron with two oxygen atoms at 1.980 Å and four at 1.946 Å for the former and two oxygen atoms at 1.980 Å and four at 1.934 Å for the latter. In *brookite,* the six oxygen atoms are located at different distances from the central Ti atoms at 1.990, 1.931, 1.923, 1.863, 1.999, and 2.052 Å. Additional strong distortions of the O–Ti–O bond angles are also present in these three compounds. The main effect visible on the Ti L_{23} edge is a systematic splitting of the L_3-e_g peak into two sub-peaks. The e_g orbitals are indeed expected to be the most sensitive probe of the octahedron distortions as they directly point toward the ligand atoms and thus experience the strongest covalent interaction.

5.2.2.3 Oxidation State

We have just highlighted the fact that the multiplet structure of both initial and final states is strongly dependent on the ground-state electronic configuration of the ion and is therefore expected to be sensitive to its oxidation state. The L_{23} near-edge fine structure has indeed been often employed as an experimental probe of the formal valence of transition metal ions in the solid state. The example of Mn L_{23} edges

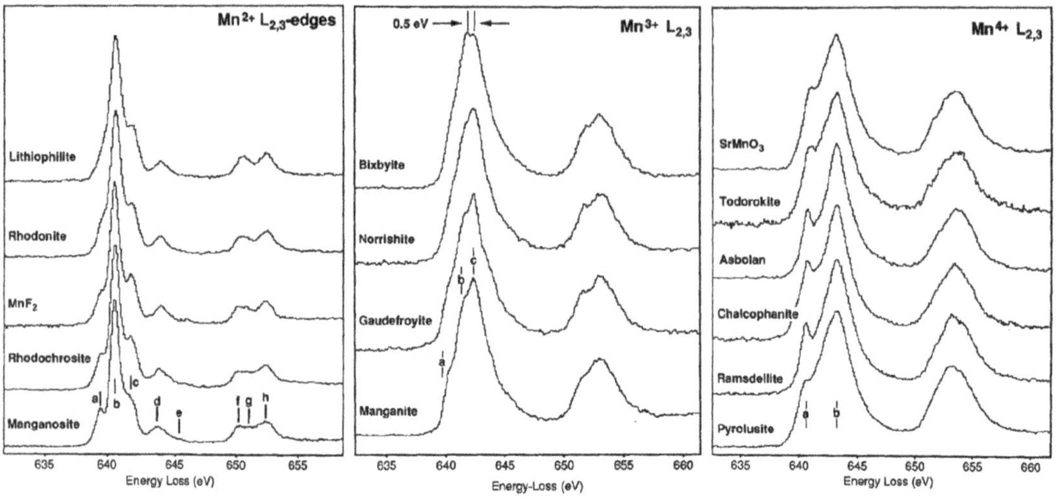

Figure 5–15. Experimental L_{23} edges of octahedrally coordinated Mn recorded in various minerals (modified from Garvie and Craven 1994). Several examples are given for each of the three most common oxidation states of Mn, namely Mn^{2+}, Mn^{3+}, and Mn^{4+}.

recorded in various minerals (Garvie and Craven 1994) is shown in Figure 5–15. This ion is commonly found under three different oxidation states: $Mn^{2+}(3d^5)$, $Mn^{3+}(3d^4)$, and $Mn^{4+}(3d^3)$. The $^6S_{5/2}$ atomic ground state of Mn^{2+} is not split under the octahedral field but projected into the $^6A_1$ cubic symmetry, corresponding to the high-spin $(t_{2g}\uparrow)^3(e_g\uparrow)^2$ electronic configuration. Mn^{3+} has four $3d$ electrons and adopts the high-spin ^5E ground state in octahedral symmetry, associated with the $(t_{2g}\uparrow)^3(e_g\uparrow)^1$ electronic configuration. The presence of a half-filled e_g shell makes this ion unstable in octahedral field and subject to strong Jahn–Teller distortion (Jahn and Teller 1937). An elongation of the octahedron, lowering the overall symmetry to D_{4h}, indeed lifts the degeneracy of the e_g states into an occupied and stabilized a_{1g} orbital and a destabilized b_{1g} orbital. Finally, Mn^{4+} possesses a stable $^4A_2$ magnetic ground state with a half-filled t_{2g} electronic shell.

The differences clearly visible in the spectra presented in Figure 5–15 reflect the large variations of the final-state multiplet structures probed in the L_{23} edges of these different ions. One can also notice that the oxidation state and the dominant octahedral (O_h) contribution to the crystal field dictate largely the shape of the edge. Crystal-specific details such as the exact value of the crystal field strength or the precise distortions lowering the symmetry of the atomic site have here only a minor influence on the spectra as can been seen within each series presented in Figure 5–15. Changes in the oxidation state induce two other important modifications of the experimental spectra:

- As we already noticed in our study of the Ba^{2+} M_{45} edge, important deviations from the statistical 3:2 branching ratio, primarily due to the electron–hole electrostatic interactions in the final state, are observed when measuring the experimental $I(M_5)/I(M_4)$ intensity ratio between the two Ba spin–orbit split white lines. Such deviations are also observed from the statistical 2:1 ratio expected for the transition metal L_{23} edges, as can be seen in Figure 5–15. The branching ratio $I(L_3)/(I(L_3) + I(L_2))$ or white-line ratio $I(L_3)/I(L_2)$ depends on both the oxidation state and the spin state of the ion in quite a complex manner (Thole and van der Laan 1988). It therefore appears as a useful parameter to quantify oxidation state ratios in mixed valence minerals (van Aken et al. 1998) and complex oxides (see Chapter 10).

- A last important difference between the divalent, trivalent, and tetravalent Mn L_{23} edges is the systematic shift of the edge onset toward higher energies – the so-called chemical shift – observed as the oxidation state increases. The qualitative physical picture behind this energy shift is that under oxidation, the transition metal atomic potential is modified by the departure of a $3d$ electron. The shielding of the nucleus is reduced and therefore leads to an increase of the core-level $2p$ binding energy. The same trend has been observed in the systematic study of chromium oxides by Theil, van Elp, and Folkmann (Theil et al. 1999).

5.2.2.4 *Spin State*

In octahedral symmetry, transition metals with four to seven electrons in their $3d$ shell may be found with at least two distinct ground states corresponding to different distributions of these electrons on the t_{2g} and e_g crystalline orbitals. A *high-spin* ground state can be thought of as an extension of the atomic Hund's rule ground state to the case of an ion in O_h symmetry: the spin alignment of the $3d$ electrons prevails over the energy cost associated with the population of high-lying e_g orbitals. On the other hand, a *low-spin* ground state corresponds to an optimized occupation of the low-lying t_{2g} orbitals by arranging two electrons of opposite spins on the same orbital. The configuration adopted by the ion depends essentially on two competing interactions, the crystal field strength 10Dq (noted Δ_{CF} hereafter) and the Stoner exchange splitting J. The $3d^6Co^{3+}$ is a well-known example of an ion that can be found in the $^1A_1$ low-spin $(t_{2g}\uparrow)^3(t_{2g}\downarrow)^3$, in the $^5T_2$ high-spin $(t_{2g}\uparrow)^3(t_{2g}\downarrow)^1(e_g\uparrow)^2$, or even in the $^3T_1$ intermediate-spin $(t_{2g}\uparrow)^3(t_{2g}\downarrow)^2(e_g\uparrow)^1$ ground states. Let us consider here only the two extreme *low-spin* and *high-spin* cases: starting from the low-spin configuration, the energy cost to promote two electrons to the e_g orbitals is equal to $2\Delta_{CF}$. If one assumes that the exchange splitting J is approximately the same for $t_{2g}-t_{2g}$, e_g-e_g, or $t_{2g}-e_g$ interactions, the ion also gains $4J$ by aligning the electron spins. The transition point from high to low spin is therefore located at $\Delta_{CF} \sim 2J$ and whether the ion is in the former or latter ground state depends strongly on the details of its local environment.

The different symmetries of the ion ground states induce differences in the final-state multiplet structure probed through the Co L_{23} edge that may be used to determine experimentally the spin state of Co^{3+} in a specific compound. In Figure 5–16, we show the Co L_{23} edges recorded in the layered cobaltite $Na_xCoO_2(x \sim 0.75)$ and in the

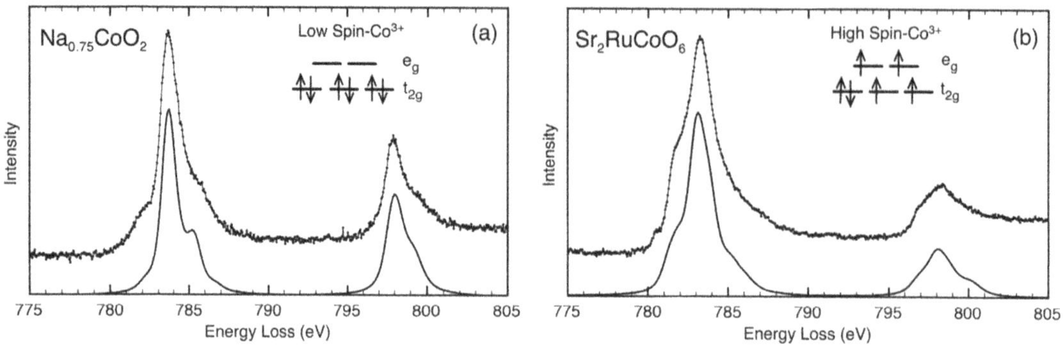

Figure 5–16. Experimental L_{23} edges of octahedrally coordinated Co^{3+}: (a) in the low-spin ${}^1A_1$ ground state as found in the layered cobaltite Na_xCoO_2(x~0.75) and (b) in the high-spin ${}^5T_2$ ground state in the perovskite Sr_2RuCoO_6.

perovskite Sr_2RuCoO_6. The former compound reveals a clear signature of low-spin Co^{3+} as can be found also in $LiCoO_2$ (de Groot et al. 1993) or $EuCoO_3$ (Hu et al. 2004). In contrast, the Co L_{23} edge recorded in Sr_2CoRuO_6 exhibits quite a different signature, especially in the shape of the L_2 edge, which is characteristic of high-spin Co^{3+} as found, for example, in Sr_2CoO_3Cl (Hu et al. 2004). The branching ratio $I(L_3)/(I(L_3)+I(L_2))$ also increases significantly from the low-spin to the high-spin compound. The experimental spectra are compared to crystal field multiplet calculations in O_h symmetry performed for a crystal field parameter 10Dq below the transition point in the case of Sr_2CoRuO_6 (high-spin ${}^5T_2$ ground state) and above in the case of $Na_{0.75}CoO_2$ (low-spin ${}^1A_1$ ground state). It should be mentioned that, strictly speaking, $Na_{0.75}CoO_2$ possesses a non-negligible amount of Co^{4+}. However, signatures recorded in mixed valence low-spin Co^{3+}/Co^{4+} (Mizokawa et al. 2001) are not significantly modified with respect to reference compounds containing only Co^{3+} ions. The same remark may be applied to Sr_2CoRuO_6 (Lozano-Gorrín et al. 2007, Potze et al. 1995). A beautiful example of low-spin Mn^{2+} recorded in $K_4Mn(CN)_6$ is given by Cramer et al. (1991) and exhibits large differences with the high-spin spectra presented in Figure 5–15.

5.2.2.5 *Charge Transfer Effects*

Up to this point, the influence of the immediate environment of the transition metal ion, i.e., the influence of its ligand coordination shell, has been accounted for in the calculations only through the addition of a crystal field potential to the atomic Hamiltonian. In this approach, the covalent interactions between the transition metal and the ligands are totally neglected, and in a certain number of cases, especially for L_{23} edges of late transition metals, it leads to a very poor description of the experimental spectra. Strong covalence may indeed induce a deep modification of the spectral features essentially through (i) the formation of small satellites and (ii) the overall contraction of the multiplet structure (de Groot 2005). This contraction may be simply reproduced in crystal field calculations by reducing the Slater integrals for Coulomb and exchange interactions. Covalence indeed tends to delocalize the transition metal 3*d* orbitals and therefore to reduce the strength of electron–electron interactions. This effect is known as the nephelauxetic effect (de Groot 1994, Lynch and Cowan 1987). An alternative approach is to extend the single $3d^n$ determinant approach of crystal field multiplet by including additional low-lying configurations of the type $3d^{n+1}\underline{L}$, $3d^{n+2}\underline{L}^2$, $3d^{n+3}\underline{L}^3, \ldots$, where $\underline{L}$ denotes a hole on the 2*p* ligand shell and allowing their mixing. Comparison with experiments has indeed shown that couplings occur predominantly with the occupied valence band, more than with the high-lying conduction states. Usually, spectral shapes are well described using only the two $3d^n$ and $3d^{n+1}\underline{L}$ configurations. One therefore needs to introduce additional parameters in the model Hamiltonian, such as the charge transfer energy $\Delta = E(3d^{n+1}\underline{L}) - E(3d^n)$ locating the energy of the charge transfer $3d^{n+1}\underline{L}$ with respect to the $3d^n$ configuration and the effective hopping parameter *t* between ligand 2*p* and transition metal 3*d* orbitals. This so-called *charge transfer multiplet* approach has given excellent results in many cases (Hu et al. 1998, Piamonteze et al. 2005), in particular to explain the presence of high-lying satellites in the spectra. Figure 5–17 reproduces the Cu L_{23} edge recorded in X-ray absorption in $CsKCuF_6$ and $La_2Li_{1/2}Cu_{1/2}O_4$ (Hu et al. 1998) together with the theoretical charge transfer multiplet calculations. In these two compounds, the Cu ions are formally trivalent with a $3d^8$ electronic configuration as, for example, in the case of Ni^{2+} in NiO (van der Laan et al. 1986).

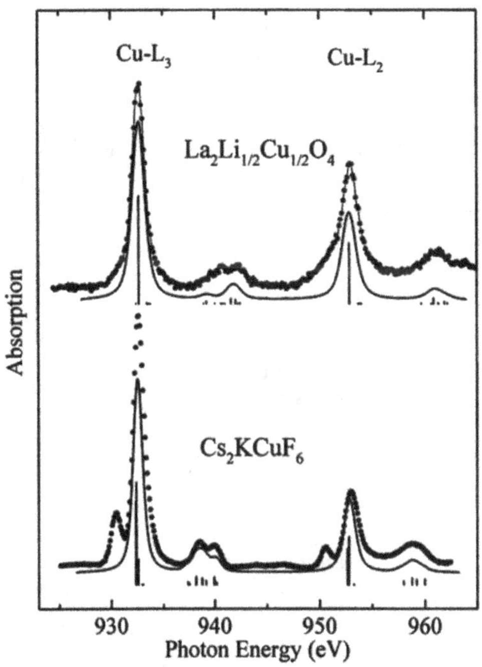

Figure 5–17. Experimental Cu L_{23} edges recorded in X-ray absorption in $CsKCuF_6$ and $La_2Li_{1/2}Cu_{1/2}O_4$ (reproduced from Hu et al. 1998).

However, the Cu^{3+} L_{23} edge recorded in these compounds exhibits a totally different shape. This is mainly due to the fact that unlike Ni^{2+} in NiO, which can be well approximated by the ionic $3d^8$ configuration, Cu^{3+} has to be thought of as a strong mixture of both $3d^8$ and $3d^9\underline{L}$ configurations in its ground state. A dominant weight on the $3d^9\underline{L}$ configuration even arises from the negative Δ usually found for trivalent Cu. In the final state, transitions to the low-lying $2p^5 3d^{10}\underline{L}$ and higher $2p^5 3d^9$ configurations lead to the formation of the sharp and intense peak at 933 eV and the satellite structures at around 936–940 eV, respectively, for the L_3 edge. This double structure, main peak + satellite, is echoed on the L_2 edge. The additional structures visible at 931 and 951 eV and not reproduced in the calculations are due to Cu^{2+} impurities.

Other advanced computational methods have been proposed in the literature aiming at an ab initio description of the transition metal L_{23} edges such as configuration interaction calculations based on cluster molecular orbitals (Ogasawara et al. 2001), multichannel multiple scattering calculations (Krüger and Natoli 2005), or Bethe–Salpeter calculations including core-hole multiplet effects (Shirley 2005).

5.3 Applications

We have seen so far the fundamental origins of the near-edge structures in energy loss spectra. As we have discussed, the technique provides essentially the same information as X-ray absorption spectroscopy with the main difference being the scattering vector dependence when bulk samples are analyzed. The other significant difference is the extent of the spatial resolution of the technique even when compared to scanning transmission X-ray microscopy (STXM) in synchrotrons. With zone plates capable of focusing X-rays, the most advanced third-generation synchrotrons offer a spatial resolution of about 13 nm (best current values) with more typical values in the range of few tens of nanometers (20–40 nm). Within the scanning transmission microscopes and the use of monochromators, high-brightness electron guns (cold field emission or modified Schottky), and aberration correctors of the probe-forming lenses one can expect to reach a spatial resolution of 0.1 nm with an energy resolution ranging between 0.15 and 0.3 eV. Indeed fast elemental mapping at atomic resolution has now been demonstrated in a number of laboratories over the last 1–2 years (Botton et al. 2010, Colliex et al. 2009, Muller 2009, Varela et al. 2009) so that even the changes in chemical composition at defects can be studied (e.g., Figure 5–18).

Still, the early demonstrations of high spatial resolution ELNES can be traced to P.E. Batson who, with a dedicated STEM, probed the Si–SiO_2 interface (Figure 5–19) (Batson 1993). In this work, the changes in the Si L_{23} edge as the electron beam is moved atomic column by atomic column from bulk Si to SiO_2 are clearly detected in two ways. First of all, there is a shift of the edge energy position, which reflects the core-level shift of the Si $2p$ levels due to the change in valence from Si^0 to Si^{4+}. Second, there is a major change in the shape of the ELNES, which

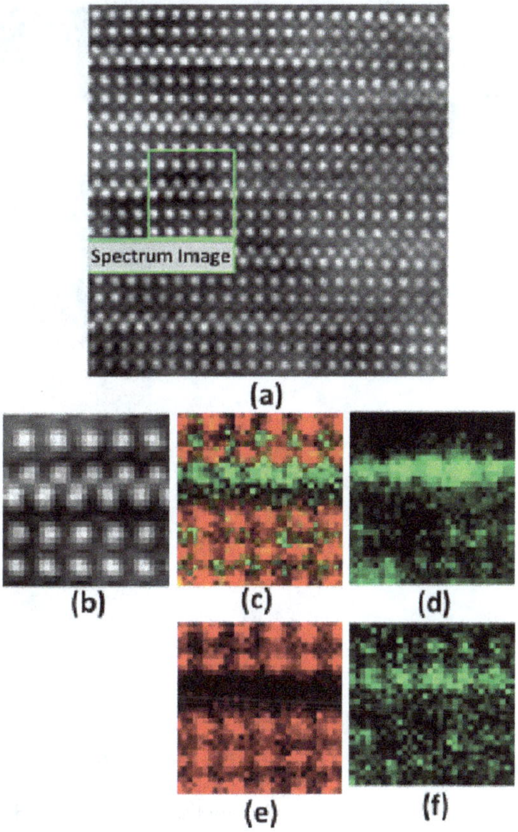

Figure 5–18. (a) HAADF image of a layered perovskite compound $BiLaTiO_{12}$ with subset area (highlighted in *green*) where EELS maps were obtained, (b) HAADF signal from the *rastered* area, (c) composite color-coded map (*red*: Ti, *green*: La), (d) La N_{45} edge signal, (e) Ti L_{23} edge signal, and (f) La M_{45} edge signal (reproduced from Gunawan et al. 2009).

reflects the change in coordination from Si in tetrahedral coordination surrounded by Si atoms to Si in SiO_4 tetrahedra and intermediate configurations as Si atoms are surrounded by increasing number of oxygen atoms (thus intermediate valence and associated core-level shift).

A corresponding complementary study was carried out with the O K edge in a gate oxide of a metal–oxide–semiconductor transistor (Figure 5–20) by Muller et al. (1999). The O K ELNES changes from the typical configuration of O in SiO_4 tetrahedra to a fine structure not observed in bulk phases. Density functional calculations were subsequently used to demonstrate the origins of this effect and attributed to the bridging oxygen sites (Neaton et al. 2000).

Changes in fine structure were also observed in early STEM work on N-doped diamond (Fallon et al. 1995). For example, the N K edge in diamond platelets demonstrates a very similar fine structure to the C K edge in bulk diamond (Figure 5–21). These results suggest that the N atoms in the platelets are substitutional and furthermore that N preferentially locates on the A centers as demonstrated subsequently with multiple scattering methods (Brydson et al. 1998). Similar results have also been observed in N-doped amorphous C films deposited by sputtering (Axén et al. 1996) and used to identify the site preference of the N atoms. Changes in the ELNES at the C K edge were also used to identify local changes in bonding in C coatings at nanometer resolution

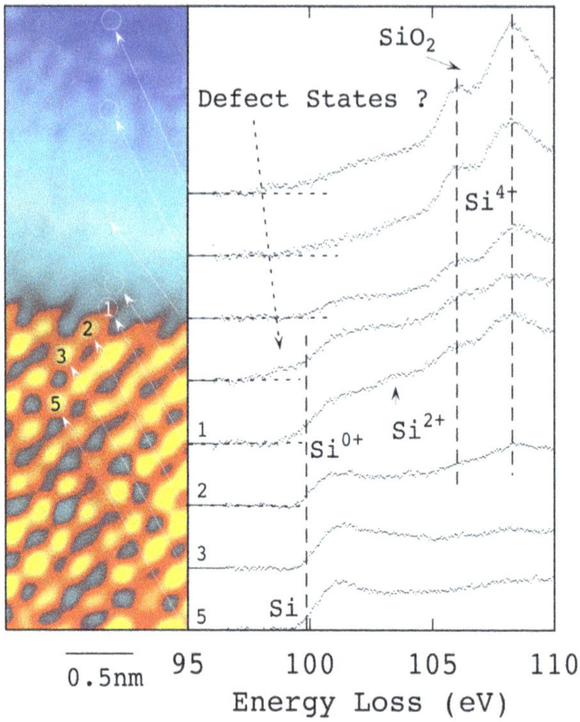

Figure 5–19. HAADF image of Si–SiO_2 interface and ELNES of the Si L_{23} edge as a function of probe position. The defect states are associated with intensity below the edge threshold corresponding to energies in the band gap of Si (figure from P.E. Batson).

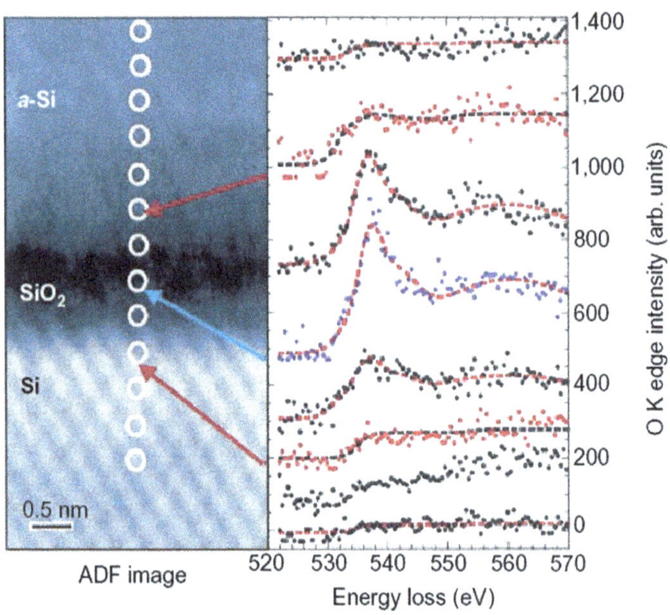

Figure 5–20. ADF image of a Si–SiO_2 interface and related O K edge ELNES as a function of atomic position (reproduced from Muller et al. 1999).

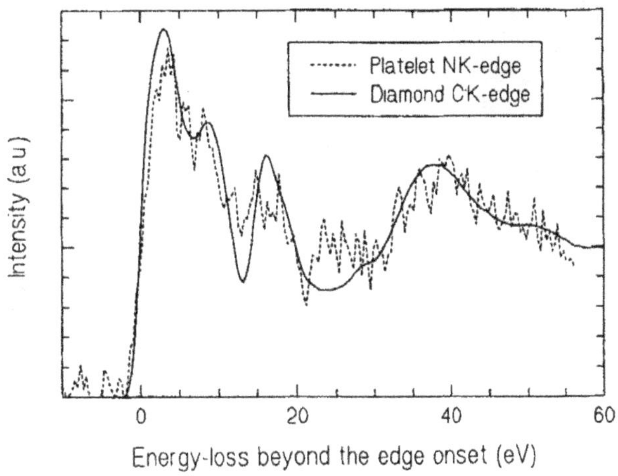

Figure 5–21. C K edge in diamond and overlapping the N K edge on an N-containing platelet. The shape of the N K edge immediately suggests that the N atoms are in a substitutional environment (reproduced from Fallon et al. 1995).

(Muller et al. 1993) through the changes in π* and σ* peaks as a function of position from the interface. These examples illustrate that, although calculations provide quantitative information on the types of electronic structure environment, simple inspection of the ELNES does already provide valuable data on the "bonding" environment of the probed atoms.

One of the major challenges in detecting the changes in bonding with high spatial resolution is the determination of statistical differences in the spectra arising from interfaces. Statistical methods were first applied in the study of the bonding changes at interfaces by Bonnet et al. (1999) with the implementation of the multivariate statistical analysis method. This technique was used to differentiate statistically significant changes in the near-edge structures in the O K edge, which appeared at the interface between Si and SiO_2 as well as the interface between SiO_2 and TiO_2. The multivariate analysis method allows the retrieval of the significant spectra (or components) that cannot be simply attributed to a linear combination of the "bulk" phases (in this case, Si and SiO_2 or the latter case of SiO_2 and TiO_2) and can be used to map the interface–spectrum contribution (Figure 5–22).

Simpler approaches, such as the multiple least square (MLS) method or nonlinear least square method, can also be used to map the location of the various "phases" having a different near-edge structure using the so-called fingerprint mapping (Arenal et al. 2008). In this case, the weight factors generated by the output of the MLS fit are used as the amplitude in a map. Examples of this work are useful to extract the distribution of the "phases" even if the arrangement is extremely complex (Figures 5–23 and 5–24). Such fitting techniques are now available in commercial software such as Gatan's Digital Micrograph™ and are thus available to all users of recent post-column spectrometers. Reference spectra used in the fitting routine can be extracted directly in the data set as given in the example shown in Figure 5–23 when they can be isolated as a single phase. In this particular example, three different phases could be identified (metallic boron, B_2O_3, and BN, this latter

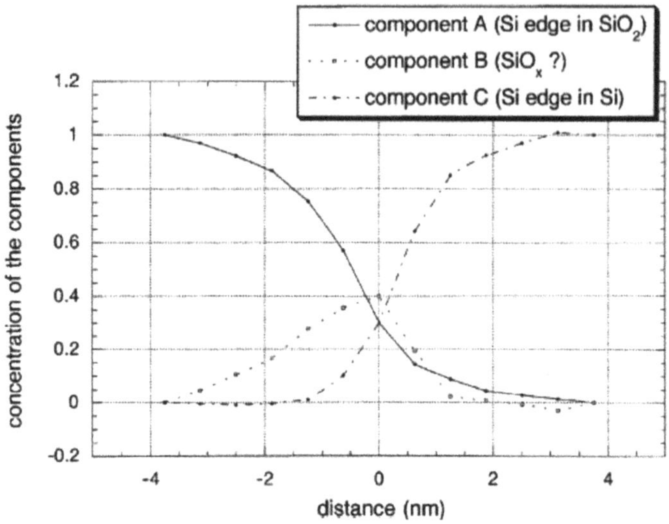

Figure 5–22. Multivariate analysis of O K edge ELNES taken at the interface between Si and SiO_2. The method provides an unbiased way to extract statistically significant spectra (identified as component B) not been the simple linear combination of the bulk phase spectra (A and C) adjacent to the interface (reproduced from Bonnet et al. 1999).

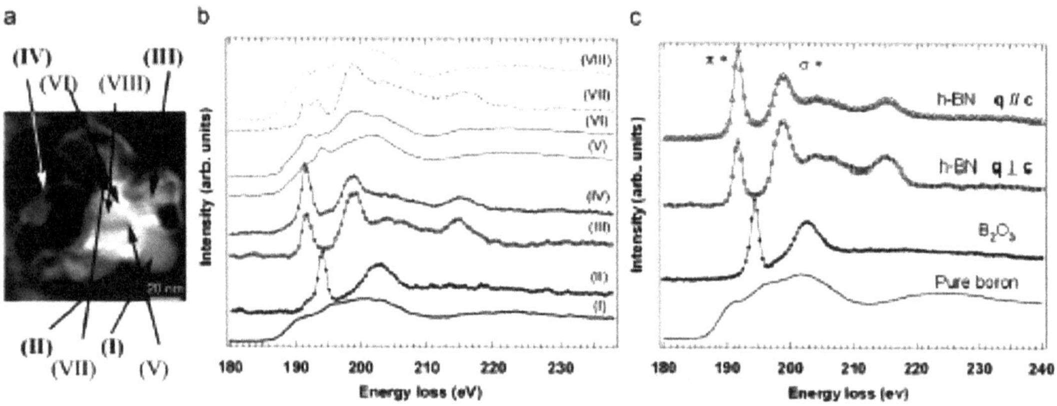

Figure 5–23. (a) STEM image of a complex $B/BN/B_2O_3$ nanostructure and (b) series of spectra extracted from different locations identified with *roman numbers* in (a). (c) Reference spectra obtained from bulk phases of metallic B, B_2O_3, and BN (two different spectra obtained for the scattering vector **q** parallel or perpendicular to the c-axis of the BN crystal; reproduced from Arenal et al. 2008).

one being an anisotropic material with different ratios of the π^* and σ^* peaks (Figure 5–23)).

Changes in ELNES were also detected at dislocation cores in GaN (Arslan et al. 2005) through which it was possible to distinguish the type of dislocation (screw or edge). In perovskites, changes in ELNES at grain boundaries (Duscher et al. 1998) and dislocation cores in perovskites (Kurata et al. 2009) were detected. For example, Duscher et al. (1998) showed changes in the O K edges in Mn-doped $SrTiO_3$

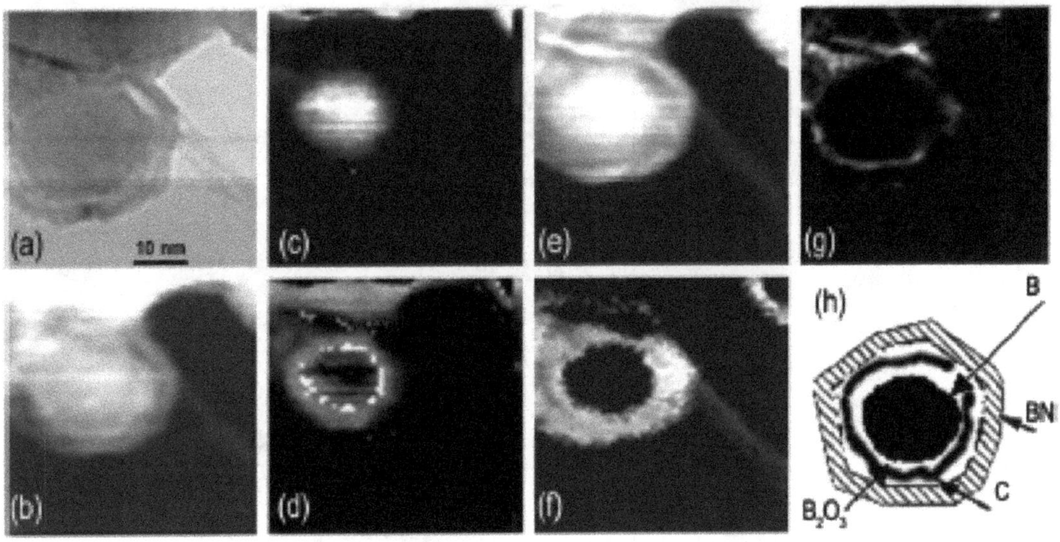

Figure 5–24. Bright-field STEM (a) and dark-field STEM (b) images of a B–N–O nanostructure with maps showing the distribution of metallic boron (c), boron oxide (d), boron nitride in the two extreme orientations (e and f), carbon (g), and composite map with the distribution of the phases (h) (reproduced from Arenal et al. 2008).

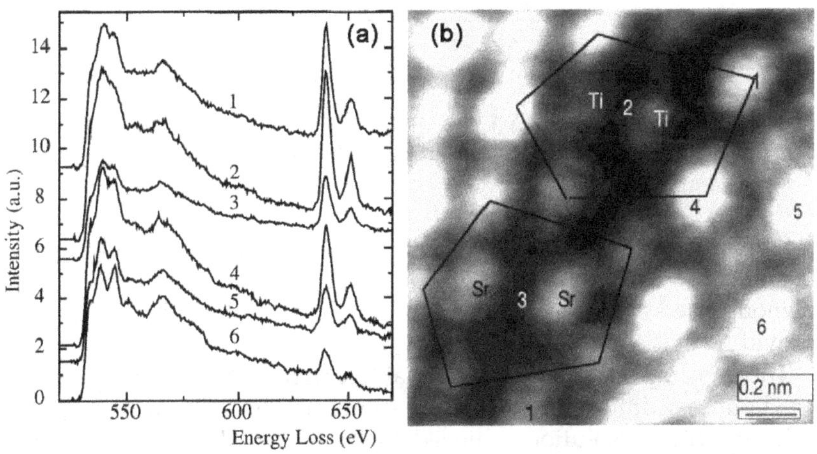

Figure 5–25. (a) EELS spectra of the O K and Mn L_{23} edges collected at a $SrTiO_3$ grain boundary. (b) HAADF image and location of points where spectra shown in (a) have been obtained (adapted from Duscher et al. 1998).

(Figure 5–25) that were consistent with the local increase in Mn content at the grain boundaries. The application of ELNES in STEM has been used to demonstrate the very clear and unambiguous changes in valence of Ti atoms in ferroelectric thin films (Muller et al. 2004). For example, the results obtained on oxygen-deficient $SrTiO_3$ show the changes of the Ti L_{23} edge from Ti^{4+} to Ti^{3+} in the regions with oxygen deficiency as well as a change in the ELNES of the O K edge indicating that the local presence of oxygen vacancies is inducing changes in the Ti valence to sustain the local charge balance (Figure 5–26).

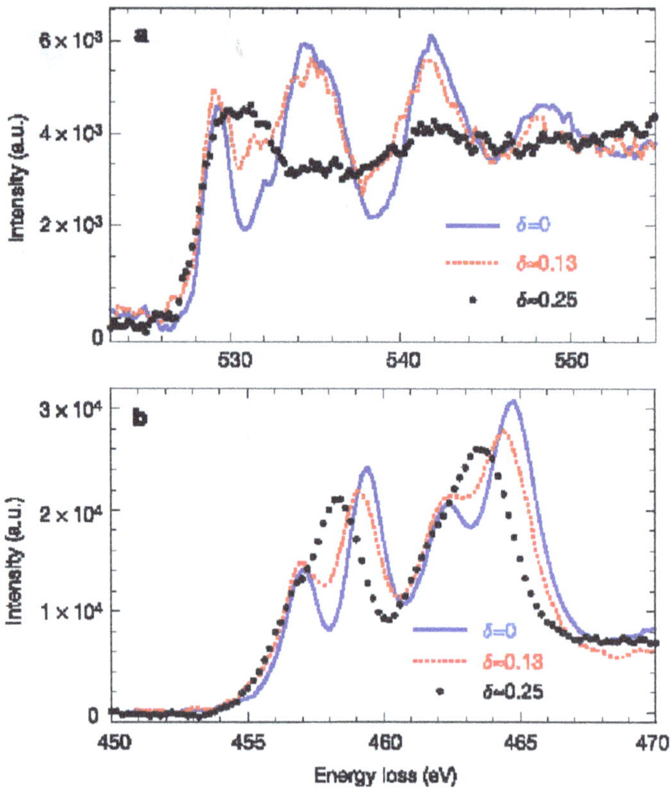

Figure 5–26. (a) O K and (b) Ti L_{23} ELNES in $SrTiO_{3-\delta}$ as a function of the oxygen vacancy content δ (reproduced from Muller et al. 2004).

Similarly, Ohta et al. (2007) have identified changes in the valence of Ti atoms in $SrTiO_3$ ultrathin films doped with Nb doping. ELNES data demonstrated the local change from Ti^{4+} to Ti^{3+} in the Nb-doped areas providing evidence of a 2D electron gas.

These examples show the potential to extract chemical environment or "bonding" information at the resolution achievable in the STEM. With the high resolution achieved today by the aberration-corrected microscopes, it is potentially possible to probe inequivalent sites in the unit cell. Recently such experiments have revealed that indeed it is possible to detect ELNES within the unit cell (Haruta et al. 2009, Lazar et al. 2010). In terms of spatial resolution of the "local" measurements, the first atomically resolved elemental maps (Bosman et al. 2007, Kimoto et al. 2007) have shown that it is possible to resolve the different atomic columns in the unit cell and that detailed calculations provide good agreement with experiments, although fine details appearing further away from the nominal position of the electron beam can also appear. In general terms, it has been shown that signals can arise further away from the precise location of the electron beam due to the Coulomb delocalization (Schenner et al. 1995, Schattschneider et al. 1999). We refer the interested reader to the detailed Chapter 6 of this book and related references dealing with imaging with inelastically scattered electrons (e.g.,

Findlay et al. 2007). It is clear, irrespective of the limitations, that atomic-resolved analysis has been experimentally demonstrated in terms of both chemical information and fine structures.

Acknowledgments We would like to acknowledge M. Couillard and P. Schattschneider for their helpful proofreading of this chapter.

References

L.J. Allen, T.W. Josefsson, Phys. Rev. B **52**, 3184 (1995)

R. Arenal, F. de la Peña, O. Stéphan, M. Walls, M. Tencé, A. Loiseau, C. Colliex, Ultramicroscopy **109**, 32 (2008)

I. Arslan, A. Bleloch, E.A. Stach, N.D. Browning, Phys. Rev. Lett. **94**, 025504 (2005)

N. Axén, G.A. Botton, H.Q. Lou, R.E. Somekh, I.M. Hutchings, Surf. Coatings Technol. **81**, 262 (1996)

P.E. Batson, Nature **366**, 727 (1993)

P.E. Batson, J. Bruley, Phys. Rev. Lett. **67**, 350 (1991)

H.A. Bethe, Annalen der Physik **5**, 325 (1930)

P. Blaha, K. Schwarz, G. Madsen, D. Kvasnicka, J. Luitz, *WIEN2k, An augmented Plane Wave Plus Local Orbitals Program for Calculating Crystal Properties* (Institute of Physical and Theoretical Chemistry, Vienna University of Technology) (2009)

M. Bosman, V.J. Keast, J.L. Garcia-Munoz, A.J. D'Alfonso, S.D. Findlay, L.J. Allen, Garcia-Munoz, Phys. Rev. Lett. **99**, 086102 (2007)

G.A. Botton, J. Electron. Spectrosc. Related Phenomena **143**, 129 (2005)

G.A. Botton, in *Encyclopedia Science of Microscopy*, eds. by J. Spence, P. Hawkes (Springer, 2007), pp. 273–405

N. Bonnet, N. Brun, C. Colliex, Ultramicroscopy **77**, 97 (1999)

G.A. Botton, S. Lazar, C. Dwyer, Ultramicroscopy **110**, 926 (2010)

G.A. Botton, C.B. Boothroyd, W.M. Stobbs, Ultramicroscopy **59**, 93 (1995)

N.D. Browning, J. Yuan, L.M. Brown, Ultramicroscopy **38**, 291 (1991)

R. Brydson, L.M. Brown, J. Bruley, J. Microsc. **189**, 137 (1998)

S.P. Cramer, F.M.F. de Groot, Y. Ma, C.T. Chen, F. Sette, C.A. Kipke, D.M. Eichhorn, M.K. Chan, W.H. Armstrong, E. Libby, G. Christou, S. Brooker, V. McKee, O.C. Mullins, J.C. Fuggle, J. Am. Chem. Soc. **113**, 7937 (1991)

C. Colliex, N. Brun, A. Gloter, D. Ihmoff, M. Kak, K. March, C. Mory, O. Stephan, M. Tencé, M. Walls, Philos. Trans. R. Soc. A **367**, 3845 (2009)

R.D. Cowan, *The Theory of Atomic Structure and Spectra* (Los Alamos Series in Basic and Applied Sciences, University of California Press, 1981)

M.T. Czyzyk, G.A. Sawatzky, Phys. Rev. B **49**, 14211 (1994)

F.M.F. de Groot, J. Electron. Spectrosc. Related Phenomena **67**, 529 (1994)

F.M.F. de Groot, Coordination Chem. Rev. **249**, 31 (2005)

F.M.F. de Groot, M. Abbate, J. van Elp, G.A. Sawatzky, Y.J. Ma, C.T. Chen, F. Sette, J. Phys.: Condens. Matter **5**, 2277 (1993)

F. de Groot, A. Kotani, in *Core Level Spectroscopy of Solids*, eds. by D.D. Sarma, G. Kotliar, Y. Tokura, Advances in Condensed Matter Science, vol. 6 (CRC Press, Taylor & Francis, Boca Raton, London, New York, 2008)

F.M.F. de Groot, J. Faber, J.J.M. Michiels, M.T. Czyzyk, M. Abbate, J.C. Fuggle, Phys. Rev. B **48**, 2074 (1993)

S.L. Dudarev, G.A. Botton, S.Y. Savrasov, C.J. Humphreys, A.P. Sutton, Phys. Rev. B **57**, 1505 (1998)

G. Duscher, N.D. Browning, S.J. Pennycook, Phys. Status Solidi (a) **166**, 327 (1998)

C. Dwyer, Ultramicroscopy **104**, 141 (2005)
C. Elsässer, S. Köstlmeier, Ultramicroscopy **86**, 325 (2001)
P.J. Fallon, L.M. Brown, J.C. Barry, J. Bruley, Philos. Mag. A **72**, 21 (1995)
S.D. Findlay, P. Schattschneider, L.J. Allen, Ultramicroscopy **108**, 58 (2007)
J.C. Fuggle, J.E. Inglesfield (eds.), in *Unoccupied Electronic States, Fundamentals for XANES, EELS, IPS and BIS*. Topics in Applied Physics, vol. 69 (Springer, Berlin, Heidelberg, New York, 1992)
L.A.J. Garvie, A.J. Craven, Phys. Chem. Minerals **21**, 191 (1994)
L. Gunawan, S. Lazar, O. Gautreau, C. Harnagea, A. Pignolet, G.A. Botton, Appl. Phys. Lett. **95**, 192902 (2009)
M. Haruta, H. Kurata, H. Komatzu, Y. Shimakawa, S. Isoda, Phys. Rev. B **80**, 165123 (2009)
C. Hébert, Micron **38**, 12 (2007)
C. Hébert, M. Kostner, P. Schattschneider, Conf. Proc. EUREM XII, Brno, 33 (2000)
C. Hébert, P. Schattschneider, H. Franco, B. Jouffrey, Ultramicroscopy **106**, 1139 (2006)
Z. Hu, G. Kaindl, S.A. Warda, D. Reinen, F.M.F. de Groot, B.G. Müller, Chem. Phys. **232**, 63 (1998)
Z. Hu, H. Wu, M.W. Haverkort, H.H. Hsieh, H.-J. Lin, T. Lorenz, J. Baier, A. Reichl, I. Bonn, C. Felser, A. Tanaka, C.T. Chen, L.H. Tjeng, Phys. Rev. Lett. **92**, 207402 (2004)
H.A. Jahn, E. Teller, Proc. Roy. Soc. London. Series A, Math. Phys. Sci. **161**, 220 (1937)
D.N. Jayawardane, C.J. Pickard, L.M. Brown, M.C. Payne, Phys. Rev. B **64**, 115107 (2001)
R.O. Jones, O. Gunnarsson, Rev. Modern Phys. **61**, 689 (1989)
D.C. Joy, A.D. Romig, Jr., J.I. Goldstein (eds.), *Principles of Analytical Electron Microscopy* (Plenum Press, New York, NY, 1986)
K. Kimoto, T. Asaka, T. Nagai, M. Saito, Y. Matusui, K. Ishizuka, Nature **450**, 702 (2007)
V.J. Keast, A.J. Scott, R. Brydson, D.B. Williams, J. Bruley, J. Microsc. **203**, 135 (2001)
V.J. Keast, A.J. Scott, M.J. Kappers, C.T. Foxon, C.J. Humphreys, Phys. Rev. B **66**, 125319 (2002)
P. Krüger, C.R. Natoli, J. Synchrotron Rad. **12**, 80 (2005)
H. Kurata, S. Isojima, M. Kawai, Y. Shimakawa, S. Isoda, J. Microsc. **236** 128 (2009)
H. Kurata, P. Wahlbring, S. Isoda, T. Kobayashi, Micron **28**, 381 (1997)
S. Lazar, Y. Shao, L. Gunawan, R. Nechache, A. Pignolet, G.A. Botton, Microsc. Microanal. **16**, 416 (2010)
R.D. Leapman, P.L. Fejes, J. Silcox, Phys. Rev. B **28**, 2361 (1983)
J.C. Le Bossé, T. Epicier, B. Jouffrey, Ultramicroscopy **106**, 449 (2006)
A.D. Lozano-Gorrín, J.E. Greedan, P. Núñez, C. González-Silgo, G.A. Botton, G. Radtke, J. Solid-State Chem. **180**, 1209 (2007)
D.W. Lynch, R.D. Cowan, Phys. Rev. B **36**, 9228 (1987)
V. Mauchamp, M. Jaouen, P. Schattschneider, Phys. Rev. B **79**, 235106 (2009)
T. Mizoguchi, I. Tanaka, M. Yoshiya, F. Oba, K. Ogasawara, H. Adachi, Phys. Rev. B **61**, 2180 (2000)
T. Mizoguchi, I. Tanaka, S. Yoshioka, M. Kunisu, T. Yamamoto, W.Y. Ching, Phys. Rev. B **70**, 045103 (2004)
T. Mizokawa, L.H. Tjeng, P.G. Steeneken, N.B. Brookes, I. Tsukada, T. Yamamoto, K. Uchinokura, Phys. Rev. B **64**, 115104 (2001)
P. Moreau, F. Boucher, G. Goglio, D. Foy, V. Mauchamp, G. Ouvrard, Phys. Rev. B **73**, 195111 (2006)

M.S. Moreno, K. Jorissen, J.J. Rehr, Micron **38**, 1 (2007)
D.A. Muller, Nat. Mater. **8**, 263 (2009)
D.A. Muller, N. Nakagawa, A.Ohtomo, J.L. Grazul, H.Y. Hwang, Nature **430**, 657 (2004)
D.A. Muller, D.J. Singh, J. Silcox, Phys. Rev. B **57**, 8181 (1998)
D.A. Muller, T. Sorsch, S. Moccio, F.H. Baumann, K. Evans-Lutterodt, G. Timp, Nature **399**, 758 (1999)
D.A. Muller, Y. Tzou, R. Raj, J. Silcox, Nature **366**, 725 (1993)
J.B. Neaton, D.A. Muller, Ashcroft, Phys. Rev. Lett. **85**, 1298 (2000)
M. Nelhiebel, P.-H. Louf, P. Schattschneider, P. Blaha, K. Schwarz, B. Jouffrey, Phys. Rev. B **59**, 12807 (1999a)
M. Nelhiebel, N. Luchier, P. Schorsch, P. Schattschneider, B. Jouffrey, Philos. Mag. B **79**, 941 (1999b)
K. Ogasawara, T. Iwata, Y. Koyama, T. Ishii, I. Tanaka, H. Adachi, Phys. Rev. B **64**, 115413 (2001)
H. Ohta, S. Kim, Y. Mune, T. Mizoguchi, K. Nomura, S. Ohta, T. Nomura, Y. Nakanishi Y. Ikuhara, M. Hirano, H. Hosono, K. Koumoto, Nat. Mater. **6**, 129 (2007)
M.P. Oxley, L.J. Allen, Phys. Rev. B **57**, 3273 (1998)
A.T. Paxton, J. Electron. Spectrosc. Related Phenomena **143**, 51 (2005)
A.T. Paxton, M. van Schilfgaarde, M. MacKenzie, A.J. Craven, J. Phys.: Condensed Matter **12**, 729 (2000)
C. Piamonteze, F.M.F. de Groot, H.C.N. Tolentino, A.Y. Ramos, N.E. Massa, J.A. Alonso, M.J. Martínez-Lope, Phys. Rev. B **71**, 020406(R) (2005)
R.H. Potze, G.A. Sawatzky, M. Abbate, Phys. Rev. B **51**, 11501 (1995)
G. Radtke A. Saúl, H.A. Dabkowska, B.D. Gaulin, G.A. Botton, Phys. Rev. B **77**, 125130 (2008)
P. Rez, J.M. MacLaren, D.K. Saldin, Phys. Rev. B **57**, 2621 (1998)
P. Schattschneider, C. Hébert, H. Franco, B. Jouffrey, Phys. Rev. B **72**, 045142 (2005)
P. Schattschneider, B. Jouffrey, M. Nelhiebel, Phys. Rev. B **54**, 3861 (1996)
P. Schattschneider, M. Nelhiebel, B. Jouffrey, Phys. Rev. B **59**, 10959 (1999)
M. Schenner, P. Schattschneider, R.F. Egerton, Micron **26**, 391 (1995)
V. Serin, C. Colliex, R. Brydson, S. Matar, F. Boucher, Phys. Rev. B **58**, 5106 (1998)
E. L. Shirley, Phys. Rev. Lett. **80**, 794 (1998)
E.L. Shirley, J. Electron. Spectrosc. Related Phenomena **144–147**, 1187 (2005)
W. Sigle, Annu. Rev. Mater. Res. **35**, 239 (2005)
I. Tanaka, T. Mizoguchi, T. Yamamoto, J. Am. Ceramic Soc. **88**, 2013 (2005)
C. Theil, J. van Elp, F. Folkmann, Phys. Rev. B **59**, 7931 (1999)
B.T. Thole, G. van der Laan, Phys. Rev. B **38**, 3158 (1988)
B.T. Thole, G. van der Laan, J.C. Fuggle, G.A. Sawatzky, R.C. Karnatak, J.-M. Esteva, Phys. Rev. B **32**, 5107 (1985)
M. Tinkham, *Group Theory and Quantum Mechanics*, International Series in Pure and Applied Physics (McGraw-Hill, 1964)
M. Varela, M.P. Oxley, W. Luo, J. Tao, M. Watanabe, A.R. Lupini, S.T. Pantelides, S.J. Pennycook, Phys. Rev. B **79**, 085117 (2009)
P.A. van Aken, B. Liebscher, V.J. Styrsa, Phys. Chem. Minerals **25**, 323 (1998)
K. van Benthem, C. Elsässer, M. Rühle, Ultramicroscopy **96**, 509 (2003)
G. van der Laan, J. Zaanen, G.A. Sawatzky, R. Karnatak, J.-M. Esteva, Phys. Rev. B **33**, 4253 (1986)
P.J.W. Weijs, M.T. Czyzyk, J.F. van Acker, W. Speier, J.B. Goedkoop, H. van Leuken, H.J.M. Hendrix, R.A. de Groot, G. van der Laan, K.H.J. Buschow, G. Wiech, J.C. Fuggle, Phys. Rev. B **41**, 11899 (1990)
D.B. Williams, C.B. Carter, *Transmission Electron Microscopy, A Textbook for Materials Science* (Plenum Press, New York, NY, 1996)

6

Simulation and Interpretation of Images

Leslie J. Allen, Scott D. Findlay and Mark P. Oxley

6.1 Introduction

Image simulation has become a standard tool in interpreting atomic-resolution scanning transmission electron microscopy (STEM) images. While new methods and approaches are continually being developed which allow for more efficient simulations under certain circumstances, the principles behind the formation of the most common kinds of STEM images are well understood. The traditional method for atomic-resolution analysis in STEM is high-angle annular dark-field (HAADF) imaging, often referred to as Z-contrast imaging as it involves collecting electrons which have, by excitation of a phonon, been scattered through large angles and thus resembles in atomic number dependence the Rutherford scattering formula (Kirkland 1988, Pennycook and Jesson 1991). For compositional analysis, electron energy-loss spectroscopy (EELS) provides more scope for accurate chemical identification and the exploration of local bonding and other electronic properties (Spence 2006). Correlating structural and chemical properties using high-resolution STEM is thus often carried out by using a Z-contrast image as a reference while recording energy-loss spectra at structurally significant points (Muller et al. 1999, Varela et al. 2004). Improvements in lens design (Batson et al. 2002, Krivanek et al. 1999) and microscope stability are greatly increasing the sensitivity at which atomic-resolution HAADF (Nellist et al. 2004, Voyles et al. 2002) and spectroscopic data can be obtained (Allen et al. 2003, Bleloch et al. 2003, Bosman et al. 2007, Kimoto et al. 2007, Varela et al. 2004). The new generation of aberration-corrected scanning transmission electron microscopes, which will admit much larger collector angles, are expected to improve still further both collector efficiency and image interpretation (Krivanek et al. 2008, Muller et al. 2008).

While one of the strengths of inelastic STEM imaging, particularly HAADF imaging, is that the images tend to be much more directly interpretable than those in conventional transmission electron microscopy (CTEM) (Varela et al. 2005), it has long been appreciated

S.J. Pennycook, P.D. Nellist (eds.), *Scanning Transmission Electron Microscopy*,
DOI 10.1007/978-1-4419-7200-2_6, © Springer Science+Business Media, LLC 2011

that there are still possible pitfalls in image interpretation. Prior to the advent of aberration correction, these included probe spreading and probe tails (Yamazaki et al. 2001). Subsequent to aberration correction, and the use of the finer STEM probes it enables, concern has focused on the elastic scattering or channelling of the probe, its spreading out through the crystal and the possibility of cross-talk effects (Dwyer and Etheridge 2003, Fitting et al. 2006, Watanabe et al. 2001). Inelastic imaging based on EELS, focusing on inner-shell ionization, allows further subtleties through the delocalized nature of the interaction and coherence effects (Bosman et al. 2007, Dwyer 2005, Oxley et al. 2005, 2007). Additionally, simulation allows for further quantification of experiments. For example, in the EELS imaging of a single impurity in the bulk, comparison with simulations allowed an estimate of its depth (Varela et al. 2004). Simulation can also be used predictively to design new experiments and assess the possible usefulness of novel imaging modes. For example, the possibility of depth sectioning to obtain 3D information with atomic resolution laterally and nanometer resolution in depth is starting to be explored in Z-contrast imaging (Borisevich et al. 2006, van Benthem et al. 2005, 2006, Wang et al. 2004) and, encouraged by preliminary simulations, the extension to EELS imaging will soon follow. Novel experimental geometries, such as atomic-resolution scanning confocal electron microscopy (SCEM) (Nellist et al. 2006), have also been proposed. A schematic of both STEM and SCEM geometries is given in Figure 6–1.

In this chapter we aim to summarize the main ideas behind some of the common simulation methods and describe how they may be used to interpret, support or extract further information from experimental images. It is not possible in such a brief overview to do justice to the variety of approaches to such simulations or the many innovative adaptations to particular problems. We describe in outline one general approach which allows for handling both the thermal scattering

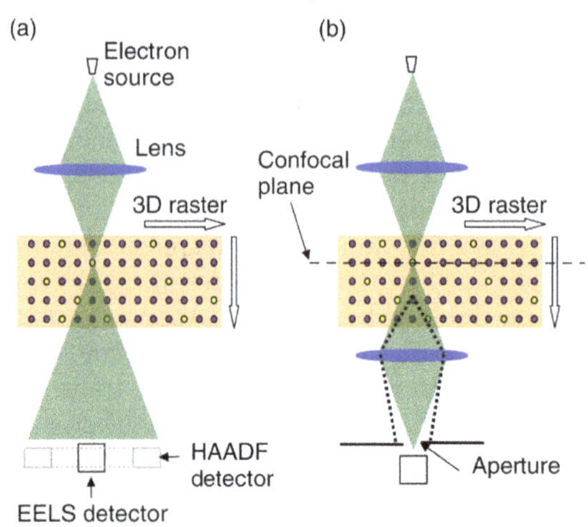

Figure 6–1. Schematic of (a) STEM and (b) SCEM geometry, allowing for depth sectioning via a 3D raster scan, reproduced from Allen et al. (2008).

that determines the HAADF images and the inner-shell ionization that determines the core-loss EELS images on a fairly equal footing, though where possible we have endeavoured to note alternative methods. We have opted to focus largely on the conceptual approach; for details of practical implementation we have tried to direct the reader to the relevant literature. We have selected case studies to demonstrate a range of the behaviour encountered in fast electron scattering through a crystal–the spreading of the probe, the role of absorption, the delocalization of the ionization interaction, etc–all of which topics have been explored before. The theory of STEM imaging has reached the point where the list of principles involved is well-known, but the complexity of the scattering in each new specimen means that one cannot always guess in advance which principles will be most important in a given instance and thus how one should best understand the image features. Image simulation enables us to answer that question.

6.2 Theoretical Background

6.2.1 Calculating the Elastic Wave Function

The current chapter will focus exclusively on signals derived from the collection of inelastically scattered electrons. However, the common feature of all detailed simulation methods of STEM imaging is the need to first calculate the elastic wave function of the fast electron probe as it scatters through the sample. Starting from the Schrödinger equation, and making the standard high-energy approximation (Humphreys 1979, van Dyck 1985), the reduced wave function[1] $\psi_0(\mathbf{R}, z, \mathbf{R}_0)$ for the ground state is governed by the equation

$$\frac{\partial}{\partial z}\psi_0(\mathbf{R}, z, \mathbf{R}_0) = \frac{\mathrm{i}}{4\pi k_0}\left\{\nabla_{\mathbf{R}}^2 + \frac{8\pi^2 m}{h^2}\left[V(\mathbf{R}) + \mathrm{i}V'(\mathbf{R})\right]\right\}\psi_0(\mathbf{R}, z, \mathbf{R}_0)\,, \tag{1}$$

where $\mathbf{r} = (\mathbf{R}, z)$ is the real space position vector, $\mathbf{R}_0$ parametrically denotes the position of the probe on the surface, $k_0 = 1/\lambda$ is the relativistic, refraction-corrected wavevector for the fast electron (λ being the corresponding wavelength) and m is the relativistically corrected electron mass. The projected potential approximation has been assumed so that the potentials $V(\mathbf{R})$ and $V'(\mathbf{R})$ do not depend on z (Bird 1989, Qin and Urban 1990). The later potential, $V'(\mathbf{R})$, is an effective absorption potential, leading to the attenuation of electron density in the elastic wave function as inelastic scattering events transfer electron density into inelastic channels (Humphreys 1979, Yoshioka 1957). In a formal derivation beginning from the Yoshioka coupled channels equations

[1] The "reduced wave function" is obtained from the full wave function by factoring out a fast oscillating phase factor of $\exp(2\pi i k_0 z)$. This is often referred to as the modulated plane wave ansatz (van Dyck 1985).

(Yoshioka 1957), the potential term in the reduced, single-channel equation has a nonlocal form, which is to say that the reduction produces an integro-differential equation with a nonlocal kernel. The most significant absorption mechanism in what follows will be thermal scattering. In that case the total absorptive potential can be well approximated by a local potential (Allen et al. 2006), and hence the form above. Later in this chapter we will encounter expressions with the nonlocal structure resulting from the formal collapse of the coupled channel equations.

The high-energy approximation has left us with a first-order differential equation in z. Consequently, the evolution of the wave function through the crystal can be determined directly from a knowledge of the wave function at the entrance surface to the specimen. In the case of STEM, the entrance surface wave function is simply the wave function of the STEM probe, which in reciprocal space may be written as

$$\begin{aligned} P(\mathbf{Q}, \mathbf{R}_0) &= A(\mathbf{Q}) \exp[-\mathrm{i}\chi(\mathbf{Q})] \exp(-2\pi \mathrm{i}\mathbf{Q} \cdot \mathbf{R}_0) \\ &\equiv T(\mathbf{Q}) \exp(-2\pi \mathrm{i}\mathbf{Q} \cdot \mathbf{R}_0)\,, \end{aligned} \tag{2}$$

where $T(\mathbf{Q})$ is the contrast transfer function of the probe-forming optics, $\mathbf{Q}$ is the coordinate reciprocal to $\mathbf{R}$ and the pupil function $A(\mathbf{Q})$ is 1 inside the probe-forming aperture and 0 elsewhere. The aberration function is defined by

$$\chi(\mathbf{Q}) = \pi\lambda\Delta f Q^2 + \frac{\pi}{2}\lambda^3 C_s Q^4 + \cdots\,, \tag{3}$$

where $Q \equiv |\mathbf{Q}|$ and for simplicity we have restricted our description to terms involving the lower order circular aberrations of defocus, Δf, and third-order spherical aberration, C_s. However, a general expression for arbitrary order aberrations (Krivanek et al. 1999) can just as easily be used.

The real space wave function is obtained from the inverse Fourier transform of Eq. (2):

$$P(\mathbf{R}, \mathbf{R}_0) = \int T(\mathbf{Q}) \exp(-2\pi \mathrm{i}\mathbf{Q} \cdot \mathbf{R}_0) \exp(2\pi \mathrm{i}\mathbf{Q} \cdot \mathbf{R}) \mathrm{d}\mathbf{Q}\,. \tag{4}$$

There are two main approaches for solving the governing equation given in Eq. (1). One is the Bloch wave method (see Humphreys 1979 for a review), which is based on determining the eigenstates of the fast electron in the sample. The total wave function is decomposed as a superposition of eigenstates, each of which can be trivially propagated individually. The evolution of the full wave function through the sample is then a consequence of the interference between the weighted superposition of Bloch states $\psi^i(\mathbf{R}, z, \mathbf{R}_0)$:

$$\psi(\mathbf{R}, z, \mathbf{R}_0) = \sum_i \alpha^i(\mathbf{R}_0) \psi^i(\mathbf{R}, z, \mathbf{R}_0)\,, \tag{5}$$

with the Bloch state amplitudes determined by the boundary conditions at the entrance surface. The Bloch states may be expanded in terms of the G_{th}-order Fourier components $C^i_{\mathbf{G}}$ as

$$\psi^i(\mathbf{R}, z, \mathbf{R}_0) = \exp(2\pi i \lambda^i z) \sum_{\mathbf{G}} C^i_{\mathbf{G}} \exp(2\pi i \mathbf{G} \cdot \mathbf{R}) \,. \qquad (6)$$

Some authors have treated the convergent STEM probe as a set of phase-linked plane waves and thus solve the problem for each individual plane wave, using the superposition principle to synthesize the full wave function if and when required (Amali and Rez 1997, Nellist and Pennycook 1999, Pennycook and Jesson 1991). We adopt the alternative approach first demonstrated in Allen et al. (2003), where the Bloch state amplitudes are determined by matching the full probe wave function at the entrance surface to the Bloch states within the crystal via the overlap integral $\int \psi^{i*}(\mathbf{R}, 0, \mathbf{R}_0) P(\mathbf{R}, \mathbf{R}_0) d\mathbf{R}$. Using Eqs. (4) and (6), we obtain for the excitation amplitude[2]

$$\alpha^i(\mathbf{R}_0) = \sum_{\mathbf{G}} C^{i*}_{\mathbf{G}} T(\mathbf{G}) \exp(-2\pi i \mathbf{G} \cdot \mathbf{R}_0) \,. \qquad (7)$$

If the probe-forming aperture is restricted to allow only the $\mathbf{G} = \mathbf{0}$ component of the incident wave function through, this reduces to the usual plane wave result of $\alpha^i(\mathbf{R}_0) = C^{i*}_{\mathbf{0}}$. Figure 6–2 shows the excitation amplitudes for a few select Bloch states in ZnS, viewed along the [110] zone axis. For simplicity we neglect absorption. Figure 6–2 shows results for a 100 keV aberration-balanced probe (first column; $\Delta f = 62$ Å, $C_s = -0.05$ mm, $\alpha = 20$ mrad) and a 100 keV aberration-free probe (second column; $\alpha = 25$ mrad) as well as a δ-function probe (third column) in which case the excitation amplitude proves to map out the Bloch state itself.

In Figure 6–2, the number below the state label for each row is the value of $|C^{i*}_{\mathbf{0}}|$, which is the magnitude of the excitation amplitude for plane wave incidence. Thus in CTEM with normal incidence, the first four Bloch states in Figure 6–2 account for 99% of the electron density. However, depending on the probe position, it is seen that in STEM these states may not be significantly excited. Additionally, state 5, which being antisymmetric has zero excitation amplitude for normal plane wave incidence, is significantly excited for certain probe positions. Indeed all available antisymmetric states are excited depending on probe position. Contributions from all states must therefore be taken into account for the correct dispersion of the probe in real space with increasing depth.

Relative magnitudes are shown below the images. State 1 may be regarded as an s-state for the zinc column, following the parlance of Buxton et al. (1978), and state 2 may be regarded as an s-state for the sulfphur column (though in both cases faint contributions should be noted on the adjacent column site). In these states especially, though

---

[2] This expression is strictly only correct in the absence of absorption. In the presence of absorption $C^{i*}_{\mathbf{g}}$ should be replaced by the appropriate element in the inverse matrix of eigenvalues (Allen and Rossouw 1989, Findlay 2005).

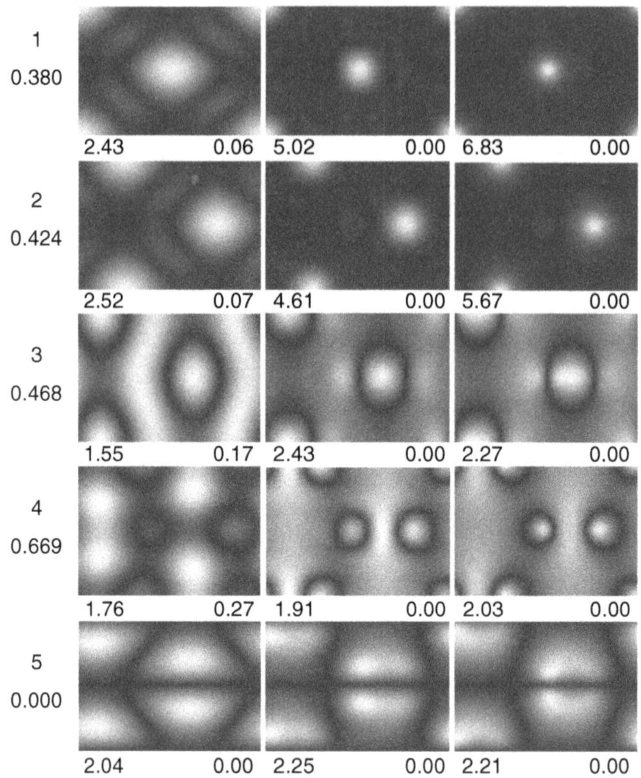

Figure 6–2. Bloch state excitation amplitudes in ZnS, viewed along the [110] zone axis, using 100 keV STEM probes. The rows are labelled with an identifying state number, below which is given the magnitude of $|C_0^{i*}|$ which would be the excitation amplitude in the case of normal plane wave incidence. The first column corresponds to an aberration-balanced probe ($\Delta f = 62$ Å, $C_S = -0.05$ mm, $\alpha = 20$ mrad) the second an aberration-free probe ($\alpha = 25$ mrad) and the third a delta-function probe. Maximum and minimum values are given below the images to provide a sense of scale.

also in the others, the variations in the excitation amplitudes with probe position become more pronounced for the finer probes. The excitation amplitudes suggest that the s-states incorporate most of the electron density when the probe is situated upon the column. Much has been made of s-state models in trying to find schemes which balance accuracy with ease of interpretation. However, it is clear from Figure 6–2 that for other probe positions it will be necessary to include many more Bloch states for an adequate description of the wave function and the signals arising from it (Allen and Rossouw 1989, Anstis et al. 2003, Cosgriff and Nellist 2007). Put another way, only for the case of plane wave incidence (in symmetrical zone axis orientation) is it true that the wave function may be adequately represented by just a few states near the top of the dispersion curve. In particular, the extent of the range of Bloch states excited is critical in assessing the spreading of the wave function about the probe location.

The equivalence of the present expression for the excitation amplitudes, Eq. (7), and the phase-linked plane wave approach has been shown rigorously elsewhere (Findlay et al. 2003). The Bloch wave method can self-consistently be applied to restricted few-beam problems, providing analytic insight. Moreover, it can be used to take great advantage of symmetry, allowing for very efficient calculations for perfect crystals (Findlay et al. 2003, Watanabe et al. 2004). But the Bloch wave method is generally implemented in reciprocal space, making it difficult to picture the effects of particular atoms. Based on matrix diagonalization using a basis of plane waves, the method further scales poorly if large or non-periodic structures are to be considered.

By contrast, the multislice method (for a review see Ishizuka (2004)) proves much more readily adaptable to non-periodic samples. Initially proposed by Cowley and Moodie in 1957, it is based on a physical optics approach, which describes the propagation of the wave function through the crystal in a series of alternating steps: phase modification, to account for the interaction with the specimen potential, and propagation as if in free space, to spatially advance the wave function a step forwards. To implement this, the crystal is divided into N slices, with thicknesses Δz_i, where i labels the slice number at depth z_i (in principle each slice may be of a different thickness, increasing the versatility of the method). The wave function incident on the first slice is simply that of the incident probe, $P(\mathbf{R}, \mathbf{R}_0)$.[3] Defining $\boldsymbol{\psi}(\mathbf{R}, z_i, \mathbf{R}_0)$ as the wave function at the top of the ith slice, we may write

$$\boldsymbol{\psi}(\mathbf{R}, z_{i+1}, \mathbf{R}_0) = [\boldsymbol{\psi}(\mathbf{R}, z_i, \mathbf{R}_0)\boldsymbol{\phi}(\mathbf{R}, z_i)] \otimes P(\mathbf{R}, \Delta z_i)\,, \tag{8}$$

where the transmission function $\boldsymbol{\phi}(\mathbf{R}, z_i)$ is given by

$$\boldsymbol{\phi}(\mathbf{R}, z_i) = \exp\{\mathrm{i}\sigma_0 \left[V(\mathbf{R}, z_i) + \mathrm{i}V'(\mathbf{R}, z_i)\right]\}\,, \tag{9}$$

in which the interaction constant $\sigma_0 = 2\pi m/h^2 k_0$. The potentials are projected over the slice i but, by retaining an explicit dependence on the slice number, different projected potentials can be used for different depths to allow for variation of the potential along the beam direction. While the elastic potential leads to only a phase change in the incident wave function, the inelastic potential leads to an attenuation of the elastic signal. The real space propagator over slice thickness Δz_i is

$$P(\mathbf{R}, \Delta z_i) = \frac{1}{\mathrm{i}\lambda\Delta z_i} \exp\left(\frac{\mathrm{i}\pi R^2}{\lambda\Delta z_i}\right)\,. \tag{10}$$

Higher order expressions have been proposed (Coene et al. 1997), but this formulation is satisfactory as long as the slices are sufficiently thin.

It should be noted that the elastic potential used in the multislice method above is thermally smeared. This broadening of the elastic

[3] The phase-linked plane wave approach could equally well be applied in the multislice method, but, following Kirkland et al. (1987), all multislice treatments of STEM seem to deal with the entire wave function.

potential results in very few electrons in the elastic signal being scattered to high angles and so does not describe directly the distribution of intensity on HAADF detectors. The thermal scattering signal on the HAADF detector must be calculated in addition to the elastic wave function, as described in the following sections. An alternative method, suitable for simulating images due to both elastic and thermal scattering within a purely elastic scattering framework, is the frozen phonon model (Loane et al. 1991, 1992).

The frozen phonon model, pioneered by Silcox and coworkers (Hillyard and Silcox 1993, Hillyard et al. 1993, Loane et al. 1991, 1992, Maccagnano-Zacher et al. 2008, Muller et al. 2001, Silcox et al. 1992), is a semi-classical model based on the fact that the time it takes a fast electron to traverse a thin specimen is much shorter than the oscillation period of an atom due to its thermal motion. Rather than using a thermally smeared potential, as the models described above do, one can perform a simulation in which each atom retains its unsmeared, sharp potential, but is displaced from its equilibrium condition in a manner consistent with the distribution of phonon modes. This is implemented in the framework of Eqs. (8), (9) and (10) by replacing the thermally smeared elastic potential in Eq. (9) with an unsmeared, displaced elastic potential. The absorptive potential, which was assumed to be entirely due to thermal diffuse scattering, is dropped completely. As different electrons see different static configurations, simulation proceeds in a number of passes, averaging the detected signal, an intensity of some sort, over a range of configurations, thus emulating the recording of many electrons which occurs in the experiments. On any given run-through, all scattering is elastic and coherent: there is no absorptive potential and hence no attenuation of the elastic flux. Because many different configurations are required for a fully converged result, a notable disadvantage of the frozen phonon method is the computation time.

Formal proofs of the equivalence between the frozen phonon model and the absorptive model are non-trivial (Wang 1998), but comparison of simulated results between the frozen phonon model and the model we shall presently describe is very good for thinner specimens[4] (Findlay et al. 2003, Klenov et al. 2007, LeBeau et al. 2008). For thicker specimens the models diverge as multiple thermal scattering becomes significant (Klenov et al. 2007), an effect which the frozen phonon model accounts for but simple application of the effective inelastic scattering potential model does not. Both these points, the initially good agreement for thin samples and the discrepancies that arise for thicker specimens, will be seen in the case study presented in Section 6.3. The frozen phonon model has become very popular, in part because

[4] We must note that the model presented here for thermal absorption and HAADF imaging follows from the Hall and Hirsch description (1965), the derivation of which involves an analytic application of the frozen phonon concept, and in light of this the agreement between the models is not surprising. But as effectively the same potential has been derived by other means which do explicitly account for the inelastic transition (Weickenmeier and Kohl 1998), the assertion is not trivial.

its conceptual underpinnings are appealingly visualisable and in part because of the ready availability of its implementation by Kirkland (2008) and the accompanying book (Kirkland 2010) describing its use and underlying theory.

6.2.2 Transition Potentials

The formal treatment of inelastic scattering in electron microscopy generally begins with a many-particle Schrödinger equation. From this starting point, Yoshioka (1957) derived a set of coupled channel equations, where the inelastic electron wave function ψ_n for each channel describes the fast electron wave function associated with an excitation of the specimen into the excited state labelled by a suitable set of quantum numbers denoted collectively by n. We shall be interested in thermal and core-loss inelastic scattering, both of which can, to a good approximation, be described on an atom-by-atom basis.[5] It follows that the inelastic transition is therefore confined about a particular depth in the crystal. The derivations of Coene and Van Dyck (1990) and Dwyer (2005) then yield that ψ_n, the inelastic wave function for the fast electron having excited the crystal from the ground state to the n th excited state, is proportional to the product of H_{n0}, a projected inelastic transition potential (Dwyer 2005), and ψ_0, the elastic wave function of the fast electron:

$$\psi_n(\mathbf{R}, z, \mathbf{R}_0) = -\mathrm{i}\sigma_n H_{n0}(\mathbf{R})\psi_0(\mathbf{R}, z, \mathbf{R}_0)\,, \tag{11}$$

where $\sigma_n = 2\pi m/h^2 k_n$ is the interaction constant for the fast electron after energy loss, in which k_n is the wave number of the fast electron resulting after an energy-loss event that excites the crystal into the nth state. The inelastic wave function thus determined may then be propagated to the exit face, and its contribution to the energy-spectroscopic diffraction pattern determined. The projected transition potential is given by

$$H_{n0}(\mathbf{R}) = \int H_{n0}(\mathbf{R}, z)\exp(2\pi \mathrm{i} q_z z)\mathrm{d}z\,, \tag{12}$$

where $H_{n0}(\mathbf{R}, z)$ is a matrix element of the type $H_{n0}(\mathbf{R}, z) = \langle n|H_{\mathrm{int}}|0\rangle$, and $q_z \approx k_0 \Delta E/2E_0$, where ΔE is the energy loss and E_0 is the incident energy. The interaction term H_{int} is the pairwise Coulomb interaction between the incident fast electron and all the particles in the crystal.

For inner-shell ionization, the projected transition matrix elements $H_{n0}(\mathbf{R})$ have been considered in detail by Dwyer (2005). The crystal

---

[5] Core loss on an atom-by-atom basis is justified because the initial bound state is meaningfully associated with a single atom, and so distinguishable from excitation of other atoms (Maslen 1987, Saldin and Rez 1987). Phonon excitation on an atom-by-atom basis is less obviously justified since it invokes an Einstein model and so neglects any correlated motion between the different atoms. For HAADF imaging, a detailed comparison was carried out by Muller et al. (2001) which showed that these two models give essentially the same predictions. We use that basis to justify the assumption.

states $|0\rangle$ and $|n\rangle$ will generally be approximated as single electron wave functions with the assumption that the ejection of the atomic electron during ionization is the only significant change to the state of the crystal.

Though the theory required to include the fine structure effects that result when the ejected electron interacts with the surrounding crystal exists (Hébert-Souche et al. 2000, Schattschneider et al. 2001), we shall neglect this effect. This is a good approximation when integrating over moderate sized energy windows and a tolerable one when looking at single energy-loss values when we are interested in the shape of the EELS image rather than its magnitude. In this approximation the appropriate basis for the ejected final states is an angular momentum basis, and Dwyer has shown that the number of final states which must be considered per atom is reasonably small for energy losses just above threshold (Dwyer 2005). The authors are not aware of transition potentials for thermal scattering having been considered specifically from this starting point, but the necessary theory exists (Weickenmeier and Kohl 1998) and the results essentially give the scattering functions of Anstis and coworkers (Anstis 1999, Anstis et al. 1996) and Rose and coworkers (Dinges and Rose 1997, Dinges et al. 1995, Hartel et al. 1996, Müller et al. 1998). The treatment of Croitoru et al. (2006) also bears some similarity to these approaches.

6.2.3 Double Channelling, Single Channelling, Nonlocal Potentials and the Local Approximation

Most simulations of EELS imaging are based on the so-called single channelling approximation (Josefsson and Allen 1996), in which further elastic scattering after the inelastic transition is neglected. Recent evidence suggests that for STEM this is not as good an approximation as previously believed (Dwyer et al. 2008), that so-called double channelling, elastic scattering after the inelastic event, can be important. In terms of the simulation theory developed thus far, double channelling involves propagating each and every inelastic wave function ψ_n in the full multislice method – i.e. allowing for elastic scattering – through the remainder of the crystal. The increased computational burden of full double channelling calculations is thus considerable. Moreover, recent developments have allowed for much larger detector collection angles than have previously been possible (Krivanek et al. 2008, Muller et al. 2008), and in this regime the single channelling approximation will hold. Therefore, as well as for consistency in making comparisons with previous work, we shall restrict our attention to the single channelling approximation unless otherwise stated. In this approximation, the contribution to the energy-spectroscopic diffraction pattern is determined by propagating all ψ_n to the exit face of the specimen via free space propagation, or rather, since we are only interested in the intensities in the diffraction plane and this additional propagation only modifies the phases of the elements in this plane, by neglecting propagation entirely. The energy-spectroscopic diffraction pattern from an event at depth z_i is thus simply the intensity of the Fourier transform of Eq. (11):

$$\psi_n(\mathbf{Q}, z_i, \mathbf{R}_0) = -\mathrm{i}\sigma_n \int H_{n0}(\mathbf{R})\psi_0(\mathbf{R}, z_i, \mathbf{R}_0) \exp(-2\pi \mathrm{i}\mathbf{Q}\cdot\mathbf{R})\, \mathrm{d}\mathbf{R} . \quad (13)$$

The recorded image[6] is obtained by multiplying by a current conversion factor k_n/k_0 (Dudarev et al. 1993), adding over all final states and integrating over some detector:

$$\begin{aligned} I(\mathbf{R}_0) &= \int_{\text{detector}} \left[\sum_{\text{final states}} \frac{k_n}{k_0} \left| \psi_n(\mathbf{Q}, z_{\text{final state}}, \mathbf{R}_0) \right|^2 \right] \mathrm{d}\mathbf{Q} \\ &= \int_{\text{detector}} \left[\sum_{n\neq 0} \frac{k_n}{k_0 t_c} \sigma_n^2 \int_0^t \left| \int H_{n0}(\mathbf{R})\psi_0(\mathbf{R}, z, \mathbf{R}_0) \right. \right. \\ &\qquad \left. \left. \exp(-2\pi \mathrm{i}\mathbf{Q}\cdot\mathbf{R})\, \mathrm{d}\mathbf{R} \right|^2 \mathrm{d}z \right] \mathrm{d}\mathbf{Q} , \end{aligned} \quad (14)$$

where we have separated the sum over final states of the atom at a particular depth from the sum over depths, the latter of which has been approximated by an integral (having introduced the repeat distance t_c).

Expanding and reordering Eq. (14)

$$\begin{aligned} I(\mathbf{R}_0) &= \int_0^t \int \int \psi_0^*(\mathbf{R}, z, \mathbf{R}_0) \left[\sum_{n\neq 0} \frac{k_n}{k_0 t_c} \sigma_n^2 H_{n0}^*(\mathbf{R}) H_{n0}(\mathbf{R}') \times \right. \\ &\qquad \left. \int_{\text{detector}} \exp\left[2\pi \mathrm{i}\mathbf{Q}\cdot(\mathbf{R}-\mathbf{R}')\right] \mathrm{d}\mathbf{Q} \right] \psi_0(\mathbf{R}', z, \mathbf{R}_0) \mathrm{d}\mathbf{R}\mathrm{d}\mathbf{R}'\mathrm{d}z \quad (15) \\ &\equiv \frac{2\pi}{hv} \int_0^t \int \int \psi_0^*(\mathbf{R}, z, \mathbf{R}_0) W(\mathbf{R}, \mathbf{R}') \psi_0(\mathbf{R}', z, \mathbf{R}_0) \mathrm{d}\mathbf{R}\mathrm{d}\mathbf{R}'\mathrm{d}z , \end{aligned}$$

where

$$W(\mathbf{R}, \mathbf{R}') = \frac{2\pi m}{h^2 t_c} \sum_{n\neq 0} \frac{1}{k_n} H_{n0}^*(\mathbf{R}) H_{n0}(\mathbf{R}') \int_{\text{detector}} \exp\left[2\pi \mathrm{i}\mathbf{Q}\cdot(\mathbf{R}-\mathbf{R}')\right] \mathrm{d}\mathbf{Q} \quad (16)$$

and $v = hk_0/m$. In the interests of obtaining an equation of the form used previously[7] (Oxley et al. 2005), the detector term and the sum over final states have been grouped into a single term, though within the

---

[6] Or, loosely speaking, cross-section expression since for high-energy electron scattering the two are related by constant factors.

[7] To make connection with previous work by the authors and collaborators (Allen and Josefsson 1995, Allen et al. 2003, 2006, Findlay et al. 2005, Oxley et al. 2005, 2007, Rossouw et al. 2003), we note that Eq. (15) has generally been evaluated in reciprocal space. Defining inelastic scattering matrix elements $\mu_{\mathbf{H},\mathbf{G}}$ via

$$\frac{2\pi}{hv} W(\mathbf{R}, \mathbf{R}') = \frac{1}{A} \sum_{\mathbf{H},\mathbf{G}} \mu_{\mathbf{H},\mathbf{G}} \exp(2\pi i\mathbf{H}\cdot\mathbf{R}) \exp(-2\pi i\mathbf{G}\cdot\mathbf{R}')$$

and substituting into Eq. (15), the nonlocal imaging expression may be rewritten as approximations made here the two terms are separable (Dwyer 2005). Ionization events from different atoms are assumed to be incoherent, and so the sum over n runs over the different atom sites as well as over the range of final states for each of these atoms. Fortunately, for a given energy loss close to the ionization threshold, the number of significant final states per atom is relatively small (Dwyer 2005).

If the range of integration in the plane is large enough (large collection aperture) then the detector integral approximates to a δ-function and we may write

$$W(\mathbf{R},\mathbf{R}') \approx 2V'(\mathbf{R})\delta(\mathbf{R}-\mathbf{R}') \equiv \frac{2\pi m}{h^2 t_c}\sum_{n\neq 0}\frac{1}{k_n}H^*_{n0}(\mathbf{R})H_{n0}(\mathbf{R})\delta(\mathbf{R}-\mathbf{R}') \,. \quad (17)$$

In this approximation, the image expression of Eq. (15) reduces to

$$I(\mathbf{R}_0) = \frac{4\pi}{hv}\int_0^t \int |\psi_0(\mathbf{R},z,\mathbf{R}_0)|^2 V'(\mathbf{R})\mathrm{d}\mathbf{R}\mathrm{d}z \,. \quad (18)$$

The local imaging expression, Eq. (18), has been derived as a special case of the nonlocal imaging expression, Eq. (15). Other, more direct derivations are possible. Ishizuka (2001) provides a succinct and elegant derivation from the multislice method using an absorptive potential. Alternatively, Eq. (18) may simply be invoked as being intuitively reasonable (Cherns et al. 1973, Pennycook and Jesson 1991). However, there is no clear route to generalize back to the nonlocal form of Eq. (15).

The use of the term "nonlocal" to describe the potential occurring in Eq. (15) requires clarification. As used here it is intended to differentiate between cases where the inelastic scattering is well described by the local approximation given in Eq. (17), for example, HAADF imaging, energy dispersive X-ray measurements and STEM EELS imaging where the detector is larger than the probe-forming semiangle and cases where the full nonlocal expression must be used, such as STEM EELS measurements with small collection angles, and even high angular resolution CTEM EELS measurements where a small detector is the norm (Allen et al. 2006). The nonlocal integral kernel was derived by Yoshioka in 1957 and hence does not represent new and unusual physics. The

---

$$I(\mathbf{R}_0) = \frac{1}{A}\sum_{\mathbf{H},\mathbf{G}} \mu_{\mathbf{H},\mathbf{G}} \int_0^t \psi_0^*(\mathbf{H},z,\mathbf{R}_0)\psi_0(\mathbf{G},z,\mathbf{R}_0)\mathrm{d}z \,.$$

The area factor A is an artefact of the assumed normalization of the wave functions, which varies widely in the literature. Here we have assumed that the integral of the intensity of the wave function over a 2D plane is dimensionless, in both real and reciprocal space forms. This contrasts with the Bloch wave formulation, where the wave function itself is often taken to be dimensionless.

The inelastic scattering matrix elements $\mu_{\mathbf{H},\mathbf{G}}$ are closely related to the mixed dynamic form factor (Kohl and Rose 1985, Schattschneider et al. 2000), the difference being that the former further incorporates information about the detector geometry.

nonlocal form will be of no surprise to readers familiar with the density matrix description of inelastic scattering (Dudarev et al. 1993, Kohl and Rose 1985, Schattschneider et al. 2000), and for implementation this form is very convenient because the full range of final states, possibly including integration over a window of different energy losses, can all be buried within the single $W(\mathbf{R}, \mathbf{R}')$ construct. But when such a form is derived from the collapse of the coupled channel equations (Allen and Josefsson 1995), and dubbed the nonlocal cross-section expression because of the mathematical form of the effective scattering potential $W(\mathbf{R}, \mathbf{R}')$ in those equations, its interpretation seems involved and indirect (Oxley et al. 2005).

From an inspection of the imaging expressions alone, the most obvious difference is that the inelastic scattering probability in the nonlocal imaging expression may depend on the phase of the wave function, while in the local imaging expression the probability of inelastic scattering is independent of the phase of the wave function. The present derivation makes clear the reason why the phase of the wave function might affect the inelastic scattering signal. The nonlocal imaging expression, Eq. (15), uses the real space elastic wave functions to describe the proportion of electrons which fall within a detector situated in the diffraction plane. Eq. (11) gives that the inelastic wave function in real space depends multiplicatively on the elastic wave function causing the transition. Therefore, the phase information in this wave function is essential in determining how the inelastic wave function will interfere with itself as it propagates to the detector plane: the phase of the elastic wave function plays a role in determining the distribution of the inelastically scattered electrons in the diffraction plane, and hence relative to the detector. It is this fact, through Eqs. (13) and (14), which means that the general inelastic imaging expression in Eq. (15) must depend on the phase of the elastic wave function. Dwyer (2005) provides a similar discussion.

In the case of EELS with a large, on-axis detector, the reason the local expression does not depend on the phase becomes equally clear: when the detector is large enough to collect all the inelastically scattered electrons, their possible redistribution is irrelevant. All that matters is the number of electrons in the inelastic channels, and this can be determined from Eq. (11) by taking the total electron density in the inelastic wave function. Because the inelastic transition is described by a multiplicative interaction, the number of electrons in the inelastic channel depends only on the electron density in the incident elastic wave function and not on its phase. The case of HAADF, where we do not collect all thermally scattered electrons, is somewhat different, and the interpretation relies more on the behaviour of the average over the detector. In mathematical terms it follows from the fast oscillatory nature of the detector integral in Eq. (16) for large momentum transfer $\mathbf{Q}$ (Muller and Silcox 1995).

If the effect of the specimen on the incident wave function is negligible, such that the wave function $\psi(\mathbf{R}, z, \mathbf{R}_0)$ can be well described by its form in free space, $P(\mathbf{R} - \mathbf{R}_0, z)$, then the integration over $\mathbf{R}$ in Eq. (18) has the form of a convolution and the expression reduces to the object function approximation (Jesson and Pennycook 1993, Loane et al. 1992)

$$I(\mathbf{R}_0) = \int_0^t |P(\mathbf{R}_0, z)|^2 \mathrm{d}z \otimes O(\mathbf{R}_0) \,, \tag{19}$$

where we have defined $O(\mathbf{R}_0) = 4\pi V'(\mathbf{R}_0)/hv$.

6.2.4 Anomalous EELS Imaging and Carbon K-Shell in SiC

In this section we will discuss how the effective nonlocality of the inelastic ionization interaction can lead to apparently anomalous results.[8] Figure 6–3 shows simulated carbon *K*-shell STEM EELS line scans in a 100 Å thick SiC sample oriented along the [011] zone axis, using a 200 keV, aberration-free probe with a 50 mrad semiangle probe-forming lens, for both a 10 and a 50 mrad detector semiangle (though note the separate vertical scales). The simulations are based on Eq. (11), in the single channelling approximation, and assume an energy loss 10 eV above the ionization threshold. The line scan is in the ⟨100⟩ direction through the SiC dumbbells. The 50 mrad detector result shows a significantly dominant, single peak on the carbon column, but a large background across the whole line scan. The 10 mrad detector results show peaks on both silicon and carbon columns, with the larger peak on the silicon, a decidedly counter-intuitive result and one which, if obtained experimentally, would initially appear quite puzzling.

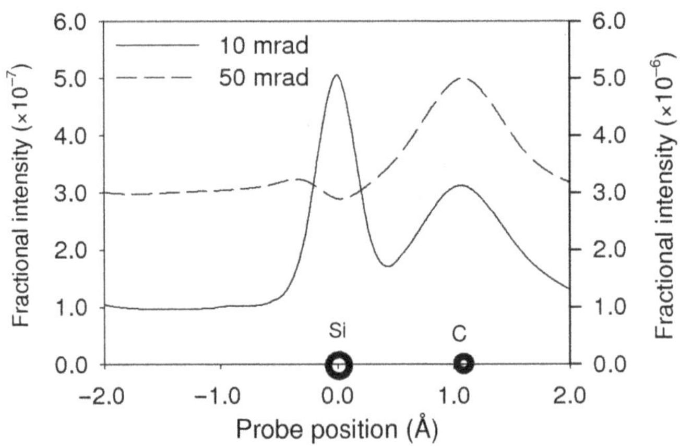

Figure 6–3. Carbon *K*-shell STEM EELS line scans in the ⟨100⟩ direction in a 100 Å thick SiC sample oriented along the [011] zone axis, using a 200 keV, aberration-free probe with a 50 mrad semiangle probe-forming lens. The images are simulated for a 10 eV energy loss above the ionization threshold. Results for both 10 and 50 mrad detector semiangles are shown with separate axes, left and right, respectively. (The units should really be "fractional intensity per eV", though to an excellent approximation one could alternatively assume the use of a 1 eV energy window.) Reproduced from Allen et al. (2008).

[8] An extended version of this investigation may be found in Allen et al. (2008).

The result for the 10 mrad detector semiangle is very similar to that presented by Oxley et al. (2005), who, for the carbon *K*-shell STEM EELS image in SiC, show significant, separate peaks on both the silicon and carbon columns, with those on the silicon column being larger.[9] The regime in which the most anomalous results arise can largely be avoided through judicious choice of experimental parameters (Allen et al. 2006, Findlay et al. 2008), most notably detector collection angle, but the highly nonintuitive result makes for a good test case to explore the possible complications which might arise in this imaging method and to demonstrate how simulations can be used to clarify what is happening.

We will present an analysis based on the transition potential model, i.e. Eq. (11). There are two aspects to an analysis of the scattering process based on transition potentials. The first aspect is to identify all the H_{n0} involved, both the different final states of each contributing atom and of the different atoms. The shape and position of these potentials relative to the elastic wave function determine the number of electrons in inelastic final states which can contribute to the imaging. The convergence with respect to the number of states and atoms included is very slow, a downside of the method as it notably increases the calculation time, but the model allows us to directly visualize the intensity in the different inelastic wave functions which is in many respects more intuitive than the somewhat rarified mathematical construction of the effective nonlocal potential. The second aspect is the proportion of electrons in each inelastic state ψ_n which contribute to the detected intensity. This is given by the fraction of the reciprocal space electron density of ψ_n which lies within the region corresponding to the detector acceptance angle.

To investigate the first aspect, Figure 6–4 shows profiles through the transition potentials with final states characterized by angular momentum quantum numbers $(l', m') = (0, 0)$, $(1, 0)$ and $(1, 1)$, a representative subset of the full range of transition potentials involved. The transition to $(1, 0)$ is the widest of the three, but its maximum value is smaller than that of the other two potentials. The transition to final state $(0, 0)$ is highly peaked on the origin, while that to $(1, 1)$ peaks off-column. Figure 6–4 also shows the probe intensity at the entrance surface for the same parameters as used in Figure 6–3. There is considerable overlap between the probe intensity and the transition potential for the $(0, 0)$ final state. Therefore, the dipole approximation, which would only allow transitions to the $l' = 1$ final state, will not be a good approximation when the probe is directly on the column. The dashed vertical line

[9] The 10 mrad case has the same shape as the test case of Oxley et al. (2005). However, since the approach based on Eq. (11) does not lend itself to integration over an energy window, a fixed energy loss of 10 eV above the threshold was chosen. This makes the units here slightly different to those in Oxley et al. (2005) since they are based on Eq. (15) where the integration over a 40 eV energy window was carried out over the effective scattering potential.

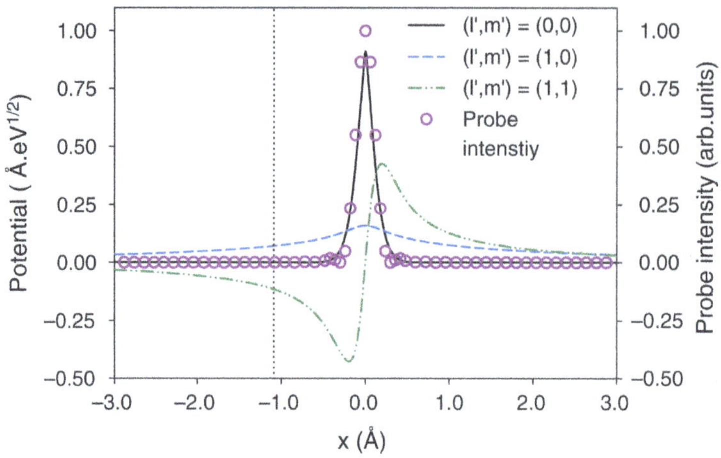

Figure 6–4. Carbon *K*-shell transition potential and probe intensity profiles. The transition potential to (0, 0) is purely real, while that to (1, 0) is purely imaginary. Only the modulus is plotted. The transition to (1, 1) is complex, and only the real part is plotted, its antisymmetric character hinting at the vortical phase ramp present in the full 2D transition potential. The *vertical dotted line* indicates where the neighbour silicon column would be in the SiC structure, reproduced from Allen et al. (2008).

in Figure 6–4 shows the distance at which, in SiC, the nearest neighbour silicon column would sit relative to the carbon atom at the origin. If the probe were at this position there would be a larger electron density in the inelastic channel for transition to the (1, 1) state than there would to the (1, 0) state.

The probe does not retain its surface form as it propagates through the crystal. Rather than considering the detailed evolution of the wave function relative to the spatial distribution of these transition potentials, we instead plot the contribution from each depth for transitions to a specific final state. Such plots are shown in Figure 6–5(a–h). Assuming a 10 mrad detector collection semiangle, Figure 6–5(a) shows the total contribution per slice from all final states, while Figures. 6–5(b), (c) and (d) shows the contribution from transitions to final states (0, 0), (1, 0) and (1, 1), respectively. The considerable spatial overlap between the probe intensity and the transition potential to final state (0, 0) is evident in the large peak at small depths in Figure 6–5(b), but this rapidly disappears beyond 20 Å as the probe diffuses through the specimen. For small depths, the contribution from transitions to (1, 1) peaks to either side of the column position, as expected from the form of the transition potential in Figure 6–4. The sum of these contributions leads to the volcano feature in the single atom case (Kohl and Rose 1985).

For larger depths, transitions to states (1, 0) and (1, 1) both give fairly consistent contributions when the probe is positioned on the carbon column but also when it is positioned on the silicon column. Indeed for transitions to the (1, 1) state, the contribution for the probe positioned on the silicon column is notably larger than that for the probe on the carbon column. This leads to the dominance of the signal from the probe

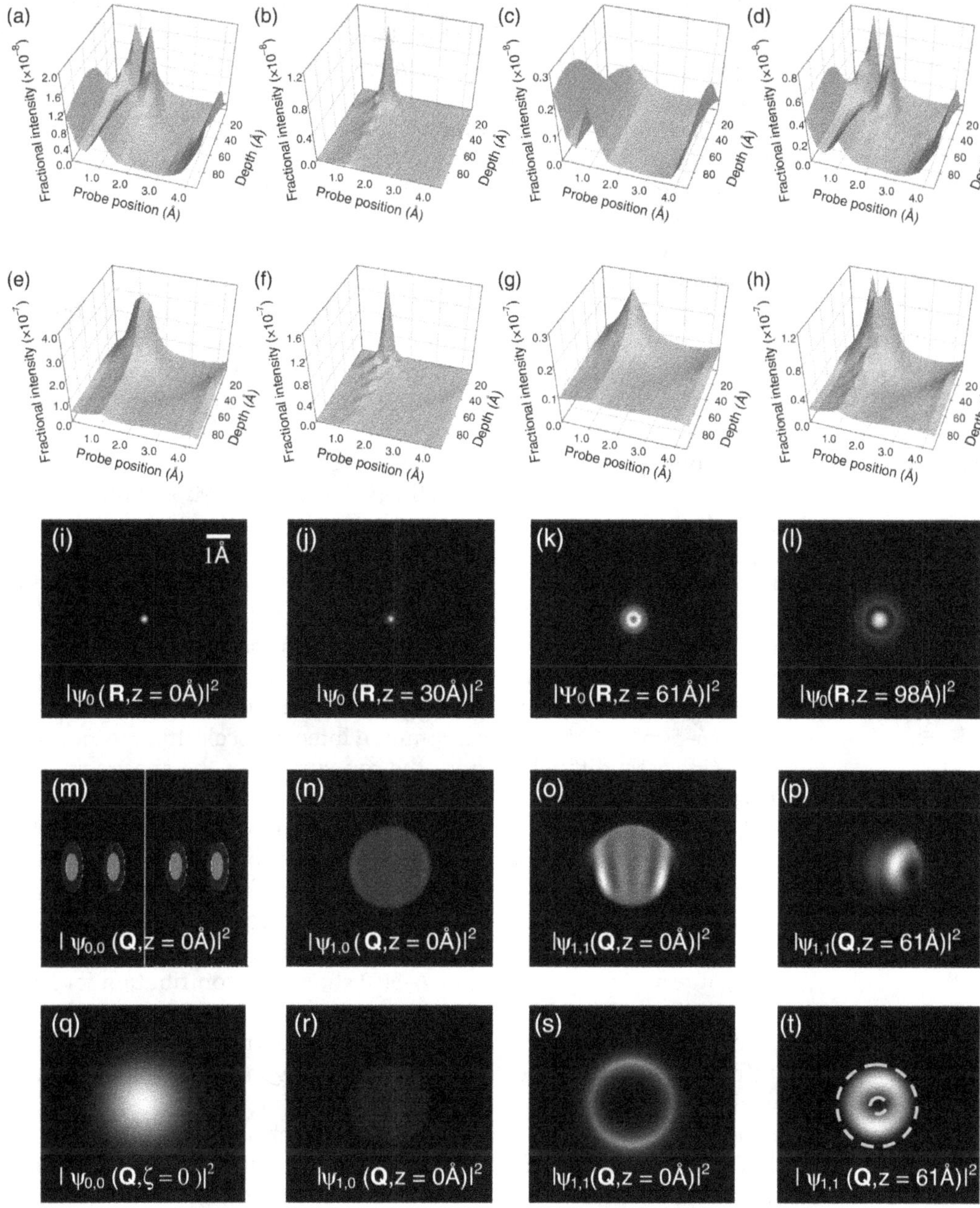

Figure 6–5. Carbon *K*-shell signal per depth for transitions to (a) all final states, (b) (0, 0), (c) (1, 0) and (d) (1, 1) for a 10 mrad detector collection semiangle; (e) all final states, (f) (0, 0), (g) (1, 0) and (h) (1, 1) for a 50 mrad detector collection semiangle. The silicon column is at the origin and the carbon column is at 1.09 Å. Real space wave function intensity for the probe on the silicon column at depths of (i) 0 Å, (j) 30 Å, (k) 61 Å and (l) 98 Å. The scale bar in (i) is applicable to (i) to (l). Diffraction pattern intensity for: (m) (0, 0), (n) (1, 0) and (o) (1, 1) at 0 Å, (p) (1, 1) at 61 Å, all for the probe on silicon; (q) (0, 0), (r) (1, 0) and (s) (1, 1) at 0 Å, (t) (1, 1) at 61 Å, all for the probe on carbon. The 10 mrad and 50 mrad detector sizes relative to the reciprocal space intensity distributions are shown by the dashed circles in (t), from which the scale for (m)–(t) can be deduced. Reproduced from Allen et al. (2008).

on the silicon column (see Figure 6–5(a), which, when integrated over depth, gives the 10 mrad detector plot in Figure 6–3).

Figure 6–5(e–h) shows equivalent plots assuming a 50 mrad detector collection semiangle. Except in so much as the contribution is large for the first few slices and decreases rapidly the deeper one goes into the crystal, a simple consequence of probe spreading, these figures are very different than those for the 10 mrad detector. In particular, nothing distinguishes the case of the probe on the silicon column to most other probe positions, and the background signal is higher and more consistent for all probe positions. It must be remembered that all that has changed is the detector size. Neither the elastic wave function nor the transition potentials in Eq. (11) have altered, and therefore neither have any of the inelastic wave functions ψ_n. All that distinguishes the form of Figure 6–5(a–d) from that of Figure 6–5(e–h) is the extent of the detector.

To better appreciate this we plot the electron distribution in the detector plane, i.e. the contribution to the (inelastic) diffraction pattern, for a few select final states and a few select atom depths. Figure 6–5(i)–(l) shows the real space intensity of the elastic wave function at the depths 0, 30, 61 and 98 Å for a probe initially on the silicon column (those for the probe on the carbon column are quite similar). These are plotted according to their individual maxima, which masks the extent of the probe diffusion. For each given distribution, the largest concentration of electron density is still strongly centred about the probe position. Figure 6–5(m)–(o) shows, on a common intensity scale, the contribution to the inelastic diffraction from the top surface of the specimen with the probe situated on the silicon column for transitions to $(0, 0)$, $(1, 0)$ and $(1, 1)$, respectively. As only the transition potentials for the latter two transitions extend out significantly to this column, only they show appreciable contributions. Because both these potentials are fairly flat in the vicinity of the silicon column, the diffraction patterns are fairly uniform disks of about 50 mrad radius – similar to the elastic diffraction pattern of the probe. Figure 6–5(p) shows the contribution for final state $(1, 1)$ from the depth of 61 Å into the crystal. While the evolution of the probe has changed its shape somewhat, the distribution in the diffraction pattern is such that a significant signal will be obtained in a small on-axis detector.

Figure 6–5(q)–(s) shows, on a common intensity scale, the contribution to the inelastic diffraction from the top surface of the specimen with the probe situated on the carbon column for transitions to $(0, 0)$, $(1, 0)$ and $(1, 1)$, respectively. The transitions to $(0, 0)$ and $(1, 0)$ are again fairly uniform, the latter being much weaker, in spite of being a "dipole-favoured" transition but quite consistent with the relative sizes of the potentials as seen in Figure 6–4. However, the complex transition potential to final state $(1, 1)$ is effectively antisymmetric (having a phase vortex) and this, combined with the rotationally symmetric probe function, leads to a doughnut-shaped diffraction pattern with very little intensity falling on the innermost region. Figure 6–5(t) shows the contribution for final state $(1, 1)$ from the depth of 61 Å into the crystal. The doughnut is again seen, which, though wider, still has very little intensity within the central region. The 10 mrad detector is shown in this figure as the inner of the two dashed circles, while the outer corresponds to 50 mrad. Now we can understand the role of the detector relative to the symmetry of the states. While there may well be more carbon *K*-shell ionization events in total when the probe is closer to the carbon column, the distribution in the diffraction pattern of the fast electrons after causing these transitions is such that the signal on a small, on-axis detector may well be larger when the probe is displaced from the column. When we go to the 50 mrad detector, we collect most of the inelastically scattered electrons. Regions of strong signal are therefore more representative of the regions where the most ionization occurs, and as such the expected peak on the carbon column is regained. The large background in this case is a consequence of the delocalized potential, the density of carbon columns and the elastic spreading of the probe.

We keep alluding to the extent to which the probe spreads, but Figure 6–5 is deceptive in this regard: it gives the signal as a function of probe position and depth, but it does not really show which atoms are contributing. To get a feel for the contributions from the nearest columns, let us simulate separately the contributions to the carbon *K*-shell EELS line scan from specific subsets of contributing columns. Figure 6–6(a) and (b) shows the results for the 10 and 50 mrad semiangle detectors, respectively.

The line scan location and column labelling are shown on the SiC structure in Figure.6–6(c). In both cases, the contribution from the

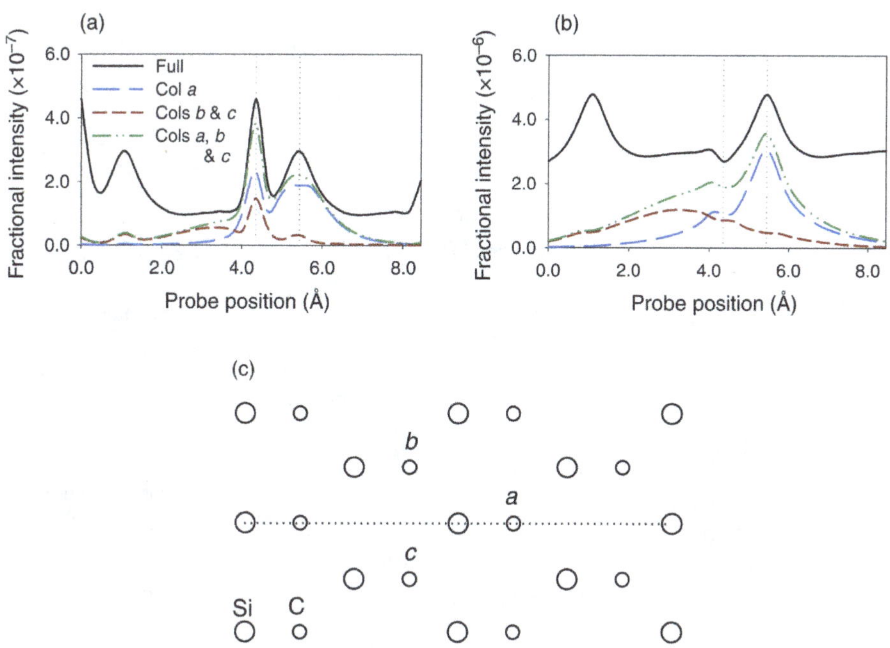

Figure 6–6. The signal strength for the (a) 10 mrad detector and (b) the 50 mrad detector, broken into contributions from select columns, with labelling corresponding to the projected structure of SiC depicted in (c), reproduced from Allen et al. (2008).

nearest column (column a) barely gives half the total contribution on the central silicon and carbon probe positions. For the smaller detector, the combination of columns a, b and c gives the correct shape at the central dumbbell, but not quite the full signal. For the large detector, the significant contribution obtained from columns other than the labelled three is quite evident; the probe spreading and background contributions are significant here.

All the simulations thus far presented in this section have been single channelling simulations, meaning, as per the discussion in the previous section, that the effects of elastic scattering on the inelastic wave functions produced by each ionization event are neglected. This approximation will be aided by the relatively thin and weakly scattering nature of this specimen. Still, for the 10 mrad detector collection semiangle there is scope for a redistribution of the inelastically scattered electrons by subsequent elastic (and thermal) scattering, which may affect the results. Figure 6–7 compares single channelling (dashed line) and double channelling (solid line) results for 4, 10, 30 and 50 mrad detector collection semiangles. For the 50 mrad detector, the results of the two calculations are indistinguishable, the single channelling approximation being valid for this large detector. The agreement gets progressively worse as the detector gets smaller, though in this case

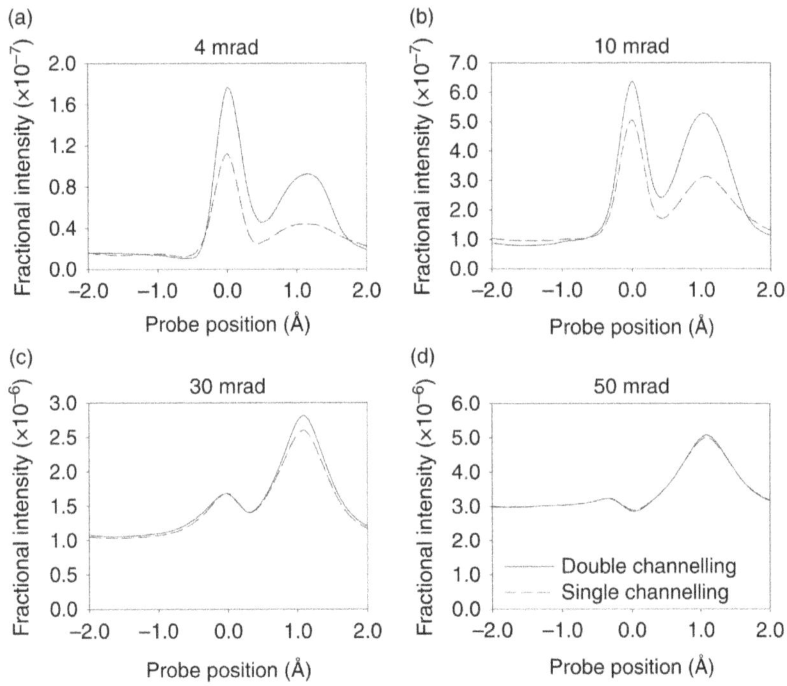

Figure 6–7. Double channelling *solid line* and single channelling *dashed line* results for the carbon K-shell STEM EELS line scans for detector collection angles of (a) 4 mrad (b) 10 mrad (c) 30 mrad and (d) 50 mrad. Other parameters are identical to Figure 6–3.

the qualitative shapes of the double and single channelling calculations are in fair agreement. For heavier scattering samples, qualitative as well as quantitative differences between the two approaches may occur (Dwyer et al. 2008). Nevertheless, the case study presented in this section serves to illustrate the principles and considerations involved in the formation of EELS images given the delocalized nature of the interaction potential.

6.2.5 Delocalization

As seen in the previous section, the long-ranged nature of the transition potentials can lead to unexpected results (in the case of small detectors) and a large background (in the case of large detectors). This raises the question of how best to characterize the delocalization of the EELS signal. Egerton has suggested the diameter in which 50% of the EELS intensity is contained (Egerton 1996):

$$(d_{50})^2 \approx \left(\frac{0.5\lambda}{\theta_E^{3/4}}\right)^2 + \left(\frac{0.6\lambda}{\beta}\right)^2, \tag{20}$$

where $\theta_E = \Delta E/2E_0$ with ΔE being the energy loss, E_0 the incident energy and β the detector semiangle. Alternative measures of delocalization are the root mean square impact parameter b_{rms} (Pennycook 1998) and the half width at half maximum (HWHM) of single atom EELS image simulations (Cosgriff et al. 2005, Kohl and Rose 1985).

In Figure 6–8 we consider the K-shell EELS signal from an isolated carbon atom for 100 keV incident electrons. Figure 6–8(a) shows the image profile for a fixed probe-forming aperture semiangle of $\alpha = 30$ mrad as a function of collection semiangle β. For small collection semiangles the signal peaks away from the atomic location and the volcano feature is clear. As the detector becomes larger the profile peaks above the atomic site, with little change in shape or intensity for β beyond approximately 50 mrad. This may be understood with reference to Figure 6–5 where it was seen that most of the inelastic wave function intensity on the diffraction plane is contained within the probe-forming aperture, and increasing the detector semiangle far beyond this does not increase the overall EELS signal significantly. In Figure 6–8(b) we plot the values of r_{50}, r_{75} and r_{90}, the radii containing 50, 75 and 90% of the EELS signal for the atomic images in (a). This is a measure of the signal in the 2D STEM EELS image, i.e. the radius describes a circle within which x% of the intensity is enclosed. We have chosen to use the radius rather than the diameter as this provides a measure of how far away from the atomic site the signal is spread, i.e. how delocalized the interaction is. The value calculated using Eq. (20) above is shown by the grey line. For other than small collection semiangles, where the volcano features exist, these plots are quite flat and there is reasonable agreement between the 50% radii calculated from Eq. (20) and that calculated directly from the simulated images. This is unsurprising given the signal is dominated by inelastic electrons within the first 30 mrad of the detector. Also plotted

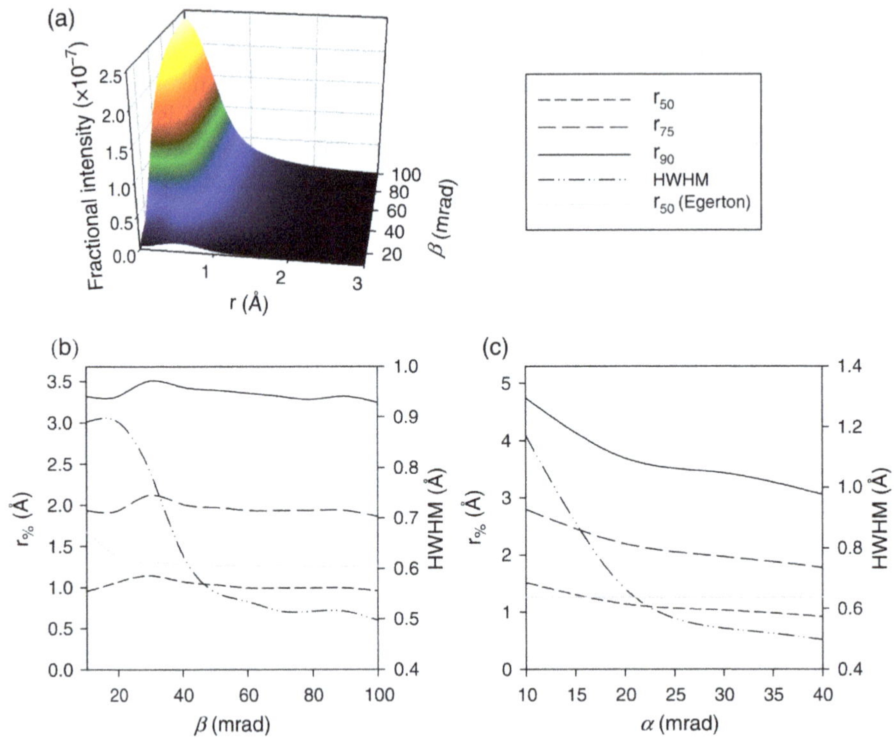

Figure 6–8. (a) Carbon *K*-shell EELS image of an isolated atom as a function of detector semiangle for 100 keV incident electrons and a 30 mrad probe-forming aperture semiangle. (b) HWHM and radii containing 50, 75 and 90% of the EELS intensity from (a). The value calculated using Eq. (20) is shown by the *grey line*. (c) Delocalization as a function of probe-forming aperture for a detector semiangle $\beta = 60$ mrad.

is the HWHM. The HWHM decreases significantly as β is increased but again levels out for $\beta > 60$ mrad. It is clear that these two measures of delocalization provide different information. The HWHM tells us how quickly the intensity drops off from an initial maximum, hence providing a measure of visibility of the signal above the background. It provides no information about how long-ranged the interaction might be. Conversely, knowing the radius within which a given amount of the intensity is contained is a measure of how long-ranged an interaction is, but does not provide useful information about how peaked (and hence visible) an isolated signal might be. A full understanding about the "delocalization" of a signal requires both pieces of information. Figure 6–8(c) plots these measures as a function of probe-forming aperture for a fixed detector of $\beta = 60$ mrad. There is a general decrease in all delocalization measures with increasing probe aperture, with the exception of that derived from Eq. (20) which includes no information about α. All measures tend to flatten out somewhat for $\alpha > 25$ mrad, indicating that ultimately delocalization depends on the interaction being measured. Indeed, the transition potentials are independent of the probe-forming aperture and those with long-range components will contribute to image formation even for an idealized point probe.

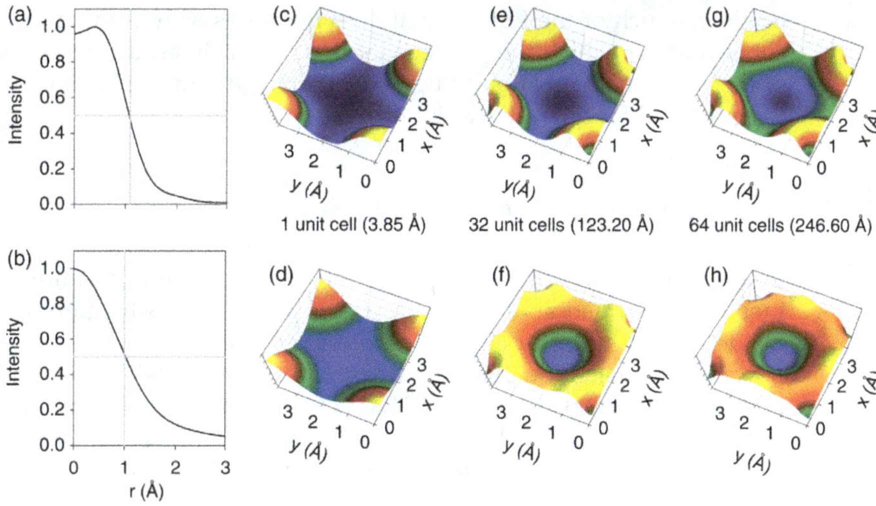

Figure 6–9. Single atom STEM EELS images based on (a) the lanthanum $M_{4,5}$ edge and (b) the lanthanum $N_{4,5}$ edge for 100 keV incident electrons with $\alpha = 20$ mrad and $\beta = 12$ mrad. The *grey vertical lines* indicate the HWHM. The evolution of images in $LaMnO_3$ in the [001] zone axis orientation is shown for the thicknesses indicated for (c), (e) and (g) for the lanthanum $M_{4,5}$ edge and (d), (f) and (h) for the lanthanum $N_{4,5}$ edge, adapted from Oxley et al. (2007).

The importance of knowing not only the HWHM but also the range of the EELS interaction is illustrated in Figure 6–9. Single atom STEM EELS images are shown for 100 keV incident electrons with a probe-forming aperture semiangle of 20 mrad and a collection semiangle of 12 mrad, based on (a) the lanthanum $M_{4,5}$ and (b) the lanthanum $N_{4,5}$ edges. The HWHMs are indicated by the vertical grey lines. Despite the difference in the edge onset values, both atomic images have similar values for the HWHM (in fact the lower lying $N_{4,5}$ edge has a slightly lower HWHM than the $M_{4,5}$ edge). However, while the lanthanum $M_{4,5}$ image has almost zero intensity 3 Å away from the atomic site, the lanthanum $N_{4,5}$ edge still has significant intensity. The effect of this long-range component is evident in Figure 6–9(c), (d) where images are calculated for each edge for atoms located in [001] zone axis oriented $LaMnO_3$. The lanthanum $N_{4,5}$ peaks have been significantly broadened by the long-range "tails" of surrounding atoms, even for a single unit cell thickness where channelling of the incident probe has played no role. In addition there is significant intensity at the center of the unit cell while the lanthanum $M_{4,5}$ image remains well defined. As the thickness of the specimen increases there is a reduction of intensity above the lanthanum columns as electrons are scattered through large angles beyond the EELS detector by the heavy lanthanum atoms. For the thickest specimen considered here (64 unit cells or 246.6 Å) the lanthanum $M_{4,5}$ image is still easily interpreted, while the lanthanum $N_{4,5}$ image now peaks above the oxygen columns.

While the discussion here has been limited to EELS chemical mapping, EELS near edge structure is often used to gain information about local bonding states. It is important to realize that in many cases the origin of much of the EELS signal does not necessarily come from the column above which the probe is located. As such, an understanding of the components that build up an EELS image from simulation is an essential tool for quantitative interpretation.

6.3 Quantitative Z-Contrast Imaging

Image simulations are innately quantitative: the image simulation methods described above give an image intensity as a fraction of the incident beam intensity. However, experiments seldom obtain sufficient information to include this fact as part of the image analysis and interpretation. The great strength of STEM HAADF imaging, that the images can often be directly and visually interpreted in terms of the projected structure (Pennycook and Jesson 1991, Varela et al. 2005), has meant that many investigations can be carried out based on qualitative analysis alone. Even when simulations are used to support the results, it has frequently sufficed to show that the qualitative image features observed are readily reproducible in simulations (Yamazaki et al. 2001). HAADF STEM imaging can be quite forgiving in this regard, as the same features can often be obtained over a wide range of thickness and defocus values (Hillyard and Silcox 1993, Hillyard et al. 1993). And, of course, in many cases this is more than sufficient. But there is a wealth of further information to be had if only the comparison can be made more quantitative. As more complex structures – defects, dopants, interfaces and the like – increasingly constitute the focus of STEM investigations, quantitative comparison is becoming more important (Carlino and Grillo 2005, Grillo et al. 2008).

Experimental intensities are seldom recorded on an absolute scale. This means that when quantitative comparisons between theory and experiment are made, a scaling factor needs to be introduced (Klenov and Stemmer 2006, Klenov et al. 2007, Kotaka et al. 2001, Watanabe et al. 2001). In such attempts it has been observed that a contrast difference tends to exist between the experimental data and simple simulations, the former having significantly less contrast than the latter[10] (Klenov and Stemmer 2006, Klenov et al. 2007). Some investigations circumvent this issue by subtracting the background, or, which amounts to the same thing, making quantitative comparisons with the data scaled between 0 and 1 (Watanabe et al. 2001). There are justifiable ways of reducing the contrast in the simulations, depending on how well the experimental set-up has been quantified. One is slight specimen tilt,

[10] This is reminiscent of a very similar problem in atomic-resolution images in CTEM, referred to as the "Stobbs factor" problem (Howie 2004, Hÿtch 1994). If the cause were to lie in some insufficiently appreciated aspect of, say, the thermal scattering, the problems in CTEM and STEM could be connected. However, though the definitive evidence has not yet been published, it seems likely that the cause of the discrepancy in CTEM is due to the modulation transfer function of the detector (Thust 2008). If so, the problems are independent, the detector modulation transfer function having no direct analogue in the STEM set-up.

which has been shown to appreciably reduce contrast before the onset of tell-tale distortions in the HAADF image (Maccagnano-Zacher et al. 2008). Incoherence, particularly spatial incoherence or finite effective source size (Batson 2006, Grillo and Carlino 2006, Klenov et al. 2007, Nellist and Rodenburg 1994, Silcox et al. 1992), is another. We will review here the findings of LeBeau et al. (2008), who have shown a truly quantitative comparison between atomically resolved experimental data and simulation for a $SrTiO_3$ single crystal over a wide range of thicknesses.

LeBeau and Stemmer (2008) have recently described a modification which can be made to standard STEM instruments using readily available equipment which allows a means of expressing the recorded HAADF intensity as a fraction of that in the incident beam, precisely what is provided by the simulations. Experiments were carried out on the FEI Titan 80–300 TEM/STEM at the University of California Santa Barbara ($E = 300$ keV, $C_s = 1.2$ mm, $\alpha = 9.6$ mrad, inner detector semiangle ~60 mrad, $\Delta f \sim 540$ Å underfocus as determined from comparison with simulation), recording 2D HAADF images of $SrTiO_3$ viewed along a $\langle 100 \rangle$ orientation for a range of specimen thicknesses. As a means of quantifying the salient features of the 2D images using only a few parameters, average values for the normalized intensities from the strontium columns, the titanium/oxygen columns and the background were extracted. EELS data were recorded to determine the local specimen thickness in each image from the low-loss spectra (Egerton 1996). Further details of the experimental preparation, operation and the image analysis can be found in LeBeau et al. (2008).

Bloch wave HAADF image simulations, based on Eq. (18), as well as frozen phonon simulations, were performed using the experimental parameters.[11] Debye–Waller factors were taken from the literature (Peng et al. 2004). The frozen phonon simulations were performed on a 1024×1024 pixel grid in which the $SrTiO_3$ unit cell was tiled in a 7×7 array. Because of the large array size, large specimen thickness and the 2D mesh of probe points, only four phonon passes were run in order to keep the simulation time manageable. For smaller thicknesses it was checked that this value, smaller than the 20 or so more traditionally advocated (Kirkland 1998), did not appreciably alter the results.

Figure 6–10(a) plots the extracted intensities described above as a function of thickness, comparing the experimental data with the simple Bloch wave and frozen phonon simulations. The Bloch wave and frozen phonon models agree to a thickness of around 200 Å, after which they diverge. This is a consequence of the effective scattering potential

[11] The outer detector semiangle for the calculations was ~240 mrad, which is likely smaller than the experimental value. However, increasing the outer angle in the calculations to 400 mrad in the Bloch wave model (the upper experimental limit given by the detector dimensions) did not significantly affect the contrast. Sampling issues prohibit frozen phonon simulations being readily attempted with the larger detector range. Hence the use of the smaller outer angle, given the Bloch wave reassurance that the difference will be small.

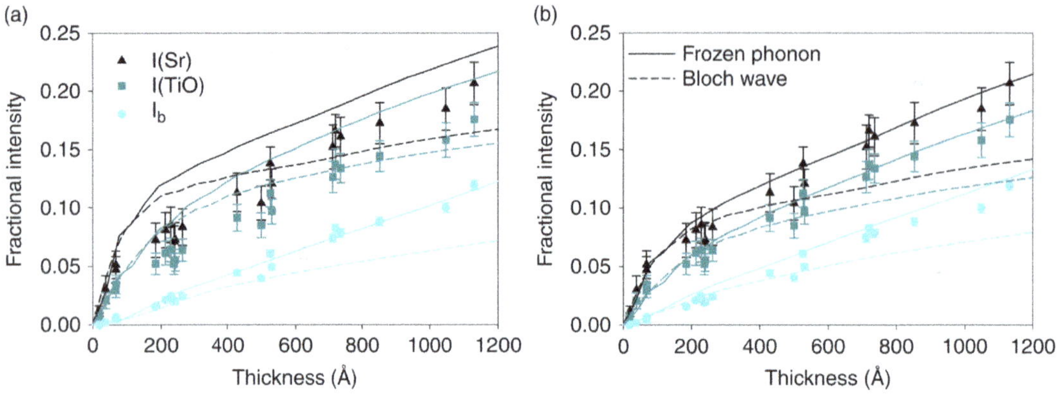

Figure 6–10. (a) Experimental *symbols* and simulated *lines* strontium and titanium/oxygen column and background intensity values for $SrTiO_3$ along ⟨100⟩ as a fraction of the incident beam intensity as a function of thickness. (b) Same as in (a) but with a 0.4 Å HWHM envelope modelling spatial incoherence in the simulations. The error bars reflect the standard deviations calculated from 400 to 600 columns for each thickness, adapted from LeBeau et al. (2008).

model not including multiple thermal scattering events, which leads to an additional increase in the recorded signal as multiple thermal scattering events tend to increase the average net scattering angle of the thermally scattered electrons. The quantitative agreement between the two simulation methods for thin regions further demonstrates that it is tolerable to use a small number of phonon configurations and probe positions in the frozen phonon calculations here. While most previous comparisons tended to subtract the background in order to obtain best agreement with experiment, it is seen in Figure 6–10(a) that the background is predicted very well by the frozen phonon model for all thickness values. Rather, we find that the contrast discrepancy with these simulations is due to the simulations overestimating the peak intensity (by a factor ~1.4).

The most obvious experimental aspect missing from simple simulations which would serve to reduce peak intensities is the effect of the finite source size (Batson 2006, Grillo and Carlino 2006, Klenov et al. 2007, Silcox et al. 1992), also referred to as spatial incoherence. This can be incorporated in STEM image simulations by convolving the resultant images with a function describing the distribution (Nellist and Rodenburg 1994), usually assumed, for lack of any clear alternative, to be Gaussian. Through trial-and-error, a Gaussian of 0.4 Å HWHM was chosen to give the best agreement, and Figure 6–10(b) shows the same comparison between experimental data and simulation where now the simulated 2D HAADF images were convolved with this Gaussian. This has had little effect on the background intensity, which was already in excellent agreement with the experiment, but has served to reduce to column intensities such that there is now good agreement between the frozen phonon method and the experiment for all thickness values considered. Such a convolution would serve to reduce the simulated intensities regardless of whether finite source size was the correct explanation or not, but the idea is strongly supported by the fact that the same effective source size distribution improves the agreement for both columns and all thickness values simultaneously. It is also telling that analysis methods which are insensitive to the effects of such a convolution with a distribution due to effective source size show significantly better agreement between simulation and experiment (Carlino and Grillo 2005, Grillo et al. 2008). Other factors which have been suggested can play a role in the contrast discrepancy, such as slight specimen misalignment (Maccagnano-Zacher et al. 2008) and plasmon scattering (Mkhoyan et al. 2008), are expected to have a thickness-dependent nature and so seem not to be playing a significant role here.

Figure 6–11 shows the 2D HAADF images for three of the thicknesses (~250, 550 and 1050 Å) at which the $SrTiO_3$ measurements were taken, with experimental data on the top row, frozen phonon simulations on the middle row and Bloch wave simulations on the bottom row. An absolute scale is used for all the data. So, for example, in the experimental data it is seen that, for the largest thickness, the thermal scattering intensity is 21% that of the incident beam when the probe is on the atomic columns but only 12% when the probe is between columns. In the two simulation rows, for each thickness, simulations not accounting for spatial incoherence are shown in the left panel while those accounting for spatial incoherence are shown in the right panel. It is clear that while fair qualitative agreement is obtained in all cases, only the convolved frozen phonon results are in good quantitative agreement with the experimental data for all three thicknesses.

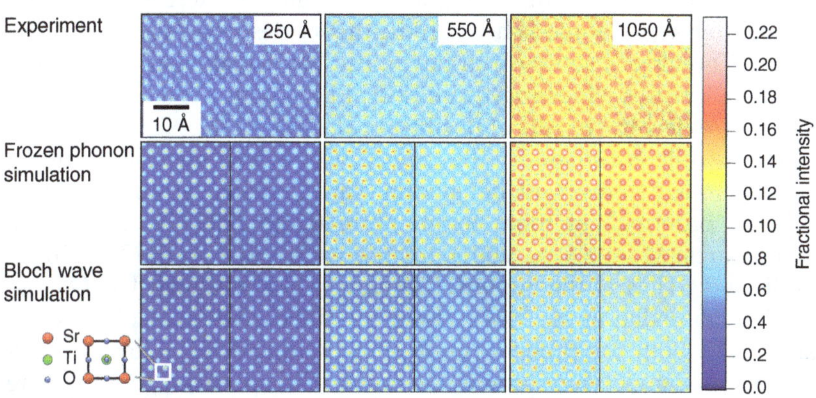

Figure 6–11. *Top row*: experimental HAADF images of $SrTiO_3$ along ⟨100⟩ with intensity variations normalized to the incident beam intensity (see scale bar on the *right*). Regions of three different thicknesses are shown. The strontium columns are the brightest and the titanium/oxygen columns are the second brightest features (see unit cell schematic on the *left*). The image of the 1050 Å region has been drift corrected. *Middle row*: frozen phonon image simulations. *Bottom row*: Bloch wave image simulations. In each case, simulations are shown without *left pane* and with convolution with a 0.4 Å HWHM Gaussian *right pane*. Adapted from LeBeau et al. (2008).

6.4 Chemical Mapping

Atomic-resolution chemical mapping, i.e. element-specific imaging via EELS, is not yet a quantitative endeavour in the sense used in the HAADF discussions in the previous section. However, improvements in aberration correction, instrument stability, detector coupling and spectrum analysis have recently made possible atomic-resolution EELS imaging in two dimensions (Allen 2008, Bosman et al. 2007, Kimoto et al. 2007, 2008, Muller et al. 2008, Okunishi et al. 2006). Previously it had only been possible to record EELS data from specific points (or averaged small regions) (Batson 1993, Browning et al. 1993, James and Browing 1999, Muller et al. 1993, 1999, Varela et al. 2004), or sometimes along a line scan (Allen et al. 2003, Oxley et al. 2007), but always with simultaneous HAADF imaging to determine the probe position relative to the structure. While simultaneous HAADF images are still recorded along with the 2D EELS images to provide support for the interpretation and maximize the information available, 2D EELS images themselves allow for new possibilities in chemical mapping analysis to the extent that bonding maps are being mooted (Muller et al. 2008). The concern with EELS images has always been whether the delocalized nature of the interaction would prevent meaningful atomic-resolution analysis from being performed (Kohl and Rose 1985, Muller and Silcox 1995). The available experimental and theoretical evidence (Allen et al. 2003, Bosman et al. 2007, Kimoto et al. 2007, 2008, Muller et al. 2008, Okunishi et al. 2006) has since been shown to support that atomic-resolution analysis is often, but not always (Kimoto et al. 2008), possible. Still, as the example of Section 6.2.4 shows, there is much scope for non-intuitive features in the images. Access to 2D EELS images allows for a good check on the theory described for their simulation. Conversely, simulations can be used to explain some of the less intuitive features which may arise in 2D EELS maps. As a case study of this, we review the results of Bosman et al. (2007).

The sample considered is $Bi_{0.5}Sr_{0.5}MnO_3$, a new material showing colossal magnetoresistant behaviour (van Tendeloo et al. 2006) and "charge ordering" at room temperature (Dörr 2006, Frontera et al. 2001, García-Muñoz et al. 2001). Images were taken on the VG HB501 dedicated STEM at the SuperSTEM facility in Daresbury, UK (Arslan et al. 2005, Falke et al. 2004), along the three zone axes $\langle 001 \rangle$, $\langle 110 \rangle$ and $\langle 111 \rangle$. Both the probe-forming semiangle and EELS detector collection semiangle were around 24 mrad. The spectrum images were acquired with the method of binned gain averaging (Bosman and Keast 2008), which improves the speed with which EELS spectra are acquired, while optimally suppressing systematic detector gain. The relatively low signal-to-noise ratio of the individual spectra prevents an accurate fit of a power-law curve, which is the conventional model to remove the background signal. The robustness of the background fit is greatly enhanced by first applying principal component analysis to remove the random spectral noise (Bosman et al. 2006). The background-subtracted oxygen and manganese intensities for all spectra were integrated over a 30 eV window above their respective thresholds to produce EELS maps.

Simulations based on Eq. (15), with a nonlocal effective scattering potential appropriate for inner-shell ionization, were performed using the Bloch wave model. As the test specimen is periodic, the Bloch wave method can be used to great effect: while the cross-section expression of Eq. (15) may not offer all the interpretive advantages of the transition potential model, it is significantly more efficient, particularly for incorporating the integration over the energy window. Structure information, including the Debye–Waller factors, was taken from Frontera et al. (2001) using the "orth 1" structure at 300 K. The cation distribution was handled via the method of fractional occupancy.[12] In the simulations, the aberration-balanced system is modelled as aberration-free within the 24 mrad probe-forming aperture semiangle. The Z-contrast images were simulated for a 60–160 mrad HAADF detector. All simulated images were convolved with a Gaussian of half-width 0.57 Å [98] to account for the finite width of the effective source (this value was selected by eye for good visual agreement of feature size). More information on the specimen preparation and experiment is given in Bosman et al. (2007). Figure 6–12 shows experimental and simulated Z-contrast images, together with EELS maps of the oxygen and manganese signal for the ⟨001⟩, ⟨110⟩ and ⟨111⟩ orientations. The specimen thicknesses, determined using low-loss spectra, are estimated at 313–358, 120–125 and 350–450 Å, respectively.

The first thing to note from Figure 6–12, the poor signal-to-noise ratio in the images for the ⟨111⟩ orientation notwithstanding, is that the visual agreement between the simulations and the experimental data is very good. The HAADF images are all directly interpretable in terms of the expected projected structure. In the ⟨001⟩ orientation, the oxygen signal, Figure 6–12(b), runs together through a combination of the delocalization of the potential and the small spacing between adjacent oxygen-bearing columns. Moreover, the signal on the Mn/O column is smaller than that on the pure oxygen columns, despite the identical oxygen densities. This is a consequence of the different scattering and absorption caused by the presence of manganese. However, the difference is too small to be evident in the experimental image. The manganese signal in the ⟨001⟩ direction, Figure 6–12(c), is directly interpretable.

More pronounced dechannelling/absorptive effects can be seen in the oxygen map of Figure 6–12(e), where the oxygen EELS signal is smallest on the Bi/Sr/O column, though admittedly the oxygen density on these columns is also only half that on the clearly visible pure oxygen columns. More intriguingly in this image, the simulations show evidence of the inclination of the MnO_6 octahedra through the alternating displacements in oxygen position when looking along horizontal

[12] It has recently been emphasized that the fractional occupancy method cannot fairly be applied in the frozen phonon model (Carlino and Grillo 2005). However, in the cross-section expression model it presents no serious inconsistencies, particularly when investigating only qualitative features rather than quantitative signals.

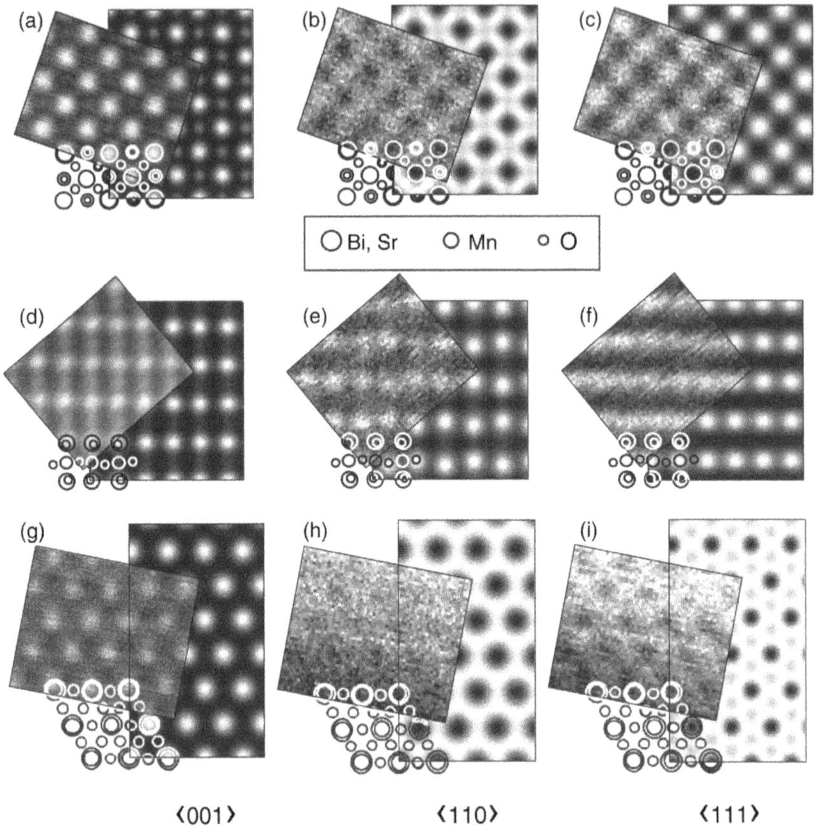

Figure 6–12. Comparison between experiments (*tilted images*) and simulations of $Bi_{0.5}Sr_{0.5}MnO_3$. For the ⟨001⟩ zone axis, (a) Z-contrast, (b) oxygen *K*-shell and (c) manganese $L_{2,3}$-shell STEM images. The simulations assume a 330 Å thick sample. For the ⟨110⟩ zone axis, (d) Z-contrast, (e) oxygen *K*-shell and (f) manganese $L_{2,3}$ STEM images. The simulations assume a 120 Å thick sample. For the ⟨111⟩ zone axis, (g) Z-contrast, (h) oxygen *K*-shell and (i) manganese $L_{2,3}$ STEM images. The simulations assume a 400 Å thick sample. The atomic structure is indicated. The EELS maps were generated by integrating the EELS spectra over a 30 eV window above the respective ionization threshold. Adapted from Bosman et al. (2007).

rows in the figure. A hint of this behaviour is also seen in the experimental data. For the manganese signal in Figure 6–12(f), the columns of atoms are difficult to separate in the horizontal direction but are clearly distinct in the vertical direction, a simple consequence of the smaller intercolumn spacing along the horizontal direction.

In the ⟨111⟩ orientation, the signal-to-noise ratio in the oxygen EELS image is too low to infer anything from the experimental data, though the structure in the simulation is what might be expected given the delocalized oxygen signal and the close column spacing. The most unexpected result, however, is the manganese EELS signal in the ⟨111⟩ orientation: the signal is a minimum on the Bi/Sr/Mn columns, the only columns containing manganese. Plotted minimum to maximum as black to white, the grey-scale images obscure the relatively low contrast

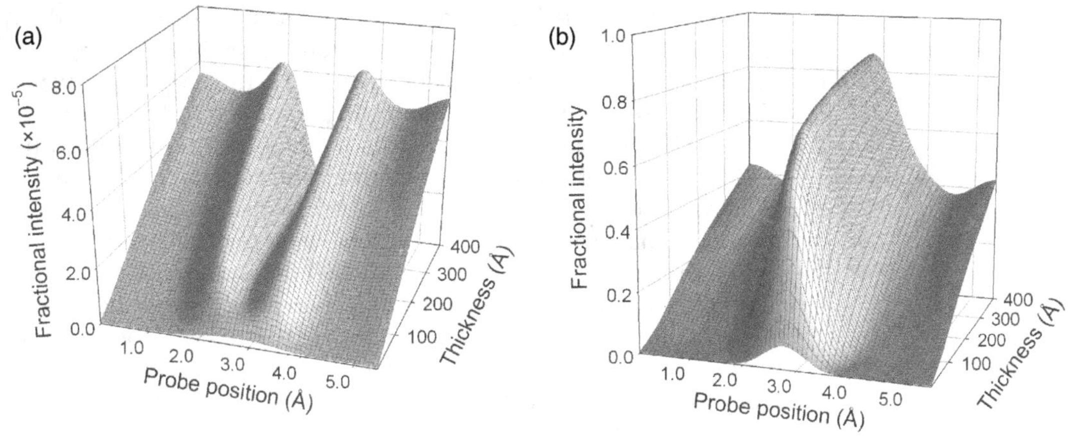

Figure 6–13. (a) Manganese EELS line scan simulation as a function of specimen thickness. With reference to the simulation and structure in Figure 6–12(i), the scan line is taken as a horizontal line passing through the manganese column, extending from an oxygen column across the manganese column to the next oxygen column. (b) The proportion of electrons lost to the elastic wave function due to thermal scattering as a function of thickness for the same line scan as in (a).

variation. Nevertheless, the effect is real. Moreover, as it is correctly reproduced in the simulations, it is not a question of some anomaly in this sample or the neglect of some important principle. Therefore this observation can be explained by exploring the dynamics in the simulations.

Figure 6–13(a) shows a simulated manganese EELS line scan from an oxygen column across a manganese column to the next adjacent oxygen column as a function of specimen thickness. It is seen that for very thin specimens the peak would be upon the Bi/Sr/Mn column, albeit a very broad one, as expected from the delocalized nature of the transition potentials. However, the signal on this column quickly saturates while that off-column continues to rise quite steadily with thickness. The cause of this rapid saturation becomes evident upon examining Figure 6–13(b), which shows the proportion of electrons removed from the elastic wave function due to TDS scattering: the absorption from the heavy Bi/Sr/Mn column is very large, rapidly attenuating the intensity of the elastic wave function capable of causing ionization events. Because the absorption is highly localized on the column sites, a probe situated off-column is not nearly so rapidly attenuated and is therefore able to interact longer with delocalized ionization transition potentials as it evolves through the sample.

In the model used here, the only electrons allowed to cause ionization events are those in the elastic wave function; electrons that undergo thermal scattering are removed from the calculation and prohibited from contributing further. This is physically unrealistic, since, despite the terminology of "absorption", the thermally scattered electrons continue to travel through the sample and will continue to interact with it. Scattering to high angles would remove electrons in the sense that they would scatter outside the on-axis detector, but not all thermal scattering is through high angles. Findlay et al. (2005) explored this question in simulation for a silver specimen, a strong absorber. The absorptive model use of the cross-section expression, Eq. (15), was compared with an implementation in which the frozen phonon wave function was used, with the phonon average applied to the EELS signal. It was shown that while there was a quantitative difference due to neglecting the contribution from thermally scattered electrons to the EELS signal, the results were still in much better qualitative agreement with the absorptive model than they were to calculations which neglected absorption. In essence, even when the number of thermally scattered electrons is large, the mutual incoherence between those electrons thermally scattered at different sites greatly reduces the possible ionization contribution of those electrons (relative to what they might have been capable of had they been, as the elastic wave function is, mutually coherent).

Recent instrument developments (Krivanek et al. 2008) have allowed a considerable increase in the angular collection range possible for EELS imaging. As described by Muller et al. (2008), this significantly increases the count rates obtainable. It also ameliorates strongly against the effects of double channelling, nonlocality and absorption as thus far described here. It should be emphasized though that while the large detector offers immunity against any redistribution of scattering caused after the primary energy-loss event, it cannot, of course, alter the effects of scattering before it. The probability of ionization will still depend on the distribution and coherence in the wave function reaching the atom to be ionized, and this depends on the characteristics of the probe and the initial scattering through the specimen. As an example of this, we review an unexpected result obtained in chemical mapping on the new, fifth-order aberration corrected, 100 keV Nion UltraSTEM at Daresbury. Using different energy windows above the $L_{2,3}$ edge in $\langle 011 \rangle$ silicon to map the position of the atomic columns we find a contrast reversal, leading to an apparent translation of the columns.

A HAADF image, Figure 6–14(a), was obtained using a 105–300 mrad annular collection range. The Nion UltraSTEM is equipped with a Gatan Enfina electron energy-loss spectrometer that was run with a collection semiangle of 67 mrad in the energy dispersive direction and 22 mrad in the non-energy dispersive direction. A 20×16 pixel subset of the full 20×20 pixels spectral data set is shown in Figure 6–14 (b) and (c) for energy-loss windows of 143–163 and 280–300 eV, respectively, subsequent to background subtraction. The power-law background fitting was sampled within a 20 eV window on the low-energy side of the silicon $L_{2,3}$ edge. Figure 6–14(d) shows a typical example of the energy-loss spectrum, indicating the background subtraction and the selected energy windows. Further details of the experiment and data processing are given in Wang et al. (2008). Simulations supporting the experimental results were carried out using the Bloch wave method.[13]

[13] Calculations predict that the contribution of electrons that excite L_1 ionizations is over an order of magnitude smaller than those that excite $L_{2,3}$ ionizations, and so the former are neglected.

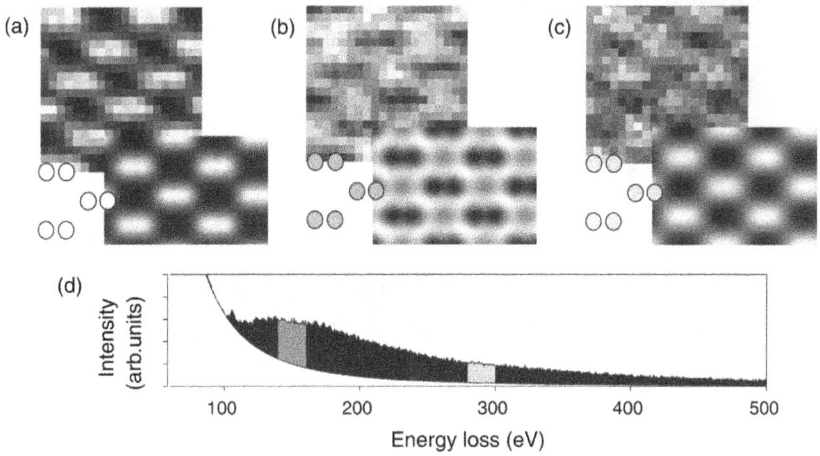

Figure 6–14. Comparison between experimental and simulated images of ⟨011⟩ silicon. (a) HAADF, (b) 143–163 eV and (c) 280–300 eV energy-filtered EELS images. The sample was estimated to be 910 Å thick, and this value was used in the simulations. The probe-forming semiangle is 24 mrad, and the probe is assumed to be aberration-free. (d) A typical acquired Si *L*-shell EELS spectrum showing the background fit curve subtraction. All images were acquired simultaneously, adapted from Wang et al. (2008).

The simulations accounted for the spatial incoherence of the probe via convolution of the image with a Gaussian of HWHM 0.6 Å, which no longer allows for resolution of the 1.4 Å silicon dumbbell spacing for the ⟨011⟩ projection in the simulated HAADF image, in agreement with the experimental results.

The projected dumbbell structure formed by adjacent columns of silicon atoms is correctly shown by the experimental and simulated HAADF images in Figure. 6–14(a). The $L_{2,3}$ EELS images for the energy windows 143–163 and 280–300 eV in Figure 6–14(b) and (c) respectively, are obtained simultaneously with the HAADF image, and thus the electron probe experiences identical scattering and absorption conditions. But while the columns for the 280–300 eV energy window image are in register with the HAADF image, correctly reflecting the known structure, the columns in the 143–163 eV energy window image have apparently been translated.

The evolution of the electron probe through the specimen is identical in both images, so the difference must be due to the variation of the ionization interaction with energy loss. The ionization probability is known to become increasingly localized with increasing energy loss (Egerton 1996). Using experimental data for this same edge, though in a thinner Si_3N_4 specimen, Kimoto et al. (2008) have demonstrated this effect in 2D EELS images. We can assess the variation in localization directly and quantitatively by exploring the dependence of the localization of the inelastic transition potential on energy loss. As per Eq. (11), the inelastic wave function is proportional to the product of the transition potential and the elastic wave function, and so the modulus squared of the transition potentials measures the strength of the

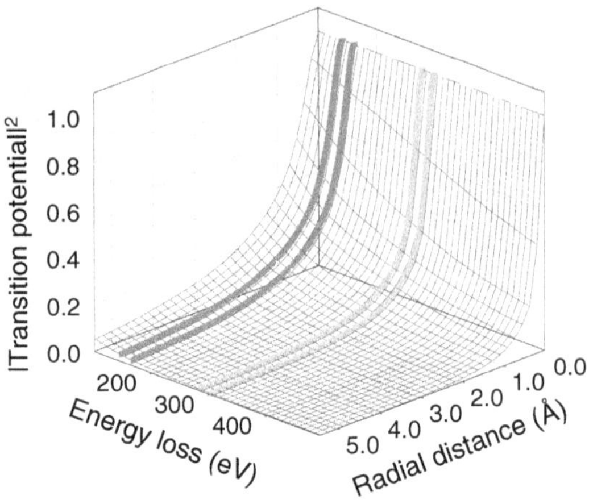

Figure 6–15. Modulus squared of the transition potentials, normalized to unity at the origin for a clear comparison of the change in shape, from initial state $l = 1$, $m_l = 0$ to final state $l' = 2$, $m_l' = 0$ as a function of energy loss for an isolated Si atom. The 143–163 and 280–300 eV energy windows are indicated, adapted from Wang et al. (2008).

transfer of electron density into the inelastic channels. Figure 6–15 shows the modulus squared of the inelastic transition potentials from an initial state with quantum numbers $l = 1$, $m_l = 0$ to a final state with quantum numbers $l' = 2$, $m'_l = 0$ as a function of energy loss for the Si $L_{2,3}$ edge. While the width of the potential over the 280–300 eV window is comparable to the inter-dumbbell spacing, over the 143–163 eV window the potential is more delocalized.

For this very thick sample, probe spreading also plays a major role. Indeed, trial calculations in which only silicon atoms in a single column are allowed to undergo ionization show that for both energy-loss ranges the ionization signal from that single column is a maximum when the probe is on that column (Wang et al. 2008), implying that the contrast reversal depends on the excitation of atoms in multiple columns, which has previously been dubbed cross-talk (see, for instance, Allen et al. 2003, Dwyer and Etheridge 2003). It was also found that if the effect of inelastic thermal scattering on the evolution of the elastic wave function was neglected then the simulations do not show a contrast reversal, implying that thermal scattering appreciably modifies the evolution of the elastic wave function, despite the generally weak scattering power of silicon atoms. That said, it is the difference in width of the interaction potential as it interacts with the spreading of the probe in this rather thick specimen that is the main effect leading to the contrast reversal.

Kimoto et al. (2008) suggested that such delocalization might prevent atomic-resolution imaging. They emphasize the idea that such grey-scale plots of EELS images should have a scale where black is the true zero level, because the hallmark of images with very delocalized interactions is very low contrast. Figure 6–14 has adopted the more traditional approach of a grey scale where black is the minimum signal, which strongly enhances the features but can be quite misleading about the contrast, which is very low in this case. As seen in Figure 6–14 and as supported by the simulations, atomic-scale features can sometimes be seen clearly above the delocalized background. But how usefully they can be interpreted may depend on some a priori assumptions about the expected structure combined with detailed simulations. The change in the EELS image contrast is a subtle competition between the elastic and inelastic scattering as a function of the probe position. General principles for how the balance between these aspects plays out remain elusive, making simulation an often essential part of atomic-resolution chemical mapping.

6.5 Imaging in Three Dimensions – Depth Sectioning

Though, as described in the previous section, 2D mapping in EELS at atomic resolution is a fairly recent development (Bosman et al. 2007, Kimoto et al. 2007, 2008, Muller et al. 2008, Okunishi et al. 2006), it is not over-reaching to consider EELS imaging in three dimensions. Spectroscopic identification of single atoms in bulk material has been achieved, and modelling the evolution of the wave function through the known supporting structure allowed an estimate of the depth of that impurity (Varela et al. 2004). We will not consider a tomographic approach, though such may well be possible (Arslan et al. 2005, Midgley and Weyland 2003). Nor shall we consider the combined use of experiment and simulation which has recently proved to be of some use in distinguishing between 3D structural candidate models for nanoparticles (Li et al. 2008). Rather, we will take advantage of the reduced depth of field which results from the increased aperture size enabled by aberration correction to carry out depth sectioning (Borisevich et al. 2006, Xin et al. 2008). This is already being done experimentally with annular dark field imaging (Borisevich et al. 2006, van Benthem et al. 2005, 2006, Wang et al. 2004). In addition to the STEM depth sectioning technique of those approaches, we will consider a confocal arrangement, an analogue of scanning confocal optical microscopy, as shown in Figure 6–1(b). This geometry has been recently achieved in an atomic-resolution transmission electron microscope with dual aberration correctors[14](Nellist et al. 2006).

Simulations can be used to explore the feasibility of depth sectioning using EELS and its extension to the novel SCEM mode. Experimental design can be explored, optimizing the use of both resources and time. As a case study in that vein, consider substituting either a carbon or aluminium impurity for a gallium atom in a crystal of GaAs. We choose the [110] zone axis to provide substantial channelling and allow the exploration of its effects. The impurity atom will be assumed to be at a depth of 152 Å within a 308 Å thick sample. We assume an aberration-free, 200 keV probe with a semiangle of 30 mrad. In addition, for the SCEM case, the image forming lens is also assumed to be aberration-free with a 30 mrad semiangle. For STEM the EELS detector aperture semiangle is taken to be 30 mrad. In that case the integrated signal in the SCEM

[14] Confocal electron microscopy at lower resolutions having been previously established (Frigo et al. 2002).

image plane would be identical to the STEM signal. The SCEM mode gains depth resolution at the expense of reducing the signal, ideally using a point detector at the same lateral position as the central axis of the incident probe – see Figure 6–1(b). The SCEM simulations below are based on a transition potential formulation, Eq. (11), where each individual inelastic wave function generated from each inelastic transition is propagated fully through the remainder of the crystal and then coherently imaged. This is what we referred to before as a double channelling calculation and the STEM calculations in this section are done in the same way to allow a fair comparison.

The STEM depth sectioning experiments carried out thus far, based on HAADF imaging (Borisevich et al. 2006, van Benthem et al. 2005, 2006, Wang et al. 2004), have been accomplished by varying the defocus value of the lens. As the confocal geometry of SCEM involves delicate alignment of electron optics, varying the defocus is not feasible once the confocal geometry has been established. Instead the specimen will be

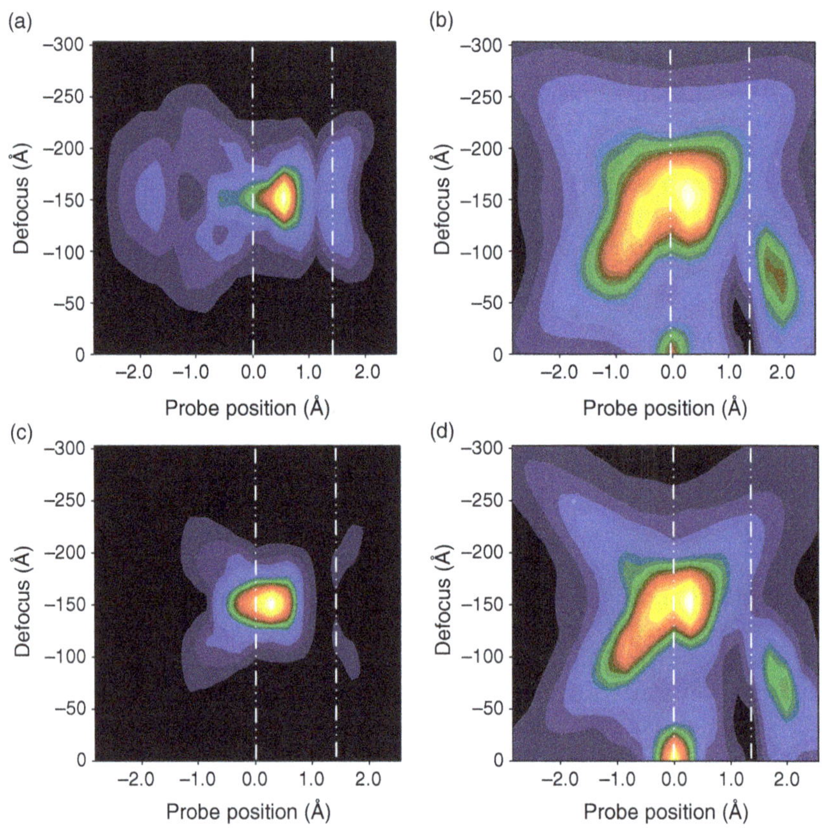

Figure 6–16. Simulated carbon *K*-shell line scans for (a) SCEM and (b) STEM geometries. Aluminium *K*-shell line scans for (c) SCEM and (d) STEM geometries. The impurity is embedded 152 Å deep, substitutionally displacing a gallium atom on column, for a 308 Å thick GaAs crystal that is oriented along the ⟨110⟩ zone axis. The gallium and arsenic column locations are indicated by the *dashed lines* (gallium on the *left*), adapted from D'Alfonso et al. (2007).

manipulated (Nellist et al. 2006). Nevertheless, in presenting our results we will use "defocus" as the axis label which corresponds to information along the beam direction, referenced to the pre-specimen lens and with a sign convention that underfocus, which corresponds to the beam waist being shifted further into the crystal, is negative.

Figures 6–16(a), (b) shows SCEM and STEM simulations for a carbon atom impurity while Figure. 6–16(c), (d) shows the SCEM and STEM results for an aluminium atom. The *K*-shell edge is monitored in each case. The lateral position of the column locations is shown by the dotted lines, with the gallium column positioned at the origin.

The carbon *K*-shell ionization produces an image which is more delocalized than that for aluminium. In Figure 6–16(a) and (b) both the SCEM and STEM images show an offset in the image intensity peak towards the arsenic column, and the increased resolution of SCEM makes this image delocalization much more pronounced. The aluminium *K*-shell ionization interaction, being more localized than the carbon *K*-shell ionization interaction, does not show as great a shift towards the arsenic column. The intensity peak around $\Delta f = 0$ Å in the STEM images is due to the probe coupling to the 1 s-like state of the gallium column (D'Alfonso et al. 2007). By coupling to the 1 s-like state, a probe focused onto the surface of the specimen can channel along the column and generate a significant number of ionization events at the depth of the dopant. The SCEM geometry suppresses this by favouring electrons which originate from the focal plane of the post-specimen lens and so this peak is not evident. The defocus HWHM is clearly smaller for the SCEM geometry than for the STEM geometry.

6.6 Summary

A range of approaches exists for the theoretical analysis and interpretation of HAADF and EELS images. The effective scattering potential formulation allows the treatment of both within the same basic framework. The transition potential formalism can be used to examine state-by-state contributions to energy-spectroscopic signals and the shapes of the individual states, which follow from the form of the probe and transition potentials, allowing insight into the form of some less intuitive features which may arise in atomic-resolution EELS imaging, at least when limitations exist on the detector collection aperture. The ability to record quantitative HAADF image and 2D EELS maps will greatly aid the interpretation of compositional information in terms of sample and chemical structure. Proof-of-principle simulations support the useful extension of these techniques to analyze the structure in three dimensions.

Acknowledgements L. J. Allen acknowledges support by the Australian Research Council. S. D. Findlay is supported by the Japanese Society for the Promotion of Science (JSPS). M.P. Oxley was supported by the Office of Basic Energy Sciences, Materials Sciences and Engineering Division, U.S. Department of Energy. We would like to thank our following collaborators for their considerable inputs into various parts of the work summarized in this chapter: G. Behan, A. L. Bleloch, M. Bosman, E. C. Cosgriff, A. J. D'Alfonso, C. Dwyer, J. L. García-Muñoz, V. J. Keast, A. I. Kirkland, J. M. LeBeau, P. D. Nellist, S. Stemmer and P. Wang.

References

L.J. Allen, New directions for chemical maps. Nat. Nanotechnol. **3**, 255–256 (2008)

L.J. Allen, S.D. Findlay, A.R. Lupini, M.P. Oxley, S.J. Pennycook, Atomic-resolution electron energy loss spectroscopy imaging in aberration corrected scanning transmission electron microscopy. Phys. Rev. Lett. **91**, 105503 (2003)

L.J. Allen, S.D. Findlay, M.P. Oxley, C.J. Rossouw, Lattice-resolution contrast from a focused coherent electron probe. Part I. Ultramicroscopy **96**, 47–63 (2003)

L.J. Allen, S.D. Findlay, M.P. Oxley, C. Witte, N.J. Zaluzec, Modelling high-resolution electron microscopy based on core-loss spectroscopy. Ultramicroscopy **106**, 1001–1011 (2006)

L.J. Allen, A.J. D'Alfonso, S.D. Findlay, M.P. Oxley, M. Bosman, V.J. Keast, E.C. Cosgriff, G. Behan, P.D. Nellist, A.I. Kirkland, Theoretical interpretation of electron energy-loss spectroscopic images. Am. Inst. Phys. Conf. Proc. **999**, 32–46 (2008)

L.J. Allen, T.W. Josefsson, Inelastic scattering of fast electrons by crystals. Phys. Rev. B **52**, 3184–3198 (1995)

L.J. Allen, C.J. Rossouw, Effects of thermal diffuse scattering and surface tilt on diffraction and channeling of fast electrons in CdTe. Phys. Rev. B **39**, 8313–8321 (1989)

A. Amali, P. Rez, Theory of lattice resolution in high-angle annular dark-field images. Microsc. Microanal. **3**, 28–46 (1997)

G.R. Anstis: The influence of atomic vibrations on the imaging properties of atomic focusers. J. Microsc. **194**, 105–111 (1999)

G.R. Anstis, S.C. Anderson, C.R. Birkeland, D.J.H. Cockayne, Computer simulation methods for the analysis of high-angle annular dark-field (HAADF) images of $Al_xGa_{1-x}As$ at high resolution. Unpublished proceedings, 15th Pfefferkorn Conference, Silver Bay, New York, 1996

G.R. Anstis, D.Q. Cai, D.J.H. Cockayne, Limitations on the s-state approach to the interpretation of sub-angstrom resolution electron microscope images and microanalysis. Ultramicroscopy **94**, 309–327 (2003)

I. Arslan, A. Bleloch, E.A. Stach, N.D. Browning, Atomic and electronic structure of mixed and partial dislocations in GaN. Phys. Rev. Lett. **94**, 025504 (2005)

I. Arslan, T.J.V. Yates, N.D. Browning, P.A. Midgley, Embedded nanostructures revealed in three dimensions. Science **309**, 2195 (2005)

P.E. Batson, Simultaneous STEM imaging and electron energy-loss spectroscopy with atomic-column sensitivity. Nature **366**, 727–728 (1993)

P.E. Batson, Characterizing probe performance in the aberration corrected STEM. Ultramicroscopy **106**, 1104–1114 (2006)

P.E. Batson, N. Dellby, O.L. Krivanek, Sub-angstrom resolution using aberration corrected electron optics. Nature **418**, 617–620 (2002)

D.M. Bird: Theory of zone axis electron diffraction. J. Electron Microsc. Tech. **13**, 77–97 (1989)

A. Bleloch, U. Falke, M. Falke, High spatial resolution electron energy loss spectroscopy and imaging in an aberration corrected STEM. Microsc. Microanal. **9**(Suppl. 3), 40–41 (2003)

A.Y. Borisevich, A.R. Lupini, S.J. Pennycook, Depth sectioning with the aberration corrected scanning transmission electron microscope. P. Natl. Acad. Sci. **103**, 3044–3048 (2006)

M. Bosman, V.J. Keast, Optimizing EELS acquisition. Ultramicroscopy **108**, 837–846 (2008)

M. Bosman, V.J. Keast, J.L. García-Muñoz, A.J. D'Alfonso, S.D. Findlay, L.J. Allen, Two-dimensional mapping of chemical information at atomic resolution. Phys. Rev. Lett. **99**, 086102 (2007)

M. Bosman, M. Watanabe, D.T.L. Alexander, V.J. Keast, Mapping chemical and bonding information using multivariate analysis of electron energy-loss spectrum images. Ultramicroscopy **106**, 1024–1032 (2006)

N.D. Browning, M.F. Chisholm, S.J. Pennycook, Atomic-resolution chemical analysis using a scanning transmission electron microscope. Nature **366**, 143–146 (1993), Corrigendum **444**, 235 (2006)

B.F. Buxton, J.E. Loveluck, J.W. Steeds, Bloch waves and their corresponding atomic and molecular orbitals in high energy electron diffraction. Philos. Mag. A **38**, 259–278 (1978)

E. Carlino, V. Grillo, Atomic-resolution quantitative composition analysis using scanning transmission electron microscopy Z-contrast experiments. Phys. Rev. B **71**, 235303 (2005)

D. Cherns, A. Howie, M.H. Jacobs, Characteristic X-ray production in thin crystals. Z. Naturforsch. A **28**, 565–571 (1973)

W. Coene, D. Van Dyck, Inelastic scattering of high-energy electrons in real space. Ultramicroscopy **33**, 261–267 (1990)

W. Coene, D. Van Dyck, M. Op de Beeck, J. Van Landuyt, Computational comparisons between the conventional multislice method and the third order multislice method for calculating high-energy electron diffraction and imaging. Ultramicroscopy **69**, 219–240 (1997)

E.C. Cosgriff, P.D. Nellist, A Bloch wave analysis of optical sectioning in aberrationcorrected STEM. Ultramicroscopy **107**, 626–634 (2007)

E.C. Cosgriff, M.P. Oxley, L.J. Allen, S.J. Pennycook, The spatial resolution of imaging using core-loss spectroscopy in the scanning transmission electron microscope. Ultramicroscopy **102**, 317–326 (2005)

J.M. Cowley, A.F. Moodie, The scattering of electrons by atoms and crystals. I. A new theoretical approach. Acta Cryst. **10**, 609–619 (1957)

M.D. Croitoru, D. Van Dyck, S. Van Aert, S. Bals, J. Verbeeck, An efficient way of including thermal diffuse scattering in simulation of scanning transmission electron microscopic images. Ultramicroscopy **106**, 933940 (2006)

A.J. D'Alfonso, S.D. Findlay, M.P. Oxley, S.J. Pennycook, K. van Benthem, L.J. Allen, Depth sectioning in scanning transmission electron microscopy based on core-loss spectroscopy. Ultramicroscopy **108**, 17–28 (2007)

C. Dinges, A. Berger, H. Rose, Simulation of TEM images considering phonon and electronic excitations. Ultramicroscopy **60**, 49–70 (1995)

C. Dinges, H. Rose, Simulation of transmission and scanning transmission electron microscopic images considering elastic and thermal diffuse scattering. Scanning Microsc. **11**, 277–286 (1997)

K. Dörr, Ferromagnetic manganites: spin-polarized conduction versus competing interactions. J. Phys. D **39**, R125–R150 (2006)

S.L. Dudarev, L.M. Peng, M.J. Whelan, Correlations in space and time and dynamical diffraction of high-energy electrons by crystals. Phys. Rev. B **48**, 13408–13429 (1993)

C. Dwyer, Multislice theory of fast electron scattering incorporating atomic inner-shell ionization. Ultramicroscopy **104**, 141–151 (2005)

C. Dwyer, J. Etheridge, Scattering of A-scale electron probes in silicon. Ultramicroscopy **96**, 343–360 (2003)

C. Dwyer, S.D. Findlay, L.J. Allen, Multiple elastic scattering of core-loss electrons in atomic resolution imaging. Phys. Rev. B **77**, 184107 (2008)

R.F. Egerton, *Electron energy-loss spectroscopy in the electron microscope*, 2nd edn. (Plenum Press, New York, NY, 1996)

U. Falke, A. Bleloch, M. Falke, Atomic structure of a (2×1) reconstructed $NiSi_2$/Si(001) interface. Phys. Rev. Lett. **92**, 116103 (2004)
S.D. Findlay, in *Theoretical aspects of scanning transmission electron microscopy*. Ph.D. thesis, The University of Melbourne, Melbourne, 2005
S.D. Findlay, L.J. Allen, M.P. Oxley, C.J. Rossouw, Lattice-resolution contrast from a focused coherent electron probe. Part II. Ultramicroscopy **96**, 65–81 (2003)
S.D. Findlay, M.P. Oxley, L.J. Allen, Modelling atomic-resolution scanning transmission electron microscopy images. Microsc. Microanal. **14**, 48–59 (2008)
S.D. Findlay, M.P. Oxley, S.J. Pennycook, L.J. Allen, Modelling imaging based on coreloss spectroscopy in scanning transmission electron microscopy. Ultramicroscopy **104**, 126–140 (2005)
L. Fitting, S. Thiel, A. Schmehl, J. Mannhart, D.A. Muller, Subtleties in ADF imaging and spatially resolved EELS: A case study of low-angle twist boundaries in $SrTiO_3$. Ultramicroscopy **106**, 1053–1061 (2006)
S.P. Frigo, Z.H. Levine, N.J. Zaluzec, Submicron imaging of buried integrated circuit structures using scanning confocal electron microscopy. Appl. Phys. Lett. **81**, 2112–2114 (2002)
C. Frontera, García-Muñoz, J.L., M.A. García-Aranda, C. Ritter, A. Llobet, M. Respaud, J. Vanacken, Low-temperature charge and magnetic order in $Bi_{0.5}Sr_{0.5}MnO_3$. Phys. Rev. B **64**, 054401 (2001)
García-Muñoz, J.L., C. Frontera, M.A. García-Aranda, A. Llobet, C. Ritter, High-temperature orbital and charge ordering in $Bi_{1/2}Sr_{1/2}MnO_3$. Phys. Rev. B **63**, 064415 (2001)
V. Grillo, E. Carlino, A novel method for focus assessment in atomic resolution STEM HAADF experiments. Ultramicroscopy **106**, 603–613 (2006)
V. Grillo, E. Carlino, F. Glas, Influence of the static atomic displacement on atomic resolution Z-contrast imaging. Phys. Rev. B **77**, 054103 (2008)
C.R. Hall, P.B. Hirsch, Effect of thermal diffuse scattering on propagation of high energy electrons through crystals. Proc. Roy. Soc. London A **286**, 158–177 (1965)
P. Hartel, H. Rose, C. Dinges, Conditions and reasons for incoherent imaging in STEM. Ultramicroscopy **63**, 93–114 (1996)
C. Hébert-Souche, P.H. Louf, P. Blaha, M. Nelhiebel, J. Luitz, P. Schattschneider, K. Schwarz, B. Jouffrey, The orientation-dependent simulation of ELNES. Ultramicroscopy **83**, 9–16 (2000)
S. Hillyard, R.F. Loane, J. Silcox, Annular dark-field imaging: Resolution and thickness effects. Ultramicroscopy **49**, 14–25 (1993)
S. Hillyard, J. Silcox, Thickness effects in ADF STEM zone axis images. Ultramicroscopy **52**, 325–334 (1993)
A. Howie, Hunting the Stobbs factor. Ultramicroscopy **98**, 73–79 (2004)
C.J. Humphreys, The scattering of fast electron by crystals. Rep. Prog. Phys. **42**, 1825–1887 (1979)
M.J. Hÿtch, W.M. Stobbs, Quantitative comparison of high resolution TEM images with image simulations. Ultramicroscopy **53**, 191–203 (1994)
K. Ishizuka, Prospects of atomic resolution imaging with an aberration-corrected STEM. J. Electron. Microsc. **50**, 291–305 (2001)
K. Ishizuka, FFT multislice method – the silver anniversary. Microsc. Microanal. **10**, 34–40 (2004)
E.M. James, N.D. Browning, Practical aspects of atomic resolution imaging and analysis in STEM. Ultramicroscopy **78**, 125–139 (1999)
D.E. Jesson, S.J. Pennycook, Incoherent imaging of thin specimens using coherently scattered electrons. Proc. R. Soc. Lond. A **441**, 261–281 (1993)

T.W. Josefsson, L.J. Allen, Diffraction and absorption of inelastically scattered electrons for K-shell ionization. Phys. Rev. B **53**, 2277–2285 (1996)

K. Kimoto, T. Asaka, T. Nagai, M. Saito, Y. Matsui, K. Ishizuka, Element-selective imaging of atomic columns in a crystal using STEM and EELS. Nature **450**, 702–704 (2007)

K. Kimoto, K. Ishizuka,Y. Matsui, Decisive factors for realizing atomic-column resolution. Micron **39**, 257–262 (2008)

E.J. Kirkland, *Advanced computing in electron microscopy* (Plenum Press, New York, NY and London, 2nd Ed. 2010)

E.J. Kirkland, (2008). http://people.ccmr.cornell.edu/~kirkland/

E.J. Kirkland, R.F. Loane, J. Silcox, Simulation of annular dark field STEM images using a modified multislice method. Ultramicroscopy **23**, 77–96 (1987)

D.O. Klenov, S.D. Findlay, L.J. Allen, S. Stemmer, Influence of orientation on the contrast of high-angle annular dark-field images of silicon. Phys. Rev. B **76**, 014-111 (2007)

D.O. Klenov, S. Stemmer, Contributions to the contrast in experimental high-angle annular dark-field images. Ultramicroscopy **106**, 889–901 (2006)

O.L. Krivanek, G.J. Corbin, N. Dellby, B.F. Elston, R.J. Keyse, M.F. Murfitt, C.S. Own, Z.S. Szilagyi, J.W. Woodruff, An electron microscope for the aberration-corrected era. Ultramicroscopy **108**, 179–195 (2008)

O.L. Krivanek, N. Dellby, A.R. Lupini, Towards sub-A electron beams. Ultramicroscopy **78**, 1–11 (1999)

H. Kohl, H. Rose, Theory of image formation by inelastic scattered electrons in the electron microscope. Adv. Electron. Electron. Phys. **65**, 173–227 (1985)

Y. Kotaka, T. Yamazaki, Y. Kikuchi, K. Watanabe, Incoherent high-resolution Z-contrast imaging of silicon and gallium arsenide using HAADF-STEM. In *Materials Research Society Symposium Proceedings*, ed. By J. Bentley, I. Petrov, U. Dahmen, C. Allen (Materials Research Society, Warrendale, PA, 2001), pp. 185–190

J.M. LeBeau, S.D. Findlay, L.J. Allen, S. Stemmer, Quantitative atomic resolution scanning transmission electron microscopy. Phys. Rev. Lett. **100**, 206101 (2008)

J.M. LeBeau, S. Stemmer, Experimental quantification of annular dark-field images in scanning transmission electron microscopy. Ultramicroscopy **108**, 1653–1658 (2008)

Z.Y. Li, N.P. Young, M. Di Vece, S. Palomba, R.E. Palmer, A.L. Bleloch, B.C. Curley, R.L. Johnson, J. Jiang, J. Yuan, Three-dimensional atomic-scale structure of size-selected gold nanoclusters. Nature **451**, 46 (2008)

R.F. Loane, P. Xu, J. Silcox, Thermal vibrations in convergent-beam electron diffraction. Acta Cryst. **A47**, 267–278 (1991)

R.F. Loane, P. Xu, J. Silcox, Incoherent imaging of zone axis crystals with ADF STEM. Ultramicroscopy **40**, 121–138 (1992)

S.E. Maccagnano-Zacher, K.A. Mkhoyan, E. Kirkland, J. Silcox, Effects of tilt on high-resolution ADF-STEM imaging. Ultramicroscopy **108**, 718–726 (2008)

V.W. Maslen, On the role of inner-shell ionization in the scattering of fast electrons by crystals. Philos. Mag. B **55**, 491–496 (1987)

P.A. Midgley, M. Weyland, 3D electron microscopy in the physical sciences: The development of Z-contrast and EFTEM tomography. Ultramicroscopy **96**, 413–431 (2003)

K.A. Mkhoyan, S.E. Maccagnano-Zacher, M.G. Thomas, J. Silcox, Critical role of inelastic interactions in quantitative electron microscopy. Phys. Rev. Lett. **100**, 025503 (2008)

D.A. Muller, B. Edwards, E.J. Kirkland, J. Silcox, Simulation of thermal diffuse scattering including a detailed phonon dispersion curve. Ultramicroscopy **86**, 371–380 (2001)

D.A. Muller, L. Fitting Kourkoutis, M. Murfitt, J.H. Song, H.Y. Hwang, J. Silcox, N. Dellby, O.L. Krivanek, Atomic-scale chemical imaging of composition and bonding by aberration-corrected microscopy. Science **319**, 1073–1076 (2008)

D.A. Muller, J. Silcox, Delocalization in inelastic scattering. Ultramicroscopy **59**, 195–213 (1995)

D.A. Muller, T. Sorsch, S. Moccio, F.H. Baumann, K. Evans-Lutterodt, G. Timp, The electronic structure at the atomic scale of ultrathin gate oxides. Nature **399**, 758–761 (1999)

D.A. Muller, Y. Tzou, R. Raj, J. Silcox, Mapping sp_2 and sp_3 states of carbon at subnanometre spatial resolution. Nature **366**, 725–727 (1993)

H. Müller, H. Rose, P. Schorsch, A coherence function approach to image simulation. J. Microsc. **190**, 73–88 (1998)

P.D. Nellist, G. Behan, A.I. Kirkland, C.J.D. Hetherington, Confocal operation of a transmission electron microscope with two aberration correctors. Appl. Phys. Lett. **89**, 124105 (2006)

P.D. Nellist, M.F. Chisholm, N. Dellby, O.L. Krivanek, M.F. Murfitt, Z.S. Szilagyi, A.R. Lupini, A. Borisevich, W.H. Sides Jr, S.J. Pennycook, Direct sub-angstrom imaging of a crystal lattice. Science **305**, 1741 (2004)

P.D. Nellist, S.J. Pennycook, Incoherent imaging using dynamically scattered coherent electrons. Ultramicroscopy **78**, 111–124 (1999)

P.D. Nellist, J.M. Rodenburg, Beyond the conventional information limit: the relevant coherence function. Ultramicroscopy **54**, 61–74 (1994)

E. Okunishi, H. Sawada, Y. Kondo, M. Kersker, Atomic resolution elemental map of EELS with a Cs corrected STEM. Microsc. Microanal. **12**(Supp. 2), 1150–1151 (2006)

M.P. Oxley, E.C. Cosgriff, L.J. Allen, Nonlocality in imaging. Phys. Rev. Lett. **94**, 203906 (2005)

M.P. Oxley, M. Varela, T.J. Pennycook, K. van Benthem, S.D. Findlay, A.J. D'Alfonso, L.J. Allen, S.J. Pennycook: Interpreting atomic-resolution spectroscopic images. Phys. Rev. B **76**, 064303 (2007)

Y. Peng, P.D. Nellist, S.J. Pennycook, HAADF-STEM imaging with sub-angstrom probes: A full Bloch wave analysis. J. Electron Microsc. **53**, 257–266 (2004)

S.J. Pennycook, Delocalization corrections for electron channeling analysis. Ultramicroscopy **26**, 239–248 (1988)

S.J. Pennycook, D.E. Jesson, High-resolution Z-contrast imaging of crystals. Ultramicroscopy **37**, 14–38 (1991)

L.C. Qin, K. Urban, Electron diffraction and lattice image simulations with the inclusion of HOLZ reflections. Ultramicroscopy **33**, 159–166 (1990)

C.J. Rossouw, L.J. Allen, S.D. Findlay, M.P. Oxley, Channelling effects in atomic resolution STEM. Ultramicroscopy **96**, 299–312 (2003)

D.K. Saldin, P. Rez, The theory of the excitation of atomic inner-shells in crystals by fast electrons. Philos. Mag. B **55**, 481–489 (1987)

P. Schattschneider, C. Hébert, B. Jouffrey, Orientation dependence of ionization edges in EELS. Ultramicroscopy **86**, 343–353 (2001)

P. Schattschneider, M. Nelhiebel, H. Souchay, B. Jouffrey, The physical significance of the mixed dynamic form factor. Micron **31**, 333–345 (2000)

J. Silcox, P. Xu, R.F. Loane, Resolution limits in annular dark field STEM. Ultramicroscopy **47**, 173–186 (1992)

J.C.H. Spence, Absorption spectroscopy with sub-angstrom beams: ELS in STEM. Rep. Prog. Phys. **69**, 725–758 (2006)

A. Thust, The Stobbs factor in HRTEM: Hunt for a phantom? in *Proceedings of 14th European Microscopy Conference*, Aachen, Germany, pp. 163–164 (2008)

K. van Benthem, A.R. Lupini, M. Kim, H.S. Baik, S.J. Doh, J. Lee, M.P. Oxley, S.D. Findlay, L.J. Allen, J.T. Luck, S.J. Pennycook, Three-dimensional imaging of individual hafnium atoms inside a semiconductor device. Appl. Phys. Lett. **87**, 34104 (2005)

K. van Benthem, A.R. Lupini, M.P. Oxley, S.D. Findlay, L.J. Allen, S.J. Pennycook, Three-dimensional ADF imaging of individual atoms by through-focal series scanning transmission electron microscopy. Ultramicroscopy **106**, 1062–1068 (2006)

D. Van Dyck, Image calculation in high-resolution electron microscopy: problems, progress, and prospects. Adv. Electron. Electron. Phys. **65**, 295–355 (1985)

G. Van Tendeloo, O.I. Lebedev, M. Hervieu, B. Raveau, Structure and microstructure of colossal magnetoresistant materials. Rep. Prog. Phys. **67**, 1315–1365 (2004)

M. Varela, S.D. Findlay, A.R. Lupini, H.M. Christen, A.Y. Borisevich, N. Dellby, O.L. Krivanek, P.D. Nellist, M.P. Oxley, L.J. Allen, S.J. Pennycook, Spectroscopic imaging of single atoms within a bulk solid. Phys. Rev. Lett. **92**, 095502 (2004)

M. Varela, A. Lupini, K. van Benthem, A. Borisevich, M. Chisholm, N. Shibata, E. Abe, S. Pennycook, Materials characterization in the aberration-corrected scanning transmission electron microscope. Annu. Rev. Mater. Res. **35**, 539–569 (2005)

P.M. Voyles, D.A. Muller, J.L. Grazul, P.H. Citrin, H.J.L. Gossmann, Atomic-scale imaging of individual dopant atoms and clusters in highly n-type bulk Si. Nature **416**, 826–829 (2002)

Z.L. Wang, The 'frozen-lattice' approach for incoherent phonon excitation in electron scattering. How accurate is it? Acta Cryst. **A54**, 460–467 (1998)

S. Wang, A.Y. Borisevich, S.N. Rashkeev, M.V. Glazgoff, K. Sohlberg, S.J. Pennycook, S.T. Pantelides, Dopants adsorbed as single atoms prevent degradation of catalysis. Nat. Mater. **3**, 143–146 (2004)

P. Wang, A.J. D'Alfonso, S.D. Findlay, L.J. Allen, A.L. Bleloch, Contrast reversal in atomic-resolution chemical mapping. Phys. Rev. Lett. **101**, 236102 (2008)

K. Watanabe, E. Asano, T. Yamazaki, Y. Kikuchi, I. Hashimoto, Symmetries in BF and HAADF STEM image calculations. Ultramicroscopy **102**, 13–21 (2004)

K. Watanabe, T. Yamazaki, Y. Kikuchi, Y. Kotaka, M. Kawasaki, I. Hashimoto, M. Shiojiri, Atomic-resolution incoherent high-angle annular dark field STEM images of Si(011). Phys. Rev. B **63**, 085316 (2001)

A. Weickenmeier, H. Kohl, The influence of anisotropic thermal vibrations on absorptive form factors for high-energy electron diffraction. Acta Cryst. **A54**, 283–289 (1998)

H.L. Xin, V. Intaraprasonk, D.A. Muller, Depth sectioning of individual dopant atoms with aberration-corrected scanning transmission electron microscopy. Appl. Phys. Lett. **92**, 013125 (2008)

T. Yamazaki, M. Kawasaki, K. Watanabe, I. Hashimoto, M. Shiojiri, Artificial bright spots in atomic-resolution high-angle annular dark field STEM images. J. Electron. Microsc. **50**, 517–521 (2001)

H. Yoshioka, Effect of inelastic waves on electron diffraction. J. Phys. Soc. Japan **12**, 618–628 (1957)

7

X-Ray Energy-Dispersive Spectrometry in Scanning Transmission Electron Microscopes

Masashi Watanabe

7.1 Introduction

Recently developed aberration correctors have brought significant improvements in materials characterization. In aberration-corrected scanning transmission electron microscopy (STEM), the incident probe dimensions can be refined significantly, and image resolution has already reached sub-angstrom levels in high-angle annular dark-field (HAADF) STEM imaging (Batson et al. 2002, Nellist et al. 2004). In addition, materials characterization at the atomic level can routinely be performed by electron energy-loss spectrometry (EELS) in aberration-corrected STEM (e.g. Varela et al. 2005). The aberration correction of the incident beam is also very useful for X-ray energy-dispersive spectrometry (XEDS) because the spatial resolution can be dramatically improved with the refined probe (Watanabe et al. 2006).

X-ray analysis in STEM with relatively high spatial resolution is a very robust and reliable approach to characterize materials. By acquiring an X-ray energy-dispersive spectrum, all major elements can be easily recognized at any measured point. Originally, X-ray analysis was applied to characterize bulk samples in electron probe microanalyzers (EPMAs). In the bulk-sample analysis in EPMAs, spatial resolution is no better than a few micrometers since all incident electrons are absorbed in the sample. Furthermore, the spatial resolution in bulk-sample analysis is degraded as incident electron energy increases, whereas such elevated electron energies are essential to generate particular X-rays for typical analysis. Obviously, the spatial resolution achievable in conventional bulk-sample analysis may not be satisfactory for the detailed characterization of advanced materials.

Several attempts have been explored to improve the spatial resolution of X-ray analysis. The use of electron-transparent thin specimens in higher kilovolt STEM is one successful approach to obtain improved

S.J. Pennycook, P.D. Nellist (eds.), *Scanning Transmission Electron Microscopy*,
DOI 10.1007/978-1-4419-7200-2_7, © Springer Science+Business Media, LLC 2011

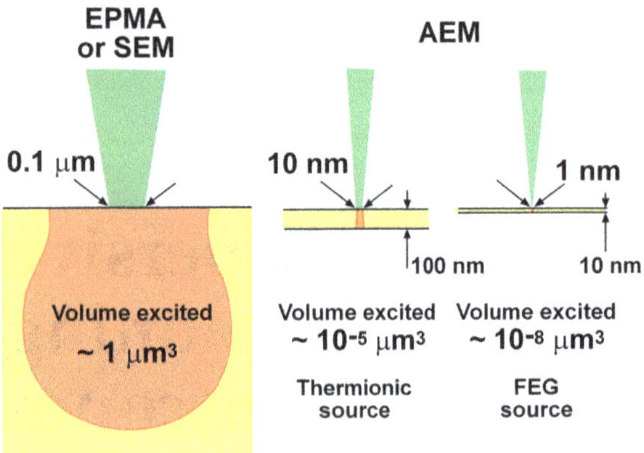

Figure 7–1. Comparison of the relative size of the analyzed volume of (**a**) a bulk sample in an EPMA, (**b**) an ~100-nm-thick specimen in a thermionic source STEM, and (**c**) an ~10-nm-thick specimen in a FEG-STEM, respectively. Due to the reduction of analyzed volume, the analytical sensitivity is degraded in the analysis of thin specimens. Reproduced from Williams and Carter (2009) with permission.

spatial resolution of X-ray analysis. In a thin specimen, the spatial resolution is governed by the incident probe size and its broadening in the specimen related to the specimen thickness. To minimize the probe broadening, thinner specimens are preferably used even for X-ray analysis. To reduce the incident probe size, higher brightness field-emission gun (FEG) sources are more popularly employed in modern instruments. With the aberration correctors, much finer incident probes are available, as mentioned later.

As the result of pursuing better spatial resolution in combination with the use of thin specimens and finer incident probes, the volume-producing X-ray signals are significantly reduced as shown in Figure 7–1, in comparison with the volumes in EPMAs and conventional STEM instruments (Williams and Carter 2009). This reduction in the analyzed volume is even more pronounced with aberration-corrected probes. In addition to the significant volume reduction, the detection of X-ray signals is even more limited by microscope column-detector configurations, which provide at most a collection angle of ~0.3 sr out of 4π sr and typically 0.15 sr in most commercialized systems. In comparison with the detection efficiency in EELS (which can be close to 100%), X-ray signal detection is very poor (only 1–3%), as shown in Figure 7–2 (Williams and Carter 2009).

The poor X-ray generation due to the reduced analyzed volume and the poor X-ray detection due to the instrument design limitation require higher probe currents than STEM imaging and EELS analysis. In operating conditions with higher probe currents, the contribution of the probe current to the incident probe formation is no longer negligible, as pointed out by Brown (1981) and Watanabe et al. (2006). Therefore, the optimum setting for X-ray analysis with higher probe currents can be very different from the setting for STEM imaging and EELS analysis. In this chapter, first the contribution of incident probe currents to probe formation will be reviewed to explore the optimum conditions of probe formation for X-ray analysis; then the benefits of aberration correction for X-ray analysis will be discussed in terms of the spatial resolution and detectability limits such as minimum mass fraction (MMF) and minimum detectable mass (MDM). X-ray analysis has been applied

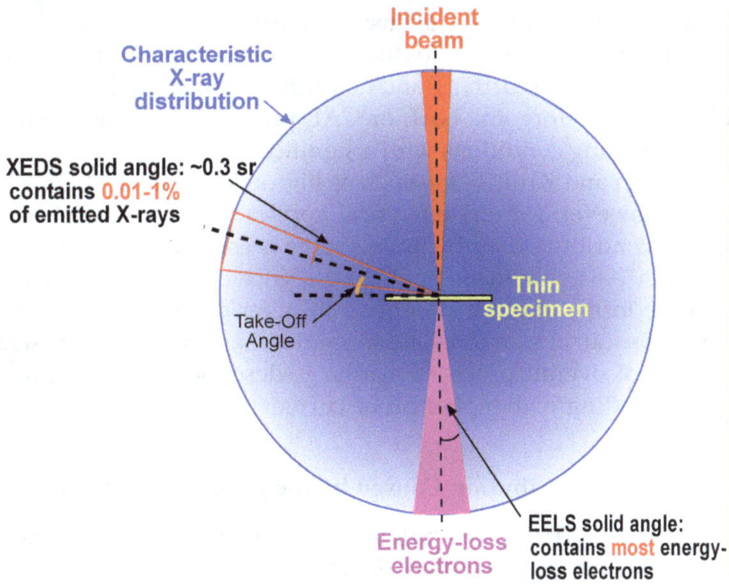

Figure 7–2. Comparison of the relative efficiency of a signal collection in XEDS and EELS. Whereas most of the energy loss electrons can be collected, only a small percentage of the generated X-ray signals are collected due to the limitation of the current design of the microscope–XEDS detector interface. Reproduced from Williams and Carter (2009) with permission.

to materials characterization in STEM for over 30 years. Recently, data acquisition and data analysis procedures including quantification associated with X-ray analysis are significantly improved. These new approaches in data acquisition and analysis are also reviewed. Applications of X-ray analysis in the aberration-corrected instruments are very limited in comparison with applications for STEM imaging and EELS analysis. Therefore, several recent applications obtained by X-ray analysis in aberration-corrected STEM including remarkable atomic-column X-ray mapping will be introduced to demonstrate the latest achievements in X-ray analysis. Finally, future prospects of X-ray analysis will be remarked on to conclude this chapter.

7.2 Optimum Instrument Settings for X-Ray Analysis

Spatial resolution in STEM imaging and analysis is directly related to the incident probe dimension, i.e., the shape and diameter containing a certain fractional current. The incident probe dimension is one of the most important factors in STEM, and hence probe formation theory has been explored intensively in previous studies. Most of the probe formation discussions are focused on the geometric aberration-limited (GAL) and/or chromatic aberration-limited (CAL) probes, which represent blurring of a point source. Note that the details about the GAL and CAL probe formation can be found in the literature (e.g. Munro 1977; Colliex and Mory 1984; Haider et al. 2000) or in other chapters.

The GAL and CAL probe dimensions do not contain any contribution of the finite source size (which is expressed through the source brightness and probe current), and no contribution of the probe current is included. Therefore, the GAL or CAL probe dimensions are useful only for operating conditions with a significantly limited probe current <20 pA (Barth and Kruit 1996), such as high-resolution HAADF-STEM imaging. However, for any analytical application by using EELS and XEDS, the contribution of the finite source is essential to probe formation (Brown 1981, Watanabe et al. 2006). Since the details of GAL and CAL probe formations are described in other chapters, the contribution of finite source size to probe formation is featured here first, and then the final overall probe formation is described for the optimum instrument setting with higher probe currents.

7.2.1 Intensity Distribution Caused by the Finite Source Size

The intensity distribution of a demagnified image of the finite source at a crossover can be adequately described by a two-dimensional (2D) Gaussian distribution (Colliex and Mory 1984, Reimer and Kohl 2009). The 2D Gaussian intensity distribution $J_S(r)$ of the demagnified source image can be expressed as (Munro 1977, Colliex and Mory 1984):

$$J_S(r) = J_{S0} \exp\left(-\frac{r^2}{r_0^2}\right) \tag{1}$$

where J_{S0} is the peak intensity of the distribution, r is the distance from the center of the distribution (radius), and r_0 is the radius of the demagnified source image. Using conventional standard deviation σ, the equation above can also be expressed as

$$J_s(r) = J_{s0} \exp\left(-\frac{r^2}{2\sigma^2}\right). \tag{2}$$

Hence, the radius of the demagnified source r_0 is equivalent to $\sqrt{2}\sigma$ in the 2D Gaussian distribution. An integral of $J_S(r)$ in the infinite range [-∞, ∞] should be equivalent to the incident probe current I_p. Therefore, the pre-factor J_{S0} can be derived as $I_p/(\pi r_0^2)$ (Colliex and Mory 1984). However, the value of the demagnified source radius r_0 is still unclear. Since it is very difficult to measure the demagnified source radius directly, an arbitrary value such as $r_0 = 0.15$ or 0.4 nm has to be chosen to calculate the intensity distribution of $J_S(r)$, depending on a selected probe current (Mory et al. 1985). Recently, the effective source diameter was determined indirectly as 0.08 nm in full width at half maximum (FWHM) of Gaussian by comparing experimental and simulated HAADF-STEM images (LeBeau et al. 2008). By matching the simulated Ronchigram to an experimentally measured one from a crystalline specimen, the effective source diameter was also determined as 0.056 nm (FWHM) (Dwyer et al. 2008). These recent approaches could be ideal to determine the effective source size in limited probe current conditions for atomic-resolution HAADF-STEM imaging. In general, the effective source size is dependent on the type of electron source and on operation conditions, e.g., the emission current that directly influences the incident probe current.

Instead of using such arbitrary values, the radius of the demagnified source image r_0 defined as $\sqrt{2}\sigma$ can be derived analytically from properties of the 2D Gaussian function. The $J_S(r)$ value becomes 10% of the peak when the distance from the origin r reaches to half the diameter at full. -width at tenth maximum (FWTM) $d_S(\mathrm{TM})$ in the 2D Gaussian distribution, i.e., 0.1 J_{S0} at $r = \pm\, d_S(TM)/2$. Therefore, the demagnified source radius r_0 can be rewritten as

$$r_0 = \left[\frac{\{d_S(\mathrm{TM})\}^2}{4\ln(10)}\right]^{1/2} = 0.330 d_S(\mathrm{TM}). \tag{3}$$

The electron source brightness remains constant at any point in the electron optical axis unless an electron energy filter is used (Reimer and Kohl 2009). Thus, the demagnified source diameter d_S can be expressed as a contributing factor to the final probe formation at the specimen position using the probe-forming aperture size α (semi-angle) and probe current after the aperture I_P (Oatley et al. 1965, Crewe et al. 1968, Crewe and Salzman 1982):

$$d_S = \frac{2}{\pi}\left(\frac{I_P}{\beta}\right)^{1/2}\alpha^{-1} \tag{4}$$

where β is the source brightness.

The above equation for the contribution of finite source size seems to be well established (Barth and Kruit 1996, Joy 1974, Vaughan 1976). However, various definitions of this diameter have been used. For example, the d_S value in Eq. (4) has erroneously been defined as the FWHM diameter in many previous studies (Joy 1974, Crewe and Wall 1970, Wells 1974, Thomas 1982, Hanai and Hibino 1984, Michael and Williams 1987, Zaluzec and Nicholls 1998). In the use of Eq. (2), the probe current value after the final probe-forming aperture and before it hits the specimen is required. The probe current is usually measured by the use of a Faraday cup, a viewing screen, or an EELS spectrometer, which collects almost 100% of the incident electrons. Thus, the definition of d_S by Eq. (4) as the FWHM diameter is clearly overestimated. In the 2D Gaussian distribution function, 90% of the total incident intensity is contained within the FWTM diameter, so the FWTM diameter d_S(FW) can be expressed from Eq. (4) as

$$d_S(\mathrm{TM}) \equiv \frac{2}{\pi}\left(\frac{0.9 I_P}{\beta}\right)^{1/2}\alpha^{-1} = 0.604\left(\frac{I_P}{\beta}\right)^{1/2}\alpha^{-1}. \tag{5}$$

Then, the demagnified source radius r_0 is given from Eq. (3) as

$$r_0 = 0.199\left(\frac{I_P}{\beta}\right)^{1/2}\alpha^{-1}. \tag{6}$$

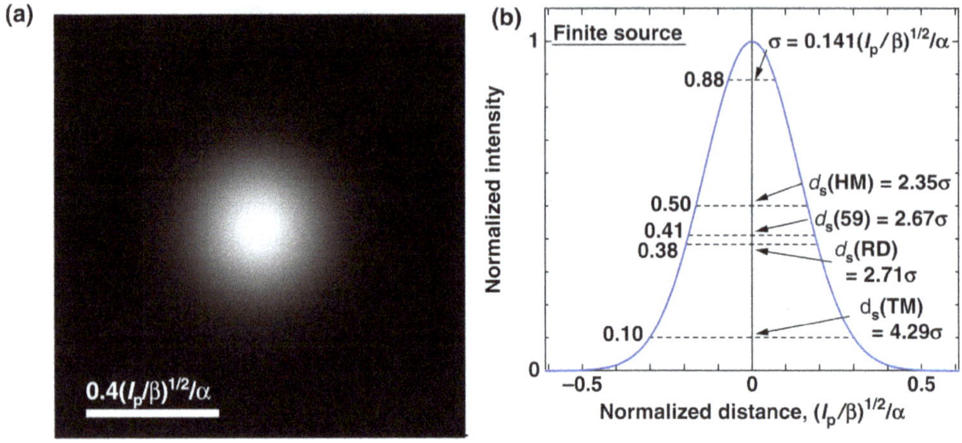

Figure 7–3. (**a**) A simulated image of the finite source contribution using Eq. (7) and (**b**) an intensity profile extracted from the image. Both the distance and the intensity are normalized by the distance parameter $(I_P/\beta)^{1/2}/\alpha$ and the peak intensity, respectively.

Finally, by substituting Eq. (6) into Eq. (1), the 2D Gaussian distribution of the demagnified source image can be derived as

$$J_S(r) = \frac{\pi \ln(10)}{0.9}\beta\alpha^2 \exp\left(-\frac{\pi^2 \ln(10)}{0.9}\frac{\beta\alpha^2}{I_P}r^2\right) = 8.038\beta\alpha^2 \exp\left(-25.251\frac{\beta\alpha^2}{I_P}r^2\right). \quad (7)$$

Therefore, the demagnified source distribution is properly linked with the probe formation parameters of β, I_P, and α.

Figure 7–3a shows a simulated image of the finite source contribution using Eq. (7) and Figure 7–3b shows an intensity profile extracted from the image, respectively. In Figure 7–3, both the distance and the intensity are normalized by the distance parameter $(I_P/\beta)^{1/2}/\alpha$ and the peak intensity, respectively. Therefore, the intensity distribution in both the image and the profile includes the contribution of the finite source under any probe formation conditions. In addition, several definitions of probe diameters proposed in the literature are shown in the extracted profile. These previously defined probe diameters and the corresponding intensity fraction (%) against the total incident intensity are summarized in Table 7–1. Furthermore, any probe diameter at a given fraction f can be derived from the 2D Gaussian function:

$$d_S(f) = \left[\frac{3.6 \ln(1-f)}{\pi^2 \ln(0.1)}\right]^{1/2} \left(\frac{I_P}{\beta}\right)^{1/2} \alpha^{-1}. \quad (8)$$

Using Eq. (8), the contribution of the finite source size can be evaluated for any probe-forming conditions.

Figure 7–4 shows the 59% diameter of the finite source contribution $d_S(59\%)$ for (a) cold and (b) Schottky electron sources in several probe current conditions, as plotted against the probe-forming semi-angle α. For the $d_S(59\%)$ calculation, the source brightness values of 1×10^{13} and

Table 7–1. Summary of several probe diameters of the finite source contribution previously defined in literature with the corresponding intensity fraction (%) against the total incident intensity.

Defined diameter	Normalized diameter in $(I_P/\beta)^{1/2}/\alpha$	Normalized diameter in σ	Relative height	Intensity fraction (%)
Standard deviation, σ	0.141	1.00	0.882	11.8
Diameter at FWHM, d_S(HM)	0.330	2.35	0.500	50.0
Diameter contains 59% of intensity, d_S(59%)	0.376	2.67	0.410	59.0
Diameter at Rayleigh distance, d_S(RD)	0.390	2.77	0.375	61.6
Source size defined in Eq. (1), $2r_0$	0.396	2.82	0.368	63.0
Diameter at FWTM, d_S(TM)	0.608	4.29	0.100	90.0

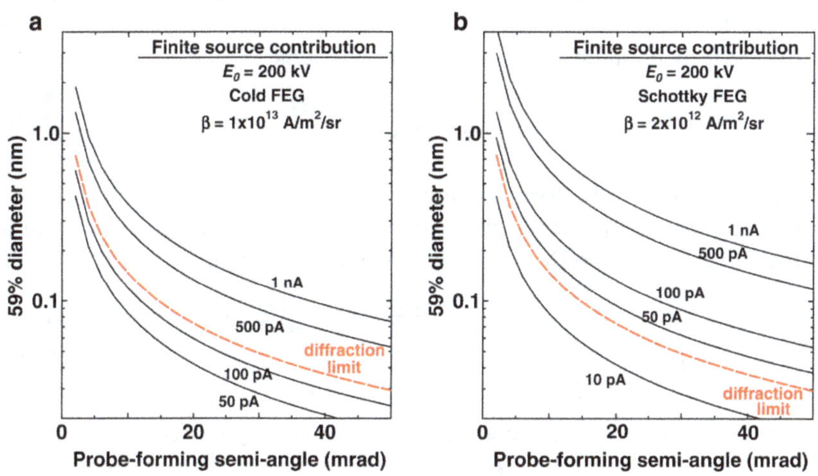

Figure 7–4. The 59% diameter of finite source contribution d_S(59%) for (**a**) cold and (**b**) Schottky electron sources in several probe current conditions, as plotted against the probe-forming semi-angle α.

2×10^{12} A/m^2/sr were used for the cold (Williams and Carter 2009) and Schottky (Tanaka et al. 2002) guns, respectively. Typical ranges of the relevant probe currents for HAADF-STEM imaging, EELS, and XEDS analyses are 0–100, 50–300, and above 300 pA, respectively. Obviously, the higher the source brightness is, the smaller the diameter becomes at the same current. In other words, with a higher brightness source, more probe current can be added in the same diameter, which is more suitable especially for X-ray analysis. For comparison, the 59% diffraction-limit diameter at 200 kV is also plotted as a dashed line in Figure 7–4a, b. When the probe current is relatively low, the contribution of the finite source is less than the diffraction limit. The diameters of the finite source contribution are superimposed on the diffraction-limited diameter at probe currents of 150 and 30 pA for the cold and Schottky sources, respectively. The contribution of the finite source size to the final probe formation will be discussed in subsequent sections.

7.2.2 Intensity Distribution of the GAL and CAL Probes

In contrast to the distribution of the finite source size, GAL and CAL probe distributions are well defined in the wave-optical treatment, which was originally developed for light optics (Born and Wolf 1999), and can be used to determine probe diameters (e.g., Munro 1977, Colliex and Mory 1984, Haider et al. 2000). In this theory, the GAL probe distribution $J_G(r)$ can be calculated as the magnitude of the wave function at the specimen position $\psi(r)$, which is obtained by the Fourier transform (FT) of the phase shift caused by geometric aberrations $\exp\{i\chi(\omega)\}$ at the front focal plane of the probe-forming lens:

$$J_G(r) = |\psi(r)|^2 = \left|\mathrm{FT}[A(\omega)\exp\{i\chi(\omega)\}]\right|^2 \tag{9}$$

where $A(\omega)$ and $\chi(\omega)$ are the probe-forming aperture function and the aberration function, respectively, defined as a function of the angle ω at the front focal plane of the probe-forming lens:

$$A(\omega) = \begin{cases} 1 & (\omega \le \alpha) \\ 0 & (\omega > \alpha) \end{cases} \tag{10}$$

where α is the probe-forming aperture semi-angle. The aberration function is given with various aberration coefficients up to the fifth order in the CEOS definition (Müller et al. 2006, Haider et al. 2008) as

$$\begin{aligned}\chi(\omega) = \mathrm{Re}\Big[\tfrac{2\pi}{\lambda}\Big(&\tfrac{1}{2}\Delta z\omega\overline{\omega} + \tfrac{1}{2}A_1\overline{\omega}^2 + B_2\omega^2\overline{\omega} + \tfrac{1}{3}A_2\overline{\omega}^3 + \tfrac{1}{4}C_S\omega^2\overline{\omega}^2 + S_3\omega^3\overline{\omega} + \tfrac{1}{4}A_3\overline{\omega}^4 \\ &+ B_4\omega^3\overline{\omega}^2 + D_4\omega^4\overline{\omega} + \tfrac{1}{5}A_4\overline{\omega}^5 + \tfrac{1}{6}C_5\omega^3\overline{\omega}^3 + S_5\omega^4\overline{\omega}^2 + R_5\omega^5\overline{\omega} + \tfrac{1}{5}A_5\overline{\omega}^6\Big)\Big]\end{aligned} \tag{11}$$

where $\overline{\omega}$ is the complex conjugate of ω, λ is the electron wave length, Δz is the defocus, and C_S and C_5 are third- and fifth-order spherical aberration coefficients, respectively. Other aberration coefficients are summarized in Table 7–2 with the corresponding designations in the NION definition (Krivanek et al. 2003).

Table 7–2. Summary of the aberration coefficients in the CEOS and NION definitions with conversion from the NION to the CEOS definitions.

Notation			Relationship	
CEOS	NION	Description	Coefficient	Angle
$C_1(\Delta z)$	$C_{1,0}$	Defocus	$C_1 = C_{1,0}$	—
A_1	$C_{1,2}$	Twofold astigmatism	$A_1 = [(C_{1,2a})^2 + (C_{1,2b})^2]^{1/2}$	$\tan^{-1}[C_{1,2a}/C_{1,2b}]$
B_2	$C_{2,1}$	Second-order axial coma	$B_2 = [(C_{2,1a})^2 + (C_{2,1b})^2]^{1/2}/3$	$\tan^{-1}[C_{2,1a}/C_{2,1b}]$
A_2	$C_{2,3}$	Threefold astigmatism	$A_2 = [(C_{2,3a})^2 + (C_{2,3b})^2]^{1/2}$	$\tan^{-1}[C_{2,3a}/C_{2,3b}]$
C_3	$C_{3,0}$	Spherical aberration	$C_3 = C_{3,0}$	
S_3	$C_{3,2}$	Star aberration	$S_3 = [(C_{3,2a})^2 + (C_{3,2b})^2]^{1/2}/4$	$\tan^{-1}[C_{3,2a}/C_{3,2b}]$
A_3	$C_{3,4}$	Fourfold astigmatism	$A_3 = [(C_{3,4a})^2 + (C_{3,4b})^2]^{1/2}$	$\tan^{-1}[C_{3,4a}/C_{3,4b}]$
B_4	$C_{4,1}$	Fourth-order axial coma	$B_4 = [(C_{4,1a})^2 + (C_{4,1b})^2]^{1/2}/5$	$\tan^{-1}[C_{4,1a}/C_{4,1b}]$
D_4	$C_{4,3}$	Three-lobe aberration	$D_4 = [(C_{4,3a})^2 + (C_{4,3b})^2]^{1/2}/5$	$\tan^{-1}[C_{4,3a}/C_{4,3b}]$
A_4	$C_{4,5}$	Fivefold astigmatism	$A_4 = [(C_{4,5a})^2 + (C_{4,5b})^2]^{1/2}$	$\tan^{-1}[C_{4,5a}/C_{4,5b}]$
C_5	$C_{5,0}$	Fifth-order spherical aberration	$C_5 = C_{5,0}$	—
S_5	$C_{5,2}$	Fifth-order star aberration	$S_5 = [(C_{5,2a})^2 + (C_{5,2b})^2]^{1/2}/6$	$\tan^{-1}[C_{5,2a}/C_{5,2b}]$
R_5	$C_{5,4}$	Rosette aberration	$R_5 = [(C_{5,4a})^2 + (C_{5,4b})^2]^{1/2}/6$	$\tan^{-1}[C_{5,4a}/C_{5,4b}]$
A_5	$C_{5,6}$	Sixfold astigmatism	$A_5 = [(C_{5,6a})^2 + (C_{5,6b})^2]^{1/2}$	$\tan^{-1}[C_{5,6a}/C_{5,6b}]$

In addition to geometric aberrations, the contribution of the chromatic aberration should be taken into account for a more accurate description of probe formation, especially for aberration-corrected conditions. The chromatic aberration can be described by the chromatic aberration coefficient (C_c), the energy spread of the electron source (ΔE), and the incident electron energy (E_0). The finite energy distribution of incident electrons causes an additional variation of the defocus in the probe formation. Therefore, the defocus is modified by the chromatic aberration as (Mory et al. 1985, Haider and Uhlemann 2000, Haider et al. 2008)

$$\Delta z_C = \Delta z + C_c \frac{E - E_0}{E_0} \tag{12}$$

where E is the electron energy within the energy distribution of the incident electron, $dN(E)/dE$. The energy distribution of the incident electron can be seen as a distribution of the zero-loss peak in an EEL spectrum, which can be assumed as a Gaussian rather than a more complicated distribution model such as a Maxwellian (Haider and Uhlemann 2000, Haider et al. 2008, Reimer and Kohl 2009). Using the GAL probe distribution $J_G(r)$ with $dN(E)/dE$, the CAL probe distribution $J_C(r)$ can be adapted into the wave-optical calculation as

$$J_C(r) = \int_{-\infty}^{+\infty} J_G(r, \Delta z_c) \frac{dN(E)}{dE} dE. \qquad (13)$$

A diameter at a given intensity fraction f from the GAL or CAL probe distribution is defined by integrating over the intensity distribution $J_G(r)$ or $J_C(r)$ (Colliex and Mory 1984, Mory et al. 1985):

$$f = \frac{1}{I_j(\infty)} \int_0^{d_j(f)/2} 2\pi r J_j(r) dr \qquad (14)$$

where $d_j(f)$ is the probe diameter at the given fraction, the subscript j represents either the GAL (G) or the CAL (C) probe diameter and $I_j(\infty)$ is the total intensity integrated to infinity, which corresponds to $\pi\alpha^2$. Traditionally, various definitions for the probe diameter were proposed.

In this manuscript, the probe diameter at $f = 0.59$, $d_j(59\%)$, has been employed to discuss the optimum instrument settings and the image resolution. The definition of $d_j(59\%)$ is originally derived from the Rayleigh criterion of intensity distribution in the aperture diffraction-limited probe, by which the image resolution can be defined as the minimum distance to distinguish the two point objects (Fertig and Rose 1979, Rose 1981, Haider and Uhlemann 2000). Figure 7–5a shows a normalized intensity distribution and Figure 7–5b shows an extracted line profile of two diffraction-limited probes in the Rayleigh criterion, respectively. The minimum distance defined by the Rayleigh criterion (the so-called Rayleigh distance d_{RC}) can be expressed as $0.61\lambda/\alpha$ in the aperture diffraction-limited condition, and the local minimum between two objects in the intensity distribution appears as 75% of the peak intensity (Fertig and Rose 1979, Rose 1981, Haider and Uhlemann 2000). The Rayleigh criterion is generally accepted as the definition of the image resolution limit, and the Rayleigh distance is equivalent to the 59% probe diameter in the aperture diffraction-limited condition. In addition, the optimum condition for probe formation derived from the $\pi/4$ limit of the phase shift is superimposed with the formation from the direct simulation of the 59% probe diameter (Watanabe and Sawada submitted). Therefore, it should be reasonable to use the $d(59\%)$ value to discuss the optimum condition for probe formation.

Figure 7–6 shows the $d(59\%)$ diameters determined from the simulated GAL distribution in conventional and aberration-corrected conditions for a 200-kV instrument, which are plotted as a function of the probe-forming semi-angle. In the conventional condition, the major limit is the third-order spherical aberration, and the calculation of the 59% diameter was performed with $C_s = 0.5$ mm, which is the best value available in commercial instruments. In contrast, the residual aberrations can be the fifth-order spherical aberration and six-fold astigmatism after complete aberration tuning in a hexapole-based corrector system (Müller et al. 2006, Sawada et al. 2009). The $d(59\%)$ diameters were calculated in the aberration-corrected condition at $C_5 = 0.5$ mm and $A_5 = 1.2$ mm, which were obtained from a comparison between experimental and simulated Ronchigrams (Watanabe and Sawada submitted). In the conventional condition, the probe diameter is minimized

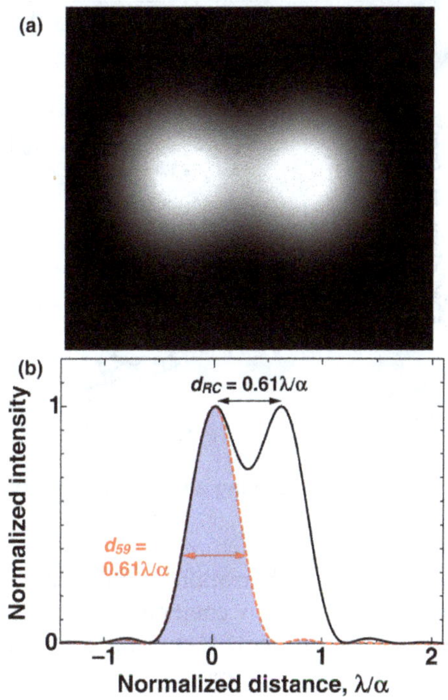

Figure 7–5. (**a**) A normalized intensity distribution and (**b**) an extracted line profile of two diffraction-limited probes in the Rayleigh criterion, which show the local minimum between two objects in the intensity distribution as 75% of the peak intensity at $0.61\lambda/\alpha$.

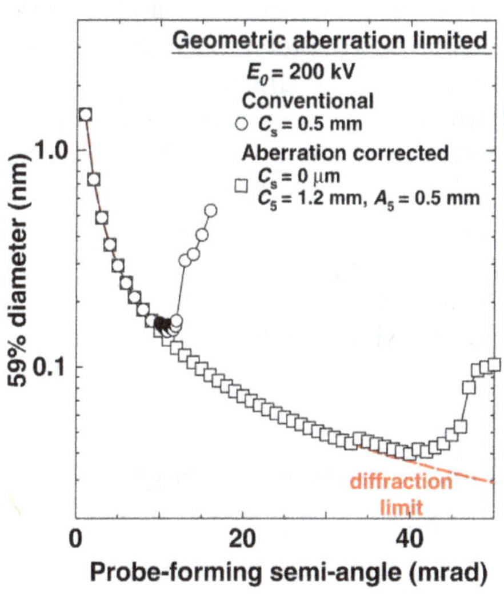

Figure 7–6. The GAL probe diameter contains 59% of the total intensity *d*(59%) simulated for the conventional condition with C_S = 0.5 mm and for the aberration-corrected condition with $C_S = 0$ μm, $C_5 =$ 1.2 mm, and $A_5 = 0.5$ mm at 200 kV, as plotted against the probe-forming semi-angle α. For comparison, the diffraction limit for 59% of the total intensity is also plotted as a solid line (Watanabe and Sawada submitted).

at α = 10.9 mrad with a defocus of −28 nm. The optimum probe-forming angle is expanded to ~40.0 mrad in the aberration-corrected condition. The optimum probe diameters determined from GAL distributions are 146 and 40 pm in the conventional and aberration-corrected conditions, respectively.

Since the optimum probe-forming angle is expanded in the aberration-corrected condition, the contribution of chromatic aberration to the probe formation may not be negligible. The CAL probe diameters were calculated based on the above GAL condition (with

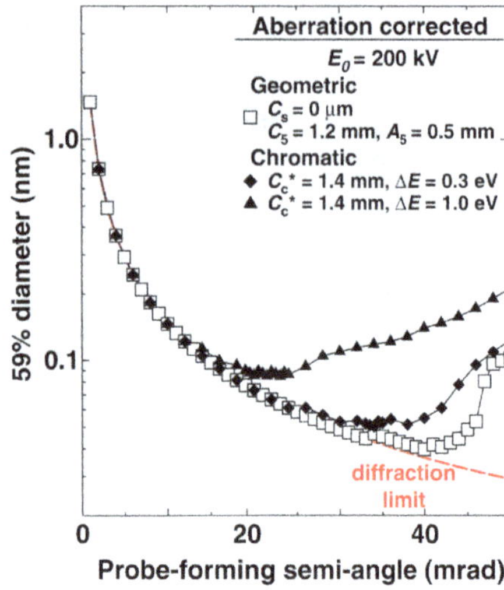

Figure 7–7. The $d(59\%)$ CAL probe diameters for the cold FEG with $\Delta E = 0.3$ eV and the Schottky FEG with $\Delta E = 1.0$ eV plotted against the probe-forming semi-angle α. For both conditions, the simulation was performed with the chromatic aberration coefficient of 1.4 mm over the optimum aberration-corrected GAL condition at 200 kV shown in Figure 7–6 ($C_S = 0$ μm, $C_5 = 1.2$ mm, and $A_5 = 0.5$ mm) (Watanabe and Sawada submitted).

dominant C_5 and A_5) with the chromatic aberration coefficient of 1.4 mm. This value of the chromatic aberration coefficient is again the best available value in a 200-kV instrument, including a slight increase of C_c due to the corrector. The probe diameter ($d(59\%)$) simulated with the energy spreads of $\Delta E = 0.3$ and 1.0 eV is shown in Figure 7–7, which is plotted against the probe-forming angle. These ΔE values represent a typical energy spread for cold and Schottky FEG sources, respectively. For comparison, the diffraction-limited (thick solid line) and the GAL probe diameters (open squares) are also plotted in Figure 7–7. The probe diameter is degraded due to the chromatic aberration and the optimum probe-forming angle shifts to lower angle, depending on ΔE, in the given chromatic aberration condition. The optimum probe diameters (and optimum angles) are 50 pm (34 mrad) and 85 pm (23 mrad) for $\Delta E = 0.3$ and 1.0 eV, respectively. Obviously, the better energy spread of the cold FEG significantly improves the fine probe formation.

7.2.3 Overall Intensity Distribution for Analysis with High Currents

As mentioned earlier, the contribution of the finite source size must be considered for probe diameters. In addition, either the GAL or the CAL intensity distribution for some applications requires higher probe currents, such as EELS and XEDS analysis. The final intensity distribution (the so-called overall intensity distribution of the extended source) is given by the convolution of either the GAL or the CAL intensity distribution with the finite source intensity distribution (Colliex and Mory 1984):

$$J_F(r) = J_j(r) \otimes J_S(r) \tag{15}$$

where $J_j(r)$ is the intensity distribution of either GAL or CAL probe (j = G or C) given in Eqs. (9) or (13), $J_S(r)$ is the intensity distribution of the finite source described in Eq. (7) and ⊗ indicates the convolution process. It should be noted that the overall (final) probe can no longer be treated as coherent illumination after the convolution step in Eq. (6) (Colliex and Mory 1984, Mory et al. 1985). From the calculated final intensity distribution, the probe diameter can be determined using Eq. (14).

As shown in Figure 7–7, the optimum probe-forming angle is ~23 mrad for the CAL probe in the instrument with the Schottky FEG (ΔE = 1.0 eV) and the chromatic aberration coefficient of 1.4 mm. In this optimum condition, the overall probe diameters contributed by the geometric and chromatic aberrations and the finite source size were simulated and plotted as a function of the probe current in Figure 7–8(a) (closed triangle). The probe diameter at 23 mrad remains almost constant up to 10 pA and then increases as the probe current increases. To maintain the probe diameter below 1 Å, the beam current must be limited below 20 pA. For comparison, the overall probe diameter at the probe-forming angle of 35 mrad is also plotted in Figure 7–8(a) (open triangles). Despite the fact that 35 mrad is not the optimum value in the presence of chromatic aberration, the probe diameter at 35 mrad is *smaller* than the probe diameter at 23 mrad if the current exceeds 100 pA, which indicates that the optimum angle of the probe-forming aperture also depends on the probe current and source brightness.

Figure 7–8(b) plots the overall probe diameter at selected currents of 30, 120, and 500 pA as a function of the probe-forming angle together with GAL and CAL probe diameters. The ranges of the relevant probe

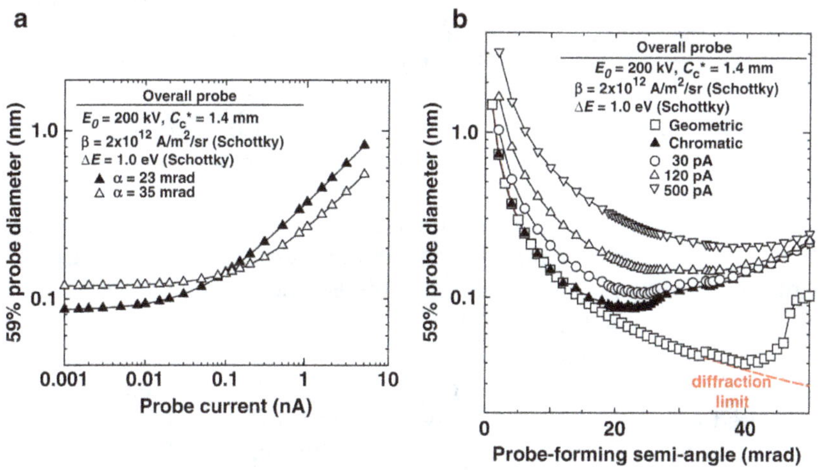

Figure 7–8. (**a**) The d(59%) diameters of the overall probe contributed by the geometric and chromatic aberrations, and by the finite source sizes at 23 mrad (*closed triangles*) and 35 mrad (*open triangles*), which are plotted as a function of the probe current in an instrument with chromatic aberration coefficient of 1.4 mm and a Schottky source. (**b**) The overall probe diameters at selected probe currents of 30, 120, and 500 pA are plotted as a function of the probe-forming angle together with GAL and CAL probe diameters in an instrument with chromatic aberration coefficient of 1.4 mm and a Schottky source (Watanabe and Sawada submitted).

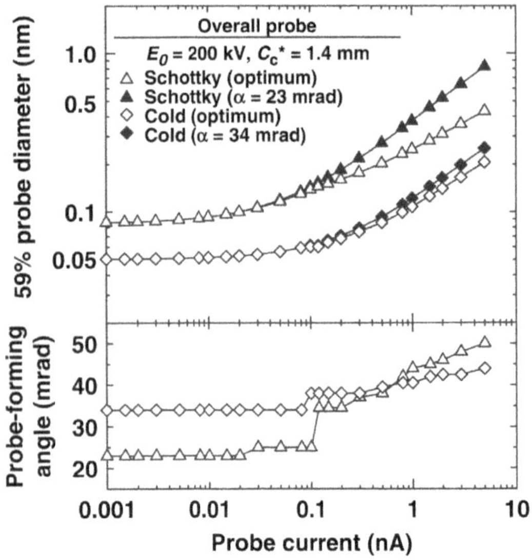

Figure 7–9. Optimum angle for the overall probe summarized as a function of the probe current with the corresponding probe diameter *d*(59%) for the Schottky and cold FEGs in a column with a chromatic aberration coefficient of 1.4 mm (Watanabe and Sawada submitted).

currents for HAADF-STEM imaging, EELS, and XEDS analyses are 0–100, 50–300, and above 300 pA, respectively. The selected currents in Figure 7–8(b) represent the typical values for these applications. The optimum angle increases from 23 mrad for the CAL probe (i.e., zero current) with an increase in the probe current, e.g., the optimum angles at 30, 120, and 500 pA are 25, 35, and 38 mrad, respectively. In addition, the probe diameter becomes almost invariant in the angular range between 30 and 40 mrad at higher probe currents. Therefore, the probe diameter (i.e., the resolution) is not very sensitive to the aperture size if a high probe current is applied.

Since the optimum probe-forming angle is also dependent on the probe current and source brightness (i.e., the finite source size), the optimum angles are summarized as a function of the probe current with the corresponding probe diameters for the Schottky and cold FEGs with the chromatic aberration coefficient of 1.4 mm (Figure 7–9). In this simulation, values for the source brightness and energy spread are 2×10^{12} A/m^2/sr and 1.0 eV for Schottky, and 1×10^{13} A/m^2/sr and 0.3 eV for cold FEGs, respectively. For comparison, the current dependence of the probe diameter at the optimum angle for the CAL probe is indicated by closed symbols. Obviously, the probe diameter becomes smaller with a larger probe-forming aperture of over 100 pA in any condition. The achievable probe diameter at 500 pA is 0.8 and 2.0 Å for the cold and Schottky source, respectively. Below 100 pA, the optimum angle remains almost the same as the angle of the CAL probe and then increases as the probe current increases. Especially for the Schottky source, the optimum angle is suddenly enlarged above 100 pA, which corresponds to the crossover point in Figure 7–8(a). In summary, below 100 pA, the optimum probe-forming aperture can be selected. Above 100 pA, a larger aperture is generally recommended for some EELS analysis and for most XEDS analysis.

Figure 7–10 shows simulated intensity distributions and extracted line profiles of the overall probes at probe currents of (a) 30 pA, (b) 120 pA, and (c) 500 pA in an aberration-corrected 200-kV instrument with a Schottky FEG. These selected probe currents are typical values for ADF-STEM imaging, EELS analysis, and XEDS analysis, respectively. The simulation was performed at optimum convergence angles for corresponding probe currents, as shown in Figure 7–9. Both the intensity distributions and the profiles are normalized with the peak intensity at 500 pA. Therefore, the peak intensities at lower current conditions are correspondingly lower (i.e., ~57% at 30 pA and ~70% at 120 pA against the peak intensity at 500 pA). In the profiles, the normalized CAL intensity profile simulated at corresponding probe formation conditions is also plotted as dashed lines. At 30 pA, the dominant contribution is chromatic/geometric aberrations to the final probe formation. However, at higher currents for analysis applications, finite source size

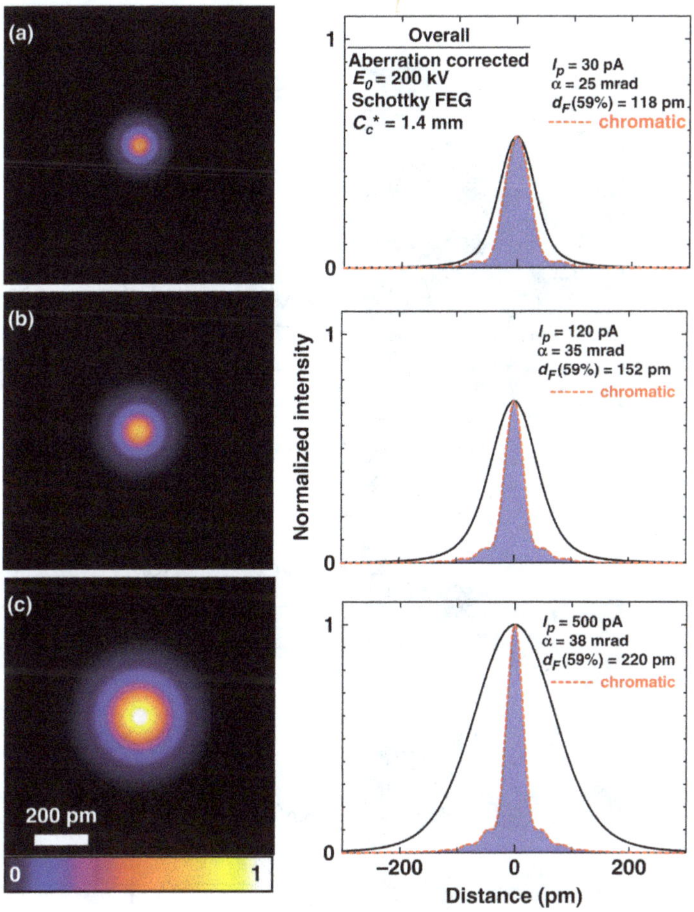

Figure 7–10. Simulated intensity distributions and extracted line profiles of the overall probes at probe currents of (**a**) 30 pA, (**b**) 120 pA, and (**c**) 500 pA in an aberration-corrected 200-kV instrument with a Schottky FEG. Normalized CAL intensity profiles simulated at corresponding probe formation conditions are also plotted as dashed lines in the profiles (Watanabe and Sawada submitted).

is the dominant factor for probe formation. Simulated $d(59\%)$ values at 30, 120, and 500 nA are 118, 152, and 220 pm, respectively.

Figure 7–11a shows HAADF-STEM images of Si<110> recorded at 30 pA, Figure 7–11b shows HAADF-STEM images of Si<110> recorded at 120 pA, and Figure 7–11c shows HAADF-STEM images of Si<110> recorded at 500 pA in an aberration-corrected 200-kV JEOL JEM-2200FS STEM with a Schottky FEG. Intensity profiles extracted from the individual images are also inserted in Figure 7–11. At 30 pA, the intensity reduction between the atomic column of the Si dumbbell reaches over 25%, which satisfies the Rayleigh criterion as described previously. However, the intensity reduction is slightly degraded at 120 pA, and no reduction can be observed at 500 pA. These experimental results are superimposed with the simulated $d(59\%)$ values shown in Figure 7–10.

In summary, the probe size simulation (including the contribution of the finite source size) is essential not only to estimate the image resolution but also to configure the optimum instrument setting for

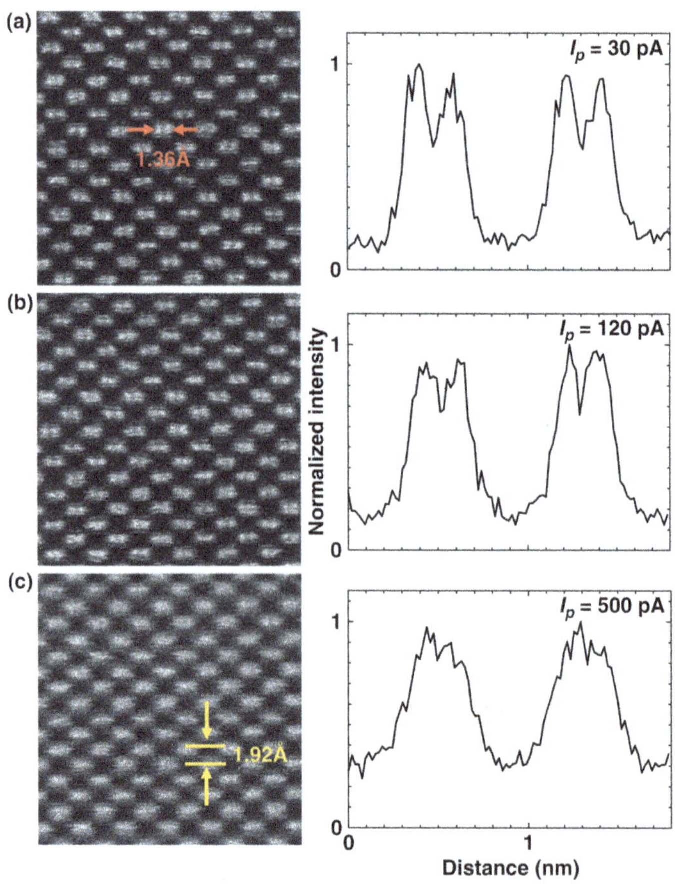

Figure 7–11. HAADF-STEM images of Si<110> recorded at (**a**) 30 pA, (**b**) 120 pA, and (**c**) 500 pA in an aberration-corrected 200-kV JEOL JEM-2200FS STEM with a Schottky FEG. Normalized intensity profiles extracted from the individual images are also inserted for comparison.

certain applications. More details about the probe formation can be found elsewhere (Watanabe and Sawada submitted).

7.3 Benefits of Aberration Correction for X-Ray Analysis

As discussed in the previous section, the incident probe dimension is significantly refined by the aberration correction in STEM. The refined incident probe dimension is generally considered to improve analytical performance. Obviously, improvements in spatial resolution for analysis are expected with the refined incident probe. Furthermore, more probe currents can be applied in a similar probe dimension after the aberration correction. Such increment of the probe current is always essential to improve analytical sensitivity (i.e., detectability limit), especially for X-ray analysis. Therefore, the analytical sensitivity may also be improved by aberration correction in STEM. In this section, the benefits of employing aberration-corrected STEM for X-ray analysis are discussed in terms of the spatial resolution and analytical sensitivity.

7.3.1 Spatial Resolution

The spatial resolution of X-ray analysis is mainly dependent on the incident probe dimension and its interaction with a thin specimen. If a crystalline thin specimen is oriented near a highly symmetric zone axis, the incident probe–specimen interaction can be localized near a single atomic column or surrounding neighbor columns (the so-called *electron channeling*). For details regarding the electron channeling, see Chapter 6. However, if the crystalline specimen is not oriented to such a highly symmetric zone axis or an amorphous thin specimen is used, the incident probe interaction with the specimen can be approximately estimated by Monte Carlo simulation.

Figure 7–12a shows electron trajectories in a 100-nm-thick pure Cu specimen at 200 kV with aberration-corrected probe simulated by a Monte Carlo code CASINO and Figure 7–12b shows electron trajectories in a 100-nm-thick pure Cu specimen at 200 kV with conventional incident probe simulated by the CASINO code (Drouin et al. 2007). In the simulation, the selected incident probe diameters are 0.4 and 1.2 nm for the aberration-corrected and conventional cases, respectively, which correspond to the $d(90\%)$ values in optimum probe-forming conditions at a current of 500 pA. The reason for employing $d(90\%)$ values over $d(59\%)$ values is that 90% of the incident electrons more accurately represents the spatial resolution of X-ray analysis since the X-ray can be generated everywhere within the electron–specimen interaction volume in a thin specimen. In the 100-nm-thick Cu specimen, the incident probes are spread significantly, relative to the incident probe diameters. More importantly, despite the difference in the incident probe diameter, the exit electron distributions (i.e., the spatial resolution of X-ray analysis) are very similar at a thickness of 100 nm in both probe-forming conditions. This similar exit probe distribution

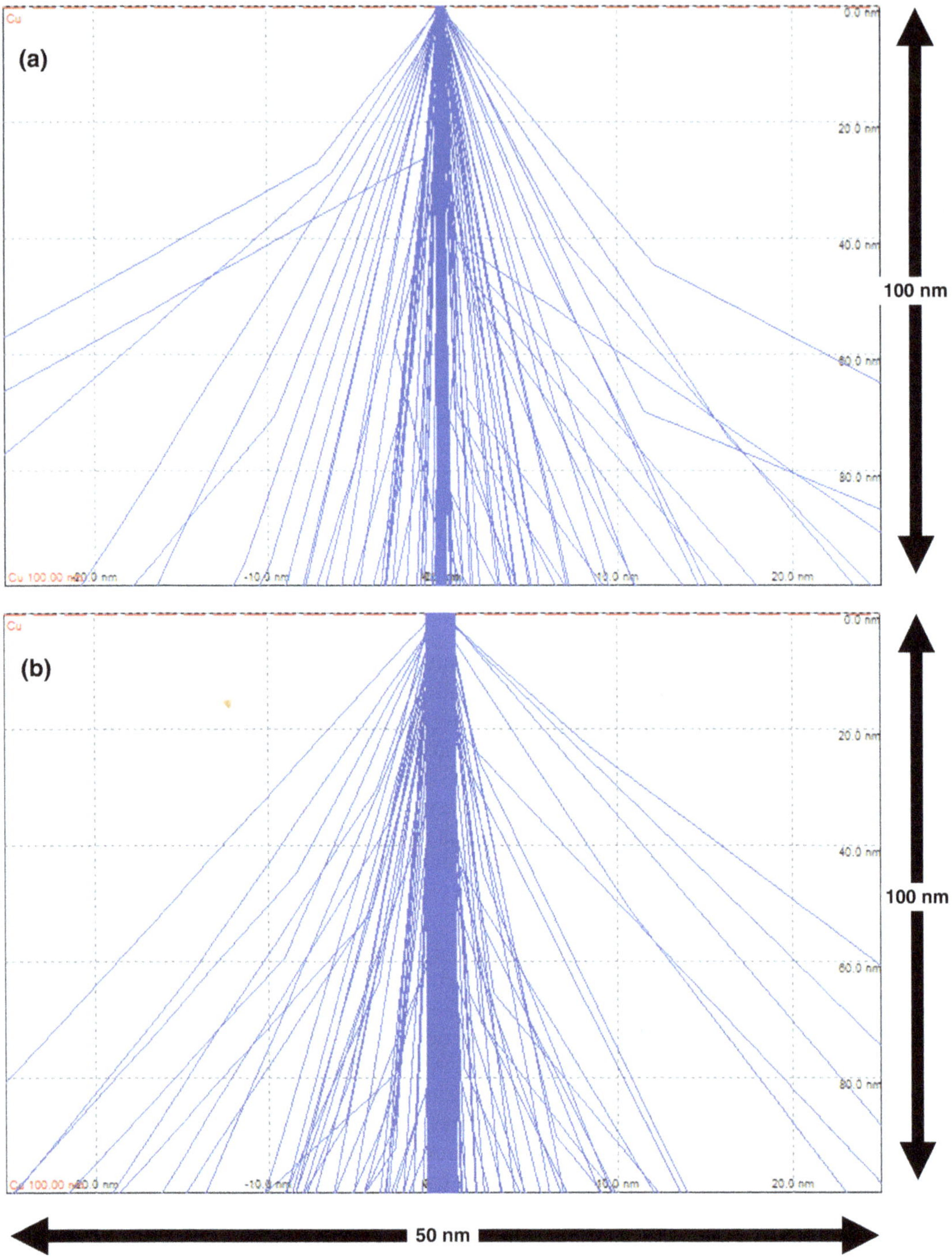

Figure 7–12. Electron trajectories in a 100-nm-thick pure Cu specimen at 200 kV with (**a**) aberration-corrected (0.4 nm) and (**b**) conventional (1.2 nm) incident probes, simulated by a Monte Carlo code CASINO (Drouin et al. 2007). Note that the incident probe sizes correspond to a 90% diameter of the total intensity instead of 59% in this simulation since an X-ray can be generated everywhere within the electron–specimen interaction volume in a thin specimen.

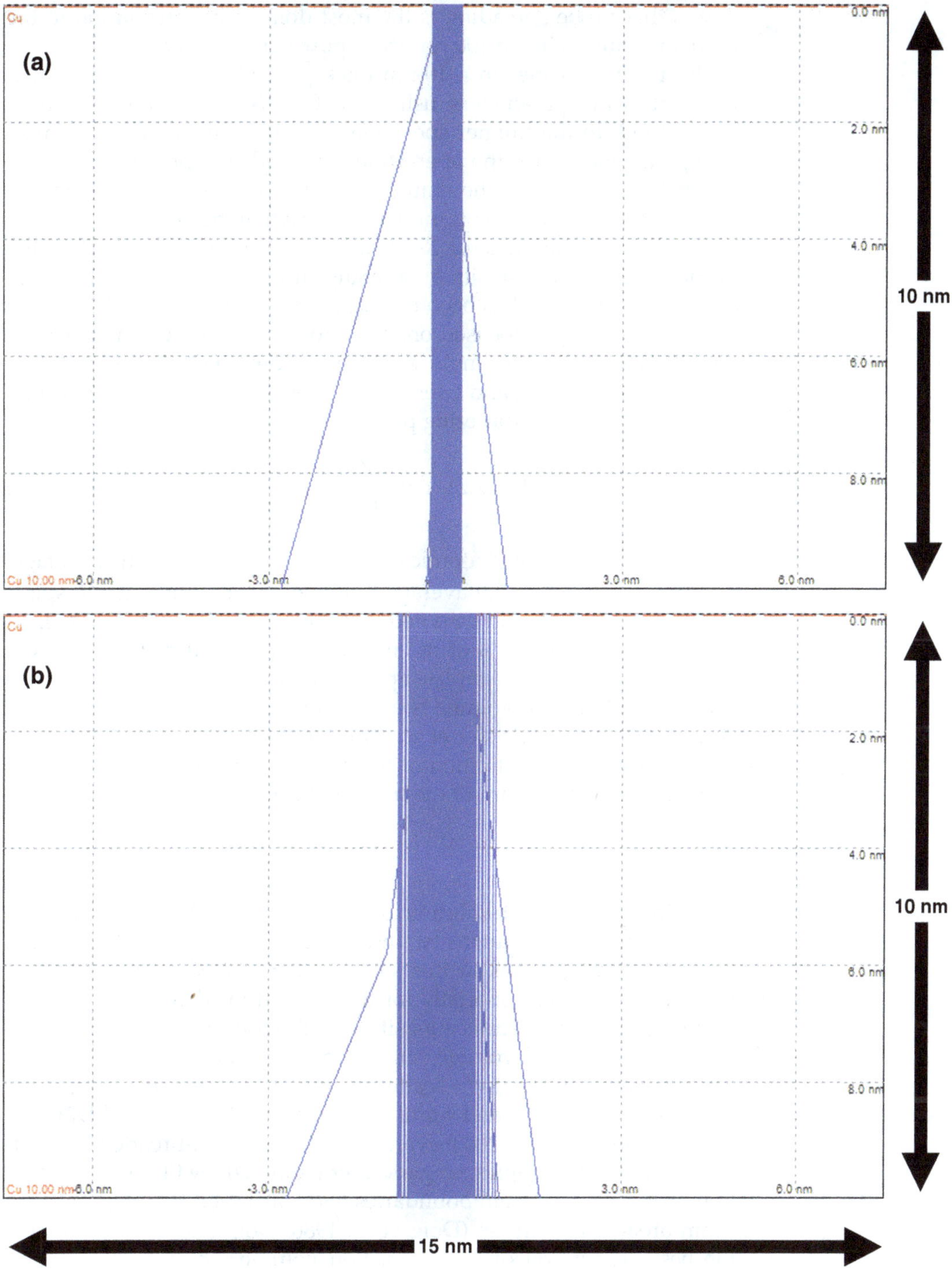

Figure 7–13. Electron trajectories in a 10-nm-thick pure Cu specimen at 200 kV in (**a**) aberration-corrected and (**b**) conventional conditions, as simulated by a Monte Carlo code CASINO (Drouin et al. 2007).

means that probe spreading is the most dominant contribution to the spatial resolution in the 100-nm-thick pure Cu specimen.

Electron trajectories in a 10-nm-thick pure Cu specimen simulated with the same parameters using the CASINO code are shown in Figure 7–13. In the thinner specimen, the probe broadening is almost negligible, and hence the aberration-corrected fine probe is the most beneficial in improving the spatial resolution of X-ray analysis since the incident probe dimension is more dominant than beam broadening.

The spatial resolution of X-ray analysis is strongly related to the incident probe size and its broadening in a thin specimen. The incident probe diameter has been discussed in detail previously. Here the probe broadening is focused on. The probe broadening can simply be described based on the single scattering model originally proposed by Goldstein et al. (1977) and later modified by Reed (1982). The modified version of the single scattering probe broadening is given as

$$b = 7.21 \times 10^5 \frac{\overline{Z}}{E_0} \left(\frac{\rho}{\overline{A}} \right)^{1/2} t^{3/2} \tag{16}$$

where $\overline{Z}$ is the averaged atomic number, E_0 is the accelerating voltage, ρ is the density, $\overline{A}$ is the averaged atomic weight, and t is the specimen thickness. It should be noted that the probe broadening described in Eq. (16) contains 90% of the incident electron intensity (Goldstein et al. 1977). One of the simplest approaches to describe incident probe–specimen interaction volume is assuming a truncated cone (Goldstein et al. 1990). Then Michael et al. (Michael et al. 1990, Williams et al. 1992) proposed the definition of the spatial resolution of analysis as a diameter at the midpoint of the truncated cone:

$$R = \frac{d(90\%) + (d(90\%)^2 + b^2)^{1/2}}{2} \tag{17}$$

Again, the spatial resolution defined in Eq. (17) contains 90% of the incident electron intensity, and the incident probe distribution is assumed homogeneous in this simplified truncated model, which is obviously incorrect, especially for the focused incident probe. In spite of neglecting the electron intensity distribution in the incident beam or in the interaction volume, this simple truncated cone approach in combination with the single scattering model provides good agreement between experimental and simulated results (Williams et al. 1992).

Keast and Williams (2000) evaluated various beam-broadening models (including the simplest single scattering model) by fitting to segregation profiles across grain boundaries and concluded that the Gaussian beam-broadening model (Doig et al. 1980, Doig and Flewitt 1982) is the best description of the interaction volume. The Gaussian beam-broadening model describes the electron intensity distribution at any depth in the specimen, which is given as (Doig et al. 1980, Doig and Flewitt 1982)

$$I(r,t) = \frac{i_p}{2\pi(\sigma^2 + \beta t^3/2)} \exp\left[-\frac{r^2}{2(\sigma^2 + \beta t^3/2)} \right] \tag{18}$$

where i_p is the total electron current in the incident beam, r is the radial distance from the center of the beam, and σ and β are terms associated with the incident probe size and its broadening, respectively, and are given as

$$\sigma = d(90\%)/4.29, \quad \beta = 500\left(\frac{4\bar{Z}}{E_0}\right)^2\left(\frac{\rho}{\bar{A}}\right). \tag{19}$$

The spatial resolution based on the Gaussian beam-broadening model was proposed by Van Cappellan and Schmitz (1992):

$$R = q\left[\sigma^2 + \beta(\kappa t)^3/2\right]^{1/2} \tag{20}$$

where q and κ are parameters which depend on the given fraction of the incident intensity: $q = 4.29$ and $\kappa = 0.68$ for the spatial resolution that contains 90% of the incident intensity (Van Cappellan and Schmitz 1992). Since a value of $q(\sigma^2 + \beta t^3/2)^{1/2}$ corresponds to the diameter at the exit surface of the probe from the specimen, for a given fraction of the total intensity, it follows that R for 90% of the incident intensity is the diameter at a depth of $0.68t$. In comparison with the simple truncated cone model, the diameter for this definition is located at a slightly deeper position.

Figure 7–14 shows the spatial resolution (90%) calculated for a Cu thin specimen in the conventional and aberration-corrected conditions at 200 kV based on the Gaussian probe-broadening model (Eq. 20), which is plotted against the specimen thickness. Again, the incident probe diameters in both conditions contain 90% of the total intensity as shown in Figure 7–14. Obviously, a finer probe size in the aberration-corrected condition provides better spatial resolution. In the aberration-corrected condition, the spatial resolution can be kept below 1 nm for specimen thicknesses up to ~30 nm in pure Cu. More importantly, the spatial resolution of X-ray analysis can reach atomic dimensions below thicknesses of 20 nm, which makes X-ray analysis competitive with EELS in terms of spatial resolution (Batson 1995,

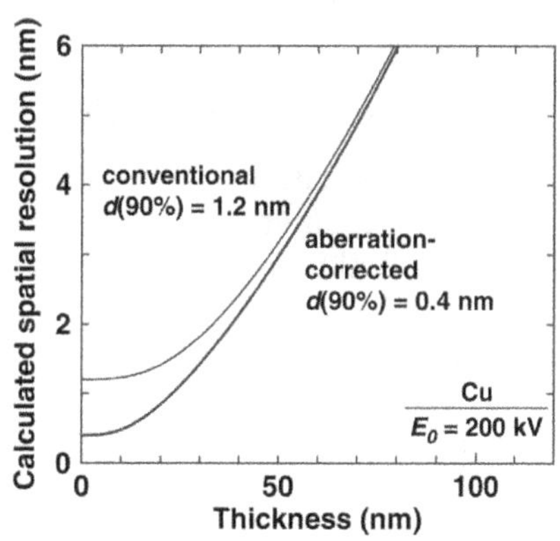

Figure 7–14. Spatial resolution (90%) calculated for a Cu thin specimen in conventional and aberration-corrected conditions at 200 kV based on the Gaussian probe-broadening model, as plotted against the specimen thickness.

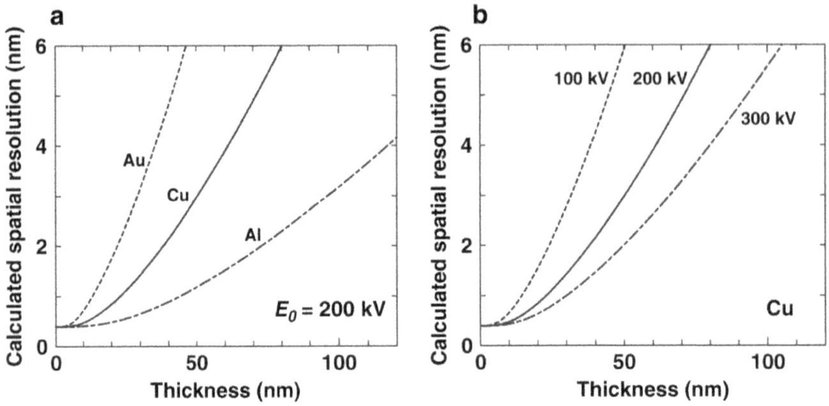

Figure 7–15. (**a**) Spatial resolution (90%) calculated for various materials (Al, Cu, and Au) in the aberration-corrected condition at 200 kV, as plotted against the specimen thickness. (**b**) The incident electron energy dependence of spatial resolution (90%) calculated for a Cu thin film, as plotted against the specimen thickness.

Browning and Pennycook 1995, Varela et al. 2005, Kimoto et al. 2007, Bosman et al. 2007, Muller et al. 2008, Varela et al. 2009).

The spatial resolutions calculated for different materials (Al, Cu, and Au) at 200 kV in the aberration-corrected condition are shown in Figure 7–15a. As described with the β term in Eq. (19), the probe broadening (and hence spatial resolution) is strongly dependent on materials, i.e., the heavier the specimen is, the more significant the incident probe is spread in the specimen due to scattering. Maximum specimen thicknesses, which maintain the spatial resolution under 1 nm, are ~40, ~30, and ~15 nm for Al, Cu, and Au at 200 kV in the aberration-corrected condition, respectively. The incident electron energy dependence of the spatial resolution is calculated for Cu in the aberration-corrected condition (Figure 7–15b). To achieve the highest spatial resolution, the higher energy is more beneficial due to the reduction of scattering. However, as long as the specimen thickness is sufficiently thin (<20 nm), the spatial resolution can be maintained within atomic dimensions. The specimen preparation can be the major limitation for ultimate analytical work.

7.3.2 Analytical Sensitivity: Minimum Mass Fraction (MMF)

Analytical sensitivity in terms of the detectability limit is another important parameter to improve in any analytical applications such as XEDS and EELS in STEM. There are mainly two definitions to describe the detectability limits in the literature: (1) the minimum mass fraction (MMF) and (2) the minimum detectable mass (MDM). The first study that addressed the issue of detection limits of a thin specimen in transmission electron microscopy was carried out by Joy and Maher (1977), who first distinguished between MMF and MDM in the paper. They described the MMF as the sensitivity of the system to identify unknown elements when they are in a strongly scattering matrix, e.g., a precipitate in a matrix. Under these conditions, the matrix bremsstrahlung governs the detection limit of the signal. The MMF was thus defined as

$$\text{MMF} = \frac{2\sqrt{(Q_\text{B}/E_\text{V})\Delta E}}{\text{P}_\text{Z}\sqrt{J\tau M_\text{B}}} \tag{21}$$

where Q_B is the bremsstrahlung ionization cross section, M_B is the mass of the matrix element (s), E_V is the energy of the characteristic peak, ΔE is the peak integration width, Pz is the count rate detected/incident electron/cm^2, τ is the counting time, and J is the incident electron current density in A/cm^2 or electrons/cm^2/s. The MMF measured in terms of a wt% can be determined from any analyzed volume of the specimen without the need to determine the exact dimensions of the probe–specimen interaction volume. To improve the MMF value, Eq. (21) implies that it is best to work in portions of the spectrum where the bremsstrahlung intensity is the lowest and the characteristic peak intensity is the highest (i.e., the highest possible peak-to-background ratio (P/B)). In fact, the study by Joy and Maher (1977) comes to the same conclusion as earlier work by Ziebold (1967), who derived a general expression for the detection limit for any analytical system with a given P/B. In the approach by Ziebold, the MMF can be expressed in essence as (Ziebold 1967)

$$\text{MMF} \propto \frac{1}{\sqrt{P(P/B)\tau}} \tag{22}$$

The relationship between the experimental parameters in Ziebold's generalized approach and the predictions of Joy and Maher's equations for AEM is clear. As mentioned above, the MMF is defined as a composition such as wt% or at%. Thus the MMF values can be determined experimentally by means of an extension of quantification procedures. In X-ray analysis for thin specimens, the MMF values can be obtained based either on the Cliff–Lorimer ratio method (Cliff and Lorimer 1975) as originally proposed by Romig and Goldstein (1979) and modified by Michael (1987) (Eq. (24)) or on the ζ-factor method (Watanabe and Williams 2006).

The best analytical sensitivity achieved in the uncorrected instrument at 300 kV was the detection of ~0.12 wt% Mn in a 15-nm-thick specimen of a Cu–Mn alloy, which corresponds to 2–3 Mn atoms in the analysis volume (Watanabe and Williams 1999). These experimental results on the detectability measurements are shown as closed circles in Figure 7–16. In this plot, the vertical axis is the ratio of the measured Mn Kα characteristic peak X-ray intensity to the minimum-required intensity for the detection of the peak, which is estimated from the corresponding background intensity (Romig and Goldstein 1979). Thus if the ratio exceeds one (drawn as a dashed line), the peak intensity is detectable with a 99% confidence limit (3σ) in this case. The details of this plot can be found in a previous paper (Watanabe and Williams 1999).

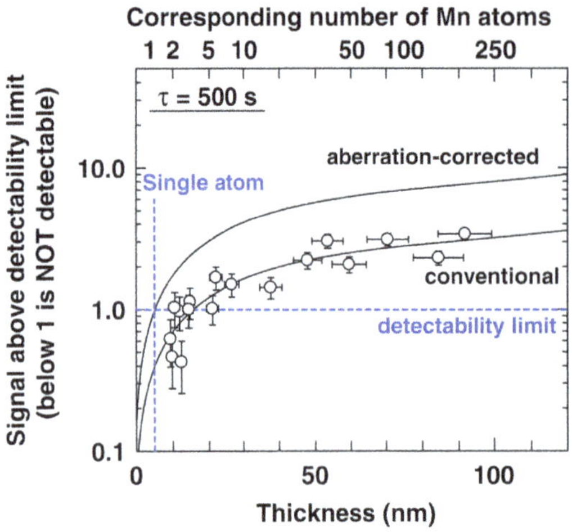

Figure 7–16. The detection limit of Mn atoms in a thin foil of Cu–0.12 wt% Mn solid solution, as experimentally obtained in the conventional HB 603-dedicated STEM at 300 kV indicated as closed circles and modified from Watanabe and Williams (1999). Under these conditions, ~ 2 Mn atoms are detectable in a 15-nm-thick region. The solid line indicates the detection limit calculated with the same probe diameter and a 10× higher probe current in the aberration-corrected condition. The detection limit (MMF) using the aberration-corrected probe can be ~3 times better than the conventional probe for the same acquisition conditions. Reproduced from Watanabe and Williams (2006) with permission.

In the aberration-corrected condition, similar peak intensities can be gathered since the probe current is not degraded by the aberration correction despite the fact that the incident probe is much more refined. In other words, the MMF is sensitive to any factors to improve the peak intensity (such as the probe current and the X-ray collection angle) and the peak-to-background ratio (e.g., the incident electron energy) but is insensitive to the incident probe dimensions and the probe–specimen interaction volume. Therefore, the MMF in the aberration-corrected condition should remain at similar levels to the conventional uncorrected condition unless *P/B* is degraded due to the installation of the aberration corrector into the column. Strictly speaking, the MMF would be slightly degraded due to the improved interaction volume (simply lack of analyzed mass). However, the degradation should be marginal as long as similar levels of probe current are applied (Watanabe and Williams 2006).

Conversely, more probe current can be added into the aberration-corrected probe in order to achieve the same probe dimension as that in the conventional uncorrected condition. In the aberration-corrected condition, at least a 10 times higher current can be applied to obtain the same probe dimension as in the conventional condition (Watanabe and Williams 2006). The MMF calculated with the 10 times higher current is plotted as a solid line in Figure 7–16. Clearly, the MMF is improved approx. three times or better with the same acquisition parameters in the aberration-corrected condition. Note that no conventional Si(Li) XEDS detector can deal with the higher X-ray intensity generated with the 10 times higher current because of the limit in throughput in X-ray acquisition (at most a few kcps). However, this situation can be offset by employing a recently developed silicon drift detector (SDD) (Bertuccio et al. 1992, Strüder et al. 1998) with which X-rays can be gathered at count rates > 100 kcps. The latest generation SDDs can be used for high-kilovolt STEMs (e.g., Zaluzec et al. 2003). Therefore, SDDs would be more ideal for X-ray analysis with higher probe currents in the aberration-corrected STEMs.

7.3.3 Analytical Sensitivity: Minimum Detectable Mass (MDM)

Another expression of the analytical sensitivity is the MDM. According to Joy and Maher (1977), the MDM is defined as the sensitivity of the system to identify unknown elements when they are "free" or in a weakly scattering matrix, e.g., a particle on a substrate or heavy atoms in low atomic number (Z) thin sections (i.e., it is reasonable to ignore the bremsstrahlung contribution to the spectrum from the matrix). Specifically, the MDM is given as (Joy and Maher 1977)

$$\mathrm{MDM} = (\mathrm{Pz}\,\tau J)^{-1} \tag{23}$$

This definition of the MDM is a convenient way to determine how many atoms are detected in the analysis volume, i.e., minimum detectable atoms (MDAs). However, it is necessary to measure the dimensions of the analyzed volume of material (Watanabe and Williams 1999). Determining the analyzed volume depends on knowledge of the specimen thickness and density, the probe size, and the amount of probe spreading through the specimen thickness. Due to difficulties in determining the probe diameter and the specimen thickness to estimate the probe spreading, the MDM or the MDA is rarely employed. Therefore, the MMF has been more widely used than the MDM, even though the MDM or the MDA of thin specimens in analytical STEM can be better than any other analysis technique except for an atom probe tomography approach, as discussed below.

The MDA values can be determined by translating from the MMF with information about the analyzed volume. Here, the MDA values for X-ray analysis are estimated in conventional and aberration-corrected instruments. First, the MMF values need to be determined, which can be performed by X-ray spectrum simulation in an NIST/NIH desktop spectrum analyzer (DTSA) (Fiori et al. 1992, Fiori and Swyt 1994, Newbury et al. 1995). The detailed procedures for MMF determination with X-ray spectrum simulation can be found in the literature (e.g., Fiori and Swyt 1989, Papworth et al. 2001, Watanabe and Williams 2006). In this study, the spectrum simulation was performed using an instrument operated at 200 kV with the XEDS detector collection angle (Ω) of 0.15 sr and a probe current (I_{p}) of 0.5 nA for the spectrum acquisition time (τ) of 100 s. The MMF values of Mn in Cu determined by X-ray spectrum simulation are plotted against the specimen thickness in Figure 7–17a. The MMF values were calculated from simulated X-ray spectra of Cu–5 wt% Mn using the Goldstein–Romig–Michael (GRM) equation (Romig and Goldstein 1979, Michael 1987):

$$\mathrm{MMF}_{\mathrm{A}} = \frac{3\sqrt{2B_{\mathrm{A}}}}{P_{\mathrm{A}} - B_{\mathrm{A}}} C_{\mathrm{A}} \tag{24}$$

where P_{A} and B_{A} are the peak and background intensities of element A in the X-ray spectrum, respectively, and C_{A} is the composition of element A in wt%. The error bars for the MMF values represent the 99% confidence limit (3σ). As shown in Figure 7–17a, the MMF value is

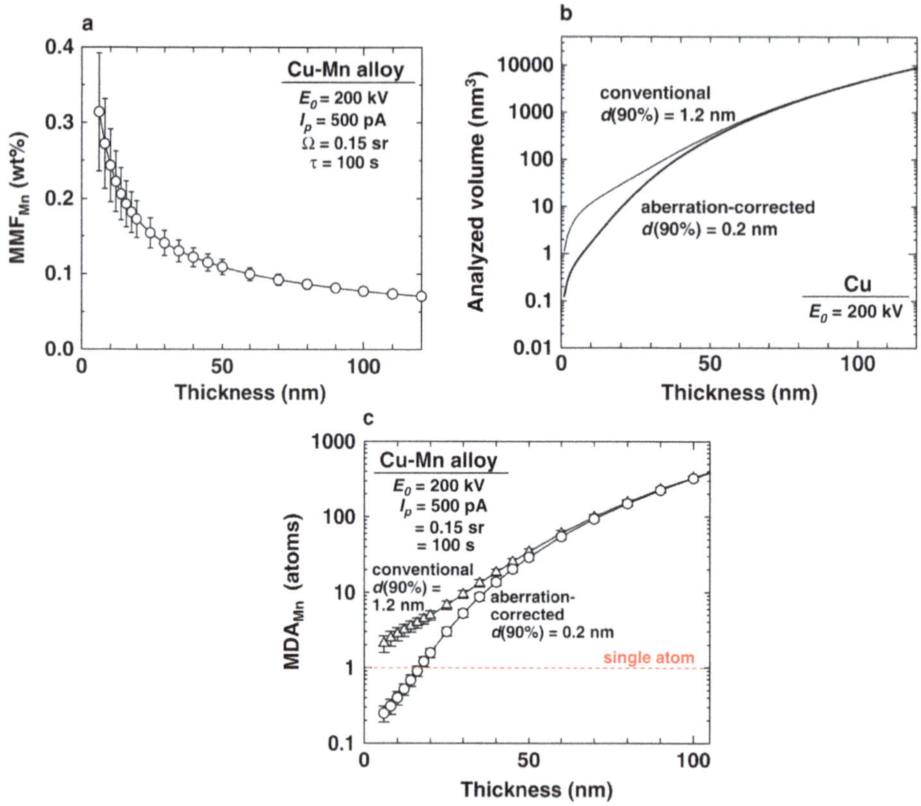

Figure 7–17. (**a**) The MMF of Mn in Cu determined by Eq. (24) with X-ray spectra of Cu–5 wt% Mn simulated by DTSA, as plotted against the specimen thickness. The spectrum simulation was performed for conditions of the operating voltage (E_0) of 200 kV with the XEDS detector collection angle (Ω) of 0.15 sr and a probe current (I_p) of 0.5 nA for the spectrum acquisition time (τ) of 100 s. The error bars of the MMF values represent the 99% confidence limit (3σ). (**b**) The analyzed volume of Cu determined using Eq. (25) for a conventional instrument with $d(90\%) = 1.2$ nm and an aberration-corrected instrument with $d(90\%) = 0.4$ nm (both operated at 200 kV), as plotted against the specimen thickness. (**c**) The MDA of Mn in Cu calculated from the MMF in (**a**) in combination with analyzed volume sizes in (**b**).

degraded as the specimen thickness is reduced simply because the net intensity from Mn decreases at a thinner region.

The MMF value needs to be translated to the MDM value in combination with the analyzed volume. In the Gaussian probe-broadening model, the analyzed volume can be given as (Van Cappellan and Schmitz 1992, Watanabe and Williams 2006)

$$V = \int_0^t \pi \left[q(\sigma^2 + \beta z^3/2)^{1/2}/2 \right]^2 \mathrm{d}z = \frac{\pi q^2}{32}\left(8\sigma^2 t + \beta t^4\right) \quad (25)$$

Figure 7–17b shows the analyzed volume of Cu using Eq. (25) for a conventional instrument (probe size, $d(90\%) = 1.2$ nm) and an aberration-corrected instrument ($d(90\%) = 0.4$ nm) both operated at 200 kV, as plotted against the specimen thickness. Obviously, the analyzed volume is significantly reduced in the aberration-corrected condition due to the smaller probe. However, this incentive to use the aberration-corrected instrument is degraded when specimens thicker than ~50 nm are analyzed. Again, thinner specimens are essential for analysis in the aberration-corrected STEM.

The analyzed volume can be transformed to the total number of atoms with basic crystallographic information. For example, Cu has an fcc structure with four atoms (i.e., the unit volume per atom can be given by $a^3/4$ (a, lattice parameter)). Then by multiplying the atomic fraction of the MMF value, the MDA can be determined. Note that the MMF value (in wt%) should be converted to the atomic fraction. The MDA of Mn in Cu translated from the MMF in both conditions is shown in Figure 7–17c as a function of the specimen thickness. Despite the degraded results of the MMF in thinner regions, the MDA value is improved as the specimen thickness decreases. According to these results, ~10 Mn atoms can be detected at 30 nm using the conventional instrument. The MDA value in the conventional instrument can be improved to a few atoms with a specimen thickness less than 20 nm. Conversely, in the aberration-corrected instrument, the MDA can be significantly improved with the smaller analyzed volume as schematically shown in Figure 7–18. This improved MDA implies that the aberration correction can improve not only the spatial resolution but also the analytical sensitivity with the refined probe dimension. In fact, the MDA may reach single-atom detection range if a specimen thinner than 20 nm is analyzed. *Single-atom detection may no longer be a dream in X-ray analysis as long as the aberration-corrected STEM instrument is properly used!* Obviously, the MDA value might be further improved if longer acquisition times are set or X-ray collection efficiency is improved with better collection capabilities (e.g., a larger collection angle). However, in the MDA, the superior advantage to be gained by using the aberration-corrected STEM instruments is simply negated if a specimen thicker than 50 nm is analyzed, as shown in Figure 7–17c. For more detailed procedures to determine MMF and MDA (or MDM), see articles by Papworth et al. (2001) and Watanabe and Williams (1999).

As a summary of improvements in the spatial resolution and the analytical sensitivity due to the aberration correction, Figure 7–19

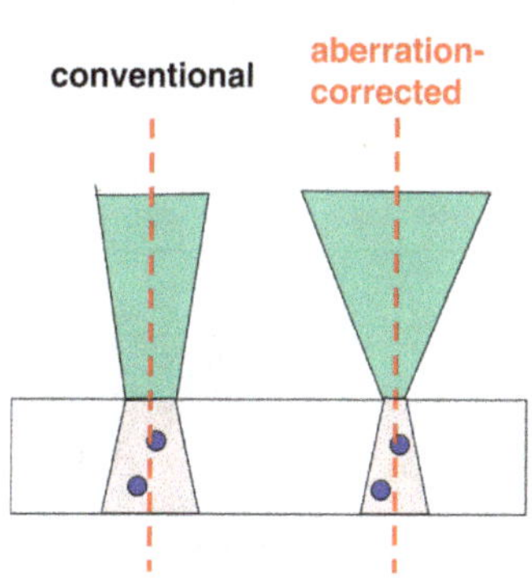

Figure 7–18. A schematic diagram to show improvement of the analyzed volume and hence MDA by the aberration-corrected refined probe over the conventional probe.

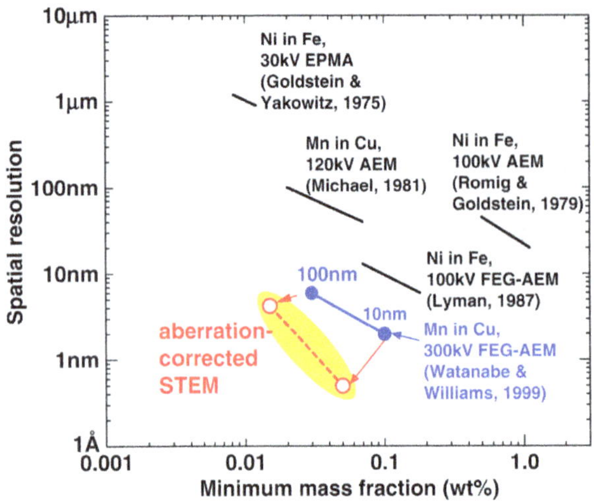

Figure 7–19. A summary of the X-ray analysis performance of several electron probe instruments with respect to the spatial resolution and the MMF, modified from Williams et al. (2002). The expected performances for X-ray analysis in the aberration-corrected STEM are added as a shadowed area, reproduced with permission.

shows a plot describing the relationship between the spatial resolution and the MMF of X-ray analysis in analytical STEM, modified from previous reports (Watanabe and Williams 2006, Williams et al. 2002). Using the aberration-corrected instruments, the spatial resolution may ultimately reach 0.4–0.5 nm for a 10-nm-thick Cu specimen with a current of 0.5 nA. The MMF can also be improved due to enhanced characteristic peak intensities because a 10 times higher probe current may be obtained in the aberration-corrected instrument while keeping the same probe size as in the uncorrected conventional STEM instrument. This improvement by the aberration correction suggests that the MDM in terms of the number of atoms detected can also be reduced by factors of 2–5 and is expected to approach or pass the single-atom level, as discussed below.

7.4 Practical X-Ray Analysis Procedures

7.4.1 X-Ray Mapping

In STEM, it has been standard practice for almost three decades to analyze individual points or measure composition changes along line profiles across specific regions of interest such as boundaries and interfaces, as summarized in Figure 7–20. However, such an approach is highly

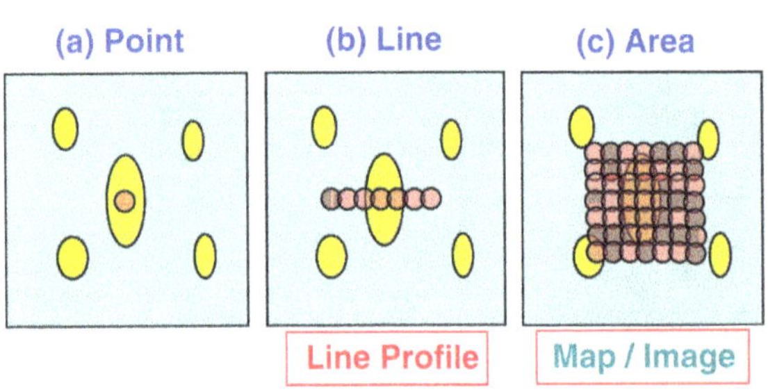

Figure 7–20. A summary of analysis modes in STEM: (**a**) point analysis, (**b**) line analysis for profiling, and (**c**) area analysis for mapping.

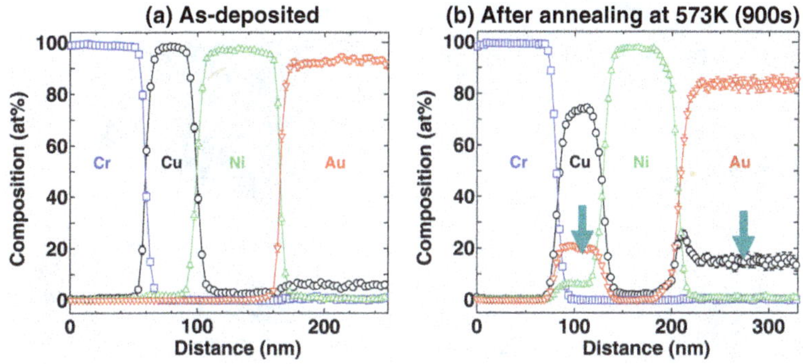

Figure 7–21. Composition line profiles of Cr, Cu, Ni, and Au determined from a Cr/Cu/Ni/Au multilayer thin specimen. (**a**) As-fabricated and (**b**) after annealing at 573 K for 900 s. Reproduced from Danylenko et al. (2003) with permission.

selective and does not reveal all the available information, since it may easily miss local two-dimensional fluctuations in composition around nano-scale features such as fine precipitates and interfaces/boundaries. For example, two composition line profiles taken from a Cr–Cu-Ni-Au multilayer thin film by X-ray analysis are compared in Figure 7–21 (a: as-fabricated and b: after annealing at 573 K for 300 s) (Danylenko et al. 2003). It should be noted that the compositions were determined by the ζ-factor method (see Section 7.4.4). The profile from the as-fabricated multilayer specimen shows that each layer suffers slight interdiffusion during the fabrication. Conversely, in the profile from the annealed film, compositions of both Cu and Au are reduced in their own layers, and $\sim$20% of Cu and Au can be seen in the Au and Cu layers, respectively. However, the concentrations of both Cu and Au are much smaller in the Ni layer. Therefore, according to the line profile, both Cu and Au seem to diffuse into one another through the Ni layer. The diffusion mechanisms of Cu and Au through the Ni layer remain unclear from the profile.

Elemental/compositional information can also be gathered by applying X-ray mapping in STEM, as shown in Figure 7–20c. Unfortunately, the X-ray mapping approach requires a longer acquisition time since X-ray maps need to be recorded sequentially at individual pixels for a sufficiently longer dwell time due to the poor efficiency of X-ray generation and detection. In Figure 7–22, composition maps of Cu (b), Ni (c), and Au (d) obtained by the XEDS approach are compared with an ADF-STEM image (a) from the annealed Cr–Cu–Ni–Au multilayer film (Danylenko et al. 2003). As shown in the Cu and Au maps, enrichments of Cu and Au are confirmed in the Au and Cu layers, respectively, and the concentrations of both Cu and Au are much lower in the Ni layer, which agrees well with the line profile shown in Figure 7–21b. However, both the Au distribution in the Cu layer and the Cu distribution in the Au layer are not homogeneous. From the Ni map, several Ni depletion regions can be seen in the Ni layer, which correspond to the grain boundaries of Ni in the ADF-STEM image. As shown in the

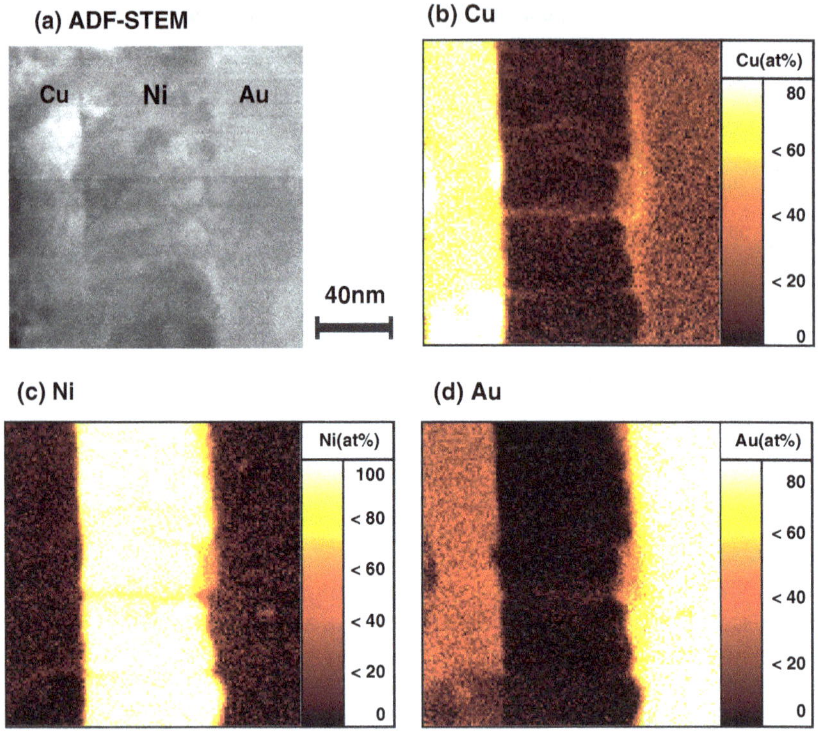

Figure 7–22 (**a**) ADF-STEM image, (**b**) Cu composition map, (**c**) Ni composition map, and (**d**) Au composition map taken from the Cr/Cu/Ni/Au multilayer thin specimen annealed at 573 K for 900 s. Reproduced from Danylenko et al. (2003) with permission.

Cu and Au maps, both Cu and Au are enriched along the same grain boundaries in the Ni layer. Thus, both Cu and Au penetrate through the Ni layer by grain boundary diffusion instead of bulk diffusion. The Cu and Au diffusion through the grain boundaries in the Ni layer cannot be identified without the use of X-ray mapping. Therefore, 2D-mapping is essential to analyze such small features and measure all local fluctuations in composition in an unbiased manner. The X-ray mapping approach is more efficiently applicable in aberration-corrected STEM since more probe current can be added into the aberration-corrected refined probe, as discussed in the previous section. Furthermore, the poor analytical sensitivity in X-ray mapping would be improved by applying spectrum imaging (SI) in combination with advanced statistical approaches, such as multivariate statistical analysis (MSA), as described below.

7.4.2 Spectrum Imaging (SI)

Conventionally, X-ray maps were recorded by setting specific energy windows for the element of interest. This approach is now obsolete due to the employment of spectrum imaging (SI), which stores a full spectrum at each pixel as schematically shown in Figure 7–23

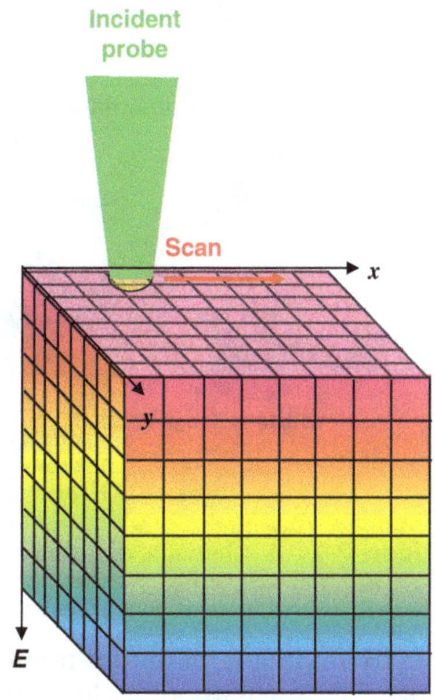

Figure 7–23 A schematic diagram of spectrum imaging acquisition in STEM, reproduced from Williams and Carter (2009) with permission.

(Jeanguillaume and Colliex 1989, Hunt and Williams 1991). The SI acquisition procedure was originally developed for the STEM–EELS approach to expand and improve conventional elemental mapping. In SI, a full spectrum is continuously recorded at each pixel. Thus, information regarding elements contained in the SI data set may not be missed even without prior knowledge of the presence of the elements; now the SI method is available not only for STEM-based EELS and XEDS but also for energy-filtering transmission electron microscopy (EFTEM).

In addition, the SI method offers post-acquisition treatments of elemental maps including regular spectral-processing methods such as background subtraction and signal deconvolution. Therefore, it is possible to map out unexpected minor elements that are not even considered for mapping beforehand if the signals from such minor elements are successfully identified. However, for characterization of elemental fluctuations in atomic-resolution elemental mapping (including SI in (S)TEM) suffers from weak signals. Furthermore, there might be many variables even in a single SI data set: some expected and others totally unexpected. It is essential to identify those variables in the data set by employing multivariate statistical analysis (MSA).

7.4.3 Multivariate Statistical Analysis (MSA)

The MSA approach is a useful family of statistics-based techniques to analyze large data sets. Principal component analysis (PCA) is one of the most popular MSA approaches and is also widely performed as a first step to other more advanced MSA approaches. The general concept of PCA is to reduce the dimensionality of an original large data

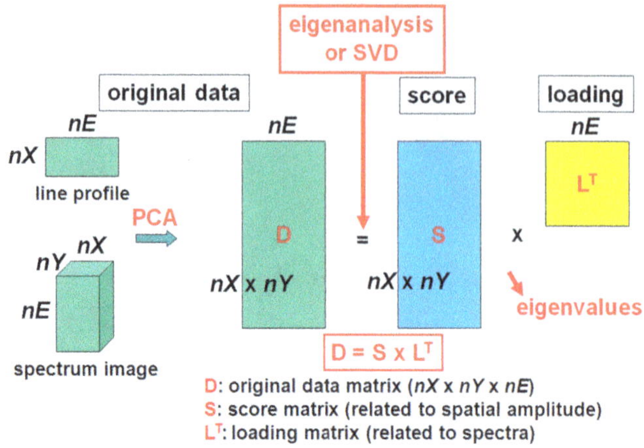

Figure 7–24 A schematic diagram of PCA data decomposition.

set by finding a minimum number of variables that describe the original data set without losing any significant information (e.g., Jolliffe 2002, Malinowski 2002). In other words, systematic deviations from the average of the dataset (such as elemental segregation to GBs) are emphasized. Since the first application to EELS by Trebbia and Bonnet (1990), PCA has been applied to segregation studies measured by line scans (e.g., Titchmarsh and Dumbill 1996, Titchmarsh 1999). This approach is useful for large complicated data such as XEDS or EELS SI data sets, as successfully demonstrated by Kotula et al. (2003).

An SI data matrix $\mathbf{D}_{((x,y),E)}$ acquired with the spatial dimensions (x, y) and the spectral (energy-channel) dimension E can be decomposed by applying PCA, as summarized in Figure 7–24:

$$\mathbf{D}_{((x,y),E)} = \mathbf{S}_{((x,y),n)} \times \mathbf{L}^{\mathrm{T}}_{(E,n)} \tag{26}$$

where $\mathbf{L}_{(E,n)}$ and $\mathbf{S}_{((x,y),\,n)}$ are called loading and score matrices, respectively. In practice, the above decomposition of a data matrix can be performed by eigenanalysis or singular-value decomposition (Jolliffe 2002, Malinowski 2002), with the singular values being equivalent to the square root of the eigenvalues. After the decomposition, each row of **L** contains a spectral feature uncorrelated to other row information, and each row of **S** represents the spatial amplitude of the corresponding loading spectrum. The superscript T of **L** indicates a matrix transpose. The individual product of each row of the loading and score matrices is called a component, and the number of the components n is mathematically equivalent to the rank of the data matrix **D**, which is equal to or less than the smaller of the numbers $(x \times y)$ and E. This decomposition process provides decomposed matrices **L** and **S** as well as eigenvalues of the data matrix. The relative magnitude of each eigenvalue indicates the amount of variance that the corresponding component contributes to the data set. In the decomposed matrices, the components are ordered from high to low according to their eigenvalues as well as the variance or information they describe.

Before decomposing the data matrix, additional mathematical pre-processing of the data sets can be useful to minimize mixing of the various spectral features in the loading spectra. The most common pre-processing routines are centering (subtracting the average values or constant values from the data set) and scaling (dividing the dataset by a constant value or a weighted average, also called weighting), and these routines can be applied to one of the data matrix directions (spatially or spectrally) or both directions (Bro and Smilde 2003). The effectiveness of such preprocessing for PCA again depends largely on the type of data set. Keenan and Kotula (2004) discussed efficient preprocessing for time-of-flight secondary ion mass spectrometry SI data sets. They concluded that the best preprocessing is a two-way scaling or weighting based on Poisson counting statistics. The justification of this method is that the experimental noise varies with the detector channel and the pixel number, e.g. the experimental noise in a minor spectral feature is less than in the energy channels of a peak with high intensity. Weighting will re-organize the data in order to normalize the distribution of the experimental noise. In this study, PCA was performed with prior weighting, as implemented with the data transformation method proposed by Cochran and Horne (1977).

Dominant features of data stored in the loading and score matrices after the decomposition are called principal components (PCs), and typically the number of the PCs is far less than the rank of the data matrix n. The number of PCs can be determined by evaluating the magnitudes of the eigenvalues. One of the most common approaches is to use the scree plot (an example for an X-ray SI data set taken from an irradiated low-alloy steel is shown in Figure 7–25 (Burke et al. 2006), which is a logarithmic plot of the eigenvalues of corresponding components against the index of the components. The magnitude of each eigenvalue indicates the amount of variance that the corresponding component contributes to the data set, i.e., the scree plot can be considered as a histogram representing the frequency. Therefore, if the eigenvalue is high, the corresponding component should be statistically significant (i.e., the corresponding component is repeated more frequently in the data set). Conversely, lower eigenvalues indicate that the corresponding components are not repeated in the data, i.e., random noise. Usually such random noise components appear as a plateau in the scree plot. In the particular scree plot shown in Figure 7–25, at least the first six components can be easily distinguished from the noise components that are expressed as a straight line in the higher index component side. The #7 component must be evaluated as to whether or not it is a PC.

For example, the results of PCA decomposition on an XEDS SI data set taken from an irradiated low-alloy steel are shown in Figure 7–26 (Burke et al. 2006). Note that the scree plot of this data set is shown in Figure 7–25. This particular XEDS SI data set was taken from data recorded with 128 × 128 pixels and 1024 energy channels for a dwell time of 400 ms in a VG HB 603-dedicated STEM operated at 300 kV. In Figure 7–26, selected pairs of loading spectra and corresponding score images of the components in this SI (a: #1, b: #2, c: #5 and d: #7) are shown. The most significant component in the data set is always

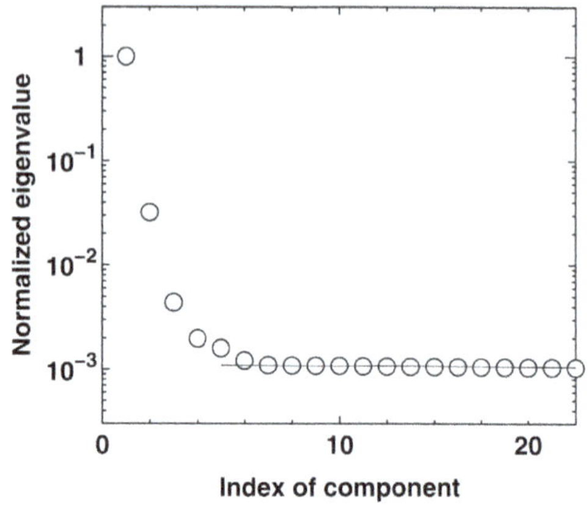

Figure 7–25 Normalized eigenvalues of the X-ray SI data set shown in Figure 7–28 taken from an irradiated low-alloy steel plotted against the component determined by PCA (scree plot). Reproduced from Burke et al. (2006) with permission.

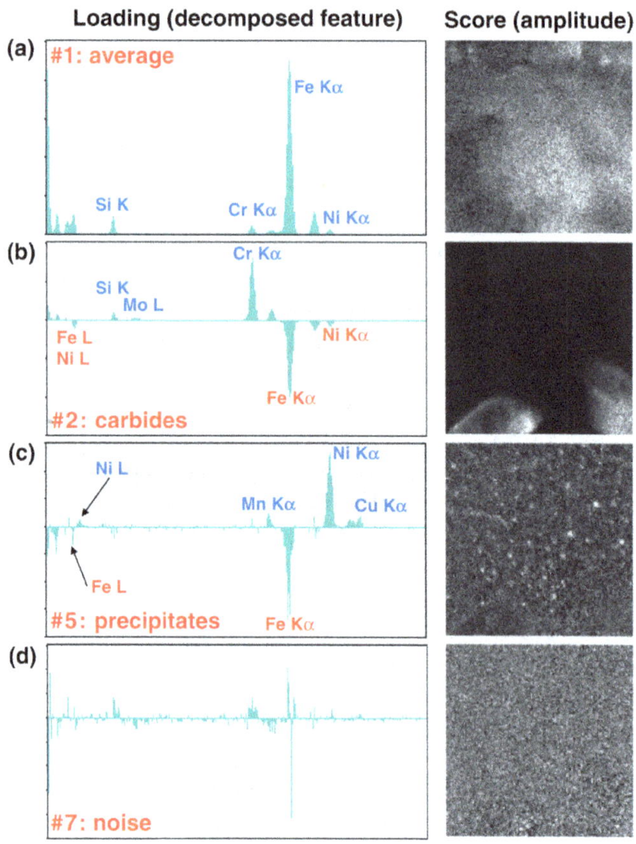

Figure 7–26. Selected pairs of the loading spectra and corresponding score images of the components for the X-ray SI data set (Figure 7–28) (**a**: #1, **b**: #2, **c**: #5 and **d**: #7), reproduced from Burke et al. (2006) with permission.

the average, and hence the loading spectrum of the #1 component (Figure 7–26a) represents the average spectrum of the X-ray SI. Any component higher than #1 indicates a significant difference from the average. Therefore, the loading spectra after the #1 component contain positive and negative peaks, which are not physically meaningful but interpretable expressions. The #2 loading spectrum has positive Cr Kα and negative Fe Kα and Ni Kα peaks. The bright regions in the corresponding score image enhance this spectral feature, and hence this component represents the carbides. The #5 loading spectrum shows positive Ni L, Mn Kα, Ni Kα, and Cu Kα peaks and a negative Fe Kα peak, and the bright regions in the score image correspond to ultrafine precipitates. Because such ultrafine precipitates were not confirmed before neutron irradiation, these precipitates can be irradiation-induced ultrafine nanoprecipitates. According to the #5 loading spectrum, ultrafine nanoprecipitates contain Ni, Mn, and Cu, which agrees with the results independently obtained by atom probe tomography (APT). The #7 component does not display specific features as expected from the scree plot (Figure 7–25). By applying PCA, PCs (i.e., dominant features) in the data set can be automatically revealed based on frequency that the features appear.

After PCs are distinguished from the random noise components in the SI data set, the original SI data set can be expressed with a limited number of PCs, α, instead of the total rank of the data matrix n (Figure 7–27):

$$\tilde{\mathbf{D}}_{((x,y),E)} = \mathbf{S}_{((x,y),\alpha)} \times \mathbf{L}^{\mathrm{T}}_{(E,\alpha)} \tag{27}$$

where $\tilde{\mathbf{D}}$ is the reconstructed data matrix with only a selected number of PCs α ($<< n$, in this case, $\alpha = 6$ out of 1024). As a result of the data reconstruction, the random noise parts can be efficiently removed from the original SI without degrading the spatial or energy resolution.

The effect of the PCA-based noise reduction is demonstrated by comparing composition maps extracted from the original SI data set used for PCA decomposition with composition maps from the PCA-reconstructed data set. Figure 7–28 shows a set of composition maps quantified from the original X-ray SI (a: Fe, b: Cr, c: Ni, d: Mn, e: Cu and f: Mo). For quantification, the ζ-factor method (described in the following section) was employed. There are two Cr-enriched regions

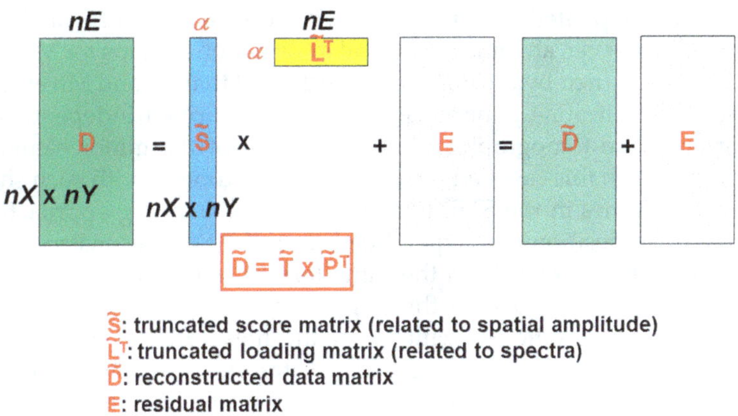

Figure 7–27. A schematic diagram of PCA data reconstruction.

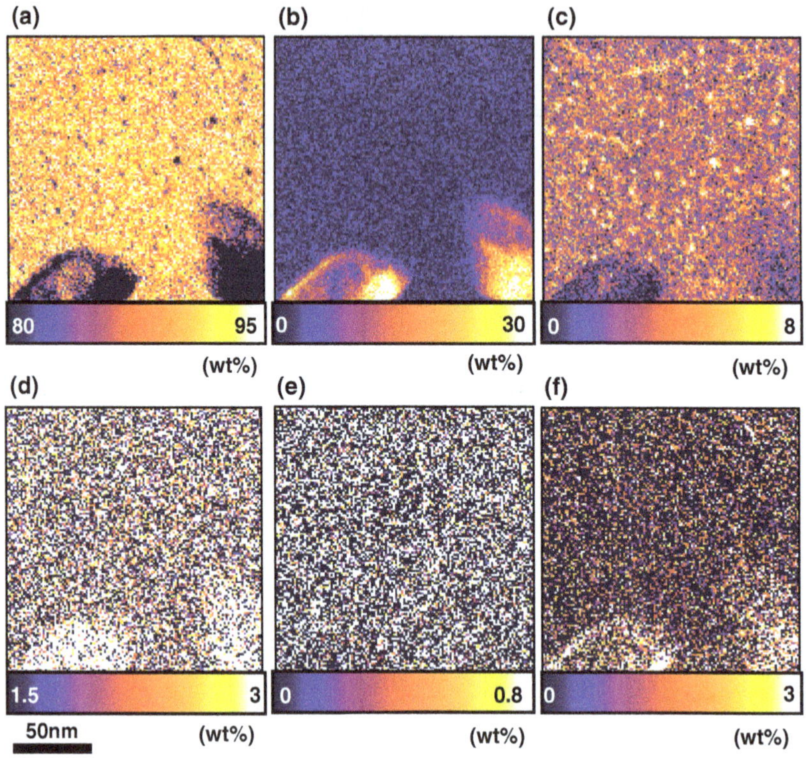

Figure 7–28. Compositional maps of the irradiated low-alloy steel quantified from the original X-ray SI: (**a**) Fe, (**b**) Cr, (**c**) Ni, (**d**) Mn, (**e**) Cu, and (**f**) Mo, reproduced from Burke et al. (2006) with permission.

at the bottom of the field of view corresponding to alloy carbides, and composition-depleted areas in the Fe and Ni maps are superimposed accordingly. The elemental enrichments are also evident in the Mn and Mo maps as well. In addition to the carbides, the localized Ni enrichments and corresponding Fe depletions can be seen in Figure 7–28; these Ni enrichments are consistent with the irradiation-induced ultrafine nanoprecipitates, which are not visible in either TEM or STEM imaging. However, any microchemical information in maps for Mn, Cu, and Mo is obscured by the noise, despite the fact that Cu and Mn enrichment of the ultrafine nanoprecipitates was confirmed independently by atom probe tomography. The distribution of the minor elements below 3 wt% in this case can never be imaged properly with such short acquisition times in the STEM-based elemental mapping approach. A series of composition maps quantified from the PCA-reconstructed SI are shown in Figure 7–29 in the same format as Figure 7–28. A comparison of Figure 7–29 with the original maps in Figure 7–28 clearly demonstrates the efficient removal of the noise. In the Fe map, the local depletions corresponding to the ultrafine nanoprecipitates are clearer. More importantly, the ultrafine nanoprecipitates are now clearly revealed in the Ni, Mn, and Cu maps. In addition, the distribution of Cr and Mo in the carbides is also clearer. These elemental distributions

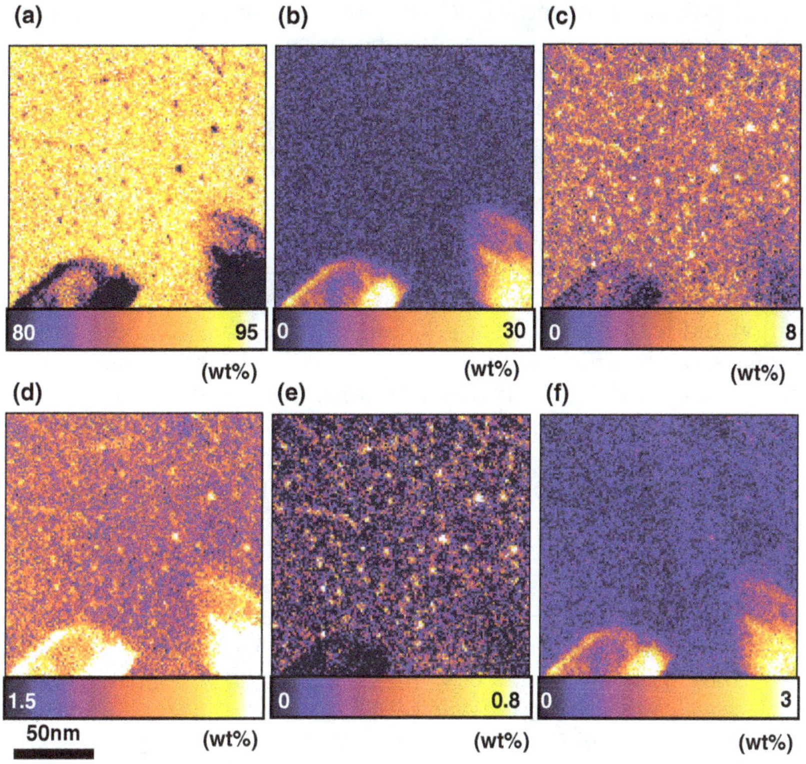

Figure 7–29. Compositional maps of the irradiated low-alloy steel from the same X-ray SI improved by the PCA reconstruction. Random noise components in the original maps were efficiently removed by applying the PCA reconstruction in the same format as Figure 7–28. Reproduced from Burke et al. (2006) with permission.

(especially for the minor constituents such as Mn, Cu, and Mo) were not clearly visible in the original maps in Figure 7–28 due to the high noise components in the unprocessed SI. However, these elemental maps, taken with relatively short acquisition times, are significantly improved by the combination of SI and PCA reconstruction. Now, small compositional changes (even below 1 wt%) are detectable in the map without degrading the high spatial resolution, which clearly demonstrates that quantitative maps of the minor elements can now be obtained using these techniques. The PCA-based noise reduction is very efficient in enhancing weak signals in SI data sets recorded at atomic resolution with a limited acquisition time.

As shown in Figures 7–28 and 7–29, the PCA-based noise reduction is very efficacious to reveal statistically significant features, which might be hidden under heavy random noise. It should be emphasized that the PCA approach is a pure statistical method. If there are minor but real components, which are not repeated in the data set frequently, they might not appear as statistically significant components after the PCA decomposition. In this case, such minor real components would be excluded by the PCA-based noise reduction. Therefore, careful evaluation of individual components especially whose corresponding eigenvalues are close to the noise level in a scree plot is essential to avoid introducing artifacts by excluding minor real components in noise reduction. Remember the PCA approach is not a dream tool. Scientists (not software or computer) should be responsible for the final judgment to determine the number of PCs in the data set.

As shown in the example, by applying PCA to SI data sets, statistically significant features that are dominant information in the data sets can be automatically extracted without prior knowledge of them. In addition, the PCA process may reveal unexpected information hidden in an SI data set. Furthermore, random noise in the data sets can be efficiently reduced by the PCA-based reconstruction. This PCA approach is also very useful for analysis of atomic-column SI data, where unexpected signal correlations might be hidden over relatively high random noise due to the short acquisition time and the small analytical volume (as shown later). Recently, the author has developed the MSA software package as a series of plug-ins (Watanabe et al. 2009) for Gatan DigitalMicrograph Suite, which is widely used to acquire/analyze EELS, XEDS, and EFTEM SI data sets. This particular MSA plug-in package has been applied to various SI data sets acquired by EELS and XEDS (Bosman et al. 2006, 2007, 2009, Herzing et al. 2008, Yaguchi et al. 2008, Watanabe et al. 2007, Varela et al. 2009). Further applications to atomic-column EELS analysis can also be found in Chapter 10. This package contains several useful utilities to deal with SI data sets more efficiently in DMS and is now available through HREM Research, Inc. More information can be found at the company's Web site or the author's Web site http://www.lehigh.edu/~maw3/research/msamain.html.

7.4.4 Quantitative X-Ray Analysis Procedures

As mentioned in previous sections, measured characteristic X-ray signals can be quantified to determine local compositions of thin specimens. For quantitative X-ray analysis of thin-foil specimens, the Cliff–Lorimer ratio method (Cliff and Lorimer 1975) is widely applied and the compositions of constituent elements (e.g., C_A and C_B, which are usually defined as the weight fraction or wt%) can be determined from the measured characteristic X-ray intensities (above background) corresponding to the elements (I_A, I_B) as

$$\frac{C_A}{C_B} = k_{AB}\left(\frac{I_A}{I_B}\right) \tag{28}$$

where k_{AB} is the Cliff–Lorimer k factor, which can be determined both theoretically and experimentally (e.g., Williams and Carter 2009, Goldstein et al. 1977). The theoretical calculation of k factors from first principles is fast and easy but may produce significant (±15–20% relative) systematic errors (e.g., Maher et al. 1981, Newbury et al. 1984). Conversely, in most cases, more accurate quantification can be performed using experimental k factors since the k factors can be determined with a relative error of $\sim \pm 1\%$ from standard thin-foil specimens of known composition (e.g., Wood et al. 1984, Sheridan 1989). However, this process for k-factor determination is tedious because standard thin specimens of known composition may not be available for the k-factor determination and many k factors need to be determined for multi-component systems, even if standards are available. Therefore, the k-factor determination limits thin-film quantification. In addition, X-ray absorption would be one of the most serious problems for quantification, even in thin specimens. Unfortunately, in order to apply the X-ray absorption correction, prior knowledge of the specimen thickness and density are required at individual analysis points (Goldstein et al. 1982). Obviously, the need for an absorption correction is a major limitation to the accurate quantitative microanalysis of thin specimens because independent measurements of the specimen thickness and density are more time consuming and would be a major problem leading to further errors in quantification.

An improved quantitative procedure for thin specimens (the ζ-factor method) has been developed to overcome these limitations (Watanabe et al. 1996, Watanabe and Williams 2003, 2006). In a thin-film specimen, the measured characteristic X-ray intensity is proportional to the mass thickness ρt and the composition C_A. Therefore, the following relationship can be established between the mass thickness and the measured X-ray intensity from element A, I_A, normalized by the composition

$$\rho t = \zeta_A \frac{I_A}{C_A D_e} \tag{29}$$

where ζ_A is a proportionality factor to connect I_A to ρt and C_A, and D_e is the total electron dose (number of electrons) during acquisition (therefore, beam current measurement is essential for this approach). Since a similar relationship to Equation (29) holds for element B, C_A, C_B, and ρt can be expressed as follows, assuming $C_A + C_B = 1$ in a binary system:

$$C_A = \frac{\zeta_A I_A}{\zeta_A I_A + \zeta_B I_B}, \quad C_B = \frac{\zeta_B I_B}{\zeta_A I_A + \zeta_B I_B}, \quad \rho t = \frac{\zeta_A I_A + \zeta_B I_B}{D_e} \tag{30}$$

Therefore, C_A, C_B, and ρt can be determined simultaneously by measuring X-ray intensities. This approach can be easily expanded to any multi-component system as long as the assumption of $\Sigma C_i = 1$ is reasonable. When some of the X-rays are absorbed and/or fluoresced significantly in the material system, absorption and fluorescence correction terms are easily incorporated into Eq. (30) since the specimen thickness information (required to calculate the correction terms) is simultaneously determined by the ζ-factor method at an individually measured point. The absorption correction term for a single X-ray line from a thin specimen can be given as (Philibert 1963)

$$A_A = \frac{(\mu/\rho)^A_{sp}\, \rho t \,\mathrm{cosec}\,\alpha}{1 - \exp[-(\mu/\rho)^A_{sp}\, \rho t \,\mathrm{cosec}\,\alpha]} \tag{31}$$

where $(\mu/\rho)^{A}_{sp}$ is the mass absorption coefficient of the characteristic X-ray line in the specimen and α is the X-ray takeoff angle. This absorption correction term is incorporated into the ζ factor by multiplying it by the corresponding X-ray intensities in Eq. (30). A complete flow chart for the quantification procedure via the ζ-factor method is summarized in Figure 7–30.

To determine the ζ factors, pure-element thin films can be used as standards, in addition to thin films with known composition, which is a major advantage and overcomes any limitations associated with thin-film standards since the pure-element thin films can be easily fabricated by evaporation, electron deposition, or sputtering and are routinely available (in contrast to the great difficulties of preparing known, multi-element, thin-film standards for k factors). In addition to the pure-element thin films, an entire set of ζ factors for K-shell X-ray lines can be estimated from a single spectrum (Watanabe and Williams 2006) generated from the National Institute of Standards and

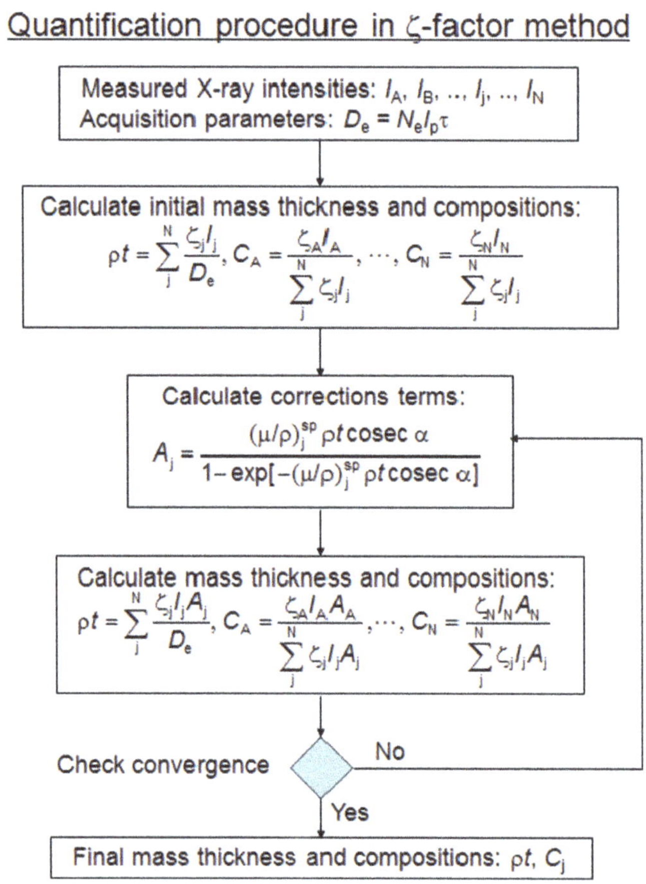

Figure 7–30. A flow chart of quantification procedure in the ζ-factor method with the X-ray absorption correction, reproduced from Watanabe and Williams (2006) with permission.

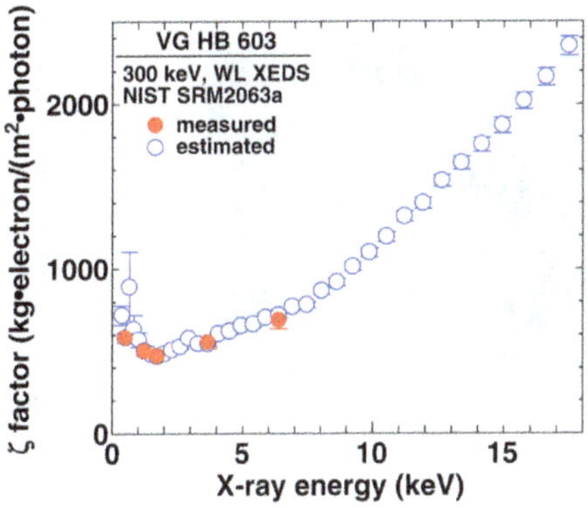

Figure 7–31. The ζ factors of K lines plotted against the X-ray energy. The open circles are the measured values from the SRM2063a glass thin film; the closed circles indicate the estimated values by parameter optimization based on the measured values, as measured in a VG HB 603. Reproduced from Watanabe and Williams (2006) with permission.

Technology (NIST) thin-film, glass, standard reference material (SRM) 2063 (Rasberry 1987) or the re-issued 2063a (Reed 1993), which was originally designed for *k*-factor determination (Steel et al. 1981, 1997). Figure 7–31 shows the ζ factors measured from a single spectrum of the SRM2063a specimen (open circles) in a VG HB 603 dedicated STEM instrument, which is plotted against X-ray energy. The closed circles in Figure 7–31 indicate a series of ζ factors estimated from the measured values by parameter optimization (Watanabe and Williams 2006).

As discussed above, there are several advantages to the ζ-factor method over the conventional Cliff–Lorimer ratio equation, mainly due to the availability of specimen thickness information which (as noted) is simultaneously determined with the compositions. The specific advantages are (1) the built-in absorption correction (which makes light-element analysis possible), (2) the calculation of spatial resolution, (3) the determination of the number of atoms in the analysis volume, (4) the determination of the analytical sensitivity (e.g., MMF or MDM), and (5) the evaluation of XEDS detector parameters. If Eq. (29) is compared with Eq. (28), the following relationship between the *k* and ζ factors can be derived:

$$k_{AB} = \frac{\zeta_A}{\zeta_B} \tag{32}$$

This relationship between *k* and ζ factors permits conversion of an existing series of the *k* factors, which have been determined in a particular instrument, into a series of ζ factors by measuring only one ζ factor, e.g., for Si K or Fe K.

To demonstrate the efficacy of the ζ-factor method in spatial resolution determination, a set of quantitative X-ray maps around an α-Fe/Fe_3P interface in a Fe–17 at% P alloy obtained in the HB 603 is shown in Figure 7–32 (a: ADF-STEM image, b: Fe composition map, c: P composition map, d: thickness map, e: map of absorption correction factor for P Kα, f: spatial resolution map, g: map of the MMF for P, h: map of a number of P atoms and i: map of the MDM for P) (Watanabe

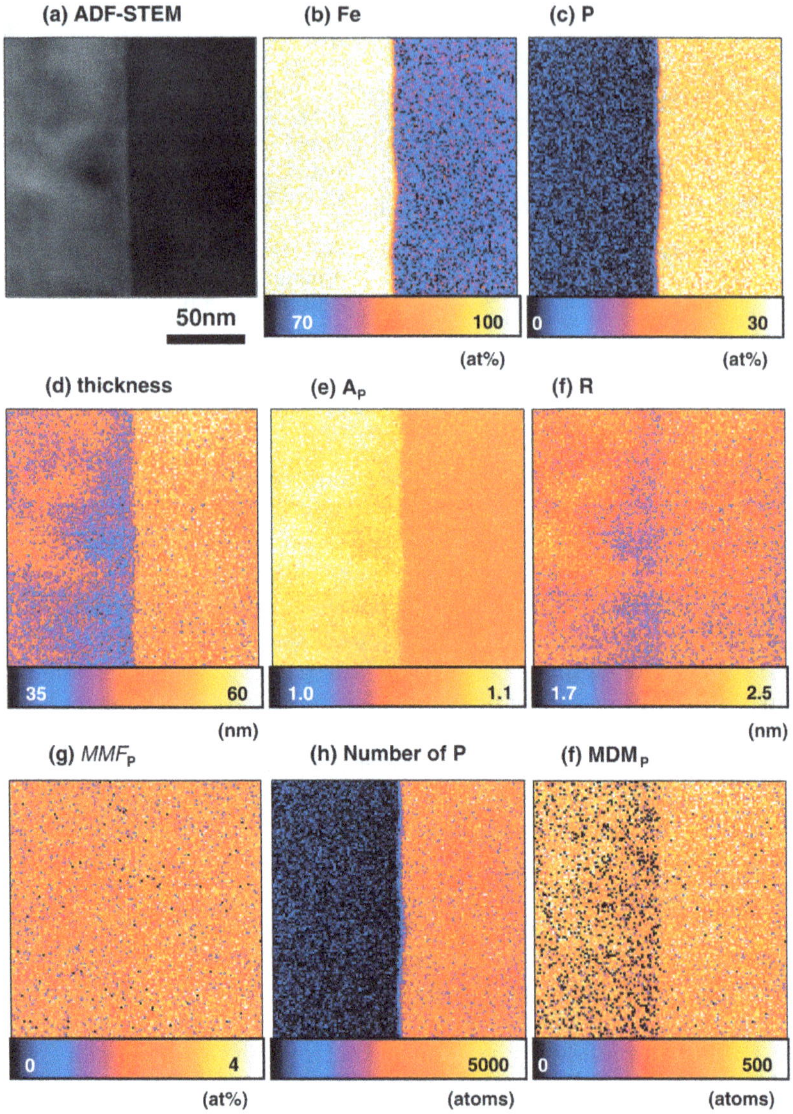

Figure 7–32. A series of X-ray maps around an α-Fe/Fe_3P interface measured in a VG HB 603. These maps were quantified by the ζ-factor method. (**a**) ADF-STEM image, (**b**) Fe map, (**c**) P map, (**d**) thickness map, (**e**) map of absorption correction factor for P Ka, (**f**) spatial resolution map, (**g**) map of the MMF for P, (**h**) map of a number of P atoms, and (**i**) map of the MDM for P. Reproduced from Watanabe and Williams (2006) with permission.

and Williams 2006). Detailed mathematical treatments and error evaluations associated with the ζ-factor method can be found elsewhere (Watanabe and Williams 2006).

7.5 Review of Recent Application of X-Ray Analysis in Aberration-Corrected STEM

Since the 1970s, X-ray analysis has been performed to characterize thin specimens in (S)TEM. There are many applications of X-ray analysis for materials characterization. Some representative applications can be found in textbooks (e.g., Joy et al. 1986, Garratt-Reed and Bell 2003, Williams and Carter 2009). As mentioned in previous sections, there are several benefits from X-ray analysis in aberration-corrected STEM, i.e., not only spatial resolution but also the analytical sensitivity can be improved significantly. In aberration-corrected STEM, however, X-ray analysis is rarely performed in comparison with EELS analysis, and hence recent applications of X-ray analysis are still very limited. Therefore, in this section, it is useful to review the current status of X-ray analysis in aberration-corrected STEM with the latest attempts, which will lead to novel, unique applications in the future.

7.5.1 Comparison of Spatial Resolution of Multilayer Analysis by EELS and XEDS

As discussed in Section 7.3.1, the primary incentive for analysis in the probe-corrected instruments is improved spatial resolution by a refined incident probe. Recently, a probe-corrected STEM was used by Fraser and his co-workers (Gençet al. 2009) to evaluate spatial resolution. The spatial resolution of XEDS and EELS analyses was evaluated by measuring concentration line profiles from Ti/Nb multilayer thin specimens in an aberration-corrected FEI TITAN 80–300 microscope operated at 300 kV. The composition profiles from a Ti/Nb = 7/9 nm multilayer specimen show good agreement between XEDS and EELS analysis. The composition line profiles from a Ti/Nb = 1.8/2.1 nm are shown in Figure 7–33 (b: XEDS and c: EELS) with an ADF-STEM image (a) (Gençet al. 2009). In contrast to the results from multilayers with wider periodicity, the XEDS composition profile from narrower multilayers is noticeably different from the EELS result. In the XEDS profile, compositions of Ti and Nb appear very mixed in both layers, whereas compositions seem mixed only in the Ti layer according to EELS results. Complementary results by atom probe tomography show good agreement with EELS results, which indicates that XEDS results can be affected by some artifact. This intermixed composition in both layers from the XEDS profile could be due to a broadening of the incident probe. The contribution of beam broadening is more serious in X-ray analysis than EELS analysis since X-ray signals can be generated from the entire analyzed volume, and there is no way to limit X-ray signals from remote regions (Williams and Carter 2009). In an EELS analysis, the spatial resolution can be controlled to some degree by the signal collection aperture (Egerton 2007).

Unfortunately, the thickness of the multilayer specimen is ~120 nm, which is not ideal to achieve the advantages of aberration-corrected fine probes for the improvement of special resolution, as discussed in

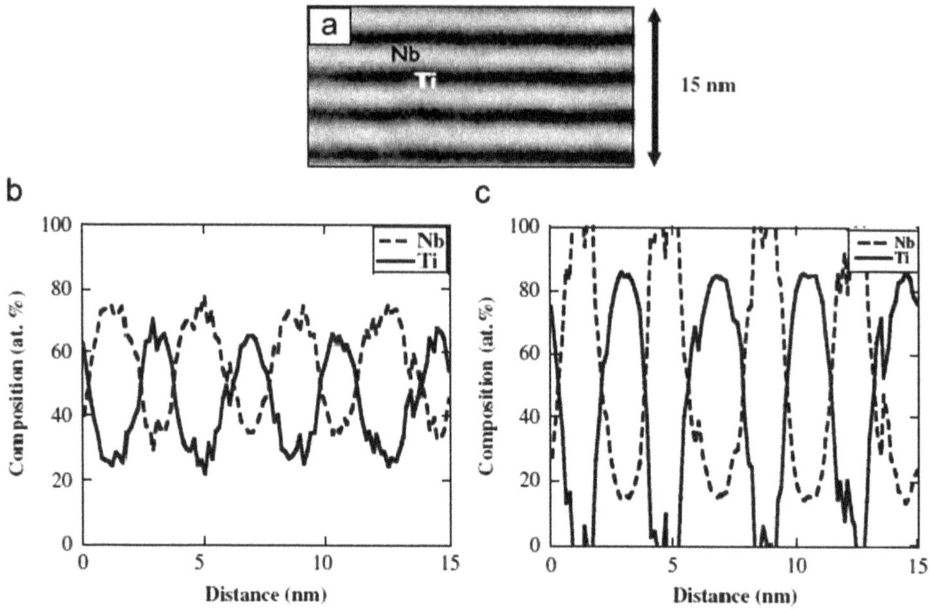

Figure 7–33. (**a**) An ADF-STEM image, (**b**) XEDS composition profile, and (**c**) EELS composition profile of a Ti/Nb multilayer specimen with a periodicity of Ti/Nb = 1.8/2.1 nm, reproduced from Genç et al. (2009) with permission.

Section 7.3.1. i.e., in thicknesses over 50 nm, beam broadening in the thin specimen would be the dominant contribution to spatial resolution. Again, thinner specimens (<50 nm) are essential to take advantage of the aberration correction, especially for high spatial resolution X-ray analysis.

7.5.2 Determination of Local Elemental Distributions in Alloy Nanoparticles

Although EELS analysis is superior to X-ray analysis in terms of spatial resolution and analytical sensitivity, X-rays could be the only practical signal with which to characterize some materials, which is certainly true when nanoparticles consisting of heavy noble metals (e.g., ,Au, Pd, Pt) are to be characterized. In fact, many catalyst systems contain metal/alloy nanoparticles with heavy elements. Therefore, X-ray analysis is the primary approach to characterize catalyst systems with noble-metal/alloy particles. Here some recent studies of metallic-alloy catalysts by X-ray analysis are reviewed.

Metal nanoparticles display fascinating optical properties due to the resonance of surface plasmons with visible light at well-defined frequencies, which are related to a number of parameters, such as nanoparticle composition (Link et al. 1999, Rodríguez-González et al. 2004) and morphology (Liz-Marzán 2004, Cao et al. 2004). Note that these plasmon resonances can also be measured by high-energy-resolution EELS with monochromated illumination, see Chapter 16, Bosman et al. (2007), Nelayah et al. (2007, 2009). For these applications,

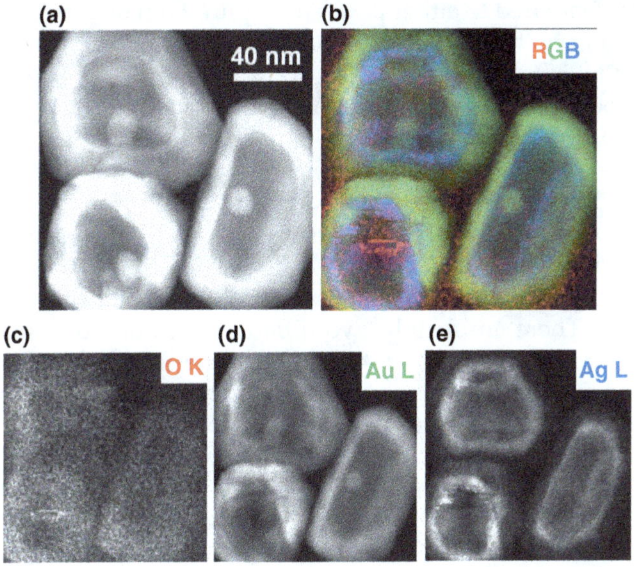

Figure 7–34. Results of X-ray SI analysis of Au/Ag/Au multishell nanoparticles. The ADF-STEM image (**a**) shows the field of view from which the X-ray SI was obtained and the RGB image (**b**) shows the relative distribution of the elements (*red,* O; *green,* Au; and *blue,* Ag), while the lower images show the respective individual maps (**c**: O, **d**: Au, and **e**: Ag). Reproduced from Rodríguez-González et al. (2005) with permission.

Au and Ag nanoparticles are traditionally studied mainly for chemical stability. By creating Au/Ag alloy nanoparticles with a controlled distribution, the plasmon resonance can be modified. However, controlling the Au/Ag distribution is particularly difficult because a perfect solid solution can be formed over the whole composition range. Recently, multilayer AuAg nanoparticles with interesting optical properties were successfully fabricated by the successive reduction of Au and Ag salts (Rodríguez-González et al. 2005). These multilayer AuAg nanoparticles were characterized by XEDS spectrum imaging in STEM in combination with PCA. A summary of the results is shown in Figure 7–34, including an ADF-STEM image of three particles (a), the individual maps of three elements (c: O, d: Au and e: Ag), and an RGB (red–green–blue) color overlay image from three elemental maps (b: red: O, green: Au, and blue: Ag). The RGB image shows the relative distribution of the elements within the particles. From the images, both the Au cores and outer shells are clearly visible. Although O is homogeneously distributed in the particles, Ag seems to be preferentially accumulated next to the outer Au shell, leaving an apparently empty space at its inner side toward the region where the central Au cores are located.

The ability to detect and visualize core–shell morphologies in bimetallic clusters has proven to be of considerable importance in the study of supported Au–Pd catalysts. These materials exhibit considerable potential for the direct production of H_2O_2 from molecular H_2 and O_2 under mild reaction conditions (Landon et al. 2002, 2003). In fact, by creating alloy Au–Pd nanoparticles, H_2O_2 production is significantly enhanced compared to either pure-Au or pure-Pd nanoparticles, which indicates that both the microstructure and the chemistry can be modified by alloying (Enache et al. 2006). Additionally, H_2O_2 production is also dependent upon support materials (Enache et al. 2006, Herzing et al. 2008). To investigate the development of the Au–Pd nanoparticle morphology more closely, the XEDS SI approach was applied, and then measured SI data sets were analyzed by PCA.

As an example, analyzed results from AuPd/Fe_2O_3 catalysts are shown in Figure 7–35 (a: ADF-STEM image, b: RGB color overlay image with red: support, green: Au and blue: Pd, c: Au and d: Pd) (Enache et al. 2006). There are mainly two different sizes of AuPd nanoparticles: smaller than 10 nm and ~40 nm. From the elemental distributions of Au and Pd, smaller nanoparticles exhibit a homogeneous random AuPd alloy. In contrast, the Pd signal occupies a larger area than the area of the Au signal in the larger particles, i.e., a Pd-enriched shell with an Au-enriched core is formed. Figure 7–36 shows a set of ADF-STEM

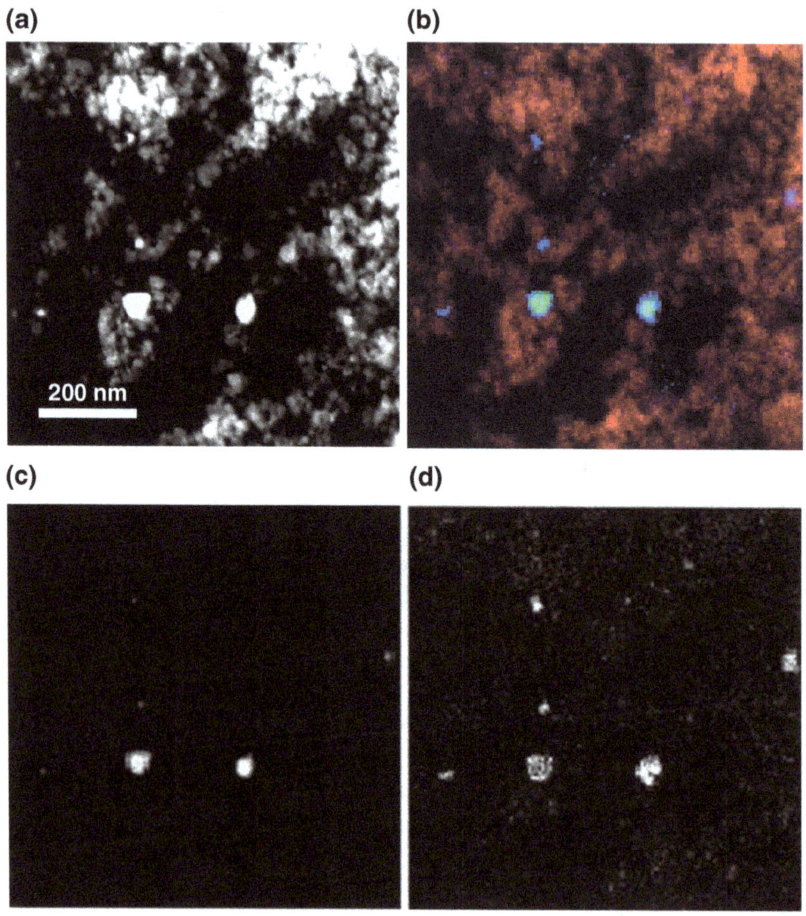

Figure 7–35 A set of X-ray maps taken from an AuPd/Fe_2O_3 catalyst: (**a**) ADF-STEM image, (**b**) RGB color overlay image (*red*, Fe_2O_3; *green*, Pd; and *blue*, Au), (**c**) Au map, and (**d**) Pd map, reproduced from Enache et al. (2006) with permission.

images, an RGB color overlay map, an Au map, and a Pd map measured from AuPd/TiO_2 (a), AuPd/Al_2O_3 (b), and AuPd/activated-carbon (c) catalysts. A similar core–shell formation of intermediate-sized AuPd nanoparticles was observed with other oxide supports such as TiO_2 and Al_2O_3. Conversely, when activated carbon is used as the support, the intermediate-scale AuPd nanoparticles form a homogeneous random alloy rather than the core–shell structure, as shown in Figure 7–36c. The formation of Pd-rich shell structures may be due to surface oxidation and consequent PdO_x oxide formation on the surface. Therefore, this process is inhibited by the reducing nature of the carbon support (Herzing et al. 2008).

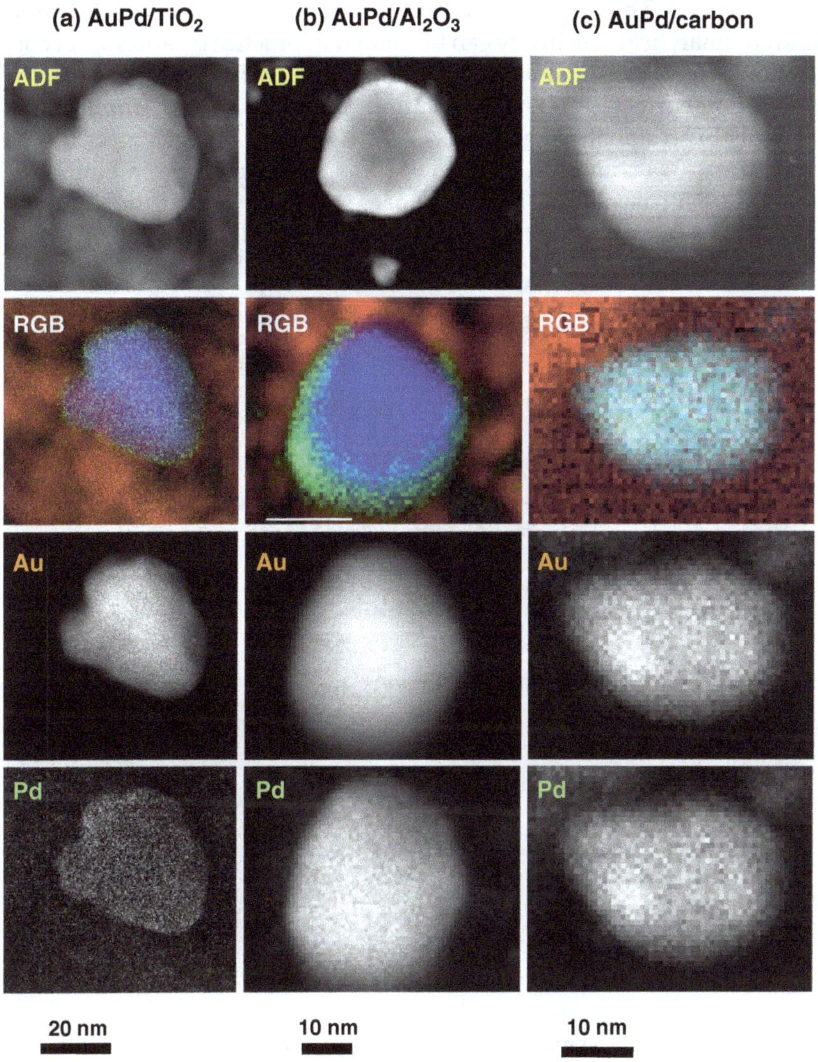

Figure 7–36. A set of ADF-STEM images, an RGB color overlay map (*red*, support; *green*, Pd; and *blue*, Au), Au map, and Pd map measured from AuPd/TiO_2 (**a**), AuPd/Al_2O_3 (**b**), and AuPd/activated-carbon (**c**) catalysts, reproduced from Herzing et al. (2008) with permission.

Obviously, the X-ray SI approach in aberration-corrected STEM is essential for investigations of noble-metal nanoparticles in combination with PCA data analysis. The approach has been employed for characterizations of other catalyst systems (e.g., Edwards et al. 2009, Zhou et al. 2009). More comprehensive descriptions of the characterization of catalyst systems by STEM can be found in Chapter 13.

7.5.3 Practical Evaluation of Improved Spatial Resolution and Sensitivity of X-Ray Analysis in Aberration-Corrected STEM

In theory, the spatial resolution and analytical sensitivity of X-ray analysis should be improved by aberration-corrected STEM, as described in Section 7.3. Thus the main question is: How much spatial resolution and/or analytical sensitivity can be gained in practice by aberration correction? Here, practical aspects of improved analytical capabilities in aberration-corrected STEM are featured based on several applications of X-ray SI (Watanabe et al. 2006).

Figure 7–37a shows an ADF-STEM image around a grain boundary (GB) in a Ni-base alloy. There is a coarse γ' precipitate at the top right

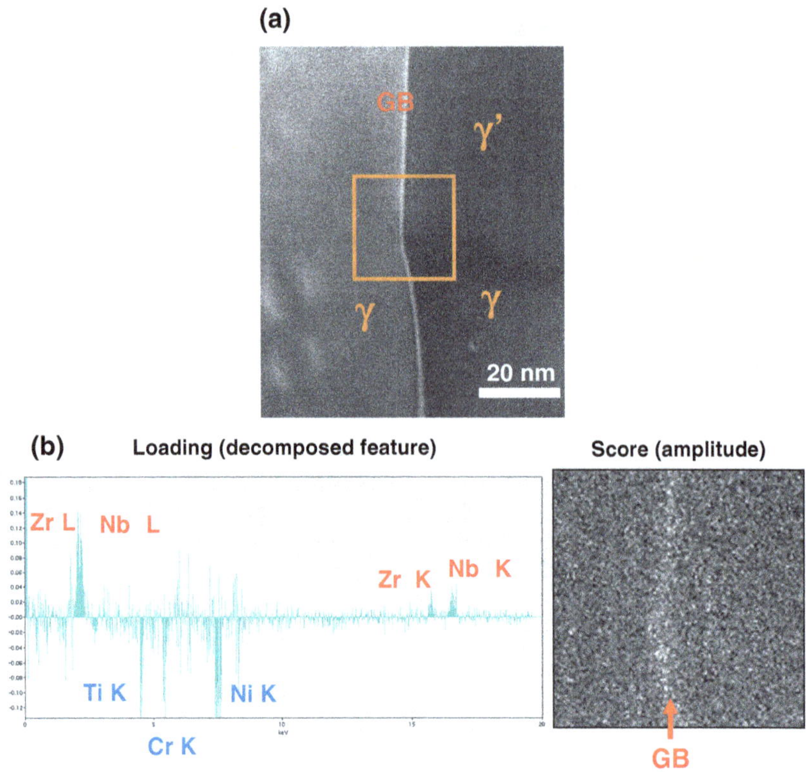

Figure 7–37. (**a**) An ADF-STEM image taken around a GB in a Ni-based alloy. An X-ray SI data set was acquired from the selected squared region in an aberration-corrected VG HB 603 STEM and (**b**) a loading spectrum and the score image of a statistically significant component extracted from the X-ray SI data set by applying PCA. Reproduced from Watanabe et al. (2006) with permission.

in this area along the GB, and hence the GB is an interphase interface between the γ/γ' phases at the top-left corner. Then this GB becomes an ordinary γ/γ at the lower region. At Lehigh University, an X-ray SI data set was acquired from the square region at the GB with 64 × 64 pixels and 1024 energy channels and a dwell time of 200 ms/pixel in the aberration-corrected VG HB 603. The actual length of the SI data set is 25 nm. X-ray SI data were analyzed by the PCA method. In Figure 7–37b, an interesting component extracted from the data set by PCA is shown, including the loading spectrum and corresponding score image. The loading spectrum shows positive K and L peaks of Zr and Nb and negative K peaks of Ti, Cr, and Ni. As described in Section 7.4.3, the spectral feature is enhanced in the corresponding score image. In this case, the brighter region in the score image is superimposed on the GB, which indicates that both Zr and Nb are enriched (i.e., cosegregated) at the GB.

The composition map of Zr quantified after the PCA noise reduction is shown in Figure 7–38a. Zr is present at a nominal 0.04 wt% (well below the detection limit!), so it has not been quantified or mapped previously in this alloy via X-ray analysis. To the authors' knowledge, there is no report of GB segregation of either Zr or Nb in this alloy. A concentration line profile extracted from the Zr map across the γ/γ' interface region is shown in Figure 7–38b. To improve counting statistics, the profile was constructed by binning 20 pixels along the interface. The error bar represents a 99% confidence limit (3σ). The error levels are relatively high due to the shorter acquisition time and much lower composition of the alloying elements, but the Zr composition at the GB can still be clearly distinguished from the Zr composition within the grains. The spatial resolution determined from these segregation profiles is as good as 0.4–0.6 nm at FWTM, which is the best spatial resolution that has ever been obtained in X-ray analysis apart from the atomic-column X-ray maps shown in Figure 7–41. Maps of the Zr enrichment and the number of Zr atoms calculated by the ζ-factor method are shown in Figure 7–38c and d, respectively. The detailed approach for extracting this information can be found in previous papers (Watanabe and Williams 2003, 2006). The Zr enrichment on the GB is 2–3 atoms/nm^2, as obtained from the enrichment map, which corresponds to 0.12–0.17 monolayer based on the atomic density of the close-packed (111) plane in this alloy. As shown in Figure 7–38d, enrichment amounts are only 20–40 Zr atoms at the GB. Such limited amounts of enriched atoms (which are highly localized in limited regions) may simply be invisible to the relatively larger probe with higher probe currents in the conventional uncorrected condition. Therefore, not only spatial resolution but also MDM is significantly improved with the aberration-corrected probes, as calculated in the previous section.

The improvement of MDM is further evaluated based on an X-ray SI data set of a Cu–0.5wt% Mn alloy. It should be noted that this alloy was previously used to evaluate the analytical sensitivity for X-ray analysis and energy-filtering TEM (Watanabe and Williams 1999, Watanabe et al. 2003). In Figure 7–39, quantified Mn maps extracted from the SI data set are shown (a raw and b: PCA reconstructed). For

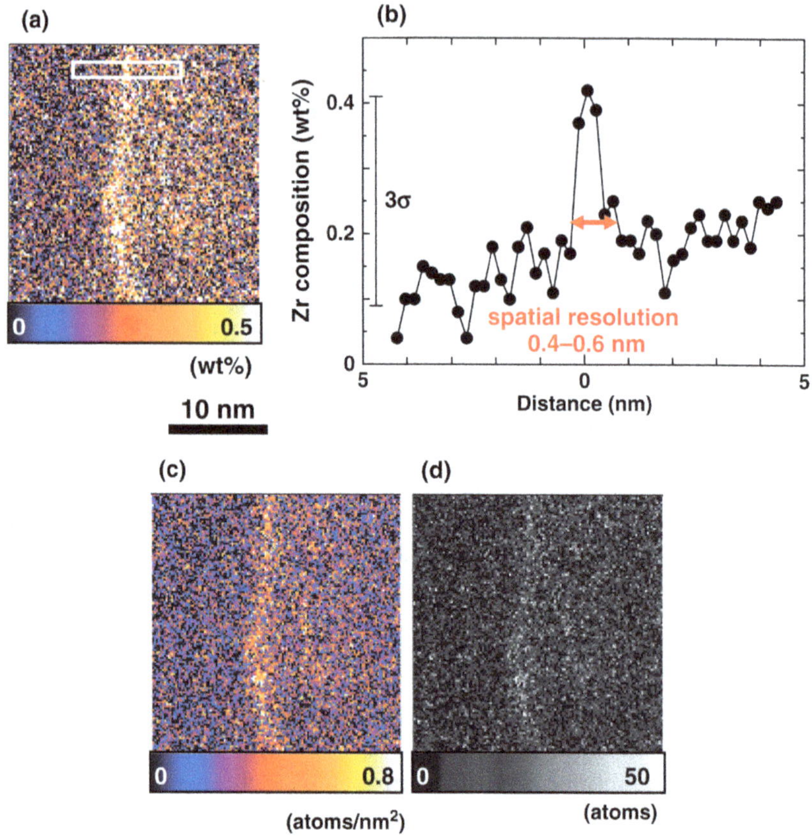

Figure 7–38 (**a**) Zr composition map quantified by the ζ-factor method after PCA noise reduction. (**b**) The Zr composition profiles across the γ/γ' interface on the GB. The error bar indicates a 99% confidence level (3σ) and the spatial resolution determined from the profiles is ~0.4–0.6 nm (FWTM). Maps of Zr enrichment (**c**) and the number of Zr atoms (**d**), reproduced from Watanabe et al. (2006) with permission.

quantification, the ζ-factor method was used and the determined thickness map is also shown in Figure 7–39c. The quantified Mn map after the PCA noise reduction exhibits ~0.5 wt%, which corresponds to the true composition of this particular specimen. These quantified results indicate that the PCA noise reduction simply removes some of the random noise from the data set but not the essential signals. A map of the number of Mn atoms is also shown in Figure 7–39d, as determined by the noise-free Mn map in combination with the thickness map with an appropriate beam–specimen interaction volume, including the beam broadening. The Mn atom map shows an average of ~2–3 atoms, which corresponds to the MDM (or MDA) value of this approach. In other words, the analytical sensitivity can be as good as 2–3 atoms by X-ray analysis in aberration-corrected STEM in combination with the PCA approach. These facts are encouraging because hitherto unsuspected details of elemental distributions may now be characterized

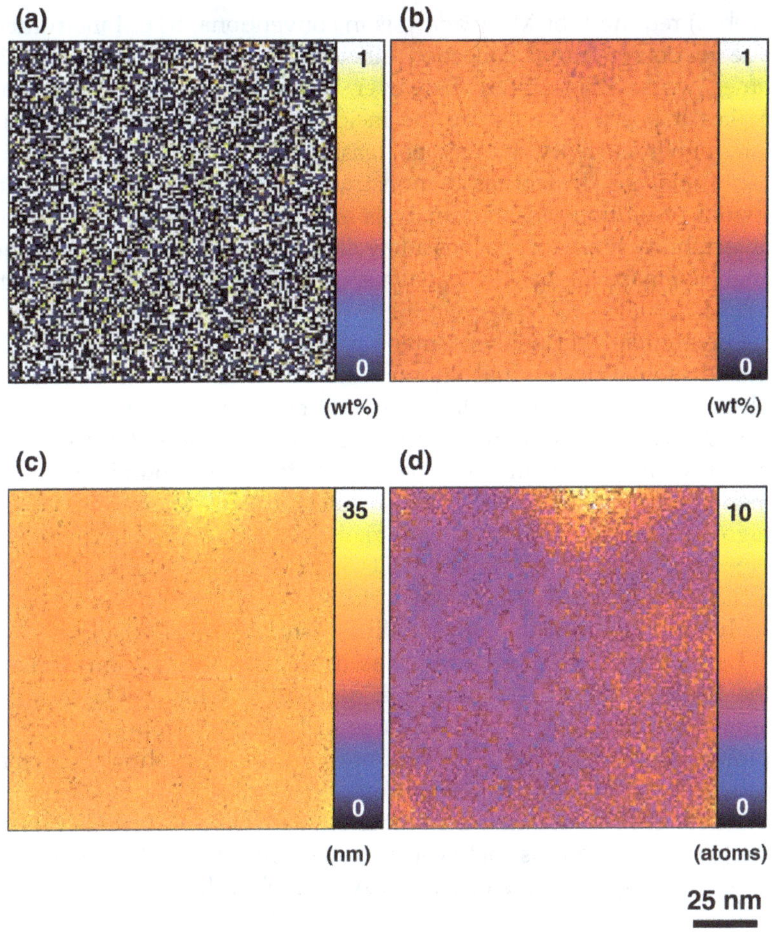

Figure 7–39 Results of X-ray SI analysis of a Cu–0.5 wt% Mn alloy: (**a**) Mn composition map determined from the original SI data set, (**b**) Mn composition map determined from the PCA noise-reduced SI data set, (**c**) thickness map, and (**d**) map of a number of Mn atoms, reproduced from Williams and Carter (2009) with permission.

in aberration-corrected STEM because of the superior quantification, analytical sensitivity, and spatial resolution.

7.5.4 Towards Atomic-Column X-ray Imaging

Atomic-column EELS analysis has routinely been performed in both conventional and aberration-corrected STEMs (Batson 1995, Browning and Pennycook 1995, Varela et al. 2005). Furthermore, atomic-resolution EELS imaging is applicable in the latest STEM with better resolution and improved stability (e.g., Kimoto et al. 2007, Bosman et al. 2007, Muller et al. 2008, Varela et al. 2009). For X-ray analysis, such atomic-column analysis or even atomic-column mapping has not even been attempted, mainly because atomic-resolution STEM images are not obtainable with the higher probe currents (and consequent large probes) required for X-ray analysis in conventional STEM instruments due to poorer signal collection efficiency as discussed above (~100 times worse than EELS). However, since the aberration correction makes it possible to reduce the incident probe size while maintaining higher currents, it may be feasible to perform atomic-column X-ray analysis. In fact, as demonstrated in the previous section, the spatial resolution of X-ray analysis is improved to ~0.4 nm and the detectability limit approaches a few atoms, which implies atomic-level analysis/mapping by X-ray analysis is feasible in aberration-corrected STEM.

For atomic-level analysis/mapping, the fine probe must be positioned above individual atomic-column sites during acquisition. Therefore, the most important key parameter in making such atomic-level X-ray analysis/mapping possible is relatively long-term stabilities not only for the instrument alone but also for the surrounding environment. Some of the latest STEM instruments might possess the required high stability to perform the atomic-level analysis and mapping. For example, the newly developed JEOL ARM-200F aberration-corrected STEM instrument is designed to perform atomic-level chemical analysis with improved instrumental stabilities (Ishikawa et al. 2009).

Using the ARM-200F, GaAs was analyzed on an atomic scale by the XEDS approach (Watanabe et al. 2010). Figure 7–40a shows an atomic-resolution ADF-STEM image of [001]-projected GaAs. In this projection, the Ga and As layers are separated, as shown schematically in Figure 7–40b (drawn by Vesta, Momma and Izumi 2008). Since the difference in the atomic number is only two between Ga (31) and As (33), the *Z* contrast between two elements may not appear unless a very thin specimen is observed. An XEDS SI data set was recorded

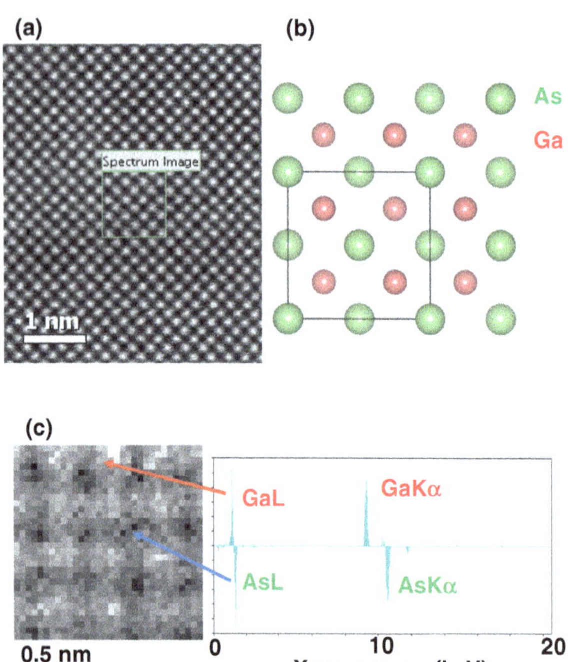

Figure 7–40. (**a**) An ADF-STEM image taken from a [001]-projected GaAs specimen. An X-ray SI data set was acquired from the selected squared region in aberration-corrected JEM ARM-200F STEM; (**b**) a schematic diagram of arrangements of Ga and As atoms in the [001]-projected GaAs, as drawn by Vesta (Momma and Izumi 2008). (**c**) A pair of loading spectrum and score image of a statistically significant component extracted from the X-ray SI data set by applying PCA, reproduced from Watanabe et al. (2010) with permission.

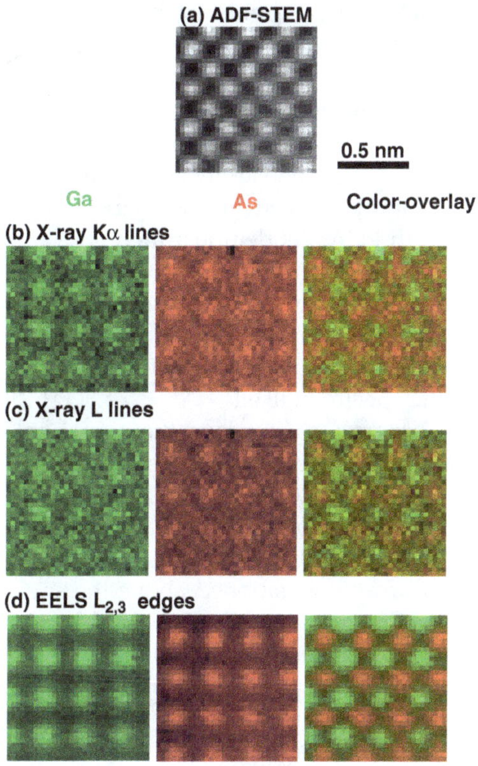

Figure 7–41. (**a**) An ADF-STEM image of the [001]-projected GaAs simultaneously recorded during SI acquisition; a set comprising a Ga map, an As map, and a color overlay map of Kα X-ray lines (**b**), a similar set for the L X-ray lines (**c**), and EELS $L_{2,3}$ edges (**d**), reproduced from Watanabe et al. (2010) with permission.

from the squared area shown in Figure 7–40a, and then PCA was performed to improve weak signals in the data set. In Figure 7–40c, the second component extracted from the data set by PCA is shown as a pair comprising the loading spectrum and the score image. The loading spectrum shows positive K and L peaks of Ga and negative K and L peaks of As. Therefore, the brighter regions in the score image must correspond to the Ga columns, whereas the darker regions correspond to the As columns. Thus, this particular component definitely shows the signal separation between Ga and As. Figure 7–41 shows a HAADF-STEM image (a) and X-ray maps of Ga Kα and As Kα lines with their color overlay image (b), X-ray maps of Ga L and As L lines with their color overlay image (c), and EELS maps of Ga $L_{2,3}$ and As $L_{2,3}$ edges with their color overlay image (c), which were also simultaneously recorded with the XEDS SI data set. Although the signal levels are still very limited in comparison with EELS results, atomic-level XEDS analysis is now possible through the combination of aberration-corrected STEM and PCA. If the detection efficiency of X-ray signals is improved, atomic-column X-ray mapping would be routinely applicable.

It should be noted that abnormal X-ray emission due to channeling effects may not be avoidable in such imaging conditions that resolve the atomic columns (Bullock et al. 1985). Therefore, for quantification of atomic-column X-ray analyses, appropriate corrections may be required. The channeling correction seems challenging but can be estimated (e.g., Allen et al. 1994, Rossouw et al. 1997).

7.6 Concluding Remarks: Future Prospects of X-Ray Analysis

X-ray analysis can be a very robust approach. As previously mentioned, there are only a few parameters to consider before acquiring an X-ray spectrum or SI data set. X-ray analysis can be significantly improved in terms of spatial resolution and analytical sensitivity in aberration-corrected STEM in combination with PCA. As demonstrated above, it is now possible to acquire atomic-resolution X-ray maps. However, in comparison to EELS, the collectable fraction of the generated X-ray signals remains extremely limited. In other words, the major limitation regarding X-ray analysis in STEM is still the poorer detection efficiency besides the major limitations on the specimen itself.

There are several attempts in progress to improve the signal collection efficiency for XEDS. Recently proposed detector configurations in a microscope are schematically summarized in Figure 7–42. Kotula et al. (2008) proposed an annular detector geometry for X-ray collection, which employs four of the latest Si-drift detectors (SDDs). As shown in Figure 7–42a, this annular SDD array is positioned above the specimen, which improves X-ray collection significantly. Expected higher count rates (which can also be a problem in conventional Si(Li) detectors) can be easily handled in the SDDs. In preliminary results of the annular SDD configuration, an input count rate over 1 Mcps can be handled, which is ~500 times higher than count rates in conventional Si(Li) detectors. The post-specimen geometry for X-ray collection was also proposed by Zaluzec (2009), as summarized in Figure 7–42b. In general, the post-specimen detector position is avoided for X-ray collection because enormously high background signals are generated by electrons scattered in forward directions (Chapman et al. 1984). According to Zaluzec, this post-specimen configuration is specifically optimized for X-ray analysis of nanoparticles since such nanoparticles would not produce significant background signals. These post-specimen detector positions were designed with the SDD to handle high count rates

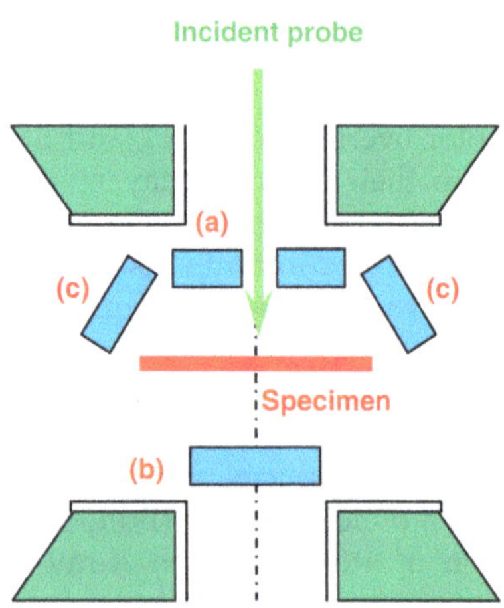

Figure 7–42. A schematic diagram of newly proposed configurations of XEDS detectors in a pole piece: (**a**) an annular SDD above the specimen (Kotula et al. 2008); (**b**) an SDD below the specimen (Zaluzec 2009); and (**c**) a combined multiple SDD system at conventional positions (von Harrach et al. 2009).

as well. The detector can be mounted with the collection angle of π sr out of the 4π sr sphere, which is at least 10 times higher than the conventional detector collection angle in STEM. It should be mentioned that both the annular and post-specimen detector configurations may suffer from back- and forward-scattered high-energy electrons, respectively. Hopefully, the issues associated with high-energy electron bombardment will soon be solved. In addition to the new detector configurations, FEI recently introduced a new microscope "Tecnai Osiris" with an integration of four SDDs in the conventional geometry, which results in a high collection angle of 0.9 sr in the microscope (at least a factor of 3 higher than with the conventional system) (von Harrach et al. 2009).

Chemical analysis at atomic-level spatial resolution with single-atom detection sensitivity is one of the ultimate goals in materials characterization. Such atomic-level materials characterization will be routinely performed (even by X-ray analysis) if the limited signal collection efficiency is improved.

Acknowledgments The author wishes to acknowledge the support of the National Science Foundation (grant DMR-0804528) and Bechtel Bettis, Inc. The author would also like to thank Prof. David Williams (currently at the Univ. of Alabama, Huntsville) for his thoughtful supervision for many years. In collaboration with Prof. Williams, the ζ-factor method and MSA plug-ins were developed. In addition, the author would like to thank Prof. Christopher Kiely, Mr. David Ackland and colleagues at Lehigh, Prof. Zenji Horita at Kyushu University, Dr. Ulrich Dahmen and colleague at National Center for Electron Microscopy, Lawrence Berkeley National Laboratory, Dr. Vicki Keast at Univ. Newcastle, Dr. Michel Bosman at Institute of Microelectronics, Dr. Kazuo Ishizuka and Mr. Kenta Yoshimura at HREM Research, Inc., Dr. Hidetaka Sawada, Mr. Eiji Okunishi and Mr. Masahiko Kanno at JEOL, Mr. Shintaro Yazuka, Mr. Toshihiro Aoki, Mr. Toshihiro Nomura, Dr. Masahiro Kawasaki, and Dr. Tom Isabell at JEOL USA, and Dr. Toshie Yaguchi at Hitachi Hitechnologies for their collaboration.

References

L.J. Allen, T.W. Josefsson, C.J. Rossouw, Interaction delocalization in characteristic X-ray emission from light elements. Ultramicroscopy **55**, 258 (1994)

J.E. Barth, P. Kruit, Addition of different contributions to the charged particle probe size. Optik **101**, 101 (1996)

P.E. Batson, Conduction bandstructure in strained silicon by spatially resolved electron energy loss spectroscopy. Ultramicroscopy **59**, 63 (1995)

P.E. Batson, N. Dellby, O.L. Krivanek, Sub-Ångstrom resolution using aberration corrected electron optics. Nature **418**, 617 (2002)

G. Bertuccio, A. Castoldi, A. Longoni, M. Sampietro, C. Gauthier, New electrode geometry and potential distribution for soft X-ray drift detectors. Nucl Instrum Methods Phys Res A **312**, 613 (1992)

M. Born, E. Wolf, *Principles of Optics*, 7th edn. (Cambridge University Press, Cambridge, 1999)

M. Bosman, V.J. Keast, J.L. García-Muñoz, A.J. D'Alfonso, S.D. Findlay, L.J. Allen, Two-dimensional mapping of chemical information at atomic resolution. Phys. Rev. Lett. **99**, 086102 (2007)

M. Bosman, V.J. Keast, M. Watanabe, A.I. Maaroof, M.B. Cortie, Mapping surface plasmons at the nanometre scale with an electron beam. Nanotechnology **18**, 165505 (2007)
M. Bosman, V.J. Keast, M. Watanabe, D.G. McCulloch, M. Shakerzadeh, E.H.T. Teo, B.K. Tay, Quantitative, nanoscale mapping of sp^2 percentage and crystal orientation in carbon multilayers. Carbon **47**, 94 (2009)
M. Bosman, M. Watanabe, D.T.L. Alexander, V.J. Keast, Mapping chemical and bonding information using multivariate analysis of electron energy-loss spectrum images. Ultramicroscopy **106**, 1024 (2006)
R. Bro, A.K. Smilde, Centering and scaling in component analysis. J. Chemometr. **17**, 16 (2003)
L.M. Brown, Scanning transmission electron microscopy: microanalysis for the microelectronic age. J. Phys. F **11**, 1 (1981)
N.D. Browning, S.J. Pennycook, Atomic-resolution electron energy-loss spectroscopy in the scanning transmission electron microscope. J. Microsc. **180**, 230 (1995)
J.F. Bullock, C.J. Humphreys, A.J.W. Mace, H.E. Bishop, J.M. Titchmarsh, in *Microscopy in Semiconductor Materials 1985*, ed. by A.G. Gullis, D.B. Holt, Crystalline effects in the analysis of semiconductor materials using Auger electrons and X-rays (Adam Hilger, Bristol, 1985), p. 405
M.G. Burke, M. Watanabe, D.B. Williams, J.M. Hyde, Quantitative characterization of nanoprecipitates in irradiated low alloy steels: advances in the application of FEG-STEM quantitative analysis to real materials. J. Mater. Sci. **41**, 4512 (2006)
E. Cao, G.C. Schatz, J.T. Hupp, Synthesis and Optical Properties of Anisotropic Metal Nanoparticles. J. Fluoresc. **14**, 331 (2004)
J.N. Chapman, W.A.P. Nicholson, P.A. Crozier, Understanding thin film X-ray spectra. J. Microsc. **136**, 179 (1984)
G. Cliff, G.W. Lorimer, The quantitative analysis of thin specimens. J. Microsc. **103**, 203 (1975)
R.N. Cochran, F.H. Home, Statistically weighted principal component analysis of rapid scanning wavelength kinetics experiments. Anal. Chem. **49**, 846 (1977)
C. Colliex, C. Mory, in *Quantitative Electron Microscopy*, ed. by J.N. Chapman, A. Craven, Quantitative aspects of scanning transmission electron microscopy. (Scottish University Summer School in Physics, Glasgow, Scotland, 1984), p. 149
A.V. Crewe, D.N. Eggenberger, J. Wall, L.M. Welter, Electron gun using a field emission source. Rev. Sci. Inst. **39**, 576 (1968)
A.V. Crewe, D.B. Salzman, On the optimum resolution for a corrected STEM. Ultramicroscopy **9**, 373 (1982)
A.V. Crewe, J. Wall, A scanning microscope with 5 Å resolution. J. Mol. Biol. **48**, 375 (1970)
M.I. Danylenko, M. Watanabe, C. Li, A.V. Krajnikov, D.B. Williams, M.A. Vasiliev, Analytical electron microscopy and Auger electron spectroscopy study of low-temperature diffusion in multilayer Cr–Cu–Ni–Au thin films. Thin Solid Films **444**, 75 (2003)
P. Doig, P.E.J. Flewitt, The detection of monolayer grain boundary segregations in steels using STEM-EDS X-ray microanalysis. Met. Trans. A **13**, 1397 (1982)
P. Doig, D. Lonsdale, P.E.J. Flewitt, The spatial resolution of X-ray microanalysis in the scanning transmission electron microscope. Philos. Mag. A **41**, 761 (1980)
D. Drouin, A.R. Couture, D. Joly, X. Tastet, V. Aimez, R. Gauvin, CASINO V2.42—A fast and easy-to-use modeling tool for scanning electron microscopy and microanalysis users. Scanning **29**, 92 (2007)

C. Dwyer, R. Erni, J. Etheridge, Method to measure spatial coherence of subangstrom electron beams. Appl. Phys. Lett. **93**, 021115 (2008)

J.K. Edwards, B. Solsona, E.N. Ntainjua, A.F. Carley, A.A. Herzing, C.J. Kiely, G.J. Hutchings, Switching off hydrogen peroxide hydrogenation in the direct synthesis process. Science **323**, 1037 (2009)

R.F. Egerton, Limits to the spatial, energy and momentum resolution of electron energy-loss spectroscopy. Ultramicroscopy **107**, 575 (2007)

D.I. Enache, J. Edwards, P. Landon, B. Solsona-Espriu, A.F. Carley, A.A. Herzing, M. Watanabe, C.J. Kiely, D.W. Knight, G.J. Hutchings, Solvent-free oxidation of primary alcohols to aldehydes using titania-supported gold-palladium catalysts. Science **311**, 362 (2006)

J. Fertig, H. Rose, On the theory of image formation in the electron microscope II. Signal-to-noise considerations on specimen resolution and quality of phase-contrast images obtained by hollow cone illumination. Optik **54**, 165 (1979)

C.E. Fiori, C.R. Swyt, in *Microbeam Analysis – 1989*, ed. by P.E. Russell, The use of theoretically generated spectra to estimate detectability limits and concentration varience in energy-dispersive X-ray microanalysis (San Francisco Press, San Francisco, CA, 1989), p. 236

C.E. Fiori, C.R. Swyt, Desk Top Spectrum Analyzer (DTSA), U.S. Patent 5299138, (1994)

C.E. Fiori, C.R. Swyt, R.L. Myklebust, NIST/NIH desk top spectrum analyzer (1992), public domain software available from the National Institute of Standards and Technology Gaithersburg, MD. http://www.cstl.nist.gov/div837/Division/outputs/DTSA/DTSA.htm

A.J. Garratt-Reed, D.C. Bell, *Energy-Dispersive X-ray Analysis in the Electron Microscope* (BIOS Scientific Publishers, Oxford, 2003)

A. Genç, R. Banerjee, G.B. Thompson, D.M. Maher, A.W. Johnson, H.L. Fraser, Complementary techniques for the characterization of thin film Ti/Nb multilayers. Ultramicroscopy **109**, 1276 (2009)

J.I. Goldstein, J.L. Costley, G.W. Lorimer, R.J.B. Reed, in *Scanning Electron Microscopy – 1977*, vol 1, ed. by O. Johari, Quantitative X-ray analysis in the electron microscope (IITRI, Chicago, IL, 1977), p. 315

J.I. Goldstein, C.E. Lyman, Z. Zhang, in *Microbeam Analysis – 1990*, ed. by J.R. Michael, P. Ingram, Spatial resolution and detectability limits in thin-film X-ray microanalysis (San Francisco Press, San Francisco, CA, 1990), p. 265

M. Haider, H. Müller, S. Uhlemann, J. Zach, U. Loebau, R. Hoeschen, Prerequisites for a C_C/C_S-corrected ultrahigh-resolution TEM. Ultramicroscopy **108**, 167, (2008)

M. Haider, S. Uhlemann, J. Zach, Upper limits for the residual aberrations of a high-resolution aberration-corrected STEM. Ultramicroscopy **81**, 163 (2000)

T. Hanai, M. Hibino, Effect of the spherical aberration on electron probe size. J. Electron Microsc. **33**, 116 (1984)

A.A. Herzing, M. Watanabe, J.K. Edwards, M. Conte, Z.-R. Tang, G.J. Hutchings, C.J. Kiely, Energy dispersive X-ray spectroscopy of bimetallic nanoparticles in an aberration corrected scanning transmission electron microscope. Faraday Discuss. **138**, 337 (2008)

J.A. Hunt, D.B. Williams, Electron energy-loss spectrum-imaging. Ultramicroscopy **38**, 47 (1991)

I. Ishikawa, E. Okunishi, H. Sawada, Y. Okura, K. Yamazaki, T. Ishikawa, M. Kawazu, M. Hori, M. Terao, M. Kanno, S. Tanba, Y. Kondo, Development of a 200 kV atomic resolution analytical electron microscope. Microsc. Microanal. **15**, Suppl. 2, 188 (2009)

C. Jeanguillaume, C. Colliex, Spectrum-image: The next step in EELS digital acquisition and processing. Ultramicroscopy **28**, 252 (1989)

I.T. Jolliffe, *Principal Component Analysis*, 2nd edn. (Springer, New York, NY, 2002)

D.C. Joy, in *Scanning Electron Microscopy/1974*, vol I, ed. by O. Johari, SEM parameters and their measurement (IIT Research Institute, Chicago, IL, 1974)

D.C. Joy, D.M. Maher, in *Scanning Electron Microscopy – 1977*, vol 1, ed. by O. Johari, Sensitivity limits for thin specimen X-ray analysis (IITRI, Chicago, IL, 1977), p. 325

D.C. Joy, A.D. Romig, Jr., J.I. Goldstein, *Principles of Analytical Electron Microscopy* (Plenum, New York, NY, 1986)

V.J. Keast, D.B. Williams, Quantification of boundary segregation in the analytical electron microscope. J. Microsc. **199**, 45 (2000)

M.R. Keenan, P.G. Kotula, Accounting for Poisson noise in the multivariate analysis of ToF-SIMS spectrum images. Surf. Interface Anal. **36**, 203 (2004)

K. Kimoto T. Asaka, T. Nagai, M. Saito, Y. Matsui, K. Ishizuka, Element-selective imaging of atomic columns in a crystal using STEM and EELS. Nature **450**, 702 (2007)

P.G. Kotula, M.R. Keenan J.R. Michael, Automated analysis of SEM X-ray spectral images: a powerful new microanalysis tool. Microsc. Microanal. **9**, 1 (2003)

P.G. Kotula, J.R. Michael, M. Rohde, Results from two four-channel Si-drift detectors on an SEM: Conventional and annular geometries. Microsc. Microanal. **14**(Suppl. 2), 116 (2008)

O.L. Krivanek, P.D. Nellist, N. Dellby, M.F. Murfitt, Z. Szilagyi, Towards sub-0.5 Å electron beams. Ultramicroscopy **96**, 229 (2003)

P. Landon, P.J. Collier, A.F. Carley, D. Chadwick, A.J. Papworth, A. Burrows, C.J. Kiely, G.J. Hutchings, Direct synthesis of hydrogen peroxide from H_2 and O_2 using Pd and Au catalysts. Phys. Chem. Chem. Phys. **5**, 1917 (2003)

P. Landon, P.J. Collier, A.J. Papworth, C.J. Kiely, G.J. Hutchings, Direct formation of hydrogen peroxide from H_2/O_2 using a gold catalyst. Chem. Commun. **18**, 2058 (2002)

J.M. LeBeau, S.D. Findlay, L.J. Allen, S. Stemmer, Quantitative Atomic Resolution Scanning Transmission Electron Microscopy. Phys. Rev. Lett. **100**, 206101 (2008)

S. Link, Z.L. Wang, M.A. El-Sayed, Alloy formation of gold–silver nanoparticles and the dependence of the plasmon absorption on their composition. J. Phys. Chem. B **103**, 3529 (1999)

L.M. Liz-Marzán, Nanometals: Formation and color. Mater. Today **7**(2), 26 (2004)

D.M. Maher, D.C. Joy, M.B. Ellington, N.J. Zaluzec, P.E. Mochel, in *Analytical Electron Microscopy – 1981*, ed. by R.H. Geiss, Relative accuracy of k-factor calculations for thin-film X-ray analysis (San Francisco Press, San Francisco, CA, 1981), p. 33

E.R. Malinowski, *Factor Analysis in Chemistry*, 3rd edn. (Wiley, New York, NY, 2002)

J.R. Michael, in *Practical Analytical Electron Microscopy in Materials Science*, 2nd edn. by D.B. Williams, (Philips Electron Optics Publishing Group, Mahwah, NJ, 1987), p. 82

J.R. Michael, D.B. Williams, A consistent definition of probe size and spatial resolution in the analytical electron microscope. J. Microsc. **147**, 289 (1987)

J.R. Michael, D.B. Williams, C.F. Klein, R. Ayer, A consistent definition of probe size and spatial resolution in the analytical electron microscope. J. Microsc. **160**, 41 (1990)

K. Momma, F. Izumi, VESTA: a three-dimensional visualization system for electronic and structural analysis. J. Appl. Crystallogr. **41**, 653 (2008)
C. Mory, M. Tence, C. Colliex, Theoretical study of the characteristics of the probe for a STEM with a field emission gun. J. Microsc. Spectrosc. Electron. **10**, 381 (1985)
D.A. Muller, L. Fitting-Kourkoutis, M. Murfitt, J.H. Song, H.Y. Wang, J. Silcox, N. Dellby, O.L. Krivanek, Atomic-scale chemical imaging of composition and bonding by aberration-corrected microscopy. Science **319**, 1073 (2008)
H. Müller, S. Uhlemann, P. Hartel, M. Haider, Advancing the hexapole C_S-corrector for the scanning transmission electron microscope. Microsc. Microanal. **12**, 442 (2006)
E. Munro, in *Proceedings of the 8th International Congress on X-Ray Optics and Microanalysis*, ed. by R. Ogilvie, D. Wittry, paper no. 19. Calculation of the combined effects of diffraction, spherical aberration, chromatic aberration and finite source size in the SEM (NBS, Washington, DC, 1977)
J. Nelayah, L. Gu, W. Sigle, C.T. Koch, I. Pastoriza-Santos, L.M. Liz-Marzán, P.A. van Aken, Direct imaging of surface plasmon resonances on single triangular silver nanoprisms at optical wavelength using low-loss EFTEM imaging. Opt. Lett. **34**, 1003 (2009)
J. Nelayah, M. Kociak, O. Stéphan, F.J. García de Abajo, M. Tencé, L. Henrard, D. Taverna, I. Pastoriza-Santos, L.M. Liz-Marzán, C. Colliex, Mapping surface plasmons on a single metallic nanoparticle. Nat. Phys. **3**, 348 (2007)
P.D. Nellist, M.F. Chisholm, N. Dellby, O.L. Krivanek, M.F. Murfitt, Z.S. Szilagyi, A.R. Lupini, A. Borisevich, W.H. Sides, Jr., S.J. Pennycook, Direct sub-angstrom imaging of a crystal lattice. Science **305**, 1741 (2004)
D.E. Newbury, R.L. Myklebust, C.R. Swyt, The use of simulated standards in quantitative electron probe microanalysis with energy-dispersive X-ray spectrometry. Microbeam Anal. **4**, 221 (1995)
D.E. Newbury, D.B. Williams, J.I. Goldstein, C.E. Fiori, in *Analytical Electron Microscopy – 1984*, ed. by D.B. Williams, D.C. Joy, Observations on the calculation of k_{AB} factors for analytical electron microscopy (San Francisco Press, San Francisco, CA, 1984), p. 276
C.W. Oatley, W.C. Nixon, R.F.W. Pease, Scanning electron microscopy. Adv. Electron. Electron Phys. **21**, 181 (1965)
A.J. Papworth, M. Watanabe, D.B. Williams, X-ray spectral simulation and experimental detection of phosphorus segregation to grain boundaries in the presence of molybdenum. Ultramicroscopy **88**, 265 (2001)
J. Philibert, in *Proceedings of the 3rd International Congress on X-ray Optics and Microanalysis*, ed. by H.H. Pattee, V.E. Cosslett, A. Engström, A method for calculating the absorption correction in electron-probe microanalysis (Academic, New York, NY, 1963), p. 379
S.D. Rasberry, Certificate of Analysis for Standard Reference Material 2063. National Bureau of Standards (now National Institute of Standards and Technology), Gaithersburg, MD (1987)
S.J.B. Reed, The single-scattering model and spatial resolution in X-ray analysis of thin foils. Ultramicroscopy **7**, 405 (1982)
W.P. Reed, Certificate of Analysis for Standard Reference Material 2063a. National Institute of Standards and Technology, Gaithersburg, MD (1993)
L. Reimer, H. Kohl, Transmission Electron Microscopy: Physics of Image Formation, 5th ed. (Springer, New York, NY, 2009)
B. Rodríguez-González, A. Burrows, M. Watanabe, C.J. Kiely, L.M. Liz Marzán, Multishell bimetallic AuAg nanoparticles: synthesis, structure and optical properties. J. Mater. Chem. **15**, 1755 (2005)
B. Rodríguez-González, A. Sánchez-Iglesias, M. Giersig, L.M. Liz-Marzán, AuAg bimetallic nanoparticles: formation, silica-coating and selective etching. Faraday Discuss. **125**, 133 (2004)

A.D. Romig Jr., J.I. Goldstein, in *Microbeam Analysis – 1979*, ed. by D.E. Newbury, Detectability limit and spatial resolution in STEM X-ray analysis: application to Fe-Ni alloys (San Francisco Press, San Francisco, CA, 1979), p. 124

H. Rose, Correction of aperture aberrations in magnetic systems with threefold symmetry. Nucl. Inst. Methods **187**, 187 (1981)

C.J. Rossouw, C.T. Forwood, M.A. Gibson, P.A. Miller, Generation and absorption of characteristic X-rays under dynamical electron diffraction conditions. Micron **28**, 125 (1997)

H. Sawada, T. Sasaki, F. Hosokawa, S. Yuasa, M. Terao, M. Kawazoe, T. Nakamichi, T. Kaneyama, Y. Kondo, K. Kimoto, K. Suenaga, Correction of higher order geometrical aberration by triple 3-fold astigmatism field. J. Electron Microsc. **58**, 341 (2009)

P.H. Sheridan, Determination of experimental and theoretical k_{ASi} factors for a 200 kV analytical electron microscope. J. Electron Microsc. Tech. **11**, 41 (1989)

E.B. Steel, R.B. Marinenko, R.L. Myklebust, Quality assurance of energy dispersive spectrometry systems. Microsc. Microanal. **3**(Suppl. 2), 903 (1997)

E.B. Steel, D.E. Newbury, P.A. Pella, in *Analytical Electron Microscopy – 1981*, ed. by R.H. Geiss, Preparation of thin-film glass standards for analytical electron microscopy (San Francisco Press, San Francisco, CA, 1981), p. 65

L. Strüder, N. Meidinger, D. Stotter, J. Kemmer, P. Lechner, P. Leutenegger, H. Soltau, F. Eggert, M. Rohde, T. Schulein, High-resolution X-ray spectroscopy close to room temperature. Microsc. Microanal. **4**, 622 (1998)

H. Tanaka, A. Sadakata, M. Watanabe, Y. Tomokiyo, N. Nishimura, M. Ozaki, Application of a JEM-2010FEF FEG-AEM for elemental analysis of microstructures in heat-resisting Cr steel. J. Electron Microsc. **51S**, S127 (2002)

L.E. Thomas, High spatial resolution in STEM X-ray microanalysis. Ultramicroscopy **9**, 311 (1982)

J.M. Titchmarsh, Detection of electron energy-loss edge shifts and fine structure variations at grain boundaries and interfaces. Ultramicroscopy **78**, 241 (1999)

J.M. Titchmarsh, S. Dumbill, Multivariate statistical analysis of FEG-STEM EDX spectra. J. Microsc. **184**, 195 (1996)

P. Trebbia, N. Bonnet, EELS elemental mapping with unconventional methods I. Theoretical basis: Image analysis with multivariate statistics and entropy concepts. Ultramicroscopy **34**, 165 (1990)

E. Van Cappellan, A. Schmitz, A simple spot-size versus pixel-size criterion for X-ray microanalysis of thin foils. Ultramicroscopy **41**, 193 (1992)

M. Varela, A.R. Lupini, K. van Benthem, A.Y. Borisevich, M.F. Chisholm, N. Shibata, E. Abe, S.J. Pennycook, Materials characterization in the aberration-corrected scanning transmission electron microscope. Annu. Rev. Mater. Res. **35**, 539 (2005)

M. Varela, M.P. Oxley, W. Luo, J. Tao, M. Watanabe, A.R. Lupini, S.T. Pantelides, S.J. Pennycook, Atomic resolution imaging of oxidation states in manganites. Phys. Rev. B **79**, 085117 (2009)

W.H. Vaughan, in *Scanning Electron Microscopy/1976*, vol I, ed. by O. Johari, The direct determination of SEM beam diameters (IIT Research Institute, Chicago, IL, 1976), p. 745

H.S. von Harrach, P. Dona, B. Freitag, H. Soltau, A. Niculae, M. Rohde, An integrated silicon drift detector system for FEI Schottky field emission transmission electron microscopes. Microsc. Microanal. **15**(Suppl. 2), 208 (2009)

M. Watanabe, D.W. Ackland, A. Burrows, C.J. Kiely, D.B. Williams, O.L. Krivanek, N. Dellby, M.F. Murfitt, Z. Szilagyi, Improvements of X-ray analytical capabilities by spherical aberration correction in scanning transmission electron microscopy. Microsc. Microanal. **12**, 515 (2006)

M. Watanabe, Z. Horita, M. Nemoto, Absorption correction and thickness determination using the ζ factor in quantitative X-ray microanalysis. Ultramicroscopy **65**, 187 (1996)

M. Watanabe, M. Kanno, D.W. Ackland, C.J. Kiely, D.B. Williams, Applications of Electron Energy-Loss Spectrometry and Energy Filtering in an Aberration-Corrected JEM-2200FS STEM/TEM. Microsc. Microanal. **13**(Suppl. 2), 1264 (2007)

M. Watanabe, M. Kanno, E. Okunishi, Atomic-resolution elemental mapping by EELS and XEDS in aberration corrected STEM. JEOL News. **45**, 8 (2010)

M. Watanabe, E. Okunishi, K. Ishizuka, Analysis of Spectrum-Imaging Datasets in Atomic-Resolution Electron Microscopy. Microsc. Anal. **23**(7), 5 (2009)

M. Watanabe, H. Sawada, Ultramicroscopy (submitted)

M. Watanabe, D.B. Williams, Atomic-level detection by X-ray microanalysis in the analytical electron microscope. Ultramicroscopy **78**, 89 (1999)

M. Watanabe, D.B. Williams, Quantification of elemental segregation to lath and grain boundaries in low-alloy steel by STEM X-ray mapping combined with the ζ-factor method. Z. Metallk. **94**, 307 (2003)

M. Watanabe, D.B. Williams, The quantitative analysis of thin specimens: a review of progress from the Cliff-Lorimer to the new ζ-factor methods. J. Microsc. **221**, 89 (2006)

M. Watanabe, D.B. Williams, Y. Tomokiyo, Comparison of detectability limits for elemental mapping by EF-TEM and STEM-XEDS. Micron **34**, 173 (2003)

O.C. Wells, in *Scanning Electron Microscopy*, Chapter 4 Electron-optical design of small-current probe-forming system (McGraw-Hill, New York, NY, 1974), p. 69

D.B. Williams, C.B. Carter, *Transmission Electron Microscopy: A Textbook for Materials Science*, 2nd edn. (Springer, New York, NY, 2009)

D.B. Williams, J.R. Michael, J.I. Goldstein, A.D. Romig Jr., Definition of the spatial resolution of X-ray microanalysis in thin foils. Ultramicroscopy **47**, 121 (1992)

D.B. Williams, A.J. Papworth, M. Watanabe, High Resolution X-ray Mapping in the STEM. J. Electron Microsc. **51S**, S113 (2002)

J.E. Wood, D.B. Williams, J.I. Goldstein, An experimental and theoretical determination of k_{AFe} factors for quantitative X-ray microanalysis in the analytical electron microscope. J. Microsc. **133**, 255 (1984)

T. Yaguchi, M. Konno, T. Kamino, M. Watanabe, Observation of three-dimensional elemental distributions of a Si device using a 360°-tilt FIB and the cold field-emission STEM system. Ultramicroscopy **108**, 1603 (2008)

Z.J. Zaluzec, Innovative instrumentation for analysis of nanoparticles: The π steradian detector. Microsc. Today **17**(4), 56 (2009)

N.J. Zaluzec, J.S. Iwanczyk, B.E. Patt, S. Barkan, L. Feng, Performance of a high count rate silicon drift X-ray detector on the ANL 300 kV advanced analytical electron microscope. Microsc. Microanal. **9**(Suppl. 2), 892 (2003)

N.J. Zaluzec, A.W. Nicholls, Experimental gun brightness measurements on a 300 kV CFEG. Microsc. Microanal. **4**(Suppl. 2), 386 (1998)

W. Zhou, E.I. Ross-Medgaarden, W.V. Knowles, M.S. Wong, I.E. Wachs, C.J. Kiely, Identification of active Zr–WO_x clusters on a ZrO_2 support for solid acid catalysts. Nat. Chem. **1**, 722 (2009)

T.O. Ziebold, Precision and sensitivity in electron microprobe analysis. Anal. Chem. **39**, 858 (1967)

8

STEM Tomography

Paul A. Midgley and Matthew Weyland

8.1 Introduction

It was in 1917 that the Austrian mathematician Johan Radon published his pioneering work on the projection of an object into a lower-dimensional space (Radon 1917). This projection or transform, now known as the Radon transform, and its inverse form the mathematical basis of tomographic techniques used today. Nearly 40 years later, in 1956, building on Radon's original paper, Bracewell (1956) proposed a method to reconstruct a 2D map of solar emission from a series of 1D 'fan beam' profiles measured by a radio telescope. In 1963, the first X-ray scanner was built (Hounsfield 1972) for medical imaging yielding remarkable 3D reconstructions of the human body; Cormack and Hounsfield, the pioneers of the technique, were awarded the Nobel Prize for Medicine in 1979. In medicine, other radiation has been used in a tomographic fashion, such as positron emission tomography (PET) (Brownell et al. 1971), ultrasound CT (Baba et al. 1989) and reconstructions from magnetic resonance imaging (Hoult 1979). In other disciplines, alternative forms of tomography were developed, for example 3D stress analysis (Hirano et al. 1995), geophysical mapping (Zhao and Kayal 2000) and non-destructive testing (Deans et al. 1983, Reimers et al. 1990).

The first examples of 3D reconstructions using electron microscopy started with the publication of three seminal papers in 1968. The first, by de Rosier and Klug (1968), determined the structure of a biological macromolecule whose helical symmetry enabled a full reconstruction to be made from a single projection (micrograph). In the second of these papers, Hoppe et al. (1968) showed how with a sufficient number of projections (images) it is possible to reconstruct fully asymmetrical systems, i.e. for objects with no internal symmetry. The third, by Hart (1968), demonstrated how the signal-to-noise ratio in a projection could be improved by using an 'average' re-projected image calculated from a tilt series of micrographs: the *polytropic montage.* Hart acknowledged the 3D information generated by such an approach without extending this to the possibility of full 3D reconstruction. A number of theoretical papers were subsequently published

S.J. Pennycook, P.D. Nellist (eds.), *Scanning Transmission Electron Microscopy*,
DOI 10.1007/978-1-4419-7200-2_8,

that discussed the limits of the Fourier techniques, used by de Rosier and Klug (Crowther et al. 1970), alternative real space reconstruction (Ramachandran and Lakshminarayanan 1971) and iterative reconstruction algorithms (Gilbert 1972a, Gordon et al. 1970).

The early successes of electron tomography were in structural biology, where 3D TEM techniques were developed to study protein structures (Unwin and Henderson 1975), viruses (Bottcher et al. 1997), ribosomes (Gabashvili 2000) and larger cellular organelles, such as the mitochondria (Marsh et al. 2001). Three basic techniques have been developed to solve structures in biology based on 3D techniques: (i) electron crystallography where a system, such as a protein, can be crystallised and then electron diffraction or high-resolution imaging can be used and the crystal structure factors derived; (ii) recording images with multiple copies of the same object of interest, such as a virus. A single micrograph can then contain many different projections (within a sub-image) of the same structure and the 3D object can be reconstructed; this is known as 'single particle' analysis; and (iii) for unique structures (such as cellular organelles) a series of images are recorded at successive tilts, reconstructing the object from the tilt series of images. The last approach is what most people would now regard as 'electron tomography'. In almost all biological microscopy, BF images are used.

The success of tomography in biological electron tomography has been mirrored in recent years in the physical sciences. Early applications included the study of block co-polymers (Spontak et al. 1988), which can assemble into a range of complex nanostructures including lamellae and gyroid networks, and the porous nature of zeolites (Koster et al. 2000). In early investigations, bright-field (BF) images were used with the approximation made that the objects were weakly scattering and non-crystalline, and as such each BF image is a reasonable approximation to a projection of the 3D structure. In 2001, it was realised that for strongly scattering, highly crystalline objects, such as those found predominantly in the physical sciences, a different form of tomography was needed. The scanning transmission electron microscope high-angle annular dark-field (STEM HAADF) image is a far 'simpler' image to interpret from crystalline materials than the equivalent BF image and has become the method of choice for most materials-based electron tomography.

Since that time, electron tomography has been applied to many classes of materials (Mobus and Inkson 2001, Spontak et al. 1996, Stegmann et al. 2003, Weyland and Midgley 2003a) and there has also been a rapid increase in the number of alternative imaging modes available for tomography, such as EFTEM-based tomography for 3D compositional mapping (Mobus et al. 2003) or holo-tomography (Twitchett et al. 2002) for mapping electric and magnetic fields. In this chapter, however, we will consider only STEM-based tomography. Elsewhere in this book there are discussions of 'confocal STEM', a through-focal approach to 3D imaging, especially for high-resolution imaging, e.g. van Benthem et al. (2005). For lower-resolution confocal STEM, especially work on thick specimens, readers should consult works by, for example, Frigo et al. (2001).

The popularity of electron tomography has evolved in part because of the availability of inexpensive, but powerful, computers not only to improve the speed of reconstruction but also to enable the electron microscope to be controlled by the computer and to allow automatic acquisition of the tilt series of images. In addition, there have been improvements in the mechanical performance of goniometers to minimise hysteresis and drift.

It is important to remember that many other tomographic methods exist in the physical sciences, for example X-ray tomography to reconstruct relatively large 3D structures, such as metallic foams (Banhart 2001) or to probe the stress in engineering structures (Marrow et al. 2004). Recent developments in Fresnel zone plate fabrication have led to X-ray tomograms using soft X-rays with a resolution of 50 nm or better (Weierstall et al. 2002). The atom probe field ion microscope (APFIM), based around a time-of-flight spectrometer and a position-sensitive detector, is able to reconstruct 3D maps of atom positions and determine each atomic species (Miller 2000). The recent development of the local electrode atom probe coupled with laser excitation enable a greater range of specimens to be studied with this technique than ever before, and to image over a far greater field of view than previously.

8.2 Tomographic Reconstruction

In Radon's paper of 1917 (Radon 1917), a transform, known now as the *Radon transform*, R, defines the mapping of a function $f(x,y)$, describing a real space object D, by a projection, or line integral, through f along all possible lines L with unit length ds so that

$$Rf = \int_L f(x,y)ds \tag{1}$$

A discrete sampling of the Radon transform is geometrically equivalent to the sampling of an experimental object by a projection or some form of transmitted signal. Given the nature of the transform, the structure of the object $f(x,y)$ can then be reconstructed from projections Rf using the inverse Radon transform. All reconstruction algorithms used in tomography are approximations of this inverse transform. The Radon transform converts real space data into 'Radon space' (l, θ), where l is the line perpendicular to the projection direction and θ is the angle of the projection. An 'image' in Radon space is called a 'sinogram'. A single projection of the object is a line at constant θ in Radon space. A series of projections at different angles will therefore sample Radon space and given a sufficient number of projections, an inverse Radon transform should reconstruct the object. In practice, the sampling of (l, θ) is always limited and any inversion will be imperfect.

An alternative, but formally related (Deans 1983), way to consider the projection of an object is to use a Fourier description. The 'central slice theorem', or 'projection-slice theorem', states that a projection of an object at a given angle in real space is a central section through the Fourier transform of that object. (This is familiar to anyone who has considered the intensities of zero-order Laue zone reflections as dependent on the *projected* potential.) By acquiring images or projections, at many different tilts, many sections of Fourier space will be sampled increasing the information available in the 3D Fourier space of the object. With a sufficiently large number of projections taken over all angles, a Fourier inversion should yield a complete description of the 3D object. Tomographic reconstruction from an inverse Fourier transform of the set of Fourier-transformed projections is known as *direct Fourier reconstruction* (Ramachandran and Lakshminarayanan 1971).

In practice, experimental data are always sampled at discrete angles and some kind of interpolation is required to fill the 'gaps' in Fourier space (Hart 1968) to avoid reconstruction artefacts. Although elegant, Fourier reconstruction methods may be difficult to implement for electron tomography where, unlike single particle reconstructions, there is often a very limited number of projections; for tomography, Fourier methods have in general been replaced by real space *backprojection* methods.

8.2.1 Backprojection

The method of backprojection is based on smearing out each image (or projection) acquired in a tilt series back into a 3D space at the angle of the original projection; this generates a 'ray' that will describe uniquely an object in the projection direction. Using a sufficient number of images (projections), from different angles, the superposition of all the backprojected 'rays' will generate the original object: this is known as *direct backprojection* (Herman 1980, Radermacher 1992), see Figure 8–1.

In general, tomographic reconstruction using backprojection relies on the premise that the intensity in the projection is some monotonic function of the physical quantity to be reconstructed (Hawkes 1992); this is known as the 'projection requirement'. In electron microscopy, there are a number of competing contrast mechanisms that could be used, only some of which, in general, fulfil the projection requirement.

Reconstructions by direct backprojection appear to be blurred because of an apparent enhancement of low frequencies and fine spatial detail reconstructed poorly. The blurring is an effect of the uneven sampling of spatial frequencies in the ensemble of original projections, as illustrated in Figure 8–2, where each of the projections is represented by a central section in Fourier space. Even if the frequency sampling in each image is uniform (the spacing of the 'data points' in the figure), this still results in proportionately higher density of data near the centre of Fourier space compared with the periphery. This undersampling of high spatial frequencies leads to a 'blurred' reconstruction. This uneven sampling can be corrected by using a weighting filter (a radially linear function in Fourier space, zero at the centre, rising to a maximum and then apodised, such that the Fourier transform has zero value at the Nyquist frequency, to avoid enhancing noise; Nyquist 1928). This

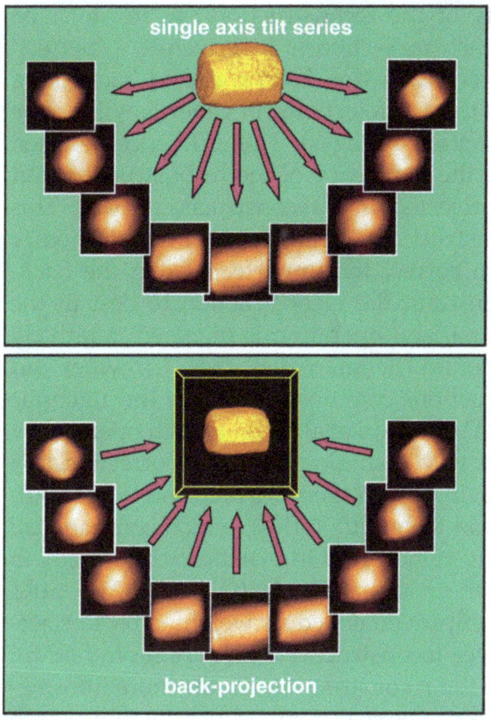

Figure 8–1. Schematic diagram showing the two-stage tomography process. In the upper figure a series of 2D images, or projections, of a 3D object is recorded about a tilt axis. In the lower figure the ensemble of images is backprojected into a 3D space to reconstruct the object.

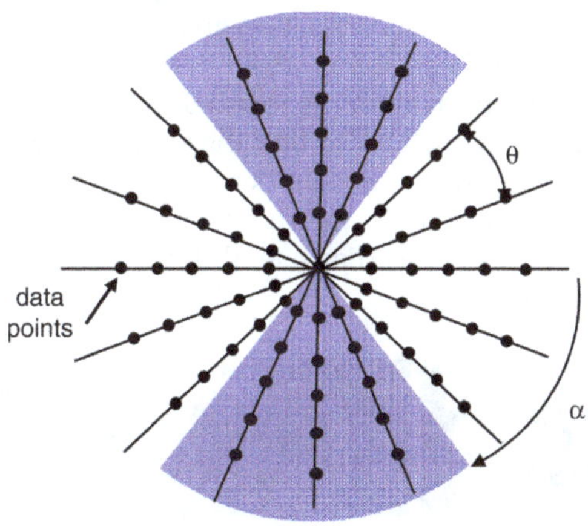

Figure 8–2. A Fourier representation of the sampling of an object by a single tilt series of images. The radial lines represent Fourier planes. Although each plane might have an even sampling of data points (*dots*), the ensemble of images leads to an under-sampling of high-frequency information and a consequent blurring of the reconstructed object. The *blue region* indicates the 'wedge' of missing information brought about by a limited tilt angle. θ is the angular sampling and α is the maximum tilt angle.

approach, known as *weighted backprojection* (Gilbert 1972b), is widely used especially in the biological community.

'Fan-shaped' artefacts and other imperfections can arise from a limited data set, a non-optimal weighting filter and if the original images have a low SNR. The quality of the reconstruction can however be improved by realising that each projection is in principle a 'perfect' reference. If the (imperfect) reconstruction is re-projected back along the original projection angles the re-projections, in general, will not be identical to the original projections (images). The difference between them will be characteristic of the deficiency of the reconstruction and this difference can be backprojected into reconstruction space. This generates a 'difference' reconstruction which can be used to modify the original reconstruction to correct the imperfections in the back-projection; in essence, we constrain the reconstruction to best fit the original projections. The comparison operation is repeated iteratively until the optimal solution is found (Bellman et al. 1971, Crowther and Klug 1971). Iterative reconstructions commonly used are the algebraic reconstruction technique (ART) (Gordon et al. 1970), which operates by comparing the reconstruction with a single projection, correcting in a single direction and then moving on to the next projection, or simultaneous iterative reconstruction technique (SIRT), which compares all the projections simultaneously rather than in isolation; SIRT tends to be more stable than ART when images are noisy (Gilbert 1972a). SIRT does not offer any resolution improvements over weighted backprojection techniques, but it can offer reconstructions that are more faithful to the original object because the reconstruction retains more information to higher frequencies. This can be seen comparing how robust each algorithm is with respect to noise, see Figure 8–3, and more quantitatively using Fourier shell correlation analysis, see the next section. The SIRT algorithm has been summarised in Figure 8–4.

There have also been attempts to use Bayesian techniques (maximum entropy methods) which try to find the simplest (least complex) reconstruction taking into account the known projections, the noise in the data, sampling artefacts in the reconstruction and the contrast limits of the original projections (Barth et al. 1989, Frank 1996, Skoglund and Ofverstedt 1996).

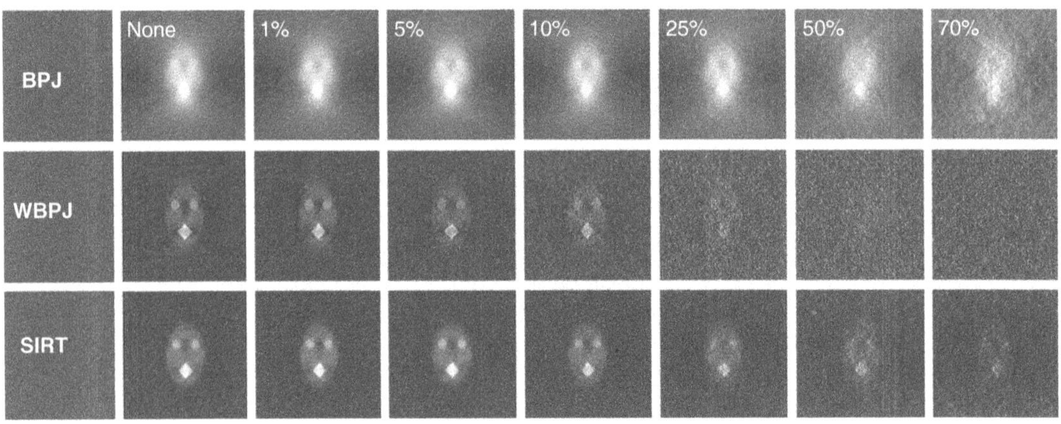

Figure 8–3. A comparison of the reconstruction fidelity of a test head phantom using three reconstruction algorithms, backprojection (BPJ), weighted backprojection (WBPJ) and simultaneous iterative reconstruction technique (SIRT). The reconstructions are shown as a function of simulated noise content (as a percentage). It is clear that the SIRT algorithm produces reconstructions that retain the highest fidelity especially at high noise levels.

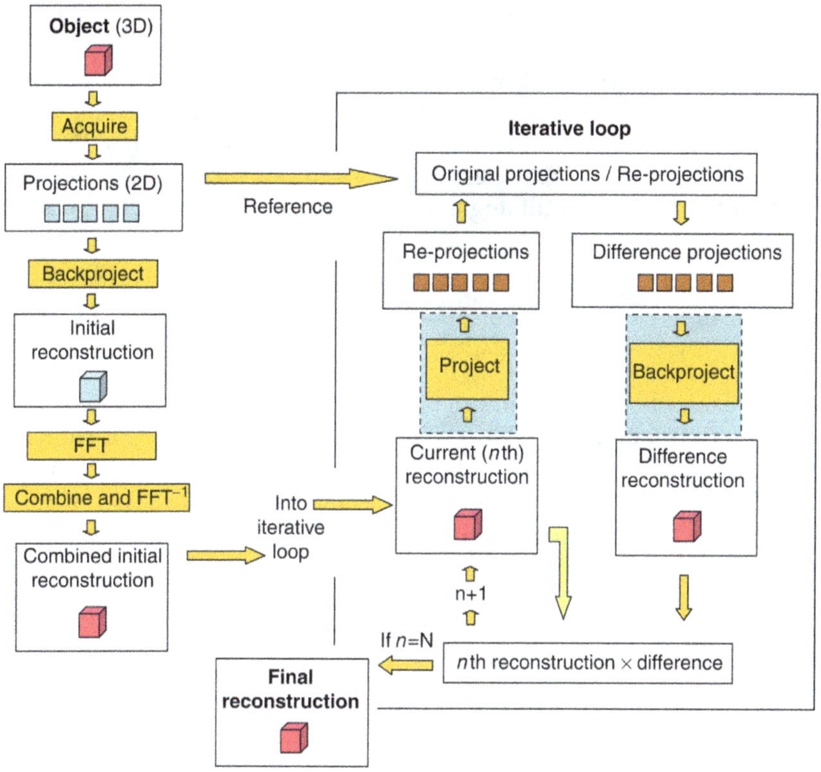

Figure 8–4. A flow diagram illustrating the SIRT algorithm.

Iterative algorithms impose constraints through the reconstruction cycle. If other a priori information is known this can be incorporated iteratively. 'Discrete tomography' considers the reconstruction object as being composed of discrete units. The discreteness can be spatial (e.g. an atomic lattice; Jinschek et al. 2008) or in terms of an object's density (i.e. the number of grey levels is discrete; Batenburg 2005). For the latter case in the extreme limit, a high-quality binary reconstruction can be found with only a handful of projections. For objects with more grey levels, more projections are required but this technique holds great promise in materials science, where very often the number of different materials (densities or grey levels) is known.

8.2.2 Resolution

As shown in Figure 8–2, to achieve a high 3D spatial resolution in a tomographic reconstruction requires the acquisition of as many projections as possible, over as wide a tilt range as possible. The relationship, first defined by Crowther (Crowther et al. 1970), between the number of projections (N) and the resolution (d) attainable can be defined as

$$d = \frac{\pi D}{N} \tag{2}$$

where D is the diameter of the reconstruction volume. Here it is assumed that the N projections are spread evenly over π radians. However, in electron tomography there is almost always an upper limit to the tilt angle, leading to a loss of resolution in the least sampled direction 'the missing wedge', and an 'elongation' of objects in that direction (usually the optic axis). An estimation of this elongation (e), as a function of the maximum tilt angle (α), is (Radermacher and Hoppe 1980)

$$e = \sqrt{\frac{\alpha + \sin\alpha\cos\alpha}{\alpha - \sin\alpha\cos\alpha}}. \tag{3}$$

For example, for a maximum tilt angle of 70°, this leads to an elongation of approximately 30%. In addition, TEM samples are often in the shape of a thin slab with far larger dimension in-plane (x, y) than in-depth (z). This can be modelled using a modified diameter (D) which is a factor of the thickness of the slab (t) and the maximum tilt angle (α) (Radermacher 1992)

$$D = t\cos\alpha. \tag{4}$$

This modification leads to a less pessimistic resolution prediction than the basic Crowther equation.

However, these equations are only estimates and the reconstruction resolution must also be governed by the noise characteristics of the data and the nature of any constraints applied. A useful approach is therefore to examine the intensity distribution of the object in Fourier space and determine whether this is above some threshold for noise. This is used in Fourier shell correlation (FSC) (Van Heel and Harauz 1986), the differential phase residual (DPR) (Frank et al. 1981, Unser et al. 1987) and the spectral signal-to-noise ratio (SSNR) method (Henderson et al. 1990), methods developed originally for determining the resolution of single particle reconstruction (Penczek 2002). As these methods rely on the analysis of the sampling and noise statistics of the reconstructed volume, accurate, or more importantly reliable, results from conventional tomographic data sets may prove difficult to determine.

With the Fourier shell correlation (FSC) method, a tilt series can be separated into two series, typically containing 'odd' and 'even' tilt images. The FSC compares the two resulting tomograms by looking at the correlation between shells of constant radius k in (3D) Fourier space, $F_n(k)$, corresponding to the two tomograms. The FSC is defined as

$$FSC(k) = \frac{\sum F_1(k)F_2^*(k)}{\sqrt{\left(\sum |F_1(k)|^2\, |F_2^*(k)|^2\right)}}. \tag{5}$$

The FSC is calculated for each shell k, and a resolution may be defined as being where the FSC declines to a certain value, often taken as 0.5, which corresponds to a SNR of 1.

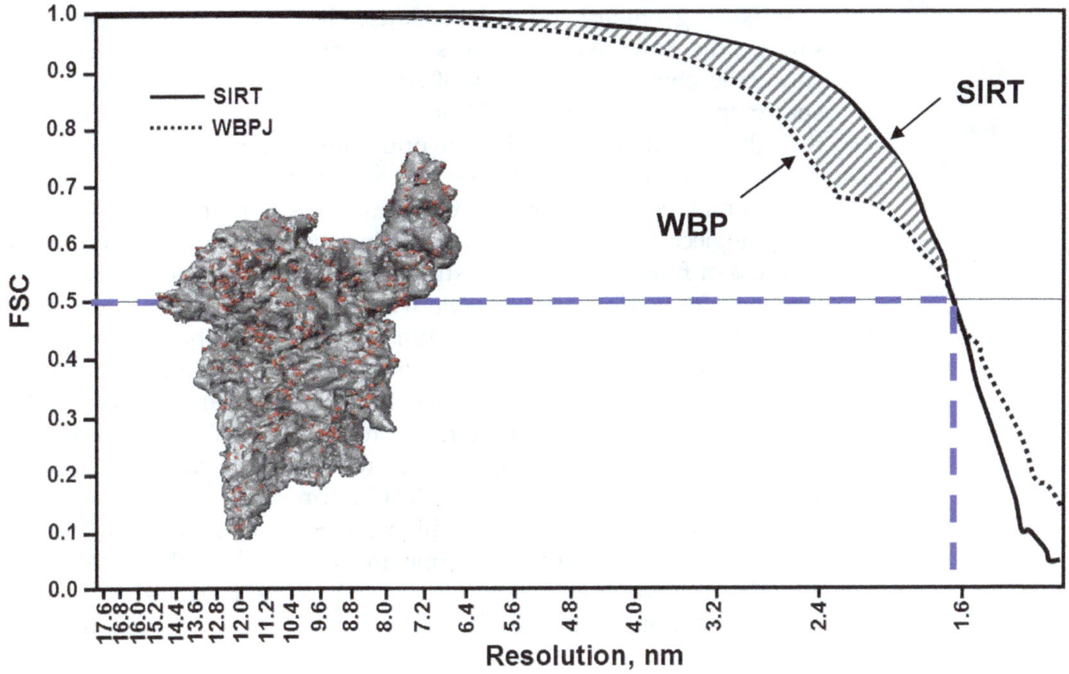

Figure 8–5. An example of Fourier shell correlation comparing the results from SIRT and WBP reconstructions for the same data set acquired from a heterogeneous catalyst. A surface render of the SIRT reconstruction of the catalyst is shown as an inset. The SIRT reconstruction retains more information at medium to high spatial frequencies and leads to a more faithful reconstruction of the original object. The 0.5 value of the FSC is often used as measure of the resolution of the reconstruction. Note both algorithms give comparable resolution measured this way for this data set.

Although there is some debate as to whether the resolution can be determined this way it is a very useful way to compare the information content present in reconstructions made with different algorithms using identical data. Figure 8–5 shows an example of FSC analysis of a tomographic reconstruction of a heterogeneous catalyst, see Section 8.6.1, made using WBP and SIRT. The analysis suggests that while the overall resolution is similar there is more 'information' (hatched area) contained in the SIRT reconstruction compared to the conventional WBP.

8.3 Practical Issues

8.3.1 Acquisition

The acquisition of a STEM tomographic tilt series is likely to take between 1 and 3 h and so it is important to ensure that the specimen suffers minimal damage or contamination during that period. The usual factors should therefore be considered to ensure the specimen and microscope are clean and that the correct voltage is chosen to minimise knock-on damage or radiolysis. Conventional TEM specimens are slab-like in geometry and their thickness chosen such that at high tilts, say 70°, the projected thickness is not so large that it leads to blurring, chromatic aberration and loss of contrast. However, of course if the specimen is too thin then the 3D information may be lost and so careful thought is needed in choosing the optimum specimen thickness.

The maximum tilt angle beyond which it will be impossible to record images may be limited by specimen thickness or by the constraints of the polepiece gap. The so-called missing wedge of information leads to a loss of fidelity in the reconstruction described previously. Modern tomography holders are available with a narrow width and profile to ensure maximum possible tilt within the confines of the polepiece gap (Weyland et al. 2004). To improve matters further, dual-axis tomography can be implemented in which two tilt series are recorded with one having a tilt axis perpendicular to the other. By combining the two series the missing wedge can be reduced to a missing pyramid, see Figure 8–6, and the amount of lost information greatly reduced for a particular tilt range. Specialist high-tilt tilt-rotate holders enable tilt series to be recorded from the same area of sample with a rotation of 90° (about the optic axis) between each tilt series. Exact 90° rotation is not necessary but an accurate knowledge of the angle is needed and this

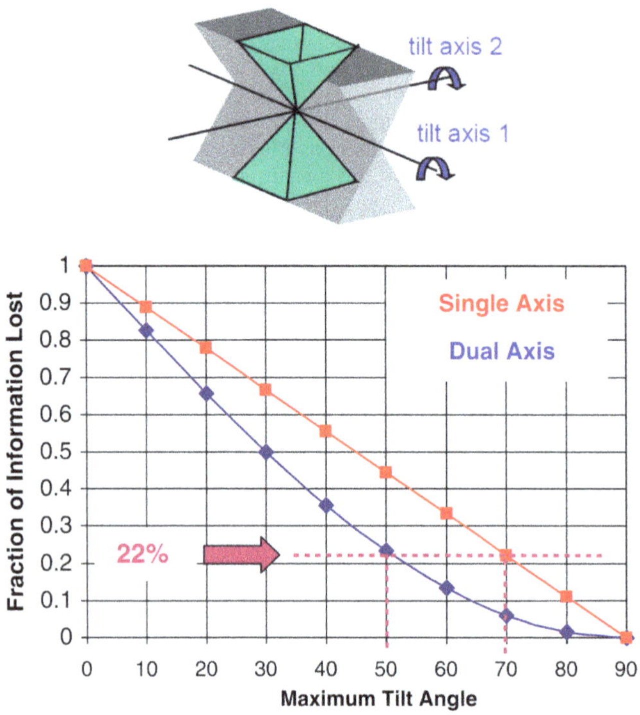

Figure 8–6. (*Top*) The missing wedge of a single tilt series (*grey*) compared to the equivalent missing wedge from a dual-axis series (*green*) using the same maximum tilt angle. (*Bottom*) A simple geometrical comparison of the information lost in a single tilt series versus a dual-axis series. As an example one can see that a dual-axis tilt series with maximum tilt angle of 50° has the same information as a single tilt series with a maximum tilt angle of 70°.

can be determined from a comparison of the two zero-degree images in each series.

Combining the data can be achieved through summing the two reconstructions (in Fourier space) or by combining iteratively the two data sets to give a single reconstruction that best fits both data sets simultaneously, using the ADA-SIRT algorithm (Tong et al. 2006). The ADA-SIRT algorithm leads to a reconstruction that is less sensitive to noise and gives the most faithful reconstruction. Image distortions brought about by scan inaccuracy in STEM mode are problematic when 'adding' the two tilt series together and must be minimised or calibrated to take the distortions into account. An example of a dual-axis reconstruction is shown in Section 8.6.

To avoid the effects of the missing wedge completely, and thus achieve the 'best' possible reconstruction, there are specimen holders available that allow complete rotation of the specimen using a separate rotation mechanism (Koguchi et al. 2001) or a combination of an internal on-axis rotation coupled with the goniometer tilting mechanism. Typical specimens that are used in such holders are needle-like or pillar-shaped and are often prepared using a FIB workstation. Examples of this will be shown later.

Even with modern goniometers, which can be pre-calibrated to account for any mechanical shift when tilted (Mastronarde 2005, Zheng 2004, Ziese et al. 2002), there will always be a small image shift when completing a tilt series acquisition. To minimise this, auto-tracking procedures have been coupled with auto-focussing algorithms to try to make the acquisition process automated. Tracking is often carried out with cross-correlation whereas focussing is corrected by measuring the image shift induced by a small beam tilt (Koster et al. 1997) for BF imaging or by maximising the image contrast in STEM HAADF mode. At the highest resolution, neither method is exact, so in practice focussing has to be optimised in a manual fashion.

Further practical issues related specifically to STEM acquisition are discussed in Section 8.4.

8.3.2 Alignment

Although the mechanical imperfections of the stage are minimised, to achieve high-quality tomographic reconstructions all projections must be aligned post-acquisition, preferably to sub-pixel accuracy, to a common tilt axis. The alignment is performed on objects that are present throughout the tilt series, and while the tilt axis may be moved to pass through any point of the reconstruction volume, the direction of the tilt axis is fixed and must be determined with high accuracy for a successful tomographic reconstruction. There are two conventional approaches used to align tomographic tilt series: tracking of fiducial markers and cross-correlation. Given adequate image contrast both techniques should lead to high alignment accuracy. The advantage of the fiducial technique is that it can determine not only spatial alignment but simultaneously the direction of the tilt axis and any secondary distortions caused by optical effects and/or beam damage (Lawrence

1992). Cross-correlation alignment however makes use of the majority of the information in an image (rather than a few selected points in the fiducial technique); it does not require the subjective selection of markers, makes no assumptions about the shape of the supporting film and is perhaps easier to implement (Frank and McEwen 1992). Another benefit to markerless alignment is that it avoids the reconstruction problems associated with high-contrast objects, such as colloidal gold, which can mask details in the reconstruction. Cross-correlation alignment however does not automatically determine the tilt axis direction and additional techniques are needed to determine this. Markerless alignment is the norm for materials science applications and will be examined in greater detail below.

8.3.2.1 Alignment Using Markers

The task of alignment can be simplified by introducing onto the specimen a dispersion of markers of known geometry, often spherical colloidal gold particles, which can be used as an alignment reference. The movement of these particles in each projection can be recorded and the tilt axis direction, relative lateral shift, magnification change and image rotation determined by a least-squares tracking of fiducial markers with comparison to a reference projection (Berriman et al. 1984, Lawrence 1983, 1992, Olins et al. 1983).

The number of fiducial markers required for alignment depends on the number of images, whether different tilt axes are used and on the accuracy of the goniometer tilt readings (leave fixed or allow to vary). In general, 15–20 markers are sufficient. Automated approaches to marker selection can be unreliable and often need manual adjustment (Brandt et al. 2001a).

8.3.2.2 Markerless Alignment

Any two projections in a tilt series, given a finely spaced angular increment, will share many common image features offset by the relative shift between the two images. The primary difference between the images will be their spatial offset which can be determined through the use of a cross-correlation algorithm (Frank and McEwen 1992); an example of such a shift determination is given in Figure 8–7. An accurate, sub-pixel, measure of the correlation can be determined by measuring the centre of mass of the central peak, or by fitting the peak with a 2D Gaussian function. Images are often 'stretched' to account for the relative tilt between correlated images and the accompanying small but sometimes significant foreshortening effect.

The quality of spatial alignment is affected strongly by the contrast and noise of the images being correlated. In order to improve the correlation, projections are often filtered to enhance image features, see Figure 8–7. Two filters are commonly used, one that enhances or suppresses particular frequencies, such as a high/band/low pass filter, and one which highlights edges (Russ 1995), such as a Sobel filter. Filtered correlation is also used during the tilt series acquisition to improve the quality of automatic tracking and focussing.

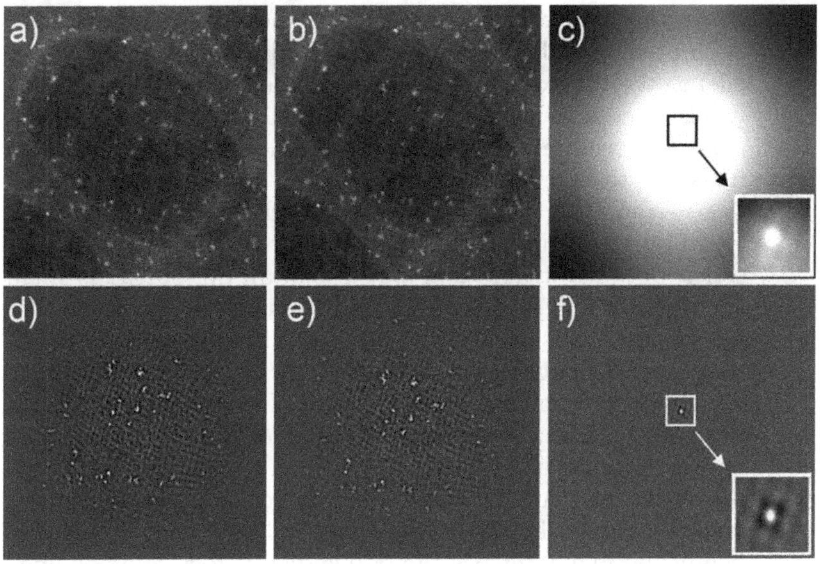

Figure 8–7. Example of cross-correlation to determine the relative shift between two images recorded 2° apart. (a), (b) Two images of a block co-polymer from a HAADF STEM tilt series. (c) Cross-correlation image between the two images, showing a broad intensity peak near the image centre. (d)–(f) The improvement in the cross-correlation peak if a high-pass filter is applied to the original images is shown.

Alignment of an image series by cross-correlation would ideally be achieved with reference to a single image. The change in shape of an object through tilt means that cross-correlation on tomographic data sets is carried out sequentially. This can lead to a build-up of errors as a consequence of small misalignments between each of the projections (Frank and McEwen 1992). In order to minimise this effect, a tilt series can be split into two parts, each using the zero-tilt image as its first reference. All alignment steps can then be applied from the zero-tilt projection out to the largest tilt in both directions, reducing the effect of cumulative shift error.

Although in theory a single alignment pass using cross-correlation should be sufficient, in practice because of the nature of rotational alignment (see below), especially without fiducial markers, and the use of apodising filters, which tend to centre-weight image features that can subtly alter correlation, more than one alignment pass is necessary. In addition the distribution of intense image features in a tilt series, especially away from the main area of interest, may demand alignment on a sub-volume in order to avoid inaccurate correlation.

While the correction of an in-plane rotation can be carried out by a number of different techniques (Frank 1981) an out-of-plane rotation (or tilt), such as induced during a tomographic tilt series, is a more difficult problem to solve. A number of approaches have been used: solution by common lines, series summation, arc minimisation and reconstruction optimisation.

Common lines: Between any two images of the same object, recorded at different tilts, there is some commonality. In the direction parallel to the tilt axis, both images and their Fourier Transforms will be identical. In the perpendicular direction, the correlation peak will be extended due to the foreshortening of image features and an estimate of the tilt angle can be made by measuring this spread. The common lines approach to axis alignment is based on measuring the trend in this spread throughout the tilt series. Between images that are close in tilt this spread is small and difficult to measure with certainty and between images that are far apart this spread can be masked by changes in object shape and/or poor noise statistics (Crowther 1971, Frank 1981). Searching for common lines, in Radon (or sinogram) space, is the basis behind a recent implementation of the common lines approach (Liu et al. 1995).

Series summation: For the single tilt axis geometry, movement of objects through the tilt series should follow a path that is perpendicular to that tilt axis and perpendicular to a 'common line'. Assuming that the spatial (x,y) alignment is close to optimal then a summed image over all, or some, of the tilt series should highlight the movement of any objects through the series (Renken and McEwen 2003). This is illustrated for an experimental tilt series in Figure 8–8. Once the tilt axis direction is determined the whole data set can be rotated to place the axis parallel to a chosen image axis. The accuracy of this approach is dependent on high-contrast image features (as often seen in HAADF images). This approach is simple and fast and so often used as a 'first guess' before more accurate, yet slower techniques are applied, such as arc minimisation.

Arc minimisation: A misalignment of the tilt axis produces an inaccurate reconstruction which manifests itself as a smearing of the reconstruction intensity into 'arcs', the direction of which depends on the direction of the misalignment away from the true axis and the degree of 'spread' reflects the magnitude of that misalignment, see Figure 8–9. These distinctive distortions can provide a way to determine the axis direction by manually aligning, in real time, 2D reconstruction slices from the whole data set.

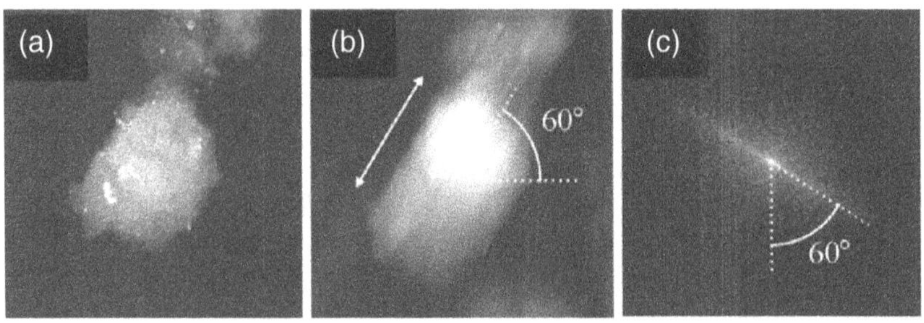

Figure 8–8. Tilt axis direction determination by series summation. (a) A single STEM HAADF image, at zero tilt, from a tilt series acquired from a Pd/C catalyst. (b) The direction of the tilt axis is determined using a tilt series summation and (c) its Fourier transform.

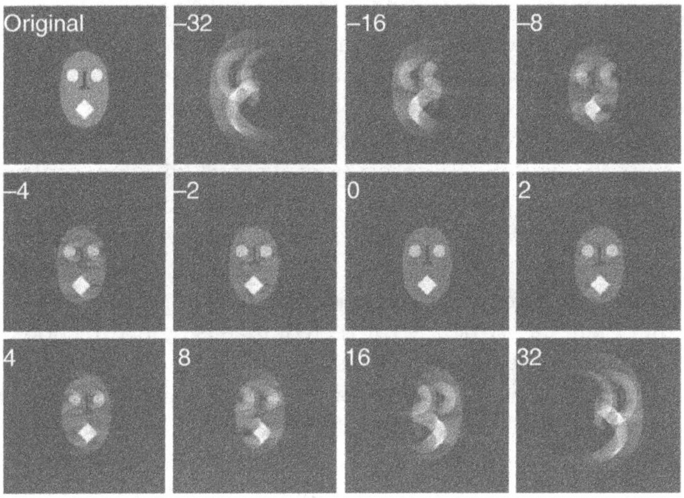

Figure 8–9. The effect of tilt axis misalignment on the reconstruction of a head phantom test object. Distinctive 'arcing' is observed whose curvature is dependent on the direction and magnitude of misalignment; the misalignment deliberately introduced here is quoted in pixels perpendicular to the tilt axis.

Ideally, this would be carried out by rotating the tilt series before reconstructing the new slice. For small angles this can be approximated by a shift of the projections in a single slice in the direction perpendicular to the tilt axis. The alignment is most easily carried out on three slices simultaneously, one at the centre of the volume and two near the edge of the volume, perpendicular to the assumed tilt axis direction. Two variables can be adjusted: an overall shift perpendicular to the axis, which shifts all slices in the same direction, and a rotation, which shifts the two outlying projections in opposite directions and leaves the centre projection unchanged.

Even though alignment by cross-correlation and some form of angular determination, by fiducial means, common lines, or observation of reconstruction, can produce good results, a more holistic automated approach to markerless alignment would be extremely useful. Several such techniques have been proposed but have yet to see broad application. These approaches include optimising the reconstruction by iterative alignment of projections and test reconstructions (Dengler 1989), a new clearer approach to which has recently been published by Winkler and Taylor (2006), or by tracking of image features based on the expected geometric relationship of a tilt series (Brandt et al. 2001b).

8.4 Visualisation

All approaches to 3D visualisation (Russ 1995) have certain drawbacks and which is chosen depends on the type of information required from the data set: sometimes a series of 2D slices through the volume can provide most information, rendering a surface in general better at revealing distinct morphologies and topography while a voxel projection may be better for visualising subtle variations and internal structure. Three-dimensional visualisation can be made more representative and informative by using some additional volumetric segmentation. Indeed, one of the most significant barriers to accurate visualisation is the difficulty of objective segmentation, especially in systems with low contrast and high noise levels. Higher-dimensional data sets, found in for example EFTEM or holographic tomography, can be displayed using colour and vectorial representations.

2D slicing: A tomographic projection is compromised by being the sum of all the structure projected in a single direction, while a slice can reveal structure unambiguously without projection artefacts. Slicing is mostly simply carried out along the three major axes of the reconstruction (x, y, z); following radiographic convention these are sometimes known as the axial, sagittal and coronal directions. However, modern software enables slices to be determined at any orientation within the reconstruction volume; inevitably this will involve 3D interpolation that may affect the fidelity of the data in that slice.

Surface rendering: Surface rendering produces a polygonal surface, normally composed of triangles, that can then be rotated and visualised from any angle and subjected to a variety of light sources, enhancing the 3D effect. A distinct advantage over volumetric visualisation is that it is much less computationally intensive and navigating, rotating, translating and zooming a surface is fast. This approach is not without its drawbacks, the largest of which is the method used to select the position of the polygonal surface. A different surface will lead to an apparently different structure and hence greatly influence any conclusions made. A common approach to rapid visualisation is to use an iso-surface. Here, a surface is generated by selecting a single threshold intensity within the data set and generating a polygon that follows that intensity. However, tomographic reconstructions often lack distinct intensity boundaries, especially from low-contrast data sets, making the choice of the iso-value rather subjective. Another limitation is that while more than one surface can be visualised using transparencies, there is no simple way of using a surface to visualise a function which is changing constantly throughout the volume.

Voxel projection: Volume rendering, or voxel projection, is in essence the re-projection, or directional summation, of the 3D data set at an arbitrary angle, and as such is analogous to the original projection operation in the microscope. However, since the volume is in silico we can modify its optical properties before projection in order to enhance the information of interest about the reconstruction. Voxels of differing intensity can be set to have specific optical characteristics such as colour, transparency or luminosity in order to differentiate them. This allows the removal of unwanted features and viewing of internal and external structures. In addition, voxel projection allows any subtle changes in object density to be visualised, which is difficult, if not impossible, by iso-surface visualisation.

Segmentation: The separation, or segmentation, of greyscale pixels, or voxels, into regions of structural, functional or chemical similarity is one of the central challenges of machine vision across multiple fields of research. The simplest and the most widely used approach, certainly in electron tomography, is manual segmentation. In each slice boundaries are drawn around each feature by hand and while this is both time-consuming and subjective, the resultant visualisations can be highly instructive. Robust automatic, or semi-automatic, segmentation methods would be of great value across a wide range of fields, including electron tomography. These can be classified broadly into those that operate globally and those that operate locally. An iso-surface is a simple form of global segmentation. However, more advanced approaches to global segmentation have been demonstrated by Frangakis and Hegerl (2002) who applied eigenvector analysis of the image grey values and their relative proximity to segment a volume. Local approaches to segmentation usually rely on the detection of changes in volumetric intensity due to some feature of interest. An early application of local segmentation to electron tomography was the use of Watershed transforms by Volkmann (2002), which treats voxel intensity as a relief map to be 'filled' and sets thresholds, the 'watersheds', above which connected areas are segmented into different volumes. Bajaj et al. (2003) have adapted another general approach to segmentation, gradient vector diffusion, to electron tomography. Gradient vector approaches are based on the local gradients within the image and are less prone to local irregularities in voxel intensity. In a related approach, anisotropic non-linear diffusion (AND), a de-noising algorithm, has proven to be successful in both the life sciences and the physical sciences (Fernandez and Li 2003, Frangakis and Hegerl 2001, Xu and Prince 1998). By reducing the noise level in the tomogram, isosurface or segmented volumes are less prone to artefacts amplified by noise content.

Once the data set has been segmented it is possible to analyse the reconstruction in a more quantitative fashion and, despite the relatively recent development of electron tomography in the physical sciences, there are now many examples of such analysis. The early study of Spontak et al. (1988) and Laurer et al. (1997) demonstrated that the lengths of repeating structures in block co-polymers measured from reconstructions agreed very closely with bulk techniques such as small-angle X-ray scattering (SAXS). In a related study, a quantitative analysis of the curvature of a styrene network (Jinnai et al. 1997, 2000) showed a statistically significant deviation from the value of the expected mean curvature, indicating the presence of packing frustration. Given a segmented volume it is also relatively straightforward to measure the size distribution and nearest-neighbour distances of structures as demonstrated by Ikeda et al. (2004) for silica inclusions in natural rubbers. The activity of a catalyst is determined in part by its loading, the mass of active particles per unit area. Electron tomography is the only technique that can provide *direct* measurement of loading for many catalysts. The measurement of particle loading of a nanoscale heterogeneous catalyst has been demonstrated by Midgley and Weyland (2003) using HAADF STEM tomography, see Section 8.6.1 for examples.

8.5 STEM HAADF Tomography

While BF imaging has dominated the life science community, the advent of electron tomography in materials science has seen the number of image modes (Midgley and Weyland 2003) used for electron tomography increase dramatically. This has arisen partly because of the need to minimise diffraction contrast, as with HAADF STEM tomography, partly because materials samples are more robust for prolonged exposure, e.g. as is needed in EFTEM tomography, and partly to be able to map in 3D other significant signals, such as electromagnetic potentials, in holographic tomography, and strain, in weak beam dark-field tomography. Here, though we concentrate on STEM tomography either used on its own or in combination with EELS (Jarausch et al. 2009) and EDX (Koguchi et al. 2001, Saghi et al. 2007) to yield 3D chemical information.

To meet the projection requirement, introduced in Section 8.2, in general crystalline materials should be studied using a tomography technique that minimises diffraction contrast. One approach, demonstrated for EFTEM tomography (Weyland and Midgley 2003b), is the use of a 'hollow cone' illumination mode (Saxton et al. 1978). In this case, the pre-specimen deflectors are used to tilt the beam at a given angle, by a few milliradians and then the beam is continuously rotated azimuthally through 360° at high speed. This results in an averaging over all diffraction conditions, thus reducing, but not eliminating, the effects of diffraction contrast. However, due to the limited accuracy of the microscope deflectors and aberrations of the objective lens, conical illumination will lead to poorer image resolution and for high-magnification studies this approach may be unsuitable. The need to meet the projection requirement led to the development of HAADF STEM tomography for use in materials science (Midgley et al. 2001). This technique has now been used to study the 3D structures of a large range of materials systems and automated acquisition software for STEM tomography is widely available through many manufacturers.

As described elsewhere in this book, images with minimal Bragg diffraction contrast can be formed in STEM mode by increasing the inner radius of an ADF detector so as to exclude Bragg-scattered beams (Howie 1979). The intensity of images collected with this detector is a function of both the atomic number (Z) of the scattering atom, approaching a Z^2 Rutherford-like relationship for a detector annulus with a large inner radius, and the projected thickness. In qualitative terms, up to certain thicknesses (see below), the HAADF signal can be considered as the reciprocal of a mass-thickness BF image, often used in life science tomography. However, at low scattering angles, electron screening will serve to reduce the Z-dependence and as a consequence, high-angle annular dark-field images show enhanced contrast over equivalent low-angle mass-thickness images. The lack of coherence in HAADF images means that such images can be considered as projections of the structure in terms of thickness and atomic number and thus to a good approximation meet the projection requirement for tomographic reconstruction.

The choice of the inner angle for the HAADF detector, θ_{HAADF}, is important and must be large enough to ensure coherent effects are minimal. For thermal diffuse scattering (TDS) to dominate the image contrast, an estimate of the correct inner angle can be obtained from $\theta_{HAADF} \geq \lambda/d_{thermal}$ (Howie 1979) where λ is the electron wavelength and $d_{thermal}$ is the amplitude of atomic thermal vibration. For example, for Si at room temperature at 200 kV, $\theta_{HAADF} > 40$ mrad. For more details about the effects of the inner angle, see elsewhere in this book.

Medium-resolution (~1 nm) STEM images, formed with a HAADF detector, are sensitive to changes in specimen composition with the intensity varying (for the most part) monotonically with composition and specimen thickness. It is well known, and described elsewhere in this book, that atomic-resolution HAADF images depend on the excitation of Bloch states and associated channelling (Kirkland et al. 1987). Such channelling effects are also present in medium-resolution STEM images and when a crystalline specimen is at or close to a major zone axis there is an increase in the STEM HAADF signal brought about by the localisation of the beam onto atomic columns (Pennycook and Nellist 1999). However, in general, strong channelling will occur very infrequently during a tilt series and will have little effect on the overall intensity distribution in the reconstruction.

The STEM probe is scanned, point by point, in a grid pattern at each point waiting (the dwell-time) while the signal is recorded on the detector. While the current density in a STEM probe is very high, the total current delivered to the specimen is similar to that for parallel TEM illumination (Rez 2003). Therefore for an equivalent dose, the STEM DF image will take longer to acquire than BF TEM because the number of electrons collected on the high-angle detector is only a small fraction of those that are incident on the specimen. While the contrast achieved by collecting at high angles is considerable, the acquisition time must be sufficiently long to overcome background noise. The relatively short focal depth of STEM can be exploited for automated focussing. A defocus series should show a clear trend in the sharpness/contrast of the image and enable the optimum focus to be attained. For indistinct or noisy features, a contrast-enhancing filter may be used. Auto-focussing in this mode will require a series of images, and that combined with the long acquisition times in STEM, results in a significant dose 'overhead' for STEM auto-focussing.

If the specimen is slab-like and tilted to high angles, it is likely that only part of the specimen will be in focus. By rotating the scan, the tilt axis of the specimen can be made perpendicular to the direction of the scan line. Given the simple geometric relationship between tilt, specimen height and defocus, it is possible, for every scan line, to adjust the beam crossover to match the change in specimen height. A focal ramp can then be applied across the image to minimise problems associated with a limited depth of field; this is known as 'dynamic focussing' and has been used for many years in scanning electron microscopy (SEM) of tilted surfaces.

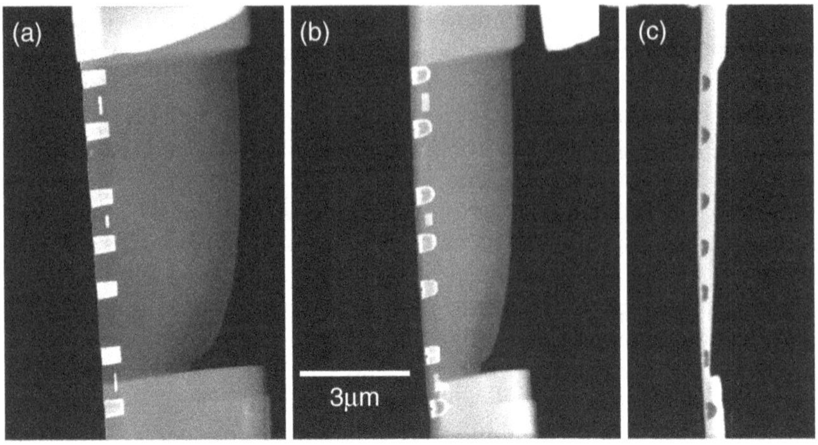

Figure 8–10. The effect of thickness and high atomic number on HAADF STEM contrast. A focussed ion beam (FIB) milled 'finger' specimen of a Si-based transistor with tungsten (W) contacts. (a) At low tilt ($\approx 0°$), the W contacts appear bright, as they are of higher atomic number than silicon. (b), (c) As the projected thickness increases the contrast reverses, so that by 74° in (c) the contacts appear dark and the silicon bright; the series of images no longer meets the projection requirement. Adapted from Dunin-Borkowski et al. (2005).

The STEM beam will diverge as it propagates through the specimen and lead to a degradation of image resolution (the 'depth of field'). Different heights inside the specimen will therefore be sampled with different resolution and what effect this has on the overall reconstruction fidelity has yet to be studied fully. For thick or massive specimens, with large average scattering angles, a significant proportion of scattering may fall outside the outer edge of the detector and lead to contrast reversals and strong deviation from monotonic behaviour. This is demonstrated in Figure 8–10 for a focussed ion beam (FIB)-prepared 'finger' specimen of a semiconductor device (Dunin-Borkowski 2005). At zero tilt, the contrast is as expected, with tungsten contacts appearing more intense than the silicon substrate. However at high tilts, and large projected thickness, the contacts appear darker as now the majority of the tungsten signal has fallen beyond the outer radius of the detector.

An early application of HAADF STEM tomography was to heterogeneous catalysts (Midgley et al. 2001) and used to visualise an ordered mesoporous silica, MCM-41, with 3 nm pore widths, embedded with nanometre-sized Pd_6Ru_6 particles. The large atomic number difference between the support and the particles led to high image contrast and both the pore geometry and particle position were reconstructed successfully, even from a somewhat limited tilt series. More recent studies on heterogeneous catalysts by HAADF tomography clearly demonstrate the power of the technique for resolving the structure of heavy active particles embedded in a lighter matrix (Thomas et al. 2004, Ward et al. 2007). The facetting of biogenic crystallites in magnetotactic bacteria was also resolved in an early study (Buseck et al. 2001) and examined further more recently (Weyland et al. 2006); see also Section 8.6. The high resolution (<1 nm^3) attainable by the technique makes it the ideal tool for examining nanoscale materials in 3D, see Section 8.6. The possibility of 3D nano-metrology using HAADF STEM has also been discussed (Ward et al. 2007, Weyland et al. 2006). ADF TEM has also been used to form images suitable for tomography and some success in reconstructing nanostructures has been reported (Bals et al. 2006).

One area that 3D metrology seems most relevant is that of semiconductor devices, and the results of Kübel et al. (2005) show that HAADF tomography may be an ideal technique for such application. The study compared reconstructions of a semiconductor device using both BF TEM and HAADF STEM; the contrast from the latter was far higher revealing porosity inside a metallic contact not visible in the BF reconstruction. BF (phase contrast) imaging is also quite insensitive to small buried objects. A study of catalysts with 3 nm pores and nanometre-sized nanoparticles (Thomas et al. 2004) shows very clearly the difference between BF TEM and HAADF STEM imaging. Figure 8–11 shows this comparison; the relative clarity of the particles in the HAADF STEM image is striking. (For aberration-corrected instruments, image delocalisation is considerably less than shown here and so phase contrast is stronger nearer the object of interest, there is less mixing of nearby signals and fine detail remains with high contrast.)

The attainable resolution of the tomographic reconstruction scales with the volume under study, following the Crowther formula (Crowther et al. 1970). It is clear that for most tilt series, where the tilt increment is typically 1°, the controlling factor for tomographic resolution will be the limited sampling rather than, for example, the image resolution or the depth of field. One effect of the convergent beam

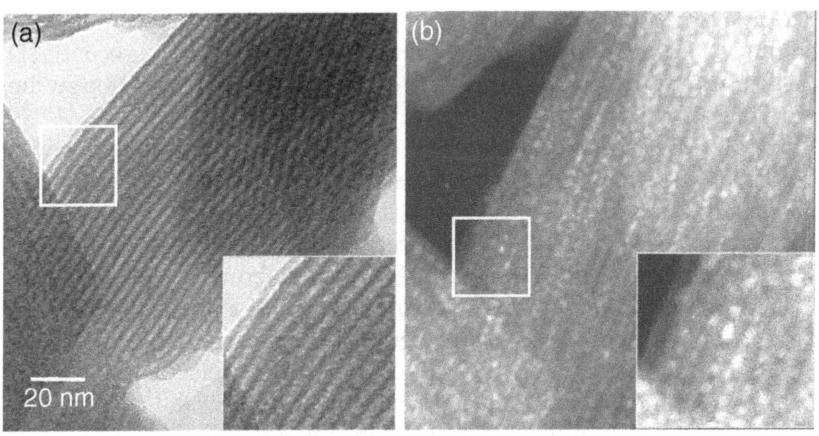

Figure 8–11. Comparison of (a) BF TEM and (b) HAADF STEM imaging of an MCM 41 mesoporous silica particle filled with bimetallic nanoparticles of $Ru_{10}Pt_2$. The nanoparticles are visible only in the HAADF STEM image. Adapted from Thomas et al. (2004).

however will be that different heights inside the specimen will be sampled with different resolution and what effect this has on the overall reconstruction has yet to be studied in detail.

Another consequence of long acquisition times is that STEM images can be distorted by specimen drift. This is a particular problem when carrying out high-magnification, atomic-resolution, STEM imaging where crystal lattice images can show significant 'shear' due to stage drift. At the lower magnifications used typically for tomography, drift should not be so severe to cause problems during acquisition.

In consideration of beam damage to the specimen, one might consider that the large current density in the STEM probe could result in severe damage. However, in specimens where beam heating is the controlling factor, STEM appears to minimise the damage caused (Midgley et al. 2001); the total dose to the specimen can be much lower than in conventional TEM imaging and the rastering nature of the scan allows the dissipation of heat (phonons) into surrounding, non-illuminated areas (Egerton et al. 2004).

8.6 Applications of STEM Tomography

8.6.1 Heterogeneous Catalysts

Some of the early work using electron tomography in materials science was in the study of materials with complex morphology, in particular polymers (Laurer et al. 1997, Spontak et al. 1988), nanostructures (Arslan et al. 2005) and catalysts (Koster et al. 2000) whose key lengthscales (tens of nanometres) made them ideal candidates for investigation by electron tomography. Early examples of catalyst study were undertaken with a BF tilt series and revealed the positions of catalytic Ag particles in a NaY-type zeolite and the porous system of an acid-leached mordenite (Arslan et al. 2005). In the first case the study revealed unambiguously that large silver particles were located on the zeolite surface and smaller ones in the interior. In the second, 2D slices through the projection revealed the internal pore position in the mordenite particle.

Shortly after these early studies, STEM HAADF tomography was used to reveal the internal structure of heterogeneous catalysts based on metallic nanoparticles distributed within disordered mesoporous siliceous (Midgley et al. 2001) and carbonaceous (Midgley et al. 2002) support structures. Surprisingly little is known about the detailed internal structure of these porous structures; conventional gas adsorption–desorption isotherms, X-ray and neutron small-angle scattering techniques yield only an average value of pore diameter and pore-size distribution and offer little insight into the nature of the connectivity of the pores. Electron tomography offers a way to elucidate such local information not possible with other techniques. The STEM HAADF signal can discriminate easily between the nanoparticles (usually of high atomic number) and the background support (usually low atomic number). Figure 8–12 shows a recent example of a reconstruction made from

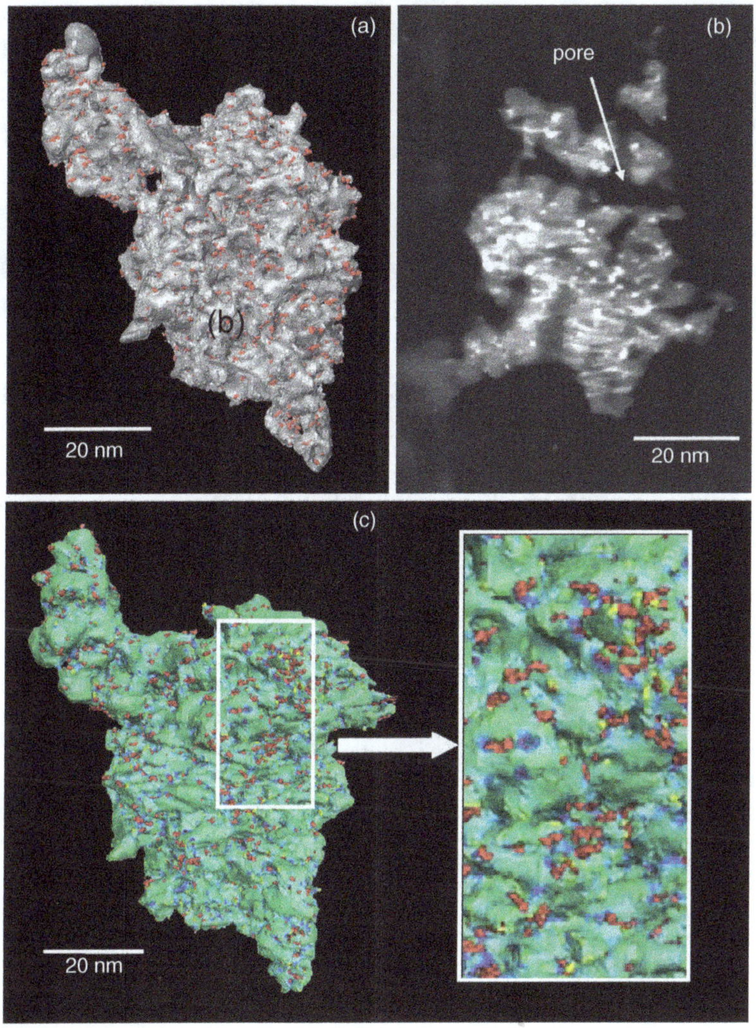

Figure 8–12. (a) Surface render of a segmented reconstruction, derived from a series of HAADF STEM images, of a heterogeneous catalyst: mesoporous silica support is shown in *grey* and Pt/Ru bimetallic nanoparticles in *red*. (b) 1-nm-thick tomographic slice of the reconstruction showing internal porosity and the internal distribution of the bimetallic nanoparticles (c) Coloured reconstruction in which the surface Gaussian curvature is colour-coded *green* and *yellow* indicating concave and convex surface areas and *blue* to indicate saddle points. The magnified inset indicates the preference of the red nanoparticles to sit at the saddle points of the surface.

a tilt series of HAADF STEM images recorded from a catalyst composed of disordered silica with Ru/Pt bimetallic nanoparticles (Ward et al. 2007). The image in Figure 8–12(a) shows a surface render with the distribution of the nanoparticles seen on the external surface of the silica. A better view of the interior can be seen from Figure 8–12(b) which shows a 1 nm slice through the interior of the catalyst revealing highly complex pore structures and the distribution of nanoparticles inside the catalyst structure. Figure 8–12(c,d) shows a similar image to

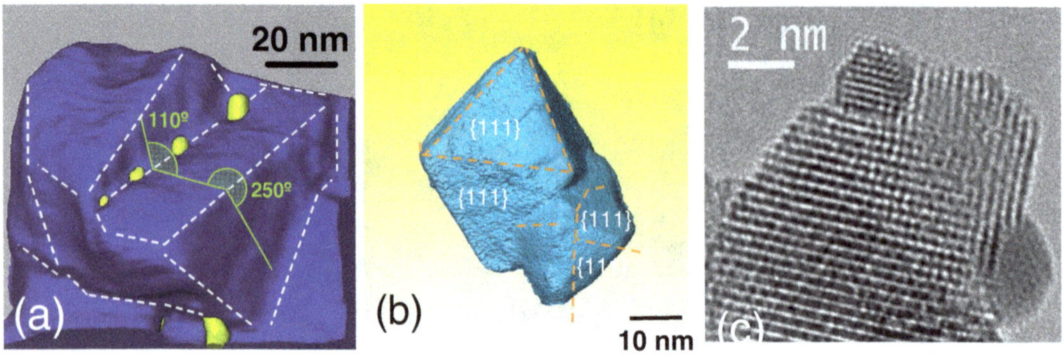

Figure 8–13. (a) Surface-rendered reconstruction of a $Au/Ce_{0.50}Tb_{0.12}Zr_{0.38}O_{2-x}$ catalyst showing the distribution of gold nanoparticles (*yellow*) on the surface of the oxide (*blue*). *Dashed lines* indicate the vertices of {111} facets. (b) Surface-rendered reconstruction of Zr-doped ceria nanoparticle showing a highly facetted structure with {111} planes prominent. (c) HREM image of the $Au/Ce_{0.50}Tb_{0.12}Zr_{0.38}O_{2-x}$ catalyst. Here the gold metal particle is simultaneously in contact with two {111} oxide facets. Adapted from Gonzalez et al. (2009).

Figure 8–12(a) but now colour coded to better describe the underlying surface curvature, illustrating here a strong preference of the nanoparticles (red) for saddle-shaped surface features (blue). Many other similar catalyst structures have been studied by STEM tomography – see, for example, the review by Friedrich et al. (2009).

The facetting of catalyst nanocrystals can play a key role in their activity and selectivity. In Figure 8–13 we show a reconstruction of heavy metal oxide catalysts based on $(Ce/Tb/Zr)O_{2-x}$ with Au nanoparticles decorating the surface (Gonzalez et al. 2009). The facetting of the oxide is dominated by {111} planes, as seen in similar reconstructions of ceria nanocrystals (Hernandez et al. 2007), see Figure 8–13(b). Here, though, we see how the Au nanoparticles are anchored preferentially in the crevices brought about by two {111} facets. The contribution of interface energy to the total energy of such nanoparticles is expected to be important and therefore maximising the interaction between the Au and oxide surface could provide additional stability. This interaction is revealed in more detail in the HREM lattice image of Figure 8–13(c).

As a final example, we consider ordered mesoporous silicas which are high-area ordered solids that allow nanoparticle bimetallic catalysts to be anchored on the walls of the support. Such structures often show high activity and selectivity but despite their enormous potential, the structure of many of these systems is still open to question. Disorder in the structure makes X-ray diffraction studies difficult and this has led to efforts to elucidate structures using electron diffraction, high-resolution imaging and electron tomography.

STEM HAADF tomography was used to reconstruct the internal structure of an ordered mesoporous silica, MCM-48, which is known to have a complex 3D porous system based on the gyroid structure (Yates et al. 2006). A series of images were recorded about a single tilt axis (of unknown crystallographic direction) and after reconstruction, the MCM-48 particle is reoriented parallel to major zone axes, as

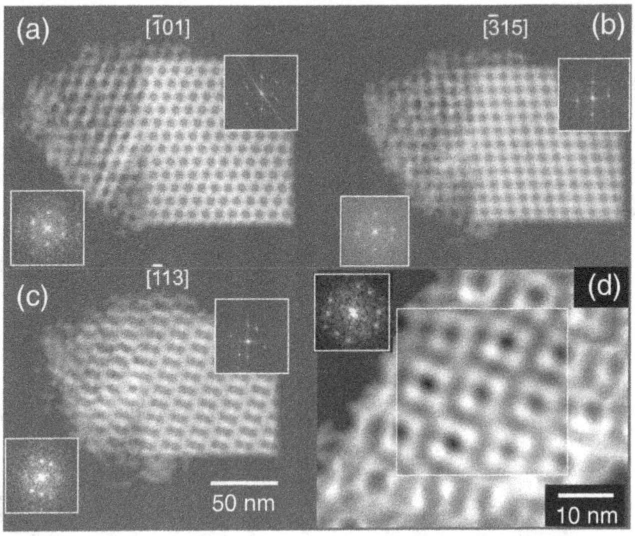

Figure 8–14. (a)–(c) A montage of tomographic voxel projections of the MCM-48 lattice shown at successive major zone axes as the reconstruction is rotated about a <112> zone axis. The *left-hand side* of each figure is the experimental image and corresponding power spectrum, and the *right-hand side* the theoretical image and its spectrum, based on a gyroid structure. (d) A reconstructed slice through an MCM-48 reconstruction, perpendicular to the <110> zone axis. A Bragg-filtered inset enhances the periodic detail. The pairs of spots in the power spectrum correspond to the {332} and {224} reflections.

seen in Figure 8–14(a–c), which shows a montage of voxel projections ('lattice images') whose zone axes are perpendicular to the <112> axis. Figure 8–14(d) though shows a slice through the MCM-48 reconstruction perpendicular to the <110> zone axis. The internal structure of the pore system is clearly revealed. The reconstruction confirmed directly the space group symmetry (Yates et al. 2006) and led to the discovery of low-angle grain boundaries with common {112} planes being present in the structure.

8.6.2 Nanostructures

8.6.2.1 Carbon Nanotubes

There remains great interest in the application of carbon nanotubes because of, for example, their high electrical and thermal conductivities, high mechanical strength, oxidation resistance and good field emission performance. The key to a better understanding of carbon nanostructures is knowledge of the structure and composition (if doped) at high spatial resolution and in three dimensions. Here we give an example of multi-walled carbon nanotubes, examined first in an undoped state and then doped with 3% nitrogen; the latter have a remarkably uniform crystallographic orientation, spacing and register between layers (Koziol et al. 2005). Figure 8–15 shows how the internal architecture of the undoped MWNTs is poorly resolved in projection but clearly revealed in tomographic slices taken parallel to the tube axis.

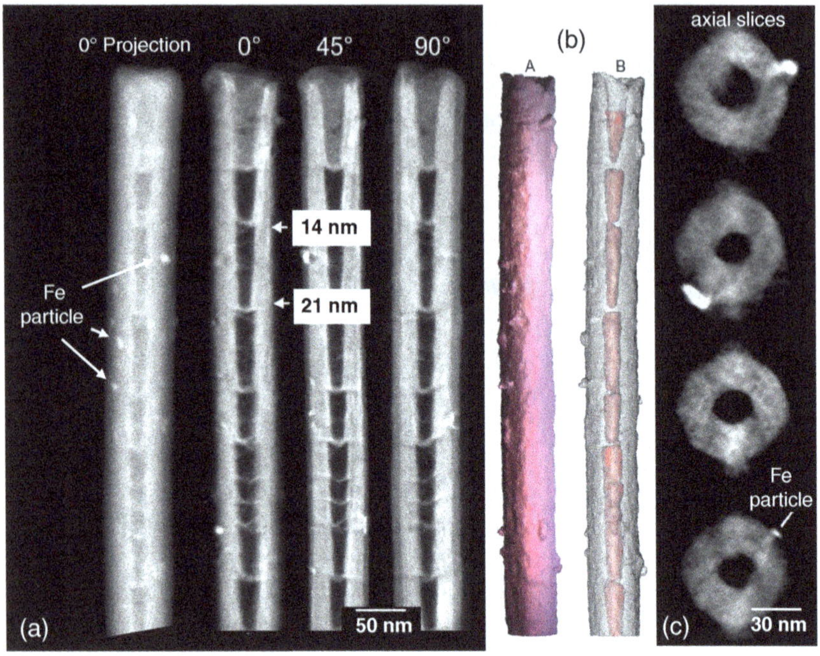

Figure 8–15. (a) A HAADF STEM image and three reconstructed slices of 2 nm thickness through a carbon multi-wall nanotube at 45° intervals, revealing the bamboo-like internal structure. (b) Surface-rendered views of (A) the external wall and (B) the bamboo cavities. (c) Single slice (2 nm thick) perpendicular to the long axis of the nanotube showing the concentric nature of the cavities with respect to the outer walls.

The tomographic reconstructions reveal the classic bamboo structure and the internal cavities are concentric with the outer graphene walls (Ducati et al. 2006). Figure 8–16 shows a montage of reconstruction slices from the nitrogen-doped nanotubes. Here the main cavities and sub-cavities are separated by a fine filamentary structure. Longitudinal slices reveal that the filaments across the tubes are surfaces made of near-constant thickness composed of a few graphene layers that seal each compartment of the tube. Axial slices reveal how the cavities are no longer concentric with the outer graphene walls and this can best be explained by some graphene layers being incomplete ending in edge-like dislocations.

8.6.2.2 Titanium Oxide Nanotubes

Titanium oxide is one of the most studied transition metal oxides because of its possible application in many fields, including photocatalysis and environmental catalysis. As with all catalytic reactions a high surface-area-to-mass ratio leads to an efficient and active catalyst. In a recent study (Hungria et al. 2009) a range of titania nanotubes were produced through a sol–gel process and examined with STEM tomography. One example, shown in Figure 8–17, illustrates the complexity of the morphology that can arise after transforming a sol–gel

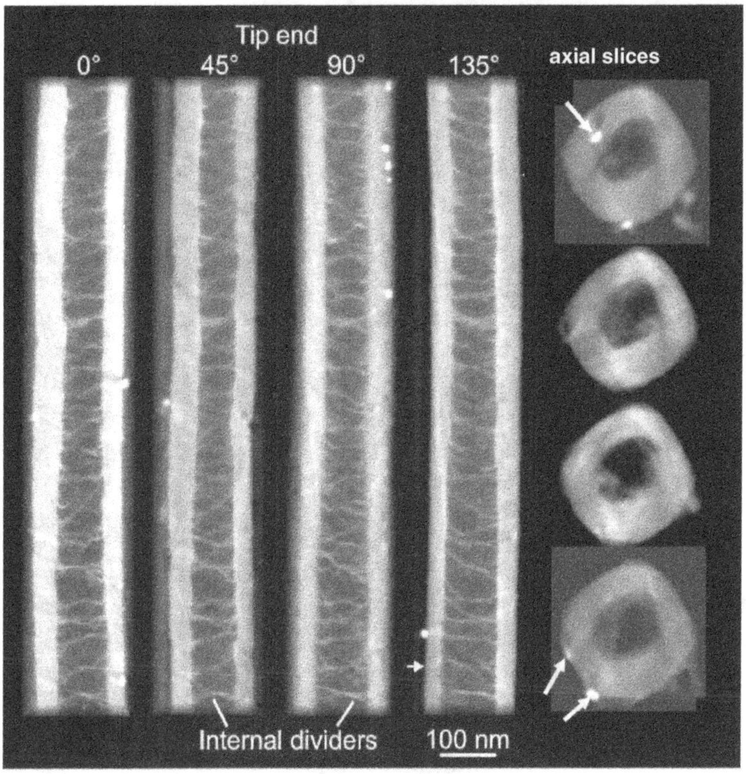

Figure 8–16. (*Left*) Four reconstructed slices of 2 nm thickness through a nitrogen-doped carbon multi-wall nanotube at 45° intervals, revealing the varying wall thickness and fine-scale internal divisions. (*Right*) Axial slices (2 nm thick) perpendicular to the long axis of the nanotube showing internal and external facetting. Iron catalyst nanoparticles are easily identified and some are arrowed in the figure.

prepared anatase nanotube into a rutile one by a series of heat treatments. Here we see as a series of slices through the reconstruction, three different rutile morphologies which reflect, among other factors, the

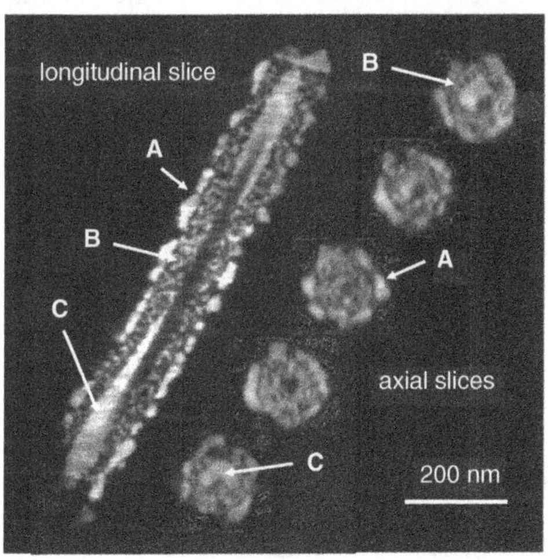

Figure 8–17. Longitudinal and axial slices through a reconstruction of a rutile nanotube showing three different crystal morphologies marked A, B and C.

thermal gradients in the nanotube as it forms rutile. Such a complex internal structure cannot be interpreted fully except through electron tomography analysis.

8.6.3 Cadmium Telluride Tetrapods – An Example of Dual-Axis Tomography

Quantum dots are used for a range of applications from sunscreens through to confocal microscopy imaging labels (Shim and Guyot-Sionnest 2000). Under certain growth conditions colloidal quantum dots with the sphalerite structure can elongate along certain crystallographic directions: 'legs' grow along <111> directions to form a structure known as a 'tetrapod'. CdTe tetrapods have been studied by dual-axis electron tomography in which STEM HAADF images were acquired about two mutually perpendicular axes ($\pm 70°$) and then combined with the ADA-SIRT algorithm (Tong et al. 2006). Figure 8–18(a) and (b) shows the two data sets reconstructed separately. The missing wedge artefacts are evident and highlighted with arrows indicating certain legs missing because of the lack of information in that direction in the acquired data set. The ADA-SIRT reconstruction seen in Figure 8–18(c) has greater fidelity and does not suffer from the missing legs seen in the single axis reconstructions (Arslan et al. 2006). The inset in Figure 8–18(c) is a perspective view of the boxed tetrapod. Figure 8–18(d) shows a 'multipod' CdTe structure formed from multiple tetrapod growths; this was also reconstructed from a dual-axis series. Dual-axis tomography is undoubtedly time-consuming but this example illustrates the significant improvement in the fidelity of reconstructions from complex nanoscale objects, crucial for high-quality metrology applications.

8.6.4 Semiconductor Devices

There has been much interest in the semiconductor industry in using electron tomography to study device structures and in particular the structure of metallic interconnects and contacts to transistor structures. A review of such work can be found in Midgley et al. (2007) but here we show two examples. First, we see how STEM tomography can reveal details about interconnects that might be missed in conventional microscopy. Figure 8–19 shows an example (Kübel et al. 2005) which shows subtle changes to a Ta barrier layer and reveals the 3D distribution of small voids in the copper interconnect. We showed previously in Figure 8–12 the problems of thick, massive contact material and how the contrast can reverse in HAADF images. In Figure 8–20 we show how this can be overcome by using an incoherent BF (IBF) STEM approach in which the BF and multiple DF discs are recorded together on a STEM detector (Ercius et al. 2006). With a large collection angle, the total signal recorded is insensitive to the contrast reversals seen previously for HAADF images and the IBF signal falls off monotonically and in an approximately exponential fashion with mass thickness.

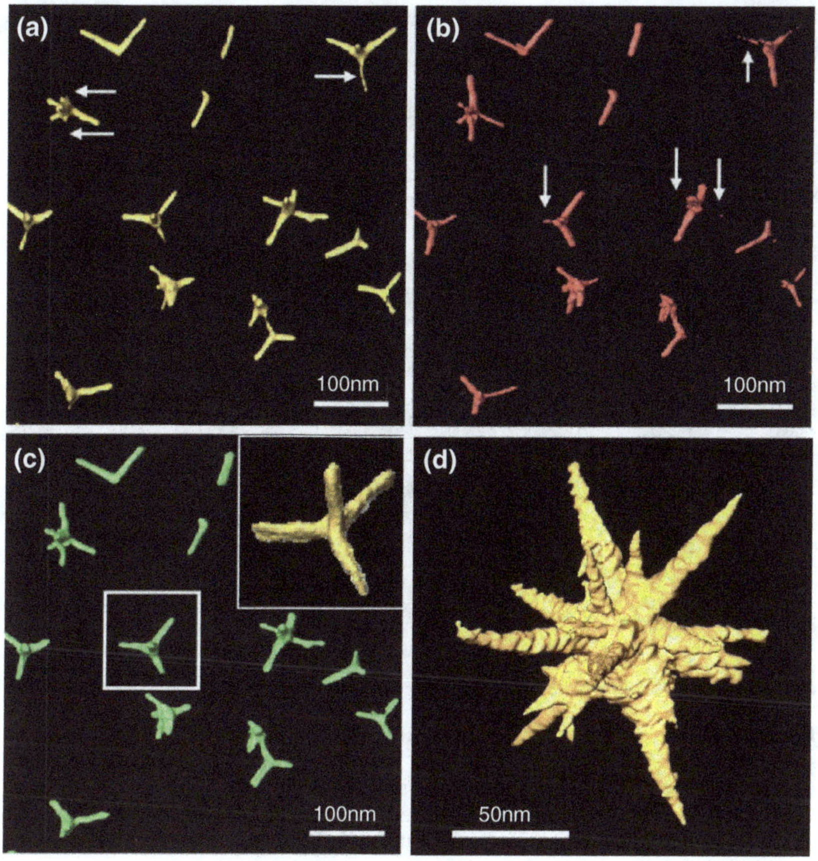

Figure 8–18. (a, b) are reconstructions from two HAADF STEM single tilt series of CdTe tetrapods showing that some of the legs of the tetrapods are missing, as indicated by arrows, due to the effects of the missing wedge; (c) is a dual-axis ADA-SIRT reconstruction of the two data sets showing no missing legs because the missing information has been greatly reduced. The inset in (c) shows one of the tetrapods in more detail. (d) An ADA-SIRT reconstruction of a CdTe 'multipod' composed of multiple tetrapod growths.

The last semiconductor example, shown in Figure 8–21, combines STEM and EELS to yield a remarkable 4D data set equivalent to 4D EFTEM tomography (Gass et al. 2006). Here at every tilt a STEM-EELS map was recorded so that 3D tomographic reconstructions could be made using chemically specific signals. Indeed in this case the different silicon signals associated with elemental silicon, oxide and nitride could be visualised separately in three dimensions (Jarausch et al. 2009). A needle-shaped sample was fabricated using the FIB so as to avoid increasing projected thickness (and thus increased plural scattering) as the sample was tilted. Similar results have been achieved combining STEM and EDX to investigate device interconnect structures (Saghi et al. 2007).

Figure 8–19. Surface render of a HAADF-STEM tomogram showing the 3D structure of copper interconnect lines; slight variations of the tantalum barrier layer (*yellow*) are seen. The copper is shown in *green*. A 3D distribution of voids (*red*) can be seen within the copper. The etch-stop layer is shown in *blue*. Taken from Kübel et al. (2005), copyright Cambridge journals, reproduced with permission.

8.6.5 Biomaterials

STEM tomography has been used to investigate a number of biomaterials including the distribution of voids in nacre (Gries et al. 2009) and the location (and toxicity) of nanotubes and nanoparticles in macrophage cells (Porter et al. 2007). Here we highlight two examples where STEM tomography has revealed 3D information on biominerals produced inside living matter.

8.6.5.1 Magnetotactic Bacteria

Magnetotactic bacteria (Blakemore 1975) are a type of bacteria in which small magnetic crystals (magnetosomes) are arranged in a chain within the cell membrane. Alignment of the magnetosomes enables the bacteria to sense the Earth's magnetic field and find a preferential direction in which to feed. Magnetotactic bacteria are found in freshwater (the magnetosomes are magnetite, Fe_3O_4) and in seawater (iron sulphide, Fe_3S_4) (Mann et al. 1990). In an early example of STEM HAADF tomography, the morphology of the magnetosomes was studied in a strain of magnetotactic bacteria called MV-1 (marine vibrios-1) (Buseck et al. 2001). In Figure 8–22 we show a reconstruction of the bacteria in which the facetting of the small magnetosomes is clearly revealed (Weyland

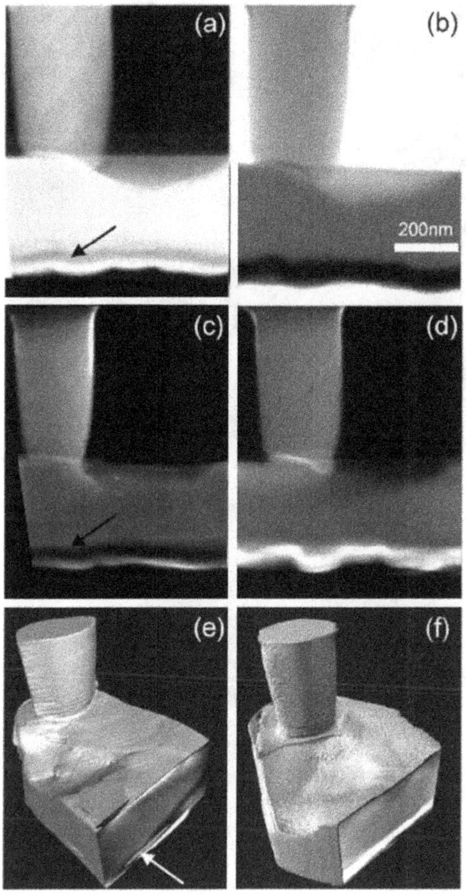

Figure 8–20. Imaging a stress void in a copper interconnect. (a) A STEM HAADF image showing a stress void in a copper interconnect void and a false void at the bottom labelled with an *arrow*. (b) An IBF image of the sample showing only the true stress void. (c) Slices from the STEM HAADF reconstruction showing the false void, but only the true void is present in (d), a slice from the IBF reconstruction. (e) The HAADF reconstruction of the Cu surface showing the false void. (f) The IBF reconstruction correctly maps the Cu surface and shows that the void is facetted near the interconnect. Adapted from Ercius et al. (2006), copyright 2006, American Institute of Physics.

et al. 2006). The fidelity of the reconstruction in this case is excellent in part because of the high tilt range (±76°). Although the facetting can be seen in the surface render, axial slices through the reconstruction give a clearer view of the facetting of the crystals and how it changes along the length of the crystal: {110} facets tend to dominate the magnetosome crystallography leading to the prismatic appearance of the crystals.

8.6.5.2 Ferritin

Interest continues to grow in the use of biomineralisation proteins for the fabrication of inorganic systems, including electronic devices, catalysts and biomimetic structures (Mann and Ozin 1996). One of the best studied examples of this type is the protein ferritin, which has a hollow shell that allows the storage of iron-rich mineral particles. The protein molecules have an outer diameter of approximately 12 nm. The structure of the core is not well established, but the mineral component is thought to be a hydrous ferric oxide with a structure similar to the mineral 'ferrihydrite' ($5Fe_2O_3.9H_2O$). Evidence of magnetite/maghemite in liver samples and brain tissues has led to the suggestion that ferritin may act as a precursor to the formation of biogenic magnetite in humans (Dobson 2001). Excess iron is toxic and can cause haemochromatosis.

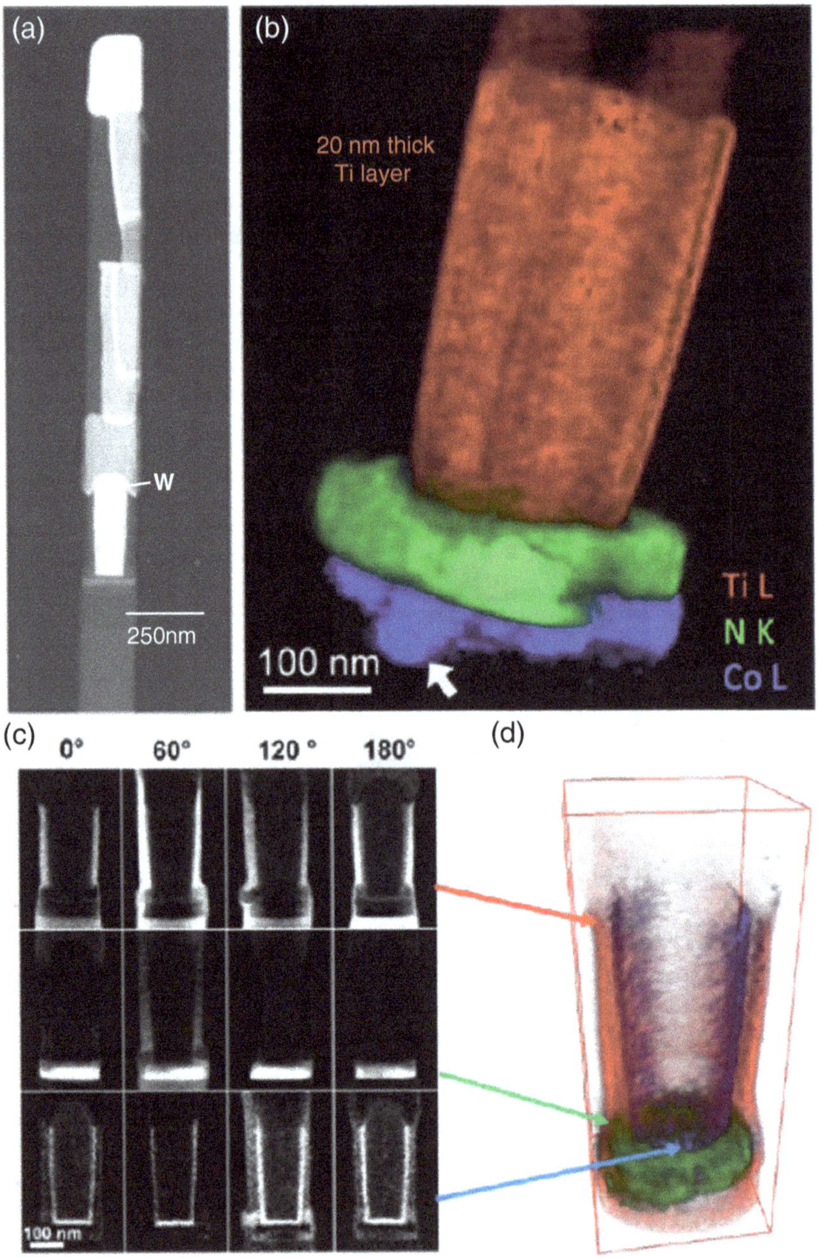

Figure 8–21. (a) A HAADF STEM image of a W-to-Si contact from a semiconductor device, prepared as a needle-shaped sample. (b) A colour composite image of the contact showing the volumetric elemental distribution maps for Ti, N and Co obtained by tomographic reconstruction from the corresponding elemental map rotation series, acquired by EELS mapping at every tilt. (c) Using MLLS fingerprint techniques over the 90–130 eV loss range, reconstructed slices are shown every 60° indicating the different state of silicon found in the contact area, *red* is silicon oxide, *green* is elemental silicon and *blue* is silicon nitride and titanium silicide phases. (d) Tomographic reconstruction of the individual states of silicon. Adapted from Jarausch et al. (2009), copyright (2009), with permission from Elsevier.

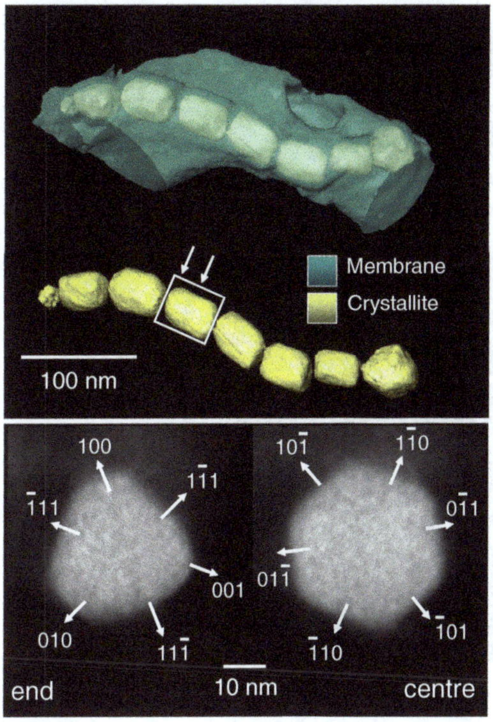

Figure 8–22. (a) A tomographic reconstruction of magnetotactic bacteria strain MV-1, shown with and without the organic 'envelope'. The facetted chain of magnetite crystals is evident. (b) Slices taken through a single magnetite crystal in the chain reveal the near-perfect hexagonal shape of the slice through the cubic crystal.

STEM HAADF tomography was used to examine unstained human pathogenic liver samples containing ferritin. In this case, the protein shell was not visualised but the mineral cores of the ferritin can be distinguished clearly. Figure 8–23(a) shows a single STEM HAADF image and Figure 8–23(b) shows a reconstruction in which the individual ferritins are evident. In Figure 8–23(b), the ferritin is surface rendered and the underlying organic medium shown as a voxel projection for clarity. A 'racetrack'-shaped region where the ferritin cores appear to be ordered is indicated by white arrows in the figure. Slices through the reconstruction show that the ferritin is well ordered in all three dimensions (Yates 2005). This ordering may be in response to an overload of particles within a membrane-bound volume, that there is some underlying template on which the ferritin can align, or that self-assembly may occur when the chemical environment in certain cell regions is different.

8.6.6 Aluminium–Germanium Alloy

STEM HAADF tomography has been used to study the morphology and distribution of precipitates in steels and alloys. One example described briefly here is that of the Al–Ge alloy system in which the limited solubility of Ge in an Al matrix leads to Ge precipitates and the production of a dispersion-hardened alloy. In the literature a great variety of precipitate morphology has been seen depending on heat treatments and small changes in composition. To better understand

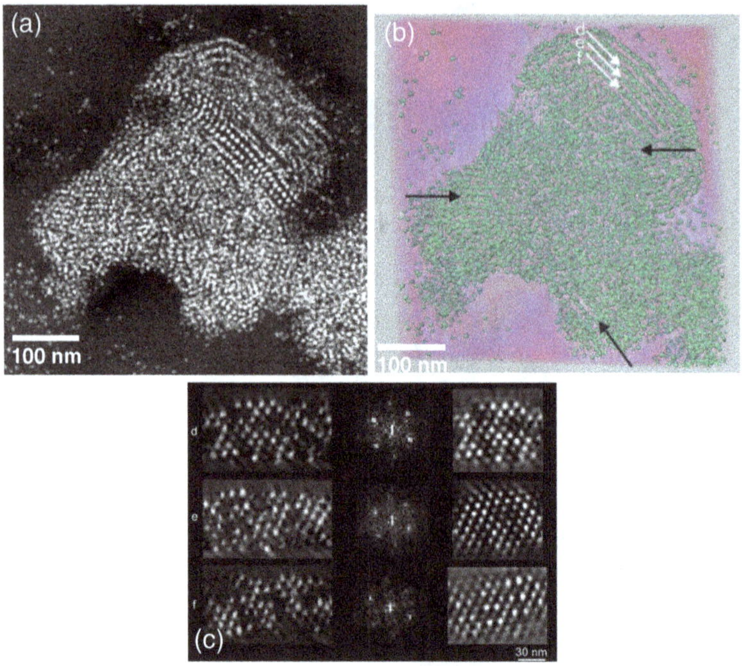

Figure 8–23. (a) A HAADF STEM image recorded from a liver section from a haemochromatosis patient showing clustering and ordering of ferritin molecules. (b) A reconstructed volume, with the ferritin cores surface rendered in *green* and the organic material displayed as a voxel projection in *pink*, oriented to show areas in which the ferritin is ordered (*white arrows*), and less obvious ordering marked by *black arrows*. (c) Single slices (nominal thickness 0.8 nm) through the planes of ordered mineral cores labelled 'd', 'e' and 'f' in (b), showing in-plane ordering, their power spectra, and corresponding Bragg filtered slices.

the system, a recent tomography study was undertaken (Kaneko et al. 2008). Figure 8–24 shows an example of a surface-rendered reconstruction of Ge precipitates in an Al–1.6 at.%Ge alloy. The variety of morphology revealed by the reconstruction is quite remarkable and has been colour-coded in the figure to distinguish the different shapes including rods, tetrahedra, triangular plates and octahedra. It is quite clear that many of these shapes would have been misidentified had only a single 2D image been recorded.

8.6.7 STEM MAADF Tomography of Dislocations

Although stereomicroscopy provides a way to visualise a 3D network of dislocations the precision with which the depth of the dislocation in the sample can be determined is limited by the small angle (5–10°) that the sample must be tilted for the stereo effect to work. In order to achieve an accurate 3D reconstruction of a dislocation network it was shown in 2006 that a weak-beam dark-field (WBDF) electron tomography approach could reveal a 3D dislocation network in GaN at far

Figure 8–24. (a) STEM HAADF reconstruction of Ge precipitates in an Al–Ge alloy. The *colours* differentiate the precipitate morphology: platelets *blue;* tetrahedra *green;* octahedra *orange;* rods *yellow;* irregular shapes *white.* Adapted from Kaneko et al. (2008).

greater spatial resolution than with X-rays and with more precision than stereomicroscopy (Barnard et al. 2006a, b).

However, in general, a WBDF tomographic reconstruction suffers from a number of artefacts which arise predominantly from oscillatory fringe contrast seen in the image of either the dislocation itself or thickness fringe contrast through the use of a wedge-shaped specimen; a persistent 'dustiness' is seen in the reconstruction within which the dislocation contrast can be lost. As the intensity and spacing of thickness fringes are determined by the deviation parameter, incoherently imaging several dark-field images simultaneously may reduce this problem. This is best configured using a STEM approach. High-angle (HAADF) scattered intensity is dominated by thermal diffuse scattering and low-angle (LAADF) is dominated by the contrast of only one or two diffracted beams (coherent contrast). In between, at medium angles (MAADF imaging), the contrast is through the effect of many diffracted beams (Sharp et al. 2008).

Figure 8–25(a) shows a comparison of a WBDF image of dislocations in deformed Si with a STEM MAADF image of the same area. The lack of thickness fringe contrast and the reduction in depth-related intensity oscillations improves greatly the dislocation visibility. Practical advantages of this STEM MAADF mode for dislocation imaging include the possibility of dynamic focussing and the availability of auto-focussing and auto-tracking routines developed for STEM HAADF tomography. However, the multi-beam image necessarily leads to a widening of the dislocation image itself and closely spaced dislocations may merge into one entity.

Figure 8–25(b) shows a STEM MAADF reconstruction of dislocations in deformed silicon (Barnard et al. 2010, Tanaka and Higashida 2004). The reconstruction shows the presence of several dislocations on the same slip plane with a precision of about 10 nm. At certain points the dislocations bunch together and cross-slip onto another {111} plane. In these cross-slip bunches, the reconstruction does not allow individual

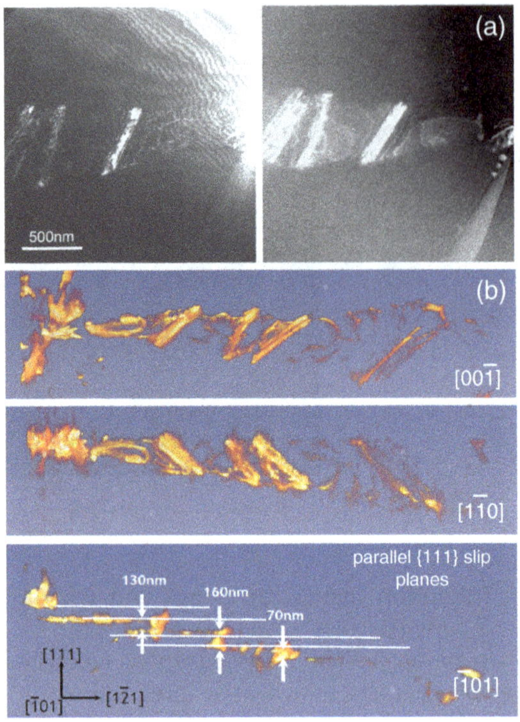

Figure 8–25. (a) On the *left* is a weak-beam dark-field image of a dislocation array induced by an indentation and annealing of silicon (recorded at **g**/3.4 **g** condition, **g** = 2–20) and on the *right* the corresponding STEM MAADF image. (b) Views of a STEM MAADF reconstruction showing parallel slip planes with approximately 100 nm separation. Taken from Barnard et al. (2010).

dislocations to be resolved but appear to lie approximately along the [–101] direction.

8.7 Summary

STEM tomography has matured into a technique that is used widely to investigate the 3D structure of many materials. The HAADF signal is ideally suited to tomographic applications involving crystalline materials as the incoherent nature of the high-angle scattering suppresses diffraction contrast that might otherwise invalidate the projection requirement. The sensitivity of the HAADF signal to changes in atomic number is also a great advantage when investigating complex materials with multiple atomic species, such as heterogeneous catalysts. Combining STEM with analytical signals such as EELS and EDX opens up the possibility of 3D chemical information mapped with high spatial resolution. By collecting multiple diffracted beams at medium scattering angles STEM tomography is optimised for defect analysis and 3D dislocation networks are revealed with clarity. There is no doubt that the use of STEM tomography will continue to grow across many disciplines. The next stage in development is likely to be in pushing the resolution towards atomic resolution, involving perhaps confocal techniques and incorporating ab initio information, using discrete tomography algorithms for example, and in pushing the technique to be a true metrology method in which reliable quantitative information can be found from reconstructions with statistical confidence.

Acknowledgements The authors thank all those with whom they have worked in this area including Jonathan Barnard, Juan-Carlos Hernandez, Ana Hungria, Joanne Sharp, Jenna Tong, John Meurig Thomas, Edmund Ward and Tim Yates.

References

I. Arslan, J.R. Tong, P.A. Midgley, Ultramicroscopy **106**, 994–1000 (2006)
I. Arslan, T.J.V. Yates, N.D. Browning, P.A. Midgley, Science **309**, 2195–2198 (2005)
K. Baba, K. Satoh, S. Sakamoto, T. Okai, S. Ishii, J. Perinat. Med. **17**, 19 (1989)
C. Bajaj, Z.Y. Yu, M. Auer, J. Struct. Biol. **144**, 132 (2003)
S. Bals, G. Van Tendeloo, C. Kisielowski, Adv. Mater. **18**, 892 (2006)
J. Banhart, Progr. Mat. Sci. **46**, 559 (2001)
J.S. Barnard, A.S. Eggeman, J. Sharp, T.A. White, P.A. Midgley, Dislocation electron tomography and precession electron diffraction – Minimising the effects of dynamical interactions in real and reciprocal space. Philos. Mag. **90**, 4711–4730 (2010)
J.S. Barnard, J. Sharp, J.R. Tong, P.A. Midgley, Science **303**, 319 (2006a)
J.S. Barnard, J. Sharp, J.R. Tong, P.A. Midgley, Phil. Mag. **86**, 4901–4922 (2006b)
M. Barth, R.K. Bryan, R. Hegerl, Ultramicroscopy **31**, 365 (1989)
K.J. Batenburg, Electron. Notes Discrete Math. **20**, 247 (2005)
S.H. Bellman, R. Bender, R. Gordon, J.E. Rowe, J. Theor. Biol. **32**, 205 (1971)
J. Berriman, R.K. Bryan, R. Freeman, K.R. Leonard, Ultramicroscopy **13**, 351 (1984)
R. Blakemore, Science, **190**, 377 (1975)
B. Bottcher, S.A. Wynne, R.A. Crowther, Nature **368**, 88 (1997)
R.N. Bracewell, Aust. J. Phys. **9**, 297 (1956)
S. Brandt, J. Heikkonen, P. Engelhardt, J. Struct. Biol. **133**, 10 (2001a)
S. Brandt, J. Heikkonen, P. Engelhardt, J. Struct. Biol. **136**, 201 (2001b)
G.L. Brownell, C.A. Burnham, B. Hoop, D.E. Bohning, in *Proceedings of the Symposium on Dynamic Studies with Radioisotopes in Medicine*, Rotterdam, 190, IAEA, Vienna, 1971
P.R. Buseck, R.E. Dunin-Borkowski, B. Devouard, R.B. Frankel, M.R. McCartney, P.A. Midgley, M. Posfai, M. Weyland, Proc. Nat. Acad. Sci. USA **98**, 13490 (2001)
R.A. Crowther, Philos. Trans. R. Soc. London, B **261**, 221 (1971)
R.A. Crowther, D.J. de Rosier, A. Klug, Proc. R. Soc. Lond. A **317**, 319 (1970)
R.A. Crowther, A. Klug, J. Theor. Biol. **32**, 199 (1971)
S.R. Deans, *The Radon Transform and Some of Its Applications* (Wiley, New York, Chichester, 1983)
J. Dengler, Ultramicroscopy **30**, 337 (1989)
D.J. de Rosier, A. Klug, Nature **217**, 130 (1968)
J. Dobson, FEBS Lett. **496**, 1–5 (2001)
C. Ducati, K. Koziol, S. Friedrichs, T.J.V. Yates, M.S. Shaffer, P.A. Midgley, A.H. Windle, Small **2**, 774–784 (2006)

R.E. Dunin-Borkowski, S.B Newcomb, T. Kasami, M.R. McCartney, M. Weyland, P.A. Midgley, Ultramicroscopy **103**, 67–81 (2005)
R.F. Egerton, P. Li, M. Malac, Micron **35**, 399 (2004)
P. Ercius, M. Weyland, D.A. Muller, L.M. Gignac, Appl. Phys. Lett. **88**, 243116 (2006)
J.J. Fernandez, S. Li, J. Struct. Biol. **144**, 152–161 (2003)
A. Frangakis, R. Hegerl, J. Struct. Biol. **135**, 239–250 (2001)
A.S. Frangakis R. Hegerl, J. Struct. Biol. **138**, 105 (2002)
J. Frank, in *Computer Processing of Electron Microscope Images*, ed. by P.W. Hawkes (Springer, New York, NY, 1981), p. 187
J. Frank, *Three-Dimensional Electron Microscopy of Macromolecular Assemblies* (Academic, San Diego, CA, 1996)
J. Frank, B.F. McEwen, in *Electron Tomography: Three-Dimensional Imaging with the Transmission Electron Microscope*, ed. by J. Frank (Plenum Press, New York, NY, 1992), p. 205
J. Frank, A. Verschoor, M. Boublik, Science **214**, 1353 (1981)
H. Friedrich, P.E. de Jongh, A.J. Verkleij, K.P. de Jong, Chem. Rev. **109**, 1613–1629 (2009)
S.P. Frigo, Z.H. Levine, N.J. Zaluzec, Appl. Phys. Lett. **81**, 2112 (2002)
I.S. Gabashvili, R.K. Agrawal, C.M.T. Spahn, R.A. Grassucci, D.I. Svergun, J. Frank, P. Penczek, Cell **100**, 537 (2000)
M.H. Gass, K.K. Koziol., A.H. Windle, P.A. Midgley, Nano Lett. **6**, 376–379 (2006)
P. Gilbert, J. Theor. Biol. **36**, 105 (1972a)
P.F.C. Gilbert, Proc. R. Soc. London B **182**, 89 (1972b)
J.C. Gonzalez, J.C. Hernandez, M. Lopez-Haro, E. del Rio, J.J. Delgado, A.B. Hungria, S. Trasobares, S. Bernal, P.A. Midgley, J.J Calvino, Angew. Chem. Int. Ed. **48**, 5313–5315 (2009)
R. Gordon, R. Bender, G.T. Herman, J. Theor. Biol. **29**, 471 (1970)
K. Gries, R. Kroger, C. Kubel, M. Fritz, A. Rosenauer, Acta Biomater. **5**, 3038–3044 (2009)
R.G. Hart, Science **159**, 1464 (1968)
P.W. Hawkes, in *Electron Tomography: Three-Dimensional Imaging with the Transmission Electron Microscope*, ed. by J. Frank (Plenum Press, New York, NY, 1992) p. 17
R. Henderson, J.M. Baldwin, T.A. Ceska, E. Beckman, F. Zemlin, K. Downing, J. Mol. Biol. **213**, 899 (1990)
G.T. Herman, *Image Reconstruction from Projections, the Fundamentals of Computerised Tomography* (Academic, New York, NY, 1980)
J.C. Hernandez, A.B. Hungria, J.A. Perez-Omil, S. Trasobares, S. Bernal, P.A. Midgley, A. Alavi, J.J. Calvino, J. Phys. Chem. **C 111**, 9001–9004 (2007)
T. Hirano, K. Usami, Y. Tanaka, C. Masuda, J. Mater. Res. **10**, 381 (1995)
W. Hoppe, R. Langer, G. Knesch, C. Poppe, Naturwissenschaften **55**, 333 (1968)
D.I. Hoult J. Magn. Reson. **33**, 183 (1979)
G.N. Hounsfield, *A Method and Apparatus for Examination of a Body by Radiation such as X or Gamma Radiation* (The Patent Office, London, England, 1972)
A. Howie, J. Microsc. **177**, 1 (1979)
A.B. Hungria, D. Eder, A.H. Windle, P.A. Midgley, Catalysis Today **143**, 225–229 (2009)
Y. Ikeda, A. Katoh, J. Shimanuki, S. Kohjiya, Macromol. Rapid Commun. **25**, 1186 (2004)
K. Jarausch, P. Thomas, D.N. Leonard, R. Twesten, C.R. Booth, Ultramicroscopy **109**, 326–337 (2009)

H. Jinnai, T. Koga, Y. Nishikawa, T. Hashimoto, S.T. Hyde, Phys. Rev. Lett. **78**, 2248–2251 (1997)

H. Jinnai, Y. Nishikawa, R. Spontak, S.D. Smith, D.A. Agard, T. Hashimoto, Phys. Rev. Lett. **84**, 518 (2000)

J.R. Jinschek, K.J. Batenburg, H.A. Calderon, R. Kilaas, V. Radmilovic, C. Kisielowski, Ultramicroscopy **108**, 589–604 (2008)

K. Kaneko, K. Inoke, K. Sato, K. Kitawaki, H. Higashida, I. Arslan, P.A. Midgley, Ultramicroscopy **108**, 210–220 (2008)

E.J. Kirkland, R.F. Loane, J. Silcox, Ultramicroscopy **23**, 77 (1987)

M. Koguchi, H. Kakibayashi, R. Tsuneta, M. Yamaoka, T. Niino, N. Tanaka, K. Kase, M. Iwaki, J. Electron. Microsc. **50**, 235 (2001)

A.J. Koster, R. Grimm, D. Typke, R. Hegerl, A. Stoschek, J. Walz, W. Baumeister, J. Struct. Biol. **120**, 276 (1997)

A.J. Koster, U. Ziese, A.J. Verkleij, A.H. Janssen, K.P. de Jong J. Phys. Chem. B **104**, 9368 (2000)

K. Koziol, M.S.P. Schaffer, A.H. Windle, Adv. Mater. **17**, 760–763 (2005)

C. Kübel, A. Voigt, R. Schoenmakers, M. Otten, D. Su, T. Lee, C. Carlsson, Microsc. Microanal. **11**, 378–400 (2005)

M.C. Lawrence, Proc. Electron. Microsc. Soc. South Africa **13**, 19 (1983)

M.C. Lawrence, in *Electron Tomography: Three-Dimensional Imaging with the Transmission Electron Microscope*, ed. by J. Frank (Plenum Press, New York, NY, 1992), p. 197

J.H. Laurer, D.A. Hajduk, J.C. Fung, J.W. Sedat, S.D. Smith, S.M. Gruner, D.A. Agard, R.J. Spontak, Macromolecules **30**, 3938 (1997)

Y. Liu, P.A. Penczek, B.F. McEwen, J. Frank, Ultramicroscopy **58**, 393 (1995)

S. Mann, G.A.Ozin, Nature **382**, 313–318 (1996)

S. Mann, N.H.C. Sparks, R.B. Frankel, D.A. Bazylinski, H.W. Jannasch, Nature **343**, 258 (1990)

T.J. Marrow, J.-Y. Buffiere, P.J. Withers, G. Johnson, D. Engelberg, Intl. J. Fatigue **26**, 717–725 (2004)

B.J. Marsh, D.N. Mastronarde, K.F. Buttle, K.E. Howell, J.R. McIntosh, Proc. Natl. Acad. Sci. USA **98**, 2399 (2001)

D.N. Mastronarde, J. Struct. Biol. **152**, 36 (2005)

P.A. Midgley, M. Weyland, Ultramicroscopy **96**, 413 (2003)

P.A. Midgley, M. Weyland, H. Stegmann, in *Advanced Tomography Methods in Materials Research and Engineering*, ed. by J. Banhart (OUP, Oxford, 2007), pp. 335–373

P.A. Midgley, M. Weyland, J.M. Thomas, P.L. Gai, E.D. Boyes, Angew. Chem. Int. Ed. **41**, 3804–3807 (2002)

P.A. Midgley, M. Weyland, J.M. Thomas, B.F.G. Johnson, Chem. Commun. **18**, 907 (2001)

M.K. Miller, *Atom-Probe Tomography: Analysis at the Atomic Level* (Kluwer/Plenum Press, New York, NY, 2000)

G. Mobus, R.C. Doole, B.J. Inkson, Ultramicroscopy **96**, 433 (2003)

G. Mobus, B.J. Inkson, Appl. Phys. Lett. **79**, 1369 (2001)

H. Nyquist, Trans. AIEE **47**, 617 (1928)

D.E. Olins, A.L. Olins, H.A. Levy, R.C. Durfee, S.M. Margle, E.P. Tinnel, S.D. Dover, Science **220**, 498 (1983)

P.A. Penczek, J. Struct. Biol. **138**, 34 (2002)

S.J. Pennycook, P.D. Nellist, in *Impact of Electron and Scanning Probe Microscopy on Materials Research*, ed. by D.G. Rickerby, G. Valdré, U. Valdré (Kluwer, Dordrecht, 1999), pp. 161–207

A.E. Porter, M. Gass, K. Muller J.N. Skepper, P.A. Midgley, M. Welland, Nat. Nanotechnol. **2**, 713–717 (2007)

M. Radermacher, in *Electron Tomography: Three-Dimensional Imaging with the Transmission Electron Microscope*, ed. by J. Frank (Plenum Press, New York, NY, 1992), pp. 91–116

M. Radermacher, W. Hoppe, In *Proceedings of the 7th European Congr. Electron Microscopy*, Den Haag, 1980

J. Radon, Ber. Verh. K. Sachs. Ges. Wiss. Leipzig, Math.-Phys. Kl. **69**, 262 (1917)

G.N. Ramachandran, A.V. Lakshminarayanan, Proc. Natl. Acad. Sci. **68**, 2236 (1971)

P. Reimers, A. Kettschau, J. Goebbels, NDT Int. **23**, 255 (1990)

C. Renken, B. McEwen, Microsc. Microanal. **9**, 1170 (2003)

P. Rez, Ultramicroscopy **96**, 117 (2003)

J.C. Russ, *The Image Processing Handbook*, 2nd edn. (CRC Press, London, 1995)

Z. Saghi, X. Xu, Y. Peng, B. Inkson, G. Möbus, Appl. Phys. Lett. **91**, 251906 (2007)

W.O. Saxton, W.K. Jenkins, L.A. Freeman, D.J. Smith, Optik **49**, 505 (1978)

J.H. Sharp, J.S. Barnard, K. Kaneko, K. Higashida, P.A. Midgley, J. Phys.: Conf. Ser. **126**, 012013 (2008)

U. Skoglund, L. Ofverstedt, J. Struct. Biol. **117**, 173(1996)

R.J. Spontak, J.C. Fung, M.B. Braunfeld, J.W. Sedat, D.A. Agard, A. Ashraf, S.D. Smith, Macromolecules **29**, 2850 (1996)

R.J. Spontak, M.C. Williams, D.A. Agard, Polymer **29**, 387 (1988)

H. Stegmann, H.H. Engelmann, E. Zschech, Microelectron. Eng. **65**, 171 (2003)

M. Tanaka, K. Higashida, J. Electron. Microsc. **53**, 353 (2004)

J.M. Thomas, P.A. Midgley, T.J.V. Yates, J.S. Barnard, R. Raja, I. Arslan, M. Weyland. Ange. Chem. Int. Ed. **43**, 6745–6747 (2004)

J. Tong, I. Arslan, P.A. Midgley, J. Struct. Biol. **153**, 55–63 (2006)

A.C. Twitchett, R.E. Dunin-Borkowski, P.A. Midgley, Phys. Rev. Lett. **88**, 238302 (2002)

M. Unser, B.L. Trus, A.C. Steven, Ultramicroscopy **23**, 39 (1987)

N. Unwin, R. Henderson, J. Mol. Biol. **94**, 425 (1975)

K. van Benthem, A.R. Lupini, M. Kim, H.S. Baik, S.J. Doh, J.H. Lee, M.P. Oxley, S.D. Findlay, L.J. Allen, J.T. Luck, S.J. Pennycook, Appl. Phys. Lett. **87**, 034104 (2005)

M. Van Heel, G. Harauz, Optik **73**, 119 (1986)

N. Volkmann, J. Struct. Biol. **138**, 123 (2002)

E.P.W. Ward, T.J.V. Yates, J.-J. Fernández, D.E.W. Vaughan, and P.A. Midgley J. Phys. Chem. C **111**, 11501–11505 (2007)

U. Weierstall, Q. Chen, J.C.H. Spence, M.R. Howells, M. Isaacson, R.R. Panepucci, Ultramicroscopy **90**, 171–195 (2002)

M. Weyland, L. Laffont, P.A. Midgley, Inst. Phys. Conf. Ser. **179 (EMAG 2003)**, 349 (2004)

M. Weyland, P.A. Midgley, Microsc. Microanal. **9**, 542 (2003a)

M. Weyland, P.A. Midgley, Microsc. Microanal. **9**, 148 (2003b)

M. Weyland, T.J.V. Yates, R.E. Dunin-Borkowski, L. Laffont, P.A. Midgley, Scripta Materialia **55**, 29–33 (2006)

H. Winkler, K.A. Taylor, Ultramicroscopy **106**, 240 (2006)

C.Y. Xu, J.L. Prince, IEEE Trans. Image Process. **7**, 359 (1998)

T.J.V. Yates, PhD thesis, University of Cambridge, 2005

T.J.V. Yates, J.M. Thomas, J.J. Fernandez, O. Terasaki, R. Ryoo, P.A. Midgley, Chem. Phys. Lett. **418**, 540–543 (2006)

D. Zhao, J.R. Kayal, Curr. Sci. **79**, 1208 (2000)

Q.X.S. Zheng, M.B. Braunfeld, J.W. Sedat, D.A. Agard J. Struct. Biol. **147**, 91 (2004)

U. Ziese, A.H. Janssen, J.L. Murk, W.J.C. Geerts, T. Krift, A.J. Verkleij, A.J. Koster, J. Microscopy **205**, 187 (2002)

9

Scanning Electron Nanodiffraction and Diffraction Imaging

Jian-Min Zuo and Jing Tao

9.1 Introduction

Electron nanodiffraction uses a nanometer-sized, or smaller, electron probe to record diffraction patterns. The small probe is formed inside a transmission electron microscope (TEM) using the same magnetic lenses of the microscope used for imaging. The probe can be as small as 1 Å, or less, in the microscopes equipped with a probe aberration corrector. The small probe, combined with the large elastic scattering cross sections of the high-energy electrons, makes electron nanodiffraction a very powerful technique for studying very small structures. The small probe is also useful for determining the local structure in materials with complicated microstructures, including interfaces. Electron nanodiffraction can also be used for imaging when it is combined with the electron beam scanning for scanning electron nanodiffraction (SEND). Information obtained from diffraction patterns recorded using a small electron probe is mapped according to the probe positions. The use of diffraction intensity to form images is called diffraction imaging. The mapping of diffraction information is one of two techniques that can be used to achieve diffraction imaging. Another diffraction imaging technique is to solve the inverse problem of image reconstruction from the recorded diffraction patterns. Diffraction imaging is not an entirely new idea. A very successful example of diffraction imaging using the scanning technique is the electron backscattering diffraction (EBSD) used for microstructure mapping in scanning electron microscopes (SEM). Image reconstruction from diffraction patterns has also been demonstrated by several groups (Chapman et al. 2006,

This manuscript has been authored by employees of Brookhaven Science Associates, LLC under Contract No. DE-AC02-98CH10886 with the U.S. Department of Energy. The publisher by accepting the manuscript for publication acknowledges that the United States Government retains a non-exclusive, paid-up, irrevocable, world-wide license to publish or reproduce the published form of this manuscript, or allow others to do so, for United States Government purposes.

S.J. Pennycook, P.D. Nellist (eds.), *Scanning Transmission Electron Microscopy*,
DOI 10.1007/978-1-4419-7200-2_9, © Springer Science+Business Media, LLC 2011

Miao et al. 1999, Robinson et al. 2001, Shapiro et al. 2005, Zuo et al. 2003). Diffraction imaging in a scanning transmission electron microscope (STEM) provides a much higher resolution. The high resolution is useful to meet the challenging characterization needs of nanostructures or complex structures in general (Billinge and Levin 2007).

Electron nanodiffraction can also be used to overcome some of the limitations of electron direct imaging. The use of high-resolution electron microscopy (HREM) imaging for structure determination is complicated by the interpretation of image contrast. Only in extremely thin samples, the HREM image contrast can be related to the sample's projected potentials. For samples of reasonable thicknesses, the image contrast is a mixture of the complex exit wave function and phases introduced by the imaging lens; image interpretation in general requires modeling of the electron scattering process and the properties of the electron imaging lens (Spence 2003). In STEM, the image contrast is less sensitive to the lens focus when an annular dark-field detector (ADF) is used. Nonetheless, a proper interpretation of ADF-STEM image contrast also requires modeling of the electron scattering process (see Chapter 2 and Chapter 6). The application of STEM imaging for quantitative structure determination is somewhat limited because of scan distortions and scan noises in a STEM. For organic materials susceptible to radiation damage, HREM imaging is often not an option because the amount of electron dose required to produce a sufficient image contrast can be larger than the material's radiation damage threshold. Electron diffraction, on the other hand, can work at low-dose situations by averaging over many unit cells for crystals. Certain structural information is also easier to obtain from diffraction patterns. For example, information about the crystal orientation, unit cell dimensions, and sample thickness can be obtained from convergent beam electron diffraction (CBED) patterns using well-established techniques (Spence and Zuo 1992). Thus, electron diffraction, in general, provides useful complementary reciprocal space information for the structure characterization of materials.

STEM can be considered as a special form of SEND. In bright-field STEM, the intensity of the direct beam is integrated over a small area detector and mapped as a function of electron probe position. In ADF-STEM, the diffraction intensity is integrated between the inner and outer cutoff angles of a circular disk-shaped detector with a hole. The ADF-STEM image intensity comes mostly from inelastic scattering that contributes mostly to the background intensities in electron diffraction patterns. In SEND, the diffraction patterns are directly recorded on a two-dimensional detector, for example, a charge-coupled device (CCD) camera. The recorded diffraction patterns are then processed either online, or offline, to obtain images, including bright-field, dark-field, or ADF-STEM images. Electron diffraction pattern processing provides the flexibility so that information in the recorded diffraction patterns, beyond the simple integrated intensities, can be extracted to form images. This last option is simply not available using the fixed STEM detectors. The tradeoff here, of course, is that one will be dealing with a far more complex data set, with four dimensions in the form of two spatial coordinates: the (x, y) in the real space and the (k_x, k_y) in the reciprocal space. It should be pointed out that direct electron diffraction pattern recording in STEM only became available recently. Diffraction pattern recording had been a particular issue in the early dedicated STEMs, which did not have the camera systems installed in the conventional TEMs. In these early STEMs, diffraction patterns were often recorded using the TV cameras; the quality of the recorded diffraction patterns was poor. In modern TEM with STEM capabilities, the same TEM cameras can be used in the STEM mode. The development of CCD cameras is also particularly useful for SEND. The CCD cameras have good dynamic ranges and the linearity for measuring electron diffraction intensities. The state-of-art CCD cameras are also capable of fast readout. This, combined with the computing power of modern computers, has improved the acquisition and processing of electron diffraction patterns.

Much of our knowledge about electron diffraction was built upon diffraction patterns recorded using fixed electron probes. There are several diffraction techniques for doing this depending on the size of the condenser aperture and the electron beam convergence angle. At very large convergence angles, diffraction patterns recorded in a STEM using a coherent electron source are known as Ronchigrams (Cowley 1979). The large convergence angle causes overlap of different diffraction beams and interferences among these beams; the resulted interference pattern is very useful for the diagnostics of the electron beam alignment and the properties of the electron probe-forming lens (see Chapter 3). The same electron probe, if imaged for a particular diffracted beam, with the help of the objective lens defocus and the selected area aperture, gives large-angle convergent beam electron diffraction (LACBED) (Spence and Zuo 1992). At medium convergence angles, the diffraction patterns recorded can be alternatively described as CBED or coherent CBED depending on whether the diffracted beams broadened by the beam convergence angle overlap and interfere with each other. At small convergence angles, the diffraction pattern recorded is similar to the selected area electron diffraction. Cowley made extensive uses of this technique for identifying the structure of small particles, interfaces, and nanotubes using a dedicated STEM (Cowley 1992, 2004).

Interpretation of electron diffraction patterns is based on diffraction theory and diffraction pattern simulations. On this aspect, the development of sophisticated simulation and data analysis algorithms, combined with the development of electron detectors, has led to a range of electron diffraction applications that simply were not possible before. These applications include crystal structure determination and structure refinement (Hovmoller et al. 2002, Jansen et al. 1998), accurate measurement of lattice parameters (Zuo et al. 1998), and electron structure factors for electron density mapping (Zuo 2004). The ability to simulate and model the electron diffraction patterns is important for these applications. The refinement technique is particularly useful for crystals with an approximately known structure; the structure, or its parameters, can be refined with high accuracy (Zuo 1998, Zuo and Spence 1991) by comparing the experimental data with simulations and by adjusting the parameters using optimization for the best fit. Multiple scattering effects can be taken into account in a refinement by using the dynamical theory to calculate the diffraction intensity. Previously, the strong electron multiple scattering effect in electron diffraction has made it difficult to use the electron intensity for crystal structure determinations in a manner similar to X-ray and neutron diffraction (Cowley 1995). In the case of accurate structure factor measurement, electron wave interference caused by multiple scattering actually enhances the electron diffraction sensitivity to the crystal potential and thickness that improves the accuracy of the electron diffraction data refinement (Zuo 2004). More recently, Zuo and his coworkers have demonstrated the use of modeling for extracting structure information from nanotubes and nanocrystals (Huang et al. 2008, Zuo et al. 2007). For many small nanostructures, electron multiple scattering effect is weak and the kinematical approximation actually becomes useful.

An early form of SEND was developed for ptychography to solve the phase problem in electron diffraction by using interference between diffraction disks in electron nanodiffraction. The concept of ptychography was first proposed by Hoppe (Hegerl and Hoppe 1970) and then further developed by Rodenburg (2008). In the original ptychography, electron diffraction patterns are recorded over an area of a crystal using a coherent probe with a diameter less than the size of the crystal unit cell, and the diffraction intensity at the middle of the overlapping disks is processed as a function of probe position to form atomic resolution images (Rodenburg 2008). The ability to invert diffraction patterns to form images has attracted considerable interest recently in the X-ray diffraction community where the lack of high-resolution imaging lens has been a major obstacle toward X-ray imaging. In electron diffraction, the additional phase introduced by the lens aberrations does not affect the diffraction intensity, and diffraction imaging by solving the phase problem provides atomic resolution imaging at diffraction-limited resolution.

In writing this chapter, we have benefited from the two earlier reviews on electron nanodiffraction by the late Professor John M. Cowley (1999, 2004). Cowley defined electron nanodiffraction as practiced in the dedicated STEMs using a focused electron probe of 1 nm in diameter or smaller (Cowley 1999). We have taken a broader view of electron nanodiffraction here to include diffraction patterns recorded with any coherent electron probes of diameter from sub-Å to tens of nanometers. This broad definition allows us to include some previous works on CBED that were categorized under electron microdiffraction but actually were performed using nanometer-sized electron probes (on the order of a few nanometers). The new definition also allows us to include recent advances in coherent electron diffraction using nanometer-sized parallel beams and their applications for nanostructure characterization.

9.2 Electron Nanodiffraction Techniques

The setup for electron nanodiffraction consists of a field emission electron gun, an illumination system with at least three lenses including the objective prefield, electron deflectors, sample, an area electron detector, and a computer system interfacing the scan coils and the electron detector. The field emission gun (FEG) provides the source brightness and the lateral coherence required for nanodiffraction. The three magnetic lens setup of the illumination system is common in modern transmission electron microscopes (TEM). The three lenses are condenser I, II, and the objective prefield, which is the part of the objective lens's (OL) magnetic field before the sample that also acts as a lens. Condenser I is a de-magnifying lens that is used to reduce the effective electron source size. The OL prefield has a short focal length, which is generally held at a constant value for the optimum microscope performance. The focal length of the condenser II can be varied continuously in the electron microscope to provide different modes of electron illumination and convergence angles. An additional condenser lens (condenser III, or condenser mini-lens as it is called in some microscopes) as shown in Figure 9–1 brings additional flexibility to the three-lens- illumination system. For example, the condenser III can be used to change the convergence angle for a focused beam with a fixed condenser aperture. Without condenser III the convergence angle is fixed by the sample position and the focal length of the objective prefield.

The cone of radiation incident on the sample, formed by the objective prefield in a TEM/STEM instrument, is controlled by the condenser

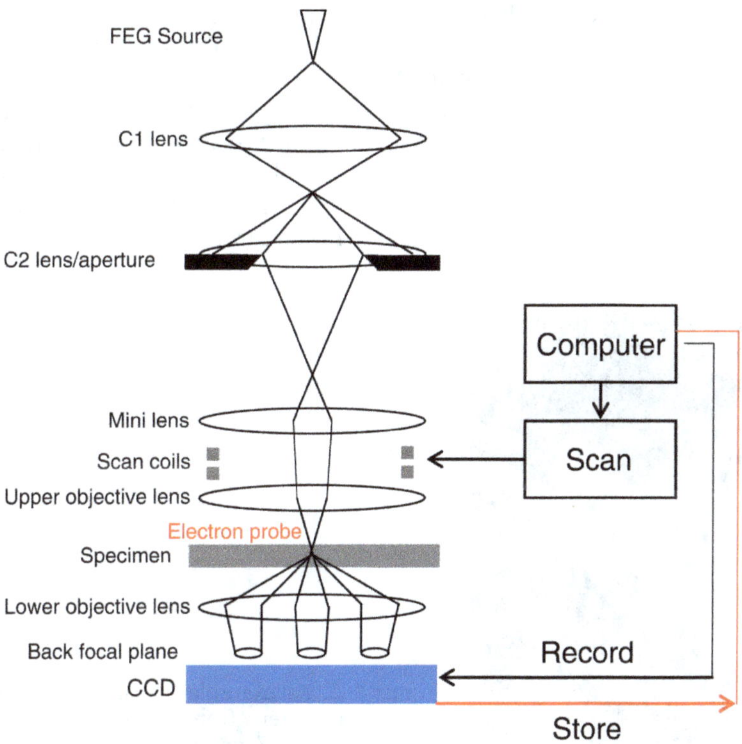

Figure 9–1. Schematic diagram of an electron illumination system for nanoprobe formation and the setup for scanning electron nanodiffraction. The projector system of the microscope between the CCD and the back focal plane is not shown. The computer system is used to position the probe, acquire and store electron diffraction patterns.

aperture and the condenser II setting. There are several electron nanodiffraction techniques that exploit different convergence angles. Nanoarea electron diffraction (NED) or nanobeam diffraction (NBD) uses a small (nanometer-sized) parallel beam illumination (Zuo et al. 2004). The small, parallel, beam is achieved by reducing the convergence angle of the condenser II crossover using condenser III and placing the crossover at the focal plane of the objective prefield, whereby it forms a parallel beam illumination on the sample. For a condenser aperture of 10 micron in diameter, the probe diameter is ~50 nm with an overall magnification factor of 1/200 in the JEOL 2010F electron microscope (JEOL, USA). The beam size in this case is much smaller than what can be achieved using a selected area aperture in a conventional TEM without a C_s corrector. In selected area electron diffraction (SAED), the illumination is spread out over a large area of sample for a parallel beam. In comparison, all electrons illuminating the sample in NED are recorded in the diffraction pattern; NED in a FEG microscope thus can be used to provide higher beam intensity for electron nanodiffraction (the probe current intensity using a 10 micron condenser II aperture in JEOL 2010F is $\sim 10^5$ e/s-nm^2) (Zuo et al. 2004). The diffraction pattern recorded in NED is similar to SAED. For crystals, the diffraction pattern consists of sharp diffraction spots. The major difference is that the diffraction volume is defined directly by the electron probe in NED. NED was originally developed for nanostructure determination (Gao et al. 2003, Zuo et al. 2004). The small beam size in NED allows the selection of an individual nanostructure and reduction of the background in the electron diffraction pattern from the surrounding materials. An example is given in Figure 9–2 for electron diffraction of a single-carbon nanotube of 3.47 nm in diameter. The nanotube was supported on holey carbon films. The diffraction pattern was recorded from a section of the tube over a hole in the carbon film.

Figure 9–2. An example of a nanoarea electron diffraction pattern from a single-wall carbon nanotube of 3.47 nm diameter. The inset at the *top-left corner* is the electron probe used to record the diffraction pattern. The nanotube is visible in the probe image. The small probe size was used to isolate a single tube for diffraction.

High spatial resolution in electron nanodiffraction is achieved using a convergent beam. The improvement comes from the reduced beam size using a focused probe at the cost of the angular resolution in the diffraction pattern. The convergent beam is obtained by placing the beam crossover after condenser II near the front focal plane of condenser III lens as shown in Figure 9–1. This gives a focused probe with the sample placed close to the back focal plane of the OL prefield. The probe size formed with a coherent illumination can be as small as few angstroms. The beam convergence angle can be changed by using different condenser apertures or changing the strength of the condenser III. At a small convergence angle, the diffraction patterns recorded consists of small disks (Cowley 2004). Figure 9–3 shows an example of the electron probe used for electron nanodiffraction in the JEOL 2010F TEM and using a 4-μm diameter condenser aperture. The full width at half maximum of the probe is 1.7 nm in this case. The probe convergence angle measured was 0.5 mrad based on the calibration using single-crystal Si particles. The convergence angle can be changed within a small range by using different condenser III lens settings in the JEOL 2010F TEM. The probe size can be reduced by changing the condenser I lens setting. For the same lens settings, the probe size and convergence angle should be fairly constant between different experiments.

The smallest probe size is obtained with a fully coherent illumination. For the STEM operations, the electron coherence is defined by the coherence length seen at the condenser aperture. According to the Zernike–van Cittert theorem, the degree of coherence between electron wave functions at two different points far away from a monochromatic electron source is given by the Fourier transform of the source intensity distribution (Cowley 1999). If we assume that the source has a uniform intensity within a circular disk, the coherence function is then given by $\lambda J_1(\pi\beta r/\lambda)/\beta r$ with J_1 as the first-order Bessel function, r the radial distance at the aperture, and β the angle sustained by the electron source

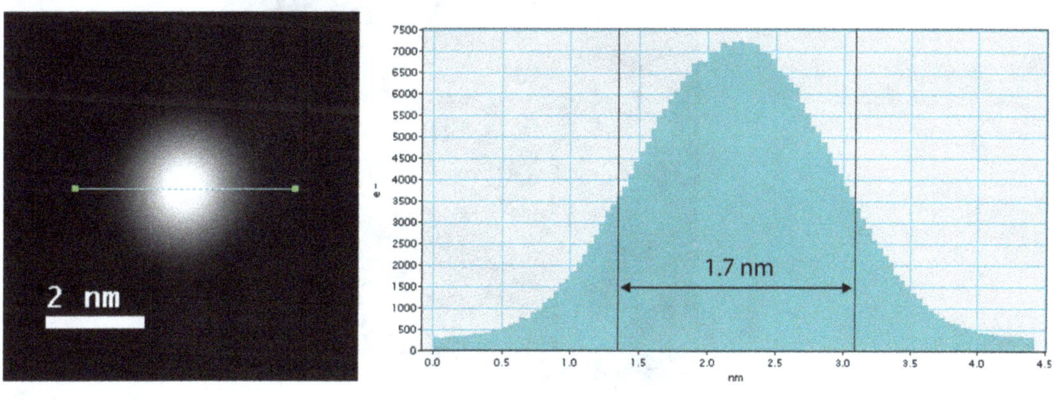

Figure 9–3. (*Left*) An image of a focused electron probe used in electron nanodiffraction. (*Right*) The intensity profile of the probe along the indicated line in the probe image. The probe was formed using a small condenser aperture with diameter ~4 microns in the CBED mode. The convergence angle is about 0.5 mrads.

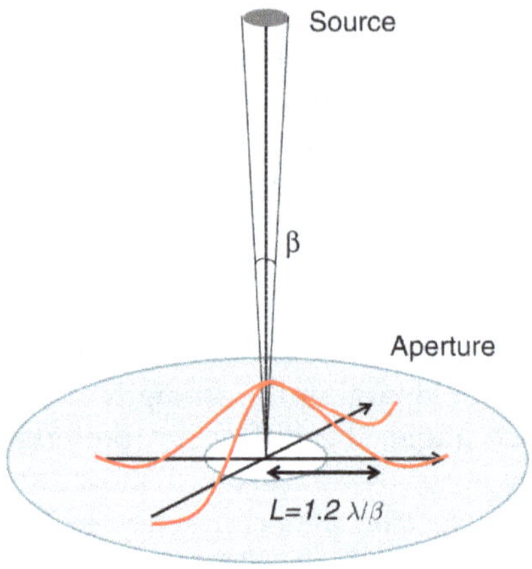

Figure 9–4. Relationship between a finite source and the lateral coherence length L at the condenser aperture. The source size is defined by its subtended angle (β) seen at the aperture.

(see Figure 9–4). The lateral coherence length L, which is often referred in the literature, is defined by r at the first zero of the J_1, which has the value of $L = 1.2\lambda/\beta$. The source seen by the condenser aperture inside a STEM is the source image formed after the condenser I lens. For a Schottky emission source, the emission diameter is between 20 and 30 nm according to Botton (2007). For a condenser aperture placed at a distance of 10 cm away from the electron source image, a factor of 10 source demagnification provides a coherence length from 100 to 150 μm.

The convergence angle of a focused probe can be changed by using different-sized condenser apertures for a fixed illumination system setting. At medium convergence angles, on the order of 10 mrads for 200 kV electrons, the diffraction pattern recorded is the same as CBED. An example of a CBED pattern is shown in Figure 9–5. The

Figure 9–5. An experimental CBED pattern recorded from silicon along the [110] zone axis orientation using a 200 kV field emission electron microscope with an in-column energy filter.

development of CBED has a much longer history than electron nanodiffraction that predates the development of electron field emission guns and STEM electron optics for forming nanometer-sized electron probes (Spence and Zuo 1992). The diameter of the electron probe formed by using a thermionic source is limited by the lateral coherence at the condenser aperture (see Figure 9–4 and the related discussions). For a LaB_6 source with an emission spot of 10 μm in diameter to achieve the same coherence length as a field emission gun requires an additional source demagnification of several hundred times, which will leave very little intensity in the electron probe. In practice, the probe size used for CBED in a TEM with a thermionic source is tens of nanometers in diameter. For this reason, CBED is commonly known as an electron microdiffraction technique. In a field emission STEM, the typical electron probe size used for CBED is on the order of 1 nm. Electron propagation in a thick crystal actually defines the sample volume that contributes to the CBED pattern (Zuo and Spence 1993).

Figure 9–5 shows an experimental CBED pattern recorded from Si along the [110] zone axis orientation using 200 kV electrons. The CBED pattern consists of large diffraction disks. The size of the disk is slightly larger than the length of the four {111} reflections seen in the [110] zone axis; this leads to a slight overlap between two neighboring disks. The electron beam convergence angle is slightly larger than 8 mrads in this case. What exemplifies a CBED pattern like Figure 9–5 is the amount of intensity information recorded in the pattern. CBED records diffraction intensity as a function of the incident beam directions, which gives additional information, including the high-order Laue zone (HOLZ) lines and the diffraction pattern symmetry. The symmetry of the [110] zone axis of the cubic silicon can be clearly seen in Figure 9–5. The amount of intensity information in a CBED pattern makes it an ideal technique for quantitative analysis of electron diffraction. The other benefit of CBED is that the diffraction pattern is taken from a small area using a focused electron probe. The smallest electron probe currently available in a aberration corrected (C_S corrected) FEG-STEM is 1 Å or less. CBED patterns recorded with such small probes, in principle, are sensitive to structural information on individual atomic columns (Zuo and Spence 1993). For the crystallographic applications, CBED patterns are typically recorded with a probe anywhere from a nanometer to a few tens of nanometers.

Using the STEM scan coils, scanning electron diffraction patterns can be recorded from an area of the sample to provide spatially resolved structural information. The double deflection STEM scan coils are placed before the OL. Scanning electron diffraction can be carried out by first selecting an area of interest and dividing this area into a number of pixels and then placing the electron probe at each of these pixels and recording diffraction patterns at these pixel locations. Diffraction pattern recording can be made on TV-rated cameras or using a CCD camera. The change in diffraction patterns can be used to map the local structural information in real space. In the simplest form, bright- and dark-field STEM images are obtained simultaneously from SEND by integrating diffraction intensities of the direct beam and diffracted beams, respectively. The advantage of SEND is that it can be used to probe structures at very fine (nanometer) scale, which is difficult to do with other scattering techniques, i.e., neutron scattering and X-ray diffraction. The scanning electron probe interacts strongly with matter that leads to local sensitivity. On the other hand, a scanning electron probe gives the structural information of the bulk material that is lacking from the other scanning probe techniques, such as atomic force microscope (AFM) and scanning tunneling microscope (STM).

9.3 Information from Diffraction Patterns

An electron diffraction pattern records the intensity of scattered electrons as a function of scattering angle. The electron scattering intensity can be further divided according to the electron energy when an energy filter is used for the electron diffraction pattern recording. The energy resolution of the energy filter used for high-energy electrons generally is not high enough to distinguish the small energy difference between elastic scattering and inelastic scattering involving phonons. A typical electron diffraction pattern consists of Bragg peaks for single crystals, or diffraction rings for nanocrystals or amorphous materials, from electron elastic scattering and a background coming mostly from inelastic scattering. The strength of the diffraction peaks relative to the background is shown in Figure 9–6 for an experimental diffraction pattern recorded from a $SrTiO_3$ crystal along the [100] zone axis at a medium thickness (several tens of nanometers) using 200 kV electrons. Structural information obtainable from electron diffraction patterns are the following:

1. The diffraction peak position can be used to measure the d-spacing of individual reflections. The combination of diffraction peak positions and their indexing can be used to determine the crystal lattice, its repeating unit cell, and the cell parameters;
2. Diffraction pattern symmetry recorded in CBED and the dynamic extinction in the form of Gjonnes–Moodie lines can be used to determine the crystal symmetry or the lack of symmetry;
3. Diffraction pattern indexing can be used to determine the crystal orientation relative to the electron beam direction;
4. The change in diffraction peak intensity and the diffuse scattering around the Bragg peaks can be used to identify structural defects.
5. For very thin samples, the Fourier transform of the diffraction pattern gives the projected inter-atomic distances. For nanocrystalline or amorphous structures, the Fourier transform of the radial diffraction intensity gives the pair distances and their distribution; the diffraction patterns can be also be used to obtain information about medium range ordering.
6. The electron diffraction intensity can be used to determine atomic positions. In cases where multiple scattering effects in the measured diffraction intensities are strong, multiple scattering must be included in order to determine the atomic structure;

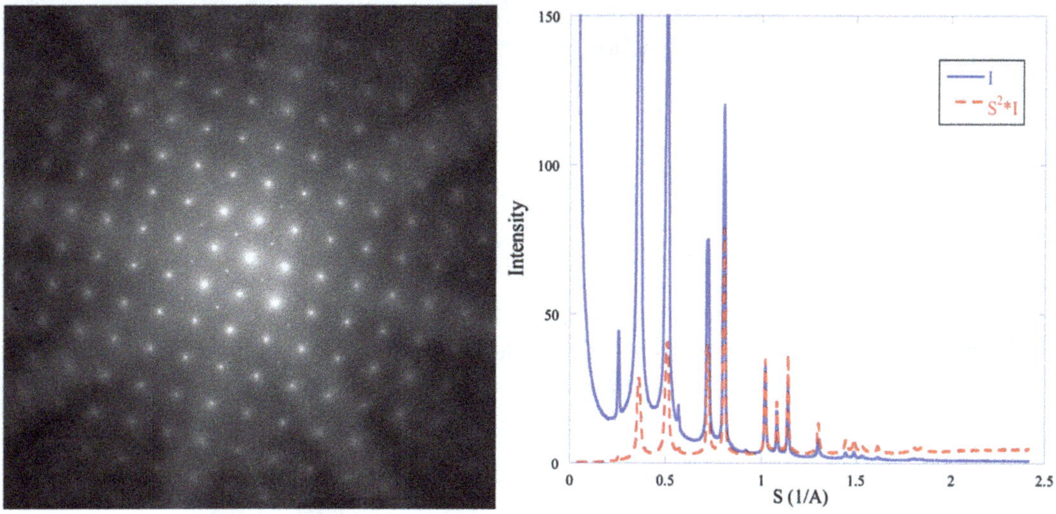

Figure 9–6. An example of electron diffraction intensity and its distribution. (*Left*) The diffraction pattern recorded on imaging plates from $SrTiO_3$ along [100] using 200 kV electrons. The display is in log scale. (*Right*) Averaged radial diffraction intensity (I) plotted against $S = 2\sin\theta/\lambda$. The *dashed line* is $S^{2*}I$, which corresponds to the intensity measured by an annular detector.

7. Accurate structure factor measurement from diffraction intensities can be used to determine the atomic thermal vibrations (the Debye–Waller factors), and crystal potential or charge density.

Procedures used to retrieve the above structural information range from relatively straightforward to highly sophisticated, involving quantitative analysis of diffraction intensities. In general, the complexity of the diffraction pattern analysis increases with the order of the above list.

The part of diffraction pattern used for structure determination resides between the very low scattering angle used for Bright-field STEM and the very high scattering angle for high-angle ADF-STEM ($S = 2.0\ 1/\text{Å}$ for the inner cutoff angle of 50 mrad for 200 kV electrons). This part contains the majority of the diffraction intensity as shown in Figure 9–6 in the form of the $S^{2*}I$ curve, which corresponds to the integrated intensity using an annular detector.

9.3.1 Kinematical Theory of Electron Nanodiffraction

A conceptual understanding of electron nanodiffraction can be developed by considering single scattering of an electron probe in the limit of very thin samples. To do this, we will consider the effect of the sample potential on the electron probe wave function $\phi_P(\mathbf{R}, z)$ placed at position $\mathbf{R}_o$. For very thin samples, the exit electron wave after passing through the sample potential field is related to the incident electron wave by the following approximation:

$$\phi(\mathbf{R}, z) \approx \phi_P(\mathbf{R} - \mathbf{R}_o, z)\left[1 + i\sigma\bar{V}(\mathbf{R})\right] \tag{1}$$

Here $\bar{V}(\mathbf{R})$ is the projected potential along the electron beam direction. The diffraction pattern records the intensity of the scattered wave as a function of the diffraction angle or scattering vector, which is obtained by Fourier transform (FT) of the scattered wave against the scattering vector **S**:

$$\begin{aligned} I(\mathbf{S}) &= |FT[\phi(\mathbf{R},z)]|^2 \approx \left|FT\left\{\phi_P(\mathbf{R}-\mathbf{R}_o,z)\left[1+i\sigma\bar{V}(\mathbf{R})\right]\right\}\right|^2 \\ &= \left|FT[\phi_P(\mathbf{R}-\mathbf{R}_o,z)] \otimes FT\left[1+i\sigma\bar{V}(\mathbf{R})\right]\right|^2. \end{aligned} \tag{2}$$

Thus, in the limit of the kinematic approximation, the electron nanodiffraction pattern recorded is simply a convolution of the probe and the complex object function of $1+i\sigma\bar{V}(\mathbf{R})$ in reciprocal (Fourier) space.

9.3.2 Electron Probe Formation

The formation of electron probes in electron nanodiffraction follows the same principle of the STEM probe formation (see Chapter 2). In STEM, the electron beam crossover from the last condenser lens is imaged by the OL. The difference between the objective lens focal length and the focal length required to image the crossover onto the sample is defined as the defocus. For a convergent beam of electrons, the lens aberrations introduce an angle-dependent phase, $\chi(\mathbf{K})$, with **K** standing for the part of the incident beam wave vector perpendicular to the optical axis of the electron lens. The phase $\chi(\mathbf{K})$ from the OL aberrations and its relation to the focused electron probe used in STEM is described in Chapter 2. For electron nanodiffraction with a defocused probe, we must also consider the electron source wave function $\phi_S(\mathbf{R})$ formed by the condenser lenses and its contribution to the electron probe. According to the image formation theory, the electron probe on the sample is an image of $\phi_S(\mathbf{R})$ magnified by the lens magnification M. The image is a convolution of $\phi_S(\mathbf{R})$ and the objective lens resolution-function $T(\mathbf{R})$:

$$\begin{aligned} \phi_P(\mathbf{R}) &= \phi_S(-\mathbf{R}/M) \otimes T(\mathbf{R}) \\ &= \int_{-\infty}^{\infty} \phi_S(-M\mathbf{K})\, A(\mathbf{K}) \exp[i\chi(\mathbf{K})] \exp(2\pi i\mathbf{K}\cdot\mathbf{R})\, d\mathbf{K}, \end{aligned} \tag{3}$$

where $A(\mathbf{K})$ is the aperture function with a value of 1 for $|\mathbf{K}| < \Theta/\lambda$ and 0 beyond with Θ standing for the beam convergence angle. The electron beam energy-spread and the chromatic aberration are neglected in Eq. (3). Equation (3) also assumes that the illuminating electron wave is perfectly coherent across the condenser aperture.

A focused electron probe on the sample is formed by placing the electron beam crossover after condenser II far away from the front focal plane of the objective lens as shown in Figure 9–7. This gives a demagnified, sharp, electron source image on the sample with the magnification of M << 1. The size of the electron probe, in this case, is largely determined by the objective lens resolution function $T(\mathbf{R})$. In reciprocal space, the demagnified electron source has a broad, spherical wave-like,

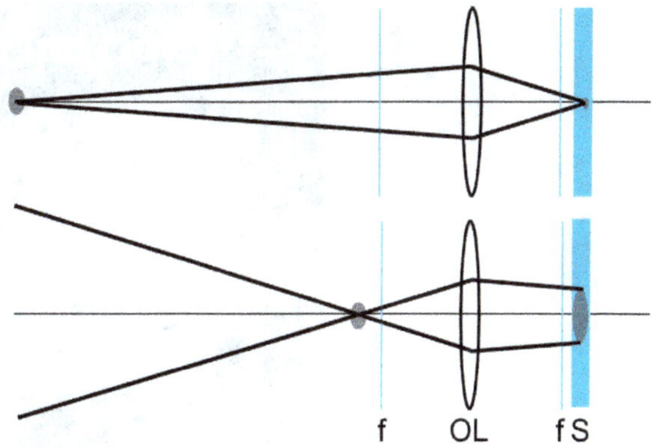

Figure 9–7. Ray diagrams comparing electron nanodiffraction with a focused probe (above) and nanoarea electron diffraction with a defocused probe.

spectrum of wave vectors. The beam convergence angle is then limited by the aperture function $A(\mathbf{K})$.

In the NED mode, the electron beam crossover is placed close to, or at, the front focal plane. The electron source in this case is magnified (M >> 1) (see Figure 9–7). The sample is also placed away from the electron source image after the OL (the image plane) near the back focal plane of the objective lens. This large underfocus must be included as a part of the lens aberration function in Eq. (3) in order to simulate the electron probe in NED (Zuo et al. 2004). The probe magnification is used to reduce the electron beam convergence angle for the parallel beam diffraction. To demonstrate this, we assume a Gaussian distribution for the electron source:

$$\phi_S(\mathbf{R}/M) = A\exp\left(-a^2R^2/M^2\right),$$

where a is one over the probe half width at A/e. The Fourier transform of this Gaussian probe after the OL is

$$\phi_S(\mathbf{K}) = \frac{A\sqrt{\pi}}{a}\exp\left[-K^2/\left(a/M\pi\right)^2\right].$$

The width of the beam in reciprocal space is reduced by a factor of 1/M. The source function in NED with its large probe magnification leads to a reduced electron beam convergence angle. The Gaussian half width of the defocused electron beam is ~ 0.05 mrad in the JEOL2010F TEM formed using a 10 μm condenser aperture. The real space probe in NED is a convolution of the magnified source with $T(\mathbf{R})$. The dominant probe features come from $T(\mathbf{R})$ as shown in Figure 9–8 for a comparison between an experimental probe and simulation based on $T(\mathbf{R})$ alone (Zuo et al. 2004).

9.3.3 Electron Nanodiffraction from an Assembly of Atoms

Once the electron probe wave function is known, electron nanodiffraction theory can be further developed by considering the details of the atomic structure. Here, we will consider an assembly of atoms in a volume several times larger than that selected by the electron probe. The

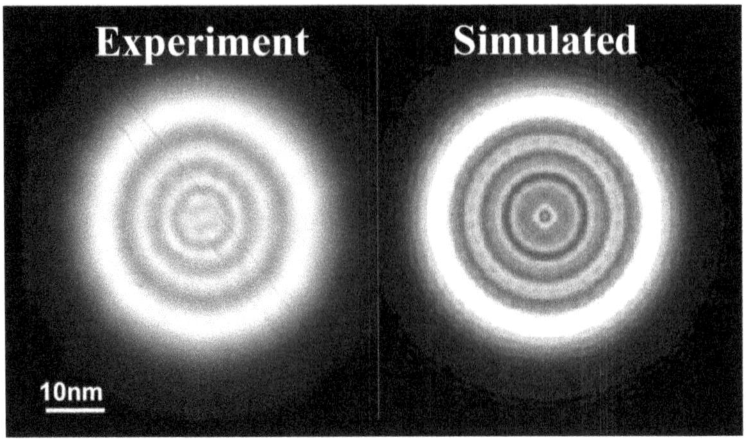

Figure 9–8. Experimental and simulated electron nanoprobe used in nanoarea electron diffraction (NED). The simulation used $C_S = 1$ mm and $\Delta f = -360$ nm, reproduced from Zuo (2004) with permission.

large volume ensures that the shape of the selected volume does not adversely affect the calculation of the diffraction patterns. Here, we will start from large to small crystals and then move to crystals with defects.

The potential of an assembly of atoms can be written as a convolution of the potential of individual atoms and their positions as described by the delta function:

$$V(\mathbf{r}) = \sum_i V_i^a(\mathbf{r}) \otimes \sum_j \delta\left(\mathbf{r} - \mathbf{R}_{i,j}\right),$$

where V^a is the atomic potential and $\mathbf{R}_{i,j}$ denotes the atomic position. The double summations are used to include different types of atoms using the subscript i and the number of the same atom using j. The Fourier transform of this potential is simply a sum of the atomic scattering factors with a phase factor dependent on the atomic position:

$$FT[V(\mathbf{r})] = \sum_i f_i(\mathbf{S})\, T_i(\mathbf{S}) \sum_j \exp\left(-2\pi i \mathbf{S} \cdot \mathbf{r}_{i,j}\right)$$

Here f is the atomic scattering factor, T_i is the thermal factor which accounts for the effect of thermal vibrations and static disorder on atomic scattering, and $\mathbf{S}$ is the scattering vector. The minus sign in the exponential is used here to be consistent with the notation for the electron wave of $\exp(2\pi i \mathbf{k} \cdot \mathbf{r})$. For a crystalline sample, the atoms are repeated in a lattice and the position of an atom is specified by the lattice vector and its position within the unit cell:

$$\mathbf{R}_{i,j} = n\mathbf{a} + m\mathbf{b} + l\mathbf{c} + \mathbf{r}_{i,j} = \mathbf{R}_{nml} + \mathbf{r}_{i,j}.$$

The lattice vector $\mathbf{R}_{nml}$ defines a point in the crystal lattice with the index of n, m, and l. The Fourier transform of the potential of a sufficiently large crystal gives the well-known reciprocal lattice formula:

$$FT\left[V\left(\mathbf{r}\right)\right] = \left\{\sum_i f_i\left(\mathbf{S}\right)T_i\left(\mathbf{S}\right)\sum_j \exp\left(-2\pi i\mathbf{S}\cdot\mathbf{r}_{i,j}\right)\right\}\cdot\left\{\sum_{h,k,l}\delta\left(\mathbf{S}-h\mathbf{a}^*-k\mathbf{b}^*-l\mathbf{c}^*\right)\right\} \tag{4}$$

Here $\mathbf{a}^*, \mathbf{b}^*, \mathbf{c}^*$ are the reciprocal lattice vectors of the crystal. The above formula has two parts indicated by the curly brackets. The first part is the structure factor:

$$F\left(\mathbf{S}\right) = \sum_i f_i\left(\mathbf{S}\right)T_i\left(\mathbf{S}\right)\sum_j \exp\left(-2\pi i\mathbf{S}\cdot\mathbf{r}_{i,j}\right). \tag{5}$$

Under the kinematical approximation, the structure factor determines the intensity of the diffracted beam. The second part of Eq. (4) defines the diffraction pattern geometry. The diffracted wave in electron nanodiffraction is given by:

$$FT\left[\phi_S\left(\mathbf{r}\right)\right]\otimes FT\left[V\left(\mathbf{r}\right)\right] = F\left(\mathbf{S}\right)\cdot\left\{FT\left[\phi_S\left(\mathbf{r}\right)\right]\otimes\sum_{h,k,l}\delta\left(\mathbf{S}-h\mathbf{a}^*-k\mathbf{b}^*-l\mathbf{c}^*\right)\right\} \tag{6}$$

For a crystal, Eq. (6) shows that the electron probe determines the shape of the diffraction peaks; the positions of the diffracted beams are determined by the reciprocal lattice and the diffraction peak intensity is determined by the structure factor. The basis for diffraction pattern geometry analysis, thus, is the crystal reciprocal lattice and the Laue diffraction condition, or the equivalent Bragg's law, for diffraction:

$$\mathbf{S} = \mathbf{k} - \mathbf{k}_o = \mathbf{g} = h\mathbf{a}^* + k\mathbf{b}^* + l\mathbf{c}^*, \tag{7}$$

where $\mathbf{k}$ and $\mathbf{k}_o$ are the incident and diffracted electron wave vectors, respectively, and $\mathbf{g}$ is a reciprocal lattice vector.

The potential of a truncated nanocrystal is the potential of a large crystal multiplied by the shape of the nanocrystal:

$$V\left(\mathbf{r}\right) = s\left(\mathbf{r}\right)\left[\sum_i V_i^a\left(\mathbf{r}\right)\otimes\sum_j\delta\left(\mathbf{r}-\mathbf{R}_{i,j}\right)\right].$$

The shape of the nanocrystal introduces an additional broadening function in the form of the shape function, $FT\left[s\left(\mathbf{r}\right)\right]$. The convolution of the shape function and the probe in reciprocal space determines the profile of the diffracted electron beam. A nanocrystalline materials can be considered as made of many nanocrystals. In this case, the diffraction pattern can be considered as a sum of diffraction patterns from individual nanocrystals. This approximation breaks down when two or more nanocrystals give the same diffracted beam. The interference between these diffracted beams results in complicated interference fringes dependent on the relative positions of the nanocrystals.

For crystals with defects, the change in crystal structure caused by a defect away from the core of the defect can be described by a lattice-dependent displacement field, $\mathbf{u}(\mathbf{R}_{nml})$. The crystal potential including this displacement then becomes

$$V(\mathbf{r}) = V_A(\mathbf{r})^* \sum_{n,m,l} \delta(\mathbf{r} - \mathbf{R}_{nml} - \mathbf{u}(\mathbf{R}_{nml})).$$

Here

$$V_A(\mathbf{r}) = \sum_i V_i^a(\mathbf{r}) \otimes \sum_j \delta\left(\mathbf{r} - \mathbf{r}_{i,j}\right)$$

is the crystal unit cell potential including contributions from different atoms inside the cell. The Fourier transform of the lattice leads to a sum of two phase terms:

$$FT\left[\sum_{n,m,l} \delta(\mathbf{r} - \mathbf{R}_{nml} - \mathbf{u}(\mathbf{R}_{nml}))\right] = \sum_{n,m,l} \exp(-2\pi i\mathbf{S}\cdot\mathbf{R}_{nml}) \exp[-2\pi i\mathbf{S}\cdot\mathbf{u}(\mathbf{R}_{nml})].$$

For small displacements, the above equation can be expanded and kept to the first-order term:

$$FT\left[\sum_{n,m,l} \delta(\mathbf{r} - \mathbf{R}_{nml} - \mathbf{u}(\mathbf{R}_{nml}))\right] \approx \sum_{n,m,l} \exp(-2\pi i\mathbf{S}\cdot\mathbf{R}_{nml}) - \sum_{n,m,l} 2\pi i\mathbf{S}\cdot\mathbf{u}(\mathbf{R}_{nml}) \exp(-2\pi i\mathbf{S}\cdot\mathbf{R}_{nml}). \tag{8}$$

The first term in Eq. (8) is the same as Eq. (4) for perfect crystals, which defines an array of diffraction peaks; the position of each peak is defined by a reciprocal lattice vector of the crystal. The second term describes the diffuse scattering around a diffraction peak. If we take the reflection as **g** and write

$$\mathbf{S} = \mathbf{g} + \mathbf{q} \; and \; \mathbf{g}\cdot\mathbf{R}_{nml} = N \tag{9}$$

with N as an integer. For $|g|>>|q|$, the diffuse scattering term can be rewritten as

$$\sum_{n,m,l} 2\pi i\mathbf{S}\cdot\mathbf{u}(\mathbf{R}_{nml}) \exp(-2\pi i\mathbf{S}\cdot\mathbf{R}_{nml}) \approx \sum_{n,m,l} 2\pi i\mathbf{g}\cdot\mathbf{u}(\mathbf{R}_{nml}) \exp\left(-2\pi i\mathbf{q}\cdot\mathbf{R}_{nml}\right). \tag{10}$$

Equation (10) is a Fourier sum of the displacements along the g direction. The intensity predicted by this equation will increase with a g^2-dependence. The atomic scattering contains the Debye–Waller factor, which describes the damping of high-angle scattering because of the thermal vibrations. The balance of these two terms results in a maximum contribution to the diffuse scattering in the diffraction pattern from deviations from an ideal crystal lattice.

In summary, for crystals, the diffraction peak position is defined by the crystal reciprocal lattice (Eq. (7)), the diffraction peak shape is given by the electron probe in reciprocal space (Eq. (6)), and the peak intensity is determined by the structure factor (Eq. (5)). For nanocrystals, the diffraction peak shape is a combination of the nanocrystal shape function and the electron probe with both in reciprocal space. For strained crystals, additional diffuse scattering around the Bragg peaks is expected (Eqs. (9) and (10)); the diffuse scattering depends on the displacement vector and also the reflection (g).

9.3.4 Convergent Beam Electron Diffraction

At medium convergence angles, the interpretation of electron nanodiffraction patterns recorded from crystals uses the same theory as for CBED. The starting point for understanding CBED is the Ewald sphere construction. Figure 9–9 shows one example. By the requirement of elastic scattering, all transmitted and diffracted beams are on the Ewald sphere. Let us take the incident beam P, which satisfies the Bragg condition for g. For an incident beam P', to the left of P, the diffracted beam also moves to the left. The difference between the incident wave and the diffracted wave is the vector **g**. The deviation of the diffracted beam away from the Bragg condition is defined by the so-called excitation error:

$$s_g = \left(k_o^2 - |\mathbf{k} + \mathbf{g}|^2\right)/2k_o. \tag{11}$$

The change in the excitation error across the CBED disk is important for understanding the rich diffraction intensity patterns often observed in CBED. To see how the excitation error changes within a CBED disk for a particular reflection, let us take the component of the wave vector **k** along a reciprocal lattice vector **g** as

$$k_g = -g/2 + \Delta.$$

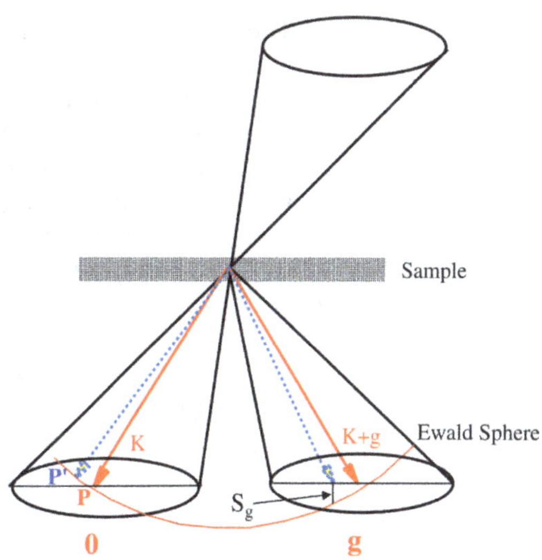

Figure 9–9. A schematic ray diagram of CBED. This figure demonstrates the variation of excitation errors at different positions of the CBED disk. The beam marked by the *full line* (P) is at the Bragg condition, while the beam marked by the *dashed line* is associated with a positive excitation error (Sg). Reproduced from Zuo (2004) with permission.

The positive Δ corresponds to a beam tilt toward **g**. Substituting k_g into Eq. (11), we have

$$s_g = \lambda\left(\left(-\frac{g}{2}+\Delta\right)^2 - \left(\frac{g}{2}+\Delta\right)^2\right)/2 = -g\Delta/(1/\lambda) \approx -g\delta\theta.$$

Here, $\delta\theta$ is the deviation angle from the Bragg condition. The excitation error given above is approximately the distance from the end of the diffracted wave vector to the Ewald sphere. The excitation error also has a sign. The S_g is negative for a positive Δ and positive for a negative Δ. Thus, an incident beam moving to the left gives a positive excitation error. Correspondingly, a beam, moving to the right of P, gives a negative excitation error. Generally, to a good approximation, the excitation error changes linearly across the CBED disk and along the direction of **g** for each diffracted beams. The slope of the change is the length of **g**. The range of excitation errors within each disk is proportional to the length of **g** and the convergence angle. Consequently, the excitation error changes much faster for a high-order Laue Zone (HOLZ) reflection than a reflection in ZOLZ close to the direct beam. The excitation error and the crystal thickness are the two parameters controlling the diffraction intensity beside the crystal potential and the underlying atomic structure. This dependence can be directly seen in the expression for the diffraction intensity in case of the two-beam approximation (without the effect of absorption), which assumes only two strong beams in the diffraction pattern with the direct beam and the diffracted beam of g:

$$I_g = |\phi_g|^2 = \frac{1}{s_g^2\xi_g^2+1}\sin^2\left(\pi t\sqrt{s_g^2+1/\xi_g^2}\right).$$

The ξ_g is the extinction distance of the diffracted beam. In CBED, the crystal thickness is approximately constant under the electron beam, and the intensity variations observed in CBED disks are mostly caused by changes in the excitation errors of the diffracted beams.

HOLZ lines are sharp lines observed in the CBED disks. They are produced by Bragg diffraction by lattice planes of high-order reflections. The rapid increase in the excitation error for a high-order reflection away from the Bragg condition results in a rapid decrease in the diffraction intensity. The maximum diffraction intensity occurs at the Bragg condition under the kinematic approximation, which appears as a straight line within the CBED disk. The position of HOLZ lines is very sensitive to small changes in lattice parameters and the local strain. The sensitivity comes from the large scattering angle. This can be seen in the case of a cubic crystal for which

$$\theta \approx g\lambda/2 = \sqrt{h^2+k^2+l^2}\lambda/2a.$$

A small change in a gives

$$\delta\theta \approx 0.5\, g\lambda\delta a/a.$$

The amount change in the Bragg angle is proportional to the length of **g**. The positions of these lines move relatively to each other when the lattice parameters change. This effect can be used for accurate measurement of lattice parameters.

The direction of a HOLZ line is normal to the reciprocal lattice vector and its position is decided by the Bragg condition. In diffraction analysis, it is useful to express the HOLZ lines using the line equations in an orthogonal zone axis coordinate system (x, y, z), with the z parallel to the zone axis direction (Zuo 1992). The x-direction can be taken along the horizontal direction of the experimental pattern and the y is normal to the x. The Bragg diffraction condition (Eq. (7)) expressed in this coordinate is given by

$$k_y = -\frac{g_x}{g_y}k_x + \frac{2g_z - g^2}{2g_y}|k_z|. \tag{12}$$

Here

$$|k_z| = \sqrt{k_o^2 - k_x^2 - k_y^2} \approx k_o = 1/\lambda.$$

The approximation holds for high-energy electrons of small wavelengths and the typical acceptance angles in electron diffraction. Within this approximation, beams that satisfy the Bragg condition form straight lines.

9.3.5 Diffraction Pattern Symmetry and Crystal Space Group

A major application of CBED is to determine the crystal symmetry. The crystal symmetry is reflected in the diffraction patterns. For example, if the crystal has a rotation axis, two diffraction patterns related by rotation should be the same. The same is true for mirror symmetry. Additional symmetries are produced in electron diffraction because of (1) the principle of reciprocity and (2) the projection along the zone axis for ZOLZ (Peng et al. 2004). The principle of reciprocity states that the intensity of the diffracted beam (B) with a source (A) is the same as the intensity detected at A with the source at B by the same scatter. The projection of crystal structure along the zone axis orientation used in observation produces a mirror symmetry at the middle of the sample which may or may not exist in the crystal. The combination of reciprocity and projection with the crystal point groups produces 31 diffraction groups, whose relationships with the 32 point groups were tabulated by Buxton (1976a). The correspondence is often not unique. The determination of crystal point groups comes down to elimination of multiple choices using the symmetry of diffraction patterns recorded along several major symmetric orientations and/or using information about the lattice determined from the diffraction pattern geometry. The diffraction pattern symmetries used in the determination are those of the whole pattern, the transmitted beam (bright-field), the diffracted beams (dark-field), and the symmetry between $+g$ and $-g$ beams. It should be emphasized that the Friedel symmetry ($I_g = I_{-g}$) is absent in electron diffraction because of dynamic scattering. The point groups can be uniquely determined by electron diffraction.

The screw and glide axes present in the crystals can be determined by observing dynamic extinction in kinematical forbidden reflections (zero structure factor due to the glide or screw axes). These reflections generally show some intensities due to electron multiple scattering. The dynamic extinction is observed when the incident beam is in the glide plane in the case of a glide; this was first reported by Gjonnes and Moodie using CBED (the extinction appears as dark lines subsequently named as the G-M lines). The dynamic extinction of a screw axis is more complicated and is described in detail in Zuo (1992).

The combination of point group determination and identification of translation symmetry allows the unique identification of space groups (Tanaka et al. 1983a, b). Both CBED and LACBED techniques can be used for this purpose. Applications of symmetry determination by CBED include phase identification and as part of the determination of unknown structures. Methods for quantifying and auto detection of the CBED symmetry can be found in references (Hu et al. 2000, Vincent and Walsh 1997).

9.3.6 Electron Diffraction Intensity

Interpretation of diffraction intensities is required for any analysis that goes beyond the measurement of the d-spacings of crystals. The diffraction intensity of ultrathin films and small nanocrystals can be approximated by summing scattering from an assembly of atoms (kinematic approximation) as shown in Section 9.3.3:

$$I(\mathbf{S}) \propto \left| \sum_{i,j} f_i(\mathbf{S}) T_i(\mathbf{S}) \exp(-2\pi i \mathbf{S} \cdot \mathbf{r}_{ij}) \right|^2, \tag{13}$$

where the summation is over individual atoms i and j, and $\mathbf{S} = \mathbf{k} - \mathbf{k}_o$ is the scattering vector. For a crystal, the sum is limited to atoms within the unit cell and $\mathbf{S} = \mathbf{k} - \mathbf{k}_o = g$. The f_i and T_i are the atomic scattering and temperature factor, respectively, and

$$T(\mathbf{S}) = \exp(-B\,|\mathbf{S}|^2\,/4) \tag{14}$$

in the case of isotropic atomic vibrations with B for Debye–Waller factor.

The kinematic approximation breaks down at a certain crystal thickness where the diffracted intensity approaches that of the incident beam. A useful rule of thumb for the validity of the kinematic approximation is

$$t < \xi_g/4,$$

where ξ_g is the extinction distance of the strongest diffracted beam in the diffraction pattern. By avoiding the strong beams using crystal rotation, the use of the kinematic approximation can be extended somewhat to thicker crystals. For some of the inorganic crystals often studied by electron microscopy such as silicon, the typical extinction distance of a strong reflection is on the order of a few tens of nanometers. For a crystal more than a few nanometers thick, multiple scattering must be taken into account for the interpretation of diffraction intensities.

There are several approaches to treat electron multiple scattering in electron diffraction (Peng et al. 2004). For perfect crystals, the Bloch wave approach is the most useful, which is based on an expansion of the electron wave function in plane waves, which was formulated first by Hans Bethe. In the Bloch wave theory, when diffraction disks do not overlap, the diffraction intensity for a point inside the diffraction disk belonging to the reflection **g** is given by

$$I_g(x,y) = |\phi_g(x,y)|^2 = \left| \sum_i c_i(x,y) C_g^i(x,y) \exp[2\pi i \gamma^i(x,y)t] \right|^2 \tag{15}$$

Here the eigenvalue γ and eigenvector C_g are obtained from diagonalizing the equation:

$$2KS_g C_g + \sum_h U_{gh} C_h = 2K\gamma C_g, \tag{16}$$

where

$$U_g = \frac{2me}{h^2} V_g = \frac{2me}{h^2} \sum_i (f_i + if_i') T_i(\mathbf{g}) \exp(-2\pi i \mathbf{g} \cdot \mathbf{r}_i) \tag{17}$$

with U_g for the electron interaction structure factor. The atomic scattering factor in Eq. (17) has an imaginary term f_i', which is included to describe the effect of inelastic scattering (absorption). Details of the evaluation of the absorption potentials can be found in references (Bird and King 1990, Peng 1997, Weickenmeier and Kohl 1991). s_g is the excitation error as defined in Section 9.3.4. The coefficients, c_i, are obtained from the first column of the inverse eigenvector matrix as determined by the incident beam boundary condition. The solution of Eq. (16) generally converges as the number of beams included in the calculation increases. In numerical calculations, the strong beams are included in the diagonalization, while weak beams can be treated by perturbation. In practice, an initial list of beams is selected using the criteria of maximum g length, maximum excitation error, and their perturbational strength. Additional criteria are used for selecting strong beams (Zuo and Weickenmeier1995).

For electron diffraction from the crystals with defects that can be characterized by a lattice-dependent displacement field, $\mathbf{u}(\mathbf{R}_{nml})$ (see Section 9.3.3), dynamic scattering of the defects can be approximately calculated by using the scattering matrix method (Hirsch et al. 1977, Sturkey 1957). In this method, the crystal is divided into parallel slices. Each slice contains a different atomic displacement. The number of slices, n, is selected to give a good approximation of the displacement field along the beam direction. The electron wave function in reciprocal space at thickness t, $\Psi = (\phi_o(t), \phi_g(t), \cdots)^T$ (T for transpose), is related to the incident wave $\Psi_o = (1, 0, \cdots)^T$ through

$$\Psi = P_n P_{n-1} \cdots P_1 \Psi_o, \tag{18}$$

where P_n stands for the scattering matrix of the nth slice of the imperfect crystal, and P can be calculated using the above Bloch wave method (Spence and Zuo 1992) using

$$P = QC\Upsilon C^{-1} Q^{-1}.$$

Here C is the Bloch wave eigenvector matrix obtained from Eq. (16), and both Q and Υ are diagonal matrices and

$$Q = \{\exp(2\pi i \mathbf{g} \cdot \mathbf{u}(z))\} \text{ and } \Upsilon = \left\{\exp\left(2\pi i \gamma^i \Delta z\right)\right\}$$

with Δz as the slice thickness. The above scattering matrix method works as long as there is a common set of reflections perpendicular to the beam direction. The number of these reflections should not be too large so the scattering matrix can be calculated and stored in a computer at a reasonable computational cost. Small changes in composition can also be included in the scattering matrix as long as the change in the lattice perpendicular to the electron beam is small. This approach is particularly useful for studying coherent interfaces, including epitaxial multilayer structures in the plane view geometry. For example, the scattering matrix method has been successfully used in simulations of CBED patterns from a buried quantum well (Jacob et al. 2008) and multilayers (Rossouw et al. 1991).

Another commonly used method for dynamic electron diffraction simulation is the multislice method developed by Cowley and Moodie and others (Cowley 1995, Ishizuka 1982). As a numerical method, multislice has the advantage that it can treat both crystals and nonperiodic structures including amorphous structures. Because of this, the multislice method is particularly suitable for electron nanodiffraction simulation. The multislice method models the forward propagation of the electron waves through successive thin slices of potentials. The basic equation is the relationship between the incident wave $\phi_n(x, y)$ and the exit wave $\phi_{n+1}(x, y)$of the *nth* slice

$$\phi_{n+1}(x, y) = \{\phi_n(x, y) \exp[i\sigma \bar{V}_n(x, y)]\} \otimes P(x, y, \Delta z_n). \tag{19}$$

The term inside the curly bracket describes a modification to the phase of the electron wave by the slice's projected potential, which is an integration of the potential over the slice thickness:

$$\bar{V}_n(x, y) = \int_{\Delta z_n} V(x, y, z)\, dz.$$

The assumption in Eq. (19) is that the change in the electron wave amplitude by the slice potential is small enough to be neglected when the slice is selected thin enough. This approximation is known as the phase-grating approximation (PGA). It should be noted that the kinematical approximation as in Eq. (1) uses the first-order term of the PGA.

The convolution in Eq. (19) describes the wave propagation over the distance of the slice thickness Δz_n. The propagation of a wave over a short distance is described by the Fresnel propagator:

$$P(x, y, \Delta z_n) = \frac{1}{\Delta z_n} \frac{1}{\lambda i} \exp\left(\pi i \frac{x^2 + y^2}{\lambda \Delta z_n}\right).$$

For electron nanodiffraction, the incident electron wave is set to the electron probe function as described in Section 9.3.2, e.g.,

$$\phi_1(x, y) = \phi_P(x, y).$$

The electron exit wave, $\phi_{\text{exit}}(x, y)$, can be obtained by applying Eq. (19) sequentially from the first to the last slices. A model for the potential can be constructed based on the approximation of the superposition of individual atomic potentials. The electron diffraction pattern recorded in far field away from the sample is the intensity of the Fourier transform of the exit wave function:

$$I(S_x, S_y) = \left| \int \int \phi_{\text{exit}}(x, y) \exp\left(-2\pi i \left(S_x x + S_y y\right)\right) dS_x dS_y \right|^2 \qquad (20)$$

The convolution used for the wave propagation in Eq. (19) can be evaluated numerically using a fast Fourier transform. In this method, the electron wave function is first multiplied by the PGA. The product is then Fourier transformed and multiplied with the Fourier transform of the Fresnel propagator. The result is then inversely transformed back to obtain the next electron wave function.

The main limitation of the multislice method is the number of atoms that can be included realistically in a simulation. The limitation comes from the atomic potential sampling considerations as illustrated in Figure 9–10. The 3-D sample potential in a multislice calculation is

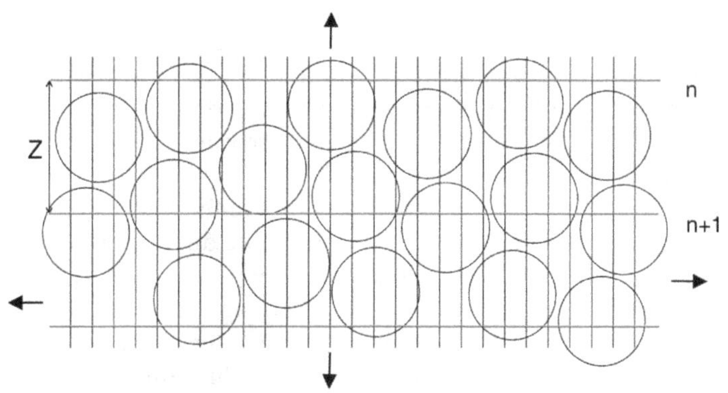

Figure 9–10. Atomic potential sampling in the multislice method. The potential is divided into slices of thickness Δz and averaged along z for each slice. Typical slice thickness is about 2 Å. The choice of slice thickness affects the numerical convergence of the calculation and the accuracy of high-order Laue zone reflections. Along the x- and y-directions, the potential is sampled in discrete points with a fixed interval or pixels.

represented in a 2-D numerical array for each slice along the beam direction. The representation of the atomic potentials requires a minimum number of sampling points. For example, a minimum of five points are required to represent the center, the size, and the gap of the atomic potential. For a 1 Å sized atom, the spacing between these points is 0.2 Å which defines a minimum pixel size in real space. A 1 k × 1 k in this case represents a sample area of 20 × 20 nm^2.

9.3.7 Inversion of Diffraction Patterns and Phase Retrieval

Inversion of electron diffraction patterns, in general, provides ultimate direct solution to the interpretation of SEND data. The diffraction pattern records the intensity of the Fourier transform of the exit wave function. Inverse Fourier transform of the diffraction pattern thus requires the phase missing in the diffraction pattern. This is known as the phase problem in diffraction. Critical procedure in the inversion of diffraction patterns is to find the phases of the diffracted waves. In crystallography, the phase problem is solved based on a priori information about the crystal structure. The priori information includes the sharply peaked atomic charge density and the periodicity of the crystal. For imaging, we must consider objects that are not perfect, infinite, crystals. In the case of electron diffraction, electron multiple scattering also leads to a complex exit wave function. These two factors complicate the solution of the phase problem for electron diffraction. The inversion of electron nanodiffraction patterns is helped by the fact that the small electron probe leads to broadened diffraction peaks, and in the case that the electron probe is smaller than the unit cell, the broadening leads to interference among the diffracted waves. This broadening effect can be seen directly in the limit of very thin samples where the diffracted wave is a convolution of the Fourier transform of the probe and the Fourier transform of the object potential (Eq. (2)).

In the case that electron nanodiffraction patterns are recorded from isolated nanostructures using a coherent and parallel beam, the object itself gives broadened diffraction peaks, which give extra information that can be used to solve the phase problem. To see this, we consider the resolution of the diffraction pattern, which is defined by the smallest spatial frequency (f). The minimal spatial frequency required to sample a finite object in the Fourier space is the reciprocal of the object size (1/S) (Sayre et al. 1998). When the resolution of the diffraction pattern is better than 1/S, the product of $(Sf)^{-n}$ with n the dimension of the image gives the oversampling ratio (Miao et al. 1999), which defines the extra information in the diffraction pattern. The smallest spatial frequency that can be recorded in an experiment is one over the coherence length (1/L). The maximum experimental oversampling ratio is then $(L/S)^n$. Recent works have shown that oversampling can be used for retrieving the missing phase (mathematically, a minimum of oversampling ratio of 2 is required for inverting diffraction patterns) (Bates 1982). The condition for an isolated object is readily met in the case of nanoparticles or other structures with distinctive boundaries that separate the object from the background. For a continuous object, Rodenburg and his co-workers proposed to use an aperture beam to isolate a part of the continuous object (Faulkner and Rodenburg 2004, Rodenburg et al. 2007). In electron nanodiffraction using a field emission source, it is reasonable to expect a fully coherent illumination across a small condenser aperture. Then, the lateral coherence is the electron wavelength divided by the convergence angle. A large coherence length can be obtained using the parallel beam produced in the NED mode. In the case of an aperture, the illumination must be fully coherent over the object. The oversampling ratio in a single diffraction pattern is at best one in this case. The requirement for oversampling is then met by scanning the aperture over an area of the object at steps smaller than the aperture size in the technique called ptychography. This technique was first proposed by Hoppe to solve the phase problem in electron microdiffraction (Hegerl and Hoppe 1970).

Figure 9–11 shows a generic algorithm for phasing the diffraction patterns. It is based on iterations between two sets of functions, the real space object function ρ_C and its modification of ρ_C^* and the Fourier transform (FT) of ρ_C, F_C and its modification, F_C^*. The modifications are represented by the two functions, f and F in Figure 9–11. The modification of f is used to change the object function toward a solution that conforms to the known constraints for the object in real space. Similarly, the modification of F is used to change the Fourier spectrum to conform to the experimental diffraction patterns. There are a number of functional forms of f and F that have been proposed; each defines a particular algorithm used for iterative phasing. Each of these algorithms relies on the available constraints that can be placed on the object and its Fourier transform. For isolated objects, a very effective constraint is the support constraint (S) where a finite region that encompasses the object is defined, and outside this region the object function is set at 0. In the Fourier space, the constraint is on the amplitudes of the complex Fourier coefficients which must agree with the square root of the measured diffraction intensities F_0^{DP} within the experimental uncertainties. The conditional step in the iterative process is typically used to compare the experimental data F_0^{DP} with the Fourier amplitude of the current estimate using a criterion function C (a typical function is the R-factor which measures the percentage of difference between the two arguments) and decide whether to end the iteration.

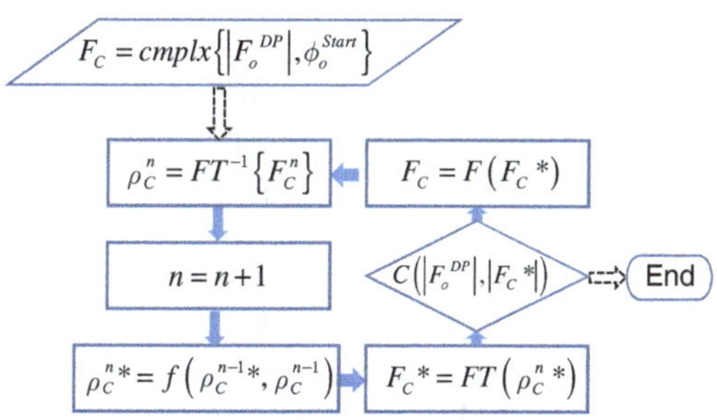

Figure 9–11. Flowchart of a generic iterative phasing procedure used to reconstruct the object from diffraction patterns.

The hybrid input and output (HIO) algorithm proposed by Fienup (1982) is a common choice for iterative phasing. This algorithm in its basic form is defined by the following modification functions:

$$f\left(\rho_C^n*, \rho_C^n\right) = \begin{cases} \rho_C^n & \text{if} \in S \\ \rho_C^n * -\beta\rho_C^n & \text{outside} \end{cases} \tag{21}$$

and

$$F\left(F_C^*\right) = F_o^{DP} \cdot F_C^* / \left|F_C^*\right|. \tag{22}$$

A modified version of HIO by Millane and Stroud uses the object symmetry and a threshold for the background noise as additional support.

For ptychography of a continuous object using a scanning aperture proposed by Rodenburg, the object function under the electron probe $\psi_{\mathrm{Pr\,obe}}(\mathbf{r})$ placed at position $\vec{R}$ is fed into a continuous object function $O_n(\mathbf{r})$ using the following modification function:

$$f\left(\rho_C^{n*}, \rho_C^n\right) = O_{n+1}(\mathbf{r})\psi_{\mathrm{Probe}}(\mathbf{r} - \mathbf{R}) \tag{23}$$

with

$$O_{n+1}(\mathbf{r}) = O_n(\mathbf{r}) + U(\mathbf{r} - \mathbf{R})\left[\rho_C^n - \rho_C^{n*}\right]. \tag{24}$$

Here U is a weight function (Rodenburg 2008). The difference between the previous object function and the feedback obtained from the diffraction patterns is weighted and used to update the continuous object function.

Examples of successful inversion of electron diffraction patterns include the work by Zuo et al. for imaging of double-wall carbon nanotubes (Zuo et al. 2003) and quantum dots and diffraction imaging of Si using an selected area aperture in aberration-corrected TEM by Morishita et al. (2008). Ptychography of a continuous object has been demonstrated for X-ray diffraction (Rodenburg et al. 2007).

9.4 Practice and Applications of Scanning Electron Nanodiffraction and Diffraction Imaging

Scanning electron nanodiffraction in the simplest form can be carried out by positioning the electron probe in the areas of interest and recording diffraction patterns from these areas. Using computer interfaces, scanning electron nanodiffraction patterns can also be acquired automatically from a predefined list of probe positions, or by scanning along a line or across an area. The automatic acquisition is achieved using the computer control of the scan coil and using a digital camera. Twesten and his co-workers at Gatan have recently developed software tools to acquire and analyze diffraction patterns based on the Gatan spectrum image (SI) framework. The diffraction patterns are stored in a stack. The tools operate in the STEM mode; diffraction pattern recording on a digital camera and annular dark-field STEM detector can be used simultaneously. The STEM imaging is useful for tracking sample drift during diffraction pattern recording. The diffraction pattern information can be used to form an image. The diffraction patterns can be indexed, measured, and analyzed based on the diffraction theory outlined in the previous section. The analysis can also be automated, although the development in this area is still lacking.

In this section, we introduce a number of examples of SEND to illustrate its practice and applications for (1) nanostructure analysis, (2) strain mapping of nanodevices, and (3) imaging of nanoscale structural phases.

9.4.1 Electron Nanodiffraction Analysis of Nanostructures

Figure 9–12 shows an example of SEND for the structure characterization of multiwall boron nitride nanotubes. The common assumption about the structure of multiwall nanotubes is that they are made of multiple, concentric, shells, and each of these shells is a cylindrical tube. Multiwall boron nanotubes, however, often show zig–zag diffraction contrast in bright-field TEM images similar to the one shown in Figure 9–12e. This type of contrast is inconsistent with the homogeneous structure one expects from concentric, cylindrical, tubes. The SEND experiment of Figure 9–12 was performed to investigate the source of the diffraction contrast. The experiment was carried out using a 50 nm parallel beam of electrons in the NED diffraction mode (see Section 9.2). A series of diffraction patterns were recorded on imaging plates (Zuo 2000) along the tube with a spacing slightly less than 50 nm. Four of these diffraction patterns are shown in Figure 9–12. The

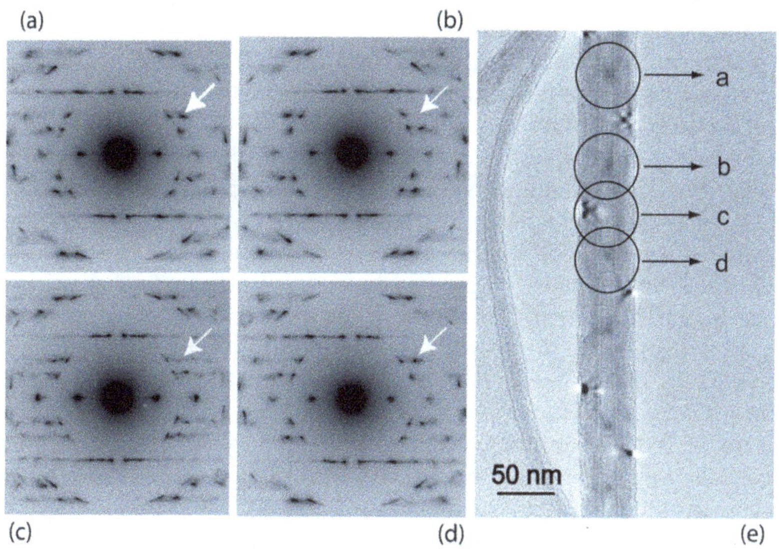

Figure 9–12. A series of electron nanodiffraction patterns (a–d) recorded from a multiwall boron nitride nanotube (e). The probe position and the size of the electron probe are indicated in (e). Reproduced from Celik-Aktas et al. (2005) with permission.

diffraction pattern at position a (see Figure 9–12e) is the same as the one recorded at d, which suggests a periodic structure associated with the zig-zag contrast. A careful examination of the diffraction patterns shows the appearance and disappearance of certain diffraction spots as the electron probe scans along the tube. One of these diffraction spots are indicated by the white arrow in Figure 9–12a–d. These diffraction spots belong to high-order zone axes of hexagonal boron nitride. Their appearance suggests the nanotube is faceted rather than a perfect cylindrical tube. The appearance and disappearance of the diffraction spots is caused by these facets coming into and out of Bragg diffraction condition. The full explanation of these diffraction patterns can be found in the Celik-Aktas et al. (2005). What Figure 9–12 shows is that SEND can be combined with electron imaging to investigate the structural origin of the image contrast.

9.4.2 Strain Mapping of Nanodevices

A major application of SEND is strain mapping and strain analysis of nanodevices. The advantages of measuring strain by diffraction are that it is not limited to thin-samples as in the case of electron imaging (Hytch et al. 2008) and it is a quantitative technique capable of very high accuracy (Armigliato et al. 2003). Strain, in general, is important in nanoscale devices because of the prevalence of surfaces and interfaces that act as strain sources. In metal–oxide–semiconductor field effect transistors (MOSFET), large device performance gain was obtained by introducing strain in the device channel (Derbyshire 2007, Thompson and Parthasarathy 2006, Thompson et al. 2005). In memory devices, it is believed that the degradation of the data retention time of the cell originates from an anomalous junction leakage current due to the presence of strain. Typical semiconductor processing also incorporates a variety of materials of different mechanical and thermal properties and strain and stress arise from these materials (Hu 1991). Strain can be measured directly from diffraction using CBED or NED.

The principle and sensitivity of the CBED technique are illustrated in Figure 9–13. A small probe of electrons is used to measure the local strain from the HOLZ lines inside the direct beam disk. The focused electron probe is about 15 Å in diameter in this case. The change in the HOLZ line positions is clearly visible as the electron probe is scanned from the substrate to the SiGe film in this case (indicated by arrows in Figure 9–13). The method for measuring the strain is by fitting the position of HOLZ lines. There are two major methods for fitting HOLZ lines. One is based on the kinematic approximation, in which HOLZ lines are approximated by straight lines (Zuo 1992) (also see Section 9.3.4). The second is based on dynamic diffraction simulations. The fitting in this case is achieved by pattern matching between the experimental and the simulated diffraction patterns (Zuo et al. 1998). Pattern matching based on the kinematic approximation is fast. It involves several tasks, including measuring the experimental lines from the recorded CBED pattern, indexing the experimental lines, and finally carrying out the fitting. These tasks can be performed through the following automated

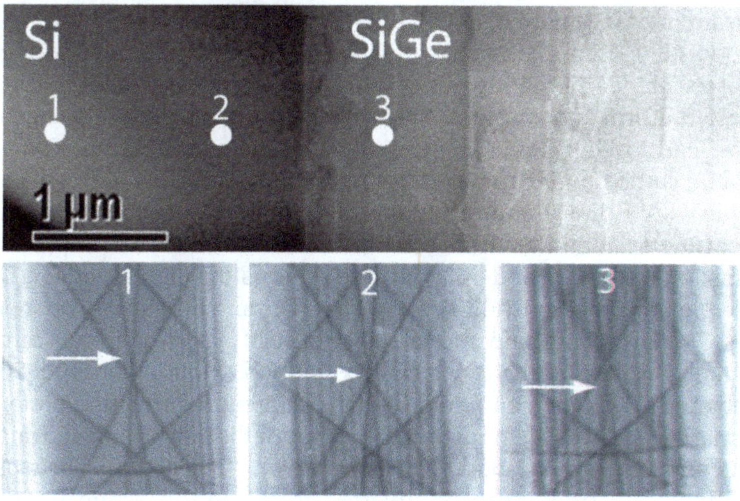

Figure 9–13. An example of CBED for detecting lattice strain. Three CBED patterns were selected from a line scan series across the interfaces for the positions of 1–3 from Si and a SiGe buffer layer grown on top of the Si. The change in the HOLZ positions is indicated by the arrows.

functions: (1) preprocessing of diffraction patterns for line detection, (2) line detection using the Hough transformation, (3) indexing the HOLZ lines, (4) defining the fitting parameters, and (5) performing the fitting. The most accurate strain measurement comes from dynamic fitting of HOLZ lines using CBED. The electron multiple scattering effects are accounted in using the Bloch wave method (Spence and Zuo 1992). Using the Si crystal as a test, Kim et al. (2004) measured the Si lattice parameter at different orientations and sample positions. The standard deviation they obtained is 0.0012 Å, which corresponds to 0.02% of the Si lattice parameter. Using this accuracy, Kim et al. (2004) examined the effect of two types of trench filling materials on strain in the shallow trench isolation structure. Their work was able to show different trench fillings leading to different stress levels, and the accuracy of their pattern matching technique at 0.02% was sufficient to detect the subtle changes in the strain for the device-related applications. The error bar of strain measurement based on the fitting of HOLZ lines using the kinematic approximation includes a systematic error due to the omission of dynamic scattering. This error can be reduced by using the so-called kinematic orientations (Buxton 1976b, Lin et al. 1989), such as the [430] and [230] zone axes in Si (Armigliato et al. 2003). The spatial resolution of strain measurement using CBED depends on several factors: the electron probe size, the probe broadening in the sample due to diffraction, and the beam direction relative to the strain axes. Theoretical consideration based on the column approximation suggests that the resolution in the zone axis orientation can be as small as a few nanometers (Zuo and Spence 1993). In the case of a thick sample, the resolution is limited by the probe broadening due to electron scattering (Chuvilin and Kaiser 2005).

Strain measurement using CBED requires relatively sharp, well-defined HOLZ lines. The sharp line is obtained where the spacing of the associated lattice plane under the illuminating electron probe is relatively uniform and constant. Large strain gradients under the electron probe lead to HOLZ line splitting. In such cases, the assumption of a constant lattice under the electron probe is no longer valid. Analysis of split HOLZ lines requires taking account of the average strain and the strain variation under the electron beam. There are two approaches to this; one is using modeling and the other is through inversion. The finite element method (FEM) can be used to model strain in nanodevices. For a given strain model, electron diffraction patterns can be simulated using the scattering matrix method (Houdellier et al. 2006, Jacob et al. 2008) and compared with experimental diffraction patterns. This approach was demonstrated by Houdellier et al. for a strained SiGe epitaxial layer (Houdellier et al. 2006). Vincent et al. proposed that the strain profile can be inverted from the diffraction intensity profile (Vincent et al. 1999). In this method, the z-dependent displacement parallel to g, R_g, is obtained directly by inverting an intensity line profile taken across the HOLZ line using an iterative phasing procedure similar to these described in Section 9.3.7. The intensity of the HOLZ line comes from the Fourier transform of the displacement if the amount of displacement is small and the kinematic approximation applies. The sample is the support here where the displacement is 0 outside the sample. In this case, the diffraction intensities measured from the CBED pattern contain the necessary phase information and can be reconstructed using iterative algorithms, such as Fienup's hybrid input and output method (see Section 9.3.7). The feasibility of using phase retrieval for measuring vertical displacements was demonstrated by Vincent et al. (1999) for a Si thin film capped with surface amorphous layers created by Ar ion milling. The method has a significant advantage since it does not require modeling. While applications of this method to nanodevices have not been demonstrated, it appears general and deserves further attention.

The use of NED, or NBD, has been proposed for strain analysis near device interfaces where the active device layer lies and the strain gradients are also large. The idea is to obtain sharp diffraction spots using a nanometer-sized parallel beam and use the diffraction spot positions and profiles to quantify the strain. The advantage of this method is that it can be used for strain analysis along the <110> zone axis. Along this orientation, the electron beam is parallel to the device interfaces; thus, it gives the highest spatial resolution for mapping strain across the interface. The <110> zone axis is difficult for CBED because of the strong dynamic effects and weak HOLZ lines, which are not visible at room temperature. The method requires a small electron probe for high spatial resolution and a small convergence angle for sharp diffraction spots required for accurate strain measurement. Armigliato and his co-workers used a 1 μm condenser aperture and the condenser minilens in the STEM/nanoprobe mode in the FEI Tecnai F20ST TEM to produce a 12-nm-sized electron probe (full width at half maximum) with a convergence angle of 0.14 mrad (Armigliato et al. 2008). They recorded diffraction patterns using a 1024 × 1024 CCD camera. The diffraction spots were sharp enough to allow the detection of a peak shift of 1 pixel, which corresponds to a 5×10^{-4} change in lattice parameters. Diffraction pattern analysis so far is limited to the measurement and analysis of the average peak positions. In the presence of large strain gradients, the diffraction peak profile is also affected by the strain as shown in Section 9.3.3. In principle, analysis of diffraction peak profiles can used to study the nature of strain and strain variation.

9.4.3 Imaging of Nanoscale Structural Phases

SEND can be used to image nanoscale structural phases using the difference in diffraction patterns between the nanophase and the matrix. This method works even in the cases where there is very little difference in atomic scattering from different atoms in the sample. Large difference in atomic scattering is required for high-angle ADF-STEM imaging. Here, we will show an example of SEND for imaging a nanoscale phase in $La_{1-x}Ca_xMnO_3$ (LCMO), where the only differences between the nanophase and the matrix are the atomic displacements. The example is taken from the work by Tao et al. (2009).

LCMO belongs to a class of manganites, which has been extensively studied recently for its magnetic field-dependent electric transport property (colossal magnetoresistance, or the CMR effect) (Salamon and Jaime 2001). The CMR effect peaks near a phase transition from an insulating high-temperature paramagnetic (PM) phase to a low temperature metallic ferromagnetic/metallic (FM) phase. During this transition, an additional phase, called charge ordering (CO), forms and disappears as the temperature is lowered (Dai et al. 2000, Zuo and Tao 2001). What's interesting about the CO phase is its size and its coincidence with the CMR effect (Kiryukhin et al. 2003, Zuo and Tao 2001). The CO phase has a correlation length of about a few nanometers. The CO phase has a superstructure with the unit cell larger than both the FM and PM phases. The characteristics of this superstructure have been observed by neutron, x-ray, and electron diffraction. What is lacking from the diffraction experiment alone is the morphology of the CO phase, its distribution, and most importantly its density.

The principle of SEND for imaging the nanometer-sized CO phase is based on detecting a diffraction signal unique to the CO phase. The larger unit cell of the CO phase (the superstructure) gives additional reflections (super reflections) in the diffraction pattern. Figure 9–14a, b are two typical electron nanodiffraction patterns recorded with the electron probe on and off the CO phase from LCMO with $x = 0.45$. The super reflections unique to the CO phase are indicated by arrows in Figure 9–14a. These super reflections are very weak compared to the lattice reflections observed in both the CO phase and the matrix. The intensity of the super reflections can be used to map the presence of the CO phase. For imaging, we used an electron probe of about 1.7 nm and scanned it across an area on the specimen. The electron nanodiffraction patterns at each probe locations in the scanning area were recorded. The

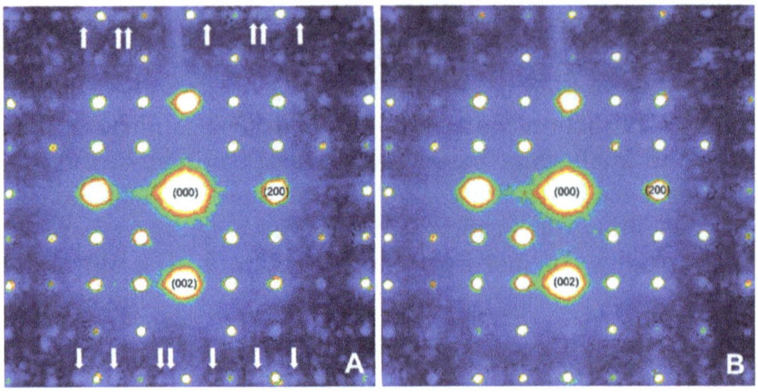

Figure 9–14. (a) An END pattern from a single CO nanocluster in the $La_{0.55}Ca_{0.45}MnO_3$ at $T = 255$ K using an electron probe of ~1.7 nm in diameter, showing the superstructure reflections indicated by the *arrows*. (b) An END pattern from the non-CO area in the $La_{0.55}Ca_{0.45}MnO_3$ at $T = 255$ K using the same electron probe as that in (a)

intensities of the super reflections were measured from the recorded diffraction patterns and mapped according to the probe position.

Figure 9–15 shows SEND intensity maps at three different temperatures as LCMO ($x = 0.45$) going through the phase transition. The color scale on the right corresponds to the ratio of the super reflection peak intensity and the surrounding background. The areas with the intensities below the background level were flattened. Each map was obtained from the same single-crystal domain in the LCMO sample ($x = 0.45$). The size of the map is 12×12 nm^2 with a scanning step of 1 nm corresponding to 12×12 recorded diffraction patterns. The SEND intensity maps in Figure 9–15 clearly show clusters of the CO phase. The average size of the CO nanoclusters measured from the maps is about 3–4 nm independent of the temperature.

It should be noted that signal/background ratio is a critical factor in making the SEND maps. The ratio is relatively small in the case of the nanometer-sized CO phase. The ability to measure weak intensities allows the direct imaging of these small structures.

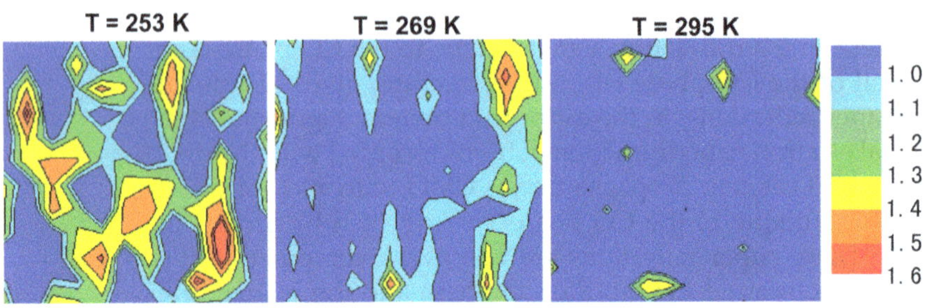

Figure 9–15. The SEND intensity maps show the evolution of the CO nanoclusters in a $La_{0.55}Ca_{0.45}MnO_3$ single-crystal domain during the phase transition, with the color scale on the right. The areas with the CO super lattice reflections intensity below the noise level were flattened to be uniformly *blue*. Reproduced from Tao et al. (2009) with permission.

While the above example is specific to the phase transition in LCMO, the principle of using diffraction signals to form images is general and can be used to extract other structural information, such as diffraction peak positions, the width and intensity to image strain, different phases and the presence of defects.

9.5 Conclusions

Scanning electron nanodiffraction and diffraction imaging are two rapidly developing techniques in electron microscopy. Both techniques are well suited for the characterization of nanostructures. The advantages of electron nanodiffraction are the small probes and the strong elastic scattering cross sections of the high-energy electrons. These advantages allow the recording of diffraction patterns from very small nanostructures, for example, a single-wall carbon nanotube. The challenge in electron nanodiffraction as in electron imaging is to relate diffraction information to the atomic structure. Since electron diffraction is not affected by the lens aberrations (except geometric distortions at large diffraction angles from the projector lens of the electron microscopes), the relationship between the electron diffraction pattern and the structure is simpler than electron imaging. In this chapter, we have outlined an electron diffraction theory based on both kinematic approximation and dynamic diffraction, which can serve as the basis for the interpretation of electron nanodiffraction patterns. We also emphasized the different electron nanoprobes that can be formed inside an electron microscope that range from a focused beam to parallel illuminations. The flexibility of the electron illumination system for forming different probes is another advantage of electron nanodiffraction. In particular, the use of parallel beams for diffraction imaging is very promising for achieving diffraction-limited resolution. We demonstrated the applications of scanning electron nanodiffraction and diffraction imaging for imaging strain, nanostructures, and defects. Since both techniques are relatively new, there is only a limited number of application examples in the literature. The examples presented in this chapter highlight the potential of electron nanodiffraction techniques in hope to stimulate further work in this area.

Acknowledgments The writing of this paper was made possible with the support by U.S. Department of Energy Grant DEFG02-01ER45923. Microscopy performed on the JEOL 2010F was carried out at the Center for Microanalysis of Materials at the Frederick Seitz Materials Research Laboratory, which is partially supported by the U.S. Department of Energy under grant DEFG02-91-ER45439. JT was supported by U.S. Department of Energy under contract DE-AC05-00OR22725 with Oak Ridge National Laboratory and by the U.S. DOE/BES under contract DE-AC02-98CH10886 with Brookhaven National Laboratory. We thank Amish Shah for Figure 9–6 and Seongwon Kim for Figure 9–13.

References

A. Armigliato, R. Balboni, G.P. Carnevale, G. Pavia, D. Piccolo, S. Frabboni, A. Benedetti, A.G. Cullis, Appl. Phys. Lett. **82**, 2172–2174 (2003)

A. Armigliato, S. Frabboni, G.C. Gazzadi, Appl. Phys. Lett. **93**, 161906 (2008)

R.H.T. Bates, Optik **61**, 247–262 (1982)

S.J.L. Billinge, I. Levin, Science **316**, 561–565 (2007)

D.M. Bird, Q.A. King, Acta Crystallographica Section A **46**, 202–208 (1990)

G. Botton, in *Science of Microscopy*, eds. by J.C.H. Spence, P.W. Hawkes, vol. I, 273–405 (Springer, New York, 2007)

B.F. Buxton, Phil. Trans. R. Soc. Lond. Ser. A-Math. Phys. Eng. Sci. **281**, 171–194 (1976a)

B. Buxton, Proc. R. Soc. Lond. Ser. A-Math. Phys. Eng. Sci. **350**, 335 (1976b)

A. Celik-Aktas, J.M. Zuo, J.F. Stubbins, C.C. Tang, Y. Bando, Acta Crystallographica Section A **61**, 533–541 (2005)

H.N. Chapman, A. Barty, S. Marchesini, A. Noy, S.R. Hau-Riege, C. Cui, M.R. Howells, R. Rosen, H. He, J.C.H. Spence, U. Weierstall, T. Beetz, C. Jacobsen, D. Shapiro, J. Opt. Soc. Am. **23**, 1179–1200 (2006)

A. Chuvilin, U. Kaiser, Ultramicroscopy **104**, 73–82 (2005)

J.M. Cowley, Ultramicroscopy **4**, 413–418 (1979)

J.M. Cowley, Electron Diffraction Techniques, vol. I, II (Oxford University Press, Oxford, UK, 1992)

J.M. Cowley, *Diffraction Physics* (Elsevier Science, NL, 1995)

J.M. Cowley, Microsc. Res. Tech. **46**, 75–97 (1999)

J. M. Cowley, Micron **35**, 345–360 (2004)

P.C. Dai, J.A. Fernandez-Baca, N. Wakabayashi, E.W. Plummer, Y. Tomioka, Y. Tokura, Phys. Rev. Lett. **85**, 2553–2556 (2000)

K. Derbyshire, Solid State Technol. **50**, 38–41 (2007)

H.M.L. Faulkner, J.M. Rodenburg, Phys. Rev. Lett. **93**, 023903 (2004)

J.R. Fienup, Applied Optics **21**, 2758–2769 (1982)

M. Gao, J.M. Zuo, R.D. Twesten, I. Petrov, L.A. Nagahara, R. Zhang, Appl. Phys. Lett. **82**, 2703–2705 (2003)

R. Hegerl, W. Hoppe, Berichte Der Bunsen-Gesellschaft Fur Physikalische Chemie **74**, 1148–1154 (1970)

P. Hirsch, A. Howie, R.B. Nicolson, D.W. Pashley, M.J. Whelan, *Electron Microscopy of Thin Crystals* (Robert E. Krieger Publishing Company, Malabar, FL, 1977)

F. Houdellier, C. Roucau, L. Clement, J.L. Rouviere, M.J. Casanove, Ultramicroscopy **106**, 951–959 (2006)

S. Hovmoller, X.D. Zou, T.E. Weirich, in *Advances in Imaging and Electron Physics*, eds. by P.W. Hawkes, P.G. Merli, G. Colestani and M. Vitlori-Antisari, vol. 123 (2002), pp. 257–289

S.M. Hu, J. Appl. Phys. **70**, R53–R80 (1991)

G.B. Hu, L.M. Peng, Q.F. Yu, H.Q. Lu, Ultramicroscopy **84**, 47–56 (2000)

W.J. Huang, R. Sun, J. Tao, L.D. Menard, R.G. Nuzzo, J.M. Zuo, Nat. Mater. **7**, 308–313 (2008)

M. Hytch, F. Houdellier, F. Hue, E. Snoeck, Nature **453**, 1086–1089 (2008)

K. Ishizuka, Acta Crystallographica Section A **38**, 773–779 (1982)

D. Jacob, J.M. Zuo, A. Lefebvre, Y. Cordier, Ultramicroscopy **108**, 358–366 (2008)

J. Jansen, D. Tang, H.W. Zandbergen, H. Schenk, Acta Crystallographica Section A **54**, 91–101 (1998)

M. Kim, J.M. Zuo, G.S. Park, Appl. Phys. Lett. **84**, 2181–2183 (2004)

V. Kiryukhin, T.Y. Koo, H. Ishibashi, J.P. Hill, S.W. Cheong, Phys. Rev. B **67**, 064421 (2003)

Y.P. Lin, D.M. Bird, R. Vincent, Ultramicroscopy **27**, 233–240 (1989)
J.W. Miao, P. Charalambous, J. Kirz, D. Sayre, Nature **400**, 342–344 (1999)
R.P. Millane, W.J. Stroud, J. Opt. Soc. Am. A **14**, 568–579 (1997)
S. Morishita, J. Yamasaki, K. Nakamura, T. Kato, N. Tanaka, Appl. Phys. Lett. **93**, 183103 (2008)
L.M. Peng, Acta Crystallogr. Section A **53**, 663–672 (1997)
L.M. Peng, S.L. Dudarev, M.J. Whelan, *High-Energy Electron Diffraction and Microscopy* (Oxford University Press, Oxford, UK, 2004)
I.K. Robinson, I.A. Vartanyants, G.J. Williams, M.A. Pfeifer, J.A. Pitney, Phys. Rev. Lett. **8719** (2001)
J.M. Rodenburg, in *Advances in Imaging and Electron Physics*, ed. by P.W. Hawkes, vol. 150 (2008), pp. 87–184
J.M. Rodenburg, A.C. Hurst, A.G. Cullis, B.R. Dobson, F. Pfeiffer, O. Bunk, C. David, K. Jefimovs, I. Johnson, Phys. Rev. Lett. **98** (2007)
C.J. Rossouw, M. Alkhafaji, D. Cherns, J.W. Steeds, R. Touaitia, Ultramicroscopy **35**, 229–236 (1991)
M.B. Salamon, M. Jaime, Rev. Mod. Phys. **73**, 583–628 (2001)
D. Sayre, H.N. Chapman, J. Miao, Acta Crystallogr. Section A **54**, 232–239 (1998)
D. Shapiro, P. Thibault, T. Beetz, V. Elser, M. Howells, C. Jacobsen, J. Kirz, E. Lima, H. Miao, A. M. Neiman, D. Sayre, Proc. Natl. Acad. Sci. USA. **102**, 15343–15346 (2005)
J.C.H. Spence, *High-Resolution Electron Microscopy* (Oxford University Press, Oxford, UK, 2003)
J.C.H. Spence, J.M. Zuo, *Electron Microdiffraction* (Plenum Press, New York, NY, 1992)
L. Sturkey, Acta Crystallogr. **10**, 858–859 (1957)
M. Tanaka, R. Saito, H. Sekii, Acta Crystallogr. Sect. A **39**, 357–368 (1983a)
M. Tanaka, H. Sekii, T. Nagasawa, Acta Crystallogr. Sect. A **39**, 825–837 (1983b)
J. Tao, D. Niebieskikwiat, M. Varela, W. Luo, M.A. Schofield, Y. Zhu, M.B. Salamon, J.M. Zuo, S.T. Pantelides, S.J. Pennycook, Phys. Rev. Letts. **103**, 097202 (2009)
S.E. Thompson, R.S. Chau, T. Ghani, K. Mistry, S. Tyagi, M.T. Bohr, IEEE Trans. Semicond. Manufact. **18**, 26–36 (2005)
S.E. Thompson, S. Parthasarathy, Mater. Today **9**, 20–25 (2006)
R. Vincent, T.D. Walsh, Ultramicroscopy **70**, 83–94 (1997)
R. Vincent, T.D. Walsh, M. Pozzi, Ultramicroscopy **76**, 125–137 (1999)
A. Weickenmeier, H. Kohl, Acta Crystallographica Section A **47**, 590–597 (1991)
J.M. Zuo, Ultramicroscopy **41**, 211–223 (1992)
J.M. Zuo, Mater. Trans. Jim **39**, 938–946 (1998)
J.M. Zuo, Microsc. Res. Tech. **49**, 245–268 (2000)
J.M. Zuo, Rep. Prog. Phys. **67**, 2053–2103 (2004)
J.M. Zuo, M. Gao, J. Tao, B.Q. Li, R. Twesten, I. Petrov, Microsc. Res. Tech. **64**, 347–355 (2004)
J.M. Zuo, T. Kim, A. Celik-Aktas, J. Tao, Zeitschrift fur Kristallographie **222**, 625–633 (2007)
J.M. Zuo, M. Kim, R. Holmestad, J. Electron. Microsc. **47**, 121–127 (1998)
J.M. Zuo, J.C.H. Spence, Ultramicroscopy **35**, 185–196 (1991)
J.M. Zuo, J.C.H. Spence, Phil. Mag. A **68**, 1055–1078 (1993)
J.M. Zuo, J. Tao, Phys. Rev. B **63**, 060407 (2001)
J.M. Zuo, I. Vartanyants, M. Gao, R. Zhang, L.A. Nagahara, Science **300**, 1419–1421 (2003)
J.M. Zuo, A.L. Weickenmeier, Ultramicroscopy **57**, 375–383 (1995)

10

Applications of Aberration-Corrected Scanning Transmission Electron Microscopy and Electron Energy Loss Spectroscopy to Complex Oxide Materials

Maria Varela, Jaume Gazquez, Timothy J. Pennycook, Cesar Magen, Mark P. Oxley and Stephen J. Pennycook

10.1 Introduction: Complex Oxides

Complex oxides exhibit a very rich set of materials properties due to the diverse interplay between structure, doping, and electronic and magnetic degrees of freedom. Collective phenomena, such as charge, orbital, oxygen vacancy, or even spin-state ordering, are quite common and underlie transport, structural, and magnetic properties including collective phenomena such as magnetism, ferroelectricity, superconductivity, metal–insulator transitions, enhanced photoconductivity, electron transfer, etc. (Bednorz and Mueller 1986, Dagotto et al. 2001, Khomskii and Low 2004, Maekawa et al. 2004). They have an enormous range of applications due to the fact that their properties change drastically as their precise composition and atomic structure are varied, and they also play a key role in both energy and information technologies. Examples include batteries, solar cells, catalysis, and miscellaneous electronic and magnetic devices (Bea et al. 2008, Bibes and Barthelemy 2007, Garcia et al. 2010, Goodenough 2003, Kilner 2008). These materials

Notice: This manuscript has been authored by UT-Battelle, LLC, under Contract No. DE-AC05-00OR22725 with the U.S. Department of Energy. The United States Government retains and the publisher, by accepting the article for publication, acknowledges that the United States Government retains a non-exclusive, paid-up, irrevocable, world-wide license to publish or reproduce the published form of this manuscript, or allow others to do so, for United States Government purposes.

S.J. Pennycook, P.D. Nellist (eds.), *Scanning Transmission Electron Microscopy*,
DOI 10.1007/978-1-4419-7200-2_10, Springer Science+Business Media, LLC 2011

show real potential toward revolutionary developments, but in many cases these will only emerge from a deeper understanding of how their novel properties arise, in particular, those that do not exist in naturally occurring materials (such as interface superconductivity or metallicity in two-dimensional electron gases). The promise posed by artificially structured and nanostructured systems based on oxides is very exciting, but harnessing the physics and structure of such thin films, interfaces, and nanostructured devices sets the ultimate limits of their potential applications.

The combination of scanning transmission electron microscopy (STEM) and electron energy loss spectroscopy (EELS) is a very powerful technique to examine the structure, properties, and chemistry of these materials with atomic resolution in real space (Browning et al. 1993, Nellist and Pennycook 1998, Nellist et al. 2004). Oxide materials, especially transition metal oxides (TMO), are ideally suited for these studies. Perovskite-based TMOs show relatively large lattice parameters, making them ideal for Z-contrast studies. Furthermore, the O 2p bands and the transition metal 3d bands are very close to the Fermi level, so the electronic properties can be probed by studying the fine structure on the O K edge and the transition metal L edge, both of which are typically well within reach of modern spectrometer optics (Egerton 1996). In this chapter we will review the state of the art of STEM-EELS techniques, which can also be combined with density functional theory (DFT) and also dynamical diffraction simulations (which are examined in detail in Chapter 6 by Allen and coworkers) to give deeper insight into the structure, chemistry, and physics of these systems. First, we will examine the current status of the analysis of oxides using aberration-corrected STEM-EELS, including limitations of the technique at its frontiers, the sensitivity to light O atoms (Section 10.2) and to isolated atoms within bulk solids (Section 10.3). Section 10.4 will review methods to measure electronic properties from EELS fine structure and then in Section 10.5 we will review a number of applications to bulk materials, defects, interfaces, and nanoscale systems.

10.2 Usefulness of High Spatial Resolution STEM-EELS Techniques to Image Light O Atoms

Aberration-corrected STEM can be used simultaneously in bright field (BF) and annular dark field (ADF) modes to generate atomic-resolution images of oxide materials in a routine fashion (Pennycook et al. 2008, 2009, Shibata et al. 2007, Varela et al. 2005). As a basic example, Figure 10–1 shows a Z-contrast and a simultaneously acquired bright field image of $SrTiO_3$ (STO) down the cubic axis obtained at 100 kV, together with a sketch of the ABO_3 generic perovskite unit cell. The heavy Sr atoms can be easily distinguished from the lighter TiO columns in the ADF image, but the pure O columns are not resolved. The simultaneous BF image shows a hint of these columns in their right positions.

Figure 10–1. Annular dark field (*left*) and simultaneous bright field (*right*) images of STO down the cubic axis, obtained in the VG Microscopes HB501UX equipped with a Nion aberration corrector. A perovskite unit cell has been highlighted on the ADF image (O columns not shown). Adapted from Varela et al. (2010).

Oxygen and other atoms such as B, C, or N have only been successfully imaged by means of electron microscopy in the last decade. By adjusting the spherical aberration coefficient of the objective lens to a negative value, O atoms were imaged for the first time, and their concentration quantified, by BF transmission electron microscopy (TEM) (Jia and Urban 2004, Jia et al. 2003). BF imaging has also been employed subsequently to study defects such as identification of O vacancies or to elucidate the structure of dislocation cores, as shown in Figure 10–2 (Shibata et al. 2007). The images unambiguously demonstrate how, while the partial dislocation cores in alumina can be nonstoichiometric due to the excess of Al or O, the total basal dislocation comprised of two such partials connected by a stacking fault chemically preserves the Al_2O_3 stoichiometry. Unfortunately, interpretation of BF images is not always straightforward. Interestingly, an imaging technique proposed decades ago has been utilized recently: annular bright field imaging (ABF), where an annular detector with a small hole in it (around 10 mrad in diameter) is used in STEM mode to produce images that resemble BF ones, e.g., atoms appear dark when in focus, but it is much more

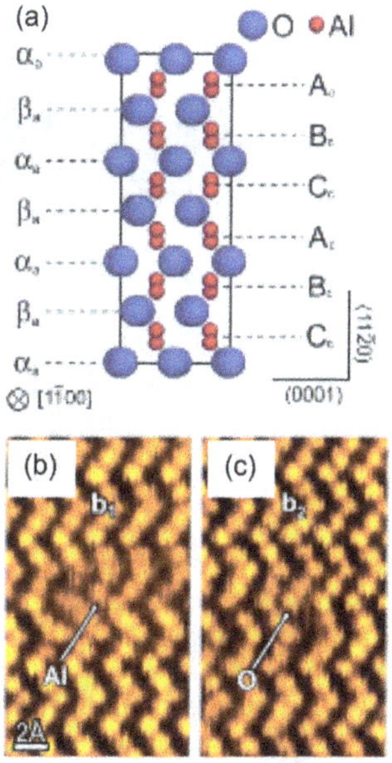

Figure 10–2. (a) Crystal structure of α-Al_2O_3, showing the Al and O columns. (b, c) BF image of the structure of dislocation cores in alumina, one terminated by an Al column (b) and the other terminated by an O column (c). Adapted from Shibata et al. (2007).

robust versus thickness or defocus changes (Findlay et al. 2009, Rose 1974). ABF imaging poses, therefore, a very interesting promise toward the study of oxide structures.

Annular dark field imaging has also been successfully employed to image and quantify O atoms (see, for example, results reviewed in Varela et al. (2005)), and even the lighter B, C, or N (Krivanek et al. 2010). But one of the most interesting possibilities regarding identification of these species arises through the combination of STEM and EELS techniques. Spectrum imaging (Hunt and Williams 1991, Jeanguillaume and Colliex 1989) in the STEM, which has been described in chapter 4 by Kociak and coworkers, allows simultaneous identification of structure, chemical composition, and also electronic properties of materials in real space and with atomic resolution (Bosman et al. 2007). EELS-based spectrum imaging can be used to visualize the O sublattice in complex oxides, and it can be done at lower voltages than, e.g., BF imaging, which can help prevent electron beam-induced damage. Figure 10–3 shows the results of the analysis of a spectrum image acquired on a crushed $LaMnO_3$ perovskite crystal down the pseudocubic [110] axis, acquired at 60 kV in a fifth order corrected Nion UltraSTEM. After principal component analysis (PCA) (Bosman et al. 2006) was used to remove random noise from the data and the background was subtracted using a power law, the lattices of Mn, La, and O can be imaged by integrating the intensity under the respective edges of interest. All these elemental maps show atomic-resolution images of their respective elements. In the O K image, the O sublattice can clearly be observed, along with the ripple associated with the distorted O octahedra, which can be unambiguously distinguished.

Interestingly, not all of the O atoms in the LMO structure are visible. The contrast associated with those sitting on the heavier La columns is

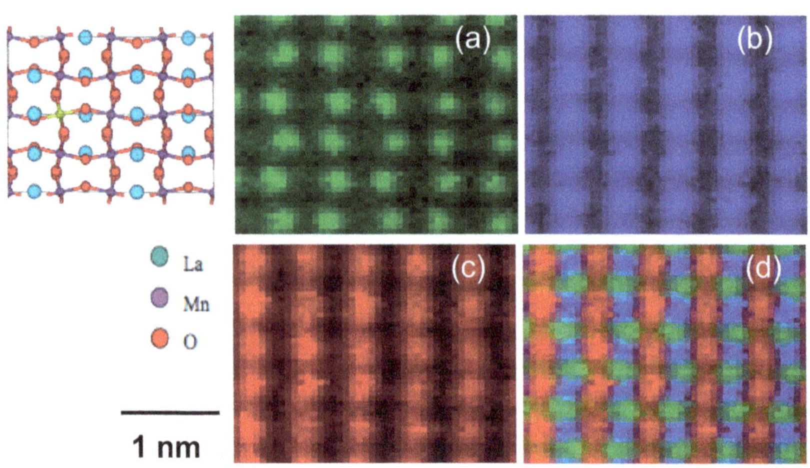

Figure 10–3. (a) Oxygen K edge image of $LaMnO_3$ down the pseudocubic [110] axis, acquired in the Nion UltraSTEM at 60 kV. (b) Simultaneously acquired Mn $L_{2,3}$ image. (c) La $M_{4,5}$ image. (d) RGB overlay of the images in (a), (b), and (c). The images have been corrected for spatial drift, and principal component analysis has been used to remove random noise from the EEL spectra.

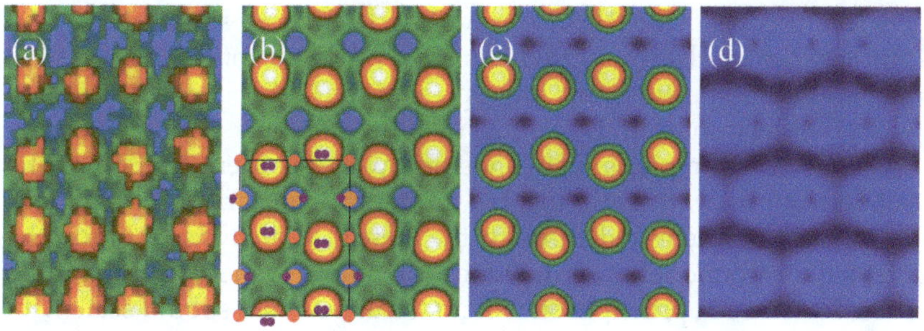

Figure 10–4. Integrated oxygen K-shell EELS signal from $LaMnO_3$ in the [010] zone axis orientation (pseudocubic [110] axis). (a) Experimental image acquired on the Nion UltraSTEM operating at 60 kV. (b) Simulated image with projected structure inset. (c) Contribution to the total image from the isolated O columns. (d) Contribution to the total image from the O atoms on the La/O columns. Adapted from Oxley et al. (2010).

lost due to high-angle scattering beyond the EELS detector, as shown by the dynamical simulations displayed in Figure 10–4. These calculations highlight the fact that interpretation of EELS spectrum images may not always be straightforward, especially in the presence of heavy atomic columns such as La (Allen et al. 2003, Findlay et al. 2005, Oxley and Allen 1998). More information on these calculations, artifacts, and interpretation is presented in Chapter 6 by Allen and coworkers.

10.3 Detection and Imaging of Isolated Impurities in Oxide Materials

The ultimate analytical sensitivity, to isolated single atoms, has been demonstrated by means of STEM-EELS. Single Sb impurities into a Si matrix were imaged for the first time by P. Voyles and coworkers (Voyles et al. 2002), while isolated La atoms in an oxide $CaTiO_3$ thin film were identified by EELS in 2004 (Varela et al. 2004). Figure 10–5 shows data from this experiment, acquired in an aberration-corrected

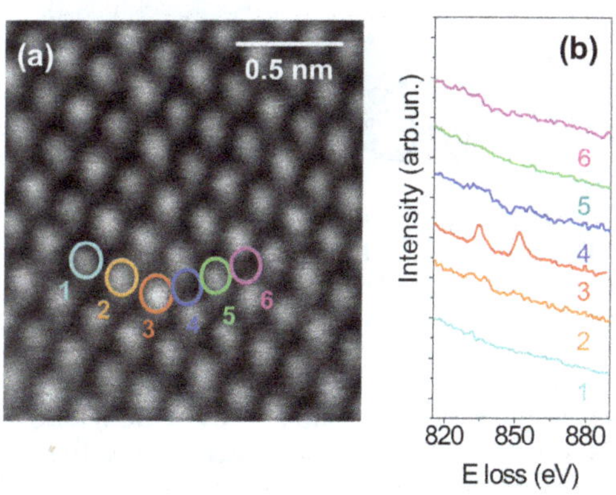

Figure 10–5. Z-contrast image of a $CaTiO_3$ thin film doped with isolated La impurities. (b) La $M_{4,5}$ lines obtained from the different atomic columns marked on (a). The spectra have been displaced vertically for clarity. Adapted from (Varela et al. 2004).

VG Microscopes HB501UX operated at 100 kV. A single, heavy La impurity can be clearly spotted thanks to its bright contrast in the ADF image. A clear spectroscopic fingerprint of the La $M_{4,5}$ edge is obtained when the electron beam is scanned over this column.

While this example is just concerned with the identification of the isolated impurity, quantifying the amount of impurities per column has also been attempted through the statistical analysis of either the intensity of atomic columns in Z-contrast images or the integrated EELS intensity (since in principle one might think that both should scale with the number of atoms per column). As reported by Voyles and coworkers (Voyles et al. 2002), for a random distribution of impurities, the probability that a given Ca column contains m La atoms is given by a binomial distribution:

$$P_{n,c}(m) = \frac{n!}{m!(n-m)!} c^m (1-c)^{n-m} \qquad (1)$$

where n is the total number of atoms in the column (i.e., directly depending on specimen thickness) and c is the concentration of La impurities in the sample. For a statistically significant number of measurements (obtainable from, for example, the analysis of a two-dimensional spectrum image), and a sufficiently thin crystal (as we will see later), the number of La atoms per column in the image can be quantified. Figure 10–6 shows an ADF image of a $Ca_{0.95}La_{0.05}TiO_3/CaTiO_3$ bilayer, obtained at 100 kV in the aberration-corrected Nion UltraSTEM. The interface between both materials is marked with a red dotted line.

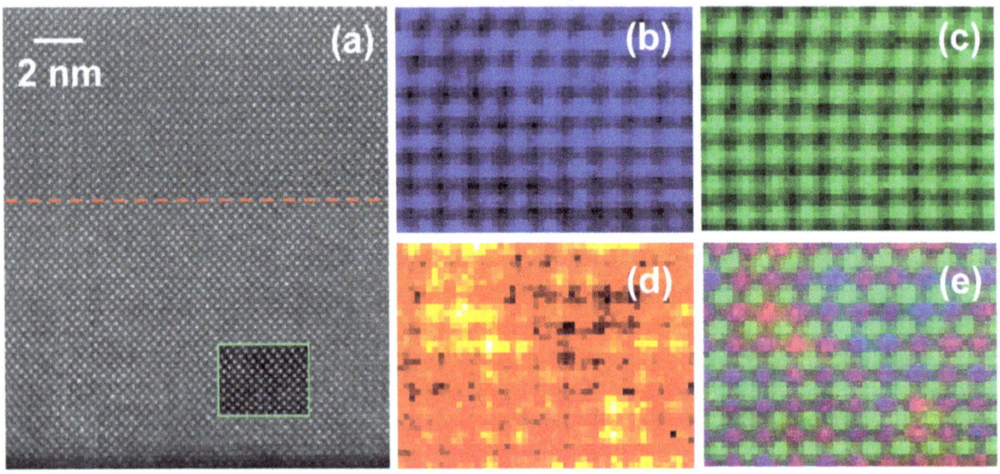

Figure 10–6. (a) Z-contrast image of a $Ca_{1-x}La_xTiO_3/CaTiO_3$ bilayer doped with $x = 0.05$ La impurities. A *red dotted line* marks the position of the interface. The inset shows the region where a spectrum image was acquired, 50 × 37 pixels in size, with a current of approximately 100 pA and an exposure time of 0.1 s per pixel. The simultaneously acquired ADF signal is overlaid in the inset. Some spatial drift is observed. (b) Ca $L_{2,3}$ map, (c) Ti $L_{2,3}$ map, and (d) La $M_{4,5}$ map, produced after noise removal using principal component analysis and background subtraction using a power law fit. The Ca and Ti maps were produced by integrating a 20 eV wide window, while for the La map a 30 eV wide window was used. (e) RGB overlay of (b, c, d), with the La map shown in *red*, the Ti map in *green*, and the Ca map in *blue*. Specimen courtesy of M. Biegalski and H. Christen from Oak Ridge National Laboratory.

A spectrum image 50 × 37 pixels in size was acquired in the area marked with a rectangle. The simultaneously acquired ADF signal is in the inset. The panels on the right show the respective Ca $L_{2,3}$, Ti $L_{2,3}$ and La $M_{4,5}$ maps. Figure 10–6(e) shows an overlay of the La (red), Ti (green), and Ca (blue maps). While the Ti intensity is relatively homogeneous through the image, the La map shows spatial variations due to the low concentration of La atoms in the specimen (5%).

By averaging and quantifying the intensity of every column in the La map, a histogram of intensities can be obtained (shown in Figure 10–7(a)). This histogram can be matched to a binomial distribution. Figure 10–7(b) shows a number of such distributions simulated for sample thicknesses (number of atoms per column) ranging from 25 to 100 in 25 atom steps. The experimental histogram has been normalized and overlaid for visual comparison. For an estimated sample thickness between 75 and 100 unit cells (approximately between 30 and 40 nm) the agreement between the experimental data and the calculated distribution is reasonably good.

With the above distribution in mind, and assuming a sample thickness of 75 atomic planes (i.e., around 30 nm), an attempt to estimate the number of La atoms per column is displayed in Figure 10–8.

Unfortunately, such a direct quantification may not be accurate due to dynamical diffraction, which causes a depth dependence of the La signal strength and, probably, some smearing of the binomial probability. For example, in the imaging conditions reported by Varela et al. (2004), when the electron probe is sitting on the Ca column (where substitutional La impurities are located) the contribution to the EELS signal initially increases due to electrostatic attraction to the column (channeling) but then decreases as elastic and inelastic scattering processes broaden the probe. As a result, the signal strength is significantly affected depending on the depth within the column where impurities are sitting (see Figure 10–9), and the atoms sitting closer to the

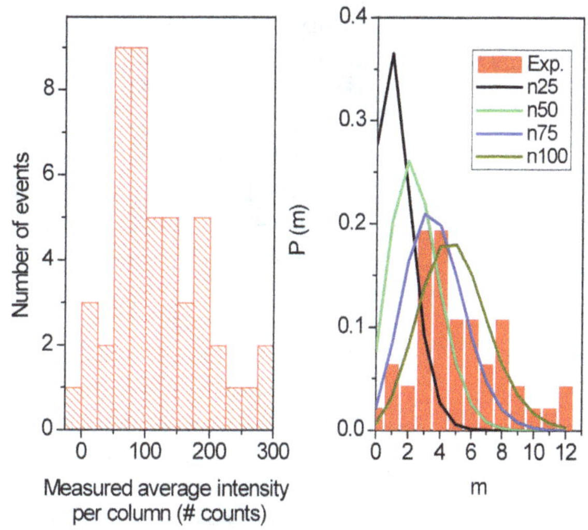

Figure 10–7. (*Left*) Histogram of column intensities, measured from the La image in Figure 10–6. (*Right*) The lines show the simulated probabilities of having *m* La atoms per column assuming a random occupation (binomial distribution) for a number of specimen thicknesses: 25 (*black*), 50 (*green*), 75 (*blue*), and 100 (*brown*) atomic planes. The scaled and normalized experimental histogram is in *red*.

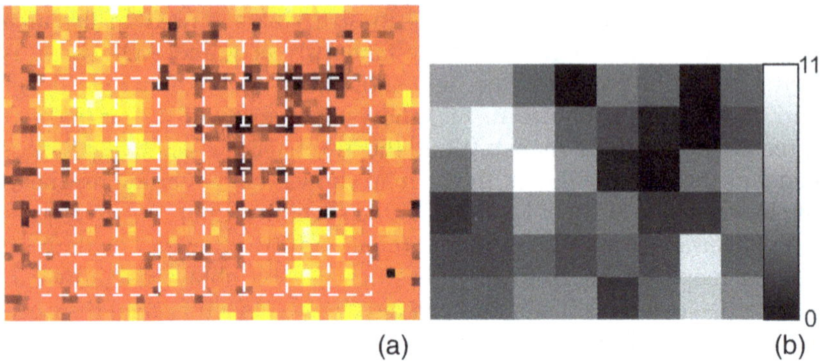

Figure 10–8. (a) Grid marked on the La image from Figure 10–6(d). (b) Estimated number of isolated La atoms per column for the atomic columns in the grid.

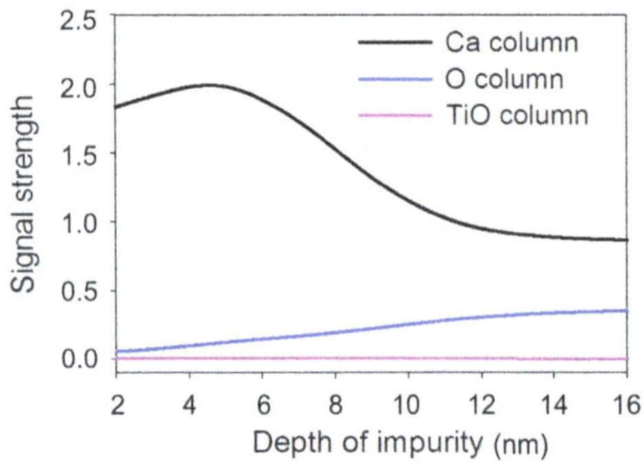

Figure 10–9. Depth dependence of the La signal strength when the electron probe is sitting on the different Ca (*black*), O (*blue*), and Ti (*pink*) columns within the unit cell. Simulations done for optical parameters of the VG Microscopes HB501UX at 100 kV, as reported in Varela et al. (2004).

entrance surface will be more easily detected. These effects get more dramatic when the probe-forming angle increases, so for the data set in Figure 10–6 the atoms near the entrance surface will contribute more strongly to the image intensity than those near the exit surface. If so, the analysis and quantification shown in Figures 10–7 and 10–8 is flawed, and more accurate simulations are needed. In summary, simple approaches may not work when using these advanced electron probes, and care should be taken when quantifying impurity concentrations from these STEM-EELS images.

In any case, EEL spectroscopic mapping around impurities allows not just identification of the elements, but also studying their electronic properties and those of the columns around them. Figure 10–10 shows a Z-contrast image obtained at 300 kV of a (3%) Co-doped TiO_2 anatase film grown on a $LaAlO_3$ substrate (Griffin-Roberts et al. 2008). At 300 kV the O columns are clearly visible in between the TiO columns. Noise removal by maximum entropy deconvolution resolves a splitting on one of the two oxygen columns in the unit cell, which appears significantly brighter than the other one: the Co atoms are sitting near this column in an interstitial configuration. Theoretical calculations (Griffin-Roberts et al. 2008) support that this is the most energetically favorable position and also show that the energy is significantly lowered by placing an O vacancy by the interstitial Co. They also demonstrate that the

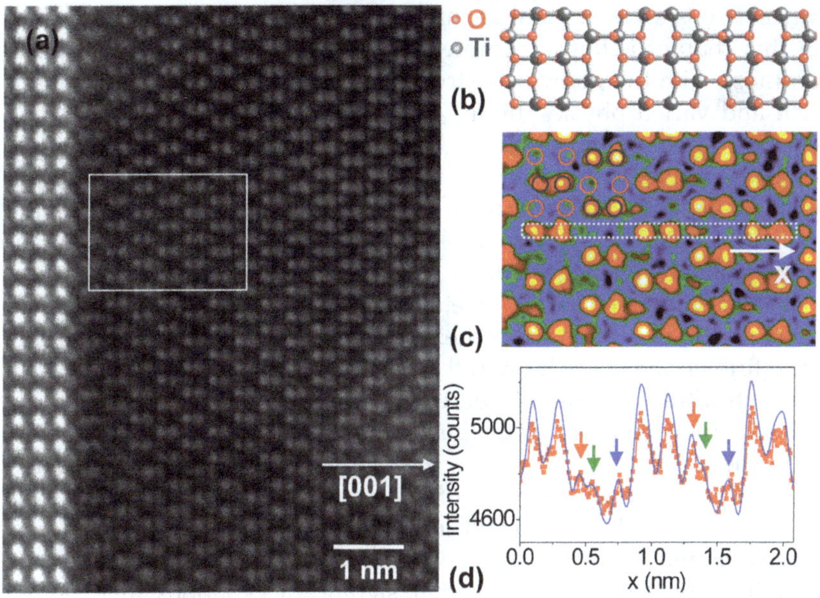

Figure 10–10. (a) Z-contrast image of a Co-doped TiO_2 anatase film grown on a $LaAlO_3$ substrate obtained in the aberration-corrected VG Microscopes HB603U. (b) Structure of the anatase phase. (c) Z-contrast image processed from the area in the inset in (a), after maximum entropy deconvolution (using the plug in for Digital Micrograph available from HREM Inc.) (d) Intensity line-trace from the box marked with a *dotted rectangle* on (c). The *solid line* comes from the deconvolved image, while the *red dots* come from the equivalent area in the raw image. *Blue arrows* mark the position of the O columns, while *red* and *green arrows* mark the position of the split Co–O columns. Adapted from Griffin-Roberts et al. (2008).

spin up minus spin down density is indeed localized around the complex, with some density being present on the vacancy and also on the neighboring TiO columns. While the next section will deal in detail with the study of electronic properties, here we will anticipate that the EELS data show that, as a result, the Ti atoms around the interstitial Co paired with the O vacancy are slightly reduced to Ti^{3+}. Thus, the combination of STEM, EELS, and DFT allows identification of the defect complex responsible for room temperature ferromagnetism in this system, which is comprised of a Co interstitial atom plus an oxygen vacancy and a number of Ti^{3+} atoms.

10.4 Measurement of Electronic Properties of Perovskites from EELS Fine Structure

So far we have mostly discussed the possibilities of STEM-EELS as a technique to image structure and chemistry, along with its limits. But in order to understand materials, their physical properties must also be measured. The electronic and physical properties of TMOs are largely determined by the metal–oxygen atomic bonds (Maekawa et al. 2004).

Since TMOs show a great number of crystal structures and the aim of this chapter is not to comprehend all of them, here we will focus on those with the perovskite structure. Perovskite compounds exhibit rich and varied physics: high-T_c superconductivity is found in some cuprates (Bednorz and Mueller 1986, Ginsberg, Tinkham 1996), while colossal magnetoresistance (CMR) is found in manganites (Dagotto et al. 2001). The basic formula of these compounds is ABO_3, with B the transition metal being surrounded by six O atoms in octahedral coordination, as shown in Figure 10–11. In perovskites, the O 2p bands and the partially occupied transition metal 3d bands are very close to the Fermi level so the electronic properties can be probed by studying the fine structure on the O K edge and the transition metal L edge, all of which are well within current EEL spectrometer optics. It is these 3d states that are largely responsible for the intriguing physics of these compounds.

In the dipole approximation, the O K near-edge structure (around 530 eV) arises from excitations of O 1 s electrons to the 2p bands, while transition metal L edges result from the excitation of 2p electrons into empty bound states (Egerton 1996). Since in TMOs those bands are heavily hybridized with each other, significant variations are expected in both edges when the occupation of the 3d bands (i.e., the metal atom valence) is changed. In perovskites, this can be easily achieved by substituting the A cation, e.g., from a divalent to a trivalent one or even a mixture of both which would result in an average mixed-valence state, as happens through the complex phase diagrams of CMR materials. These oxides have a chemical formula of $A_xB_{1-x}MnO_3$, where A is a trivalent cation (La, Nd, Bi, Pr) and B a divalent cation (Sr, Ca, Ba). The resulting mixed-valence state within the Mn sublattice produces a complex electronic structure. The common view so far is that a fraction $1-x$ Mn ions per unit cell are in a +4 formal oxidation state, with a $3d^3$ electronic configuration ($t_{2g}{}^3e_g{}^0$) while the rest, being Mn^{3+}, have a $3d^4$ configuration ($t_{2g}{}^3e_g{}^1$) (Dagotto et al. 2001, Maekawa et al. 2004). Hence

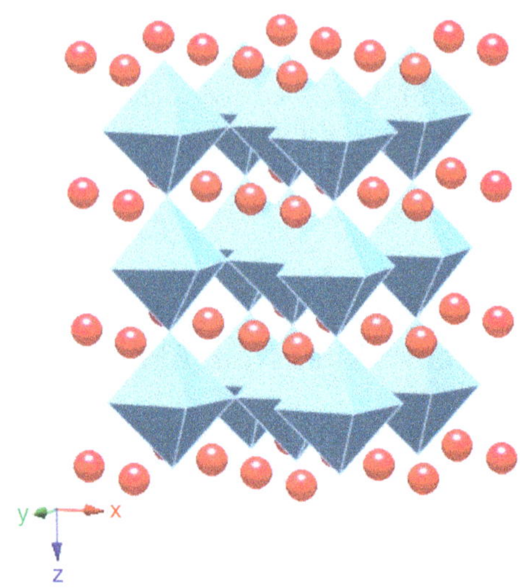

Figure 10–11. Schematic of the perovskite structure. The O octahedra are shown in *light blue*, with the B transition metal enclosed within. The A sites are shown as *red* atoms.

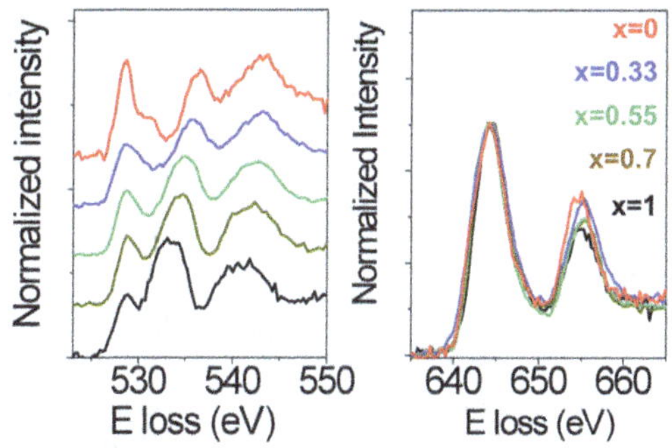

Figure 10–12. O K (*left*) and Mn $L_{2,3}$ (*right*) edges for a set of $La_xCa_{1-x}MnO_3$ crystals oriented down the pseudocubic [100] axis for x values of 1 (*black*), 0.7 (*brown*), 0.55 (*green*), 0.33 (*blue*), and 0 (*red*). The O K edges have been displaced vertically for clarity, while the Mn L edges have been normalized and aligned for direct visual comparison. Adapted from Varela et al. (2009).

the Mn oxidation state can be tuned continuously by adjusting the relative A/B cation ratio. For example, in the $La_xCa_{1-x}MnO_3$ system, the Mn oxidation state ranges from +3 to +4 when x values are increased from 0 to 1, tuning the 3d e_g band occupation from 1 down to 0 electrons. As a result, both the Mn $L_{2,3}$ and the O K edges vary, as shown in Figure 10–12 for values of $x = 1$, 0.7, 0.55, 0.33, and 0, respectively (Varela et al. 2009).

Both the O K and the Mn L fine structures show significant changes when the Mn oxidation state varies. For the O K edge the most prominent variations are observed at the prepeak feature near the edge onset. This behavior is to be expected, since there is a major hybridization with the Mn 3d bands near this energy (Luo et al. 2009). It has been theoretically demonstrated that the variation in the prepeak's intensity with doping is controlled by the orbital occupancy of the majority-spin Mn 3d states, while its width is controlled by crystal field splitting (Luo et al. 2009). Experimentally, the prepeak intensity increases gradually as the 3d e_g band gets emptier and emptier, while the separation in energy between the prepeak and the next peak, below 535 eV, increases. Both these parameters can be quantified by fitting two Gaussian curves to both peaks and extracting their areas and positions. The normalized prepeak intensity along with the peak separation for the compounds in Figure 10–12 are plotted versus nominal Mn oxidation state in Figure 10–13. Both of them show a linear dependence with nominal Mn valence, providing a calibration that can be extrapolated to quantify other datasets. Identical behavior is found when quantifying the L_{23} intensity ratio from the Mn $L_{2,3}$ edge. This parameter can be quantified in a number of ways, most of which involve background subtraction and continuum correction. For Mn oxides, the $L_{2,3}$ ratio is known to increase when the oxidation state of Mn decreases (Krivanek and Paterson 1990, Kurata and Colliex 1993, Kurata et al. 1993, Rask et al. 1987). Figure 10–13 shows the values of $L_{2,3}$ ratios extracted from the data in Figure 10–12. The data points can be adjusted to a linear fit, which can also be used as external calibration for other EELS data sets.

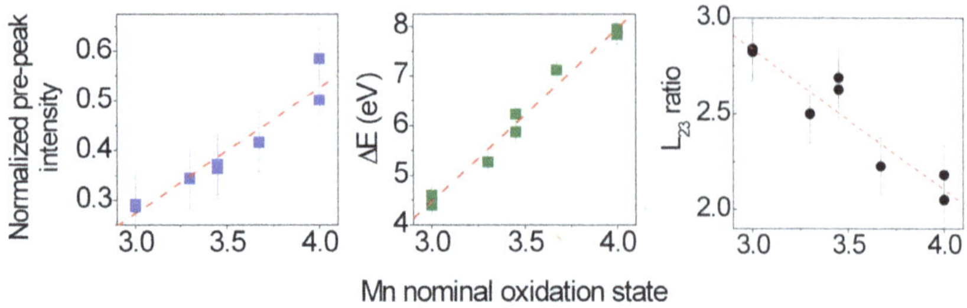

Figure 10–13. O K edge normalized prepeak intensity (*left*), peak separation, ΔE (*center*), and L_{23} intensity ratio (*right*) versus nominal Mn oxidation state for the set of $La_xCa_{1-x}MnO_3$ crystals Adapted from Varela et al. (2009).

Figure 10–14. Ti $L_{2,3}$ and O K edges from STO (*blue*) and LTO (*red*), obtained in the aberration-corrected VG Microscopes HB501UX. Adapted from Garcia-Barriocanal et al. (2010).

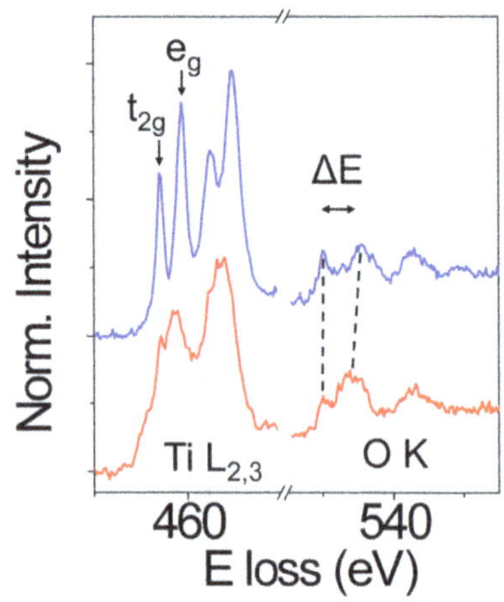

However, L_{23} ratio quantification is subject to larger error bars, mainly due to the unknown contribution of continuum states.

Similar behaviors are found for other TMOs (Van Aken and Leibscher 2002) and also perovskites such as titanates or cobaltites. Figure 10–14 shows the O K and Ti $L_{2,3}$ edges of $SrTiO_3$ (STO) and $LaTiO_3$ (LTO), where Ti atoms are in the +4 ($3d^0$ electronic configuration) and +3 ($3d^1$) oxidation states, respectively. As in the manganites, the separation between the prepeak and the second peak decreases for lower oxidation state. The $L_{2,3}$ edge of Ti atoms shows a richer fine structure. For these compounds, especially $SrTiO_3$, the crystal field splitting is large enough to be resolved by EELS, and both the L_3 and the L_2 lines exhibit splittings associated with the t_{2g} and the e_g bands (Abbate et al. 1991, Brydson et al. 1989, Kurata et al. 1993). There are several ways that these reference spectra can be used to quantify Ti oxidation states in EELS data sets: one is to use them as a reference basis for a multiple linear least squares fit, as proposed by Ohtomo et al. (2002). Another is to measure the relative t_{2g}/e_g line intensity ratio, as suggested by (Abbate et al. 1991). A third method proposed by Garcia-Barriocanal et al. (2010) is to use again the O K edge peak separation, ΔE, assuming a linear dependence between this parameter and Ti oxidation state as in the manganites. The values obtained for Ti oxidation states from the analysis of EELS images using this approach have been demonstrated to be consistent with the aforementioned MLLS fit (Garcia-Barriocanal et al. 2010).

Another interesting family of perovskites which allows the probing of one extra degree of freedom, spin, is the cobaltites, with a formula of $A_xB_{1-x}CoO_3$. Co ions in perovskites exhibit a competition between the crystal field splitting and Hund's-rule exchange energy in the 3d states, which ultimately determines the spin state of the individual Co ions (Imada et al. 1998, Khomskii and Low 2004). It is widely accepted that Co^{3+} ions (e.g., in $LaCoO_3$) display a $3d^6$ configuration which is susceptible to change with temperature from low spin (LS) $t_{2g}{}^6e_g{}^0$ to intermediate spin (IS) $t_{2g}{}^5e_g{}^1$ or high spin (HS) $t_{2g}{}^4e_g{}^2$ states (Klie et al. 2007, Korotin et al. 1996, Podlesnyak et al. 2006, Raccah and Goodenough 1967, Senaris-Rodriguez and Goodenough 1995). While the dynamics involved in such spin transitions and whether the IS or the HS is the first excited state remain controversial (He et al. 2009, Korotin et al. 1996, Phelan et al. 2006, Raccah and Goodenough 1967), it is established that the transition actually takes place. With this idea in mind, the bulk parent compound, $LaCoO_3$, has been used to test the sensitivity of EELS to the spin state of Co in this compound. Klie et al. (2007) reported changes in the O K edge prepeak intensity in $LaCoO_3$ when cooling the compound through the 100 K spin transition, with a higher prepeak intensity found for the LS state (see Figure 10–15).

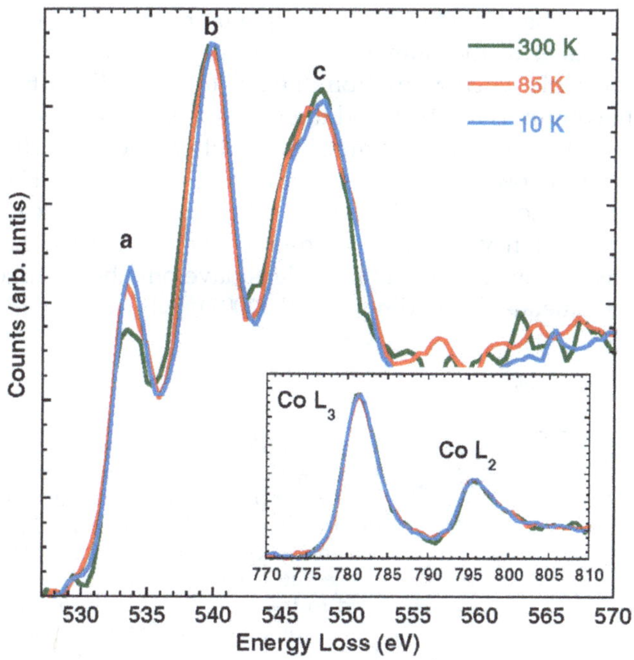

Figure 10–15. EEL spectra including both the O K edge and the Co $L_{2,3}$ edge (*inset*) acquired on $LaCoO_3$ crystals tilted down the pseudocubic [100] axis at 10 K (*light blue*), 8 5 K(*red*), and 300 K (*green*). Data acquired in a JEOL3000F. Extracted from Klie et al. (2007).

Meanwhile, the Co L_{23} ratio stayed constant, as one would expect from the fact that the Co average oxidation state does not change in the process. Therefore, changes that occur in the O K pre-edge feature, while the Co L fine structure remains unchanged, are to be looked for when using this technique to study spin states.

Further complexity is introduced into these systems with chemical doping. It is believed that Sr doping into the $La_{1-x}Sr_xCoO_3$ (LSCO) system generates a mixed valence Co^{3+}/Co^{4+} system where the IS is stabilized due to the strong hybridization between the Co 3d bands and the O 2p bands (Podlesnyak et al. 2008). In the absence of spin transitions, changes occurring in EEL spectra of LSCO compounds with different values of x can be directly associated with variations in electronic structure. Figure 10–16 shows a set of O K edges and Co $L_{2,3}$ edges for a number of LSCO samples with $x = 0, 0.15, 0.3, 0.4$, and 0.5. The O K edges show clear changes with x (Sr content). All of them exhibit three main peaks, a prepeak feature around 529 eV, which is attributed to the hybridization of the O 2p with Co 3d states, and two more peaks near 535 and 542 eV that are related to the bonding of the O 2p with the La 5d (or the Sr 4d) and the Co 4sp bands, respectively (Abbate et al. 1991). As in other perovskites, the peak related to the Co–O hybridization grows in intensity and width for higher Sr doping, accompanied by a gradual increase of the peak separation between the prepeak and the adjacent central peak, ΔE. Interestingly, the L_{23} intensity ratio for these Co oxides follows a trend opposite to the one observed in manganites: its value increases for higher x values. This behavior could be related to the experimental observation that the L_{23} ratio increases with the number of electrons in the 3d bands across the periodic table when going from the $3d^0$ toward $3d^5$ configuration, where a maximum is reached, and then decreases toward the $3d^{10}$ configuration within the transition metal row (Mitterbauer et al. 2003, Riedl et al. 2007, Sparrow et al. 1984, Waddington et al. 1986). Hence, for atoms such as Co one would expect the L_{23} ratio to increase when the d band occupancy decreases as observed. Unfortunately, theoretical simulations of transition metal $L_{2,3}$ edges have only been successfully achieved in metals (Ankudinov et al. 2003), while for oxides only

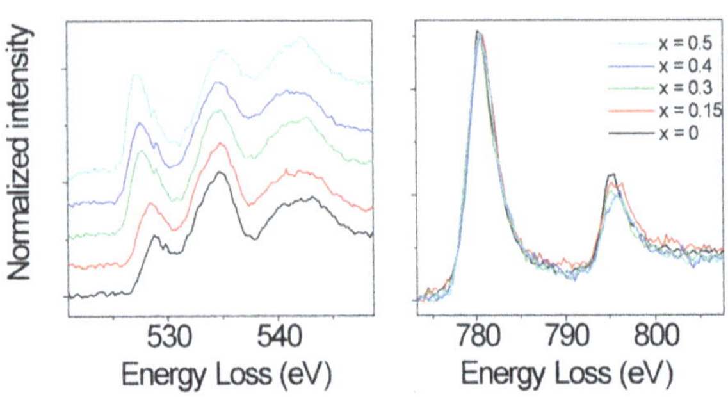

Figure 10–16. (*Left*) O K edges and (*right*) Co $L_{2,3}$ edges after background subtraction from LSCO samples with increasing values of Sr doping ($x = 0, 0.15, 0.3, 0.4, 0.5$). The O K data have been displaced vertically for clarity, while the Co L spectra have been normalized to the L_3 line intensity to enable direct visual comparison of the data. Notice that the L_2 line around 794 eV decreases in intensity with Sr content. Data from the aberration-corrected VG Microscopes HB501UX operated at 100 kV. Samples courtesy of Professor C. Leighton's group at the University of Minnesota (J. Gazquez et al. in preparation).

multiplet-based calculations (Cramer et al. 1991, de Groot 1994, 2005) have been able to fully reproduce actual L_{23} ratio values across the full range of oxidation states. In the case of cobaltites, understanding of the $L_{2,3}$ lines is further obscured by the fact that cobaltites tend to accommodate more and more O vacancies as x is increased, but this is just another example of an intriguing problem that remains open in this exciting field.

10.5 Applications

So far we have discussed the capabilities of aberration-corrected STEM-EELS in the framework of complex oxide materials, including quantification of structural, chemical, and electronic properties. In what follows we will review a number of applications to these systems, including both bulk materials and low dimensionality systems such as thin films, interfaces, and nanoparticles.

10.5.1 Atomic Resolution Measurement of Electronic Structure in High-T_c Superconductors

High-T_c superconducting materials are considered those with critical temperatures above 30–35 K. Given the promise they pose regarding both new physics and also applications, over a hundred thousand research articles have been devoted to their study since the discovery of high-T_c superconductivity (HTCS) in cuprates in 1986 (Bednorz and Mueller 1986, Wu et al. 1987). As a result, our understanding of the phenomenon has improved, although the underlying physical mechanism remains elusive. Understanding the materials electronic structure is a key task toward unraveling HTCS, since it is widely believed that hole states at the top of the valence band are the ones involved in superconductivity (Ginsberg 1989, Tinkham 1996). Around the Fermi level there is a very strong mixing of the Cu 3d bands and the O 2p bands. Therefore, the analysis of the Cu $L_{2,3}$ edge near 930 eV and, especially, the O K edge around 530 eV allows relevant information regarding the holes responsible for HTCS in these compounds to be extracted from the analysis of the fine structure (Nücker et al. 1988). As an example of the early work on this front, Browning et al. (1993) analyzed the fine structure of the O K edge in a series of $YBa_2Cu_3O_{7-\delta}$ (YBCO) samples with changing O contents. Removal of small amounts of O in YBCO causes the material to go into the underdoped regime and for the superconducting properties to degrade and, eventually, disappear (Nagaoka et al. 1998). The fingerprint in the O K edge spectra for the hole reduction can be seen as a reduction in intensity of the O K edge prepeak, as shown in Figure 10–17.

Thanks to aberration correction in the STEM, these studies can be carried out in real space, atomic plane by atomic plane, allowing the density of holes to be mapped with atomic resolution. Figure 10–18 shows a Z-contrast image of YBCO recorded at 100 kV. The unit cell

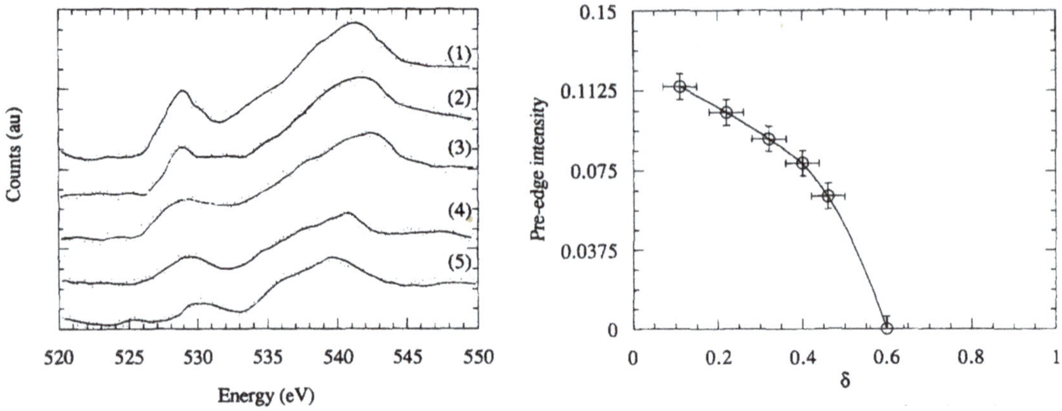

Figure 10–17. (*Left*) O K edge spectra obtained from a series of YBCO samples with oxygen contents of 6.89, 6.78, 6.67, 6.60, and 6.54, respectively, from *top* to *bottom*. (*Right*) Normalized prepeak intensity using a three-Gaussian fit as a function of O concentration for the aforementioned series of YBCO samples. Adapted from Browning et al. (1992).

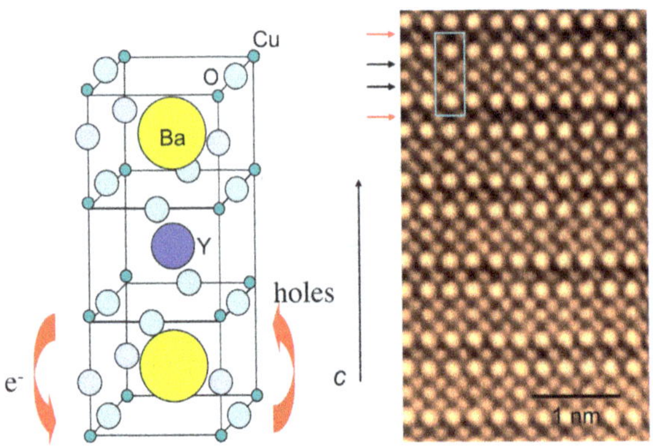

Figure 10–18. Z-contrast image of YBCO acquired in the aberration-corrected VG Microscopes HB501UX at 100 kV, with the electron beam perpendicular to the [001] axis. The unit cell (*sketched on the left*) has been marked with a *blue rectangle*. *Black arrows* point to the CuO_2 planes on the image, while *red arrows* show the CuO chains.

has been marked with a blue rectangle. The superconducting condensate is believed to take place on the CuO_2 planes of the structure thanks to a transfer of electrons to the CuO chains (Ginsberg 1989, Tinkham 1996), as shown in the image.

Figure 10–19 shows a series of O K edge spectra produced from a linescan acquired as the electron beam was scanned along the c direction of YBCO (marked with a white arrow on the image). The prepeak feature, marked with a blue arrow, oscillates up and down as the electron beam is scanned over CuO_2 planes and CuO chains, respectively. Figure 10–19(b) shows the average spectra from those positions: planes in black and chains in red, with a clear decrease of the prepeak on the CuO chains. These spectroscopic data indeed confirm that the O 2p holes widely believed to give rise to the superconducting condensate are confined to the CuO_2 planes.

Analysis of the Cu $L_{2,3}$ edge could give further proof of the transfer of electrons to the CuO chains. When Cu is reduced in an oxide, there is a chemical shift of the edge onset by a few electronvolts along with a decrease of the L_{23} intensity ratio. The Cu $L_{2,3}$ edge can also be acquired

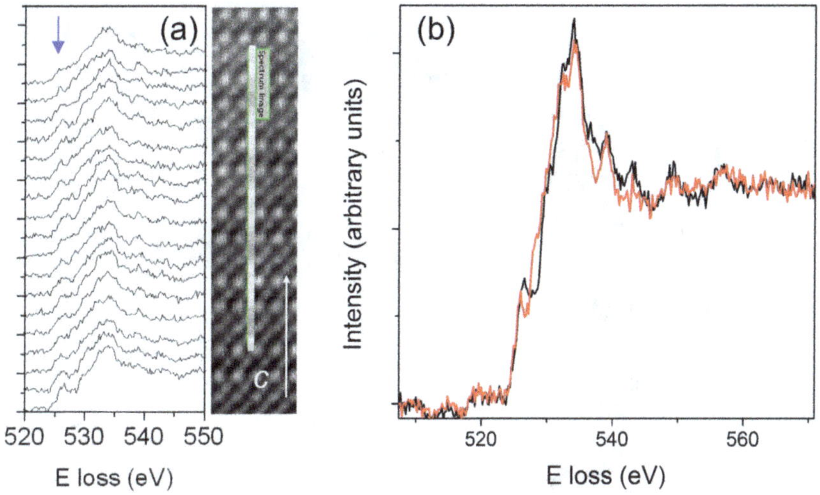

Figure 10–19. (a) O K edge spectra produced from a linescan along the region marked on the Z-contrast image, parallel to the YBCO *c* axis. Data acquired in the aberration-corrected VG Microscopes HB501UX at 100 kV. (b) Average O K edges from the CuO_2 planes (*black*) and CuO chains (*red*), normalized to the continuum for presentation purposes. The spectra have been averaged laterally to decrease random noise.

as the electron beam is scanned parallel to the *c* direction, an example being shown in Figure 10–20. In this case, a split structure is observed at both the L_3 (around 930 eV) and the L_2 (around 950 eV) lines. An enhanced view of the average L_3 line is displayed in Figure 10–20(b). The magnitude of the splitting is a few electronvolts, consistent with a chemical shift due to variation in oxidation state. Thus, the most straightforward way to interpret these data is that the Cu L edges measured here result from a mixture of Cu in two distinct valence states. These data can then be interpreted as follows: on the CuO chains two different types of Cu species are detected, which we will call Cu^A and Cu^B. Cu^B is responsible for the second peak of the splitting in Figure 10–20(b), and indicates a reduced valence with respect to Cu^A. In the CuO_2 planes, this peak is essentially absent. This behavior is seen consistently, as shown by the linescans in Figure 10–20(c). The Cu atoms on the CuO chains are of course expected to be reduced compared to those in the CuO_2 planes, but comparison to bulk standards could show quantitatively if there were any additional changes due to the transfer of holes as indicated in Figure 10–18(a), which is widely believed to be responsible for superconductivity in the CuO_2 planes.

This example comprised a bulk-like superconducting system, but the possible applications are endless and, in some cases, unique. Mapping of electronic properties and holes around defects such as dislocation cores in Ca-doped YBCO grain boundaries via STEM-EELS has clarified the mechanism for increased Jc observed in these systems (Hammerl et al. 2000). STEM-EELS combined with DFT in these Ca-doped dislocation cores has demonstrated how the inhomogeneous strain around the

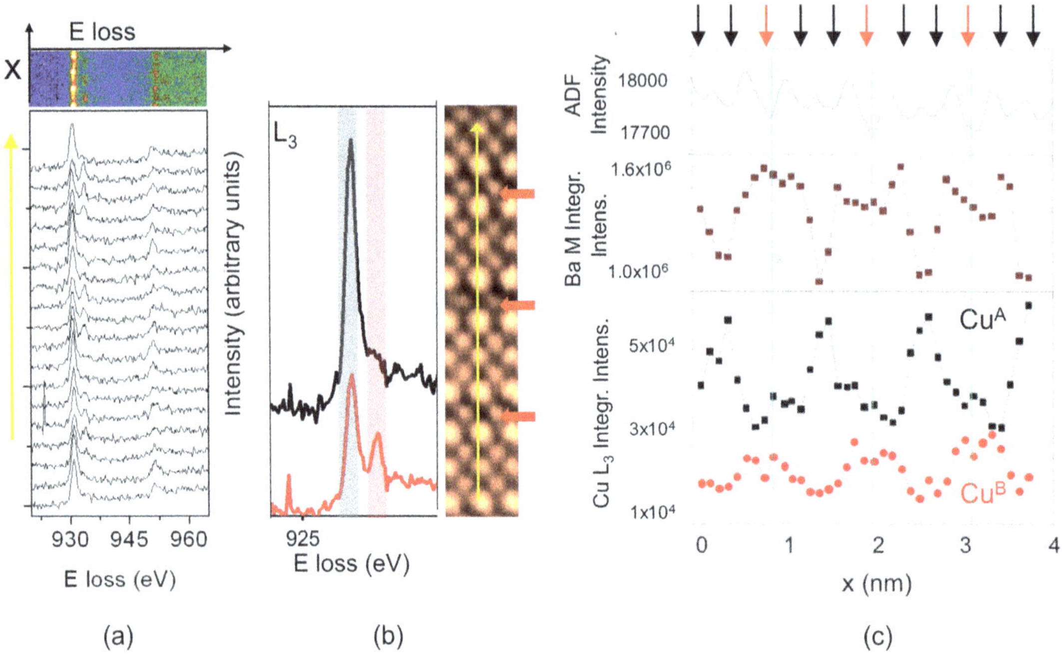

Figure 10–20. (a) Cu $L_{2,3}$ spectra acquired as the electron beam is swept parallel to the YBCO *c* axis. (b) Averaged L_3 line from the location of the CuO_2 planes (*black*) and the CuO chains (*red*). (c) The *top panel* shows the ADF signal, while the *medium panels* show the integrated intensity under the Ba $M_{4,5}$ edge, both acquired simultaneously with the Cu $L_{2,3}$ edge in (a). These allow the location of the CuO_2 planes and CuO chains along the scan. *Blue arrows* mark the CuO chains, and a *yellow arrow* shows the direction of the linescan. (c) Integrated intensities under the first (*in black*) and second (*in red*) peaks of the split L_3 line using windows such as the ones marked in (b). Data acquired in the aberration-corrected VG Microscopes HB501UX at 100 kV. The spectra have been averaged laterally to reduce noise.

core (compressive on one end and extensive on the other end) inhibits the formation of O vacancies, therefore causing a passivation of the grain boundary and the reported increase of Jc (Klie et al. 2005). The methods of prepeak intensity quantification along with Cu $L_{2,3}$ edge analysis demonstrated in this section can be utilized for further understanding of other superconducting systems, allowing not only the hole concentration but also the local atomic and electronic structure changes to be unraveled.

10.5.2 Column-Dependent Fine Structure: Atomic Resolution Measurement of Oxidation States in Manganites

With the advent of aberration correction, electron probes with sub-angstrom full widths at half maximum have been made readily available for atomic column EELS. This development opens up a new level of sensitivity when probing the electronic properties of oxides at the atomic scale. These materials have complex structures, and, within a given chemical species, not all atomic columns in a given projection may be equivalent. This behavior may be due to structural, electronic, or other reasons, and different ordering phenomena may ensue such as

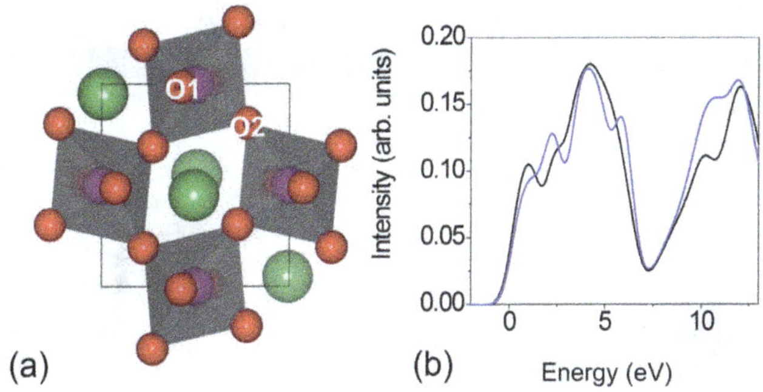

Figure 10–21. (a) Sketch of the LMO structure extracted from Rodriguez-Carvajal et al. (1998) with the O1 and O2 positions highlighted. (b) O K edge simulations for O1 (*black*) and O2 (*blue*) using density functional theory with the Z+1 approximation and 1 eV broadening. Adapted from Varela et al. (2009).

charge ordering, spin-state ordering. An example can be found in Jahn–Teller ordered perovskites such as $LaMnO_3$ (LMO) (Maekawa et al. 2004). In these compounds the O octahedra are distorted, and orbital ordering develops at low temperatures. Figure 10–21 shows a sketch of the LMO pseudocubic unit cell. Two different types of O species can be clearly distinguished: the apical O atoms and the equatorial ones. We will call them O1 and O2, respectively.

O1 and O2 have quite distinct environments. O2 is the Jahn–Teller active bond, with Mn–O2 bond lengths of 0.197 nm (short) and 0.2178 nm (long). The Mn–O1 bond length is 0.1968 nm. Simulations of the O K edge for both O species are shown in Figure 10–21(b). The pre-edge fine structure is quite different for both types of atoms, suggesting that EEL spectra from these two O positions should show different features. The prepeak position is shifted to higher energies for O2, and its intensity is slightly higher. Experimentally, these changes in prepeak intensity and also peak separation, ΔE, should be detectable. And indeed, experimental data show an oscillation of the peak separation ΔE when the electron beam is scanned along the pseudocubic [110] direction of crushed LMO samples, see Figure 10–22 (Varela et al. 2009). The oscillation is small, though, since some mixing from different columns leading to smearing of the changes is to be expected due to dechanneling of the probe.

While data such as that shown in Figure 10–22 suggest that indeed the fine structure, of the O K edge in this case, changes with the relative position of the probe and the different atomic columns, the data are noisy. Without tools to remove random noise like principal component analysis (PCA) (Bosman et al. 2006), it is not possible to get an interpretation (without degrading the spatial resolution). After noise has been removed the oscillation denoted by the red line suggests that the ripple in Figure 10–22 is only possible if the O1 atoms in this LMO sample are located in the Mn–O column of the scan, and the four pure O column neighbors are composed of O2 atoms. Thanks to PCA, inequivalent O species can be distinguished from the EELS data. However, if we try to quantify the Mn oxidation state from these scans acquired with atomic

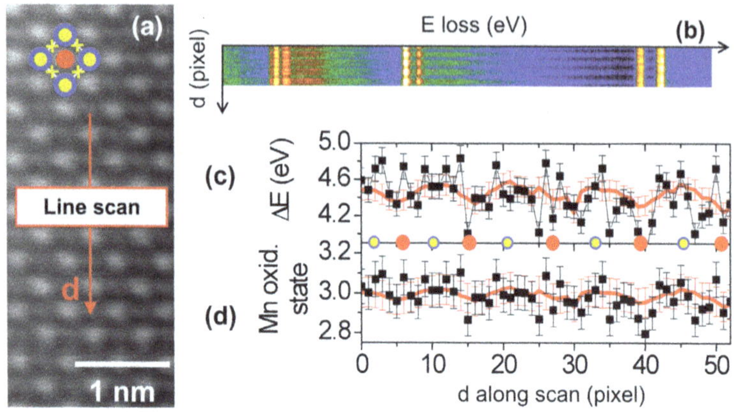

Figure 10–22. (a) Z-contrast image of an LMO crystal down the pseudocubic axis, obtained in the aberration-corrected VG Microscopes HB501UX operated at 100 kV. A cubic unit cell is highlighted (La in *red*, O in *yellow*, Mn in *blue*). (b) EELS linescan along the [110] pseudocubic direction marked with a *red arrow* in (a), including the O K, Mn $L_{2,3}$, and La $M_{4,5}$ edges. (c) ΔE extracted from the analysis of the O K edge fine structure. *Black dots* derive from the analysis of raw EELS data, while the *red line* was extracted from the analysis of the same data treated with PCA to remove random noise. The approximate position of the different atomic columns in the scan is marked. (d) Mn oxidation state derived from the ΔE measurement. Adapted from Varela et al. (2009).

size electron probes, an oscillation with an amplitude near 0.03 valence units is obtained (most likely, the influence of dynamical diffraction on the EELS signal). The values obtained on top of the Mn–O columns are in better agreement with the expected nominal ones (+3) pointing to the conclusion that oxidation states should be measured on columns, but more work is needed on this front.

Similar conclusions can be obtained for the analysis of 2D spectrum images such as the one shown in Figure 10–3. The O K edge map from Figure 10–3(a) is reproduced in Figure 10–23 and compared to a ΔE 2D map calculated from the same dataset. A clear fingerprint of the atomic lattice is observed in the peak position map, denoting fine structure changes along the lines of those predicted by theory from column

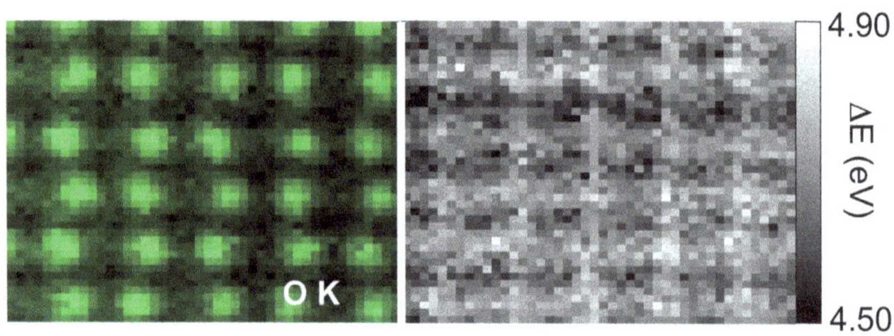

Figure 10–23. O K edge image for LMO down the [110] pseudocubic direction (*left*), along with the peak separation ΔE map showing a contrast due to the dependence of the fine structure on the atomic lattice (*right*). Data acquired at 60 kV on the aberration-corrected Nion UltraSTEM.

to column. These are the first steps toward identification in real space of inequivalent O atoms over long (a few unit cells) lateral distances, and to generalize, a new capability for spatial mapping of electronic structure.

10.5.3 Sensitivity of EELS to Spin: Spatial Mapping of a Spin-State Superlattice in Cobaltite Thin Films

As mentioned before, EELS is not only sensitive to electronic structure but, under certain conditions, to the spin state of atoms (Klie et al. 2007). Therefore, aberration-corrected STEM-EELS can be used to spatially map the spin state of certain systems with atomic resolution. In this section, we will describe the application of this technique to an oxygen-deficient cobaltite oxide thin film where the combination of ordered O vacancies and epitaxial strain in the system can stabilize a spin-state superlattice (Gazquez et al. 2010). When $LaCoO_3$ is doped with Sr, an oxygen-deficient perovskite phase is stabilized, characterized by an ordering of such vacancies (Ito et al. 2002, Wang and Yin 1998). When the $La_{0.5}Sr_{0.5}Co_3O_{3-\delta}$ (LSCO) composition is grown in the form of thin films, nanodomains of ordered oxygen vacancies arise, with a geometry connected to the release of epitaxial strain (Klenov et al. 2003, Torija et al. 2008). The structural relaxation due to the inclusion of ordered vacancies causes an elongation and a distortion of the perovskite unit cell along the *c*-axis, hence accommodating the mismatch with the substrate. This superstructure is responsible for every other Co–O plane showing a dimmer contrast in Z-contrast images such as the one shown in Figure 10–24(a). Figures 10–24(b) and (c)

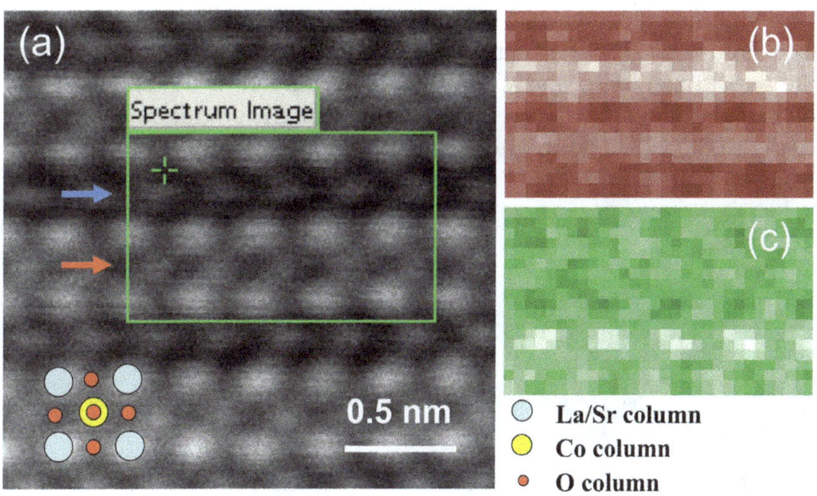

Figure 10–24. (a) Z-contrast image of LSCO [100] thin film showing the domain structure, acquired in the VG Microscopes HB501UX at 100 kV. The *rectangle* highlights the window for the acquired spectrum image. (b) Co $L_{2,3}$ edge and (c) O K edge EELS maps. Adapted from Gazquez et al. (2010).

show the atomic-resolution maps of the Co $L_{2,3}$ and O K edges, respectively, derived from a spectrum image acquired on the area marked in Figure 10–24(a). The EELS maps were generated by integrating the EELS spectra over a 30 eV window above the respective ionization threshold. Previously, PCA was applied to remove the spectral random noise (Bosman et al. 2006). Although some spatial drift and noise are present in the chemical maps they replicate the crystal lattice. The brighter Co–O plane on the Z-contrast image shows a slightly dimmer contrast in the Co $L_{2,3}$ image. This is due to the fact that more electrons are scattered to high angles when the beam is on this plane, so the overall incoming intensity into the EEL spectrometer is reduced. However, in the O K edge image, this same plane shows an enhanced signal relative to the dark stripe. Hence, one can conclude that the O content is severely reduced along the dark Co–O stripe.

The Co 3d and the O 2p states in cobaltite perovskites lie close to the Fermi level; as a result the unoccupied part of these states can be investigated by exciting transitions from Co 2p and O 1 s levels. Thus, by examining the fine structure of the Co L and O K edges along different Co–O planes from the spectrum image acquired before, one can study how the distribution of oxygen vacancies affects the local electronic properties of those Co atoms. The intensity ratio of the L_3 and L_2 lines of the Co $L_{2,3}$ edge (the $L_{2,3}$ ratio) is known to correlate with the oxidation state of Co (Abbate et al. 1992, Wang and Yin 1998). The Co $L_{2,3}$ ratio remains unchanged when shifting the electron beam from the bright to the dark Co–O stripe (Gazquez et al. 2010). This fact implies that there is no change in Co valence along the different Co–O planes. The Co valence can also be quantified from the $L_{2,3}$ ratio, which shows that Co atoms are close to a Co^{2+} state.

On the other hand, the O K edge depicted in Figure 10–25 shows a very significant decrease in the prepeak intensity along the dark stripe (see Figure 10–25). This pre-edge feature is related to the filling of the hybridized O 2p and Co 3d states (Abbate et al. 1992). While the observed change of the prepeak intensity might be due to a different filling of the Co 3d band and a changing population of the O 2p bands from dark to bright Co–O stripes, the Co $L_{2,3}$ ratio does not change

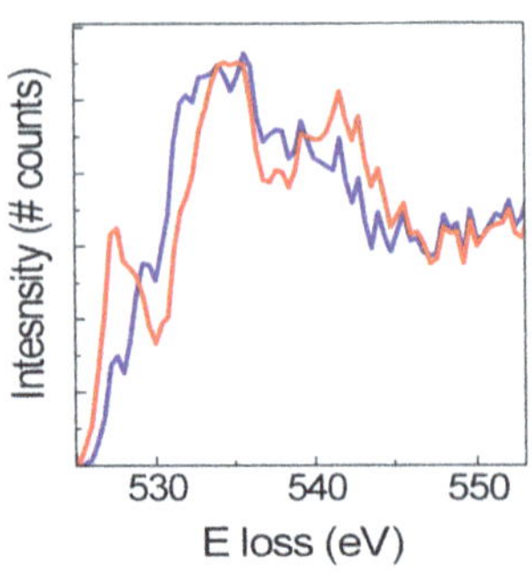

Figure 10–25. Average O K edges along the dark (*in blue*) and along the bright (*in red*) Co–O planes of the superstructure. Adapted from Gazquez et al. (2010).

significantly from plane to plane, which rules out this scenario. Klie and coworkers reported a similar behavior in $LaCoO_3$ (Klie et al. 2007), where the higher prepeak intensity was found to correspond to the lowest spin state. The results shown here point to different spin states in the Co atoms along the bright and the dark stripes (Gazquez et al. 2010). In view of this capability, EELS in combination with Z-contrast imaging provides a unique tool to measure spin-state transitions with atomic resolution, a key necessity in systems that exhibit nanoscale features such as domains, interfaces, or defects, where other techniques do not have the necessary spatial resolution.

10.5.4 Interface Quantification: Structure, Roughness, Interdiffusion, and Electronic Properties

By far, within the field of complex oxides, one area that is benefiting greatly from the applications of aberration-corrected STEM-EELS is the study of interface regions. When a material is grown on top of another one (be it a substrate or another thin layer in a heterostructure), the interface region acts as mediator between the properties of both: phenomena such as charge transfer due to mismatch of Fermi level, proximity effects, interface dipoles or charge localization, interface magnetism, exchange bias, or even new exotic phenomena such as the 2D metallic electron gases reported in some insulator–insulator oxide interfaces depend directly on the structure and morphology of the interface and also on the defects present (Ashcroft and Mermin 1976, Gonzalez et al. 2008, Harrison 1989, Herranz et al. 2007, Howe 1997, Kittel 1956, Mannhart and Schlom 2010, Ohtomo and Hwang 2004, Okamoto and Millis 2004, Nogues and Schuller 1999, Sefrioui et al. 2003). Therefore, its structure, chemistry, and the nature of the atomic bonding often determine the macroscopic properties of the whole system. Understanding and quantifying these issues is of the utmost importance when attempting to harness the different systems' physical properties.

While average diffraction techniques can accurately estimate interface structures and disorder-related parameters in the case of periodic structures, such as superlattices (Fullerton et al. 1992), the study of isolated systems (e.g., buried interfaces) is often easier through probes that can look "at the thing" in real space. And atomic-resolution STEM combined with EELS is one of the most powerful ones (Browning et al. 1993; Kimoto et al. 2007; Muller et al. 2008, Varela et al. 2006, Verbeeck et al. 2010). In this section we will review a few examples of studies of not only interface structure, but also chemistry and bonding with atomic resolution when these techniques are combined.

A nice example can be found in the study of cuprate–ferrite interfaces. Figure 10–26 shows a pair of simultaneously acquired BF and ADF images of a $Sm_2CuO_4/LaFeO_3$ (SCO/LFO) superlattice grown on a $SrTiO_3$ substrate. The interfaces between the layers and also with

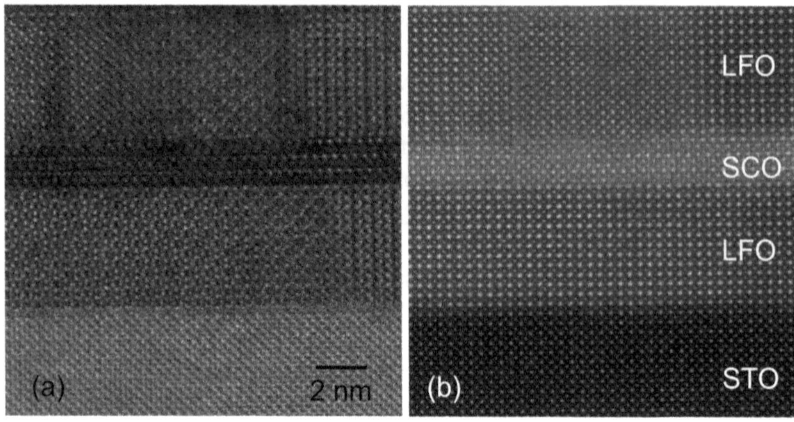

Figure 10–26. (a) BF and (b) high-angle ADF images of a $Sm_2CuO_4/LaFeO_3$ (SCO/LFO) superlattice grown on a $SrTiO_3$ (STO) substrate, acquired at 100 kV in the aberration-corrected Nion UltraSTEM. Sample courtesy of F.Y. Bruno and J. Santamaria, Universidad Complutense de Madrid, Spain.

the substrate are coherent, and the growth is epitaxial. While the interfaces are mostly flat, occasional one atomic plane high interface steps are observed between the SCO and the LFO layers. Interestingly, the BF images are more sensitive to changes in the LFO layer structure. While the ADF images always show a perovskite pseudocubic-like contrast, the BF images clearly demonstrate that there are lateral nanoscale domains within the LFO layers with different pseudocubic orientations. Some of the domain walls seem to be pinned to interface steps, suggesting that the domains arise during growth perhaps to compensate epitaxial mismatch, and help relax strain.

While in some cases the atomic plane stacking sequence at the interface can be studied from ADF images, EEL spectrum images are most useful for this purpose, since they allow acquisition of images where the EEL spectrum within every pixel can include a number of different absorption edges. Hence, several elemental maps can be produced simultaneously and then compared to each other in order to identify the atomic plane stacking sequence. Figure 10–27 shows an example for an SCO/LFO interface from a superlattice such as the one depicted in Figure 10–26. The inset marks the approximate region where a spectrum image was acquired, producing the simultaneous O, Fe, La, and Sm maps included in the figure. All of them show the atomic-resolution contrast related to their respective elemental lattices. The atomic planes can be counted one by one, and having also in mind the ADF images, the interface stacking sequence can be clearly identified as a ...–FeO_2–LaO– termination on the LaO side facing –CuO_2–SmO–... planes from the cuprate.

Not just the identity of the atoms by the interface but also the width of the region's chemical extent can be estimated from the analysis of both ADF images and EELS profiles. The interface shown in Figure 10–27 is atomically sharp, but this is not always the case. When interfaces are rough or chemically mixed, a careful statistical analysis of the images

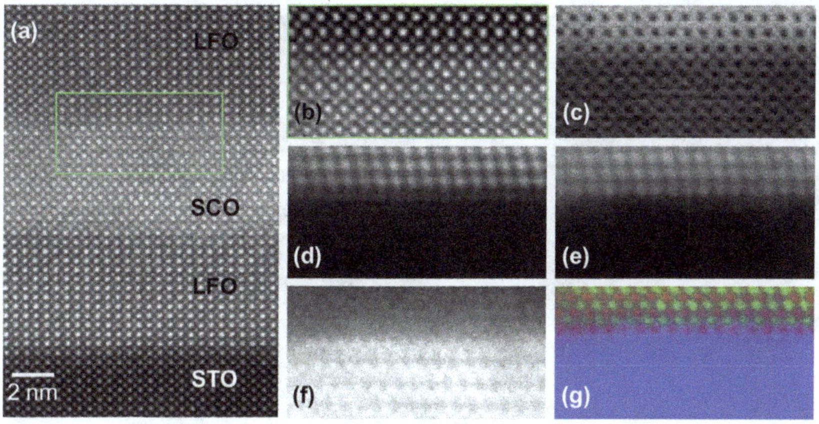

Figure 10–27. (a) ADF image of another SCO/LFO superlattice grown on an STO substrate. The *green box* marks the approximate area where a spectrum image was acquired at 100 kV in the Nion UltraSTEM with an exposure time of 30 ms per pixel. (b) Simultaneously acquired ADF signal. Minor spatial drift is observed. (c, d, e, f) O K, Fe $L_{2,3}$, La $M_{4,5}$, and Sm$M_{4,5}$ edge images, respectively. (g) RGB map produced by overlaying the Fe (*green*), La (*red*), and Sm (*blue*) images. All EELS images produced by integrating the signal under the respective edges after background subtraction using a power law and also after applying PCA to the raw data to remove random noise. Sample courtesy of F.Y. Bruno and J. Santamaria, Universidad Complutense de Madrid, Spain.

is needed to quantify the degree of disorder. Several methods have been proposed in the literature (Luysberg et al. 2009, Van Aert et al. 2009). An example is included in Figure 10–28, which shows an analysis by STEM-EELS of a $DyScO_3/SrTiO_3$ (DSO/STO) interface studied in an aberration-corrected FEI Titan 80-300 operated at 300 kV (from Luysberg et al. (2009)). In this work, the width of the interfaces was estimated from the ADF images for the large Sr and Dy cations and from the EELS signal for the lighter Ti and Sc. This work showed that the interfaces are not atomically sharp, but a couple of atomic planes wide, due to chemical intermixing. The interfaces were found to be symmetrically widened irrespective of their being at the top or bottom position. This interdiffusion was suggested to arise to compensate charge neutrality in these polar interfaces. These measurements are also subject to a number of artifacts that might obscure interpretation. One of the most important ones may be, again, beam broadening due to dechanneling of the STEM probe, which can lead to smearing of the interface width. In this sense, and unless the specimen thickness is small (below 5–10 nm thick) the estimations of interface widths either by ADF or by EELS must be taken as an upper estimate of the interface width.

Lastly, we will review a brief example on how to measure electronic properties across oxide interfaces from the EEL spectra fine structure as described in the previous sections. Since effects such as the aforementioned beam broadening may affect the measurements, it is desirable to use as many methods as possible to quantify the spectra and cross check

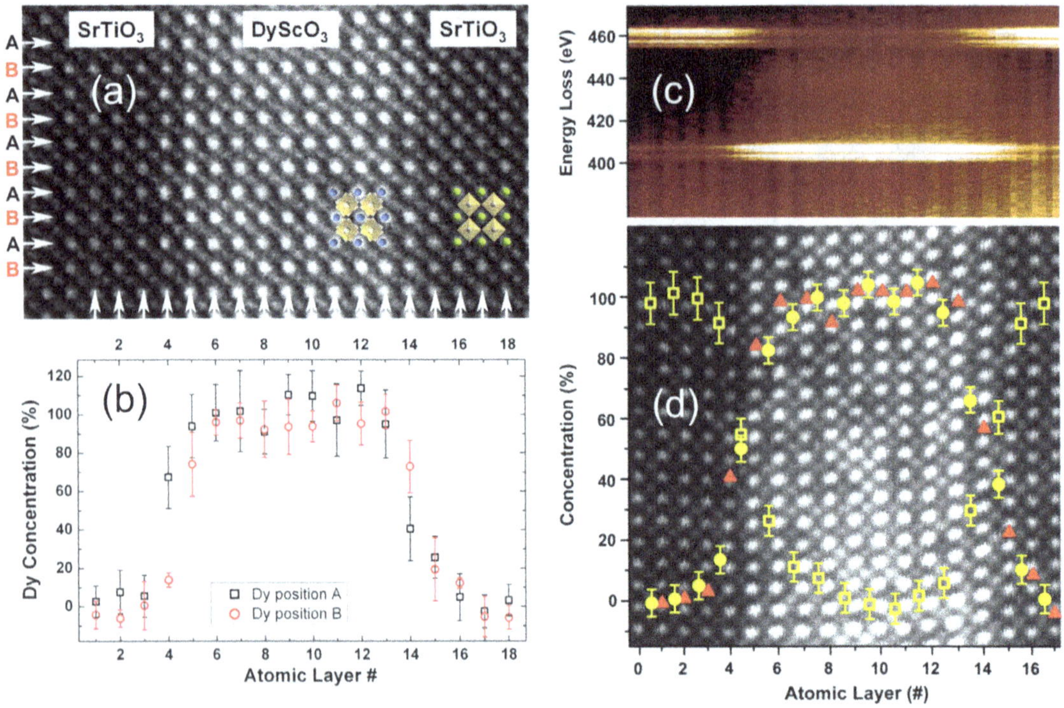

Figure 10–28. HAADF (*left*) and EELS (*right*) analysis of a $DyScO_3/SrTiO_3$ interface by M. Luysberg and coworkers (Luysberg et al. 2009). (a) HAADF image of the interface, with the DSO and STO unit structures marked. (b) Dy concentration evaluated from the image on (a). For each Dy layer the concentrations of positions A and B are displayed. (c) EELS image showing the $L_{2,3}$ edges of Sc and Ti. (d) Sc and Ti concentrations across the interfaces, estimated from the spectroscopic data in (c). Reproduced from Luysberg et al. (2009) with permission.

their results for consistency. Figure 10–29 shows the result of a STEM-EELS study of a set of $LaMnO_3/SrTiO_3$ (LMO/STO) superlattices, a system that has attracted a lot of interest recently (Garcia-Barriocanal et al. 2010, Shah et al. 2010, Zhai et al. 2010). Neither LMO nor STO are materials that are conducting or ferromagnetic in the bulk, and still, when combined in the form of superlattices, not only an intense ferromagnetism but also unexpected metallicity may arise in the system (Garcia-Barriocanal et al. 2010). Large changes in these properties in LMO/STO samples grown by high oxygen pressure sputtering can be tuned through changes in the LMO/STO relative thickness ratio. Perhaps the most peculiar feature in this particular system is that it is symmetric: all interfaces in the system, regardless of them being the top or the bottom one in each layer, show the same atomic structure: a LaO plane from the manganite facing a TiO_2 plane from the titanate, as detected by atomic-resolution STEM-EELS (Garcia-Barriocanal et al. 2010). This symmetry dopes the system with an extra LaO plane per LMO layer. Since the nominal oxidation states of the atoms in this plane are $La^{3+}O^{2-}$, the additional LaO plane can be envisioned as an electron donor giving one electron into the system per LMO layer.

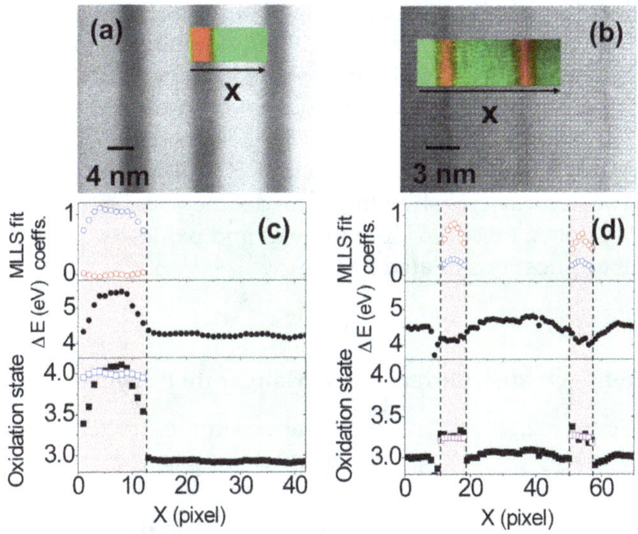

Figure 10–29. (a, b) Z-contrast images of a LMO_{17}/STO_{12} and a LMO_{17}/STO_2 superlattice, respectively. The LMO layers show the brighter contrast. The inset shows the region where a spectrum image was acquired, and displays the integrated Ti $L_{2,3}$ (*red*) and Mn $L_{2,3}$ (*green*) signals. (c, d) Analysis of the transition metal valences for the spectrum images in (a) and (b), respectively, averaged laterally. The *upper panel* shows the multiple linear least squares fit coefficients when the EEL spectra from the STO layer are fitted to the reference LTO (*red dots*) and STO (*blue dots*) spectra shown in Figure 10–14. The *middle panel* shows the value of the peak separation parameter, ΔE, across the spectrum image. The *lower panel* depicts the oxidation state of the transition metal (Mn in LMO and Ti in STO) calculated from both the MLLS fit and the ΔE parameter. The figure is color coded for clarity: *red* to mark the STO layers and *green* for the LMO layers. Data from the aberration-corrected VG Microscopes HB501UX, operated at 100 kV. Adapted from Garcia-Barriocanal et al. (2010).

The oxidation state of the transition metal in these superlattices has been derived from the measurement of the value of the peak separation at the O K edge, ΔE, and in the STO layers also by fitting the Ti $L_{2,3}$ edges to the reference spectra of bulk STO and LTO shown in Figure 10–14. The Mn oxidation state is close to +3 in all samples. But the Ti oxidation state varies with changing layer thickness ratio. As shown in Figure 10–29, the additional electron stays in the STO layer for samples with ultrathin STO layers such as a LMO_{17}/STO_2 superlattice, rendering the oxidation state of Ti close to +3.25. This reduction in the Ti oxidation state is detected by both the MLLS fit and the ΔE methods. As an aside, we note that the gradual drop in ΔE across the interface is, most likely, not an electronic effect, but a result of beam broadening since the O lattice is continuous across the interface.

These samples exhibit ferromagnetism with a magnetic moment per Mn close to the bulk value of the equivalent parent compound, and they are metallic perhaps thanks to the reduced Ti. For samples with thicker STO layers Ti is in a +4 oxidation state. Here, a slightly reduced value of the Mn oxidation state suggests that the additional electron may be accommodated within the LMO layer. However, the changes are too close to the error bars (a few percentage points for this data set) to draw conclusions. It is worth mentioning that these samples with thick LMO and STO layers exhibit insulating behavior along with a weak ferromagnetism. In summary, this example shows how electronic properties such as oxidation states can be measured across interfaces using the calibration methods described in Section 10.4. The next couple of sections will be dedicated to showing more detailed analysis of other oxide interface systems: metal–oxide interfaces and oxide–oxide interfaces of interest in applications related to energy.

10.5.5 Metal–Oxide Interfaces: Au–Manganite Bilayers

Another interesting application of aberration-corrected STEM-EELS relates to the study of interface interactions between nanomaterials, where this technique can provide unique structural and chemical information about epitaxy, chemical diffusion, charge transfer phenomena, or electronic structure of the interfaces. A good example is the analysis of the proximity effects of surface Au nanoparticles in the magnetic properties of $La_{0.67}Sr_{0.33}MnO_3$ (LSMO) thin films. Macroscopic characterization techniques have demonstrated that the effect of a thin (2 nm nominal thickness) Au capping was a sharp decrease of the magnetization and Curie temperature of LSMO films with thicknesses below 8 nm, even when an ultrathin $SrTiO_3$ (STO) spacing layer was deposited between Au and LSMO to avoid direct contact. In the latter, the effect was diminished but still present (Brivio et al. 2010). The fact that the existence of an intermediate layer does not destroy the effect suggested that the phenomena had electrostatic origin. Aberration-corrected STEM-EELS helped to unveil the microscopic origin of this exotic phenomenon. In order to do so, several samples were studied: a reference 6 nm LSMO thin film, a Au capped 6 nm thick LSMO, and the same stack including a 2 nm thick STO spacer, all of them grown on STO (100) substrates.

ADF images such as the ones in Figure 10–30 reveal a good epitaxial growth and the existence of structurally defect-free atomically sharp interfaces between the flat thin film layers in all samples. The Au capping layer, nominally a few nm thick, grows forming a discontinuous sheet of dispersed nanoparticles adopting quasi-spherical shapes with sizes ranging between 3 and 10 nm approximately. No preferential orientation of the Au nanoparticles was found.

Spectrum images were used to study the chemical and electronic nature of the layers and the interfaces. Figure 10–31 shows the distribution of chemical elements in the system along a linescan perpendicular to the film growth extracted from EELS spectra for La $M_{4,5}$, Mn $L_{2,3}$, Ti $L_{2,3}$, and O K edges. The interfaces are less than 1 nm thick, which is compatible with atomically sharp interfaces but beam broadening effects are smearing the interface chemical width. The data suggest that the interfaces are sharp and there is negligible chemical intermixing. A small offset between the La and Mn profiles is due to the different termination of the LSMO layer in the upper (right) and lower (left) interfaces.

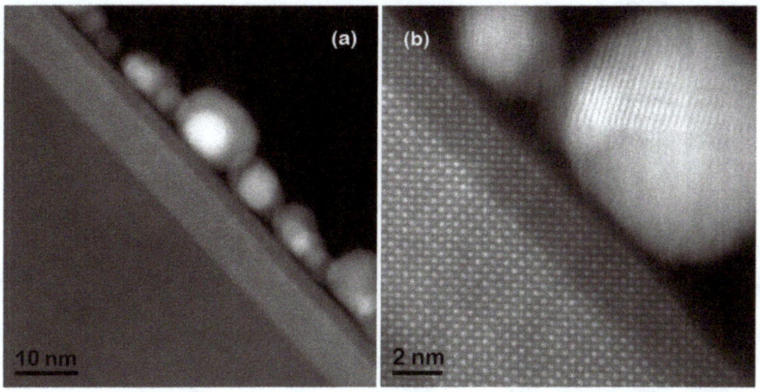

Figure 10–30. (a) ADF STEM image of the Au/2 nm STO/6 nm LSMO/STO thin film. (b) Details of the interface between LSMO and STO, and STO and Au particles. Data from the aberration-corrected VG Microscopes HB501UX operated at 100 kV and equipped with an Enfina spectrometer and a Nion aberration corrector. Adapted from Brivio et al. (2010).

Figure 10–31. Chemical profiles of the Au/2 nm STO/6 nm LSMO/STO thin film extracted from a spectrum image linescan perpendicular to the interfaces below the Au nanoparticle superimposed over the ADF STEM image as a reference. PCA was used to subtract the random noise from the EEL spectra. Adapted from Brivio et al. (2010).

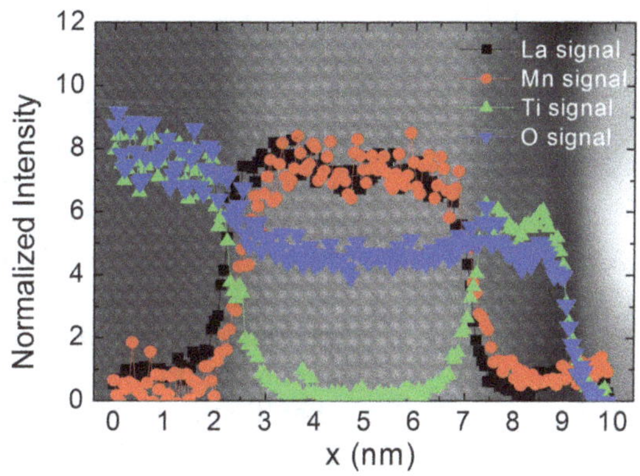

Whereas the LSMO film is LaO terminated at the interface with the substrate, the LSMO layer termination consists of an MnO atomic layer on the STO spacer side.

To get a deeper insight into the microscopic origin of the degradation of the magnetic properties, the Mn oxidation state has been analyzed by measuring again the peak separation between the prepeak and the main peaks of the O K edge (Varela et al. 2009). The result is illustrated in Figure 10–32, where a map of the Mn oxidation state is plotted. The noise is significantly high, so the Mn valence values have been averaged along the direction parallel to the substrate to reduce the statistical error. The result is an averaged profile of the Mn oxidation state as a function of depth, which gives an average value of $+3.15 \pm 0.04$, which is significantly reduced when compared to the nominal +3.33 expected for this composition (which is consistent with the experimental value measured under the same conditions for a reference LSMO film of the same thickness, $+3.28 \pm 0.04$, Brivio et al. 2010). Furthermore, in the case of the Au–LSMO sample without STO spacer, the average

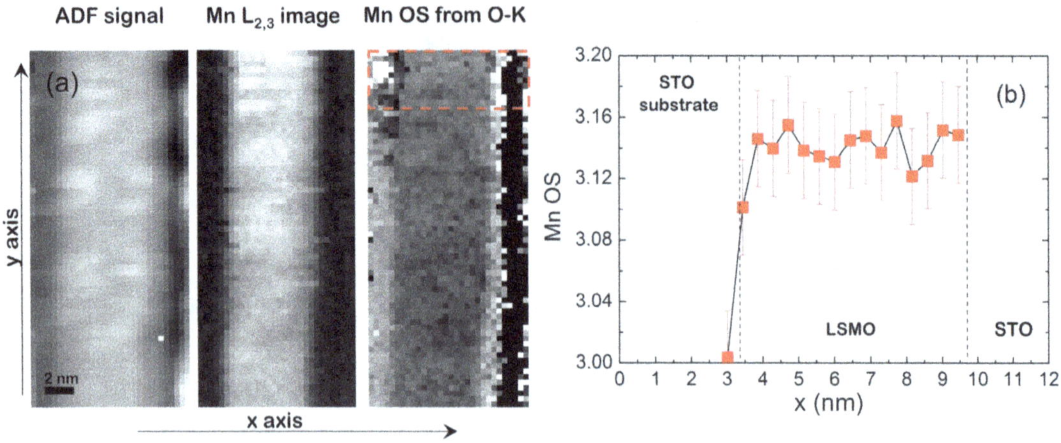

Figure 10–32. STEM-EELS spectrum image of the Au/2 nm STO/6 nm LSMO/STO thin film. (a) 2D maps of the ADF signal, of the Mn $L_{2,3}$ edge signal, and of the Mn oxidation state extracted from the energy difference of the main peak and the prepeak of the O K edge. (b) Mn oxidation state as a function of depth (x) calculated by statistically averaging Mn oxidation state values along the y axis in the *red dashed* box. Adapted from Brivio et al. (2010).

Mn valence value decreases down to $+2.92 \pm 0.05$. Therefore, we can conclude that the deposition of the Au nanoparticle layer induces a dramatic decrease of the Mn oxidation state, which is somewhat attenuated when the intermediate STO layer is present. Other experimental techniques support the decrease of the Mn oxidation state observed by STEM-EELS: nuclear magnetic resonance experiments reveal the disappearance of the Mn^{4+} site-related peak when compared to the double exchange-related Mn peak. Furthermore, DFT calculations have shown that the presence of the Au–LSMO interface does not cause any sizable change in the electronic properties of LSMO (Brivio et al. 2010). In consequence, a model for the Au-driven deoxygenation of the LSMO layer has been proposed to explain this drastic decrease of the Mn valence, an effect that disappears when the thickness of the LSMO and/or STO spacers is increased (Brivio et al. 2010).

10.5.6 Oxide Interfaces for Energy Applications: Colossal Ionic Conductivity in Strained YSZ–STO Superlattices

Hydrogen-based fuel cells convert chemical energy directly into electricity with water as the sole by-product. Their ability to cleanly produce electricity makes them an attractive alternative to fossil fuel-based energy sources. Hydrogen, however, is typically not freely available. Without a completely clean means of producing H, the efficiency of the fuel cell determines how clean a source of power it is. Solid oxide fuel cells (SOFCs), named for their solid oxide electrolytes, are the most efficient currently under development. The electrolytes must conduct O ions from cathode to anode while remaining electrically insulating. Yttria-stabilized zirconia $(Y_2O_3)_x(ZrO_2)_{1-x}$ (YSZ) is the most commonly used electrolyte, but requires operating temperatures of at least 800°C for sufficiently high O conductivity (Goodenough 2003, Kilner 2008, Ormerod 2003, Steele 2001). The need for such high temperatures has hampered the application of SOFCs, and a major research effort has been devoted to a search for new materials with enhanced ionic conductivity at lower temperatures.

The greatest improvements in O ion conductivity have been achieved in YSZ/STO multilayers, which have recently been found to exhibit up to eight orders of magnitude greater ionic conductivity than bulk YSZ near room temperature (Garcia-Barriocanal et al. 2008). The discovery, dubbed colossal ionic conductivity, brought intense interest due to the potential to achieve high-efficiency SOFCs operating at low temperatures. The ionic conductance was found to scale with the number of interfaces, but to be virtually independent of the YSZ layer thickness from 1 to 30 nm, indicating that the majority of the ionic conduction occurs near the interfaces. The contribution of electronic conductivity was determined to be three to four orders of magnitude lower, and experiments also ruled out a protonic contribution to the conductance (Garcia-Barriocanal et al. 2009).

Both X-ray and STEM analysis of the multilayers with thin 1–30 nm YSZ layers showed them to be highly coherent, with the YSZ lattice rotated 45° to that of the STO. Figure 10–33 shows low- and high-magnification Z-contrast images of a YSZ/STO superlattice with 1-nm-thick YSZ layers and nine repeats, from which it can be seen that the layers are continuous and flat over long distances. From the high-magnification image it can be seen that the cation lattices are perfectly coherent across the interface, indicating that the YSZ is strained a large 7%. At larger thicknesses, however, the YSZ was seen to have

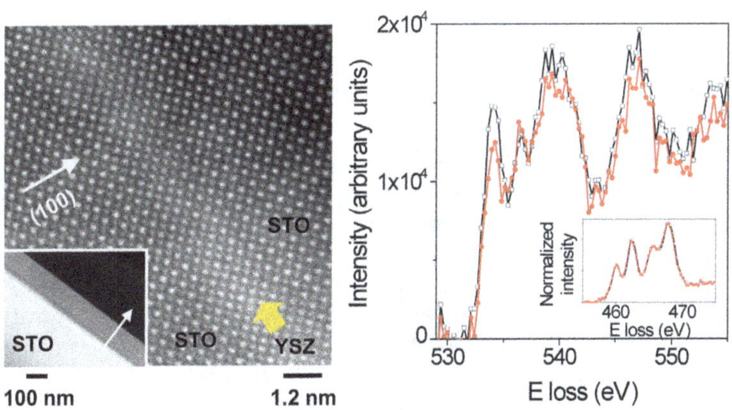

Figure 10–33. (*left*) Z-contrast STEM image of the STO/YSZ interface of the $[YSZ_{1\ nm}/STO_{10\ nm}]_9$ superlattice, obtained in the VG Microscopes HB603U microscope. A *yellow arrow* marks the position of the YSZ layer. (*Inset*) Low-magnification image obtained in the VG Microscopes HB501UX column. In both cases a *white arrow* indicates the growth direction. (*right*) EEL spectra showing the O K edge obtained from the STO unit cell at the interface plane (*red circles*) and 4.5 nm into the STO layer (*black squares*). (*Inset*) Ti $L_{2,3}$ edges for the same positions, same color code. Reproduced from Garcia-Barriocanal et al. (2008).

relaxed, releasing the strain. The conductance of the sample was found to be three orders of magnitude lower than those with strained YSZ, suggesting that strain is vital to colossal ionic conductivity.

EELS provides further insight into the nature of the interface. Figure 10–33 displays spectra taken from the interface plane and from the center of the STO layer. No significant change in the Ti fine structure is seen, indicating that a Ti^{4+} configuration is predominant at the interface plane, consistent with the lack of electronic conductivity.

To further probe the origins of colossal ionic conductivity, the information from the STEM and X-ray experiments was used as a starting point for DFT calculations (Pennycook et al. 2010). To determine the effect of the 7% strain, finite-temperature quantum molecular dynamics simulations for both strained and unstrained zirconia were performed at various temperatures. Using simulated annealing to explore the structure between 2,500 and 0 K, it was found that the O sublattice completely changes when strained. Below 1,000 K the O atoms are ordered in zigzags. At higher temperatures, the O sublattice becomes increasingly disordered, appearing completely amorphous above 2,000 K. The disordered O atoms were found to be far more mobile than the O atoms in the unstrained structure. As the experiments pointed to an interfacial mechanism, DFT simulations were also performed for an YSZ–STO multilayer. With a 1 nm YSZ thickness, it was found that the YSZ O atoms became as disordered near room temperature as that of the strained bulk zirconia at 2,000 K. Also, the interfacial O atoms complete the O octahedra around the interfacial Ti atoms as in bulk STO, in agreement with the EELS measurements of a 4+ Ti oxidation state. These O positions are at odds with the positions desired by the YSZ. It seems it is this O sublattice incompatibility which perturbs the YSZ O atoms into disorder. This is the key to the low-temperature colossal ionic conductivity. The estimated ionic conductivity of the strained multilayer YSZ was six orders of magnitude greater than that of bulk unstrained YSZ, giving strong theoretical evidence that the origin of colossal ionic conductivity is a combination of strain and O sublattice mismatch (Pennycook et al. 2010).

10.6 Conclusions

This chapter has been dedicated to review the status of aberration-corrected STEM-EELS applied to complex oxides, especially those with the perovskite structure. We have shown how challenges that were unthinkable a decade ago have been tackled and solved: light atom imaging has become a doable task using different imaging modes (BF, ABF, ADF, EELS), and even single atoms can be detected and quantified, and the properties of the surrounding matrix can be studied. Atomic-resolution spectrum imaging allows elemental maps to be produced with atomic resolution and fine structures and the underlying electronic properties to be studied as a function of probe position. Different inequivalent species can be distinguished, and the properties of defects, interfaces, and nanoparticles can be studied with great detail.

Unfortunately, and due to limited space, not every possible exciting result or highlight has been gathered here. But we hope the reader will get the flavor of this exciting field. Aberration correction really has revolutionized the study of these materials, opening up an endless number of possibilities. Most of the examples shown here and reported in the literature are more a proof of principle than a complete summary deeply understood. Now, more than ever, "more work is needed."

Acknowledgments The authors are most grateful to all of our colleagues and collaborators who made this work possible through the years. We cannot name all of them here, but a special word of gratitude goes to R. Sanchez, A.R. Lupini, M.F. Chisholm, J.T. Luck, W.H., Sides, W. Luo, S.T. Pantelides, J. Santamaria, J. Garcia-Barriocanal, Z. Sefrioui, C. Leon, A. Rivera-Calzada, N. Shibata, Y. Ikuhara, T. Mizoguchi, K.M. Krishnan, K. Griffin-Roberts, S.N. Rashkeev, D.G. Mandrus, H.M. Christen, M. Biegalski, M.A. Torija, M. Sharma, C. Leighton, J. Tao, R. Bertacco and his group, M. Watanabe, R. Klie, L.J. Allen, S.D. Findlay, P.D. Nellist, and of course O.L. Krivanek, N. Dellby, M. Murfitt, and everybody at Nion Co. Research at ORNL sponsored by the Materials Sciences and Engineering Division, Office of Science, US DOE, and research at Universidad Complutense supported by the European Research Council Starting Investigator Award.

References

M. Abbate, F.M.F. de Groot, J.C. Fuggle, A. Fujimori, Y. Tokura, Y. Fujishima, O. Strebel, M. Domke, G. Kaindl, J. van Elp, G.A. Sawatzky, M. Sacchi, N. Tsuda, Soft-x-ray-absorption studies of the location of extra charges induced by substitution in controlled-valence materials. Phys. Rev. B **44**, 5419–5422 (1991)

M. Abbate, F.M.F. de Groot, J.C. Fuggle, A. Fujimori, O. Strebel, F. Lopez, M. Domke, G. Kaindl, G.A. Sawatzky, M. Takano, Y. Takeda, H. Eisaki, S. Uchida, Controlled-valence properties of $La_{1-x}Sr_xFeO_3$ and $La_{1-x}Sr_xMnO_3$ studied by soft-x-ray absorption spectroscopy. Phys. Rev. B **46**, 4511–4519 (1992)

L.J. Allen, S.D. Findlay, M.P. Oxley, C.J. Rossouw, Lattice-resolution contrast from a focused coherent electron probe. Part I. Ultramicroscopy **96**, 47–63 (2003)

A.L. Ankudinov, A.I. Nesvizhskii, J.J. Rehr, Dynamic screening effects in x-ray absorption. Phys. Rev. B **67**, 115120 (2003)

N.W. Ashcroft, N.D. Mermin, *Solid State Physics* (Saunders College, 1976)

H. Bea, M. Gajek, M. Bibes, A. Barthelemy, Spintronics with multiferroics. J. Phys.: Condens. Matter. **20**, 434221 (2008)

J.G. Bednorz, K.A. Mueller, Possible high TC superconductivity in the Ba-La-Cu-O system. Z. Phys. B **64**, 189–193 (1986)

M. Bibes, A. Barthelemy, Oxide spintronics. IEEE Trans. Electron Dev. **54**, 1003–1023 (2007)

M. Bosman, M. Watanabe, D.T.L. Alexander, V.J. Keast, Mapping chemical and bonding information using multivariate analysis of electron-energy-loss spectrum images. Ultramicroscopy **106**, 1024–1032 (2006)

M. Bosman, V.J. Keast, J.L. Garcia-Muñoz, A.J. D'Alfonso, S.D. Findlay, L.J. Allen, Two-dimensional mapping of chemical information at atomic resolution. Phys. Rev. Lett. **99**, 086102 (2007)

S. Brivio, C. Magen, A.A. Sidorenko, D. Petti, M. Cantoni, M. Finazzi, F. Ciccacci, De R. RENzi, M. Varela, S. Picozzi, R. Bertacco, Effects of Au nanoparticles on the magnetic and transport properties of $La_{0.67}Sr_{0.33}MnO_3$ ultrathin layers. Phys. Rev. B **81**, 094410 (2010)

N.D. Browning, N.D. Chisholm, S.J. Pennycook, Atomic-resolution chemical analysis using a scanning transmission electron microscope. Nature **366**, 143–146 (1993)

N.D. Browning, J. Yuan, L.M. Brown, Determination of local oxygen stoichiometry in $YBa_2Cu_3O_{7-\delta}$ by electron energy loss spectroscopy in the scanning transmission electron microscope. Physica C **202**, 12–18 (1992)

R. Brydson, H. Sauer, W. Engel, J.M. Thomas, E. Zeitler, N. Kosugi, H. Kuroda, Electron energy loss and X-ray absorption spectroscopy of rutile and anatase: a test of structural sensitivity. J. Phys.: Condens. Matter. **1**, 797–812 (1989)

S.P. Cramer, F.M.F. de Groot, Y. Ma, C.T. Chen, F. Sette, C.A. Kipke, D.M. Eichhorn, M.K. Chan, W.H. Armstrong, E. Libby, G. Christou, S. Brooker, V. McKee, O.C. Mullins, J.C. Fuggle, Ligand field strengths and oxidation states from manganese L-edge spectroscopy. J. Am. Chem. Soc. **113**, 7937–7940 (1991)

E. Dagotto, T. Hotta, A. Moreo, Colossal magnetoresistant materials: the key role of phase separation. Phys. Rep. **344**, 1–153 (2001)

R.F. Egerton, *Electron Energy Loss Spectroscopy in the Electron Microscope* (Plenum Press, New York, NY, 1996)

S.D. Findlay, M.P. Oxley, S.J. Pennycook, L.J. Allen, Modelling imaging based on core-loss spectroscopy in scanning transmission electron microscopy. Ultramicroscopy **104**, 126–140 (2005)

S.D. Findlay, N. Shibata, H. Sawada, E. Okunishi, Y. Kondo, T. Yamamoto, Y. Ikuhara, Robust atomic resolution imaging of light elements using scanning transmission electron microscopy. Appl. Phys. Lett. **95**, 191913 (2009)

E.E. Fullerton, I.K. Schuller, H. Vanderstraeten, Y. Bruynseraede, Structural refinement of superlattices from x-ray diffraction. Phys. Rev. B **45**, 9292–9310 (1992)

V. Garcia, M. Bibes, L. Bocher, S. Valencia, F. Kronast, A. Crassous, X. Moya, S. Enouz-Vedrenne, A. Gloter, D. Imhoff, C. Deranlot, N.D. Mathur, S. Fusil, K. Bouzehouane, A. Barhelemy, Ferroelectric control of spin polarization. Science **327**, 1106–1110 (2010)

J. Garcia-Barriocanal, A. Rivera-Calzada, M. Varela, Z. Sefrioui, E. Iborra, C. Leon, S.J. Pennycook, J. Santamaria, Colossal ionic conductivity at interfaces of epitaxial ZrO_2:Y_2O_3/$SrTiO_3$ heterostructures. Science **321**, 676–680 (2008)

J. Garcia-Barriocanal, A. Rivera-Calzada, M. Varela, Z. Sefrioui, E. Iborra, C. Leon, S.J. Pennycook, J. Santamaria, Response to the comment on "Colossal ionic conductivity at interfaces of epitaxial ZrO_2:Y_2O_3/$SrTiO_3$ heterostructures". Science **324**, 465 (2009)

J. Garcia-Barriocanal, F.Y. Bruno, A. Rivera-Calzada, Z. Sefrioui, N.M. Nemes, M. Garcia-Hernández, J. Rubio-Zuazo, G.R. Castro, M. Varela, S.J. Pennycook, C. Leon, J. Santamaría, "Charge leakage" at $LaMnO_3$/$SrTiO_3$ interfaces. Adv. Mater. **22**, 627–632 (2010)

J. Gazquez, W. Luo, M.P. Oxley, M. Orange, M.A. Torija, M. Sharma, C. Leighton, S.T. Pantelides, S.J. Pennycook, M. Varela, In preparation (2010)

D.M. Ginsberg, *Physical Properties of High Tc Superconductors* I (World Scientific, Singapore, 1989) and subsequent volumes

I. Gonzalez, S. Okamoto, S. Yunoki, A. Moreo, E. Dagotto, Charge transfer in heterostructures of strongly correlated materials. J. Phys.: Condens. Matter. **20**, 264002 (2008)

J.B. Goodenough, Oxide-ion electrolytes. Annu. Rev. Mater. Res. **33**, 91–128 (2003)

K. Griffin-Roberts, M. Varela, S. Rashkeev, S.T. Pantelides, S.J. Pennycook, K.M. Krishnan, Defect-mediated ferromagnetism in insulating Co-doped anatase TiO_2 thin films. Phys. Rev. B **78**, 014409 (2008)

F.M.F. de Groot, X-ray absorption and dichroism of transition metals and their compounds. J. Electron. Spectrosc. Relat. Phenom. **67**, 529–622 (1994)

F. de Groot, Multiplet effects in X-ray spectroscopy. Coord. Chem. Rev. **249**, 31–63 (2005)

G. Hammerl, A. Schmehl, R.R. Schulz, B. Goetz, H. Bielefeldt, C.W. Schneider, H. Hilgenkamp, J. Mannhart, Enhanced supercurrent density in polycrystalline $YBa_2Cu_3O_{7-\delta}$ at 77 K from calcium doping of grain boundaries. Nature **407**, 162–164 (2000)

W. Harrison, *Electronic Structure and the Properties of Solids* (Dover, New York, NY, 1989)

C. He, H. Zheng, J.F. Mitchell, M.L. Foo, R.J. Cava, C. Leighton, Low temperature Schottky anomalies in the specific heat of $LaCoO_3$: defect-stabilized finite spin states. Appl. Phys. Lett. **94**, 102514 (2009)

G. Herranz, M. Basletic, M. Bibes, C. Carretero, E. Tafra, E. Jacquet, K. Bouzehouane, C. Deranlot, A. Hamzic, J.-M. Broto, A. Barthelemy, A. Fert, High mobility in $LaAlO_3/SrTiO_3$ heterostructures: origin, dimensionality, and perspectives. Phys. Rev. Lett. **98**, 216803 (2007)

J.H. Howe, Interfaces in Materials: Atomic Structure, Thermodynamics and Kinetics of Solid-Vapor, Solid-Liquid and Solid-Solid Interfaces (Wiley, New York, NY, 1997)

J.A. Hunt, D.B. Williams, Electron energy-loss spectrum-imaging. Ultramicroscopy **38**, 47–73 (1991)

M. Imada, A. Fujimori, Y. Tokura, Metal insulator transitions. Rev. Mod. Phys. **70**, 1039–1263 (1998)

Y. Ito, R.F. Klie, N.D. Browning, T.J. Mazanec, Atomic resolution analysis of the defect chemistry at domain boundaries in Brownmillerite type strontium cobaltite. J. Am. Ceram. Soc. **85**, 969–976 (2002)

C. Jeanguillaume, C. Colliex, Spectrum-image: the next step in EELS digital acquisition and processing. Ultramicroscopy **28**, 252–257 (1989)

C.L. Jia, M. Lentzen, K. Urban, Atomic resolution imaging of oxygen in perovskite ceramics. Science **299**, 870 (2003)

C.L. Jia, K. Urban, Atomic resolution measurement of oxygen concentration in oxide materials. Science **303**, 2001–2004 (2004)

D.I. Khomskii, U. Low, Superstructures at low spin-high spin transitions. Phys. Rev. B **69**, 184401 (2004)

K. Kimoto, T. Asaka, T. Nagai, M. Saito, Y. Matsui, K. Ishizuka, Element-selective imaging of atomic columns in a crystal using STEM and EELS. Nature **450**, 702–704 (2007)

J.A. Kilner, Ionic conductors: feel the strain. Nat. Mater. **7**, 838–839 (2008)

C. Kittel, *Introduction to Solid State Physics* (Wiley, New York, NY, 1956)

D.O. Klenov, W. Donner, B. Foran, S. Stemmer, Impact of stress on oxygen vacancy ordering in epitaxial $La_{0.5}Sr_{0.5}CoO_{3-\delta}$ thin films. Appl. Phys. Lett. **82**, 3427–3429 (2003)

R.F. Klie, J.P. Buban, M. Varela, A. Franceschetti, C. Jooss, Y. Zhu, N.D. Browning, S.T. Pantelides, S.J. Pennycook, Enhanced current transport at grain boundaries in high-Tc superconductors. Nature **435**, 475–478 (2005)

R.F. Klie, J.C. Zheng, Y. Zhu, M. Varela, J. Wu, C. Leighton, Direct measurement of the low temperature spin-state transition in $LaCoO_3$. Phys. Rev. Lett, **99**, 047203 (2007)

M.A. Korotin, S.Yu. Ezhov, I.V. Solovyev, V.I. Anisimov, D.I. Khomskii, G.A. Sawatzky, Intermediate-spin state and properties of $LaCoO_3$. Phys. Rev. B, **54**, 5309–5316 (1996)

O.L. Krivanek, J.H. Paterson, ELNES of 3d transition-metal oxides I. Variations across the periodic table. Ultramicroscopy **32**, 313–318 (1990)

O.L. Krivanek, M.F. Chisholm, V. Nicolosi, T.J. Pennycook, G.J. Corbn, N. Dellby, M.F. Murfitt, C.S. Own, Z.S. Szilagyi, M.P. Oxley, S.T. Pantelides, S.J. Pennycook, Atom-by-atom structural and chemical analysis by annular dark field electron microscopy. Nature **464**, 571–574 (2010)

H. Kurata, C. Colliex, Electron-energy-loss core-edge structures in manganese oxides. Phys. Rev. B **48**, 2102–2108 (1993)

H. Kurata, E. Lefevre, C. Colliex, R. Brydson, Electron-energy-loss near-edge structures in the oxygen K-edge spectra of transition-metal oxides. Phys. Rev. B **47**, 13763–13768 (1993)

W. Luo, M. Varela, J. Tao, S.J. Pennycook, S.T. Pantelides, Electronic and crystal-field effects in the fine structure of electron energy-loss spectra of manganites. Phys. Rev. B. **79**, 052405 (2009)

M. Luysberg, M. Heidelmann, L. Houben, M. Boese, T. Heeg, J. Schubert, M. Roeckerath, Intermixing and charge neutrality at $DyScO_3/SrTiO_3$ interfaces. Acta Mater. **57**, 3192–3198 (2009)

S. Maekawa, T. Tohyama, S.E. Barnes, S. Ishihara, W. Koshibae, G. Khaliullin, *Physics of Transition Metal Oxides* (Springer, Berlin, Germany, 2004)

J. Mannhart, D.G. Schlom, Oxide interfaces – an opportunity for electronics. Science **327**, 1607–1611 (2010)

C. Mitterbauer, G. Kothleitner, W. Grogger, H. Zandbergen, B. Freitag, P. Tiemeijer, F. Hofer, Electron energy-loss near edge structures of 3d transition-metal oxides recorded at high energy resolution. Ultramicroscopy **96**, 469–480 (2003)

D.A. Muller, L. Fitting-Kourkoutis, M. Murfitt, J.H. Song, H.Y. Hwang, J. Silcox, N. Dellby, O.L. Krivanek Atomic-scale chemical imaging of composition and bonding by aberration-corrected microscopy. Science **319**, 1073–1076 (2008)

T. Nagaoka, Y. Matsuda, H. Obara, A. Sawa, T. Terashima, I. Chong, M. Tacaño, M. Suzuki, Hall anomaly in the superconducting state of high-Tc cuprates: universality in doping dependence. Phys. Rev. Lett. **80**, 3594–3597 (1998)

P.D. Nellist, S.J. Pennycook, Subangstrom resolution by underfocused incoherent transmission electron microscopy. Phys. Rev. Lett. **81**, 4156–4159 (1998)

P.D. Nellist, M.F. Chisholm, N. Dellby, O.L. Krivanek, M.F. Murfitt, Z.S. Szilagyi, A.R. Lupini, A.Y. Borisevich, W.H. Sides, S.J. Pennycook, Direct sub-angstrom imaging of a crystal lattice. Science **305**, 1741 (2004)

J. Nogues, I.K. Schuller, Exchange bias. J. Magn. Magn. Mater. **192**, 203–232 (1999)

N. Nücker, J. Fink, J.C. Fuggle, P.J. Durham, W.M. Temmerman, Evidence for holes on oxygen sites in the high-Tc superconductors $La_{2-x}Sr_2CuO_4$ and $YBa_2Cu_3O_{7-y}$. Phys. Rev. B **37**, 5158–5163 (1988)

A. Ohtomo, D.A. Muller, J.L. Grazul, H.Y. Hwang, Artificial charge-modulation in atomic-scale perovskite titanate superlattices. Nature **419**, 378–380 (2002)

A. Ohtomo, H.Y. Hwang, A high-mobility electron gas at the $LaAlO_3/SrTiO_3$ heterointerface. Nature **427**, 423–426 (2004)

S. Okamoto, A.J. Millis, Electronic reconstruction at an interface between a Mott insulator and a band insulator. Nature **428**, 630–633 (2004)

R.M. Ormerod, Solid oxide fuel cells. Chem. Soc. Rev. **32**, 17–28 (2003)

M.P. Oxley, L.J. Allen, Delocalization of the effective interaction for inner-shell ionization in crystals. Phys. Rev. B **57**, 3273–3282 (1998)

M. Oxley, H. Chang, A. Borisevich, M. Varela, S. Pennycook, Imaging of Light Atoms in the Presence of Heavy Atomic Columns. Microsc. Microanal. **16**, 92 (2010)

J.H. Paterson, O.L. Krivanek, ELNES of 3d transition-metal oxides II. Variations with oxidation state and crystal structure. Ultramicroscopy **32**, 319–325 (1990)

S.J. Pennycook, M.F. Chisholm, A.R. Lupini, M. Varala, K. van Benthem, A.Y. Borisevich, M.P. Oxley, W. Luo, S.T. Pantelides, Materials application of aberration-corrected scanning transmission electron microscopy. Adv. Img. Electron. Phys. **153**, 327–384, (2008)

S.J. Pennycook, M.F. Chisholm, A.R. Lupini, M. Varela, A.Y. Borisevich, M.P. Oxley, W.D. Luo, K. van Benthem, S.-H. Oh, D.L. Sales, S.I. Molina, J. Garcia-Barriocanal, C. Leon, J. Santamaria, S.N. Rashkeev, S.T. Pantelides, Aberration-corrected scanning transmission electron microscopy: from atomic imaging and analysis to solving energy problem. Phil. Trans. R. Soc. A **367**, 3709–3733 (2009)

T.J. Pennycook, M.J. Beck, K. Varga, M. Varala, T.J. Pennycook, S.T. Pantelides, Origin of colossal ionic conductivity in oxide multilayers: interface induced sublattice disorder. Phys. Rev. Lett. **104**, 115901 (2010)

D. Phelan, D. Louca, S. Rosenkranz, S.-H. Lee, Y. Qiu, P.J. Chupas, R. Osborn, H. Zheng, J.F. Mitchell, J.R.D. Copley, J.L. Sarrao, Y. Moritomo, Nanomagnetic droplets and implications to orbital ordering in $La_{1-x}Sr_xCoO_3$. Phys. Rev. Lett. **96**, 027201 (2006)

A. Podlesnyak, S. Streule, J. Mesot, M. Medarde, E. Pomjakushina, K. Conder, A. Tanaka, M.W. Haverkort, D.I. Khomskii, Spin-state transition in $LaCoO_3$: direct neutron spectroscopic evidence of excited magnetic states. Phys. Rev. Lett. **97**, 247208 (2006)

A. Podlesnyak, M. Russino, A. Alfonsov, E. Vavilova, V. Kataev, B. Buchner, Th. Strassle, E. Pomjakushina, K. Conder, D.I. Khomskii, Spin-state polarons in lightly hole doped $LaCoO_3$. Phys. Rev. Lett. **101**, 247603 (2008)

P.M. Raccah, J.B. Goodenough, First-order localized-electron collective-electron transition in $LaCoO_3$. Phys. Rev. **155**, 932–943 (1967)

J.H. Rask, B.A. Miner, P.R. Buseck, Determination of manganese oxidation states in solids by electron energy-loss spectroscopy. Ultramicroscopy **21**, 321–326 (1987)

T. Riedl, T. Gemming, W. Gruner, J. Acker, K. Wetzig, Determination of manganese valency in $La_{1-x}Sr_xMnO_3$ using ELNES in the (S)TEM. Micron **38**, 224–230 (2007)

J. Rodriguez-Carvajal, M. Hennion, F. Moussa, A.H. Moudden, L. Pinsard, A. Revcolevschi, Neutron-diffraction study of the Jahn-Teller transition in stoichiometric $LaMnO_3$. Phys. Rev. B **57**, R3189–R3192 (1998)

H. Rose, Phase-contrast in scanning-transmission electron-microscopy. Optik **39**, 416–436 (1974)

Z. Sefrioui, D. Arias, M. Villegas, V. Peña, W. Saldarriaga, P. Prieto, C. Leon, J.L. Martinez, J. Santamaría, Ferromagnetic/superconducting proximity effect in LCMO/YBCO superlattices. Phys. Rev. B **67**, 214511 (2003)

M.A. Senaris-Rodriguez, J.B. Goodenough, $LaCoO_3$ revisited. J. Solid State Chem. **116**, 224–231 (1995)

S.B. Shah, Q.M. Ramasse, X. Zhai, J.G. Wen, S.J. May, I. Petrov, A. Bhattacharya, P. Abbamonte, J.N. Eckstein, J.-M. Zuo, Probing interfacial electronic structures in atomic layer $LaMnO_3$ and $SrTiO_3$ superlattices. Adv. Mater. **22**, 1156–1160 (2010)

N. Shibata, M.F. Chisholm, A. Nakamura, S.J. Pennycook, T. Yamamoto, Y. Ikuhara, Nonstoichiometric dislocation cores in α-alumina. Science **316**, 82–85 (2007)

T. Sparrow, B. Williams, C. Rao, J. Thomas, L_3/L_2 white-line intensity ratios in the electron energy-loss spectra of 3d transition-metal oxides. Chem. Phys. Lett. **108**, 547–550 (1984)

B.C.H. Steele, A. Heinzel, Materials for fuel-cell technologies. Nature **414**, 345–352 (2001)

M. Tinkham, *Introduction to Superconductivity* (McGraw-Hill, New York, NY, 1996)

M.A. Torija, M. Sharma, M.R. Fitzsimmons, M. Varela, C. Leighton, Epitaxial $La_{0.5}Sr_{0.5}CoO_3$ thin films: structure, magnetism, and transport. J. Appl. Phys. **104**, 023901 (2008)

S. Van Aert, J. Verbeeck, R. Erni, S. Bals, M. Luysberg, D. Van Dyck, G. Van Tendeloo, Quantitative atomic resolution mapping using high-angle annular dark field scanning transmission electron microscopy. Ultramicroscopy **109**, 1236–1244 (2009)

P. Van Aken, B. Leibscher, Quantification of ferrous/ferric ratios in minerals: new evaluation schemes of Fe L_{23} electron energy-loss near edge spectra. Phys. Chem. Minerals **29**, 188–200 (2002)

M. Varela, A.R. Lupini, H.M. Christen, N. Dellby, O.L. Krivanek, P.D. Nellist, S.J. Pennycook, Spectroscopic identification of single atoms within a bulk solid. Phys. Rev. Lett. **92**, 095502 (2004)

M. Varela, A.R. Lupini, K. van Benthem, A.Y. Borisevich, M.F. Chisholm, N. Shibata, E. Abe, S.J. Pennycook, Materials characterization in the aberration corrected scanning transmission electron microscope. Annu. Rev. Mater. Res. **35**, 539–569 (2005)

M. Varela, T. Pennycook, W. Tian, D. Mandrus, S. Pennycook, V. Peña, Z. Sefrioui, J. Santamaria, Atomic scale characterization of complex oxide interfaces. J. Mater. Sci. **41**, 4389–4393 (2006)

M. Varela, W. Luo, J. Tao, M.P. Oxley, M. Watanabe, A.R. Lupini, S.T. Pantelides, S.J. Pennycook, Atomic resolution imaging of oxidation states in manganites. Phys. Rev. B. **79**, 085117 (2009)

M. Varela, J. Gazquez, A.R. Lupini, J.T. Luck, M.A. Torija, M. Sharma, C. Leighton, M.D. Biegalski, H.M. Christen, M. Murfitt, N. Dellby, O. Krivanek, S.J. Pennycook, Applications of aberration corrected scanning transmission electron microscopy and electron energy loss spectroscopy to thin oxide films and interfaces. Int. J. Mater. Res. **101**, 21–26 (2010)

J. Verbeeck, S. Bals, A.N. Kratsova, D. Lamoen, M. Luysberg, M. Hiujben, G. Rijnders, A. Brinkman, H. Hilgenkamp, D.H.A. Blank, G. Van Tendeloo, Electronic reconstruction at n-type $SrTiO_3/LaAlO_3$ interfaces. Phys. Rev. B **81**, 085113 (2010)

P.M. Voyles, D.A. Muller, J.L. Grazul, P.H. Citrin, H.-J.L.Grossman, Atomic scale imaging of individual dopant atoms and clusters in highly n-type bulk Si. Nature **416**, 826–829 (2002)

W.G. Waddington, P. Rez, I.P. Grant, C.J. Humphreys, White lines in the $L_{2,3}$ electron-energy-loss and x-ray absorption spectra of 3d transition metals. Phys. Rev. B **34**, 1467–1473 (1986)

Z.L. Wang, J.S Yin, Co valence and crystal structure of $La_{0.5}Sr_{0.5}CoO_{2.25}$. Philos. Mag. B **77**, 49–65 (1998)

M.K. Wu, J.R. Ashburn, C.J. Torng, P.H. Hor, R.L. Meng, L. Gao, Z.J. Huang, Y.Q. Wang, C.W. Chu, Superconductivity at 93 K in a new mixed-phase Y-Ba-Cu-O compound system at ambient pressure. Phys. Rev. Lett. **58**, 908–910 (1987)

X. Zhai, C.S. Mohapatra, A.B. Shah, J.-M. Zuo, J.N. Eckstein, Adv. Mater. **22**, 1136–1139 (2010)

11

Application to Ceramic Interfaces

Yuichi Ikuhara and Naoya Shibata

11.1 Introduction

The mechanical and electronic properties of ceramics are strongly influenced by the atomic structure of grain boundaries and interfaces (Ikuhara 2001, Sakuma et al. 2000, Sutton and Ballufi 1995). On the other hand, grain boundary and interface structures themselves are sensitive to the grain boundary character. Therefore, it is important that we investigate the relationship between grain boundary structure and its character so that we can understand how grain boundary structure affects the intrinsic properties of ceramics. In this chapter, the importance of grain boundary character is briefly described, and typical examples are introduced for a small angle grain boundary (Read 1953) and a coincidence site lattice (CSL) grain boundary (Kronberg and Wilson 1949). Like the grain boundaries of metals, the grain boundaries of ceramics can be described as either dislocation boundaries or CSL boundaries. However, the atomic structures in ceramics are pretty complicated, compared with simple metals.

Sutton (Sutton and Vitek 1983) proposed the idea of structural units to describe grain boundary atomic structure. The concept is that a grain boundary generally consists of some structural units, and it has been successfully applied to grain boundaries in various ceramics. It therefore will be briefly covered in this chapter. Doping impurities into ceramics is a useful way to control grain boundary properties. The impurities often segregate along the grain boundary, changing the intrinsic properties. A typical example of this is shown for a small amount of impurity-doped alumina which has high creep resistance (Ikuhara et al. 2001). Covalent-bonded ceramics such as Si_3N_4 and SiC are known as hard sintering materials, and therefore sintering additives are usually used for the sintering. In this case, an amorphous film with a thickness of about 1 nm is frequently formed along the grain boundaries. The chemical composition and bonding state in the film are considered to determine the high-temperature mechanical properties of such ceramics (Clarke 1987, Ikuhara et al. 1987, Shibata et al. 2004). Amorphous grain boundaries are also briefly discussed in this chapter.

S.J. Pennycook, P.D. Nellist (eds.), *Scanning Transmission Electron Microscopy*,
DOI 10.1007/978-1-4419-7200-2_11, © Springer Science+Business Media, LLC 2011

Hetero-interfaces are always formed in ceramic composites and thin films. The hetero-interface structure is of wide interest not only from a fundamental point of view but also because of its importance in many modern materials which are composites consisting of two or more phases. The significance of the interface lies in the fact that many of the properties of structural or electronic composites depend sensitively on it. The adhesion between two materials is basically made by atomic interactions across the interface, and its strength is ultimately determined by the strength of interfacial bonds between the atoms of two constituent materials (Sutton and Ballufi 1995, Ikuhara and Pirouz 1998, Kohyama 1999, 2002). Hence, the adhesion between two materials is inherently related to the structure of, and defects in, the interface between them. The strength and fracture properties of a structural composite are in turn related to adhesion. Besides its practical importance, the structure of a hetero-interface is important from a fundamental viewpoint because of the desire to understand how nature accommodates the mismatches across the contact plane of two translationally periodic structures. Interfacial dislocations play an important role here and are described in detail for various hetero-interfaces in this chapter.

So far, transmission electron microscopy (TEM), in particular, high-resolution electron microscopy (HRTEM) has been a useful technique for studying the atomic structure of grain boundaries and was used to investigate various kinds of grain boundaries in many kinds of ceramics. However, among recent nano-characterization technologies, there has been remarkable progress by scanning transmission electron microscopy (STEM) utilizing the spherical aberration (Cs) corrector (Haider et al. 1998). The technique enables us not only to identify the atomic structures but also the location of dopants segregated at grain boundaries and interfaces. A STEM image is formed by the scattered electrons in each probe position collected by the annular dark field (ADF) detector at the bottom of the sample, displayed on the monitor in synchronism with the scanning probe (Nellist et al. 2004, Pennycook and Jesson 1990, 1991). An atomic-resolution image can be obtained by focusing the electron probe down to below the atomic column interval. The advantages of this method are as follows: there is no inversion of the image contrast with defocusing and change in the sample thickness, and thus the positions of the atomic columns can be determined directly from the image. These excellent characteristics are very useful to determine the complicated atomic structures in the grain boundaries and interfaces in ceramic materials.

11.2 Grain Boundary Structure

11.2.1 Grain Boundary Character

Grain boundary character can be described in terms of the relative orientation relationship between two crystals and the orientation of the boundary plane. Geometrically speaking, there are nine degrees of freedom that we must consider to exactly describe the grain boundary character (Sutton and Ballufi 1995). Consider a grain boundary with the plane normal to the vector *P*, in which one crystal is rotated around the rotation axis *n* with respect to the other crystal. In this case, there are totally five macroscopic parameters because two degrees of freedom are given for selecting the rotation axis *n*, one degree of freedom is given for rotation angle θ, and the additional two degrees of freedom are given for selecting *P*. The remaining four are microscopic parameters and are introduced from atomic structure relaxation at the grain boundary. That is to say, three parameters for a rigid body translation of one crystal referred to the other and one parameter for the sequential periodicity at the boundary plane, which can be observed by HRTEM and STEM.

A grain boundary can be classified into a small angle boundary or a large angle boundary depending on the degree of rotation angle. Although it varies a little by material, the angle of a small angle grain boundary is generally limited to 10–15° which is close to the point where dislocation cores begin to overlap (Bandon 1966). On the other hand, among the angles of large angle grain boundaries, there are some specific angles at which two adjacent grains are well matched geometrically. A grain boundary having such an angle is called a coincidence site lattice (CSL) grain boundary (Ranganathan 1966, Sutton and Ballufi 1995), and generally the energy is low and its structure is considered to be stable. A CSL grain boundary is expected to be mechanically strong, and therefore has been used to design grain boundary controlled materials.

We often use the terms tilt boundary and twist boundary to describe grain boundary character. A tilt boundary has the plane parallel to the rotation axis *n*, while a twist boundary has the plane perpendicular to *n*. A grain boundary that falls in between these two is called a mixed grain boundary, which comprises both tilt and twist components. Whether a grain boundary becomes a tilt or twist grain boundary depends on the location of the grain boundary plane even if the orientation of the two crystals is exactly the same.

11.2.2 Low-Angle Grain Boundary

A low-angle grain boundary compensates the misorientation angle by introducing a periodic array of dislocations. Figure 11–1 shows a schematic view of a simple low-angle grain boundary (Ikuhara, 2009). If the edge dislocation with Burgers vector b is periodically arranged at intervals of h along the low-angle grain boundary with a misorientation tilt angle 2θ, the following relationship is obtained between θ, h, and b (Frank 1951):

$$2\theta = \tan^{-1} b/h = b/h. \tag{1}$$

Figure 11–2 shows a typical HRTEM image obtained from a GB dislocation in a $[1\bar{1}00]$ 2° tilt grain boundary in $\alpha\text{-}Al_2O_3$. In this case, a pair of lattice discontinuities were observed in the HRTEM image. It was found from the Burgers circuits in the figure that the total edge component of the lattice discontinuities is $1/3[11\bar{2}0]$, which corresponds to the Burgers vector of a perfect basal dislocation that is expected to be

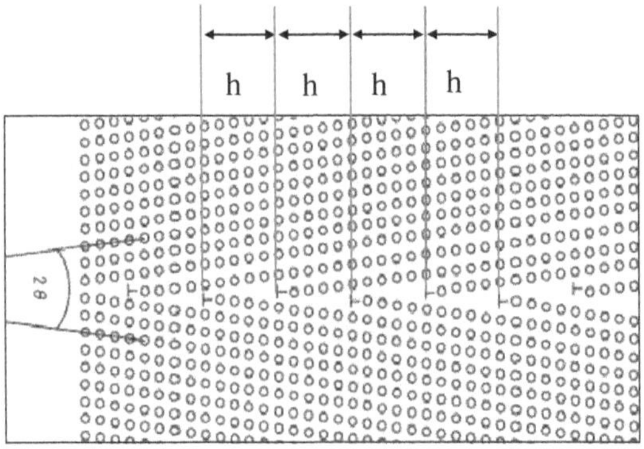

Figure 11–1. Schematic of a simple low-angle grain boundary (misorientation angle θ, dislocation spacing h). Reproduced from Ikuhara (2009) with permission.

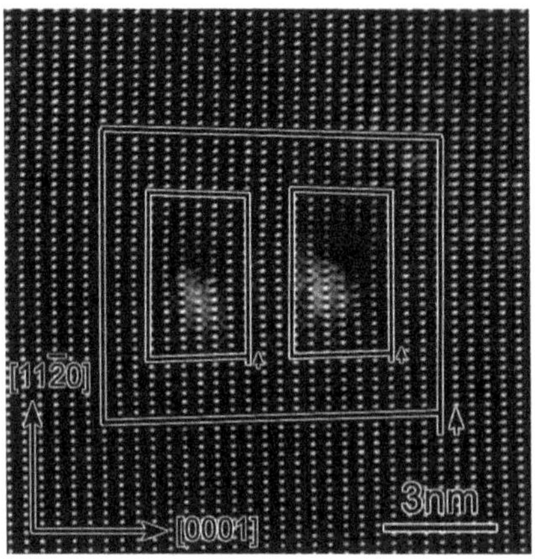

Figure 11–2. A typical HRTEM image of partial dislocations in a [1$\bar{1}$00] 2° low-angle grain boundary in α-Al_2O_3. It is seen from the Burgers circuit that the total size of the lattice discontinuity is 1/3[11$\bar{2}$0], which is the perfect translation vector of the corundum structure and that the size of each lattice discontinuity is 1/6[11$\bar{2}$0]. Reproduced from Nakamura et al. (2002) with permission.

formed for compensating the misorientation angle between two adjacent crystals. It is known that a basal dislocation in α-Al_2O_3 dissociates into two partial dislocations (Mitchell et al. 1976, Nakamura et al. 2002) according to Eq. (2):

$$1/3 <11\bar{2}0> \rightarrow 1/3 <10\bar{1}0> + 1/3 <01\bar{1}0>. \quad (2)$$

The size of the edge component of each lattice discontinuity in Figure 11–2 is 1/6[11$\bar{2}$0] and corresponds to the projection of the two partial dislocations with $\boldsymbol{b} = 1/3[10\bar{1}0]$ or $\boldsymbol{b} = 1/3[01\bar{1}0]$. It is thus considered that the lattice discontinuities form a pair due to the dissociation of the perfect basal dislocation. It is considered that this dislocation structure is exactly the same as the basal dislocation structure in the crystal lattice. We recently applied STEM techniques to quantitatively determine the dislocation core structures for the present specimen.

11.2.3 Dislocation Core Structures of α-Al_2O_3

The core structures of dislocations are critical to the electronic, optical, and mechanical properties of a wide range of materials. In complex crystals such as oxides, either cation or anion columns (or both) can be the terminating atomic columns even with the same dislocation character, i.e., characteristic displacement vectors called Burgers vectors, ***b***. Thus, the detailed knowledge of dislocation core structures and compositions is of critical importance to understand the dislocations in ionic crystals.

The inherent structural complication of alpha alumina (α-Al_2O_3) has led to conflicting models for dislocation glide (Bilde-Sørensen et al. 1996, Kronberg 1957). Slip on the (0001) basal plane is reported to be the dominant deformation system at elevated temperatures (Lagerlöf et al. 1994), and thus important for understanding the high-temperature mechanical behavior. Kronberg first proposed a basal dislocation slip model based on structurally related hexagonal metals (Kronberg 1957). Slip was assumed to occur between Al and O basal plane layers. In order to maintain the normal octahedral coordination of the oxygen to aluminum sites, Kronberg proposed the synchroshear mechanism, where two shears cooperatively operate in different directions on adjacent atomic planes. This mechanism has been shown to operate in the Laves phase compound $HfCr_2$ (Chisholm et al. 2005). Later, Bilde-Sørensen et al. (1996) argued that the slip between Al and O planes would require charge transport. Alternatively, they proposed that dislocation slip would occur along the midplane on the puckered Al (cation) layer. They argued that this choice of the slip plane allows the moving dislocations to carry no net charge. However, there has been no direct observation of the dislocation core structures in α-Al_2O_3. Here, aberration-corrected STEM is used to directly observe dislocation core structures in α-Al_2O_3.

Figure 11–3(a) illustrates a ball and stick model of α-Al_2O_3 in the $<1\bar{1}00>$ projection. From this viewing direction, Al and O atom sites are arranged as distinct columns. The stacking sequence of α-Al_2O_3 along

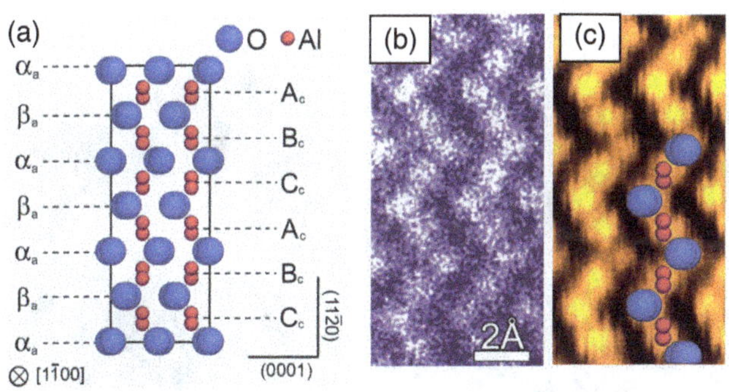

Figure 11–3. (a) Unit cell of α-Al_2O_3 viewed from the $<1\bar{1}00>$ direction. The structure of α-Al_2O_3 consists of alternating Al and O planes along the <0001> direction, whose stacking sequence is $\cdots\alpha_a A_c \beta_a B_c \alpha_a C_c \beta_a A_c \alpha_a B_c \beta_a C_c \alpha_a \cdots$. From the $<1\bar{1}00>$ projection, we can distinguish individual Al and O columns. (b) and (c) are simultaneously obtained HAADF and bright field STEM images of α-Al_2O_3 viewed from the $<1\bar{1}00>$ direction. Comparing the HAADF and bright field STEM images, the bright spots in the bright field image are found to directly correspond to the position of atomic columns in the present experimental conditions. Reproduced from Shibata et al. (2005) with permission.

the <0001> direction consists of 12 alternating cation and anion basal layers. The cation layers are slightly puckered along the <0001> direction. Bilde-Sørensen et al. (1996) proposed basal slip occurred between these shifted Al sites.

Figures 11–3(b) and 11–1(c) show simultaneously obtained HAADF and bright field STEM images of α-Al_2O_3 viewed along the <1$\bar{1}$00> direction (Shibata et al. 2007). The Z-contrast image obtained using the HAADF detector is an incoherent image (Nellist et al. 2004, Pennycook and Jesson 1990, 1991); it is essentially a map of the scattering power of the specimen. There is a direct correspondence between the features in the specimen and their image. On the other hand, the phase-contrast image obtained using a small bright field detector has coherent image characteristics, which is comparable to the parallel beam high-resolution TEM. The contrast is influenced by focus of the objective lens and specimen thickness, orientation, and scattering power. This makes phase-contrast images of unknown structures difficult to directly interpret. However, this sensitivity can be exploited to provide much greater contrast variations than can be obtained from the Z-contrast images of low atomic number elements (such as Al and O). The simultaneously recorded Z- and phase-contrast images (Figure 11–3(b) and (c)) of the $< 1\bar{1}00 >$ projection of α-Al_2O_3 reveal parallel kinked lines of spots that reflect the arrangement of alternating O and Al columns along the <0001> direction. Under the conditions used to obtain these two images, the bright features of the Z-contrast image are seen to correspond to the bright features in the bright field image. These bright features identify the positions of the two oxygen columns and the puckered Al column in each segment of the kinked lines seen in this projection.

Figure 11–4(a) shows a typical bright field STEM image of a basal dislocation core in α-Al_2O_3 (Shibata et al. 2007). The line direction of the dislocation is parallel to the observing direction, so that the core

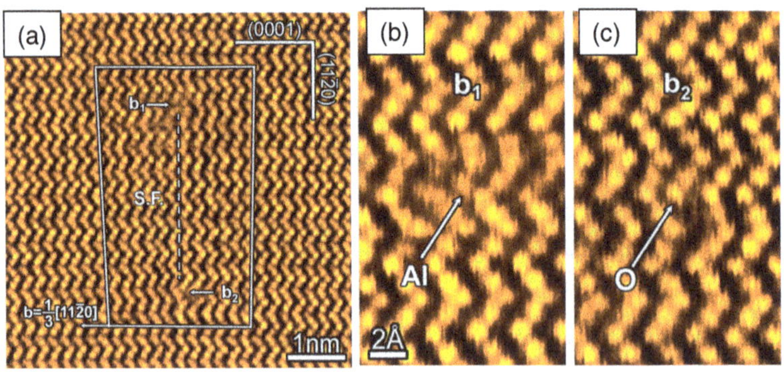

Figure 11–4. (a) Typical bright field STEM image of a basal dislocation core observed from the <1$\bar{1}$00> direction. (b) and (c) are bright field STEM images of the upper and lower partial dislocation cores, respectively. It is clearly resolved that the upper partial core is terminated by an Al column, while the lower partial core is terminated by an O column. High-resolution electron micrographs of (a) 10° small angle tilt grain boundary in Al_2O_3 and (b) Burgers circuit around the dislocation in (a). Reproduced from Shibata et al. (2005) with permission.

structure is set at an "edge-on" condition. In this condition, bright spots in the image correspond to the position of atomic columns as determined from the simultaneous HAADF-STEM image. It is clearly seen that at room temperature this dislocation is not a single perfect dislocation ($\mathbf{b} = 1/3 <11\bar{2}0>$), but has an extended structure with partials ($\mathbf{b} = 1/3 <10\bar{1}0>$ and $\mathbf{b} = 1/3 <01\bar{1}0>$) connected by a $\{11\bar{2}0\}$ stacking fault, consistent with the previous reports of dislocation observations after basal slip deformation (Lagerlöf et al. 1984, Mitchell et al. 1976, Nakamura et al. 2002). The dislocation is considered to be driven by the reduction of strain energy and suppressed by increasing stacking-fault energy.

The bright field STEM images of each dissociated partial core are shown in Figure 11–4(b) and (c). The arrows in the images indicate the respective position of extra half plane termination of the two partials. The images reveal that the upper core is terminated between vertices of the kinked line of atoms at an Al column position, while the lower core is terminated at a vertex of the kinked line of atoms at an O column position. These observations clearly show both dislocation core terminations and, thus, indicate that the slip planes of the dislocations are located between the Al and O atomic planes. Moreover, the partials terminated by Al or O column indicate that the partial cores are locally not stoichiometric. Contrary to the common assumption, non-stoichiometric core structures actually exist in an ionic crystal. While each partial dislocation core is non-stoichiometric, the total dislocation preserves Al_2O_3 stoichiometry. Based on these observations, a new atomic-scale basal slip mechanism of α-Al_2O_3 has been proposed (Shibata et al. 2007). The possibility for atomic-scale characterization of dislocation core structures will significantly assist our understanding of dislocation activity and related properties in complex materials.

11.2.4 Coincidence Site Lattice (CSL) Theory and Structural Units

As mentioned above, in the case of a single-phase polycrystal, a concept that has often been used to predict the orientation relationship (OR) between two adjacent grains is the "coincidence site lattice" (CSL) (Kronberg and Wilson 1949). The concept of CSL is based on the fact that certain rotations about an axis bring a lattice into partial self-coincidence. The common lattice sites (c.l.s.) then form a larger lattice known as a CSL. The CSL can also be considered a lattice of coincidence sites in the composite lattice obtained by interpenetration of two single lattices. Figure 11–5 shows interpenetration of two simple cubic lattices by means of rotation about the <001> axis (Bollmann 1970). The points of overlap correspond to CSL points indicated by white circles. Like any other lattice, a CSL can be defined in terms of a unit cell, the volume of which is proportional to the density of lattice sites; the higher the c.l.s. density is, the smaller the volume of the CSL unit cell becomes. Conventionally, the volume of the CSL unit cell is normalized with respect to the volume of the lattice unit cell and the result, which is necessarily an integer, n, is denoted by $\Sigma = n$. Each Σ defines a particular orientation relationship between the two lattices; thus, for two

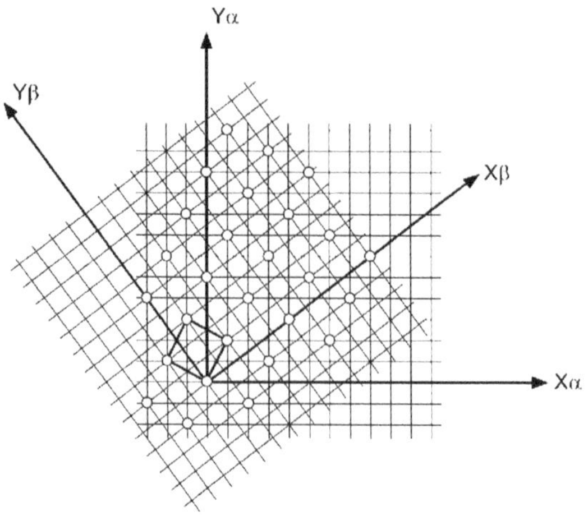

Figure 11–5. CSL plot of a $\Sigma 5$ grain boundary in the simple cubic lattice, in which one lattice is rotated around the <001> axis by 36.52° with respect to the other lattice. Reproduced from Bollmann et al. (1970) with permission.

lattices in parallel orientation, $\Sigma = 1$ and the c.l.s. density is a maximum. The angle θ, which forms the CSL by rotating around the $[hkl]$ axis, can be obtained from the following equation:

$$\theta = 2\tan^{-1}(Ry/x), \tag{3}$$

where $R^2 = h^2 + k^2 + l^2$, and x and y are integers. In this case, Σ can be expressed as $\Sigma = x^2 + Ry^2$. In the case of a cubic unit cell, the particular orientations at which low values of Σ occur can be deduced from simple geometrical considerations (Ranganathan 1966); thus $\Sigma = 3, 9, 11, 17$ when one lattice is rotated around the <110> axis of the other lattice through angles 70.53°, 38.94°, 50.48°, and 86.63°, respectively.

A particular grain boundary can be obtained by passing a plane (hkl) through the interpenetrating composite. The grain boundary is then termed Σ (hkl). Relaxation of atoms at the two sides of the boundary, or slight rigid-body translations of one grain with respect to the other (which destroys the CSL), can lower the energy of the system. Numerous calculations have shown that the higher the density of the coincident sites, i.e., the lower the Σ and the more closely packed the plane (hkl) is, the lower the energy of the grain boundary becomes.

The CSL model is just deduced from the geometry of two adjacent crystals. It is, however, a periodic atomic configuration that must be directly related to the grain boundary energy. As the model for treating this, we have the structural unit model proposed by Sutton (Sutton and Vitek 1983), which is widely accepted in this field at present (Sutton and Ballufi 1995). This idea is that a grain boundary can be constructed by the combination of some kinds of structural units. In the case of a CSL boundary, the boundary consists of a couple of stable structural units, and the strain of each structural unit is small, so the grain boundary energy is low. On the other hand, a large angle grain boundary, which is composed of distorted structural units, shows high grain boundary energy. Recent HREM studies and first-principles grain boundary calculations have confirmed that the structural unit model is appropriate. In the next section, it will be shown that CSL grain boundaries in ceramics also consist of structural units.

11.2.5 CSL Grain Boundary in ZrO_2

In this section, an example of the presence of structural units in a $\Sigma9$ [110] symmetric tilt grain boundary of zirconia ceramics is demonstrated. In order to predict the grain boundary atomistic structures theoretically, systematic lattice statics calculations performed using the GULP program code have an advantage (Gale 1997). It has been well demonstrated that lattice statics calculations are an effective method for predicting the stable grain boundary structures in many kinds of ceramic materials. In the calculation, the atomistic interactions are described by a potential function which divides the interatomic forces into long-range interactions (described by Coulomb's Law and summed by the Ewald method) and short-range interactions treated by a pairwise function using the Buckingham potential. The potential parameters used in this study were taken from the literature report by Lewis and Catlow (1985). The lattice energies were calculated by summing all the potentials of constituent ions in the calculation cells. The grain boundary excess energies were estimated by subtracting the calculated lattice energy for the single crystal cell with the same number of ions from the calculated lattice energy for the cell including the grain boundary. The calculated energies were then evaluated as a function of the translation states, and the atomic configurations with local energy minima were subsequently selected as the equilibrium structures.

Figure 11–6 shows the thus obtained lowest energy grain boundary structure, indicating that the structural units are periodically located

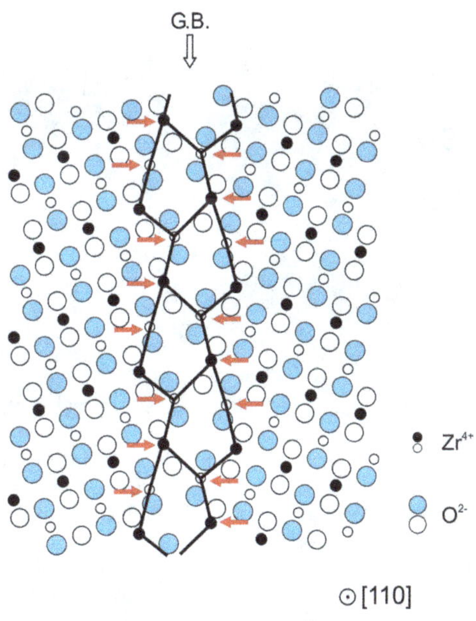

Figure 11–6. The lowest energy structure of the $\Sigma = 9$, {221} grain boundary obtained by lattice statics calculations. Note that a high density of cation sites with sevenfold coordination is formed in the structural unit along the boundary, as indicated by the small solid arrows. Reproduced from Shibata et al. (2004) with permission.

along the boundary (Shibata et al. 2004). The grain boundary excess energy was calculated to be 3.01 J/m^2, and the lowest energy grain boundary structure had a rather large boundary expansion of 1.00 Å, resulting in the formation of slightly large free volumes at the core of the boundary. The open spaced structures at the boundary core in a real material may give rise to channeling contrast on the HRTEM image where there are no atomic columns. Such contrast would make the interpretation of the experimental HRTEM image complicated and would require extensive image simulation of various kinds of possible structure models to determine the real atomistic core structures. To avoid such ambiguity, atomic-resolution high-voltage electron microscopy (ARHVEM) was applied to directly determine the atomic columns of the cation sublattice at the present grain boundary. Cross-sectional ARHVEM observations were then carried out to directly image the atomic column structure of the boundary. In this case, the thickness of the sample was controlled to be as thin as about 4 nm, and the image was taken near the Scherzer defocus of about –38 nm, so that the atomic columns can be imaged as black dots reflecting their potentials.

Figure 11–7(a) shows the ARHVEM image of the $\Sigma = 9, \{221\}$ grain boundary (Shibata et al. 2004). As can be seen in the image, the black dots were imaged slightly elongated in the [001] directions. This is because the anion sites are located very close to the cation sites in this direction. Since the open spaced cation sublattice structure can be directly observed in the micrograph, this boundary is confirmed to have a periodic array of asymmetrical structural units along the grain boundary, as indicated by the solid lines in Figure 11–7(b). Figure 11–7(c) shows the calculated ARHVEM image based on the predicted model as shown in Figure 11–6. The simulated image approximately agrees with the experimental image, as far as the periodic array of asymmetrical structural units. The cation sites with different coordination

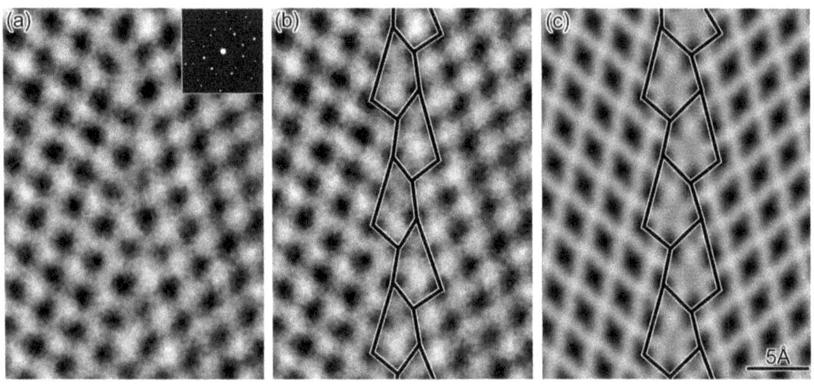

Figure 11–7. Experimental ARHVEM image of the $\Sigma = 9, \{221\}$ grain boundary. In this condition, the *black dots* in the image correspond to the position of the cation column at the boundary. (b) Asymmetric structural units drawn as *solid lines* on the experimental image in (a). (c) The simulated ARHVEM image using the structure model is shown in 11–6. The image simulation was performed for a defocus value of –38 nm and a film thickness of 4 nm. Reproduced from Shibata et al. (2005) with permission.

states are accumulated along the calculated $\Sigma = 9$ grain boundary structure, indicated by the solid arrows in Figure 11–6. These sites have sevenfold coordination of oxygen ions and almost maintain the cubic polyhedra. The density of the coordination deficient sites is considered to be related to the grain boundary energy in zirconia ceramics.

11.2.6 CSL Grain Boundary in $SrTiO_3$

Perovskite oxides, such as $SrTiO_3$ and $BaTiO_3$, exhibit a large variety of electrical properties, and thus they have been utilized for electrical devices. Since some properties of the perovskite oxides arise only in polycrystalline materials, it is considered that grain boundaries play an important role for their properties. Thus, understanding the peculiar atomic structures and defect energetics at the interface is indispensable for further developments and applications of perovskite oxides. In this section, $SrTiO_3$ is selected as a model sample and the [001](310)$\Sigma5$ and [001](210)$\Sigma5$ grain boundaries are investigated by high-resolution HAADF-STEM to find the relationships among the atomic arrangements, electronic structures, and defect energetics. For investigating the relationships quantitatively, the first-principles projector augmented wave (PAW) method as implemented in Vienna *Ab Initio* Simulation Package (VASP) code was used (Kresse and Furthmüller 1996). In this case, a three-dimensional rigid-body translation of one grain with respect to the other was taken into account to obtain the stable grain boundary structure. After obtaining the stable structure, the vacancy formation energy at possible sites in the vicinity of the grain boundary was systematically calculated.

Figure 11–8 shows (a) HAADF-STEM image, (b) most stable atomic arrangements, (c) strains, and (d) defect energetics of the $SrTiO_3$ [001](310)$\Sigma5$ grain boundary (Imaeda et al. 2008). In the calculation, the rigid-body translations of one grain with respect to the other were fully considered. It is seen that the calculated most stable structure well reproduces the experimental image (Figure 11–8(a), (b)). By analyzing the calculated structure, it was found that the structural distortions, strains, and dangling bonds are present mainly at the grain boundary core (Figure 11–8(c)). On the other hand, although the vacancy formation energy depends on the atomic site, the defect energetics at the grain boundary was found to be similar to that in the bulk. It was also found that the Ti vacancy is more sensitive to structural distortions than Sr and O vacancies (Figure 11–8(d)). This would be caused by the difference in the bonding character of Ti–O and Sr–O. Figure 11–9 shows (a) the most stable calculated structure and (b) the HAADF-STEM image of [001](210)$\Sigma5$ grain boundary. The calculated atomic positions of Sr and Ti–O columns of the [001](210)$\Sigma5$ grain boundary are shown as white circles in the figure, respectively. As can be seen, the shape of the structural unit formed on this GB is different from that in [001](310)$\Sigma5$ grain boundary, indicating that the shape of the structural unit is influenced by the grain boundary plane although the relative orientation is the same. Through this study, the atomic structures of the $\Sigma5$ grain

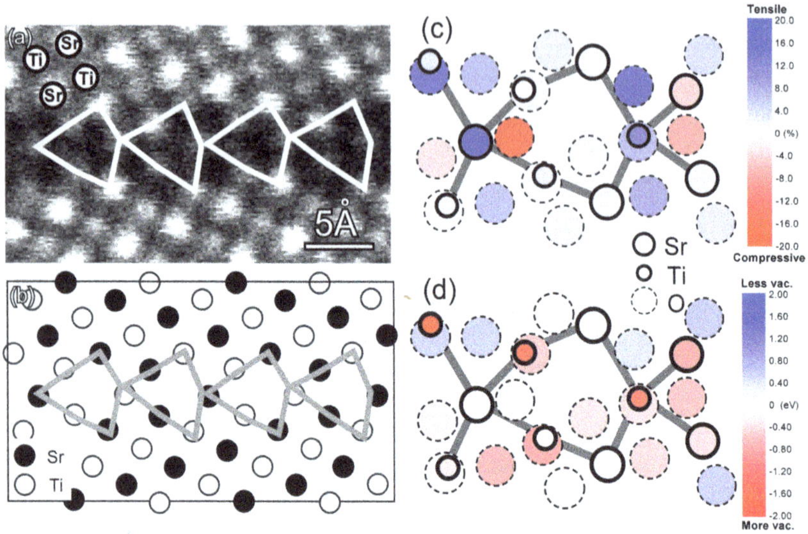

Figure 11–8. (a) HAADF-STEM image, (b) calculated most stable structure, (c) strains, and (d) defect energetics of $SrTiO_3$ [001](310)Σ5 grain boundary. Only Sr and Ti columns are shown in (b) to compare the theoretically obtained structure with the experimental results. Reproduced from Imaeda et al. (2008) with permission.

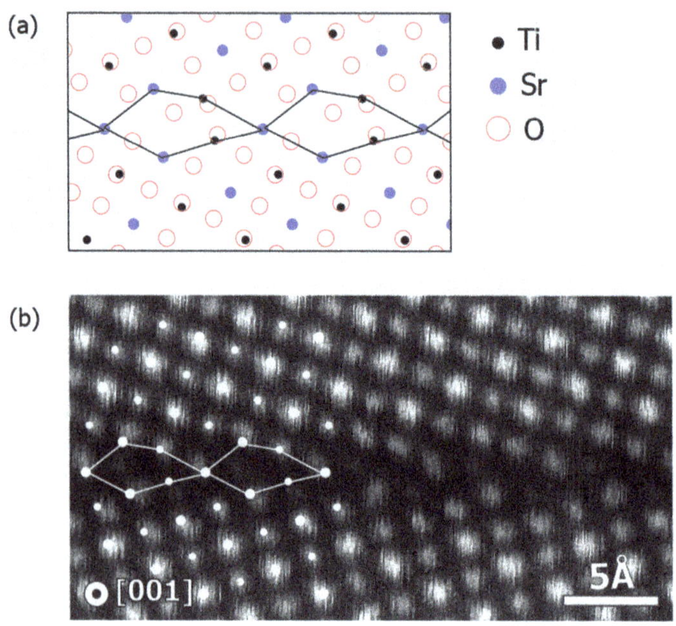

Figure 11–9. (a) Most stable atomic structures obtained by theoretical calculation and (b) the HAADF-STEM image of [001](210)Σ5 grain boundary.

boundaries of $SrTiO_3$ can be fully determined, and the characteristic electronic structures and defect energetics of those grain boundaries can be identified (Imaeda et al. 2008).

11.3 Grain Boundary Segregation

11.3.1 McLean's Concept

The explanation above is for grain boundaries at which two pure crystals are directly joined without any impurities or secondary phase. However, in the case of polycrystals, impurities often segregate at grain boundaries, which affect the properties of materials. For example, phosphorus is well known to segregate in iron along the grain boundaries, where it makes the chemical bonding state at the boundaries brittle (Losch 1979). Generally, the amount of segregation is small in low energy boundaries and large in high energy boundaries. McLean considered grain boundary segregation on the basis of thermodynamics and proposed the following equation (McLean 1957):

$$\frac{X_b}{(X_b^0 - X_b)} = \frac{X_c}{(1 - X_c)} \exp\left(\frac{Q}{RT}\right), \qquad (4)$$

where X_c, X_b, and $X_b{}^0$ are the solute concentration in a grain, in a grain boundary, and the saturated solute concentration in a grain boundary. Q represents the difference in strain energy that arises when solute atoms exist in a grain and a grain boundary. Grain boundary segregation is closely related to the properties of ceramics (Ikuhara et al. 1997), and a typical example is rare-earth cation doped alumina ceramics (Yoshida et al. 1998, 1999). In the next section, it will be demonstrated that the dopants segregated at a grain boundary can be directly observed by the HAADF-STEM technique.

11.3.2 Grain Boundary of Y-Doped Alumina Ceramics

α-Alumina (Al_2O_3) is one of the most important structural ceramics for high-temperature applications, and in particular, its creep behavior has been extensively studied so far. It has been known that a small amount of lanthanide ions are effective in improving the creep behavior in Al_2O_3 ceramics (Cho et al. 1997, Yoshida et al. 1998, 1999, 2002). Lanthanide ions have larger ionic radii than Al ions and tend to segregate at the grain boundaries, which are considered to retard the grain boundary sliding during creep. A number of mechanisms for the lanthanide-dopant effect have been proposed so far. It is thought that the dopants improving Al_2O_3 creep behavior affect the structure and chemistry of grain boundaries in Al_2O_3, but the mechanism of the dopant effect in Al_2O_3 has not been clarified yet in detail. In order to clarify the dopant mechanism, grain boundary atomic structures including the dopant should be directly observed for well-defined specimens. For this purpose, bicrystals are advantages to directly study the dopant effects. This is because grain boundary characters in bicrystals can be controlled (Gemming et al. 2003, Ikuhara et al. 1999), and their atomic structures can be systematically analyzed in combination with STEM.

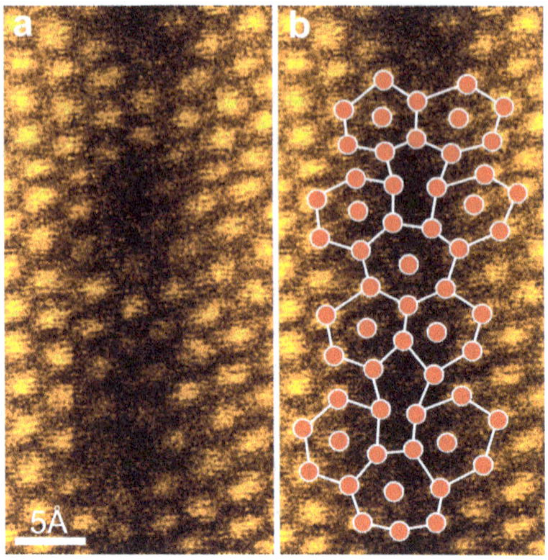

Figure 11–10. (a) HAADF-STEM image of a pristine Σ31 [0001] tilt grain boundary in alumina. (b) Same image with overlay to illustrate the aluminum atomic column arrangement in the structural units. Note the large open structure of the 7-membered ring unit, which is nearly periodic along the grain boundary. Reproduced from Buban et al. (2006) with permission.

In this section, the atomistic mechanism of Y-doping is described, using the HAADF-STEM technique to determine the GB structure on a bicrystal pair (Ikuhara et al. 1999, Matsunaga et al. 2003). The HAADF-STEM technique is especially well suited for understanding the role of heavy impurities such as Y in a bicrystal composed of much lighter ions, here Al and O. Figure 11–10(a) shows a HAADF-STEM image of an undoped ∑31 grain boundary in Al_2O_3 (Buban et al. 2006). Bright spots in the image correspond to atomic columns of Al (columns of oxygen do not scatter strongly enough to be seen in the image). The schematic overlay (Figure 11–10(b)) clearly illustrates the presence of periodic structural units along the boundary plane. A notable feature of the grain boundary structure is the presence of a 7-membered ring of Al ions leading to a large open structure. For the Y-doped Σ31 grain boundary, the Z-contrast image is shown in Figure 11–11(a). The

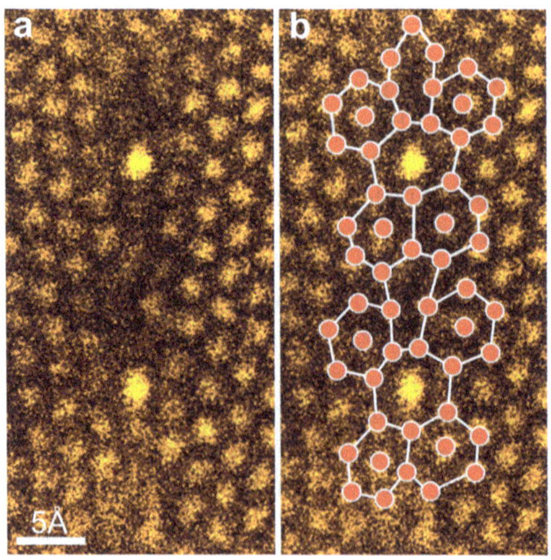

Figure 11–11. (a) HAADF-STEM image of Y-doped Σ31 [0001] tilt grain boundary in alumina. (b) Same image with overlay to illustrate the atomic column arrangement. The two brightest columns indicate the presence of the heavy Y ions. These Y-containing columns are found right at the center of the 7-membered ring unit. Reproduced from Buban et al. (2006) with permission.

most striking feature is the unusually bright columns that lie periodically along the boundary plane, indicating the presence of Y. Using nano-probe energy-dispersive spectroscopy (EDS) in the STEM, Y was confirmed to be confined to the boundary plane, consistent with the direct observation. Figure 11–11(b) shows a structural schematic of the Y-doped Al_2O_3 grain boundary superimposed on the image. Here, the structural units observed at the Y-doped boundary closely resemble the units found in the undoped case, suggesting that Y does not alter the basic grain boundary structure on length scales of more than ~0.1 nm. Instead it appears that Y^{3+} simply replaces Al^{3+} on the specific site of the cation sublattice. The Y-containing columns are found at the center of the 7-membered ring, periodically along the grain boundary. Y was only rarely detected at other sites, suggesting that Y preferentially segregates to cation sites in the center of the 7-membered ring.

In order to understand the chemical and bonding environment at the grain boundary, ab initio calculations can provide accurate information on the local atomic bonding and charge distributions. To this end, high precision ab initio calculations were again performed using the VASP code (Kresse and Furthmüller 1996). A large periodic supercell with 700 atoms containing two oppositely oriented grain boundaries was constructed using the structure obtained from static calculations. First, the supercell was constructed based on the experimentally obtained structure and fully relaxed to obtain the most accurate grain boundary structure for the undoped case. The results showed that further relaxations occurred. However, these relaxations were relatively small (less than 0.1 nm), yielding a grain boundary structure that still matched the structure observed in the STEM image. Next, assuming that all four distinct Al sites were substituted with Y ions in the column at the center of the 7-membered ring – the location that was observed in the STEM image – the structure was fully relaxed, and again, the final structure matched well with the corresponding experimental image.

Changes in the bonding character between the Y–O bonds and the Al–O bonds can be best illustrated by plotting the charge density maps. Figure 11–12(a) and (b) shows charge density maps along the (0001) plane for the undoped case and Y-doped case, respectively. Due to the complexity of the grain boundary structure in Al_2O_3, cations and anions do not lie on the same (0001) plane. Therefore, we have carefully selected appropriate (0001) planes, which are close to a cation belonging to the center column of the 7-membered ring such that the charge densities of the neighboring oxygen ions can also be clearly seen. To facilitate visualization, the locations of the cation columns are indicated schematically. The charge density map for the undoped grain boundary (Figure 11–12(a)) shows the presence of sharp nodes between the oxygen charge densities and the charge density from the Al ion in the center of the 7-membered ring. In contrast, the Y-doped grain boundary charge density map (Figure 11–12(b)) shows that the oxygen electron densities are elongated toward the Y ion, indicating a stronger covalency for the Y–O bonds. It can be seen that Y at the center column interacts considerably with the surrounding oxygen ions. This should result in a much stronger grain boundary, which explains why Y-doped grain

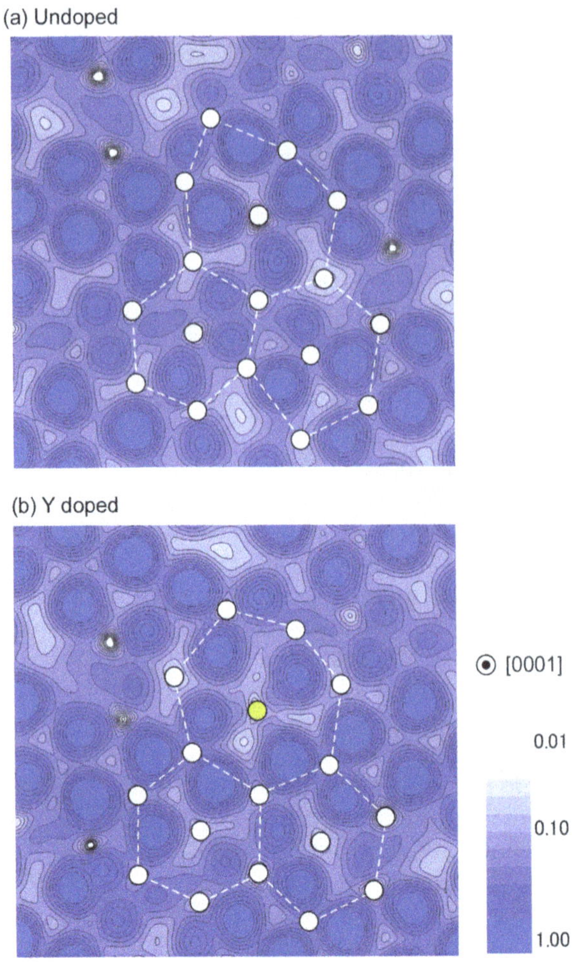

Figure 11–12. A charge density map for the undoped grain boundary (a) is shown for a (0001) plane near an Al ion in the middle of the 7-membered ring, where the charge density from the neighboring O ions (appearing as graduated *blue spots*) can be easily seen. A charge density map for the Y-doped grain boundary (b) is displayed using a similar (0001) plane. Here, one can see the elongation of the O charge density toward the Y ion in the center of the 7-membered ring indicating covalent type bonding. Note: *White circles* indicate the location of Al ions while the *yellow circle* indicates the location of the Y column. Reproduced from Buban et al. (2006) with permission.

boundaries can have such a large increase in creep resistance despite the fact that only a small amount of Y is present.

11.3.3 Grain Boundaries in Pr-Doped ZnO Varistors

It is known that zinc oxide (ZnO) ceramics show highly nonlinear current–voltage characteristics. Due to these electrical properties, ZnO ceramics are used as varistors in electronic devices (Clarke 1999). Doping of praseodymium (Pr) or bismuth in ZnO ceramics is a well-known method to obtain high nonlinearity in current–voltage characteristics, and the presence of the dopants at ZnO grain boundaries is thought to be an important point (Clarke 1999, Mukae et al. 1977). In order to understand the microscopic origin of the properties, the detailed structure of ZnO grain boundaries including the location and distribution of dopants should be known on the atomic scale. Results on the atomic arrangement and location of Pr at ZnO grain boundaries (Sato et al. 2006, 2007) are introduced in this section. Grain boundaries with properly controlled orientation relationships and boundary planes were fabricated within ZnO bicrystals, thereby direct observations of the atomic arrangements were enabled. HAADF-STEM was used for the grain boundary observations. The incoherent imaging nature and atomic-number-dependent contrast of HAADF-STEM led to straightforward understanding of the atomic arrangements and location of heavy elements such as Pr at ZnO grain boundaries.

Figure 11–13 shows HAADF-STEM images of a Pr-doped ZnO grain boundary. For this grain boundary, the <0001> axes of both crystals are parallel to each other and the rotation angle is ~21.8° about the $<1\bar{1}00>$ axis to give an orientation relationship with a value of $\sum 7$. The boundary plane is parallel to the $\{12\bar{3}0\}$ plane of both crystals. This selection of the orientation relationship and the boundary plane yields a short structural periodicity and common low-index axis for both crystals, which are well suited for atomic-resolution electron microscopy observations and calculations of atomic arrangement that are often performed under three-dimensional periodic boundary conditions. The electron is incident parallel to the <0001> axes of both crystals in the HAADF-STEM images. Therefore, Zn and O are aligned in the same columns and the position of the columns appears as bright spots. Their location was clearly observed not only in the bulk crystal but also at the grain boundary. It was found that two ZnO crystals are directly bonded at the atomic scale at the grain boundary. This grain boundary has a structural periodicity as noted above, which is indicated by

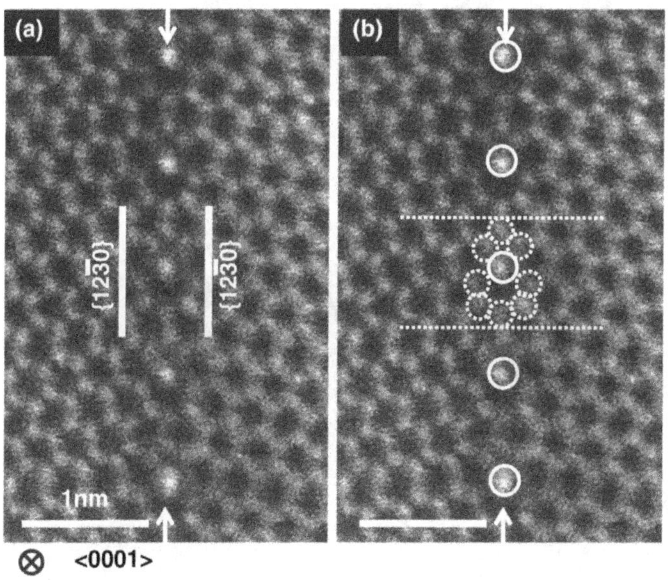

Figure 11–13. (a) HAADF-STEM image of Pr-doped ZnO [0001] $\Sigma 7$ tilt grain boundary. Electron incident direction is parallel to the <0001> axes of the ZnO crystals, and boundary planes are parallel to $\{12\bar{3}0\}$ planes. (b) The same image with an overlay of a structural unit is shown by a set of circles. *Arrows* and *dotted lines* indicate the grain boundary planes and the structural periodicity along the boundary plane. *Solid circles* indicate positions of much higher intensities suggesting the presence of Pr at these columns. Reproduced from Sato et al. (2006) with permission.

dotted lines. The atomic arrangement within one period, the structural unit, can be understood with the set of circles shown in Figure 11–13(b). In addition, the intensities at some particular columns are higher than those in the bulk crystal, showing the presence of heavier elements at these columns, as shown by solid circles in Figure 11–13(b). Since Pr has a higher atomic number ($Z = 59$) than Zn does ($Z = 30$), the higher intensities show the presence of Pr at these columns. Pr atoms preferentially occupy specific columns in the structural units and periodically appear along the boundary plane. The preference of Pr's location will be discussed later. It was found that the structural unit of the Pr-doped ZnO grain boundary (sets of circles in Figure 11–13(b)) is similar to that of the undoped ZnO boundary (sets of circles in Figure 11–14(a)) (Oba et al. 2004, Sato et al. 2006, 2007). This suggests that the atomic arrangement of the ZnO $\Sigma 7$ tilt GB did not change significantly with the doping of Pr, and Pr simply substitutes for Zn at these columns in the structural units.

In order to obtain further insight, the stable atomic arrangement of the Pr-doped ZnO grain boundary was simulated by first-principles band-structure methods with the VASP code (Kresse and Furthmüller 1996). For the simulation of the Pr-doped ZnO boundary, Zn marked

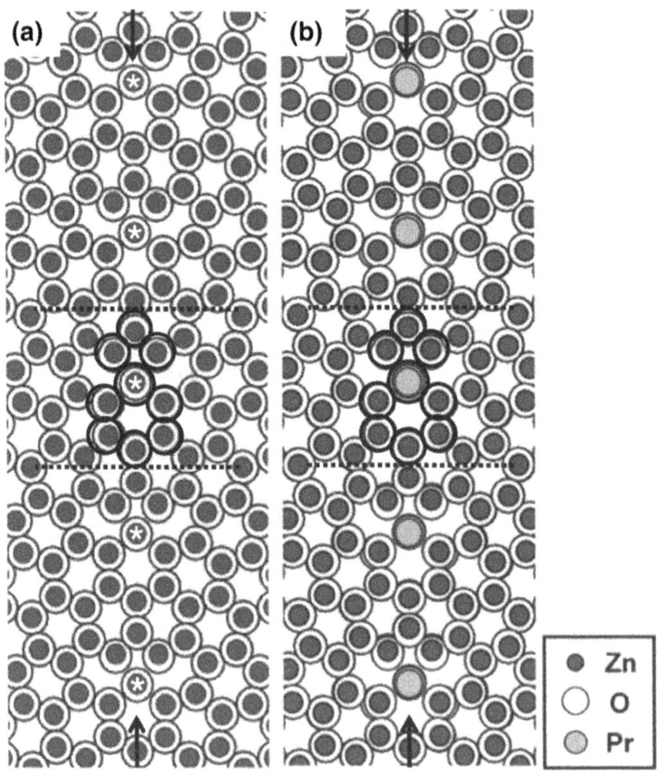

Figure 11–14. Optimized atomic arrangements of (a) the undoped and (b) the Pr-doped ZnO [0001] $\Sigma 7$ tilt grain boundaries. Both of the arrangements are viewed along <0001> directions. *Asterisks* in (a) indicate the locations of atomic columns corresponding to higher intensities in STEM images (Figure 11–13). Reproduced from Sato et al. (2007) with permission.

with asterisks was replaced with Pr, and the atomic arrangements were optimized. Figure 11–14(b) shows the optimized atomic arrangement of the Pr-doped ZnO grain boundary. The atomic arrangement did not change significantly during the structural optimization, keeping the structural unit similar. The optimized atomic arrangement agrees with the HAADF-STEM image in Figure 11–13, supporting the validity of the atomic arrangement in Figure 11–14(b) from both the experimental and theoretical sides. On this structural basis, formation energies of acceptor-like point defects such as V_{Zn} and O_i at ZnO grain boundaries were calculated (Sato et al. 2006, 2007). It was found that the formation energies of these defects were lower for the Pr-doped case, and thus, it was suggested that facilitation of defect formation is the key role of Pr-doping for the generation of nonlinear current–voltage characteristics (Sato et al. 2006, 2007).

Another topic to be mentioned here is why Pr selects specific columns of the ZnO grain boundary. An important insight was obtained from intensive inspection of the undoped boundary. Figure 11–15 shows a map of interatomic distances with neighboring O at respective Zn sites of the undoped ZnO boundary. It was found that the interatomic distances are locally longest at the Zn sites marked with asterisks in Figure 11–15. These Zn sites correspond to the asterisked Zn sites in Figure 11–14(a), showing that Pr selectively substitutes Zn having the

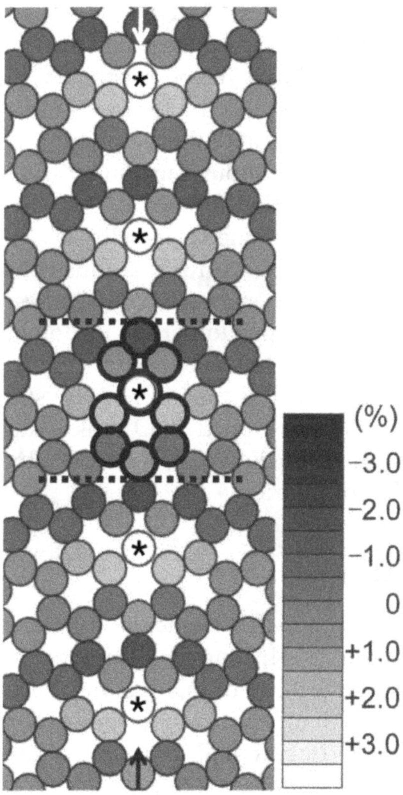

Figure 11–15. Map of interatomic distances with neighboring O atoms at respective Zn sites of the undoped ZnO [0001] Σ7 tilt grain boundary. Position and gray shading of circles show the location of Zn sites and difference of the averaged interatomic distances from those in bulk crystal. *Asterisks* indicate the Zn sites with locally longest interatomic distances. Reproduced from Sato et al. (2007) with permission.

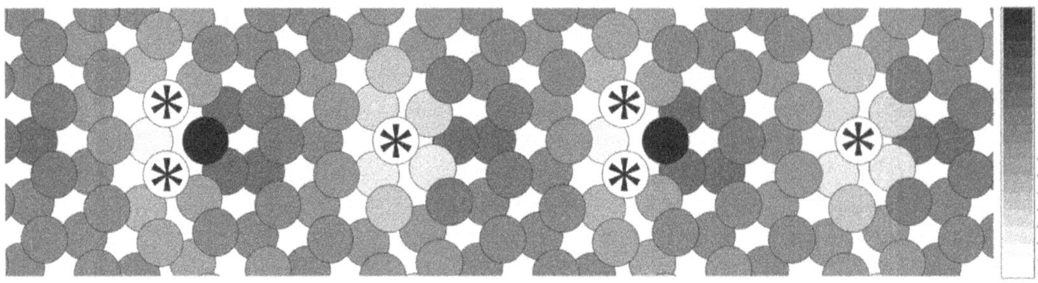

Figure 11–16. (a) Interatomic distance at individual Zn sites in the undoped ZnO [0001] $\Sigma 49$ grain boundary. Location and gray shading of the circles denote position of the Zn sites and their interatomic distances, respectively. *Asterisks* indicate the Zn sites with locally longest interatomic distances. Reproduced from Sato et al. (2009) with permission.

longest interatomic distances. This is probably because Pr has larger ionic radius than Zn does, and therefore, Pr would prefer atomic sites with relatively larger space due to longer interatomic distances. In other words, Pr must segregate to sites with larger space. Figure 11–16 shows the interatomic distance at individual Zn sites at the undoped ZnO [0001] $\Sigma 49$ grain boundary, which has a different orientation from the $\Sigma 7$ boundary. The location and gray color of the circles denote the position of Zn sites and their interatomic distances, respectively. Asterisks indicate the Zn sites with locally longest interatomic distance, indicating that the atomic site has a unique alternating configuration with one and two possible Pr sites. Figure 11–17 shows HAADF-STEM image of the Pr-doped ZnO [0001] $\Sigma 49$ grain boundary obtained with the incident electron beam parallel to the [0001] axis for both crystals, in which the grain boundary atomic arrangement and the locations of Pr are clearly observed. It was found that Pr appeared alternately in one and two atomic columns periodically along the boundary, suggesting that there is selectivity on the atomic-level locations of the Pr. This configuration is consistent with the theoretically obtained sites with locally longest interatomic distance as shown in Figure 11–16. It is thus concluded that Pr is selectively segregated at the sites with the longest interatomic distance. Furthermore, detailed inspection of the Pr-doped ZnO grain boundary revealed that Pr–O bonds tend to have coordination and electronic structures more similar to those in Pr_2O_3 crystal bulk, when compared with Pr–O bonds in Pr-doped ZnO bulk (Sato et al. 2009). This would be another important insight for the reason why Pr prefers specific Zn sites in the grain boundaries.

11.3.4 3D Observation of Y-Doped Al_2O_3

HAADF-STEM is well suited to identifying heavy elements in lighter surroundings (Choi et al. 2009, Nellist et al. 2004, Pennycook and Jesson 1990, 1991, Voyles et al. 2002). As we have shown, this fact has been used to explore the detailed configuration of impurity atoms segregated at grain boundaries in ceramic materials (Buban et al. 2006, Sato et al. 2006, Voyles et al. 2002), where it is known that such dopant addition significantly changes the physical properties of the material. However,

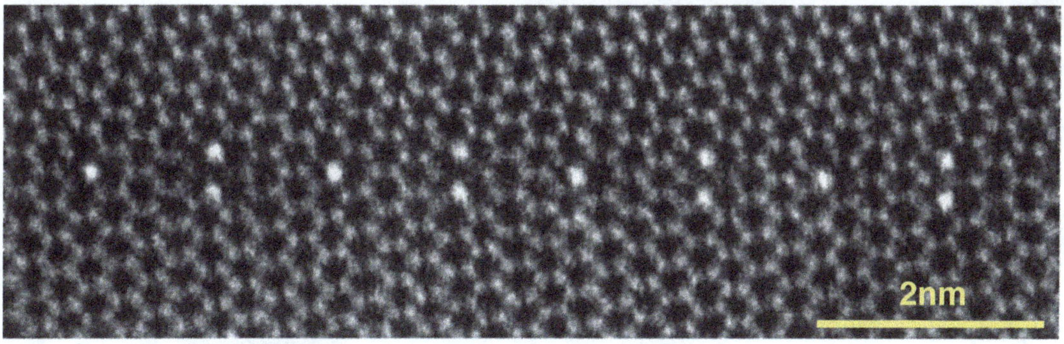

Figure 11–17. HAADF-STEM image of a Pr-doped ZnO [0001] $\Sigma 49$ grain boundary. The image is observed along the [0001] common axis for both crystals. Reproduced from Sato et al. (2009) with permission.

at best, a single image only gives a clear indication of the projected structure along one direction. The complexity of interface structure makes it desirable to seek further information, such as the distribution of dopant atoms on the interface plane. Recent advances in image quantification may provide one route to this goal (LeBeau et al. 2008, Molina et al. 2009, Robb and Craven 2008), and depth sectioning in aberration-corrected microscopes may provide another (van Benthem et al. 2005).

However, the most direct approach is to observe the specimen from multiple directions, such as cross-sectional and plan-view images, provided the different images can be interpreted equally well. A spherical aberration corrector allows the formation of an atomically fine probe on the crystal surface, but, depending on the specimen orientation, it becomes important to describe in detail how the probe spreads through the sample in order to best interpret the experimental images. Therefore we explore the effects of sample orientation, defocus selection, and probe spreading on atomic-resolution HAADF-STEM imaging using a Y doped alumina (α-Al_2O_3) grain boundary as a test case, in order to explore the way to highlight the 3D positioning of Y atoms buried in the boundary. It will be demonstrated here that the aberration-corrected HAADF-STEM is a powerful method to directly highlight individual dopant atoms on buried crystalline interfaces (Shibata et al. 2009).

As a model system, an Y doped α-Al_2O_3 grain boundary was prepared by diffusion bonding of two single crystals in the $\Sigma 13$ orientation relationship (Azuma et al. 2010, Fabris and Elsässer 2001). Y was added to the interface before diffusion bonding, as described elsewhere (Azuma et al. 2010, Matsunaga et al. 2003). Figure 11–18(a) shows schematically the bicrystal fabricated in this study. The bicrystallography of the interface between the top and bottom crystals is summarized as follows: $(10\bar{1}4)_{top} || (10\bar{1}4)_{bottom}$, $[1\bar{2}10]_{top} || [\bar{1}2\bar{1}0]_{bottom}$, and $[\bar{2}021]_{top} || [\bar{2}021]_{bottom}$. The interface normal is parallel to the high index $<50\bar{5}4>$ directions in both the top and bottom crystals. Figure 11–18(b) shows typical atomic-resolution HAADF-STEM images of the interface projected along the $<1\bar{2}10>$ and $<\bar{2}021>$ directions. The doped Y atomic columns are clearly imaged with very strong contrast along

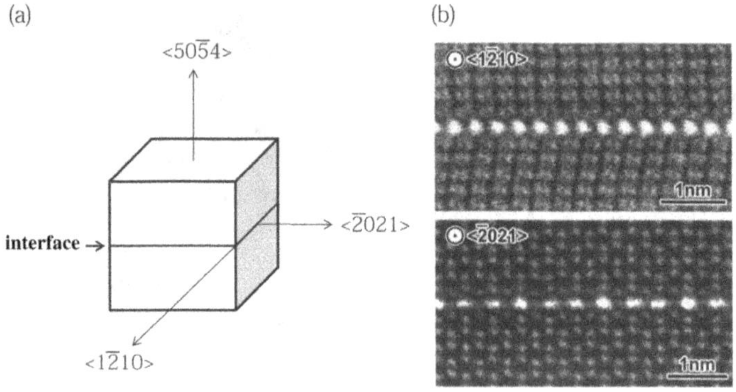

Figure 11–18. Schematic illustration and two cross-sectional HAADF-STEM images of the Y-doped $\Sigma 13$ grain boundary of α-Al_2O_3. (a) Schematic illustration of the α-Al_2O_3 bicrystal fabricated in this study, with the orientation relationship between the top and bottom crystals indicated. Y atoms are artificially doped in the boundary plane. (b) HAADF-STEM images of the Y-doped grain boundary observed from two orthogonal directions parallel to the interface plane. Reproduced from Shibata et al. (2009) with permission.

the boundary and form a monatomic layer structure in the core of the boundary.

Figure 11–19 shows a high-resolution HAADF-STEM image of the Y doped $\Sigma 13$ grain boundary observed from the <50$\bar{5}$4> plan-view direction perpendicular to the interface plane (Shibata et al. 2009). In this plan-view image, the interface Y atoms are visible as strong image intensity spots due to their much larger atomic number compared with Al and O atoms. These spots are periodically arrayed along the <$\bar{2}$021> direction and accompany a weak background contrast elongated along the same direction. This striped background corresponds to the projected image of Al atomic planes of the matrix

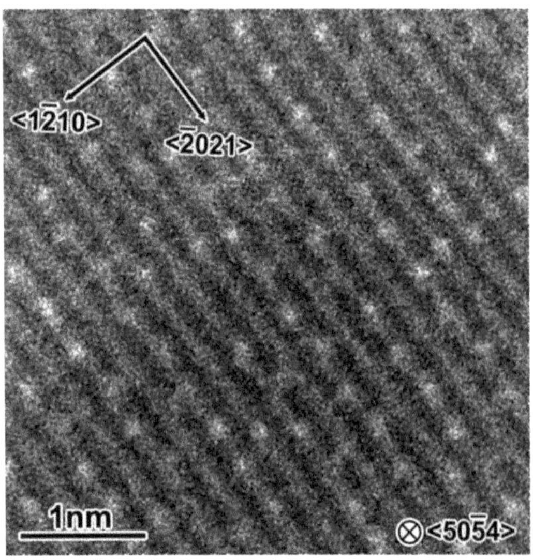

Figure 11–19. A plan-view HAADF-STEM image of the Y-doped $\Sigma 13$ grain boundary observed from the < 50$\bar{5}$4 > plan-view direction. The interface Y atoms are visible as strong image intensity spots on weak background contrast elongated along the < $\bar{2}$021 > direction. Reproduced from Shibata et al. (2009) with permission.

α-Al_2O_3. It is found that the Y atom occupancy is not uniform on these stripes; Y atoms are much more densely situated on every second stripe. The plan-view image thus directly reveals the two-dimensional ordered array of Y atoms on the buried interface plane.

To theoretically support the visibility of the Y atoms in the $\Sigma 13$ α-Al_2O_3 grain boundary, multislice STEM image simulations were performed (Allen et al. 2003, Shibata et al. 2009). Figure 11–20(a) shows a schematic of the model pristine $\Sigma 13$ α-Al_2O_3 grain boundary used in the simulations. Figure 11–20(b) shows simulated intensity profiles, projected along the <1$\bar{2}$10> direction, with the electron beam in vacuum and focused on the interface plane between the crystals. The intensity profiles are almost identical, which suggests that the atomic-resolution capability of the probe in vacuum is preserved inside the light α-Al_2O_3 crystal. Figure 11–20(c) shows simulated plan-view images of pristine and Y doped $\Sigma 13$ grain boundaries. Corresponding structure models viewed from the <50$\bar{5}$4> direction are also shown, and the Y position is indicated. In the pristine case, the periodic striped contrast parallel to the <$\bar{2}$021> direction is due to the periodic Al-containing atomic planes. In the Y doped case, in addition to the stripes due to the Al-containing atomic planes, we further see the bright contrast spot corresponding to the interface Y atom. This result is in good agreement with our experimental images.

Since the striped background originates from the Al atomic planes in the matrix α-Al_2O_3 crystals, we performed image filtering to remove it and so improve the visibility of the Y atom arrangement (Shibata et al. 2009). In this case, a simple FFT-masking method was used to remove the background stripes from the original image. Figure 11–21 shows

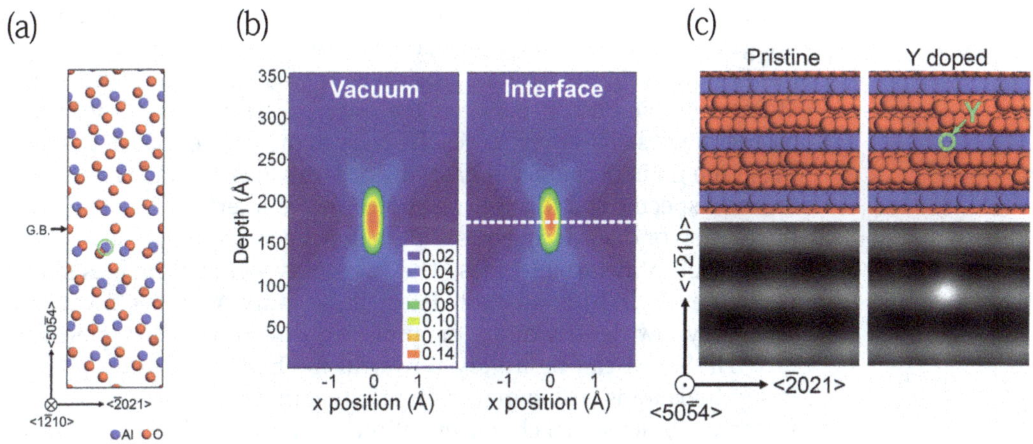

Figure 11–20. HAADF-STEM image simulation of the plan-view Y-doped $\Sigma 13$ grain boundary. (a) Atomic structure model of the $\Sigma 13$ grain boundary used for the STEM image simulations. (b) Simulated intensity profiles of the electron beam, projected along the <1$\bar{2}$10> direction, in vacuum and focused on the interface plane between the crystals. The thickness of the sample is assumed to be 355 Å. (c) Simulated HAADF-STEM images of the pristine and Y-doped $\Sigma 13$ grain boundary observed from the <50$\bar{5}$4> direction. Reproduced from Shibata et al. (2009) with permission.

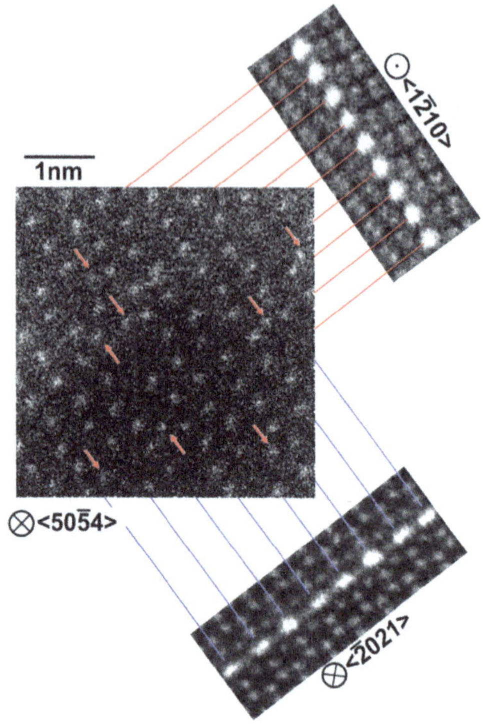

Figure 11–21. Filtered plan-view HAADF-STEM image highlighting the two-dimensional positioning of the interface Y atoms. The filtered image was obtained by a simple FFT-masking method to remove the background stripes from the original image shown in 11–19. The image is displayed with a nonlinear intensity scale to highlight bright features. The two-dimensional ordered array of individual Y atoms is now more clearly evident. Reproduced from Shibata et al. (2009) with permission.

the filtered version of the image in Figure 11–19. The two-dimensional ordered array of individual Y atoms is now more clearly evident. The comparison of Y atom positioning between the filtered plan-view image and the two cross-sectional images is also shown in the figure. The Y atom array along the <1$\bar{2}$10> and <$\bar{2}$021> directions corresponds well to the Y atomic column positions observed in the two cross-sectional images. In addition, there are some clear deviations of the Y atom positioning from the ordered positions. It was sometimes seen that Y atoms were shifted in between the ordered array sites, as indicated by the arrows. The present results thus demonstrate that Y positioning at the grain boundary is basically ordered in two dimensions based on their specific stable atom positions on the interface, but that some fluctuation or disordering exists, perhaps due to trapping and/or overflowing onto metastable sites. The two cross-sectional images shown in Figure 11–18(b) could not detect these stray Y atoms because of their very low density along the projected directions. Thus, plan-view HAADF-STEM can be a very powerful method for directly imaging individual atoms within materials, bringing us a crucial step toward the full three-dimensional characterization of interface atomic structures.

11.4 Amorphous Grain Boundary

11.4.1 Equilibrium Thickness of Amorphous Layer

In the case of ceramics sintered with sintering additives, an amorphous film is frequently formed along grain boundaries. For example, Si_3N_4

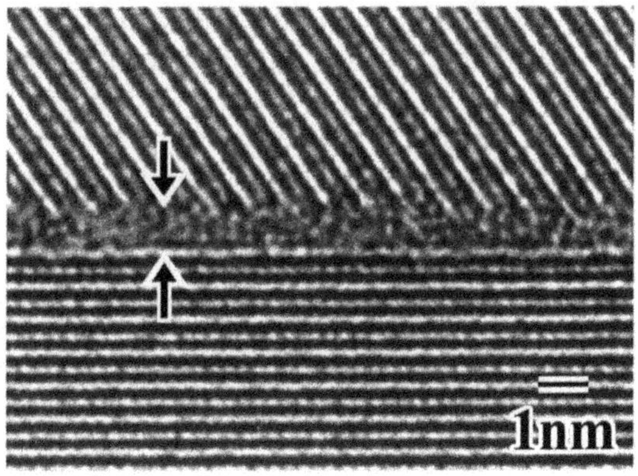

Figure 11–22. HREM image of a grain boundary in a Si_3N_4 sintered body containing Y_2O_3–Al_2O_3 additives, indicating an amorphous film with a thickness about 1 nm is present at the grain boundary. Reproduced from Ikuhara et al. (1996) with permission.

is usually sintered with metal oxides, and the added oxides form an amorphous film along the boundaries. High-temperature properties of Si_3N_4 are actually influenced by the composition and chemical bonding state of the amorphous film (Clarke 1987, Ikuhara et al. 1988, 1996, Kleebe 1997, Tanaka et al. 1994). Figure 11–22 shows a HREM image of a grain boundary in a Si_3N_4 sintered body containing Y_2O_3–Al_2O_3 additives (Ikuhara et al. 1996). It is obvious that an amorphous film with the thickness of about 1 nm is formed along the grain boundary. Clarke theorized that the equilibrium thickness of the amorphous layer is determined by balancing the van der Waals force between two adjacent grains and the steric force of the amorphous layer (Clarke 1987). According to his theory, van der Waals force is related to the dielectric properties of the grain and amorphous layer, and the steric force is dependent on the chemical composition of the thin amorphous film. As a result, the equilibrium thickness h can be expressed as the following equation:

$$\frac{H}{6\pi h^3} = a\eta_0^2 \frac{1}{\sinh^2(h/2\xi)}, \tag{5}$$

where H is related to the Hamaker constant, η_0 is a factor derived from the free energy difference between the amorphous film with and without structural ordering, and ξ is the correlation length in the amorphous film. The details to derive this equation are reported elsewhere (Clarke 1987), but this equation actually well explains some experimental results obtained for Si_3N_4 ceramics (Kleebe 1997). However, there are still some questions as to whether van der Waals force effectively works between two grains 1 nm apart. On the other hand, an amorphous-like film is often observed even in ceramics sintered without sintering additives, particularly in ceramics having covalent bonds, such as SiC (Ikuhara et al. 1987, 1988, Tsurekawa et al. 1995). If the surface of the starting powders is oxidized, the oxide layer can form an amorphous film during sintering even without additives. The amorphous film in hot isostatically pressed Si_3N_4 without sintering additives is in this category (Tanaka et al. 1994), but the amorphous-like layer in well-defined, high-purity materials is very much different (Ikuhara et al. 1988, Tsurekawa et al. 1995). This layer is not composed of impurities, and the layer has been confirmed to consist of the same elements as the grain interior by nano-probe EDS (Ikuhara et al. 1987). It is, therefore, believed that the amorphous-like structure appears in order to relax the atomic configuration and reduce the high energy of the grain boundary. In other words, a high energy grain boundary is extended to form a relaxed structure with some thickness to reduce its high energy. We previously proposed the concept of "the extended grain boundary" (Ikuhara et al. 1987, 1988, Tsurekawa et al. 1995) and the amorphous-like grain boundary in high-purity ceramics is likely such a boundary. The width of the layer is considered to depend on the grain boundary character. The presence of the extended grain boundary is also reasonably predicted by first-principles calculations (Kohyama 1999, 2002).

11.4.2 Amorphous Films in Si_3N_4 Grain Boundaries

Silicon nitride (Si_3N_4) ceramics are used in numerous applications because of their ability to overcome the inherent brittleness of ceramics through reinforcing microstructures with whisker-like grains (Chen et al. 1993). However, the formation of such anisotropic grains is very sensitive to the dopant cations used in the sintering additives (Becher et al. 2005, Hoffmann et al. 2000, Satet and Hoffmann 2004). Understanding the underlying atomistic mechanisms of these dopant effects is a key to designing high-performance Si_3N_4 ceramics through microstructure optimization.

In general, Si_3N_4 ceramics are fabricated through liquid phase sintering, where oxynitride amorphous intergranular films (IGFs) with thickness on the nanometer scale are formed between grains (Kleebe 1997). Dopant atoms are expected to reside in IGFs and control the anisotropic grain growth essential for the formation of a whisker-like grain morphology. However, the atomic-level details about how the dopants are distributed within the IGF have been extremely difficult to assess, due to its very small thickness (i.e., < 2 nanometers) and its amorphous nature.

It has been well demonstrated that HAADF-STEM is very powerful to directly observe dopant atoms within the nanometer thick IGFs (Shibata et al. 2004, Winkelman et al. 2005, Ziegler et al. 2004). Figure 11–23(a) is a HAADF-STEM image of the IGF region in La doped Si_3N_4 ceramics (Shibata et al. 2004). The grain on the right is aligned with the [0001] projection of β-Si_3N_4, so that the $(10\bar{1}0)$ prismatic boundary plane is set at an "edge on" condition. In the image, bright spots inside the grains correspond to the Si columns, and the bright vertical band at the center of the image indicates the position of the IGF, which is confirmed by the amorphous-like contrast in the BF-STEM image of the same region. The estimated IGF thickness is about 1 nm in both images. The strong bright contrast in the IGF is due to the presence of atoms with high atomic number, in this case, La ($Z = 57$). Notice that the highest intensity in the HAADF image is observed at the IGF/grain interfaces and a minimum occurs around the center of

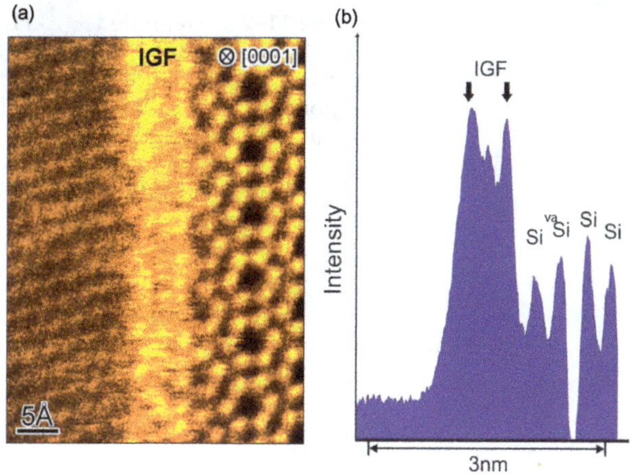

Figure 11–23. (a) HAADF-STEM image of an IGF in a La-doped Si_3N_4 ceramic. (b) Image intensity profile across the IGF is shown in (a). Reproduced from Shibata et al. (2004) with permission.

the IGF. Figure 11–23(b) shows the image intensity profile across the boundary summed along Figure 11–23(a). It is clearly seen that the maximum intensities appear at both IGF/crystalline interfaces, resulting in a bimodal intensity distribution across the IGF. The strong zone of image intensity along the right edge of the IGF is located where the terminal Si columns in the right-hand grain would have existed, suggesting that this zone represents the first cation layer of the glass attached to the β-Si_3N_4 terminating surface.

Figure 11–24 is a magnified HAADF-STEM image of the IGF/prismatic crystalline interface. The β-Si_3N_4 lattice structure is superimposed on the image. Within the interfacial zone, La atoms are readily observed as bright spots (denoted by red arrows). Note that the positions of the La atoms are shifted from those expected for Si atoms for a continuation of the β-Si_3N_4 structure; these expected positions are shown by open green circles. These La sites are in good agreement with the calculated stable positions of La at a N-terminated prismatic plane using first-principles cluster calculations (Shibata et al. 2004). The present result clearly indicates that the anisotropic grain growth should be related to the strong preferential segregation of La to the prismatic planes, retarding the grain growth effectively on these specific planes.

Figure 11–25 shows typical HAADF-STEM images of Lu, Gd, and La-doped Si_3N_4 ceramics (Shibata et al. 2005). Each grain on the right side is aligned along the [0001] projection of β-Si_3N_4 with the smooth $(10\bar{1}0)$ prismatic plane imaged edge-on. The very bright spots in the IGF represent high atomic number atoms (e.g., $Z = 71$ (Lu), 64 (Gd), 57 (La), confirmed by energy-dispersive X-ray spectroscopy (EDS) microanalysis). Each periodic bright spot along the edge of the right-hand interface (indicated by the red arrows) corresponds to a concentration of dopant atoms attached to the prism surfaces of grains. In each case, rare earth attachment at the anion terminated prismatic surface forms the first cation layer in the IGF. Comparison of the interface images reveals that, in these three dopant systems, a distinct sequence of surface occupations is observed by Z-contrast STEM. The difference in surface occupancies in each dopant system is also confirmed by first-principles

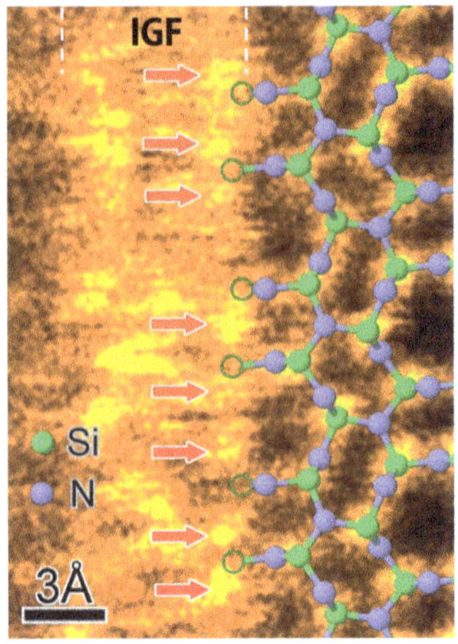

Figure 11–24. Magnified HAADF-STEM image of the IGF/crystal interface. *Green* and *blue circles* correspond to Si and N, respectively. The *arrows* indicate the position of La atoms attached to the prismatic surface of the Si_3N_4 grain. Reproduced from Shibata et al. (2004) with permission.

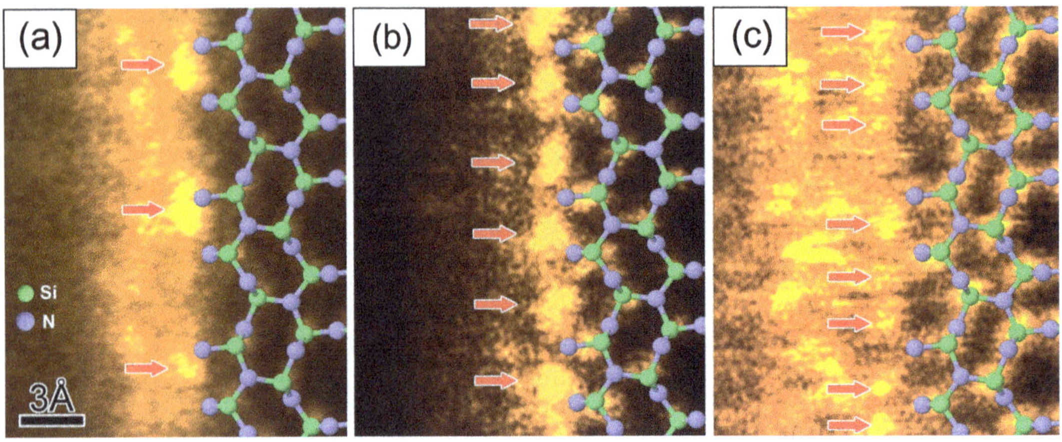

Figure 11–25. HAADF-STEM images of the IGF/crystal interfaces in (a) Lu, (b) Gd, and (c) La systems, respectively. *Arrows* indicate the positions of the rare earth atoms attached to the prismatic crystal surface. Reproduced from Shibata et al. (2004) with permission.

calculations (Painter et al. 2008). It is found that the segregation energy of these dopants to the prismatic β-Si_3N_4 surface is the key parameter to determine the anisotropic grain growth behavior in doped Si_3N_4 ceramics. Thus, the combination of Z-contrast STEM with theory should provide the basis for designing very tough microstructures in silicon nitride ceramics from the atomistic dimensions.

11.5 Hetero-Interface Structures

11.5.1 Coherent and Incoherent Interfaces

To describe the interface structure between two crystals, and the accommodation by misfit dislocations of the lattice mismatch between them, one should, in general, distinguish between systems with small and large lattice mismatches. During epitaxial growth of a film on a substrate which is only slightly lattice mismatched, growth normally takes place in a coherent fashion, which leaves the film homogeneously strained and commensurate with the substrate (van der Merwe 1950). The resulting strain energy increases the total energy of the film compared to a relaxed, unstrained, film. As the thickness of the strained film increases, the strain energy of the film increases proportionately. In other words, in this strained, pseudo-morphic film, the strain can be described by an array of (fictitious) "coherency dislocations" each of which has a very small Burgers vector and is separated from a neighboring one by one lattice spacing along the interface (Figure 11–26(a)) (Ikuhara and Pirouz 1998, Ikuhara et al. 1994, 1995, Olson and Cohen 1979). Thus the strain energy in the film can be considered to be the sum total of the strain energy of coherency dislocations.

When the film thickness, h, reaches a critical value $h = h_C$, it becomes energetically favorable for (real) "misfit" or "anti-coherency" dislocations to be introduced at the interface, to accommodate the lattice mismatch and relax the strained film (Figure 11–26(b)) (van der Merwe 1963). This process is equivalent to a cancellation of the elastic field of coherency dislocations by anti-coherency dislocations (Olson and Cohen 1979) and usually takes place in a gradual fashion. In the intermediate stages, at $h > h_C$ but before complete relaxation, part of the

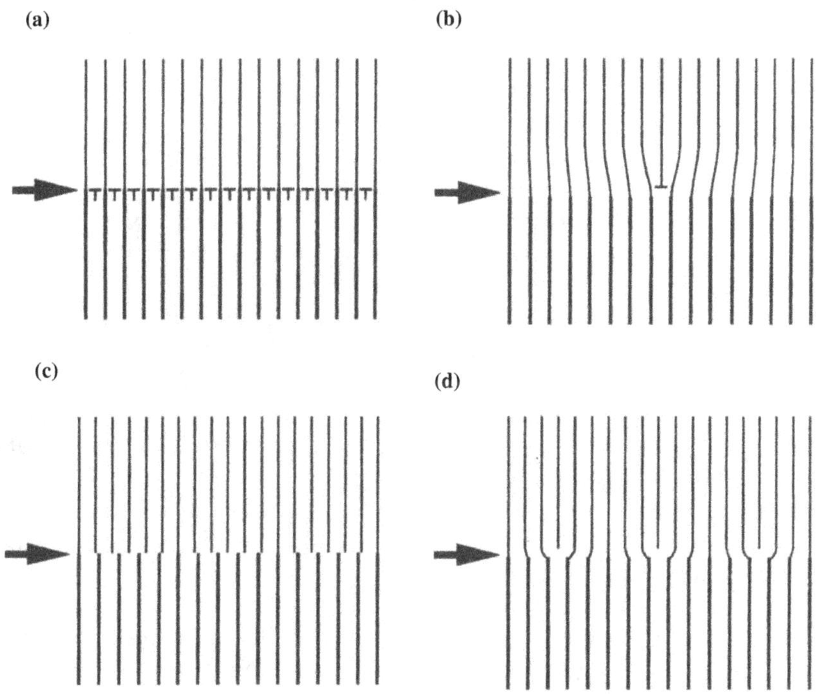

Figure 11–26. Various models of a hetero-interface. (a) coherent with coherency dislocations, (b) semicoherent with misfit or anti-coherency dislocations, (c) incoherent, and (d) pseudo-semicoherent with geometrical misfit dislocations. Reproduced from Ikuhara and Pirouz (1998) with permission.

strain is accommodated elastically by coherency dislocations while the remainder is accommodated by misfit, anti-coherency dislocations. In this intermediate stage, the anti-coherency dislocations cancel out only a fraction of the coherency dislocations, and thus the film is relaxed only partially; the film is still elastically strained and contains a fraction of coherency dislocations. As the film thickness increases further, the proportion of elastic accommodation of strain (i.e., the fraction of coherency dislocations) decreases while the proportion accommodated by misfit (anti-coherency) dislocations increases. During this process, the spacing between misfit dislocations decreases. Eventually, when all the strain is accommodated by misfit dislocations, all the coherency dislocations have been canceled out by anti-coherency dislocations; the film is now completely relaxed and the interface is "semicoherent" with a constant spacing of misfit dislocations. Among the different mechanisms by which a misfit dislocation can be formed is the nucleation of a dislocation half-loop at the film surface and its glide on a slip plane of the epilayer to the interface (Matthews 1979). The segment of the half-loop at the interface constitutes a misfit dislocation, and its two ends are connected to the film surface by "threading" dislocations. The misfit dislocation segment can expand its length by the expansion of the two threading dislocations on their (common) slip plane in the epilayer. The formation of a semicoherent interface is in general valid only for misfit parameters less than 4–5%. However, when the misfit parameter is larger than 4–5%, the critical thickness becomes of the order of atomic dimensions and the whole concept of a critical thickness loses meaning, and incoherent interfaces form (Figure 11–26(c)). Thus, when the lattice mismatch is large (>4–5%), the stages of complete or partial coherency do not exist and the interface may be expected to be incoherent with no continuity between the lattice planes on the two sides of the interface.

A truly incoherent interface implies a complete lack of interfacial bonding between the opposing atoms. In that case, there is no interface adhesion and the two constituent crystals simply fall apart. If there is indeed some interfacial adhesion, even though it may be very weak, there must be some localized bonding between the atoms at the two sides of the interface. Assuming that the elastic constants on one side of the interface are much higher than on the other side, then usually a displacement of atoms in the vicinity of the interface takes place in the softer component. As a result, there will be an appearance of coherency along some planes on the two sides of the interface separated by what we call "geometrical" misfit dislocations (or "mismatch dislocations") which separate these "pseudo-semicoherent" planes (Figure 11–26(d)) (Ikuhara and Pirouz 1998, Ikuhara et al. 1994). The difference between misfit dislocations (with an invariant Burgers vector) and mismatch dislocations (with a mismatch vector that depends on the particular Burgers circuit drawn around the dislocation line) manifests itself in the fact that different mismatch dislocations can be observed in HREM along different viewing directions. The only restrictions, which are set by the requirements of HREM observation, are that (i) the corresponding parallel planes on the two sides of the interface are parallel to the incident electron beam and (ii) the spacing between these planes is within the resolution of the electron microscope. With this reservation in mind, we shall continue to use the term "misfit dislocation" except that we qualify the term by adding the word "geometrical" to distinguish it from a proper "dislocation" with a definable Burgers vector (Ikuhara and Pirouz 1998, Ikuhara et al. 1994). The terms "mismatch dislocation" and "mismatch vector" will be also employed to define such a line feature. It should also be added that even for the Burgers vector of a proper interfacial dislocation, the definition is not very clear; it generally requires transformation of both lattices to a common coordinate system (or a common lattice) and using a DSC (displacement shift complete)vector to define it (Balluffi et al. 1982). According to Bollmann (1970), a Moiré pattern is closely related to the dislocation network at the interface. In fact, a Moiré pattern from two overlapping mismatched lattices ideally represents the position of these mismatch dislocations and forms a Wigner–Seitz cell. In the next section, a couple of examples will be shown for the incoherent interface, the coherent interface, and the size dependence of the coherent–incoherent transition.

11.5.2 SiC/Ti_3SiC_2 Coherent Interface

Silicon carbide (SiC) is a very promising semiconductor to succeed Si in next generation electronic devices especially for high power and frequency applications due to its various unusual intrinsic properties. One of the key technological issues currently limiting device processing is the fabrication of robust and low-resistance Ohmic contacts, which allows higher current driving, faster switching speed, and less power dissipation (Porter and Davis 1995). Most studies to date on obtaining this Ohmic contact have focused on the deposition of TiAl-based metals (Tsukimoto et al. 2004a), the only materials currently available to yield significantly lower contact resistance. The formation of the Ohmic contact has been attributed empirically to a functional interface between SiC and Ti_3SiC_2 generated after annealing (Tsukimoto et al. 2004b), which presumably serves as a primary current-transport pathway to lower the Schottky barrier formed between the metals and the semiconductor. However, an understanding of the role of this SiC/Ti_3SiC_2 interface on the mechanism whereby a Schottky barrier becomes Ohmic has not yet been well developed. It is not even clear how the two materials atomically bond together due to the complexity associated with the study of a buried interface. The local chemistry and bonding at the interface, which deeply affect the physical properties, are also hardly accessible as a result of the unknown atomic structure. Detailed knowledge on the atomic and electronic structure and their impact on electronic properties is essential to elucidate the mechanism and for better device design and performance control.

In this section, the atomic-scale structure of the 4H–SiC/Ti_3SiC_2 interface in Ohmic contact to p-SiC is determined and related to properties by combining HAADF-STEM and first-principles calculation. Figure 11–27(a) depicts a HAADF-STEM image of the SiC(0001)/Ti_3SiC_2(0001) interface. The interface is atomically abrupt and coherent without transition regions, contaminants, or remaining

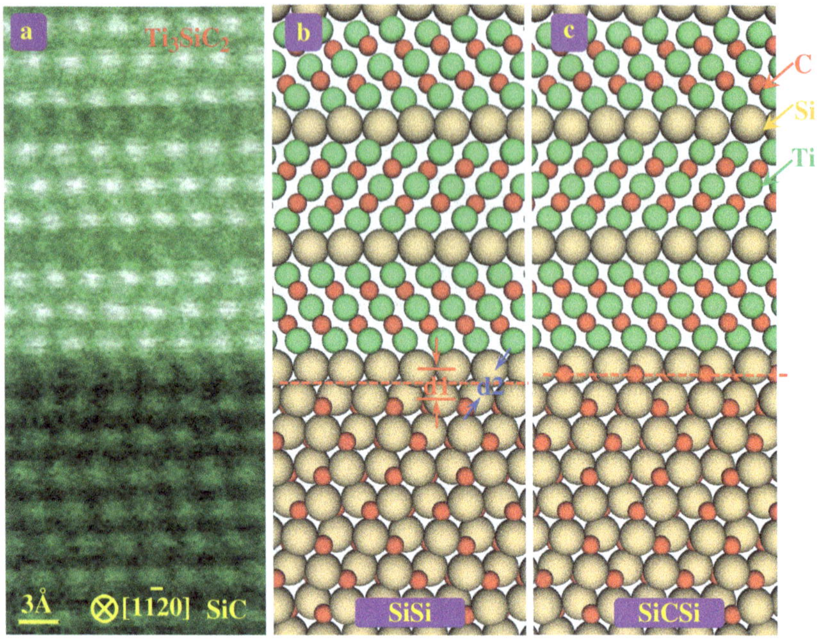

Figure 11–27. (a) HAADF-STEM image of the 4H–SiC/Ti_3SiC_2 interface in the Ohmic contact sample. Optimized SiC(0001)/Ti_3SiC_2(0001) interface models (b) without interfacial C atoms (SiSi) and (c) with interfacial C atoms (SiCSi). The distance between interfacial Si/Si layers is represented by d_1 and that between interfacial Si/Si atoms projected onto the plane of the figure by d_2. The interface is indicated by a *horizontal dotted line*. Reproduced from Wang et al. (2009) with permission.

reaction layers, which means that we have successfully deposited epitaxial Ti_3SiC_2 on the SiC substrate via the annealing technique. Bright spots in the image correspond to atomic columns of Ti, while the darker spots represent Si columns. However, columns of carbon do not scatter strongly enough to be visualized in this image, making it incomplete. To complement the image so as to relate the atomic structure to properties on the atomic scale, we also performed first-principles calculations, taking into account all of the 96 possible interface models (Wang et al. 2009). Of all the models, the image can be intuitively fitted by two models shown in Figure 11–27(b) and (c). The difference between these two models is that their local environments surrounding the interface are remarkably different, as carbon unseen by the STEM is trapped at the SiCSi interface (Figure 11–27(c)) but not at the SiSi interface (Figure 11–27(b)).

To investigate which interface is more likely and understand the interfacial bonding nature, the energy and electronic structure were calculated using VASP (Kresse and Furthmüller 1996) within density functional theory (DFT). The SiC(0001)/Ti_3SiC_2(0001) interface was modeled by a symmetric SiC slab connected to a symmetric Ti_3SiC_2 slab (Wang et al. 2009). To form coherent interfaces, the in-plane lattice constants of Ti_3SiC_2 were expanded by 0.63% to match those of the harder SiC. All atoms were fully relaxed until the force on each atom converged to less than 0.05 eV/Å. As for transport calculations, we used a state-of-the-art quantum transport technique: the fully self-consistent nonequilibrium Green's function method combined with DFT, which was implemented in the Atomistix Toolkit code (Brandbyge et al. 2002, Wang et al. 2007). The two-probe transport model consists of a sandwich system, $Ti/Ti_3SiC_2/SiC/Ti_3SiC_2/Ti$, wherein the SiC/Ti_3SiC_2 interfaces could be either SiSi or SiCSi, whereas other interfaces are maintained identical for the two systems (Wang et al. 2009). In this sense, the difference between the two systems is mainly due to the distinct SiC/Ti_3SiC_2 interfaces.

Calculations of adhesion energies (W_{ad}) of the two interfaces reveal that the SiCSi interface is more favored by having larger W_{ad} (6.81 J/m^2 for the SiCSi and 1.62 J/m^2 for the SiSi) (Wang et al. 2009). Next, we calculated the optimal distances of interfacial Si–Si planes ("d1" in Figure 11–27(b)) and interfacial Si–Si atoms projected onto the plane of the figure ("d2" in Figure 11–27(b)) and found that the distances for the SiCSi ($d_1 = 2.53$ Å; $d_2 = 2.81$ Å) are very close to the experimental values obtained from the STEM image ($d_1 = 2.5$ Å; $d_2 = 2.8$ Å), while those for the SiSi interface($d_1 = 2.13$ Å; $d_2 = 2.31$ Å) deviate severely from the experimental values. Therefore, the interface with C is inferred to better match the HAADF-STEM image both qualitatively and quantitatively. The interface with C was confirmed to be of less Schottky character by calculating the p-type Schottky barrier height from the difference between the Fermi level and the valence band top of the bulk SiC region in the supercell. The SiCSi interface has a Schottky barrier height of 0.60 eV, lower than that of the SiSi interface (1.05 eV) (Wang et al. 2009). The strong adhesion in the SiCSi can be explained by the mixed covalent–ionic nature of its interfacial bonds, as most of the charges are localized on interfacial C with distortions directed toward interfacial Si (Figure 11–28(b)). This is remarkably different from the interfacial bonds in the SiSi showing a clearly covalent nature (11–28(a)). The partial ionic character in the SiCSi, together with its considerable charge transfer, generates a large dipole shift, which lowers the electrostatic potential of interfacial Si in Ti_3SiC_2 relative to the SiC, thus reducing the SBH.

11.5.3 Interface Structure of $SrTiO_3$/Nb–$SrTiO_3$/$SrTiO_3$ Superlattices (Coherent Interface)

Figure 11–29 shows a HAADF-STEM image of the $SrTiO_3$/Nb-doped $SrTiO_3/SrTiO_3$ superlattice film (Ohta et al. 2007), which has a perfectly coherent interface. As described above, since the image intensity in a HAADF-STEM image is roughly proportional to the square of Z, it is recognized that Sr ($Z = 38$) columns are observed brightly compared with the Ti ($Z = 22$) columns. In this case, Nb-doped $SrTiO_3$ layers deposited at every 24 unit cells are observed as stripe contrast (Figure 11–29(a)). Figure 11–29(b) and (c) shows the magnified HAADF-STEM image around the Nb-doped $SrTiO_3$ layer and a line profile of

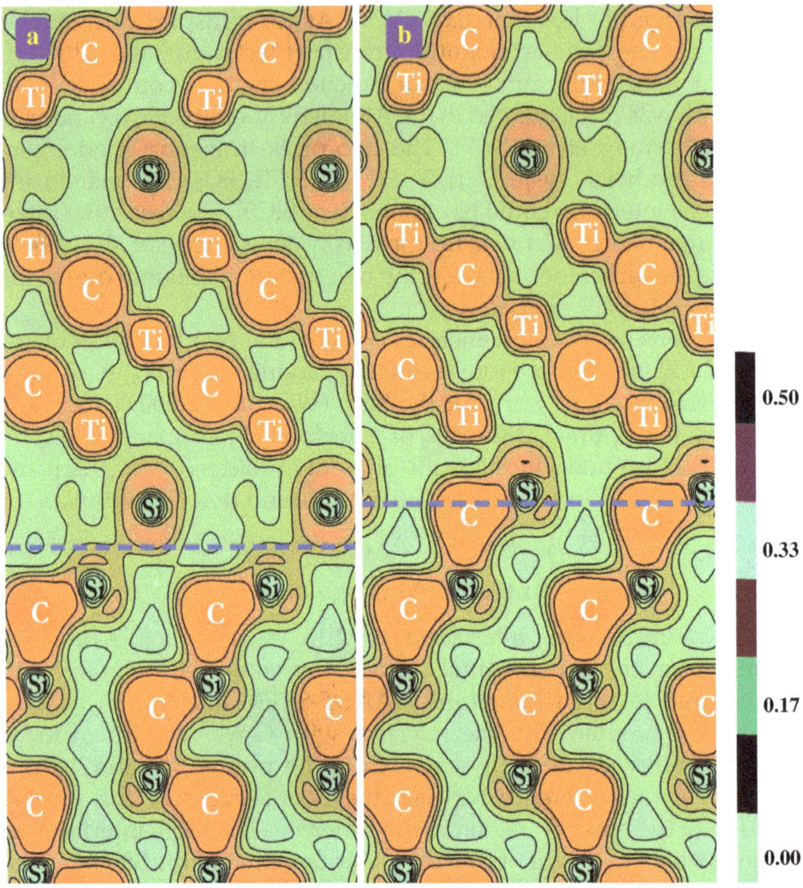

Figure 11–28. Contour plots of charge densities for (a) SiSi and (b) SiCSi interfaces taken along the $(11\bar{2}0)$ plane. The interface is represented by a *horizontal line* and the atoms that intersect the contour plane are *labeled*. The magnitude of charge is denoted by a scale on the *right*. Reproduced from Wang et al. (2009) with permission.

the image intensity of the Sr atomic row and the Ti atomic row in the same region. From these figures, it is recognized that the image intensity does not change in the Sr atomic row; however, the image intensity becomes high in the Ti atomic row at the Nb-doped $SrTiO_3$ layer. Taking into consideration that the atomic number of Nb is 41, it is considered that Nb occupies the Ti sites by substitution. On the other hand, since the atomic numbers of Nb and Sr are close, whether Nb exists in the Sr sites or not cannot be judged only by the contrast of the HAADF-STEM image. Then, the solubility energy of Nb was calculated by the first-principles PAW (Projector Augmented Wave) method. It is then clarified that the solubility energy of Nb to the Sr sites is 7.6 eV higher than to the Ti sites. This result also shows the solubility of Nb in the Ti sites.

Figure 11–30 shows spectra of Ti-$L_{2,3}$ ELNES (Energy Loss Near Edge Structure) obtained from the $SrTiO_3$ layer and the Nb-doped $SrTiO_3$ layer in the $SrTiO_3$/Nb-doped $SrTiO_3$/$SrTiO_3$ superlattice. The upper

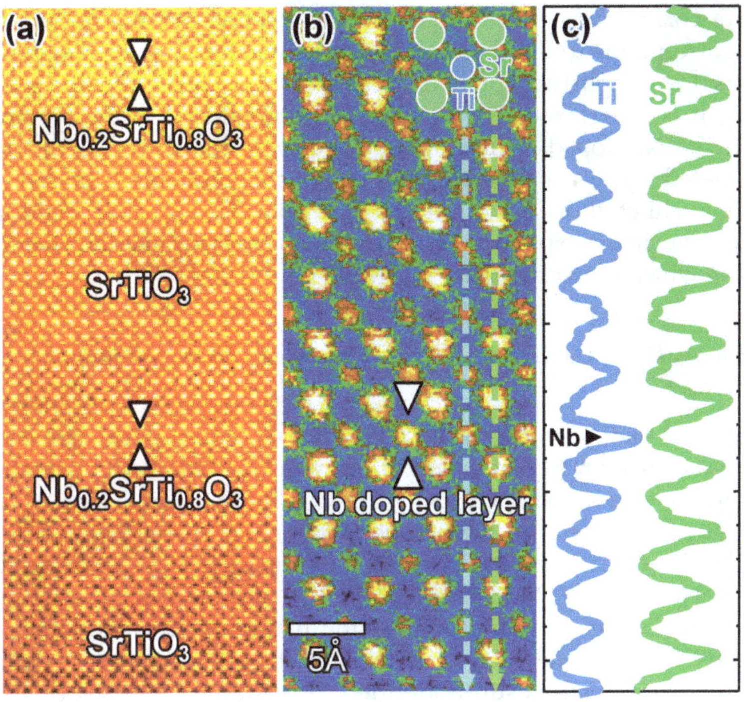

Figure 11–29. (a) HAADF-STEM image of $SrTiO_3/Nb–SrTiO_3$ superlattice, (b) magnified image of the region (a) and (c) image intensity profile of Ti and Sr layers. Reproduced from Ohta et al. (2009) with permission.

Figure 11–30. (*Top*) Experimental and (*bottom*) theoretically calculated Ti–$L_{2,3}$ ELNES obtained from the $SrTiO_3$ layer and the Nb-doped $SrTiO_3$ layer in the $SrTiO_3$/Nb-doped/$SrTiO_3$ superlattice. The calculations were made by the first-principles relativistic multi-electron method. Reproduced from Ohta et al. (2009) with permission.

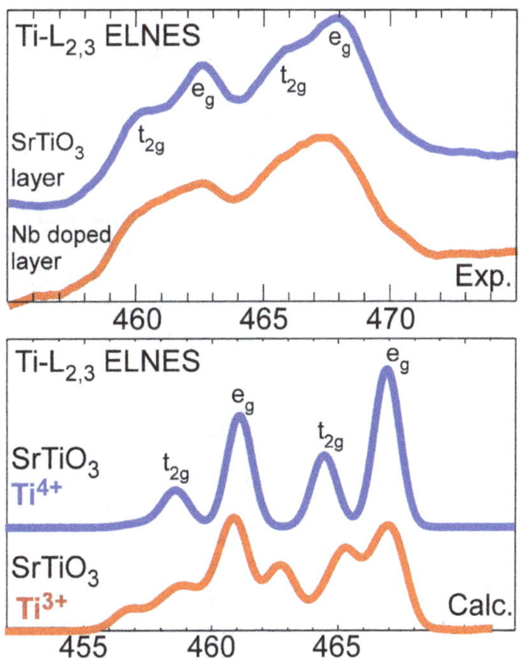

figure corresponds to the experimental spectra, and the bottom figure corresponds to the theoretically calculated spectra. Although four peaks (t_{2g}, e_g split) are apparent in the spectra obtained from the $SrTiO_3$ layer, it is recognized that the peaks are broadened in the spectrum from the Nb-doped $SrTiO_3$ layer. Comparing the theoretical ELNES calculated by the first-principles relativistic multi-electron method (Mizoguchi et al. 2006) with the experimental ELNES, it is found that the change in the experimental spectra is due to the transition from Ti^{4+} to Ti^{3+} which is accompanied by the Nb doping.

11.5.4 Au/TiO_2 Catalyst Interface (Coherent and Incoherent Interface)

Gold (Au) nanoparticles dispersed on metal oxide supports are active catalysts for a variety of chemical reactions, including CO oxidation, propylene epoxidation, and the water gas shift reaction (Haruta 1997). It has been reported that the unique catalytic activities strongly depend on the size of Au nanoparticles and the properties of metal oxide supports (Haruta 1997, Valden et al. 1998), suggesting the importance of gold–support oxide interactions at the nanometer regime. Extensive studies have been devoted to understand the origin of these activities, but the mechanism is still under debate because of the lack of structural information on the active gold on the support oxide surfaces. It is thus essential to characterize the atomic structures of nanosized Au on metal oxide surfaces, in order to truly understand the origin of the unique catalytic activities. It has been well demonstrated that atomic-resolution STEM is a powerful method for directly characterizing metal cluster catalysts on supports at subnanometer dimensions (Nellist and Pennycook 1996, Rashkeev et al. 2007). Here, Au/TiO_2 (110) model catalysts were fabricated and atomic-resolution plan-view STEM imaging was carried out, in order to directly observe the atomic structures of Au nanoparticles on the TiO_2 (110) surface.

A commercially available rutile TiO_2 (110) substrate was thinned down by mechanical polishing followed by ion bombardment to obtain electron transparent TEM samples. The TEM samples were annealed in air at 973 K for 30 min to produce atomically flat (110) surfaces, in accordance with previous reports (Nakamura et al. 2005, Rashkeev et al. 2007, Shibata et al. 2008). High-purity gold (99.95%) was deposited on the TEM sample surface by vacuum evaporation at room temperature.

Figure 11–31(a) shows a HAADF-STEM image of a rutile TiO_2 crystal observed in the <110> projection (Shibata et al. 2009). There are two types of bright spots with different image intensities as indicated by the arrows. These bright spots correspond to the two different Ti-containing columns as shown schematically in Figure 11–31(b). The brighter spots correspond to the atomic columns with both Ti and O atoms (Ti–O columns), while the darker spots correspond to the atomic columns with body-centered Ti atoms only (Ti-only columns), as confirmed by HAADF-STEM simulation.

Figure 11–32(a) shows a typical plan-view HAADF-STEM image of Au nanoparticles on the TiO_2(110) surface (Shibata et al. 2009).

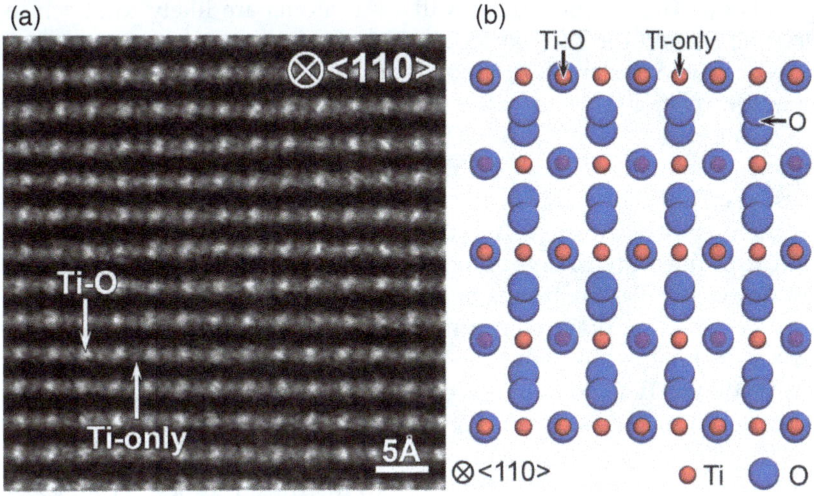

Figure 11–31. (a) Atomic-resolution HAADF-STEM image of a rutile TiO_2 single crystal observed from the <110> direction. (b) The corresponding crystal structure model. Reproduced from Shibata et al. (2009) with permission.

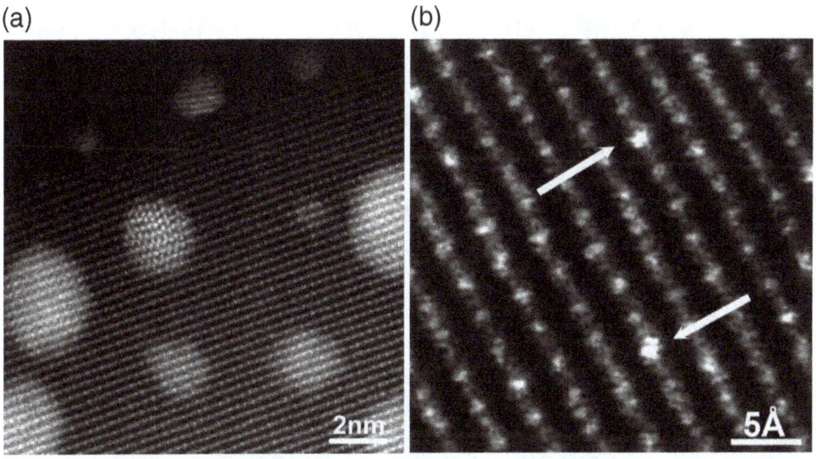

Figure 11–32. (a) Typical HAADF-STEM image of Au nanoparticles deposited on a TiO_2 (110) surface. (b) HAADF-STEM image of single Au atoms on the TiO_2 (110) surface. Reproduced from Shibata et al. (2009) with permission.

It is clearly seen that the HAADF-STEM detects the presence of Au islands on the TiO_2 surface by the strong Z-dependent image contrast (Z: Au = 79, Ti = 22, O = 8). The atomic structure of both Au nanoislands and the TiO_2 support is simultaneously resolved. It is found that the projected sizes of Au islands are in the range of 1 ~ 5 nm with the present deposition condition. In addition, single Au atoms are observed to attach on the TiO_2 surface. Figure 11–32(b) shows single Au atoms attached to specific sites on the TiO_2 (110) surface, as indicated by the arrows. The strong image intensity from the Au single atoms is only found at the positions on top of the Ti–O columns in the [110] projection. This result indicates that Au atoms are likely to attach to specific sites on the TiO_2 surface.

When the Au island size becomes less than 3 nm in size, unique orientation relationships are observed between Au nanoislands and the TiO_2 support (Shibata et al. 2009). Figure 11–33(a)–(e) shows typical atomic-resolution HAADF-STEM plan-view images of Au nanoparticles on a TiO_2 (110) surface, arranged in order of projected particle size. In the case of very small Au islands ($<\sim$2 nm), as in Figure 11–33(a) and (b), the strong image intensities corresponding to Au atoms are found on the Ti–O columns and on the O columns along the trough between the Ti-containing columns (along the arrows in the magnified image in Figure 11–33(f)). The strong preference of Au atom attachments on the Ti–O columns is consistent with the Au single atom case. The present results suggest that Au atoms in the small Au islands also preferentially attach to specific sites on the TiO_2 surface, and therefore form an epitaxial coherent-type hetero-interface (Ikuhara et al. 1995, Wolf 1992). These epitaxial Au islands are not only small in size, but are also very thin (a few atomic layers) as estimated from the STEM image intensity profiles. If the Au atoms on the Ti–O columns are in the first Au atomic layer, Au atoms in the trough between the Ti-containing columns (on top of O columns) can be considered as second layer atoms, because nearest Au–Au distances would be too close to each other (less than 80% of the stable Au–Au interatomic distance in bulk Au) if these Au atoms were on the same first atomic layer. This proposed Au bilayer stacking sequence is similar to the (110) stacking ($\{110\}_{Au}//\{110\}_{TiO_2}$, $\langle 100\rangle_{Au}$ // $\langle 110\rangle_{TiO_2}$) or (100) stacking ($\{100\}_{Au}//\{110\}_{TiO_2}$, $\langle 110\rangle_{Au}$ // $\langle 110\rangle_{TiO_2}$) of fcc Au on TiO_2 (110), despite the extremely large lattice mismatch (>20%) in one direction.

On the other hand, if the Au particle size becomes larger ($>\sim$3 nm), as in Figure 11–33(c)–(e), various orientation relationships exist between the Au nanoparticles and the TiO_2 substrate. Frequently observed Moiré fringes suggest that the lattices are gradually displaced or rotated from each other. These results indicate that the Au/TiO_2 (110) interface is not a coherent interface, but an incoherent-type hetero-interface (Ikuhara et al. 1994, Wolf 1992), which is often found in large-mismatched metal/oxide hetero-interface systems (Ikuhara et al. 1994, Matsunaga et al. 2006). Figure 11–34 shows a histogram of the formation of coherent or incoherent interfaces as a function of the projected Au nanoparticle size. Although the thickness of the Au nanoparticles along the beam direction is difficult to precisely estimate, our systematic observations strongly suggest that there is a structural transition at the Au/TiO_2 (110) interface for Au particles around 2–3 nm in size. The interface structural transition (coherent to incoherent) depending on Au particle size has been also reproduced in DFT calculations (Shibata et al. 2009). These calculations also predict that the electronic structures of deposited Au islands can be strongly altered by the changes in interface structures. The existence of unique nanoscale interface atomic structures is an essential factor in understanding the origin of the size-dependent catalytic activity of Au nanoparticles supported on metal oxides.

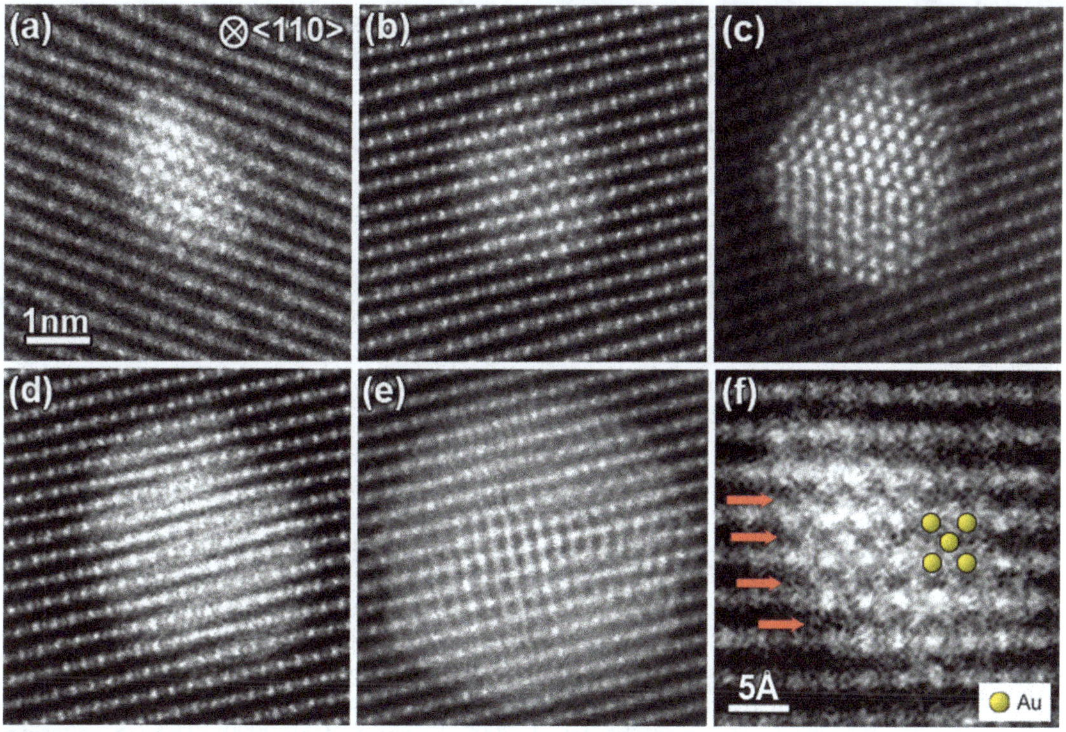

Figure 11–33. (a)–(e) HAADF-STEM images of Au nanoparticles arranged in order of projected particle size. The lattice coherency between Au nanoparticles and the TiO_2 substrate clearly changes according to the Au particle size ((a) and (b) are coherent, but (c)–(e) are incoherent). (f) Magnified image of the epitaxial Au structure is shown in (a). Reproduced from Shibata et al. (2009) with permission.

Figure 11–34. A histogram of the formation of coherent or incoherent interfaces as a function of Au nanoparticle lateral size estimated from the HAADF-STEM images. Reproduced from Shibata et al. (2009) with permission.

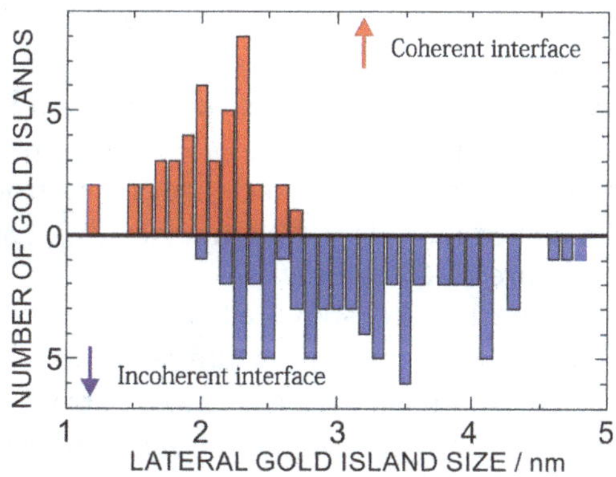

11.6 Another Application to Ceramics

11.6.1 Ordered Structures of Ca in $Ca_{0.33}CoO_2$ Thin Films

Ca_xCoO_2, a kind of layered cobalt oxide, has attracted increasing research interest because of its high potential for thermoelectric (TE) devices and other electronic and magnetic applications (Sugiura et al. 2006, Takada et al. 2003, Wang et al. 2003). It consists of alternately stacked sheets of Ca ions and CoO_2^- layers along the c-axis direction. The content of Ca ions can vary over a wide range and forms some ordered structures at certain concentrations, leading to correspondingly different physical properties (Sugiura et al. 2006, Roger et al. 2007). Two well-defined ordered structures, a $\sqrt{3}a \times \sqrt{3}a$ hexagonal superstructure and a $2a \times \sqrt{3}a$ orthorhombic superstructure, have been reported to exist in Ca_xCoO_2 bulk, corresponding to $x = 0.33$ and $x = 0.5$, respectively (Yang et al. 2006). However, it has been found that the $\sqrt{3}a \times \sqrt{3}a$ hexagonal superstructure in a $Ca_{0.33}CoO_2$ thin film can be transformed to the $2a \times \sqrt{3}a$ orthorhombic superstructure without changing the Ca concentration by post-annealing in air, accompanied by a metal–insulator transition (Sugiura et al. 2009). Direct observation of the ordered structures of Ca ions at the atomic scale can provide valuable information for clarifying this controversy on crystal structure and understanding the relationship between cation ordering and physical properties.

In this section, the two ordered structures of Ca ions and vacancies in $Ca_{0.33}CoO_2$ thin films are directly observed and differentiated by HAADF-STEM. The $Ca_{0.33}CoO_2$ thin films were fabricated by the topotactic ion-exchange treatment using a $Na_{0.7}CoO_2$ epitaxial film as a precursor, which was grown on the c (0001) plane of α-Al_2O_3 by the reactive solid-phase epitaxy method. The as-prepared $Ca_{0.33}CoO_2$ thin film was annealed at 400°C for 1 h in air to change the arrangement of Ca^{2+} ions (Ohta et al. 2005, Sugiura et al. 2009). The Ca ordering in the intercalation plane of the $\sqrt{3}a \times \sqrt{3}a$ (as-prepared thin film) and $2a \times \sqrt{3}a$ (annealed thin film) superstructures are illustrated in Figure 11–35(a) and (b), respectively. The black lines represent the Co lattice and the red balls are Ca ions. It can be clearly seen that these two superstructures are easily discriminated by selected-area diffraction (SAD) along the $[1\bar{1}00]$ zone axis and by HAADF-STEM along the $[11\bar{2}0]$ zone axis. Particularly, when these superstructures are observed from the $[11\bar{2}0]$ direction, the contrast of Ca ions can be imaged by HAADF-STEM, and in the $2a \times \sqrt{3}a$ superstructure they take the form of closely spaced pairs, which is different from the uniformly distributed Ca ions in the $\sqrt{3}a \times \sqrt{3}a$ superstructure, as illustrated by dashed lines in Figure 11–35(a) and (b).

Figure 11–35(c) shows a SAD pattern of the as-prepared $Ca_{0.33}CoO_2$ film when the electron beam is parallel to the $[1\bar{1}00]$ zone axes. The superlattice reflections (marked by arrows) at 1/3 and 2/3 of $[11\bar{2}0]$ are due to the Ca ordering and clearly indicate the presence of the $\sqrt{3}a \times \sqrt{3}a$ superstructure. On the other hand, Figure 11–35(d) shows a SAD pattern obtained from the annealed $Ca_{0.33}CoO_2$ film along the $[1\bar{1}00]$ zone axis. The superlattice reflections (marked by arrows) reveal the $2a \times \sqrt{3}a$ superstructure of Ca ions, consistent with the in-plane XRD results (Huang et al. 2008).

Figure 11–36(a) and (b) shows typical HAADF-STEM images of the $\sqrt{3}a \times \sqrt{3}a$ hexagonal and $2a \times \sqrt{3}a$ orthorhombic superstructures, respectively. It can be clearly seen that the thin films are composed of alternate stacking of bright layers (Co) and in between dark layers (Ca). Because the atomic number of oxygen ($Z = 8$) is much smaller

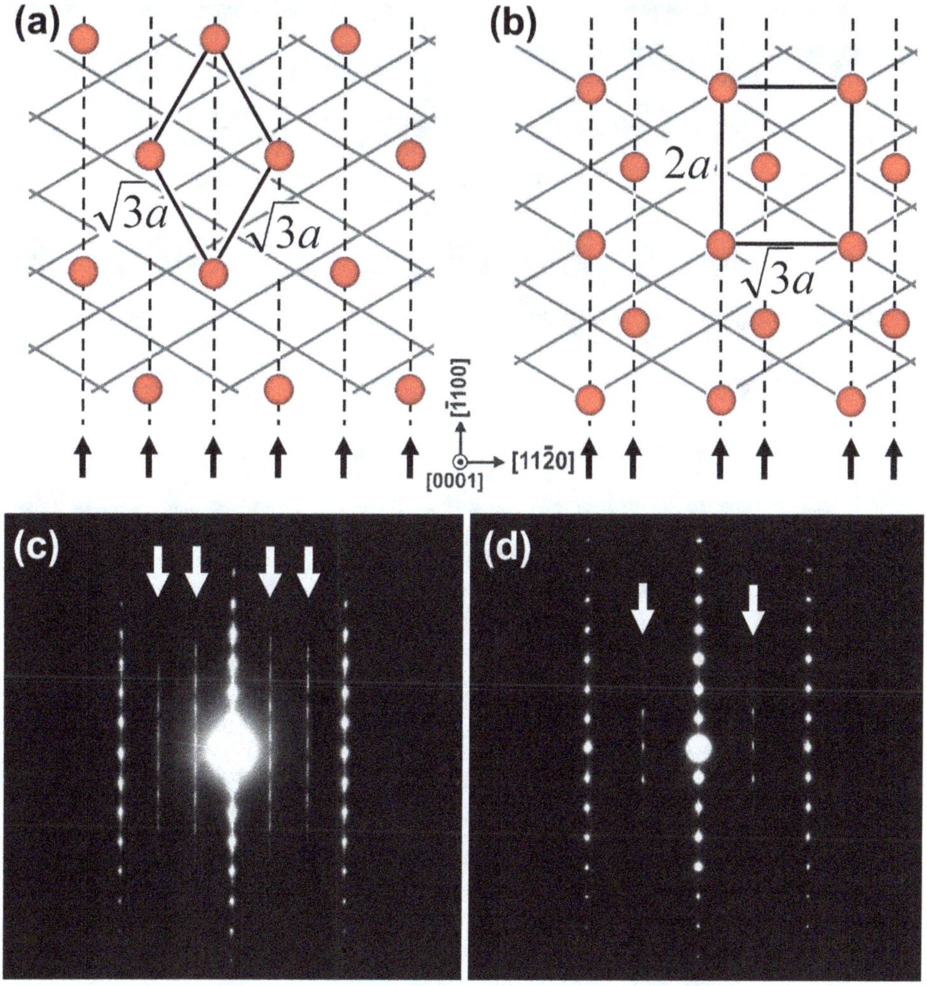

Figure 11–35. Schematics of Ca ordering in the intercalation plane of (a) the $\sqrt{3}a \times \sqrt{3}a$ hexagonal superstructure and (b) the $2a \times \sqrt{3}a$ orthorhombic superstructure. The *black lines* represent the Co lattice and the *red balls* are Ca ions. The *dashed lines* indicate the Ca columns along the [11$\bar{2}$0] zone axis. Selected-area electron diffraction patterns of (c) the $\sqrt{3}a \times \sqrt{3}a$ superstructure and (d) the $2a \times \sqrt{3}a$ superstructure along the [1$\bar{1}$00] zone axes. *Arrows* indicate the superlattice reflections from Ca ordering. Reproduced from Huang et al. (2008) with permission.

than Co (Z = 27), the oxygen columns cannot be seen in this imaging condition. The HAADF-STEM image shown in Figure 11–36(b) exhibits pairs of Ca ions in the $2a \times \sqrt{3}a$ orthorhombic superstructure, which is easily distinguished from the uniformly distributed Ca ions in the $\sqrt{3}a \times \sqrt{3}a$ hexagonal superstructure shown in Figure 11–36(a), when observed along the [11$\bar{2}$0] zone axes. In order to get more detailed information about the local structure of Ca layers, an intensity profile was taken along the arrow in Figure 11–36(b). From the intensity of each Ca column, the Ca pairs can be identified, as shown in Figure 11–37. On the other hand, the intensities of Ca columns are not uniform but fluctuate with a relatively large amplitude, which indicates that there

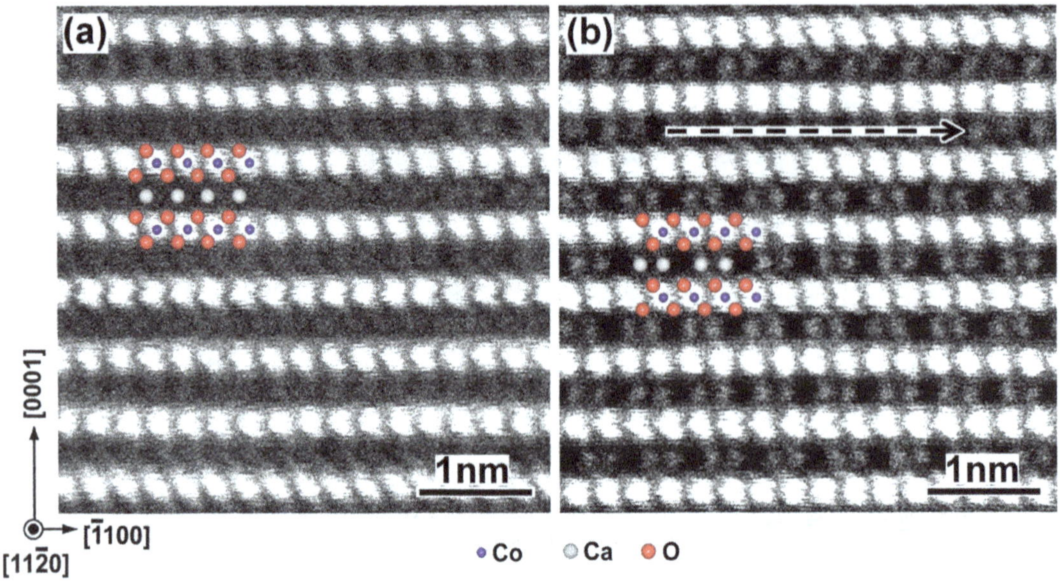

Figure 11–36. HAADF-STEM images of (a) the $\sqrt{3}a \times \sqrt{3}a$ hexagonal superstructure and (b) the $2a \times \sqrt{3}a$ orthorhombic superstructure along the [11$\bar{2}$0] zone axes. Insets are the corresponding crystal structures. Reproduced from Huang et al. (2008) with permission.

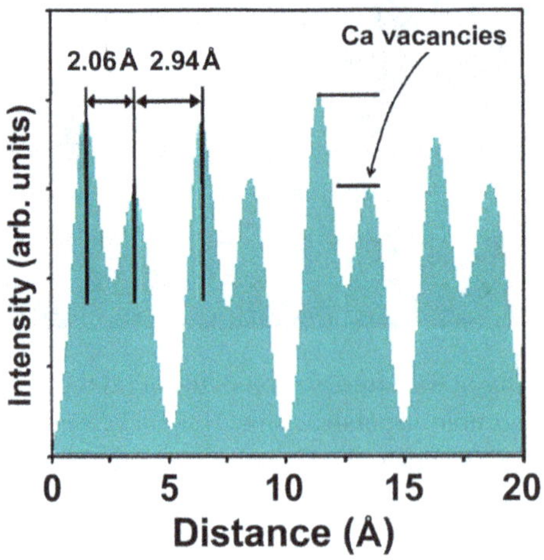

Figure 11–37. Intensity profile of Ca layer along the arrowed line in Figure 11–36(b). The Ca sites showing lower intensity contain Ca vacancies. Reproduced from Huang et al. (2008) with permission.

are different numbers of Ca^{2+} ions along the electron beam direction in each Ca column. It might be that some Ca vacancies have formed in the $Ca_{0.33}CoO_2$ thin film. In fact, the $2a \times \sqrt{3}a$ orthorhombic superstructure has cation sites of $x = 0.5$ (Yang et al. 2006). However, the concentration of Ca in the present sample is $x = 0.33$, which can only partly occupy the Ca sites in the $2a \times \sqrt{3}a$ orthorhombic superstructure, leaving many Ca vacancies in the structure (Huang et al. 2008).

Furthermore, we found that the darker Ca columns always appear at the Ca sites which lie in between two adjacent cobalt ions. It requires extra energy for occupation relative to the other Ca sides due to the relatively large electrostatic repulsion between the adjacent Ca^{2+} and Co^{3+}/Co^{4+} ions. This gives a reasonable explanation for the preferred sites of Ca vacancies (Huang et al. 2008). Thus, the HAADF-STEM observations have confirmed the existence of Ca vacancies in the $2a \times \sqrt{3}a$ orthorhombic superstructure, important information for understanding the physical properties of these layered cobalt oxides.

11.6.2 Determination of Li Ion Sites in $LiFePO_4$ Crystals

The crystal structure of $LiFePO_4$ has two kinds of octahedral interstitial sites: edge-sharing "*M1*" for Li and corner-sharing "*M2*" for Fe. Control of the cation occupancy between *M1* and *M2* sites in $LiFePO_4$ has been an important issue for understanding crystallographic stability under the various chemical environments and improving the mass transport behavior as a cathode material for Li ion batteries (Henderson et al. 1996, Papike and Cameron 1976). So far, most of the analytical investigations on cation disordering, the so-called anti-site exchange, were based on neutron powder diffraction. However, such a macroscopic diffraction method can only show the overall distribution of the anti-site defects between the *M1* and *M2* sites. In this regard, direct atomic-level observation of the anti-site defects is essential to precisely understand the local distribution of the defects in the crystal lattice. To this end, HAADF-STEM equipped with an aberration corrector was unprecedentedly applied to visualize the lithium columns containing anti-site iron.

Figure 11–38(a) shows a HAADF-STEM image in the [010] projection of a $LiFePO_4$ crystal where the two-dimensional atom arrays are superimposed. Li (yellow sphere) and Fe (red sphere) are located in the *M*1 and *M*2 sites, respectively, forming an ordered orthorhombic olivine structure in a space group of *Pnma*. The periodically arrayed bright spots in Figure 11–38(a) directly represent Fe and P columns. Even though the projected distance between Fe and P, 1.26 Å, lies below the resolution limit of the aberration-corrected STEM, the Fe and P columns are not obviously discriminated and instead are represented as a single oval-shaped feature in this case, consistent with image simulations.

Although each atom column for Li shows no contrast at all because of its low atomic number, Figure 11–38(a) does reveal a visible contrast for some of the Li columns, which means that Fe atoms must occupy the Li sites as anti-site defects. Judging from the fact that the overall *M* site exchange between the Li and Fe sites is as low as 1% based on the neutron diffraction result (Chung et al. 2008a), such a remarkable variation in *M1* sites in Figure 11–38(a) implies that the exchange defects in the Li sites are localized in a preferential orientation. Such an orientation-dependent array of anti-site defects is verified by observing from two directions; while the HAADF image from the [010] projection as in Figure 11–38(a) clearly demonstrates the presence of anti-site iron cations in some of the lithium sites, almost no detectable white contrast in the lithium sites for the anti-site defects is found for HAADF images taken from the [001] projection as shown in Figure 11–38(b). As the intensity of each atomic column in the HAADF images is critically dependent on the average *Z*, the absence of observable intensities in the lithium columns in the [001] projection directly indicates

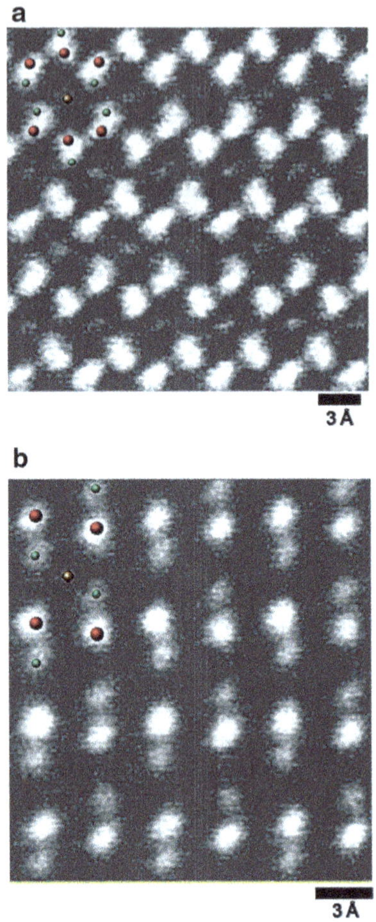

Figure 11–38. HAADF-STEM images in (a) [010] and (b) [001] projections and superimposed atomic array indicating the locations of each atom [Li (*yellow*), Fe (*red*), and P (*green*)]. The image in the [001] projection demonstrates that some of the lithium columns have a bright contrast with significant intensity, while maintaining an ordered arrangement of the iron–phosphorous contours. No lithium columns with visible intensity are observed in HAADF images obtained in the [001] projection. Reproduced from Chung et al. (2008a) with permission.

the strong preferential arrangement of anti-site iron cations along the [010] direction.

The *M*1 sites are edge-sharing octahedral interstitials and thus the distance between the neighboring cations in the *M*1 sites is shorter than that in the corner-sharing *M*2 sites, which accordingly leads to the distorted shape of the oxygen octahedra for the *M*1 sites so as to minimize the electrostatic repulsion between the cations. If some of Fe^{2+} ions occupy the *M*1 sites instead of Li^+, the induced additional electrostatic repulsion due to their higher valence state eventually results in a structural instability. Therefore, the local aggregation of disordered Fe in the Li sites would not be thermodynamically favorable, and therefore the cation disordering could be suppressed by annealing at a higher temperature or possibly for a longer period of time (Chung et al. 2008b). This presumption was experimentally examined through annealing the samples at different temperatures, as shown in Figure 11–39. The HAADF image in Figure 11–39(a) shows a detectable contrast for some of the *M1* sites (red arrows) in the sample annealed at 600°C. By contrast, any visible intensity in the Li columns was practically undetectable in the 800°C-annealed sample, as represented in Figure 11–39(b). To clarify that the bright contrast of the visible Li

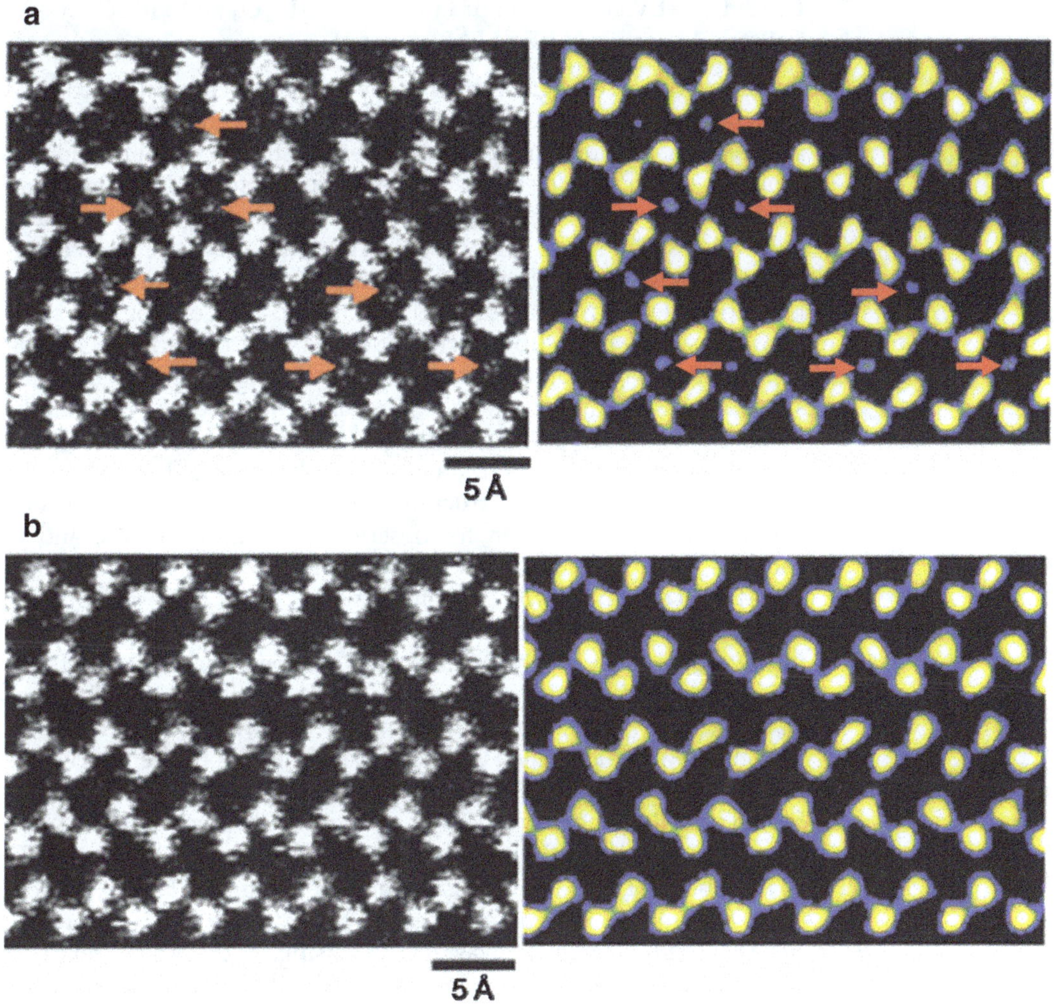

Figure 11–39. HAADF-STEM images of $LiFePO_4$ crystals along a [010] zone axis. (a) The images clearly show that a significant number of Li columns have a bright contrast for a sample annealed at 600°C (*red arrows*) while maintaining the ordered arrangement of the Fe–P contours. (b) No Li columns with visible intensity are observed when annealing at 800°C. Corresponding deconvoluted images are provided in color on the *right*. Reproduced from Chung et al. (2008b) with permission.

columns does not originate from statistical noise, a deconvolution processing technique was carried out on the raw HAADF images. Each deconvoluted image is shown in color, consistently confirming the anti-site defects in the Li columns.

Assuming that the iron cations are randomly distributed in a $LiFePO_4$ crystal, they may block lithium transport along the fastest diffusion path. On the other hand, such blockage by anti-site iron ions can be avoided and further lithium transport can be even enhanced if they are localized into just a few columns; in particular, by forming a one-dimensional passage along the [010] axis for lithium transport. Therefore, this atomic-scale analysis suggests that the distribution of anti-site defects in $LiFePO_4$ can be modified to improve the lithium intercalation/deintercalation process.

11.6.3 Direct Visualization of Li Ions in $LiMn_2O_4$ and $LiCoO_2$ Cathode Materials (ABF STEM Imaging)

Ceramics, in which light elements are the main constituent components, are important as energy and environment-related materials. The properties of ceramics are thus dependent on the atomic sites and distribution of the light elements in the microstructures. It is, therefore, necessary to directly observe light elements in ceramics to understand the origin of their functional properties. Recently, we have reported that annular bright field (ABF)-STEM imaging, whereby an annular detector is positioned within the bright field region in an atomic-resolution scanning transmission electron microscope, is a very powerful technique to produce images showing both light and heavy element columns simultaneously (Findlay et al. 2009). In this section, the ABF STEM technique is demonstrated to directly observe Li and O ions in $LiCoO_2$ and $LiMn_2O_4$ battery materials (Ikuhara et al. 2010, Huang et al. 2011).

According to the crystal structure of $LiMn_2O_4$ spinel, the [110] zone axis is the best orientation for observation because Li, Mn, and O columns are separated from each other. Figure 11–40(a) shows a typical HAADF-STEM image with the detection angle of 92–228 mrad obtained from a $LiMn_2O_4$ particle viewed along the [110] zone axis. The diamond arrangement of the Mn columns can be clearly observed, but the contrast of O and Li columns is invisible. Figure 11–40(b) shows an ABF-STEM image, using a 6–25 mrad detection angle, of a $LiMn_2O_4$ particle viewed along the [110] zone axis. The Li ions are clearly visible together with the O and Mn columns in the image, as indicated by the model structure overlaid in the figure. The image contrast was also confirmed by ABF image simulations. Figure 11–41 shows ABF-STEM image of a $LiCoO_2$ crystal projected along the $[11\bar{2}0]$ direction, indicating that Li ion columns can be clearly observed between two oxygen layers. This image was obtained using an 8–25 mrad detection angle, and oxygen and Co columns are also observed simultaneously with good signal-to-noise ratio. The sites of respective ion columns are consistent with the atomic model structure overlaid on the figure. These results indicate that ABF techniques in Cs-corrected STEM will be very useful to directly characterize light elements in any ceramic materials.

11.6.4 Direct Visualization of Fluorine Dopants in Iron Arsenide Superconductor (STEM EELS Mapping)

The STEM EELS mapping technique is another useful method to characterize ceramics. Remarkable progress has recently been made in atomic-scale chemical mapping based on electron-energy-loss spectroscopy (EELS) (Kimoto et al. 2007, Muller et al. 2008, Varela et al. 2004). Combined with the Cs-corrected HAADF-STEM image, the method can be a powerful tool for the investigation of microscopic modification of materials which are responsible for various materials properties. In this section, the recently discovered $LaFeAsO_{1-x}F_x$ superconductor is shown as a model system for STEM EELS mapping.

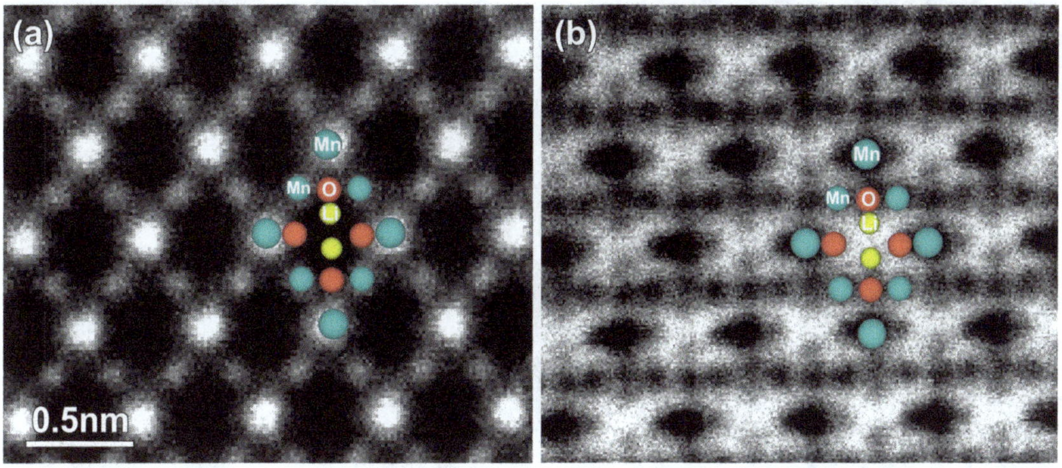

Figure 11–40. (a) HAADF-STEM and (b) ABF-STEM images of $LiMn_2O_4$ powder observed along the [110] zone axis. Reproduced from Huang et al. (2011) with permission.

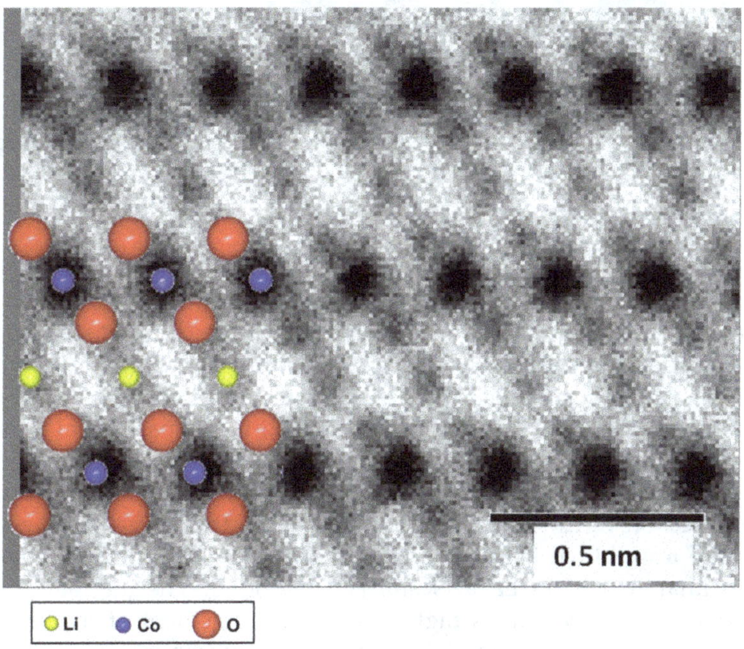

Figure 11–41. (a) ABF-STEM image of a $LiCoO_2$ crystal projected along the $[11\bar{2}0]$ direction, indicating that Li ions can be clearly observed between oxygen layers. Reproduced from Ikuhara et al. (2010) with permission.

Although the fluorine ion doping is critical for the superconductivity, so far no microscopic observations have been made on the dopant state in this system because the fluorine is believed to be substituted for the oxygen sites and invisible to conventional imaging techniques. Figure 11–42 shows a HAADF-STEM image of the fluorine-doped LaFeAsO observed along the [100] zone axis. In HAADF images, the

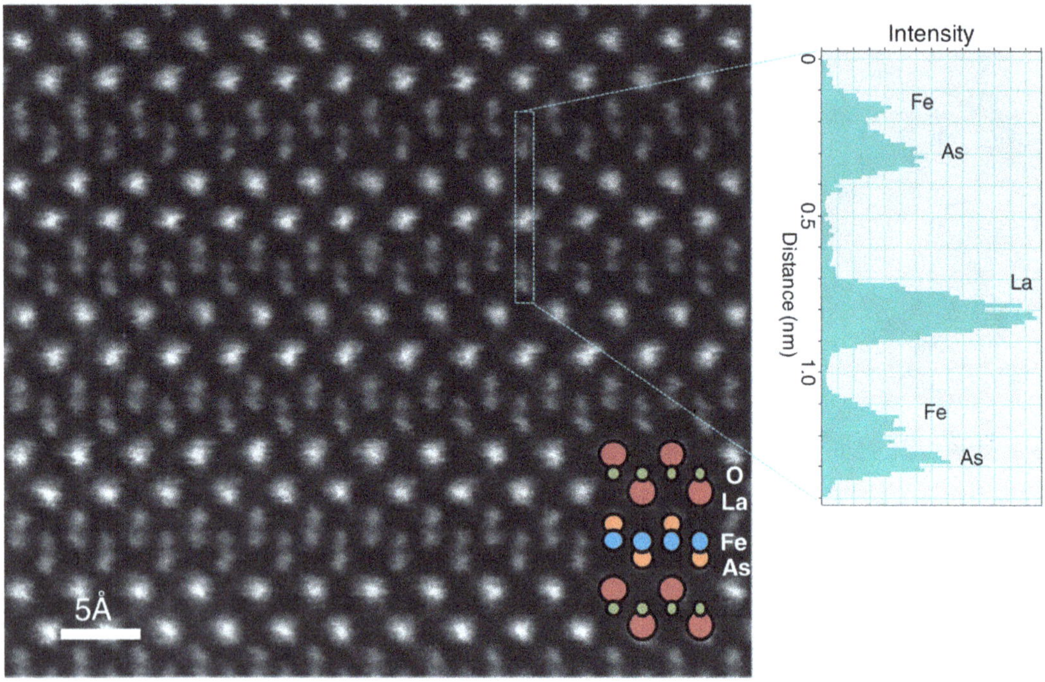

Figure 11–42. HAADF-STEM image of the $LaFeAsO_{0.9}F_{0.1}$ compound observed along the [100] zone axis. Reproduced from Tohei et al. (2009) with permission.

layered structure of the crystal is clearly imaged by Z-contrast as bright zig-zag spots (La) and arrays of dumbbells (Fe–As) in between. Oxygen and fluorine are not visible in HAADF images at the present observation condition. To reveal the hidden fluorine dopants, EELS measurements were performed with an electron probe focused to atomic size to detect the location of the doped fluorine. Spectroscopic imaging based on EELS allows imaging of lighter atoms such as fluorine that are invisible in HAADF images. Figure 11–43 shows EELS spectrum imaging of the $LaFeAsO_{0.9}F_{0.1}$ compound. The mapping with La-$M_{4,5}$ and Fe-$L_{2,3}$ edges highlights the arrangement of the zig-zag lanthanum ions and the straight arrangement of iron ions, which coincides with the atomic sites in the HAADF image (Figure.11–43(b)–(d)). The most remarkable observation is the direct imaging of the fluorine ion dopants. The spatial distribution of fluorine ions, which is undetectable by HAADF imaging, is clearly shown in the spectroscopic image from the fluorine K edge (Figure 11–43(e)). The intensity of the fluorine signal increases at the middle of the lanthanum zig-zag layer, proving the fluorine substitution into the oxygen sites. The present result demonstrates the power of combined observation of Z-contrast imaging by HAADF and spectroscopic imaging by EELS for investigating dopants in materials at the atomic scale (Tohei et al. 2009).

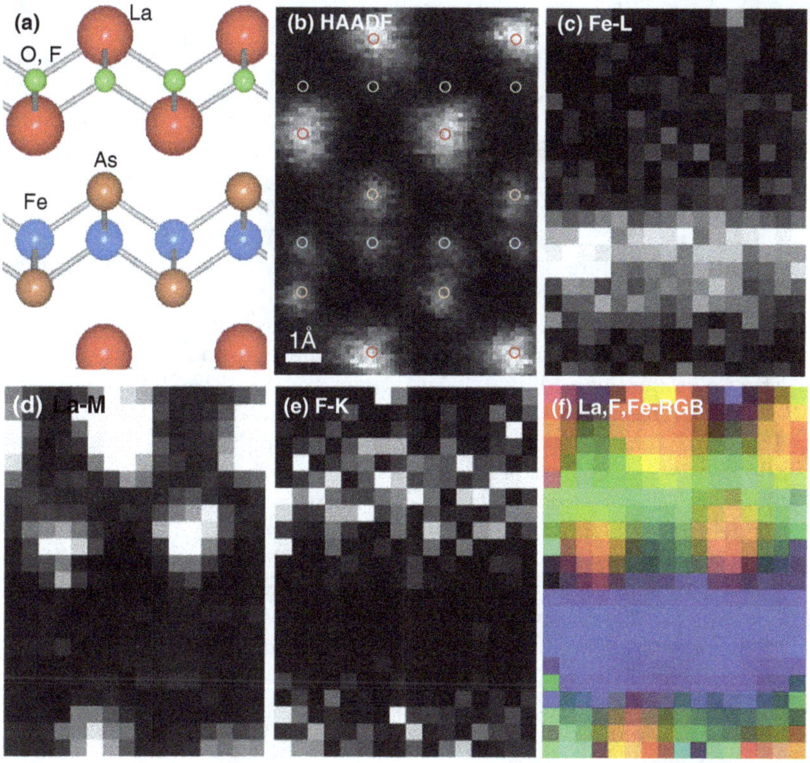

Figure 11–43. Atomic structure and STEM EELS spectrum imaging of fluorine-doped LaFeAsO. Reproduced from Tohei et al. (2009) with permission.

11.7 Conclusion

In this chapter, geometrical treatments of grain boundaries and hetero-interfaces were reviewed together with experimental results for ceramics mainly obtained by STEM. Throughout the manuscript, the emphasis was on how the grain boundary and interface characters are related to their atomic structure, chemical composition, and chemical bonding state, which are characterized by TEM/STEM techniques. The grain boundary character is an important concept even in ceramics, and some grain boundaries in ceramics can be described as dislocation boundaries and CSL boundaries. The idea of the structural unit is very useful in considering grain boundary atomic structures and has been successfully applied to ceramics. In addition to the structural features, chemistry is also crucial for considering grain boundaries in ceramics. Grain boundary segregation and the formation of an amorphous layer fall into this category, and their chemical properties strongly affect the bulk properties of ceramics. The hetero-interface can be classified into coherent, semicoherent, incoherent, and pseudo-semicoherent interfaces, which can be characterized by STEM imaging techniques. It is thus concluded that STEM is a very powerful experimental technique that allows us to ascertain the atomic structure and chemistry of grain boundaries and hetero-interfaces in ceramics.

References

L.J. Allen, S.D. Findlay, M.P. Oxley, C.J. Rossouw, Lattice-resolution contrast from a focused coherent electron probe. Part I. Ultramicroscopy **96**, 47 (2003)

S. Azuma, N. Shibata, S.D. Findlay, T. Mizoguchi, T. Yamamoto, Y. Ikuhara, HAADF STEM observations of a Σ13 grain boundary in α-Al_2O_3 from two orthogonal directions. Phil. Mag. Lett. **90**, 539 (2010).

R.W. Balluffi, A. Brokman, A.H. King, CSL DSC lattice model for general crystal-crystal boundaries and their line defects. Acta Met. **30**, 1453 (1982)

D.G. Bandon, Structure of high-angle grain boundaries. Acta Metall. **8**, 1221 (1966)

P.F. Becher, G.S. Painter, M.J. Lance, S. Ii, Y. Ikuhara, Direct observations of debonding of reinforcing grains in silicon nitride ceramics sintered with yttria plus alumina additives. J. Am. Ceram. Soc. **88**, 1222 (2005)

J.B. Bilde-Sørensen, B.F. Lawlor, T. Geipel, P. Pirouz, A.H. Heuer, K.P.D. Lagerlöf, On basal slip and basal twinning in sapphire (α-Al_2O_3). 1. Basal slip revisited. Acta Mater. **44**, 2145 (1996)

W. Bollmann, *Crystal Defects and Crystalline Interfaces* (Springer-Verlag, Berline-Heidelberg-New York, 1970)

M. Brandbyge, J.-L. Mozos, P. Ordejón, J. Taylor, K. Stokbro, Density-functional method for nonequilibrium electron transport. Phys. Rev. B **65**, 165401 (2002)

J.P. Buban, K. Matsunaga, J. Chen, N. Shibata, W.Y. Ching, T. Yamamoto, Y. Ikuhara, Grain boundary strengthening in alumina by rare earth impurities. Science **311**, 212 (2006)

I.-W. Chen, P.F. Becher, M. Mitomo, G. Petzow, T.-S. Yen (eds.), Silicon nitride ceramics scientific and technological advances. MRS Proc. **287** (Mater. Res. Soc. Pittsburgh, PA) (1993)

M.F. Chisholm, S. Kumar, P. Hazzledine, Dislocations in complex materials. Science **307**, 701 (2005)

J. Cho, M.P. Harmer, M. Chan, J.M. Rickman, A.M. Thompson, Effect of yttrium and lanthanum on the tensile creep behavior of aluminum oxide. J. Am. Ceram. Soc. **80**, 1013 (1997)

S.Y. Choi, S.Y. Chung, T. Yamamoto, Y. Ikuhara, Direct determination of dopant site selectivity in ordered perovskite $CaCu_3Ti_4O_{12}$ polycrystals by aberration-corrected STEM. Adv. Mater. **21**, 885 (2009)

D.R. Clarke, On the equilibrium thickness of intergranular glass phases in ceramic materials. J. Am. Ceram. Soc. **70**, 15 (1987)

D.R. Clarke, Varistor ceramics. J. Am. Ceram. Soc. **82**, 485 (1999)

S.-Y. Chung, S.-Y. Choi, T. Yamamoto, Y. Ikuhara, Atomic-scale visualization of antisite defects in $LiFePO_4$. Phys. Rev. Lett. **100**, 125502 (2008a)

S.-Y. Chung, S.-Y. Choi, T. Yamamoto, Y. Ikuhara, Orientation-dependent arrangement of antisite defects in lithium iron (II) phosphate crystals. Angew. Chem. Int. Ed. **47**, 1 (2008b)

S. Fabris, C. Elsässer, Σ13 $(10\bar{1}4)$ twin in α-Al_2O_3: A model for a general grain boundary. Phys. Rev. B **64**, 245117 (2001)

S.D. Findlay, N. Shibata, H. Sawada, E. Okunishi, Y. Kondo, T. Yamamoto, Y. Ikuhara, Robust atomic resolution imaging of light elements using scanning transmission electron microscopy. Appl. Phys. Lett. **95**, 191913 (2009)

F. C. Frank, Crystal dislocations – elementary concepts and definitions. Phil. Mag. **42**, 809 (1951)

J.D. Gale, GULP: A computer program for the symmetry-adapted simulation of solids. J. Chem. Soc. Faraday Trans. **93**, 629 (1997)

T. Gemming, S. Nufer, W. Kurtz, M. Ruhle, Structure and chemistry of symmetrical tilt grain boundaries in α-Al_2O_3: I, bicrystals with "clean" interface. J. Am. Ceram. Soc. **86**, 581 (2003)

M. Haider, H. Rose, S. Uhlemann, B. Kabius, K. Urban, Towards 0.1 nm resolution with the first spherically corrected transmission electron microscope. J. Electron. Microsc. **47**, 395 (1998)

M. Haruta, Size- and support-dependency in the catalysis of gold. Catal. Today **36**, 153 (1997)

C.M.B. Henderson, K.S. Knight, S.A.T. Redfern, B.J. Wood, High-temperature study of octahedral cation exchange in olivine by neutron powder diffraction. Science **271**, 1713 (1996)

M.J. Hoffmann, H. Gu, R.M. Cannon, Influence of the interfacial properties on the microstructural development and properties of silicon nitride ceramics. Interfacial Eng. Optimized Properties II **586**, 65–74 (Mater. Res. Soc. Warrendale, PA, 2000)

R. Huang, Y.H. Ikuhara, T. Mizoguchi, S.D. Findlay, A. Kuwabara, C.A.J. Fisher, H. Moriwake, H. Oki, T. Hirayama, Y. Ikuhara, Oxygen-vacancy ordering at surfaces of lithium manganese (III, IV) oxide spinel nanoparticles. Angewandte Chemie, in press (2011)

R. Huang, T. Mizoguchi, K. Sugiura, H. Ohta, K. Koumoto, T. Hirayama, Y. Ikuhara, Direct observations of Ca ordering in $Ca_{0.33}CoO_2$ thin films with different superstructures. Appl. Phys. Lett. **93**, 181907 (2008)

Y. Ikuhara, Grain boundary and interface structures in ceramics. J. Ceram. Soc. Jpn. **109**, S110 (2001)

Y. Ikuhara, Nanowire design by dislocation technology. Prog. Mater. Sci. **54**, 770 (2009)

Y. Ikuhara, R. Huang, S.D. Findlay, T. Mizoguchi, N. Shibata, T. Hirayama, T. Yamamoto, H. Oki, New approach to characterize ceramics by cs-corrected STEM-Three dimensional observation and light elements visialization-AMTC Letters, vol. 2, p. 14 (2010) (ISSN 1882-9465)

Y. Ikuhara, H. Kurishita, H. Yoshinaga, Grain boundary and high temperature strength of sintered SiC. J. Ceram. Soc. Jpn. Inter. Ed. **95**, 592 (1987)

Y. Ikuhara, H. Kurishita, H. Yoshinaga, Grain boundary structure and mechanical properties of covalent-bonded ceramics. in *Proceedings of the 2nd International Conference on Composition Interfaces*, Cleveland, Elsevier Science Publisher, 1988, p. 673

Y. Ikuhara, P. Pirouz, High resolution transmission electron microscopy studies of metal/ceramics interfaces. Microsc. Res. Tech. **40**, 206 (1998)

Y. Ikuhara, P. Pirouz, A.H. Heuer, S. Yadavalli, C.P. Flynn, Structure of V-Al_2O_3 interfaces grown by molecular-beam epitaxy. Philos. Mag. A. **70**, 75 (1994)

Y. Ikuhara, P. Pirouz, S. Yadavalli, C.P. Flynn, Structure of V-MGO and MGO-V interfaces. Philos. Mag. A **72**, 179 (1995)

Y. Ikuhara, H. Suzuki, T. Suzuki, High-temperature deformation and fracture behavior of Al_2O_3-Y_2O_3 doped silicon nitride. Mater. Trans. JIM **37**, 430 (1996)

Y. Ikuhara, P. Thavorniti, T. Sakuma, Solute segregation at grain boundaries in superplastic SiO_2-doped TZP. Acta Mater. **45**, 5275 (1997)

Y. Ikuhara, T. Watanabe, T. Saito, H. Yoshida, T. Sakuma, Atomic structure and chemical bonding state of sapphire bicrystal. Mater. Sci. Forum **294**, 273 (1999)

Y. Ikuhara, H. Yoshida, T. Sakuma, Impurity effects on grain boundary strength in structural ceramics. Mater. Sci. Eng. **A319-321**, 24–30 (2001)

M. Imaeda, T. Mizoguchi, Y. Sato, H.S. Lee, S.D. Findlay, N. Shibata, T. Yamamoto, Y. Ikuhara, Atomic structure, electronic structure, and defect energetics in [001](310)Σ5 grain boundaries of $SrTiO_3$ and $BaTiO_3$. Phys. Rev. B **78**, 245320 (2008)

K. Kimoto, T. Asaka, T. Nagai, M. Saito, Y. Matsui, K. Ishizuka, Element-selective imaging of atomic columns in a crystal using STEM and EELS. Nature **450**, 702 (2007)

H.-J. Kleebe, Structure and chemistry of interfaces in Si_3N_4 ceramics studied by transmission electron microscopy. J. Ceram. Soc. Jpn. **105**, 453 (1997)
M. Kohyama, Atomic structures and properties of grain boundaries. Solid State Phys. **34**, 803 (in Japanese) (1999)
M. Kohyama, Computational studies of grain boundaries in covalent materials, Mode. Simul. Mater. Sci. Eng. **10**, R31 (2002)
M.L. Kronberg, Plastic deformation of single crystals of sapphire – basal slip and twinning. Acta Metall. **5**, 507 (1957)
G. Kresse, J. Furthmüller, Efficient iterative schemes for ab initio total-energy calculations using a plane-wave basis set. Phys. Rev. B. **54**, 11169 (1996)
M.L. Kronberg, F.H. Wilson, Secondary recrystallization in copper. Met. Trans. **185**, 501 (1949)
K.P.D. Lagerlöf, A.H. Heuer, J. Castaing, J.P. Rivière, T.E. Mitchell, Slip and twinning in sapphire (α-Al_2O_3). J. Am. Ceram. Soc. **77**, 385 (1994)
K.P.D. Lagerlöf, T.E. Mitchell, A.H. Heuer, J.P. Riviere, J. Cadoz, J. Castaing, D.S. Phillips, Stacking-fault energy in sapphire (α-Al_2O_3). Acta Metall. **32**, 97 (1984)
J.M. LeBeau, S.D. Findlay, L.J. Allen, S. Stemmer, Quantitative atomic resolution scanning transmission electron microscopy. Phys. Rev. Lett. **100**, 206101 (2008)
G.V. Lewis, C.R.A. Catlow, Potential models for ionic oxides. J. Phys. C: Solid State Phys. **18**, 1149 (1985)
W. Losch, New model of grain-boundary failure in temper embrittled steel. Acta Metall. **27**, 1885 (1979)
K. Matsunaga, H. Nishimura, H. Muto, T. Yamamoto, Y. Ikuhara, Direct measurements of grain boundary sliding in yttrium-doped alumina bicrystals. Appl. Phys. Lett. **82**, 1179 (2003)
K. Matsunaga, T. Sasaki, N. Shibata, T. Mizoguchi, T. Yamamoto, Y. Ikuhara, Bonding nature of metal/oxide incoherent interfaces by first-principles calculations. Phys. Rev. B **74**, 125423 (2006)
J.W. Matthews, *Epitaxial Growth, Part B*, ed. by F.R.N. Nabarro (North-Holland, Amsterdam, 1979)
D. McLean, *Grain Boundaries in Metals* (Oxford University Press, Oxford, 1957)
T.E. Mitchell, B.J. Pletka, D.S. Phillips, A.H. Heuer, Climb dissociation of dislocations in sapphire (α-Al_2O_3). Phil. Mag. **34**, 441 (1976)
T. Mizoguchi, J.P. Buban, K. Matsunaga, T. Yamamoto, Y. Ikuhara, First-principles study on incidence direction, individual site character, and atomic projection dependences of ELNES for perovskite compounds. Ultramicroscopy **106**, 92 (2006)
S.I. Molina, D.L. Sales, P.L. Galindo, D. Fuster, Y. González, B. Alén, L. González, M. Varela, S.J. Pennycook, Column-by-column compositional mapping by Z-contrast imaging. Ultramicroscopy **109**, 172 (2009)
K. Mukae, K. Tsuda, I. Nagasawa, Non-ohmic properties of ZnO-rare earth metal oxide Co_3O_4 ceramics. Jpn. J. Appl. Phys. **16**, 1361 (1977)
D. A. Muller, L.F. Kourkoutis, M. Murfitt, J.H. Song, H.Y. Hwang, J. Silcox, N. Dellby, O.L. Krivanek, Atomic-scale chemical imaging of composition and bonding by aberration-corrected microscopy. Science **319**, 1073 (2008)
R. Nakamura, N. Ohashi, A. Imanishi, T. Osawa, Y. Matsumoto, H. Koinuma, Y. Nakato, Crystal-face dependences of surface band edges and hole reactivity, revealed by preparation of essentially atomically smooth and stable (110) and (100) n-TiO_2 (rutile) surfaces. J. Phys. Chem. B **109**, 1648 (2005)

A. Nakamura, T. Yamamoto, Y. Ikuhara, Direct observation of basal dislocation in sapphire by HRTEM. Acta. Mater. **50**, 101 (2002)

P.D. Nellist, M.F. Chisholm, N. Dellby, O.L. Krivanek, M.F. Murfitt, Z.S. Szilagyi, A.R. Lupini, A.Y. Borisevich, W.H. Sides, S.J. Pennycook, Direct sub-angstrom imaging of a crystal lattice. Science **305**, 1741 (2004)

P.D. Nellist, S.J. Pennycook, Direct imaging of the atomic configuration of ultradispersed catalysts. Science **274**, 413 (1996)

F. Oba, H. Ohta, Y. Sato, H. Hosono, T. Yamamoto, Y. Ikuhara, Atomic structure of [0001]-tilt grain boundaries in ZnO: A high-resolution TEM study of fiber-textured thin films. Phys. Rev. B **70**, 125415 (2004)

H. Ohta, S.W. Kim, Y. Mune, T. Mizoguchi, K. Nomura, S. Ohta, T. Nomura, Y. Nakanishi, Y. Ikuhara, M. Hirano, H. Hosono, K. Koumoto, Giant thermoelectric Seebeck coefficient of two-dimensional electron gas in $SrTiO_3$. Nat. Mater. **6**, 129 (2007)

H. Ohta, S. Kim, S. Ohta, K. Koumoto, M. Hirano, H. Hosono, Reactive solid-phase epitaxial growth of $Na_x CoO_2$ (x similar to $_{0.83}$) via lateral diffusion of Na into a cobalt oxide epitaxial layer. Cryst. Growth Des. **5**, 25 (2005)

G.B. Olson, M. Cohen, Interphase-boundary dislocations and the concept of coherency. Acta Met. **27**, 1907 (1979)

G.S. Painter, F.W. Averill, P.F. Becher, N. Shibata, K. van Benthem, S.J. Pennycook, First-principles study of rare earth adsorption at β-Si_3N_4 interfaces. Phys. Rev. B **78**, 214206 (2008)

J.J. Papike, M. Cameron, Crystal-chemistry of silicate minerals of geophysical interest. Rev. Geophys. Space Phys. **14**, 37 (1976)

S.J. Pennycook, D.E. Jesson, High-resolution incoherent imaging of crystals. Phys. Rev. Lett. **64**, 938 (1990)

S.J. Pennycook, D.E. Jesson, High-resolution Z-contrast imaging of crystals. Ultramicroscopy **37**, 14 (1991)

L.M. Porter, R.F. Davis, A critical review of ohmic and rectifying contacts for silicon-carbide. Mater. Sci. Eng. B **34**, 83 (1995)

S. Ranganathan, On geometry of coincidence-site lattices. Acta Cryst. **21**, 197 (1966)

S.N. Rashkeev, A.R. Lupini, S.H. Overbury, S.J. Pennycook, S.T. Pantelides, Role of the nanoscale in catalytic CO oxidation by supported Au and Pt nanostructures. Phys. Rev. B. **76**, 035438 (2007)

W.T. Read, *Dislocations in Crystals* (McGraw-Hill, New York, NY, 1953)

P.D. Robb, A.J. Craven, Column ratio mapping: A processing technique for atomic resolution high-angle annular dark-field (HAADF) images. Ultramicroscopy **109**, 61 (2008)

M. Roger, D.J.P. Morris, D.A. Tennant, M.J. Gutmann, J.P. Goff, J.-U. Hoffmann, R. Feyerherm, E. Dudzik, D. Prabhakaran, A.T. Boothroyd, N. Shannon, B. Lake, P.P. Deen, Patterning of sodium ions and the control of electrons in sodium cobaltate. Nature **445**, 631 (2007)

T. Sakuma, L. Shepard, Y. Ikuhara (eds.), *Grain Boundary Engineering in Ceramics-from Grain Boundary Phenomena to Grain Boundary Quantum Structures*, Ceram. Trans, vol. 118 (The American Ceramic Society, Columbus OH, 2000)

R.L. Satet, M.J. Hoffmann, Grain growth anisotropy of beta-silicon nitride in rare-earth doped oxynitride glasses. J. Eur. Ceram. Soc. **24**, 3437 (2004)

Y. Sato, T. Yamamoto, Y. Ikuhara, Atomic structures and electrical properties of ZnO grain boundaries. J. Am. Ceram. Soc. **90**, 337 (2007)

Y. Sato, T. Mizoguchi, N. Shibata, T. Yamamoto, T. Hirayama, Y. Ikuhara, Atomic-scale segregation behavior of Pr at a ZnO [0001] $\Sigma 49$ tilt grain boundary. Phys. Rev. B **80**, 094114 (2009)

Y. Sato, T. Mizoguchi, N. Shibata, M. Yodogawa, T. Yamamoto, Y. Ikuhara, Role of Pr segregation in acceptor-state formation at ZnO grain boundaries. Phys. Rev. Lett. **97**, 106802 (2006)

N. Shibata, M.F. Chisholm, A. Nakamura, S.J. Pennycook, T. Yamamoto, Y. Ikuhara, Nonstoichiometric dislocation cores in α-alumina. Science **316**, 82 (2007)

N. Shibata, S.D. Findlay, S. Azuma, T. Mizoguchi, T. Yamamoto, Y. Ikuhara, Atomic-scale imaging of individual dopant atoms in a buried interface. Nat. Mater. **8**, 654 (2009)

N. Shibata, A. Goto, S.-Y. Choi, T. Mizoguchi, S.D. Findlay, T. Yamamoto, Y. Ikuhara, Direct imaging of reconstructed atoms on TiO_2 (110) surfaces. Science **322**, 570 (2008)

N. Shibata, A. Goto, K. Matsunaga, T. Mizoguchi, S.D. Findlay, T. Yamamoto, Y. Ikuhara, Interface structures of gold nanoparticles on TiO_2 (110). Phys. Rev. Lett. **102**, 136015 (2009)

N. Shibata, F. Oba, T. Yamamoto, Y. Ikuhara, Structure, energy and solute segregation behaviour of [110] symmetric tilt grain boundaries in yttria-stabilized cubic zirconia. Phil. Mag. **84**, 2381 (2004)

N. Shibata, G.S. Painter, R.L. Satet, M.J. Hoffmann, S.J. Pennycook, P.F. Becher, Rare-earth adsorption at intergranular interfaces in silicon nitride ceramics: Subnanometer observations and theory. Phys. Rev. B **72** 140101(R) (2005)

N. Shibata, S.J. Pennycook, T.R. Gosnell, G.S. Painter, W.A. Shelton, P.F. Becher, Observation of rare-earth segregation in silicon nitride ceramics at subnanometre dimensions. Nature **428**, 730 (2004)

K. Sugiura, H. Ohta, Y. Ishida, R. Huang, T. Saito, Y. Ikuhara, K. Nomura, H. Hosono, K. Koumoto, Structural transformation of Ca-arrangements and carrier transport properties in $Ca_{0.33}CoO_2$ epitaxial films. Appl. Phys. Express **2**, 035503 (2009)

K. Sugiura, H. Ohta, K. Nomura, M. Hirano, H. Hosono, K. Koumoto, High electrical conductivity of layered cobalt oxide $Ca_3Co_4O_9$ epitaxial films grown by topotactic ion-exchange method. Appl. Phys. Lett. **89**, 032111 (2006)

A.P. Sutton, R.W. Ballufi, *Interfaces in Crystalline Materials* (Oxford University Press, Oxford, 1995)

A.P. Sutton, V. Vitek, On the structure of tilt grain-boundaries in cubic metals. 1. Symmetrical tilt boundaries. Phil. Trans. R. Soc. Lond. **A309**, 1 (1983)

K. Takada, H. Sakurai, E. Takayama-Muromachi, F. Izumi, R.A. Dilanian, T. Sasaki, Superconductivity in two-dimensional CoO_2 layers. Nature **422**, 53 (2003)

I. Tanaka, H.-J. Kleebe, M.K. Cinibulk, J. Bruley, D.R. Clarke, M. Ruhle, Calcium-concentration dependence of the intergranular film thickness in silicon-nitride. J. Am. Ceram. Soc. **77**, 911 (1994)

T. Tohei, T. Mizoguchi, H. Hiramatsu, Y. Kamihara, H. Hosono, Y. Ikuhara, Direct imaging of doped fluorine in $LaFeAsO_{1-x}F_x$ superconductor by atomic scale spectroscopy. Appl. Phys. Lett. **95**, 193107 (2009)

S. Tsurekawa, S. Nitta, H. Nakashima, H. Yoshinaga, Grain-boundary structures in silicon-carbide – verification of the extended boundary concept. Interface Sci. **3**, 75 (1995)

S. Tsukimoto, K. Nitta, T. Sakai, M. Moriyama, M. Murakami, Correlation between the electrical properties and the interfacial microstructures of TiAl-based ohmic contacts to p-type 4H-SiC. J. Electron. Mater. **33** 460 (2004b)

S. Tsukimoto, T. Sakai, M. Murakami, Electrical properties and microstructure of ternary Ge/Ti/Al ohmic contacts to p-type 4H-SiC. J. Appl. Phys. **96**, 4976 (2004a)

M. Valden, X. Lai, D.W. Goodman, Onset of catalytic activity of gold clusters on titania with the appearance of nonmetallic properties. Science **281**, 1647 (1998)

K. van Benthem, A.R. Lupini, M. Kim, H.S. Baik, S. Doh, J.-H. Lee, M.P. Oxley, S.D. Findlay, L.J. Allen, J.T. Luck, S.J. Pennycook, Three-dimensional imaging of individual hafnium atoms inside a semiconductor device. Appl. Phys. Lett. **87**, 034104 (2005)

J.H. van der Merwe, On the stresses and energies associated with intercrystalline boundaries. Proc. Phys. Soc. Lond. **A63**, 616 (1950)

J.H. van der Merwe, Crystal interfaces. Part I. Semi-infinite crystals. J. Appl. Phys. **34**, 117 (1963)

M. Varela, S.D. Findlay, A.R. Lupini, H.M. Christen, A.Y. Borisevich, N. Dellby, O.L. Krivanek, P.D. Nellist, M.P. Oxley, L.J. Allen, S.J. Pennycook, Spectroscopic imaging of single atoms within a bulk solid. Phys. Rev. Lett. **92**, 095502 (2004)

P.M. Voyles, D.A. Muller, J.L. Grazul, P.H. Citrin, H.-J.L. Gossman, Atomic-scale imaging of individual dopant atoms and clusters in highly *n*-type bulk Si. Nature **416**, 826 (2002)

Z. Wang, T. Kadohira, T. Tada, S. Watanabe, Nonequilibrium quantum transport properties of a silver atomic switch. Nano Lett. **7**, 2688 (2007)

Y. Wang, N.S. Rogado, R.J. Cava, N.P. Ong, Spin entropy as the likely source of enhanced thermopower in $Na_xCo_2O_4$. Nature **423**, 425 (2003)

Z.C. Wang, M. Saito, S. Tsukimoto, Y. Ikuhara, Interface atomic-scale structure and its impact on quantum electron transport. Adv. Mater. **21**, 4966 (2009)

Z. Wang, S. Tsukimoto, M. Saito, Y. Ikuhara, SiC/Ti_3SiC_2 interface: Atomic structure, energetics, and bonding. Phys. Rev. B **79**, 045318 (2009)

G.B. Winkelman, C. Dwyer, T.S. Hudson, D. Nguyen-Manh, M. Döblinger, Three-dimensional organization of rare-earth atoms at grain boundaries in silicon nitride. Appl. Phys. Lett. **87**, 061911 (2005)

D. Wolf, in *Materials Interfaces: Atomic-Level Structure and Properties*, eds. by D. Wolf, S. Yip (Chapman & Hall, London, 1992), pp. 1–57

H.X. Yang, Y.G. Shi, X. Liu, R.J. Xiao, H.F. Tian, J.Q. Li, Structural properties and cation ordering in layered hexagonal Ca_xCoO_2. Phys. Rev. B **73**, 014109 (2006)

H. Yoshida, Y. Ikuhara, T. Sakuma, High-temperature creep resistance in rare-earth-doped, fine-grained Al_2O_3. J. Mater. Res. **13**, 2597 (1998)

H. Yoshida, Y. Ikuhara, T. Sakuma, High-temperature creep resistance in lanthanoid ion-doped polycrystalline Al_2O_3. Phil. Mag. Lett. **79**, 249 (1999)

H. Yoshida, Y. Ikuhara, T. Sakuma, Grain boundary electronic structure related to the high temperature creep resistance in polycrystalline Al_2O_3. Acta Metall. **50**, 2955 (2002)

A. Ziegler, J.C. Idrobo, M.K. Cinibulk, C. Kisielowski, N.D. Browning, R.O. Ritchie, Interface structure and atomic bonding characteristics in silicon nitride ceramics. Science **306**, 1768 (2004)

12

Application to Semiconductors

James M. LeBeau, Dmitri O. Klenov and Susanne Stemmer

12.1 Introduction

Scanning transmission electron microscopes capable of achieving atomic resolution became widely available in the late 1990s with the development of Schottky field emitters capable of providing a stable, finely focused electron probe (James and Browning 1999, James et al. 1998). This transformed scanning transmission electron microscopy (STEM) from being a tool used mainly in research into a major instrument for the metrology of semiconductor-based structures and devices in a multi-billion dollar industry. In particular, STEM addressed an urgent need for high-spatial resolution physical characterization in silicon industry, where devices were being scaled to ever-smaller dimensions. For example, the gate length in silicon field effect transistors is expected to drop below 30 nm in the near future. In parallel with device scaling, new materials were being introduced into semiconductor technology, such as high-permittivity (k) gate dielectrics and metal gates in silicon devices. Exploration of the structure and chemistry of interfaces in these highly scaled device structures required characterization methods with a spatial resolution approaching the Ångstrom level. For example, today's silicon gate stacks are comprised of multiple layers, some less than 1 nm in thickness, making STEM one of the most powerful tools to characterize these layers. As discussed in other chapters, high-angle annular dark-field (HAADF or Z-contrast) imaging in STEM offers excellent atomic number (Z) sensitivity and can normally be directly interpreted in terms of atom column positions regardless of specimen thickness or the defocus condition. In addition, STEM allows for high-spatial resolution electron energy-loss spectroscopy (EELS) and energy-dispersive x-ray spectroscopy (EDS). In silicon transistor development, these techniques have been widely employed to characterize the composition, atomic and electronic structure of high-permittivity (k) oxides and their interfaces with gate materials and the silicon channels (Busch et al. 2002, Craven et al. 2005, Diebold et al. 2003, Foran et al. 2005, Muller et al. 1999, Wilk and Muller 2003) and to detect single dopant and impurity atoms (Klenov et al. 2006, Oh et al. 2008, van Benthem et al. 2006, Voyles et al. 2002).

S.J. Pennycook, P.D. Nellist (eds.), *Scanning Transmission Electron Microscopy*,
DOI 10.1007/978-1-4419-7200-2_12,

In this chapter, we first present several examples from the literature that illustrate the application and impact of STEM in semiconductor research and technology. There is insufficient space in this chapter to cover even a fraction of the extensive literature reporting on STEM studies of semiconductors; however, we hope that a few specific examples can serve to illustrate the unique capabilities of STEM. This chapter will focus mainly on STEM imaging. We discuss how HAADF-STEM has been applied to directly determine the atomic structure and chemistry of interfaces and nanoscale structures in semiconductor research. We discuss the potential for future developments of STEM in semiconductor characterization, such as how unique aspects of the STEM image formation process can be useful for interpretation of local strain and in tomography. Finally we discuss how semiconductors, because of their high perfection, have played a significant role in developing an understanding of STEM images, leading in turn to further advances in the analysis of semiconductors.

12.2 Analysis of Semiconductor Interfaces

STEM imaging has played a major role in the scientific understanding of the structure and composition of semiconductor interfaces and those between semiconductors and dissimilar materials, such as dielectrics and metals. For example, the analysis of non-periodic features, such as interface reconstructions, is non-trivial in conventional high-resolution transmission electron microscopy (HRTEM) and requires considerable effort to correctly determine the structure using iterative approaches of comparing interface model-based simulations with experiments (Thust et al. 1996). In contrast, the incoherent nature of the imaging process in HAADF-STEM often allows for direct determination of the interface atomic structure without the need to employ preconceived structure models (Nellist and Pennycook 1998). Furthermore, the *Z*-sensitivity of HAADF-STEM images allows for identification of atomic columns based on their atomic number, thus providing additional information aiding in the correct interpretation of interface structures (Diebold et al. 2003). Even in HAADF-STEM, however, comparisons between simulation and experiment are needed if quantitative compositional information is to be extracted from the image intensities (LeBeau et al. 2008). Such comparisons are facilitated by recently developed methods to place experimental STEM image intensities on an absolute scale for direct comparisons with simulations (LeBeau and Stemmer 2008, Pizarro et al. 2008). These methods also allow for quantitative analysis of the background signal (the intensity measured between the atom columns), which contains important information such as the local thickness of the sample. While model-based approaches have been used to study the composition at interfaces (Van Aert et al. 2009), care must be taken to properly consider the effects of modified thermal diffuse scattering at interfaces and defects. In particular, it is generally not known how Debye–Waller factors change at interfaces and around defects relative to their bulk values. A change in contrast at an interface is therefore not always due to a compositional change as has also been demonstrated for strained interfaces and around doping atoms (Grillo 2009, Grillo et al. 2008, Perovic et al. 1993, Yu et al. 2004, 2008). In the following sections we discuss examples of the successful application of STEM to determine the atomic structure and probe the chemical changes that occur at interfaces involving semiconducting materials.

12.2.1 Dielectric/Semiconductor Interfaces

The integration of high-*k* gate dielectrics into highly scaled Si technology allowed manufacturers to scale the gate length to the 45 nm generation of silicon transistors in 2007. The development of high-*k* gate dielectric stacks required investigations of the stability and interface quality at the atomic scale. For example, initial attempts to integrate high-*k* gate dielectrics were fraught with problems due to interface reactions with the polycrystalline-Si gate electrode due to reducing deposition conditions (Stemmer 2004), which was finally solved with the introduction of metal gate electrodes (Chau et al. 2004). Another problem was phase separation of some of the early candidates for high-*k* gate dielectrics, in particular the hafnium and zirconium silicates (Stemmer et al. 2003). Today's high-*k*s of choice are Hf-based oxides and multiple interfaces play a role in the performance and properties of high-*k* gate stacks, such as the Si/SiO_2, SiO_2/high-*k*, and high-*k*/metal gate interfaces. While many issues could be identified in conventional HRTEM studies, HAADF-STEM and EELS were particularly suited to study the precise chemistry and electronic structure of ultrathin (< 1 nm) SiO_2-like interface layers that are omnipresent even in high-*k* stacks (Agustin et al. 2006, Muller et al. 1999). For example, unlike in HRTEM, single heavy atoms such as Hf can be easily identified in HAADF-STEM images of amorphous layers, as shown in Figure 12–1 (Agustin et al. 2005), and these images can be used to study interdiffusion and reactions in layers that are less than 1 nm in thickness. With aberration-corrected STEM, the improved depth of field allowed for the position of Hf atoms in an amorphous SiO_2 layer to be determined in three dimensions (Borisevich et al. 2006, van Benthem et al. 2006). Further, HAADF-STEM imaging established that Hf atoms do not diffuse in properly densified SiO_2 layers (Agustin et al. 2005, Klenov et al. 2006).

More recently, high-*k* gate oxides have also been investigated for field effect transistors with III–V semiconductor channels, such as $In_xGa_{1-x}As$ and GaAs, which have higher electron mobilities than Si and would thus potentially allow for further device scaling. Here, the advantages of HAADF-STEM are that low-Z interfacial layers can be much more easily detected than in HRTEM. In the example shown in Figure 12–2, the HRTEM image shows an apparently abrupt interface between GaAs and HfO_2. In contrast, the HAADF image clearly shows the presence of an interfacial layer that is amorphous and composed of low-Z material such as the native semiconductor oxides.

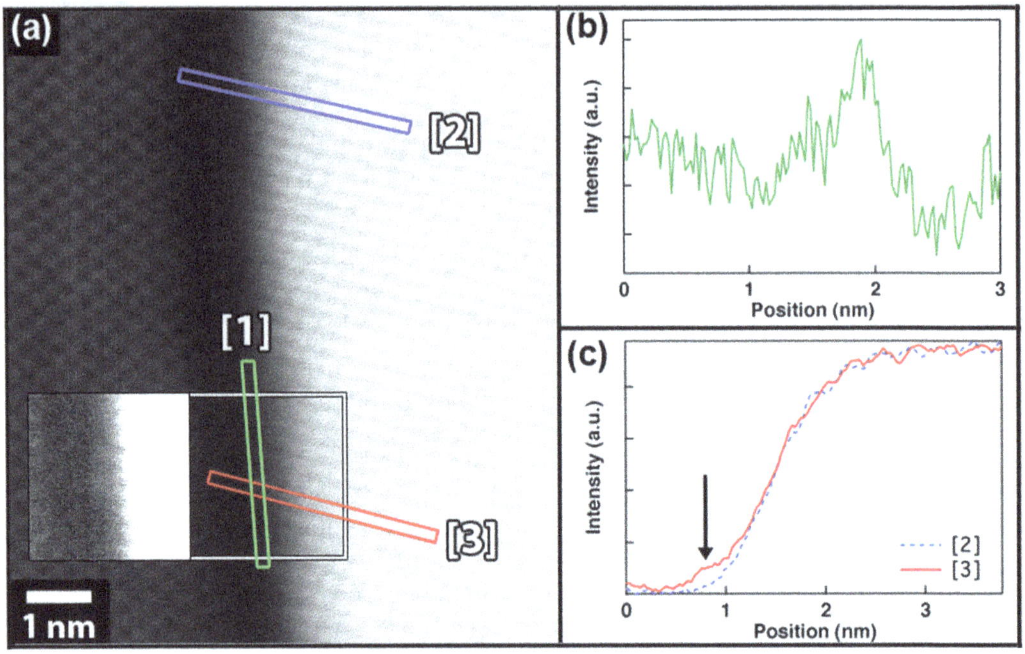

Figure 12–1. (a) HAADF-STEM image of a single Hf atom protruding from a crystalline HfO_2 film (bright layer on *right*) into an interfacial SiO_2 layer (dark amorphous layer). The crystalline Si substrate is to the *left*. The inset shows a magnified portion of the interface with the contrast adjusted to that the Hf atoms are visible in the SiO_2 near the HfO_2 layer. Detection of the Hf atoms is possible due to the strong atomic number contrast between the Hf and the SiO_2 layer. Intensity line profiles were taken across the image in (a) and positions one to three are shown in (b) and (c). After Agustin et al. (2005).

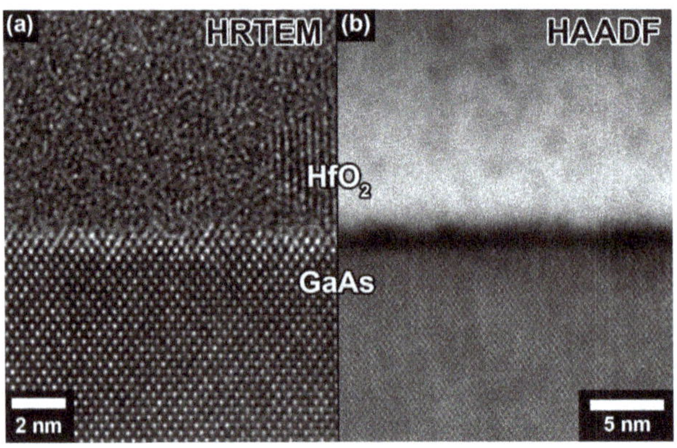

Figure 12–2. (a) HRTEM image of an HfO_2/GaAs interface showing no discernable interface layer. (b) HAADF image from the same region showing that a low-Z amorphous oxide layer (*dark band*) is formed at the interface. Furthermore, density variations are also visible in the HfO_2 film.

In addition to amorphous or polycrystalline oxides, epitaxial dielectrics are also of interest for future semiconductor devices. One of the major limitations of HRTEM in the analysis of crystalline, epitaxial interfaces is its inability (under most practical conditions) to directly interpret the image intensities in terms of the interface atom column

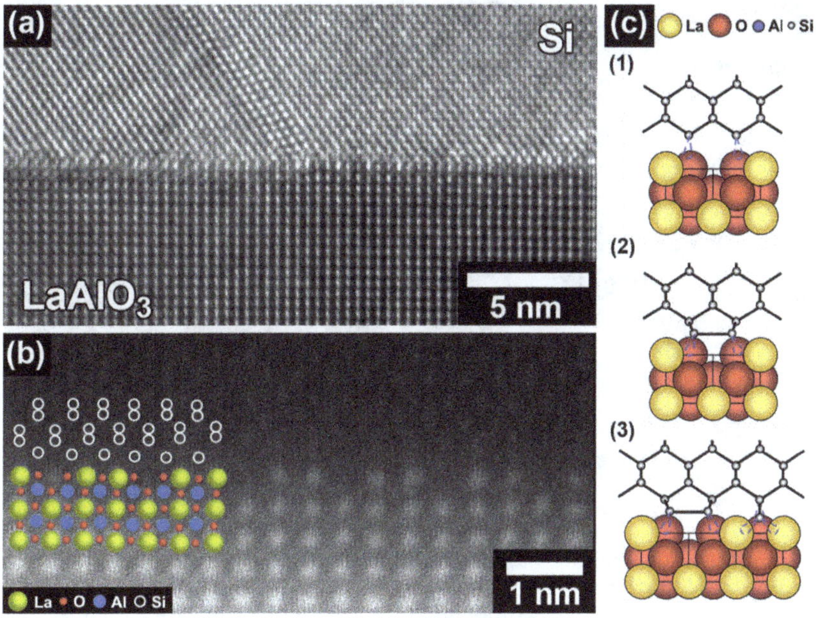

Figure 12–3. (a) HRTEM and (b) HAADF-STEM images from an epitaxial $LaAlO_3$/Si interface showing an interface reconstruction where every third La column is missing at the interface. (c) Interface models based on the HAADF-STEM images. After Klenov et al. (2005).

positions due to dynamical scattering and delocalization (Haider et al. 1998, Spence 2003). These limitations are particularly severe for non-aberration-corrected HRTEM images, as shown in Figure 12–3a for a $LaAlO_3$/Si interface (Klenov et al. 2005). Inspection of the HRTEM image would lead to the conclusion that the interface is atomically abrupt and does not reconstruct. The HAADF-STEM image of the same $LaAlO_3$/Si interface reveals a different picture. As can be seen from Figure 12–3b, the interface is reconstructed such that every third La column is missing in the terminating LaO layer at the interface. Closer inspection of the HRTEM image reveals some variation of contrast along the interface, but without the HAADF-STEM image it is unlikely that the correct interface atomic structure could have been guessed from this image. From HAADF-STEM images along two mutually perpendicular directions, it was possible to construct possible models of the three-dimensional structure as shown in Figure 12–3c, which could were then used for a detailed investigation of the interface with density functional theory calculations (Först et al. 2005).

12.2.2 Metal/Semiconductor Interfaces

Metal/semiconductor interfaces are ubiquitous in electronic devices, where they serve as Ohmic or Schottky contacts. More recently they have also attracted attention for novel spintronic devices. The interface atomic structure directly determines the electrical properties of these interfaces, such as the Schottky barrier height (Tung 1993).

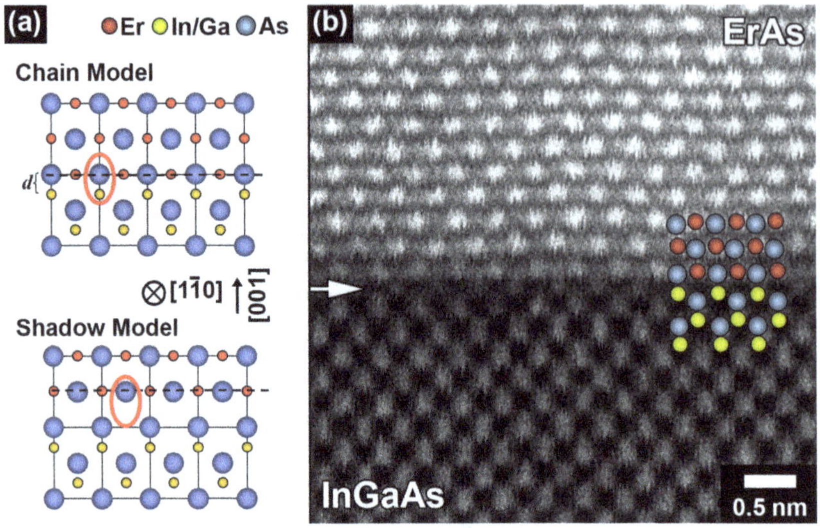

Figure 12–4. (a) Interface models for the ErAs/GaAs interface corresponding to two different terminations of the semiconductor (Lambrecht et al. 1998). ErAs has the rock salt structure and GaAs has the zinc blende structure. The As-sublattice is continuous across the interface. (b) HAADF-STEM image of the interface between the ErAs and GaAs. The observed interface structure is directly determined to correspond to the chain model. After Klenov et al. (2005).

STEM has the potential to play a pivotal role in understanding these interfaces, because the interface structure can be directly determined. For instance, two alternative configurations had been suggested for interfaces between ErAs, a semi-metal, and the semiconductor GaAs (Lambrecht et al. 1998). These models are shown in Figure 12–4a and have been denoted the chain and shadow models, respectively. These two models are easily distinguished in HAADF-STEM, as shown in Figure 12–4b (Klenov et al. 2005). From this image, the observed structure can be directly identified as corresponding to the chain model. In addition, close inspection of the interplanar spacing at the interface indicated that an increase in the Ga and As column spacing occurs at the interface (Klenov et al. 2005).

Metal/semiconductor interface structures determine the performance of novel spintronic devices, such as their spin injection efficiency (Schultz et al. 2009). An important interface for such devices is the one formed between ferromagnetic metals, such as Fe, and semiconductors, such as GaAs (Hanbicki et al. 2002, Zega et al. 2006). By using HAADF-STEM, shown in Figure 12–5, the Fe/GaAs interface structure was recently determined (LeBeau et al. 2008). Unlike in HRTEM, surface steps at the interface can be easily identified, see Figure 12–5a. This can help to eliminate confusion in the interpretation due to overlapping regions. When viewed along the [1$\bar{1}$0] zone axis, the interface appeared abrupt without any additional contrast between adjacent GaAs dumbbells at the interface. Images acquired along [110], however, clearly showed additional intensity between As columns at the interface. By

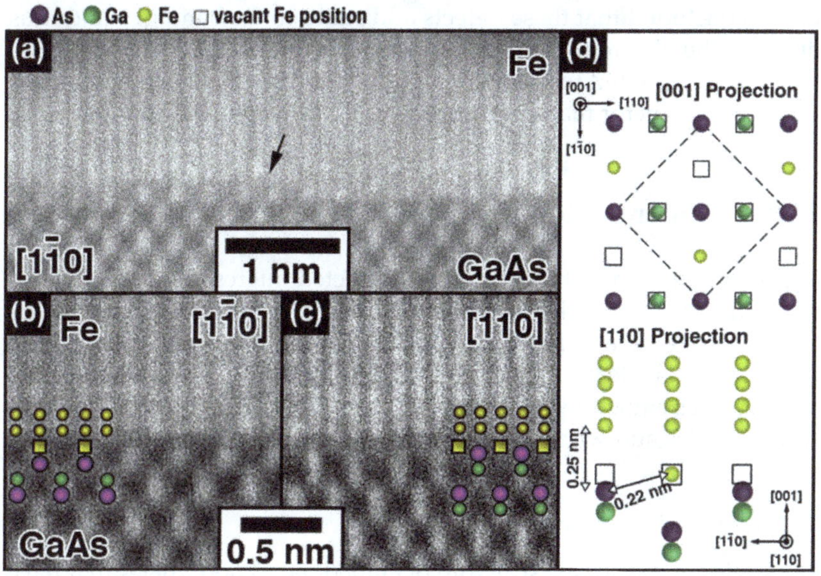

Figure 12–5. (a) A interface step (see *arrow*) is clearly revealed in an HAADF-STEM image of a Fe/GaAs interface. (b, c) HAADF-STEM image of the Fe/GaAs viewed down two perpendicular directions. The location of the atomic columns is shown in the overlay. (d) Model of the interface based on the column positions and intensities identified from the HAADF images. After LeBeau et al. (2008).

using the Z-dependence of the signal, the reduced intensity of these additional columns relative to the Fe film implies that they are half-filled Fe columns. Based on this information, a model of the interface structure could be constructed (Figure 12–5d) in which Fe bonds to every other As pair at the surface.

Other STEM studies that have explored the semiconductor/metal interfaces have included Au/GaAs (Morgan et al. 2009) and silicide/silicon interfaces (Falke et al. 2004, 2005).

12.2.3 Semiconductor/Semiconductor Interfaces and Defects

Many STEM explorations of semiconductor interfaces have focused on the defects that occur at interfaces. One of the first of such studies investigated the nature of dislocation structures in at the CdTe/GaAs interface (McGibbon et al. 1995). Others have explored the configuration of dopant atoms at Si grain boundaries (Chisholm et al. 1998). Defect analysis using HAADF-STEM of epitaxial GaAs/Si interfaces is another important example. Because of the high carrier mobility of GaAs, the integration of III–V materials into traditional Si technology is of great interest to the electronics industry (Chau 2008). Direct observation of these interfaces has helped to explain the nature of the misfit interface dislocations. Specifically, Z-contrast images of the GaAs/Si interface show 90° and 60° dislocations that are not reconstructed (Lopatin et al. 2002). In combination with EELS and DFT calculations, it was concluded that these defects add states to the band gap and thus degrade the electronic properties (Lopatin et al. 2002). HAADF-STEM has also contributed to the understanding of dislocations in wide-band gap semiconductor materials, such as GaN (Xin et al. 1998).

12.3 Semiconductor Nanostructures

Semiconductor nanostructures have attracted interest for a wide range of fields including the biological sciences, sensing devices, opto-electronics, and alternative energy. Scanning transmission electron microscopy is ideally suited for the exploration of nanoscale structures such as embedded quantum dots and freestanding semiconductor nanoparticles and wires.

12.3.1 Embedded Nanostructures

Determining the atomic structure of embedded nanoparticles that have a crystal structure that is different from the surrounding matrix is nearly impossible in HRTEM because of the complex nature of the image formation process due to overlap, non-planar interfaces, and unknown size and shapes of the particles. In contrast, HAADF-STEM imaging has been shown to be able to directly determine the structure of embedded nanoscale structures. One such example is that of ErAs nanoparticles embedded in an epitaxial $In_{0.53}Ga_{0.47}As$ layer. This composite system is of interest for thermoelectric applications because the ErAs particles scatter phonons, decreasing the thermal conductivity, and dope the semiconductor, increasing the electrical conductivity (Kim et al. 2006). While bulk ErAs has the rock salt structure, it is conceivable that very small ErAs particles may adopt the crystal structure of the zinc blende $In_{0.53}Ga_{0.47}As$ host. As shown in Figure 12–6, ErAs nanoparticles are readily observed in the HAADF-STEM images as the Er atoms have much higher Z than the surrounding matrix (Klenov et al. 2005). In addition, the structure of the embedded particles is readily determined.

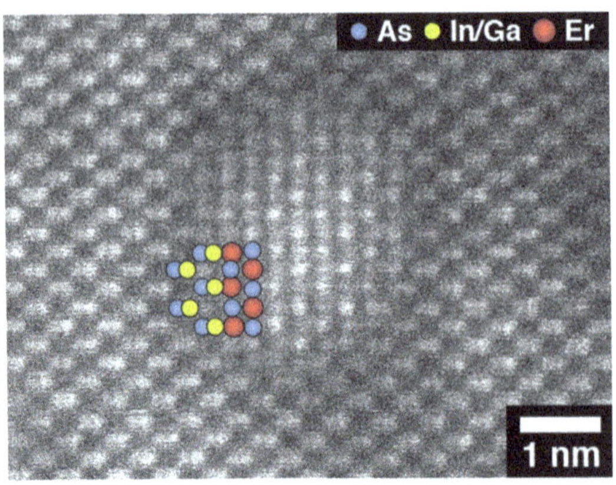

Figure 12–6. HAADF-STEM micrograph of an ErAs particle embedded in an $In_{0.53}Ga_{0.47}As$ matrix. The overlay shows the atomic positions obtained directly from the image, showing that the particle has the cubic rock salt structure. After Klenov et al. (2005).

From the image, it is clear that the As sublattice is continuous across the matrix/particle interface and the crystal structure of the particles is rock salt even though it is embedded in a zinc blende matrix (Klenov et al. 2005).

Other examples of embedded nanostructures where HAADF-STEM provided information not obtainable by other techniques include determination of the boundaries of Er clusters in Si (Kaiser et al. 2002) and the investigation of strain around InAs/InP quantum wires (Molina et al. 2006, 2007). HAADF-STEM has also been widely used for electron tomography of nanostructures because of the huge improvement of mass-contrast relative to HRTEM imaging (Inoue et al. 2008, Midgley et al. 2006). A challenge in the study of embedded nanostructures by HAADF-STEM is the simulation of the image contrast even for relatively small nanoscale structures. The requirement of large supercells precludes the use of slow frozen phonon simulations. Instead, parallel computer techniques or simplifications are being developed to tackle computing time-intensive simulations (Pizarro et al. 2008).

Another advantage of STEM in the imaging of embedded nanostructured semiconductors is its ability to readily control the degree of phase contrast within the images by varying the inner and outer angle of the annular dark-field (ADF) or bright-field detector. For bright-field STEM, interference effects dominate the image contrast if the selected collection angle is sufficiently small. Through reciprocity, a bright-field STEM image approaches an HRTEM image (LeBeau et al. 2009). Conversely, by increasing the bright-field detector angle, the resulting image can be made incoherent. This approach has been successfully applied to bright-field STEM tomography of Cu interconnects in a Si semiconductor device, where interference effects would otherwise have dominated (Ercius et al. 2006). For ADF-STEM, dechanneling of the electron probe is caused by lattice strain, for example, that surrounding lattice-mismatched particles in a matrix. For a sufficiently small ADF detector inner angle, these effects become apparent and manifest as areas of bright intensity in images. An example is shown in Figure 12–7, where a bright halo surrounds GeMn nanoparticles in the low-angle ADF (LAADF) image whereas no such contrast is seen in the corresponding HAADF image (Bougeard et al. 2009, Li et al. 2007). This information can help in understanding the role of strain in the properties of nanocomposites or at interfaces (Yu et al. 2004).

12.3.2 Freestanding Nanostructures

Nanowires and particles have been an area of significant interest in recent years. An example is shown in Figure 12–8, which shows semiconducting SnO_2 nanowires that are functionalized with Pd catalyst nanoparticles (Kolmakov et al. 2005). Although the Pd nanoparticles, which are located on the surface of the nanowires, are visible in HRTEM (Figure 12–8a), the strong Z-contrast in HAADF-STEM makes them much more easily to detect for analysis of their size and distribution (Figure 12–8b) (Kolmakov et al. 2005). In particular, HAADF-STEM

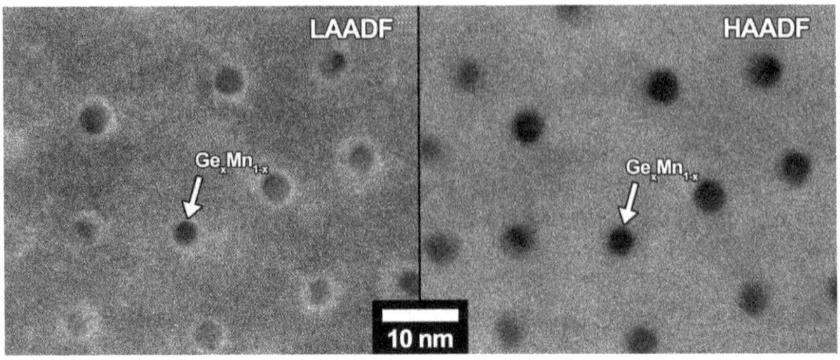

Figure 12–7. Plan-view STEM images of a sample containing Mn-rich particles in a Ge matrix. (*left*) LAADF, inner semiangle 23 mrad where strain is observed as bright halos around the particles, (*right*) HAADF, inner semiangle 65 mrad. After Bougeard et al. (2009).

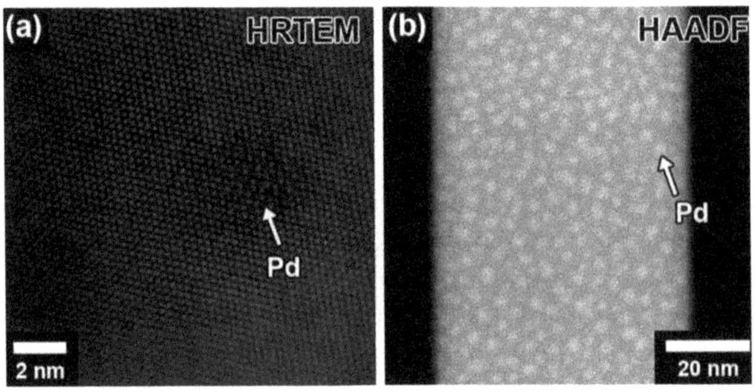

Figure 12–8. HRTEM (a) and HAADF-STEM (b) images of SnO_2 nanowires covered with Pd catalyst particles. After Kolmakov et al. (2005).

images allowed for quantification of the coverage of the nanoparticles (Kolmakov et al. 2005).

Aberration-corrected HAADF-STEM has recently been used to detect single atoms of Au on the surface of and within Si nanowires. By studying the frequency at which various Au atoms appear at different defects, it has been shown that Au atoms can occupy both substitutional and interstitial sites (Oh et al. 2008). Furthermore, throughfocus series in aberration-corrected STEM can be used to show that Au atoms can lie within the core of the wire, rather than at the surface (Allen et al. 2008). Core–shell nanostructures have also been imaged using HAADF-STEM (Tambe et al. 2008). The high-spatial resolution of aberration-corrected STEM has allowed for detailed morphological and structural studies of core–shell CdSe/ZnS nanocrystals (Kadavanich et al. 2000, Rosenthal et al. 2007). A three-dimensional picture of particle could be determined based on the thickness-dependent signal and facets (Kadavanich et al. 2000, Pennycook et al. 2009). Furthermore, because aberration correction allows for an increase in the bright-field detector collection aperture, images of nanoparticles can be near-simultaneously acquired in both bright-field and HAADF-STEM imaging modes. In addition, bright-field STEM imaging allows for more ready determination of nanoparticles boundaries with a loss of single atom detection (Rosenthal et al. 2007).

12.4 Semiconductors as Test Structures and Model Systems

Semiconductors are ideal model systems for understanding image formation in STEM and for the testing of theoretical models of electron scattering because of their low defect densities, high purity, relative ease of TEM sample preparation and because they already have been well characterized by other methods due to their technological importance. For example, resolving the Si dumbbells (0.136 nm separation) when imaged along the <110> zone axis was a standard test of STEM image resolution. For aberration-corrected microscopes, this test has now shifted to the <112> zone axis, with dumbbells separated by 0.078 nm (Nellist et al. 2004). Silicon was also one of the first materials used to test the frozen phonon multislice method for simulations of HAADF-STEM images (Kirkland et al. 1987). Accurate knowledge of the inelastic mean-free path for Si has allowed for determining the modification of the scattered electron distribution due to plasmon scattering and accurate thickness determination with EELS (Mkhoyan et al. 2008).

For HAADF-STEM, knowledge of the Debye–Waller factors of each atomic site is particularly important for simulations that correctly incorporate the effects of thermal diffuse scattering on image intensities (Hillyard and Silcox 1995, LeBeau et al. 2008, 2009). In complex crystal structures that have different Debye–Waller factors for each atom site, and/or heavy element single crystals, it has been shown that the contrast of columns containing different atomic species in HAADF cannot be directly interpreted in terms of Z-number differences (LeBeau et al. 2009). Semiconductors, such as Si, have played an important role in experimentally testing the validity of different models that account for phonon scattering. For Si, experimentally determined phonon density of states are available (Jian et al. 1993) and have allowed for comparisons of semi-classical frozen phonon simulations using the Einstein model of uncorrelated atom motion with models using detailed phonon dispersion curves (Muller et al. 2001). From these simulations, it was found that the inclusion of correlated motion of the atoms does not significantly alter the contrast of images based on high-angle scattering.

12.5 Summary

The use of STEM in semiconductor research played a crucial role in enabling the continued scaling of devices for silicon technology and in the fundamental research of new semiconducting materials and structures. In these areas, the major advantages of HAADF-STEM, namely providing images that are directly interpretable in terms of atom column locations and its excellent Z-number sensitivity, have been used to directly determine interface and embedded nanoscale structures. As aberration-corrected microscopes are being introduced the investigation of semiconductors is likely going to be one of the biggest benefactors. Due to continued increase in the spatial resolution combined with EELS analysis, mapping chemical bonding on the atomic scale has become possible (Muller et al. 2008). For semiconductors this will likely mean that STEM will be able to provide answers to some of long-standing questions regarding the nature of bonding around defects.

Acknowledgments Melody Agustin provided the image shown in Figure 12–1 of this chapter. We would like to also thank the following collaborators for providing samples: Joshua Zide, Jeramy Zimmerman, Art Gossard, Andrei Kolmakov, Martin Moscovitz, Hao Li, Darrell Schlom, Paul McIntyre, Dominique Bougeard, Jacob Hooker, Qi Hu, and Chris Palmstrøm. We thank the U.S. National Science Foundation (Grant No. DMR-0804631) and the Department of Energy for support (Grant No. DE-FG02-06ER45994). J.M.L. also thanks the U.S. Department of Education for support under the GAANN program (Grant No. P200A07044).

References

M.P. Agustin, G. Bersuker, B. Foran, L.A. Boatner, S. Stemmer, J. Appl. Phys. **100**, 024103 (2006)

M.P. Agustin, L.R.C. Fonseca, J.C. Hooker, S. Stemmer, Appl. Phys. Lett. **87**, 3 (2005)

J.E. Allen, E.R. Hemesath, D.E. Perea, J.L. Lensch-Falk, Z.Y. Li, F. Yin, M.H. Gass, P. Wang, A.L. Bleloch, R.E. Palmer, L.J. Lauhon, Nat. Nano. **3**, 168 (2008)

A.Y. Borisevich, A.R. Lupini, S.J. Pennycook, Proc. Nat. Acad. Sci. USA **103**, 3044 (2006)

D. Bougeard, N. Sircar, S. Ahlers, V. Lang, G. Abstreiter, A. Trampert, J.M. LeBeau, S. Stemmer, D.W. Saxey, A. Cerezo, Nano Lett. **9**, 3743 (2009)

B.W. Busch, O. Pluchery, Y.J. Chabal, D.A. Muller, R.L. Opila, J.R. Kwo, E. Garfunkel, MRS Bull. **27**, 206 (2002)

R. Chau, Solid State Technol. **51**, 30 (2008)

R. Chau, S. Datta, M. Doczy, B. Doyle, J. Kavalieros, M. Metz, Electron Dev. Lett. **25**, 408 (2004)

M.F. Chisholm, A. Maiti, S.J. Pennycook, S.T. Pantelides, Phys. Rev. Lett. **81**, 132 (1998)

A.J. Craven, M. MacKenzie, D.W. McComb, F.T. Docherty, Microelectron. Eng. **80**, 90 (2005)

A.C. Diebold, B. Foran, C. Kisielowski, D.A. Muller, S.J. Pennycook, E. Principe, S. Stemmer, Microsc. Microanal. **9**, 493 (2003)

P. Ercius, M. Weyland, D.A. Muller, L.M. Gignac, Appl. Phys. Lett. **88**, 243116 (2006)

U. Falke, A. Bleloch, M. Falke, S. Teichert, Phys. Rev. Lett. **92**, 116103 (2004)

M. Falke, U. Falke, A. Bleloch, S. Teichert, G. Beddies, H.J. Hinneberg, Appl. Phys. Lett. **86**, 203103 (2005)

B. Foran, J. Barnett, P.S. Lysaght, M.P. Agustin, S. Stemmer, J. Electron Spectrosc. Relat. Phenom. **143**, 149 (2005)

C.J. Först, K. Schwarz, P.E. Blöchl, Phys. Rev. Lett. **95**, 137602 (2005)

V. Grillo, Ultramicroscopy **109**, 1453 (2009)
V. Grillo, E. Carlino, F. Glas, Phys. Rev. B **77**, 054103 (2008)
M. Haider, H. Rose, S. Uhlemann, E. Schwan, B. Kabius, K. Urban, Ultramicroscopy **75**, 53 (1998)
A.T. Hanbicki, B.T. Jonker, G. Itskos, G. Kioseoglou, A. Petrou, Appl. Phys. Lett. **80**, 1240 (2002)
S. Hillyard, J. Silcox, Ultramicroscopy **58**, 6 (1995)
T. Inoue, T. Kita, O. Wada, M. Konno, T. Yaguchi, T. Kamino, Appl. Phys. Lett. **92**, 031902 (2008)
E.M. James, N.D. Browning, Ultramicroscopy **78**, 125 (1999)
E.M. James, N.D. Browning, A.W. Nicholls, M. Kawasaki, Y. Xin, S. Stemmer, J. Electron Microsc. **47**, 561 (1998)
W. Jian, Z. Kaiming, X. Xide, Solid State Commun. **86**, 731 (1993)
A.V. Kadavanich, T.C. Kippeny, M.M. Erwin, S.J. Pennycook, S.J. Rosenthal, J. Phys. Chem. B **105**, 361 (2000)
U. Kaiser, D.A. Muller, J.L. Grazul, A. Chuvilin, M. Kawasaki, Nat. Mater. **1**, 102 (2002)
W. Kim, J. Zide, A. Gossard, D. Klenov, S. Stemmer, A. Shakouri, A. Majumdar, Phys. Rev. Lett. **96**, 045901 (2006)
E.J. Kirkland, R.F. Loane, J. Silcox, Ultramicroscopy **23**, 77 (1987)
D.O. Klenov, D.C. Driscoll, A.C. Gossard, S. Stemmer, Appl. Phys. Lett. **86**, 111912 (2005)
D.O. Klenov, T.E. Mates, S. Stemmer, Appl. Phys. Lett. **89**, 041918 (2006)
D.O. Klenov, D.G. Schlom, H. Li, S. Stemmer, Jap. J. Appl. Phys. **44**, L617 (2005)
D.O. Klenov, J.M. Zide, J.D. Zimmerman, A.C. Gossard, S. Stemmer, Appl. Phys. Lett. **86**, 241901 (2005)
A. Kolmakov, D.O. Klenov, Y. Lilach, S. Stemmer, M. Moskovits, Nano Lett. **5**, 667 (2005)
W.R.L. Lambrecht, A.G. Ptukhov, B.T. Hemmelman, Solid State Commun. **108**, 361 (1998)
J.M. LeBeau, A.J. D'Alfonso, S.D. Findlay, S. Stemmer, L.J. Allen, Phys. Rev. B **80**, 174106 (2009)
J.M. LeBeau, S.D. Findlay, L.J. Allen, S. Stemmer, Phys. Rev. Lett. **100**, 206101 (2008)
J.M. LeBeau, S.D. Findlay, X. Wang, A.J. Jacobson, L.J. Allen, S. Stemmer, Phys. Rev. B **79**, 214110 (2009)
J.M. LeBeau, Q.O. Hu, C.J. Palmstrom, S. Stemmer, Appl. Phys. Lett. **93**, 121909 (2008)
J.M. LeBeau, S. Stemmer, Ultramicroscopy **108**, 1653 (2008)
S. Lopatin, S.J. Pennycook, J. Narayan, G. Duscher, Appl. Phys. Lett. **81**, 2728 (2002)
A.P. Li, C. Zeng, K. van Benthem, M.F. Chisholm, J. Shen, S.V.S.N. Rao, S.K. Dixit, L.C. Feldman, A.G. Petukhov, M. Foygel, H.H. Weitering, Phys. Rev. B **75**, 201201 (2007)
A.J. McGibbon, S.J. Pennycook, J.E. Angelo, Science **269**, 519 (1995)
P.A. Midgley, M. Weyland, T.J.V. Yates, I. Arslan, R.E. Sunin-Borkowski, J.M. Thomas, J. Microsc. **223**, 185 (2006)
K.A. Mkhoyan, S.E. Maccagnano-Zacher, M.G. Thomas, J. Silcox, Phys. Rev. Lett. **100**, 025503 (2008)
S.I. Molina, T. Ben, D.L. Sales, J. Pizarro, P.L. Galindo, M. Varela, S.J. Pennycook, D. Fuster, Y. Gonzalez, L. Gonzalez, Nanotechnology **17**, 5652 (2006)
S.I. Molina, M. Varela, D.L. Sales, T. Ben, J. Pizarro, P.L. Galindo, D. Fuster, Y. Gonzalez, L. Gonzalez, S.J. Pennycook, Appl. Phys. Lett. **91**, 143112 (2007)
D.G. Morgan, Q.M. Ramasse, N.D. Browning, J. Electron. Microsc. **58**, 223 (2009)

D.A. Muller, T. Sorsch, S. Moccio, F.H. Baumann, K. Evans-Lutterodt, G. Timp, Nature **399**, 758 (1999)
D.A. Muller, B. Edwards, E.J. Kirkland, J. Silcox, Ultramicroscopy **86**, 371 (2001)
D.A. Muller, L.F. Kourkoutis, M. Murfitt, J.H. Song, H.Y. Hwang, J. Silcox, N. Dellby, O.L. Krivanek, Science **319**, 1073 (2008)
P.D. Nellist, M.F. Chisholm, N. Dellby, O.L. Krivanek, M.F. Murfitt, Z.S. Szilagyi, A.R. Lupini, A. Borisevich, W.H. Sides Jr., S.J. Pennycook, Science **305**, 1741 (2004)
P.D. Nellist, S.J. Pennycook, J. Microsc. **190**, 159 (1998)
S.H. Oh, K.V. Benthem, S.I. Molina, A.Y. Borisevich, W. Luo, P. Werner, N.D. Zakharov, D. Kumar, S.T. Pantelides, S.J. Pennycook, Nano Lett. **8**, 1016 (2008)
S.J. Pennycook, M.F. Chisholm, A.R. Lupini, M. Varela, A.Y. Borisevich, M.P. Oxley, W.D. Luo, K. van Benthem, S.-H. Oh, D.L. Sales, S.I. Molina, J. García-Barriocanal, C. Leon, J. Santamaría, S.N. Rashkeev, S.T. Pantelides, Philos. Trans. Royal Soc. A **367**, 3709 (2009)
D.D. Perovic, C.J. Rossouw, A. Howie, Ultramicroscopy **52**, 353 (1993)
J. Pizarro, P.L. Galindo, E. Guerrero, A. Yanez, M.P. Guerrero, A. Rosenauer, D.L. Sales, S.I. Molina, Appl. Phys. Lett. **93**, 153107 (2008)
S.J. Rosenthal, J. McBride, S.J. Pennycook, L.C. Feldman, Surf. Sci. Rep. **62**, 111 (2007)
B.D. Schultz, N. Marom, D. Naveh, X. Lou, C. Adelmann, J. Strand, P.A. Crowell, L. Kronik, C.J. Palmstrom, Phys. Rev. B **80**, 201309 (2009)
J.C.H. Spence, *High-Resolution Electron Microscopy* (Oxford Science Publications, New York, 2003)
S. Stemmer, J. Vac. Sci. Technol. B **22**, 791 (2004)
S. Stemmer, Y.L. Li, B. Foran, P.S. Lysaght, S.K. Streiffer, P. Fuoss, S. Seifert, Appl. Phys. Lett. **83**, 3141 (2003)
M.J. Tambe, S.K. Lim, M.J. Smith, L.F. Allard, S. Gradečak, Appl. Phys. Lett. **93**, 151917 (2008)
A. Thust, W.M.J. Coene, M. Op de Beeck, D. Van Dyck, Ultramicroscopy **64**, 211 (1996)
R.T. Tung, J. Vac. Sci. Technol. B **11** (1993)
S. Van Aert, J. Verbeeck, R. Erni, S. Bals, M. Luysberg, D.V. Dyck, G.V. Tendeloo, Ultramicroscopy **109**, 1236 (2009)
K. van Benthem, A.R. Lupini, M.P. Oxley, S.D. Findlay, L.J. Allen, S.J. Pennycook, Ultramicroscopy **106**, 1062 (2006)
P.M. Voyles, D.A. Muller, J.L. Grazul, P.H. Citrin, H.-J.L. Gossmann, Nature **416**, 826 (2002)
G.D. Wilk, D.A. Muller, Appl. Phys. Lett. **83**, 3984 (2003)
Y. Xin, S.J. Pennycook, N.D. Browning, P.D. Nellist, S. Sivananthan, F. Omnes, B. Beaumont, J.P. Faurie, P. Gibart, Appl. Phys. Lett. **72**, 2680 (1998)
Z. Yu, D.A. Muller, J. Silcox, J. Appl. Phys. **95**, 3362 (2004)
Z. Yu, D.A. Muller, J. Silcox, Ultramicroscopy **108**, 494 (2008)
T.J. Zega, A.T. Hanbicki, S.C. Erwin, I. Zutic, G. Kioseoglou, C.H. Li, B.T. Jonker, R.M. Stroud, Phys. Rev. Lett. **96**, 196101 (2006)

13

Nanocharacterization of Heterogeneous Catalysts by Ex Situ and In Situ STEM

Peter A. Crozier

13.1 Introduction to Heterogeneous Catalysts

Heterogeneous catalysts are an important set of materials that increase the rate of chemical reactions. The increase in the reaction rate can be many orders of magnitude and depends on the degree to which the activation energy of the reaction can be lowered. Heterogeneous catalysts have traditionally played major roles in fields such as fuel processing, chemical synthesis and polymer production (van Santen et al. 1999). More recently they have played increasing roles in environmental technology, energy production and materials synthesis. The field is undergoing a significant expansion and it is widely recognized that, for substantial progress, it is necessary to develop an atomic-level understanding of the interaction between the catalyst and the reactants/products in order to design novel catalytic materials that can address the problems of the 21st century.

The catalyst functions by providing an alternative reaction pathway to create products from reactants and, in general, the activation energies of the fundamental steps in the catalysed pathway are lower than the steps in the uncatalysed path (van Santen and Niemantsverdriet 1995). For a heterogeneous catalyst, this is accomplished through four essential processes. Firstly, reactant molecules are adsorbed onto the surface of the catalytic nanoparticles. These adsorbates undergo bond rearrangement to form a reaction intermediate which often involves dissociation of the reactants. The reaction intermediates interact on the surface to create the desired product molecule and these product molecules are then desorbed from the surface returning the catalyst to its original configuration. Changing the structure and composition of the surface can dramatically alter the activation energies of the elementary steps in the process resulting in an increase or a decrease in the reaction rate. Moreover, competing low-energy reaction pathways may lead to the creation of different product molecules lowering the selectivity of the catalysts. To achieve high selectivity in the chemical

S.J. Pennycook, P.D. Nellist (eds.), *Scanning Transmission Electron Microscopy*,
DOI 10.1007/978-1-4419-7200-2_13,

conversion process, the surface of the catalyst must be controlled so that the adsorbate interactions are directed along the correct reaction pathway.

The nanoparticle surface essentially choreographs this sequence of events and may itself undergo dynamic changes to facilitate the surface reaction. The specific reactions that take place on the surface are controlled by the atomic structure and composition of the surface as well as the electronic properties of the surface. An ongoing goal in the field of heterogeneous catalysts is to elucidate the relationship between these catalytically relevant surface structures and catalyst activity and selectivity for a particular chemical reaction. The main role of electron microscopy is to provide a detailed description of the underlying catalyst at length scales down to the atomic level. Heterogeneous catalysts are not really a class of materials in the traditional sense. The exact catalyst formulation depends very much on the application of interest. They may be composed of metals, ceramics or combinations of materials. For example, noble metals are often used in catalytic converters to control automobile emissions, oxides are used in many reactions where redox chemistry is necessary and sulphides play a critical role in desulphiding fossil fuels.

Although the materials choices are very different, for each application there are a number of characteristics associated with superior catalyst functionality. Firstly, they usually have high specific surface areas to maximize contact with the reactants. Many catalysts are composed of porous aggregates of nanoparticles or their crystal structures form "internal surfaces" such as those found in zeolites or mesoporous materials. For practical applications, the high surface areas associated with these materials are critical to achieve high rates of converting reactants into products. The catalyst may be composed of more than one material giving rise to nanoscale composites. For example, in supported metal catalysts, metal nanoparticles are dispersed over a high surface area support. In some cases, the support functions purely as a passive surface on which the metal nanoparticles sit. In other cases, the support may modify the character of the metal or it may play an active role in the surface chemistry for one or more of the intermediate steps in the reaction. Such catalysts may be described as bifunctional because several different surface reactions may be necessary to create the product. For example, reducible oxides like titania and ceria are often used together with noble metals for redox reactions. In additional to providing high surface area support for the metal catalyst, these oxides can essentially act as a buffer for storage and release of oxygen from the crystal lattice at different points in a reaction pathway. Similarly, more than one metal may also be present on the support to provide additional functionality giving rise to bimetallic or even trimetallic catalysts.

The catalytically active surfaces should be stable under reactor conditions over extended periods of time. This means that particle sintering should be controlled, the catalyst should be resistant to poisons and undesirable phase transformations should not take place during operation. As society's utilization of catalysts becomes more widespread, it becomes increasingly important to control costs. There is growing incentive to discover new catalytically active materials that do not involve expensive noble metals or materials synthesis processes.

Electron microscopy is the only atomic resolution tool that can allow the location, morphology and shape of individual nanoparticles dispersed on an irregular high surface area support to be directly observed. For this reason, electron microscopy is a critical tool for characterization of heterogeneous catalysts. Atomic resolution phase contrast microscopy in a conventional transmission electron microscopy has been extensively used to study the structure and morphology of a wide variety of catalysts at the atomic level (e.g. Datye 2003, Gontard et al. 2007, Thomas and Gai 2004, Yacaman et al. 2002). It can provide vital information on the interior and surface structure of the nanoparticles on many heterogeneous catalysts. Scanning transmission electron microscopy provides several additional techniques which make it particularly well suited for the characterization of certain classes of catalyst. Z-contrast imaging is a powerful technique for locating heavy metal particles and atoms on low atomic number supports. This particular combination occurs frequently in supported metal catalysts and consequently Z-contrast STEM has become a vital tool for investigating these systems. The ability to perform high spatial resolution chemical analysis in a STEM with both energy-dispersive X-ray spectroscopy (EDX) and electron energy loss spectroscopy (EELS) plays an increasingly important role in catalyst characterization. This not only allows the composition of individual nanoparticles to be explored but also permits interfacial interactions (e.g. metal–support interactions) to be studied. With the advent of aberration-corrected STEM, atomic resolution spectroscopic information can now be acquired from single atoms (Varela et al. 2004) and atomic resolution two-dimensional chemical images can be acquired in reasonable times (at least in favourable cases) (Kimoto et al. 2007, Muller et al. 2008).

This chapter is laid out to illustrate some of the successful approaches to catalyst characterization using STEM. Section 13.2 describes the vital role that STEM imaging plays in the characterization of supported metal catalysts. Z-contrast imaging has proven to be extremely useful for this class of catalysts, especially for the common case of high atomic number noble metal particles or atoms on a high-surface-area, low-atomic-number support. Section 13.3 describes work on the application of high spatial resolution EDX and EELS to determine the nanochemistry of catalyst particles. Sections 13.4 and 13.5 deal with the emerging technique of tomography and the re-emerging technique of nanodiffraction, two approaches that show great promise for elucidating catalyst structure. Many of the catalytically relevant structures may form only in the conditions present in a chemical reactor, i.e. in the presence of reactant/product gases at operating temperature. Consequently there is increasing recognition of the importance of characterizing catalytic materials in situ of under *operando* conditions. Section 13.6 describes a limited amount (mostly the author's work) of in situ STEM work conducted under reactive gas conditions. This is a vitally important area in catalyst characterization and it is expected to grow in the years ahead. The final section looks ahead and discusses some of the challenges and opportunities which may be addressed in the future in the application of STEM to heterogeneous catalysts.

13.2 STEM Imaging of Heterogeneous Catalysts

In a supported metal catalyst, a large percentage of the metal atoms are on the surface of the nanoparticles and thus available to interact with surrounding gas or liquid phases. This morphology is favoured in many applications where expensive precious metals like platinum, rhodium or palladium are used because most of the metal atoms are available to facilitate catalysis and are not wasted in subsurface sites. To develop a fundamental understanding of the structure–property relations of supported metal catalysts, it is necessary to have a detailed understanding of the metal dispersion, nanoparticle shape and composition. This is especially true for expensive noble metal catalysts where there is economic pressure to ensure that every metal atom is directly contributing to the catalytic reaction, thus allowing the amount of metal used in the catalyst formulation to be reduced. However, interference from the underlying crystalline support often hinders attempts to locate small metal particles with bright-field TEM. Indeed from a historical point of view, the need to identify the location of metal particles on supported metal catalysts was a significant motivation for the development of high-angle annular dark-field STEM imaging – the so-called Z-contrast imaging approach. This imaging technique plays a major role in characterizing supported metal catalysts and is the primary focus of this section.

13.2.1 Z-Contrast Imaging of Supported Metal Catalysts

Early work by Crewe and co-workers had demonstrated that dark-field STEM techniques could be employed to image individual heavy atoms on thin carbon films (Crewe et al. 1970). The initial approach involved taking a ratio of the elastic-to-inelastic signals recorded simultaneously on different detectors. Later work by the same group showed that the elastic signal recorded on an annular detector could also yield images of individual heavy atoms on light element supports (Isaacson et al. 1976, Wall et al. 1974). The ratio approach was applied with some success to supported metal catalyst by the Cavendish group (Treacy et al. 1978). However the ratio method was found to be unsuitable for crystalline supports because of the strong diffraction contrast contribution to the resulting signal (Donald and Craven 1979). Further work by the same group demonstrated that using a high-angle annular detector significantly suppressed diffraction contrast from crystalline supports yielding a powerful technique for detecting metal particles in supported metal catalysts (Pennycook 1981, Treacy 1981, Treacy et al. 1980).

High-angle annular dark-field imaging is now used routinely to determine particle size distributions and metal dispersions in supported metal catalysts (Datye 2003, 2006, Liu 2004, 2005, Liu and

Cowley 1990, Rice and Bradley 1994, Treacy and Rice 1989). Particle size distributions can then be converted to metal dispersion and surface areas for comparison with catalytic data. Figure 13–1a and b shows typical Z-contrast images from two noble metal catalysts. Figure 13–1a is of Pt nanoparticles approximately 3 nm in diameter uniformly distributed over the carbon support. The sample was synthesized with standard wet chemical techniques and is typical of many commercial catalysts (Crozier et al. 1999). To prepare the sample for STEM analysis, the C powder was embedded in a resin which was then cured and ultramicrotomed into thin sections of about 50 nm in thickness. The contrast difference between the Pt and the C is striking making it relatively easy to determine a size distribution from the data. Figure 13–1b is of Rh/γ-Al_2O_3 and in this case, the contrast of the heavier crystalline support is more pronounced and the signal from the lighter metal is weaker but it is still easy to make out the location of the metal nanoparticles.

Figure 13–2a shows a higher resolution image from a similar catalyst subject to a different aging treatment. In this case, the resulting Rh particles are considerable smaller and on the thinner isolated alumina particles, metal particles down to about 0.5 nm can be easily resolved. The images for Figures 13–1 and 13–2 were recorded on a JEOL 2010F equipped with an objective lens with a spherical aberration coefficient C_s of 0.5 mm and no aberration correction. Such an instrument has an optimum probe size of about 1.5 Å which should be sufficient to resolve individual metal atoms. Figure 13–3 is an image of different Rh/Al_2O_3 catalyst which was believed to have near atomically dispersed Rh. The image is fairly typical of those obtained from a real catalyst where the support may show significant intensity variation due to thickness changes. The bright specks on the image are the atoms or small atomic clusters on the oxide support, although the contrast patches vary significantly in both intensity and spatial extent.

These images are fairly disappointing in comparison to the images obtained from well-oriented crystalline material on the same instrument. For example, Figure 13–4 was recorded on the same instrument under identical conditions from the (110) projection of Si. In this projection, the so-called silicon dumb-bells with a separation of 0.135 nm are clearly resolved with high contrast and good signal-to-noise. Several factors contribute to the less dramatic appearance of the images of individual metal atoms from catalyst. The signal from atomic-scale features in images from crystals does not come from single atoms but from a column of many atoms (typically 50–100 atoms in length). Moreover, for many crystals in zone-axis orientations, electron channelling (Hillyard and Silcox 1995, Pennycook and Jesson 1990) plays a major role in concentrating the incident electron beam along these atomic columns. This tightly bound s-like state strongly scatters electrons to the high-angle detector. Both these effects combine to increase the signal to noise of atomic-scale features from a single crystal sample.

For a metal atom sitting on the surface of a support particle, we have almost the exact opposite situation. The scattered signal from a single atom is much weaker than that generated by a column of atoms. With a typical convergence semi-angle of about 10 mrad (for non-aberration

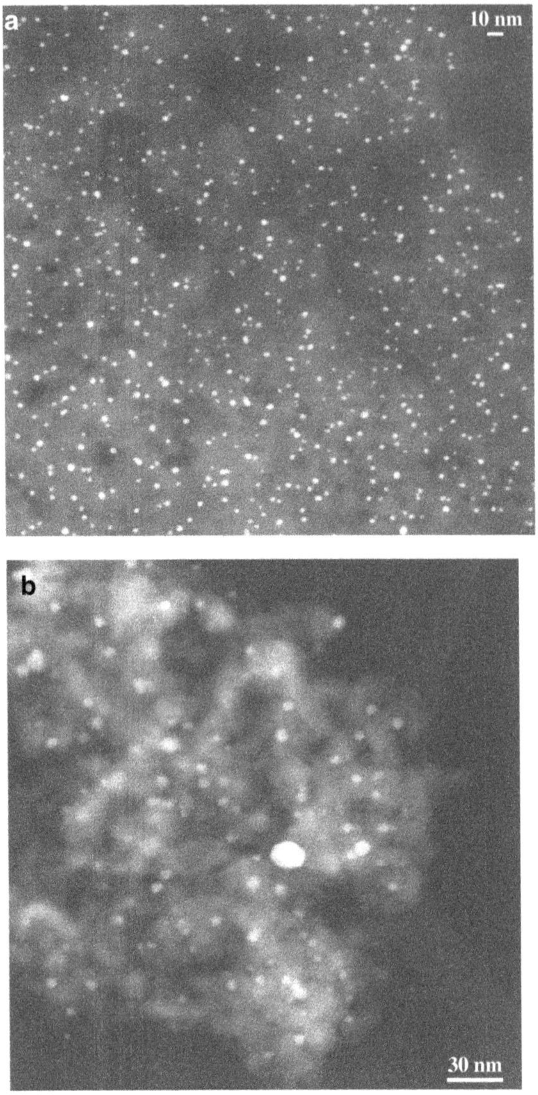

Figure 13–1. (a) Z-contrast image of Pt/C catalyst. The Pt nanoparticles are supported on a high surface are a active carbon support. (b) Z-contrast image from Rh γ-Al_2O_3 catalyst showing presence of Rh metal particles in size range of 3–30 nm.

corrected instruments), image blurring of a few angstroms or more may exist between the entrance and the exit side of a support particle that is 200 Å or more in thickness. Essentially substantial differences in focus are present between the two surfaces. Moreover, on a typical nanoscale catalyst support particle, entrance and exit surfaces will in general not be perpendicular to the beam direction so that only one point or line on the entrance surface can be at the optimum defocus. Finally, the contribution to the background comes from the support particles and this is usually comparable to or much larger than the signal from the atom (see Figure 13–3b). These effects combine to give significant variations in the contrast from atoms and atomic clusters over extended areas of the support.

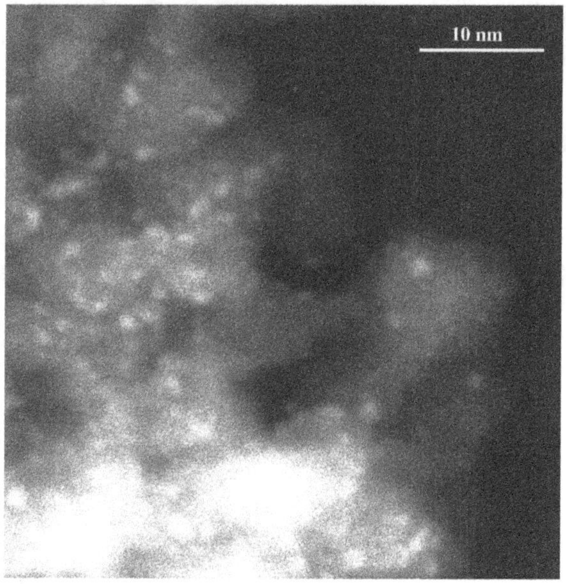

Figure 13–2. Z-contrast image from Rh/γ-Al_2O_3 catalyst showing presence of Rh metal particles in size range of 0.5–3 nm.

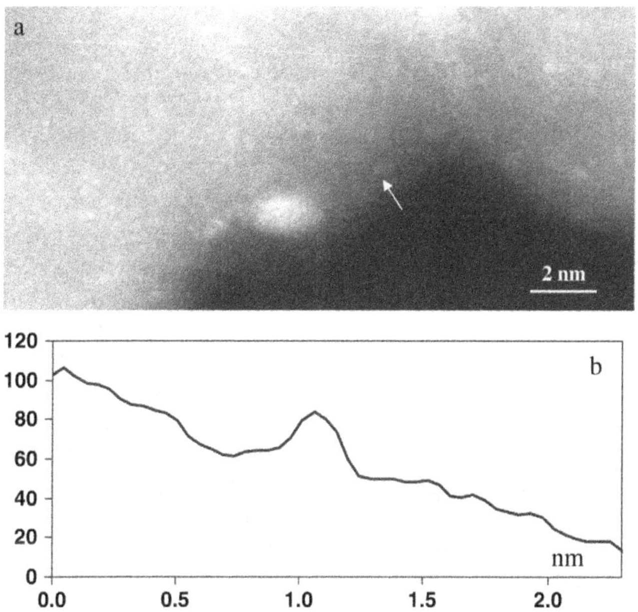

Figure 13–3. (a) Z-contrast image from Rh/Al_2O_3 catalyst where Rh is atomically dispersed (see atom at *arrow*) on the γ-Al_2O_3. (b) Linescan through arrowed region of (a) showing intensity peaks from metal atoms on Al_2O_3 background.

Several methods have been developed to determine the number of atoms in metal clusters by removing the substrate background and integrating the remaining image intensity. Treacy and Rice (1989) were the first to determine the number of atoms in the clusters in supported metal catalysts. The method relied on comparing the measured cluster size with the image intensity to generate a calibration curve. With this

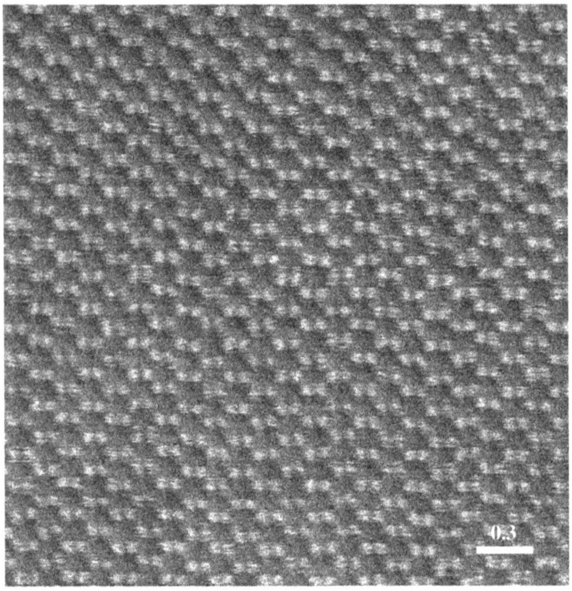

Figure 13–4. Z-contrast image from Si (110) projection under same conditions as those used to record Figure 13–3.

approach they claimed to be able to detect three atom clusters of Pt on 20 nm of γ-Al_2O_3. Singhal et al. (1997) used a similar approach to study Re clusters supported on thin sheets of graphite. Variations in the image intensity versus particle size can also be used to deduce information about the 2D and 3D character of supported metal particles (Menard et al. 2006, Yang et al. 2003). A method involving the use of blurring has recently been developed to provide a tool for quantifying annular dark-field images from supported metal clusters when particle stability may limit the signal-to-noise ratio of the image (Okamoto et al. 2008).

Nellist and Pennycook (1996) successfully imaged individual atoms and small clusters of Pt and Rh on a γ-Al_2O_3 support. The recent development of aberration-corrected STEM has dramatically improved the sensitivity for atom detection on high surface area catalyst supports. The smaller probe size gives a substantial increase in the signal-to-background ratio for detection of single atoms and effectively makes it easier to achieve higher signal-to-noise ratios for reliable atom detection (Blom et al. 2006). Figure 13–5 is a typical example of locating atomically dispersed Pt on a γ-Al_2O_3 with aberration correction (Blom et al. 2006). In this case, individual atoms show up clearly on the oxide support and demonstrate the dramatic impact that aberration correction can have on this type of measurement. Detailed information on metal cluster configurations is now possible as demonstrated by the investigation of anomalous Pt–Pt bond distances on γ-Al_2O_3 (Sohlberg et al. 2004). The Z-contrast approach was also used to show that bilayer Au clusters about 0.5 nm in diameter may be responsible for high CO oxidation activity at room temperature (Herzing et al. 2008).

Many other exciting insights into the structure of supported metal catalyst are now possible as the following examples show. The detection of surface segregation becomes very convincing with enhanced

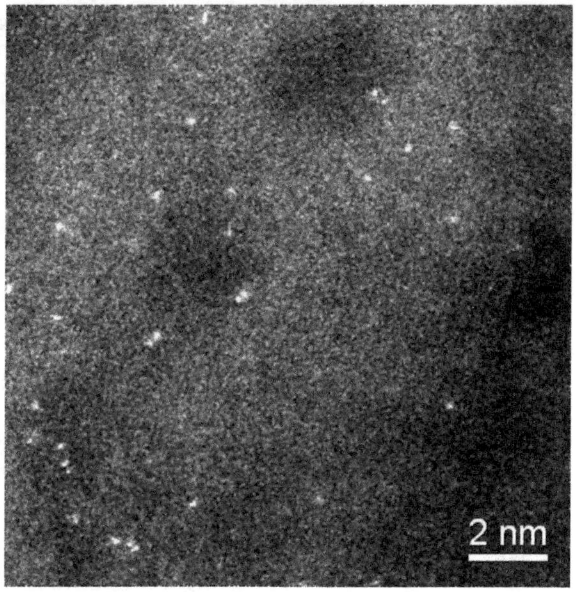

Figure 13–5. Z-contrast image of individual Pt atoms and Pt dimers on γ-Al_2O_3 support recorded with aberration-corrected STEM (from Blom et al. 2006, reproduced with permission).

resolution and contrast of aberration correction as demonstrated by the work on Au–Pd bimetallic particles (Ferrer et al. 2008). Identification of the location of single metal atoms relative to the unit cell of the support becomes much easier (Wang et al. 2004). The ability to use Z-contrast image intensity to suggest structural models is now being exploited in more complex oxide catalysts. In recent work on MoVNbTeO, Z-contrast imaging was employed to explore the occupancy of heptagonal channels in two different catalyst preparation methods (Pyrz et al. 2008).

13.2.2 Signal-to-Background Ratios and Single-Atom Visibility in Catalysts

The most dramatic Z-contrast images of metal atoms in catalysts have been obtained from the higher atomic number metals like Pt and Au because they give images with higher signal-to-background ratios. It is useful to explore the variations in peak-to-background ratio for atom detection in typical catalyst samples with atomically dispersed metal under different experimental conditions. Such a calculation could be carried out rigorously using multislice methods (see Chapters 6) for typical combinations of metals and supports. This would be somewhat tedious, however, by employing a number of simplifying assumptions, the overall trends in metal atom sensitivity for some typical metal–oxide combinations of importance to heterogeneous catalysis can be explored.

We begin by considering the overall signal change that occurs in the Z-contrast image when a monolayer of metal is placed on a thin oxide support. Figure 13–6 shows the geometry of a metal layer m sitting on top of a support s of thickness t. The signal scattering into the high

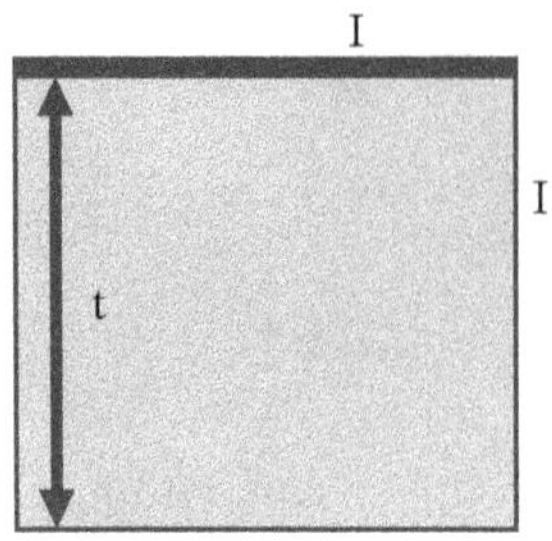

Figure 13–6. Schematic representation of metal monolayer on substrate of thickness t. Signal arriving at annular detector from metal atom and substrate are I_m and I_s, respectively.

angle detector can be written as I_m and I_s for the metal layer and the support, respectively. The ratio of the signal from the metal layer to the support signal, which we will call the signal-to-background ratio (SBR), can then be written as

$$\mathrm{SBR} = \frac{I_m}{I_s} \quad (1)$$

Assuming a thin film, the signal strength for scattering from an atom on the entrance surface of a substrate is given by

$$I_m = I_o N_m \sigma_m \quad (2)$$

where I_o is the number of incident electrons striking the atom, σ_m is the cross section for scattering into the annular dark-field detector and N_m is the number of atoms per unit area. For the background, we can write a similar expression:

$$I_s = I_o N_s \sigma_s \quad (3)$$

where the symbols have the same meaning for the background atoms. For high-angle scattering, the cross sections increase with Z^n, where Z is the atomic number of the scattering atom and n is typically between 1.5 and 2. For typical oxide supports, it is convenient to consider the support as being composed of "average atoms" with an average atomic number Z_{av} determined from the formula unit (i.e. Al_2O_3 or SiO_2). Assuming that the number density of the metal surface layer is the same as the number density on the support surface and setting $n \sim 1.7$ (Kirkland 1998), Eq. (1) can be written as

$$\mathrm{SBR} = \frac{1}{tN_V^{1/3}} \left(\frac{Z_m}{Z_{av}} \right)^{1.7} \quad (4)$$

where N_V is the number of support atoms per unit volume. This equation is plotted in Figure 13–7 as a function of metal atomic number for three common catalyst supports, carbon, alumina and ceria, and two typical thicknesses, 5 and 20 nm.

Figure 13–7 shows several important aspects of metal detection in catalyst. Firstly, there is an enormous advantage if the metal atoms are imaged on a light thin substrate such as carbon. For a given metal atom, the SBR scales inversely with both substrate thickness and with ${Z_{av}}^{1.7}$. For example, the SBR for a Au monolayer on 5 nm of carbon is almost

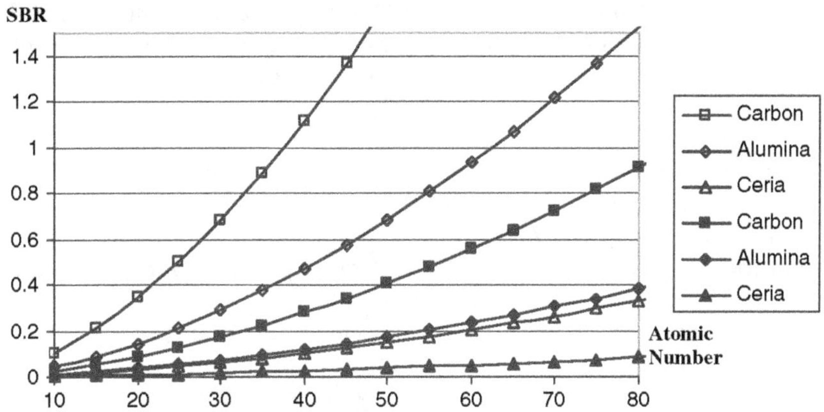

Figure 13–7. Signal-to-background ratio (SBR) for metal monolayer of atomic number Z on three different supports. *Open symbols* are for support thickness of 5 nm and *solid symbols* for support thickness of 20 nm.

4 (this contributed to the early success of the STEM pioneers in demonstrating single-atom detection). However, Figure 13–7 also shows that as the support thickness and the average atomic number increase, the SBR drops significantly. For example, the SBR for Au on 20 nm of ceria is less than 0.1. This simple monolayer approach seems to be reasonably accurate at predicting the SBR for the Rh atoms shown in Figure 13–3. The line scan through the arrowed atoms has an experimental SBR of about 0.6 which agrees with the values from Figure 13–7 of 0.6 for Rh on 2 nm of Al_2O_3. (This level of agreement is somewhat fortuitous in this case and will be discussed in the following section.) Figure 13–7 explains why it is often so difficult to detect 3d transition metals and other lighter noble metals such as Rh or Pd on the heavier supports such as zirconia or ceria. Of course it could be argued that the problem of detecting metal atoms when the SBR is low can simply be addressed by increasing the counting statistics. However, this is often precluded by radiation damage, contamination and sample stability. In practice, high SBR is often necessary for successful atom detection.

13.2.3 Probe Size Consideration for Single-Atom Visibility

The Z-contrast imaging signal is proportional to the sharp part of the projected potential. Classically, this arises from impact parameter arguments where large-angle scattering is mainly associated with electrons passing close to the atom (i.e. small-impact parameter events). This dimension is significantly smaller than the atomic size (which for many catalytically important metals typically is ~1.3–1.5 Å). For a 200-kV electron scattering from a heavy atom through an angle of 40 mrad or more, impact parameter arguments show that an area of diameter 0.3 Å or less will contribute to the high-angle signal (Reimer 1985). Similar dimensions arise in wave mechanical formulations of electron scattering. For example, a significant contribution to the Z-contrast image arises from tightly bound 1 s-like states of width approximately

0.2–0.3 Å (Nellist and Pennycook 1999). Lupini et al. explored a real-space representation of the high-angle annular dark-field detector function to derive an equivalent and correspondingly narrow point spread function (Lupini and Pennycook 2003).

A set of multislice calculations using Kirkland codes (Kirkland 1998) were undertaken to investigate the intensity distribution in Z-contrast images of a single atom for different microscope aberrations. The atom was placed at the centre of a 1 nm × 1 nm unit cell and the image calculated with the defocus Δf and convergence angles α optimized to minimize the size of the atom image. Calculations were undertaken for 200-kV incident electrons for three catalytically important metals (Cr, Rh and Pt) and similar results were obtained. Figure 13–8 shows images and intensity line scans through two images of a Rh atom from a microscope with almost no aberrations (C_S = 0.001 mm) and a typical non-corrected instrument with C_S = 0.5 mm. (The images were normalized to ensure that the integrated intensities from each atom were identical.) The full-width half maximum for the Rh peak in the image with no aberration is about 0.35 Å, a value which agrees with impact parameter and Bloch wave dimension described above. The full-width half maximum of the peak in the image from the non-aberration corrected microscope was about 1.4 Å. The experimentally measured width for the Rh atom in Figure 13–3 is a bit larger measuring about 2 Å possibly due to instabilities in the catalyst sample or focus drifts.

It is useful to explore how the probe size will affect the visibility of a metal atom on a typical catalyst support particle. If we assume the support may be in a random orientation (or disordered) and the projected potential will be smeared out and thus the background

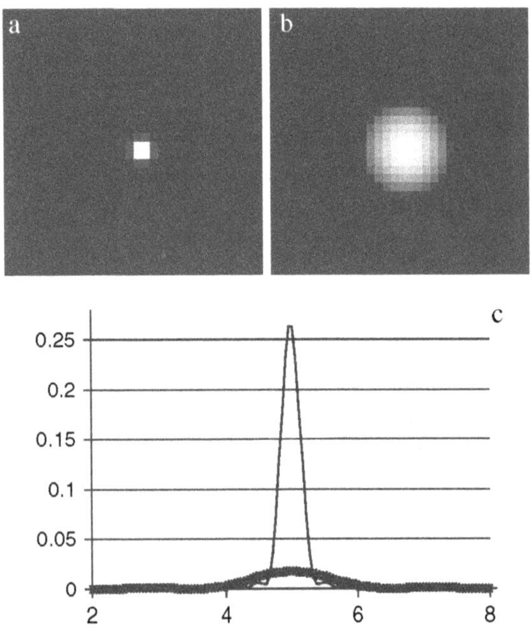

Figure 13–8. (a) and (b) Simulated Z-contrast images and (c) line profiles through Z-contrast images of a Rh atom from microscope with no aberrations (a, *solid line*) and microscope with C_S = 0.5 mm (b, *hashed curve*). (Calculation parameter for: *solid curve*: C_S = 0.001 mm, α = 40 mrad, Δf = 5 nm. *Solid curve*: C_S = 0.5 mm, α = 10 mrad, Δf = 300 nm). Image width = 5 Å.

signal will not be strongly affected by the size of the probe. However for a metal atom sitting on the entrance side of the support, the appearance of the atom in the Z-contrast image will be a convolution between the probe distribution and the sharp part of the projected potential. The magnitude of the peak will depend of the size of the probe relative to the width of the scattering potential. In the monolayer calculations of Figure 13–7, the spacing between the atoms on the surface is about 2 Å. This is comparable with the probe size on non-aberration-corrected instruments and simulations suggest that the peak-to-background ratio from individual atoms will lie within a factor of 2 for these monolayer curves. This is why the predictions of Figure 13–7 are consistent with the experimentally measured values of Figure 13–3.

Figure 13–8 suggests that the peak-to-background ratio can increase by almost a factor of 10 if the aberrations are eliminated. The peak-to-background curves for this ideal situation for single atoms on carbon, alumina and ceria supports are shown in Figure 13–9. The calculations suggest that lighter atoms should be visible on thin supports and that medium and heavy atomic number atoms should be visible on thicker supports of higher average atomic number. Of course such a machine does not exist at present (this corresponds to a probe size of 0.3 Å) and sample stability may prevent such large peak-to-background ratios from being achieved in some cases. For the current generation of aberration-corrected instruments, the peak-to-background ratio will lie somewhere between Figures 13–7 and 13–9 depending on the degree of correction. Recent experimental results have already demonstrated that, with aberration correction, individual Cr atoms on γ-alumina can be easily detected (Borisevich et al. 2007). This discussion also demonstrates why aberration correction is so important for supported metal catalyst research. For Z-contrast imaging, the improvement in atom visibility is directly related to the improved peak-to-background ratio. The visibility may improve until the electron probe size is comparable with the sharp part of the scattering potential (~ 0.3 Å).

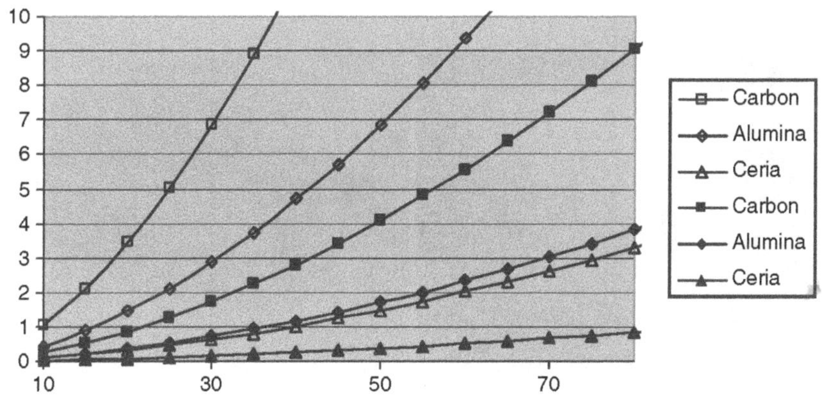

Figure 13–9. Peak-to-background ratio (SBR) versus metal atomic number for STEM with no aberrations on three different supports. *Open symbols* are for support thickness of 5 nm and *solid symbols* for support thickness of 20 nm.

The *detection limits* are ultimately determined by the signal-to-noise ratio. High peak-to-background ratios make detection easier to achieve but does not tell us what the actual detection limit is for a given measurement. This can be calculated by considering the beam current, acquisition time, elastic scattering cross section and support thickness. However, for individual metal atoms on a support, calculations based purely on Poisson statistics are often of limited value because in practice the electron doses required to achieve a predicted detection limit may not be practical because of sample instability (Batson 2008). It is often more valuable to experimentally determine the maximum dose that a sample can take without undergoing significant change. The experimentally determined variance in a small region of the background δB can then be used to determine the lowest detectable peak-to-background ratio in an image. Assuming that the signal from the atom, I_{min}, has to be greater than twice the background noise

$$I_{min} = 2\delta B$$

gives a minimum detectable SBR of $2\delta B/B$. The curves of Figures 13–7 and 13–9 can then be employed to determine if the metal atom of interest will be detectable within the image.

13.2.4 Bright-Field STEM Imaging

While Z-contrast imaging is the preferred technique for locating metal particles on supported metal catalyst, there is still value in phase contrast bright-field STEM. The power of Z-contrast imaging to suppress diffraction and phase contrast also makes it insensitive to easily differentiating between crystalline and amorphous phases. This is illustrated in Figure 13–10 which shows a Z-contrast and phase contrast STEM image from Pd particles on alumina recorded on a JEOL 2010F. The Z-contrast image is rather featureless and provides very little information about the metal particle and its relationship to the support. However, the phase contrast image immediately shows that the particle

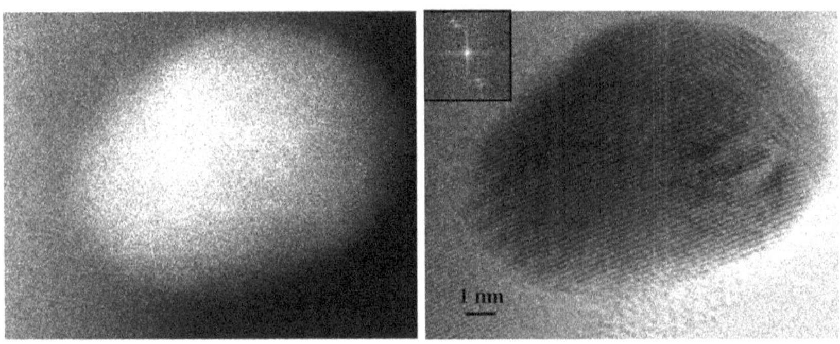

Figure 13–10. Z-contrast image (*left*) and bright-field STEM image (*right*) of Pd particle on Al_2O_3 support. *Inset* is Fourier transform of bright-field image showing parallel sets of diffraction spots from metal particle and support (see text).

is crystalline, has a well-defined orientation relationship with the substrate (which is also crystalline in this region) and the metal particle is partially covered with an amorphous layer. This is a powerful reminder of the value of different imaging modes in STEM (and TEM) and that selection of an appropriate imaging technique depends very much on the type of information that is required. Ideally one would record pairs of Z-contrast and phase contrast STEM images simultaneously so that important nanostructural information is not overlooked.

13.3 High Spatial Resolution Chemical Analysis

Many catalysts exhibit significant nanoscale heterogeneity both in local composition and in bonding so nanospectroscopy is an essential tool for investigating these variations. A major advantage of STEM is the ability to employ nanospectroscopy to determine the elemental composition and oxidation states of the components of the catalysts and to explore the interfacial phenomena that characterize many metal–support interactions. Moreover, with aberration-corrected STEM, local chemistry can be measured at the 1 Å level making it possible (at least in principle) to determine the composition and bonding of individual atoms on nanoparticle surfaces. For example, there is an enormous effort underway to develop completely new nanocatalysts based on inexpensive metal combinations to replace noble metal systems. However, there are still significant problems associated with understanding and controlling the metal distributions within bimetallic nanoparticles. Aberration-corrected STEM nanospectroscopy will allow the location of metal atoms on the surface of oxide supports to be mapped out at different stages of the synthesis and aging process. The spectroscopic data will provide fundamental benchmark information that will allow us to develop a complete atomic-level understanding of the relationship between structure, composition, bonding and activity. This will point the way for designing new supported bimetallic particles that are uniquely suited to catalytic applications. Both energy-dispersive X-ray spectroscopy (EDX) and electron energy loss spectroscopy (EELS) play an important role in determining the local nanochemistry of catalysts.

13.3.1 Energy-Dispersive X-ray Spectroscopy in Catalysis

Energy-dispersive X-ray spectroscopy is a powerful and versatile tool for determining the elemental composition of catalytic nanoparticles. Figure 13–11 is a Z-contrast image recorded from a used Zeigler-Natta polypropylene catalyst after approximately 100 s exposure to ethylene. The catalyst consists of a $TiCl_4/MgCl_2$ procatalyst which is then activated in situ with triethyl aluminate – $Al(C_2H_5)_3$. The catalyst is known to undergo fragmentation as a result of localized mechanical stress during the polymerization process. The ADF image of Figure 13–11

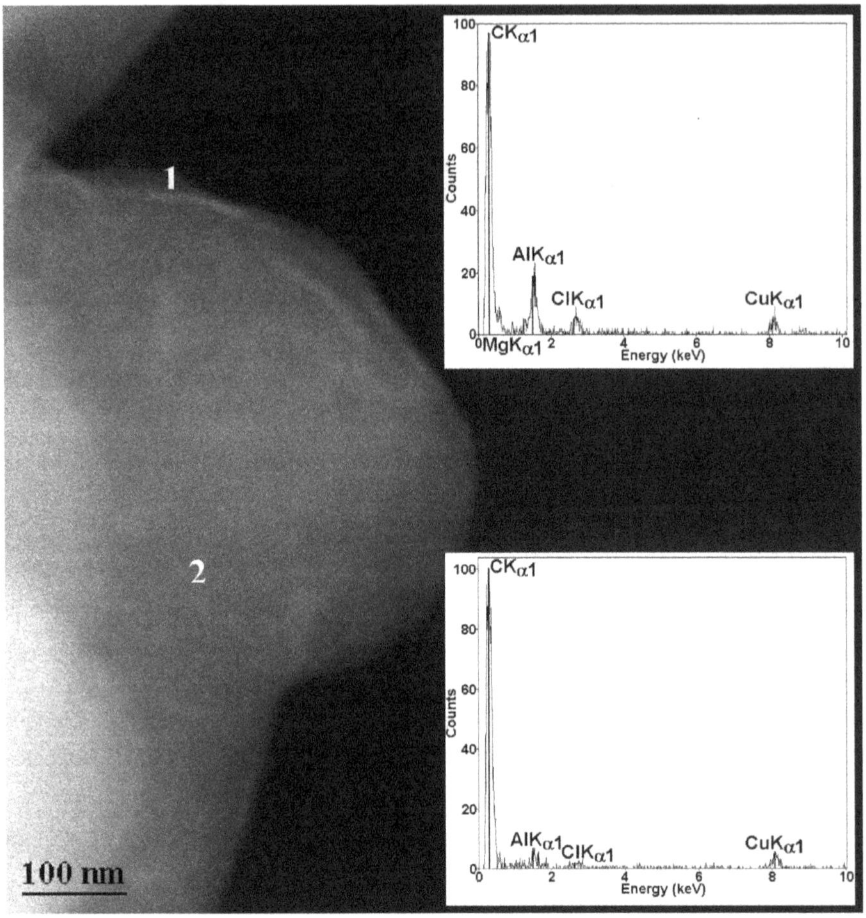

Figure 13–11. Z-contrast image of the Ziegler-Natta catalyst particle after exposure to ethylene. Insets are EDX spectra recorded in points 1 and 2 (from Oleshko et al. 2002, reproduced with permission).

shows light contrast patches and lamellae embedded in a darker contrast matrix. The EDX analysis shows a strong carbon signal associated with the matrix demonstrating that this component is a polypropylene globule. EDX nanoanalysis from a fragment (point 1) shows it to be chlorinated cocatalyst (Al Kα at 1.49 keV and Cl Kα series at 2.62 keV). Surprisingly, only traces of Mg and Ti were occasionally found in the fragments. This suggests that during the early stage of polymerization, the active sites are not floating on the surface of the growing polymer but instead are located at the interface between the polymer and the catalysts with monomer diffusing through the polymer layer (Oleshko et al. 2002).

A goal for some bimetallic catalysts is to make nanoparticles that are identical. However, this rarely occurs in practice for real catalysts and STEM EDX plays an important role in determining the nature of the compositional variations between metal particles. The average composition of a bimetallic particle is often dependent on particle size due to

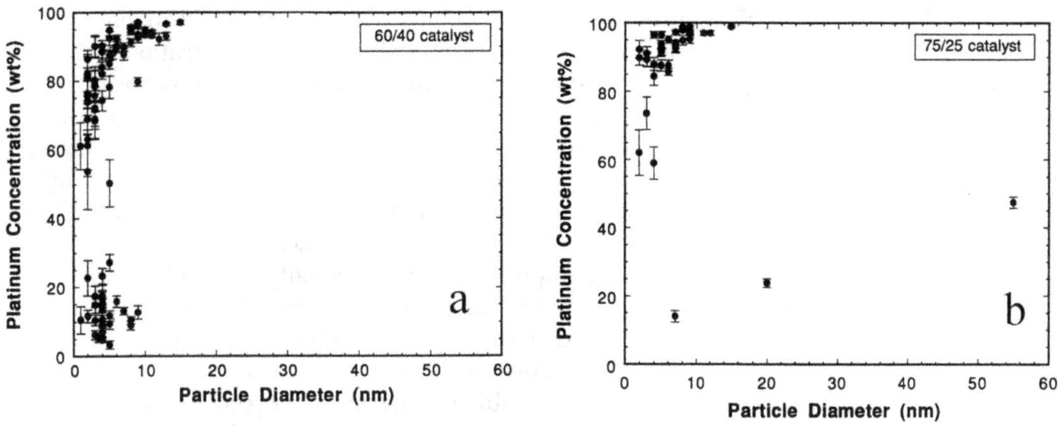

Figure 13–12. Composition versus particle size for individual Pt–Rh particles supported on γ-Al_2O_3. (a) Pt was impregnated first and (b) Rh was impregnated first (from Lyman et al. 1995, reproduced with permission).

preferential diffusion of one metal to the surface or other catalyst preparation effects. Lyman and co-workers have used EDX nanospectroscopy to explore this phenomenon by studying the composition versus particle size for several different supported bimetallic catalysts (Bednarova et al. 2002, Lyman et al. 1995, 2000, Prestvik et al. 1998). This approach is illustrated in Figure 13–12 which shows the composition versus size for Pt–Rh/alumina catalysts prepared in two different ways. For Figure 13–12a, the catalyst was prepared by first impregnating and calcining with a Pt precursor followed by Rh impregnation and reduction. For Figure 13–12b, the reverse procedure was employed, i.e. the catalyst was first impregnated and calcined with the Rh precursor followed by Pt impregnation and reduction. In both cases, the alloy particles separate into two different phase populations corresponding to Pt-rich and Rh-rich particles. Impregnating with Rh first results in the formation of rather large Rh-rich particles and only a very small fraction of Rh is incorporated into the smaller Pt-rich particles. Impregnating first with Pt gives a smaller range of particle sizes but a bimodal composition profile consisting of Pt-rich and Rh-rich particles. This work demonstrates the dramatic effect that different catalyst preparation methods can have on the resulting bimetallic nanoparticles.

One problem with the application of EDX to small nanoparticles is the very small X-ray signal that is generated from nanometre-sized particles. Characteristic X-rays are emitted isotropically making efficient collection challenging in current electron microscopes. The Lehigh group has optimized a 300-kV dedicated STEM (VG HB603) for EDX analysis by increasing the peak-to-background ratio in the spectra and increasing the solid angle of collection through simultaneous collection from two detectors (Lyman et al. 1994). The resulting solid angle of collection is 0.3 sr giving excellent EDX performance compared to other STEM systems. However, despite the improvements in the Lehigh system, the total fraction of X-rays collected remains a little over 2% and typical acquisition times to obtain reasonable statistics from nanometre-sized particles are in the range of 50–200 s. While aberration correction can increase the total beam current to improve signal-to-noise ratio, it will also increase the rate at which the nanoparticle is destroyed. A remaining challenge for EDX analysis is the need to develop instrumentation which permits a much higher fraction of the emitted X-rays to be detected. Significant improvements in the collection solid angle are possible with the silicon drift detector (Zaluzec 2009).

The presence of compositional heterogeneity within bimetallic particles can also be explored using spectral mapping techniques. EDX elemental mapping of catalytic nanoparticles can provide information about the distribution of elements in nanoparticles. The much higher beam current available in aberration-corrected instruments has increased interest in the approach for systems where radiation damage is not a limitation. The Lehigh group has taken the lead in this area by having their VG HB603 fitted with a NION aberration corrector (Watanabe et al. 2006). The combination of the high current in the aberration-corrected probe along with the high X-ray collection efficiency allows 128×128 spectrum images to be acquired in about 1 h. They have recently used this approach to explore the nanoscale compositional variations in AuAg and AuPd bimetallic particles (Herzing et al. 2008). Details can be found in Chapter 7.

13.3.2 Electron Energy Loss Spectroscopy in Catalysis

Electron energy loss spectroscopy is another powerful technique for determining the physico-chemical properties of catalytic materials on the nanometre scale. It can be employed to provide information not only on elemental composition but also on the electronic structure of materials (for a recent review, see Egerton 2009). It has not been as widely used in catalyst characterization (Egerton 2002) as in other areas of materials partly because the signal-to-noise problems discussed for EDX are also present for EELS analysis of small particles. In the EELS case, a greater fraction of the energy loss signal is collected but the signal-to-background ratio of the relevant characteristic signal is much lower compared to EDX. Moreover, many of the common noble metals (like Pt and Au) have useful ionization edges only at high energy losses where the signal is weak or (for example, Pd and Rh) have ionization edges with delayed maxima which exacerbate the poor signal-to-background ratio.

In spite of these issues, STEM EELS is competitive with EDX for exploring compositional fluctuation in catalyst particles when suitable ionization edges are available. For example ceria zirconia oxide is important in automotive three-way catalyst (Shelef and McCabe 2000, Trovarelli 2002) and has potential applications in solid oxide fuel cells (Mogensen 2002, Mogensen et al. 2000, 2004). The exact nature of the catalytic activity in this system is still being investigated but it may be influenced by nanoscale compositional heterogeneity. Figure 13–13 shows the variation in the Ce/Zr elemental ratio across a nanoparticle. This profile was generated by performing an EELS linescan,

Figure 13–13. (a) Z-contrast image and (b) EELS line scan profile from individual nanoparticle of $Ce_{0.5}Zr_{0.5}O_2$. *Arrow* indicates direction of line scan (from Wang et al. 2006, reproduced with permission).

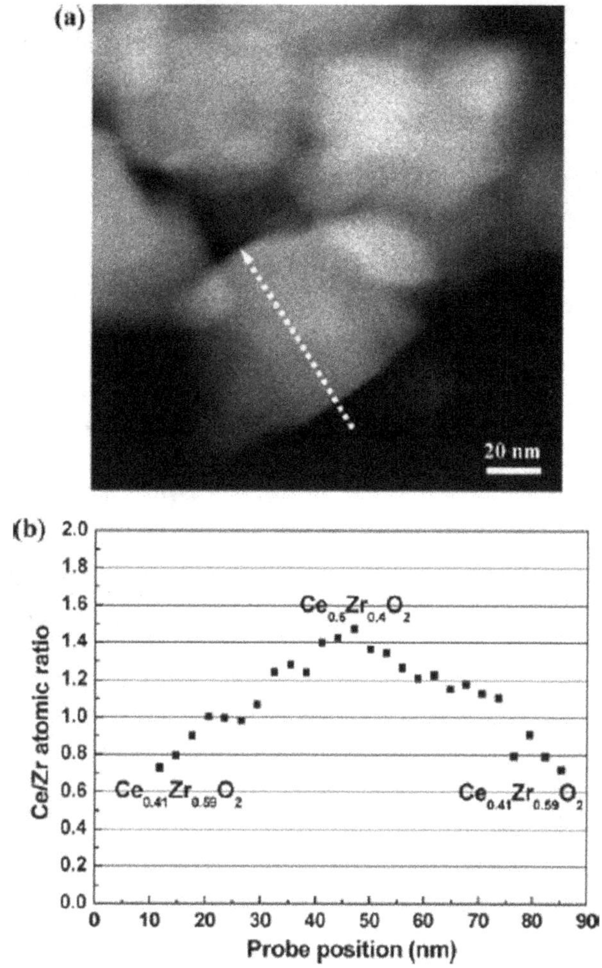

i.e. acquiring a series of energy loss spectra every couple of nanometres along the line drawn on the image of Figure 13–13a. The profile clearly shows that the particle has a core-shell structure with a Ce-rich centre. This nanospectroscopic characterization can be correlated with catalyst synthesis methods and redox activity and plays a vital role in developing a thorough understanding of the functionality of this system (Wang et al. 2006, 2008).

STEM EELS is the only method which can provide simultaneous information on the oxidation state, bonding and composition in catalytic materials with subnanometre resolution. Changes in the shape of the near-edge structure and chemical shifts in ionization edges (see Chapter 5) can be directly interpreted in terms of changes in oxidation state and bonding with resolution on the order of an angstrom on aberration-corrected STEMs. Browning and co-workers have used a combination of STEM EELS and Z-contrast imaging to investigate bonding at the metal–support interface support for several catalysts (Klie et al. 2002, Sun et al. 2002a). For example, by examining chemical shifts at the metal–support interface of Pd /γ-Al_2O_3, they were able to show that there is a size-dependent chemical shift in the position of the oxygen K-edge (Sun et al. 2002b). They interpret this shift in terms of an electron transfer from the alumina support to metal particles of 10 nm or larger in size. The same group investigated the effect of reducing conditions on the compositional heterogeneity of CuPd particles (Sun et al. 2002c). They found preferential segregation of Pd to the surface when reduction is carried out at temperatures below 500°C, whereas uniform alloy particles were formed when the reduction temperature is increased to 800°C.

In supported metal catalysts, it is often important to know if a metal species is present in the oxide or the metallic state. This is not easy to determine from EDX because there is often a strong oxygen signal from the underlying oxide support. EELS is ideally suited to this because the ionization edge shapes can be substantially modified by the presence of an oxygen–metal bond (see Section 13.6 and Sun et al. 2002a). Even for noble metals which are not always considered to be ideal edges for EELS analysis, this approach can often work. Figure 13–14 shows a series of EELS spectra recorded from Rh particles supported on alumina. The Rh M_{45} edge (not shown) and M_2 and M_3 edges do not show a large change when the particle transforms from metal to oxide. However, the oxygen K-edge associated with the formation of oxides of Rh has a lower edge onset compared to the O K-edge associated with the alumina support. Thus the shape of the O K-edge is a reliable indicator of the oxidation state of the Rh.

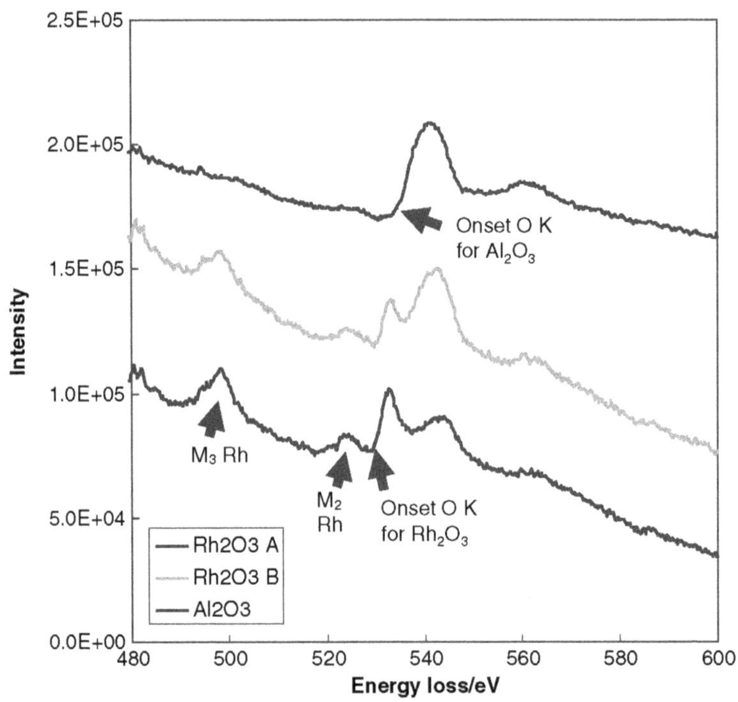

Figure 13–14. Energy loss spectra for two Rh_2O_3 particles A and B and from the alumina substrate (spectra courtesy of Duncan Alexander).

STEM EELS is clearly a powerful tool for exploring the nanochemistry of catalytic materials but it is also somewhat limited by the weak signal strength associated with nanometre particles. Aberration-corrected probes not only increase the beam current but also increase the radiation damage rate. Improved electron detectors and coupling between the spectrometer and the microscope can improve the signal-to-noise ratio in the spectrum. One advantage of EELS of small nanoparticles is that plural scattering effects are not significant, so working at lower accelerating voltages may offer advantages in terms of less radiation damage and higher inelastic scattering cross sections (Egerton 1996). The improved resolution of the monochromator will also open up new opportunities for probing local bonding on the surface of catalytic particles.

13.4 Tomography of Catalysts

In general, single-shot imaging techniques yield only a projection of the structure of the catalyst and do not directly provide information about the 3D structure of the material. Since catalysts are high surface area materials, they typically contain high degrees of porosity and it would be extremely useful to be able to correlate the 3D porosity with the nanostructure. Moreover, understanding the nanoparticle shape and surface structure is critical if we are to develop a deeper understanding of structure–property relations for heterogeneous catalysts. Tomographic techniques provide a powerful tool for generating 3D nanoscale descriptions of heterogeneous catalysts. A complete description of this approach is described elsewhere in this book (see Chapter 8). The discussion here will focus primarily on the current status with regard to applications to catalysis.

Several STEM approaches have been adopted to generate 3D representations of catalysts and nanoparticles. They all employ Z-contrast imaging because the incoherent nature of the high-angle annular dark-field signal means that, at least for thin samples, the signal intensity is linearly proportional to sample thickness (or mass thickness). One successful approach for nanometre resolution work involves reconstructing the 3D objects from a series of Z-contrast images taken at regular tilt intervals (Midgley and Weyland 2003). To achieve a reconstruction free from artefacts with nanometre resolution, a large number of images must be recorded ($\sim$100) over a wide tilt range (ideally 180°). This approach has been successfully employed to map out the distribution of metal particles in mesoporous silicates and carbons (Moreno et al. 2006, Wikander et al. 2007). It has also been successful in exploring the distribution of Co species over alumina in Fischer–Tropsch catalysts (Arslan et al. 2008).

The development of aberration-corrected STEM has stimulated other approaches for extracting 3D information from catalysts. The high convergence angles that are necessary for forming sub-angstrom focused probes result in a depth of focus of only a few nanometres. This opens up a confocal type of approach to tomography in which a through-focal series of Z-contrast images is recorded in which different sections of the sample are in focus and the full 3D object is then reconstructed (Borisevich et al. 2006). This depth sectioning certainly has atomic resolution in the lateral dimension and should be capable of determining the position of small clusters and individual atoms in the third direction to within several nanometres.

A final STEM approach for extracting atomic-resolution 3D information from small nanoparticles relies simply on the linear relationship between the intensity of the Z-contrast image and the number of atoms under the beam. This is not really a true tomography method because it employs only a single image of the particle. However, in an atomic-resolution, Z-contrast image of a small nanoparticle, the intensity in each column is linearly dependent on the number of atoms in the column (Li et al. 2008). The incoherent nature of scattering means that 3D cluster models may be constructed directly from the image intensities and it has been used to provide a detailed representation of an Au cluster on a carbon support film. The method should be effective for providing an atomic-level representation of monometallic particles but it will be more problematic for bimetallic particles or partially oxidized particles because the image intensity will be affected by the atom species. However, one advantage of the method is that it does not require that multiple STEM images are recorded. For nanoparticles and atomic clusters which are unstable under STEM irradiation, this may be the only practical method that works.

Regardless of the strengths and weaknesses of the various approaches to 3D reconstructions, it is clear that continued development of these methods will have a major impact on catalyst nanocharacterization. These recent examples demonstrate that it may soon be possible to map out the location and identity of each atom in catalytic particles at least in favourable cases. This will allow us to determine the nature of the defect sites on the cluster surface and correlate this with catalytic activity.

13.5 Nanodiffraction

STEM nanodiffraction is a somewhat neglected technique in catalyst characterization. It was pioneered by Cowley and co-workers (Pan et al. 1990) mainly as a method for determining particle structure and topotactic relationships between the particle and the underlying support. The convergent beam patterns obtained with the 0.3-nm probe showed a variety of effects including spot splitting that depended on the position of probe in particle, lens current and particle orientation (Pan et al. 1989). This made interpretation difficult and the technique was not pursued much in catalysis research. Recent Zuo and co-workers (2004) developed the technique of coherent nano-area electron diffraction in which near-parallel illumination approximately 5 nm in diameter is used to record diffraction patterns from nanoparticles. This pattern is more directly interpretable than the early form of nanodiffraction and offers the possibility of determining coordination-dependent surface relaxations for metal nanoparticles. If applicable to supported metal catalyst, such an approach may offer new insights into the active sites on surfaces of metal nanoparticles (Huang et al. 2007, 2008). See Chapter 9 for more information.

13.6 In Situ Environmental STEM (ESTEM)

A major goal of microscopy studies on catalysts is to elucidate the structure–property relations. In general, the catalyst operates in a liquid or a gas environment at some temperature and the relevant structure is that assumed by the system under these so-called reactor conditions. Conventional microscopy characterization under high vacuum conditions may provide results which are either misleading or difficult to interpret unambiguously. To solve this problem, STEM should ideally be performed under conditions identical to those found in the reactor. Moreover, it would be highly desirable to measure the activity and selectivity of the catalyst simultaneously so that the complete structure–property relation can be determined. This approach gives rise to the so-called *operando* methods first pioneered for Raman spectroscopy (Bañares and Wachs 2002). At present, true *operando* microscopy characterization is possible only where the products can be directly imaged for applications such as catalytic growth of polymers and nanotubes (e.g. see Oleshko et al. 2002, Ross et al. 2005, Sharma and Iqbal 2004, Sharma et al. 2009). For example with nanotubes/wires, the kinetics of the growth process relates directly to the catalyst activity and the type of wire growing relates directly to the catalyst selectivity. However, for many gas (or liquid)-phase reactions, the reactant and the product gases are not directly visible with microscopy techniques.

With in situ environmental observations we are able to observe the catalyst under near-reactor conditions, i.e. at temperature in a reactive gas environment, but the catalyst activity/selectivity is not simultaneously measured for gaseous products. Environmental STEM (ESTEM) is a powerful approach for understanding the nanoscale structural and chemical modifications taking place, under near-reactor conditions, in catalytic nanoparticles. It can avoid many common disadvantages of the ex situ characterization techniques, such as the change of nanostructure and chemistry that may occur when the catalyst is removed from the reactor and/or exposed to air. It can also avoid changes in the catalyst that may take place due to exposure to vacuum.

13.6.1 Approaches to In Situ Electron Microscopy

The earliest motivation for controlling the gas environment inside an electron microscope dates back to early efforts in the 1930s to examine biological samples in their hydrated state (Marton 1935) and for the study and control of contamination (Stewart 1934, Watson 1947, 1948). This so-called environmental electron microscopy was more extensively developed during the 1970s and a comprehensive review on environmental TEM and other in situ techniques from that time can be found in the book by Butler and Hale (1981). While all of this early work was focused on conventional broad beam TEM, there is not really any fundamental difference in the design approaches for TEM and STEM.

A critical requirement for all gas systems compatible with STEM is that they maintain a high gas pressure around the sample and high vacuum around the field-emission electron source. This can be accomplished in three ways:

(a) Window method – gas or liquid is confined around the sample region by using thin electron transparent windows, e.g. thin amorphous carbon or SiN films.
(b) Differential pumping – a pressure difference is maintained by installing small apertures above and below the sample area and using additional pumping.
(c) Gas injection system – an injection needle is placed near the sample surface and gas is allowed to flow from the tip of the needle into the sample area.

The gas injection approach has been used extensively in applications where localized CVD depositions are required (Kohlmann et al. 1991, Matsui and Ichihashi 1988). This approach is advantageous because it introduces a relatively small volume of gas into the system and usually does not require extra pumping capacity. One disadvantage is that the pressure varies across the sample with the distance from the injection point, although it does not vary by much within a distance of several micrometres. For electron beam-induced deposition employing low precursor gas pressure, this approach seems to work very well for nanoscale patterning over a small area (Mitsuishi et al. 2003). However, for catalyst analysis, it is often necessary to image many areas because of statistical variations in the catalyst powder. The differential gas pressure across the sample may result in changes in microstructure because of non-uniform exposure to the gas environment complicating data interpretation. However, this should not be a problem for well-defined nanocatalysts.

In the window method, thin electron transparent membranes are employed to separate the high-pressure atmosphere around the sample from the vacuum in the rest of the microscope. This is usually accomplished via a custom-built holder which incorporates windows and a gas inlet system and atomic resolution has been demonstrated on catalysts (Giorgio et al. 2006, Parkinson 1989). The windowed design has the advantage of being able to handle high gas pressures (depending upon the strength and the thickness of the window) and can be used with any electron microscope. Recent application of MEMS technology has resulted in the development of a windowed cell that has shown atomic resolution of 0.18 nm at pressures above 1 atm and temperatures up to 500°C (Creemer et al. 2008). These exciting results open the way for performing controlled atmosphere experiments at pressures at or above 1 atm. They can also handle wet samples and are often called "wet cell" sample holders (de Jonge et al. 2009). The main disadvantage of the window method is that the additional scattering from the amorphous structure of the window films may interfere with imaging and spectroscopy. Moreover, windowed holders are usually limited to one axis of tilt and there is a risk of membrane rupture, so additional precautions may be necessary to protect the field emission source.

Differentially pumped cells have a long history of development for ETEM. For a short review of their history and development the reader is referred to the article by Sharma and Crozier (2005). Modern differential pumping systems are designed after the basic principles outlined by Swann and Tighe (1972). In this type of cell, a series of differential pumping stages are employed to create a large pressure difference between the sample area and the rest of the microscope. This is accomplished by inserting a series of differential pumping apertures into the microscope column and adding additional pumping capacity to remove the gas that leaks through the differential pumping apertures. The first atomic resolution with this approach was achieved by the Oxford group on a modified JEOL 4000 (Doole et al. 1991) and was continuously developed through the 1990 s often involving substantial modification of the objective lens pole pieces (Boyes and Gai 1997, Lee et al. 1991, Sharma and Weiss 1998, Yao et al. 1991).

13.6.2 Environmental STEM

Working in collaboration with scientists at Haldor Topsoe, FEI designed and commercialized a differential pumped system and demonstrated that this system was compatible with the Schottky field emission electron source of a Philips CM 300 FEG-TEM (Hansen and Wagner 2000). The modified column was shown to permit cell pressures of up to 10 mbar without adversely affecting the performance of the field emission source. These modifications were incorporated into the FEI Tecnai F20 TEM/STEM and the first commercial system was delivered to Arizona State University in the 2002 (Sharma et al. 2003). A photograph of the ASU system is shown in Figure 13–15 and a schematic diagram of part of the differential pumping system in Figure 13–16. The microscope column has been modified to add three sets of differential pumping apertures for three pumping stages between the gun valve and the viewing chamber. The first two sets of apertures are placed within the upper and lower objective lens pole pieces and most of the gas leaking through this first set of apertures is pumped out using a turbo-molecular pump. The second stage is between the condenser aperture and the selected area aperture and is pumped using a molecular drag pump. An additional ion pump is employed for the last stage before the gun valve. There is a control box to open and close various pneumatic valves and thus the microscope can be switched between ESTEM and STEM mode easily. The gas inlet is controlled by a set of shut-off valves and pressure is regulated using a fine needle valve. The first publication on catalysts performed with the field

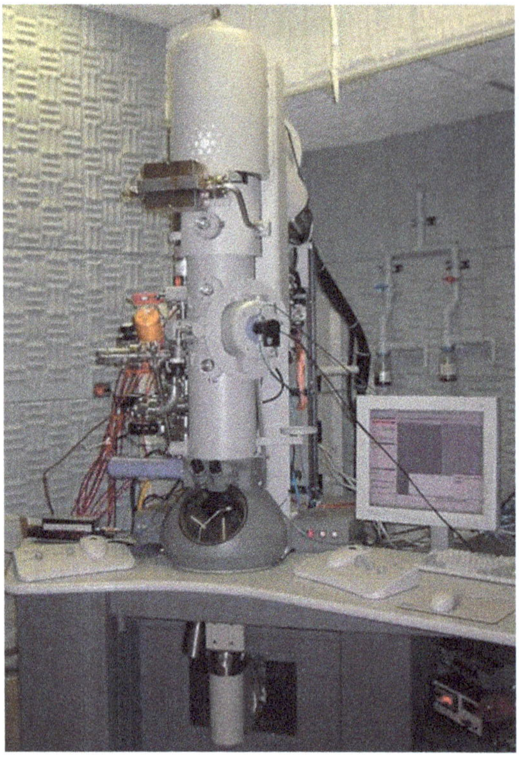

Figure 13–15. Photograph of FEI Tecnai F20 ESTEM at ASU.

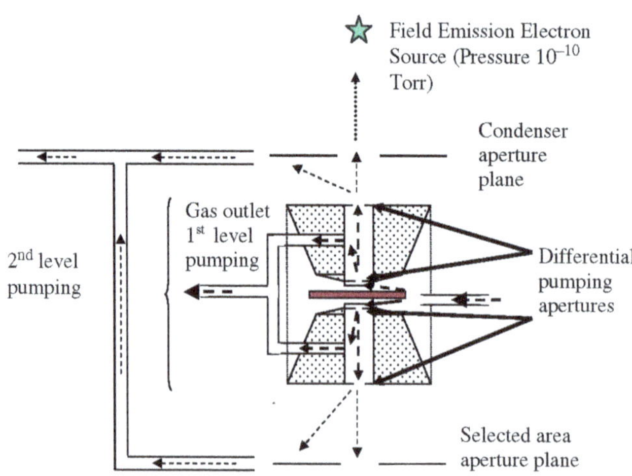

Figure 13–16. Schematic diagram of first and second level of pumping in FEI Tecnai F20 ESTEM.

emission ESTEM at high pressures was conducted on CoRu bimetallic particles and demonstrated nanometre resolution Z-contrast images and electron energy loss spectroscopy at 400°C in 1 Torr of H_2 and N_2 (Li et al. 2006b).

13.6.3 Issues for In Situ Environmental STEM

Several issues do arise when the environmental cell is combined with STEM capability. The differential pumping systems add to the cost of an instrument because of the modification to the column architecture, the additional pumping capability and modifications to the vacuum control logic. However, the resulting system is very robust and can achieve pressures of up to 20 mbar without damaging the FEG. The differential pumping ESTEM takes standard side entry holders, so heating, cooling and double tilting capability are not compromised. The Schottky FEG may be preferable to a cold FEG because the high operating temperature is more compatible with higher gun pressures that inevitably results when there is gas in the cell. The upper limit on the pressure is usually specified for gases that can be efficiently pumped like N_2 or O_2. Gases like H_2 are more difficult to pump and the upper pressure limit in the cell typically 2–4 times lower. The operator can close the gun valve and increase the pressure in the column to several hundred torr. This can be helpful for reactions that may need a high-pressure impulse for initiation. The pressure in the cell is then reduced and the gun valve re-opened to permit sample observation.

In principle, the maximum allowable pressure could be raised by increasing the number of pumping stages and/or decreasing the size of the differential pumping apertures. On our system, the differential pumping apertures restrict the convergence and collection semi-angles to about 50 mrad at 200 kV. This should not affect the large convergence angles needed to form small electron probes even on aberration-corrected STEMs. The upper differential pumping apertures could be reduced by about a further factor of 2 without significantly affecting the probe-forming capability. However, the lower differential pumping apertures cut all scattering above 50 mrad. Essentially, this makes it impossible to perform high-angle annular dark-field imaging and consequently there is a significant degradation of the contrast and signal-to-noise ratio in the Z-contrast image. For Z-contrast imaging we typically collect scattering over the range of 40–50 mrad which balances atomic number contrast with signal-to-noise ratio.

The gas present on the column represents an additional channel for electron scattering which could affect both the probe formation process and the collected signal. The effect on the entrance side of the sample has been discussed extensively in the environmental SEM literature and is often referred to as the "skirt" effect (Thiel and Toth 2005). These authors show that the resolution can degrade by 1–2 nm in 10 Torr of water vapour. However, the much higher accelerating voltages employed in STEM dramatically decrease this effect and we routinely observe subnanometre resolution in gas pressures of a few torr.

It is useful to calculate the mean free path λ for electron scattering in a column of gas. The mean free path can be written in terms of gas pressure P and temperature T as

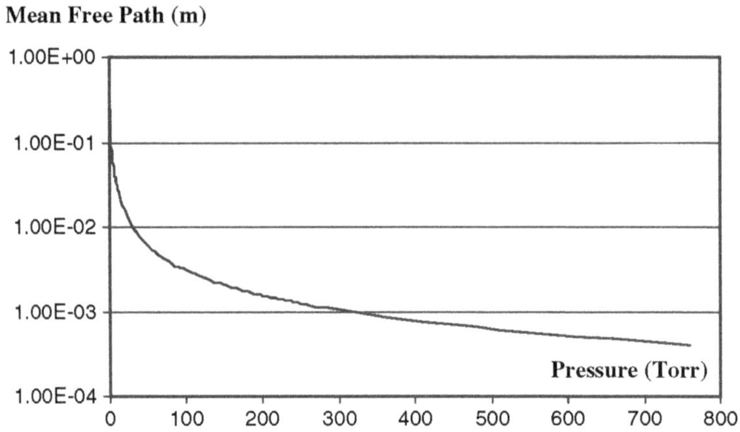

Figure 13–17. Mean free path for 200-kV electron in O_2 atmosphere as a function of pressure.

$$\lambda = \frac{RT}{PN_A\sigma}$$

where R is the gas constant, N_A is the Avogadro's number and σ is the electron scattering cross section. For light elements, the electron scattering is dominated by inelastic scattering and we take the total inelastic scattering cross sections at 200 kV to be 1×10^{-22} m^2 for O_2 dropping to about 0.5×10^{-22} m^2 for H_2 (Inokuti et al. 1981). The mean free path for scattering in an O_2 atmosphere at room temperature is plotted as a function of pressure in Figure 13–17. At 1 Torr the mean free path is about 30 cm dropping to about 400 μm at 760 Torr (1 atm). The mean free paths increase by a factor of 2 in an H_2 atmosphere.

For good STEM performance, the ESTEM must be designed to keep the probability of electron–gas scattering small while achieving high pressures around the sample. The probability of gas scattering can then be determined from knowledge of the length of the gas column L. The probability P_e of an electron being scattered at least once on passing through a column of gas of length L is then given by

$$P_e = 1 - \exp(-L/\lambda)$$

In our Tecnai F20, the high-pressure path length is confined to the pole piece gap of 0.54 cm. The pressure in the first stage of the differential pumping is usually a factor of 10^4 smaller than the cell pressure allowing us to neglect additional gas scattering in the rest of the column.

To evaluate the effect on the probe formation process we assume that the sample lies in the middle of the pole piece, so the important gas scattering length is about 0.27 cm. The probability P_e is plotted as a function of pressure for the Tecnai F20 in Figure 13–18. Notice that even at 10 Torr, the probability of scattering is less than 10%, so the gas should not significantly affect either the probe size or the probe current. By 100 Torr, the probability of scattering is on the order of 60%

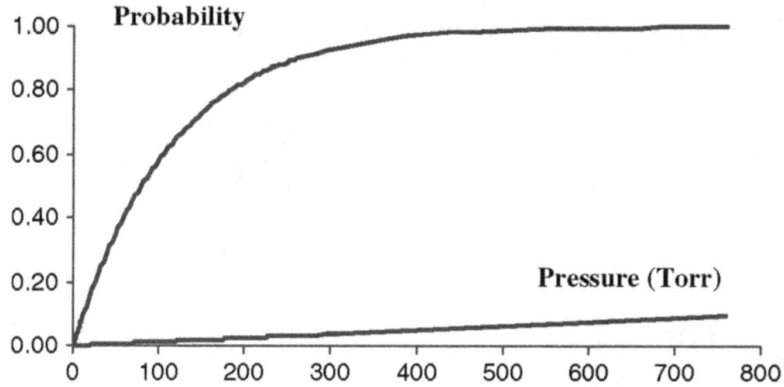

Figure 13–18. Probability of electron being scattered once by column of O_2 gas as a function of pressure. *Upper curve* – gas path length 0.27 cm; *lower curve* – gas path length 40 μm.

and we would expect to see a drop in STEM performance at these pressures or higher pressures. This suggests that adding additional stages of differential pumping may not be advantageous with this cell design because the electron path length through the high pressure region is too long. This demonstrates the advantage of a suitably designed window cell for high-pressure applications. For example, in the MEMS cell of Creemer et al, the minimum spacing between the two windows is only 40 μm. The probability of gas scattering in such a cell is also plotted in Figure 13–18 and shows that even at 1 atm, the probability of gas scattering is only 10%. In such a cell, the pressure could be increased by another order of magnitude before the scattering probability reaches 50%.

The windowed cell has clear advantages for high-pressure work but the reliability of the current generation of membranes has yet to be demonstrated. The high-pressure results published so far were obtained by inserting the holder into an ETEM equipped with a differential pumping system to protect the FEG in the event of membrane rupture. For STEM applications, the rupture problem may be more severe especially if the membrane is irradiated with a focused electron probe in the reactive gas environment. Recent work shows that combining a focused electron probe with a reactive gas atmosphere can provide a very efficient method for hole drilling and phase transformation in thin films (Crozier 2007). The effect depends not only on the membrane material and current density but also on the ambient atmosphere. This is demonstrated in Figure 13–19 showing results of irradiating a 30-nm-thick Si_3N_4 membrane with a 0.7-nm probe with a beam current of 0.2 nA. Initial experiments in the high vacuum of a FEI Tecnai F20 E-STEM (10^{-7} Torr) demonstrated that the Si_3N_4 films were relatively unreactive during exposures of up to 10 s with the focused electron beam. However, introduction of 0.5 Torr of either N_2 or H_2 resulted in immediate localized etching and hole drilling during electron irradiation. This shows that even relatively inert materials and gases can become highly reactive in the presence of a high-energy, focused electron beam. In this case, the gas molecules are ionized in the vicinity

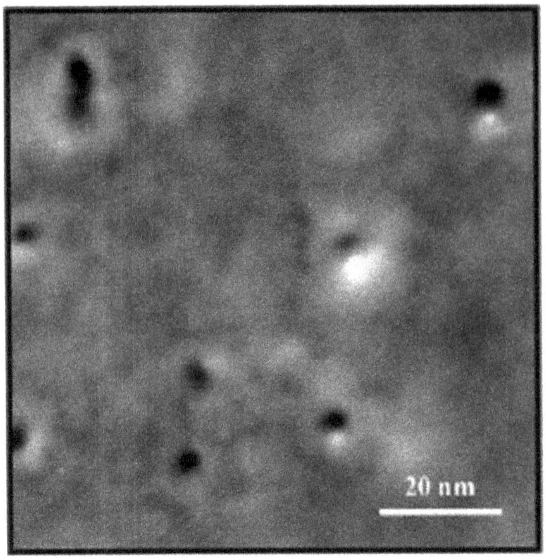

Figure 13–19. Z-contrast STEM image of randomly placed holes created in a Si_3N_4 substrate using focused electron beam etching in a H_2 atmosphere (holes are *dark*). H_2 pressure was kept close to 0.5 Torr and the electron beam current and diameter were 0.2 nA and 0.7 nm, respectively (from Crozier, 2007, reproduced with permission).

of the surface becoming highly reactive and capable of etching with relative ease. A completely different type of reaction takes place in an oxidizing gas. For example with H_2O, the Si_3N_4 gets locally converted to SiO_x during focused electron irradiation (Crozier 2007).

The risk of rupture may depend on the way in which the sample is loaded into the holder. If the catalyst particles are loaded directly onto the window material, the electron beam will necessarily be focused just above or below the membrane during STEM increasing the probability of electron–membrane–gas reactions. Several emerging hot-stage designs involve placing catalyst particles directly onto MEMS-fabricated membranes which are then directly heated (Allard et al. 2008, Creemer et al. 2008). Such a design can yield remarkable control and stability for high-temperature experiments. Figure 13–20 is a Z-contrast STEM image from a Pt particle on a γ-Al_2O_3 support recorded at 700°C under high-vacuum conditions (Allard et al. 2008, 2009). Individual Pt atoms can be easily resolved on the support in the vicinity of the nanoparticle. It is possible that at high temperature, rapid diffusion processes may anneal out any damage from the electron beam and reduce the chance of a membrane failure, although this diffusion may also change the surface of the catalyst. However, it is clear that the windowed cell offers great promise for STEM analysis of catalysts under high-pressure and high-temperature conditions and should be aggressively pursued.

For many in situ experiments, it is desirable to follow the evolution of the catalyst particles at atomic resolution in real time in order to get information on phase transformation processes. Many STEM imaging systems are rather slow and show significant distortions in the raster if scanned at TV rates. For improved time resolution during in situ experiments, future systems should be able to acquire and store image data at least at TV rates. Rapid electron diffraction and energy loss spectroscopy may also be exploited to study changes in the structure and

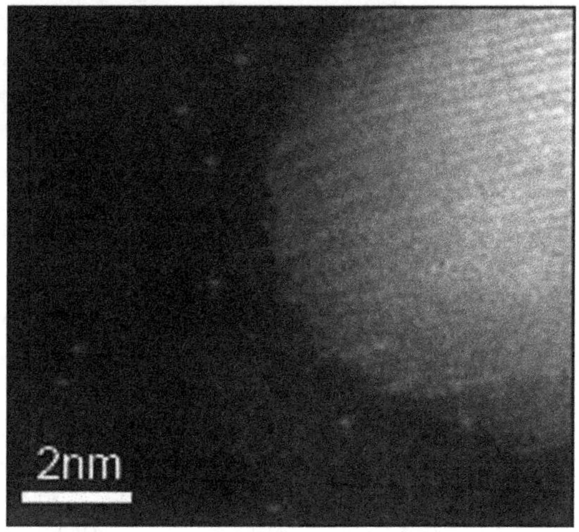

Figure 13–20. Image of Pt particle on Al_2O_3 support at nominal temperature of 700°C showing individual Pt atoms dispersed on the surface of support (from Allard et al. 2009, reproduced with permission).

chemistry of catalytic nanoparticles under reactive gas conditions. The desire to follow the transformation on a molecular timescale has stimulated interest in developing ultrafast DTEM techniques (Kim et al. 2008).

13.6.4 In Situ Studies on Catalyst

Baker and co-workers were the first to extensively use ETEM to study heterogeneous catalysts. His first work focused on the growth of carbon filaments using Ni-based catalysts (Baker et al. 1972). Catalytic gasification and filamentous growth of carbon remained common themes for many of Baker's publications in the 1980's and 1990's (Baker and Chludzinski 1980, 1986, Baker et al. 1973, 1985, 1987). He also worked on a wide range of metal catalysts studying the influence of gaseous environments on particle shape and metal–support interactions (Baker and Rodriguez 1994, Derouane et al. 1984, Dumesic et al. 1986, Simoens et al. 1984, Upton et al. 1993). Gai and Boyes have also developed ETEM and applied it extensively to a wide number of different heterogeneous catalysts (Gai 1983, 1999, Gai and Boyes 1992, 1997). Our group at ASU has been active studying various catalytic processes at atomic level under reaction conditions (Crozier and Datye 2000, Crozier et al. 1998, 2008, Li et al. 2005, 2006a, b, 2009, Liu et al. 2004, 2005, Oleshko et al. 2000, 2001, 2002, Sharma and Crozier 1999, 2005, Sharma et al. 2004, Wang et al. 2006, 2008) as has the group at Haldor Topsoe (Hansen et al. 2001, 2002; Simonsen et al. 2008, Vesborg et al. 2009). Very little work has been published on ESTEM imaging and analysis of catalysts (or any other material system) under reactive gas pressures. The author has been applying ESTEM to study supported bimetallic catalyst and ceria-based oxide supports. Examples from the work on bimetallic catalysts are presented to illustrate the power of the ESTEM approach for catalyst characterization.

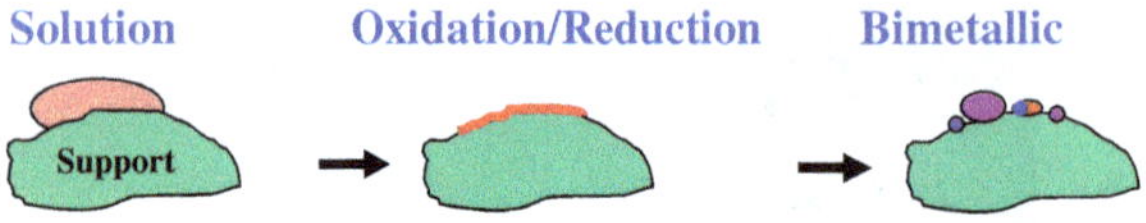

Figure 13–21. Schematic diagram illustrating solution impregnation synthesis of supported metal nanoparticles.

13.6.5 In Situ Preparation and Evolution of Supported Bimetallic Catalysts

The catalytic properties of bimetallic nanoparticles are directly related to their structure and composition which may be controlled by kinetics during catalyst preparation or thermodynamics during operation. For example, even though Cu has a lower surface free energy compared to Ni and should, based on thermodynamics, preferentially occupy surface sites (Sinfelt 1977), core-shell structures with Ni surface enrichment (Wu et al. 1996) or Cu surface enrichment (Wang and Baker 2004) have been prepared. This indicates that the kinetics associated with the catalyst preparation processes can play a major role in determining the final structure of the bimetallic particles. Classical impregnation techniques have been remarkably successful in preparing a wide variety of supported metal catalysts. Classical impregnation consists of two stages (see schematic diagram of Figure 13–21):

(i) Loading metal precursor salts onto a high surface area oxide support
(ii) Thermal decomposition of the metal salts in oxidizing and/or reducing atmospheres leading to the formation of metal nanoparticles.

The thermal treatment not only guarantees decomposition of the precursor to metal but also may result in the particles being anchored on the support and thus resistant to sintering during high-temperature operation. Although simple in conception, there are many outstanding questions about the adsorption, dissociation, diffusion, nucleation, particle growth and phase transformation processes taking place during catalyst preparation. In bimetallic systems, the interaction of the metal species with each other and the support further complicates the process. There is wide recognition of the importance of understanding and controlling the preparation protocols to improve the performance of the final catalysts.

We have used ESTEM to investigate the nanoscale dynamic processes taking place on supported metal catalysts with applications to energy production. In one set of experiments, we investigated the evolution of metallic particles in the $Co/\gamma\text{-}Al_2O_3$ and $CoRu/\gamma\text{-}Al_2O_3$ Fischer Tropsch catalyst (Belambe et al. 1997, Iglesia et al. 1993). We also investigated the $CuNi/TiO_2$ system which is important for water gas-shift reactions and partial oxidation of methane (Huang et al. 2006, Liu and Liu 1999). For both the systems, the metal precursor was formed by mixing two metal salts together in solution and co-impregnating into the oxide support. For the CuNi system, titanium dioxide powder (Degussa

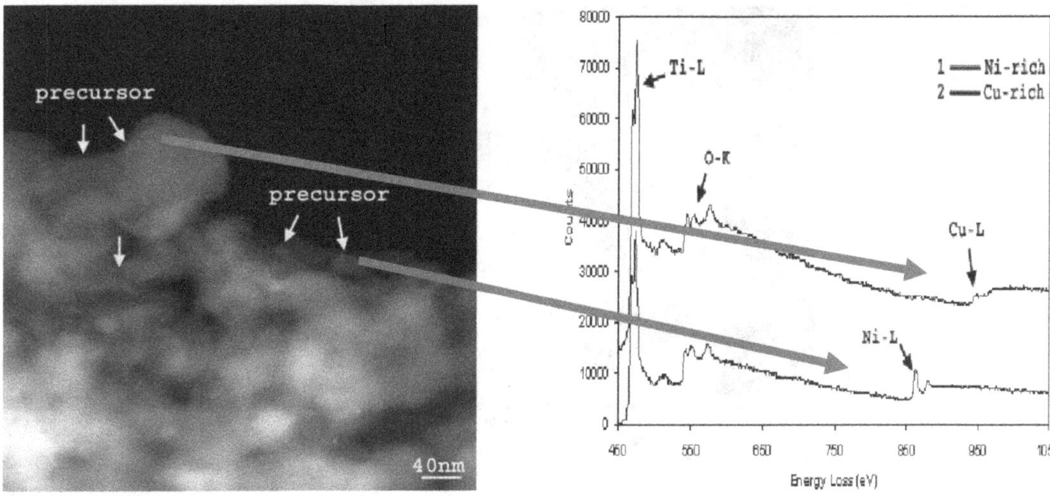

Figure 13–22. Z-contrast image and associated energy loss spectra from individual NiCu precursor patches on titania support showing compositional heterogeneity. Spectrum 1 is Ni rich and spectrum 2 is Cu rich.

P-25) was impregnated with the desired amount of Ni and Cu salt solutions. For the CoRu system, Co and Ru salts were impregnated into γ-alumina.

One interesting observation that we made in both precursor/support systems is that there is a strong tendency for the two salts to phase separate when they dry on the high surface area support. Figure 13–22 shows a Z-contrast image and energy loss spectra from two individual precursor patches for the NiCu system. The spectra show that each patch is either Ni or Cu rich indicating that the precursors mostly separated on the titania support, forming individual Ni-rich and Cu-rich nanopatches with sizes around 1–10 nm. We observed a similar result for the Co and Ru distribution for the alumina support (Li et al. 2006b). This shows that the efflorescing species is not a homogeneous mixture of the two metal salts and the drying process generates a molecularly separate but, on a nanometre scale, intermingled Ni and Cu precursor domains.

In situ Z-contrast images of Figure 13–23 show the evolution of the nanoparticle nucleation process during the reduction of the metal precursors in H_2 at 400°C for both Co/alumina (Figure 13–23a and b) monometallic and CoRu/alumina bimetallic catalysts (Figure 13–23c and d). The catalyst precursors contain many diffuse patches with slightly brighter contrast as indicated by the arrows in Figure 13–23a and c. During reduction, it is predominantly these regions that directly transform to large, Co-rich nanoparticles (shown in Figure 13–23b and d). The evolution and final morphology of the Co and CoRu catalyst appears to be similar. The alumina grains in contact with the precursor offer multiple nucleation sites for nanoparticle formation and the initial nucleation gives rise to a very fine dispersion of 1–3 nm particles. This very fine initial dispersion of metal particles undergoes a

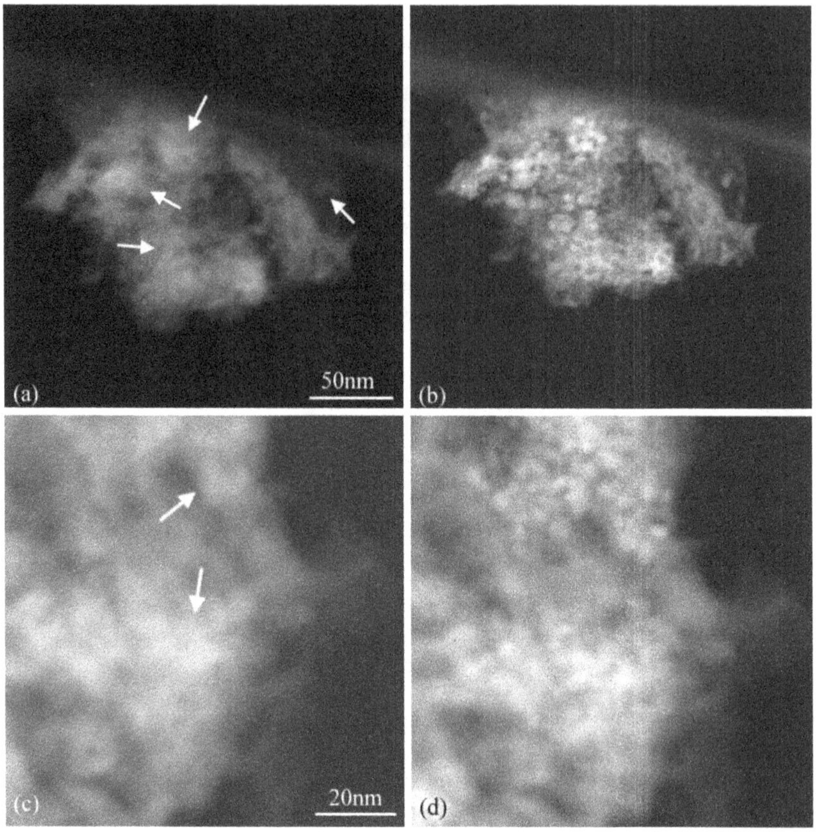

Figure 13–23. In situ Z-contrast images in Co/alumina system: precursor (a) before reduction and (b) after 120 min reduction; in situ Z-contrast images in CoRu/alumina system: precursor (c) before reduction and (d) after 17 min reduction. The reduction of the catalyst precursor was performed in situ under 1 Torr of 10% H_2 and 90% N_2 at 400°C (from Li et al. 2006b, reproduced with permission).

rapid coarsening as reduction proceeds. The non-uniformity of the precursor dispersion correlates directly with the non-uniform distribution of Co nanoparticles in the final catalyst. To improve the metal dispersion in high loading catalysts, e.g. the Co/Al_2O_3 system, it is necessary to develop protocols that can provide better precursor dispersion on the high surface area supports.

In situ EELS nanoanalysis can be used to follow the phase transformations taking place during the reduction process in the monometallic and bimetallic systems. For the pure Co system, EELS analysis shows that cobalt nitrate transforms to CoO_x at 400°C in hydrogen as shown in Figure 13–24. Note that in this case, the oxygen signal associated with Co is easily distinguished from the alumina support because the former gives rise to a peak (labelled 1st peak) at the onset of the oxygen K-edge. In situ imaging also showed that the nanoparticles contain lattice spacings of 0.25 nm corresponding to a stable intermediate phase of CoO. Other researchers have also found that CoO was the dominant cobalt oxide phase during temperature-programmed reduction on

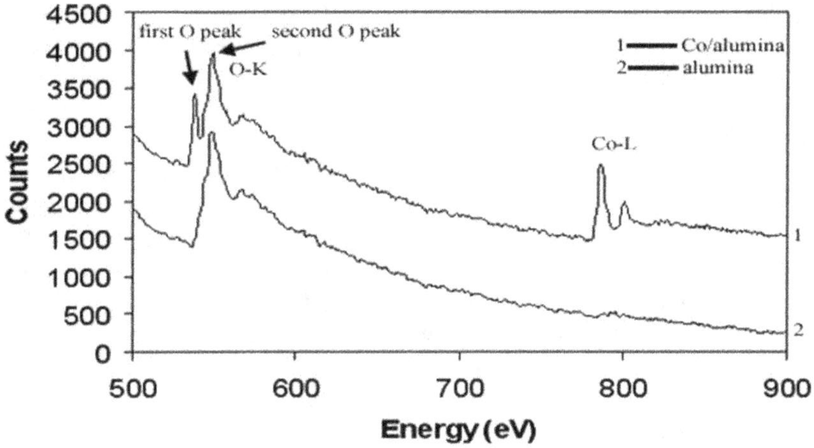

Figure 13–24. In situ EELS nanoprobe analysis from the nanoparticles of Co/Al_2O_3 system after 120 min reduction in the ESTEM at 400°C under 1 Torr of 10% H_2/90% N_2 (from Li et al. 2006b, reproduced with permission).

a Co/alumina catalyst (Zhang et al. 1999). No cobalt metal particles were found in our study even after 2 h of reducing the Co/alumina precursor materials in a hydrogen-containing gas. This is consistent with previous observations that also found that at 400°C, much longer reducing times (i.e., normally more than 9 h) and/or higher pressures are needed to completely reduce the cobalt precursor to cobalt metal in the Co/alumina system (Jacobs et al. 2002, van de Loosdrecht et al. 1997).

The reduction products are more diverse when Ru is added to the precursor. In situ energy loss spectra were acquired from 40 individual particles of differing sizes and a typical image and representative spectra are shown in Figure 13–25. Spectrum 4 was acquired from a particle of about 8 nm in diameter and shows the presence of only Co $L_{2,3}$-edge and oxygen K-edge and the sharp threshold O peak indicates that all

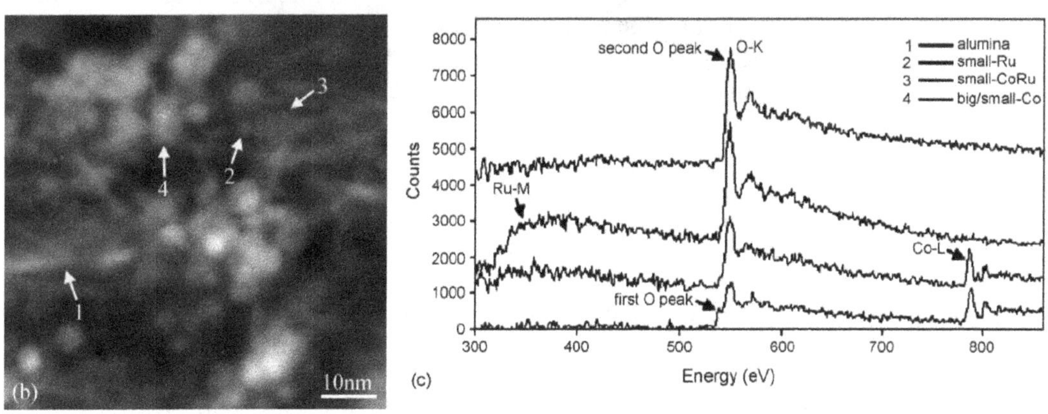

Figure 13–25. (a) In situ Z-contrast image and (b) in situ EELS nanoprobe analysis from the four labelled nanoparticles after 120 min reduction in the ESTEM at 400°C under 1 Torr of 10% H_2/90% N_2 (from Li et al. 2006b, reproduced with permission).

the large particles are oxides. Spectra 2 and 3 were acquired from small particles with sizes around 2 nm. The spectra acquired from small particles showed either only a Ru edge (i.e., indicated as spectrum 2) or both Ru and Co peaks (i.e., indicated as spectrum 3) but not oxygen pre-peak.

The ESTEM analysis of smaller CoRu particles shows that Co is mostly in the metallic form, indicating that the addition of Ru helps to enhance the reducibility of the particles. The presence of these additional small metallic particles in the CoRu/alumina catalyst increases the reducibility of the surface metal atoms, thus enhancing catalyst activity. Other studies have shown that the enhanced catalytic properties in Ru-promoted Co/Al_2O_3 catalysts may be attributed to the enhanced catalyst properties of small metal nanoparticles present in the system (Jacobs et al. 2002). The Ru-promoted Co/alumina catalyst was found to have a lower reduction temperature compared to the non-promoted catalyst (i.e. about 100°C) (Kogelbauer et al. 1996). It was suggested that the Ru promoter appears to enhance the reducibility of the small metal particles (Kogelbauer et al. 1996, Takeuchi et al. 1989). This is because noble metals can activate hydrogen and become a source for hydrogen spillover to cobalt oxides, therefore promoting its reduction at lower temperature.

The NiCu system shows a similar effect where Cu plays a promotional role in reducing NiO at low temperatures (Li et al. 2009). EELS analysis in the ESTEM shows that copper nitrate will reduce to Cu metal in about 1 Torr of hydrogen at 300°C, whereas Ni nitrate will reduce only to NiO under identical conditions. In this case, the white lines on both the Ni and Cu L_{23} edges were utilized to determine the oxidation state of the two species. This is consistent with the previously published results on reduction of these two precursors using more traditional conditions and techniques (e.g. Li et al. 1998, Naghash et al. 2005, 2006). This shows that the trends observed in the STEM environmental reactor chamber are consistent with higher pressure conditions for these particular systems.

Finally it is important to emphasize the role of the controlled atmosphere in understanding the intermediate and final products that form during the preparation of bimetallic catalysts. Ex situ methods in which the samples are exposed to air during transfer to the microscope would result in partial oxidation of the metal components making interpretation of oxidation state data by EELS problematic. Following the entire process in a controlled atmosphere is critical to avoid interpretation ambiguities.

13.7 Future Challenges, Developments and Opportunities

At present, the main role of microscopy is to map out the surface structures that form under reaction conditions and use this together with other experimental or theoretical methods to deduce the reaction rates for each surface structure or motif. Continued development of aberration correction has already made a major impact on mapping out metal atoms and clusters on supports. Some problems with radiation damage under focused probe (e.g. migrating atoms; for example see Batson 2008) are problematic for relating catalyst structure and properties. However, the new generation of aberration-corrected STEMs are now capable of operating at 60 kV or below and this has already shown dramatic reductions in radiation damage in many systems (Dellby et al. 2008). Since atom motion is most likely caused by knock-on damage processes, this problem should be reduced or mostly eliminated by working at low voltage. Low-voltage operation offers many advantages for characterization of nanomaterials and nanocatalysts. One motivation for going to higher accelerating voltage was driven in part by the difficulties of preparing suitable thin samples. However, sample thickness is less of a limiting factor in nanocatalysts because the small dimensions guarantee thin areas at least around the edges of aggregates. Indeed for smaller objects, low-voltage operation is advantageous, provided that ionization damage is not limiting, because of the much higher scattering cross sections for both imaging and spectroscopy signals.

For many catalysts, especially those that react with air or those that function under strongly reducing conditions, in situ observation will be critical for correct interpretation. STEM observations under reactive gas conditions and at operating temperature will be required to map out the relevant surface structures for many catalytic materials. Indeed a major contribution of microscopy is the identification of the nanostructures formed on high surface area nanocatalysts under reaction conditions. Considerable progress has been made in this field but more work is required to develop higher pressures to reproduce the conditions present in many industrial reactors.

The continued evolution of EELS and EDX coupled with aberration-corrected STEM has already opened up the path for elemental mapping of a catalyst at or close to the atomic level. This now makes it possible to map not just the primary species on the catalyst but also promoter species. Promoters can make a dramatic difference in the activity and selectivity of a catalyst but in many cases their functional mechanism is not well understood. Monochromated high-resolution EELS may open up a whole new area of bond mapping of the catalyst surface especially in the presence of adsorbates. Moreover, adding additional techniques to the conventional set of imaging and spectroscopy tools would further strengthen our ability to understanding the fundamental process taking place during catalysis. For example, secondary electron imaging of catalytically active surfaces under reactive gas conditions could provide information of the location of adsorbates with large residence times (site blockage). Variations in the coverage and character of absorbates may significantly modulate the local secondary electron emission, providing a further powerful tool to explore catalyst structure and absorbate characteristics.

The ultimate challenge in catalyst characterization is correlating reaction pathways with the local catalyst surface structure. There may be many possible pathways associated with different structures on the catalyst surface and we ideally want to increase the surface structure associated with the faster pathway. For microscopy it is very challenging to directly observe the chemical reaction taking place at atomic locations on the surface of the catalyst. The development of aberration-corrected STEM and associated techniques has essentially eliminated spatial resolution as a constraint. However, the scattering strength from surface adsorbates is rather weak and the residence times on the surface may be rather short especially for the kinetically important reaction pathways (these are the catalytic pathways). The combination of ultrafast techniques coupled with aberration correction and in situ capability would address this concern. Such an instrument is not available at the present time nor is it likely to become available in the near future. However, such a dream machine would open up a whole new area of characterization where adsorbate structures can be directly observed at local sites on the catalyst surface.

Acknowledgements I would like to thank colleagues and former students for contributions, collaborations and discussions over the years from many who have contributed to the work present in this chapter including Renu Sharma, Karl Weiss, Vladimir Oleshko, Peng Li, Ruigang Wang, Jingyue Liu and Nabin Nag. Financial support is acknowledged from Dow Chemical Company, Monsanto Company and the National Science Foundation (NSF-CTS-0306688 and NSF-CBET-0553445) and the Department of Energy for funding the ESTEM (DOE # AAD-0-30621-01). I also gratefully acknowledge access to the instrumentation in John M. Cowley Center for High Resolution Electron Microscopy in the LeRoy Eyring Center for Solid State Science at Arizona State University.

References

L.F. Allard, W.C. Bigelow, D.P. Nackashi, J. Damiano, S.E. Mick, A new paradigm for ultra-high-resolution imaging at elevated temperature. Microsc. Microanal. **14**(Suppl. 2), 792–793 (2008)

L.F. Allard, W.C. Bigelow, M. Jose-Yacaman, D.P. Nackashi, J. Damiano, S.E. Mick, A new MEMS-based system for ultra-high-resolution imaging at elevated temperatures. Microsc. Res. Tech. **72**, 208–215 (2009)

I. Arslan, J.C. Walmsley, E. Rytter, E. Bergene, P.A. Midgley, Toward three-dimensional nanoengineering of heterogeneous catalysts. J. Am. Chem. Soc. **130**, 5716–5719 (2008)

R.T.K. Baker, M.A. Barber, P.S. Harris, F.S. Feates, R.J. White, Nucleation and growth of carbon deposits from the nickel catalyzed decomposition of acetylene. J. Catal. **26**, 51–62 (1972)

R.T.K. Baker, J.J. Chludzinski, Filamentous carbon growth on nickel-iron surfaces: The effect of various oxide additives. J. Catal. **64**, 464–478 (1980)

R.T.K. Baker, J.J. Chludzinski, In-situ electron microscopy studies of the behavior of supported ruthenium particles. 1. The catalytic influence on graphite gasification reactions. J. Phys. Chem. **90**, 4730–4734 (1986)

R.T.K. Baker, J.J. Chludzinski, J.A. Dumesic, Filamentous carbon growth on nickel surfaces treated with titanium dioxide: Migration of titania and ramifications of strong metal–support interactions. J. Catal. **93**, 312–320 (1985)

R.T.K. Baker, J.J. Chludzinski, C.R.F. Lund, Further studies of the formation of filamentous carbon from the interaction of supported iron particles with acetylene. Carbon **25**, 295–303 (1987)

R.T.K. Baker, P.S. Harris, R.B. Thomas, R.J. Waite, Formation of filamentous carbon from iron, cobalt and chromium catalyzed decomposition of acetylene. J. Catal. **30**, 86–95 (1973)

R.T.K. Baker, N.M. Rodriguez, A review of the use of in situ electron microscopy techniques for the study of iron-based catalysts for coal conversion processes. Energy Fuels **8**, 330–340 (1994)

M.A. Bañares, I.E. Wachs, Molecular structures of supported metal oxide catalysts under different environments. J. Raman Spectrosc. **33**, 359 (2002)

P.E. Batson, Motion of Au atoms on carbon in the aberration corrected STEM. Microsc. Microanal. **14**, 89–97 (2008)

L. Bednarova, C.E. Lyman, E. Rytter, A. Holmen, Effect of support on the size and composition of highly dispersed Pt–Sn particles. J. Catal. **211**, 335–346 (2002)

A.R. Belambe, R. Oukaci, J.G.J. Goodwin, Effect of pretreatment on the activity of a Ru-promoted Co/Al2O3 Fischer Tropsch catalyst. J. Catal. **166**, 8–15 (1997)

D.A. Blom, L.F. Allard, S. Mishina, O'M.A. Keefe, Early results from an aberration-corrected JEOL 2200FS STEM/TEM at Oak Ridge National Laboratory. Microsc. Microanal. **12**, 483–491 (2006)

D.A. Blom, S.A. Bradley, W. Sinkler, L.F. Allard, Observation of Pt atoms, clusters and rafts on oxide supports, by sub-Angstrom Z-contrast imaging in an aberration-corrected STEM/TEM. Microsc. Microanal. **12**(Suppl. 2), 50–51 (2006)

A.Y. Borisevich, A.R. Lupini, S.J. Pennycook, Depth sectioning with the aberration-corrected scanning transmission electron microscope. Proc. Natl. Acad. Sci. **103**, 3044–3048 (2006)

A.Y. Borisevich, S. Wang, S.N. Rashkeev, M. Glazoff, S.J. Pennycook, S.T. Pantelides, Dual nanoparticle/substrate control of catalytic dehydrogenation. Adv. Mater. **19**, 2129–2133 (2007)

E.D. Boyes, P.L. Gai, Environmental high resolution electron microscopy and applications to chemical science. Ultramicroscopy **67**, 219–232 (1997)

P. Butler, K. Hale, Dynamic experiments in the electron microscope. Prac. Methods Electron Microsc., North Holland **9**, 239–308 (1981)

J.F. Creemer, S. Helveg, et al., Atomic-scale electron microscopy at ambient pressure. Ultramicroscopy **108**, 993–998 (2008)

A.V. Crewe, J. Wall, J. Langmore, Visibility of single atoms. Science **168**, 1338–1340 (1970)

P.A. Crozier, Nanoscale oxide patterning with electron–solid–gas reactions. Nanoletters **7**, 2395–2398 (2007)

P.A. Crozier, A.K. Datye, Direct observation of reduction of PdO to Pd metal by in situ electron microscopy. Stud. Surf. Sci. Catal. **130**, 3119–3124 (2000)

P.A. Crozier, R. Sharma, A.K. Datye, Oxidation and reduction of small palladium particles on silica. Microsc. Microanal. **4**, 278–285 (1998)

P.A. Crozier, S.-C. Tsen, J. Liu, C. Lopez Cartes, J.A. Perez-Omil, Factors affecting the accuracy of lattice spacing determination by HREM in nanometre-sized Pt particles. J. Electron Microsc. **48**(Suppl.), 1015–1024 (1999)

P.A. Crozier, R. Wang, R. Sharma, In situ environmental TEM studies of dynamic changes in cerium-based oxide nanoparticles during redox processes. Ultramicroscopy **108**, 1432–1440 (2008)

A.K. Datye, Electron microscopy of catalysts: recent achievements and future prospects. J. Catal. **216**, 144–154 (2003)

A.K. Datye, Particle size distributions in heterogeneous catalysts: What do they tell us about the sintering mechanism? Catal. Today **111**, 59–67 (2006)

N. de Jonge, D.B. Peckys, G.J. Kremers, D.W. Piston, Electron microscopy of whole cells in liquid with nanometer resolution. Proc. Natl. Acad. Sci. USA. **106**, 2159–2164 (2009)

N. Dellby, M. Murfitt, O.L. Krivanek, M. Kociak, K. March, M. Tence, C. Colliex, Atomic-resolution STEM at 60 kV primary voltage. Microsc. Microanal. **14**(Suppl 2), 136–137 (2008)

E.G. Derouane, J.J. Chludzinski, R.T.K. Baker, Direct observation of wetting and spreading of copper particles on magnesium oxide. J. Catal. **85**, 187–196 (1984)

A.M. Donald, A.J. Craven, A study of grain boundary segregation in Cu–Bi alloys using STEM. Philos. Mag. A **39**, 1–11 (1979)

R.C. Doole, G.M. Parkinson, J.M. Stead, Inst. Phys. Conf. Ser. **119**, 157–160 (1991)

J.A. Dumesic, S.A. Stevenson, R.D. Sherwood, R.T.K. Baker, Migration of nickel and titanium oxide species as studied by in situ scanning transmission electron microscopy. J. Catal. **99**, 79–87 (1986)

R.F. Egerton, *Electron Energy-Loss Spectroscopy in the Electron Microscope*, 2nd edn. (Plenum Press, New York, NY, 1996)

R.F. Egerton, Application of electron energy-loss spectroscopy to the study of solid catalysts. Top. Catal. **21**, 185–190 (2002)

R.F. Egerton, Electron energy-loss spectroscopy in the TEM. Rep. Prog. Phys. **72**, 016502 (2009)

D. Ferrer, D.A. Blom, L.F. Allard, S. Mejia, E. Perez-Tijerina, M. Jose-Yacaman, Atomic structure of three-layer Au/Pd nanoparticles revealed by aberration-corrected scanning transmission electron microscopy. J. Mater. Chem. **18**, 2442–2446 (2008)

P.L. Gai, Philos. Mag. **48**, 359–371 (1983)

P.L. Gai, Environmental high resolution electron microscopy of gas-catalyst reactions. Top. Catal. **8**, 97–113 (1999)

P.L. Gai, E.D. Boyes, Catal. Rev. Sci. Eng. **34**, 1–54 (1992)

P.L. Gai, E.D. Boyes, in *In Situ Microscopy in Materials Research*, ed. by P.L. Gai (Kluwer, Dordrecht, 1997), pp. 123–146

S. Giorgio, S. Sao Joao, S. Nitsche, D. Chaudanson, G. Sitja, H.C.R., Environmental electron microscopy (ETEM) for catalysts with a closed E-cell with carbon windows. Ultramicroscopy **106**, 503–507 (2006)

L.C. Gontard, L.Y. Chang, C.J.D. Hetherington, A.I. Kirkland, D. Ozkaya, R.E. Dunin-Borkowski, Aberration-corrected imaging of active sites on industrial catalyst nanoparticles. Ange. Chem. Int. Ed. **46**, 3683–3685 (2007)

L. Hansen, J.B. Wagner, in *Proceedings of the 12th European Congress on Electron Microscopy*, (Brno, 2000) 537–538

P.L. Hansen, J.B. Wagner, S. Helveg, J.R. Rostrup-Nielsen, B.S. Clausen, H. Topsoe, Atom resolved imaging of dynamic shape changes in supported copper nanocrystals. Science **295**, 2053 (2002)

T.W. Hansen, J.B. Wagner, P.L. Hansen, S. Dahl, H. Topsoe, C.J.H. Jacobson, Science **294**, 1508–1510 (2001)

A.A. Herzing, C.J. Kiely, A.F. Carley, P. Landon, G.J. Hutchings, Identification of active gold nanoclusters on iron oxide supports for CO oxidation. Science **321**, 1331–1335 (2008)

A.A. Herzing, M. Watanabe, J.K. Edwards, M. Conte, Z.R. Tang, G.J. Hutchings, C.J. Kiely, Energy dispersive X-ray spectroscopy of bimetallic nanoparticles in an aberration corrected scanning transmission electron microscope. Faraday Discussions, 138, 337–351 (2008)

S. Hillyard, J. Silcox, Detector geometry, thermal diffuse scattering and strain effects in ADF STEM imaging. Ultramicroscopy **58**, 6–17 (1995)

T. Huang, T. Yu, S. Jhao, Weighting variation of water–gas shift in steam reforming of methane over supported Ni and Ni–Cu catalysts. Ind. Eng. Chem. Res. **45**, 150–156 (2006)

W.J. Huang, B. Jiang, R.S. Sun, J.M. Zuo, Towards sub-Å atomic resolution electron diffraction imaging of metallic nanoclusters: A simulation study of experimental parameters and reconstruction algorithms. Ultramicroscopy **107**, 1159–1170 (2007)

W.J. Huang, R. Sun, J. Tao, L.D. Menard, R.G. Nuzzo, J.M. Zuo, Coordination-dependent surface atomic contraction in nanocrystals revealed by coherent diffraction. Nat. Mater. **7**, 308–313 (2008)

E. Iglesia, S.L. Soled, R.A. Fiato, G.H. Via, Bimetallic synergy in cobalt–ruthenium Fischer–Tropsch synthesis catalysts. J. Catal. **143**, 345–368 (1993)

M. Inokuti, J.L. Dehmer, T. Baer, J.D. Hanson, Oscillator strength moments, stopping power and total inelastic scattering cross sections of all atoms through strontium. Phys. Rev. A **23**, 95–109 (1981)

M.S. Isaacson, J. Langmore, N.W. Parker, D. Kopf, M. Utlaut, The study of the adsorption and diffusion of heavy atoms on light element substrates by means of the atomic resolution STEM. Ultramicroscopy **1**, 359–376 (1976)

G. Jacobs, P.M. Patterson, Y. Zhang, T. Das, J. Li, B.H. Davis, Fischer–Tropsch synthesis: deactivation of noble metal-promoted Co/Al_2O_3 catalysts. Appl. Catal. A **233**, 215–226 (2002)

J.S. Kim, T. LaGrange, B.W. Reed, M.L. Taheri, M.R. Armstrong, W.E. King, N.D. Browning, G.H. Campbell, Imaging of transient structures using nanosecond in situ TEM. Science **321**, 1472–1475 (2008)

K. Kimoto, T. Asaka, T. Nagai, M. Saito, Y. Matsui, K. Ishizuka, Element-selective imaging of atomic columns in a crystal using STEM and EELS. Nature **450**, 702–704 (2007)

E.J. Kirkland, *Advanced Computing in Electron Microscopy* (Plenum Press, New York, NY, 1998)

R.F. Klie, M.M. Disko, N.D. Browning, Atomic scale observations of the chemistry at the metal–oxide interface in heterogeneous catalysts. J. Catal. **205**, 1–6 (2002)

A. Kogelbauer, J.G. Goodwin Jr., R. Oukaci, Ruthenium promotion of Co/Al_2O_3 Fischer–Tropsch catalysts. J. Catal. **160**, 125–133 (1996)

K.T. Kohlmann, M. Thiemann, W.H. Brünger, E-beam induced X-ray mask repair with optimized gas nozzle geometry. Microelectron. Eng. **13**, 279–282 (1991)

T.C. Lee, D.K. Dewald, J.A. Eades, I.M. Robertson, H.K. Birnbaum, An environmental cell transmission electron microscope. Rev. Sci. Instrum. **62**, 1438–1444 (1991)

Y. Li, J. Chen, L. Chang, Y. Qin, The doping effect of copper on the catalytic growth of carbon fibers from methane over a Ni/Al_2O_3 catalysts prepared from Feitknecht compound precursor. J. Catal. **178**, 76–83 (1998)

P. Li, J. Liu, N. Nag, P.A. Crozier, Atomic-scale study of in-situ metal nanoparticle synthesis in a Ni/TiO_2 system. J. Chem. Phys. B **109**, 13883–13890 (2005)

P. Li, J. Liu, N. Nag, P.A. Crozier, Dynamic nucleation and growth of Ni nanoparticles on high surface area titania. Surface Sci. **600**, 693–702 (2006a)

P. Li, J. Liu, N. Nag, P.A. Crozier, In situ synthesis and characterization of Ru promoted Co/Al_2O_3 Fischer–Tropsch catalysts. Appl. Catal. A **307**(2), 212–221 (2006b)

P. Li, J. Liu, N. Nag, P.A. Crozier, In situ preparation of Ni-Cu/TiO_2 bimetallic catalysts. J. Catal. **262**, 73–82 (2009)

Z.Y. Li, N.P. Young, M. DiVece, S. Palomba, R.E. Palmer, A.L. Bleloch, B.C. Curley, R.L. Johnson, J. Jiang, J. Juan, Three-dimensional atomic-scale structure of size-selected gold nanoclusters. Nature **451**, 46–48 (2008)

J.Y. Liu, *Advanced Electron Microscopy Characterization of Nanostructured Heterogeneous Catalysts*, Microscopy and Microanalysis, **10**, 55–76 (2004)

J.Y. Liu, Scanning transmission electron microscopy and its application to the study of nanoparticles and nanoparticle systems. J. Electron Microsc. **54**, 251–278 (2005)

J. Liu, J.M. Cowley, High-angle ADF and high-resolution SE imaging of supported catalysts clusters. Ultramicroscopy **34**, 119–128 (1990)

R.-J. Liu, P.A. Crozier, C.M. Smith, D.A. Hucul, J. Blackson, G. Salaita, In-situ electron microscopy studies of the sintering of palladium nanoparticles on alumina during catalyst regeneration processes. Microsc. Microanal. **10**, 77–85 (2004)

R.-J. Liu, P.A. Crozier, C.M. Smith, D.A. Hucul, J. Blackson, G. Salaita, A new contact sintering mechanism associated with regeneration of Pd/Al_2O_3 hydrogenation catalyst. Appl. Catal. A **282**, 111–121 (2005)

Y. Liu, D.Z. Liu, Study of bimetallic Cu–Ni/g-Al_2O_3 catalysts for carbon dioxide hydrogenation. Int. J. Hydrogen Energy **24**, 351–354 (1999)

A.R. Lupini, S.J. Pennycook, Localization in elastic and inelastic scattering. Ultramicroscopy **96**, 313–322 (2003)

C.E. Lyman, J.I. Goldstein, D.B. Williams, D.W. Ackland, S. von Harrach, N. A.W., P.J. Statham, High-performance X-ray detection in a new analytical electron microscope. J. Microsc. **176**, 85–98 (1994)

C.E. Lyman, R.E. Lakis, H.G. Stenger, X-ray emission spectrometry of phase separation in Pt–Rh nanoparticles for nitric oxide reduction. Ultramicroscopy **58**, 25–34 (1995)

C.E. Lyman, R.E. Lakis, H.G. Stenger Jr., B. Tøtdal, R. Prestvik, Analysis of alloy nanoparticles. Microchim. Acta **132**, 301–308 (2000)

L. Marton, Bull. Acad. R. Belg. Cl. Sci **21**, 553 (1935)

S. Matsui, T. Ichihashi, In situ observation on electron-beam-induced chemical vapor deposition by transmission electron microscopy. Appl. Phys. Lett. **53**, 842–844 (1988)

L.D. Menard, F.T. Xu, R.G. Nuzzo, J.C. Yang, Preparation of TiO_2-supported Au nanoparticle catalysts from a Au-13 cluster precursor: Ligand removal using ozone exposure versus a rapid thermal treatment. J. Catal. **243**, 64–73 (2006)

P.A. Midgley, M. Weyland, 3D electron microscopy in the physical sciences: The development of Z-contrast and EFTEM tomography. Ultramicroscopy **96**, 413–431 (2003)

K. Mitsuishi, M. Shimojo, M. Han, K. Furuya, Electron-beam-induced deposition using a subnanometer sized probe of high-energy electrons. Appl. Phys. Lett. **83**, 2064–2066 (2003)

M. Mogensen, in *Ceria-Based Electrodes. Catalysis by Ceria and Related Materials*, ed. by A. Trovarelli (Imperial College Press, London, 2002), pp. 453–481

M. Mogensen, D. Lybye, N. Bonanos, P.V. Hendriksen, F.W. Poulsen, Factors controlling the oxide ion conductivity of fluorite and perovskite structured oxides. Solid State Ionics **174**, 279–286 (2004)

M. Mogensen, N.M. Sammes, G.A. Tompsett, Physical, chemical and electrochemical properties of pure and doped ceria. Solid State Ionics **129**, 63–94 (2000)

M.S. Moreno, M. Weyland, P.A. Midgley, J.F. Bengoa, M.V. Cagnoli, N.G. Gallegos, A.M. Alvarez, S.G. Marchetti, Highly anisotropic distribution of iron nanoparticles within MCM-41 mesoporous silica. Micron **37**, 52–56 (2006)

D.A. Muller, L.F. Kourkoutis, M. Murfitt, J.H. Song, H.Y. Hwang, J. Silcox, N. Dellby, O.L. Krivanek, Atomic-scale chemical imaging of composition and bonding by aberration-corrected microscopy. Science **319**, 1073–1076 (2008)

R. Naghash, T.H. Etsell, S. Xu, XRD and XPS study of Cu–Ni interactions on reduced copper–nickel–aluminum oxide solid solution catalyst. Chem. Mater. **18**, 2480–2488 (2006)

R. Naghash, S. Xu, T.H. Etsell, Coprecipitation of nickel–copper–aluminum takovite as catalyst precursors for simultaneous production of carbon nanofibers and hydrogen. Chem. Mater. **17**, 815–821 (2005)

P.D. Nellist, S.J. Pennycook, Direct imaging of the atomic configuration of ultradispersed catalysts. Science **274**, 413–415 (1996)

P.D. Nellist, S.J. Pennycook, Incoherent imaging using dynamically scattered coherent electrons. Ultramicroscopy **78**, 111–124 (1999)

N.L. Okamoto, B.W. Reed, S. Mehraeen, A. Kulkarni, D.G. Morgan, B.C. Gates, N.D. Browning, Determination of nanocluster sizes from dark-field scanning transmission electron microscopy images. J. Phys. Chem. C **112**, 1759–1763 (2008)

V. Oleshko, P. Crozier, R. Cantrell, A. Westwood, In situ and ex situ study of propylene polymerization with a $MgCl_2$-supported Ziegler-Natta catalyst. Stud. Surf. Sci. Catal. **130**, 935–940 (2000)

V. Oleshko, P.A. Crozier, R. Cantrell, A. Westwood, In-situ real-time environmental TEM of gas phase Ziegler-Natta catalytic polymerization of propylene. J. Electron Microsc. **51**(Suppl.), S27–S39 (2002)

V.P. Oleshko, P.A. Crozier, R.D. Cantrell, A.D. Westwood, In-situ and ex-situ study of gas phase propylene polymerization over a high activity $TiCl_4$–$MgCl_2$ heterogeneous Zeigler-Natta catalyst. Macromol. Rapid Commun. **22**, 34–40 (2001)

M. Pan, J.M. Cowley, J.C. Barry, Coherent electron microdiffraction from small metal particles. Ultramicroscopy **30**, 385–394 (1989)

M. Pan, J.M. Cowley, I.Y. Chan, Study of highly dispersed Pt in Y-zeolites by STEM and electron microdiffraction. Ultramicroscopy **34**, 93–101 (1990)

G.M. Parkinson, Catal. Lett. **2**, 303 (1989)

S.J. Pennycook, Study of supported ruthenium catalysts by STEM. J. Microsc. **124**, 15–22 (1981)

S.J. Pennycook, D.E. Jesson, High-resolution incoherent imaging of crystals. Phys. Rev. Lett. **64**, 938–941 (1990)

R. Prestvik, B. Tøtdal, C.E. Lyman, A. Holmen, Bimetallic particle formation in Pt–Re/Al_2O_3 reforming catalysts revealed by energy-dispersive X-Ray spectrometry in the analytical electron microscope. J. Catal. **176**, 246–252 (1998)

W.D. Pyrz, D.A. Blom, N.R. Shiju, V.V. Guliants, T. Vogt, D.J. Buttrey, Using aberration-corrected STEM imaging to explore chemical and structural variations in the M1 phase of the MoVNbTeO oxidation catalyst. J. Phys. Chem. C **112**, 10043–10049 (2008)

L. Reimer, *Scanning Electron Microscopy* (Springer-Verlag, Berlin, 1985)

S.B. Rice, S.A. Bradley, Imaging supported metal catalysts: the case for annular dark-field microscopy. Catal. Today **21**, 71–82 (1994)

F.M. Ross, J. Tersoff, M.C. Reuter, Sawtooth faceting in silicon nanowires. Phys. Rev. Lett. **95**, 4 (2005)

R. Sharma, P.A. Crozier, In situ electron microscopy of CeO_2 and CeO_2–ZrO_2 reduction. Inst. Phys. Conf. Ser. **61**, 569–572 (1999)

R. Sharma, P.A. Crozier, in *Environmental Transmission Electron Microscopy in Nanotechnology Handbook of Microscopy for Nanotechnology*, eds. by N. Yao, Z.L. Wang (Kluwer, New York, NY, 2005), pp. 531–563

R. Sharma, P.A. Crozier, Z.C. Kang, L. Eyring, Observation of dynamic nanostructural and nanochemical changes in ceria-based catalysts during in situ reduction. Philos. Mag. **84**, 2731–2747 (2004)

R. Sharma, P.A. Crozier, R. Marx, K. Weiss, An environmental transmission electron microscope for in-situ observation of chemical processes at the nanometer level. Microsc. Microanal. **9**(Suppl. 02), CD 912–913 (2003)

R. Sharma, Z. Iqbal, In situ observations of carbon nanotube formation using environmental transmission electron microscopy. Appl. Phys. Lett. **84**, 990–992 (2004)

R. Sharma, E. Moore, P. Rez, M.M.J. Treacy, Site-specific fabrication of Fe particles for carbon nanotube growth. Nano Lett. **9**, 689–694 (2009)

R. Sharma, K. Weiss, Microsc. Res. Tech. **42**, 270–280 (1998)

M. Shelef, R.W. McCabe, Twenty-five years after introduction of automotive catalysts: what next? Catal. Today **62**, 35–50 (2000)

A.J. Simoens, R.T.K. Baker, D.J. Dwyer, C.R.F. Lund, R.J. Madon, A study of the nickel–titanium oxide interaction. J. Catal. **86**, 359–372 (1984)

S.B. Simonsen, S. Dahl, E. Johnson, S. Helveg, Ceria-catalyzed soot oxidation studied by environmental transmission electron microscopy. J. Catal. **255**, 1–5 (2008)

J.H. Sinfelt, Catalysis by alloys and bimetallic clusters. Accounts Chem. Res. **10**, 15–20 (1977)

A. Singhal, J.C. Yang, J.M. Gibson, STEM-based mass spectroscopy of supported Re clusters. Ultramicroscopy **67**, 191 (1997)

K. Sohlberg, S. Rashkeev, A.Y. Borisevich, S.J. Pennycook, S.T. Pantelides, Origin of anomalous Pt–Pt distances in the Pt/alumina catalytic system. Chemphyschem **5**(12), 1893–1897 (2004)

R.L. Stewart, Insulating films formed under electron and ion bombardment. Phys. Rev. **45**, 488–490 (1934)

K. Sun, J. Liu, N. Nag, N.D. Browning, Studying the metal–support interaction in Pd/gamma-Al_2O_3 catalysts by atomic-resolution electron energy-loss spectroscopy. Catal. Lett. **84**, 193–199 (2002a)

K. Sun, J. Liu, N.K. Nag, N.D. Browning, Atomic scale characterization of supported Pd–Cu/gamma-Al_2O_3 bimetallic catalysts. J. Phys. Chem. B **106**, 12239–12246 (2002b)

K. Sun, J.Y. Liu, N.D. Browning, Direct atomic scale analysis of the distribution of Cu valence states in Cu/-gamma-Al_2O_3 catalysts. Appl. Catal. B Environ. **38**, 271–281 (2002c)

P.R. Swann, N.J. Tighe, Proc. 5th Eur. Reg. Cong. Electron Microsc. 436 (1972)

K. Takeuchi, T. Matsuzaki, H. Arakawa, T. Hanaoka, Y. Sugi, Synthesis of C2-oxygenates from syngas over cobalt catalysts promoted by ruthenium and alkaline earths. Appl. Catal. A **48**, 149–157 (1989)

B.L. Thiel, M. Toth, Secondary electron contrast in low-vacuum/environmental scanning electron microscopy of dielectrics. J. Appl. Phys. **97**, 051101 (2005)

J.M. Thomas, P.L. Gai, *Electron Microscopy and the Materials Chemistry of Solid Catalysts. Advances in Catalysis*, vol. 48. (Elsevier, San Diego, 2004), pp. 171–227

M.M.J. Treacy, Imaging with Rutherford scattered electrons in the scanning transmission electron microscope. Scann. Electron Microsc. **1**, 185–197 (1981)

M.M.J. Treacy, A. Howie, S.J. Pennycook, Z contrast of supported catalysts particles on the STEM. Inst. Phys. Conf. Ser. **52**, 261–264 (1980)

M.M.J. Treacy, A. Howie, C.J. Wilson, Z-contrast of platinum and palladium catalysts. Philos. Mag. A **38**, 569–585 (1978)
M.M.J. Treacy, S.B. Rice, Catalyst particle sizes from Rutherford scattered intensities. J. Microsc. **156**, 211–234 (1989)
A. Trovarelli, *Catalysis by Ceria and Related Materials* (Imperial College Press, London, 2002)
B.H. Upton, C.C. Chen, N.M. Rodriguez, R.T.K. Baker, In situ electron microscopy studies of the interaction of iron, cobalt, and nickel with molybdenum disulfide single crystals in hydrogen. J. Catal. **141**, 171–190 (1993)
J. van de Loosdrecht, M. van der Haar, A.M. van der Kraan, A.J. van Dillem, J.W. Geus, Preparation and properties of supported cobalt catalysts for Fischer–Tropsch synthesis. Appl. Catal. A **150**, 365–376 (1997)
R.A. van Santen, J.W. Niemantsverdriet, *Chemical Kinetics and Catalysis* (Plenum Press, London, 1995)
R.A. van Santen, P.W.N.M. van Leeuwen, J.A. Moulijn, *Catalysis: An Integrated Approach* (Elsevier, Amsterdam, 1999)
M. Varela, S.D. Findlay, A.R. Lupini, H.M. Christen, A.Y. Borisevich, N. Dellby, O.L. Krivanek, P.D. Nellist, M.P. Oxley, L.J. Allen, S.J. Pennycook, Spectroscopic imaging of single atoms within a bulk solid. Phys. Rev. Lett. **92**, 095502 (2004)
P.C.K. Vesborg, I. Chorkendorff, I. Knudsen, O. Balmes, J. Nerlov, A.M. Molenbroek, B.S. Clausen, S. Helveg, Transient behavior of Cu/ZnO-based methanol synthesis catalysts. J. Catal. **262**, 65–72 (2009)
J. Wall, J. Langmore, M.S. Isaacson, A.V. Crewe, Scanning transmission electron microscopy at high resolution. Proc. Natl. Acad. Sci. **71**, 1–5 (1974)
H. Wang, R.T.K. Baker, Decomposition of methane over a Ni–Cu–MgO catalyst to produce hydrogen and carbon nanofibers. J. Phys. Chem. B **108**, 20273–20277 (2004)
R. Wang, P.A. Crozier, R. Sharma, J.B. Adams, Nanoscale heterogeneity in ceria zirconia with low-temperature redox properties. J. Phys. Chem. B **110**, 18278–18285 (2006)
R. Wang, P.A. Crozier, R. Sharma, J.B. Adams, Measuring the redox activity of individual catalytic nanoparticles in cerium-based oxides. Nanoletters **8**, 962–967 (2008)
S.W. Wang, A.Y. Borisevich, S.N. Rashkeev, M.V. Glazoff, K. Sohlberg, S.J. Pennycook, S.T. Pantelides, Dopants adsorbed as single atoms prevent degradation of catalysts. Nat. Mater. **3**, 274–274 (2004)
M. Watanabe, D.W. Ackland, A. Burrows, C.J. Kiely, D.B. Williams, O.L. Krivanek, N. Dellby, M.F. Murfitt, Z. Szilagyi, Improvements in the X-Ray Analytical Capabilities of a Scanning Transmission Electron Microscope by Spherical-Aberration Correction. Microscopy and Microanalysis, **12**, 515–526 (2006).
J.H.L. Watson, An effect of electron bombardment upon carbon black. J. Appl. Phys. **18**, 153–161 (1947)
J.H.L. Watson, Specimen contamination in electron microscopes. J. Appl. Phys. **19**, 110–111 (1948)
K. Wikander, A.B. Hungria, P.A. Midgley, A.E.C. Palmqvista, K. Holmberg, J.M. Thomas, Incorporation of platinum nanoparticles in ordered mesoporous carbon. J. Colloid Interf. Sci. **305**, 204–208 (2007)
S. Wu, C. Zhu, W. Huang, Properties of polymer supported Ni–Cu bimetallic catalysts prepared by solvated metal atom impregnation. Chin. J. Polym. Sci. **14**, 217–224 (1996)

M.J. Yacaman, J.A. Ascencio, S. Tehuacanero, M. Marin, Modem applications of electron microscopy to catalysis. Top. Catal. **18**, 167–173 (2002)

J.C. Yang, S. Bradley, J.M. Gibson, The oblate morphology of supported PtRu5 on carbon black. Mater. Charact. **51**, 101–107 (2003)

N. Yao, G.E. Spinnler, R.A. Kemp, D.C. Guthrie, R.D. Cates, C.M. Bolinger, *Proceedings of the 49th Annual; Meeting of Microscopy Society of America*, San Francisco Press, 1028–1029 (1991)

N.J. Zaluzec, Detector solid angle formulas for use in X-ray energy dispersive spectrometry. Microsc. Microanal. **15**, 93–98 (2009)

Y. Zhang, D. Wei, S. Hammache, J.G. Goodwin Jr., Effect of water vapor on the reduction of Ru-promoted Co/Al_2O_3. J. Catal. **188**, 281–290 (1999)

J.M. Zuo, M. Gao, J. Tao, B.Q. Li, R. Twesten, I. Petrov, Coherent nano-area electron diffraction. Microsc. Res. Tech. **64**, 347–355 (2004)

14

Structure of Quasicrystals

Eiji Abe

14.1 Introduction

Quasicrystals are aperiodic solids that exhibit rotational symmetries incompatible with conventional periodic lattice order, e.g., icosahedral symmetry in three dimensions and tenfold symmetry in the plane. A revolutionary announcement of the first quasicrystal was made for a thermodynamically metastable phase in a rapidly solidified Al–Mn alloy (Shechtman et al. 1984). At present, not only metastable phases but also stable quasicrystalline phases (Dubost et al. 1986, Ohashi and Spaepen 1987, Tsai et al. 1987) are available in a variety of metallic alloys (Tsai 1999), and certain stable phases can be grown into a single grain several millimeters (Figure 14–1a (Fisher et al. 1998)) or even centimeters in size. It would appear obvious, therefore, that quasicrystals can represent a free energy minimum state at a given temperature. Some of the highly perfect quasicrystalline materials exhibit a striking diffraction pattern, see Figure 14–1b, c. There are a large number of diffraction peaks, which are aperiodically arranged and located at the ideal positions being consistent with a fivefold symmetry. Besides, a remarkable fact is that the peak sharpness appears to be comparable to that from nearly perfect crystals such as silicon, as evidenced by high-resolution synchrotron diffraction experiments.

These diffraction features, in particular the sharp diffraction peaks represented by delta-functions, which had been believed to be possible only for periodic crystals, can no longer be explained according to a classical framework of incommensurate crystals. Shortly after their discovery, quasicrystal structure was discussed in relation to a rather disordered/imperfect state of solids. Representative early models are an icosahedral glass model (Stephens and Goldman 1986) that assumes only a short-range icosahedral order distributed randomly to form the solid, and the so-called Pauling's model (1986) that employs multiply twinned configurations of giant cubic crystals to generate a pseudo-fivefold symmetry pattern. One intuitively notices that neither of these configurations can be responsible for generating the sharp Bragg peaks. Nowadays we interpret this unique long-range aperiodic order as true quasiperiodicity (Levine and Steinhardt 1984), which is not a simple

S.J. Pennycook, P.D. Nellist (eds.), *Scanning Transmission Electron Microscopy*,
DOI 10.1007/978-1-4419-7200-2_14, © Springer Science+Business Media, LLC 2011

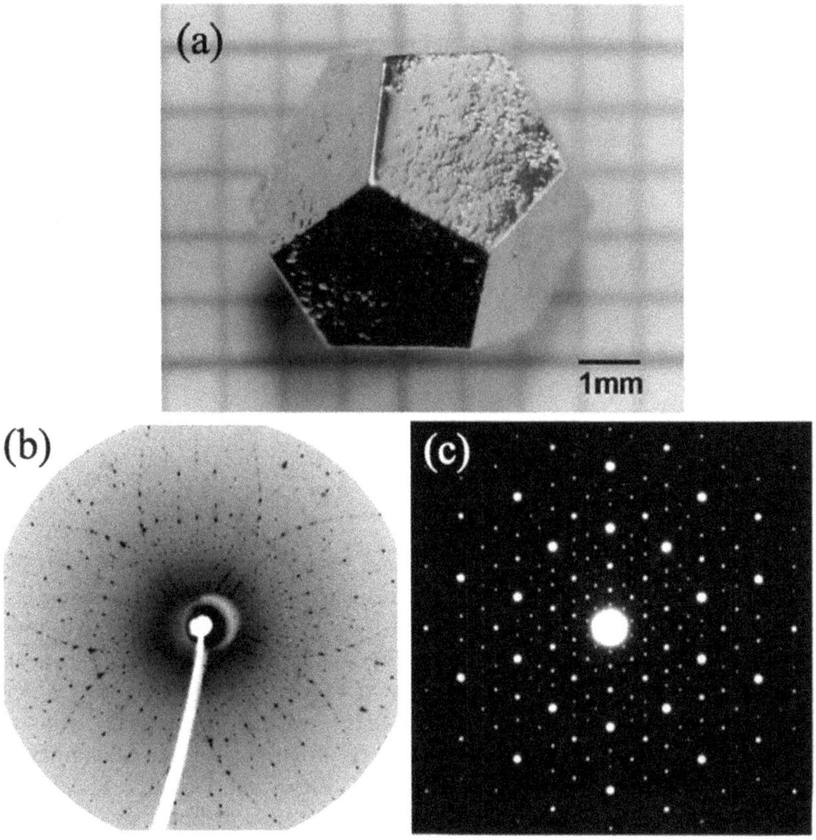

Figure 14–1. (a) Dodecahedral single grain of $Zn_{56.8}Mg_{34.6}Ho_{8.7}$ icosahedral quasicrystal successfully grown from the melt via slow cooling method (courtesy of Ian Fisher and Paul Canfield (Fisher et al. 1998)). (b) Transmission X-ray Laue photograph obtained from Cd_6Yb icosahedral quasicrystal. (c) Electron diffraction pattern taken along the fivefold symmetry axis of Zn_6Mg_3Ho icosahedral quasicrystal.

periodic arrangement of a unit cell as a normal crystal, but instead is composed of an array of two length-scales as represented by the Fibonacci sequence (Figure 14–2). This is a precisely defined sequence that is able to generate the delta-function diffraction peaks and account for the experimental observations, as described below. In time, the discovery of quasicrystals led to a redefinition of the term "crystal" to mean "*any solid giving essentially discrete diffraction peaks,*" as declared by the International Union of Crystallography in 1991. Microscopic unit cells that repeat periodically are not necessary any longer for a material to be called a crystal, and within the family of crystals we now distinguish between periodic and aperiodic solids based on their diffraction features. We particularly emphasize that the term quasicrystal is a short form for "quasiperiodic crystal," and never means an imperfect, pseudo-crystal as a layperson might guess from the expression. It is important to recognize that quasicrystals indeed represent a well-ordered condensed state of matter that is now generally accepted

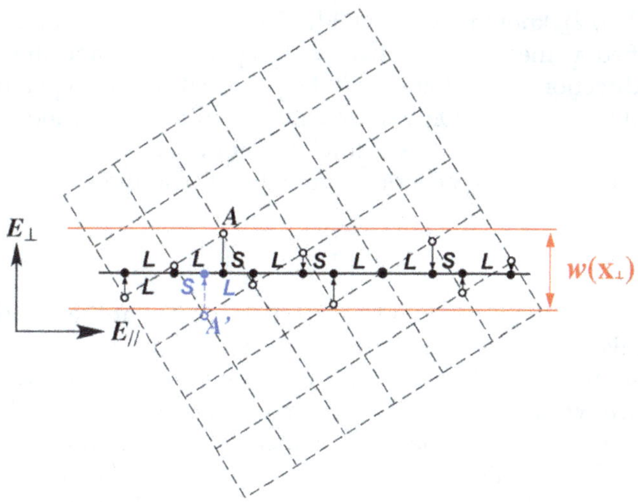

Figure 14–2. Generation of a one-dimensional (1D) quasiperiodic order from a two-dimensional (2D) square lattice. By projecting the 2D lattice points (*open circles*) contained within an acceptance window bounded by the *red lines* along the $E_\perp$ direction, a 1D quasiperiodic lattice (*solid circles*) is obtained along the $E_{//}$ direction. This is a Fibonacci sequence of the two length-scales *L* and *S* (adapted from Abe et al. (2004) with permission).

as a new form of solid; a third phase following the classical crystalline (periodic) and amorphous (random) solids.

14.1.1 Hyperspace Crystallography

Similar to the manner of periodic crystals, the structure of quasicrystals can be described by a combination of a quasiperiodic lattice (quasilattice) and its atomic decorations. The quasilattice, a fundamental framework responsible for generating sharp Bragg peaks, is constructed by a set of two or more unit cells. Diffraction intensity calculations of any quasicrystal model structures are carried out by the hyperspace crystallography (Janssen 1986, Yamamoto 1996), a mathematical recipe that treats a quasicrystal as a periodic structure embedded in a hyperspace, e.g., a cubic lattice defined in six-dimensions generates the icosahedral quasilattice in three dimensions. Below we briefly describe the concept of hyperspace crystallography, by reference to the generation of one-dimensional quasiperiodic order from a two-dimensional square lattice (Figure 14–2).

In the hyperspace crystallography, a density function periodically distributed in hyper-dimensions is described as $\rho^h(\mathbf{r}_{//}, \mathbf{r}_\perp)$, where $\mathbf{r}_{//}$ and $\mathbf{r}_\perp$ correspond to the components along the real dimensions $E_{//}$ and the extra dimensions $E_\perp$, respectively; the latter $E_\perp$ is a complementary space orthogonal to $E_{//}$.

In order to generate a quasiperiodic structure in the $E_{//}$ space, the $\rho^h(\mathbf{r}_{//}, \mathbf{r}_\perp)$ of a cubic lattice (hypercubic) is projected onto $E_{//}$ along $E_\perp$, with the condition that the slope of the hypercubic lattice with respect to the $E_{//}$ direction is $-1/\tau$, where τ is an irrational number

(= (1 + 5)/2) known as the golden mean. In Figure 14–2, this is exemplified as the relation between the square lattice and the rectilinear $E_{//}$ direction. Note that not all the potentials in the hyperspace are subjected to projection. Only those within a finite distance along $E_{\perp}$ are selected by an "acceptance window" to filter the $\rho^h\left(\mathbf{r}_{//}, \mathbf{r}_{\perp}\right)$, projecting them onto $E_{//}$ space to generate a potential distribution,

$$\rho\left(\mathbf{r}_{//}\right) = \int d\mathbf{r}_{\perp}\rho^h\left(\mathbf{r}_{//}, \mathbf{r}_{\perp}\right) \cdot w(\mathbf{r}_{\perp} - \mathbf{R}_{\perp}), \quad (1)$$

where $w(\mathbf{r}_{\perp} - \mathbf{R}_{\perp})$ is the window function located at the position $\mathbf{R}_{\perp}$. Again, this is schematically understood by Figure 14–2. By projecting the two-dimensional lattice points (open circles) contained within the acceptance window bounded by the red-lines along the $E_{\perp}$ direction, a one-dimensional quasiperiodic sequence (solid circles) is obtained along the $E_{//}$ direction as *LLSLSLLSL...*; this is a Fibonacci sequence constructed by the two length-scales *L* and *S*. All in all, the procedure includes *cutting* the hyperspace crystal and then *projecting* its potential onto the real dimension space; hence this is referred to as "cut-and-projection" method.

Next, we derive the structure factor of the projected structure $\rho\left(\mathbf{r}_{//}\right)$. The periodic potential in the hyperspace $\rho^h\left(\mathbf{r}_{//}, \mathbf{r}_{\perp}\right)$ can be decomposed into a Fourier series by using a structure factor of the hyperspace lattice f_G^h,

$$\rho^h\left(\mathbf{r}_{//}, \mathbf{r}_{\perp}\right) = \sum_G f_G^h \exp(i\mathbf{G}_{//} \cdot r_{//} + i\mathbf{G}_{\perp} \cdot \mathbf{r}_{\perp}), \quad (2)$$

where $\mathbf{G}_{//}$ and $\mathbf{G}_{\perp}$ represent the components of the reciprocal vectors $\mathbf{G}$ of the hyperspace lattice in the real dimensions and extra dimensions, respectively. From Eqs. (1) and (2), we obtain the potential distribution $\rho\left(\mathbf{r}_{//}\right)$ and the corresponding structure factor f_G that is observable in the real dimension space,

$$\rho\left(\mathbf{r}_{//}\right) = \sum_G f_G \exp(i\mathbf{G}_{//} \cdot \mathbf{r}_{//}), \quad (3)$$

where

$$f_G = f_G^h \exp(i\mathbf{G}_{\perp} \cdot \mathbf{R}_{\perp}) \int d\mathbf{x}_{\perp} w(\mathbf{x}_{\perp}) \cdot \exp(i\mathbf{G}_{\perp} \cdot \mathbf{x}_{\perp}), \quad (4)$$

where $\mathbf{x}_{\perp} \equiv \mathbf{r}_{\perp} - \mathbf{R}_{\perp}$. In Eq. (4), the latter integral part corresponds to a Fourier transform of the window function, which is sufficiently spread over the real dimension space $E_{//}$ but with a limited thickness along the complementary space $E_{\perp}$ (shown as a "band" in Figure 14–2). Therefore, the diffraction peaks are expected to be significantly streaked along the $E_{\perp}$ direction due to the shape effect of the window, while retaining their delta-function-like distributions along $E_{//}$ when projected and observed in the $E_{//}$ space. It should be remembered, for those who are familiar with electron diffraction, that this scheme is definitely analogous to the diffraction from a thin crystal; diffraction peaks will be elongated along the thickness direction (described according to an "excitation error" parameter) due to the crystal-shape effect, the occurrence of which is well explained by the Laue function.

To sum up, from Eqs. (3) and (4), it is known that we observe Bragg reflections at $\mathbf{q} = \mathbf{G}_{//}$ whose intensities are proportional to $|f_G|^2$ for the (projected) quasiperiodic structure. Here, it is noteworthy that a summation of $\mathbf{G}$ is taken over the entire reciprocal vectors that extend into the hyperspace, and consequently the reciprocal lattice points are, in principle, densely distributed in the $\mathbf{G}_{//}$ space. However, the observable structure factor f_G is dependent on $\mathbf{x}_{\perp}$ that represents the distance from the window position (Eq. (4)); therefore, the diffraction intensity becomes significantly weaker for the reflections with larger $G_{\perp}$components. These characteristics derived from the hyperspace descriptions of the quasiperiodic order fairly account for the experimental diffraction pattern of highly perfect quasicrystalline materials, for which a large number of weak reflections appear even at low-q ranges (e.g., see Figure 14–1b and c).

14.1.2 Local Isomorphism

As described above, by supposing a hidden periodic structure it becomes possible to calculate the diffraction intensity of any model structure of a quasicrystal. But, unlike the structure determination of crystals, in the case of quasicrystals a unique structural solution cannot be easily obtained solely based on standard X-ray diffraction experiments. This is due to the local isomorphic nature of the quasilattice. To illustrate this feature, we shall look again at the one-dimensional quasiperiodic sequence in Figure 14–2. When the window position (or the square lattice) is translated along the $E_{\perp}$ direction, some external lattice points will move into the acceptance window and remove some of the original points outside of the window. As an example, if the blue lattice point A' comes into the window instead of the original A lattice point, a new local sequence *LSL* (shown by blue) is generated instead of the original *LLS* sequence. According to this prescription, several quasiperiodic arrangements of *S* and *L* can be generated when the window function is uniformly translated along the $E_{\perp}$ direction; all of the resultant *S*–*L* arrangements belong to a class of local isomorphism. Cut-and-projection of the five-dimensional hypercubic lattice generates a two-dimensional quasiperiodic lattice. Figure 14–3 shows locally isomorphic quasilattices that are constructed by an identical set of tiles (fat and thin rhombi) (Ishihara and Yamamoto 1988), which are generated by differing the window position $\mathbf{R}_{\perp}$.

Concerning the structure factor of these quasilattices, it should be noted in particular that the window position $\mathbf{R}_{\perp}$ just alters a phase term in Eq. (4). Therefore, each of these gives rise to identical diffraction intensity distributions, as shown in the bottom of Figure 14–3 (Ishihara and Yamamoto 1988) (namely a phase problem along the $E_{\perp}$ direction). Multiple structural solutions intrinsically occur from the diffraction data for a quasicrystal structure analysis, similar to the case known as *homometric structures* in normal crystals.

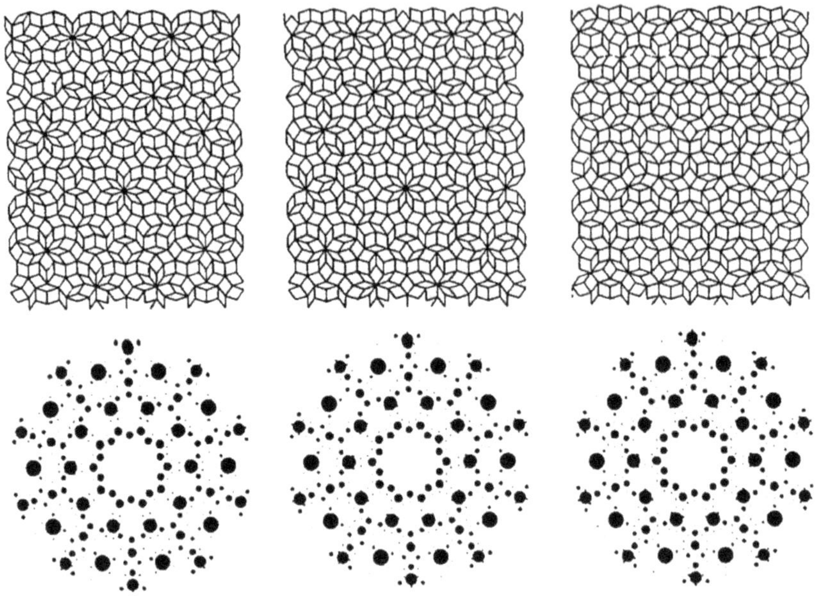

Figure 14–3. Examples of two-dimensional quasiperiodic lattices constructed by an identical set of tiles, fat- and skinny-rhombic tiles in the Penrose tiling (Reproduced from Ishihara and Yamamoto 1988 with permission). These quasilattices belong to a class of local isomorphism, generating the identical structure factor shown below.

"Where are the atoms?"—this has been the fundamental key question since it was used as the title of an early article (Bak 1986). Atomic-resolution electron microscope investigations are unique in this regard, as they directly provide the (local) atomic arrangements even for aperiodic solids. Electron diffraction is also available to probe the average features seen in an image, in a comparable manner to X-ray diffraction. In early structural studies of quasicrystals, high-resolution phase-contrast imaging provided evidence that their structures are indeed represented by a combination of clusters and the particular quasilattice – direct observation is able to distinguish one of the patterns in Figure 14–3. Nevertheless, there have been still significant information lacking concerning the local atomic configurations of quasicrystals. In real quasicrystalline material, are the clusters always placed at the ideal quasiperiodic positions? Like the "unit-cell" in normal crystals, do all the clusters have almost identical atomic configurations in accordance with ideal quasiperiodicity? What types of local disorder/defects do they possess as characteristic for quasicrystals? These local structure issues are especially important for quasicrystals, since they have direct influence on their thermodynamic stability, which in turn is important for understanding why quasicrystals form. Yet these issues have not been clarified sufficiently because of experimental limitations of standard X-ray techniques and conventional high-resolution electron microscopy. We describe below some recent insights made into these critical issues, which have been provided through direct structural observations by advanced scanning transmission electron microscopy.

14.2 Local Symmetry of Quasicrystals

Decagonal quasicrystals (Bendersky 1985) are the planar realization of a quasiperiodic order, see Figure 14–4. Decagonal structure is described as a periodic stack of quasiperiodic layers and is composed of decagonal columnar clusters as a building unit. Because of their two-dimensional character, quasiperiodic planar arrangements of atoms can be directly addressed through high-resolution electron microscope observations viewing along the tenfold-symmetry axis. Individual decagonal clusters appear as decagons in the projected images, so that their packing, or tiling, may be directly observed (Hiraga 1998). We note that for icosahedral quasicrystals, the atomic images are the projections of three-dimensional quasiperiodic structures, and unfortunately individual icosahedral clusters cannot be distinguished (e.g., see Abe et al. (2001)).

Previous high-resolution phase-contrast observations, mostly made on Al-transition metal (TM) decagonal alloys, provided evidence that their structures are quasiperiodic arrangements of decagonal clusters (to be described later in Figure 14–6). The picture emerging from these studies is that the quasicrystal can be viewed as a cluster aggregate, for which the basic atomic clusters are supposed to have the same point symmetry of the corresponding quasicrystalline phases. For the decagonal quasicrystals, all models assumed either tenfold or five-fold symmetric atomic configurations within the clusters; typical and well-known examples are the large decagonal clusters with a diameter of 2 nm, commonly found in Al–Ni–Co, Al–Cu–Co, and Al–Pd–Mn decagonal alloys (Beeli and Horiuchi 1994, Hiraga 1998, Tsuda et al. 1996). These cluster symmetries, in particular the tenfold rotation axis, were originated not from confident experimental evidence, but rather

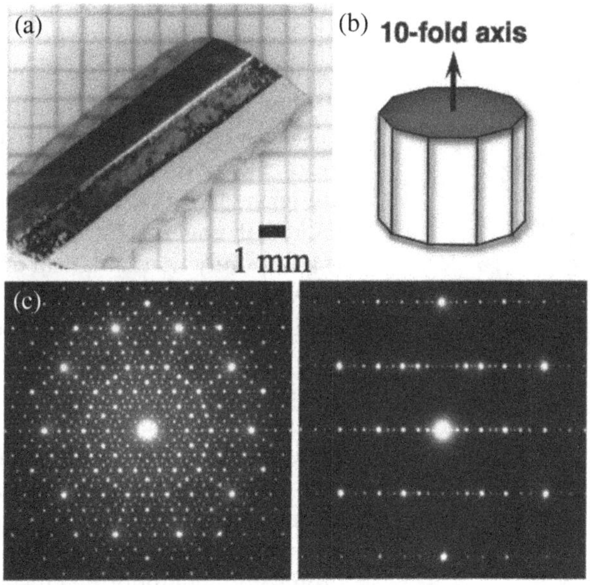

Figure 14–4. (a) Single grain of $Al_{56.8}Ni_{34.6}Co_{8.7}$ decagonal quasicrystal (reproduced from Fisher et al. 1999 with permission). Decagonal structure is an example of the two-dimensional quasicrystals, whose entire symmetry is represented by a regular prism with the corresponding symmetry. (b) Schematic representation of the decagonal symmetry. (c) Electron diffraction patterns taken along the tenfold (*left*) and twofold (*right*) symmetry axis of the $Al_{72}Ni_{20}Co_8$ decagonal quasicrystal, one of the best quasiperiodic ordered materials available today.

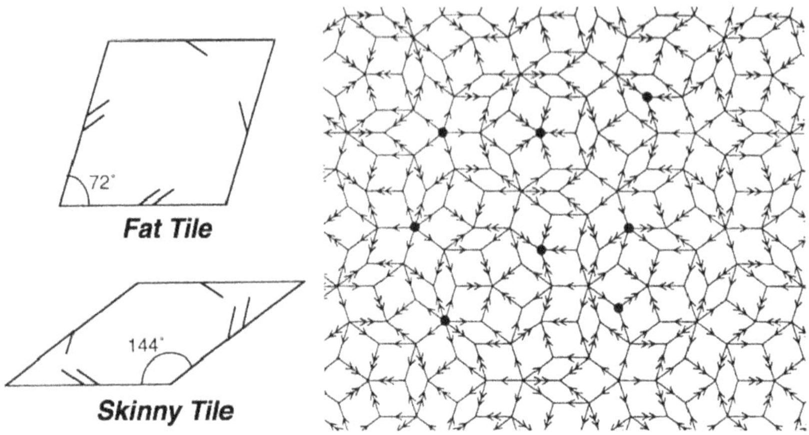

Figure 14–5. Penrose tiling, the first aperiodic tiling constructed by two types of rhombic tiles. The entire pattern is generated by a local joint rule referred to as "matching rule," which requires each of the tiles to complete types and directions of the arrowheads on the tile edges, as shown in the figure. The pattern appears to be complicated, but there are only eight local environments (vertex symmetry), as indicated by the black dots.

from the notion that the symmetry of every component cluster should directly reflect the entire symmetry seen in the diffraction pattern; in other words, that the microscopic and macroscopic symmetry should be the same, similar to the unit cell concept for a regular crystal. Hence there still remained ambiguities in terms of detailed atomic distributions within the clusters in terms of true cluster symmetry.

Some decagonal quasicrystalline samples were found to have a cluster arrangement similar to a Penrose pattern (1974), a planar tiling composed of two different rhombic tiles with matching rules (Figure 14–5). The matching rule is a strict mathematical rule that forces the tiles to join uniquely into a perfect quasiperiodic pattern. We note that even being constructed by the same set of rhombic tiles, no matching rule is applied for the variant quasiperiodic patterns shown in Figure 14–3. In this sense, the Penrose pattern is definitely a unique tiling with a strict local rule, and accordingly the variant patterns of local isomorphism are termed *generalized* Penrose patterns. The matching rule implies a trick that governs the local growth of quasicrystals, though the rule is still purely mathematical and does not provide any physical insight on why the atoms should favor such a complicated structure. It cannot explain how quasicrystals arise as a minimum free energy state against competing periodic crystals.

As an alternative to the two-tile Penrose tiling (or its subset tiling composed of multiple shapes of tiles), Burkov (1991, 1992) described the quasiperiodic pattern in a broader sense that discards the matching rule, describing the model structure of Al–Cu–Co as a random packing of decagon clusters having tenfold symmetry. The clusters are allowed to overlap with their neighbors, in the sense that they partially share atoms with neighboring clusters. In random packing, there are no rules that force the clusters into a unique arrangement, and hence many possible configurations appear due to a large degree of freedom on how to join or overlap the neighboring decagons—many degenerate ways of packing are an unavoidable consequence of this model. This situation is supported by the scenario that *configurational entropy* might be an important factor causing quasicrystals to be more stable than competing crystals. The so-called random tiling model would give significant contributions of configurational entropy (Henley 1991) and seems to be consistent with the fact that the stable quasicrystalline phases discovered so far occur only at high-temperatures (Tsai 1999), and most transform into periodic structures at lower temperatures. Besides, in a random-packing picture, resultant structures may appear to have a considerable amount of chemical disorder—most atomic sites are mixed with constituent atoms. Atomic disorder is, of course, another contribution to configurational entropy (Joseph et al. 1997).

14.2.1 Symmetry-Breaking of Internal Clusters

Striking features of cluster symmetry and their arrangement have become apparent through investigations of decagonal $Al_{72}Ni_{20}Co_8$ (electron diffraction patterns are shown in Figure 14–4). This material exhibits the highest quasicrystalline structural perfection available today, as confirmed from microscopic (Joseph et al. 1997, Ritsch et al. 1996) to macroscopic scales (Abe et al. 2000), and is reproducible as a single phase by annealing at 1170 K followed by water-quenching. Therefore, the decagonal $Al_{72}Ni_{20}Co_8$ is an excellent candidate to investigate the intrinsic features of a quasiperiodic structure. The insights have come from the use of Z-contrast STEM (Pennycook and Boatner 1988, Pennycook and Jesson 1990, 1991), an alternative to classical phase-contrast TEM; see the different appearances of the contrast in Figure 14–6.

The first application of the STEM technique to the quasicrystal (Saitoh et al. 1997, Steinhardt et al. 1998, Yan and Pennycook 1998) provided immediate insights into the veiled structural details, which were difficult to figure out solely based on phase-contrast imaging. First, a significant breakthrough to emerge is a breaking of the tenfold symmetry within the 2 nm decagonal cluster, see Figure 14–6b. The brightest spots in the Z-contrast image represent atomic columns of Ni or Co (Ni and Co are neighboring elements in the periodic table and are not distinguishable). As indicated by arrows, in the cluster center they are clearly not arranged in a tenfold form, but appear to show only mirror symmetry. Secondly, such local broken-symmetries in every decagon are found to be not in random orientations but in accord with a perfect quasiperiodic pattern (Figure 14–7a)—the pattern can be well represented by the novel form of decagon packing proposed by Gummelt (1996) (Figure 14–7b).

In Gummelt's construction plan, decagons are not tenfold symmetric, and they overlap with their neighbors according to a well-defined overlap rule (Figure 14–7b) that is equivalent to the Penrose matching rule. Therefore, the overlap rule forces the decagons into a perfect

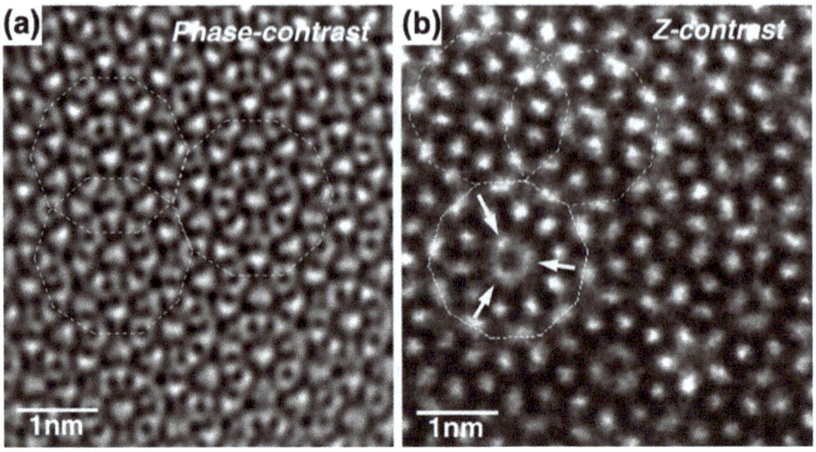

Figure 14–6. Atomic-resolution (a) phase-contrast TEM and (b) Z-contrast STEM images taken along the tenfold symmetry axis of the $Al_{72}Ni_{20}Co_8$ decagonal quasicrystal. Decagonal clusters of 2 nm across are shown by dashed lines in both images. The image (a) was obtained from the very thin region (~60 Å; near-edge of a cleavage grain) and under nearly Scherzer defocus (~45 nm for the JEM-4000EX with Cs = 1.0 mm). The image (b) was taken from a slightly thicker region than that of (a), approximately ~120 Å thick evaluated by EELS plasmon method.

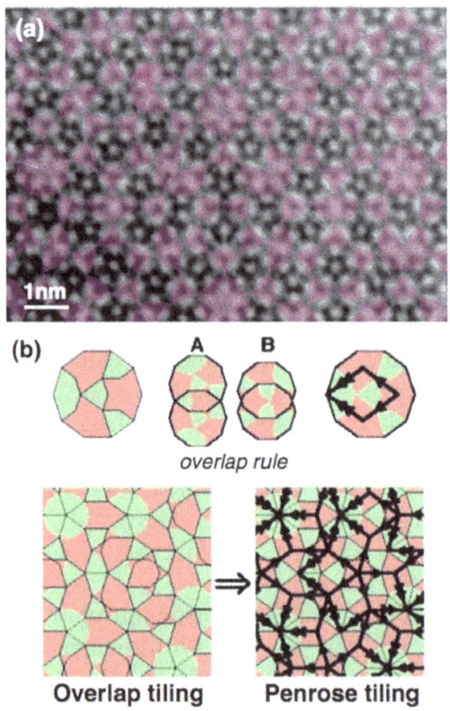

Figure 14–7. (a) Long-range structure of the decagonal $Al_{72}Ni_{20}Co_8$ can be well represented by the novel decagon tiling (Gummelt 1996) that is equivalent to a perfect Penrose tiling. (b) Using the decagons marked so as to break their tenfold-symmetry, a quasiperiodic tiling can be forced if the decagons are permitted to overlap only if shaded regions overlap, limiting the possible overlaps to just two ways, A-type and B-type (overlap rule). By inscribing a fat rhombus within each decagon, a decagon-overlapping tiling (*left*) is converted into a Penrose tiling with the matching rule (*right*) (adapted from Steinhardt et al. 1998 with permission).

quasiperiodic arrangement. This remarkable mathematical proof has led to a physically plausible picture for the origin of quasicrystals. A subsequent but important proof by Steinhardt and Jeong (1996) has shown that the overlap rule realizes the condition that the density of the decagons is maximized. Suppose the atomic configuration within a decagon, i.e., the atomic cluster in the form of Gummelt's decagon (Figure 14–8 (Abe et al. 2000)), is energetically favored. Then, quasicrystals occur as a consequence of simple energetics, following "density maximization of energy minimized clusters." The fact that the $Al_{72}Ni_{20}Co_8$ structure appears to be the realization of a unique packing of symmetry-breaking clusters (Gummelt's decagon) therefore suggests that the phase is dominantly stabilized by energy; if there were significant entropy contributions, a considerable amount of random structural disorder including site mixing by Al and Ni/Co atoms (chemical disorder), or deviations from an ideal tiling, would be observed.

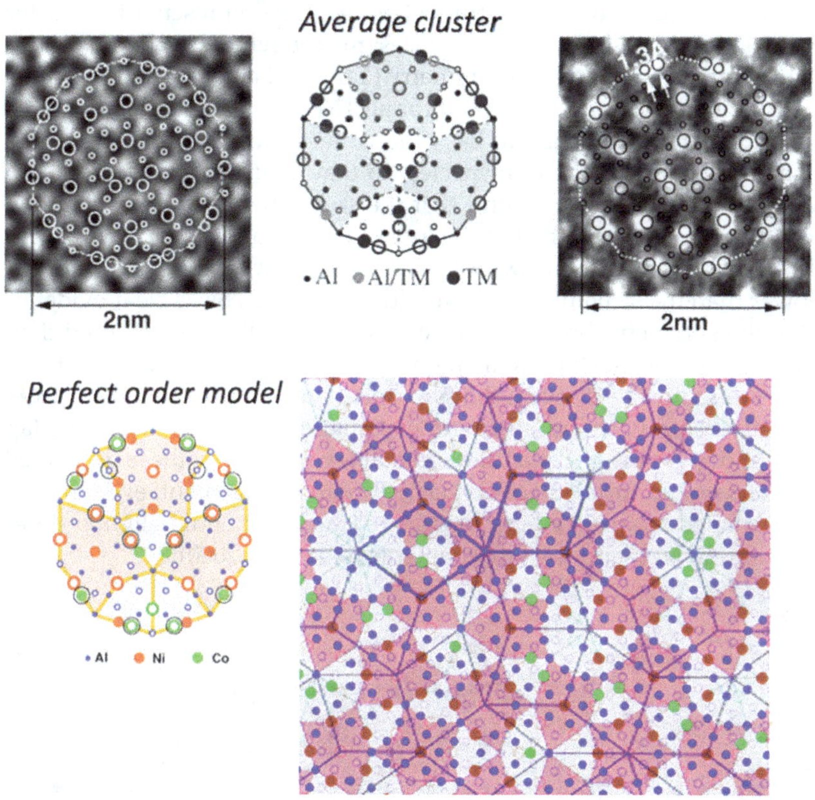

Figure 14–8. Atomic model of the decagonal $Al_{72}Ni_{20}Co_8$; the structure has two distinct atomic layers, and *solid* and *open circles* represent different levels along the tenfold axis, $c = 0$ and $c = 1/2$, respectively. A perfect quasiperiodic atomic order (*below*) can be constructed from the decagonal cluster properly decorated according to Gummelt's motif (adapted from Abe et al. 2000 with permission). Average decagonal cluster (*above*), derived by averaging over the local variations that occur in the perfect decoration (*below*), fairly well explains both the phase-contrast and Z-contrast images.

On these bases, it is concluded that the decagonal clusters have intrinsic broken symmetry that is a built-in aspect of the atomic decoration; the symmetry breaking is definitely not a consequence of random chemical and occupational (vacancy) disorder, as likely to occur for Burkov's random packing scenario. As shown in Figure 14–8, symmetry-breaking atomic decorations on the 2 nm decagonal cluster in fact provide a remarkably better fit to TEM/STEM images on some details, such as triangular arrangements of Al atoms around the cluster center and closely spaced (~1.3 Å) pairs of transition metal (TM) atoms, which did not exist in any previous tenfold symmetric models.

14.2.2 Cluster Symmetries and Stability

Strong supporting evidence on the cluster stability comes from a first-principles total energy calculation (Mihalkovic et al. 2002, Yan and Pennycook 2001), which demonstrates that the symmetry-breaking cluster is energetically more favored than competing symmetric-based configurations. Shown in Figure 14–9 are three representative cluster models for three different symmetries: mirror symmetry, fivefold rotation symmetry, and tenfold symmetry (Yan and Pennycook 2001) (in the calculations, Ni atoms were used for all TM atoms). Note that the three structures in the calculations have the same unit cell dimension and the same number of each atom species, so that their total energies can be directly compared. It is found that the structure with broken symmetry, the mirror symmetry, has the lowest energy, about 12 eV/(unit cell) lower than the structure with tenfold symmetry, and 5 eV/(unit cell) lower than the structure with fivefold symmetry. This is clear evidence that the symmetry-breaking cluster is energetically more favored than any symmetric clusters. Remember that most of the decagons identified in Figure 14–7a reveal the symmetry-breaking feature; the broken symmetry provides an atomistic explanation for Gummelt's overlap rule and gives insight at a fundamental level into why these decagonal clusters would form a perfect quasiperiodic arrangement, even though the detailed atomic configurations in each cluster could slightly differ depending on their local environments (Abe et al. 2000).

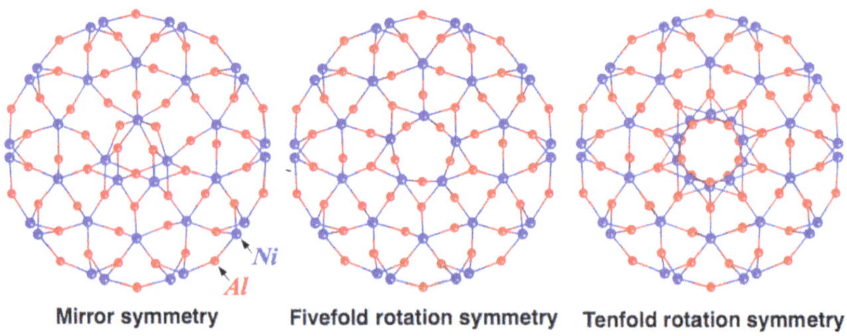

Figure 14–9. Different atomic decorations of the decagonal cluster for three different symmetries (reproduced from Yan and Pennycook 2001 with permission): mirror symmetry, fivefold rotation symmetry, and tenfold symmetry.

Decagonal $Al_{72}Ni_{20}Co_8$ has turned out to be a quasiperiodic intermetallic compound with nearly perfect atomic order—it is definitely not comparable to order–disorder type alloys, which take the form of a disordered solid solution at high temperature to account for significant entropy contributions. Good chemical order between the Al and TM, directly confirmed by the Z-contrast imaging, seems to be consistent with the fact that the present highly perfect Al–Ni–Co structure occurs only for a narrow composition range, within a few atomic percent for both the Al and TM contents (Goedecke et al. 1998). If the structure could tolerate a considerable amount of chemical disorder, essential for a random-packing model, then the single-phase region would extend to a much wider composition range at high-temperatures; this is evidently not the case. In this sense, the $Al_{72}Ni_{20}Co_8$ compound is close to its ideal stoichiometry, being tuned in favor of structural energy. This may well be explained by an optimized average valence electron concentration per atom (*e/a*) (Tsai 1999), which is known as Hume-Rothery's empirical rule (Hume-Rothery 1926) that concerns structural stability of ordered alloys in terms of Brillouin-zone/Fermi-surface interactions. When the composition deviates from the ideal stoichiometry, Al–Ni–Co alloys form several types of less ordered quasicrystalline phases (Ritsch et al. 1998) with diffraction patterns that show a strong diffuse background and broadened Bragg peaks, direct signs of significant disorder. For such disordered quasicrystals, their *average* structures may be well described by random-packing of clusters (Burkov 1991, 1992). The apparent high-symmetry of these patterns is then a result of averaging over the local random disorder. Structural variations of the basic cluster (at some conditions, fivefold symmetric decagonal clusters occur for the decagonal Al–Ni–Co alloys (Hiraga et al. 2001, Yamamoto et al. 2005)) will then simply depend on the alloy composition and annealing temperature, in the same way as ordinary crystalline alloys behave, and can be depicted through equilibrium phase diagrams. Nevertheless, we now see that the best quasicrystalline sample appears to be a highly ordered intermetallic compound with only a minimal amount of disorder on the atomic scale, within the resolution limit of the STEM imaging. The fundamental reason for the existence of such well-ordered quasicrystals is that their structure is energetically favored. Only rarely do practical experimental conditions allow the perfect structure to be realized, and therefore most quasicrystalline phases do not reach that degree of perfection. The situation seems to be quite analogous to that of normal crystals, i.e., highly perfect, almost defect-free single crystals can only be grown for a limited range of materials under carefully controlled conditions.

14.3 Atomistic Fluctuations in Quasicrystals

As described in the introduction, the striking features of quasicrystals—aperiodic order with non-crystallographic symmetry but being nevertheless compatible with sharp Bragg peaks—can be reasonably explained according to the hyperspace crystallography, which is given as a generalized form of normal crystallography by extending its framework into n-dimensions. Suppose that the quasicrystalline order we view at our (observable) dimension is a "projected shadow" of a hyperspace lattice. Then, the relevant quasicrystals should involve an intrinsic degree of freedom along the extra dimensions, even though they are hidden to us. On this basis, we expect an extra elastic degree of freedom for quasicrystals termed a phason. We will never be able to detect anything directly even if the hyper-lattice fluctuates along the hidden dimensions. However, such fluctuations, if any, indeed leave traces in the observable dimensions as a discontinuous change of local structures in quasicrystals, as described later. As a consequence, the lattice dynamics of quasicrystals can be described by a combination of lattice vibrations (phonons) and atomic fluctuations (phasons). In this section, we describe phasonic fluctuations in the highly perfect quasicrystalline decagonal $Al_{72}Ni_{20}Co_8$, through successful observations of a local Debye–Waller factor anomaly by annular dark-field STEM imaging. Meanwhile, the results provided the first direct, real-space imaging of a local thermal vibration anomaly in a solid (Abe et al. 2003).

14.3.1 Phason—Extra Degree of Freedom

Within the context of hyperspace n-dimensional ($n \geq 4$) crystallography (Janssen 1986, Yamamoto 1996), potential distributions in the quasicrystal structure can be expressed by a Fourier series (density waves) based on the structure factor of the hyperspace lattice f_G^h, which basically has the dual components for the real dimensions ($\mathbf{G}_{\parallel}$) and the extra dimensions ($\mathbf{G}\perp$), as described in Eq. (2). Therefore, their phase term ϕ_G can be written as

$$\phi_G = \mathbf{G}_{\parallel} \cdot u + \mathbf{G}_{\perp} \cdot v, \tag{5}$$

where the $\boldsymbol{u}$ and $\boldsymbol{v}$ represent the displacement along $\mathbf{G}_{\parallel}$ and $\mathbf{G}\perp$, respectively. Here, $\boldsymbol{v}$ defines an extra degree of freedom and causes a unique elastic property specific to quasicrystals (Bak 1985, Levine et al. 1985). The $\boldsymbol{u}$ corresponds to the same phonons as for normal crystals, while the $\boldsymbol{v}$ provides an exotic property termed phasons. It should be noted here that, before the discovery of quasicrystals, a similar extra degree of freedom had been already discussed for incommensurate structures (Cowley and Bruce 1978), for which the additional modulation vector was also supposed according to a four-dimensional crystallography (Figure 14–10, left). Nevertheless, the phason in quasicrystals provides intriguing behaviors that were not expected for the ordinary incommensurate structures. As schematically represented in Figure 14–10, for the incommensurate structures the relevant hyperspace potentials are taken to be continuous with a periodic modulation along the extra dimension. Hence their lattice dynamics at the observable-dimensions appear to be a (continuous) displacive mode when the hyper-lattice fluctuates along the extra dimensions. This is basically analogous to phonon behavior. On the other hand, for ideal quasicrystals the corresponding hyperspace potentials are "discrete" segments (Yamamoto

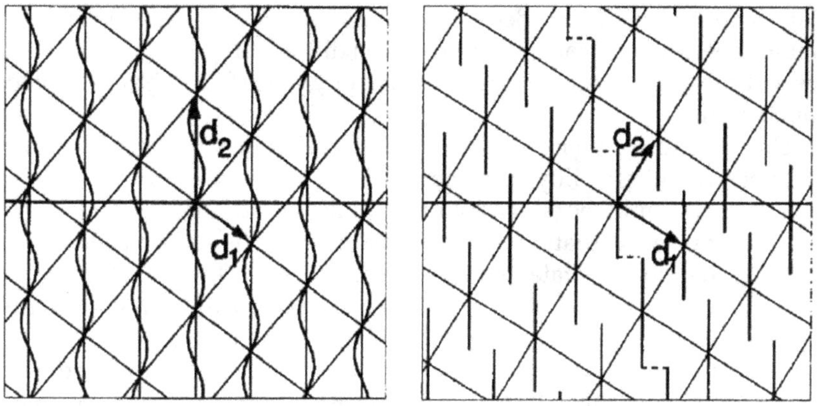

Figure 14–10. Hyperspace potential distributions shown by "atomic surfaces" that represent scattering objects decorating the hyper-lattice (reproduced from Yamamoto 1996 with permission). Atomic surfaces lie along the extra dimensions (*vertical direction*), and the corresponding aperiodic structures are generated in the real-space dimensions (*horizontal direction*) at the intersections. This procedure, referred to as the "section method", is mathematically equivalent with the "cut-and-projection method" described in Figure 14–2. Atomic surfaces are given as discrete segments for an ideal quasicrystal (*right*), while they are continuously modulated for an incommensurate aperiodic crystal (*left*).

1996), the phasons in quasicrystals will show up as being not continuous displacements but discrete jumps/flips between the two distinct sites. This flipping behavior can be understood by referring again to Figure 14–2. Suppose that the hyper-lattice (square-lattice) is elastically excited along the $E_{\perp}$ direction to give a non-uniform translation of the lattice points, which essentially represents a phason excitation. This may turn out to switch *S* and *L* at some places, flipping a local sequence *LLS* into the sequence *LSL* (shown in blue), as exemplified in Figure 14–2. Displacive distortion along $E_{//}$, of course, corresponds to a phonon in quasicrystals.

In the real quasicrystal structure, primary phason effects are often manifested as fluctuations or occupational disorder at pairs of atomic sites that are separated by less than a typical interatomic distance. These closely spaced dumbbell sites can be indeed generated due to a phason degree of freedom for hyperspace potentials, and hence they are termed phason-flip sites (details will be described later). For the phason-flip sites, one site should be vacant when the other is occupied due to simple geometrical frustrations; the atom may be hopping (Coddens and Steurer 1999, Dolinsek et al. 1999) between these sites. Consequently, the lattice dynamics of quasicrystals can be described by a combination of lattice vibrations (phonons) and atomic fluctuations or jumps (phasons). Note that such atomistic fluctuations at the local phason-flip sites do not destroy the long-range quasiperiodic order, but instead give rise to significant diffuse scattering in the same manner as thermal vibrations (or phonons) do (de Boissieu et al. 1995, Ishii 1992,

Jaric and Nelson 1988). Because of this, even for the best quasicrystalline specimen revealing delta-function diffraction peaks, we may still expect substantial diffuse scattering that originates from phason-related fluctuations.

Understanding phason-related atomic behaviors is critical for the thermodynamic stability of quasicrystals, whatever their origin maybe; the context could be a random-tiling-like realization (Henley 1991) where phason fluctuations are essential to provide entropic stabilization, or an unlocked state (Jeong and Steinhardt 1993) of an intrinsically energy-minimized perfect quasicrystal. In either case, phason fluctuations are expected to be significant above some critical temperature. So far, numerous experimental measurements on phason-related disorders/dynamics have been attempted by various experimental techniques (Bancel 1989, Colella et al. 2000, Edagawa et al. 2000, Francoual et al. 2003, Zeger et al. 1999); however, there has been no direct evidence on *where* such localized atomic fluctuations take place in the real-space quasicrystal structure. This is because the local disorder/fluctuation effects are all averaged in the X-ray and other diffraction patterns, and therefore it is difficult to specify the local source for diffuse scattering. We use STEM to tackle this issue.

14.3.2 Thermal Diffuse Scattering in STEM

Annular dark-field (ADF) STEM provides the atomic-resolution images by effectively illuminating each atomic column one-by-one as a finely focused electron probe is scanned across the specimen, generating an intensity map from the annular detector (Figure 14–11). To a good approximation, the resultant atomic image is interpreted assuming independent scattering from individual atomic columns, and hence the observed intensity distribution ($I(\mathbf{R})$) can be simply described by a convolution between a probe-intensity function ($P(\mathbf{R})$) and a scattering object function ($O(\mathbf{R})$) (*incoherent imaging*) (Pennycook and Jesson 1990, 1991):

$$I(\mathbf{R}) = O(\mathbf{R}) \otimes P(\mathbf{R}). \tag{6}$$

There are two major effects that cause Eq. (6) to be a valid approximation. (1) *Electron channeling*: the fast incidence-electron propagates along the atomic columns with strong channeling effects, which are well described by 1*s* Bloch-state excitation through dynamical diffraction. (2) *High-pass filtering by the detector*: the annular detector set at high-angle ranges captures the 1*s*-dominant scattering by effectively filtering the other Bloch-state excitations (e.g., 2*s*-state). Details of these ingenious ideas of ADF-STEM, based on advantageous use of dynamical electron diffraction, are described in the relevant chapters in this book (see Chapter 2 or Chapter 6). Accordingly, in Eq. (6), $O(\mathbf{R})$ represents the columnar scattering cross-section that contributes to the annular detector.

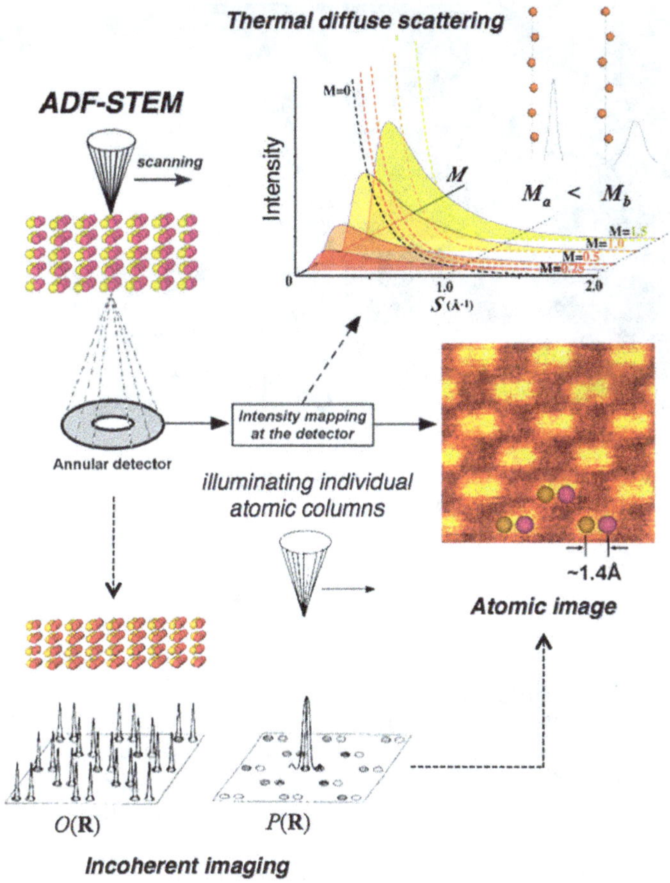

Figure 14–11. Schematic drawing of atomic-resolution ADF-STEM. Within the incoherent imaging approximation, the resolution is primarily determined by the size of the convergent electron beam. The intensity of electrons reaching the annular detector is dominated by thermal diffuse scattering (TDS), which is phenomenologically described by an absorptive atomic form factor, $f'_{HA}(M, s)$, given in Eq. (7). TDS intensity depends strongly on atomic number (Z), so that ADF-STEM provides enhanced chemical contrast (Z-contrast). Note that the TDS intensity at given detector angle ranges is also sensitive to the Debye–Waller factor (M) values of individual atomic sites (or columns); M-dependent changes of the TDS angular distribution are shown upper-right, which are calculated according to σ_{TDS} (Eq. (8)) for an aluminum atom. Adapted from Abe et al. (2004) with permission.

At high-angle scattering, the Bragg intensities are significantly attenuated, and instead the background diffuse scattering becomes dominant, see the scattering ranges larger than ~ 1 Å^{-1} in Figure 14–12b. The origin of this diffuse scattering at high-angles is mostly due to a thermal vibration of atoms (i.e., phonon scattering events), and hence it is denoted as thermal diffuse scattering (TDS). TDS intensities at given detector ranges can be estimated either through elastic scattering based on the frozen phonon model (Loane et al. 1991, Muller et al. 2001), or inelastic (quasi-elastic) scattering described by an absorptive potential added as imaginary component to the normal electrostatic potential (Bird and King 1990, Hashimoto et al. 1962, Weickenmeier

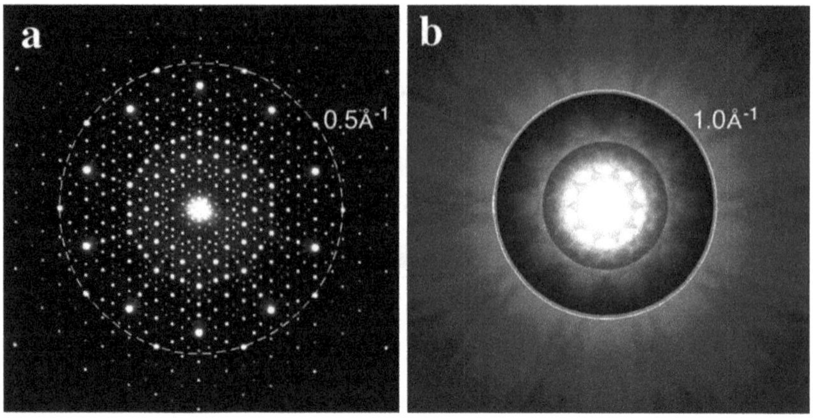

Figure 14–12. Electron diffraction patterns of the decagonal quasicrystal $Al_{72}Ni_{20}Co_8$, taken by (a) parallel beam and (b) convergent beam illumination. In (b), the intensity distributions at low *s* ranges are adjusted to show the Bragg reflection disks.

and Kohl 1991). Note that both methods fairly well estimate the TDS intensities necessary for ADF-STEM. In the following, we describe TDS characteristics through the inelastic scattering model.

The imaginary potentials phenomenologically lead to loss of electrons and instead give rise to the TDS as a counter-part. Angular distribution of the TDS is well represented by the absorptive atomic form factor f' described as (Weickenmeier and Kohl 1991),

$$f'(s, M) \propto \int f(|\mathbf{s}'|) f(|\mathbf{s} - \mathbf{s}'|)[1 - \exp\{-2M(s'^2 - \mathbf{s} \cdot \mathbf{s}')\}] d^2 s', \qquad (7)$$

where f is the atomic form factor for elastic scattering. In the absorptive description, the form factor is given as a function of s ($=\sin\theta / \lambda$, θ is a scattering angle, λ is the electron wave-length) and M, which is a Debye–Waller (DW) factor defined by the mean-square thermal vibration amplitude, $<u^2>$, of the atoms. We note that expression (7) does not provide any fine details (Loane et al. 1991, Muller et al. 2001) of the TDS, such as Kikuchi lines indeed observed in Figure 14–12b. However, for ADF-STEM imaging we are concerned with the TDS intensity integrated over the annular detector, and hence the effect of the Kikuchi lines will become negligible when the detector covers scattering ranges sufficiently larger than the widths of the Kikuchi lines (Pennycook and Jesson 1990, 1991). Furthermore, the high-angle TDS is dominated by multi-phonon scattering, which makes an Einstein model of independently vibrating atoms valid enough for the estimation of the scattering intensity reaching to the detector. In these situations, Eq. (7) may be further simplified for high-angle scattering, and the TDS cross section (σ_{TDS}) can be derived as a high-angle approximation of f' (Pennycook and Jesson 1990, 1991),

$$f'_{HA}(M, s) \approx \sigma_{\mathrm{TDS}} \propto \int_{\mathrm{detector}} f^2(s) \left[1 - \exp(-2M \cdot s^2)\right] d^2\mathbf{s}, \qquad (8)$$

where the integration is carried out over the detector angle range. With this simplified cross-section description, each atom is supposed to be a δ-function source for TDS generation, whose intensity only depends on the detector angle range (δ-function approximation (Pennycook and Jesson 1990, 1991). Accordingly, $O(\mathbf{R})$ (Eq. (6)) for each atomic column can be directly related to the sum of σ_{TDS} within the relevant column, and the resultant image intensity ($I(\mathbf{R})$) after illumination by the probe ($P(\mathbf{R})$) is thus dependent on σ_{TDS}.

Since the σ_{TDS} is proportional to the square of $f(s)$ (Eq. (8)), the ADF-STEM is well-known for its atomic number-dependent contrast referred to as Z-contrast. Here, we pay attention to the fact that the σ_{TDS} (or $f'_{HA}(M, s)$) is a function of M; the TDS intensity in given detector ranges differs depending on the M values, as shown at the upper-right hand side in Figure 14–11. This means that ADF-contrast is also sensitive to the DW factors at individual atomic sites (columns). With this in mind, we attempt to detect a local thermal vibration anomaly in the quasicrystal through the M-sensitive ADF-STEM imaging.

14.3.3 Local Debye–Waller Factor Anomaly

As described in the legend to Figure 14–10, the hyperspace lattice is decorated by the so-called *atomic surfaces* that represent the hyperspace potentials. In order to convert these hypothetic potentials finally into the atomic arrangement in the real quasicrystal, the atomic surfaces will be divided into small partitions, each of which is allocated to the individual atomic sites of the quasicrystal (Yamamoto 1996) (details will be described later with Figure 14–18b). Given that the relevant atomic surfaces are discrete in form (Figure 14–10, right), a phason fluctuation causes the edges of the atomic surfaces to be smeared. This means that the phason effects appear to be not equivalent for all the atomic sites, but localized at the particular sites related to the edges of the atomic surfaces. Therefore, the phason fluctuation may cause a local anomaly of the DW factor at the corresponding real-space atomic sites. Note that this phasonic perturbation does not destroy the long-range quasiperiodic order, but causes a reduction in the Bragg intensity and generates some additional diffuse scattering, in the same manner as the TDS generation.

The phason has no counterpart in periodic crystals, and hence the DW factor of a quasicrystal may have both phonon and phason contributions (Ishii 1992, Jaric and Nelson 1988), written as

$$M = M_{\text{phonon}} + M_{\text{phason}} \tag{9}$$

Below we describe successful direct observations of DW factor anomalies at the specific atomic sites in a quasicrystal (Abe et al. 2003), based on in situ high-temperature as well as angle-resolved STEM experiments.

In the ADF-STEM image of the highly perfect decagonal $Al_{72}Ni_{20}Co_8$ taken at room temperature (300 K: Figure 14–13a), Z-contrast highlights the transition metal (TM: Ni or Co) positions relative to the Al due to

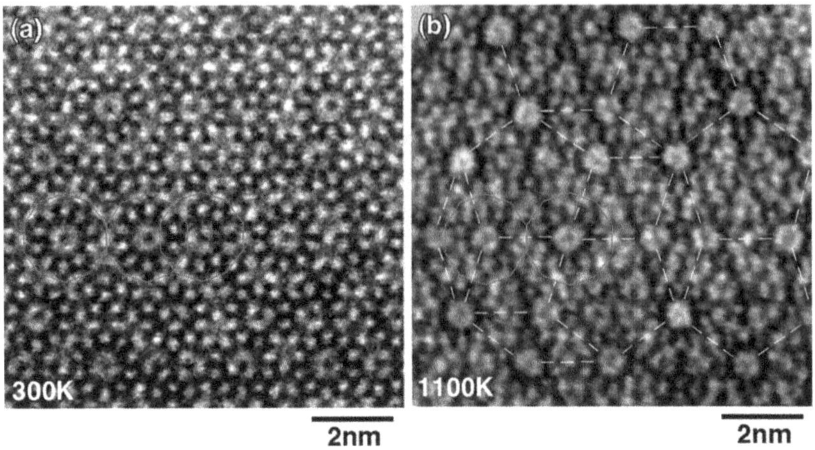

Figure 14–13. ADF-STEM images of the decagonal $Al_{72}Ni_{20}Co_8$. These were taken at (a) 300 K and (b) 1100 K by collecting the electrons scattered at angles approximately between 45 and 100 mrad ($0.9 \leq s \leq 2.0$) with a 200 kV-STEM. Contrast differences between (a) and (b) are due to the different temperatures. By connecting the center of the 2 nm decagonal clusters (*red*) that reveal significant temperature-dependent contrast changes, a pentagonal quasiperiodic lattice (*yellow*) with an edge length of 2 nm can be drawn in (b). Adapted from Abe et al. (2003), with permission.

the $f^2(s)$ dependence of the contrast (Eq. (8)). But when the sample is heated and held at a temperature of approximately 1100 K within the microscope, we find a remarkable change in the relative contrast, see Figure 14–13b compared to Figure 14–13a. A significant enhancement in the image intensity appears at some specified places, which can be well represented by the pentagonal Penrose tiling with an edge-length of 2 nm (note that the pentagonal Penrose tiling is uniquely related to Gummelt's overlap tiling (Figure 14–7b) by connecting the specified decagons separated by 2 nm). Considering that the present highly perfect quasicrystalline phase is obtained as a quenched-in high-temperature configuration (water-quenched after annealing at 1100 K), the image of Figure 14–13b is showing the "true face" of the decagonal $Al_{72}Ni_{20}Co_8$, in the sense that it is indeed at its equilibrium situation. Interestingly, such contrast anomaly is also observed even at room temperature when the ADF image is formed by relatively low-angle scattering, as shown in Figure 14–14.

These angular-dependent as well as temperature-dependent anomalous ADF contrast effects can naturally be attributed to local anomalies of the DW factor, as expected through Eq. (8). In the images of Figures14–13 and 14–14, it is found that the DW-factor contrast occurs at cores of the decagonal clusters; representative angular-dependent and temperature-dependent features are summarized in Figure 14–15a–d. Viewing carefully the interior contrast of the clusters, significant increase in intensities occurs at the positions indicated by arrowheads, which correspond to the Al sites by referring to the structural model (Figure 14–15e). The intensity profiles confirm that these Al sites,

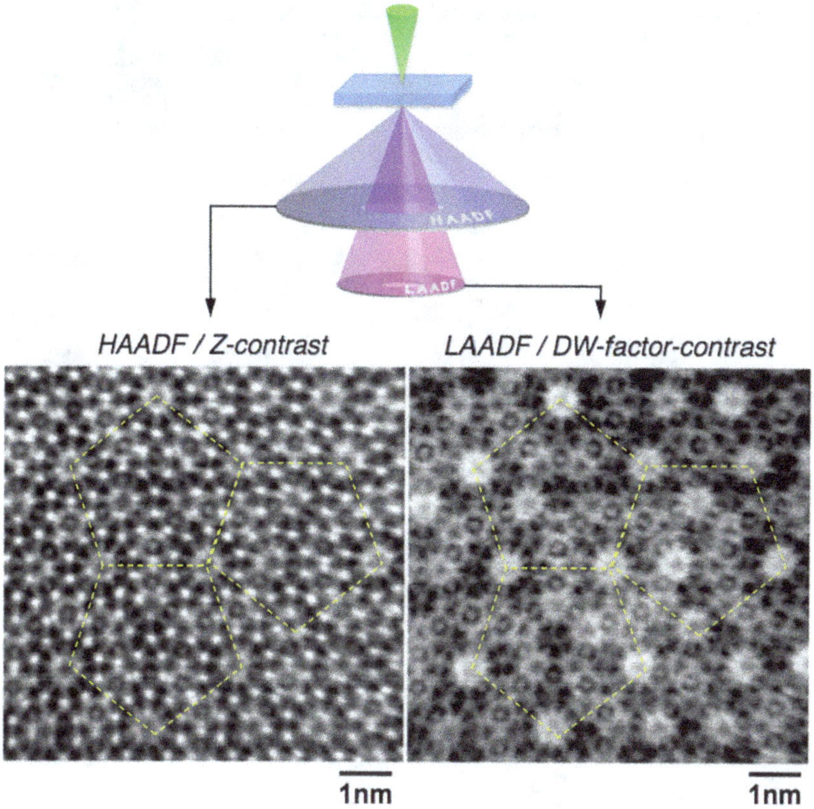

Figure 14–14. ADF-STEM images of the decagonal $Al_{72}Ni_{20}Co_8$, taken with an aberration-corrected 100 kV-STEM (Cs-corrected VG501). These images are simultaneously obtained with the different annular detectors, which covered high-angle scattering (HAADF: inner-angle ~50 mrad) and low-angle scattering (LAADF: inner-angle ~30 mrad), respectively. Occurrences of anomalous contrast seen in LAADF (*right*) are formed in the same manner as those observed during the in situ high-temperature observation (Figure 14–13b), as represented by the 2-nm-scale pentagonal quasiperiodic lattice (*yellow*).

denoted as Alα, in fact reveal stronger intensity at 1100 K than that at 300 K. It is also noticed that at the room temperature condition, the intensity at the Alα sites ($I_{Al\alpha}$) is originally stronger than that of the other Al sites (Figure 14–15b). Besides, as confirmed by Figure 14–15a–c, the $I_{Al\alpha}$ shows significant scattering angle dependence, while the intensities of the other Al and TM atoms do not (I_{TM}/I_{Al} is almost constant). On these bases, the local DW factor anomaly is predominantly attributed to the Alα sites; if the DW factor effect is equivalent for all the Al sites, neither the temperature dependence nor angle dependence of the $I_{Al\alpha}$ is observed.

On the basis of the above discussions, the significant increase in the $I_{Al\alpha}$ can be interpreted in terms of atomic vibration amplitudes, $\langle u^2 \rangle$, for the observations at both 300 and 1100 K. Before describing details of the DW effects, it is worth remembering that electron scattering is, in principle, not able to distinguish whether the displacement is due

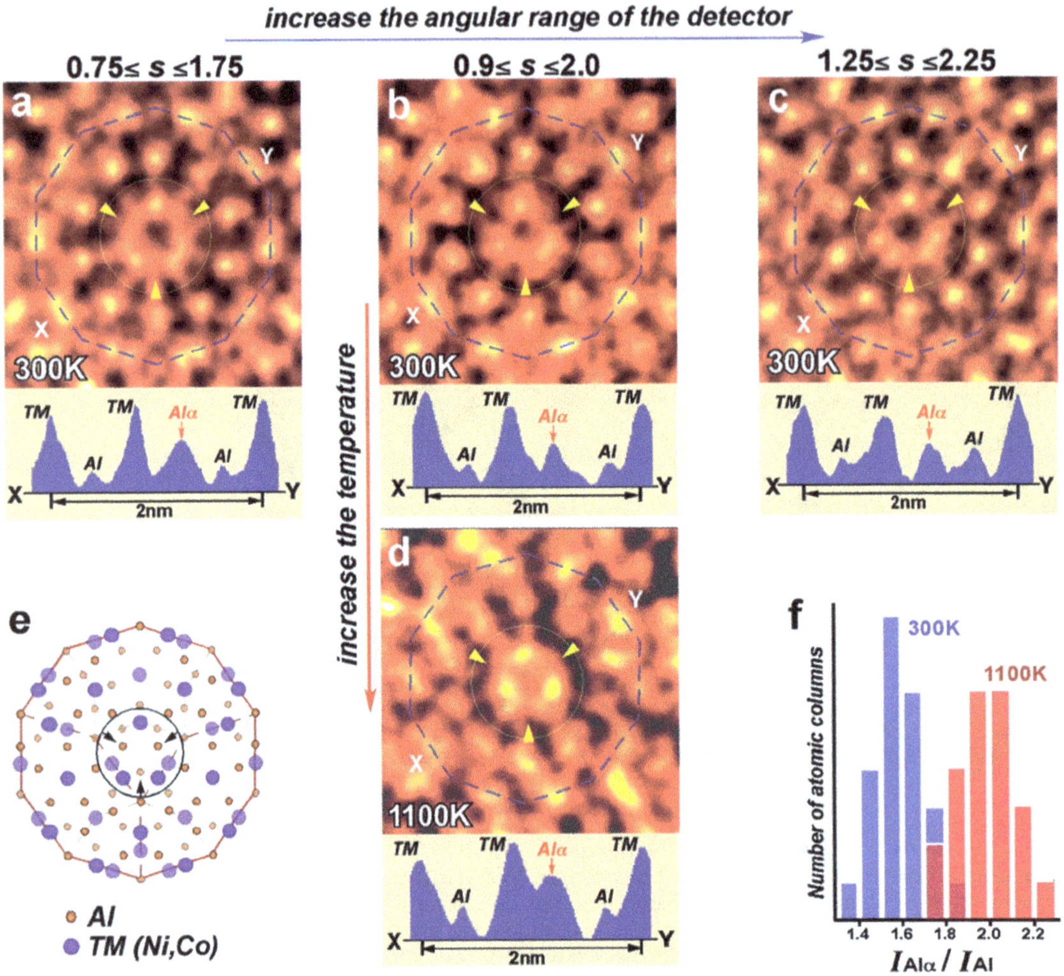

Figure 14–15. Decagonal cluster of about 2 nm across, a structural unit of the $Al_{72}Ni_{20}Co_8$. The clusters observed (a–c) at 300 K with different angular ranges of the detector and (d) at 1100 K, together with the corresponding line-profiles across X–Y. Images (a) and (c) were obtained by the 300 kV STEM with detector angle ranges of about 30–70 mrad ($0.75 < s < 1.75$) and 50–90 mrad ($1.25 < s < 2.25$), respectively. Images of (b) and (d) are enlargements of a part of Figure 14–13a, b, respectively. Viewing carefully the images (a–d), one notices that the contrast changes significantly at the core of the cluster, particularly at the positions indicated by arrowheads, which correspond to the Alα sites denoted in the line-profiles. (e) Structural model of the $Al_{72}Ni_{20}Co_8$, derived by averaging all possible atomic positions in the perfect order model (same as that shown in Figure 14–8). *Bold* and *semi-transparent dots* denote the atoms at layers $c = 0$ and $c = 1/2$, respectively. (f) Histograms of intensity ratios between the Alα and Al atomic columns, $I_{Al\alpha}/I_{Al}$, measured for the decagonal clusters located at the 2-nm-scale pentagonal Penrose lattice (Figure 14–13a, b). Adapted from Abe et al. (2003), with permission.

to dynamic motion or static distortions, see Figure 14–16. In the STEM imaging of individual atomic columns, mean broadening of their projected potentials is directly related to the M values in Eq. (7) and hence affect the ADF intensity. Since the electron scattering occurs much faster ($> \sim 10^{-15}$ s) than the thermal vibration period (an order of $\sim 10^{-13}$ s), even the dynamically vibrating atoms are recognized as stationary

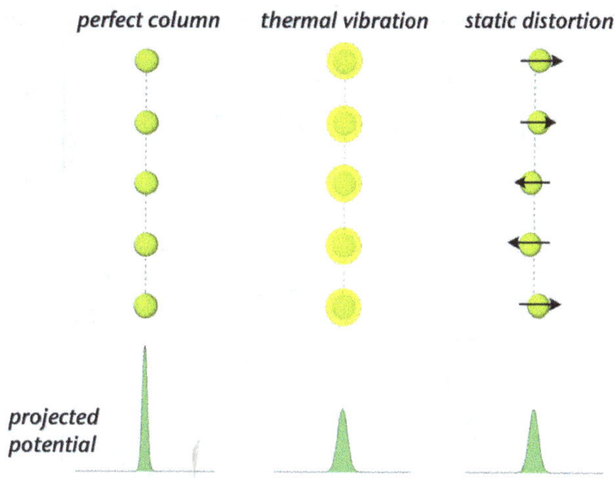

Figure 14–16. Schematic illustrations of a *mean* Debye–Waller factor of the atomic columns.

distorted columns (Figure 14–16, right) by electrons at each scattering event (Loane et al. 1991) (i.e., the dynamical potential broadening (Figure 14–16, center) can be described by time-averaging of a large number of distorted configurations). Because of this, both the thermally diffused and the statically distorted atomic columns are reasonably described by the relevant DW factor, M, whose values are directly related with the full-width half-maximum of the projected potentials of the columns.

Now we come back to the issue of the DW factor anomaly in the $Al_{72}Ni_{20}Co_8$ quasicrystal. It is natural to assume that the present anomalous DW factor at the Alα site is not due to static column distortions but mostly dynamically vibrating effects, since the $I_{Al\alpha}$ indeed reveals significant temperature dependence. Suppose that the static effects are dominant for the anomalous $I_{Al\alpha}$ at 300 k; then, static transverse displacements of the Alα atoms cause relaxations for the nearest-neighbor TM atoms. In this case, these TM atoms are also distorted and expected to reveal anomalous contrast compared to the other TM atoms; however, they do not show any significant angle-dependent contrast change ($I_{\rm TM}/I_{\rm Al}$ is almost constant in Figure 14–15a–c). Note that $I_{\rm TM}/I_{\rm Al}$ does not change remarkably even after elevating the temperature (Figure 14–15d), indicating that the temperature effects on the DW factor of these Al and TM atoms are almost equivalent and small as compared with that of the Alα. Thus, we normalize the $I_{Al\alpha}$ with reference to the I_{Al} at each temperature, and Figure 14–15f shows the histograms of $I_{Al\alpha}/I_{Al}$ distributions, in which the $I_{Al\alpha}/I_{Al}$ peaks at around ~1.5 and ~2.0 for the observations at 300 K and 1100 K, respectively.

Both temperature- and angular-dependences of the $I_{Al\alpha}$ are fairly well explained by the $\sigma_{\rm TDS}$ for different $\langle u^2\rangle$ values, see the $\sigma_{\rm TDS}^{Al}$–$\langle u^2\rangle$ curves calculated based on Eq. (8) (Figure 14–17). In Figure 14–17, the bold curve ($0.9 \leq s \leq 2.0$) explains the temperature-dependent change by assuming the appropriate $\langle u^2\rangle$ values for the standard Al (~0.4×10^{-4} nm^2; green line), Alα at 300 K (~0.9×10^{-4} nm^2; green line),

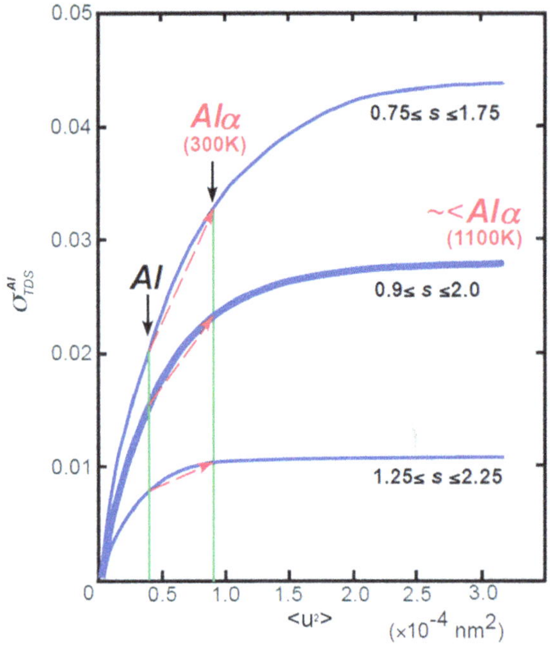

Figure 14–17. Estimates of the thermal diffuse scattering cross section of Al ($\sigma_{\mathrm{TDS}}^{Al}$ defined by Eq. (8)) for several detector angle ranges, as a function of mean-square thermal vibration amplitude ($<u^2>$). These are calculated using the atomic form factor ($f(s)$) in Weickenmeier and Kohl (1991), by assuming an isotropic Debye–Waller factor ($M = 8\pi^2 < u^2>$). DW factor of the standard Al is assumed to be not temperature dependent. Adapted from Abe et al. (2003), with permission.

and Alα at 1100 K ($>\sim 1.0\times 10^{-4}$ nm^2); these generate the $\sigma_{\mathrm{TDS}}^{Al\alpha}/\sigma_{\mathrm{TDS}}^{Al}$, relevant to $\boldsymbol{I}_{Al\alpha}/\boldsymbol{I}_{Al}$, as being approximately ~1.5 and ~2.0 for 300 and 1100 K, respectively. We note that these $<u^2>$ values also explain well the angle-dependent changes of $\mathbf{I}_{Al\alpha}/\mathbf{I}_{Al}$ observed at 300 K, see red arrows in Figure 14–17. Although the $<u^2>$ at 1100 K cannot be determined because σ_{TDS} saturates at large $<u^2>$ values, we nevertheless emphasize that the observed anomalous contrast at the Alα site can be clearly correlated with differences in $<u^2>$, demonstrating the first direct observation of a local vibration anomaly in a solid.

We here note that the $<u^2>$ values estimated above are based on the simplified δ-function approximation (σ_{TDS}) and hence may be regarded as only tentative. More accurate $<u^2>$ values may be available by fitting the observed intensity with those estimated based on a full absorptive potential f'_{HA} (M, s), which can be incorporated into a dynamical diffraction calculation (e.g., multi-slice simulations (Ishizuka 2002)). For such purpose, a significant signal-to-noise ratio improvement of the experimental dataset is essential at both room- and elevated-temperatures, and perhaps a state-of-the-art aberration-corrected STEM is capable of meeting this challenge. In any case, although the present $<u^2>$ values in Figure 14–17 are semi-quantitative, it is noteworthy that significantly large $<u^2>$ at high temperature is consistent with the X-ray diffuse scattering measurement of the same sample of $Al_{72}Ni_{20}Co_8$, in which the overall $<u^2>$ is suggested to be on the order of $\sim 1.0\times 10^{-3}$ nm^2 (at 1100 K) when compared with a uniform harmonic vibration (Abe et al. 2003) (the local anomalies of $<u^2>$ presently observed may imply significant anharmonicity at the Alα site). Further supporting evidence comes from molecular dynamics simulations of the structure at 1000 K (Henley et al. 2002), which predicted the occurrence of DW factor anomalies for the Al atoms at the core of the 2 nm decagonal cluster.

The present observation of the local DW factor anomaly implies some intriguing physics of quasicrystals. It is quite interesting to note that the Alα atoms, located at the center of the decagonal clusters that are on the 2-nm-scale pentagonal quasiperiodic lattice, are shown to be generated from the edge of the atomic surfaces (α in Figure 14–18a) within the hyperspace structural description (Takakura et al. 2001). Therefore, the present local DW factor anomaly can be attributed to significant fluctuations of the atomic surfaces in specific edge regions, which, in turn, indicate an occurrence of phasonic fluctuations realized at the relevant Alα sites in the real quasicrystal. This phenomenon is reasonably interpreted as a perturbation of quasiperiodic order that can be described through M_{phason} (Eq. (9)). It is worthwhile to note that some Al atoms within the 2 nm cluster are also placed in similar local environment to that of the Alα atoms—see the "kite" tiles drawn in Figure 14–15e. However, they are symmetrically (crystallographically) not equivalent to the Alα site and hence may not reveal any significant anomalies (i.e., atomic configuration in the kite tiles can be slightly different depending on their local-neighbor environment). The occurrence of DW factor anomalies at the Alα site is probably induced by the presence of the phason-flip atomic sites, denoted as β in Figure 14–18, which are separated by less than a typical interatomic distance. These α and β sites cannot be occupied simultaneously, and the β sites could act as vacancies by providing an effective space for relaxation. Consequently, it is reasonably presumed that the Alα atoms reveal a significant anisotropic DW factor, as illustrated in Figure 14–18b, although the STEM resolution (~1.5 Å) was not sufficient to check directly this anisotropic shape. Molecular-dynamics structural simulations at 1000 K provided supporting evidence (Takakura et al. 2001), which indicates an anisotropic DW factor identical to that shown in Figure 14–18b.

We have demonstrated that the quasicrystal may intrinsically possess a phason degree of freedom based on successful STEM observations of the real structure of $Al_{72}Ni_{20}Co_8$ (Figure 14–19). The idea of extra dimensions is a compelling, universal issue in physics (Randall 2007), and the phason in quasicrystals is definitely one of the relevant issues.

14.4 Point Defects in Quasicrystals

In the highly perfect quasicrystalline $Al_{72}Ni_{20}Co_8$, we indeed find phason-related localized fluctuations, which are observed as a DW factor anomaly at the particular Al sites. Under the high-temperature equilibrium condition, it is natural to consider that the local vibration anomalies enhance the short-range diffusion of atoms located around the cluster center with the help of a phason-site–mediated transport mechanism, causing occasional diffusion jumps of not only Al but also TM atoms into the phason-related site. When these local atomic fluctuations show long-range correlations, phason dynamics modes may well be characterized by long wavelengths of the order of ~100 nm

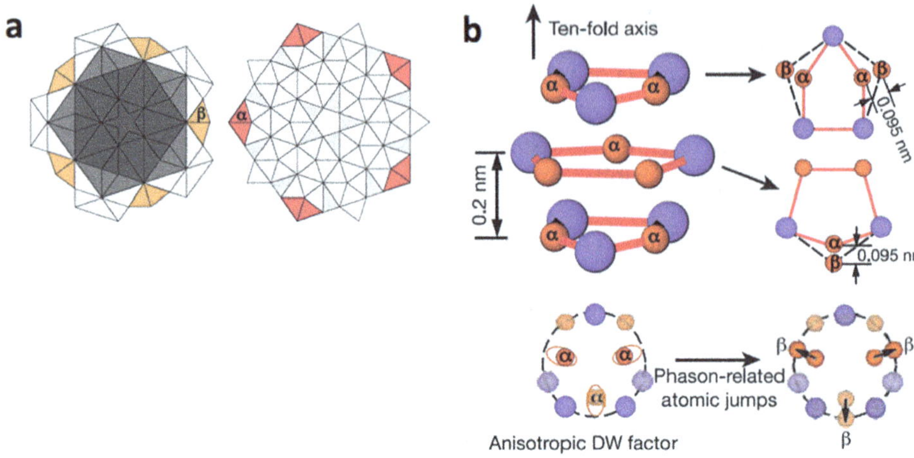

Figure 14–18. (a) Hyperspace description of the decagonal $Al_{72}Ni_{20}Co_8$ structure. Atomic surfaces placed at (1/5, 1/5, 1/5, 1/5, 1/4) (*left*) and (2/5, 2/5, 2/5, 2/5, 1/4) (*right*) in the five-dimensional decagonal lattice with successful partitions assigned by Al and TM, where the gray level represents the concentration of the TM atoms at relevant sites (see Takakura et al. (2001) for details). Extra β portions (*yellow*) are added to the edges of the atomic surfaces to generate phason-flip atomic sites with respect to the α sites. (b) Pentagonal columnar atomic configuration around the center of the 2 nm cluster (encircled region of Figure 14–15e). Phason-related α and β sites are on the quasiperiodic atomic plane, separated by approximately 0.95 Å. Possible anisotropy of the Debye–Waller factor of the Al atoms at the α sites and occurrence of atomic jumps into the β sites are described in the projected atomic positions shown below, where the semitransparent atoms represent those at a different level along the tenfold axis. Reproduced from Abe et al. (2003), with permission.

Figure 14–19. Experimental approaches to the quasicrystal-hypercrystal structures: diffraction versus microscopy. Direct observations of a real quasicrystal structure enable it to be converted into the relevant hyper-dimension structure, an entirely opposite way to diffraction analysis. This provides an interesting testing ground whether or not there exists an intrinsic phason degree of freedom in the quasicrystal.

(Francoual et al. 2003) and slow relaxation times, several tens of seconds (Edagawa et al. 2000, Francoual et al. 2003) due to the diffusive nature of a phason excitation. The diffusive phason modes are believed to be truly typical of quasiperiodic order. In this sense, the local thermal vibration anomaly (anharmonicity?) may play a role in precursory phenomena that systematically connect fast vibration motion (Coddens et al. 1999) with slow diffusional dynamics (Francoual et al. 2003).

With this scenario of diffusion-based phason dynamics, each of the mode-driven local configurations at a given moment can be quenched-in due to its slow diffusive nature. Accordingly, the phasonic atomic motions are expected to leave traces of localized point defects at particular atomic sites, which appear as chemical and occupational disorders in a sample quenched from the equilibrium high-temperature.

Therefore, quantitative evaluation of these phason-related point defects becomes an intriguing issue to understand the local, primary origin of the diffusive phason modes (if any). For this purpose, we use an aberration-corrected STEM to identify the local atomic occupations at individual atomic sites in the decagonal $Al_{72}Ni_{20}Co_8$, and here we briefly describe preliminary results to demonstrate its promising performance.

14.4.1 Imaging with Aberration-Corrected STEM

STEM incoherent imaging provides an ideal intensity map that can be directly inverted to the specimen structure (Eq. (6)), with significantly less artifacts compared with phase-contrast imaging (Pennycook and Jesson 1990). As described earlier, incoherent imaging becomes possible only when the ADF detector is set to a sufficiently high-angle scattering range where the detector captures mostly thermal diffuse scattering (TDS). TDS intensity is significantly dependent on the atomic number Z (Eq. (8)), and therefore the Z-contrast is an unavoidable consequence of incoherent imaging.

While Z-contrast STEM led to remarkable progress for the quasicrystal structure analysis (Abe et al. 2004) by selectively highlighting the (relatively) heavy atom positions, it was quite difficult to detect the light atom position and fractionally occupied weak sites simultaneously. Recently, the spherical aberration (Cs) correction (Haider et al. 1998) of the objective lens has been successful in converging the beam into the sub-Å scale (Batson et al. 2002, Dellby et al. 2001), providing a much sharper/brighter electron probe (P (**R**)) and hence improving remarkably the STEM resolution (Nellist et al. 2004, Varela et al. 2005).

Figure 14–20 shows incoherent Z-contrast images of the decagonal $Al_{72}Ni_{20}Co_8$, obtained before and after the Cs-correction, respectively. The improved resolution can be immediately noticed for the brightest dots representing the Ni/Co atoms, which appear to be remarkably sharper than those in the non-Cs-corrected image. In addition, by looking carefully at the Cs-corrected image, many weak dots also appear between the brightest dots; these are reasonably attributed to the Al sites, the smallest Z constituent in the $Al_{72}Ni_{20}Co_8$ compound. This is due to an improved signal-to-noise ratio with increasing probe current and an effective suppression of secondary-maxima that existed in the uncorrected probe tails (see Figure 14–20 bottom), allowing the light atom positions, even those sitting nearby to heavy atom sites, to be visible. Details are clearly seen in the enlarged images in Figure 14–21, where the improved resolution made possible by the Cs-correction is also demonstrated for different accelerating voltages. In Figure 14–21, the Al sites are hardly recognized in the non-Cs-corrected image, but they gradually emerge with the higher resolution provided by increasing the voltage; a large number of weak sites are distinctly detected by the best resolution of 300 kV-STEM. It is obvious from Figure 14–20 that even a small difference of the achievable resolution (with ideal conditions, a full-width at half-maximum of $P(\mathbf{R})$ can be calculated as ~0.8 and ~0.5 Å for 200 and 300 kV, respectively) certainly maximizes the

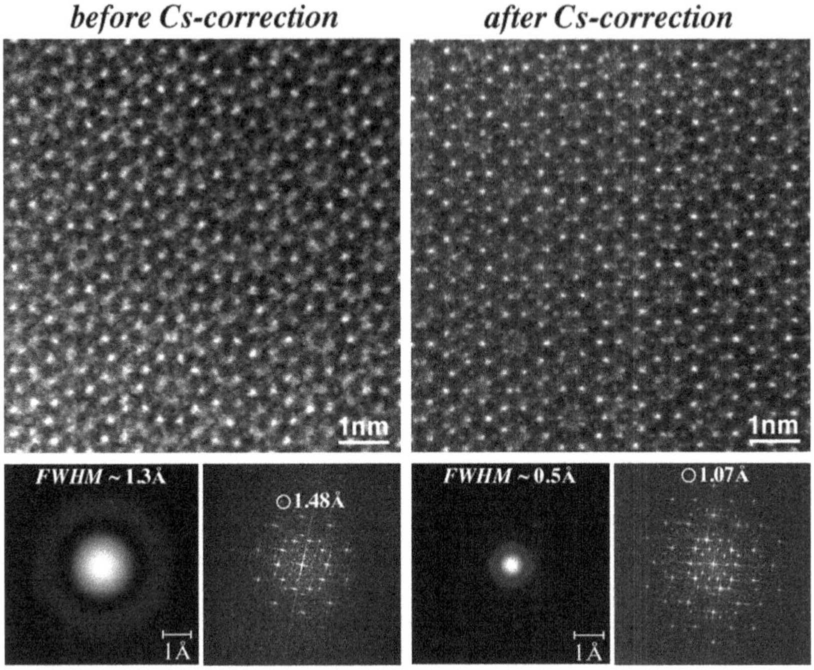

Figure 14–20. Atomic-resolution Z-contrast images of the decagonal quasicrystal $Al_{72}Ni_{20}Co_8$ taken by 300 kV-STEM (VG-HB603U) before (*left*) and after (*right*) the aberration correction. Inner-angle of the detector was set approximately at 40 mrad for both images. The corresponding calculated probe functions ($P(\mathbf{R})$) and the Fourier transform patterns are shown at the *bottom*.

structure information of the complex quasicrystal, for which both the strong and weak atomic sites are densely distributed in the projected structure.

Within the incoherent approximation represented by Eq. (6), atomic structures, $O(\mathbf{R})$, can be directly addressed by a simple deconvolution procedure using $P(\mathbf{R})$ known for given equipment parameters. Deconvolution using a maximum entropy (ME) algorithm gives a safest, least possible structure that fits the experimental image. Figure 14–22 shows the result after ME-deconvolution on the Cs-corrected STEM image (Figure 14–20), allowing the weak spots to emerge clearer by effective reduction of background noise. Here it is noteworthy that even applying the ME-deconvolution, the weak peaks never showed up from the non-Cs-corrected STEM image. Only when sufficient signal-to-noise ratio is available with the Cs-corrected sharper/brighter probe, does the ME-deconvolution work effectively to reveal the weak peaks that represent relatively light atomic sites (Taniguchi and Abe 2008).

The Cs-corrected ultrahigh-resolution STEM is now able to reveal both the heavy and light atom positions simultaneously even under the incoherent Z-contrast condition. From a direct comparison of the experimental (ME-deconvoluted) image with the simulation based on

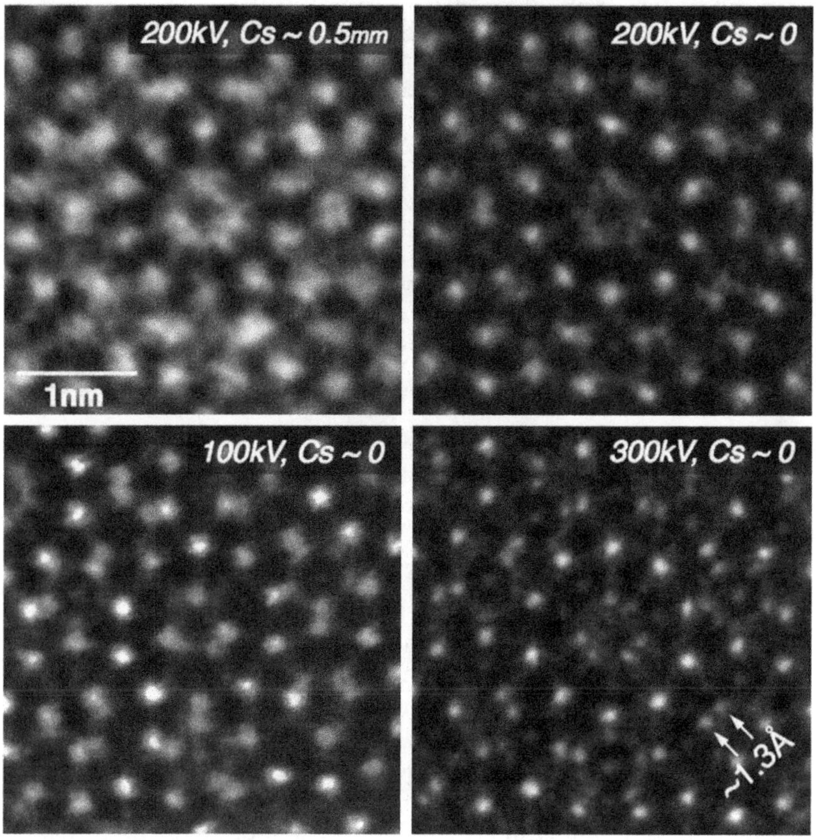

Figure 14–21. Enlarged views of the 2 nm cluster of the decagonal $Al_{72}Ni_{20}Co_8$ obtained by several types of STEM with different accelerating voltages: non-Cs-corrected 200 kV-STEM image (*top left*, JEM-2010F with Cs~0.5 mm) and Cs-corrected STEM images at 100 kV (*bottom left*, VG-501), 200 kV (*top right*, JEM-2100F), and 300 kV (*bottom right*, VG-HB603U). The dumbbell of Ni/Co sites separated by approximately 1.3 Å, similar to the Si dumbbell in the [110] projection, is distinctly resolved in the 300 kV-STEM image, as indicated by *arrows*.

the perfect ordered model (Figure 14–22), it is obvious that significant point defects occur even in the highly ordered decagonal $Al_{72}Ni_{20}Co_8$. One immediately notices that a large number of extra weak peaks occur in the ME-deconvoluted experimental map, including the phason-flip Al sites separated by 0.95 Å (α–β in Figure 14–18b) such as indicated by arrows in Figure 14–22. Note that these phason sites are fractionally occupied by nature (these sites are too close to be occupied simultaneously), demonstrating that such weak potential sites can even be detectable. Furthermore, looking carefully at the intensities of individual atomic sites, they are definitely not limited to just two types of columns, i.e., fully occupied either with Al or with TM atoms in the perfect order model (Figure 14–22, right). This suggests that there indeed occurs chemical disorder (Al/TM mixed occupations) at a considerable number of atomic sites.

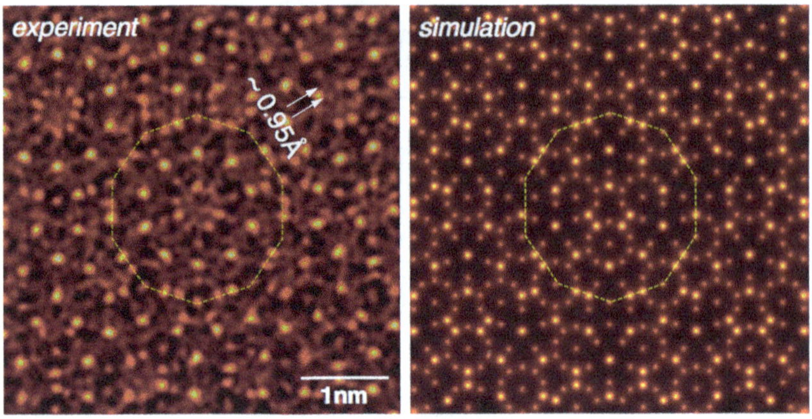

Figure 14–22. (*Left*) Maximum-entropy deconvolution map obtained from the Cs-corrected STEM image of Figure 14–20. (*Right*) Projected potential (σ_{TDS} in Eq. (8)) distributions calculated based on the perfect order model of the decagonal $Al_{72}Ni_{20}Co_8$ ("perfect order" means that there is neither fractionally occupied site nor chemically mixed Al/TM sites in the structure).

Taking advantages of the amplitude-sensitive Z-contrast image, we are attempting quantitative evaluations of the atomic occupation at individual atomic sites based on intensity-fitting. A preliminary finding is that the occurrence of point defects, well-localized around the DW factor anomaly sites, seems to be governed by the underlying local symmetry of the Penrose tiling. These remarkable defect configurations can be possibly interpreted as due to a phasonic perturbation which, in turn, implies an underlying "phason mode." We are now able to draw some intriguing physics based on the state-of-the-art electron microscopy observations. Details will soon be described in a forthcoming article.

14.5 Summary

The discovery of quasicrystals provided a paradigm shift in solid state physics since it had long been assumed, though never strictly proven, that the best and most stable long-range order should be realized in the form of a periodic solid constructed by repeating unit cells. Quasicrystals are now established as a well-ordered form of solids, and we have made significant progress in understanding their microscopic details based on STEM observations, including some answers to the key question "Where are the atoms?" Further veiled insights into this unique solid promise to be uncovered by aberration-corrected STEM, which now provides remarkable performance not only for imaging but also for spectroscopy; elucidating local electronic states will certainly lead to deeper understanding of why quasicrystals form.

Acknowledgments I would be grateful to A. P. Tsai, S. J. Pennycook, K. Saitoh, H. Takakura, P. J. Steinhardt, H.-C. Jeong, and Y. Yan for collaborations, on which the present chapter is based. I would also thank T. J. Sato, M. Widom,

C. L. Henley, M. Miharcovic, W. Steurer, M. de Boissieu, A. Yamamoto, N. Tanaka, and K. Ishizuka for valuable comments and discussions. I acknowledge support from the CREST-JST and SORST-JST (1996–2001, 2002–2007 Project leader: A. P. Tsai), a Grant-in-Aid for Scientific Research of Priority Areas "Atomic Scale Modification" from the Ministry of Education, Science, Sports and Culture of Japan, and by Kazato Research Foundation.

References

H. Abe, Y. Matsuo, H. Saitoh, T. Kusawake, K. Ohshima, H. Nakao, Jpn. J. Appl. Phys. **39**, L1111 (2000)
E. Abe, S.J. Pennycook, A.P. Tsai, Nature **421**, 347 (2003)
E. Abe, K. Saitoh, H. Takakura, A.P. Tsai, P.J. Steinhardt, H.-C. Jeong, Phys. Rev. Lett. **84**, 4609 (2000)
H. Abe, H. Saitoh, T. Ueno, H. Nakao, Y. Matsuo, K. Ohshima, H. Matsumoto, J. Phys. Soc. Jpn. **72**, 1828 (2003)
E. Abe, H. Takakura, A.P. Tsai, J. Electron. Microsc. **50**, 187 (2001)
E. Abe, Y. Yan, S.J. Pennycook, Nat. Mater. **3**, 759 (2004)
P. Bak, Phys. Rev. Lett. **54**, 1517 (1985)
P. Bak, Phys. Rev. Lett. **56**, 861 (1986)
P.A. Bancel, Phys. Rev. Lett. **63**, 2741 (1989)
P.E. Batson, N. Dellby, O.L. Krivanek, Nature **418**, 617 (2002)
C. Beeli, S. Horiuchi, Phil. Mag. B **70**, 215 (1994)
L. Bendersky, Phys. Rev. Lett. **55**, 1461 (1985)
D.M. Bird, Q.A. King, Acta Cryst. A **46**, 202 (1990)
S. Burkov, Phys. Rev. Lett. **67**, 614 (1991); J. Phys. **2**, 695 (1992)
G. Coddens, S. Lyonnard, B. Hennion, and Y. Calvayrac, Phys. Rev. Lett. **83**, 3226 (1999)
G. Coddens, W. Steurer, Phys. Rev. B **60**, 270 (1999)
R. Colella, Y. Zhang, J.P. Sutter, S.N. Ehrlich, S.W. Kycia, Phys. Rev. B **63**, 014202-1 (2000)
R.A. Cowley, A.D. Bruce, J. Phys. C **11**, 3577, 3591, 3609 (1978)
M. de Boissieu, M. Boudard, B. Hennion, R. Bellissent, S. Kycia, A. Goldman, C. Janot, M. Audier, Phys. Rev. Lett. **75**, 89 (1995)
N. Dellby, O.L. Krivanek, P.D. Nellist, P.E. Batson, A.R. Lupini, J. Electron. Microsc. **50**, 177 (2001)
J. Dolinsek, T. Apih, M. Simsic, J.M. Dubois, Phys. Rev. Lett. **82**, 572 (1999)
B. Dubost, J.-M. Lang, M. Tanaka, P. Sainfort, M. Audier, Nature **324**, 48 (1986)
K. Edagawa, K. Suzuki, S. Takeuchi, Phys. Rev. Lett. **85**, 1674 (2000)
I.R. Fisher, Z. Islam, A.F. Panchula, K.O. Cheon, M.J. Kramer, P.C. Canfield, A.I. Goldman, Phil. Mag. B **77**, 1601 (1998)
I.R. Fisher, M.J. Kramer, Z. Islam, A.R. Ross, A. Kracher, T. Wiener, M.J. Sailer, A.I. Goldman, P.C. Canfield, Phil. Mag. B **79**, 425 (1999)
S. Francoual, F. Livet, M. de Boissieu, F. Yakhou, F. Bley, A. Létoublon, R. Caudron, J. Gastaldi, Phys. Rev. Lett. **91**, 225501 (2003)
T. Goedecke, M, Scheffer, R. Luck, S. Ritsch, C. Beeli, Z. Metallkd. **89**, 687 (1998)
P. Gummelt, Geometriae Dedicata 62, 1 (1996)
M. Haider, S. Uhlemann, E. Schwan, H. Rose, B. Kabius, K. Urban, Nature **392**, 768 (1998)
H. Hashimoto, A. Howie, M.J. Whelan, Proc. R. Soc. London Ser. A, **269**, 80 (1962).
C.L. Henley, in *Quasicrystals: The State of the Art*, ed. by D. DiVincenzo, P.J. Steinhardt (World Scientific, Singapore, 1991) pp. 429–524
C.L. Henley, M. Mihalkovic, M. Widom, J. Alloys Compd. **342**, 221 (2002)

K. Hiraga, in *Advances in Imaging and Electron Physics*, ed. by P.W. Hawkes (Academic, London, 1998), pp. 37–98
K. Hiraga, T. Ohsuna, W. Sun, K. Sugiyama, Mater. Trans. **42**, 2354 (2001)
W. Hume-Rothery, J. Inst. Met. **36**, 295 (1926)
Y. Ishii, Phys. Rev. B **45**, 5228 (1992)
K.N. Ishihara, A. Yamamoto, Acta Crystallogr. A **44**, 508 (1988)
K. Ishizuka, Ultramicroscopy **90**, 71 (2002)
T. Janssen, Acta Crystallogr. A **42**, 261 (1986)
M.V. Jaric, D.R. Nelson, Phys. Rev. B **37**, 4458 (1988)
H.-C. Jeong, P.J. Steinhardt, Phys. Rev. B **48**, 9394 (1993)
D. Joseph, S. Ritsch, C. Beeli, Phys. Rev. B **55**, 8175 (1997)
D. Levine, T.C. Lubensky, S. Ostlund, S. Rawaswamy, P.J. Steinhardt, Phys. Rev. Lett. **54**, 1520 (1985)
D. Levine, P.J. Steinhardt, Phys. Rev. Lett. **53**, 2477 (1984)
R.F. Loane, P. Xu, J. Silcox, Acta Cryst. A **47**, 267 (1991)
W. Ohashi, F. Spaepen, Nature **330**, 555 (1987)
M. Mihalkovic, I. Al-Lehyani, E. Cockayne, C.L. Henley, N. Moghadam, J.A. Moriarty, Y. Wang, M. Widom, Phys. Rev. B **65**, 104205 (2002)
D.A. Muller, B. Edward, E.J. Kirkland, J. Silcox, Ultramicroscopy **86**, 371 (2001)
P.D. Nellist, M.F. Chisholm, N. Dellby, O.L. Krivanek, M.F. Murfitt, Z.S. Szilagyi, A.R. Lupini, A. Borisevich, W.H. Sides Jr., S.J. Pennycook, Science **305**, 1741 (2004)
L. Pauling, Nature **317**, 512 (1986)
S.J. Pennycook, L.A. Boatner, Nature **336**, 565 (1988)
S.J. Pennycook, D.E. Jesson, Ultramicroscopy **37**, 14 (1991); Phys. Rev. Lett. **64**, 938 (1990)
R. Penrose, Bull. Inst. Math. App. **10**, 266 (1974)
L. Randall, Phys. Today **60**, 80 (2007)
S. Ritsch, C. Beeli, H.-U. Nissen, T. Godecke, M, Scheffer, R. Luck, Phil. Mag. Lett. **74**, 99 (1996)
S. Ritsch, C. Beeli, H.-U. Nissen, T. Godecke, M, Scheffer, R. Luck, Phil. Mag. Lett. **78**, 67 (1998)
K. Saitoh, K. Tsuda, M. Tanaka, K. Kaneko, A.P. Tsai, Jpn. J. Appl. Phys. **36**, L1400 (1997)
D. Shechtman, I. Blech, D. Gratias, J.W. Cahn, Phys. Rev. Lett. **53**, 1951 (1984)
P.J. Steinhardt, H.-C. Jeong, Nature **382**, 431 (1996)
P.J. Steinhardt, H.-C. Jeong, K. Saitoh, M. Tanaka, E. Abe, A.P. Tsai, Nature **396**, 55 (1998)
P.W. Stephens, A.I. Goldman, Phys. Rev. Lett. **56**, 1168 (1986), ibid. **57** 2331 (1986)
H. Takakura, A. Yamamoto, A.P. Tsai, Acta Crystallogr. A **57**, 576 (2001)
S. Taniguchi, E. Abe, Philos. Mag. **88**, 1949 (2008)
A.P. Tsai, in *Physical Properties of Quasicrystals*, ed. by Z M. Stadnik (Springer, Heidelberg, 1999), pp. 5–50
A.P. Tsai, A. Inoue, T. Masumoto, Jpn. J. Appl. Phys. **26**, L1505 (1987)
K. Tsuda, Y. Nishida, K. Saitoh, M. Tanaka, A.P. Tsai, A. Inoue, T. Masumoto, Phil. Mag. A **74**, 697 (1996)
M. Varela, A.R. Lupini, K. van Benthem, A. Borisevich, M.F. Chisholm, N. Shibata, E. Abe, S.J. Pennycook, Annu. Rev. Mater. Res. **35**, 539 (2005)
A. Weickenmeier, H. Kohl, Acta Crystallogr. A **47**, 590 (1991)
A. Yamamoto, Acta Crystallogr. A **52**, 509 (1996)
A. Yamamoto, H. Takakura, E. Abe, Phys. Rev. B **72**, 144202-1 (2005)
Y. Yan, S.J. Pennycook, A.P. Tsai, Phys. Rev. Lett. **81**, 5145 (1998)
Y. Yan, S.J. Pennycook, Phys. Rev. Lett. **86**, 1542 (2001)
G. Zeger, D. Plachke, H.D. Carstanjen, H.-R. Trebin, Phys. Rev. Lett. **82**, 5273 (1999)

15

Atomic-Resolution STEM at Low Primary Energies

Ondrej L. Krivanek, Matthew F. Chisholm, Niklas Dellby and Matthew F. Murfitt

15.1 Introduction

The scanning transmission electron microscope (STEM) is now able to produce electron probes as small as 1 Å at 60 keV, a primary energy that is low enough to avoid direct knock-on damage in materials made of light atoms such as graphene. The difference between high-energy and low-energy imaging of these materials is remarkable: at 100 keV, i.e., above the knock-on threshold, holes are drilled at a rapid pace; at 60 keV, i.e., below the threshold, one can observe the same patch of graphene for minutes to hours without any substantial changes. A similarly large reduction in radiation damage is expected below the knock-on threshold energy in all materials that do not suffer from ionization damage.

In a previous paper (Krivanek et al. 2010b), we have called STEM operation at primary energies lower than the knock-on threshold "gentle STEM." There are no regular inter-atomic distances smaller than 1.2 Å not involving hydrogen, and 1 Å resolution therefore means that gentle STEM is now able to resolve near-neighbor atoms in light materials without altering the sample structure. This is an important advance and the main topic of this chapter.

Many researchers have contributed to the advance. The full history of the STEM is reviewed in depth elsewhere in this volume (Chapter 1); here we only note the major mileposts encountered along the way. Crewe's cold field emission STEM (Crewe et al. 1968b, Crewe 2009) was the key development, because it showed the wealth of results that can

---

Notice: This manuscript has been authored by UT-Battelle, LLC, under Contract No. DE-AC05-00OR22725 with the U.S. Department of Energy. The United States Government retains and the publisher, by accepting the article for publication, acknowledges that the United States Government retains a non-exclusive, paid-up, irrevocable, world-wide license to publish or reproduce the published form of this manuscript, or allow others to do so, for United States Government purposes.

S.J. Pennycook, P.D. Nellist (eds.), *Scanning Transmission Electron Microscopy*,
DOI 10.1007/978-1-4419-7200-2_15,

be obtained when a small probe of electrons is focused on a thin sample and several signals, such as those due to elastically and inelastically scattered electrons, are collected simultaneously.

Once their new STEM was working well, Crewe's group progressed quickly onto imaging of single heavy atoms—a significant first in electron microscopy (Crewe et al. 1970)—and also to a demonstration of the power of Electron Energy Loss Spectroscopy (EELS) carried out in the STEM (Isaacson and Johnson 1975). Their work inspired two Cornell workshops on Analytical Electron Microscopy organized by Silcox and colleagues (Fraser et al. 1976, Fejes 1978), which spearheaded the adoption of the STEM technique by a large number of researchers, and defined the agenda for the development of the STEM over the next two decades.

It is useful to remember that Crewe's work was preceded by the development of the STEM concept by von Ardenne (1938, 1940, 1985), and a demonstration by Zworykin, Hillier, and Snyder (1942) that a cold field emitter can be a useful source of electrons. But these developments were "ahead of their time:" the detectors, the electronics, the ultra-high vacuum technology, and the computers necessary for a practical STEM did not yet exist.

The resolution of the Chicago STEM ultimately reached 2.4 Å at 40 keV, limited by the spherical aberration coefficient C_S, even though the coefficient had been reduced to the low value of $C_S = 0.3$ mm. To improve the resolution further, either the spherical aberration had to be eliminated, with the help of an aberration corrector, or the operating voltage had to be raised significantly, so as to decrease the electron wavelength λ. Crewe's lab embarked on both approaches, without a crowning success. In aberration correction, they were probably themselves "ahead of their time," because computers were not up to the task of analyzing and correcting aberrations automatically in the 1970s. Raising the operating energy significantly may have needed more funding than they were able to secure.

The STEM field grew much wider from the late 1970s on. Others began to make significant contributions (Hawkes 2009a), e.g., by producing a STEM capable of 1.2 Å resolution at 300 keV (von Harrach 2009). But to approach 1 Å-level resolution at operating energies <200 keV, a working aberration corrector was needed.

The history of aberration correction has been reviewed several times recently; the review by Hawkes (2009b) stands out for its comprehensiveness. Here we provide a brief account emphasizing the STEM-related developments. The first proof-of-principle corrector for an electron probe instrument was built by Deltrap (1964a, b). It was based on Archard's (1955) quadrupole–octupole form of Scherzer's (1947) aberration corrector. Deltrap showed that combined quadrupole–octupoles acting on an electron beam can make spherical aberration in a probe-forming instrument zero or negative. However, he did not demonstrate an improved resolution. We now know that abandoning cylindrical symmetry together with the imperfect nature of quadrupoles (which are much harder to construct with micron-level precision than round lenses) produces parasitic aberrations of many different orders, and that if these aberrations are not corrected along with the spherical aberration, the resolution is likely to become worse than it was before the correction attempt. Deltrap's corrector did not have enough flexibility to null the parasitic aberrations, and the task would not have been an easy one without modern style computer control even if it did.

The correctors built in Crewe's lab were not as successful as Deltrap's advance, and no correction of aberrations was demonstrated at all. This was probably at least partly due to the fact that in the first corrector built by Beck and Crewe (1976), whose electron-optical design was similar to a C_S corrector designed by Thomson (1968), the magnetic circuit was built from permendur, whose high remanence and hysteresis made it unsuitable for producing the low-magnitude fields that needed to be controlled with high precision in a corrector. But Crewe's group contributed greatly to the theoretical understanding of corrector optics: the concepts of a sextupole corrector and of a fifth order-optimized corrector were introduced by them (Beck 1979, Crewe and Kopf 1980, Shao 1988).

The next attempt at correcting spherical aberration in a STEM was by Krivanek, Dellby, and coworkers, with a quadrupole–octupole corrector (Krivanek et al. 1997). Their work demonstrated that with a sufficient number of computer-controlled auxiliary optical elements and newly developed aberration-diagnosing software, parasitic aberration could be mastered and the resolution of the STEM the corrector was built into could be improved relative to its pre-corrector value. This group then moved onto a second-generation corrector design (Krivanek et al. 1999, Dellby et al. 2001), which became the first commercially available corrector, and which allowed a more optimized STEM to reach directly interpretable sub-Å resolution for the first time in electron microscopy (Batson et al. 2002).

At about the same time and on a parallel track, a sextupole corrector design due to Rose (1990) was developed into a practical aberration corrector for the TEM by Rose's students and collaborators (Haider et al. 1998). The same group also developed a practical aberration corrector for an SEM (Zach and Haider 1995), using a Rose (1971) design whose principles were similar to a C_C probe corrector designed and built by Hardy (1967).

With aberration correction thus attained for all the three principal types of electron microscopes, further progress in aberration-corrected STEM has been rapid. Resolution records have been bested repeatedly (Nellist et al. 2004, Sawada et al. 2007, 2009, Erni et al. 2009), the improvements coming principally from the use of higher primary energies (and thus smaller electron wavelength λ). Sextupole+round lens STEM correctors have been shown to be capable of similar probe-forming performance as quadrupole+octupole ones (Müller et al. 2006), the resolution of both types being typically limited by uncorrected chromatic aberration C_C that comes primarily from the objective lens.

The new instruments made possible many new types of materials studies. Elemental mapping by EELS with atomic resolution in an aberration-corrected STEM (Bosman et al. 2007, Muller et al. 2008) has greatly improved on similar mapping in an uncorrected STEM (Kimoto et al. 2007) and has in fact become a standard tool. Single atoms with large EELS cross sections are now routinely detected (Varela et al.

2004, Suenaga et al. 2009), and the 2-D variation in electronic bonding can be studied with atomic resolution (Muller et al. 2008, Fitting-Kourkoutis 2010). The availability of atomic resolution at 60–80 keV primary energy, i.e., low enough to avoid significant knock-on damage in light-Z materials such as graphene and BN (Zobelli et al. 2007), has produced a wealth of results in these materials (Alem et al. 2009, Girit et al. 2009, Jin et al. 2009, Meyer et al. 2009, Krivanek et al. 2010a, b). Several meetings and conference sessions have been devoted to aberration correction and its results, many review articles have been written (e.g., Muller 2009), and several compendia of aberration-corrected work that include a large proportion of STEM results have been published (e.g., Hawkes 2008).

In this chapter, we focus on gentle STEM at primary energies lower than 100 keV. We review the factors that determine the resolution and the STEM probe current, show examples of investigations that have now become possible, and discuss likely future developments.

15.2 Microscope Performance at Low Primary Energies

Predicting the attainable resolution has become more complicated in the aberration-corrected era. Spherical aberration no longer dominates, and the STEM resolution limit typically comes from one or more of the following:

(a) chromatic aberration (if not corrected)
(b) higher-order geometric or mixed geometric-chromatic aberrations
(c) finite brightness of the electron source
(d) finite size of the atoms and scattering delocalization effects
(e) statistical noise in the images
(f) instrumental instabilities.

Formulas for evaluating contributions (a)–(c) and (e) numerically are given in Krivanek et al. (2008a). The principal resolution-limiting influences are discussed below.

15.2.1 Optimizing the Resolution at Low Primary Energies

Despite the large number of factors that need to be considered, the practical strategy for obtaining the smallest possible STEM probe is usually very simple. The electron optics of the microscope is optimally adjusted (tuned), and the result of the tuning is checked by observing an experimental shadow image (Ronchigram, see for instance Krivanek et al. 2009a) of an amorphous sample.

At the optimum adjustment close to zero defocus, the central part of the Ronchigram from a very thin amorphous sample is featureless. In thicker amorphous samples for which the defocus cannot be zero throughout the sample depth, the central part of the Ronchigram shows weak large-scale structures whose scale is uniform. Outside the central region, there is contrast even in very thin samples, and a change of the structural scale in thicker samples. These arise because aberrations have distorted the electron wavefront. The parts of the wavefront that deviate from the ideal spherical wavefront shape by more than $\pm\lambda/4$, where λ is the electron wavelength, do not contribute electrons to the probe maximum and are therefore best excluded by an aperture. The aperture is similar in its function to the objective aperture used in fixed-beam TEMs. It is usually not located inside the objective lens itself, and in dedicated STEMs it is therefore called the "virtual" objective aperture (VOA). In TEM/STEM instruments which already have an objective aperture after the sample, the probe-defining aperture is usually called the "condenser aperture," or the "illumination aperture."

We denote the largest probe semi-angle admitted by the maximum aperture allowed by the aberrations present in the optical system by α_o. The minimum attainable size of the probe on the sample is given by the diffraction limit due to the aperture as

$$d_o = 0.61\ \lambda/\alpha_o \tag{1}$$

Excluding the aberrated parts of the wavefront by an aperture means that the image formation in a non-corrected and an aberration-corrected STEM is very similar: the wavefront used to form the electron probe is not seriously distorted by aberrations in either case, and the main difference between the two types of microscopes is that the illumination aperture is larger in the aberration-corrected case.

It is useful to note that the chosen illumination semi-angle may be decided by other considerations than optimizing the attainable resolution. For example, a smaller illumination angle may be selected because it provides a greater depth of focus, and stronger channeling down atomic columns. The smaller angle will also allow the signal due to small angle scattering to be collected more efficiently, or spot-type diffraction patterns to be acquired instead of convergent-beam ones. This type of operation will become more common when aberrations are overcome to such an extent that they are no longer limiting. Much of the discussion presented here is also applicable when the illumination semi-angle is set to a smaller value than α_o.

With a sufficiently precise knowledge of the relevant aberration parameters, α_o and d_o can of course be calculated, and detailed expressions that consider geometric aberrations up to seventh order as well as the first-order chromatic aberration are given in Krivanek et al. (2008a). In practice, in a well-tuned, C_S-corrected STEM with minimized or corrected fifth-order aberrations operating at 200 keV and below, the smallest attainable probe size is usually determined by the chromatic aberration. To reach the smallest probe, the illumination aperture is then set to a half-angle given by (Krivanek et al. 2008a):

$$\alpha_o = \alpha_{chrom} = 1.2\ (\lambda/(C_c\delta E/E_o))^{1/2}, \tag{2}$$

where C_c is the coefficient of chromatic aberration, δE the full width at half maximum (FWHM) of the energy spread of the primary beam, and E_o the primary energy. The minimum attainable probe size is then given by

$$d_{o\,chrom} = 0.5(\lambda\ C_c\ \delta E/E_o)^{1/2}, \tag{3}$$

Three other useful formulas for determining the smallest probe size attainable in the presence of dominant geometric aberration $C_{5,4}$, $C_{5,6}$, or $C_{7,8}$ (4- and 6-fold astigmatism of fifth order and 8-fold astigmatism of seventh order) are as follows (Krivanek et al. 2008a):

$$d_{o\,C5,4} = 0.44\, C_{5,4}{}^{1/6}\lambda^{5/6}, \quad (4)$$

$$d_{o\,C5,6} = 0.64\, C_{5,6}{}^{1/6}\lambda^{5/6}, \quad (5)$$

$$d_{o\,C7,8} = 0.61\, C_{7,8}{}^{1/8}\lambda^{7/8}. \quad (6)$$

Equations (1), (2), (3), (4), (5), and (6) do not consider the influence of the finite size of the electron source that must be projected onto the sample if the probe is to contain a non-zero current. For a useful probe current, i.e., a source demagnification value that is not infinite, the probe size is broadened to

$$d_{\text{probe}} = \left(d_o^2 + d_{\text{source}}^2\right)^{1/2}, \quad (7)$$

where d_{source} is the diameter of the source projected onto the sample (assuming perfect (Gaussian) optics). d_{source} depends on the selected probe current I_p and the source brightness as

$$d_{\text{source}} = 2\left(I_p/B_n V_o^*\right)^{1/2} / \left(\pi\, \alpha_o\right), \quad (8)$$

where B_n is the "normalized" (or "reduced") brightness $B_n = B/V_o{}^*$, B is the source brightness defined as $B = I_p/(A\ \Omega)$ in which A is the area of the virtual electron source as it appears when looking back at the source from just after the electron accelerator and Ω the solid angle of the electron beam emanating from the virtual source that ends up contributing to the probe, and $V_o{}^*$ is the relativistically corrected accelerating voltage. ($V_o{}^* = V_o\,(1 + eV_o/2m_ec^2)$ where V_o is the accelerating voltage, e is the electron charge, m_e the electron rest mass, and c the velocity of light.) α_o appears in this expression because the probe current is proportional to the solid angle of illumination.

Typical values of the normalized brightness are $B_n = 1{\times}10^8\ \text{A}/(\text{m}^2\ \text{sr V})$ for cold field emission guns (CFEGs), and $B_n = 2{\times}10^7\ \text{A}/(\text{m}^2\ \text{sr V})$ for Schottky guns. $B_n = 7.7{\times}10^9\ \text{A}/(\text{m}^2\ \text{sr V})$ has been reported for a nanotip CFEG (Qian et al. 1993), and we often measure B_n of $2{\times}10^8\ \text{A}/(\text{m}^2\ \text{sr V})$ for the VG 100 keV CFEG. Freitag et al. (2008) have reported $B_n = 1{\times}10^8\ \text{A}/(\text{m}^2\ \text{sr V})$ for what is probably a small tip diameter Schottky source. The "typical" brightness values used here are thus on the low side of what is possible in a record-breaking instrument, but they correspond to what is readily reachable in most CFEG and Schottky guns presently in use.

A key situation arises when $d_o = d_{\text{source}}$. The source size projected into the plane of the probe is then equal to the diffraction limit due to the illumination aperture at the probe. Since the diffraction limit defines the coherence width at the probe, the projected source size and the coherence width are therefore identical in this situation. The current corresponding to a source whose size is equal to the coherence width is appropriately called *the coherent current* I_c. It is given by

$$I_c = B\,A_c\,\Omega_c \tag{9a}$$

$$I_c = \left(\pi^2/4\right) B_n V_o^* d_o^2 \alpha_o^2 \tag{9b}$$

where A_c and Ω_c are the projected source area and the solid angle corresponding to the coherent current, and in (9b) we are using the fact that when $I_p = I_c$, $d_{source} = d_o$.

Combining (9b) with (1) with the expression relating the electron wavelength to its energy ($\lambda = h/(2m_eE_o^*)^{1/2}$) gives I_c in terms of the normalized brightness B_n:

$$I_c = \left(\pi^2 0.61^2 h^2/\,(8m_e e)\right) B_n = 1.4\ \times\ 10^{-18} B_{n,} \tag{10}$$

where h is the Planck constant, m_e the rest mass of the electron, e the electron charge, $E_o^* = eV^*$ the relativistically corrected primary energy, and SI units are used for all the quantities (and I_c is therefore in amperes). The coherent current values corresponding to the normalized brightness given above are $I_c = 1.4 \times 10^{-10}$ A (=0.14 nA) for the typical CFEG and $I_c = 2.8 \times 10^{-11}$ A (= 0.028 nA) for the typical Schottky gun.

The coherent current is a characteristic property of the source, and, in the absence of brightness-reducing effects such as Coulomb interactions or high-frequency instrumental instabilities, it remains the same throughout the illumination column: at the source, in the condenser section, at the final probe incident on the sample. It has the same value before and after the acceleration of the electrons, and it is independent of the size of the illuminating aperture. If the aperture size is decreased by 2× while everything else is kept the same, the range of angles accepted from the source is of course decreased by 2×. But the size of the diffraction limit projected back onto the source then increases 2×. The same coherent probe current (and the same degree of coherence) can thus be obtained from a virtual source that is 2× as large spatially but 2× narrower in angle than before, by adjusting the condenser lenses of the microscope appropriately.

Reformulating Eq. (7) to have I_p and d_o as its two main variables, while remembering that the probe current is proportional to the square of the source size and that when $d_{source} = d_o$, $I_p = I_c$, leads to a simple fundamental expression for the probe size with a non-zero probe current:

$$d_{probe} = \left(1 + I_p/I_c\right)^{1/2} d_o. \tag{11}$$

Figure 15–1 shows the dependence of d_{probe} on I_p for the CFEG case with $I_c = 0.14$ nA, and the Schottky case with $I_c = 0.028$ nA. When the probe current I_p is equal to the coherent current I_c, the probe size becomes $\sqrt{2}\times$ larger than in the zero-current limit, and this is also indicated in the figure.

The graph can be made universal for all values of I_c simply by rescaling the abscissa to show the probe current value as a fraction of I_c rather than as an absolute current. This scaling is implemented at the top of Figure 15–1, for the CFEG curve.

Since using the full coherent probe current does cause a 41% broadening of the probe compared to the diffraction limit, those aiming for

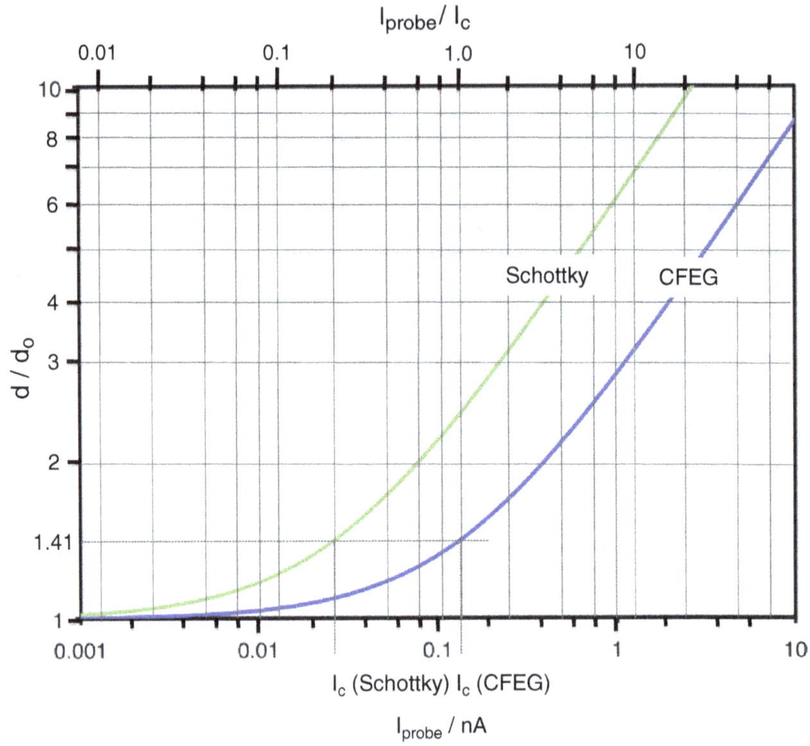

Figure 15–1. Probe size as a function of the probe current for the typical CFEG and Schottky electron guns, calculated for B_n(CFEG) = 1×10^8 A/(m^2 sr V) and B_n(Schottky) = 2×10^7 A/(m^2 sr V).

the highest resolution typically only use probe currents of 0.1–0.3 I_c. On the other hand, those needing to optimize the signal-to-noise ratio in noisy spectra often run with the source demagnified less strongly, e.g., at a beam current of $3I_c$, at which point the probe is 2× as wide as the diffraction limit, and is mostly incoherent. Increasing the beam current to $3I_c$ gives about 0.5 nA current for the typical CFEG and 0.1 nA for the typical Schottky gun.

When operating in the large-current regime, the illumination aperture can usually be opened up a little to provide an even bigger beam current, without the additional aberration-caused broadening affecting the probe more than the increased source size. This adjustment is capable of giving up to about a 2× increase in the probe current before the additional aberration-caused broadening becomes objectionable.

Counteracting the above is the fact that because the solid angle of the beam emitted from the source that ends up contributing to the final probe on the sample grows at higher beam currents, the diameter of the used part of the beam emitted from the tip then becomes larger. With a wider beam traversing them, the contributions of the gun and the condensers to the total aberrations of the optical system then increase in importance.

The increase is a steep one: the contribution of an optical element j, in which the beam has a diameter D_j, to the total "aberration budget" increases as D_j^2 for chromatic aberration, and as D_j^4 for spherical aberration. At some value of the beam current, typically around 5–10 I_c for the CFEG, the gun and condenser aberrations begin to dominate the total aberrations of the probe-forming system. The aberration corrector then needs to worry more about correcting the aberrations arising in the gun and the condensers than those arising in the final probe-forming lens. The exact point at which this happens depends on the details of the gun optics, and a lens that restricts the spread of the beam coming out of the gun (a "gun lens," e.g., Venables and Cox 1987) is useful for delaying the onset of the effect.

In the case of a STEM whose optical performance is limited by chromatic aberration, Eqs. (3) and (11) lead to a practical expression for the probe size:

$$d_{\mathrm{probe}} = 5.5 \times 10^5 \left(\left(1 + I_{\mathrm{p}}/I_{\mathrm{c}}\right) C_{\mathrm{c}}\delta E\right)^{1/2} / E_{\mathrm{o}}^{*3/4}, \quad (12)$$

where d_{probe} is in pm, I_{p} and I_{c} can use any unit of current as long as it is the same for both of them, but are most conveniently specified in nA, C_{c} is in mm, and δE and E_{o}^* are in eV. In our experience, the equation predicts the attainable probe size in a C_{c}-limited STEM with about ±10% accuracy in the low-current regime ($I_{\mathrm{p}}<I_{\mathrm{c}}$), and within about ±30% in the high-current regime up to about 10 I_{c} (in a CFEG). The larger uncertainty of the high-current regime arises because it is subject to the complicating factors already mentioned: sometimes being able to increase the size of the illuminating aperture slightly without greatly worsening the resolution and sometimes needing to decrease the aperture size because of extra gun and condenser aberrations, and also because the coherent current can be tricky to measure and is typically not known very accurately. Beyond about 10 I_{c}, the gun and condenser aberrations dominate the optics of probe forming in a CFEG STEM, and Eq. (12) no longer applies.

Equation (12) shows that in C_{c}-limited STEMs, the probe size for a given probe current I_{p} and given primary energy E_{o}^* depends on just three parameters: the coherent current I_{c}, the chromatic aberration coefficient C_{c}, and the energy spread δE. This means that for most present-day aberration-corrected STEMS, two out of the three parameters that determine the attainable probe size depend on the characteristics of the electron source.

Despite the central role played by the electron source, there has been much less emphasis on improving its performance than on aberration correction, and the performance has not progressed greatly since the first consistent use of the cold field emission gun by Crewe et al. (1968a) over 40 years ago. This may change in the future, and it is therefore useful to review the present status of high-brightness electron sources.

15.2.2 Cold Field Emission and Schottky Sources

High-brightness electron sources used by present-day STEMs come in two principal varieties: cold field emission guns (CFEGs) and field-assisted thermionic emission sources called Schottky guns. Figure 15–2 shows a schematic comparison of the emission mechanism from a cold

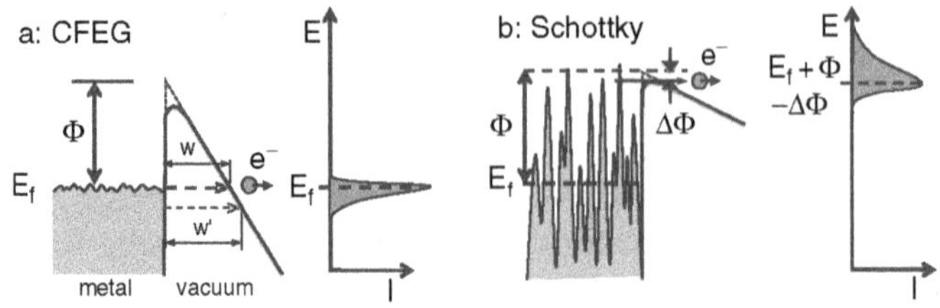

Figure 15–2. Comparison of the cold field electron emission mechanism with the field-assisted thermionic (Schottky) emission mechanism.

field emission gun (Crewe et al. 1968a) and a Schottky gun (Swanson and Schwind 2009).

In the CFEG, the electric field at the surface of the emission tip is typically about 10× higher than in the Schottky gun (~1 V/Å vs. ~0.1 V/Å). This decreases the width w of the potential barrier due to the metal's work function Φ so much that electrons at the Fermi energy level E_f are able to tunnel through the barrier. Electrons of slightly lower energy than E_f are able to tunnel through the barrier too, but the tunneling distance w' for them is longer, which significantly reduces the tunneling probability. The tunneling of these electrons causes a "low-energy tail" of the emission peak (see Figure 15–2). With an emission tip at room temperature, some electrons have energies slightly greater than E_f and these are able to tunnel out too. This causes a high-energy tail, which is typically less extended in the CFEG than the low-energy tail, giving a characteristically asymmetric zero loss peak in energy spectra.

In the Schottky gun, the electric field at the tip is normally too weak for the electrons at the Fermi energy to be able to tunnel out, and the electrons therefore go over the top of the barrier. The barrier is lowered slightly, by ΔΦ, by the applied electric field; this is known as the "Schottky effect". The electrons are able to go over the top of the barrier because of the extra energy supplied to them by the elevated temperature of the tip. A picturesque way of viewing this is that the Fermi sea is rough, and some of its "waves" are able to "splash" over the workfunction barrier. The energy of the Schottky emission peak is therefore displaced by about Φ relative to the CFEG peak. There is a finite probability that an electron will be excited to an energy some distance above the workfunction barrier, and this causes a high-energy tail of the Schottky emission peak. Electrons excited to just below the workfunction barrier are able to tunnel out of the tip, and this causes the low-energy tail of the Schottky distribution.

Lowering the work function Φ lowers the energy for both types of guns. Two different mechanisms are involved. In the CFEG, lower value of Φ means that the same tunneling width w is reached at a lower applied electric field. The decreased gradient of the potential then causes the width of the tunneling barrier to increase faster for energies lower than E_f, resulting in a faster decay of the tunneling tail. In the Schottky gun, the temperature required to excite electrons over a lower workfunction barrier is lower, and this decreases the width of the tail of the energy distribution extending above the barrier. The workfunction is typically lowered in the Schottky gun case, by building up a layer of ZrO or a similar oxide covering a (100) W tip (Swanson and Schwind 2009). Attempts to do the same for the CFEG (e.g., Batson 1987) have so far not resulted in a practical electron source.

The maximum attainable brightness of both the CFEG and the Schottky gun is chiefly determined by electron–electron Coulomb interactions, which modify the electron trajectories and thereby increase the size of the virtual electron source the electrons appear to come from. Without the interactions, one could simply increase the extraction voltage (in the CFEG case) or the extraction voltage in combination with the tip heating (in the Schottky case) and thereby extract more electrons from the tip and increase the brightness with no clear upper limit. In reality, the Coulomb interactions cause an increase in the virtual source size and hence a drop-off in the brightness (current per unit source area per unit solid angle) at total emission currents greater than about 5 μA in the W (310) CFEG (Bacon private communication, 2008), and at larger total currents for blunter Schottky sources (which typically emit in the forward direction from a larger total area than CFEG sources). The Coulomb interaction is much reduced once the electrons have been accelerated, and the faster acceleration in the stronger extraction field gives the CFEG an edge over Schottky in the ultimate brightness it can attain.

For all electron sources, there is another important aspect beyond the brightness and energy spread: how stable is the emitted current? With the W (310) and (111) tips used in normal CFEG guns, contamination of the tip typically causes an increase in the work function and thus a drop in the emitted current (Crewe et al. 1968a). Furthermore, the adsorbed contaminants tend to be mobile, and their rearrangement causes short-term changes in the tip geometry and work function. The current emitted from a "contaminated" tip therefore fluctuates more than the current emitted from a "clean" tip.

When a tungsten cold field emission tip becomes contaminated, it is readily restored to the clean condition by brief heating (flashing) at a high temperature, typically by passing a current through the tip-supporting wire loop for about a second. The clean tip then once again becomes gradually covered by adsorbates, and the emission decays. The time over which the emission current decays to one half of the starting value is usually denoted by $t_{1/2}$. In actual electron guns, $t_{1/2}$ varies widely, from less than a minute to several hours or even several days (Martin et al. 1960).

When $t_{1/2}$ is greater than about 30 min, the usual practice is to use the gun in its "clean" state and to flash the tip as often as is required to keep it clean. This way of operating a CFEG appears to have been introduced by Vacuum Generators (VG). It is also used with the Nion CFEG (Bacon et al. 2010). We call this type of gun a "clean" cold field emission gun. Its advantages are increased brightness and reduced energy spread relative to a gun with a higher workfunction, and greatly reduced emission noise. Being able to operate a CFEG in this way requires that the vacuum near the tip be $<10^{-10}$ torr even when emitting, and ideally in the 10^{-12} torr to the low 10^{-11} torr regimes. (The vacuum levels given here are those measured near the vacuum pump. At the tip, where incorporating a vacuum gauge would be more problematic, and where contamination due to the gauge would have a strong effect on the tip cleanliness, the vacuum is probably about 10× worse.)

In CFEG guns with vacuum levels worse than about 1×10^{-10} torr, the emission current typically decays much too fast for the gun to be usable in the freshly flashed "clean" state. The gun is then operated with the emission occurring with the adsorbate layers present, rather than from a clean tungsten tip. The extraction voltage is raised as needed to get a usable emission current, sometimes repeatedly to compensate for the current drop-off due to additional adsorbates. We call this type a "dirty" field emission gun. Its advantage is that vacuum requirements are reduced. Its disadvantages are that the emission current fluctuates much more than with clean tips, and that the gun brightness is typically reduced compared to the "clean" state.

The Schottky gun can operate in a poorer vacuum of around 10^{-9} torr, because arriving contaminants are continuously desorbed from the hot tip, which is in a state of dynamic equilibrium. Because of the high temperature, the mobility of the adsorbants is high and the changes in the work function are therefore much faster than they would be in a CFEG. Furthermore, the tip radius is typically larger in the Schottky gun than in a CFEG, and the emitting area that contributes to the final probe is therefore also larger. The high adsorbant mobility and the larger emitting area lead to a more stable emission current than in the dirty CFEG case.

The electric field at the tip in the Schottky gun is weaker than the field that would be needed to obtain field emission of electrons with the Fermi energy by tunneling. If tunneling of electrons with the Fermi energy did occur in the Schottky gun, electrons of many different energies would be able to go through the barrier, and the energy width of the emitted beam would grow to several electron volts (Swanson and Martin 1975, Swanson and Schwind 2009). This needs to be avoided in a practical source of electrons, and a Schottky gun used in an electron microscope is therefore never run at an applied field high enough to allow electrons of the Fermi energy to tunnel out of the tip. The standard Schottky gun therefore behaves very differently from a CFEG: the Schottky emission current goes to zero when the heating is turned off. Its advantages are that it can operate in a considerably poorer vacuum than a CFEG, and that its emission current remains stable over long periods of time. Its disadvantages are a wider energy spread than achievable with CFEG, typically of the order of 0.5–1.0 eV, lower brightness, and shorter lifetime: the Schottky emitter typically runs out of the pool of Zr needed to replenish its ZrO coating after about a year of continuous operation, whereas a well run clean CFEG emitter typically lasts more than 3 years.

The Schottky gun is nowadays often referred to as a field emission gun (FEG), almost certainly following a practice coined by a marketing department. It is ironic and regrettable that many scientists have started to employ this terminology, i.e., to call the Schottky gun by an acronym based on an emission mechanism that the gun *must minimize in order to operate as an optimized electron source.*

15.2.3 Practical Probe-Forming Performance

Figure 15–3 shows the influence of chromatic aberration of $C_c = 1.3$ mm combined with an energy spread $\delta E = 0.35$ eV, and of geometric aberrations of $C_{5,4} = 50$ mm and $C_{7,8} = 50$ mm, on the theoretical probe size. The assumed probe current is $0.25\ I_C$, which is equal to 35 pA for a typical CFEG. The theoretical curves were computed using Eqs. (4), (6), (11), and (12) above. $C_{5,4} \sim 50$ mm is typically the most important geometric aberration of Nion-corrected VG columns (Dellby et al. 2001, Batson 2009), and $C_{7,8} \sim 50$ mm is the most important geometric aberration of a well-tuned Nion UltraSTEM™ (Krivanek et al. 2008b). The chromatic aberration and the energy spread of both these instruments correspond to the ones used in the graph.

The attainable probe size is plotted as a function of the primary energy. In the C_s-corrected VG columns, the uncorrected fifth order aberrations limit the probe size more than the chromatic aberration above about 50 keV. In the C_s- and C_5-corrected UltraSTEM, the chromatic aberration dominates at all primary energies up to 200 keV (and it would dominate up to 500 keV if a microscope with similar aberration performance was operated at those energies). The strong dependence of the C_c-limited probe size on the primary energy means that it is about 5× longer at 20 keV than at 200 keV. This underscores the need for developing practical C_c correctors as the next step in aberration-corrected STEM, particularly for STEMs designed to operate at low primary energies.

The performance predicted by Figure 15–3 is largely borne out in practice. For instance, we have been able to resolve sample spacings of 1.09 Å at 60 keV with a 30 pA probe (Krivanek et al. 2010a, b) and 0.8 Å at 100 keV. With our new 200 keV CFEG (Bacon et al. 2010), we should be able to reach 0.5 Å. We have now recorded spacings of 0.6 Å, and we should be able to progress to 0.5 Å with more precise tuning.

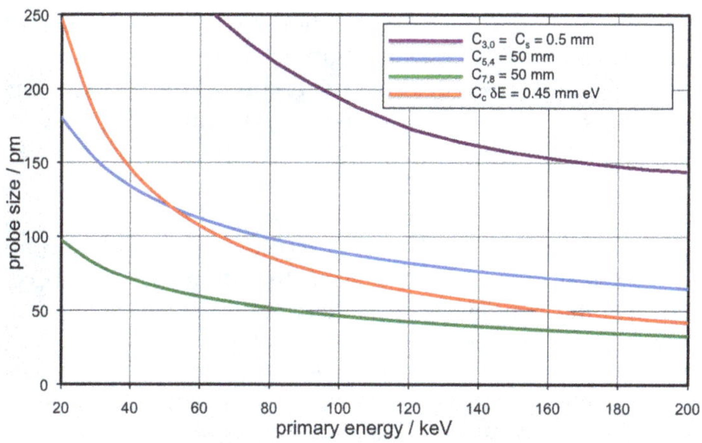

Figure 15–3. Theoretical probe size at various primary energies in a non-corrected STEM with $C_S = 0.5$ mm compared to the probe size in aberration-corrected STEMs of indicated parameters. Probe current $I_p = 0.25\ I_c$ for all the cases.

In experimental practice, there are of course many additional components of the total system that need to be optimized in addition to the aberrations, the gun brightness and the energy spread. These include

(i) the high voltage power supply, which must be stable enough so that the energy spread of the primary electrons does not increase significantly, thereby worsening the effect of the chromatic aberration,
(ii) the tuning of aberrations, which must be accurate enough so that mistuned aberrations (including defocus and astigmatism) do not worsen the resolution,
(iii) the power supplies for the optics, which must be stable over the medium term so that the tuning remains unchanged over time periods long enough to find areas of interest on the sample and record data from them, and stable over the short term so that the probe is not deflected randomly,
(iv) the microscope suspension, which must provide sufficient isolation from floor vibrations,
(v) the mechanical rigidity of the microscope column, which must be high enough so that the optical elements do not shift with respect to each other, thereby causing a change of the aberrations and/or probe drift,
(vi) the sample stage, which must be free of vibrations and drift,
(vii) the shielding of the microscope column and of the electronics, which must be good enough to keep out external disturbances such as stray magnetic fields, cell phone transmissions, and acoustic noise,
(viii) the water cooling of the lenses, which must not introduce vibrations or thermal drift,
(ix) the microscope room, which must be acoustically quiet, free of floor vibrations and stray magnetic fields, and have a stable temperature,
(x) the post-sample detector-coupling optics, which must be able to bring the right signals to the right detectors,
(xi) the detectors, which must be fast and sensitive enough to record scattering events with good detective quantum efficiency,
(xii) the vacuum of the microscope, which must be high and clean enough so that contamination and sample etching are avoided.

Failure to meet these requirements results in the resolution becoming worse, the data becoming noisier, the atomic images becoming "squiggly," or the sample being destroyed prematurely.

The stability requirements are rather high, but are now being attained. Figure 15–4 illustrates this with a small portion of an experimental image of a filled nanotube shown in six versions: as recorded (a), and with intentionally added random probe displacements simulating probe "wiggles" of 0.05 Å r.m.s. (b) to 1.0 Å r.m.s. (f).

The smooth nanotube outer wall seen in profile on the right side of the images is especially sensitive to the added disturbances. Only the image with 0.05 Å added wiggles looks essentially the same as the

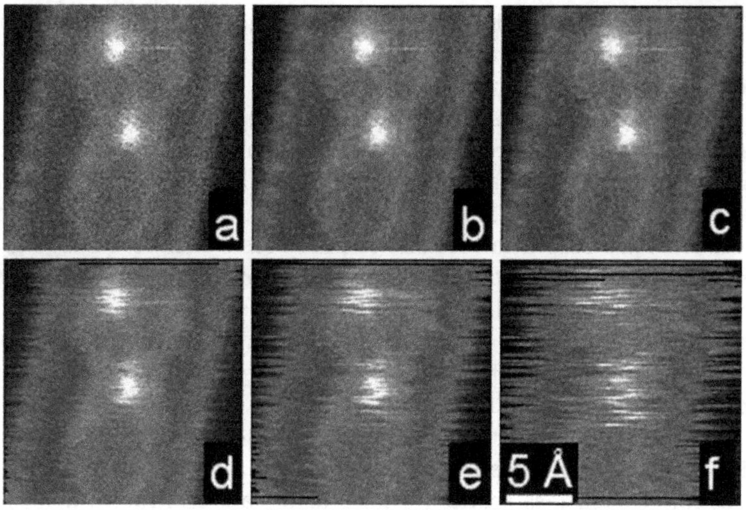

Figure 15–4. Images of a small part of a nanotube filled with nanopods filled with single Er atoms, with various amounts of probe "wiggles" added artificially, recorded with the Nion UltraSTEM at 60 keV. (**a**) No added wiggles, (**b**) 0.05 Å r.m.s, (**c**) 0.1 Å, (**d**) 0.25 Å, (**e**) 0.5 Å, and (**f**) 1.0 Å. Sample courtesy Dr. K. Suenaga, AIST.

original image (a), indicating that the actual level of the microscope instabilities was 0.05–0.10 Å r.m.s. (Krivanek et al. 2010c).

Not meeting stability requirements (iii–ix) typically introduces image streaks that are similar to those shown in Figure 15–4. It is a considerable achievement that overall stabilities better than 0.1 Å (10 pm) r.m.s. are now being reached in practice in instruments such as the Nion UltraSTEM™ (Krivanek et al. 2008b).

15.2.4 HAADF and EELS Image Resolution

STEM images are obtained by scanning the probe over the sample. They can be described as a convolution of an ideal image (which would be obtained with an infinitesimally small probe) with the actual probe. Obtaining good experimental resolution therefore requires a small probe *and* a sharp ideal image, i.e., an image formed by an electron–sample interaction that is sufficiently well localized.

Electrons that form high-angle annular dark field (HAADF) images originate from Rutherford scattering by the deep potential well surrounding the atomic nucleus. Ideal HAADF images of single atoms, i.e. images that are free of the influence of aberrations, source size broadening and of resolutions limits due to noise, would show the potential well rather than the electron orbitals. The full width at half maximum (FWHM) of the central peak of the image would therefore be very small—typically less than 0.3 Å in diameter for heavier atoms (Kirkland 1998, Batson 2006). The size would be partly due to the finite dimension of the potential well and partly due to the thermal vibration of the atomic nucleus.

With probe sizes greater than 1 Å, the broadening effect due to the finite size of atoms can usually be neglected and the ADF resolution taken as equal to the probe size. With probe sizes smaller than 1 Å, however, this approximation may produce significant errors. Beck and Crewe (1975) showed already 35 years ago that a C_s-corrected 100 keV STEM using an illumination half-angle of 30 mr will produce ADF images of C atoms that are 0.9 Å wide even though the probe size is 0.8 Å. The ADF resolution for sub-Å probe sizes therefore needs to be worked out either by summing the squares of the probe size and the atom size, or by a full calculation of the expected images.

Inelastic scattering at energy losses of several electron volts and higher originates from the interaction of the incident fast electrons with the sample's electrons (Egerton 1996), and this interaction is much more spread out than Rutherford scattering. The resolution in EELS maps is determined by the probe size plus the spatial spreading of inner shell loss scattering (and also statistical noise, discussed in Krivanek et al. 2008a). The spreading is called "delocalization." For the aberration-corrected STEM case, with large incident illumination and EELS collection angles, the delocalization can be described by an approximation for the diameter d_{50} of the area that contains 50% of the scattering events given by Egerton (2006):

$$d_{50} \sim 0.4\lambda/\theta_E^{3/4} = 0.7\,\lambda/(\Delta E/E_o^*)^{3/4} = 0.5\,\mathrm{h}\,\mathrm{m_e}^{-1/2}\Delta E^{-3/4}E_o^{*1/4}, \quad (13a)$$

where θ_E is the characteristic scattering angle used in EELS theory (Egerton 1996) and ΔE is the energy loss. Recent work shows that because the angular distribution of the inner shell loss scattering has rather wide tails and is better approximated by a Lorentzian distribution than a Gaussian one, the pre-factor in (13a) may need to be lower (Egerton private communication, 2010). Our experimental results (see below) have been matched well using a pre-factor of 0.5 in the second form of the expression, i.e.,

$$d_{50} = 0.5\lambda/(\Delta E/E_o^*)^{3/4}. \quad (13b)$$

Inner shell loss scattering is of course a lot more complex than the description provided by (13a) and (13b). EELS images of single atoms computed using full quantum-mechanical treatment (Chapter 6; Cosgriff et al. 2005, Kohl and Rose 1985) typically have sharper maxima than described by (13) and long tails, and they can also have, for certain combinations of illumination and collection angles, a dip in the center, resulting in a "volcano-like" appearance. The results of these calculations, however, do not appear to have been generalized into a form that can be used for all the various inner shell loss edges normally employed for analysis. Egerton's approximation provides guidance for what spatial resolution to expect for normal STEM-EELS imaging at any edge energy, and this makes it very useful. The approximation has been verified experimentally at 60 keV for energy losses of 170 and 285 eV (Krivanek et al. 2010b), and it also matches the result of the full quantum-mechanical treatment to within about 30% for the large

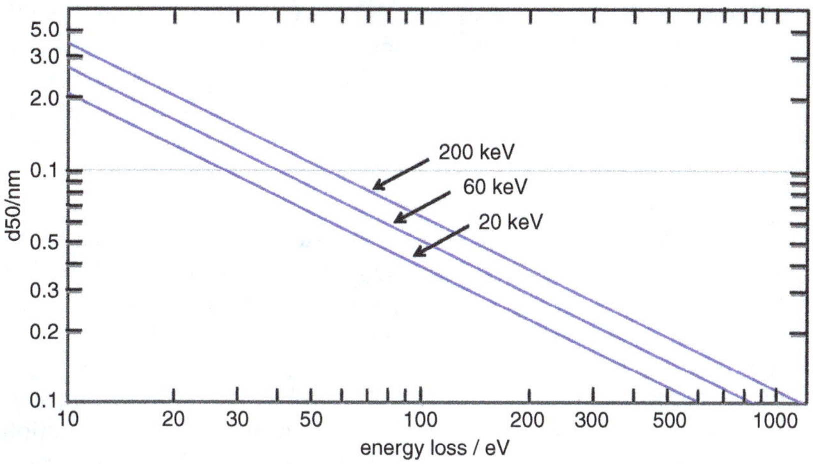

Figure 15–5. EELS delocalization calculated according to Egerton's formula as a function of the energy loss, for three different primary energies.

illumination angle/large collection angle geometry typically used for STEM-EELS (see Allen et al.'s Figure 6.8).

Figure 15–5 shows d_{50} as a function of the energy loss computed using (13b) for three primary energies: 20, 60, and 200 keV. Taking 2 Å as a convenient benchmark for "atomic" resolution, this level is achieved at energy losses greater than 460 eV at 200 keV, 340 eV at 60 keV, and 240 eV at 20 keV. The precise values of the energy losses of course depend on the pre-factor used in (13), but the predicted dependencies on the energy loss and the primary energy are a characteristic consequence of Egerton's approximation. The dependence of the delocalization on the primary energy (d_{50} α $E_o{}^{*1/4}$) is not fully supported by a full quantum-mechanical treatment (Pennycook private communication, 2010b) and it has not been verified experimentally. But if it exists even in a reduced form, it will provide another motive for operating at lowered primary energies.

The resolution expected in EELS maps is given by

$$d_{\text{EELS}} = (d_{\text{probe}}^2 + d_{50}^2)^{1/2}. \tag{14}$$

Figure 15–5 makes it clear that with a probe size of the order of 0.1 nm (1 Å), d_{50} typically limits the spatial resolution of EELS maps much more than d_{probe}. In order to improve the EELS mapping resolution for a particular element, d_{50} would need to be lowered at a given energy loss. A possible way to do this without changing the primary energy may be to concentrate on large-angle EELS scattering events (Muller and Silcox 1995), such as those that occur at energies considerably above an edge threshold and give rise to the Bethe ridge in the angular scattering distribution (Egerton 1996). This can be done for instance by using an annular EELS entrance aperture that prevents electrons scattered by low angles from entering the spectrometer. Its effectiveness remains to be tested experimentally.

When both the illumination and EELS collection angles are small, the EELS delocalization becomes much larger (Chapter 6; Cosgriff et al. 2005, Egerton 2006). In CTEM EELS imaging both the angles usually are small, because the CTEM illumination is quasi-parallel, and the chromatic aberration of its post-sample imaging optics blurs the inelastic image severely except when the selected EELS energy range and the collected scattering angles are small. In the STEM, the EELS collection optics does not focus the electrons into an image and is therefore much less affected by aberrations. With EELS coupling optics capable of accepting an angular range larger than the probe cone without worsening the EELS energy resolution, both the illumination and collection angles are usually large: the illumination angle is made large in order to optimize the probe size and current, and the collection angle is made large in order to optimize the collection efficiency. This is why the resolution of STEM-EELS maps is usually considerably higher than the resolution of energy-filtered TEM (EFTEM) maps.

It is worth noting that even though the formula for delocalization shows that atomic-resolution EELS elemental mapping is not possible with very low-energy losses of the order of a few tens of electron volt, the literature contains many experimental energy-filtered images that have been recorded at low losses and yet appear to show atomic resolution. This is readily explained by the fact that low loss, high spatial resolution EELS images can result from double scattering: low angle inelastic scattering that provides a new "primary beam" with the selected energy, plus high-angle elastic scattering that gives the fine structures seen in the image. The resultant images are not indicative of the sample composition. In a material with two elements that give low-energy edges of similar energies, energy-filtering to select one edge and then the other will produce substantially identical images even if the two types of atoms occupy atomic sites that project into different places in the image.

15.2.5 Image Acquisition and Processing

When imaging thin low Z materials at low primary energies, a slightly different strategy is called for than when imaging heavier and thicker samples at higher energies. First, the high-angle scattering from low-Z atoms not being particularly abundant, it is typically best to select a lower ADF cut-off semi-angle of 50–60 mr rather than the 80–90 mr one would use for heavier atoms at 60 keV primary energy. This gives an increase in the image signal of about 2×. We call the corresponding imaging mode medium-angle annular dark field (MAADF). A slight increase in non-linearity is expected in MAADF images compared to HAADF ones, whereby two atoms lying on top of each other give more than two times the signal of a single atom. We have tested for the effect by comparing the intensities of MAADF images of aligned single and double layers of graphene and BN. The experiments showed that for 50–60 mr lower cut-off angle, the non-linearity is smaller than the statistical noise typically present in typical images.

Second, the dark field detector gain has to be increased (by increasing the MAADF detector's PMT voltage) so that the signal from a single B or C atom is reliably detected above the background detector noise. We usually increase the gain to a level whereby the signal from about 10 graphene layers will saturate the detector, and then stay in thin sample areas where the saturation is not a concern.

Third, to get a good signal-to-noise ratio in the images of single light atoms, the exposure level (the electron dose per unit sample area) usually needs to be higher than would be needed for imaging thicker and heavier materials. Raising the probe current would worsen the resolution, and this leaves just one useful option: increasing the time the probe spends scanning over each atom. Instead of raising the per-pixel dwell time, we usually raise the time per atom by decreasing the pixel size. The resultant images are greatly oversampled, with each atom occupying an area of 10 × 10 or even 20 × 20 pixels, but the oversampled images contain useful information on the exact atomic position and on the atomic movement, as is shown for instance in Krivanek et al. (2010b).

The STEM MAADF images obtained with the Nion UltraSTEM at 60 keV typically resolve individual atoms in graphene and monolayer boron nitride, in which the nearest neighbor spacings are 1.42 and 1.45 Å, respectively. However, the image intensity does not go to zero in the center of the hexagonal rings in these structures, and it decays over several Ångströms away from the specimen edge. Both these effects are due to an extended "tail" of the electron probe, which is typically much stronger in actual imaging experiments than the tails of probes modeled by Gaussian profiles. The tail creates a "background fog" in MAADF images and decreases their clarity. It also contributes extra intensity to the images of the nearest neighbors of each atom and thereby makes the MAADF intensity of each individual atom depend on how many neigbors it had, and what their atomic numbers were.

It is important to note that the effects of the probe tail are well visible in MAADF images because of the quantitative nature of dark field imaging, in which the intensity in the vacuum next to the sample goes to zero, and thus provides a baseline to which the intensity in the center of the hexagons can be compared. In bright field (BF) phase-contrast imaging, by comparison, the intensity in the vacuum is 1. A tail in a BF image of an atom results in a slight change in the image pattern and a reduction of the overall phase contrast. But there is no readily visible change in the DC level of any image area, and the contrast is easily boosted back up, rendering the reduction nearly invisible.

A second undesirable effect in the as-recorded highly oversampled MAADF images is that by spreading the signal from each atom over many pixels, the signal per pixel is reduced, and the statistical noise increased artificially. However, the extra statistical noise is occurring at spatial frequencies higher than the spatial frequencies of sample details captured in the image and can therefore be readily filtered out. Provided that the noise introduced by the detector at every pixel is negligible, which is the case for well-designed and properly adjusted MAADF detectors, the image with the high-frequency noise filtered out will then have no extra noise compared to an image acquired at a sampling frequency corresponding to the spatial frequency of the filtering.

Both the above effects can be corrected by a simple Fourier-filtering procedure described by Krivanek et al. (2010a, b). The procedure is illustrated in Figure 15–6, which shows a part of an experimental 1 k × 1 k MAADF image of a graphene monolayer plus its multiple layer surroundings, acquired at 60 keV with about 50 pA beam current, at 64 μs per each 0.12 Å wide pixel, i.e., at about 1.4×10^6 $e^-/Å^2$. The figure also shows the filtering steps and the end result.

The filtering amounts to convolving the image with

(a) a broad Gaussian, whose width corresponds to the experimental resolution, and which therefore filters out the artificial statistical noise occurring at spatial frequencies higher than the highest actual sample frequencies captured in the image, and
(b) a negative Gaussian, whose width corresponds to the width of the probe tail, and whose intensity matches the intensity of the probe tail. The negative Gaussian causes the central dip of the filter and

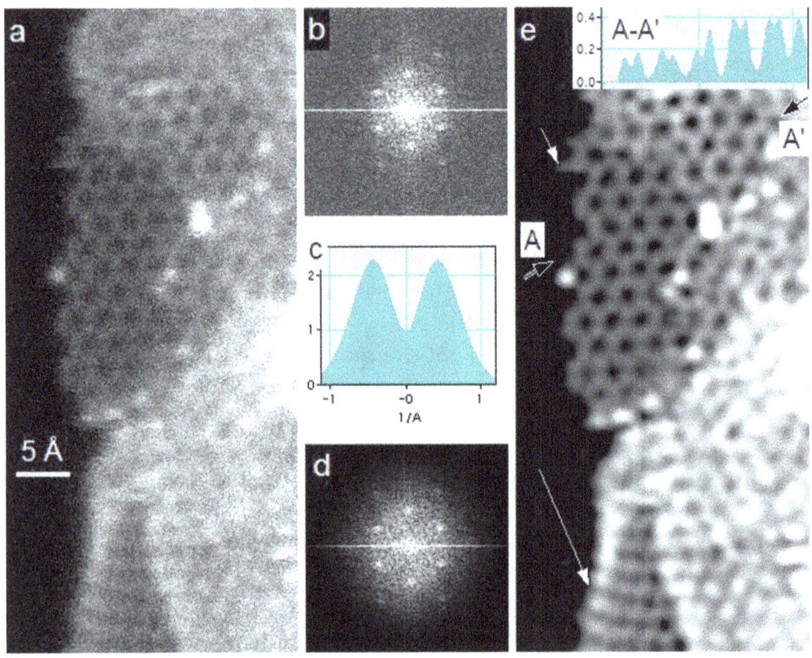

Figure 15–6. MAADF images of graphene illustrating the Fourier-filtering procedure designed to remove probe tails and artificially introduced statistical noise. (**a**) As-recorded image, (**b**) fast Fourier transform (FFT) of image, (**c**) profile through the applied Fourier filter, (**d**) filtered FFT obtained by multiplying (b) with (c), and (**e**) filtered image obtained by an inverse FFT. *Black arrows* in (e) mark the direction of profile A–A′ shown as an insert in the image. Sample courtesy Dr. V. Nicolosi, Oxford U. (Krivanek et al. (2010b), *Ultramicroscopy*, by permission).

amounts to subtracting the experimental probe tail contribution from the image, i.e., to "de-fogging" the image.

The shapes of the probe and of the probe tail are typically not known exactly, and they vary from image to image and especially from one autotuning operation to the next. This is the reason for choosing a particularly simple filtering procedure, in which the probe tail is greatly reduced compared to the unfiltered image, even though it is not subtracted exactly. Because the filtering is rotationally symmetric and has no sharp cut-offs that might cause "ringing" in the processed image, the probability of creating misleading artifacts out of random noise is small.

Figure 15–6(e) includes a profile through the filtered image, taken along the line A–A′, which starts in vacuum, crosses a monolayer of graphene, and ends in a double layer. The profile traverses the centers of the graphene hexagons, where it drops to about 10% of the single atom intensity. In unprocessed images, the intensity in the center of the hexagons was 50–70% of the single atom intensity, and this provided a reliable measure of the strength of the tail at 1.42 Å from the probe center. We avoided subtracting the probe tail completely, which would have produced unphysical negative intensities in the centers of some of the hexagons, and also in some places along the sample edge.

It is interesting to note that the second graphene layer was aligned over the first layer in A–A stacking in the sampled area, even though the normal stacking in double-layer graphene is A–B, in which atoms in the second layer lie over the centers of hexagons in the first layer. However, the second layer was probably pinned by amorphous carbon and hydrocarbons present around the edges of the monolayer, and was therefore not in an equilibrium configuration. Reassuringly, the intensity recorded in the double layer for atoms aligned on top of each other is about 2× the intensity of single carbon atoms.

Two other interesting details in the image are marked by white arrows. The short white arrow points to a location that probably had a single carbon atom dangling off the graphene edge, but which ran away while the probe was scanning over it. This can be seen in the corresponding place of the unsmoothed image (a), in which there is an extra intensity off the graphene edge that is cut-off abruptly, from one scan line to the next. The long white arrow marks a monolayer graphene sheet that curled over at the edge, thereby creating a shape resembling one quarter of a complete nanotube. Many other interesting details of graphene, single-layer BN, and nanotube structures are shown and discussed in the next section.

15.3 Graphene, Carbon Nanotubes, and Monolayer BN

Graphene and BN samples were prepared by liquid phase exfoliation of bulk graphite and BN powders in N-Methyl-2-Pyrrolidone (NMP, Hernandez et al. 2008). Full details of the sample preparation are given in Krivanek et al. (2010a).

The exfoliation produced graphite and BN flakes with small monolayer regions in various locations at or near the flake edges. The size of the regions varied. Smaller monolayer areas of around 5 × 5 to 30 × 30 hexagons surrounded by thicker regions were typically the most stable under the beam and were therefore preferred for observation.

15.3.1 Graphene: Lattice Defects and Adatoms at the Graphene Edge

Figure 15–7 shows the central portion of the image of Figure 15–6 at higher magnification (Figure 15–7(b)), and the same part of the sample imaged about a minute earlier (Figure 15–7(a)). Both the images were processed by the Fourier filter and are displayed slightly non-linearly, in order to make the monolayer graphene clearly visible without saturating the images of impurity adatoms.

An impurity atom at the graphene edge stayed in its place, which was only about 50% probable, as could be seen by observing, with the same electron dose, the mobility of impurity edge atoms at other locations in the same sample. The structure of the edge underwent major modifications. In the left image, a variety of atomic arrangements is seen at the edge: two 5-fold rings (indicated by single white arrows), a single dangling carbon atom (indicated by a double arrow), a distorted "armchair" (in which a complete carbon hexagon sits right at the sample's edge) just above the bottom 5-fold ring, and some atoms that were moving and left streaks behind. In the right image, the edge terminates in four regular armchairs. The rearrangement required the addition of just one carbon atom below the impurity atom and the removal of one carbon atom above the impurity atom. The armchair-terminated edge is similar to graphene edges imaged by bright field phase-contrast TEM (Girit et al. 2009), but the observations of 5-fold rings at graphene's edge and of a single dangling carbon atom appear to be new.

Many carbon hexagons are seen to be somewhat distorted, and the distortion of the same hexagon is typically different in the two

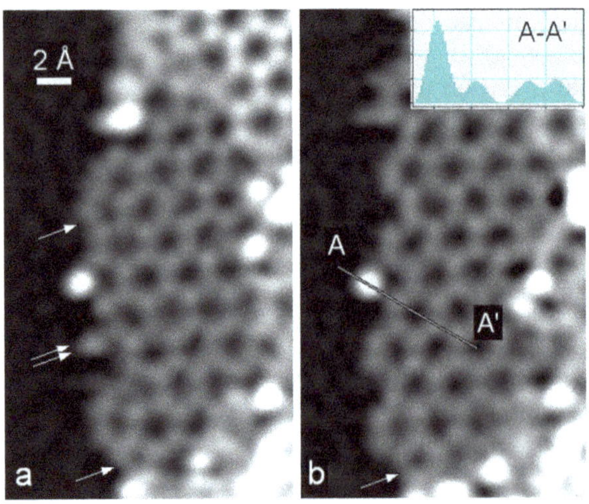

Figure 15–7. MAADF images of monolayer graphene taken about a minute apart. Image (b) is a higher magnification version of Figure 15–6(e). The *single arrows* in (a) and (b) point to 5-fold rings at the graphene's edge, and the *double arrow* in (a) points to a single atom of carbon dangling off the graphene edge. A–A′ profile through the impurity atom at the edge is shown as an insert in (b). Nion UltraSTEM, 60 keV, sample courtesy Dr. V. Nicolosi, Oxford University (Krivanek et al. (2010b), *Ultramicroscopy*, by permission).

images. As discussed in Krivanek et al. (2010b), there were three principal causes for the distortions: (a) statistical noise, which randomly enhanced different parts of the spread-out atomic images, and thus caused the smoothed images of individual atoms to shift randomly from frame to frame, (b) sample movement, which translated into the displacement of some parts of the image but not others, and (c) real distortions present in the carbon sheet, plus apparent distortions caused by the fact that the sheet was not aligned perpendicular to the beam and was probably also slightly buckled. The best way to separate the random distortions from the real ones is to image the same area in a sequence of images. The two images shown here indicate that most of the distortions in the present case were of the random kind. In stable samples the random distortions grow smaller at larger electron doses, and our practical experience (Krivanek et al. 2010a) indicates that they can be kept as small as about 0.1 Å if the dose is increased about 4× relative to the one used for the images of Figures 15–6 and 15–7.

There were several impurity adatoms, which gave much stronger contrast than the carbon atoms. Adatoms on the right side of the images were located on top of the graphene sheet and were moving frequently, and this made their analysis difficult. The single adatom at the graphene's edge was stationary and formed the apex of a 5-fold ring, with larger separation from its neighbors than the apex atom in the carbon-only 5-fold ring seen just above the adatom in Figure 15–7a. A profile through the adatom (insert in (b)) showed its intensity as 3.6× larger than that of the C atom images. Using the $I = a\ Z^{1.64}$ dependence of the atomic intensity I on the atomic number Z that we have measured experimentally (Krivanek et al. 2010a, see also Figure 15–14 below) on images of B, C, N, and O atoms obtained under essentially the same conditions as here, gave $Z_{\text{impurity}} = 6 \times 3.6^{1/1.64} = 13.1$, and we therefore tentatively identified the atom as aluminium. However, the extrapolation to $Z = 13$ based on experimental data obtained for $Z = 5$–8 is a stretch, and it is therefore possible that the impurity atom was Mg or Si, or even Na or P. EELS was tried on similar intensity impurity atoms in the vicinity, but it was not conclusive: the atoms were not strongly attached and tended to run away under the beam.

Figure 15–8 shows a time sequence of MAADF images of the edge of a graphene monolayer that was decorated by many adatoms, recorded as a sequence of images each one of which took 8 s to record. The top row shows unprocessed images, which provide useful information about atomic movement. The bottom row shows smoothed and tail-subtracted Fourier-filtered images, which provide clearer information about the sample structure. The adatom intensities were similar to the adatom whose profile is shown in Figure 15–7b, and they were therefore probably also Al. Once more, trying to identify the adatoms by EELS resulted in them running away, without providing useful EELS data.

Arrows mark various interesting features in Figure 15–8. The double arrow in (a) marks an adatom that came and went while the beam was scanning over its general area, resulting in short streaks in the

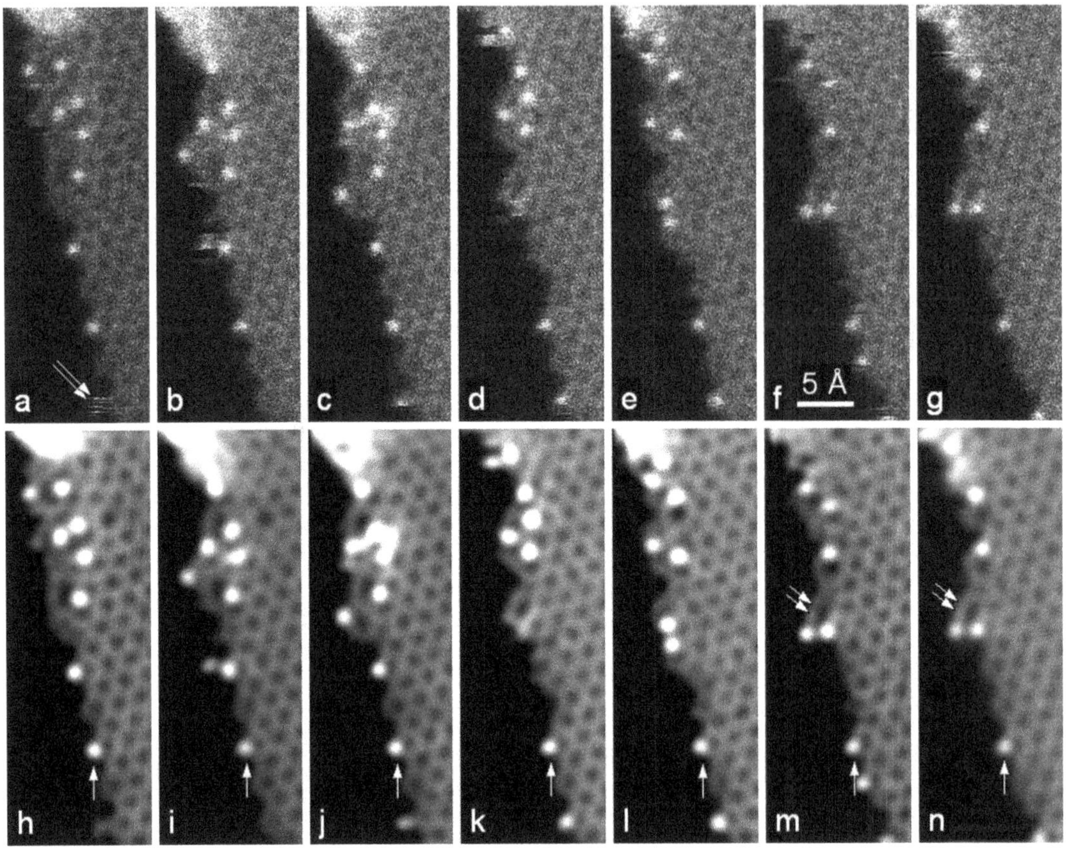

Figure 15–8. Time sequence of MAADF images of a graphene edge decorated by several adatoms, most of which were rather mobile. *Top row*: unprocessed images; *bottom row*: smoothed and tail-subtracted images. Nion UltraSTEM, 60 keV, sample courtesy Dr. V. Nicolosi, Oxford University.

image. This adatom was absent in the next image (b), came back in (c) but jumped off as the probe was scanning over it, was back in (d) and (e), and gone again in (f) and (g). The single arrows in (h)–(n) mark an adatom that remained stationary throughout the sequence. The double arrows in (m) and (n) mark a single chain of C atoms, about 3 Å long, terminating in a single adatom.

The bottom half of the portion of the graphene edge shown in the image was relatively stable, with the armchair termination dominant. The top half was much more mobile, and had 5- and 7-fold rings of carbon that came and went. The whole sequence illustrates the detailed nature of the studies of the dynamics of low-Z materials that have now become possible.

Figure 15–9 shows a pair of MAADF images of monolayer graphene recorded about 1 min apart, some distance away from the sample edge. Both show four 7-fold carbon rings (marked by white circles) and 5-fold rings (marked in by white crosses in (a)).

The atomic arrangement for the 7-fold rings and the associated 5-fold rings is related to a Stone–Wales defect (Saito et al. 1998, Suenaga et al.

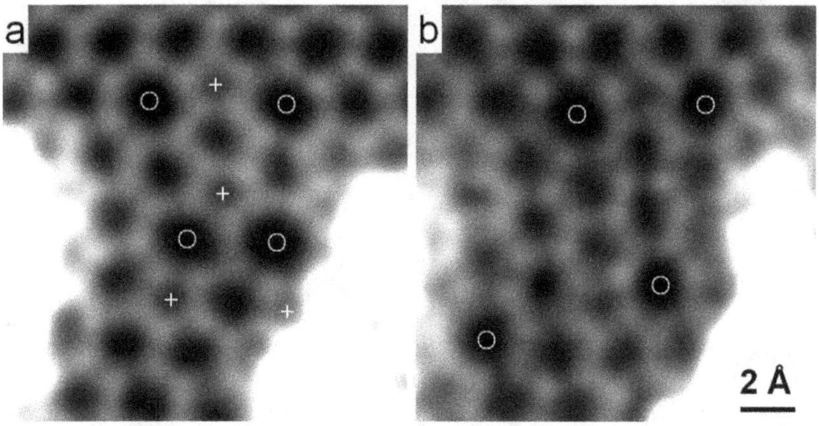

Figure 15–9. Defects in monolayer graphene imaged 1 min apart. MAADF, 60 keV. White circles mark 7-fold carbon rings, and white crosses mark 5-fold rings. Nion UltraSTEM, sample courtesy Dr. V. Nicolosi, Oxford University.

2007). In (b), two of the 7-fold rings have moved to different places, and the atomic arrangement has grown more complicated. Overlayers at bottom left and bottom right are only an additional layer thick, but they appear as saturated white in the present images, whose contrast has been adjusted to show the monolayer clearly.

15.3.2 Single-Wall Nanotubes Imaged with Atomic Resolution

Nanotubes are essentially graphene sheets rolled up into tubes. The orientation of the rolled-up sheet determines the nanotube chirality and helical pitch, which in turn determines the conducting properties of the nanotube (Saito et al. 1998). Figure 15–10a shows a MAADF image of a single-wall nanotube obtained at 60 keV and processed using the noise and tail-subtracting Fourier filter.

The nanotube displays an interesting periodic structure, with a longitudinal periodicity of 31 Å, but the image does reveal the nanotube structure clearly. However, a Fourier transform of the nanotube gives two sets of mirror-related reflections (insert in (a)). Masking one set followed by an inverse FFT produces the image (b), which is simply either the front or the back half of the nanotube. Masking the other set produces an image of the complementary half. Determining the chiral angle (Hashimoto et al. 2004) of the nanotube helix becomes rather easy with the two halves of the nanotube separated in this way. Determining the polarity of the nanotube's helical pitch should also be possible, for instance by tilting the illuminating beam by 10 mr or more and observing the resultant shift between the top and bottom halves. It is also interesting to note that the nanotube is slightly deformed, with a shape that conforms to the shape of an irregular nanotube pressing against it from the right side, but remaining about 3.6 Å away.

The front–back separation for a nanotube has been done before, on a bright field image (Suenaga et al. 2007). The work of Hashimoto et al.,

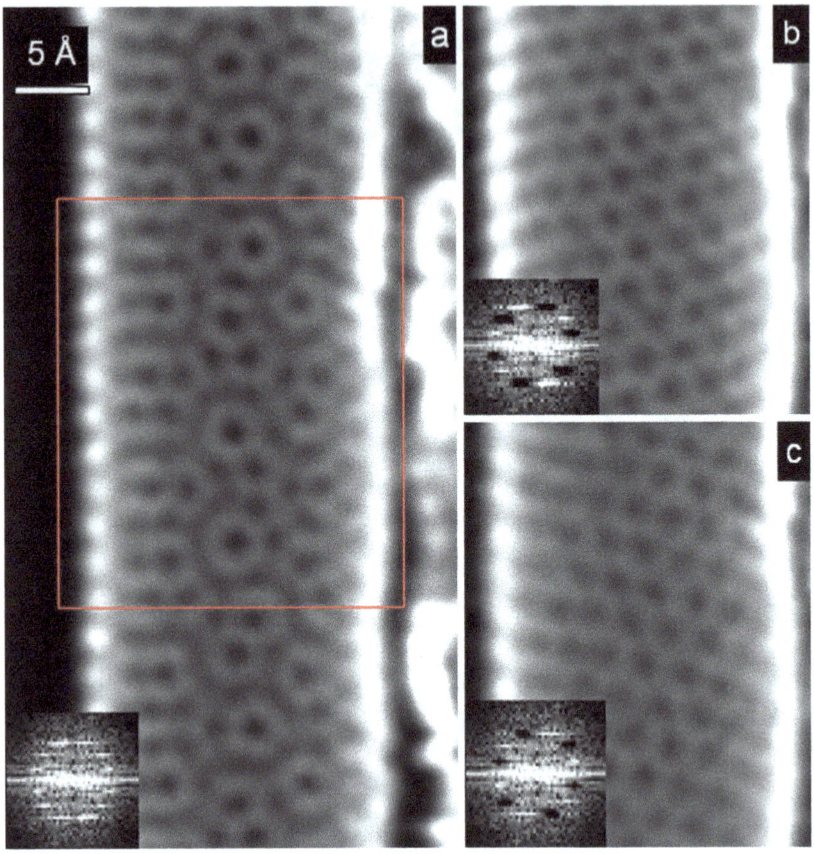

Figure 15–10. (**a**) MAADF image of a single-wall carbon nanotube obtained at 60 keV, with the diffractogram shown in an insert. (**b**) One half of the nanotube from the area marked by the *red rectangle* in (a), obtained by Fourier filtering that masked 50% of the nanototube's reflections while admitting the other 50% of the reflections plus the diffuse scattering. (**c**) The other half of the nanotube. Nion UltraSTEM, sample courtesy Dr. David Geohegan, ORNL.

Suenaga et al., and our work, which was done at a lower energy and with better resolution, show that nanotubes can now be imaged atom-by-atom, and defects and impurities present in them can be identified with clarity. Perfect single-wall carbon nanotubes are also very suitable as containers for holding molecules of unknown structures, since their contribution to the observed image of the molecule can be subtracted away quite precisely.

15.3.3 Monolayer BN: Distinguishing B from N and Identifying Impurities

Monolayer hexagonal BN is similar to grahene, but with three boron and three nitrogen atoms making up each 6-atom ring. Unlike graphene, monolayer BN is an insulator and could therefore potentially be used for separating graphene electronic devices (Geim 2009) incorporated in a continuous graphene–BN sheet. Figure 15–11 shows the

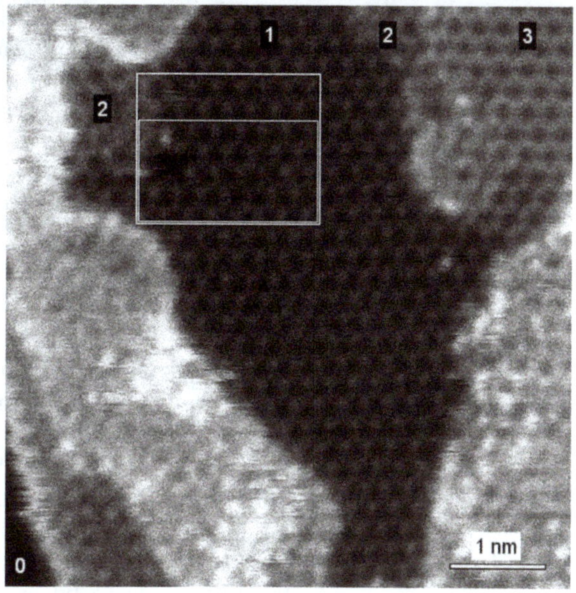

Figure 15–11. MAADF image of BN. Different numbers in the image mark the number of BN layers in that image area. "0" indicates the vacuum beyond the sample edge. The *white rectangles* show image areas studied in greater detail. Nion UltraSTEM, 60 keV, sample courtesy Dr. V. Nicolosi, Oxford University (Krivanek et al. (2010a) supplementary materials, *Nature*, by permission).

central part of an unprocessed 1 k × 1 k MAADF image of a BN monolayer, acquired with a per-pixel dwell time of 64 μs and a pixel size of 0.12 Å.

The monolayer area is surrounded by double, triple, and thicker layers of BN as well as less ordered overlayers that included hydrocarbons, as revealed by a strong C K-edge in their EEL spectra.

Figure 15–12 compares an unprocessed image of the sample region outlined by the smaller white rectangle in Figure 15–11 to an

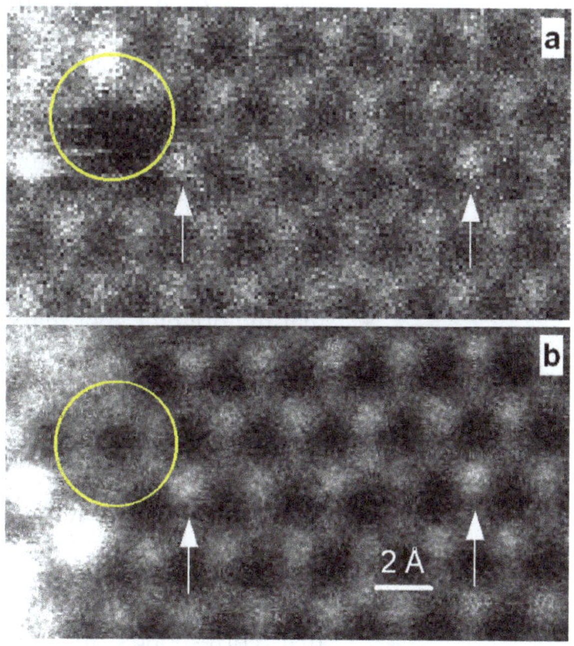

Figure 15–12. Comparison of the BN area containing the hole (**a**) with the same area imaged about 2 min later, at a higher magnification (**b**). (Krivanek et al. (2010a) supplementary materials, *Nature*, by permission).

unprocessed image of the same region taken 2 min later. The magnification was twice as high for the second image, in which each pixel was only 5.9 pm (0.059 Å) wide. The pixel dwell time was the same for both the images. This meant that there were 4× as many electrons per $Å^2$ in the higher magnification image, and its statistical image noise was therefore significantly reduced.

A hexagonal pattern of bright spots is clearly visible in both the images, with three of the spots in each hexagon considerably brighter than the other three spots. This is exactly what is expected in ADF images of monolayer BN. The brighter spots correspond to the heavier nitrogen, and the darker ones to the lighter boron.

Several departures from the regular bright spot—dark spot pattern—are visible in Figure 15–12. Two spots that are brighter than the spots corresponding to nitrogen are indicated by white arrows. They are only just brighter than the nitrogen spots, and their location did not change from one image to the next. This indicates that they are due to heavier substitutional atoms, probably oxygen, incorporated into the BN lattice. They provide fiducials that allow individual atoms in the BN lattice to be followed from one image to the next in the sequence of several images we recorded from this area.

Some image spots located on the left side of the images are considerably brighter than the spots due to substitutional atoms. These spots mostly occur in different locations in the two images. They are almost certainly due to mobile impurity adatoms on the BN surface.

A hole in the sample seen in Figure 15–12(a) is marked by a small yellow circle. There were atoms moving around in the hole, and the motion produced horizontal white streaks about 1.5 Å long that are visible inside the hole. The streak lying at about 7 o'clock within the circle is two scan lines wide, meaning that an atom arrived at this location while the beam was scanning nearby, and left 66–198 ms later, 66 ms being the line scan interval (i.e., it stayed for one whole scan line interval, plus two unknown portions of line intervals). The streak at 9 o'clock is only one scan line wide, meaning that the atom departed 1–132 ms after its arrival.

The same yellow circle is also shown in Figure 15–12b. The hole is now filled, but the atoms within it deviate from the bright–dark pattern of spots in the rest of the image: their intensity is roughly the same. It seems likely that carbon atoms available in the hydrocarbon deposits next to the hole on the left side of the image have filled the hole up, and that the brighter spot marked by the left white arrow was an oxygen atom that lodged itself in the BN monolayer as a part of the same hole-filling process.

The hole and the left oxygen atom were absent in an image recorded even earlier. Another small hole, roughly where the right oxygen atom is, was seen briefly in an earlier image and then filled up. This suggests that the substitutional impurities seen in Figure 15–12b were incorporated in the BN sheet following hole creation by the electron beam, and their subsequent filling by mobile adatoms traveling over the sheet.

Figure 15–13 shows an area of the original image of Figure 15–12b that corresponds to the larger rectangle in Figure 15–11. The

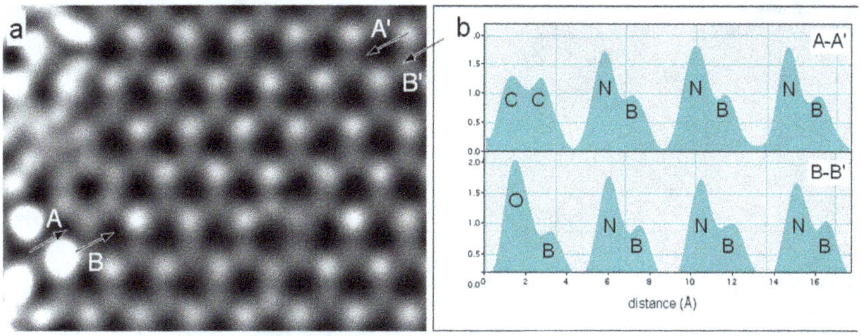

Figure 15–13. (**a**) Filtered version of the MAADF image of the BN monolayer area containing atomic substitutions. (**b**) Profiles through marked locations in (a). (Krivanek et al. (2010a), *Nature*, by permission).

experimental data have been smoothed and de-fogged using the double Gaussian filter, and also corrected for a small scan distortion of about 0.4 Å amplitude.

The strength of the negative Gaussian component of the filter was adjusted so that the intensity at the center of the BN hexagons (which was ~50% in the unprocessed image), became close to zero. This guaranteed that the intensity contribution of individual atoms to their nearest neighbor sites, which are the same distance away from the atoms as the centers of the hexagons, was also reduced to zero. In other words, the spurious contributions that the tails of the images of the nearest neighbors would have made to each atomic image have been subtracted by the procedure.

Profiles A–A′ and B–B′ shown in Figure 15–13b therefore portray the correct intensities, rather than intensities altered by a probe tail. They show a consistent pattern of peaks of alternating intensity, with the higher peaks corresponding to nitrogen atoms and the lower ones to boron. There are three significant deviations from the pattern: two peaks whose intensity is about half-way between the N and B peaks in profile A–A′, and a single peak in profile B–B′, whose intensity is significantly higher than the N peaks.

The most plausible explanation is the one already given: the intermediate peaks are due to C atoms, and the high one due to an O atom. Without a quantitative statistical analysis, however, atomic assignments such as these are subject to an unquantified statistical uncertainty. The appropriate way to quantify the assignments is to compute a histogram showing the distribution of the atom intensities (Isaacson et al. 1979, Voyles et al. 2002) for all the atoms in a given area, and to use the histogram to determine the probability that the atomic assignments were made correctly.

Figure 15–14a shows a histogram of all the image peaks within the monolayer area of the corrected image of Figure 15–13a.

The histogram separates into four distinct peaks, showing that we selected the illumination dose (6×10^6 electrons per Å^2) just right: 2× fewer electrons would have resulted in enough additional statistical noise to cause the peaks to overlap significantly, 2× more would have

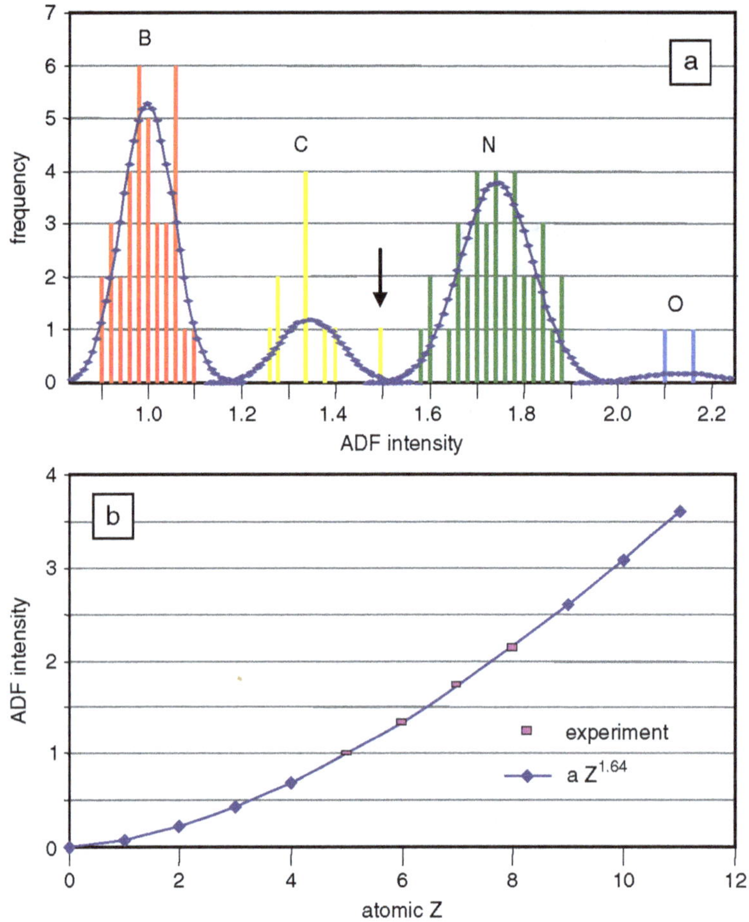

Figure 15–14. (**a**) Histogram of the peak intensities in the monolayer area of Figure 15–13(a), (**b**) Plot of the histogram peak positions as a function of atomic number Z, together with the best fit of $I = a\ Z^{1.64}$. The uncertainty of the experimental points in (b) is indicated by the height of the small rectangles. (Krivanek et al. (2010a), *Nature*, by permission).

produced a better separation of the peaks, but may have caused extra damage to the sample. The B and N distributions are modeled by Gaussians whose widths were extrapolated from the B and N distributions. The centers of the Gaussian peaks correspond to the measured centers of the four distributions. The intensity of the image has been normalized so that the center of the B peak is at 1.0.

Figure 15–14b shows the dependence of the average atomic peak intensity, i.e., the centers of the histogram peaks, on the assigned atomic number *Z*, plus a theoretical fit using an $I = a\ Z^{1.64}$ model. The fit is excellent, passing within 1σ of all the centers of the experimental data rectangles, whose heights correspond to 2σ for the experimental points. An exponent of 1.64 is about what is expected for MAADF imaging on theoretical grounds (Hartel et al. 1996, Treacy 1982).

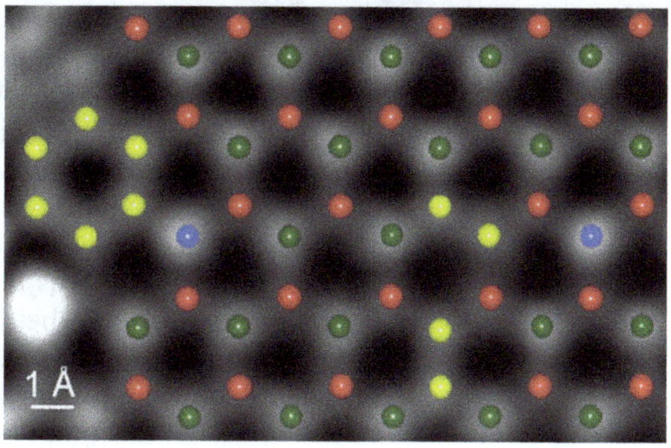

Figure 15–15. MAADF image of monolayer BN shown in Figure 15–13a, with a DFT-relaxed atomic model corresponding to the atomic types derived from the observed intensities shown on top of the image. Boron = *red,* carbon = *yellow,* nitrogen = *green,* and oxygen = *blue.* DFT model courtesy T.J. Pennycook. (Krivanek et al. (2010a), *Nature,* by permission).

There was one exception to the clear separation: the arrowed bar in the valley between the C and the N peaks. The intensity of the corresponding atom's peak was 3 standard deviations from the center of the C peak, and 5.6 standard deviations from the center of the N peak. This means that the atom was likely to be carbon at 94% confidence level. For all the other atoms, the probability of having made the correct assignment was >99%.

Figure 15–15 shows the resultant atomic model superimposed on the experimental image. The oxygen atoms substituted for nitrogen atoms, singly, whereas the carbon atoms substituted for boron and nitrogen atoms in pairs. The paired substitution avoided an energy penalty due to the unbalanced charge distribution associated with a single substitutional carbon atom in BN, and is therefore not surprising.

The substitutional atoms created small in-plane distortions in the BN lattice next to them. In particular, the O atoms pushed their nearest neighbors away by about 0.1 Å. This is most readily seen for the O atom next to the C hexagon: the C atom nearest to the oxygen is pushed toward the center of the carbon ring. The stability of the substitutions was verified by density-functional theory (DFT) calculations (Krivanek et al. 2010a). The calculations also confirmed the lattice distortions caused by the substitutional atoms, although the amplitude of the distortions predicted by DFT was about 50% smaller than the distortions observed experimentally.

Going beyond the single-layer BN, the image area on the lower left side of Figure 15–13a contains three bright spots, whose intensity is a good match for sodium atoms sitting over N atoms in the BN layer. The image area to the left of and above the carbon hexagon in Figure 15–13a shows a disordered second layer lying mostly over a continuation of the BN layer. It provides tantalizing glimpses of a disordered 3D structure, some of whose atoms we are able to place. However, we have not been able to model the entire structure. These kinds of investigations may well become more fruitful when the low-energy STEM resolution improves further, as discussed in Section 15.4.

It is useful to note that there have been several previous attempts to distinguish boron atoms from nitrogen atoms in monolayer BN using bright field phase-contrast imaging (Alem et al. 2009, Jin et al. 2009, Meyer et al. 2009), but that none of them has succeeded in being able to identify a particular atom in a single image as being either B or N. The reason is that the bright field scattering strength of the two types of atoms is very similar (Meyer et al. 2009), which makes it difficult to distinguish them without extensive averaging to improve the signal-to-noise ratio, either over several images, or many atomic sites, or both.

The "traditional wisdom" in electron microscopy used to be that annular dark field imaging in the STEM was excellent for imaging heavier atoms, but that it was not a suitable technique for imaging single light atoms because of their small cross sections. Imaging light atoms by ADF STEM is indeed more difficult than imaging heavy atoms. But the results shown here demonstrate that aberration correction has made incoherent ADF imaging of light atoms readily possible, and that this technique enjoys the standard advantages of incoherent dark field techniques: better resolution than axial BF imaging in the same instrument, quantitative results, and simple interpretation.

An instructive example illustrating the quantitative nature of incoherent ADF imaging is shown in Figure 15–16, which looks very much like an image of monolayer graphene: all the atomic maxima have about the same intensity. The image was one of the first atomic-resolution MAADF images we recorded from BN at 60 keV, and initially we were mostly producing images just like this one. This was rather confusing: we expected images in which the boron and nitrogen intensities differed substantially, i.e., images similar to those shown in Figures 15–11, 15–12, and 15–13. The explanation was not long in coming: we were looking at a double BN layer. The sample region we initially examined had many more double than single-layer areas, a property that we have seen repeatedly in BN samples, and most of the images we took were from double and thicker areas. The BN stacking is A–A′—boron atoms in the second layer lie over nitrogen atoms in

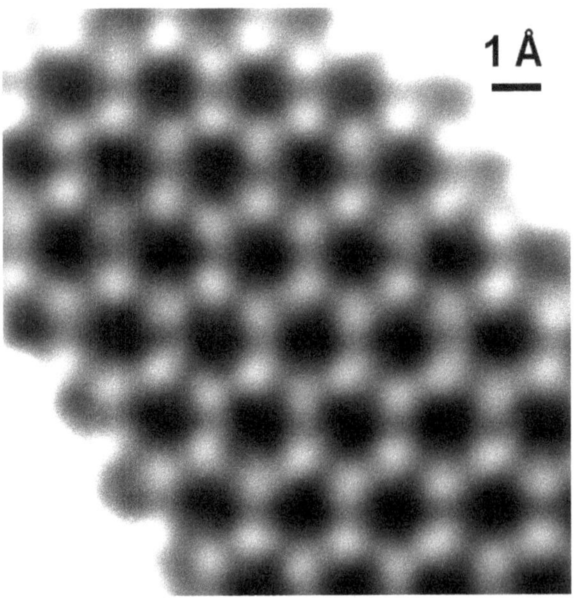

Figure 15–16. A smoothed, tail-subtracted and distortion-corrected MAADF image of double-layer BN. 60 keV, 64 μs per each 0.05 Å wide pixel. The image looks very similar to an image of monolayer graphene, except in one crucial aspect: its average intensity is double that of monolayer graphene. Nion UltraSTEM, sample courtesy Dr. V. Nicolosi, Oxford University.

the first layer and vice versa, and graphene-like contrast is therefore expected for the double layer.

The simple measurement which demonstrated that this was happening was a line profile spanning from the area shown in Figure 15–16 to a thicker area recorded in the same micrograph. Instead of the 1:2 ratio of average intensities that we expected, we obtained a ratio of 2:3. This showed clearly that the thin area was in fact thicker than a monolayer. Had we been looking at our sample using bright field phase-contrast imaging, no such quantitative tool allowing us to determine how many layers we were looking at would have been available.

15.3.4 EELS of Single Heavy Atoms

Our attempts to record EEL spectra from light impurity adatoms (Z~13) resting on graphene and monolayer BN have been largely unsuccessful so far. The chief reason is that the EELS cross sections are typically 100×–1000× smaller than the MAADF ones, and obtaining an EEL spectrum with a good signal-to-noise ratio therefore requires that the electron probe spends much longer over each atom. Adatoms are not strongly bound to graphene and BN, and nearly always run away while an EEL spectrum or a spectrum image is being recorded.

The situation is much more favorable for EELS if the impurity atom is confined, for instance when it is inside a nanopod, which can itself be inside a nanotube. Such samples are now being produced, sometimes with small molecules being confined in this way instead of single atoms (e.g., Suenaga et al. 2000, Koshino et al. 2007, Liu et al. 2007).

Figure 15–17 shows an MAADF image of single Er atoms inside C_{82} nanopods stuffed inside a single-wall nanotube, plus Er EEL spectra and an Er EELS map extracted from a spectrum image (Krivanek et al. 2010b).

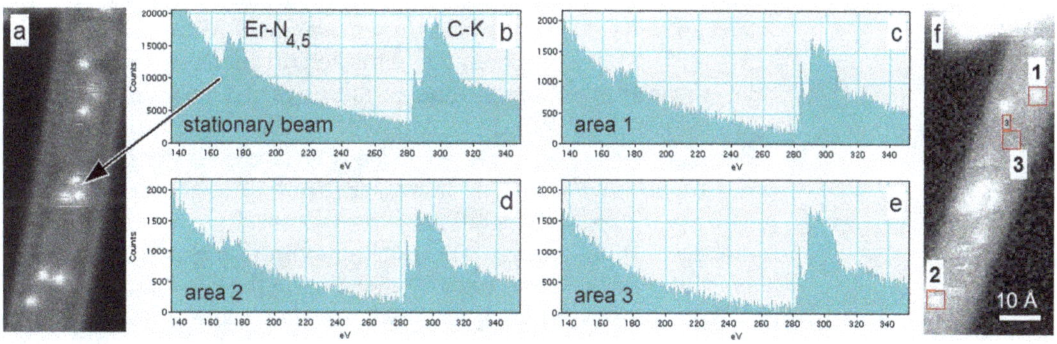

Figure 15–17. (**a**) An unprocessed MAADF image of a single-wall carbon nanotube filled with C_{82} nanopods, which originally contained one Er atom each. (**b**) EEL spectrum recorded with the STEM probe placed between the three Er atoms indicated by the arrow, at an acquisition time of 1 s. (**c–e**) EEL spectra extracted from the areas in spectrum image (**f**) marked by *small red rectangles* 1→c, 2→d, 3→e), (f) post-Er $N_{4,5}$ energy slice through a spectrum image recorded with 9 ms per each 0.5 × 0.5 Å pixel. (**c**) and (**d**) originate from single Er atoms, (**e**) originates from the carbon nanotube only. Nion UltraSTEM, 60 keV, sample courtesy Dr. K. Suenaga, AIST. (Krivanek et al. 2010b, *Ultramicroscopy*, by permission).

The Er atoms are readily visible in the MAADF image, which was obtained with about the same beam current (50 pA), a shorter per-pixel time (10 μs) and larger pixel size (0.12 Å) compared to the settings we normally use for imaging graphene and monolayer BN. The spectra show good signal-to-noise ratios, more than adequate for identifying the single Er atoms.

The spatial resolution predicted by Eqs. (13b) and (14) for a 1.4 Å probe and a 170 eV energy loss (the energy of the Er $N_{4,5}$ edge threshold) at 60 keV primary energy is 3.6 Å. This is in good agreement with the data shown in Figure 15–17. It is much larger than the probe size. Even with the carbon K-edge energy of 285 eV, the EELS resolution at 60 keV was a relatively poor 2.6 Å (Krivanek et al. 2010b).

15.4 Current State of the Art and Future Directions

15.4.1 Present Status

The observations shown here document several important points:

(1) The signal-to-noise ratio (SNR) in ADF images of individual atoms as light as boron is now high enough to allow the atoms to be imaged clearly, and to determine the chemical types of the atoms by their ADF intensity. The ADF discrimination capability is at its best when the atoms are non-overlapping, as in the BN example shown here, but it should also be applicable to heavier atoms lying on lighter thin substrates.

(2) Our measurements showed that the dependence of the MAADF image intensity I of a single atom on its atomic number Z goes as $I = a\,Z^n$, where a is a constant and the measured exponent n is 1.64. This was determined for an illumination semi-angle of 35 mr and an inner ADF cut-off angle semi-angle of 55 mr, for the range $5 \leq Z \leq 8$. The exponent is expected to increase toward 2 for larger ADF inner angles, and to fall toward 1.3 for smaller inner angles (Treacy 1982). It is also expected to vary with Z, but the dependence is not expected to be a strong one, with n decreasing by about 5% as Z is increased from 10 to 90 (Treacy 1982). Moreover, the ADF signal also shows a weak dependence on the electron orbital configuration of the atom (Langmore et al. 1973, Humphreys 1979, see curve D of figure 5 in the Humphreys paper).

The one experimental result we have been able to find that compares the ADF intensities of a heavy and a light atom (Wall 1979) gives the surprising result $I_{\text{uranium}}/I_{\text{carbon}} = 9 \pm 4$, measured in a 40 keV STEM. For $n = 1.5$, the ratio should be $92^{1.5}/6^{1.5} = 60$, yet the measurement appears to have stood the test of time so far. A ratio of 9 corresponds to an exponent of just 0.8. If it is correct, our understanding of useful approximations that model the ADF scattering process is clearly rather limited, and it is high time we revisited the subject.

(3) In order to estimate how well the atomic identification by ADF intensity is likely to work for heavier atoms, we assume for now that an explanation for the above discrepancy will be found and that for sufficiently high collection angles, the exponent n will be confirmed to remain within about 10% of 1.64 across the whole periodic table. Using the histogram method illustrated by Figure 15–14a, the separation of the peaks for adjacent elements will then increase as $\Delta I/\Delta Z \sim 1.64\ a\ Z^{0.64}$. The width of the histogram peaks in an image whose SNR is limited by the finite statistics due to the limited electron dose. The width will increase as $\sqrt{I} = a^{0.5}\ Z^{0.82}$, which means that the relative separation of the histogram peaks (the absolute separation divided by the width of the peaks) for adjacent elements will decrease as $Z^{-0.18}$. This is a rather weak dependence: the relative separation of Pt (Z = 78) and Au (Z = 79) histogram peaks will be about $(78/6)^{-0.18}$ = 63% of the relative separation of C and N peaks. Provided that no influences other than the image shot noise limit the precision with which individual atomic intensities can be measured, distinguishing isolated Pt atoms from Au ones with high confidence level should thus be possible with an electron dose that is $(1/0.63)^2$ = 2.5× times higher than the one used in the present work for distinguishing B, C, N, and O atoms, i.e., about 2×10^7 electrons per Å^2. Distinguishing iridium (Z = 77) from Au should be possible with a slightly smaller dose than used here.

The above derivation clearly depends on the value of the exponent n being known across the periodic table and points out the importance of measuring n experimentally.

(4) Extending the SNR considerations to lighter elements shows that individual atoms of all elements down to H present in monolayer samples should be identifiable by ADF imaging with a slightly smaller dose than the one used in this work, provided that they remain stationary while the electron beam is scanning over them. Unfortunately, hydrogen or helium atoms are not likely to remain stationary at the high doses used for this kind of imaging and identification. They will therefore continue to be hard to detect directly, even though their detection has become possible in principle.

(5) The high SNR of MAADF imaging makes it possible to distinguish whether an individual atom was in place for each pixel in an atomic image spanning an area consisting of 100 or more pixels, for atoms as light as carbon. This is allowing atomic motions to be studied on a time scale corresponding to the per-pixel dwell time, in our case 10 μs (for single Er atoms) and 64 μs (for single carbon atoms). (This is shown more clearly in Krivanek et al. 2010b, c.)

(6) Heavy atoms in nanotubes and Z ~ 13 atoms on graphene were seen to be mobile even when the beam was not directly over them. The atoms tended to be more stationary when the electron dose was smaller, and this suggests that the beam had to be in the general vicinity in order for the atoms to move. Previous studies of atomic motion with Crewe's original 30 keV STEM (Isaacson et al.

1977, Crewe 1979) have suggested that much of the atomic motion is thermal in origin, but the higher primary energy used here (60 keV rather than 30 keV) may have made the beam-induced effects more important. In the future, when probe correctors of chromatic aberration start being more widely available, 30 keV may become a common operating energy.

(7) Delocalization of inelastic scattering makes EELS images of single atoms larger than the diameter of the electron probe. It now limits the spatial resolution of EELS mapping rather more than the electron-optical performance of the STEM. An accurate correlation between theoretical and experimental values for the delocalization is still being worked on, as is a verification of whether the delocalization is dependent on the primary energy as predicted by Egerton's approximation.

(8) An electron energy loss spectrum from a single atom with a suitable EELS edge can now be collected with good SNR in a few ms, provided that the edge has a large cross section and the atom remains stationary during the acquisition. Nevertheless, there are several factors that make atomic identification by EELS more complementary than competitive with atomic identification using MAADF imaging:
    (a) much higher doses need to be used for the EELS, making it likely that the atoms of interest will run away,
    (b) as described in point (7), the spatial resolution of the EELS elemental map is typically much worse than the probe size. For energy losses smaller than about 300 eV, it is typically not good enough to resolve the nearest neighbors in closely packed materials,
    (c) EELS edges suitable for elemental mapping, i.e., edges with energies between about 100 and 2000 eV, with well-defined thresholds and sufficiently large cross sections, are only available for about half the elements in the periodic table. This means that EELS mapping cannot become a general technique applicable to all atomic species. EELS mapping therefore needs to be supplemented either by ADF imaging or by other spectroscopic techniques such as energy-dispersive X-ray spectroscopy (EDXS).

(9) Holes were made in a BN monolayer away from its edges by a 60 keV beam, even though this energy was below the theoretical knock-on displacement threshold of 78 keV. Two explanations appear possible.

First, an intermediary agent may be able to transfer more energy from an incident electron to a B or N atom that can be transferred by a direct electron–B or electron–N collision. Hydrogen could be such an agent: it is known to be able to lower the knock-on threshold energy of its neighbors by acting as an impedance-matching medium, whereby a fast electron impacts the proton that constitutes the hydrogen nucleus, and the proton immediately impacts an atom in the lattice. This mechanism can transfer a quantum of energy to a lattice atom that is 3.7× higher that the maximum amount that can be transferred directly from the fast electron (Bond et al. 1987). Hydrogen was likely present as a migrating adatom species on the BN monolayer surface. The cross section for this double process is likely to be small, and the frequency of the event will depend on how many migrating hydrogen atoms there are.

Second, an electronic transition may be responsible for the ejection of the initial atom, e.g., a double or even higher ionization.

Given the fact that hole creation was a rare event, and that it was seen to occur mostly in areas close to the hydrocarbon overlayers, the first explanation seems more likely. Work at different primary energies and with different types of light-Z materials is likely to clarify the mechanism in the future.

(10) In order to improve the resolution of ADF images taken at low primary energies, the chromatic aberration limit will have to be overcome. Implementing the correction may allow atomic resolution to be reached at even lower primary energies, and this may improve the EELS spatial resolution.

15.4.2 Future Directions

In pre-aberration correction days, a STEM that could form a probe whose size was 50 λ (1.3 Å at 200 keV) was "top of the line." With aberration correction, we have progressed to probe sizes of the order of 20 λ (1 Å at 60 keV). In order to progress to 0.5 Å probe size at 60 keV, or to 1 Å at 20 keV, the probe size will have to come down to around 10 λ. This is a tall order even by aberration correction standards, but it should be reachable with chromatic correction, and a further reduction of instrumental instabilities. Practical C_C correctors for STEMs are now being developed (Haider et al. 2009, Krivanek et al. 2009b, Zach 2009), and it will be interesting to see how well they succeed in improving the spatial resolution available at low primary energies.

The principal goal of making the probe size smaller is to concentrate the signal from each atom into a smaller area. This will result not only in an improved ability to resolve atoms lying close to each other, but also in a better signal-to-noise ratio for individual atoms at a given electron dose. At resolutions better than 1 Å, we will be taking advantage of the fact that ADF STEM basically images the atomic nucleus, which is very much smaller than the electron orbitals around it. The potential ADF resolution is therefore considerably higher than the resolution of techniques that image the outer electron orbitals, such as STM and AFM.

The main reason for lowering the operating energy further will be to avoid radiation damage and other sample instabilities to such an extent that complex structures can be analyzed with irradiation doses large enough to allow atom-by-atom imaging and analysis. This will be especially important for materials such as graphene and BN, in which there appears to be no major ionization damage and thus no major damage mechanism left when knock-on damage is eliminated. A secondary reason will be to try to improve the resolution of EELS elemental mapping, provided of course that it is confirmed that the resolution does improve at lower primary energies.

The analysis that we have been able to perform on monolayer BN with impurities shows that in favorable circumstances, every single atom in a small area of a sample can now be resolved and individually identified. The challenge posed by this success is to see whether the approach can be made applicable to small molecules of unknown structure. This will amount to extending the tantalizing glimpses of 3-D sample structures that we saw in the thicker sample areas of Figures 15–11 and 15–13 to a full 3-D characterization of non-periodic structures, such as molecules of unknown shapes.

Some molecules may be able to withstand the high dose we have used here, but many others will require doses of less than 100 electrons/Å^2. The best way to achieve 3-D atomic resolution will then probably be to adapt the low-dose techniques developed for determining macromolecular structures (e.g., Frank 2006) to ADF imaging. The molecules can probably be supported on hydrophilic substrates such as monolayer graphene oxide (Pantelic et al. 2010), and they could also be encapsulated in nanotubes. The image contribution from these kinds of support structures can in principle be modeled and subtracted perfectly (apart from an increase in statistical noise), rendering them nearly invisible.

Imaging a large number of identical (and well separated) molecules of random orientations may then lead to an atomically resolved 3D structure of the molecule, at illumination doses low enough to avoid serious damage, even in biological molecules containing hydrocarbon chains. The task should be made easier by using chemical information about the molecule to narrow down the search among candidate structures. Since we often know the amino acid sequence making up a particular protein, but do not know the precise structure of the protein, this kind of capability should find a very wide range of applications.

The sample temperature may need to be lowered for this work, without sacrificing the sample stability. This is, however, a problem that has been solved several times before, as much of biological microscopy is carried out only at low temperatures.

Another promising avenue for the capabilities demonstrated here will be to analyze and track the motion of individual atoms of various species. Our improving ability to image single atoms has brought many insights into catalysis (e.g., Rashkeev et al. 2007) in which individual atoms such as Au, Pt, and Ru exhibit markedly different catalytic properties compared to atomic aggregates. This kind of ability can now be extended to lighter atoms, and it is a safe bet that it will lead to new insights into both natural and man-made materials.

Determining which atom is which using the atom's ADF intensity would be a lot more precise if one had the ability to put down atomic markers of known species. This could take the form of a simple evaporator or an ion beam deposition system, preferably in-situ or designed so that it is easy to go back and forth between the deposition system and the microscope. We plan on constructing such a system and using it to measure experimentally the scattering cross sections for atoms across the periodic table.

15.5 Conclusion

David Cockayne has recently remarked that with aberration correction, it is as if a veil of fog has finally lifted from the things we look at with electron microscopes. Aberration-corrected gentle STEM is making a major contribution to the lifting of the fog, which will undoubtedly clear up even more in the future. The dreams of electron microscopy pioneers such as Ruska, Scherzer, Gabor, and Crewe about being able to see the atomic structure of matter with an electron microscope are thus being realized, step by step. It is an exciting time to be active in this field, and to be helping to advance it further.

Acknowledgment We are grateful to Steve Pennycook and the Oak Ridge National Laboratory for the use of Nion UltraSTEM after its installation, to David Geohegan, Valeria Nicolosi, and Kazu Suenaga for the provision of samples, to George Corbin, Chris Own, James Woodruff, and Zoltan Szilagyi for their part in the design and construction of the Nion UltraSTEM, and to Neil Bacon, Phil Batson, David Cockayne, Ray Egerton, David Muller, Peter Nellist, Steve Pennycook, Tim Pennycook, Peter Rez, John Spence and Mike Treacy for useful discussions. Research at Oak Ridge National Laboratory (MFC) was sponsored by the Materials Sciences and Engineering Division of the U.S. Department of Energy.

References

N. Alem, R. Erni, C. Kisielowski, M.D. Rossell, W. Gannett, A. Zettl, Atomically thin hexagonal boron nitride probed by ultrahigh-resolution transmission electron microscopy. Phys. Rev. **B80**, 155425 (2009)

G.D. Archard, Two new simplified systems for the correction of spherical aberration in electron lenses. Proc. R. Soc. **B68**, 156–164 (1955)

M. von Ardenne, in *The Beginnings of Electron Microscopy. Advances in Electronics and Electron Physics*, ed. by P.W. Hawkes, Suppl. 16 (Academic, London, 1985), pp. 1–21

M. von Ardenne, Das Elektronen-Raster Mikroskop (The scanning electron microscope). Z. Tech. Phys **19**, 407–416 (1938)

M. von Ardenne, *Elektronen Übermikroskopie* (Springer, Berlin, 1940)

N.J. Bacon, G.J. Corbin, N. Dellby, P. Hrncirik, O.L. Krivanek, M.F. Murfitt, Z.S. Szilagyi, A High-Performance 200 kV Cold Field Emission Electron Source. *Proceedings 12th International Microscopy Congress* (IFSM, Rio de Janeiro, 2010) pp. 240–241

P.E. Batson, Low-workfunction field-emission source for high-resolution EELS. Proceedings 45th EMSA Meeting, ed. by G.W. Bailey (San Francisco Press, San Francisco, CA, 1987), pp. 132–133

P.E. Batson, Characterizing probe performance in the aberration corrected STEM. Ultramicroscopy **106**, 1104–1114 (2006)

P.E. Batson, Control of parasitic aberrations in multipole optics. J. Electron. Microsc. **58**, 123–130 (2009)

P.E. Batson, N. Dellby, O.L. Krivanek, Sub-Ångström resolution using aberration corrected electron optics. Nature **418**, 617–620 (2002)

V. Beck, Hexapole spherical-aberration corrector. Optik **53**, 241–245 (1979)

V. Beck, A.V. Crewe, High resolution imaging properties of the STEM. Ultramicroscopy **1**, 137–144 (1975)

V. Beck, A.V. Crewe, A quadrupole–octupole corrector for a 100 keV STEM. Proc. 32nd EMSA Meeting, 578–579 (1976)

G.M. Bond, I.M. Robertson, F.M. Zeides, H.K. Birnbaum, Sub-threshold electron irradiation damage in hydrogen-charged aluminium. Phil. Mag. A **55**, 669 (1987)

M. Bosman, V.J. Keast, J.L. García-Muñoz, A.J. D'Alfonso, S.D. Findlay, L.J. Allen, Two-dimensional mapping of chemical information at atomic resolution. Phys. Rev. Lett. **99**, 086102 (2007)

E.C. Cosgriff, M.P. Oxley, L.J. Allen, S.J. Pennycook, The spatial resolution of imaging using core-loss spectroscopy in the scanning transmission electron microscope. Ultramicroscopy **102**, 317–326 (2005)

A.V. Crewe, Direct imaging of single atoms and molecules using the STEM. Proceedings 47th Nobel Symposium, Chemica Scripta **14**, 17–20 (1979)

A.V. Crewe, in *Cold Field Emission and the Scanning Transmission Electron Microscope*, ed. by P.W. Hawkes. Advances in Imaging and Electron Physics, vol. 159. (Elsevier, Amsterdam, 2009), pp. 1–61

A.V. Crewe, D. Kopf, A sextupole system for the correction of spherical aberration. Optik **5**, 1–10 (1980)

A.V. Crewe, D.N. Eggenberger, J. Wall, L.M. Welter, Electron gun using a field emission source. Rev. Sci. Inst. **39**, 576–583 (1968a)

A.V. Crewe, J. Wall, L.M. Welter, A high-resolution scanning transmission electron microscope. J. Appl. Phys. **39**, 5861–5868 (1968b)

A.V. Crewe, J. Wall, J. Langmore, Visibility of single atoms. Science **168**, 1338–1340 (1970)

N. Dellby, O.L. Krivanek, P.D. Nellist, P.E. Batson, A.R. Lupini, Progress in aberration-corrected scanning transmission electron microscopy. J. Electron. Microsc. **50**, 177–185 (2001)

J.H.M. Deltrap, Correction of spherical aberration with combined quadrupole-octopole units. Proc. EUREM-3 (Prague), 45 (1964a)

J.H.M. Deltrap, Correction of spherical aberration of electron lenses. Ph.D. dissertation, University of Cambridge, 1964b

R.F. Egerton, *EELS in the Electron Microscope*, 2nd edn. (Plenum Press, New York, NY, 1996)

R.F. Egerton, Limits to the spatial, energy and momentum resolution of electron energy-loss spectroscopy. Ultramicroscopy **107**, 575–586 (2006)

R. Erni, M.D. Rossell, C. Kisielowski, U. Dahmen, Atomic-resolution imaging with a sub-50-pm electron probe. Phys. Rev. Lett. **102**, 096101 (2009)

P.L. Fejes (ed.), *Proceedings of Cornell Specialist Workshop in Analytical Electron Microscopy, Cornell University*, July 1978

L. Fitting-Kourkoutis, From imaging individual atoms to atomic resolution 2D mapping of bonding. In *Proceedings 12th International Microscopy Congress* (IFSM, Rio de Janeiro, 2010), pp. 354–355

J. Frank, *Three-Dimensional Electron Microscopy of Macromolecular Assemblies* (Oxford University Press, Oxford, 2006)

H.L. Fraser, R.H. Geiss, M.S. Isaacson, D.C. Joy, D.M. Maher, J. Silcox (eds.), *Proceedings of Cornell Specialist Workshop in Analytical Electron Microscopy*, Cornell Univeristy, August 1976

B. Freitag, G. Knippels, S. Kujawa, P.C. Tiemeijer, M. Van der Stam, D. Hubert, C. Kisielowski, P. Denes, A. Minor, U. Dahmen, First Performance Measurements and Application Results of a New High Brightness Schottky Field Emitter for HR-S/TEM at 80–300 kV Acceleration Voltage, ed. by M. Luysberg, K. Tillmann, T. Weirich. European Microscopy Congress, vol. 1: Instrumentation and Methods (2008), pp. 55–56

A.K. Geim, Graphene: Status and prospects. Science **324**, 1530–1534 (2009)
Ç.Ö. Girit, J.C. Meyer, R. Erni, M.D. Rossell, C. Kisielowski, L. Yang, C.-H. Park, M.F. Crommie, M.L. Cohen, S.G. Louie, A. Zettl, Graphene at the edge: Stability and dynamics. Science **323**, 1705–1708 (2009)
M. Haider, P. Hartel, H. Müller, S. Uhlemann, J. Zach, Current and future aberration correctors for the improvement of resolution in electron microscopy. Phil. Trans. R. Soc. A **367**, 3665–3682 (2009)
M. Haider, S. Uhlemann, E. Schwan, H. Rose, B. Kabius, K. Urban, Electron microscopy image enhanced. Nature **392**, 768–769 (1998)
D.F. Hardy, Combined Magnetic and Electrostatic Quadrupole Electron Lenses. Ph.D. dissertation, University of Cambridge, 1967
H.S. von Harrach, in *Cold Field Emission and the Scanning Transmission Electron Microscope*, ed. by P.W. Hawkes. Advances in Imaging and Electron Physics, vol. 159 (Elsevier, Amsterdam, 2009), pp. 287–323
P. Hartel, H. Rose, C. Dignes, Conditions and reasons for incoherent imaging in STEM. Ultramicroscopy **63**, 93–114 (1996)
A. Hashimoto, K. Suenaga, A. Gloter, K. Urita, S. Iijima, Direct evidence for atomic defects in graphene layers. Nature **430**, 870–873 (2004)
P.W. Hawkes (ed.), *Aberration-Corrected Electron Microscopy*. Advances in Imaging and Electron Physics, vol. 153 (Elsevier, Amsterdam, 2008)
P.W. Hawkes (ed.), *Cold Field Emission and the Scanning Transmission Electron Microscope*. Advances in Imaging and Electron Physics, vol. 159 (Elsevier, Amsterdam, 2009a)
P.W. Hawkes, Aberration correction past and present. Phil. Trans. R. Soc. A **367**, 3637–3664 (2009b)
Y. Hernandez, V. Nicolosi, M. Lotya, F.M. Blighe, Z. Sun, S. De, I.T. McGovern, B. Holland, M. Byrne, Y. Gun'ko, J. Boland, P. Niraj, G. Duesberg, S. Krishnamurthy, R. Goodhue, J. Hutchison, V. Scardaci, A.C. Ferrari, J.N. Coleman, High yield production of graphene by liquid phase exfoliation of graphite. Nat. Nanotechnol. **3**, 563–568 (2008)
C.J. Humphreys, The scattering of fast electrons by crystals. Rep. Prog. Phys. **42**, 1825–1887 (1979)
M. Isaacson, D. Johnson, Microanalysis of light-elements using transmitted energy-loss electrons. Ultramicroscopy **1**, 33–52 (1975)
M. Isaacson, D. Kopf, M. Utlaut, N.W. Parker, A.V. Crewe, Direct observation of atomic diffusion by scanning transmission electron microscopy. Proc Natl. Acad. USA. **74**, 1802–1806 (1977)
M. Isaacson, D. Kopf, M. Ohtsuki, M. Utlaut, Atomic imaging using the dark field annular detector in the STEM. Ultramicroscopy **4**, 101–104 (1979)
C. Jin, F. Lin, K. Suenaga, S. Iijima, Fabrication of a freestanding boron nitride single layer and its defect assignments. Phys. Rev. Lett. **102**, 195505 (2009)
K. Kimoto, T. Asaka, T. Nagai, M. Saito, Y. Matsui, K. Ishizuka, Nature **450**, 702–704 (2007)
E.J. Kirkland, *Advanced Computing in Electron Microscopy* (Plenum Press, New York and London, 1998), p. 69
H. Kohl, H. Rose, Theory of image formation by inelastically scattered electrons in the electron microscope. Adv. Imaging Electron. Phys. **65**, 173–227 (1985)
M. Koshino, T. Tanaka, N. Solin, K. Suenaga, H. Isobe, E. Nakamura, Imaging of single organic molecules in motion. Science **316**, 85 (2007)
O.L. Krivanek, M.F. Chisholm, V. Nicolosi, T.J. Pennycook, G.J. Corbin, N. Dellby, M.F. Murfitt, C.S. Own, Z.S. Szilagyi, M.P. Oxley, S.T. Pantelides, S.J. Pennycook, Atom-by-atom structural and chemical analysis by annular dark-field electron microscopy. Nature **464**, 571–574 (2010a). See also the on-line supplementary information available at http://www.nature.com/nature/journal/v464/n7288/suppinfo/nature08879.html

O.L. Krivanek, G.J. Corbin, N. Dellby, B.F. Elston, R.J. Keyse, M.F. Murfitt, C.S. Own, Z.S. Szilagyi, J.W. Woodruff, An electron microscope for the aberration-corrected era. Ultramicroscopy **108**, 179–195 (2008b)

O.L. Krivanek, N. Dellby, R.J. Keyse, M.F. Murfitt, C.S. Own, Z.S. Szilagyi, in *Advances in Imaging and Electron Physics*, ed. by P.W. Hawkes (Academic Press, London, 2008a), pp. 121–155

O.L. Krivanek, N. Dellby, A.R. Lupini, Towards sub-Å electron beams. Ultramicroscopy **78**, 1–11 (1999)

O.L. Krivanek, N. Dellby, M.F. Murfitt, in *Handbook of Charged Particle Optics*, 2nd edn., ed. by J. Orloff (CRC Press, Boca Raton, 2009a), pp. 601–640

O.L. Krivanek, N. Dellby, M.F. Murfitt, M.F. Chisholm, T.J. Pennycook, K. Suenaga, V. Nicolosi, Gentle STEM: ADF imaging and EELS at low primary energies. Ultramicroscopy **110**, 935–945, (2010b)

O.L. Krivanek, N. Dellby, M.F. Murfitt, Z.S. Szilagyi, M.F. Chisholm, K. Suenaga, Slow and fast atomic motion observed by aberration-corrected STEM. In *Proceedings MSA meeting (Portland)*, Microscopy and Microanalysis **16** (Suppl. 2), 70–71 (2010c)

O.L. Krivanek, N. Dellby, A.J. Spence, R.A. Camps, L.M. Brown, Aberration correction in the STEM, in *Proceedings 1997 EMAG meeting*, ed. by J.M. Rodenburg, (Institute of Physics Conference Series vol. 153, 1997), pp. 35–40

O.L. Krivanek, J.P. Ursin, N.J. Bacon, G.J. Corbin, N. Dellby, P. Hrncirik, M.F. Murfitt, C.S. Own, Z.S. Szilagyi, High-energy-resolution monochromator for aberration-corrected scanning transmission electron microscopy/electron energy-loss spectroscopy. Phil. Trans. R. Soc. A **367**, 3683–3697 (2009b)

J.P. Langmore, J. Wall, M.S. Isaacson, The collection of scattered electrons in dark field electron microscopy. Optik **38**, 335–350 (1973)

Z. Liu, K. Yanagi, K. Suenaga, H. Kataura, S. Iijima, Imaging the dynamic behaviour of individual retinal chromophores confined inside carbon nanotubes. Nat. Nanotechnol. **2**, 422–425 (2007)

E.E. Martin, J.K. Trolan, W.P. Dyke, Stable, high density field emission cold cathode. J. Appl. Phys. **31**, 782–789 (1960)

J.C. Meyer, A. Chuvilin, G. Algara-Siller, J. Biskupek, U. Kaiser, Selective sputtering and atomic resolution imaging of atomically thin boron nitride membranes. Nano Lett. **9**, 2683–2689 (2009)

D.A. Muller, Structure and bonding at the atomic scale by scanning transmission electron microscopy. Nat. Mater. **8**, 263–270 (2009)

D.A. Muller, L. Fitting-Kourkoutis, M.F. Murfitt, J.H. Song, H.Y. Hwang, J. Silcox, N. Dellby, O.L. Krivanek, Atomic-scale chemical imaging of composition and bonding by aberration-corrected microscopy. Science **319**, 1073–1076 (2008)

D.A. Muller, J. Silcox, Delocalization in inelastic electron scattering. Ultramicroscopy **59**, 195–213 (1995)

H. Müller, S. Uhlemann, P. Hartel, M. Haider, Advancing the hexapole C_S-corrector for the scanning transmission electron microscope. Microsc. Microanal. **12**, 442–455 (2006)

P.D. Nellist, M.F. Chisholm, N. Dellby, O.L. Krivanek, M.F. Murfitt, Z.S. Szilagyi, A.R. Lupini, A. Borisevich, W.H. Sides, S.J. Pennycook, Direct sub-angstrom imaging of a crystal lattice. Science **305**, 1741–1742 (2004)

R.S. Pantelic, J.C. Meyer, U. Kaiser, W. Baumeister, J.M. Plitzko, Graphene oxide: a substrate for optimizing preparations of frozen-hydrated samples. J. Struct. Biol. **170**, 152–156 (2010)

W. Qian, M. Scheinfein, J.C.H. Spence, Brightness measurement of nanometer sized field emission electron sources. J. Appl. Phys. **73**, 7041–7045 (1993)

S.N. Rashkeev, A.R. Lupini, S.H. Overbury, S.J. Pennycook, S.T. Pantelides, Role of the nanoscale in catalytic CO oxidation by supported Au and Pt nanostructures. Phys. Rev. B **76**, 035438 (2007)

H. Rose, Abbildungseigenschaften sphärisch korrigierter elektronenoptischer Achromate. Optik **33**, 1–24 (1971)

H. Rose, Outline of a spherically corrected semiaplanatic medium-voltage transmission electron-microscope. Optik **85**, 19–24 (1990)

R. Saito, G. Dresselhaus, M.S. Dresselhaus, *Physical Properties of Carbon Nanotubes* (World Scientific Publishing, Singapore, 1998)

H. Sawada, F. Hosokawa, T. Kaneyama, T. Ishizawa, M. Terao, M. Kawazoe, T. Sannomiya, T. Tomita, Y. Kondo, T. Tanaka, Y. Oshima, Y. Tanishiro, N. Yamamoto, K. Takayanagi, Achieving 63 pm resolution in scanning transmission electron microscope with spherical aberration corrector. Jpn. J. Appl. Phys. **46**, L568–L570 (2007)

H. Sawada, Y. Tanishiro, N. Ohashi, T. Tomita, F. Hosokawa, T. Kaneyama, Y. Kondo, K. Takayanagi, STEM imaging of 47-pm-separated atomic columns by a spherical aberration-corrected electron microscope with a 300-kV cold field emission gun. J. Electron. Microsc. **58**, 357–361 (2009)

O. Scherzer, Sphärische und chromatische Korrektur von Elektronen-linsen, Optik **2**, 114–132 (1947)

Z. Shao, On the fifth order aberration in a sextupole corrected probe forming system. Rev. Sci. Instrum. **59**, 2429–2437 (1988)

K. Suenaga, Y. Sato, Z. Liu, H. Kataura, T. Okazaki, K. Kimoto, H. Sawada, T. Sasaki, K. Omoto, T. Tomita, T. Kaneyama, Y. Kondo, Visualizing and identifying single atoms using electron energy-loss spectroscopy with low accelerating voltage. Nat. Chem. **1**, 415–418 (2009)

K. Suenaga, M. Tence, C. Mory, C. Colliex, H. Kato, T. Okazaki, H. Shinohara, K. Hirahara, S. Bandow, S. Iijima, Element-selective single atom imaging. Science **290**, 2280–2282 (2000)

K. Suenaga, H. Wakabayashi, M. Koshino, Y. Sato, K. Urita, S. Iijima, Imaging active topological defects in carbon nanotubes. Nat. Nanotechnol. **2**, 358–360 (2007) (see also the supplementary materials)

L.W. Swanson, N.A. Martin, Field electron cathode stability studies: Zirconium/tungsten thermal-field cathode. J. Appl. Phys. **46**, 2029–2050 (1975)

L.W. Swanson, G.A. Schwind, Review of ZrO/W Schottky Cathode, in *Handbook of Charged Particle Optics*, 2nd edn., ed. by J. Orloff (CRC Baton Rouge, 2009), pp. 1–28

M.G.R. Thomson, The aberrations of quadrupole electron lenses. Ph.D. Dissertation, University of Cambridge, 1968

M.M.J. Treacy, Optimistic atomic number contrast in annular dark field images of thin films in the scanning transmission electron microscope. J. Microsc. Spectrosc. Electron. **7**, 511–523 (1982)

M. Varela, S.D. Findlay, A.R. Lupini, H.M. Christen, A.Y. Borisevich, N. Dellby, O.L. Krivanek, P.D. Nellist, M.P. Oxley, L.J. Allen, S.J. Pennycook, Spectroscopic imaging of single atoms within a bulk solid. Phys. Rev. Lett. **92**, 095502 (2004)

J.A. Venables, G. Cox, Computer modeling of field emission gun scanning electron microscope columns. Ultramicroscopy **21**, 33–46 (1987)

P.M. Voyles, D.A. Muller, J.L. Grazul, P.H. Citrin, H.-J.L. Gossman, Atomic-scale imaging of individual dopant atoms and clusters in highly n-type bulk Si. Nature **416**, 826–829 (2002)

J.S. Wall, Limits on visibility of single heavy atoms in the scanning transmission electron microscope – an experimental study. Proc. 47th Nobel Symp. Chemica Scripta **14**, 271–278 (1979)
J. Zach, Chromatic correction: a revolution in electron microscopy? Phil. Trans. R. Soc. A **367**, 3699–3707 (2009)
J. Zach, M. Haider, Aberration correction in a low-voltage SEM by a multipole corrector. Nucl. Instr. Meth. A **363**, 316–325 (1995)
A. Zobelli, A. Gloter, C.P. Ewels, G. Seifert, C. Colliex, Electron knock-on cross section of carbon and boron nitride nanotubes. Phys Rev. B **75**, 245402 (2007)
V.K. Zworykin, J. Hillier, R.L. Snyder, A scanning electron microscope. A.S.T.M. Bull. No. **117**, 15–23 (1942)

16

Low-Loss EELS in the STEM

Nigel D. Browning, Ilke Arslan, Rolf Erni and Bryan W. Reed

16.1 Introduction

The main goal of this chapter is to introduce the concept of "low-loss" or "valence loss" electron energy loss spectroscopy (VEELS) in the STEM. Much of the discussion will assume that the microscope is aligned to form the optimum probe size (as described in other chapters in this book) with only special attention being drawn to the monochromator and how its use modifies the electron optics of the microscope (i.e., how the probe is formed). VEELS is traditionally described by energy loss processes that are seen in the 0–50 eV region of the spectrum (Figure 16–1) and processes that are typically characterized as collective excitations. These collective oscillations can provide key insights into optical and electronic properties that are fundamentally different from the composition and structure information that is typically extracted from core-loss spectra. Here we will provide a basic physical model for these collective excitations that allows materials properties to be interpreted from experimental spectra acquired in the STEM. As the low-loss region of the experimental spectrum is dominated by the zero-loss peak, experimental considerations needed to acquire optimal spectra and subtract the zero-loss peak will also be discussed. Finally, a few examples of the use of monochromated VEELS to provide insights into materials properties will be discussed.

16.2 The Physics of the Low-Loss Spectrum

In this section, the main principles behind the formation of a VEEL spectrum will be discussed. Particular attention will be paid to the differences in spectra obtained from thick vs thin samples and to the effects of surfaces and interfaces on the observed spectrum. Guidelines in the use of VEELS to study materials properties will also be developed that indicate where a simple bulk interpretation is valid and where more complex modeling is required.

S.J. Pennycook, P.D. Nellist (eds.), *Scanning Transmission Electron Microscopy*,
DOI 10.1007/978-1-4419-7200-2_16, © Springer Science+Business Media, LLC 2011

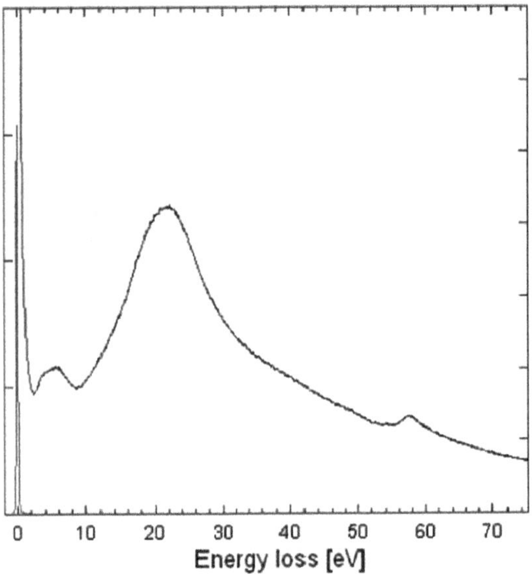

Figure 16–1. Low-loss spectrum of Fe_2O_3 showing the typical energy range and features of a low-loss EEL spectrum (reproduced from Erni et al. (2005) with permission).

16.2.1 Basic Concepts

Let us consider what sorts of material excitations exist in the low-loss region and how they might be modeled. The low-loss energy range is typically too high for phonons (although this is becoming less true for monochromated systems) and too low for inner-shell excitations. This means that so far as processes measured by low-loss EELS are concerned, the ions can usually be considered to be standing still, being neither ionized nor moved significantly in the process. So low-loss EELS is essentially measuring the response of the valence electrons—or, to be more precise, the combined response of the valence electrons and the electromagnetic field. This distinction is important. For example, in the case of a surface plasmon (SP), the excitation is actually a combination of a surface charge density wave and an electromagnetic wave, and much of the energy is stored in the vacuum near the surface of the material. Such modes can be excited in an "aloof" manner, i.e., with the probe electron never touching the material. If we thought of the excitation as being confined within the material, aloof excitation would be difficult to understand in intuitive, classical terms.

For many EELS experiments, a quasiclassical dielectric formalism captures most of the physics (Echenique et al. 1987, Egerton 1996, Ritchie 1957, 1981, Ritchie and Howie 1988, Rivacoba et al. 2000, Wang 1996, Zabala et al. 2001). In this formalism, a material's valence electron response is encoded in the complex dielectric function $\varepsilon(\omega)$ (or, in models including spatial dispersion, $\varepsilon(\boldsymbol{k},\omega)$). Once the geometry is specified, classical electrodynamics suffices to calculate the frequency spectrum of the modes excited by a probe electron. Quantum mechanics only appears in the identification of a discrete energy loss E with the temporal frequency via the equation $E = \hbar\omega$. The quantum mechanics and internal dynamics of the material are hidden within the dielectric function. The probability of exciting a given mode then comes straight out of Maxwell's equations for the classical electromagnetic field, explicitly solving for the response of the material to a passing electrical charge. This remarkably simple model successfully describes an enormous variety of experimental EELS measurements. It was developed in the early years of EELS (Echenique et al. 1987, Raether 1977, Ritchie 1957, 1981, Ritchie and Howie 1988), and recent years have seen relatively little fundamental development of the theory. Instead, most of the recent applications of the theory have had more of a computational flavor, applying the model to specific geometries and materials in order to assist in the interpretation of experimental spectra (Erni and Browning 2008, García de Abajo and Aizpurua 1997, García de Abajo and Howie 1998, García de Abajo and Sáenz 2005, Mkhoyan et al. 2007, Reed et al. 1999, Stöger-Pollach 2008, Stöger-Pollach and Schattsneider 2007, Ugarte et al. 1992). The model can be expanded (still within the realm of classical electromagnetism) by adding magnetic properties, anisotropy, and spatial dispersion. The formalism may need some modifications for nano-size effects (e.g., modified band gaps and plasma frequencies from quantum confinement and increased lifetime broadening from surface scattering) and the fundamentally quantum nature of the electron–solid interaction (Reed et al. 1999, Ritchie 1981, Ritchie and Howie 1988, Rivacoba et al. 2000, Stöckli et al. 1997, Ugarte et al. 1992, Wang 1996, Zabala et al. 2001). But in many practical cases such corrections can be neglected or easily accounted for, and the essentially classical result holds to a very good approximation.

And yet, despite the simplicity of the fundamental concepts (comprising nothing more than Maxwell's equations), the range of excitations is great and the interpretation of low-loss EELS can be quite tricky. This is because, depending on the geometry and the materials, there are quite a few ways to get a resonant coupling between a probe electron and a material's valence electrons. The spectrum can include peaks from surface plasmons, bulk plasmons, interband transitions, guided light and whispering gallery modes, and Cerenkov effects. The strengths, widths, shapes, and positions of all of these peaks carry different kinds of information about the sample. In principle, low-loss EELS is sensitive to effective carrier densities, band gaps, refractive indices, the shape and orientation of surfaces and interfaces, thin surface coatings, and (in some cases) magnetic properties and crystal orientation. Moreover, the spectrum depends on these parameters not only along the actual path of the probe electron but at distances of up to several nanometers in all directions. It also depends on the acceleration voltage and the convergence and collection angles used in the experiment. This is the double-edged sword of EELS; the spectrum is sensitive to everything. So while in principle the spectrum contains an enormous amount of information about the sample, it can be quite challenging to pull out a specific piece of information. Even deceptively simple tasks like measuring a dielectric function or a semiconductor band gap can be undermined by relativistic and surface effects (Erni and Browning 2008, Gu et al. 2007, Mkhoyan et al. 2007, Stöger-Pollach 2008, Stöger-Pollach and Schattsneider 2007, Zhang et al. 2008).

Fortunately in practice the problem is not as difficult as it may seem. While there are certainly pitfalls in the interpretation of spectra, these pitfalls are for the most part well understood, and solutions are available in the literature. By careful choice of experimental parameters, many confounding effects can be minimized. Undesired peaks can be suppressed (e.g., by choice of probe placement), and many of the more complicated effects can be neglected provided that a few simple rules are observed. For example, retardation effects can typically be neglected for nanoparticles much smaller than the corresponding free-space wavelength of light ($\lambda = 2\pi c/\omega$) (Ugarte et al. 1992), and probe electron coherence effects can usually be neglected if the collection angle is significantly larger than the convergence angle (Ritchie and Howie 1988). In borderline cases, sophisticated computational tools (García de Abajo and Aizpurua 1997, García de Abajo and Howie 1998) can be applied to ensure that the interpretation of spectra is correct.

One of the most common pitfalls is to confuse the bulk energy loss function $\mathrm{Im}(-1/\varepsilon)$ (discussed below) with the far more complicated and geometry-dependent total loss function. For nanostructured materials such as nanotubes, quantum dots, and metamaterials, the spectrum can be so dominated by surface and interface modes that the bulk loss function only accounts for a small fraction of the loss spectrum. The bulk loss function completely misses the modes most interesting for technological applications of such materials, including surface plasmon, exciton, and guided light modes. Fortunately, this pitfall is rapidly gaining recognition as nanostructured materials gain prominence.

16.2.2 Theoretical Background

As we have stated, the complex dielectric function $\varepsilon(\omega)$ (or $\varepsilon(\boldsymbol{k},\omega)$) is the main material property that governs the low-loss spectrum. Let us consider some typical $\varepsilon(\omega)$ curves for a generic metal and a generic semiconductor (Figure 16–2). While each material will have its own peculiar band structure, many materials will look qualitatively similar to one of these two graphs (although transition metals, e.g., may show additional complications from d-orbital effects). The metal (Figure 16–2a) roughly follows the response curve of a simple Drude model plasma, characterized by a plasma frequency ω_p and scattering time τ, while the semiconductor also approximates this behavior for frequencies well above the band gap (Figure 16–2b). The semiconductor exhibits a resonance at the band gap (with $\mathrm{Im}\varepsilon$ reaching a peak), and at low frequencies it acts like a simple dielectric, with a positive, almost purely real ε approaching a constant as ω approaches 0. For frequencies well above the band gap, there is not much difference in the dielectric behavior of metals and semiconductors; both act like simple plasmas, at least qualitatively.

Let us consider the Drude model in more detail. The dielectric function is given by Jackson (1975)

$$\varepsilon(\omega) = 1 - \frac{\omega_p^2}{\omega^2 + i\omega/\tau}. \tag{1}$$

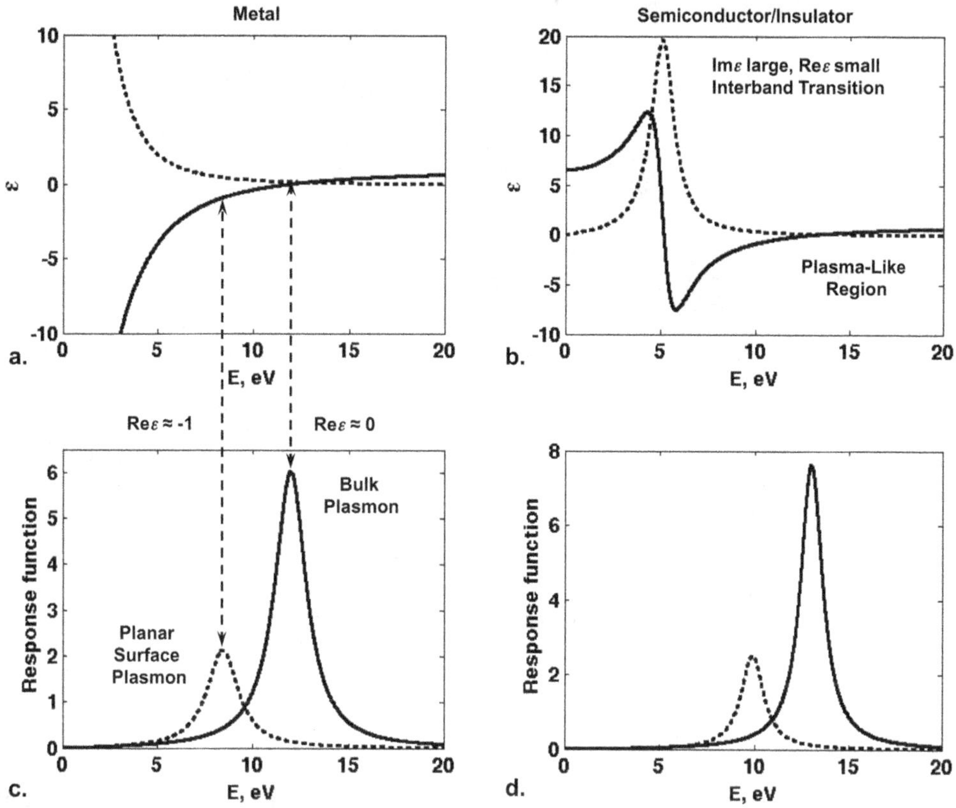

Figure 16–2. Typical complex dielectric and EELS response functions for (a, c) a metal and (b, d) a dielectric with a band gap ~5 eV. (a, b) Real (*solid*) and imaginary (*dashed*) parts of the dielectric functions. (c, d) The bulk loss function $\mathrm{Im}(-1/\varepsilon)$ (*solid*) and approximate planar surface loss function $\mathrm{Im}(-1/(\varepsilon+1))$ (dashed) for the functions in (a) and (b). These loss functions are missing contributions from retardation, spatial dispersion, and the geometry-dependent effects that occur in nanostructured materials.

The Drude model considers the ions to form a stationary, uniform, positive charge density. So long as the valence electrons in the bulk are also stationary, uniform, and of the same charge density, the system will be at equilibrium (Figure 16–3a). If a group of electrons is displaced and then released (Figure 16–3b), leaving behind an uncompensated positive charge, this creates a local electrical dipole moment and electric field that tends to restore the electrons to uniform density. The result is a plasma oscillation, with the group of electrons shifting back and forth at the characteristic frequency ω_p, gradually losing energy on the time scale τ to various damping processes (e.g., coupling between the electrons and the ion lattice). Such an oscillation, viewed quantum mechanically, is a bulk plasmon. A fast electron passing through the plasma (Figure 16–3c) will repel valence electrons as it does so. This sets up charge density waves that propagate outward. This is an intuitive classical picture of how bulk plasmons are created in an EELS experiment. The mobile charge in a metal moves so as to cancel out any applied electric field, but the finite electron mass and density

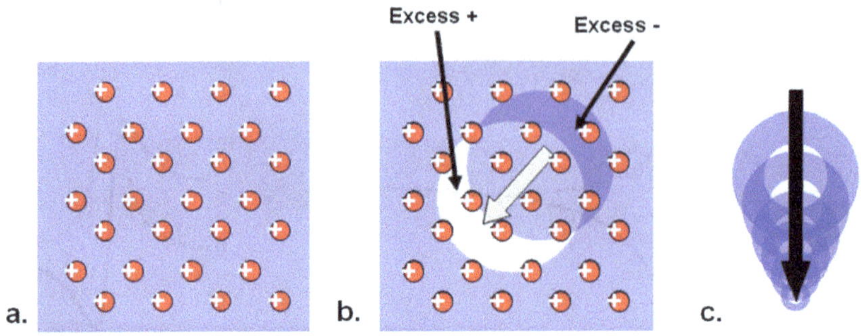

Figure 16–3 A conceptual schematic of a plasma oscillation or plasmon. (**a**) A metal at equilibrium, with a background of fixed positive charges and a sea of mobile electrons (*light gray*). The bulk macroscopically averaged positive charge density is equal in magnitude to the macroscopically averaged negative charge density. (**b**) When some of the electrons are collectively displaced, regions of excess positive and negative charge produce a local dipole moment and associated electric field. This results in a restoring force on the electrons (*block arrow*) that tends to eliminate the imbalance and restore the equilibrium state. Since the force is proportional to the mobile charge density and to the electron displacement, this yields simple harmonic motion at a well-defined frequency (the plasma frequency) that scales with the square root of the mobile charge density. (**c**) An intuitive picture of how a high-speed electron (*black arrow*) repels electrons in its vicinity, setting up charge density waves that propagate outward.

means that the response is not instantaneous but rather takes a characteristic time $\sim 1/\omega_{\mathrm{p}}$, which in the Drude model scales as the inverse square root of the mobile charge density. The corresponding characteristic frequency shows up as an energy loss peak at $E = \hbar\omega_p$ with a lifetime-broadened width of order $\Delta E = \hbar/\tau$. Thus the bulk plasmon peak carries information about both the density and the characteristic scattering time of the mobile valence electrons.

A semiconductor can be modeled by positing a similar dielectric function but with a finite number of real resonant frequencies ω_j with lifetimes τ_j and oscillator strengths f_j (the sum of which must be 1) (notation adapted from Jackson [1975, p. 285]),

$$\varepsilon(\omega) = 1 + \omega_p^2 \sum_j \frac{f_j}{\omega_j^2 - \omega^2 - i\omega/\tau}. \tag{2}$$

These resonances typically correspond to band gaps or, more generally, to peaks in the joint electron density of states for allowed direct interband transitions. Equation (1) is a special case of Eq. (2) with a single resonant frequency $\omega_j = 0$. There are of course much more sophisticated models of the dielectric function; we have only used these models as intuitive illustrations. For example, the Lindhard model includes spatial dispersion and a more fundamentally quantum mechanical view of the solid (Egerton 1996 and references therein). Modern quantum chemistry has yielded sophisticated computational tools for calculating band structures and dielectric response, and these can also be applied to EELS.

For an electron traveling at high speed through an extended, homogeneous material, the approximate energy loss probability per unit length, energy, and solid angle is given by the differential cross section

$$\frac{\partial^3 P}{\partial z \partial \Omega \partial E} = \frac{1}{\pi^2 a_0 m v^2} \frac{1}{\theta^2 + \theta_E^2} \mathrm{Im} \frac{-1}{\varepsilon}, \tag{3}$$

where a_0 is the Bohr radius, m is the electron mass, v is the probe electron velocity, θ is the scattering angle, and $\theta_E = E/(\gamma m v^2)$ is a characteristic scattering angle with γ the relativistic dilation factor (Egerton 1996, p. 150). The dominant material-dependent factor in this function is $\mathrm{Im}(-1/\varepsilon)$ (plotted as solid curves in Figure 16–2c and d), which is often called the bulk loss function.

Assuming that we are in a regime where the bulk loss function dominates (i.e., the surface-to-volume ratio is small and relativistic retardation effects are either insignificant or easily removed from the spectrum), low-loss EELS provides a way to measure the complex dielectric function throughout the visible to far-ultraviolet regime. Even though $\mathrm{Re}(-1/\varepsilon)$ cannot be directly measured by EELS, there are strict constraints on any complex material response function. These constraints derive from causality, i.e., an effect cannot—in any reference frame—precede its cause. For a frequency-dependent function like $1/\varepsilon(\omega)$, this means that the real and imaginary parts must be linked via a Kramers–Kronig transform (Egerton 1996, Jackson 1975); knowing one of these two functions allows the other to be computed. There are many pitfalls in the procedure (Stöger-Pollach 2008, Zhang et al. 2008), but when performed correctly, the Kramers–Kronig analysis provides a measurement of $\varepsilon(\omega)$ over a much wider range of ω than is possible for most optical measurement systems.

Next, we will discuss the modes that are not properly modeled by the bulk loss function $\mathrm{Im}(-1/\varepsilon)$. To start, consider the dielectric function for a typical semiconductor in Figure 16–2b. At low frequencies, this function is essentially that of a dielectric, with ε very nearly purely real and greater than 1. This means that the electrons are bound at low frequencies (so we do not see the plasma-like behavior of nearly free electrons), and also that the refractive index $n = \varepsilon^{1/2}$ itself is essentially real and greater than 1. In other words, below the band gap frequency, ordinary electromagnetic waves can propagate in the material at a phase speed c/n. The bulk loss function $\mathrm{Im}(-1/\varepsilon)$ (Figure 16–2d) is quite small in this regime, which would suggest that no such waves will be generated by the probe electron if this function were capturing all of the loss mechanisms. Yet there are at least two mechanisms to excite such waves that are not captured by the bulk loss function.

The first such mechanism is that of Cerenkov, whereby a cone of radiation is generated by a charged particle passing through the material faster than c/n. This can be a very important effect with high accelerating voltages (i.e., probe electrons moving at a large fraction of c) and high dielectric constants. In terms of the Maxwell equations, this mechanism arises from the coupling of the electric and magnetic fields through the time derivative terms, which are responsible for the fact that electromagnetic effects propagate at finite speed. Thus the Cerenkov effect is described as a relativistic or retardation effect. The growing importance of semiconductor band gap measurements in nanoparticles, coupled with the improved visibility of extreme low-loss (<2–3 eV) excitations in EELS through monochromation, has inspired a large number of recent publications discussing the Cerenkov effect (Erni and Browning 2008, Gu et al. 2007, Mkhoyan et al. 2007, Stöger-Pollach and Schattsneider 2007, Stöger-Pollach 2008, Zhang et al. 2008). Usually this effect is regarded as a nuisance, since it can obscure the semiconductor band gap and carries little information of its own.

The second mechanism is via so-called guided light modes (Raether 1977), which are exactly the same modes that carry information over many kilometers of optical fiber (Figure 16–4a) and thus are relevant for communications and optoelectronic applications. Inside the dielectric, electromagnetic waves can propagate with real wave vectors, reflecting

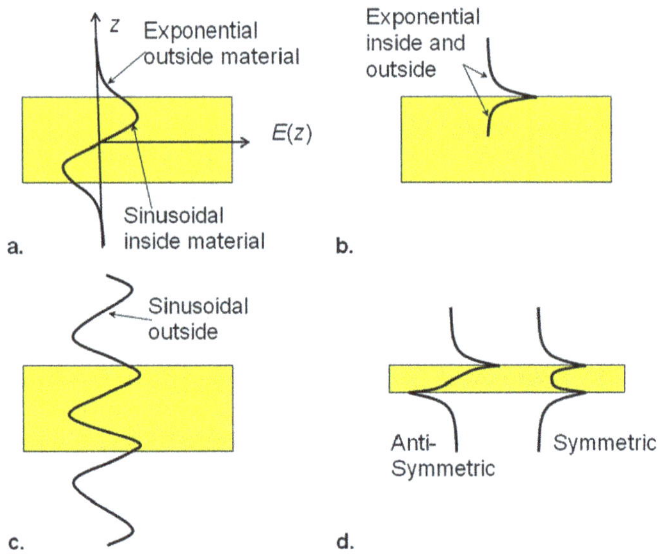

Figure 16–4. Various electromagnetic modes associated with surfaces, with schematic indications of the electric field amplitudes as a function of position z. The complex vector amplitudes are complicated by the elliptical polarization of the modes, which include both in-plane and normal components whose complex amplitude ratios are different inside the material than they are outside (see (Raether 1977) for more details). (**a**) Guided light mode. Inside the material, the mode propagates as an ordinary wave with a mostly real wave vector. Outside the material, total internal reflection produces an evanescent field that decays exponentially (i.e., with an imaginary normal wave vector component). (**b**) A non-radiative surface plasmon, with evanescent waves both inside and outside the material. This mode has too much in-plane momentum to decay directly to a free photon, so it remains bound to the surface. However, it can still couple to free photons via various mechanisms such as surface roughness scattering. (**c**) A radiative surface plasmon, with a real wave vector outside the material. This mode is not bound to the surface like a non-radiative mode is. (**d**) Coupling of surface plasmon modes when two surfaces are close enough for the evanescent waves to overlap. The energy eigenstates are now linear combinations of the modes on each surface.

off the material surface. If the reflection angles are large enough, all of the energy gets reflected back into the dielectric in the well-known phenomenon of total internal reflection. Evanescent waves, demanded by the boundary conditions implied by Maxwell's equations, decay exponentially into the vacuum. In other words, the normal component $k_\perp$ of the wave vector is purely imaginary outside the material, so that the usual $\exp(i\boldsymbol{k}\cdot\boldsymbol{r})$ expression implies an exponential decay in the normal direction. Whispering gallery modes (Hyun et al. 2008) are a very closely related phenomenon, representing essentially the same physics in a different geometry (i.e., quantized modes on a localized resonator rather than a band of related modes in an extended waveguide).

The guided light and whispering gallery modes are representative of a class of similar excitations that arise from solutions of Maxwell's equations in the presence of one or more surfaces where the dielectric function changes discontinuously (i.e., interfaces and vacuum-exposed surfaces) (Raether 1977, Ritchie 1973, Rivacoba et al. 2000, Wang 1996). The typical guided light mode occurs in a frequency regime in which all of the refractive indices are essentially real and greater than or equal to 1. Yet Figure 16–1 shows that $n = \varepsilon^{1/2}$ can also be less than 1, or even complex or purely imaginary, for a metal or a semiconductor measured above its band gap frequency. The advent of metamaterials even suggests that n (in an appropriate spatial average (García de Abajo and Sáenz 2005)) can be real and negative. This opens up other possibilities for the solutions of Maxwell's equations. When ε is negative, it is possible for $k_\perp$ to be purely imaginary both inside and outside the surface, so that the wave is evanescent on both sides of the surface (Figure 16–4b). This surface-bound wave is a surface plasmon (SP) or interface plasmon (IP), which combines a surface charge density wave with a surface-bound electromagnetic wave. Neglecting retardation, for an isolated, planar surface the condition $\mathrm{Re}(\varepsilon_\mathrm{A}+\varepsilon_\mathrm{B}) = 0$ (where A and B are the two materials) determines a necessary resonance condition for an SP or IP mode, with ε_A and ε_B the dielectric functions on either side of the interface. In the case of an SP, one of these is vacuum ($\varepsilon_\mathrm{B} = 1$), and the condition is $\mathrm{Re}(\varepsilon_\mathrm{A}) = -1$. Getting a strong resonance also requires that $\mathrm{Im}(\varepsilon_\mathrm{A})$ be relatively small. Adding a dielectric surface layer (thus increasing ε_B) tends to shift the energy downward.

More generally, the probability of exciting a given surface plasmon mode depends on the position of the probe and the geometrical shape of the mode (specified, e.g., by the in-plane component $\boldsymbol{k}_{||}$ for a planar surface, or by a longitudinal wave vector k_z and azimuthal mode number m for a cylinder). The function can be fairly complicated (Echenique et al. 1987, Rivacoba et al. 2000, Stöger-Pollach 2008, Ugarte et al. 1992, Zabala et al. 2001) but is typically dominated by a first-order pole at the spatiotemporal resonance condition for that mode. In practical terms, the most important material-dependent factor for SP and IP losses is dominated by a pole that scales as $\mathrm{Im}(-1/(\varepsilon + p))$, with p a dimensionless quantity that depends on the geometry of the mode. Neglecting retardation, an isolated, planar surface yields a mode with $p = 1$ (plotted in Figure 16–1c, d), while the Mie mode (i.e., dipole mode) on a sphere has $p = 2$ (Raether 1977). Bulk plasmon responses follow a similar functional form with $p = 0$. When two surfaces or interfaces are close together (at a distance $\sim 1/|k_\perp|$ or less), the modes on each surface will couple, resulting in superposed modes with properties quite different from those on each isolated surface (Raether 1977) (Figure 16–4d).

There are also other possibilities for the ε dependence of the surface loss function. For example, a monopole surface mode on a cylindrically symmetric nanowire produces a response that scales roughly as $\mathrm{Im}(\varepsilon)$ (Reed et al. 1999), which is large in the vicinity of a direct resonance such as a band gap or other interband transition (the ω_j from Eq. (2)). This enhancement of the direct interband transition peaks is not limited to cylinders but also occurs in spheres and seems to be a generic property associated with nanometer-scale surface geometry. Thus, in nanoscale geometries, one can directly measure the direct interband transitions (and thus the band gaps), provided confounding effects such as Cerenkov radiation be avoided.

So far we have neglected the dependence of the SP energy on the wave vector $\boldsymbol{k}$ (or, to be more precise, the in-plane component $\boldsymbol{k}_{||}$ of this wave vector for this surface-bound oscillation). Two physical effects modify the SP resonant frequency as a function of $\boldsymbol{k}_{||}$ (i.e., its dispersion relation, see Figure 16–5), namely spatial dispersion and retardation (Raether 1977, Ritchie 1973). The effects of spatial dispersion are fairly simple, although precise calculation from first principles can be challenging. As the magnitude $\boldsymbol{k}_{||}$ becomes comparable to a characteristic wave vector for the material's band structure (e.g., the Fermi wave vector k_F), the dielectric response of the material can no longer be taken to be purely local. That is, the polarization at one point in space can be affected by the electric field a finite distance away. The result is an upward curvature of the dispersion curves at large $\boldsymbol{k}_{||}$. This can potentially produce a mild asymmetry in the plasmon peaks in the energy loss spectrum, although the effect is typically very small because of the rapid decay of inelastic cross sections at larger wave vectors.

Now let us consider retardation. At small $\boldsymbol{k}_{||}$, the wavelength of the mode approaches the free-space wavelength of light at the same frequency (the light line $\omega = ck$ in Figure 16–5). This means that the

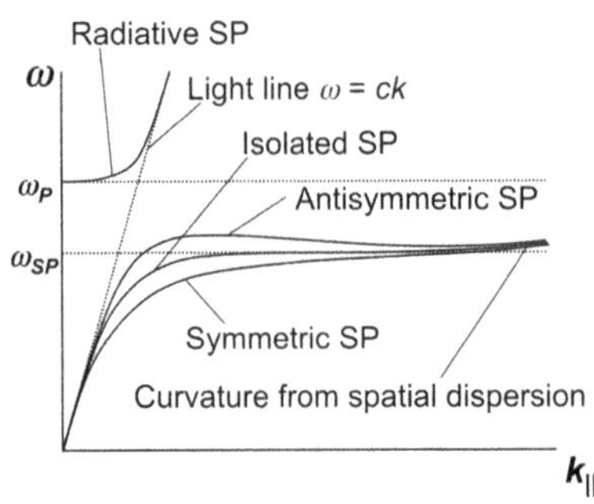

Figure 16–5. Example dispersion curves for surface plasmons. Most non-radiative SP modes have a frequency close to ω_{SP}, which is $\sim 0.71\omega_P$ for the Drude model. Polaritonic coupling (i.e., retardation effects) changes the curves in the vicinity of the light line. The "Isolated SP" curve corresponds to the mode in Figure 16–4b, the "Radiative SP" to Figure 16–4c, and the symmetric and antisymmetric mode curves to Figure 16–4d. Spatial dispersion (exaggerated on the scale of the figure) causes a slight curvature apparent at larger wave vectors.

SP can resonantly couple to a free-space photon mode. The result is a mode called a surface plasmon-polariton (SPP), which acts like a superposition of an SP and a photon. As a result, the dispersion relation splits into two branches, called radiative and non-radiative, both of which approach the light line asymptotically. The non-radiative branch approaches zero energy as the wavelength approaches infinity, with a well-defined phase speed in the limit, somewhat analogous to an acoustic phonon mode. The radiative branch is peculiar in that $k_\perp$ is actually real in the vacuum. So, unlike the non-radiative SPs which are bound to the surface, the radiative SPs are more like a resonant coupling between a free photon and the surface of the material (Figure 16–4c).

These dispersion relations can all be calculated from ordinary classical electrodynamics, assuming that $\varepsilon(\boldsymbol{k},\omega)$ is known. The polaritonic coupling that produces the interesting behavior near the light line comes out of the time-derivative terms in Maxwell's equations—the same terms that are in part responsible for Cerenkov radiation and other relativistic effects. Neglecting these terms means neglecting relativistic retardation, i.e., the coupling between electric and magnetic fields is ignored, and the field at any given time is given by Coulomb's law as applied to the spatial distribution of charge at that instant, regardless of distance. This so-called electrostatic approximation often works quite well for nanoparticles (Ugarte et al. 1992) but completely fails to predict Cerenkov radiation, polaritons, and radiative SPs.

All of these excitations have some nonlocal aspects in their properties and their response to an electron probe. The probe electron produces a long-range electromagnetic field that can shake valence electrons at distances of several nanometers. The dispersion relations of many of the electromagnetic modes depend on the geometry some distance away from the point of excitation. Many of these modes have significant amplitude in the vacuum and can be strongly excited from outside the material. The ability to set up a coherent Cerenkov radiation cone depends on there being a sufficiently large volume of bulk-like material. The material itself has built-in length scales such as excitonic radii and the Fermi wavelength.

A localized, generic excitation driven by an electron passing by it at a distance b typically produces an EELS excitation probability that decays as $\exp(-2b\omega/v)$. This approximate functional form (often better approximated as a K-type Bessel function) arises in both quasiclassical and quantum mechanical models and is also consistent with experimental measurements (Muller and Silcox 1995, Ritchie and Howie 1988, Rivacoba et al. 2000). Because of this "dynamic screening" effect, the spatial resolution for low energy excitations can be quite poor, typically on the nanometer scale, and it gets worse at lower energies: For a 200 keV electron traveling at $v = 2.08 \times 10^8$ m/s, a 2 eV excitation has a 1/e-fold decay length of $v/(2\omega) = 34$ nm. Fortunately, this only applies to the electrons scattered at very small angles, and high spatial resolution can still be attained in dark-field EELS even at quite low values of energy loss (Ritchie and Howie 1988).

This dynamic screening effect also highlights a subtlety in low-loss EELS spectrum imaging, which produces real-space images corresponding to various excitation modes (Bosman et al. 2007, Nelayah et al. 2007). Because the dynamic screening effect depends on the speed v of the electron probe, such images cannot be strictly regarded as images of the electromagnetic fields associated with each mode. The apparent shape will actually be somewhat different for a 100 keV probe than for a 200 keV probe; the image that is produced results from a combination of the properties of the sample and those of the probe. In other words the image is a map, not of the mode itself, but of the ability of the probe electron to excite the mode as a function of position. Fortunately, this effect is well understood and can be dealt with using available modeling techniques (García de Abajo and Aizpurua 1997, Rivacoba et al. 2000, Zabala et al. 2001).

Table 16–1 summarizes the valence electron excitations that are typically seen in low-loss EELS measurements, along with rough guidelines for when each is likely to be important. We have not discussed phonon excitation, which is rarely visible in TEM-EELS because the energy scale of phonons is typically much smaller than the energy resolution. But this may change as monochromated EELS systems become more common and the energy resolution is pushed to smaller fractions of an electron volt. For the present, though, the terms "low-loss EELS" and "valence EELS" are almost synonymous.

In summary, low-loss EELS provides a measure of a probe electron's ability to excite all manner of excitations involving valence electrons and the coupled electromagnetic field. Besides providing a method of measuring the complex dielectric function of a very small volume of material, this technique also specifically excels at quantifying the very modes (especially excitons, surface plasmons, and waveguide modes) that are essential to present-day and emerging applications of nanostructures. The spatial resolution is often relatively poor by TEM standards (on the nanometer scale) but can be improved through dark-field techniques. The essential physics underlying all modern models of low-loss EELS was established some decades ago and can usually be interpreted in very nearly classical terms.

Table 16–1. Excitations frequently encountered in low-loss EELS, together with typical conditions under which each type of excitation may be highly visible.

Excitation	Condition on ε	Sample conditions
Bulk plasmon	Re$\varepsilon \sim 0$, Imε small	Relatively thick
Surface or interface plasmon	Re$\varepsilon \sim -1$ to -2, or $-\mathrm{Re}(\varepsilon_{\mathrm{surface}})$, Im$\varepsilon$ small	Relatively thin, or probe near an edge or interface
Cerenkov radiation	Reε large	Relatively thick
Direct interband transition	Imε large	Usually nanoscale
Guided light mode	Re$\varepsilon > 1$, Imε small	Resonance depends on geometry

16.3 Experimental Considerations

As described in detail in the previous section, the interaction of an incident electron beam with a thin film of material leads to characteristic energy losses reflected in the energy distribution of the transmitted electrons. The energy-loss distribution of the electron beam after transmission is directly related to the dielectric response of the material and can thus be analyzed in order to derive dielectric properties of the material under investigation. The main advantage of performing energy-loss spectroscopy in an electron microscope is the spatial resolution with which energy-loss spectra, and thus dielectric information such as band gap information, can be collected. One approach to map dielectric information at high energy and high spatial resolution is based on employing energy-filtered imaging (EFI) in combination with a monochromated broad-beam illumination (Sigle et al. 2008). Another approach, on which the present instrumentation section focuses is based on scanning transmission electron microscopy where the ability to position a small electron probe on a point of interest provides the spatial resolution of the energy-loss spectrum that is recorded while the electron probe is kept on the point of interest. Hence, low-loss electron energy-loss spectroscopy in scanning transmission mode is performed similarly to the more common STEM/EELS experiments that are often employed to probe core-loss absorption edges at highest spatial resolution (see, e.g., Browning et al. (1993)). An annular dark-field detector is used for STEM imaging while the forward scattered beam, which contains elastically and inelastically scattered electrons, is analyzed in an electron energy-loss spectrometer. STEM imaging allows for precisely positioning the electron probe for point analyses and/or spectrum mapping.

The main instrumental requirements to perform low-loss electron energy-loss spectroscopy in scanning transmission mode include thus a transmission electron microscope equipped with an electron source of small inherent energy spread, typically a field emission microscope, an illumination system with probe-forming optics, a scanning unit, and a spectrometer which can be either an in-column or a post-column electron energy-loss spectrometer. Optional but for many applications beneficial are an illumination aberration corrector and a gun electron monochromator.

16.3.1 Energy Resolution

The information content of low-loss electron energy-loss spectra depends on the energy resolution of the spectrum. It is in the nature of the technique that low-loss electron energy-loss spectroscopy aims at measuring energy losses that lead to spectral features close to the zero-loss peak (ZLP) of the electron energy-loss spectrum. Low-loss features are thus superimposed on the low-energy tail of the ZLP. Hence, the width of the ZLP not only imposes a limit on the resolvable spectral fine structure, but also on the signal-to-background ratio at low energy losses. A narrow ZLP with a rapidly decaying low-energy tail is thus of fundamental importance in the experimental setup of low-loss experiments. The main characteristic of the ZLP depends either on the energy spread of the emitted electrons, i.e., on the type of electron source used, or on the availability of a gun electron monochromator. Secondary effects that can alter the characteristics of the ZLP are high-voltage instabilities, residual aberrations of the spectrometer, the point spread function of the recording device, and environmental instabilities, such as electromagnetic stray fields, the power supply system, and acoustic and low-frequency vibrations (Erni and Browning 2005).

The usual figure of merit of the ZLP is its full width at half maximum (FWHM), which typically is used to describe the energy resolution of the electron energy-loss spectrum. The comparison of the ZLP of a thermally assisted Schottky field emission microscope with the ZLPs of a cold field-emission microscope and a monochromated Schottky field emission microscope in Figure 16–6 reveals that the monochromated setup enables an energy resolution of better than 100 meV which is substantially smaller than the non-monochromated Schottky field emitter (520 meV) and the cold field emitter (330 meV). As mentioned above, besides the FWHM of the ZLP, it is the decay of the low-energy tail of the ZLP that is of fundamental importance for the signal-to-background ratio of spectral information close to the ZLP. Hence, an alternative figure of merit of the ZLP has been suggested which corresponds to the

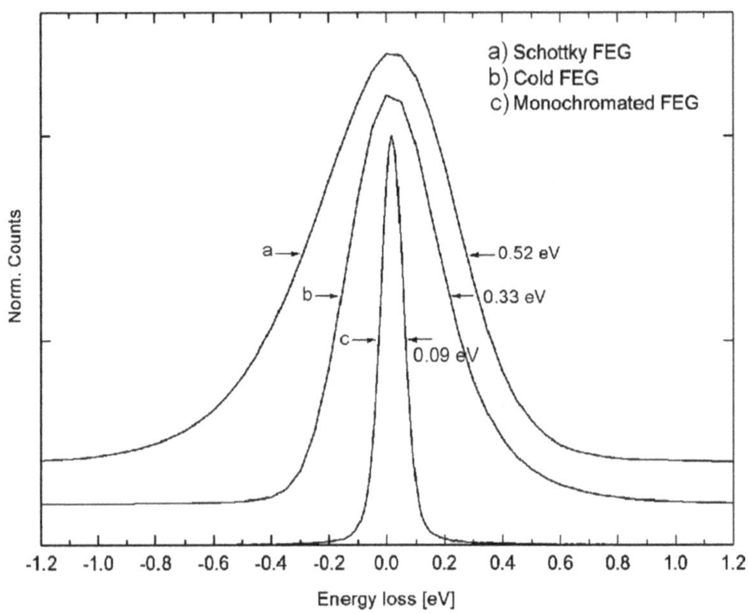

Figure 16–6. Comparison of the ZLPs of a Schottky field emission microscope (200 kV), a cold field-emission microscope (100 kV), and a monochromated electron beam in a Wien-filter type monochromated Schottky field emission microscope (80 kV). The FWHM of the zero-loss peaks indicated in each case provides a measure for the energy resolution. The decay of the low-energy tail (*right hand side*) of the ZLPs defines the signal-to-background ratio of the low-loss signal.

energy loss where the intensity of the tail drops to 1/1000 of the maximum intensity of the ZLP (Kimoto et al. 2005). This critical energy loss would then reflect the lowest detectable energy loss feature. Due to similar tail characteristics, for both the ZLPs of the Schottky and cold field emitter microscopes shown in Figure 16–6, this critical energy loss is around 1.5–2.0 eV. And for the monochromated spectrum in Figure 16–6, it is ~0.8 eV. It has to be pointed out that the energy loss at 1000th of the maximum is not a simple measure, since it is very sensitive to residual spectrometer aberrations, the point spread function of the detector and particularly to shot noise that would, however, not directly impact the low-loss signal. Although deconvolution and sophisticated background subtraction methods can be applied to extract spectral features of low signal-to-background ratio from the tail of the ZLP (Nelayah et al. 2007, Reed and Sarikaya 2002), it is the combination of energy resolution and this critical energy loss which reveals the main benefit of a monochromated beam setup; spectral features can directly be recorded with high signal-to-background ratio down to an energy loss of ~1 eV.

Two types of gun electron monochromators have been developed and brought to application: an electrostatic Omega-filter type monochromator (Benner et al. 2004) and Wien-filter type monochromators (Terauchi et al. 1999, Tiemeijer 1999). Both types of monochromators, located in front of the accelerator, disperse the electrons emitted by the tip according to their energy. An energy slit inserted into the energy dispersion plane selects electrons with a narrow energy distribution. Compared to the intrinsic asymmetry of the ZLPs of a Schottky or cold field emission source, the energy slit ensures that the emission characteristic of the tip is not translated into the shape of the ZLP. Hence, apart from the improved energy spread, another advantage of a monochromated electron beam is its symmetric ZLP. This circumstance simplifies modeling of the ZLP in the data processing.

16.3.2 Monochromated Electron Probe

The presence of an electron monochromator can challenge the probe forming optics. This is particularly true for the case of a Wien-filter type monochromator which, in general, leads to an enlarged virtual source size at the dispersion plane where the energy slit is inserted. Increased demagnification in the condenser lens system is necessary to enable an effective source size projected onto the specimen that is comparable to the unfiltered case. Nonetheless, for both designs of electron monochromators, i.e., Wien filter and Omega filter, it has been shown that the formation of an atomic size electron probe is feasible (Erni et al. 2008, Walther et al. 2006). However, as pointed out above, an electron monochromator reduces the energy spread of the beam by filtering. Reducing the energy spread of a Schottky field emission gun down to a value of 100–200 meV implies that roughly 70–80% of the initial beam current is lost. Assuming comparable probe sizes for the monochromated and the non-monochromated case, the loss in beam current that comes with the monochromator can be counterbalanced if the monochromated instrument is equipped with an illumination aberration corrector that allows for increasing the beam convergence angle. This of course only applies for an experimental setup where the probe size is limited by coherent aberrations rather than by insufficient demagnification of the (virtual) electron source. Provided that the effective source size projected onto the specimen plane is unchanged, the increased beam convergence angle enlarges the probe current by a factor of 4–9, counterbalancing the loss in current that is caused by the monochromator. It has been shown that by employing a beam convergence angle of ~20 mrad in a 300 kV monochromated Schottky field emission microscope, an aberration-corrected electron probe is feasible that enables a spatial resolution in STEM of better than 0.14 nm with an energy resolution of 130 meV (FWHM). The resulting beam current of more than 50 pA exceeds the requirements of low-loss experiments by a factor of about 5. As a rule of thumb it can be stated that the probe current of a monochromated and aberration-corrected electron probe is comparable to the probe current that is achievable with a non-monochromated and non-aberration-corrected instrument, provided that other parameters, such as lateral probe size and initial source brightness, remain unchanged.

16.3.3 Data Acquisition

A critical point of the data acquisition in low-loss electron energy-loss spectroscopy is the dynamic range of the detector. The dynamic range of the detector should allow for simultaneously recording the ZLP and the low-loss signal with sufficient signal-to-noise ratio (Erni et al. 2005). Information about the ZLP is essentially needed for the data processing and cannot simply be neglected. PEELS and slow-scan charge-coupled device (CCD) cameras commonly have a dynamic range of 14 or 16 bits. Typically and particularly true for thin samples, the intensity of the ZLP exceeds the actual low-loss signal by about three orders of magnitude. Hence, with standard electron detectors employing a single acquisition mode, it is impossible to record a "noise-free" low-loss signal of a thin sample without saturating the detector with the ZLP. There are two approaches to solve this "detection problem." Using a short exposure time in cumulative acquisition mode to access the full dynamic range of the spectrum is one way. In this way, the dynamic range is enlarged by multiple acquisitions. This approach has the benefit that the ZLP is directly contained as part of the low-loss data. However, the disadvantage is that through the summation of the individual "noisy" spectra recorded with short exposure times the noise in the resulting summed spectrum is amplified. Particularly, correlated noise is being amplified and can even lead to artifacts in the summed spectrum.

Another practical solution to the acquisition problem makes use of two consecutive exposures of alternating energy range; after an exposure of an energy window, which does not contain the ZLP, and which is optimized for recording "noise-free" the low-loss signal, the spectrum is shifted by a defined energy shift (~1–3 eV) to move the ZLP into the energy window and a short acquisition is made (~0.05 s) which is optimized to record the ZLP without saturating the detector (Dorneich et al. 1998). This two-step acquisition procedure takes about 3–5 s, i.e., a period of time where spectrum drift is usually not critical. However, although the low-loss spectrum can be spliced with the ZLP, the disadvantage of this acquisition method is that the ZLP is not directly contained in the low-loss data. This can complicate the data processing. In any case, both acquisition methods provide the level of data needed for the analysis, firstly by allowing a precise energy-scale calibration, secondly to have the low-loss signal "noise-free" without saturating the detector, and thirdly to have the information about the basic shape of the ZLP necessary for data processing. Both acquisition methods have pros and cons that basically need to be weighted according to the data that need be extracted from the low-loss spectrum.

16.4 Interpreting Spectra

The previous sections of this chapter have examined the theory behind low-loss EELS and the experimental aspects of acquiring the spectrum. In this section we examine the interpretation of experimental spectra by considering a set of examples that highlight particular features of the low-loss spectrum. The key to applying the theory for low-loss EELS to experimental spectra is to evaluate the experimental conditions that are being used (including all specimen and microscope parameters) to start with a basic picture of which modes are likely to be excited in the specimen being examined. From this, simplifications can be made to the general theory described in Section 16.2, which allow you to focus on a particular property.

In most cases, STEM specimens will be thin foils/nanostructures, which are typically below 100 nm in thickness (this is not true in all cases, as the final example in this section demonstrates). As discussed earlier, if a charged particle moves from one dielectric medium to another, i.e., into a thin sample, it can excite collective oscillations of surface (or interface) electrons. Surface (or interface) plasmons are longitudinal waves of the surface (or interface) charge density that run along the boundary. For a thin sample transmitted by fast electrons, the excitation of the upper and the lower surface can be coupled. This leads to thickness-dependent excitation modes of surface electrons and to the corresponding energy losses in the transmitted electron beam. The impact of these surface losses is not a priori negligible.

In practical situations we also have to remember that a fast electron interacting with a solid can also be impacted by retardation effects. For materials with a high dielectric constant (real part ε_1), retardation of the incident electrons can lead to the emission of Cerenkov radiation and to the corresponding energy losses in the transmitted beam. For many semiconductor materials ε_1 is large enough such that Cerenkov radiation can in principle be emitted at acceleration voltages common in (S)TEM (100–300 kV). The probability of the emission of Cerenkov radiation increases with increasing ε_1. Hence, Cerenkov losses are typically peaked on the energy-loss axis where the real part of the dielectric function ε_1 is maximal. Cerenkov absorption features thus fall exactly in the spectral region where band structure and band gap information is contained. A simple picture estimating the materials where Cerenkov radiation is likely to be important is shown in Figure 16–7.

One key issue that needs to be considered in addition to the plot in Figure 16–7 is the thickness of the specimen. The ~100 nm thickness of the thin films commonly used in (S)TEM limits the emission of Cerenkov radiation and the appearance of Cerenkov losses in VEELS. In the case of very thin foils, the Cerenkov light cone cannot be built up and no Cerenkov radiation can be emitted—even if the bulk condition for the emission of Cerenkov radiation given above is fulfilled. As a rule of thumb, if the thickness of the foil is smaller than the wavelength of the Cerenkov radiation, the emission of Cerenkov radiation is inhibited and Cerenkov losses are absent in VEELS. Although the thickness restriction is not strictly formulated here, it shows that the probability of Cerenkov losses for typical TEM foils (< 100 nm) is small. Similar to

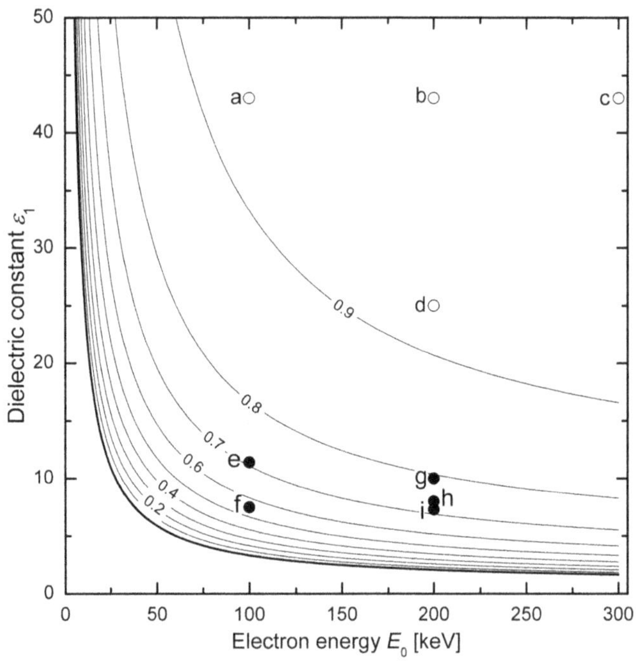

Figure 16–7. Contour plot for the probability of retardation effects (reproduced from Erni and Browning (2008) with permission). The *bold solid line* corresponds to the condition for the emission of Cerenkov radiation; for pairs of E_0 and ε_1 that are on the left or below this line, retardation effects are not possible. The *open circles* indicate measurements where strong retardation effects were observed, whereas the *full circles* indicate measurements where retardation effects were marginal. Points (a), (b), and (c) are values for Si from the literature (Mkhoyan et al. 2007, Stöger-Pollach et al. 2006); point (d) is GaAs (Stöger-Pollach et al. 2006), point (e) is for $SrTiO_3$ from van Benthem et al. (2001), point (f) is a result for AlN from Dorneich et al. (1998), point (g) is for CdSe from Erni and Browning (2008), point (h) is for InN from Jinschek et al. (2006), and point (i) is for GaN and Si_3N_4 from Erni and Browning (2008). The transition between strong and weak retardation effects is in the range of probability = 0.9.

small particles, very thin TEM foils do not show bulk Cerenkov losses in VEELS.

Surface, interface, and finite-size effects, potentially related to the retardation of the electron, can be important. Which excitation modes are feasible depends on the magnitude of the real part ε_1 of the material's dielectric function. If an electron passes through a metal whose dielectric function ε_1 goes through 0, radiative surface plasmons can be excited. Energy losses caused by the excitation of a radiative surface plasmon are superimposed on the volume plasmon loss and show, however, a different θ dependency. The energy losses related to the radiative surface plasmons are equal to or larger than the energy losses caused by the excitation of the volume plasmon. Radiative surface plasmons decay by emitting light. Non-radiative surface plasmons can be excited by transmitting electrons if the material's dielectric function ε_1 becomes smaller than –1. The corresponding energy losses show a characteristic dispersion; the energy losses are smaller than the losses related to the actual surface plasmon mode $E_S(\theta)$, approaching however E_S for large θ. If the real part ε_1 of the material's dielectric function becomes larger than 1, retardation effects can occur.

If the primary electron energy is large enough, electrons can then excite radiative or non-radiative guided light modes. For materials with high dielectric constants, non-radiative guided light modes have to be considered. A surface (or interface) plasmon corresponds to an excitation mode of the surface charge density, leading to a longitudinal electromagnetic wave that propagates with a given phase velocity parallel to the boundary. The maximum amplitude of the electromagnetic wave associated with a surface plasmon is located at the surface. Compared to a surface plasmon, a guided light mode involves collective excitations of electrons inside the foil. The component parallel to the foil normal of the electromagnetic field associated with a guided light mode is a standing wave that shows one (or more) amplitude maximum (maxima) within the foil, whereas at the boundary of the material the amplitude is small. The electromagnetic wave propagates with a given phase velocity parallel to the foil. A surface (or interface) plasmon is determined by the boundary configuration, whereas a guided light mode is essentially determined by the finite thickness (or finite size) of the sample. If the condition for total internal reflection is not fulfilled, a guided light mode can decay by emitting light. Similar to a radiative surface plasmon, a guided light mode is then called radiative. Guided modes can also be excited through coupling to Cerenkov radiation. If the opening angle of the Cerenkov light cone is larger than the angle of total internal reflection, no Cerenkov radiation can be emitted. In such cases, the retardation radiation is confined in the sample at guided mode frequencies.

16.4.1 Large, Strongly Peaked Dielectric Function

The prototypical example of a material with a large, strongly peaked dielectric function is Si (Figure 16–8a). Figure 16–8b shows a thickness series of experimental low-loss EEL spectra from Si recorded at 300 kV,

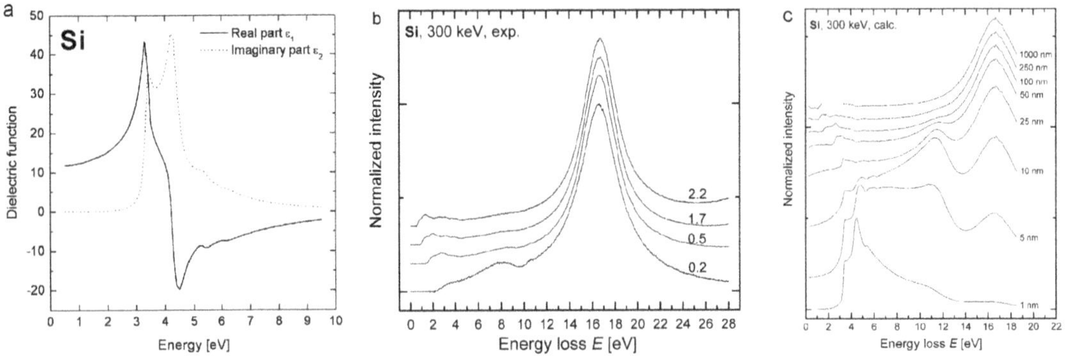

Figure 16–8. (**a**) Dielectric function for Si. (**b**) EEL spectra from Si recorded at 300 keV using a probe semi-convergence angle of 19 mrad and a collection angle of 3.8 mrad. The foil thickness is given in units of the inelastic mean free path λ_{in}; 0.2, 0.5, 1.7, 2.2. All spectra are normalized and for clarity shifted along the y-axis. (**c**) Series of spectra calculated for the foil thickness are indicated in each case (reproduced from Erni and Browning (2008) with permission).

the thickness is given in units of the inelastic mean free path λ_{in}, where the inelastic mean free path λ_{in} is ~130 nm (Erni and Browning 2008). The effective collection angle for these experimental results is 3.8 mrad and the probe semi-convergence angle is 19 mrad. A calculated thickness series of low-loss EEL spectra from Si is shown in Figure 16–8c, where the calculations include volume, surface, and retardation effects discussed in Section 16.2.

The real part of the dielectric function of Si is sharply peaked at 3.3 eV with a value exceeding 43. The condition for the emission of Cerenkov radiation is thus fulfilled for electrons with an energy exceeding $eV_c = 5$ keV. Figure 16–8 illustrates how the Cerenkov losses and the energy losses that are due to the excitation of the guided light modes impact the low-loss EEL spectra. For foil thicknesses exceeding ~25 nm, the calculated spectra in Figure 16–8c reveal a broad absorption feature between 1.5 and 4.5 eV. Except for the Si spectrum of 0.2 λ_{in} thickness, this spectral signature is also observable in the experimental spectra shown in Figure 16–8b. For very thin foils, there seems to be a mismatch between calculated and experimental spectra of Si. This mismatch is likely caused by the oxidized surface of the Si sample used for the measurements. The calculated as well as the experimental series of spectra in Figure 16–8 gives the impression that the low-loss retardation absorption feature below ~5 eV moves toward lower energy losses with increasing foil thickness. This behavior has also been shown by Stöger-Pollach et al. (2006).

However, it is not the Cerenkov-loss peak that moves toward lower energies. Figure 16–9 illustrates in more detail the thickness dependency of the absorption features in the spectrum by calculating the scattering probabilities for an energy loss of 3 eV. Apart from a surface mode observable below 0.01 mrad, labeled A, two maxima, B and C, can be identified. The relative intensity of absorption feature B decreases with increasing foil thickness. Furthermore, it moves to lower θ values. The position of feature C remains unchanged. The relative intensity of

Figure 16–9. Calculated scattering probabilities for a fixed energy loss E of 3 eV as a function of the scattering angle θ for 200 keV electrons transmitting a Si film. Curve (a) shows the volume contribution of a 100 nm thick film. Curve (b) shows the volume contribution of a 100 nm thick film including retardation effects. The other curves are full calculations of the energy loss. The foil thickness is indicated in each case. The numbers running from +3.0 to –4.5 indicate the shifts on the y-axis (reproduced from Erni and Browning (2008) with permission).

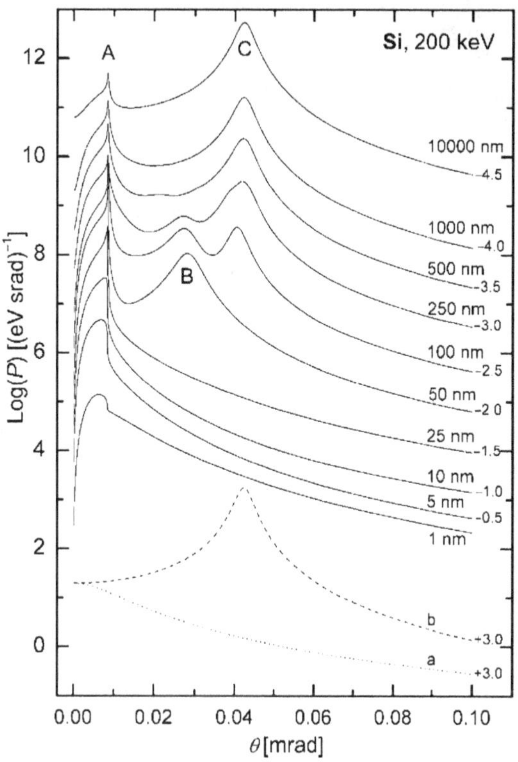

peak C increases with increasing foil thickness. Peak B can be associated with a guided light mode, whereas the invariant feature C corresponds to the Cerenkov-loss peak and is not visible for foil thicknesses below ~100 nm.

From the results shown in Figure 16–9, we can see that for foil thicknesses smaller than 250 nm, the absorption feature below 5 eV is dominated by energy losses that are due to the excitation of a guided light mode. Only for foil thicknesses above 250 nm, the Cerenkov-loss peak becomes dominant. The low-loss absorption feature then remains stationary. The seeming movement of the Cerenkov-loss peak from higher to lower energies between 50 and 250 nm foil thickness is in fact a transition between two different retardation absorption features: the guided light mode and the actual Cerenkov-loss peak. In terms of interpreting low-loss spectra for Si, this complex interplay between different retardation effects makes a simple identification of the band-gap and the bulk dielectric constant difficult without a high degree of care in the experiment to track the thickness and a detailed set of theoretical simulations.

16.4.2 Small, Smoothly Varying Dielectric Function

Gallium nitride (GaN) is a material where the dielectric function is smoothly varying and with a low maximum value. Figure 16–10a shows that the real part of the dielectric function for GaN is peaked

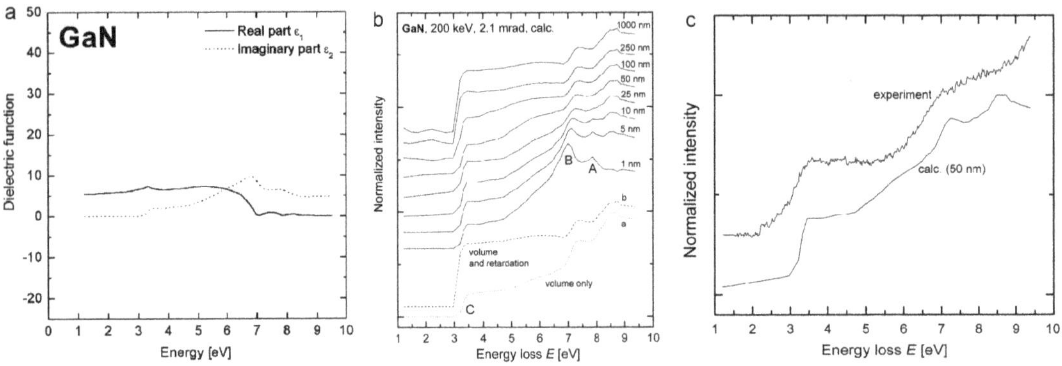

Figure 16–10. (**a**) Dielectric function for GaN (**b**) simulations of the low-loss spectrum and (**c**) experimental spectrum compared to the simulation (reproduced from Erni and Browning (2008) with permission).

at 3.3 eV with a value of 7.3. This means that $eV_C \approx 40$ keV for the emission of Cerenkov radiation and for 200 keV electrons the condition is fulfilled for energy losses smaller than 6.8 eV. However, the band-gap energy of GaN has been measured reliably using low-loss EELS independently by several groups using different data analysis methods (Brockt and Lakner 2000, Gutierrez-Sosa et al. 2003, Jinschek et al. 2006, Lazar et al. 2003). The reason for this can be clearly seen in Figure 16–10b, where calculated low-loss spectra of GaN are shown for three different cases; considering bulk or volume losses only, considering bulk and retardation losses and thirdly bulk, surface and retardation losses.

The intensity onset C in spectrum (i) of Figure 16–10b contains non-retarded volume contributions only and accurately reflects the band-gap signal of GaN. Including bulk retardation, see curve (ii) in Figure 16–10b, the intensity onset is shifted toward lower energies. However, if apart from the bulk retardation surface contributions are taken into account as well, the impact of the volume retardation on the band-gap signal becomes negligible for foil thicknesses smaller than 100 nm (Figure 16–10b). The spectra 1–50 nm reveal the proper band-gap signal, comparable to the "volume-only" case shown in curve (i). Only for foil thicknesses exceeding 100 nm, bulk retardation starts to interfere with the band-gap signal. Hence, it can be stated that provided that the thickness of the foil is below ~100 nm, retardation effects do not alter the band-gap signal as observed in VEEL spectra of GaN.

However, retardation effects not only alter the intensity onset C for foils exceeding 100 nm in thickness, they also impact the spectral area between the peaks A and B. With increasing foil thickness the intensity between 3.5 and 7 eV increases, clearly deviating from the "volume-only" case shown in curve (i). The spectra calculated for 25 and 50 nm foil thickness show the closest similarity to the "volume-only" spectrum (i). Apart from the retardation effects, surface effects modulate the low-loss EEL spectra of GaN for foils thinner than ~25 nm; peak A disappears and peak B moves to higher energy losses with increasing foil thickness. For foils exceeding ~25 nm in thickness, peaks A and B represent the bulk absorption feature as observable in the "volume-only" spectrum of curve (i). From this it can be concluded that although the correct value for the band-gap energy of GaN can be extracted from low-loss EEL spectra for sample thicknesses lower than 100 nm, to obtain a reasonable measure for the dielectric function thicknesses between 25 and 50 nm is required. For foils thinner than ~25 nm, surface effects result in a thickness-dependent modulation of the bulk absorption feature. The selection of the correct thickness for low-loss spectra from GaN can be seen from the comparison of the experimental spectrum with a simulation in Figure 16–10c. The experimental spectrum is taken from a GaN film with a thickness of 0.42 in units of the inelastic mean free path λ_{in}(~101 nm), with an effective collection angle of 2.1 mrad and a probe semi-convergence angle of ~20 mrad.

16.4.3 Measuring the Band-Gap

CdSe is another material with a smoothly varying dielectric constant that allows us to measure the band-gap of the material directly (Erni and Browning 2007). The band-gap for bulk CdSe has been determined optically by Rabani et al. to be 1.75(±0.5) eV (Rabani et al. 1999). For CdSe nanoparticles in excess of 15 nm in size, quantum confinement effects should be non-existent and the band-gap will have the bulk value (Troparevsky 2003). Figure 16–11a shows a low-loss EEL spectrum acquired at 200 kV with an energy resolution measured from the FWHM of the zero-loss peak of 0.15 eV obtained from a spherical CdSe nanoparticle with a diameter of about 30 nm. The position of the particle on the edge of the carbon mesh made it possible to have no measurable contribution of the carbon support. Figure 16–11b shows the background-corrected spectrum (raw) and the spectrum after applying a 21-point Savitzky–Golay filter (second order polynomial) to reduce the noise (Erni and Browning 2007).

In Figure 16–11c, the first derivative of the filtered spectrum and the corresponding Lorentz fit function (multiple peak fit) are shown. The first positive peak of this fit function corresponds to the inflexion point of the intensity onset. This first inflexion point at energy E_{IP} can be related to the band gap E_G by $E_G \approx E_{IP} - 0.5 \times \text{FWHM}$, where the FWHM is the full width at half maximum of the Lorentz function fitted to the first peak of the spectrum's first derivative. The error of the energy-gap measurement given by the uncertainty of the first peak position in the VEEL spectrum becomes then 0.5×FWHM of the first Lorentz function. The limited energy resolution of the measurement (~0.15 eV) contributes to the error of the measurements too. It must, however, be emphasized that the inflexion point of an intensity onset (similar to a peak position) can be determined with a distinctly higher precision than the actual energy resolution. For this and for the fact that the energy resolution affects the width of the first Lorentz peak as well, the contribution of the limited energy resolution was explicitly taken into account by half the value of the energy resolution. The total experimental error of the energy gap of a QD amounts then to 0.5×FWHM(LF) + 0.5×FWHM(ZLP), where FWHM(LF) and FWHM(ZLP) are the full width at half maximum of the first Lorentz

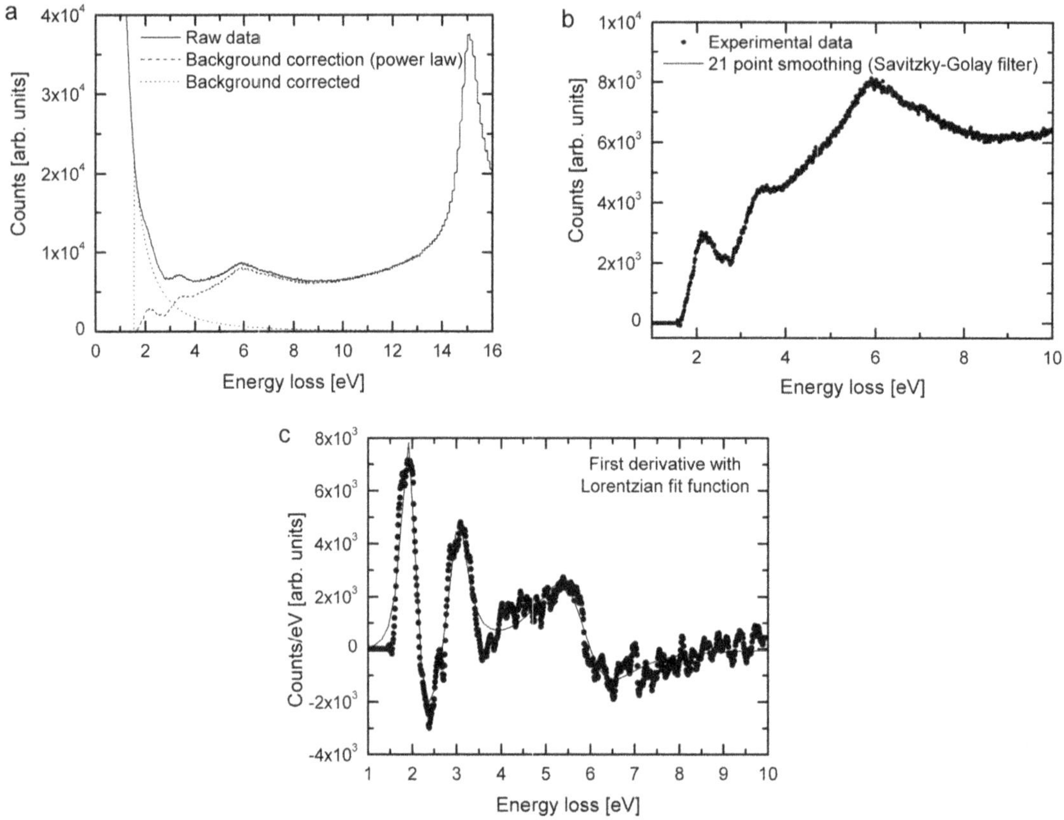

Figure 16–11. (**a**) VEELS energy-gap analysis of a 30 nm CdSe nanoparticle: the VEEL spectrum of the CdSe particle (*full line*), the background-corrected VEEL spectrum (*dashed line*), and the power-law background model (*dotted line*). The energy-resolution is 0.18 eV given by the FWHM of the zero-loss peak (not shown). (**b**) Background corrected VEEL spectrum (*dots*) and the smoothed spectrum (*full line*) using a Savitzky–Golay filter. (**c**) The first derivative of the smoothed spectrum (*dots*) and the multiple-peak Lorentz fit function (*full line*). The first peak of the Lorentz fit has its maximum at 1.94 eV and a FWHM of 0.42 eV, which results in a band gap of the nanoparticle of 1.7(±0.1) eV (reproduced from Erni and Browning (2007) with permission).

function and the zero-loss peak, respectively. The Lorentzian fit of the first peak of the derivative has its maximum at 1.94 eV with a FWHM of 0.42 eV. This gives a band-gap energy E_G of 1.7(±0.1) eV. This result is in good agreement with band-structure calculations and band-gap measurements of bulk CdSe which predict a band gap of 1.7–1.8 eV. Such results show that under the correct experimental conditions, band gaps can be accurately and readily determined from low-loss EELS.

16.4.4 Surfaces

We now briefly show an example of surface effects in EELS of nanostructured materials. As we have discussed, the bulk loss function Im(−1/ε) captures the material dependence of the EELS signal only in cases where surface and retardation effects can be neglected, and furthermore, the dielectric functions for bulk materials may not match those of nanostructured materials because of quantum confinement,

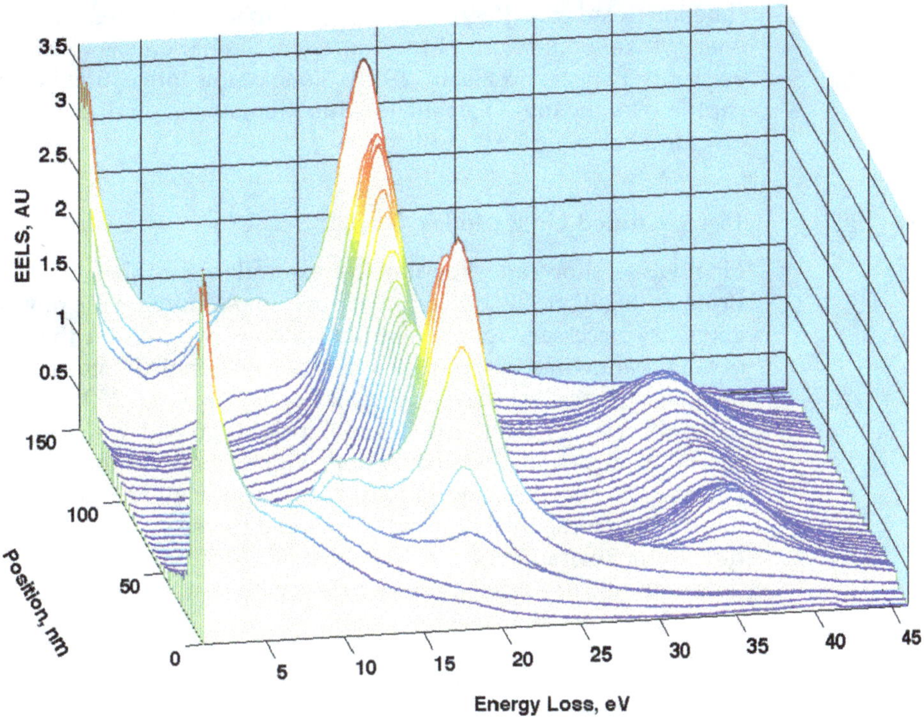

Figure 16–12. Line scan across a 130 nm diameter Si cylinder showing the very different spectra as a function of the amount of material intersected by the beam. (reproduced from Reed et al. (1999) with permission).

surface scattering, and other size effects. Figure 16–12 shows a series of energy loss spectra obtained from a line scan across an isolated, suspended silicon wire 130 nm in diameter (Reed et al. 1999). The wire is roughly centered on the position axis, so that the spectra at the extreme ends (near 0 and 150 nm) are aloof, those at ~10 and ~140 nm are just grazing the surface, and those in between are penetrating bulk material. Thus this single plot shows spectra over quite a large range of thicknesses. The bulk spectra are very simple, consisting of peaks at integer multiples of the bulk plasmon energy ~17 eV. As expected, the 130 nm thick region is more than one mean free path thick for the 100 keV electron energy.

As the beam reaches and passes the edge of the material, the bulk plasmon peaks disappear very rapidly and are replaced with surface plasmon peaks at ~11 eV. Theory suggests that for this large diameter, these peaks include a mix of many azimuthal mode numbers *m* (Reed et al. 1999) and also that an apparent shift to ~8 eV in the aloof mode may be due in part to retardation effects (Moreau et al. 1997). In principle the aloof spectrum should also include a direct interband transition peak at ~5 eV, but for such a large diameter this peak cannot be clearly separated from the background. This peak is clearly visible when the material diameter is reduced to a 3.5 nm hemispherical tip, and moreover, the spectra at the tip suggest that quantum confinement and surface scattering effects may be altering the effective dielectric function (Reed et al. 1999); the peaks are broader, stronger, and at higher energy than would normally be expected. Similar effects were reported earlier by Batson and Heath (1993). None of this interesting behavior is captured by the Im(−1/ε) bulk loss function, which predicts no peaks of any kind below ~17 eV in silicon.

16.4.5 Guided Light Modes

Finally, we show an experimental example of guided light modes. Bringing together the discussion in previous sections on the physics and geometry necessary for guided light modes, we present the example of GaN nanowires of two different cross-sectional geometries (hexagonal and triangular) (Arslan et al. 2008, Hersee et al. 2006), compared to a bulk (planar) geometry. As previously discussed, GaN has a small, smoothly varying dielectric function, which allows for the extraction of information such as band gap, dielectric function, and surface effects depending on the thickness of the material traversed by the probe electron. Cerenkov radiation is not problematic for sufficiently thin samples (less than ~50 nm), so any intensity below the band gap of GaN (3.4 eV) can be interpreted to arise from another source, in this case guided light modes.

Before discussing the details of the data, it is important to discuss the location of the electron beam for each sample, and the way the data are presented. For the bulk case, the electron beam goes *through* 50 nm of material. For the hexagonal and triangular nanowires, the electron beam is *aloof*, ~1 nm away from the surface of the wire (i.e., the beam is in vacuum). The three spectra have all been background subtracted using a power law in the Digital Micrograph software. This is not intended as a rigorous analysis procedure; it is performed here merely to enhance the visibility of features that arise from the sample and not from the tails of the zero-loss peak. Although the beam traverses different amounts of material for different cases, the comparison is a stunning example of the excitation of guided light modes, their dependence on geometry, and the uniformity required of that geometry.

Figure 16–13 shows STEM images of the two nanowires used in this study, along with the corresponding EELS spectra compared to the spectrum from the bulk geometry. The characteristics of the hexagonal nanowire are a uniform hexagonal diameter of ~450 nm. The triangular nanowire is tapered, having a diameter varying from ~50 nm to ~300 nm along one nanowire. For this nanowire, spectra were taken with the probe placed near the surface of the nanowire at regions with diameters ranging from ~50 nm up to ~100 nm with no difference in the resulting spectra. The presented spectrum is a representative spectrum at a location of ~50 nm. The dots in the images show the locations from which spectra were taken. The bulk sample is a planar geometry with a thickness of ~50 nm. The spectra are all normalized to the peak at ~3.8 eV. No vertical displacements have been performed; all spectra are plotted directly on top of each other.

What is immediately obvious is the strong set of peaks present in the hexagonal nanowire, which has been confirmed by theory to arise from

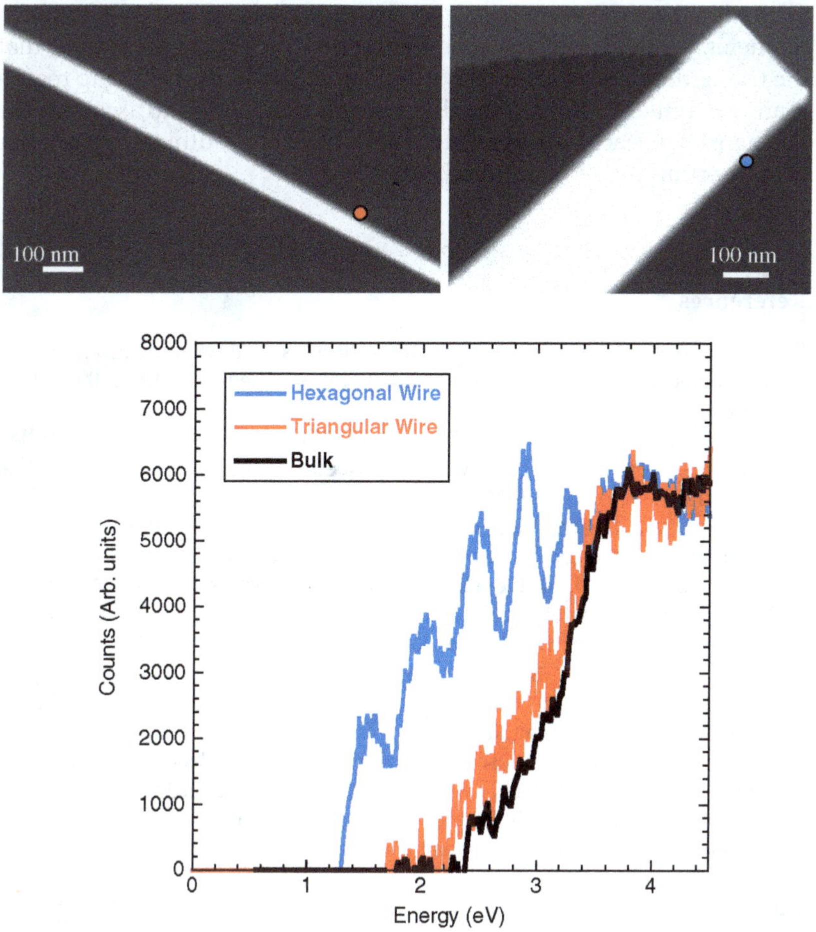

Figure 16–13. The two STEM images show the locations from which EELS spectra were taken for a triangular nanowire (*left*) and a hexagonal nanowire (*right*). These spectra are compared to a "bulk" sample. The hexagonal nanowire shows sharp peaks in the band gap region which are due to the excitation of guided light modes by the electron beam.

guided light modes (Arslan et al. 2009). This is due to the very uniform geometry of the nanowire, with the number of peaks dependent on the diameter of the wire (larger diameter, more peaks). This is typical of optical waveguides, which (depending on the shape, diameter, and dielectric properties) will carry one or more sets of propagating modes, each mode having a different transverse spatial profile and a different minimum "cutoff" frequency. This frequency varies inversely with the diameter. Although detailed calculations were not performed for the triangular nanowire, the diameter scaling predicts that, at a diameter of 100 nm, even the lowest cutoff frequency is well above the band gap. In other words, this diameter is too small for GaN to function as an optical waveguide at all. Furthermore, even at the larger diameters, the tapered shape should cause the waveguide properties (cutoff frequencies and impedances for each mode) to vary along the length, resulting in modes that are more localized than they would be for a uniform diameter. While it appears (consistently for every point measured) that the triangular nanowires do have more intensity in the band gap region than the "bulk" geometry, the difference is not enough to ascertain the mechanism. Clearly, the geometry of the object is of utmost importance in supporting guided light modes.

References

I. Arslan, A.A. Talin, G.T. Wang, Three-dimensional visualization of surface defects in core-shell nanowires. J. Phys. Chem. C **112**, 11093–11097 (2008)

I. Arslan, J.K. Hyun, R. Erni, M.N. Fairchild, S.D. Hersee, D.A. Muller, Using electrons as a high-resolution probe of optical modes in individual nanowires. Nano Lett. **9**, 4073–4077 (2009)

P.E. Batson, J.R. Heath, Electron-energy-loss spectroscopy of single silicon nanocrystals—the conduction-band. Rev. Lett. **71**, 911–914 (1993)

G. Benner, E. Esser, M. Matijevic, A. Orchowski, P. Schlossmacher, A. Thesen, M. Haider, P. Hartel, Performance of monochromized and aberration-corrected TEMs. Microsc. Microanal. **10**(Suppl. 2), 108–109 (2004)

M. Bosman, V.J. Keast, M. Watanabe, A.I. Maaroof, M.B. Cortie, Mapping surface plasmons at the nanometre scale with an electron beam. Nanotechnology **18**, 1–5 (2007)

G. Brockt, H. Lakner, Nanoscale EELS analysis of dielectric function and bandgap properties in GaN and related materials. Micron **31**, 435–440 (2000)

N.D. Browning, M.F. Chisholm, S.J. Pennycook, Atomic-resolution chemical analysis using a scanning transmission electron microscope. Nature **366**, 143–146 (1993)

A.D. Dorneich, R.H. French, H. Müllejans, S. Lughin, M. Rühle, Quantitative analysis of valence electron energy-loss spectra of aluminium nitride. J. Microsc. **191**, 286–296 (1998)

P.M. Echenique, J. Bausells, A. Rivacoba, Energy-loss probability in electron microscopy. Phys. Rev. B **35**, 1521–1524 (1987)

R.F. Egerton, *Electron Energy-Loss Spectroscopy in the Electron Microscope*, 2nd edn. (Plenum Press, New York, NY, 1996)

R. Erni, N.D. Browning, Valence electron energy-loss spectroscopy in monochromated scanning transmission electron microscopy. Ultramicroscopy **104**, 176–192 (2005)

R. Erni, N.D. Browning, Quantification of the size-dependent band gap of individual quantum dots. Ultramicroscopy **107**, 267–273 (2007)

R. Erni, N.D. Browning, The impact of surface and retardation losses on valence electron energy-loss spectroscopy. Ultramicroscopy **108**, 84–99 (2008)

R. Erni, N.D. Browning, Z.R. Dai, J.P. Bradley, Analysis of extraterrestrial particles using monochromated electron energy-loss spectroscopy. Micron **36**, 369–379 (2005)

R. Erni, S. Lazar, N.D. Browning, Prospects for analyzing the electronic properties in nanoscale systems by VEELS. Ultramicroscopy **108**, 270–276 (2008)

F.J. García de Abajo, J. Aizpurua, Numerical simulation of electron energy loss near inhomogeneous dielectrics. Phys. Rev. B **56**, 15873–15884 (1997)

F.J. García de Abajo, A. Howie, Relativistic electron energy loss and electron-induced photon emission in inhomogeneous dielectrics. Phys. Rev. Lett. **80**, 5180–5183 (1998)

F.J. García de Abajo, J.J. Sáenz, Electromagnetic surface modes in structured perfect-conductor surfaces. Phys. Rev. Lett. **95**, 233901 (2005)

L. Gu, V. Srot, W. Sigle, C. Koch, P. van Aken, F. Scholtz, S.B. Thapa, C. Kirchner, M. Jetter, M. Rühle, Band-gap measurements of direct and indirect semiconductors using monochromated electrons. Phys. Rev. B **75**, 195214 (2007)

A. Gutierrez-Sosa, U. Bangert, A.J. Harvey, C. Fall, R. Jones, Energy loss spectroscopy of dislocations in GaN and diamond: a comparison of experiment and calculations. Diam. Rel. Mater. **12**, 1108–1112 (2003)

S.D. Hersee, X.Y. Sun, X. Wang, The controlled growth of GaN nanowires. Nano Lett. **6**, 1808–1811 (2006)

J.K. Hyun, M. Couillard, P. Rajendran, C.M. Liddell, D.A. Muller, Measuring far-ultraviolet whispering gallery modes with high energy electrons. Appl. Phys. Lett. **93**, 243106 (2008)

J.D. Jackson, *Classical Electrodynamics*, 2nd edn. (Wiley, New York, NY, 1975)

J.R. Jinschek, R. Erni, N.F. Gardner, A.Y. Kim, C. Kisielowski, Local indium segregation and bang gap variations in high efficiency green light emitting InGaN/GaN diodes. Solid State Comm. **137**, 230–234 (2006)

K. Kimoto, G. Kothleitner, W. Grogger, Y. Matsui, F. Hofer, Advantages of a monochromator for bandgap measurements using electron energy-loss spectroscopy. Micron **36**, 185–189 (2005)

S. Lazar, G.A. Botton, M.-Y. Wu, F.D. Tichelaar, H.W. Zandbergen, Materials science applications of HREELS in near edge structure analysis and low-energy loss spectroscopy. Ultramicroscopy **96**, 535–546 (2003)

K.A. Mkhoyan, T. Babinec, S.E. Maccagnano, E.J. Kirkland, J. Silcox, Separation of bulk and surface-losses in low-loss EELS measurements in STEM. Ultramicroscopy **107**, 345–355 (2007)

P. Moreau, N. Brun, C.A. Walsh, C. Colliex, A. Howie, Relativistic effects in electron-energy-loss-spectroscopy observations of the Si/SiO_2 interface plasmon peak. Phys. Rev. B **56**, 6774–6781 (1997)

D.A. Muller, J. Silcox, Delocalization in inelastic scattering. Ultramicroscopy **59**, 195–213 (1995)

J. Nelayah, M. Kociak, O. Stephan, F.J. García de Abajo, M. Tence, L. Henrard, D. Taverna, I. Pastoriza-Santos, L.M. Liz-Marzan, C. Colliex, Mapping surface plasmons on a single metallic nanoparticle. Nat. Phys. **3**, 348–353 (2007)

E. Rabani, B. Hetenyi, B.J. Berne, L.E. Brus, Electronic properties of CdSe nanocrystals in the absence and presence of a dielectric medium. J. Chem. Phys. **110**, 5355–5369 (1999)

H. Raether in *Surface Plasma Oscillations and Their Applications*. Edited by G. Hass, M.H. Francombe and R.W. Hoffman, Physics of Thin Films, vol 9 (Academic, New York, NY, 1977), pp. 145–261

B.W. Reed, M. Sarikaya, Background subtraction for low-loss transmission electron energy-loss spectroscopy. Ultramicroscopy **93**, 25–37 (2002)

B.W. Reed, J.M. Chen, N.C. MacDonald, J. Silcox, G.F. Bertsch, Fabrication and STEM/EELS measurements of nanometer-scale silicon tips and filaments. Phys. Rev. B **60**, 5641–5652 (1999)

R.H. Ritchie, Plasma losses by fast electrons in thin films. Phys. Rev. **106**, 874–881 (1957)

R.H. Ritchie, Surface plasmons in solids. Surf. Sci. **34**, 1–19 (1973)

R.H. Ritchie, Quantal aspects of the spatial resolution of energy-loss measurements in electron microscopy: I. Broad-beam geometry. Phil. Mag. A **44**, 931–942 (1981)

R.H. Ritchie, A. Howie, Inelastic scattering probabilities in scanning transmission electron microscopy. Phil. Mag. A **58**, 753–767 (1988)

A. Rivacoba, N. Zabala, J. Aizpurua, Image potential in scanning transmission electron microscopy. Prog. Surf. Sci. **65**, 1–64 (2000)

W. Sigle, L. Gu, C. Koch, V. Srot, J. Nelayah, P.A. van Aken, Application of monochromated electrons in EELS. Microsc. Microanal. **14**(Suppl. 2), 134–135 (2008)

T. Stöckli, J.-M. Bonard, P.-A. Stadelmann, A. Châtelain, EELS investigation of plasmon excitations in aluminum nanospheres and carbon nanotubes. Z. Phys. D **40**, 425–428 (1997)

M. Stöger-Pollach, Optical properties and bandgaps from low loss EELS: Pitfalls and solutions. Micron **39**, 1092–1110 (2008)

M. Stöger-Pollach, P. Schattsneider, The influence of relativistic energy losses on bandgap determination using valence EELS. Ultramicroscopy **107**, 1178–1185 (2007)

M. Stöger-Pollach, H. Franco, P. Schattsneider, S. Lazar, B. Shaffer, W. Grogger, H.W. Zandbergen, Cerenkov losses: A limit for bandgap determination and Kramers–Kronig analysis. Micron **37**, 396–402 (2006)

M. Terauchi, M. Tanaka, K. Tsuno, M. Ishida, Development of a high energy resolution electron energy-loss spectroscopy microscope. J. Microsc. **194**, 203–209 (1999)

P.C. Tiemeijer, Operation modes of a TEM monochromator. Inst. Phys. Conf. Ser. **161**, Section 5, 191–194 (1999)

M.C. Troparevsky, L. Kronik, J.R. Chelikowsky, Optical properties of CdSe quantum dots. J. Chem. Phys. **119**, 2284–2287 (2003)

D. Ugarte, C. Colliex, P. Trebbia, Surface- and interface-plasmon modes on small semiconducting spheroids. Phys. Rev. B **45**, 4332–4343 (1992)

K. van Benthem, R.H. French, W. Sigle, C. Elsässer, M. Rühle, Valence electron energy loss study of Fe-doped $SrTiO_3$ and a $\Sigma 13$ boundary: electronic structure and dispersion forces. Ultramicroscopy **86**, 303–318 (2001)

T. Walther, E. Quandt, H. Stegmann, A. Thesen, G. Benner, First experimental test of a new monochromated and aberration-corrected 200 kV field-emission scanning transmission electron microscope. Ultramicroscopy **106**, 963–969 (2006)

Z.L. Wang, Valence electron excitations and plasmon oscillations in thin films, surfaces, interfaces, and small particles. Micron **27**, 265–299 (1996)

N. Zabala, E. Ogando, A. Rivacoba, F.J. García de Abajo, Inelastic scattering of fast electrons in nanowires: A dielectric formalism approach. Phys. Rev. B **64**, 205410 (2001)

L. Zhang, R. Erni, J. Verbeeck, G. van Tendeloo, Retrieving the dielectric function of diamond from valence electron energy-loss spectroscopy. Phys. Rev. B **77**, 195119 (2008)

17

Variable Temperature Electron Energy-Loss Spectroscopy

Robert F. Klie, Weronika Walkosz, Guang Yang and Yuan Zhao

17.1 Introduction

For more than a decade, high-resolution Z-contrast imaging in combination with electron energy-loss spectroscopy (EELS) has been used as an important tool for microstructural evaluations of interfaces and defects in materials that are important for a variety of technological applications (Browning et al. 1993, Muller et al. 1993, Batson 1993). In the published literature, one can find many examples, where atomic-resolution Z-contrast imaging and EELS have been used to address important materials science issues, ranging from ultra-thin gate oxides in semiconductor devices (Muller et al. 1999, Green et al. 2001, Klie et al 2003) to hetero-interfaces in ceramics (Shibata et al. 2004, Ziegler et al. 2004, 2006, Winkelman et al. 2004, Walkosz et al. 2008, Ohtomo et al. 2002), grain boundaries in high-T_c superconductors (Klie et al. 2005, McGibbon et al. 1994, Kim et al. 2000, Browning et al. 1999) and heterogeneous catalysts (Qi et al. 2001, Sun et al. 2002 a-c, Klie et al. 2002). Recently, the introduction of aberration-corrected scanning transmission electron microscopes (STEMs) (Krivanek et al. 2003) has further pushed the limits of Z-contrast imaging and EELS to single-atom sensitivity, (Varela et al. 2004), sub-Å spatial resolution (Yang et al. 2008), and atomic-resolution spectrum imaging (Muller et al. 2008). However, these measurements are all conducted at room temperature and in equilibrium with the surrounding environment to assure the stability of the sample holder and to minimize the drift during the EELS acquisition time, typically of the order of 1–3 s. These conditions might not always be ideal for understanding the behavior of materials or structures that exhibit interesting properties at temperatures above or below 300 K.

For more than three decades, in situ heating experiments have been successfully performed under vacuum or even under controlled gas conditions in the transmission electron microscope (TEM) (Packan and Braski 1970, Braski 1970, Baker 1979, Heinemann et al. 1975). In these experiments, the samples are typically mounted on a carbon-coated TEM grid and then loaded into the heating element of the specimen

S.J. Pennycook, P.D. Nellist (eds.), *Scanning Transmission Electron Microscopy*,
DOI 10.1007/978-1-4419-7200-2_17, © Springer Science+Business Media, LLC 2011

holder (Boyes and Gai 1997, Gai 1999a, b, Sharma 2001, 1998). For dedicated environmental TEMs, the specimen can also be directly loaded inside the objective lens pole piece, which means that the sample cannot be easily removed and has to remain inside the microscope column. Another approach to heating holders involves the use of a helical wire heater that is coated with a thin carbon film to support the sample material. The film and the sample are then directly heated using a current flowing through the wire, which allows for rapid heating and cooling rates, but is associated with large drift rates. Nevertheless, specimen holders using this approach enable lattice imaging in the TEM at temperatures up to 1300 K (Kamino et al 2005a-c).

In this chapter, we will review recent experiments that demonstrate high spatial resolution for Z-contrast imaging combined with EELS in the temperature range between 10 and 700 K. While many of the results presented here are not obtained using aberration-corrected STEMs, the techniques and approaches described here can be directly applied to aberration-corrected instruments. In addition to reviewing these results, we will also address some fundamental issues related to high-resolution variable temperature EELS, and suggest further improvements to the instrumentation and analysis software to deal with some of the problems that are unique to variable temperature EELS in a STEM.

The remaining chapter is organized as follows: In Section 17.2, we describe the different specimen holders that were used for the studies described in Section 17.3. In Section 17.3 we describe a series of in situ heating and cooling EELS experiments including a high-temperature study of $SrTiO_3$ tilt grain boundaries (Section 17.3.1), a variable temperature study of $LaCoO_3$ (Section 17.3.2), and in situ heating experiments in $Ca_3Co_4O_9$ (Section 17.3.3).

17.2 Methods and Instrumentation

The variable temperature EELS results described here were obtained using the JEOL 3000F at Brookhaven National Laboratory (BNL) and the JEOL 2010F at the University of Illinois at Chicago (UIC) operated at 300 and 200 KV, respectively. Both microscopes are equipped with a Schottky field-emission source, an ultra high-resolution objective lens pole piece (URP), a high-angle annular dark-field detector, and a post-column Gatan imaging filter. The microscope and spectrometer were set up for a convergence angle (α) of 13 mrad (at 300 KV) and 15 mrad (at 200 KV) to achieve a probe size of between 1.4 and 2.0 Å for imaging and spectroscopy, respectively. The high-angle annular dark-field detector inner angle was chosen at $\sim 3\alpha$ and the spectrometer collection angle (β) was 28 mrad at 300 KV and 38 mrad at 200 KV.

17.2.1 Heating Holders

The in situ heating experiments utilize the Gatan Model 652 double tilt heating stage (see Figure 17–1). This holder can be used with the

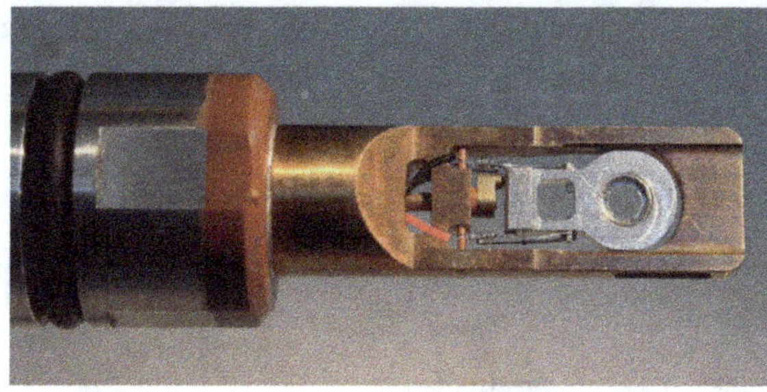

Figure 17–1. Tip of the Gatan double tilt heating stage (Model 652).

JEOL ultra-high resolution (URP) pole piece within a tilt range of ±10° in the x- and y-tilt directions. As shown in Figure 17–1, the sample is mounted inside a heating element at the tip of the specimen holder. This mini-furnace can be tilted around two axes, but extreme care is required to assure that the furnace does not come in contact with the objective lens pole piece during the heater operation. The specimen furnace contains a miniature, encapsulated heater, which is attached to the two terminal posts in the specimen tip. In this heating stage, the furnace body, the anti-welding washers, and the hexring to mount the sample are all made of tantalum. A SmartSet Hot Stage Controller is used for temperature control of the heating stage, which provides a variable temperature control in the range between 293 and 1273 K. If the holder is operated above 773 K, the water recirculation system must be connected to prevent heat transfer through the rod of the stage. The turbulent flow of water through the recirculation system makes it impossible to achieve atomic-resolution Z-contrast images for temperatures higher than 773 K. The furnace suspension system has been designed to minimize thermal expansion of the stage so that the total specimen movement while heating or cooling to 773 K is less than 5 μm, whereas the drift rate at operating conditions is less than 5 nm/min. This means that for each desired temperature, the sample must stabilize for at least 1 h before any measurement can be taken. Thus, in situ time-resolved atomic-scale measurements are nearly impossible. However, after stabilization, the experiments shown here indicate that the drift rate can be even lower than expected, at times as low as 1 nm/min. The oxygen partial pressure near the specimen location in the microscope column is $\sim 5 \times 10^{-8}$ Pa during the experiment, which means that samples experience a highly reducing environment at temperatures as high as 700 K.

In recent years, a number of MEMS-based heating holders have become available, which promise to dramatically increase the temperature and spatial stability compared to traditional heating stages. These stages will become particularly interesting in combination with aberration-corrected STEM, where sub-Å spatial resolution can now be achieved (Nellist et al. 2004). These MEMS-based holders can potentially overcome a number of performance problems associated with standard heating stage technologies, allowing, for example, rapid temperature cycling with $\Delta T > 1000\,\mathrm{K}$ in less than 1 ms with negligible spatial drift (Allard et al. 2009). The advantage of these MEMS-based

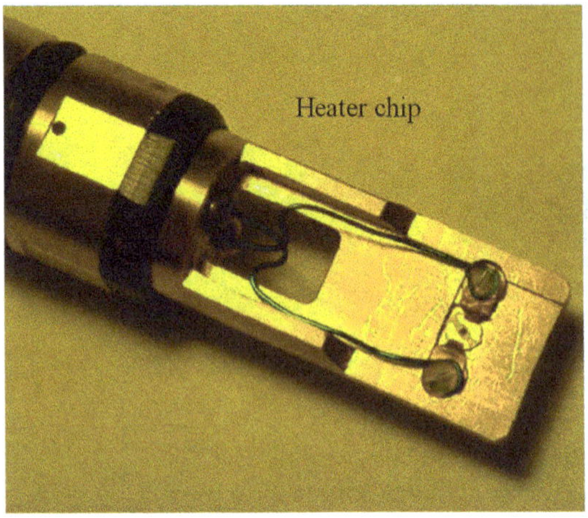

Figure 17–2. Tip of prototype specimen holder for a JEOL 200 KV STEM/TEM, showing heater chip clamped into place, with electrical leads connected. (Image courtesy of L.F. Allard (2009).)

designs (see for example Figure 17–2) compared to traditional heating stages is that only a 150 nm thick, 500 μm^2 freestanding membrane made from a conductive ceramic is heated (compared to a 3 mm lacey carbon-coated Cu grid in the Gatan 652 tilt heating stage). While a large heating region can be advantageous, due to the likely temperature uniformity of the sample area and can be calibrated using thermocouples to give a measure of the temperature experienced by the specimen, the disadvantage is that such a large volume makes it impossible to achieve rapid heating and cooling rates. Furthermore, the image drift rates due to thermal expansion/contraction of the material in the large heating area cannot be neglected. For more information on MEMS-based heating stages, see for example Allard et al. (2009).

17.2.2 Cooling Holders

The cooling experiments are performed using the Gatan 636LHe double tilt cryo-holder (Figure 17–3), which allows atomic-resolution imaging and spectroscopy at temperatures as low as 86 K using liquid nitrogen as a cryogenic. Similarly to the Gatan double tilt heating stage, the sample is mounted at the tip of the holder using two tantalum washers and a hexring. The sample sits inside the cooling element, which is directly connected to the dewar and is cooled via thermal coupling. Such a setup allows for a tilt range in x- and y-directions similar to that reported for the heating stage (i.e., $\pm 10°$). The double-sided dewar sits outside the microscope column and can be refilled during the cooling experiments. The temperature is measured close to the specimen and no significant specimen heating due to the electron beam exposure was noticed. In order to minimize the thermal drift and increase the stability of the sample, the specimen is usually kept at temperature around 86 K for several hours before any imaging or spectroscopy can be performed. In addition, an electrical current can be applied to warm up the specimen holder to a specific temperature. For experiments that utilize liquid He as a coolant, the inner dewar, which is directly connected to the metal rod cooling the tip of the holder, is filled with liquid He, while

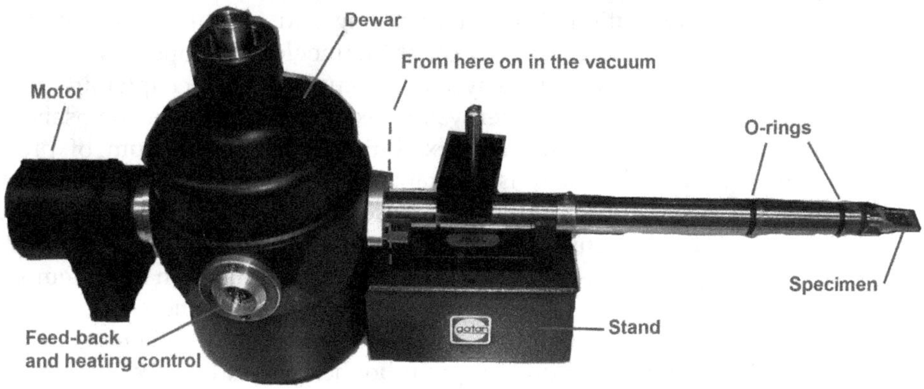

(a)

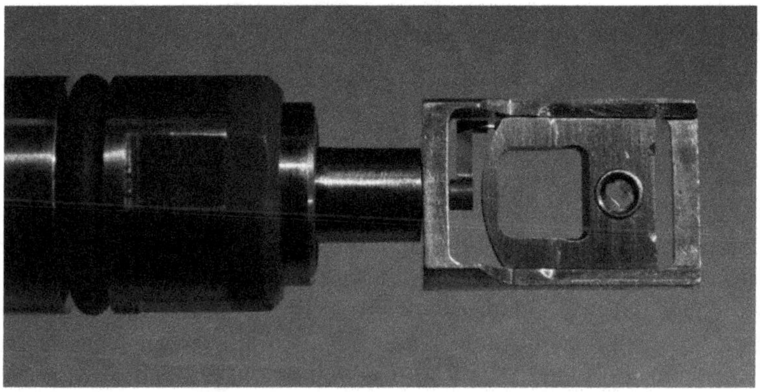

(b)

Figure 17–3. Pictures of the (**a**) Gatan 636LHe double tilt cooling stage and (**b**) tip of the Gatan 636 double tilt cooling stage.

the outer dewar is filled with liquid N_2. While this setup allows for higher thermal stability and lower drift rates at temperatures around 10 K, the liquid He in the inner dewar only lasts for 10 min before completely boiling off, which means that long-time thermal stability cannot be achieved.

17.3 Results

In this section, we will describe several experimental results using variable temperature EELS and high-resolution Z-contrast imaging. Section 17.3.1 describes an in situ heating study of tilt grain boundaries in $SrTiO_3$ and the effects of oxygen vacancy diffusion in grain boundaries at elevated temperature. Section 17.3.2 describes an in situ heating and cooling study of $LaCoO_3$, where the effects of a spin state transition of the Co^{3+} ions can be directly measured using EELS. Finally, the thermoelectric $Ca_3Co_4O_9$ will be examined in Section 17.3.3 using high-resolution Z-contrast imaging and in situ heating to study the effects of charge transfer on its thermoelectric properties.

All of the materials systems discussed in this chapter are perovskite oxide based, since perovskite oxides have proven themselves to be extremely versatile and exhibiting a broad spectrum of properties. In particular, there is substantial interest in exploiting their functionality for superconductivity, ferromagnetism, ferroelectricity, magneto resistance, ionic conductivity, and as dielectrics, and many of these properties are exhibited at temperatures other than room temperature. To develop a fundamental understanding of the structure–property relationships of perovskite oxide systems, we will therefore study their behavior in the bulk at grain boundaries and hetero-interfaces in the temperature range between 10 and 724 K.

17.3.1 $SrTiO_3$ Grain Boundaries

$SrTiO_3$ represents a relatively simple and ideal perovskite oxide system, in which the detailed investigation of defects and grain boundaries can be directly applicable to structurally and chemically more complex perovskite materials. In addition, $SrTiO_3$ itself displays a wide range of physical and chemical properties, such as superconductivity, catalytic activity, ferroelectricity, and semiconductivity. These properties are directly influenced by the presence of grain boundaries, and a fundamental understanding of the grain boundary mechanism might lead to further applications for $SrTiO_3$. Therefore, much can be learned about the fundamental structure–property relationships of perovskite oxide grain boundaries and hetero-interfaces by understanding the atomic-scale properties of low-angle tilt grain boundaries in $SrTiO_3$.

While the bulk structure and properties of $SrTiO_3$ are now well understood, there is still considerable debate as to the origin of many widely observed grain boundary properties. Many theories to explain the microscopic properties of grain boundaries have introduced generic grain boundary states that lead to the formation of a double Schottky barrier (Kliewer and Koehler 1965). In these models, the boundary states are induced by the presence of immobile charges on the grain boundary plane, which are compensated for by an opposite space charge in a depletion layer on either side of the boundary (Kliewer and Koehler 1965). Several high-resolution transmission electron microscopy (HRTEM) studies have suggested amorphous phases or cation interstitials to be the origin of the charge imbalance in the boundary plane (Denk et al. 1997, Chiang et al. 1990). More recently, the correlation between the structural and the local electronic properties of $SrTiO_3$ grain boundaries was obtained by the combination of Z-contrast imaging and electron energy-loss spectroscopy (EELS) in the scanning transmission electron microscope (STEM) (McGibbon et al. 1996, Browning and Pennycook 1996, Klie and Browning 2000, McGibbon et al. 1994, Klie et al. 2003). In these studies, it was found that [001] tilt grain boundaries contained characteristic sequences of structural units that did not contain any inter-granular grain boundary phases (McGibbon et al. 1994, Kim et al. 2000). However, these structural units did contain reconstructions that were proposed to give rise to the local states responsible for the electronic behavior of the grain boundary (Browning et al. 1999). Self-consistent ab initio density functional calculations of these units suggest that the behavior is more subtle than previously proposed. In particular, it has been shown that it is energetically favorable for there to be an excess of oxygen vacancies in these units, and in the case of units centered on the Ti sub-lattice, a Ti excess (Kim et al. 2000). Such non-stoichiometry leads to the formation of a highly donor-doped, or n-type, region at the boundary rather than the formation of a Schottky barrier.

Two sets of experiments were performed on a 58° [001] tilt misorientation bicrystal purchased from Shinkosha; each set consists of a series of EELS spectra acquired both from the bulk and from the grain boundary. The first set is taken at room temperature, while the second set was acquired at 724 K after the sample was heated and stabilized for at least 3 h. The sample was not exposed to air between the two experiments, assuring that the changes are attributable only to the heating experiments.

Figure 17–4 shows a high-resolution Z-contrast image of $SrTiO_3$ [001] bulk taken with the aberration-corrected VG HB603U. In this image, as well as in the images taken from the 58° tilt grain boundaries in $SrTiO_3$, the atomic columns with the highest image intensity represent the Sr columns which form a square lattice, while the TiO columns are shown as the less bright spots in the center of the Sr square. Figures 17–5 and 17–6 show Z-contrast images of the $SrTiO_3$ 58° [001] tilt grain boundary taken at room and elevated (724 K) temperatures. The positions of the Sr and TiO atomic columns are highlighted at the interface by black and white circles. Pure oxygen columns are not visible in these images, due to the small scattering amplitude of light elements at large angles. This symmetric grain boundary exhibits the same structural units as

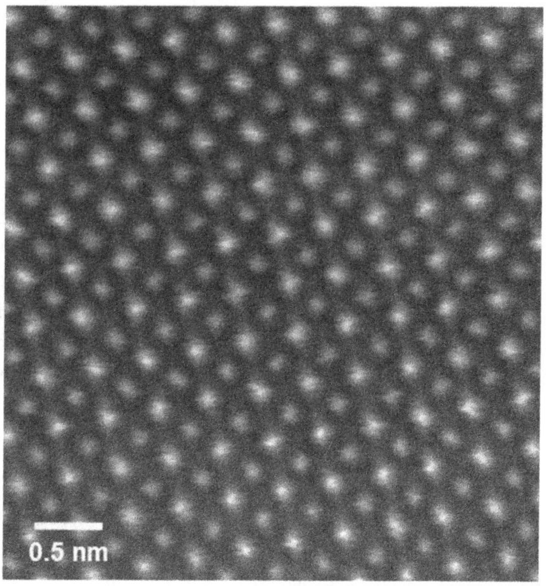

Figure 17–4. High-resolution Z-contrast image of $SrTiO_3$ [001] bulk, acquired using an aberration-corrected VG HB603U with a convergence angle $\alpha = 24$ mrad and a detector inner angle $\beta = 100$ mrad.

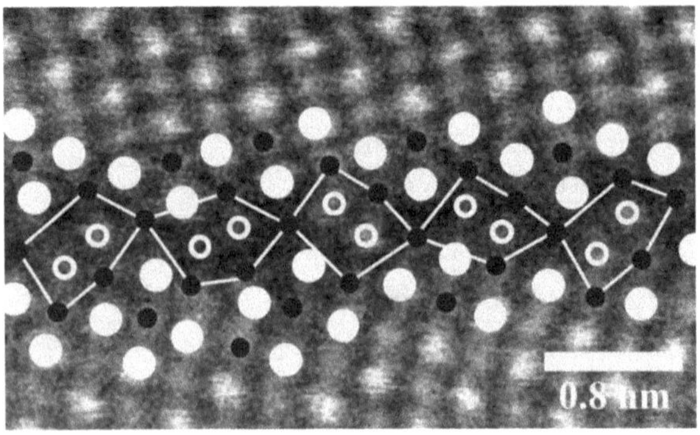

Figure 17–5. Z-contrast image of the 58°[001] tilt grain boundary at room temperature. Reproduced from Klie and Browning (2000) with permission.

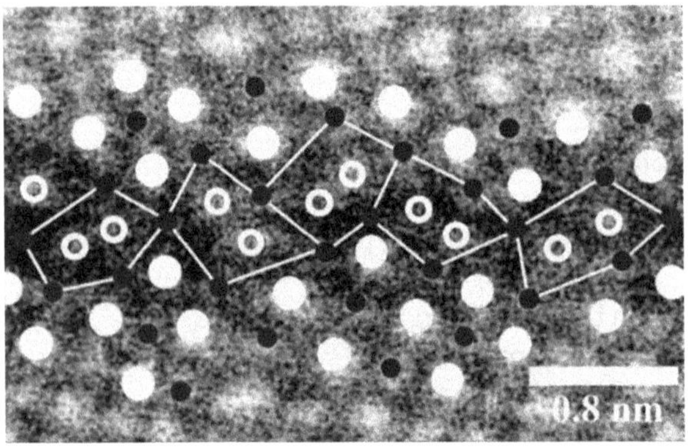

Figure 17–6. Z-contrast image of the 58°[001] tilt grain boundary at 724 K. Reproduced from Klie and Browning (2000) with permission.

reported earlier (McGibbon et al. 1994, 1996, Browning and Pennycook 1996, Klie et al. 2003). Here, Sr atomic columns are located in the center of the grain boundary resulting in only partially occupied Sr atomic columns. The location of these partially occupied atomic Sr columns is indicated by the white rings. At 724 K (see Figure 17–6), the grain boundary exhibits the same structural cation arrangement as it does at room temperature, apart from the broadening of the atomic columns, which is due to thermal vibrations. As can be seen from the image taken after 3 h of heating to 724 K, the spatial drift is comparable to that at room temperature, and no additional noise due to the heating operation has been observed.

Figure 17–7a, b shows two spectra taken from the bulk and the center of the grain boundary at room temperature and 724 K. Each spectrum shown in Figure 17–7a, b represents the sum of 7–14 individual spectra that were added and background subtracted prior to normalizing the intensity to the continuum interval 30 eV before the onset of the O *K*-edge (532 eV). The spectra show the Ti *L*-edges, containing two peaks, the Ti L_3 peak at ~461 eV and the L_2 peak at ~467 eV, and the O *K*-edge at 532 eV, which exhibits three distinct peaks in the bulk labeled *a*, *b*, and *c*. The energy scale for all the spectra shown in this section is normalized to the onset of the O *K*-edge which is located at 532 eV. Therefore, any shift in the position of the Ti *L*-edge will always be measured with respect to the O *K*-edge position. The spectra labeled "bulk" were taken from an area of the bicrystal at least 20 nm away from the

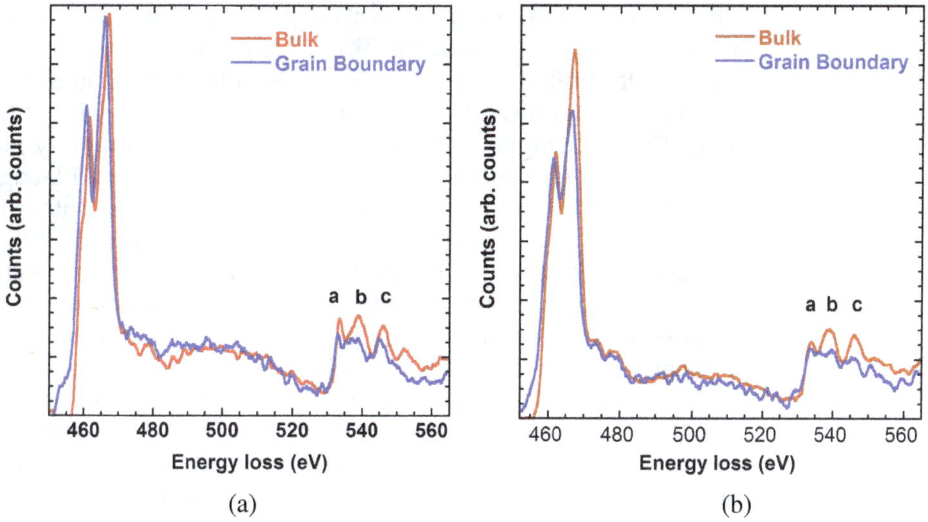

Figure 17–7. (**a**) Electron energy-loss spectrum from the bulk and the grain boundary showing the shift in the Ti *L*-edges, the lower count rate under the O *K*-edge and the non-existent near-edge structure at the boundary. (**b**) Comparison of EEL spectra from the bulk and the grain boundary at 724 K. Reproduced from Klie and Browning (2000) with permission.

grain boundary to minimize the influence of the grain boundary on the near-edge fine structure of both the Ti *L*-edge and the O *K*-edge.

At room temperature, the most obvious difference between the spectra taken in the bulk and from the grain boundary (Figure 17–7a) is the decrease in intensity of O *K*-edge spectrum in the grain boundary. Additionally, there is a small change in the O *K*-edge fine structure, with peak *c* being much more visible than peaks *a* and *b* in the boundary. The reduction of peaks *a* and *b* and fewer counts under the O *K*-edge suggest a destruction of long-range order and the presence of excess oxygen vacancies at the grain boundary (Browning and Moltaji 1998). Additional evidence for the presence of oxygen vacancies comes from an energy shift of the Ti *L*-edge s by 1.4 ± 0.2 eV down in energy and an increase in the Ti L_3/L_2 intensity ratio by 3.5% (both indicative of a lowering of the Ti valence to compensate for oxygen vacancies) (Sankararaman and Perry 1992). The integrated Ti:O intensity ratio is also increased by 25% at the grain boundary, again suggesting an increased presence of oxygen vacancies.

After heating the $SrTiO_3$ bicrystal for 3 h at 724 K, we measured the same series of images and spectra at similar locations in the bulk and at the grain boundary. Figure 17–7b shows two background subtracted spectra that contain 7–14 summed spectra, normalized to the onset at 532 eV and the intensity 30 eV before the O *K*-edge. Both spectra show the Ti *L*-edges and the O *K*-edge which again exhibits three peaks labeled *a*, *b*, and *c*. While these three peak are still clearly visible in the bulk at 724 K, the O *K*-edge pre-peaks shows a decrease in intensity. Moreover, the fine structure of the grain boundary spectrum at 724 K shows a much greater change than in the unheated sample and now resembles the hydrogenic edge expected for an isolated atom.

Furthermore, the spectra taken at the grain boundary at 724 K exhibit an increase in the Ti:O and the Ti L_3/L_2 intensity ratios by 30% with respect to the bulk (Figure 17–7b). This is a larger effect than is observed at room temperature and strongly suggests that after 3 h of in situ heating to 724 K, the number of oxygen vacancies in the boundary plane is increasing. There is again a shift down in energy of the Ti *L*-edges by $\sim 1 \pm 0.2$ eV, which indicates a decrease in the Ti valence. Interestingly, but maybe not all that surprising, there is an increase in the Ti:O ratio in the bulk by 3.5% over the unheated case, indicating that there are also oxygen vacancies introduced into the bulk of the grains as a result of the in situ reduction. It has been previously shown that the O *K*-edge pre-peak intensity in $SrTiO_3$ and related oxides can be directly correlated to the oxygen vacancy concentration, and the observed decrease in the O *K*-edge pre-peak intensity in the bulk at 724 K further confirms the increased concentration of oxygen vacancies as far as 20 nm from the grain boundary. Such vacancies may explain the smaller shift in the edge onset of the Ti *L*-edge since we would now expect Ti^{3+} to also be present in the bulk of the grains.

In addition to the spectra taken from the bulk and the center of the grain boundary, a set of EELS spectra was acquired as a function of distance from the $SrTiO_3$ grain boundary. At each position up to 14 spectra were summed up for an improved signal-to-noise ratio. From the acquired core-loss spectra containing the Ti *L*-edges and the O *K*-edge, it is now possible to measure the oxygen content of $SrTiO_3$ with respect to its distance from the boundary. After the core-loss spectra (similar to the ones shown in Figure 17–7a) are background subtracted and corrected for the differences in sample thickness, the Ti:O ratio is calculated at each location. The Ti intensity is integrated from 455 to 480 eV and the O intensity from 530 to 560 eV. The difference of this ratio with respect to the bulk value as a function of distance from the interface gives the oxygen vacancy concentration profile shown in Figure 17–8. A Lorentzian function is fitted to the data at both room temperature and 724 K. It has to be noted here that the profile of the $SrTiO_3$ grain boundary heated to 724 K is taken after 20 h of in situ heating in an effort to achieve equilibrium conditions. Furthermore, the EELS data at the grain boundary are only taken from one side of the grain boundary. The data shown in Figure 17–8 are mirrored at the grain boundary plane under that assumption of a symmetric oxygen vacancy concentration profile.

From the fit through the experimental data, the area under the curve and the full width at tenth maximum (FWTM) can be determined. Prior to the in situ heating experiment, the FWTM of the 58° tilt grain boundary is 16 ± 1.6 Å, while the area under this plot is $1.035 \pm 0.1 \frac{\text{vacancies}}{\text{Å}}$. For both profiles, the bulk vacancy concentration is used to normalize the profile and all the measured changes are with respect to this bulk concentration. After 20 h of in situ heating, the grain boundary width (defined as the full width at tenth maximum) widens to 25.6 ± 2.6 Å and there are now 18% more oxygen vacancies at the interface than in the bulk. This value is nearly half of that measured after 3 h of heating (Figure 17–7b), but the total integrated number of oxygen

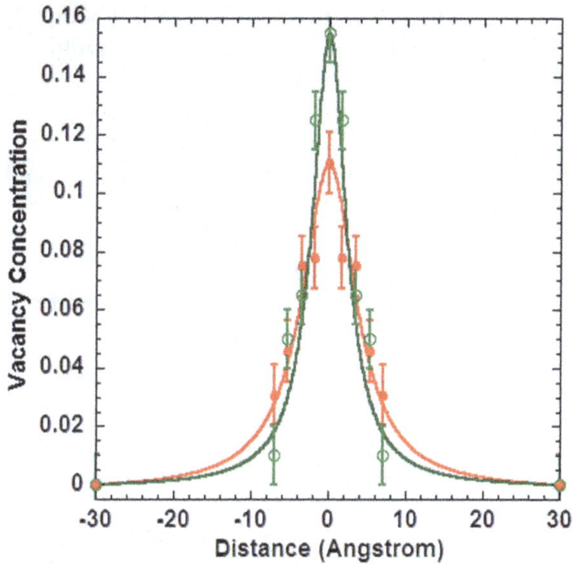

Figure 17–8. Oxygen vacancy concentration profiles taken at room temperature prior to heating and at 724 K after in situ heating for 20 h. Please note that the profiles were only taken from one side of the interface and then mirrored at the origin.

vacancies is increased with respect to the room temperature results to $1.194 \pm 0.1 \frac{\text{vacancies}}{\text{Å}}$.

These experimental results show that atomic column-resolved Z-contrast imaging and EELS can be achieved in $SrTiO_3$ grain boundaries at temperatures as high as 724 K after stabilizing for at least 3 h. The spatial drift at 724 K was comparable to that at room temperature, and no additional noise was measured as the result of the heating operation. Furthermore, we could demonstrate that the 58° tilt grain boundary in $SrTiO_3$ contains an excess of oxygen vacancies that are compensated for by a decrease in the Ti valence. As the reducing conditions are increased in the microscope by heating the sample to 724 K, the number of oxygen vacancies at the boundary is increased. However, the cation arrangement of the boundary plane is not changed significantly during the reduction process. We also did not observe any electron beam damage during the heating experiments, and the increased O vacancy concentration profile was not the result of radiation damage. It was restored close to the profile prior to the heating experiment after keeping the $SrTiO_3$ sample for an additional 5 h in the microscope column at room temperature after the completion of the heating experiment.

17.3.2 Spin State Transitions in $LaCoO_3$

The perovskite oxide $LaCoO_3$ has been studied intensely over the last 40 years due to its unique magnetic behavior and its related nonmetal–metal transitions (see for example Heikes et al. 1964, Raccah and Goodenough 1967, Medarde et al. 2006). Specifically, two broad transitions in the magnetic susceptibility of $LaCoO_3$ are of interest, the first one occurring at 50–90 K when $LaCoO_3$ undergoes a gradual transition from a non-magnetic to a paramagnetic semiconductor, followed by a second transition at 500–600 K, that coincides with a semiconductor-to-metal transition. While these two transitions have been attributed to spin-state transitions of the Co^{3+} ion spins (Heikes et al. 1964, Raccah and Goodenough 1967, Abbate et al. 1993, Korotin et al. 1996), the underlying electronic structure and spin states have not yet been fully understood. Goodenough (1958), Raccah and Goodenough (1967), and Senaris-Rodriguez and Goodenough (1995) first interpreted these magnetic transitions as spin-state transitions of the Co^{3+} ions from a low-spin state (LS) to a high spin state (HS) due to the close values of the intra-atomic exchange energy (J_H) and the crystal field splitting (10Dq) at the Co^{3+} sites. Thus, depending on the relative values of the J_H and 10Dq, either the LS with $t_{2g}^6 e_g^0$ resulting in $S = 0$ or the HS with $t_{2g}^4 e_g^2$ resulting in $S = 2$ is suggested to be more stable. While this model can explain the high-temperature transition in $LaCoO_3$, several different models for the Co^{3+} spin state in the temperature regime between 80 and 500 K and the associated transition at 80 K have been proposed in the past (Korolin et al. 1996, Senaris-Rodriguez and Goodenough 1995, Asai et al. 1989).

One popular model is the mixed spin state of the Co^{3+} ions, where the population of the HS state is increased with increasing temperature, resulting in a stable LS–HS spin-state array (ratio between LS and HS of 1:1). Many spectroscopic studies, including photoemission spectroscopy (XPS), and X-ray absorption spectroscopy (XAS) have been reported investigating these spin-state models in $LaCoO_3$ (Medarde et al. 2006, Abbate et al. 1993, Saitoh et al. 1997, Moodenbaugh et al. 2000). However, LDA+U calculations of the total energy for different spin states in $LaCoO_3$ raised questions about the existence of a mixed LS–HS state, and moreover predicted the occurrence of an intermediate spin state (IS, $S = 1$) with $t_{2g}^5 e_g^1$ of the Co^{3+} ion. Korotin and coworkers (Korotin et al. 1996) have shown that this IS state is energetically comparable to the LS state, and much more stable than the HS state, due to the larger O 2p-Co 3d hybridization, as well as orbital-ordering effects. These calculations also predict changes in the total DOS between the LS and the IS states, due to the increased filling of the e_g-band and the splitting of the spin-up and spin-down spin projections.

In Section 17.3.2.1, we will describe in situ cooling experiments of bulk $LaCoO_3$ at 10 K to quantify the Co^{3+} ion spin-state transition in $LaCoO_3$. We will further demonstrate that high-resolution TEM imaging and EELS can be achieved at temperatures as low as 10.4 K. We will also show the results of in situ heating experiments of bulk $LaCoO_3$ at temperatures up to 700 K in an effort to quantify the high-temperature Co^{3+} spin-state transition. In Section 17.3.2.2, we will explore the effects of interfacial strain on the low-temperature spin-state transition in epitaxially strained $LaCoO_3$ thin films.

17.3.2.1 Bulk $LaCoO_3$

The $LaCoO_3$ powder samples were fabricated by the conventional solid-state reaction of La_2O_3 and Co_3O_4 powders at 1000°C for 1 week (for more details see English et al. 2002). Susceptibility and resistivity measurements as a function of temperature in a magnetic field of $H = 50$ kOe and $H = 0$, respectively, show a clear anomaly in the

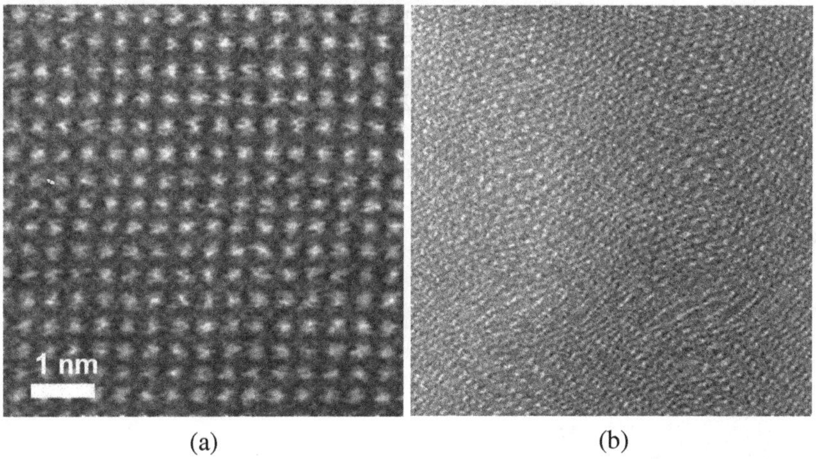

Figure 17–9. Z-contrast image of $LaCoO_3$ [2$\bar{2}$1] at (**a**) 85 K showing the square lattice of La atoms. (**b**) at 10 K. The random drift makes it impossible to achieve atomic resolution at 10 K. Reproduced from Klie et al. (2007) with permission.

magnetic as well as in the electronic behavior at 80 K (English et al. 2002). The powder of $LaCoO_3$, with an average grain size of more than 1 μm, was crushed and dispersed on a lacey carbon film and subsequently heated to 100°C in air to reduce specimen contamination prior to the TEM studies. Next, we utilize the Gatan LHe stage to perform a series of in situ cooling experiments to 85 and 10 K. High-resolution TEM and Z-contrast images, EELS spectra, and diffraction patterns were taken prior to the cooling experiment at room temperature, at 85 and 10 K, respectively. The entire cooling experiment (i.e., 300 K > T > 10.4 K) was performed without removing the sample from the microscope column and exposing it to air. Therefore, we can assume that the sample remains unchanged, unless altered by the electron beam.

Figure 17–9a shows a atomic-resolution Z-contrast image of $LaCoO_3$ [2$\bar{2}$1] taken at 85 K. The images shows the square lattice of La atomic columns, while neither the Co-O nor the pure O columns are visible in this micrograph, due to the reduced specimen stability resulting from the sample cooling. Moreover, the overall specimen drift can be seen as a slight bending of the vertical lines of La atoms. Figure 17–9b shows a Z-contrast image taken at 10.4 K. It can be clearly seen in this image that the specimen stability is no longer sufficient to allow atomic-resolution imaging in Z-contrast imaging mode, but conventional HRTEM images can still be obtained under these conditions (Figure 17–10).

The diffraction patterns taken at room temperature and at 10.4 K are shown superimposed in the inset in Figure 17–10. It can be seen directly that no significant structural transition occurs between 300 and 10.4 K that could explain the observed changes in the susceptibility and resistivity. We confirm that the space group for $LaCoO_3$ remains $R\bar{3}c$ over the entire temperature range measured here and only a small decrease in the lattice parameter from 300 to 10.4 K can be seen, consistent with earlier neutron powder diffraction data (Radaelli and Cheong 2002).

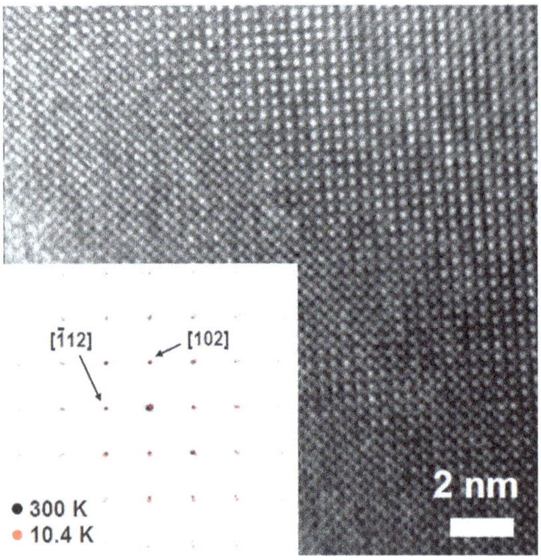

Figure 17–10. High-resolution TEM image of $LaCoO_3$ [$2\bar{2}1$] at 10 K. Insert: Diffraction pattern at 300 K and 10 K superimposed. Reproduced from Klie et al. (2007) with permission.

EELS spectra (Figure 17–11) of the O *K*-edge at 300, 86, and 10 K were taken from the grain shown in Figures 17–9a and exhibit three main peaks labeled *a*, *b*, and *c*. The O *K*-edge pre-peak (peak *a* in Figure 17–11) decreases notably above 86 K, while the peaks labeled *b* and *c* remain unchanged. The Co L_3 and L_2-edges (see inset in Figure 17–11) do not exhibit any change in either the white-line intensity ratio or the edge onset. Finally, the integrated intensity ratio of the Co *L*-edge and the O *K*-edge (not shown here) remains constant within the margin of error as a function of temperature, indicating that the stoichiometry of $LaCoO_3$ grain does not change during the cooling experiment.

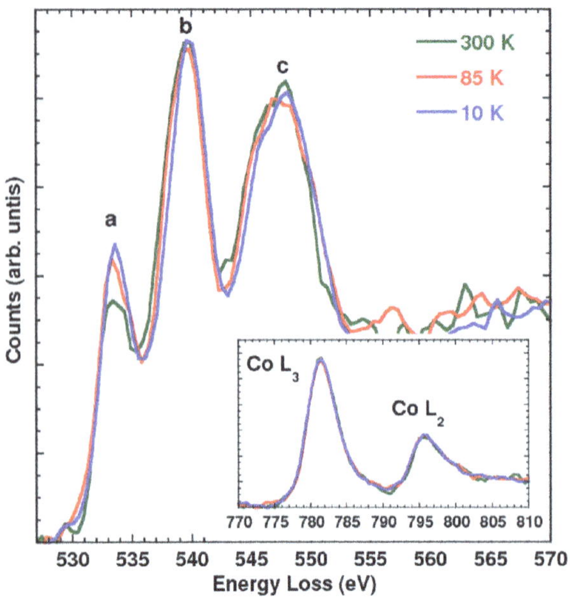

Figure 17–11. EELS spectrum of the O *K*-edge at 300, 85 and 10 K. The Co *L*-edge at these temperatures is shown in the insert. Reproduced from Klie et al. (2007) with permission.

In order to fully understand why the Co *L*-edge remains unchanged and the O *K*-edge pre-peak decreases during the Co^{3+} ion spin-state transition, we have to take a closer look at the origin of the different peaks in both the Co *L*-edge and the O *K*-edge. Several studies have shown the correlation between the Co *L*-edge energy onset (and intensity ratio) and the local Co valence (Abbate et al. 1993, 1992, Wang et al. 2000, Ito et al. 2002, Riedl et al. 2006). Since neither the Co *L*-edge fine structure nor its energy onset changes in the temperature range studied here, we conclude that the Co *L*-edges, measured with an energy resolution of 0.8 eV, are not sensitive to the Co^{3+} spin state. Recent XAS measurements (Haverkort et al. 2006) showed small changes in the Co *L*-edge fine-structure related to the Co^{3+} spin-state transition. However, these changes could not be detected by EELS.

Focussing now on the O *K*-edge, it was previously shown that the pre-edge feature is related to the filling of the hybridized O 2*p* and Co 3*d* states (Abbate et al. 1993, de Groot et al. 1989). Furthermore, the peak at 540 eV (peak *b*) is commonly attributed to the La 5*d* band, while the peak at 548 eV stems from the Co 4*sp* bands (Abbate et al. 1993). Therefore, the O *K*-edge features reveal that the bonding between the O 2*p* with the La 5*d* and the Co 4*sp* bands remains unchanged during the in situ cooling experiment. However, the electronic structure of the hybridized Co 3*d* O 2*p* bands changes with the onset of the spin-state transition of the Co^{3+} ions.

First-principles GGA and LDA+U calculations (Anisimov et al. 1993, Hohenberg and Kohn 1965, Parr and Yang 1989) are used to simulate the near-edge fine structure of the O *K*-edge and total energy of $LaCoO_3$ for three different spin states. We have calculated the projected density of unoccupied states for $LaCoO_3$ in the LS state that can be found at low temperatures ($T < 80\,K$), for an IS state and for a HS state. The EELS spectra were calculated using the `TELNES.2` package included in the `WIEN2K` code (Blaha et al. 2001), a full-potential linear augmented plane-wave (FLAPW) plus local-orbitals method within DFT. In order to simulate the spectra for the different spin states (e.g., $S = 0,\ 1$, and 2) of $LaCoO_3$, the fixed-spin-moment calculation was used to constrain the total spin magnetic moment (Klie et al. 2007).

Figure 17–12 shows the resulting EELS spectra of the O *K*-edge near-edge fine-structure as determined by our DFT calculations and then broadened by 1.0 eV. In all three simulated spectra, three peaks can be seen in the O *K*-edge. While the peaks labeled *b* and *c* are very similar for all three spin states, our calculations show a clear difference in the intensity and energy position of the O *K*-edge pre-peak (*a*) for $S = 1$ and $S = 2$. The O *K*-edge pre-peak energy is shifted toward lower energies by 1.2 eV from the LS state to the HS state, while the pre-peak in the IS spectrum shows only a small shift in energy and a decrease in the pre-peak intensity compared to the LS spectrum. Simply changing the lattice parameter of $LaCoO_3$, but not the spin of the Co^{3+} ions, results in only minor changes in the O *K*-edge fine structure that would not have been detectable in the experimental spectra.

Thus, our DFT calculations predict that the low-temperature spin-state transition from a LS into an IS state should be resolved by core-loss

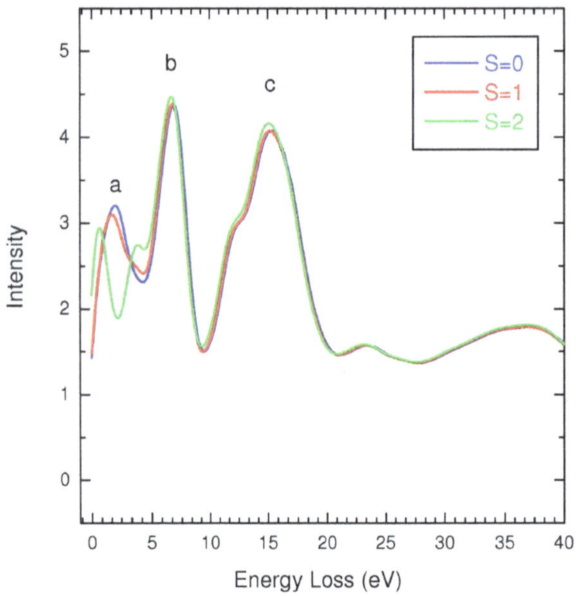

Figure 17–12. Calculated EELS spectrum of the O *K*-edge broadened by 1.0 eV for the low, intermediate, and high spin states. Reproduced from Klie et al. (2007) with permission.

EELS of the O *K*-edge. The experimental EELS spectra (Figure 17–11) clearly show that the O *K*-edge pre-peak changes as a function of temperature, and comparing the experimental spectra with the DFT calculations, it is obvious that the changes in the O *K*-edge pre-peak for $S = 2$ are not observed in the temperature range between 10 and 300 K. Moreover, the observed decrease in the O *K*-edge pre-peak intensity without any measurable chemical shift shows that the spin-state transition at 80 K in $LaCoO_3$ occurs from a LS to a IS state of the Co^{3+} ions. A mixed LS–HS state can be excluded since no chemical shift or significant broadening of the pre-peak was measured. It should be noted that the spectra of the O *K*-edge (Figure 17–11) show a small peak at 529.5 eV at both 10 and 300 K that might indicate the existence of a HS state even at 10 K. However, this peak remains within the noise level and its position might be purely coincidental.

We have shown that atomic-resolution Z-contrast imaging and EELS can be achieved at temperatures as low as 85 K. While we do not show direct evidence of atomic column-resolved EELS spectra in this study, it should be noted that the Z-contrast image in Figure 17–9a was taken with an acquisition time of 2.1 s and shows a spatial drift of less than 1 unit cell (~4 Å), which means that the drift per 1 s (acquisition time for a core-loss EELS spectrum) should be less than 2 Å, which will be sufficient to distinguish neighboring Co sites (~4 Å) in perovskite oxides. This example further shows that the high spatial drift at temperatures below 85 K due to the small volume of the LHe dewar makes it impossible to achieve atomic-resolution Z-contrast images or spectra.

In a second set of experiments, the $LaCoO_3$ sample is heated to 700 K in an attempt to measure the changes in the near-edge fine structure of the O *K*-edge and the Co *L*-edge during the semiconductor-to-metal transitions that has been reported at 500–600 K. Associated with this

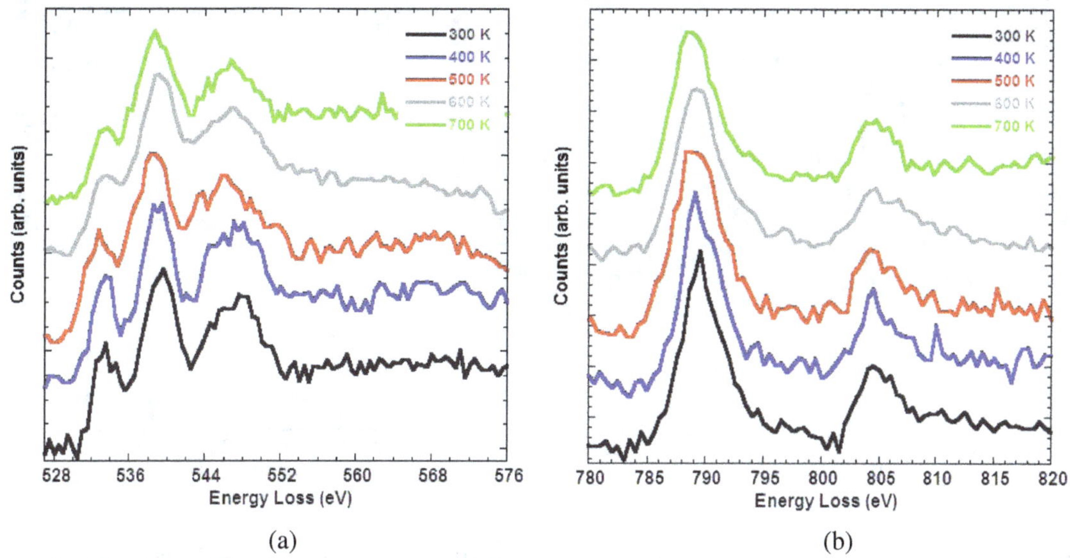

Figure 17–13. EELS spectra of $LaCoO_3$ [$2\bar{2}1$] as a function of temperature. (**a**) The O *K*-edge and (**b**) the CoO_2 at 300, 400, 500, 600, and 700 K.

transition has also been a spin state transition to a high spin-state (HS, $S = 2$). However, when heating an oxide sample to temperatures as high as 700 K in a highly reducing environment, such as a TEM column, particular attention has to be paid to the loss of oxygen during the heating experiment. In Section 17.3.1, we have shown that atomic-resolution in situ heating experiments can be conducted using $SrTiO_3$ without inducing a significant amount of electron beam damage. While $SrTiO_3$ is a very stable material for STEM analysis, materials such as $LaCoO_3$ might be highly susceptible to the loss of oxygen when exposed to the electron beam at high temperatures.

Figure 17–13a, b shows the O *K*-edge and Co *L*-edge of $LaCoO_3$ as a function of temperature. It can be seen immediately that there are no significant changes in the peak positions in either the O *K*-edge or the Co *L*-edge. However, the O *K*-edge pre-peak shows a decrease in its intensity at elevated temperature. We have used three Gaussian functions to fit the O *K*-edge fine structure and extracted the relative O *K*-edge pre-peak intensity. As shown in Figure 17–14a, the area under the Gaussian function that was fitted to the O *K*-edge pre-peak can now be measured directly, relative to the total area under the fitted function. We find that the O *K*-edge pre-peak intensity remains constant at 300 and 400 K, but shows a significant decrease above 500 K (see Figure 17–14b).

Next, the Co valence was calculated using the functional relation between the Co *L*-edge ratio and the numerical value of the Co valence state established by Wang et al. (2000). The Co L_3,L_2 -ratio was determined by measuring the intensity of the L_3 and L_2 peaks of the second derivative of the original spectrum (so as to be insensitive to changes in the specimen thickness) (Botton et al. 1995). The second derivative

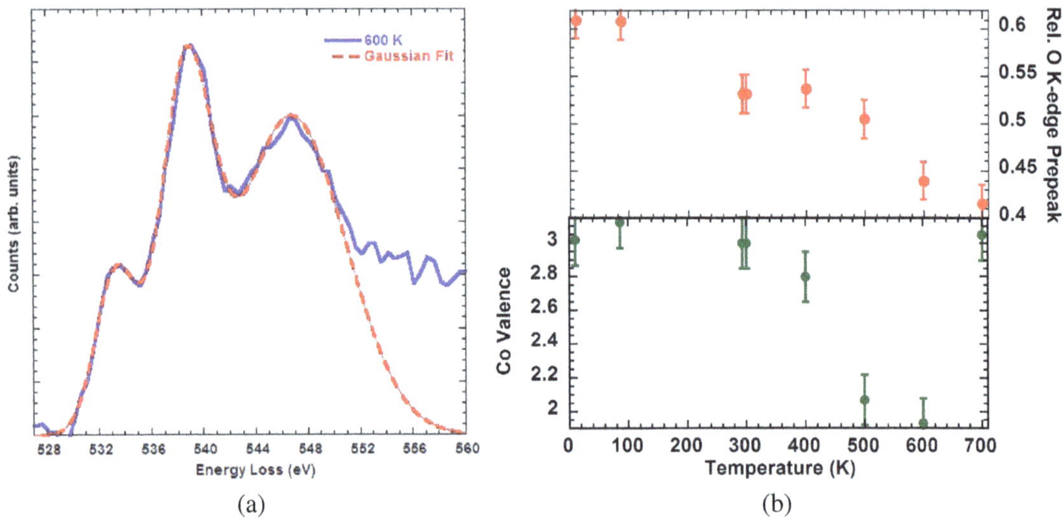

Figure 17–14. (**a**) O *K*-edge in $LaCoO_3$ taken at 600 K with three Gaussian functions fit. (**b**) Summary of the Co valence and the O *K*-edge pre-peak intensity as a function of temperature.

spectrum was obtained using the numerical filter available in the Gatan DigitalMicrograph program, with a positive and a negative window width of 4.1 and 1.7 eV, respectively (Kundmann et al. 1990, Klie and Browning 2002). As shown in Figure 17–14b, we find that the Co valence in $LaCoO_3$ is 3+ and constant within the experimental error in the temperature range between 10.4 and 400 K. At 500 and 600 K, the Co valence drops to 2+; at 700 K a valence state of Co^{3+} is measured again. Meanwhile, the Co/O ratio as measured by integrating the Co and O intensity in the background-subtracted spectra remains constant within the experimental error bars in the temperature range between 10 and 700 K.

The results of the heating experiments demonstrate some of the difficulties that can occur during in situ heating experiments inside a highly reducing environment. While the measured Co/O ratio indicates stoichiometric $LaCoO_3$ throughout the entire heating series, the Co *L*-edges show a significant decrease in the Co valence state that could be caused by a local loss of oxygen under the electron beam. Furthermore, the decrease in the O *K*-edge pre-peak intensity indicates that the Co e_g orbital is being filled, potentially as a result of an increased amount of O vacancies. Therefore, one needs to be mindful of the effects of in situ reduction during in situ heating experiments, in particular in ceramic oxide materials, since this change in the local stoichiometry can mask the effects that were intended to be measured at elevated temperature.

In summary, we have shown that the thermally excited spin-state transition in $LaCoO_3$ occurs from the LS state to the IS state and can be directly quantified using the low-temperature EELS near-edge fine structure of the O *K*-edge. Thus, we have shown that the O *K*-edge pre-peak provides an ideal fingerprint for identifying the different spin states of the Co^{3+} ions in $LaCoO_3$ and related compounds.

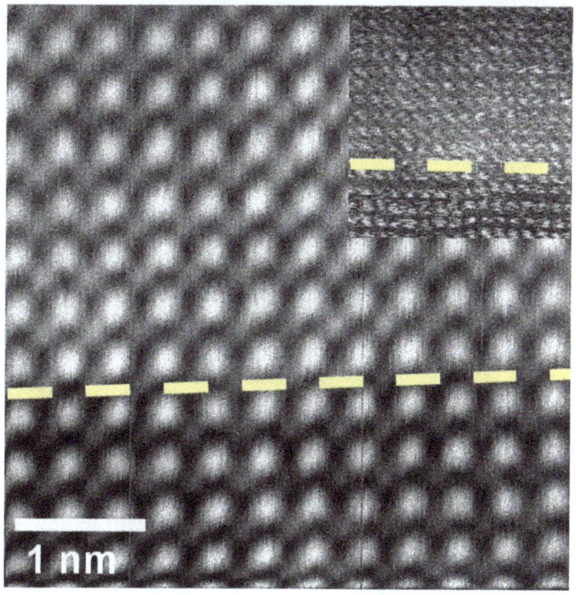

Figure 17–15. Atomic-resolution Z-contrast image of epitaxial $LaCoO_3$ [100] (*top*) on $LaAlO_3$ [100] (*bottom*) using the aberration-corrected VG 501. The *bright spots* indicate the La atomic columns and the *dashed line* indicates the position of the interface. Insert: Z-contrast image taken at 94.5 K using the JEOL2010F showing that the atomic structure of the $LaCoO_3$ film has not changed. Reproduced from Klie et al. (2010) with permission.

17.3.2.2 Single-Crystal $LaCoO_3$ Thin Film

In Section 17.3.2.1, we have shown that by using low-temperature EELS the spin-state transition of the Co^{3+} ions can be directly measured and quantified by using the O *K*-edge pre-peak intensity (Klie et al. 2005). Using first-principles DFT calculations, Korotin et al. (1996) have suggested that by changing the $LaCoO_3$ lattice parameter, different Co^{3+} ion spin states can be stabilized even at low temperatures, suggesting that ferromagnetic ordering in $LaCoO_3$ can be achieved (Yan et al. 2004, Fuchs et al. 2007).

Here, we will examine 30-nm-thick film of fully stoichiometric $LaCoO_3$ (001) grown by molecular beam epitaxy on $LaAlO_3$ (001). The lattice mismatch between the $LaAlO_3$ support (a = 3.789 Å) and the pseudo-cubic unit cell of $LaCoO_3$ (ac = 3.805 Å) should result in a $\varepsilon = -0.42\%$ lattice compression of the $LaCoO_3$ film. Figure 17–15 shows an atomic-resolution Z-contrast image taken with an aberration-corrected VG HB 501 of the $LaCoO_3$ [001] film on the $LaAlO_3$ [001] support. The $LaCoO_3$ film can be seen on the top of Figure 17–15 as the brighter part, due to higher atomic number of Co (Z = 27) compared to Al (Z = 13). The Z-contrast image does not show any defects at the interface, which appears to be atomically abrupt. Diffraction patterns of the $LaCoO_3$ films and the $LaAlO_3$ support taken at room temperature further confirm the single-crystal structure of the $LaCoO_3$ film and the eitaxial relationship between the $LaAlO_3$ film, where the $LaCoO_3$ film is epitaxially strained to match the lattice parameter of the support.

Next, the sample is cooled to 94.5 K using the in situ cooling stage, without prior exposing the sample to air. Diffraction patterns of the $LaCoO_3$ films and the $LaAlO_3$ support (Figure 17–16a) taken at room temperature do not exhibit any additional peaks due to the $LaCoO_3$ thin

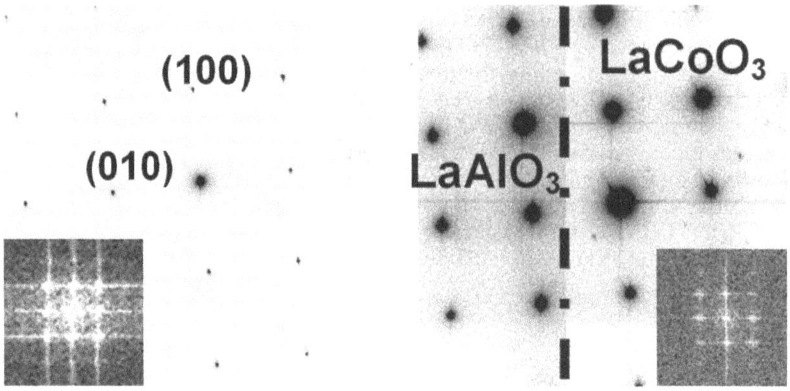

(a) (b)

Figure 17–16. Electron diffraction patterns of the $LaAlO_3$ [001] and the 30 nm $LaCoO_3$ [001] film (**a**) at room temperature and (**b**) at 94.5 K taken with the JEOL2010F. This diffraction pattern consists of two different parts that are superimposed. The *left side* corresponds to a diffraction pattern taken from the $LaAlO_3$ support, while the *right side* is taken from the film. The image intensity has been enhanced to show the low-intensity satellite peaks. The inserts in (**a**) and (**b**) are fast-Fourier transforms (FFTs) of the Z-contrast images taken at 300 and 94.5 K. Reproduced from Klie et al. (2010) with permission.

film, indicating that the film is single crystal and epitaxially strained to the lattice parameter of the $LaAlO_3$ support. Upon cooling, two additional satellite peaks appear around every major diffraction spot at low temperature (Figure 17–16b). Such superstructure peaks are usually associated with either cation, orbital, or charge ordering (Hong et al. 2007, Jooss et al. 2007). However, Z-contrast images taken at 94.5 K (Figure 17–15 insert) do not exhibit any obvious change in the crystal structure or the interfacial morphology (Klie et al. 2010).

Figures 17–17 and 17–18 show background-subtracted EELS spectra of the O *K*-edge and Co *L*-edge taken from the $LaCoO_3$ film at both room temperature and 94.5 K with an acquisition time of 5 s. Figure 17–17 shows the O *K*-edge at room temperature and 94.5 K exhibiting the typical near-edge fine structure that was previously reported for poly-crystalline and powder $LaCoO_3$ samples. The three main peaks of the O *K*-edge can be separated into the pre-peak (labeled peak *a*) and the two main peaks (peak *b* at 540 eV and peak *c* at 548 eV). As shown in the previous section, the pre-peak feature stems from the filling of the hybridized O 2*p* and Co 3*d* states, while the second peak *b* of the O *K*-edge has been attributed to the La 5*d* band and the third peak *c* is due to transition into the Co 4*sp* bands (Abbate et al. 1993). Therefore, the similarity of the O *K*-edge fine structure of the $LaCoO_3$ thin film with previously studied $LaCoO_3$ powder samples shows that the 30 nm thin film is fully stoichiometric.

In the previous section, we have also shown that the O *K*-edge pre-peak intensity increases significantly upon cooling below 86 K, associated with a Co^{3+} ion spin-state transition from an intermediate spin state ($S = 1$) at 300 K to the low spin state ($S = 0$) at low

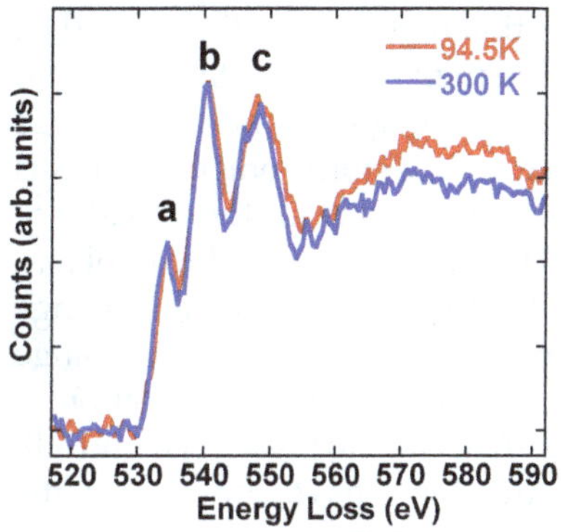

Figure 17–17. Electron energy-loss spectra (EELS) of the O K-edge at room temperature and 94.5 K. Reproduced from Klie et al. (2010) with permission.

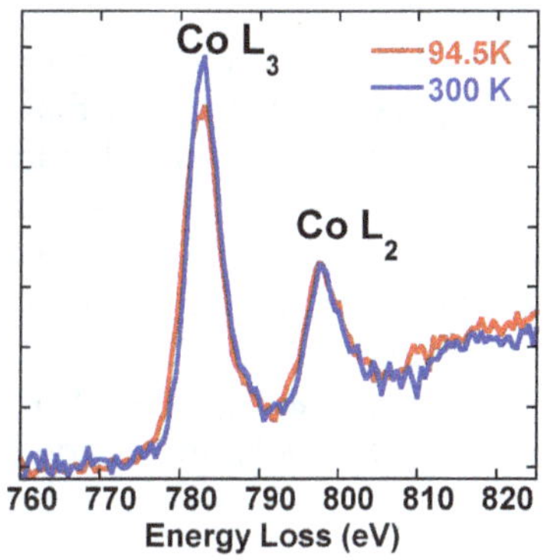

Figure 17–18. Electron energy-loss spectra (EELS) of the Co *L*-edges at room temperature and 94.5 K. Reproduced from Klie et al. (2010) with permission.

temperatures. However, such an increase in the O *K*-edge pre-peak intensity is not observed in the $LaCoO_3$ thin film samples at 94.5 K. Moreover, neither of the other two peaks (peaks *b* and *c*) exhibits any change in intensity or energy position upon cooling to 94.5 K. This suggests that the epitaxially strained thin film sample has not undergone the Co^{3+} ion spin-state transition that is observed in unstrained $LaCoO_3$.

The Co *L*-edges as a function of temperature are shown in Figure 17–18. The Co L_3 - peak stems from transition of the Co $2p^{\frac{3}{2}}$ to the Co 3*d* states, while the L_2-peak stems from transitions of the Co $2p^{\frac{1}{2}}$ to the Co 3*d* states. Using the relationship between the Co *L*-edge and the Co valence state, we find that the local Co valance at room temperature is Co^{3+} indicating a fully stoichiometric $LaCoO_3$ thin film. However, upon cooling to 94.5 K the Co L_2-edge shows a considerable increase in intensity, resulting in a decreased L_3/L_2 ratio, which would indicate a decrease in the Co valence to $Co^{2.6+}$. Since we did not observe any significant change in the crystal structure of the $LaCoO_3$ film (see Figure 17–16a), this decrease in the Co valence state could only be caused by the creation of O vacancies during the cooling process or as a result of beam damage. As mentioned above, while the O K-edge pre-peak intensity is a very sensitive measure of the O vacancy concentration, no change of the O *K*-edge fine structure has been observed, suggesting that the reason for the difference in the Co L_3/L_2 ratio at 94.5 K is not due to increased concentration of O vacancies.

It has been long known that the transition metal white-line ratio (i.e., L_3/L_2 ratio) does depend not only on the metal valence state, but also on the local O stoichiometry and the local magnetic moment (Pease et al. 2001). We have shown above that the O stoichiometry remains unchanged during the cooling experiment, which means the Co valence state should remain the same as well. Therefore, one possible explanation for the observed change in the Co L_3/L_2 ratio could be the change in the magnetic moment of the Co^{3+} ions. This means that the change in the Co L-edge intensity measures a magnetic ordering transition in epitaxially strained $LaCoO_3$ thin films that was not observed in bulk $LaCoO_3$ samples. Magnetization measurements have shown that the strained $LaCoO_3$ film on $LaAlO_3$ undergo a ferromagnetic transition at low temperature.

One might still argue that the change in the Co L_3/L_2 ratio stems from the creation of O vacancies during the cooling experiments, and that, by coincidence, the increase of the O K-edge pre-peak intensity due to the Co^{3+} ion spin-state transition to a low spin state is compensated by the decrease of the O K-edge pre-peak as a result of an increased O vacancy concentration. To conclusively prove that the change of the Co L_3/L_2 ratio measures the ferromagnetic transition in the $LaCoO_3$ thin films other TEM techniques, such as electron holography or electron magnetic circular dichroism (Schattschneider et al. 2006), are needed (Klie et al. 2010).

17.3.3 $Ca_3Co_4O_9$

Layered cobaltate materials have been the focus of many recent studies due to the wide variety of electrical, magnetic, and structural properties they exhibit. One of these properties is the two-dimensional superconductivity found in water-intercalated Na_xCoO_2 (Takada et al. 2003), where the superconducting CoO_2 sheets are separated by insulating layers of Na ions. Another outstanding property of layered cobaltates is the large thermoelectric power in materials such as $NaCo_2O_4$ (Terasaki et al. 1997), $(CaOH)_{1.14}CoO_2$ (Shizuya et al. 2007, Isobe et al. 2007), $Ca_3Co_4O_9$ (Masset et al. 2000), and $(Bi_2Sr_2O_4)_xCoO_2$ (Funahashi and Matsubara 2001). The crystal structure of all these materials is very similar, with a CdI_2-type conducting CoO_2 layer that is separated by an insulating rocksalt-type structure with n layers, where $n = 1$ for $NaCo_2O_4$, $n = 2$ for $(CaOH)_{1.14}CoO_2$, and so on. These layered structures exhibit a low electrical resistivity ρ and thermal conductivity κ, resulting in a figure of merit ZT ($ZT = \frac{S^2T}{\rho\kappa}$) comparable to that of traditional intermetallic thermoelectric materials such as Bi_2Te_3, $CoSb_3$, while exhibiting superior thermal stability.

Since its discovery in 1997 (Terasaki et al. 1997), the high thermoelectric power in $NaCo_2O_4$ has been attributed to a number of different mechanisms, including the large effective mass of the charge carriers due to the strong correlations in the CoO_2 subsystem, the spin degree freedom of charge carriers, the occurrence of a mixed valence state in the CoO_2 layer, or a pseudo gap. Similar mechanisms have also been suggested for the other layered thermoelectric cobalt oxide materials. It is interesting to note here that the structure of the CoO_2 layer remains nearly unchanged (Grebille et al. 2007) while the thermoelectric power of the different layered cobaltate compounds increases as the thickness of the insulating rocksalt layer increases from 100 μV/K at 300 K in $NaCo_2O_4$ (Terasaki et al. 1997) to 140 μV/K at 300 K in Pb- and Ca-doped $(Bi_2Sr_2O_4)_xCoO_2$ (Funahashi and Matsubara 2001). Therefore, the insulating rocksalt layer must play a crucial role in the high thermoelectric power of these misfit-layered compounds.

Among the different layered cobaltate systems, the $Ca_3Co_4O_9$ stands out as the only system containing one cation with nominally different oxidation states, namely Co^{2+} in the rocksalt buffer layers (Ca_2CoO_3) and Co^{4+} in the octahedral CoO_2 layers, which makes it an ideal system for studying effects such as charge transfer, orbital ordering, and spin-state transitions on the material's thermoelectric behavior. The structure of $Ca_3Co_4O_9$ has been reported to be monoclinic with two misfit-layered subsystems, a distorted rocksalt-type Ca_2CoO_3 layer sandwiched between two CdI_2-typed CoO_2 layers along the c-axis. Both subsystems share the same lattice parameters with $a = 4.8339$ Å, $c = 10.8436$ Å, and $\beta = 98.14°$, but along the b-axis the incommensurate structure results in $b_1 = 2.8238$ Å for the CoO_2 subsystem and $b_2 = 4.5582$ Å for the Ca_2CoO_3 subsystem (Miyazaki et al. 2002). The triangular CoO_2 layer consists of edge-sharing oxygen octahedra, and recent studies have shown that CoO_2 is a metal near a Mott transition with unit-cell parameters of $a = b = 2.806$ Å (de Vaulx et al. 2007). Therefore, the CoO_2 subsystem in $Ca_3Co_4O_9$ is subject to compressive strain in the a-axis direction, and several studies have shown that increasing the compressive strain will further increase the thermoelectric power (Matsubara et al. 2002, Xu 2002, Hu et al. 2005).

It has also been suggested that the occurrence of a mixed Co valence state in the CoO_2 layers and the transition of different Co ion spin states play a crucial role in understanding the high thermoelectric properties of $Ca_3Co_4O_9$. Previous studies on Co valences estimated the Co valence to be +3.5 in the CoO_2 layers and +2.8 in CoO layers based on the measured average bond length (Lambert et al. 2001). Moreover, previous powder X-ray and neutron diffraction experiments have suggested a strong undulation of O atomic sites in the CoO_2 layers and a strong displacive modulation of both the Co and O sites in the rocksalt subsystem (Miyazaki et al. 2002, Muguerra et al. 2008).

In this section, we explore the high thermoelectric properties of $Ca_3Co_4O_9$ at room and elevated temperatures using aberration-corrected Z-contrast imaging and atomic column-resolved electron energy-loss spectroscopy (EELS) as well as in situ heating experiments to 500 K. At room temperature, the atomic-resolution STEM images and EELS spectra were obtained using an aberration-corrected VG HB 501 dedicated STEM (Varela et al. 2005), the TEAM instrument (FEI Titan 300 KV TEM/STEM) located at the National Center for Electron Microscopy (NCEM), and the JEOL2200FS at Brookhaven National Laboratory (BNL). The in situ heating experiments were conducted in the JEOL2010F at UIC. Particular attention was paid to the effects of electron irradiation on the sample materials, to assure that all the results reported here are not attributable to electron beam damage.

Figure 17–19a, b shows atomic-resolution Z-contrast images of $Ca_3Co_4O_9$ in the [010] and [100] orientations acquired with the aberration-corrected FEI Titan and JEOL2200FS, respectively. Each micrograph clearly exhibits four distinct layers of varying brightness. The CoO_2 layer can be seen as the brightest layer followed by the CaO, CoO, and CaO layer, respectively. The incommensurate structure of $Ca_3Co_4O_9$ is visible only in the [100] orientation (Figure 17–19b), where the misfit-layered structure is clearly shown. It is interesting to note here that while the atomic columns in the CoO_2 and CaO layers can be clearly resolved in both Z-contrast images, the atomic columns in the CoO layers appear blurred in the [010] orientation (see Figure 17–19a). This is not due to insufficient spatial resolution. On the contrary, the electron probe size in the FEI Titan is calculated to be about 0.8 Å using Haider's d59 criterion (Haider et al. 2000) (or close to a full width at half maximum, FWHM, of 0.5 Å). Z-contrast image simulations of $Ca_3Co_4O_9$ using such an electron probe size (Figure 17–19c) show that the CoO columns should be clearly resolved in an undistorted $Ca_3Co_4O_9$ structure. Therefore, we conclude that the Co and O sites in the Ca_2CoO_3 layers exhibit a large undulation along the *b*-axis.

Atomic column-resolved electron energy-loss spectra (acquired using the VG HB 501) of the different layers in the $Ca_3Co_4O_9$ unit cell are shown in Figure 17–20. Figure 17–20a shows the O *K*-edge spectrum acquired from the CoO_2, the CaO, and the CoO layers. The near-edge fine structure of the O *K*-edge spectra can be divided into two regions, the pre-peak from ~526 to ~533 eV and the main peak from ~533 to ~549 eV. The pre-peak region contains a dominant peak at 530 eV (labeled *A*) and a small satellite peak at 528.5 eV (labeled A_1) in the CoO_2 spectrum and 532.5 eV (labeled A_2) in the CoO spectrum. By comparing the experimental spectra from the CoO_2 and the CoO layers with EELS spectra from similar materials, such as $LaCoO_3$ (see Section 17.3.2), we find that peak *A* stems from transitions from the O 1s into the hybridized O 2*p*-Co 3*d* orbitals (De Groot et al. 1989). In this layered cobaltate material the Co t_{2g} states are further split into a_{1g} and $e_{g'}$ orbitals due to the rhombohedral distortion of the CoO_2 layer, with the a_{1g} orbital at a higher energy than the $e_{g'}$ (Singh 2000, Wu et al. 2005). Therefore, peak A_1 in the CoO_2 spectrum stems from the transitions to hybridized O 2*p* and Co^{4+} a_{1g} states, while no such peak is observed in the CoO spectrum. The peak A_2 in the CoO spectrum is characteristic of a Co^{3+} oxidation state due to transitions to hybridized O 2*p*-Co^{3+} e_g states (Valkeapaa et al. 2007), which indicates the presence of Co^{3+} ions in the CoO layers. Finally, the first peak of the main O *K*-edge (peak *B*) has been shown to originate from transitions to hybridized O 2*p*-Ca 4*sp* orbitals, while peak *C* has been attributed to transitions to the hybridized O 2*p*-Co 4*sp* band (Moltaki et al. 2000, Gu and Ceh 2000). Therefore, the high intensity of peak *C* in the CoO_2 and CoO spectra indicates strong Co-O bonding, while the high intensity of peak *B* in the CaO and CoO spectra shows high Ca-O bonding.

Figure 17–20b shows the Co L-edge from the CoO_2 and the CoO, respectively. By using the relationship between the Co L_3/L_2 -ratio and the Co valence reported by Wang et al. (2000), we find that a mixed

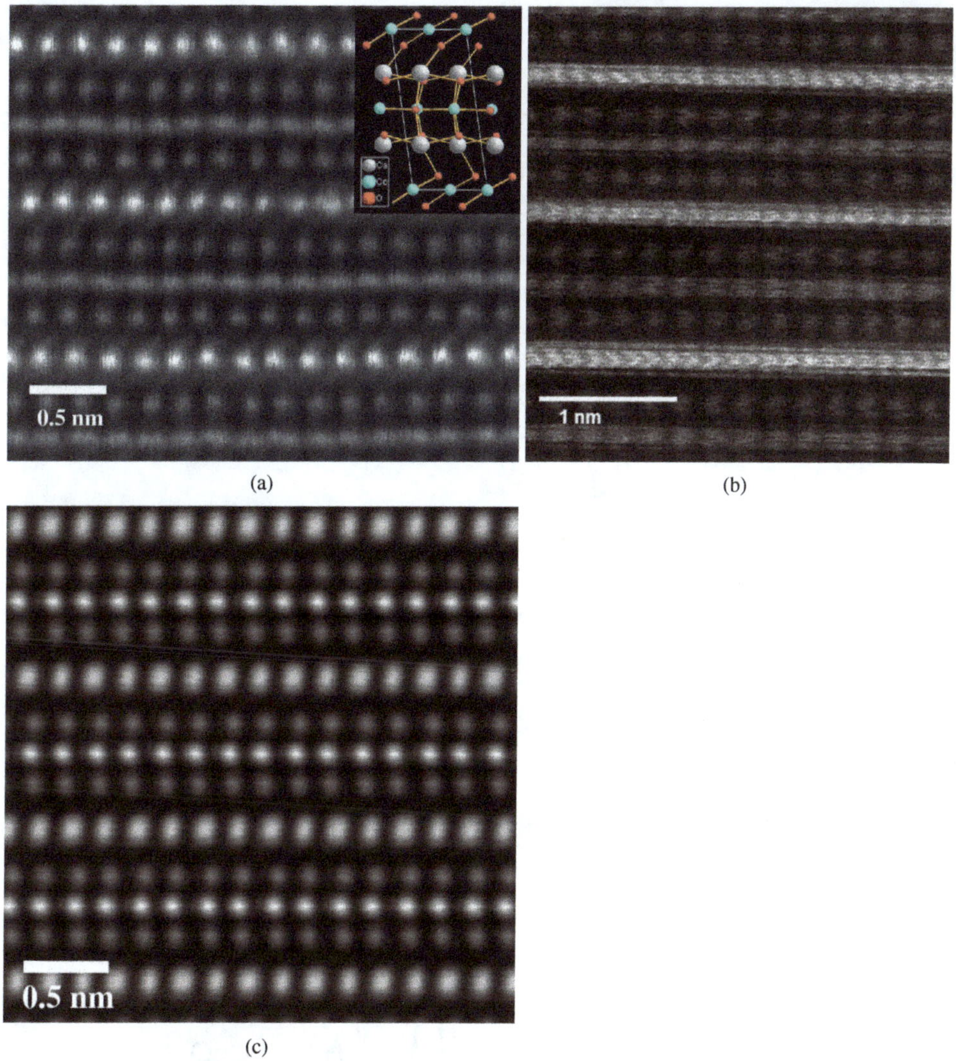

Figure 17–19. (**a**) Atomic-resolution Z-contrast image of $Ca_3Co_4O_9$ in the [010] orientation. The brightest atomic columns show the Co atoms in the CoO_2 layer with the adjacent O atoms clearly visible. The inset shows a model of the $Ca_3Co_4O_9$ unit cell in the same orientation. (**b**) Atomic-resolution Z-contrast image of $Ca_3Co_4O_9$ in the [100] orientation. (**c**) Calculated Z-contrast image of $Ca_3Co_4O_9$ [010] showing that the CoO column in the middle of the rocksalt Ca_2CoO_3 should be resolved clearly in the experimental image. Reproduced from Yang et al. (2008) with permission.

valence state exists in the CoO_2 layers with a nominal Co valence of 3.5+, while the valence in the CoO layers is 3.0+. This measured Co valence state is in good agreement with previous estimates. However, compared to the expected valence state of Co for charge-neutral CoO and CoO_2 layers, we find the CoO layer in the rocksalt C_2CoO_3 layer to be positively charged ($(CoO)^{1+}$), while the hexagonal CoO_2 layer is negatively charged ($(CoO_2)^{0.5-}$). By preserving the overall charge neutrality of both layers, holes are now transferred from the CoO to the

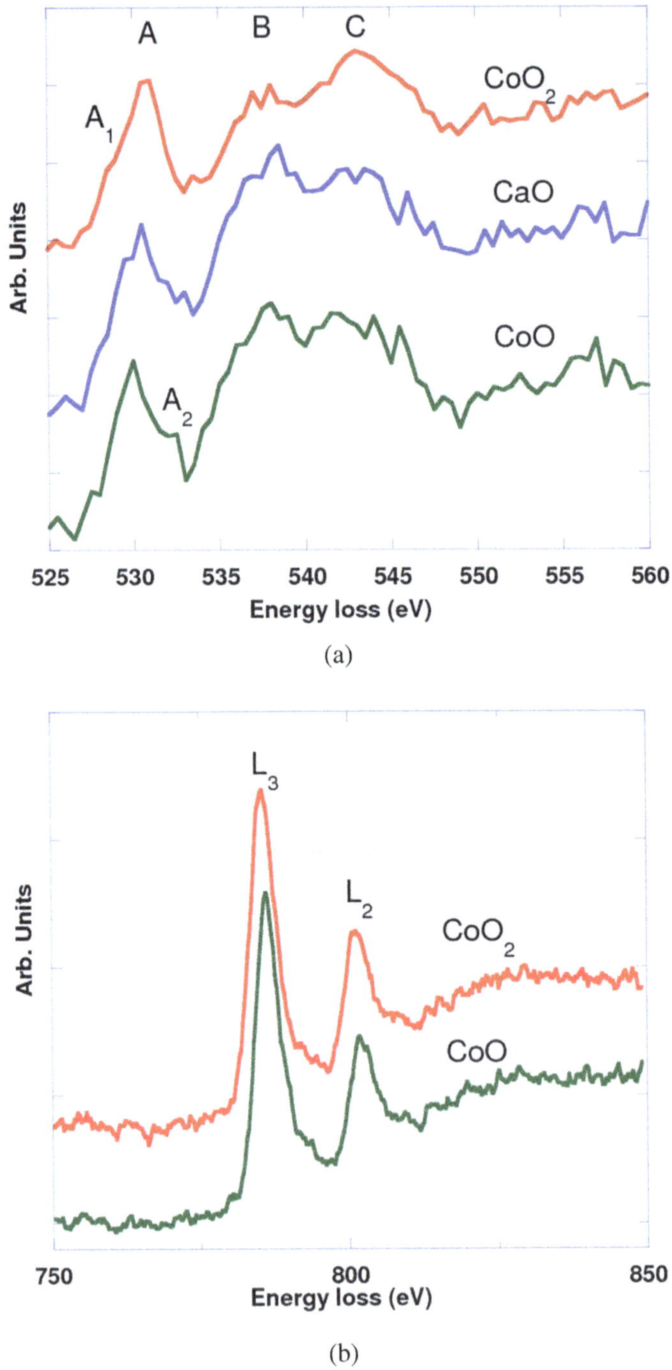

Figure 17–20. (**a**) O *K*-edge spectra of different layers in $Ca_3Co_4O_9$, the energy scale is calibrated to the Ca L_3-edge onset, while the intensity is normalized to the pre-peak *A* intensity; (**b**) Co *L*-edges for the different Co-O layers showing the Co L_3 and the L_2 *white lines*. The spectra are normalized to the Co L_3 peak intensity. All experimental spectra are averaged over three individual spectra. Reproduced from Yang et al. (2008) with permission.

CoO_2 layer, resulting in the high concentration of mobile holes measured in the CoO_2 layer. It has been previously shown that such a hole transfer is essential for the thermoelectric effect, since it not only provides the necessary mobile charge carriers, but the existence of a half-filled band (or the existence of particle–hole symmetry) as in the case for Co^{4+} in the CoO_2 layers, will result in a zero thermoelectric power (Seebeck coefficient) (Beni and Coll 1975). The observed hole transfer will thus remove the orbital degeneracy (t_{2g} splits into a_{1g} and $e_{g'}$ orbitals, as observed in Figure 17–20a), thereby explaining the non-zero thermopower in $Ca_3Co_4O_9$.

17.3.3.1 High-Temperature Study of $Ca_3Co_4O_9$

High-temperature studies of the thermopower in $Ca_3Co_4O_9$ as a function of temperature have revealed an abrupt change in the magnetic susceptibility and electrical resistivity at ~420 K, which has been suggested to be due to a transition of the Co-ion spin state (Masset et al. 2000). However, the effect of this transition on the mixed Co valence state or the previously measured charge transfer within the $Ca_3Co_4O_9$ remains unclear. Therefore, it is entirely possible that a change in the charge transfer, and thus the mixed Co valence state in the CoO_2 layers or a structural transition, could be responsible for the observed change in the magnetic and electrical properties of $Ca_3Co_4O_9$ at ~420 K. We utilize high-temperature Z-contrast imaging in combination with electron energy-loss spectroscopy (EELS) in a scanning transmission electron microscope (STEM) to quantify the lattice structure of $Ca_3Co_4O_9$, the Co valence state, and the occurrence of a Co ion spin-state transition in the temperature regime between 300 and 500 K.

Figure 17–21 shows a Z-contrast image acquired at 500 K of $Ca_3Co_4O_9$ [010] from a region similar to that shown in Figure 17–19a. While the drift of the heating stage appears to be comparable to that reported in Section 17.3.1, we find additional high-frequency noise that decreases the contrast in the image at 500 K. As this noise strongly decreased the clarity of the image, post-image processing was performed by removing all components in the image that pertain to white noise in a statistical manner and reconstruct a noise-reduced image with

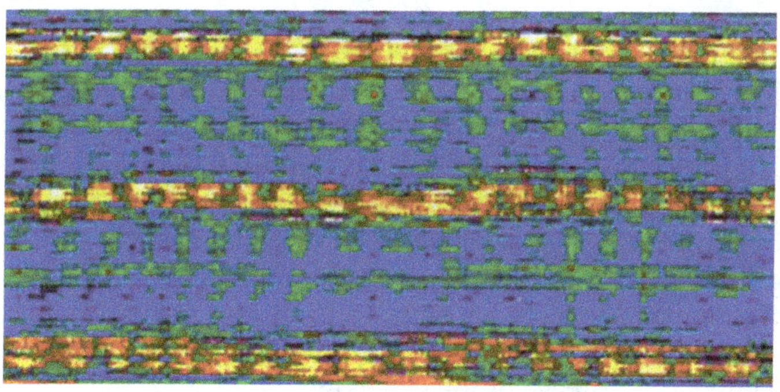

Figure 17–21. Atomic-resolution Z-contrast image of $Ca_3Co_4O_9$ in the [010] orientation at 500 K. Due to the inevitable environmental vibrations, the image has low signal-to-noise ratio. Reproduced from Yang et al. (2009) with permission.

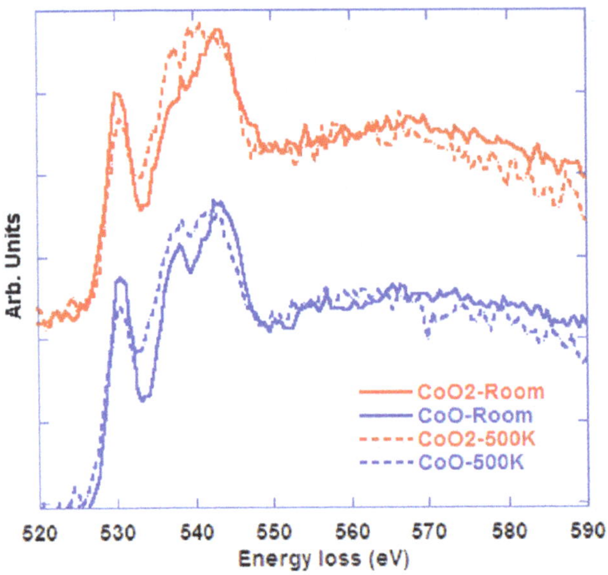

Figure 17–22. O *K*-edge spectra of CoO_2 and CoO layers at 500 K (*dotted line*), compared with spectra at room temperature (*solid line*). Reproduced from Yang et al. (2009) with permission.

the essential information only (Chatfield and Collins 1984). Compared to the room temperature measurements, there is no observable structure change at 500 K image despite its low signal-to-noise ratio.

Figure 17–22 shows the O *K*-edge spectra obtained from the CoO and CoO_2 layers at 500 K (dotted line), compared to the spectra acquired at room temperature (solid line). Similar to the spectra shown in Figure 17–20a, the O *K*-edge in the CoO layer contains three distinct peaks, while the O *K*-edge in the CoO_2 layers shows only two peaks at room temperature. It can be seen that upon heating $Ca_3Co_4O_9$ to 500 K, the peak positions of the O *K*-edge spectra do not change in the CoO or the CoO_2 layers. The most obvious change in the spectra taken at 500 K is the decrease in the pre-peak intensity compared to room temperature. As mentioned earlier, this decrease in the O *K*-edge pre-peak intensity can be either due to a change in the local Co valence or due to a spin-state transition.

To further distinguish these effects, we have measured the local Co valence at room temperature and 500 K. Figure 17–23 shows the EELS spectra of Co L_3 -and L_2 -edges taken from the CoO and CoO_2 layers. The solid lines are the Co *L*-edge spectra at room temperature while the dotted lines are taken at 500 K. It can be seen here that both the Co L_3-and L_2-edges obtained at 500 K are slightly broader than those at room temperature; however, the Co L_3/L_2 ratio remains unchanged during the in situ heating process, which suggests that Co valence state has not changed during the in situ heating experiments. Therefore, we can exclude the presence of oxygen vacancies as a possible explanation for the intensity decrease of the O *K*-edge pre-peak. Hence, the abrupt change in the magnetic and electrical properties of $Ca_3Co_4O_9$ at 420 K is not associated with a change in the charge transfer between the CoO_2 and the CoO layers.

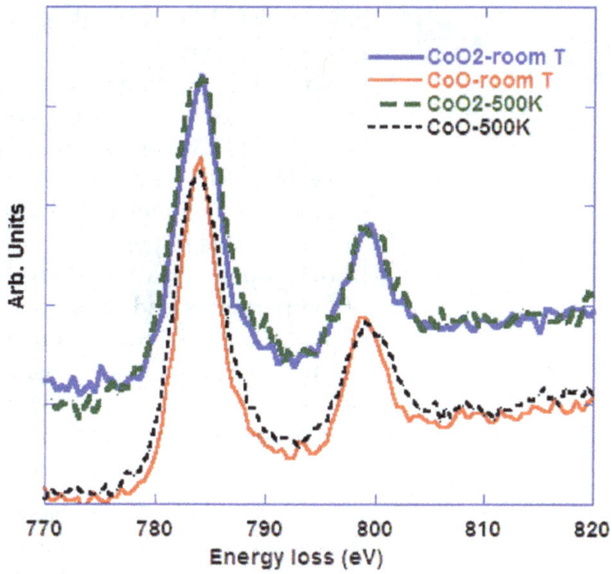

Figure 17–23. Comparison of Co *L*-edge spectra of room temperature (*solid line*) at 500 K (*dotted line*). Reproduced from Yang et al. (2009) with permission.

In summary, atomic-resolution Z-contrast imaging and EELS study of the misfit-layered thermoelectric, $Ca_3Co_4O_9$, were performed at room temperature and 500 K. We find that charge transfer from the rocksalt reservoir layers to the CoO_2 layers is responsible for the high thermoelectric properties of $Ca_3Co_4O_9$. Above the transition temperature of 420 K, we did not find any change in crystal structure or the charge transfer in $Ca_3Co_4O_9$ that could account for the abrupt change in magnetic susceptibility and electrical resistance. However, we measure a decrease in the O K-edge pre-peak intensity upon heating the materials to 500 K, which could indicate a spin-state transition of the Co^{3+} ions, similar to that reported in Section 17.3.2.1 for bulk $LaCoO_3$ from low spin to an intermediate spin state.

17.4 Conclusions

We have shown that high-resolution Z-contrast imaging and EELS can be achieved in the temperature range between 86 and 700 K. Using currently available double-tilt heating and cooling stages for ultra high-resolution objective lens pole pieces, we have shown that the stability of the sample is sufficient to achieve a spatial resolution for spectroscopy of less than 2 Å. However, at temperatures below 86 and above 750 K, the stability of these holders decreases significantly, thereby making it impossible to achieve high resolution. Novel, MEMS-based heating stages promise to improve the spatial and thermal stability at temperatures up to 1000 K. However, there are currently no side-entry cooling stages available that can provide sufficient stability at liquid helium temperature. On the other hand, liquid He cooling inside the microscope column (e.g., in a dedicated liquid He (S)TEM) could provide significantly improved sample stability. The authors are not aware of any results showing high spatial resolution EELS results using such instrumentation.

Now that aberration correction has become well established in modern Z-contrast imaging and energy-loss spectroscopy, the field of analytical STEM has to move beyond trying to achieve the smallest possible probe size and begin to characterize materials in a larger array of environments, including variable temperatures and gas pressures. Therefore, it is imperative that new instrumentation and methods will be developed that allow atomic-resolution spectroscopy in a wide range of temperatures and pressures. The current holder technology has not kept pace with the revolution in electron optics, but the increasing interest in measuring phase transitions or ordering phenomena at elevated or cryogenic temperatures will, without a doubt, drive a new revolution in holder technology and analysis methods to match the attainable spatial and energy resolution of state-of-the-art STEM instrumentation.

Acknowledgments The authors would like to thank Drs. A.W. Nicholls, Q. Ramasse, Q. Li, Y. Zhu, M. Varela, C. Leighton, and C.H. Ahn. This research was in part funded by a National Science Foundation CAREER award (Grant No. DMR-0846748).

References

M. Abbate, J.C. Fuggle, A. Fujimori, L.H. Tjeng, C.T. Chen, R. Potze, G.A. Sawatzky, H. Eisaki, S. Uchida, Electronic-structure and spin-state transition of $LaCoO_3$. Phys. Rev. B **47**, 16,124–16,130 (1993)

M. Abbate, F.M.F. de Groot, J.C. Fuggle, A. Fujimori, O. Strebel, F. Lopez, M. Domke, G. Kaindl, G.A. Sawatzky, M. Takano, Y. Takeda, H. Eisaki, S. Uchida, Controlled-valence properties of $La_{1-x}Sr_xFeO_3$ and $La_{1-x}Sr_xMnO_3$ studied by soft-X-ray absorption spectroscopy. Phys. Rev. B **46**, 4511 (1992)

L.F. Allard, W.C. Bigelow, M. Jose-Yacaman, D.P. Nackashi, J. Damiano, S.E. Mick, A new MEMS-based system for ultra-high-resolution imaging at elevated temperatures. Microsc. Res. Tech. **72**, 208–215 (2009)

V.I. Anisimov, I.V. Solovyev, M.A. Korotin, M.T. Czyzyk, G.A. Sawatzky, Density-functional theory and NiO photoemission spectra. Phys. Rev. B **48**, 16, 929–16,934 (1993)

K. Asai, P. Gehring, H. Chou, G. Shirane, Temperature-induced magnetism in $LaCoO_3$. Phys. Rev. B **40**, 10,982–10,985 (1989)

R.T.K. Baker, In-situ electron microscopy studies of catalyst particle behavior. Catal. Rev. Sci. Eng. **19**, 161–209 (1979)

P.E. Batson, Simultaneous STEM imaging and electron energy-loss spectroscopy with atomic-column sensitivity. Nature **366**, 727–728 (1993)

G. Beni, C.F. Coll, Thermoelectric power in half-filled bands. Phys. Rev. B **11**(2), 573–576 (1975)

P. Blaha, K. Schwarz, G. Madsen, D. Kvasnicka, J. Luitz, WIEN2k, An Augmented Plane Wave + Local Orbitals Program for Calculating Crystal Properties. Technische Universität Wien (2001)

G.A. Botton, C.C. Appel, A. Horsewell, W.M. Stobbs, Quantification of the EELS near-edge structures to study Mn doping in oxides. J. Microsc. Oxf. **180**, 211–216 (1995)

E.D. Boyes, P.L. Gai, Environmental high resolution electron microscopy and applications to chemical science. Ultramicroscopy **67**, 219–232 (1997)

D.N. Braski, High-vacuum evaporation stage for an electron microscope. J. Vac. Sci. Technol. **7**, 164–168 (1970)

N.D. Browning, J.P. Buban, H.O. Moltaji, G. Duscher, S.J. Pennycook, R.P. Rodrigues, K. Johnson, V.P. Dravid, The influence of atomic structure on the formation of electrical barriers at grain boundaries in $SrTiO_3$. Appl. Phys. Lett. **74**, 2638–2640 (1999)

N.D. Browning, M.F. Chrisholm, S.J. Pennycook, Atomic resolution chemical analysis using a scanning transmission electron microscope. Nature **366**, 143–146 (1993)

N.D. Browning, H.O. Moltaji, J.P. Buban, Investigation of three-dimensional grain-boundary structures in oxides through multiple-scattering analysis of spatially resolved electron-energy-loss spectra. Phys. Rev. B **58**, 8289–8300 (1998)

N.D. Browning, S.J. Pennycook, Direct experimental determination of the atomic structure at internal interfaces. J. Phys. D-Appl. Phys. **29**, 1779–1798 (1996)

C. Chatfield, A.J. Collins, *Introduction to Multivariate Statistical Analysis* (Chapman and Hall, New York, NY, 1984)

Y.M. Chiang, T. Takagi, Grain-boundary chemistry of barium-titanate and strontium-titanate. 1. High-temperature equilibrium space-charge. J. Am. Ceram. Soc. **73**, 3278–3285 (1990)

F.M.F. de Groot, M. Grioni, J.C. Fuggle, J. Ghijsen, G.A. Sawatzky, H. Petersen, Oxygen 1*s* X-ray-absorption edges of transition-metal oxides. Phys. Rev. B **40**, 5715–5723 (1989)

I. Denk, J. Claus, J. Maier, Electrochemical investigations of $SrTiO_3$ boundaries. J. Electrochem. Soc. **144**, 3526–3536 (1997)

S.R. English, J. Wu, C. Leighton, Thermally excited spin-disorder contribution to the resistivity of $LaCoO_3$. Phys. Rev. B **65**, 220407 (2002)

D. Fuchs, C. Pinta, T. Schwarz, P. Schweiss, P. Nagel, S. Schuppler, R. Schneider, M. Merz, G. Roth, H.v. Lohneysen, Ferromagnetic order in epitaxially strained $LaCoO_3$ thin films. Phys. Rev. B **75**, 144402 (2007)

R. Funahashi, I. Matsubara, Thermoelectric properties of Pb- and Ca-doped $(Bi_2Sr_2O_4)_xCoO_2$ whiskers. Appl. Phys. Lett. **79**, 362–364 (2001)

P.L. Gai, Environmental high resolution electron microscopy of gas-catalyst reactions. Topics Catal. **8**, 97–113 (1999)

P.L. Gai, Probing selective oxidation catalysis under reaction conditions by atomic scale environmental high resolution electron microscopy. Curr. Opin. Solid State Mater. Sci. **4**, 63–73 (1999)

J.B. Goodenough, An interpretation of the magnetic properties of the perovskite-type mixed crystals. J. Phys. Chem. Solids **6**, 287–297 (1958)

D. Grebille, H. Muguerra, O. Perez, E. Guilmeau, H. Rousseliere, R. Funahashi, Superspace crystal symmetry of thermoelectric misfit cobalt oxides and predicted structural models. Acta Crystallogr. Sect. B-Struct. Sci. **63**, 373–383 (2007)

M. Green, E. Gusev, R. Degraeve, E. Garfunkel, Ultrathin (< 4 nm) SiO_2 and Si-O-N gate dielectric layers for silicon microelectronics: Understanding the processing, structure, and physical and electrical limits. J. Appl. Phys. **90**(5), 2057–2121 (2001)

H. Gu, M. Ceh, Indirect EELS imaging reaching atomic scale – CaO planar faults in $CaTiO_3$. Ultramicroscopy **78**, 221–231 (1999)

M. Haider, S. Uhlemann, J. Zach, Upper limits for the residual aberrations of a high-resolution aberration-corrected STEM. Ultramicroscopy **81**, 163–175 (2000)

M.W. Haverkort, Z. Hu, J.C. Cezar, T. Burnus, H. Hartmann, M. Reuther, C. Zobel, T. Lorenz, A. Tanaka, N.B. Brookes, H.H. Hsieh, H.J. Lin, C.T.

Chen, L.H. Tjeng, Spin state transition in $LaCoO_3$ studied using soft X-ray absorption spectroscopy and magnetic circular dichroism. Phys. Rev. Lett. **97**, 176405 (2006)

R.R. Heikes, R.C. Miller, R. Mazelsky, Magnetic and electrical anomalies in $LaCoO_3$. Physica **30**, 1600–1608 (1964)

K. Heinemann, D.B. Rao, D.L. Douglass, Oxide nucleation on thin-films of copper during in-situ oxidation in an electron microscope. Oxidation Met **9**, 379–400 (1975)

P. Hohenberg, W. Kohn, Inhomogeneous electron gas. Phys. Rev. **136**, B864–B871 (1965)

B. Hong, H.L. Li, L.S. Ling, S. Tan, Y. Ying, L. Pi, Y.H. Zhang, Charge ordering phase submerged in ferromagnetic ordering phase in the $Nd_{0.5}Sr_{0.5}MnO_3$ system. EPL **80**, 37002 (2007)

Y.F. Hu, W.D. Si, E. Sutter, Q. Li, In situ growth of c-axis-oriented $Ca_3Co_4O_9$ thin films on Si(100). Appl. Phys. Lett. **86**, 082103 (2005)

M. Isobe, M. Shizuya, E. Takayama-Muromachi, Crystal structure and physical properties of a misfit-layered cobaltite $(CaOH)_{1.14}CoO_2$. J. Magn. Magn. Mater. **310**, E269–E271 (2007)

Y. Ito, R.F. Klie, N.D. Browning, T.J. Mazanec, Atomic resolution analysis of the defect chemistry and microdomain structure of brownmillerite-type strontium cobaltite. J. Am. Ceram. Soc. **85**, 969–976 (2002)

C. Jooss, L. Wu, T. Beetz, R.F. Klie, M. Beleggia, M.A. Schofield, S. Schramm, J. Hoffmann, Y. Zhu, Polaron melting and ordering as key mechanisms for colossal resistance effects in manganites. Proc. Natl. Acad. Sci. USA **104**, 13,597–13,602 (2007)

T. Kamino, T. Yaguchi, M. Konno, T. Hashimoto, In situ high temperature TEM observation of interaction between multi-walled carbon nanotube and in situ deposited gold nano-particles. J. Electron Microsc. **54**, 461–465 (2005)

T. Kamino, T. Yaguchi, M. Konno, A. Watabe, T. Marukawa, T. Mima, K. Kuroda, H. Saka, S. Arai, H. Makino, Y. Suzuki, K. Kishita, Development of a gas injection/specimen heating holder for use with transmission electron microscope. J. Electron Microsc. **54**, 497–503 (2005)

T. Kamino, T. Yaguchi, T. Sato, T. Hashimoto, Development of a technique for high resolution electron microscopic observation of nano-materials at elevated temperatures. J. Electron Microsc. **54**, 505–508 (2005)

C.S. Kim, M. Kim, J.K. Furdyna, M. Dobrowolska, S. Lee, H. Rho, L.M. Smith, H.E. Jackson, E.M. James, Y. Xin, N.D. Browning, Evidence for 2*d* precursors and interdiffusion in the evolution of self-assembled CdSe quantum dots on ZnSe. Phys. Rev. Lett. **85**, 1124–1127 (2000)

R.F. Klie, M. Bellegia, Y. Zhu, J.P. Buban, N.D. Browning, Atomic scale model of the grain boundary potential in perovskite oxides. Phys. Rev. B **68**, 214,101 (2003)

R.F. Klie, N.D. Browning, Atomic scale characterization of oxygen vacancy segregation at $SrTiO_3$ grain boundaries. Appl. Phys. Lett. **77**, 3737–3739 (2000)

R.F. Klie, N.D. Browning, Atomic scale characterization of vacancy ordering in oxygen conducting membranes. Microsc. Microanal. **8**, 475–486 (2002)

R.F. Klie, N.D. Browning, A.R. Chowdhuri, C.G. Takoudis, Analysis of ultrathin SiO_2 interface layers in chemical vapor deposition of Al_2O_3 on Si by in situ scanning transmission electron microscopy. Appl. Phys. Lett. **83**, 1187–1189 (2003)

R.F. Klie, J.P. Buban, M. Varela, A. Franceschetti, C. Jooss, Y. Zhu, N.D. Browning, S.T. Pantelides, S.J. Pennycook, Enhanced current transport at grain boundaries in high-T_c superconductors. Nature **435**, 475–478 (2005)

R.F. Klie, M.M. Disko, N.D. Browning, Atomic scale observations of the chemistry at the metal-oxide interface in heterogeneous catalysts. J. Catal. **205**, 1–6 (2002)

R.F. Klie, T. Yuan, M. Tanase, G. Yang, Q. Ramasse, Direct measurement of ferromagnetic ordering in biaxially strained $LaCoO_3$ thin films. Appl. Phys. Lett. **96**, 082510 (2010)

R.F. Klie, J.C. Zheng, Y. Zhu, M. Varela, J. Wu, C. Leighton, Direct measurement of the low-temperature spin-state transition in $LaCoO_3$. Phys. Rev. Lett. **99**, 047203 (2007)

K.L. Kliewer, J.S. Koehler, Space charge in ionic crystals. I. General approach with application to NaCl. Phys. Rev. **140**, A1226–A1240 (1965)

M.A. Korotin, S.Y. Ezhov, I.V. Solovyev, V.I. Anisimov, D.I. Khomskii, G.A. Sawatzky, Intermediate-spin state and properties of $LaCoO_3$. Phys. Rev. B **54**, 5309–5316 (1996)

O.L. Krivanek, P.D. Nellist, N. Dellby, M.F. Murfitt, Z. Szilagyi, Towards sub-0.5 Å electron beams. Ultramicroscopy **96**, 229–237 (2003)

M.K. Kundmann, X. Chabert, K. Truong, O.L. Krivanek, *EL/P Software for Macintosh II Computer* (Gatan, Pleasanton, CA 1990)

S. Lambert, H. Leligny, D. Grebille, Three forms of the misfit layered cobaltite $[Ca_2CoO_3][CoO_2]_{1.62}$ – a 4*d* structural investigation. J. Solid State Chem. **160**, 322–331 (2001)

A.C. Masset, C. Michel, A. Maignan, M. Hervieu, O. Toulemonde, F. Studer, B. Raveau, J. Hejtmanek, Misfit-layered cobaltite with an anisotropic giant magnetoresistance: $Ca_3Co_4O_9$. Phys. Rev. B **62**, 166–175 (2000)

I. Matsubara, R. Funahashi, M. Shikano, K. Sasaki, H. Enomoto, Cation substituted $(Ca_2CoO_3)_xCoO_2$ films and their thermoelectric properties. Appl. Phys. Lett. **80**, 4729–4731 (2002)

M.M. McGibbon, N.D. Browning, M.F. Chisholm, A.J. McGibbon, S.J. Pennycook, V. Ravikumar, V.P. Dravid, Direct determination of grain-boundary atomic-structure in $SrTiO_3$. Science **266**, 102–104 (1994)

M.M. McGibbon, N.D. Browning, A.J. McGibbon, S.J. Pennycook, The atomic structure of asymmetric [001] tilt boundaries in $SrTiO_3$. Philos. Mag. A **73**, 625–641 (1996)

M. Medarde, C. Dallera, M. Grioni, J. Voigt, A. Podlesnyak, E. Pomjakushina, K. Conder, T. Neisius, O. Tjernberg, S.N. Barilo, Low-temperature spin-state transition in $LaCoO_3$ investigated using resonant X-ray absorption at the Co-K edge. Phys. Rev. B **73**, 054424 (2006)

Y. Miyazaki, M. Onoda, T. Oku, M. Kikuchi, Y. Ishii, Y. Ono, Y. Morii, T. Kajitani, Modulated structure of the thermoelectric compound $[Ca_2CoO_3]_{0.62}CoO_2$. J. Phys. Soc. Jpn. **71**, 491–497 (2002)

H.O. Moltaki, J.P. Buban, J.A. Zaborac, N.D. Browning, Simulating the oxygen K-edge spectrum from grain boundaries in ceramic oxides using the multiple scattering methodology. Micron **31**, 381–99 (2000)

A.R. Moodenbaugh, B. Nielsen, S. Sambasivan, D.A. Fischer, T. Friessnegg, S. Aggarwal, R. Ramesh, R.L. Pfeffer, Hole-state density of $La_{1-x}Sr{-}CoO_{3-\delta}$ $(0 \leq x \leq 0.5)$ across the insulator/metal phase boundary. Phys. Rev. B **61**, 5666–5671 (2000)

H. Muguerra, D. Grebille, F. Bouree, Disordered misfit $[Ca_2CoO_3][CoO_2]_{1.62}$ structure revisited via a new intrinsic modulation. Acta Crystallogr. Sect. B-Struct. Sci. **64**, 144–153 (2008)

D. Muller, T. Sorsch, S. Moccio, F. Baumann, K. Evans-Lutterodt, G. Timp, The electronic structure at the atomic scale of ultrathin gate oxides. Nature **399**, 758–761 (1999)

D.A. Muller, L.F. Kourkoutis, M. Murfitt, J.H. Song, H.Y. Hwang, J. Silcox, N. Dellby, O.L. Krivanek, Atomic-scale chemical imaging of composition

and bonding by aberration-corrected microscopy. Science **319**, 1073–1076 (2008)
D.A. Muller, Y. Tzou, R. Raj, J. Silcox, Mapping sp^2 and sp^3 states of carbon at sub-nanometre spatial resolution. Nature **366**, 725–727 (1993)
P.D. Nellist, M.F. Chisholm, N. Dellby, O.L. Krivanek, M.F. Murfitt, Z.S. Szilagyi, A.R. Lupini, A. Borisevich, W.H. Sides, S.J. Pennycook, Direct sub-angstrom imaging of a crystal lattice. Science **305**, 1741 (2004)
A. Ohtomo, D.A. Muller, J.L. Grazul, H.Y. Hwang, Artificial charge-modulation in atomic-scale perovskite titanate superlattices. Nature **419**, 378–380 (2002)
N.H. Packan, D.N. Braski, Electron microscope in situ annealing study of voids induced by irradiation in aluminum. J. Nuclear Mater. **34**, 307–314 (1970)
G.G. Parr, W.T. Yang, *Density-Functional Theory of Atoms and Molecules* (Oxford University Press, New York, NY, 1989)
D.M. Pease, A. Fasihuddin, M. Daniel, J.I. Budnick, Method of linearizing the 3*d* L_3/L_2 white line ratio as a function of magnetic moment. Ultramicroscopy **88**, 1–16 (2001)
C. Qi, T. Akita, M. Okumura, M. Haruta, Epoxidation of propylene over gold catalysts supported on non-porous silica. Appl. Catal. A **218**, 81–89 (2001)
P.M. Raccah, J.B. Goodenough, First-order localized-electron $\leftrightarrows$ collective-electron transition in $LaCoO_3$. Phys. Rev. **155**, 932 (1967)
P.G. Radaelli, S.W. Cheong, Structural phenomena associated with the spin-state transition in $LaCoO_3$. Phys. Rev. B **66**, 094408 (2002)
T. Riedl, T. Gemming, K. Wetzig, Extraction of EELS white-line intensities of manganese compounds: Methods, accuracy, and valence sensitivity. Ultramicroscopy **106**, 284–291 (2006)
T. Saitoh, T. Mizokawa, A. Fujimori, M. Abbate, Y. Takeda, M. Takano, Electronic structure and magnetic states in $La_{1-x}Sr_xCoO_3$ studied by photoemission and X-ray-absorption spectroscopy. Phys. Rev. B **56**, 1290–1295 (1997)
M. Sankararaman, D. Perry, Valence determination of titanium and iron using electron-energy loss spectroscopy. J. Mater. Sci. **27**, 2731–2733 (1992)
P. Schattschneider, S. Rubino, C. Hebert, J. Rusz, J. Kunes, P. Novak, E. Carlino, M. Fabrizioli, G. Panaccione, G. Rossi, Detection of magnetic circular dichroism using a transmission electron microscope. Nature **441**, 486–488 (2006)
M.A. Senaris-Rodriguez, J.B. Goodenough, Magnetic and transport-properties of the system $La_{1-x}Sr_xCoO_{3-\delta}$ $(0 \leq x \leq 0.5)$. J. Solid State Chem. **118**, 323–336 (1995)
R. Sharma, Design and applications of environmental cell transmission electron microscope for in situ observations of gas-solid reactions. Microsc. Microanal. **7**, 494–506 (2001)
R. Sharma, K. Weiss, Development of a TEM to study in situ structural and chemical changes at an atomic level during gas-solid interactions at elevated temperatures. Microsc. Res. Tech. **42**, 270–280 (1998)
N. Shibata, S. Pennycook, T. Gosnell, G. Painter, W. Shelton, P. Becher, Observation of rare-earth segregation in silicon nitride ceramics at subnanometre dimensions. Nature **428**, 730–733 (2004)
M. Shizuya, M. Isobe, Y. Baba, T. Nagai, M. Osada, K. Kosuda, S. Takenouchi, Y. Matsui, E. Takayama-Muromachi, New misfit-layered cobalt oxide $(CaOH)_{1.14}CoO_2$. J. Solid State Chem. **180**, 249–259 (2007)
D.J. Singh, Electronic structure of $NaCo_2O_4$. Phys. Rev. B **61**, 13,397–13,402 (2000)
K. Sun, J. Liu, N.D. Browning, Correlated atomic resolution microscopy and spectroscopy studies of $Sn(Sb)O_2$ nanophase catalysts. J. Catal. **205**, 266–277 (2002a)

K. Sun, J. Liu, N. Browning, Direct atomic scale analysis of the distribution of Cu valence states in $Cu/\delta - Al_2O_3$ catalysts. Appl. Catal. B **38**, 271–281 (2002b)

K. Sun, J. Liu, N. Nag, N. Browning, Studying the metal-support interaction in $Pd/\gamma - Al_2O_3$ catalysts by atomic-resolution electron energy-loss spectroscopy. Catal. Lett. **84**, 193–199 (2002c)

K. Takada, H. Sakurai, E. Takayama-Muromachi, F. Izumi, R.A. Dilanian, T. Sasaki, Superconductivity in two-dimensional CoO_2 layers. Nature **422**, 53–55 (2003)

I. Terasaki, Y. Sasago, K. Uchinokura, Large thermoelectric power in $NaCo_2O_4$ single crystals. Phys. Rev. B **56**, 12, 685–12,687 (1997)

M. Valkeapaa, Y. Katsumata, I. Asako, T. Motohashi, T.S. Chan, R.S. Liu, J.M. Chen, H. Yamauchi, M. Karppinen, Charge compensation and oxidation in $Na_xCoO_{2-\delta}$ and $Li_xCoO_{2-\delta}$ studied by XANES. J. Solid State Chem. **180**, 1608–1615 (2007)

M. Varela, S.D. Findlay, A.R. Lupini, H.M. Christen, A.Y. Borisevich, N. Dellby, O.L. Krivanek, P.D. Nellist, M.P. Oxley, L.J. Allen, S.J. Pennycook, Spectroscopic imaging of single atoms within a bulk solid. Phys. Rev. Lett. **92**, 95502 (2004)

M. Varela, A.R. Lupini, K. van Benthem, A.Y. Borisevich, M.F. Chisholm, N. Shibata, E. Abe, S.J. Pennycook, Materials characterization in the aberration-corrected scanning transmission electron microscope. Ann. Rev. Mater. Res. **35**, 539–569 (2005)

C. de Vaulx, M.H. Julien, C. Berthier, S. Hébert, V. Pralong, A. Maignan, Electronic correlations in CoO_2, the parent compound of triangular cobaltates. Phys. Rev. Lett. **98**, 246402 (2007)

W. Walkosz, R.F. Klie, S. Ogut, A. Borisevich, P.F. Becher, S.J. Pennycook, J.C. Idrobo, Atomic resolution study of the interfacial bonding at $Si_3N_4/CeO_{2-\delta}$ grain boundaries. Appl. Phys. Lett. **93**, 053104 (2008)

Z.L. Wang, J. Bentley, N.D. Evans, Valence state mapping of cobalt and manganese using near-edge fine structures. Micron **31**, 355–362 (2000)

G. Winkelman, C. Dwyer, T. Hudson, D. Nguyen-Manh, M. Doblinger, R. Satet, M. Hoffmann, D. Cockayne, Arrangement of rare-earth elements at prismatic grain boundaries in silicon nitride. Philos. Mag. Lett. **84**, 755–762 (2004)

W.B. Wu, D.J. Huang, J. Okamoto, A. Tanaka, H.J. Lin, F.C. Chou, A. Fujimori, C.T. Chen, Orbital symmetry and electron correlation in Na_xCoO_2. Phys. Rev. Lett. **94**, 146,402–146,404 (2005)

G.J. Xu, Thermoelectric properties of the Bi- and Na- substituted $Ca_3Co_4O_9$ system. Appl. Phys. Lett. **80**, 3760–3762 (2002)

J.Q. Yan, J.S. Zhou, J.B. Goodenough, Ferromagnetism in $LaCoO_3$. Phys. Rev. B **70**, 014402 (2004)

G. Yang, Q. Ramasse, R.F. Klie, Direct measurement of charge transfer in thermoelectric $Ca_3Co_4O_9$. Phys. Rev. B **78**, 153109 (2008)

G. Yang, Q. Ramasse, R.F. Klie, Direct measurement of Co-ion spin state transitions in $Ca_3Co_4O_9$ using variable-temperature electron energy-loss spectroscopy. Appl. Phys. Lett. **94**, 093112 (2009)

A. Ziegler, J. Idrobo, M. Cinibulk, C. Kisielowski, N.D. Browning, R. Ritchie, Atomic-resolution observations of semicrystalline intergranular thin films in silicon nitride. Appl. Phys. Lett. **88**, 041919 (2006)

A. Ziegler, J.C. Idrobo, M.K. Cinibulk, C. Kisielowski, N.D. Browning, R.O. Ritchie, Interface structure and atomic bonding characteristics in silicon nitride ceramics. Science **306**, 1768–1770 (2004)

18

Fluctuation Microscopy in the STEM

Paul M. Voyles, Stephanie Bogle and John R. Abelson

18.1 Introduction

Fluctuation electron microscopy (FEM) is a technique for measuring structure in amorphous materials using electron nanodiffraction. Since its invention by Treacy and Gibson (1996, 1997), it has been applied to a variety of materials, including amorphous semiconductors (Chen et al. 2004, Cheng et al. 2001, 2002, Gerbi et al. 2003, Gibson and Treacy 1997, Gibson et al. 1998, Johnson et al. 2004, Voyles et al. 2001a), oxides (Ho et al. 2003, Kisa et al. 2006), other covalent network materials (Kwon et al. 2007, Zhao et al. 2009), and metals (Hwang et al. 2007, Li et al. 2003, Stratton et al. 2005, Wen et al. 2007). Treacy et al. (2005) recently reviewed the state of the field.

Although FEM was originally developed using dark-field imaging in the TEM, there are substantial advantages to using nanodiffraction in a STEM instead. Voyles and Muller did the first FEM in the STEM experiments (Voyles and Muller 2002), although the possibility (like many other aspects of electron nanodiffraction) was foreseen by John Cowley (2001, 2002). We will start by describing the problem FEM was developed to solve, measuring medium-range order in amorphous materials, and why FEM experiments make progress where so many other techniques founder. We will then review the two theoretical models that have been developed to describe FEM and then discuss the advantages and disadvantages of using STEM for FEM. The last major topic is a review of current STEM FEM experimental results and discoveries. We conclude with a short discussion of the outlook for STEM FEM and some thoughts about future directions. We will discuss inorganic materials almost exclusively, although there are parallel problems in molecular and polymeric glasses.

18.1.1 Measuring Medium-Range Order

Structurally, amorphous materials have no long-range periodicity, which is demonstrated by a lack of Bragg peaks in diffraction measurements. They must, however, have some short-range order, in the form of well-defined nearest-neighbor distances at a minimum, since they

S.J. Pennycook, P.D. Nellist (eds.), *Scanning Transmission Electron Microscopy*,
DOI 10.1007/978-1-4419-7200-2_18,

are still comprised of atoms. Materials may be structurally amorphous without exhibiting a glass transition in their heat capacity or viscosity; amorphous silicon is one example. All glassy materials that we are aware of are amorphous.

Structure between the extremes of short-range order (SRO) and long-range order (LRO) is called, for lack of a better term, medium-range order (MRO). (Various other terminologies have been proposed (Elliot 1989), but not widely adopted.) Materials with strongly directional bonding can have regular arrangements of nearest-neighbor atoms with well-defined angles between bonds, but that is still SRO. MRO can be defined as structure starting at the third coordination shell and extending up to the length scale at which Bragg peaks appear in diffraction. In many inorganic materials, this corresponds to a physical length scale starting at 0.7–1.0 nm, depending on the nearest-neighbor distance, and extending up to ~3 nm, that corresponds in turn to clusters of ten to several hundred atoms.

The nature of MRO, or indeed whether it even exists, is a difficult question to answer. The most influential models for the structure of amorphous materials, the continuous random network for covalently bonded solids (Zachariasen 1932), and the dense random packed model for metallic bonding (Bernal 1964) have no MRO. They have just enough SRO to satisfy the bonding constraints of the atoms and no other structure. That said, there are a variety of physical phenomena in amorphous materials that involve collective action of groups of atoms at the MRO scale or electron wavefunctions or phonons localized at MRO length scales. In amorphous silicon, MRO structure influences the vibrational density of states, and hence the Raman spectrum (Sokolov and Shebanin 1990, Voyles et al. 2001b), and the localized electronic states responsible for the Urbach tails near the band edges (Pan et al. 2008) and for hopping conduction (Nakhmanson et al. 2001). In amorphous metals, plastic deformation may be mediated by "shear transformation zones" (Argon 1979), which are analogous to nanoscopic, MRO-size dislocations which nucleate but cannot propagate without a supporting crystal lattice (see Schuh et al. 2007 for a review). In both covalent networks and metals, MRO may play a controlling role in crystallization (Kwon et al. 2007, Stratton et al. 2005). Finally, MRO in glass-forming liquids around the glass transition temperature plays a central role in the Adam and Gibbs (1965) and Kivelson et al. (1995) atomistic models of the glass transition, but not in some other models (Bengtzelius et al. 1984, Leutheusser 1984).

Wide-angle diffraction with x-rays or neutrons is the primary tool for measurements of the structure of amorphous materials (Fischer et al. 2006). LRO has a clear signature as sharp peaks in the structure factor, $S(k)$, especially at low divergence angle sources like synchrotrons. SRO can be measured from the two-atom position distribution function $g_2(r)$ (or related functions like the pair correlation function or the radial distribution function), which is related to the Fourier transform of $S(k)$. As discussed in Section 18.2.1, $g_2(r)$ has limited sensitivity to MRO. Element-specific two-body function measurements like resonant x-ray scattering, extended x-ray absorption fine structure, or partial pair distribution functions from isotope substitution in neutron diffraction can extend the physical length scale covered by the two-body function, but they only measure a subset of the structure. The equivalent electron scattering techniques have also been employed (Cockayne 2007).

In the early days of high-resolution phase-contrast TEM imaging, there were a number of reports of images of "microcrystallites" in amorphous thin films (Howie et al. 1973, Rudee and Howie 1972). They unfortunately turned out to be an artifact of the imaging process: if the phases of similar micrographs were scrambled, destroying any information about the sample, some of the apparent microcrystallites remained (Krivanek et al. 1976). The images were strongly affected by artifacts associated with contrast delocalization and the oscillatory nature of the phase-contrast transfer function (CTF). Similar dubious identifications of MRO-scale structures in HRTEM images have persisted (e.g., Tsu et al. (1997)). C_S-corrected HRTEM removes the oscillations in the CTF and contrast delocalization (Haider et al. 1998), making it much more reliable for characterizing nanoscale order in amorphous materials. There have been some successes, primarily in characterizing dilute, well-ordered inclusions, such as Si nanocrystals in amorphous silicon (Perrey et al. 2004) and icosahedral symmetry, possibly quasicrystalline nanocrystals in a bulk metallic glass (Hirata et al. 2007). Van Dyck has shown, however, that even with a perfect microscope, the size of the atoms will cause too much overlap in two-dimensional projection to simply measure the position of the all atoms in the sample (Van Dyck et al. 2003); measuring all the atom positions will require an atomic-resolution tomography technique. Ultimately, any EM technique will be limited by the need to remain below the electron dose that displaces too many atoms, which may be difficult if the displacement energy distribution in amorphous materials is broad (Stratton et al. 2006).

Fluctuation microscopy is based on two key insights by Treacy and Gibson. The first is that there is more information to be had from electron diffraction at *moderate* resolution than at the highest possible resolution. The kinematic diffracted intensity from a perfect crystal at an exact Bragg condition scales as the square of the number of atoms in the crystal through perfect constructive interference. Including an entire ordered cluster of atoms inside the probed volume therefore results in a higher signal to background than sampling only a fraction of the cluster with a very small probe. The advantage of having some spatial resolution (instead of none as in x-ray or neutron diffraction) is that a nanodiffraction measurement samples fewer copies of the ordered cluster. With a probe diameter of ~1 nm and a TEM sample thickness of ~30 nm, an electron nanodiffraction pattern arises from on the order of 1,000 atoms, corresponding to 1–100 MRO clusters, depending on their size and density. A standard diffraction experiment with no spatial resolution averages over at least millions of clusters, which makes them very difficult to distinguish, especially if they are not perfectly structurally identical due to disorder and if they are randomly oriented.

Treacy and Gibson's second insight was that it was necessary to quantify the statistics of scattering from many regions. One of the distinguishing factors between actual order and imaging artifacts or chance correlations of atoms in projection is that actual order will exhibit higher diffracted intensity. More fundamentally, however, amorphous materials sampled 1,000 atoms a time which will show significant variability from place to place because of their disordered nature: almost any structure can be found once if the experimenter searches hard enough. Some measure of structural *significance* is therefore required. In FEM, that comes from examining the statistics of an ensemble of many nanodiffraction measurements, using the normalized variance of the intensity

$$V(\mathbf{k}, Q) = A \frac{\int I^2(\mathbf{r}, \mathbf{k}, Q)\, d^2\mathbf{r}}{\left(\int I(\mathbf{r}, \mathbf{k}, Q)\, d^2\mathbf{r}\right)^2} - 1. \tag{1}$$

The fundamental measured data set is $I(\mathbf{r}, \mathbf{k}, Q)$, the diffracted intensity I from position $\mathbf{r}$ on the sample, into a diffraction vector $\mathbf{k}$, measured with a probe of momentum spread Q, set by the convergence angle.[1] The real-space resolution is then proportional to $1/Q$. The integrals are over the total area A sampled in a set of many nanodiffraction measurements. V is often reported as $V(k)$, where $k = |\mathbf{k}|$ and Q is constant for a particular set of experiments. This statistical approach is essential to FEM and distinguishes it from the electron nanodiffraction techniques in Chapter 9, which solve a single structure completely.

The utility of the variance $V(k, Q)$ for characterizing MRO is captured qualitatively in Figure 18–1. In Figure 18–1a, the structure is without MRO, such as a maximally disordered CRN or DRP structure. A typical nanodiffraction pattern from this material has weak, fuzzy rings arising from SRO, which are broken into small dots called "speckles" by the small number of atoms in the samples. The size of the speckles is set by the probe Q. If the probe is moved to another position, the position and intensity of the speckles change, because the specific configuration of atoms under the probe is different, but the fluctuations are small. In Figure 18–1b, the sample consists of MRO-sized, ordered regions which exhibit Bragg diffraction. If the probe lands on an ordered region oriented on a Bragg condition, there is a high-intensity disk at a particular orientation in the nanodiffraction pattern. If the probe lands on an ordered cluster oriented between Bragg conditions, there is a particularly dark spot in the nanodiffraction pattern due to destructive interference between the scatterings from the atoms in the cluster.

For k corresponding to a Bragg peak in the structure factor of the clusters, V for Figure 18–1b is larger than V for Figure 18–1a. $V(k)$ also has

---

[1] We use the electron diffraction definitions for $\mathbf{k}$ and Q, $|\mathbf{k}| = \sin(\theta)/\lambda$, as does most of the FEM literature. In Treacy et al. (2005) and subsequent publications, Treacy and co-workers have adopted the x-ray scattering notation, in which the diffraction vector is $|\mathbf{q}| = 2\pi \sin(\theta)/\lambda$ and the probe convergence half angle wave vector is $\mathbf{K}$.

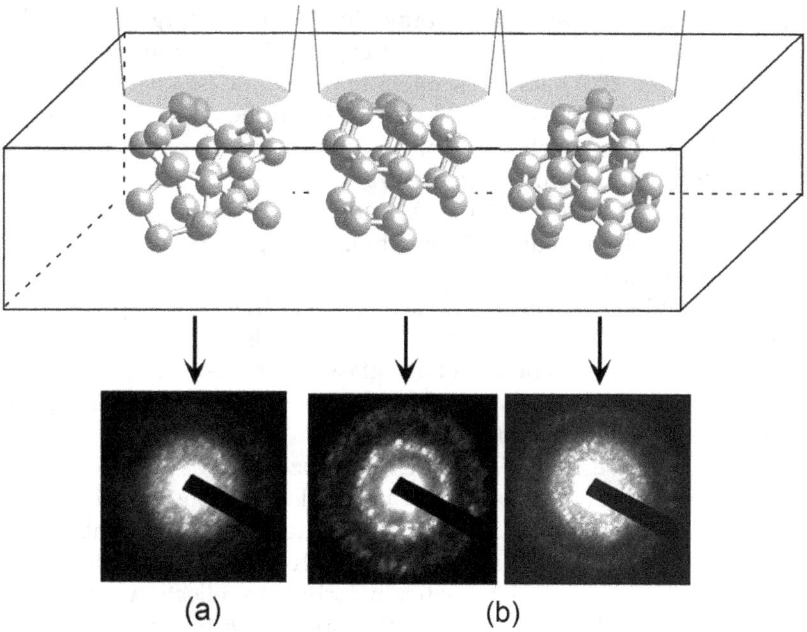

Figure 18–1. Qualitative picture of why $V(k)$ is useful for MRO: (**a**) Nanodiffraction from a uniformly random sample shows small fluctuations with position and little structure in k. (**b**) Nanodiffraction from a sample containing small-ordered clusters varies strongly with position and has significant structure in k. Adapted from Voyles et al. (2000a).

more structure, as k passes through maxima and minima in the cluster structure factor. $V(k)$ for the sample without MRO is closer to flat. This qualitative picture also shows the importance of moderate spatial resolution. Nanodiffraction with a very small probe, $R \leq$ the nearest-neighbor distance, will be similar from both samples, since it captures primarily SRO. For R near the cluster size, V is sensitive to the MRO.

Since its invention, FEM has been applied to a variety of different systems. Experiments on semiconductors deposited as amorphous thin films or amorphized by ion implantation have found that a "paracrystalline" structure of small, strained, crystal-like regions (Treacy et al. 1998) is ubiquitous. Amorphous silicon (Cheng et al. 2001, 2002, Gerbi et al. 2003, Khare et al. 2004, Nakhmanson et al. 2001, Treacy et al. 1998, Voyles et al. 2000b, 2001a, b), germanium (Gibson and Treacy 1997, Treacy et al. 1998), and diamond-like and graphite-like amorphous carbons (Chen et al. 2004, Johnson et al. 2004) all contain paracrystallites. Treacy developed a description of these regions as "topologically crystalline" (Treacy et al. 2000), which means that they have the same arrangement of bonds as the corresponding crystal, although they may be heavily strained, distorted, or defective. These regions may be frustrated crystal proto-nuclei formed during film deposition (Gerbi et al. 2003) or the ultra-rapid quenching of a region melted by an ion implantation thermal spike (Cheng et al. 2002). In amorphous silicon, paracrystallites can lead to localized electronic states (Nakhmanson et al. 2001).

Proto-nuclei have also been found in Al-based marginal metallic glasses (Gibbons et al. 2006, Stratton et al. 2005, Stratton and Voyles 2007, Wen et al. 2007) and chalcogenide optical phase-change materials (Kwon et al. 2007) using FEM. The MRO in the Al-based glasses is crystal-like as in amorphous semiconductors. Its presence or absence is correlated with a primary crystallization reaction which results in a high density (up to 10^{20} m^{-3}) of Al nanocrystals 10–50 nm in diameter in an amorphous matrix (Foley et al. 1996, Stratton et al. 2005). In chalcogenides, the presence of proto-nuclei strongly effects the speed of the crystallization reaction, which controls the switching speed of optical memories (Kwon et al. 2007). FEM has also been used to study the structure of various bulk metallic glasses (Hruszkewycz et al. 2008, Hufnagel et al. 2002, Hwang et al. 2007, Li et al. 2003, Wen et al. 2009), but with less definitive results.

Zhao et al. have used FEM to characterize various nanoscale carbon fullerenes and nanotubes either in conglomerations of carbon soot or as they occur naturally in the mineral shungite (Zhao et al. 2009). Fullerenes are detectable because the curvature of the graphite sheets breaks the condition for a kinematically forbidden reflection, giving the fullerenes a large V signal at that k with respect to an amorphous carbon background. A few amorphous oxides (Ho et al. 2003, Kisa et al. 2006) and oxide glasses (Ryan and Pantano 2007) have also been studied.

18.2 Theoretical Models

18.2.1 Higher Order Atom Position Distribution Functions

Treacy and Gibson developed an imaging theory for FEM based on higher order atom position distribution functions (Gibson et al. 2000, Treacy and Gibson 1993, 1996, Treacy et al. 1998). Voyles has described it in detail (Voyles 2001) and Treacy has reviewed it (Treacy et al. 2005) in terms of dark-field TEM FEM, so we will summarize it here in terms of nanodiffraction in the STEM. The two modes are formally equivalent by the optical reciprocity of TEM and STEM, so the equations are unchanged. The difference is in the aspect of the microscope represented by the various quantities.

The kinematic diffracted wave from a monatomic ensemble of atoms at positions $\{\mathbf{r}_i\}$, illuminated by a plane wave with wavevector $\mathbf{k}_i$, into a wavevector $\mathbf{k}_f$ is

$$\psi\left(\mathbf{k}_f - \mathbf{k}_i\right) = i\lambda f\left(\left|\mathbf{k}_f - \mathbf{k}_i\right|\right) \sum_j e^{-2\pi i(\mathbf{k}_f - \mathbf{k}_i)\mathbf{r}_j}, \tag{2}$$

where f is the atomic scattering factor. The diffracted intensity $|\psi(\mathbf{k}_f - \mathbf{k}_i)|^2$ is

$$I\left(\mathbf{k}_f - \mathbf{k}_i\right) = \lambda^2 f^2\left(\left|\mathbf{k}_f - \mathbf{k}_i\right|\right) \sum_{j,l} e^{2\pi i(\mathbf{k}_f - \mathbf{k}_i)\mathbf{r}_{jl}}, \tag{3}$$

where $\mathbf{r}_{jl} = \mathbf{r}_j - \mathbf{r}_l$. For nanodiffraction with a convergent probe at normal incidence to position $\mathbf{r}$ on the sample, we integrate Eq. (2) over the probe wave function, $P(\mathbf{k}, \mathbf{r})$:

$$\psi\left(\mathbf{r}, \mathbf{k_f}, Q\right) = i\lambda \int_{-\infty}^{\infty} d^2\mathbf{k_i} f\left(\left|\mathbf{k_f} - \mathbf{k_i}\right|\right) P\left(\mathbf{r}, \mathbf{k_i}, Q\right) \sum_j e^{-2\pi i(\mathbf{k_f}-\mathbf{k_i})\mathbf{r_j}}. \quad (4)$$

As described in Chapter 1, for a probe-forming aperture that subtends a wave vector magnitude Q, $P\left(\mathbf{k}, \mathbf{r}\right) = e^{2\pi i \mathbf{k}\cdot\mathbf{r}} e^{i\chi(\mathbf{k})} \Theta\left(k - Q\right)$, where the Heaviside function Θ is 1 for $k = |\mathbf{k}| < Q$ and 0 for $k > Q$. $\chi\left(\mathbf{k}\right)$ is the wave aberration function of the microscope, described in detail in Chapter 15. As described in Section 18.4, FEM requires probes that are relatively large in real space, with a diameter >1.0 nm, which corresponds to small Q. For a modern STEM, even without an aberration corrector, $\chi\left(\mathbf{k}\right) \sim 0$ over that size aperture and $P\left(\mathbf{k}\right) \simeq e^{2\pi i \mathbf{k}\cdot\mathbf{r}} \Theta\left(k - Q\right)$. In addition, f changes very little over the small range in wave vector $\mathbf{k_i}$, so we will assume that $f\left(\left|\mathbf{k_i} - \mathbf{k_f}\right|\right) \simeq f\left(k_f\right)$. These approximations make

$$\psi\left(\mathbf{r}, \mathbf{k_f}, Q\right) \simeq i\lambda f\left(k_f\right) \sum_j e^{-2\pi i \mathbf{k_f}\mathbf{r_j}} \int_{-\infty}^{\infty} d^2\mathbf{k_i} \Theta\left(k_i - Q\right) e^{2\pi i \mathbf{k_i}(\mathbf{r}-\mathbf{r_j})}. \quad (5)$$

The integral is then simply the real-space probe wave function, $a_Q(\mathbf{r})$, which is also the point-spread function of the measurement. Without aberrations, $a_Q(\mathbf{r})$ is an Airy function:

$$a_Q\left(\mathbf{r}\right) = \frac{Q}{\sigma} J_1\left(2\pi\sigma Q\right), \quad (6)$$

where $\boldsymbol{\sigma}$ is the two-dimensional projection of $\mathbf{r}$ perpendicular to the beam direction, $\sigma = |\boldsymbol{\sigma}|$, and J_1 is the first-order Bessel function of the first kind. $a_Q(\mathbf{r})$ is more complicated if we include aberrations and has a different form entirely for other means of forming the probe, such as real-space aperture method discussed in Section 18.4. To make later equations integrable, we make the approximation of a Gaussian probe:

$$a_Q\left(\mathbf{r}\right) = 2\pi Q^2 e^{-2\pi Q^2 \sigma^2}. \quad (7)$$

The probe nanodiffraction intensity is therefore

$$I\left(\mathbf{r}, \mathbf{k}, Q\right) = \lambda^2 f^2\left(k\right) \sum_{j,l} a_Q\left(\mathbf{r} - \mathbf{r}_j\right) a_Q^*\left(\mathbf{r} - \mathbf{r}_l\right) e^{2\pi i \mathbf{k}\cdot\mathbf{r}_{jl}}. \quad (8)$$

To construct the variance from Eq. (1), we need the first and second moments of Eq. (8). The first moment, the mean intensity, is

$$\left\langle I\left(\mathbf{k}, Q\right)\right\rangle = \frac{1}{A} \int_A d^2\mathbf{r} I\left(\mathbf{r}, \mathbf{k}, Q\right) = \frac{\pi^2 Q^2 f^2\left(k\right) \lambda^2}{A} \sum_{j,l} e^{2\pi i \mathbf{k}\cdot\mathbf{r}_{jl}} e^{-2\pi Q^2 \sigma^2}. \quad (9a)$$

The sum over pair vectors $\mathbf{r}_{\mathbf{jl}}$ can be replaced with a integral over the atom pair distribution function $g_2(\mathbf{r}_{12})$ (Chandler 1987). There are a variety of related definitions for the pair function; we use the definition from Cusack (1987). This definition excludes the one-body, $j = l$, terms from the sum in Eq. (8), so

$$\langle I(\mathbf{k}, Q)\rangle = \pi^2 Q^2 f^2(k)\,\lambda^2 \rho t\left(1 + \rho \int d^3\mathbf{r}_{12} g_2(\mathbf{r}_{12})\, e^{2\pi i\mathbf{k}\cdot\mathbf{r}_{12}} e^{-\pi Q^2 \sigma_{12}}\right). \tag{9b}$$

The equivalent expression for parallel illumination, based on Eq. (3), is widely employed to interpret large-area diffraction measurements.

The second moment of nanodiffraction intensity is

$$\begin{aligned}\left\langle I^2(\mathbf{k}, Q)\right\rangle &= \frac{1}{A}\int_A d^2\mathbf{r} I^2(\mathbf{r}, \mathbf{k}, Q) \\ &= \frac{2\pi^3 Q^6 f^4(k)\lambda^4}{A}\sum_{j,l,m,n} e^{2\pi i\mathbf{k}\cdot(\mathbf{r}_{\mathbf{jl}} - \mathbf{r}_{\mathbf{mn}})} e^{-\frac{1}{2}\pi^2 Q^2\left(\sigma_{jl}^2 + \sigma_{jm}^2 + \sigma_{jn}^2 + \sigma_{lm}^2 + \sigma_{\mathrm{ln}}^2 + \sigma_{mn}^2\right)},\end{aligned} \tag{10}$$

which can also be rewritten in terms of distribution functions. First, we must divide the quadruple sum over atom positions into unique-order terms. There is one type of one-body term, when $j = l = m = n$. There are two types of two-body terms, one when $(j = l) \neq (m = n)$ (and two other permutations of j, l, m, and n), and one when $(j = l = m) \neq n$. There is one type of three-body term, when $j \neq l \neq (m = n)$ and one type of four-body term, when $j \neq l \neq m \neq n$. (All the permutations are given explicitly in Voyles 2001). The result is

$$\begin{aligned}\left\langle I^2(\mathbf{k}, Q)\right\rangle &= 2\pi^3\lambda^4 f^4(k)\, Q^6 \rho t \\ &\times\Big\{1 + \rho \int g_2(\mathbf{r}_{12})\left[\left(2 + e^{-4\pi i\mathbf{k}\cdot\mathbf{r}_{12}}\right) e^{-2\pi^2 Q^2\sigma_{12}^2} + 4e^{-2\pi i\mathbf{k}\cdot\mathbf{r}_{12}} e^{-\frac{3}{2}\pi^2 Q^2\sigma_{12}^2}\right] d\mathbf{r}_{12} \\ &+ \rho^2 \int g_3(\mathbf{r}_{12}, \mathbf{r}_{13})\left[4e^{-2\pi i\mathbf{k}\cdot\mathbf{r}_{12}} + e^{2\pi i\mathbf{k}\cdot(\mathbf{r}_{12} - 2\mathbf{r}_{13})} + e^{-2\pi i\mathbf{k}\cdot(\mathbf{r}_{12} - 2\mathbf{r}_{13})}\right] e^{-\frac{1}{2}\pi^2 Q^2\left[\sigma_{12}^2 + 2\sigma_{13}^2 + 2|\sigma_{13} - \sigma_{12}|^2\right]} d\mathbf{r}_{12} d\mathbf{r}_{13} \\ &+ \rho^3 \int g_4(\mathbf{r}_{12}, \mathbf{r}_{13}, \mathbf{r}_{14}) e^{-2\pi i\mathbf{k}\cdot(\mathbf{r}_{12} + \mathbf{r}_{13})} e^{-\frac{1}{2}\pi^2 Q^2\left[\sigma_{12}^2 + \sigma_{13}^2 + \sigma_{14}^2 + |\sigma_{12} - \sigma_{13}|^2 + |\sigma_{12} - \sigma_{14}|^2 + |\sigma_{13} - \sigma_{14}|^2\right]} d\mathbf{r}_{12} d\mathbf{r}_{13} d\mathbf{r}_{14}\Big\}.\end{aligned} \tag{11}$$

Equations (9) and (11) could be substituted into Eq. (1) to yield an expression for $V(\mathbf{k}, Q)$ in terms of the distribution functions. Unfortunately, the resulting expression has not been inverted to yield the distribution functions directly from $V(\mathbf{k}, Q)$ data. It does offer a more quantitative explanation of why $V(\mathbf{k}, Q)$ contains more information than a large-area diffraction measurement: $V(\mathbf{k}, Q)$ depends on higher order atom distribution functions, $g_3(\mathbf{r}_{12}, \mathbf{r}_{13})$ and $g_4(\mathbf{r}_{12}, \mathbf{r}_{13}, \mathbf{r}_{14})$. Figure 18–2 is a qualitative explanation of why higher order correlation functions contain more information about MRO than $g_2(\mathbf{r}_{12})$. In an isotropic medium, $g_2(r)$ effectively counts the number of atoms that sit in a shell with inner radius r and outer radius $r+dr$ centered on the average sample atom. As r increases, the number of ways that atoms can be packed inside the sphere of radius r that put an atom somewhere in the shell also increases, and eventually $g_2(r)$ only reflects the surface area

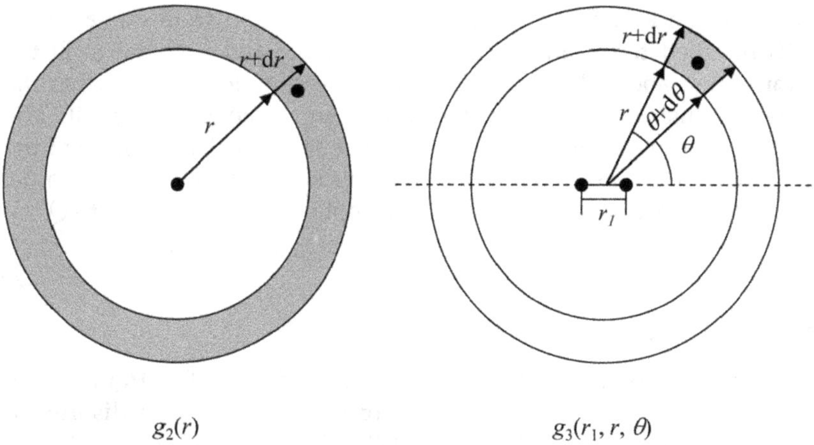

Figure 18–2. The search volume for the pair distribution function $g_2(r)$ and the three-atom distribution function $g_3(r_1, r, \theta)$ in an isotropic sample, showing why $g_3(r_1, r, \theta)$ retains more information about MRO (reproduced from Voyles and Abelson (2003) with permission).

of the sphere and average atom density. $g_3(r_1, r, \theta)$, on the other hand, has a pair of atoms at the origin, separated by a distance r_1. The vector between them defines an axis, and the search volume for atoms is a strip a distance r to $r+dr$ away, at an angle θ to $\theta+\Delta\theta$ with respect to the axis. The search volume stays much smaller than for $g_2(r)$ as r increases, so $g_3(r_1, r, \theta)$ retains useful information up to MRO length scales. $g_4(r_1, r_2, r, \theta)$ can be thought of as a pair of atoms at the origin separated by r_1 and another pair a distance r away separated by r_2, with an angle θ between the pair vectors. It is therefore sensitive to sets of aligned pairs of atoms, which is naturally connected to diffraction. Subsets of $g_4(r_1, r_2, r, \theta)$ calculated from computer models can be found in Voyles et al. (2000a).

One path forward from Eq. (11) is to make a parameterized *ansatz* for the distribution functions and then develop an expression for the parameters from the data. Gibson et al. (2000) took this approach with the *ansatz* of a Gaussian decay of the four-body correlation function,

$$g_4(\mathbf{r}_1, \mathbf{r}_2, \mathbf{r}) = G_4(\mathbf{r}_1, \mathbf{r}_2)\, e^{-r^2/2\Lambda^2}, \tag{12}$$

where $G_4(\mathbf{r}_1, \mathbf{r}_2)$ contains all the other unknown information in the full four-body function. They substituted Eq. (12) into Eq. (11) and neglected the lower order constant, g_2 and g_3 terms. In principle, the g_2 term could be calculated from large-area diffraction data and subtracted off, and Gibson et al. believe the g_3 term to be smaller than the g_4 term (Gibson et al. 2000). With some other approximations (Gibson et al. 2000),

$$V(\mathbf{k}, Q) \simeq \frac{\Lambda^3 Q^2}{1 + 4\pi^2 Q^2 \Lambda^2}\, \beta(\mathbf{k}), \tag{13}$$

in which the **k** and Q dependencies are separated from one another. $\beta(\mathbf{k})$ is called the "pair persistence function," and it contains all the information about the type of order in the sample, such as what the ordered cluster is and how the atoms are arranged within it. The Q-dependent term determines Λ, the characteristic length scale of the structural order. Experimentally, one fixes **k**, in principle at any value, and extracts Λ from the slope and intercept of a straight line fit to Q^2/V vs. Q^2.

18.2.2 An Order/Disorder Composite

Stratton and Voyles recently presented a model of FEM from a composite structure consisting of small, ordered regions in a disordered matrix (Stratton and Voyles 2007, 2008). The model was motivated by experiments and computer modeling which suggest that several classes of amorphous materials, including amorphous semiconductors (Chen et al. 2004, Treacy et al. 1998) (Si, Ge, and C) and amorphous metals (Stratton et al. 2005), have this type of structure. The model is less general than the distribution functions model, which applies to any structure, but it has the advantage that the sample structure is completely defined by the cluster's structure, volume fraction Φ, and size distribution. The model connects the measured V directly to these sample structural parameters. The model presented here contains some additional refinements by F. Yi and P.M. Voyles to be published in more detail elsewhere.

The model sample is divided in N_c columns, the size of the microscope resolution R laterally, and the thickness t vertically, and the integrals in Eq. (1) are replaced by the corresponding sums over the intensities I_i from the columns. The probe wave function only defines the size of the column, so it is 1 inside the column and 0 outside. I_i is determined by the number of nanocrystals in the column N_i and the nanocrystal sizes. We make the very simple scaling approximation that the diffracted intensity from a crystal of volume V is $\gamma(\rho V)^2$, where ρ is the atom number density of the sample (assumed to be the same in the crystalline and disordered phases), and $\gamma = \lambda^2 f^2(k)$. A given crystal does not necessarily lie entirely inside a single column, so the diffracted intensity it contributes to a particular column is $\gamma(\rho\chi V)^2$, where χ is the fraction of the volume of the crystal inside a particular column. The disordered material between the columns is more like a gas, for which the scattered intensity scales linearly, as $\gamma\rho V$. These considerations together make

$$I_i\left(k_{hkl}\right) = \gamma\rho R^2 t + \gamma \sum_{j=1}^{N_i} A_j\left(k_{hkl}\right)\left[\left(\rho\chi_j \frac{\pi}{6} d_j^3\right)^2 - \rho\chi_j \frac{\pi}{6} d_j^3\right]. \qquad (14)$$

$A_j(k_{hkl})$ is 1 if nanocrystal j is oriented such that one of the $\{hkl\}$ family of Bragg conditions is excited and 0 otherwise; the model is only valid for scattering into a Bragg condition. The first term is scattering from the disordered phase. Inside the sum over nanocrystals, the first term is the scattering from the crystals and the second term is the scattering from the disordered phase that the crystals replace.

The first moment of the nanodiffraction intensity distribution is therefore

$$\langle I\,(k_{hkl})\rangle = \gamma\rho R^2 t + \frac{\gamma}{N_C}\sum_{i=1}^{N_C}\sum_{j=1}^{N_i} A_j\,(k_{hkl})\left[\left(\rho\chi_j\frac{\pi}{6}d_j^3\right)^2 - \rho\chi_j\frac{\pi}{6}d_j^3\right]. \quad (15)$$

If the number of columns is large and the size, orientation, and position of the nanocrystals are uncorrelated, we can replace the sum over nanocrystals in each column with the expectation values, denoted $\langle\ \rangle$, of the various quantities inside the sum times the expectation value of the number of crystals $\langle N\rangle$:

$$\langle I\rangle = \gamma\rho R^2 t + \gamma\,\langle N\rangle\,\langle A_{hkl}\rangle\left[\left(\frac{\rho\pi}{6}\right)^2\left\langle\chi^2\right\rangle d^6 - \frac{\rho\pi}{6}\,\langle\chi\rangle\,d^3\right]. \quad (16)$$

For simplicity, we have assumed that all the nanocrystals are the same diameter, d. Models for, e.g., a Gaussian distribution of diameters have also been developed (F. Yi and P.M. Voyles to be published). If the nanocrystals are randomly oriented, $\langle A_{hkl}\rangle$ is determined by the acceptance angle $\Delta\theta$ about the perfect Bragg condition and the multiplicity of the $\{hkl\}$ family of planes, M_{hkl}. $\Delta\theta$ contains contributions from the probe convergence angle, the size of the detector pixels in the nanodiffraction pattern, and the finite size of the nanocrystals (Freeman et al. 1977). For typical FEM experimental conditions (in TEM or STEM), the finite nanocrystal size is by far the largest contribution (Stratton and Voyles 2008), so we will approximate $\Delta\theta \approx d_{hkl}/d$. To a good approximation, $\langle A_{hkl}\rangle \approx M_{hkl}d_{hkl}/4d$ (Stratton and Voyles 2008), although this overestimates $\langle A_{hkl}\rangle$ for $\Delta\theta > 100$ mrad (Stratton and Voyles 2008). For convenience, we define $C_{hkl} \equiv M_{hkl}d_{hkl}/4$, a property of a particular crystal structure, with a typical magnitude of 0.25 nm. Expectation values of χ^n must be evaluated numerically as a function of d/R, as shown in Figure 18–3. Note that $\langle\chi^2\rangle \neq \langle\chi\rangle^2$.

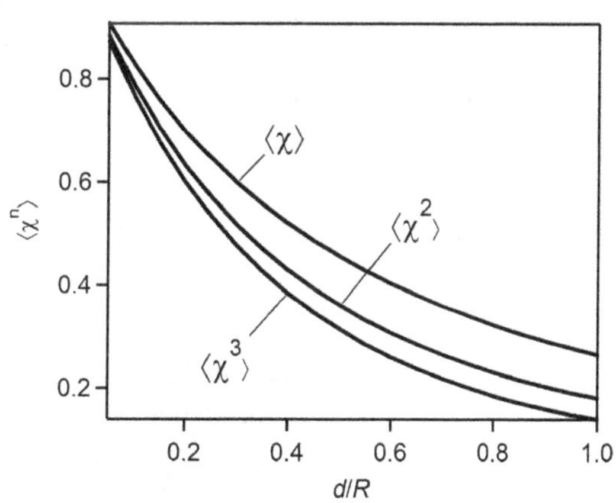

Figure 18–3. Expectation values of χ, the average fraction of the volume a nanocrystal which lies inside a column, as a function of the ratio of the diameter of the crystal d to the size of the column R, calculated assuming that the nanocrystals are randomly distributed.

The second moment of the intensity distribution is

$$\left\langle I^2\right\rangle = \frac{1}{N_C}\sum_{i=1}^{N_c}\left\{\gamma\rho R^2 t + \gamma\sum_{j=1}^{N_i} A_j\left(k_{hkl}\right)\left[\left(\rho\chi_j\frac{\pi}{6}d_j^3\right)^2 - \rho\chi_j\frac{\pi}{6}d_j^3\right]\right\}^2. \quad (17)$$

The sum over columns can again be replaced with expectation values, this time of squared quantities. Because A is either 0 or 1, for self-terms in the squared intensity from the nanocrystal, $\langle A_jA_j\rangle = \langle A^2\rangle = C_{hkl}/d$. However, the cross-terms $\langle A_jA_k\rangle$ are a joint probability of two different nanocrystals being simultaneously on the same Bragg condition, so $\langle A_jA_j\rangle = \langle A\rangle\langle A\rangle = C_{hkl}^2/d^2$. Similarly, the self-terms depend on $\langle N\rangle$, but the cross-terms depend on $\langle N^2\rangle - \langle N\rangle$. Therefore

$$\begin{aligned}\frac{\langle I^2\rangle}{\gamma^2} = {} & \left(\rho R^2 t\right)^2 + C_{hkl}\langle N\rangle\left[\left(\tfrac{\rho\pi}{6}\right)^4\left\langle\chi^4\right\rangle d^{11} - 2\left(\tfrac{\rho\pi}{6}\right)^3\left\langle\chi^3\right\rangle d^8 + \left(\tfrac{\rho\pi}{6}\right)^2\left\langle\chi^2\right\rangle d^5\right] \\ & + C_{hkl}^2\left(\left\langle N^2\right\rangle - \langle N\rangle\right)\left[\left(\tfrac{\rho\pi}{6}\right)^2\left\langle\chi^2\right\rangle d^5 - \left(\tfrac{\rho\pi}{6}\right)\langle\chi\rangle d^2\right]^2 \\ & + 2\rho R^2 t C_{hkl}\langle N\rangle\left[\left(\tfrac{\rho\pi}{6}\right)^2\left\langle\chi^2\right\rangle d^5 - \left(\tfrac{\rho\pi}{6}\right)\langle\chi\rangle d^2\right].\end{aligned} \quad (18)$$

If the nanocrystals are randomly distributed in space,

$$\langle N\rangle = \frac{6\Phi R^2 t}{\pi\langle\chi\rangle d^3}, \quad (19)$$

which is just the volume of a column divided by the average volume of one nanocrystal. If N_i is large enough (Stratton and Voyles 2008)

$$\left\langle N^2\right\rangle - \langle N\rangle^2 = \frac{6\Phi R^2 t}{\pi\langle\chi\rangle d^3}\left(1 - \frac{6\Phi}{\pi}\right). \quad (20)$$

Substituting Eqs. (19) and (20) into Eqs. (16) and (18) and then into Eq. (1) yields, after some simplification, an expression for the variance in terms of the experimental parameters R and k_{hkl} and sample parameters ρ, Φ, d, and C_{hkl}.

$$V = \frac{\left[\left(\frac{\rho\pi}{6}\right)^4\left\langle\chi^4\right\rangle d^{11} - 2\left(\frac{\rho\pi}{6}\right)^3\left\langle\chi^3\right\rangle d^8 + \left(\frac{\rho\pi}{6}\right)^2\left\langle\chi^2\right\rangle d^5\right] - \frac{6C_{hkl}\Phi}{\pi}\left[\left(\frac{\rho\pi}{6}\right)^2\left\langle\chi^2\right\rangle d^5 - \left(\frac{\rho\pi}{6}\right)\langle\chi\rangle d^2\right]^2}{R^2 t\left\{\frac{\rho^2\pi\langle\chi\rangle d^3}{6C_{hkl}\Phi} + \frac{6C_{hkl}\Phi}{\pi\langle\chi\rangle d^3}\left[\left(\frac{\rho\pi}{6}\right)^2\left\langle\chi^2\right\rangle d^5 - \left(\frac{\rho\pi}{6}\right)\langle\chi\rangle d^2\right]^2 + 2\rho\left[\left(\frac{\rho\pi}{6}\right)^2\left\langle\chi^2\right\rangle d^5 - \frac{\rho\pi}{6}\langle\chi\rangle d^2\right]\right\}}. \quad (21)$$

Figure 18–4 shows $V(d, \Phi)$ computed for Al nanocrystals at k_{200} and $R = 1.6$ nm. The key prediction of this model is that the d and Φ dependences are quite different: V increases monotonically with d, but goes through a broad maximum as a function of Φ. The maximum occurs because V measures spatial variability, not absolute scattered intensity. As the structure becomes saturated with crystals, the variability of the structure eventually decreases. Because V depends more weakly on Φ above the maximum, this model of the FEM signal is most useful for relatively dilute ordered regions, with $\Phi < \sim 10\%$. This may mean that

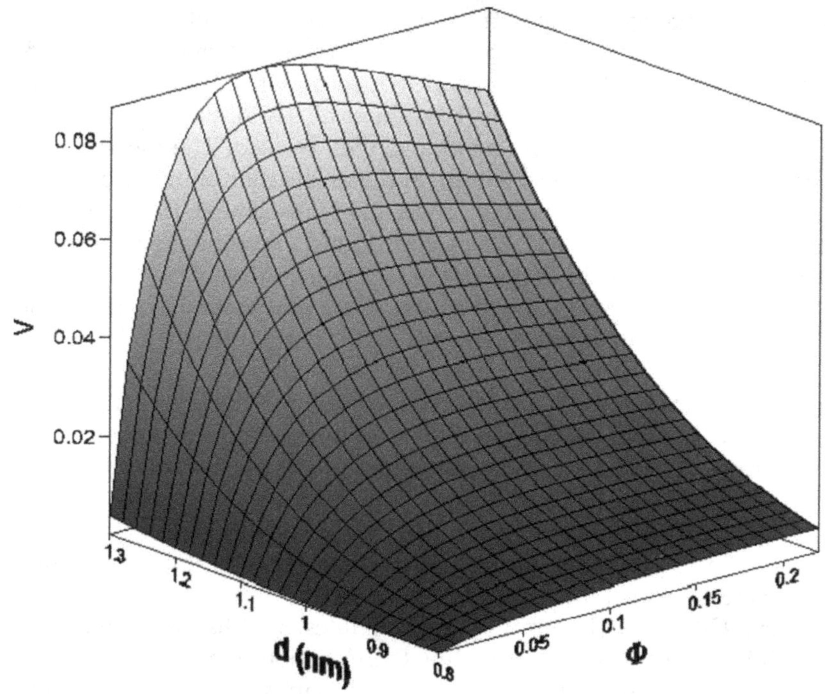

Figure 18–4. $V(d, \Phi)$ calculated from Eq. (21) for Al nanocrystals at k_{200}. $\rho = 60$ atoms/nm^3, $C_{200} = 0.303$ nm, $R = 1.6$ nm, and $t = 60$ nm.

FEM is relatively ineffective at characterizing pervasive MRO of the type suggested, for example, for some metallic glasses (Miracle 2003, Sheng et al. 2006, Wen et al. 2009). The distribution functions model would certainly be a better tool for interpreting FEM data from such a structure than Eq. (21).

In principle, it is possible to extract d and Φ from Eq. (21), if the structure of the ordered regions and the thickness of the sample are known. V from two different k_{hkl} with different C_{hkl} gives a system of non-linear equations for d and Φ which can be solved numerically. Models with additional parameters like a Gaussian size distribution for the nanocrystals require additional data points. In practice, this procedure is not robust with TEM FEM data at a single resolution. As discussed in Section 18.3, the magnitude of TEM FEM V is suppressed and subject to systematic errors from sample thickness fluctuations. These difficulties are reduced in STEM FEM. STEM FEM also has the potential to systematically vary R. A fit of $V(1/R^2)$ should yield a more robust estimate for the remainder of Eq. (21) than a single data point. Equation (21) predicts the same behavior of $V(R)$ as the distribution function Λ *ansatz* result in Eq. (13) for $1/Q >> \Lambda$, which is equivalent to $R >> d$. However, the characteristic length Λ does not distinguish the effects of d and Φ.

18.2.3 Computational Models

A semi-quantitative interpretation of FEM data for amorphous silicon has been obtained by forward simulating $V(k, Q)$ from a family of realistic paracrystalline structural models (Bogle et al. 2007). The CRN matrix was synthesized using the Wooten, Weiner, and Weaire (WWW) bond-switching algorithm which maintains fourfold coordination as required for silicon (Wooten et al. 1985). Paracrystals were inserted into the matrix by hand and then relaxed. The models have realistic bond angle, bond length, and coordination number distributions. Nakhmanson et al. (2001) and Nakhmanson et al. (2001) showed that they correctly predict the electronic and vibrational properties of a-Si and yield peaks in $V(k)$ at the same k as experiment.

To study the effect of paracrystallite size d and volume fraction Φ on the variance, 13 computational cells of 1,000 atoms each, 2.7 nm on a side, were constructed. The grain sizes in the models range from 1.10 to 2.54 nm in diameter and the ordered volume fractions range from 12 to 40%. In order to reproduce the film thickness used in experiment (~20 nm), seven cells are stacked along the beam direction to give a total simulated thickness of 19 nm. Simulations were performed assuming kinematic diffraction with a flat Ewald sphere (Dash et al. 2003). Preliminary simulations including dynamical scattering give similar results for these samples (see Section 18.3.1).

For a CRN model with no paracrystallites, $V(k)$ has weak but non-negligible peaks at $\mathbf{k}$ corresponding to diffraction from the {111} planes and a combination of the {220} and {311} planes. Previous FEM simulations from CRNs did not show these peaks (Treacy et al. 1998, Voyles et al. 1999), probably because the models were too small; the variation in the number of atoms probed compared to the thickness of the model is large for a small model, resulting in a large background signal (for a discussion of the contribution of thickness variation to $V(k)$, see Section 18.3.1). The non-zero variance from a CRN sets a modest lower limit on the ability of FEM to detect dilute paracrystalline content, which we estimate at <0.1 vol% for 2 nm diameter paracrystals or <5% for a ~1 nm diameter.

Using this family of paracrystalline-Si models at a single FEM resolution, Bogle et al. determined that the ratio of the first and second variance peaks affords the size d of the ordered regions, and the magnitude of the variance affords a semi-quantitative measure of the volume fraction Φ (Bogle et al. 2007), as shown in Figure 18–5. Bogle et al. then used the same family of models to calculate characteristic length using three simulated resolutions (1.1, 1.3, and 1.7 nm) for a variety of Φ values (Bogle et al. 2009). The characteristic length Λ increases monotonically with d as shown in Figure 18–6. The magnitude is small (less than a Si–Si bond length for the CRN at $\Phi = 0$!), but Λ is a correlation function decay length, not simply connected to a nanostructural feature size. In fact, in this range of parameters, Λ does not follow a simple geometric expectation, such as $\Lambda \propto$ diameter $\propto N^{1/3}$. The origin of the observed dependence is still under investigation. Λ depends more weakly on Φ than N.

Although the paracrystalline structure is a highly plausible model, it has not been proven as a unique explanation for the FEM data. A new method for synthesizing computer models with less interference from the simulator, experimentally constrained molecular relaxation

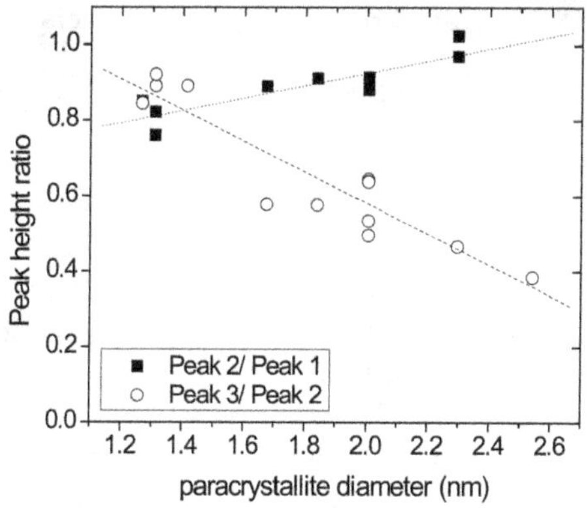

Figure 18–5. Peak height ratios (peak 2/peak 1 and peak 3/peak 2) in $V(k)$ simulated at 1.1 nm resolution for monodisperse paracrystalline-Si models of various paracrystallite volume fraction and diameter. *Dotted lines* are linear least square fits (reproduced from Bogle et al. (2007) with permission).

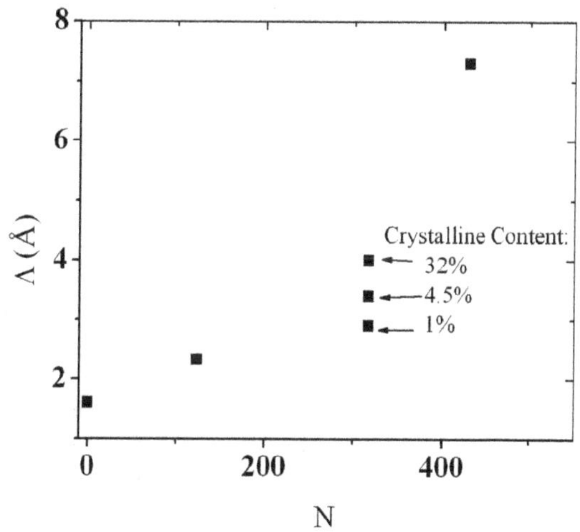

Figure 18–6. Λ extracted from FEM simulations from paracrystalline-Si models with different sizes and densities of paracrystals.

(ECMR), may be useful in identifying other possible forms of MRO in some systems (Biswas et al. 2004b). ECMR is a hybrid approach: it creates and optimizes a model structure in order to fit all sets of experimental data that are used as input through a reverse Monte Carlo approach (McGreevy 2001) incorporating physically realistic bonding constraints (Biswas et al. 2004a), and it uses approximate first-principles energies to relax the trial structures.

18.3 Fluctuation Microscopy in the TEM and STEM

FEM in the STEM and in the TEM fundamentally measure the same $I(\mathbf{r}, \mathbf{k}, Q)$ data set, by the principle of reciprocity of TEM and STEM. However, there are experimental advantages to using STEM nanodiffraction for FEM experiments. These include greater coherence, reduced influence of the detector point-spread function and chromatic aberration, and a greater flexibility in the size of the probed volume.

18.3.1 Data Collection and Artifacts

One difference between TEM and STEM is in how the data set is sampled, as shown in Figure 18–7. In STEM, the data set is a series of nanodiffraction patterns from different positions, so the samples in **k** are collected in parallel on an imaging detector. The spatial samples in **r** are collected serially by moving the probe from position to position on the sample. The Q samples are collected serially by changing the probe lens configuration as described in Section 18.4. In TEM, the data set is a series of dark-field images, so the **r** samples are collected in parallel on the imaging detector, the **k** samples are collected serially by tilting the beam, and the Q samples are collected serially by changing the objective aperture size. The parallel collected dimension of the data is typically sampled with $O(10^6)$ points, for example, on a 1024×1024 pixel CCD camera. The serial dimension is sampled by $O(10^1)$ data points for **k** in TEM and by $O(10^3)$ data points for **r** in STEM. On a typical TEM, there are at most four values of Q available, controlled by the number of objective apertures, and often only one or two of them are small enough to be useful for FEM. As discussed in Section 18.4 below, more values of Q covering a much larger range are available in the STEM and 4–10 might be used in a typical experiment.

Dense, parallel sampling in **k** has the advantage of enabling measurements of $V(\mathbf{k})$. All TEM FEM experiments to date have measured $V(k)$, either in the form of a hollow-cone dark-field images which average **k** over direction at constant magnitude or in the form of tilted dark-field images which make a one-dimensional trace through the two-dimensional **k** space. $V(\mathbf{k})$ data should in principle be sensitive to anisotropic MRO and having the entire $I(\mathbf{k})$ data set opens up new possible experiments, such as measuring the autocorrelation (Rodenburg and Rauf 1990) or variance (Hruszkewycz et al. 2008) of the intensity around a ring in the nanodiffraction pattern at constant k.

Serial sampling in **k** in TEM FEM has the advantage that the exposure time of each dark-field image can be adjusted to maintain a constant number of electrons per spatial sample (pixel) in each image. This is important because there is a Poisson noise contribution to V given by Voyles and Muller (2002)

$$V_P(\mathbf{k}, Q) = \frac{A}{\int I(\mathbf{r}, \mathbf{k}, Q)\, d^2\mathbf{r}}. \tag{22}$$

Adjusting the exposure time keeps V_P constant in TEM FEM, but it varies strongly with k in STEM FEM. The average scattered intensity scales as the atomic scattering factor, which decreases as $1/k^4$, so V_P can become sizeable at moderate to large k. V_P also has some structure in k, since it varies inversely with the peaks and valleys in $\langle I(k)\rangle$.

Poisson noise also contributes to the uncertainty in $V(\mathbf{k}, Q)$. The uncertainty, however, scales as the total number of electrons at a particular **k** and Q in the entire data set, instead of the number of electron scattered from a single position. The fractional uncertainty from Poisson noise is (Voyles and Muller 2002),

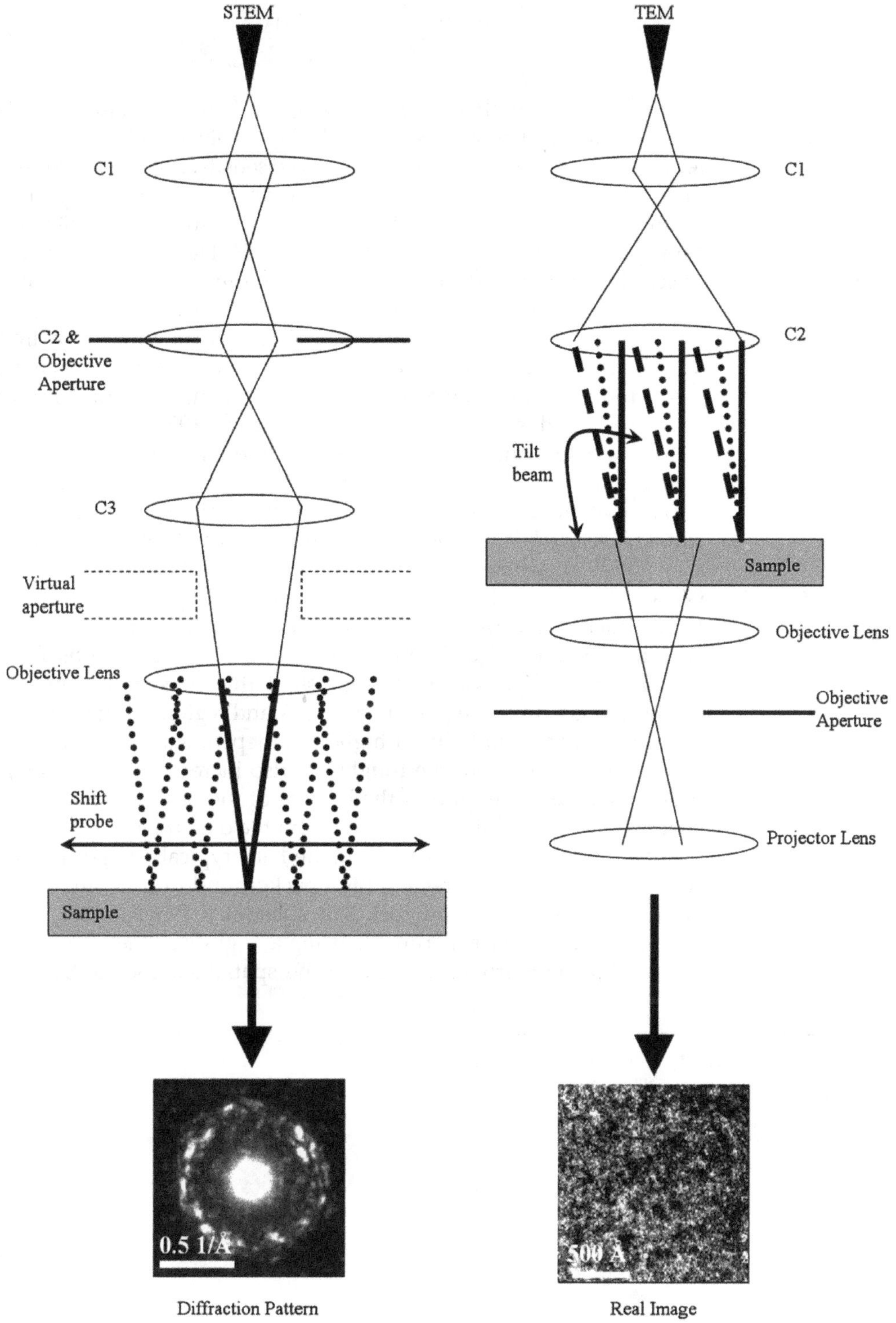

Figure 18–7. Illustration of the different means of collecting the $I(\mathbf{r}, \mathbf{k}, Q)$ data set for FEM using STEM nanodiffraction and dark-field TEM imaging (reproduced from Stratton and Voyles (2007) with permission).

$$\frac{\delta V_P(\mathbf{k},Q)}{V(\mathbf{k},Q)} = \frac{2A\sqrt{2}}{N\int I(\mathbf{r},\mathbf{k},Q)\,d^2\mathbf{r}'} \tag{23}$$

where N is the number of spatial samples. Parallel, denser sampling in $\mathbf{r}$ in TEM FEM compared to STEM FEM substantially reduces this term, in principle making TEM FEM more sensitive to small differences in V. Equations (22) and (23) also show that FEM can be applied to beam-sensitive samples: the absolute offset due to Poisson noise can in principle be subtracted from the measured V (Fan et al. 2007) and the uncertainty can be reduced by acquiring more spatial samples at low dose, as long as the number of counts is above the detector noise.

Artifacts in $V(k, Q)$ can also be caused by non-idealities in the sample. Spatial variation in the projected sample thickness is a common problem, which can arise from imperfections in sample thinning, wedged-shaped samples (Gibson and Treacy 1997), surface roughness, or a distribution of internal voids. Differences in thickness create changes in the overall magnitude of $I(\mathbf{r}, \mathbf{k}, Q)$ which contributes variance at all values of k. Small values of roughness ($\sim$ 1–2 nm) contribute a nearly *k-independent* background that can be measured at high k (> 1.4 Å^{-1}), where diffraction contrast is suppressed (Treacy and Gibson 1993) and subtracted from the variance (Voyles 2001). However, larger thickness variations give rise to a *k-dependent* background, as shown in Figure 18–8. Figure 18–8 shows $V(k)$ simulations from a paracrystalline-Si model with a constant thickness of 19 nm and a model that is a mix of regions 11 nm thick and regions 19 nm thick. $V(k)$ from the rough sample has a higher k-independent background than the smooth sample, but the roughness also increases the total magnitude of $V(k)$ and the ratio of the heights of the peaks. These kinds of changes cannot be reliably removed from the data (Bogle 2009).

Some early FEM data on a-Si exhibited an atypically large and broad second peak and an extremely high background due to film agglomeration during growth on rock salt substrates (Voyles 2001, Voyles et al. 2001a). (Newer experiments using a-C grids as the substrate did not exhibit these artifacts.) Because the spatial samples in STEM are

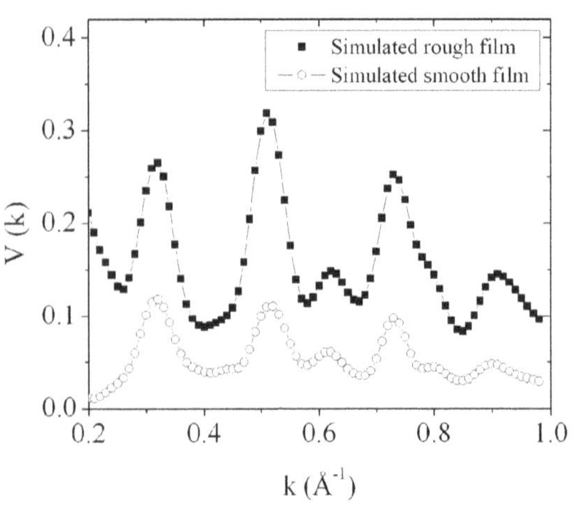

Figure 18–8. Simulated variance for a smooth paracrystalline-Si film of thickness of 19 nm and a rough paracrystalline film with a mixed thickness of 11 and 19 nm.

acquired serially, patterns from areas of the sample that all are of the same thickness can be selected by using only patterns for which the high-k intensity falls within certain limits to calculate the variance. The resulting spatial samples may or may not be contiguous, but that does not affect the result. In general, however, it is advisable to evaluate the surface roughness using AFM prior to FEM analysis. Carbon contamination on the sample surface affects the variance in a similar way. A thin, uniform surface layer of carbon or oxide is relatively unimportant. However, non-uniform carbon contamination can be quickly created in the STEM by stopping the electron beam, leading to artifacts that are more difficult, although not impossible, to correct.

18.3.2 Coherence and Chromatic Aberration

Voyles and Muller compared $V(k)$ measured on the same sample by STEM FEM and TEM FEM at comparable resolution (Voyles and Muller 2002). The result, shown in Figure 18–9, is that $V(k)$ differs in magnitude only, by a constant factor of two in this case. The increased magnitude of $V(k)$ in STEM FEM is caused by a decrease in experimental imperfections of poor coherence and chromatic aberration.

High coherence in TEM imaging is achieved by using an objective aperture semiangle β than is much larger than the illumination semiangle α (see, e.g., Kirkland (1998) for a compact discussion). This is routinely achieved in high-resolution TEM, but it is not the case in FEM. In FEM, β must be small, since it is proportional to Q, and $R \propto 1/Q$ must be relatively large. The value of α must be large to achieve a high enough intensity in the dark-field image to avoid prohibitively long exposure times. For example, on the LaB_6-source LEO 912 TEM used for FEM in Stratton et al. (2005) and Stratton and Voyles (2007), typically $\alpha = 2$–4 mrad and $\beta = 1.3$ mrad and the coherence is not high. A field-emission gun (FEG) TEM might make it possible to achieve α somewhat less than β, but the requirement for a small objective aperture will prevent high coherence from being achieved. In STEM, the probe coherence is the relevant quantity and high coherence can be

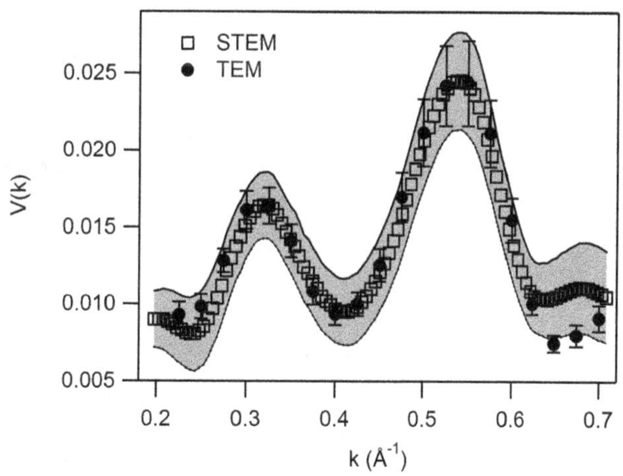

Figure 18–9. $V(k)$ measured by STEM FEM and TEM FEM on the same a-Si sample. The TEM FEM data have been multiplied by two and offset by 0.009 to match the STEM FEM data. The gray band in the error interval about the STEM FEM data points (reproduced from Voyles and Muller (2002) with permission).

readily achieved. As discussed in Section 18.4, the probe must be minimally convergent for its size and the source size must be small. A FEG STEM is required.

Better coherence increases V (Stratton and Voyles 2007). With imperfect coherence, the intensity from an ordered cluster scales less strongly with the cluster size, making the excursion away from the mean intensity smaller and reducing V. In TEM language, poor coherence reduces the image contrast, and V is simply a quantitative measure of contrast. As a result, the STEM FEM $V(k)$ is anywhere from two to nine times higher than the TEM FEM V measured from similar samples (Stratton and Voyles 2007, Voyles and Muller 2002). This increased signal compensates for the reduced sensitivity of STEM FEM due to the smaller number of spatial samples.

The effects of chromatic aberration are reduced in STEM vs. TEM FEM. As discussed in Chapter 15, the relevant energy spread for chromatic aberration for STEM is the energy spread of the source, which for a Schottky FEG STEM is ~0.7 eV. Chromatic aberration can be significant for sub-Ångstrom, aberration-corrected probes, but not for the much less convergent, much larger probes required for FEM. The energy spread relevant to chromatic aberration in TEM FEM is imposed by plasmon losses in the sample, which is typically ~10 eV. The objective aperture semiangle in TEM FEM is still very small, so chromatic aberration effects do not significantly change the spatial resolution; they may, however, reduce the contrast of the image and thus the variance V. In STEM, chromatic blurring from plasmon losses will reduce the $\mathbf{k}$ resolution of the nanodiffraction patterns, but the reduction in resolution is not significant compared to the already large width of the features imposed by the convergent probe and the size broadening due to the small structure features being measured. STEM FEM may be affected by the angular distribution of inelastic scattering. Minimizing this effect using energy filtering is crucial to using electron diffraction to accurately measure the structure factor $S(k)$ (Cockayne and McKenzie 1988), but the effect of energy filtering has not been investigated for STEM or TEM FEM.

Similarly, the STEM FEM V is less effected by the detector point-spread function (PSF) or modulation transfer function. The PSF of a CCD camera can extend over tens of pixels (Zuo 1996), blurring the image, which will reduce V in TEM FEM. Image processing has been used to reduce these effects (Voyles 2001, Voyles et al. 1999). In STEM FEM, that blurring takes place in reciprocal space, so it reduces the spatial variance V significantly less. As shown in Section 18.2.1, the spatial PSF is just the probe wave function, and it is easy to separate the spatial intensity samples by many multiples of the probe size to completely prevent cross talk.

18.3.3 Comparison to Simulation

Simulations of $V(k)$ have always been systematically larger in magnitude than experiment. Although it is not fully understood, much of this discrepancy appears to be the joint result of computational models that afford a maximum estimate of the variance and experiment that may be

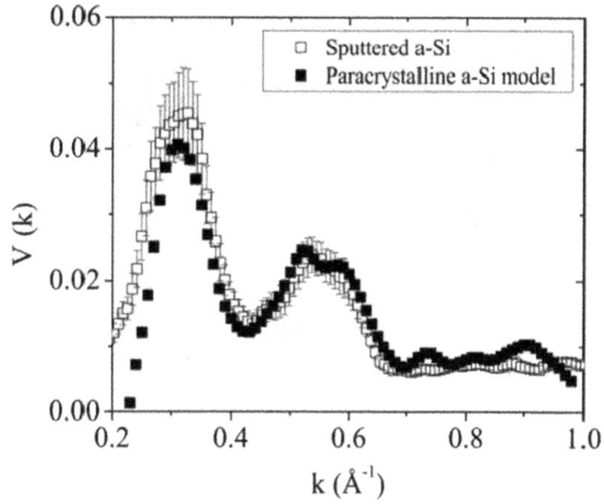

Figure 18–10. Plot of simulated $V(k)$ for a paracrystalline model and measured $V(k)$ from an a-Si film sputtered at 230°C. The experimental variance was multiplied by 5 and an offset of 0.017 was subtracted from the simulated variance (reproduced from Bogle (2009) with permission).

degraded by less than ideal coherence, multiple scattering, and other factors. In earlier work, the heights of the simulated variance peaks were as much as 10 times larger than the experimental values (Treacy et al. 1998, Voyles et al. 2001b). A significant portion of that discrepancy was traced to the small size of the computational models (Treacy et al. 1998, Voyles et al. 1999), which can also lead to spurious peaks in $V(k)$ (Treacy et al. 1998, Voyles 2001). Recent models with thickness similar to experiment (Section 18.2.3) match the data better, to within a factor of 2–5 in overall magnitude, and essentially perfectly in shape, as shown in Figure 18–10.

Almost all of the simulations to date have been performed either in the phase-grating approximation (Nakhmanson et al. 2001, Treacy et al. 1998, Voyles et al. 2001b) or with kinematic diffraction in two-dimensional projection, which is equivalent to assuming a flat Ewald sphere (Dash et al. 2003, Khare et al. 2004, Stratton et al. 2005). Zhao et al. recently emphasized the need for a full kinematic diffraction treatment in their study of curved fullerenes, and the same argument may apply here (Zhao et al. 2009). Bogle et al. (2009) recently examined the possible effects of multiple scattering using a Bloch wave dynamical diffraction approach. In that work, 16 nm thick models of paracrystalline a-Si were converted into *pseudo* a-Ge simply by changing the atomic number from 14 to 32, but not performing any relaxations. Simulated $V(k)$ for the *pseudo* a-Ge is shown in Figure 18–11. For kinematic scattering, the a-Si and *pseudo* a-Ge produced exactly the same variance curve; for multiple scattering, the ratio of peak heights in $V(k)$ changed for the a-Ge but not for the a-Si. Future work will need to consider the effect of multiple scattering on the measured $V(k)$ and calculated Λ for samples with heavy atoms.

18.4 Probes for FEM

Various aspects of the formation of electron probes have been covered in Chapters 1, 9, and 15. FEM requires probes that are 1 nm or larger, which is large by the standards of many of the other applications in

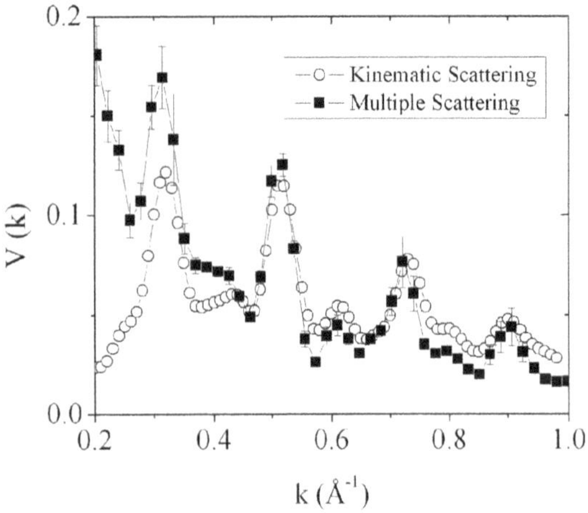

Figure 18–11. Simulated variance for kinematic vs. multiple scattering for 16 nm thick *pseudo* a-Ge.

this volume. Continuously variable probe size is desirable for variable resolution FEM, and the probes must have high coherence. High coherence requires a field-emission gun imaged with large demagnification into a probe whose size is defined by a small objective aperture. The dramatic advantage of STEM vs. TEM is that the STEM uses a virtual objective aperture: there are lenses between the physical aperture and the strong probe-forming objective lens, as shown in Figure 18–12. By changing the excitation of those lenses, the probe convergence angle and probe size can be changed for the same physical aperture. A virtual aperture allows essentially continuously variable probe size, within the limits imposed by the focal lengths and physical distances between the lenses.

All STEM FEM experiments so far have used a Fourier space aperture (FSA), which is the mode used for high-resolution STEM as described in Chapter 2 and Section 2.1. In this mode, the physical aperture is imaged onto the objective lens front focal plane, as shown in Figure 18–12a. Q is proportional to the physical aperture radius, and the probe is the Fourier transform of the demagnified aperture image. For the aberration-free Airy function probe of Eq. (6), the probe size can be defined by the Rayleigh resolution criterion, $R = 0.61/Q$. Voyles and Muller created probes 0.8–5.0 nm in diameter in this mode (Voyles and Muller 2002) on a STEM with two condenser lenses by making large changes to the objective lens focal length. However, these probes were impractical for routine use because changing probe size required extensive realignment of the microscope scan and projector systems. Bogle et al. used a STEM with two full-strength condenser lenses and one condenser mini-lens to create probes 1.2–4 nm in diameter at a constant objective lens current (Bogle et al. 2009), as shown in Figure 18–13. On that STEM, smaller probes (down to ~0.2 nm) are easily created. Larger probes sizes are inaccessible due to the limited minimum focal length of the mini-lens. Newer commercial STEMs have three full-strength condenser lenses, which should extend the range of probes to larger sizes.

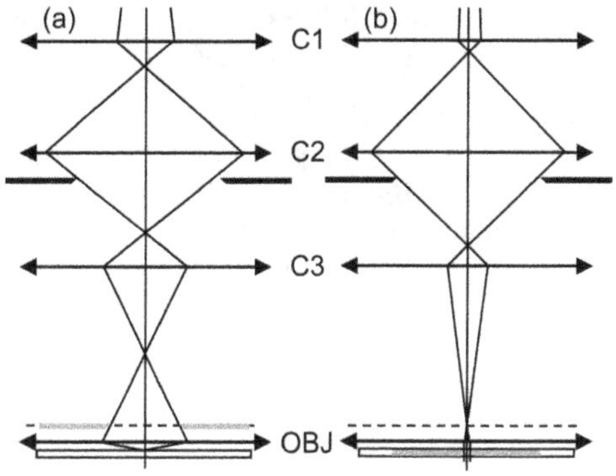

Figure 18–12. Two different optical configurations for probes in STEM: (a) Fourier space aperture mode, in which the physical aperture is imaged into the objective lens front focal plane; (b) real-space aperture mode, in which the physical aperture is imaged onto the sample. The *black* aperture is the physical aperture. The *light gray* aperture is its image.

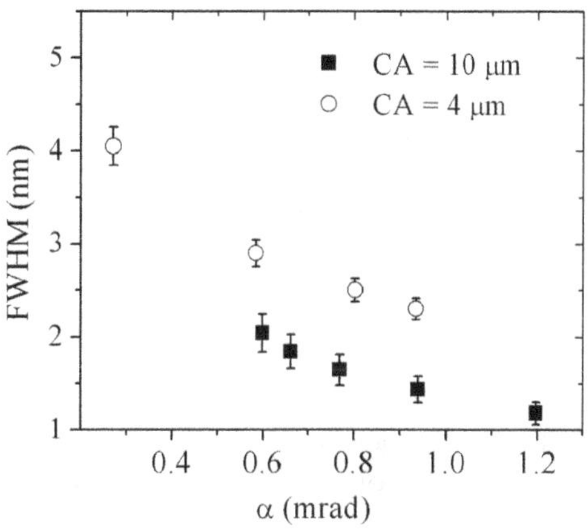

Figure 18–13. Measured probe size as a function of convergence angle for a set of probes created on a JEOL 2010F STEM by varying the condenser minilens excitation. "CA" is the physical diameter of the condenser aperture.

Real-space aperture (RSA) probes, used for nanoarea electron diffraction as described in Chapter 9, could also be used for FEM. In this mode, shown in Figure 18–12b the focal length of the condenser lenses is changed substantially to place the image of the physical aperture in the object plane of the objective lens, and the probe is a demagnified image of the aperture. The probe size is thus directly proportional to the aperture size, and Q can be estimated from the Rayleigh criterion. Zuo has used this mode on a two full-strength/one mini-lens STEM to produce a probe 50 nm in diameter (Zuo et al. 2003). In RSA mode, creating larger probes is simply a matter of using a larger aperture, although the coherence may suffer. The ability to create smaller probes is limited by the maximum possible demagnification set by the focal length of the lenses.

Because FSA probes have a sharp cutoff in Fourier space, the probe wave function exhibits ringing in real space, as in Eq. (6). RSA probes, on the other hand, have a sharp cutoff in real space and thus ringing in Fourier space, as has been demonstrated experimentally by Dwyer et al. (2007). Because RSA probes have large diameters and very small Q, the extent of this ringing may be experimentally unimportant, but it has not yet been incorporated into the FEM theoretical models.

As discussed in Chapter 15, a hexapole probe aberration corrector adds a substantial number of lenses to the STEM column before the sample. Because the convergence angles for FEM probes are small, the

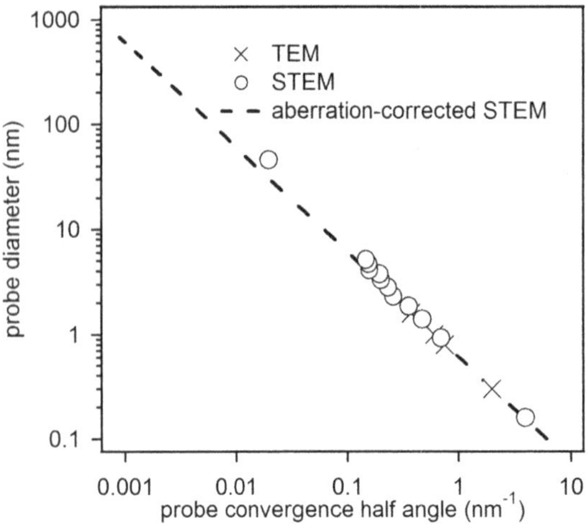

Figure 18–14. Current and future probe sizes for FEM. *Crosses* are published TEM results, *circles* <10 nm are FSA probes from Voyles and Muller (2002) and Bogle et al. (2009). The *circle* at 50 nm diameter is an RSA from Zuo et al. (2004). The *dashed line* is based on calculations for an aberration-corrected STEM with three full-strength condenser lenses.

negative C_S of the hexapole elements is not needed, but the additional round lenses in the coupling and transfer systems of the corrector can be used to provide additional demagnification of the virtual aperture in either RSA or FSA mode (Dwyer et al. 2007). We estimate that on an FEI Titan STEM with three full-strength condenser lenses and a CEOS hexapole aberration corrector, by combining FSA and RSA modes and four condenser aperture sizes, it will be possible to generate high-coherence probes with diameters running from 0.1 to at least 750 nm, continuously. The smallest probes, $\leq$4 nm, can be generated only in FSA mode, and the largest probes, $\geq$300 nm, only in RSA mode, but in between there is the possibility for comparison between the two modes. The current results and future prospects for probes for FEM, in TEM and STEM, are summarized in Figure 18–14.

18.5 STEM FEM Experiments

STEM FEM has proven to be a robust technique in our laboratories. Most of the STEM FEM experiments to date have examined vapor-deposited a-Ge (Voyles and Muller 2002) and a-Si (Bogle et al. 2009, Voyles and Muller 2002) films, although some data on amorphous metals are also available (Stratton and Voyles 2007). The data reported here were acquired on a JEOL 2010F STEM using 1.2–1.8 nm probes formed with the 10 μm condenser aperture and 2.8–3.8 nm probes formed with the 4 μm condenser aperture from Figure 18–13.

Figure 18–15 shows variable resolution FEM data acquired on a-Si films grown under significantly different conditions, magnetron sputtering of a silicon target vs. high-pressure plasma-enhanced chemical vapor deposition of silane. The latter had been hypothesized to afford greater MRO because nanoclusters are generated in the gas phase and impinge on the growth surface (Bogle et al. 2009). Figure 18–16 shows a fit of this data at $k = 0.32$ Å^{-1} to the form predicted by Eq. (13). The data are an excellent fit to a line, with a Pearson's R of 0.98. The extracted

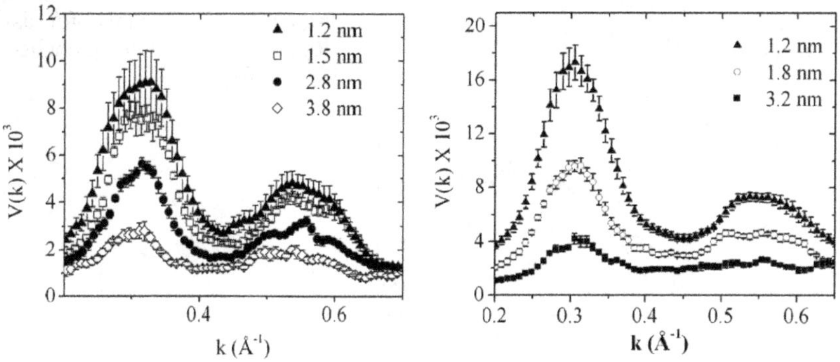

Figure 18–15. VR-FEM data for a magnetron sputtered a-Si (*left*) and high-pressure PECVD a-Si (*right*) (reproduced from Bogle et al. (2009) with permission).

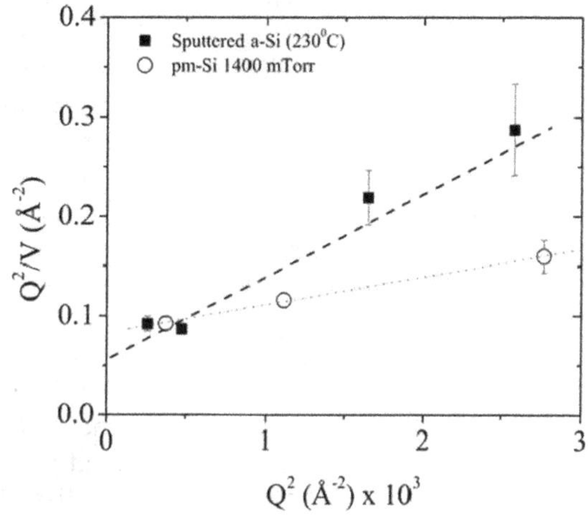

Figure 18–16. Λ analysis applied to variable resolution FEM data from two amorphous silicon samples, one deposited by sputtering and one by plasma-enhanced chemical vapor deposition under "polymorphous" silicon deposition conditions. Λ is determined from the slope and intercept of the fit lines.

characteristic lengths are 0.6 and 0.3 nm for the two samples, respectively. However, the magnitude of the variance from the PECVD sample is larger. Based on the results in Section 18.2.3, we interpret this to indicate that the PECVD sample has a higher volume fraction of smaller ordered regions. This result raises the question of whether ordered clusters created in the gas phase can retain their size or configuration upon bonding to the growth surface or upon incorporation into the bulk. Significantly, the same analysis using $k = 0.53$ and 0.61 Å yielded the same values for the characteristic lengths, within experimental uncertainty, which was not the case for previous experiments (Voyles and Muller 2002). The previous experiments had contained artifacts unrelated to MRO (mostly due to thickness variations) and the characteristic length could not be determined reliably.

STEM FEM has also been used to examine the MRO in amorphous chalcogenide films, such as $Ge_2Sb_2Te_5$, in the as-deposited, melt-quenched and annealed states (Kwon et al. 2007). These materials are used in phase-change memory devices, where the incubation time prior to crystallization, and thus the switching speed of the device,

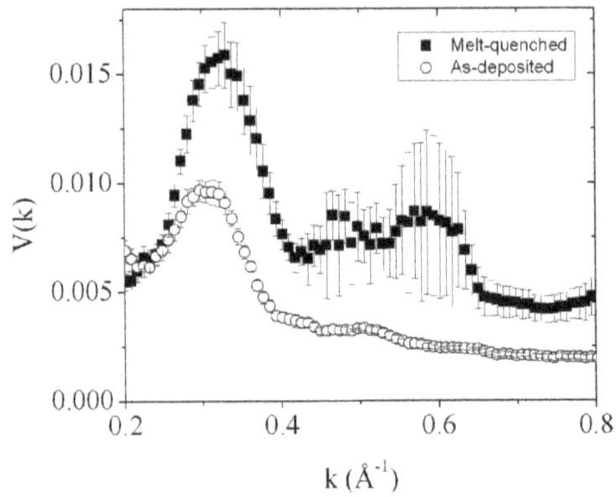

Figure 18–17. Variance for as-deposited vs. melt-quenched $Ge_2Sb_2Te_5$.

depends on the thermal history of the sample. It had been proposed that, in the case of $Ge_2Sb_2Te_5$, melt-quenching creates a large population of nuclei because the incubation time becomes very short (Lee 2006). Figure 18–17 shows $V(k)$ for $Ge_2Sb_2Te_5$ films in the as-deposited and melt-quenched states. The characteristic lengths derived from similar data at other resolutions are 0.3 and 0.9 nm, respectively, consistent with the deduction from crystallization kinetics that the MRO had been enhanced by melt quenching. It was further shown that the melt-quenched sample contained a dilute population of ordered regions that were not always resolved in HRTEM (~3 nm). One way to identify such large ordered regions from the $I(\mathbf{r}, \mathbf{k}, Q)$ data is to compute the azimuthal variance $V(\theta)$ of individual nanodiffraction patterns. Nanocrystals give rise to unusually high values.

Changes in the MRO upon thermal annealing are generally subtler, not associated with such large ordered objects as in the previous case. In recent studies of another chalcogenide composition, AgInSbTe, these changes mapped consistently onto the incubation time for crystallization and have been interpreted in terms of the coarsening of the distribution of subcritical nuclei in the material (Lee et al. 2009). With a low-temperature anneal, the variance increased (and corresponding time to crystallization decreased). Upon melt quenching, the MRO in this alloy was still comparable to the MRO in the as-deposited state, indicating that the formation rate of ordered regions is considerably slower than in $Ge_2Sb_2Te_5$. These studies indicate that STEM FEM is a uniquely powerful means to analyze the development of subcritical nuclei in amorphous materials.

18.6 Future Directions

We have already mentioned several possible new directions in the sections above: STEM FEM on an aberration-corrected STEM will make it possible to perform variable resolution FEM over a very wide range of resolution R. This in turn will provide a more robust data set for measuring the correlation length of MRO using the theory of Section 18.2.2 or the characteristic size of ordered regions using the theory of Section

18.2.1. The rich $V(\mathbf{k})$ data set available from STEM FEM makes it possible to measure angular correlations at constant k as a function of polar angle, using, for example, the average angular autocorrelation function (Rodenburg 1999) or the angular variance, $V(\theta)$, as mentioned above for chalcogenide samples (Hruszkewycz et al. 2008).

Fan et al. (2005, 2007) have recently implemented fluctuation microscopy measurements using x-ray probe nanodiffraction instead of electrons. In order to achieve good coherence, they used fairly soft 1.83 keV x-rays and a simple pinhole to define a probe ~1 μm in diameter. The longer wavelengths and larger probes of x-rays make them potentially useful for studying systems like polymers and molecular glasses in which the structural building blocks and the ordering length scales are substantially larger than in the inorganic materials we have discussed so far. Figure 18–18 shows some proof-of-principle fluctuation x-ray microscopy (FXM) measurements on a disordered packing of 277 nm diameter latex spheres, which is roughly analogous to a dense random packed glass. By using different pinholes to change the probe size, they extract an ordering length scale of 1.1 to 1.4 μm for these samples from the Q^2/V vs. Q^2 analysis.

Steady advances in x-ray focusing optics have yielded smaller and smaller x-ray probes. Probes 100 nm in diameter have been achieved on the same beamline Fan et al. used for FXM, and probes as small as 10 nm may be possible in the near future. This will extend the reach of FXM to smaller structural units and shorter length scales. It also creates extensive overlap in probe sizes with aberration-corrected STEM instruments, raising the possibility of quantitative comparison between scattering with the two types of radiation.

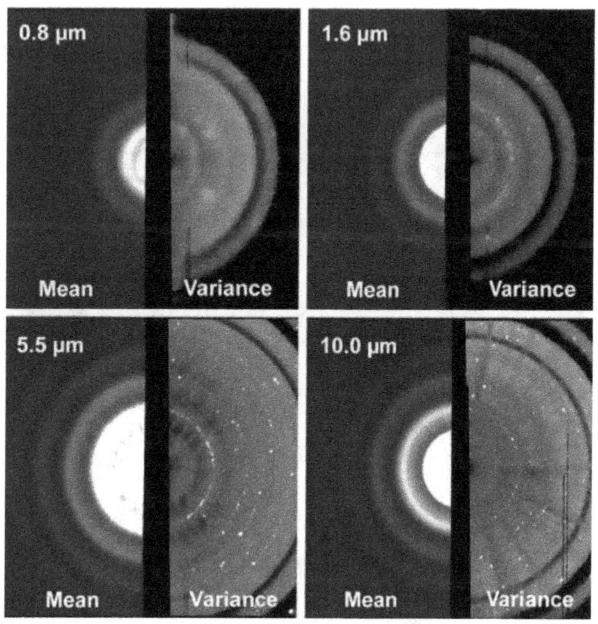

Figure 18–18. Fluctuation x-ray microscopy data from a random compact of polystyrene beads. The beads are 277 nm in diameter, and the data sets are acquired with different probe sizes (reproduced from Fan et al. 2007 with permission).

In addition to organic glasses, FEM may also find application characterizing packaged nanostructures. Nanostructures in service are often packaged in amorphous organic or inorganic hosts, for example, optically active nanoparticles in a SiO_2 host (Walters et al. 2005). While bare nanoparticles are straightforward to characterize using high-resolution TEM, STEM, or scanned probe microscopies, those techniques are difficult to apply within a host.

Acknowledgments The authors thank Jian-Min Zuo, Bong-Sub Lee, and Feng Yi for helpful discussions, and gratefully acknowledge the support of the U.S. National Science Foundation (DMR-0605890).

References

G. Adam, J.H. Gibbs, On the temperature dependence of cooperative relaxation properties in glass-forming liquids. J. Chem. Phys. **43**, 139–146 (1965).

A.S. Argon, Plastic deformation in metallic glasses. Acta Metall. **27**, 47–58 (1979).

U. Bengtzelius, W. Goetze, et al., Dynamics of supercooled liquids and the glass transition. J. Phys. C: Solid State Physics **17**, 5915–5934 (1984).

J.D. Bernal, The Bakerian lecture, 1962: the structure of liquids. Proc. Roy. Soc. Lond. A **280**(1382), 299–322 (1964).

P. Biswas, R. Atta-Fynn, et al., Reverse Monte Carlo modeling of amorphous silicon. Phys. Rev. B **69**, 195207 (2004a).

P. Biswas, D.N. Tagen, et al., The inclusion of experimental information in first principles modelling of materials. J. Phys. Condens. Matter **16**, S5173–S5182 (2004b).

S.N. Bogle, *Quantifying Nanoscale Order in Amorphous Materials via Fluctuation Electron Microscopy. Materials Science and Engineering* (University of Illinois, Champaign-Urbana, IL, 2009).

S.N. Bogle, L.N. Nittala, et al., Size analysis of nanoscale order in amorphous materials by variable-resolution fluctuation electron microscopy. Ultramicroscopy **110**, 1273–1278 (2010).

S.N. Bogle, P.M. Voyles, et al., Quantifying nanoscale order in amorphous materials: simulating fluctuation electron microscopy of amorphous silicon. J. Phys. Cond. Mat. **19**, 455204 (2007).

D. Chandler, *Introduction to Modern Statistical Mechanics* (Oxford University Press, Oxford, 1987).

X. Chen, J.P. Sullivan, et al., Fluctuation microscopy studies of medium-range ordering in amorphous diamond-like carbon films. Appl. Phys. Lett. **84**, 2823–2825 (2004).

J.-Y. Cheng, J.M. Gibson, et al., Quantitative analysis of annealing-induced structure disordering in ion-implanted amorphous silicon. J. Vac. Sci. Tech. A **20**(6), 1855–1859 (2002).

J.-Y. Cheng, J.M. Gibson, et al., Observation of structural order in ion-implanted amorphous silicon. J. Mat. Res. **16**, 3030–3033 (2001).

D.J.H. Cockayne, The study of nanovolumes of amorphous materials using electron scattering. Ann. Rev. Mater. Res. **37**, 159–187 (2007).

D.J.H. Cockayne, D.R. McKenzie, Electron diffraction analysis of polycrystalline and amorphous thin films. Acta Cryst. **A44**, 870–878 (1988).

J.M. Cowley, STEM imaging with a thin annular detector. J. Electron Microsc. **50**, 147–155 (2001).

J.M. Cowley, Electron nanodiffraction methods for measuring medium-range order. Ultramicroscopy **90**(2–3), 197–206 (2002).

N. Cusack, *The Physics of Structurally Disordered Matter* (A. Hilger, 1987).

R.K. Dash, P.M. Voyles, et al., A quantitative measure of medium-range order in amorphous materials from transmission electron micrographs. J. Phys Condens Matter **15**(31), S2425–S2435 (2003).

C. Dwyer, A.I. Kirkland, et al., Electron nanodiffraction using sharply focused parallel probes. Appl. Phys. Lett. **90**, 151104 (2007).

S.R. Elliot, The structure of amorphous hydrogenated silicon and its alloys: A review. Adv. Phys. **38**, 1–88 (1989).

L. Fan, I. McNulty, et al., Fluctuation microscopy – a tool for examining medium-range order in noncrystalline systems. Nucl. Instr. Meth. Phys. Res. B **238**(1–4), 196–199 (2005).

L. Fan, D. Paterson, et al., Fluctuation X-ray microscopy: a novel approach for the structural study of disordered materials. J. Microsc. **225**(1), 41–48 (2007).

H.E. Fischer, A.C. Barnes, et al., Neutron and x-ray diffraction studies of liquids and glasses. Rep. Prog. Phys. **69**, 233–299 (2006).

J.C. Foley, D.R. Allen, et al., Analysis of nanocrystal development in Al-Y-Fe and Al-Sm glasses. Scripta Mat. **35**, 655–660 (1996).

L.A. Freeman, A. Howie, et al., Bright field and hollow cone dark field electron microscopy of palladium catalysts. J. Microsc. **111**, 165–178 (1977).

J.E. Gerbi, P.M. Voyles, et al., Control of medium-range order in amorphous silicon via ion and neutral bombardment. Appl. Phys. Lett. **82**(21), 3665–3667 (2003).

J.E. Gerbi, P.M. Voyles, et al., Increasing medium-range order in amorphous silicon with low-energy ion bombardment. Appl. Phys. Lett. **82**(21), 3665–3667 (2003).

P.C. Gibbons, Y.T. Shen, et al., Intermediate-range order in amorphous metal alloys. Phil. Mag. **86**(3–5), 293–298 (2006).

J.M. Gibson, M.M.J. Treacy, Diminished medium-range order observed in annealed amorphous germanium. Phys. Rev. Lett. **78**, 1074–1077 (1997).

J.M. Gibson, M.M.J. Treacy, et al., Atom pair persistence in disordered materials from fluctuation microscopy. Ultramicroscopy **83**(3–4), 169–178 (2000).

J.M. Gibson, M.M.J. Treacy, et al., Structural disorder induced in hydrogenated amorphous silicon by light soaking. Applied Physics Letters **73**(21), 3093–3095 (1998).

M. Haider, S. Uhlemann, et al., Electron microscopy image enhanced. Nature **392**, 768–769 (1998).

A. Hirata, Y. Hirotsu, et al., Direct imaging of local atomic ordering in a Pd–Ni–P bulk metallic glass using C_S-corrected transmission electron microscopy. Ultramicroscopy **107**, 116–123 (2007).

M.Y. Ho, H. Gong, et al., Morphology and crystallization kinetics in HfO_2 thin films grown by atomic layer deposition. J. Appl. Phys. **93**(3), 1477–1481 (2003).

A. Howie, O.L. Krivanek, et al., Interpretation of electron micrographs and diffraction patterns of amorphous materials. Philosophical Magazine **27**(1), 235–255 (1973).

S.O. Hruszkewycz, T. Fujita, et al., Selected area nanodiffraction fluctuation electron microscopy for studying structural order in amorphous solids. Scripta Mat. **58**, 303–306 (2008).

T.C. Hufnagel, C. Fan, et al., Controlling shear band behavior in metallic glass through microstructural design. Intermetallics **10**, 1163–1166 (2002).

J. Hwang, H. Cao, et al., in *Nanometer-Scale Structure of a Zr-Based Bulk Metallic Glass. Bulk Metallic Glasses*, eds. by J. Schoers, R. Busch, N. Nishiyama, M. Li (Fall Materials Research Society Meeting, Boston, MA, 2007).

J.A. Johnson, J.B. Woodford, et al., Insights into "near-frictionless carbon films". J. Appl. Phys. **95**, 7765–7771 (2004).

S.V. Khare, S.M. Nakhmanson, et al., Evidence from atomistic simulations of fluctuation electron microscopy for preferred local orientations in amorphous silicon. Appl. Phys. Lett. **85**(5), 745–747 (2004).

E.J. Kirkland, in *Advanced Computing in Electron Microscopy* (Plenum Press, New York, NY, 1998).

M. Kisa, T.K. Minton, et al., Homogeneous silica formed by the oxidation of Si(100) in hyperthermal atomic oxygen. J. Spacecraft Rockets **43**, 431–435 (2006).

D. Kivelson, S.A. Kivelson, et al., A thermodynamic theory of supercooled liquids. Physica A **219**(1–2), 27–38 (1995).

O.L. Krivanek, P.H. Gaskell, et al., Seeing order in 'amorphous' materials. Nature **262**, 454–457 (1976).

M.H. Kwon, B.S. Lee, et al., Nanometer-scale order in amorphous $Ge_2Sb_2Te_5$ analyzed by fluctuation electron microscopy. Appl. Phys. Lett. **90**(2) (2007).

B. Lee, in *Optical and Electronic Properties, Nanoscale Structural Order, and Transformation Kinetics of Phase Change Materials* (University of Illinois at Urbana-Champaign, 2006).

B.S. Lee, G.W. Burr, et al., Observation of the role of subcritical nuclei in crystallization of a glassy solid. Science **326**, 980–984 (2009).

E. Leutheusser, Dynamical model of the liquid-glass transition. Phys. Rev. A **29**, 2765–2773 (1984).

J. Li, X. Gu, et al., Using fluctuation microscopy to characterize structural order in metallic glass. Microsc. Microanal. **9**, 509–515 (2003).

R.L. McGreevy, Reverse Monte Carlo modelling. J. Phys. Condens. Matter **13**, R877–R913 (2001).

D.B. Miracle, On the universal model for medium-range order in amorphous metal structures. J. Non-Cryst. Sol. **317**, 40–44 (2003).

S.M. Nakhmanson, P.M. Voyles, et al., Realistic models of paracrystalline silicon. Phys. Rev. B **63**(23), 235207 (2001).

Y. Pan, F. Inam, et al., Atomistic origin of Urbach tails in amorphous silicon. Phys. Rev. Lett. **1000**, 206403 (2008).

C.R. Perrey, S. Thompson, et al., Observation of Si nanocrystals in a/nc-Si: H films by spherical-aberration corrected transmission electron microscopy. J. Non-Cryst. Solids **343**(1–3), 78–84 (2004).

J.M. Rodenburg, in *Measurement of Higher-Order Correlation Functions in Amorphous Materials Via Coherent Microdiffraction. EMAG99 Proceedings* (Institute of Physics Conference Serials, 1999).

J.M. Rodenburg, I.A. Rauf, in *A Cross-Correlation Measure of Order in 'Amorphous' Indium Oxide. EMAG-MICRO 89* (Institute of Physics, London, UK, 1990).

M.L. Rudee, A. Howie, Structure of amorphous Si and Ge. Philos. Mag. **25**(4), 1001–1007 (1972).

J.V. Ryan, C.G. Pantano, Medium-range order in silicon oxycarbide glass by fluctuation electron microscopy. J. Phys. Cond. Mat. **19**, 455205 (2007).

C.A. Schuh, T.C. Hufnagel, et al., Mechanical behavior of amorphous alloys. Acta Mat. **55**, 4067–4100 (2007).

H.W. Sheng, W.K. Lou, et al., Atomic packing and short-to-medium-range order in metallic glasses. Nature **439**, 419–425 (2006).

A.P. Sokolov, A.P. Shebanin, Structural order and optical properties of amorphous silicon. Sov. Phys. Semicon.: 720. REPLACE WITH: A.P. Sokolov, A.P. Shebanin, O.A. Golikova, M.M. Mezdrogina (1991) Structural order in amorphous silicon and its alloys: Raman spectra and optical gap J. Non-Cryst. Sol. **137–138**, 99–102 (1990).

W.G. Stratton, J. Hamann, et al., Aluminum nanoscale order in amorphous $Al_{92}Sm_8$ measured by fluctuation electron microscopy. Appl. Phys. Lett. **86**, 141910 (2005).

W.G. Stratton, J. Hamann, et al., Electron beam induced crystallization of amorphous Al-based alloys in the TEM. Intermetallics **14**, 1061–1065 (2006).
W.G. Stratton, P.M. Voyles, Comparison of fluctuation electron microscopy theories and experimental methods. J. Phys. Cond. Mat. **19**, 455203 (2007).
W.G. Stratton, P.M. Voyles, A phenomenological model of fluctuation electron microscopy for a nanocrystal/amorphous composite. Ultramicroscopy **108**, 727–736 (2008).
M.M.J. Treacy, J.M. Gibson, Coherence and multiple scattering in "Z-contrast" images. Ultramicroscopy **52**(1), 31–53 (1993).
M.M.J. Treacy, J.M. Gibson, Variable coherence microscopy: a rich source of structural information from disordered systems. Acta Cryst. A **52**(2), 212–220 (1996).
M.M.J. Treacy, J.M. Gibson, et al., Fluctuation microscopy: a probe of medium range order. Rep. Prog. Phys. **68**, 2899–2944 (2005).
M.M.J. Treacy, J.M. Gibson, et al., Paracrystallites found in evaporated amorphous tetrahedral semiconductors. J. Non-Cryst. Sol. **231**(1–2), 99–110 (1998).
M.M.J. Treacy, P.M. Voyles, et al., Schläfli cluster topological analysis of medium range order in paracrystalline amorphous semiconductor models. J. Non-Cryst. Sol. **266**, 150–155 (2000).
D.V. Tsu, B.S. Chao, et al., Effect of hydrogen dilution on the structure of amorphous silicon alloys. Appl. Phys. Lett. **71**, 1317–1319 (1997).
D. Van Dyck, S. Van Aert, et al., Is atomic resolution transmission electron microscopy able to resolve and refine amorphous structures? Ultramicroscopy **98**, 27–42 (2003).
P.M. Voyles, Fluctuation electron microscopy of medium-range order in amorphous silicon. Physics. Urbana-Champaign, IL, University of Illinois (2001).
P.M. Voyles, J.R. Abelson, Medium-range order in amorphous silicon measured by fluctuation electron microscopy. Solar Energy Mater. Solar Cells **78**(1–4), 85–113 (2003).
P.M. Voyles, J.E. Gerbi, et al., Absence of an abrupt phase change from polycrystalline to amorphous in silicon with deposition temperature. Phys. Rev. Lett. **86**(24), 5514–5517 (2001).
P.M. Voyles, J.M. Gibson, et al., Fluctuation microscopy: a probe of atomic correlations in disordered materials. J. Electron Microsc. **49**(2), 259–266 (2000).
P.M. Voyles, D.A. Muller, Fluctuation microscopy in the STEM. Ultramicroscopy **93**(2), 147–159 (2002).
P.M. Voyles, M.M.J. Treacy, et al., in *Experimental Procedures and Data Analysis for Fluctuation Microcopy. Advances in Materials Problem Solving with the Electron Microscope*. (Materials Research Society, Boston, MA, 1999).
P.M. Voyles, M.M.J. Treacy, et al., in *Comparative Fluctuation Microscopy Study of Medium-Range Order in Hydrogenated Amorphous Silicon Deposited by Various Methods. Amorphous and Heterogeneous Silicon Thin Films 2000*. (Materials Research Society, San Francisco, CA, 2000).
P.M. Voyles, N. Zotov, et al., Structure and physical properties of paracrystalline atomistic models of amorphous silicon. J. Appl. Phys. **90**(9), 4437–4451 (2001).
R.J. Walters, G.I. Bourianoff, et al., Field-effect electroluminescence in silicon nanocrystals. Nature Materials **4**(2), 143–146 (2005).
J. Wen, Y.Q. Cheng, et al., Distinguishing medium-range order in metallic glasses using fluctuation electron microscopy: A theoretical study using atomic models. J. Appl. Phys. **105**, 043519 (2009).
J. Wen, H.W. Yang, et al., Fluctuation electron microscopy of Al-based metallic glasses: effects of minor alloying addition and structural relaxation on

medium-range structural homogeneity. J. Phys. Condens. Matter **19**, 455211 (2007).

F. Wooten, K. Winer, et al., Computer-generation of structural models of amorphous Si and Ge. Phys. Rev. Lett. **54**(13), 1392–1395 (1985).

W.H. Zachariasen, The atomic arrangement in glass. J. Am. Chem. Soc. **54**, 3841–3851 (1932).

G. Zhao, P.R. Buseck, et al., Medium-range order in molecular materials: fluctuation electron microscopy for detecting fullerenes in disordered carbons. Ultramicroscopy **109**, 177–188 (2009).

J.M. Zuo, Electron detection characteristics of slow-scan CCD camera. Ultramicroscopy **66**, 21–33 (1996).

J.M. Zuo, M. Gao, et al., Coherent nano-area electron diffraction. Microsc. Res. Tech. **64**, 347–355 (2004).

J.M. Zuo, I. Vartanyants, et al., Atomic resolution imaging of a carbon nanotube from diffraction intensities. Science **300**, 1419–1421 (2003).

Index

S.J. Pennycook, P.D. Nellist (eds.), *Scanning Transmission Electron Microscopy*,
DOI 10.1007/978-1-4419-7200-2, © Springer Science+Business Media, LLC 2011

www.ingramcontent.com/pod-product-compliance
Lightning Source LLC
Chambersburg PA
CBHW080325050326
40679CB00005B/508
9781441971999